FOURTH EDITION

Physical Chemistry

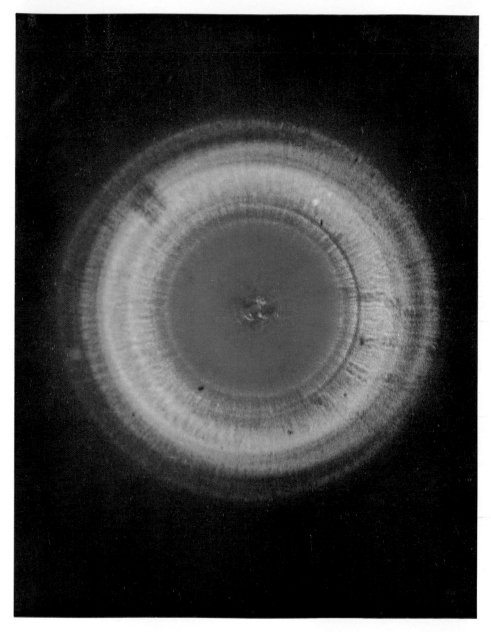

A stimulated Raman spectrum from benzene. An intense beam from a ruby laser was focused into a cell containing liquid benzene. The scattered light was recorded photographically on a plate behind the cell. The wavelength of the laser light was 694.3 nm, giving the deep red color at the center of the picture. The wavelength of the Raman scattered light was shifted to both lower and higher values, but the latter lie in the infrared and were not photographed. The successive circles of colored light, red, orange, yellow, and green, show the Raman shifts to lower wavelengths, corresponding to addition to the laser frequency of one, two, three, or four multiples of a fundamental vibration frequency of the benzene molecule. This picture is from the work of R. W. Terhune and P. D. Maker at the Research Laboratories of the Ford Motor Company.

FOURTH EDITION

Physical Chemistry

Walter J. Moore
Professor of Physical Chemistry
The University of Sydney

PRENTICE-HALL, INC., Englewood Cliffs, New Jersey

10 9 8 7 6

ISBN: 0-13-665968-3

PRENTICE-HALL INTERNATIONAL, INC., London
PRENTICE-HALL OF AUSTRALIA, PTY. LTD., Sydney
PRENTICE-HALL OF CANADA, LTD., Toronto
PRENTICE-HALL OF INDIA PRIVATE LIMITED, New Delhi
PRENTICE-HALL OF JAPAN, INC., Tokyo

Preface

Oor universe is like an e'e
Turned in, man's benmaist hert to see,
And swamped in subjectivity.

But whether it can use its sicht
To bring what lies withoot to licht
To answer's still ayont my micht.

Hugh MacDiarmid
1926*

Like the last edition, this fourth edition of *Physical Chemistry* is the result of a substantial rewriting of the entire book. About 22 years ago, in the preface to the first edition, I said that this book was not designed to be a collection of facts, but rather an introduction to ways of thinking about the world. Actually this edition was written under the title *Foundations of Physical Chemistry*, and such a title expresses quite well the basic intention of the book. I have tried to emphasize critical discussions of definitions, postulates, and logical operations. The concepts of physical chemistry today are transient states in the progress of the science. The historical background in the book is intended to help the student reach this understanding, without which science becomes static and comparatively uninteresting.

For some students of physical chemistry the use of mathematics remains a major difficulty. We try to convince students that the scientist must learn mathematics while he studies science. It is neither necessary nor desirable to learn "pure mathematics" first and then to apply it to scientific problems. The level of mathematical difficulty in this edition is somewhat higher than previously, but as a compensation, more careful discussions of mathematical details have been given. Nevertheless, many students would find it worthwhile to acquire one of several excellent books on mathematics for the physical sciences, references to which are made in the text.

In this edition, the order of the subject matter has been changed in order to bring statistical mechanics into the text as early as possible, and then to use its methods in subsequent discussions. Examination of current textbooks of general chemistry and physics (universal prerequisites for the study of physical chemistry)

---

*From "The Great Wheel" by Hugh MacDiarmid (C. M. Grieve) in *A Drunk Man Looks at the Thistle* (Edinburgh: Wm. Blackwood & Sons, 1926).

v

indicates that almost all contain sufficient atomic physics and elementary quantum theory to serve as an adequate foundation for the principles of statistical mechanics as given in my Chapter 5.

I have tried to follow the recommendations on nomenclature and units of the International Union of Pure and Applied Chemistry, except for the retention of the *atmosphere* as a unit of pressure, a relic of nonsystematic units, which also should disappear in due course. Probably within a decade, the SI system of units will be in general use by all scientists.*

There are not many worked out numerical problems in the text, but Professors William Bunger of Indiana State University and Theodore Sakano of Rose–Hulman Polytechnic Institute have prepared a manual of solutions to all the problems at the ends of chapters. In my experience students will learn most quickly if they obtain this manual as a companion to the text.

It is always a pleasant duty to thank my confreres who have contributed so generously with illustrations, corrections, and suggestions to improve the book. So many people have helped that I am sure to forget to mention some, but these also have my thanks. The publishers wisely enlisted Thomas Dunn to provide a general analysis of the book and Jeff Steinfeld to make a critical reading of the manuscript. Walter Kauzmann was a continual source of help, both in the extensive comments he sent me and in the excellent material I found in his clearly written books. By devious pathways, an exegesis of the third edition by George Kistiakowsky fell into my hands, which provided many valuable clarifications.

Apart from these major efforts on the total book, much work on individual chapters was done by Peter Langhoff, Edward Bair, Donald McQuarrie, Robert Mortimer, John Bockris, Donald Sands, Edward Hughes, John Ricci, John Griffith, Dennis Peters, Ludvik Bass, Albert Zettlemoyer and Dieter Hummel (who has made a German translation). Lucky is the author who has such good neighbors as these. Acknowledgments to scientists who sent illustrations are included in the text. At Prentice-Hall, Albert Belskie, Editor for Chemistry, was a solid source of support and good counsel at all times.

With all this help one may wonder why the book is still so far from an ideal state. The answer must have something to do with the fact that we are not working closer to absolute zero.† As always, I shall welcome comments from readers and try to correct all the mistakes that they will find.

W.J.M.

---

*M. A. Paul, "The International System of Units (SI)—Development and Progress," *J. Chem. Doc. 11*, 3 (1971).

†A concise summary of thermodynamics has been given: (1) The First Law says you can't win; the best you can do is break even. (2) The Second Law says you can break even only at absolute zero. (3) The Third Law says you can never reach absolute zero.

Contents

10   Electrochemistry: Ionics   420

11   Interfaces   475

12   Electrochemistry—Electrodics   520

13   Particles and Waves   570

14   Quantum Mechanics and Atomic Structure   614

15   The Chemical Bond   670

9. Proof of Bragg Reflection   10. Fourier Transforms and Reciprocal Lattices
11. Structures of Sodium and Potassium Chlorides   12. The Powder Method
13. Rotating Crystal Method   14. Crystal-Structure Determinations   15.
Fourier Synthesis of a Crystal Structure   16. Neutron Diffraction   17.
Closest Packing of Spheres   18. Binding in Crystals   19. The Bond Model
20. Electron-Gas Theory of Metals   21. Quantum Statistics   22. Cohesive
Energy of Metals   23. Wave Functions for Electrons in Solids   24. Semi-
conductors   25. Doping of Semiconductors   26. Nonstoichiometric Com-
pounds   27. Point Defects   28. Linear Defects: Dislocations   29. Effects
Due to Dislocations   30. Ionic Crystals   31. Cohesive Energy of Ionic
Crystals   32. The Born–Haber Cycle   33. Statistical Thermodynamics of
Crystals: Einstein Model   34. The Debye Model   *Problems*

19   Intermolecular Forces and the Liquid State   902

1. Disorder in the Liquid State   2. X-Ray Diffraction of Liquid Structures
3. Liquid Crystals   4. Glasses   5. Melting   6. Cohesion of Liquids—
The Internal Pressure   7. Intermolecular Forces   8. Equation of State and
Intermolecular Forces   9. Theory of Liquids   10. Flow Properties of
Liquids   11. Viscosity   *Problems*

20   Macromolecules   928

1. Types of Polyreactions   2. Distribution of Molar Masses   4. Light
Scattering—The Rayleigh Law   5. Light Scattering by Macromolecules   6.
Sedimentation Methods: The Ultracentrifuge   7. Viscosity   8. Stereochemistry of Polymers   9. Elasticity of Rubber   10. Crystallinity of Polymers   *Problems*

1

Physicochemical Systems

Nosotros (la indivisa divinidad que opera en nosotros) hemos soñado el mundo.
Lo hemos soñado resistente, misterioso, visible, ubicuo en el espacio y firme en
el tiempo; pero hemos consentido en su arquitectura tenues y eternos intersticios
de sinrazón, para saber que es falso.

Jorge Luis Borges
1932*

On the planet Earth the processes of evolution created neural networks called
brains. Reaching a certain degree of complexity, these networks generated electrical
phenomena in space and time called *consciousness*, *volition*, and *memory*. The brains
in some of the higher primates, *genus Homo*, devised a medium called *language*
to communicate with one another and to store information. Some of the human
brains persistently sought to analyze the input signals received from the world
in which they had their existence. One form of analysis, called *science*, proved to be
especially effective in correlating, modifying, and controlling the sensory input data.

Most of the structure of brains was laid down in conformity with information
coded into the base sequence of the DNA molecules of the genetic material. Additional structuring was caused by a relatively uniform experience during their
periods of growth and maturation. Thus, heredity and early environment combined
to produce adult brains with rather stereotyped capabilities for analysis and
communication.

Language was effective in communications that dealt with the content of
sensory input data, but it did not allow the brains to talk about themselves or their
relation to the world without breakdowns into paradox or contradiction. In
particular, although it was possible to find thousands of books filled with results
of science, to observe thousands of men at work in the fields of science, and to
experience the earthshaking effects of science, it was not possible to explain in
words what science was or even the mechanism by which it operated. Different
views on these questions were eloquently put forth from time to time.

*From "La Perpetua Carrera de Achilles e la Tortuga" in *Discusion* (Buenos Aires: M. Gleizer,
1932). "We (the undivided divinity that operates within us) have dreamed the world. We have
dreamed it resistant, mysterious, visible, ubiquitous in space and firm in time; but we have allowed
into its architecture tenuous and eternal interstices of unreason to let us understand that it is false."

1. What Is Science?

According to one view, called *conventionalism*, the human brains created or invented certain beautiful logical structures called *laws of nature* and then devised special ways, called *experiments*, of selecting sensory input data so that they would fit into the patterns ordained by the laws. In the conventionalist view, the scientist was like a creative artist, working not with paint or marble but with the unorganized sensations from a chaotic world. Scientific philosophers supporting this position included Poincaré,* Duhem,† and Eddington.‡

A second view of science, called *inductivism*, considered that the basic procedure of science was to collect and classify sensory input data into a form called *observable facts*. From these facts, by a method called *inductive logic*, the scientist then drew general conclusions which were the laws of nature. Francis Bacon, in his *Novum Organum* of 1620, argued that this was the only proper scientific method, and at that time his emphasis on observable facts was an important antidote to medieval reliance on a formal logic of limited capabilities. Bacon's definition accords most closely with the layman's idea of what scientists do, but many competent philosophers have also continued to support the essentials of inductivism, including Russell§ and Reichenbach.||

A third view of science, called *deductivism*, emphasized the primary importance of theories. According to Popper,♯ "Theories are nets cast to catch what we call 'the world': to rationalize, to explain, and to master it. We endeavor to make the mesh ever finer and finer." According to the deductivists, there is *no* valid inductive logic, since general statements can never be proved from particular instances. On the other hand, a general statement can be *disproved* by one contrary particular instance. Hence, a scientific theory can never be proved, but it can be disproved. The role of an experiment is therefore to subject a scientific theory to a critical test.

The three philosophies outlined by no means exhaust the variety of efforts made to capture science in the web of language. As we are studying the part of science called *physical chemistry*, we should pause sometimes (but not too often) to ask ourselves which philosophic school we are attending.

---

*Henri Poincaré, *Science and Hypothesis* (New York: Dover Publications, Inc., 1952).

†Pierre Duhem, *The System of the World*, 6 Vols. (Paris: Librarie Scientifique Hermann et Cie., 1954).

‡Arthur Stanley Eddington, *The Philosophy of Physical Science* (Ann Arbor, Mich.: University of Michigan Press, 1958).

§Bertrand Russell, *Human Knowledge, Its Scope and Limits* (New York: Simon and Schuster, Inc., 1948).

||Hans Reichenbach, *The Rise of Scientific Philosophy* (Berkeley: University of California Press, 1963).

♯Karl R. Popper, *The Logic of Scientific Discovery* (New York: Harper Torchbooks, 1965).

2. Physical Chemistry

There appear to be two reasonable approaches to the study of physical chemistry. We may adopt a synthetic approach and, beginning with the structure and behavior of matter in its finest known state of subdivision, gradually progress from electrons to atoms to molecules to states of aggregation and chemical reactions. Alternatively, we may adopt an analytical treatment and, starting with matter or chemicals as we find them in the laboratory, gradually work our way back to finer states of subdivision as we require them to explain experimental results. This latter method follows more closely the historical development, although a strict adherence to history is impossible in a broad subject whose different branches have progressed at different rates.

Two main problems have been primary concerns of physical chemistry: the question of the position of chemical equilibrium, which is the principal problem of chemical thermodynamics; and the question of the rate of chemical reactions, which is the field of chemical kinetics. Since these problems are ultimately concerned with the interactions of molecules, their complete solutions should be implicit in the mechanics of molecules and molecular aggregates. Therefore, molecular structure is an important part of physical chemistry. The discipline that allows us to bring our knowledge of molecular structure to bear on the problems of equilibrium and kinetics is found in the study of statistical mechanics.

We shall begin our study of physical chemistry with thermodynamics, which is based on concepts common to the everyday world. We shall follow quite closely the historical development of the subject, since usually more knowledge can be gained by watching the construction of something than by inspecting the polished final product.

3. Mechanics: Force

The first thing that may be said of thermodynamics is that the word itself is evidently derived from *dynamics*, which is a branch of mechanics dealing with matter in motion.

Mechanics is founded on the work of Isaac Newton (1642–1727), and usually begins with a statement of the well-known equation

$$\mathbf{F} = m\mathbf{a}$$

with

$$\mathbf{a} = \frac{d\mathbf{v}}{dt} = \frac{d^2\mathbf{r}}{dt^2} \tag{1.1}$$

The equation states the proportionality between a vector quantity $\mathbf{F}$, called the *force* applied to a particle of matter, and the acceleration $\mathbf{a}$ of the particle, a vector in the same direction, with a proportionality factor m, called the mass. Equation (1.1) may also be written

$$\mathbf{F} = \frac{d(m\mathbf{v})}{dt} \tag{1.2}$$

where the product of mass and velocity is called the *momentum*.

In the International System of Units (SI), the unit of mass is the kilogram* (kg), the unit of time is the second† (s), and the unit of length is the metre‡ (m). The SI unit of force is the *newton* (N).

Mass might be introduced in Newton's Law of Gravitation,

$$F = \frac{Gm_1 m_2}{r_{12}^2}$$

which states that there is an attractive force between two masses proportional to their product and inversely proportional to the square of their separation. If this gravitational mass is to be the same as the inertial mass of (1.1), the proportionality constant

$$G = 6.670 \times 10^{-11} \text{ m}^3 \cdot \text{s}^{-2} \cdot \text{kg}^{-2}$$

The weight of a body, W, is the force with which it is attracted toward the earth, and may vary slightly over the earth's surface, since the earth is not a perfect sphere of uniform density. Thus

$$W = mg$$

where g is the acceleration of free fall in vacuum.

In practice, the mass of a body is measured by comparing its weight by means of a balance with that of known standards ($m_1/m_2 = W_1/W_2$).

4. Mechanical Work

In mechanics, if the point of application of a force $\mathbf{F}$ moves, the force is said to *do work*. The amount of work done by a force $\mathbf{F}$ whose point of application moves a distance dr along the direction of the force is

$$dw = F \, dr \tag{1.3}$$

If the direction of motion of the point of application is not the same as the direction of the force, but at an angle θ to it, we have the situation shown in Fig. 1.1.

The component of $\mathbf{F}$ in the direction of motion is $F \cos \theta$, and the element of work is

$$dw = F \cos \theta \, dr \tag{1.4}$$

If we choose a set of Cartesian axes XYZ, the components of the force are

---

*Defined by the mass of the international prototype, a platinum cylinder at the International Bureau of Weights and Measures at Sèvres, near Paris.

†Defined as duration of 9 192 631 770 periods of the radiation corresponding to the transition between the two hyperfine levels in the ground state of the cesium-133 atom.

‡Defined as the length equal to 1 650 763.73 wavelengths in vacuum of the radiation corresponding to the transition between the levels $2p_{10}$ and $5d_5$ of the krypton-86 atom.

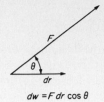

FIG. 1.1 Definition of differential element of work.

F_x, F_y, F_z and

$$dw = F_x \, dx + F_y \, dy + F_z \, dz \tag{1.5}$$

For the case of a force that is constant in direction and magnitude, (1.3) can be integrated to give

$$w = \int_{r_0}^{r_1} F \, dr = F(r_1 - r_0)$$

An example is the force acting on a body of mass m in the earth's gravitational field. Over distances that are short compared to the diameter of the earth, this $F = mg$. To lift a body against earth's gravitational attraction we must apply to it an external force equal to mg. What is the work done on a mass of 1 kg when it is lifted a distance of 1 m?

$$w = mgr_1 = (1)(9.80665)(1) \text{ kg} \cdot \text{m} \cdot \text{s}^{-2} \cdot \text{m} = 9.80665 \text{ kg} \cdot \text{m}^2 \cdot \text{s}^{-2}$$
$$= 9.80665 \text{ newton metre (N} \cdot \text{m)} = 9.80665 \text{ joule (J)}$$

An application of (1.3) in which the force is not constant is to the stretching of a perfectly elastic spring. In accord with the law of Hooke, 1660, *ut tensio sic vis*: the restoring force is directly proportional to the extension,

$$F = -\kappa r \tag{1.6}$$

where κ is called the *force constant* of the spring. Hence, the work dw done on the spring to extend it by dr is

$$dw = \kappa r \, dr$$

Suppose the spring is stretched by a distance r_1,

$$w = \int_0^{r_1} \kappa r \, dr = \frac{\kappa}{2} r_1^2$$

The work done on the spring is taken by convention to be positive.

In the general case, we can write the integral of (1.5) as

$$w = \int_a^b (F_x \, dx + F_y \, dy + F_z \, dz) \tag{1.7}$$

The components of the force may vary from point to point along the curve followed by the mass point. They are functions of the space coordinates x, y, z: $F_x(x, y, z)$, $F_y(x, y, z)$, and $F_z(x, y, z)$. It is evident that the value of the integral depends upon the exact path or curve between the two limits a and b. It is called a *line integral*.

5. Mechanical Energy

In 1644, René Descartes declared that in the beginning God imparted to the universe a certain amount of motion in the form of vortices, and this motion endured eternally and could be neither increased nor decreased. For almost a century after the death of Descartes, a great controversy raged between his followers and those of Leibniz on the question of whether motion was conserved. As often happens, lack of precise definitions of the terms used prevented a meeting of minds. The word *motion* then usually designated what we now call *momentum*. In fact, the momentum in any given direction is conserved in collisions between elastic bodies.

In 1669, Huygens discovered that if he multiplied each mass m by the square of its velocity v^2, the sum of these products was conserved in all collisions between elastic bodies. Leibniz called mv^2 the *vis viva*. In 1735, Jean Bernoulli asked himself what happened to the *vis viva* in an inelastic collision. He concluded that some of it was lost into some kind of *vis mortua*. In all mechanical systems operating without friction, the sum of *vis viva* and *vis mortua* was conserved at a constant value. In 1742, this idea was also clearly expressed by Emilie du Châtelet, who said that although it was difficult to follow the course of the *vis viva* in an inelastic collision, it must nevertheless be conserved in some way.

The first to use the word "energy" appears to have been d'Alembert in the French *Encyclopédie* of 1785. "There is in a body in movement an effort or *énergie*, which is not at all in a body at rest." In 1787, Thomas Young called the *vis viva* the "actual energy" and the *vis mortua* the "potential energy." The name "kinetic energy" for $\frac{1}{2}mv^2$ was introduced much later by William Thomson.

We can give these developments a mathematical formulation starting with (1.3). Let us consider a particle at position r_0 and apply to it a force $F(r)$ that depends only on its position. In the absence of any other forces, the work done on the body in a finite displacement from r_0 to r_1 is

$$w = \int_{r_0}^{r_1} F(r)\, dr \tag{1.8}$$

The integral over distance can be transformed to an integral over time:

$$w = \int_{t_0}^{t_1} F(r)\frac{dr}{dt}\, dt = \int_{t_0}^{t_1} F(r)v\, dt$$

Introducing Newton's Law of Force, (1.1), we obtain

$$w = \int_{t_0}^{t_1} m\frac{dv}{dt}v\, dt = m \int_{v_0}^{v_1} v\, dv$$

$$w = \tfrac{1}{2}mv_1^2 - \tfrac{1}{2}mv_0^2 \tag{1.9}$$

The kinetic energy is defined by

$$E_k = \tfrac{1}{2}mv^2$$

Hence,

$$w = \int_{r_0}^{r_1} F(r)\, dr = E_{k1} - E_{k0} \tag{1.10}$$

The work done on the body equals the difference between its kinetic energies in the final and the initial states.

Since the force in (1.10) is a function of r alone, the integral defines another function of r which we can write as

$$F(r) \, dr = -dU(r)$$

or

$$F(r) = -\frac{dU(r)}{dr} \tag{1.11}$$

Thus, (1.10) becomes

$$\int_{r_0}^{r_1} F(r) \, dr = U(r_0) - U(r_1) = E_{k1} - E_{k0}$$

or

$$U_0 + E_{k0} = U_1 + E_{k1} \tag{1.12}$$

The new function $U(r)$ is the *potential energy*. The sum of the potential and the kinetic energies, $U + E_k$, is the total mechanical energy of the body, and this sum evidently remains constant during the motion. Equation (1.12) has the form typical of an *equation of conservation*. It is a statement of the mechanical principle of the *conservation of energy*. For example, the gain in kinetic energy of a body falling in a vacuum is exactly balanced by an equal loss in potential energy.

If a force depends on velocity as well as position, the situation is more complex. This would be the case if a body falls, not in a vacuum, but in a viscous fluid, such as air or water. The higher the velocity, the greater is the frictional or viscous resistance opposed to the gravitational force. We can no longer set $F(r) = -dU/dr$, and we can no longer obtain an equation such as (1.11), because the mechanical energy is no longer conserved. From the dawn of history it has been known that the frictional dissipation of energy is attended by the evolution of something called *heat*. We shall see later how it became possible to include heat among the ways of transforming energy, and in this way to obtain a new and more inclusive principle of the conservation of energy.

It may be noted that whereas the kinetic energy is zero for a body at rest, there is no naturally defined zero of potential energy. Only differences in potential energy can be measured. Sometimes, however, a zero of potential energy is chosen *by convention*; an example is the choice $U(r) = 0$ for the gravitational potential energy when two bodies are an infinite distance apart.

6. Equilibrium

The ordinary subjects for chemical experimentation are not individual particles of any sort but more complex *systems*, which may contain solids, liquids, and gases. A *system* is a part of the world separated from the rest of the world by definite boundaries. The world outside the system is called its *surroundings*. If the boundaries of the system do not permit any change to occur in the system as a consequence of a change in the surroundings, the system is said to be *isolated*.

The experiments that we perform on a system are said to measure its *properties,* these being the attributes that enable us to describe it with all requisite completeness. This complete description is said to define the *state* of the system.

The idea of predictability enters here. Having once measured the properties of a system, we expect to be able to predict the behavior of a second system with the same set of properties from our knowledge of the behavior of the original. When a system shows no further tendency to change its properties with time, it is said to have reached a *state of equilibrium.* The condition of a system in equilibrium is reproducible and can be defined by a set of properties, which are *functions of the state,* i.e., which do not depend on the history of the system before it reached equilibrium.*

A simple mechanical illustration will clarify the concept of equilibrium. Figure 1.2(a) shows three different equilibrium positions of a box resting on a table. In

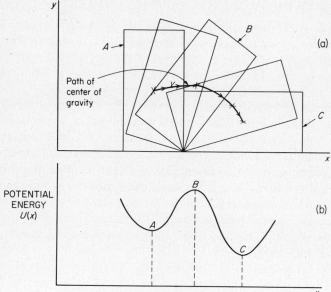

FIG. 1.2 An illustration of mechanical equilibrium.

both positions A and C, the center of gravity of the box is lower than in any slightly displaced position, and if the box is tilted slightly it will tend to return spontaneously to its original equilibrium position. The gravitational potential energy of the box in positions A or C is at a minimum, and both positions represent *stable equilibrium* states. Yet it is apparent that position C is more stable than position A, and a certain tilt of A will suffice to push it over into C. In position A, therefore, the box is said to be in *metastable equilibrium.* Position B is also an equilibrium position, but it is a state of *unstable equilibrium,* as anyone who has

*The specification of the state of a system that is not in equilibrium is a more difficult problem. It will require a larger number of variables and sometimes will not even be possible in practice.

tried to balance a chair on two legs will agree. The center of gravity of the box in *B* is higher than in any slightly displaced position, and the tiniest tilt will send the box into either position *A* or *C*. The potential energy at a position of unstable equilibrium is a maximum, and such a position could be realized only in the absence of any disturbing forces.

These relations may be presented in more mathematical form by plotting in Fig. 1.2(b) the potential energy of the system as a function of the horizontal position of the center of gravity in Fig. 1.2(a). Positions of stable equilibrium are seen to be minima in the curve, and the position of unstable equilibrium is represented by a maximum. Positions of stable and unstable equilibrium alternate in this way in any system. For an equilibrium position, the slope dU/dr of the curve for U vs. displacement is equal to zero, and one may write the equilibrium condition as

$$\left(\frac{dU}{dr}\right)_{r=r_0} = 0$$

Examination of the second derivative will indicate whether the equilibrium is stable or unstable:

$$\left(\frac{d^2U}{dr^2}\right) > 0 \quad \text{stable}$$

$$\left(\frac{d^2U}{dr^2}\right) < 0 \quad \text{unstable} \qquad .$$

Although these considerations have been presented in terms of a simple mechanical model, similar principles will be found to apply in the more complex physicochemical systems that we shall study. In addition to purely mechanical changes, such systems may undergo temperature changes, changes of state of aggregation, and chemical reactions. The problem of thermodynamics is to discover or invent new functions that will play the role in these more general systems which the potential energy plays in mechanics.

7. The Thermal Properties of Matter

To specify precisely the state at equilibrium of a substance studied in the laboratory, we must give the numerical values of certain of its measured properties. Since there are equations that give relations between properties, it is not necessary to specify the values of each and every property to define exactly the state of a substance. In fact, if we ignore external fields of force (gravitational, electromagnetic) and take a gas or a liquid as the substance under consideration,* the exact specification of its state requires the values of only a few quantities. At present, we shall confine the problem to individual, pure substances, for which no composition variables are needed. To specify the state of a pure gas or liquid, we may first of all state the mass *m* of the substance. There are many properties of the substance that might be measured to define its state. In particular, we focus attention on three thermodynamic variables: pressure *P*, volume *V*, and temperature θ. If any two of these are

---

*The properties of solids may depend in a rather complicated way on direction.

fixed, we find as a matter of experimental fact that the value of the third will also be fixed, because of the existence of a relation between the variables. In other words, of the three variables of state, P, V, θ, only two are independent variables. Note particularly that we may describe the state of the substance entirely in terms of the two mechanical variables P and V, and not use the thermal variable θ at all.

The use of the pressure P as a variable to describe the state of a substance requires some care. In Fig. 1.3, consider a fluid contained in a cylinder with a

FIG. 1.3 Definition of pressure in a fluid, neglecting gravitational field in the fluid. The force F represented by the weight includes also the force due to the earth's atmosphere.

frictionless piston. We can calculate the pressure on the fluid by dividing the force on the piston by its area. ($P = \text{force/area}$). At equilibrium, this pressure will be uniform throughout the fluid, so that on any specified unit area in the fluid there will be a force P. A pressure is thus a stress that is uniform in all directions.

In this analysis, we neglect the effect of the weight of the fluid itself. If we included the weight, there would be an extra force per unit area, increasing with the depth of the fluid and equal to the weight of the column of fluid above the given section. In the subsequent analysis, this effect of the weight will be neglected and we shall consider the pressure of a volume of fluid to be the same throughout. This simplification is what we mean by "ignoring the gravitational field."

If the fluid is not in equilibrium, we can still speak of the external pressure P_{ex} on the piston, but this is clearly not a property of the state of the fluid itself. Until equilibrium is restored, the pressure may vary from point to point in the fluid and we cannot define its state by a single pressure P.

The properties of a system can be classified as *extensive* or *intensive*. Extensive properties are additive; their value for the whole system is equal to the sum of their values for the individual parts. Sometimes they are called *capacity factors*. Examples are the volume and the mass. Intensive properties, or *intensity factors*, are not additive. Examples are temperature and pressure. The temperature of any small part of a system in equilibrium is the same as the temperature of the whole.

Before we use the temperature θ as a physical quantity, we should consider how it can be measured quantitatively. The concept of *temperature* evolved from sensual perceptions of hotness and coldness. It was found that these perceptions could be correlated with the readings of *thermometers* based on the volume changes of liquids. In 1631, the French physician Jean Rey used a glass bulb having a stem partly filled with water to follow the progress of fevers in his patients. In 1641, Ferdinand II, Grand Duke of Tuscany and founder of the Accademia del Cimento of Florence, invented an alcohol-in-glass *thermoscope* to which a scale was added by marking equal divisions between the volumes at "coldest winter cold" and "hottest summer heat." A calibration based on two fixed points was introduced in 1688 by Dalencé, who chose the melting point of snow as $-10°$ and the melting point of butter as $+10°$. In 1694, Renaldi took the boiling point of water as the upper fixed point and the melting point of ice as the lower. To make the specification of these fixed points exact, we must add the requirements that the pressure is maintained at 1 atmosphere (atm), and that the water in equilibrium with ice is saturated with air. Apparently Elvius, a Swede, in 1710, first suggested assigning the values $0°$ and $100°$ to these two points. They define the *centigrade scale*, officially called the *Celsius scale*, after a Swedish astronomer who used a similar system.

8. Temperature as a Mechanical Property

The existence of a temperature function can be based on the fact that whenever two bodies are separately brought to equilibrium with a third body, they are then found to be in equilibrium with each other.

We may choose one body (1) to be a pure fluid whose state is specified by P_1 and V_1 and call it a *thermometer*, and use some property of the state of this body $\theta_1(P_1, V_1)$ to define a temperature scale. When any second body (2) is brought into equilibrium with the *thermometer*, the equilibrium value of $\theta_1(P_1, V_1)$ measures its temperature.

$$\theta_2 = \theta_1(P_1, V_1) \tag{1.13}$$

Note that the temperature defined and measured in this way is defined entirely in terms of the mechanical properties of pressure and volume, which suffice to define the state of the pure fluids. We have left our sensory perceptions of hotness and coldness and reduced the concept of temperature to a mechanical concept.

A simple example of (1.13) is a liquid thermometer, in which P_1 is kept constant and the volume V_1 is used to measure the temperature. In other cases, electrical, magnetic, or optical properties can be used to define the temperature scale, since in every case the property θ_1 can be expressed as a function of the state of the pure fluid, fixed by specifying P_1 and V_1.

9. The Spring of the Air and Boyle's Law

The mercury barometer was invented in 1643 by Evangelista Torricelli, a mathematician who studied with Galileo in Florence. The height of the column under atmospheric pressure may vary from day to day over a range of several centimetres of mercury, but a standard *atmosphere* has been *defined* as a pressure unit equal to 101 325 newtons per square metre ($N \cdot m^{-2}$). Workers in the field of high pressures often use the *kilobar* (kbar), $10^8 \ N \cdot m^{-2}$. At low pressures, one often uses the *torr*, which equals atm/760.*

Robert Boyle and his contemporaries often referred to the pressure of a gas as the spring of the air. They knew that a volume of gas behaved mechanically like a spring. If you compress it in a cylinder with a piston, the piston recoils when the force on it is released. Boyle tried to explain the springiness of the air in terms of the corpuscular theories popular in his day. "Imagine the air," said he, "to be such a heap of little bodies, lying one upon another, as may be resembled to a fleece of wool. For this . . . consists of many slender and flexible hairs, each of which may indeed, like a little spring, be still endeavoring to stretch itself out again." In other words, Boyle supposed that the corpuscles of air were in close contact with one another, so that when air was compressed, the individual corpuscles were compressed like springs. This hypothesis was not correct.

In 1660, Boyle published the first edition of his book *New Experiments, Physico-Mechanical, Touching the Spring of the Air, and its Effects*, in which he described observations made with a new vacuum pump he had constructed. He found that when the air surrounding the reservoir of a Torricellian barometer was evacuated, the mercury column fell. This experiment seemed to him to prove conclusively that the column was held up by the air pressure. Nevertheless, two attacks on Boyle's work were immediately published, one by Thomas Hobbes, the famous political philosopher and author of *Leviathan*, and the other by a devout Aristotelian, Franciscus Linus. Hobbes based his criticism on the "philosophical impossibility of a vacuum." ("A vacuum is nothing, and what is nothing cannot exist.") Linus claimed that the mercury column was held up by an invisible thread, which fastened itself to the upper end of the tube. This theory seemed quite reasonable, he said, for anyone could easily feel the pull of the thread by covering the end of the barometer tube with his finger.

In answer to these objections, Boyle included an appendix in the second edition of his book, published in 1662, in which he described an important new experiment. He used essentially the apparatus shown in Fig. 1.4. By addition of mercury to the open end of the J-shaped tube, the pressure could be increased on the gas in the closed end. Boyle observed that as the pressure increased, the volume of the gas proportionately decreased. The temperature of the gas was almost constant during these measurements. In modern terms, we would therefore state Boyle's result thus: *At constant temperature, the volume of a given sample of gas varies inversely as the pressure.* In mathematical terms, this becomes $P \propto 1/V$ or

---

*The *torr* differs from the conventional mmHg by less than 2 parts in 10^7.

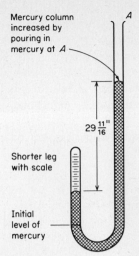

Mercury column
increased by
pouring in
mercury at A

$29\frac{11}{16}''$

Shorter leg
with scale

Initial
level of
mercury

FIG. 1.4 Boyle's J-tube, as used in an experiment in which the volume of gas was halved when the pressure was doubled.

$P = C/V$ where C is a constant of proportionality. This relation is equivalent to

$$PV = C \quad \text{(at constant } \theta \text{)} \tag{1.14}$$

Equation (1.14) is known as *Boyle's Law*. It is followed quite closely by many gases at moderate pressures, but the real behavior of gases may deviate greatly, especially at higher pressures.

10. The Law of Gay-Lussac

The first detailed experiments on the variation with temperature of the volumes of gases at constant pressures were those published by Joseph Gay-Lussac from 1802 to 1808. Working with the "permanent gases," such as nitrogen, oxygen, and hydrogen, he found that the different gases all showed the same dependence of V on θ.

His results can be put into mathematical form as follows. We define a gas temperature scale by assuming that the volume V varies linearly with the temperature θ. If V_0 is the volume of a sample of gas at 0°C, we have

$$V = V_0(1 + \alpha_0\theta) \tag{1.15}$$

The coefficient α_0 is a *thermal expansivity* or *coefficient of thermal expansion.*[*] Gay-Lussac found α_0 approximately equal to $\frac{1}{267}$, but in 1847, Regnault, with an improved experimental procedure, obtained $\alpha_0 = \frac{1}{273}$. With this value, (1.15) may be written as

$$V = V_0\left(1 + \frac{\theta}{273}\right)$$

This relation is called the *Law of Gay-Lussac*. It states that a gas expands by $\frac{1}{273}$ of its volume at 0°C for each degree rise in temperature at constant pressure.

[*]In Section 1.13, we define a somewhat different thermal expansivity, α.

Careful measurements revealed that real gases do not obey exactly the laws of Boyle and Gay-Lussac. The variations are least when a gas is at high temperature and low pressure. Furthermore, the deviations vary from gas to gas; for example, helium obeys closely, whereas carbon dioxide is relatively disobedient. It is useful to introduce the concept of the *ideal gas*, a gas that follows these laws exactly. Since gases at low pressures—i.e., low densities—obey the gas laws most closely, we can often obtain the properties of ideal gases by extrapolation to zero pressure of measurements on real gases.

Figure 1.5 shows the results of measurements of α_0 on different gases at suc-

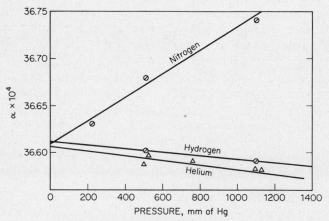

FIG. 1.5 Extrapolation of thermal expansion coefficients to zero pressure.

cessively lower pressures. Note that the scale is greatly expanded so that the maximum differences do not exceed about 0.5%. Within experimental uncertainty, the value found by extrapolation to zero pressure is the same for all these gases. This is the value of α_0 for an ideal gas. The consensus of the best measurements gives

$$\alpha_0 = 36.610 \times 10^{-4} \; °C^{-1}$$

or

$$\frac{1}{\alpha_0} = 273.15°C \pm 0.02°C = T_0$$

Thus, the Law of Gay-Lussac for an ideal gas may be written

$$V = V_0\left(1 + \frac{\theta}{T_0}\right) \tag{1.16}$$

It is now possible and most convenient to define a new temperature scale with the temperature denoted by T and called the *absolute temperature*. The unit of temperature on this scale is called the kelvin, K.* Thus,

$$T = \theta + T_0$$

In terms of T, the Law of Gay-Lussac, (1.16), becomes

*It is read, for example, simply as "200 kelvin" or "200 K."

$$V = V_0\left[1 + \frac{(T - T_0)}{T_0}\right]$$

$$V = \frac{V_0 T}{T_0}$$

(1.17)

Extensive work at very low temperatures, from 0 to 20 K, revealed serious inconveniences in the definition of a temperature scale based on two fixed points. Despite the most earnest attempts, it was evidently impossible to obtain a measurement of the ice point (at which water and ice are in equilibrium under 1 atm pressure) accurate to within better than a few hundreths of a kelvin. Many years of careful work yielded results from 273.13 to 273.17 K.

Therefore, in 1954, the Tenth Conference of the International Committee on Weights and Measures, meeting in Paris, defined a temperature scale with only one fundamental fixed point, and with an arbitrary choice of a universal constant for the temperature at this point. The point chosen was the *triple point of water*, at which water, ice, and water vapor are simultaneously in equilibrium. The temperature of this point was set equal to 273.16 K. The value of the ice point then became 273.15 K. The boiling point of water is not fixed by convention, but is simply an experimental point to be measured with the best accuracy possible.

11. Definition of the Mole

In accord with the latest recommendations of the International Union of Pure and Applied Chemistry, we shall consider the *amount of substance n* to be one of the basic physicochemical quantities. The SI unit of amount of substance is the *mole* (abbreviated *mol*). The mole is the amount of substance of a system containing as many elementary units as there are carbon atoms in exactly 0.012 kg of carbon-12. The elementary unit must be specified and may be an atom, a molecule, an ion, an electron, a photon, etc., or a specified group of such entities. Examples are as follows:

1. One mole of $HgCl$ has a mass of 0.23604 kg.
2. One mole of Hg_2Cl_2 has a mass of 0.47208 kg.
3. One mole of Hg has a mass of 0.20059 kg.
4. One mole of $Cu_{0.5}Zn_{0.5}$ has a mass of 0.06446 kg.
5. One mole of $Fe_{0.91}S$ has a mass of 0.08288 kg.
6. One mole of e^- has a mass of 5.4860×10^{-7} kg.
7. One mole of a mixture containing 78.09 mol % N_2, 20.95 mol % O_2, 0.93 mol % Ar, and 0.03 mol % CO_2, has a mass of 0.028964 kg.

12. Equation of State of an Ideal Gas

Any two of the three variables P, V, and T suffice to specify the state of a given amount of gas and to fix the value of the third variable. Equation (1.14) is an expression for the variation of P with V at constant T, and (1.17) is an expression for the variation of V with T at constant P.

$$PV = \text{const.} \quad (\text{constant } T)$$

$$\frac{V}{T} = \text{const.} \quad (\text{constant } P)$$

We can readily combine these two relations to give

$$\frac{PV}{T} = \text{const.} \tag{1.18}$$

It is evident that this expression contains both the others as special cases.

The next problem is the evaluation of the constant in (1.18). The equation states that the product PV divided by T is always the same for all specified states of the gas; hence, if we know these values for any one state we can derive the value of the constant. Let us take this reference state to be an ideal gas at 1 atm pressure (P_0) and 273.15 K. The volume under these conditions is 22 414 cm³·mol⁻¹ (V_0/n). According to Avogadro's Law (Sect. 4.2), this volume is the same for all ideal gases. If we have n moles of ideal gas in the reference state, (1.18) becomes

$$\frac{PV}{T} = \frac{P_0 V_0}{T_0} = \frac{(1 \text{ atm})(n)(22 \text{ 414 cm}^3)}{273.15 \text{ K}} = 82.057n \text{ cm}^3 \cdot \text{atm} \cdot \text{K}^{-1}$$

$$\frac{PV}{T} = nR$$

The constant R is called the *gas constant per mole*. We often write this equation as

$$PV = nRT \tag{1.19}$$

Equation (1.19) is called the *equation of state of an ideal gas* and is one of the most useful relations in physical chemistry. It contains the three gas laws: those of Boyle, Gay-Lussac, and Avogadro.

We first obtained the gas constant R in units of cm³·atm·K⁻¹ mol⁻¹. Note that cm³·atm has the dimensions of energy. Some convenient values of R in various units are summarized in Table 1.1.

TABLE 1.1
Values of the Ideal Gas Constant R in Various Units

(SI):	$J \cdot K^{-1} \cdot mol^{-1}$	8.31431
	$cal \cdot K^{-1} \cdot mol^{-1}$	1.98717
	$cm^3 \cdot atm \cdot K^{-1} \cdot mol^{-1}$	82.0575
	$l \cdot atm \cdot K^{-1} \cdot mol^{-1}$	0.0820575

Equation (1.19) allows us to calculate the molar mass M of a gas from measurements of its density. If a mass of gas m is weighed in a gas density bulb of volume V, the density $\rho = m/V$. The amount of gas $n = m/M$. Therefore, from (1.19),

$$M = \frac{RT\rho}{P}$$

13. The Equation of State and *PVT* Relationships

If P and V are chosen as independent variables, the temperature of a given amount n of a pure substance is some function of P and V. Thus, if $V_m = V/n$,

$$T = f(P, V_m) \tag{1.20}$$

For any fixed value of T, this equation defines an *isotherm* of the substance under consideration. The state of a substance in thermal equilibrium can be fixed by specifying any two of the three variables, pressure, molar volume, and temperature. The third variable can then be found by solving (1.20). Equation (1.20) is a general form of the *equation of state*. If we do not care to specify a particular independent variable, it could be written as $g(P, V_m, T) = 0$; for example, $(PV - nRT) = 0$.

Geometrically considered, the state of a pure fluid in equilibrium can be represented by a point on a three-dimensional surface, described by the variables P, V, and T. Figure 1.6(a) shows such a *PVT* surface for an ideal gas. The isothermal lines connecting points at constant temperature are shown in Fig. 1.6(b), projected on the *PV* plane. The projection of lines of constant volume on the *PT* plane gives the *isochores* or *isometrics* shown in Fig. 1.6(c). Of course, for a nonideal gas, these would not be straight lines. Constant pressure lines are called *isobars*.

The slope of an isobaric curve gives the rate of change of volume with temperature at the constant pressure chosen. This slope is therefore written $(\partial V/\partial T)_P$. It is a partial derivative because V is a function of the two variables T and P. The fractional change in V with T is α, the *thermal expansivity*.

$$\alpha = \frac{1}{V}\left(\frac{\partial V}{\partial T}\right)_P \tag{1.21}$$

Note that α has the dimensions of T^{-1}.

In a similar way, the slope of an isothermal curve gives the variation of volume with pressure at constant temperature. We define β, the *isothermal compressibility* of a substance, as

$$\beta = \frac{-1}{V}\left(\frac{\partial V}{\partial P}\right)_T \tag{1.22}$$

The negative sign is introduced because an increase in pressure decreases the volume, so that $(\partial V/\partial P)_T$ is negative. The dimensions of β are those of P^{-1}.

Since the volume is a function of both T and P, a differential change in volume can be written*

$$dV = \left(\frac{\partial V}{\partial T}\right)_P dT + \left(\frac{\partial V}{\partial P}\right)_T dP \tag{1.23}$$

Equation (1.23) can be illustrated graphically by Fig. 1.7, which shows a section

---

*W. A. Granville, P. F. Smith, W. R. Longley, *Elements of Calculus* (Boston: Ginn & Company, 1957), p. 445. The total differential of a function of several independent variables is the sum of the differential changes that would be caused by changing each variable separately. This is true because a change in one variable does not influence the change in another independent variable.

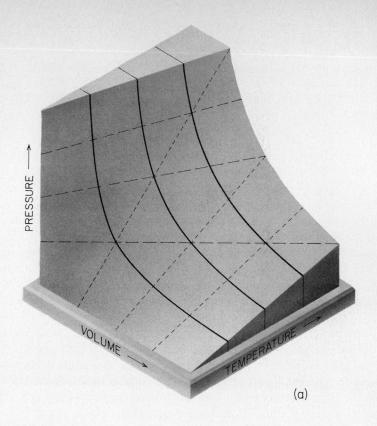

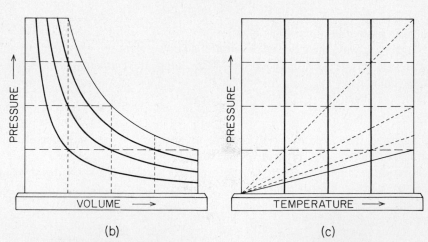

(b)

(c)

FIG. 1.6 (a) *PVT* surface for an ideal gas. The solid lines are isotherms, the dashed lines are isobars, and the dotted lines are
isometrics. (b) Projection of the *PVT* surface on *PV* plane,
showing isotherms. (c) Projection of the *PVT* surface on
the *PT* plane, showing isometrics. [After F. W. Sears, *An
Introduction to Thermodynamics* (Cambridge, Mass.: Addison-Wesley, 1950).]

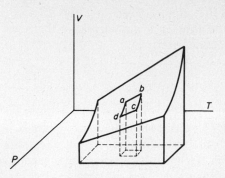

FIG. 1.7 The surface $V(P, T)$ to illustrate the partial derivatives,

$$\left(\frac{\partial V}{\partial T}\right)_P \text{ and } \left(\frac{\partial V}{\partial P}\right)_T$$

of the PVT surface plotted with V as the vertical axis. The area $abcd$ represents an infinitesimal element of surface area, cut out from the surface by planes parallel to the VT plane and the VP plane. Suppose we start with a state of the gas specified by point a, corresponding to values of the state variables specified by V_a, P_a, and T_a. Now suppose we change both P and T by infinitesimal amounts to $P + dP$ and $T + dT$. The new state of the system is represented by the point c. The change in V is

$$dV = V_c - V_a = (V_b - V_a) + (V_c - V_b)$$

We see, however, that $V_b - V_a$ is the change in V that would occur if the P is kept constant and only the temperature changed. The slope of the line ab is, therefore,

$$\operatorname*{Lim}_{\substack{T \to 0 \\ (P \text{ const.})}} \frac{V_b - V_a}{T_b - T_a} = \left(\frac{\partial V}{\partial T}\right)_P$$

The infinitesimal change $V_b - V_a$ is, therefore, $(\partial V/\partial T)_P \, dT$. In the same way, we can see that $V_c - V_b$ is $(\partial V/\partial P)_T \, dP$. Hence, the total change in V becomes the sum of these partial changes, as shown in (1.23). For such infinitesimal changes as shown in Fig. (1.7), it makes no difference which partial change is considered first.

We can derive an interesting relation between the partial differential coefficients. By solving (1.23), we have

$$dP = \frac{1}{(\partial V/\partial P)_T} \, dV - \frac{(\partial V/\partial T)_P}{(\partial V/\partial P)_T} \, dT$$

But also, by analogy with (1.23), which is a general form,

$$dP = \left(\frac{\partial P}{\partial V}\right)_T dV + \left(\frac{\partial P}{\partial T}\right)_V dT$$

The coefficients of dT must be equal, so that

$$\left(\frac{\partial P}{\partial T}\right)_V = \frac{-(\partial V/\partial T)_P}{(\partial V/\partial P)_T} = \frac{\alpha}{\beta} \tag{1.24}$$

The variation of P with T for any substance can therefore be readily calculated if we know α and β. An interesting example is suggested by a common laboratory accident, the breaking of a mercury-in-glass thermometer by overheating. If a thermometer is exactly filled with mercury at 50°C, what pressure will be developed within the thermometer if it is heated to 52°C? For mercury, under these conditions, $\alpha = 1.8 \times 10^{-4}°C^{-1}$, $\beta = 3.9 \times 10^{-6}$ atm^{-1}. Therefore, $(\partial P/\partial T)_V = \alpha/\beta$ = 46 atm·°C^{-1}. For $\Delta T = 2°C$, $\Delta P = 92$ atm. It is not surprising that even a little overheating will break the usual thermometer.

14. *PVT* Behavior of Real Gases

The pressure, volume, temperature (*PVT*) relationships for gases, liquids, and solids would preferably all be succinctly summarized in the form of equations of state of the general type of (1.20). Only in the case of gases has there been much progress in the development of these state equations. They are obtained by correlation of empirical *PVT* data, and also from theoretical considerations based on atomic and molecular structure. These theories are farthest advanced for gases, but more recent developments in the theory of liquids and solids give promise that suitable state equations may eventually be available in these fields also.

The ideal gas equation $PV = nRT$ describes the *PVT* behavior of real gases only to a first approximation. A convenient way of showing the deviations from ideality is to write, for the real gas,

$$PV = znRT \qquad (1.25)$$

The factor z is called the *compressibility factor*. It is equal to PV/nRT. For an ideal gas, $z = 1$, and departure from ideality will be measured by the deviation of the compressibility factor from unity. The extent of deviations from ideality depends on the temperature and pressure, so that z is a function of T and P. Some compressibility-factor curves are shown in Fig. 1.8; these are determined from

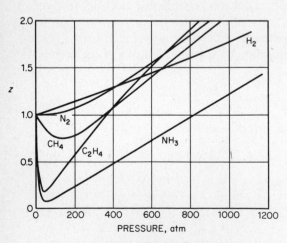

FIG. 1.8 Compressibility factors $z = PV/nRT$ at 0°C.

experimental measurements of the volumes of the substances at different pressures. (The data for NH_3 and C_2H_4 at high pressures pertain to the liquid substances.)

15. Law of Corresponding States

Let us consider a liquid at some temperature and pressure at which it is in equilibrium with its vapor. This equilibrium pressure is called the *vapor pressure* of the liquid. The liquid will be more dense than the vapor, and if we have a sample of the substance in a closed transparent tube, we can see a meniscus between liquid and vapor, indicating the coexistence of the two phases. Above a certain temperature, however, called the critical temperature T_c, only one phase exists, no matter how great the pressure applied to the system. A substance above its T_c is said to be in the *fluid state*.

The pressure that would just suffice to liquefy the fluid at T_c is the *critical pressure* P_c. The molar volume occupied by the substance at T_c and P_c is its *critical volume* V_c. These critical constants for various substances are collected in Table 1.2.

TABLE 1.2
Critical Point Data and van der Waals Constants

Formula	$T_c(K)$	P_c(atm)	V_c(cm$^3\cdot$mol^{-1})	$10^{-4}a$ (cm$^6\cdot$atm$\cdot$mol^{-2})	b(cm$^3\cdot$mol^{-1})
He	5.3	2.26	61.6	3.41	23.7
H$_2$	33.3	12.8	69.7	24.4	26.6
N$_2$	126.1	33.5	90.0	139	39.1
CO	134.0	34.6	90.0	149	39.9
O$_2$	154.3	49.7	74.4	136	31.8
C$_2$H$_4$	282.9	50.9	127.5	447	57.1
CO$_2$	304.2	72.8	94.2	359	42.7
NH$_3$	405.6	112.2	72.0	417	37.1
H$_2$O	647.2	217.7	55.44	546	30.5
Hg	1735.0	1036.0	40.1	809	17.0

The ratios of P, V, and T to the critical values P_c, V_c, and T_c are called the *reduced* pressure, volume, and temperature. These reduced variables may be written

$$P_R = \frac{P}{P_c}, \qquad V_R = \frac{V}{V_c}, \qquad T_R = \frac{T}{T_c} \qquad (1.26)$$

In 1881, van der Waals pointed out that to a fairly good approximation, especially at moderate pressures, all gases would follow the same equation of state in terms of reduced variables, P_R, T_R, V_R—i.e., $V_R = f(P_R, T_R)$. He proposed to call this rule the *Law of Corresponding States*. If this "law" were true, the *critical ratio* $P_c V_c/RT_c$ would be the same for all gases. Actually, as you can readily confirm from the data in Table 1.2, the ratio varies from about 0.2 to 0.33 for the common gases.

Chemical engineers and other workers actively concerned with the properties of gases at elevated pressures have prepared extensive and useful graphs to show the variation of the compressibility factor z in (1.25) with P and T, and they have found that to a good approximation, even at fairly high pressures, z appears to be a universal function of P_R and T_R,

$$z = f(P_R, T_R) \tag{1.27}$$

This rule is illustrated in Fig. 1.9 for a number of different gases, where $z = PV/nRT$ is plotted at various reduced temperatures, against the reduced pressure. At these moderate pressures, the fit is good to within about 1%.

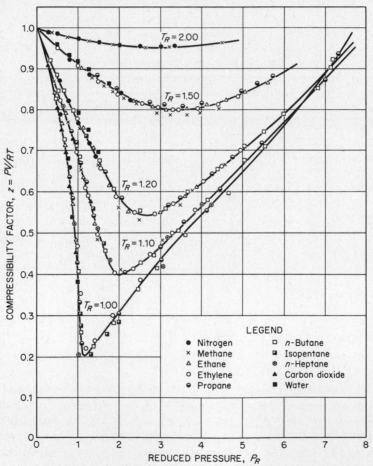

FIG. 1.9   Compressibility factor as function of reduced state variables.
[Gouq-Jen Su, *Ind. Eng. Chem.*, *38*, 803 (1946).]

16. Equations of State for Gases

If the equation of state is written in terms of reduced variables, as $F(P_R, V_R) = T_R$, it is evident that it contains at least two independent constants, characteristic of the gas in question—e.g., P_c and V_c. Many equations of state, proposed on semi-empirical grounds, serve to represent the PVT data more accurately than does the ideal gas equation. Several of the best known of these also contain two added constants. For example, the equation of van der Waals,

$$\left(P + \frac{n^2 a}{V^2}\right)(V - nb) = nRT \tag{1.28}$$

and the equation of D. Berthelot,

$$\left(P + \frac{n^2 A}{TV^2}\right)(V - nB) = nRT \tag{1.29}$$

The van der Waals equation provides a reasonably good representation of the PVT data of gases in the range of moderate deviations from ideality. For example, consider the following values in liter atmospheres of the PV product for 1 mol of carbon dioxide at 40°C, observed experimentally and as calculated from van der Waals' equation. We have written $V_m = V/n$ for the volume per mole.

P, atm	1	10	50	100	200	500	1100
PV_m, obs.	25.57	24.49	19.00	6.93	10.50	22.00	40.00
PV_m, calc.	25.60	24.71	19.75	8.89	14.10	29.70	54.20

The constants a and b are evaluated by fitting the equation to experimental PVT measurements, or from the critical constants of the gas. Some values for van der Waals' a and b are included in Table 1.2. Berthelot's equation is somewhat better than van der Waals' at pressures not much above 1 atm, and is preferred for general use in this range.

17. The Critical Region

The behavior of a gas in the neighborhood of its critical region was first studied by Thomas Andrews in 1869 in a classic series of measurements on carbon dioxide. Results of determinations by A. Michels of these PV isotherms around the critical temperature of 31.01°C are shown in Fig. 1.10.

Consider the isotherm at 30.4°C, which is below T_c. As the vapor is compressed, the PV curve first follows AB, which is approximately a Boyle's Law isotherm. When the point B is reached, a meniscus appears and liquid begins to form. Further compression then occurs at constant pressure until the point C is reached, at which all the vapor has been converted into liquid. The curve CD is the isotherm of liquid carbon dioxide, its steepness indicating the low compressibility of the liquid.

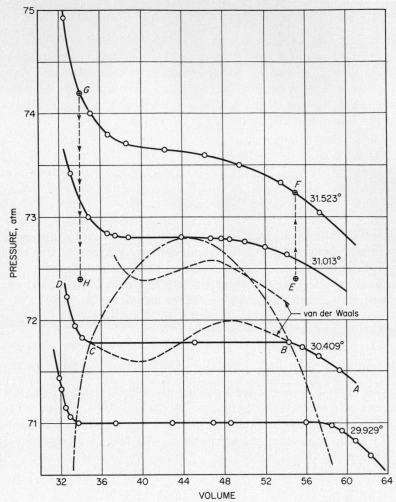

FIG. 1.10 Isotherms of carbon dioxide near the critical point. [Michels, Blaisse, and Michels, *Proc. Roy. Soc. A. 160*, 367 (1937).]

As isotherms are taken at successively higher temperatures, the points of discontinuity *B* and *C* are observed to approach each other gradually, until at 31.01°C they coalesce, and no appearance of a second phase is observable. This isotherm corresponds to the critical temperature of carbon dioxide. Isotherms above this temperature exhibit no formation of a second phase no matter how great the applied pressure.

Above the critical temperature, there is no reason to draw any distinction between liquid and vapor, since there is a complete *continuity of states*. This may be demonstrated by following the path *EFGH*. The vapor at point *E*, at a temperature below T_c, is warmed at constant volume to point *F*, above T_c. It is then compressed along the isotherm *FG*, and finally cooled at constant volume along *GH*. At the point *H*, below T_c, the carbon dioxide exists as a liquid, but at no point along this path are two phases, liquid and vapor, simultaneously present. One must conclude that the transformation from vapor to liquid occurs smoothly and continuously.

18. The van der Waals Equation and Liquefaction of Gases

The van der Waals equation provides a reasonably accurate representation of the *PVT* data of gases under conditions that deviate only moderately from ideality. When we apply the equation to gases in states that depart greatly from ideality, we do not obtain a quantitative representation of the data, but we still get an interesting qualitative picture. A typical example is shown in Fig. 1.10, where the van der Waals isotherms, drawn as dashed lines, are compared with the experimental isotherms for carbon dioxide in the neighborhood of the critical point. The van der Waals equation provides an adequate representation of the isotherms for the homogeneous vapor and even for the homogeneous liquid.

As might be expected, the equation cannot represent the discontinuities arising during liquefaction. Instead of the experimental straight line, it exhibits a maximum and a minimum within the two-phase region. We note that as the temperature gradually approaches the critical temperature, the maximum and the minimum gradually approach each other. At the critical point itself, they have merged to become a point of inflection in the *PV* curve. The analytical condition for a maximum is that $(\partial P/\partial V)_T = 0$ and $(\partial^2 P/\partial V^2)_T < 0$; for a minimum, $(\partial P/\partial V)_T = 0$ and $(\partial^2 P/\partial V^2)_T > 0$. At the point of inflection, both the first and the second derivatives vanish, $(\partial P/\partial V)_T = 0 = (\partial^2 P/\partial V^2)_T$.

According to van der Waals' equation, therefore, the following three equations must be satisfied simultaneously at the critical point $(T = T_c,\ V = V_c,\ P = P_c)$ for 1 mol of gas, $n = 1$:

$$P_c = \frac{RT_c}{V_c - b} - \frac{a}{V_c^2}$$

$$\left(\frac{\partial P}{\partial V}\right)_T = 0 = \frac{-RT_c}{(V_c - b)^2} + \frac{2a}{V_c^3}$$

$$\left(\frac{\partial^2 P}{\partial V^2}\right)_T = 0 = \frac{2RT_c}{(V_c - b)^3} - \frac{6a}{V_c^4}$$

When these equations are solved for the critical constants, we find

$$T_c = \frac{8a}{27bR}, \qquad V_c = 3b, \qquad P_c = \frac{a}{27b^2} \tag{1.30}$$

Values for the van der Waals constants and for *R* can be calculated from these equations. We prefer, however, to consider *R* as a universal constant, and to obtain the best fit by adjusting *a* and *b* only. Then (1.30) would yield the relation $P_c V_c/T_c = 3R/8$ for all gases.

In terms of the reduced variables of state, P_R, V_R, and T_R, one obtains from (1.30),

$$P = \frac{a}{27b^2} P_R, \qquad V = 3bV_R, \qquad T = \frac{8a}{27Rb} T_R$$

The van der Waals equation then reduces to

$$\left(P_R + \frac{3}{V_R^2}\right)\left(V_R - \frac{1}{3}\right) = \frac{8}{3} T_R \tag{1.31}$$

A reduced equation of state similar to (1.31) can be obtained from an equation of state containing no more than three arbitrary constants, such as a, b, and R, provided it has an algebraic form capable of giving a point of inflection. Berthelot's equation is often used in the following form, applicable at pressures of the order of 1 atm:

$$P_R V_R = nR'T_R\left[1 + \frac{9}{128}\frac{P_R}{T_R}\left(1 - \frac{6}{T_R^2}\right)\right] \tag{1.32}$$

where $R' = R(T_c/P_c V_c)$.

19.   Other Equations of State

To represent the behavior of gases with greater accuracy, especially at high pressures or near their condensation temperatures, we must use expressions having more than two adjustable parameters. Consider, for example, a *virial equation* similar to that given by Kammerlingh–Onnes in 1901:

$$\frac{PV}{nRT} = 1 + \frac{B(T)n}{V} + \frac{C(T)n^2}{V^2} + \frac{D(T)n^3}{V^3} + \cdots \tag{1.33}$$

Here, B, C, D, etc., which are functions of temperature, are called the second, third, fourth etc., *virial coefficients*. Figure 1.11 shows the second virial coefficients B of several gases over a range of temperature. This B is an important property in theoretical calculations on imperfect gases.*

The virial equation can be extended to as many terms as are needed to represent the experimental PVT data to any desired accuracy. It may also be extended to mixtures of gases, and in such cases gives important data on the effects of intermolecular forces between the same and different molecules.†

One of the best of the empirical equations is that proposed by Beattie and Bridgeman.‡ It contains five constants in addition to R, and fits the PVT data over a wide range of pressures and temperatures, even near the critical point, to within 0.5%. An equation with eight constants, which is based on a reasonable model of

---

*See, for example, T. L. Hill, *Introduction to Statistical Thermodynamics* (Reading, Mass.: Addison-Wesley Publishing Co., Inc., 1960), Ch. 15.

†An example is the system methane-tetrafluoromethane, studied by D. R. Douslin, R. H. Harrison, R. T. Moore, *J. Phys. Chem.*, *71*, 3477 (1967). This paper illustrates the continuing interest in experimental work on fundamental properties of gases (see Table 1.3).

‡J. A. Beattie and O. C. Bridgeman, *Proc. Am. Acad. Arts. Sci.*, *63*, 229 (1928).

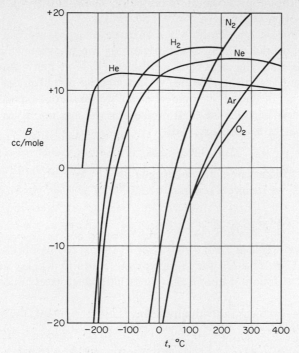

B
cc/mole

t, °C

FIG. 1.11   The second virial coefficients B of several gases as functions of temperature.

dense fluids, has been described that reproduces the isotherms in the liquid region quite well.*

20.   Mixtures of Ideal Gases

A mixture can be specified by stating the amount of each component substance that it contains, $n_1, n_2, \ldots, n_j \ldots$. The total amount of all components is

$$n = \sum_{j=1}^{c} n_j$$

The composition of the mixture is then conveniently described by stating the *mole fraction* X_j of each component,

$$X_j = \frac{n_j}{n} = \frac{n_j}{\sum n_j} \tag{1.34}$$

Another method of specifying composition is the *concentration*,

$$c_j = \frac{n_j}{V} \tag{1.35}$$

*M. Benedict, G. W. Webb, L. C. Rubin, *J. Chem. Phys.*, *10*, 747 (1942).

The SI unit of concentration is the mole per cubic metre, but the mole per cubic decimetre is more often used.*

If the system under consideration is a mixture of gases, we can define the *partial pressure P_j* of any particular component as the pressure which that gas would exert if it occupied the total volume all by itself. If we know the concentration of a component gas in the mixture, we can find its partial pressure from its *PVT* data or equation of state. For nonideal gases, we would not expect the sum of these partial pressures to equal the total pressure of the mixture. Even if each gas individually behaves as an ideal gas with

$$P_j = c_j RT \tag{1.36}$$

it is possible that specific interactions between unlike gases would cause $\sum P_j$ to differ from P. Thus, we require a separate definition of an *ideal gas mixture* as one in which

$$P = P_1 + P_2 + \cdots + P_c = \sum P_j \tag{1.37}$$

This is called *Dalton's Law of Partial Pressures*. It simply represents a special kind of gas mixture. It is called a *law* for the historical reason that many gas mixtures at ordinary T and P follow it about as well as individual gases follow the ideal gas law.

In the event that each gas individually behaves as an ideal gas, the system is an *ideal mixture of ideal gases*:

$$P = RT(c_1 + c_2 + \cdots + c_c) = RT \sum c_j$$

$$P = \frac{RT}{V} \sum n_j$$

Since

$$P_j = \frac{RT}{V} n_j$$

we have

$$P_j = X_j P \tag{1.38}$$

The partial pressure of each gas in an ideal mixture of ideal gases is equal to its mole fraction times the total pressure.

21. Mixtures of Nonideal Gases

The *PVT* behavior of a mixture of gases at any given constant composition can be determined just as that for a single pure gas. The data can then be fitted to an equation of state. When such data are obtained for mixtures of different compositions, we find that the parameters of the state equations depend on the composition of the mixture.

The virial equation is most suitable for representing the *PVT* properties of gas

---

*The *liter* (*l*) is defined as 10^{-3} m^3 or 1 dm^3. A solution with a concentration of (e.g.) 1.63 mol·dm^{-3} is often called a *1.63 molar solution*.

mixtures, since theoretical relations between the coefficients can be obtained from statistical thermodynamics. For instance, the second virial coefficient B_m of a binary gas mixture is

$$B_m = X_1^2 B_{11} + 2X_1 X_2 B_{12} + X_2^2 B_{22} \qquad (1.39)*$$

where X_1 and X_2 are mole fractions of the two components. The coefficient B_{12} represents the contribution to the second virial coefficient that is due to specific interactions between unlike gases. Table 1.3 is an example of precise data on second virial coefficients.

TABLE 1.3
Second Virial Coefficients for Equimolar Mixtures of CH_4 and CF_4

K	$B_1(CH_4)$ $(cm^3 \cdot mol^{-1})$	$B_2(CF_4)$ $(cm^3 \cdot mol^{-1})$	B_{12} $(cm^3 \cdot mol^{-1})$
273.15	-53.35	-111.00	-62.07
298.15	-42.82	-88.30	-48.48
323.15	-34.23	-70.40	-37.36
348.15	-27.06	-55.70	-28.31
373.15	-21.00	-43.50	-20.43
423.15	-11.40	-24.40	-8.33
473.15	-4.16	-10.10	$+1.02$
523.15	$+1.49$	$+1.00$	8.28
573.15	5.98	9.80	14.10
623.15	9.66	17.05	18.88

22. The Concepts of Heat and Heat Capacity

The experimental observations that led to the concept of *temperature* also led to that of *heat*, but for a long time students did not clearly distinguish between these two concepts, often using the same name for both, *calor* or *caloric*.

The beautiful work of Joseph Black on *calorimetry*, the measurement of heat changes, was published in 1803, four years after his death. In his *Lectures on the Elements of Chemistry*, he pointed out the distinction between the intensive factor, *temperature*, and the extensive factor, *quantity of heat*. Black showed that equilibrium required an equality of *temperature* and did not imply that there was an equal "quantity of heat" in different bodies.

He then proceeded to investigate the capacity for heat or the amount of heat needed to increase the temperature of different bodies by a given number of degrees.

> It was formerly a supposition that the quantities of heat required to increase the heat of different bodies by the same number of degrees were directly in proportion to the quantity of matter in each. . . . But very soon after I began to think on this subject (Anno 1760) I perceived that this opinion was a mistake, and that the quantities of heat which different kinds of matter must receive to reduce them to an equilibrium with one another, or to raise their temperatures by an equal number of

*J. E. Lennard-Jones and W. R. Cook, *Proc. Roy. Soc.* (*London*), *A 115*, 334 (1927).

degrees, are not in proportion to the quantity of matter of each, but in proportions widely different from this, and for which no general principle or reason can yet be assigned.

In explaining his experiments, Black assumed that heat behaved as a substance, which could flow from one body to another but whose total amount must always remain constant. This idea of heat as a substance was generally accepted at that time. Lavoisier even listed *caloric* in his "Table of the Chemical Elements." In the particular kind of experiment often done in *calorimetry*, heat does, in fact, behave much like a weightless fluid, but this behavior is the consequence of certain special conditions. Consider a typical experiment: A piece of metal of mass m_2 and temperature T_2 is introduced into an insulated vessel containing a mass m_1 of water at temperature T_1. We impose the following conditions: (1) the system is isolated from its surroundings; (2) any change in the container itself can be neglected; (3) there is no change such as vaporization, melting, or solution in either substance, and no chemical reaction. Under these strict conditions, the system finally reaches a new temperature T, somewhere between T_1 and T_2, and the temperatures are related by an equation of the form

$$c_2 m_2 (T_2 - T) = c_1 m_1 (T - T_1) \tag{1.40}$$

Here, c_2 is the *specific heat* of the metal and $c_2 m_2 = C_2$ is the *heat capacity* of the mass of metal used. The corresponding quantities for the water are c_1, and $c_1 m_1 = C_1$. The specific heat is the heat capacity per unit mass.

Equation (1.40) has the form of an equation of conservation like (1.12). Under the restrictive conditions of this experiment, it is permissible to consider that heat is conserved, and flows from the hotter to the colder substance until their temperatures are equal. The flow of heat is

$$q = C_2 (T_2 - T) = C_1 (T - T_1) \tag{1.41}$$

A more exact definition of heat will be given in the next chapter.

The *unit of heat* was originally defined in terms of such an experiment in calorimetry. The *gram calorie* was the heat that must be absorbed by 1 g of water to raise its temperature 1°C. It followed that the specific heat of water was 1 cal per °C.

More careful experiments showed that the specific heat was itself a function of the temperature. It therefore became necessary to redefine the calorie by specifying the range over which it was measured. The standard was taken to be the 15° calorie, probably because of the lack of central heating in European laboratories. This is the heat required to raise the temperature of 1 g of water from 14.5 to 15.5°C. Finally, another change in the definition of the calorie was found to be desirable. Electrical measurements are capable of greater precision than calorimetric measurements. The Ninth International Conference on Weights and Measures (1948) therefore recommended that the *joule* (*volt coulomb*) be used as the unit of heat.* The SI unit of heat capacity is the joule per kelvin $(J \cdot K^{-1})$.

---

*The *calorie*, however, is still popular among chemists, and the National Bureau of Standards uses a *defined calorie* equal to exactly 4.1840 J. The Bureau plans to discontinue use of the calorie in its publications in 1972.

The heat capacity, being a function of temperature, should be defined precisely only in terms of a differential heat flow dq and temperature change dT. Thus, in the limit, (1.41) becomes

$$dq = C\,dT \quad \text{or} \quad C = \frac{dq}{dT} \tag{1.42}$$

23. Work in Changes of Volume

In our discussion of the transfer of heat, we have so far carefully restricted our attention to the simple case in which the system is isolated and is not allowed to interact mechanically with its surroundings. If this restriction does not apply, the system may either do work on its surroundings or have work done on itself. Thus, in certain cases, only a part of the heat added to a substance causes its temperature to rise, the remainder being used in the work of expanding the substance. The amount of heat that must be added to produce a certain temperature change depends on the exact process by which the change is effected.

A differential element of work was defined in (1.3) as $dw = F\,dr$, the product of a force and the displacement of its point of application when both have the same direction. Figure 1.3 showed a simple thermodynamic system, a fluid confined in a cylinder with a movable piston that is assumed to be frictionless. The external pressure on the piston of area A is $P_{ex} = F/A$. If the piston is displaced a distance dr in the direction of the force F, the element of work is

$$dw = \frac{F}{A} \cdot A\,dr = P_{ex}\,A\,dr = P_{ex}\,dV$$

This is the work done by the force. In mechanics, the work is always associated with the force. It does not matter what the force acts upon—a mass point, a collection of mass points, or a continuous body or system. Given the forces and the displacements of their points of application, we can calculate the work.

In thermodynamics, however, we focus attention upon the system (a definite enclosed part of the world) and its surroundings (the rest of the world). We speak of the work done *on* the system, and the work done *by* the system on its surroundings. We adopt the international convention that work done *on* the system is positive and work done *by* the system is negative.* Therefore, we write for the *work done on the system*,

$$dw = -P_{ex}\,dV \tag{1.43}$$

Since dV is negative for a compression, the work done by an external force acting on the system is positive in accord with our convention.

Note that the calculation of the work requires that we know the *external pressure* P_{ex} on the system. It does not require that the system be in equilibrium with this external pressure. If the pressure is *kept constant* during a finite compres-

---

*Some authors adopt the opposite convention for historical reasons, but our convention is in accord with the *general acquisitive convention* of thermodynamics that regards all signed quantities from the point of view of the system: You stand inside the system and call anything you see coming into the system *positive*.

sion from V_1 to V_2, we can calculate the work done on the fluid by integrating (1.43):

$$w = \int_{V_1}^{V_2} - P_{ex}\, dV = -P_{ex} \int_{V_1}^{V_2} dV = -P_{ex}(V_2 - V_1) = -P_{ex}\, \Delta V \quad (1.44)$$

If a finite change in volume is carried out in such a way that the external pressure is known at each successive state of expansion or compression, we can plot the process on a graph of P_{ex} vs. volume V. Such a plot is called an *indicator diagram*; an example is shown in Fig. 1.12(a). The work done by the system is equal to the area under the curve.

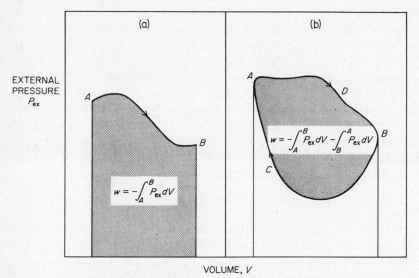

FIG. 1.12   Indicator diagrams for *PV* work. (a) A general process from *A* to *B*. (b) A cyclic process *ADBCA*.

It is evident that the work done in going from point *A* to point *B* in the $P_{ex} - V$ diagram depends upon the particular path that is traversed. Consider, for example, two alternate paths from *A* to *B* in Fig. 1.12 (b). More work will be done in going by the path *ADB* than by the path *ACB*, as is evident from the greater area under curve *ADB*. If we proceed from state *A* to state *B* by path *ADB* and return to *A* along *BCA*, we shall have completed a *cyclic process*. The net work done by the system during this cycle is seen to be equal to the difference between the areas under the two paths, which is the shaded area in Fig. 1.12(b).

In thermodynamic discussion, we must always be careful to define carefully what we mean by the system and by its surroundings. In the discussion of Fig. 1.3, we tacitly assumed that the piston was weightless and that it operated without friction. Thus, the system was the gas, and the piston and cylinder were treated as idealized boundaries that could be neglected in our consideration of work terms. Suppose, on the other hand, we had a real cylinder with a creaky piston generating considerable friction with the cylinder walls. Then we should need to specify carefully whether the piston and cylinder were to be included in the system or as part of the surroundings. We could do a lot of work on the piston, of which only a fraction would be done on the gas, the rest being dissipated as frictional heat by the creaky piston.

If each successive point along the $P_{ex} - V$ curve is an equilibrium state of the system, we have the very special case that P_{ex} always equals P, the pressure of the fluid itself. The indicator curve then becomes an equilibrium curve for the system. Such a case is shown in Fig. 1.13. Only when equilibrium is maintained can the work be calculated from functions of the state of the substance itself, P and V.

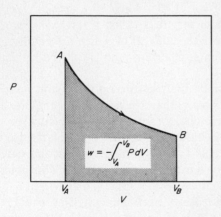

FIG. 1.13 Indicator diagram for the work done on a system consisting of a fluid in equilibrium, with an external pressure $P_{ex} = P$.

24. General Concept of Work

In the mechanical systems described, the work was always formulated as the product of two terms, an intensive factor, which is a generalized force, and an extensive factor, which is a generalized displacement. Such a formulation applies also to nonmechanical work.

In physical chemistry, we are often interested in changes carried out in electrical cells. We shall postpone a detailed description of such systems until Chapter 11, but mention now that in the case of electrical work, the generalized force becomes the electromotive force (emf) E of the cell, and the generalized displacement be-

TABLE 1.4
Examples of Work

Intensive factor	Extensive factor	Element of work dw
Tension f	Distance l	$f \, dl$
Surface tension γ	Area A	$\gamma \, dA$
Pressure P	Volume V	$-P \, dV$
Electromotive force E	Charge Q	$E \, dQ$
Magnetic field $\mathscr{H}$	Magnetization M	$\mathscr{H} \, dM$

comes the charge dQ transferred through the external circuit as the cell discharges [$dQ < 0$]. The element of work done on the cell is $E \, dQ$.

Similarly, in the magnetic case, the intensive factor is the magnetic field strength $\mathscr{H}$. If this acts on a substance to produce a magnetization dM in the direction of the field, the work done on the substance is $\mathscr{H} \, dM$.

We can thus summarize the various examples of work in Table 1.4. Work done on a system is always taken as *positive*.

25. Reversible Processes

The path followed in the PV diagram of Fig. 1.13 belongs to a special class, of great importance in thermodynamic arguments. It is called a *reversible path*. A reversible path is one connecting intermediate states all of which are equilibrium states. A process carried out along such an equilibrium path is called a *reversible process*.

For example, to expand a gas reversibly, the pressure on the piston must be released so slowly, in the limit infinitely slowly, that at every instant the pressure everywhere within the gas volume is exactly the same and is just equal to the opposing pressure on the piston. Only in this case can the state of the gas be represented by the variables of state, P and V.* In geometric terms, the state is represented by a point in the PV plane. The line joining such points is a line joining points of equilibrium.

Consider the situation if the piston were drawn back suddenly. Gas would rush in to fill the space, pressure differences would be set up throughout the gas volume, and even a state of turbulence might ensue. The state of the gas under such conditions could no longer be represented by the two variables, P and V. Indeed, an enormous number of variables would be required, corresponding to the many different pressures at different points throughout the gas volume. Such a rapid expansion is a typical *irreversible process*; the intermediate states are no longer equilibrium states.

Reversible processes are never realizable in actuality because they must be carried out infinitely slowly. All processes that occur naturally are therefore irreversible. The reversible path is the limiting path that is reached as we carry out an irreversible process under conditions approaching more and more closely equilibrium conditions. We can define a reversible path exactly and calculate the work done in moving along it, even though we can never carry out an actual change reversibly. The conditions for reversibility can, however, be closely approximated in certain experiments.

In Fig. 1.13, the change from A to B can be carried out along different reversible paths, of which only one is drawn. These different paths are possible because the volume V is a function of the temperature T, as well as of the pressure P. If one particular temperature is chosen and held constant throughout the process, only

---

*We can represent an irreversible path on the indicator diagram by plotting P_{ex} vs. V. Only in the reversible case does $P_{ex} = P$, the state property of the substance itself.

one reversible path is possible. Under such an *isothermal condition*, the work done on the system in going from A to B via a path that is reversible is the *minimum work* possible for the particular temperature in question.* This is true because in the reversible case the expansion takes place against the maximum possible opposing force, which is one exactly in equilibrium with the driving force.

PROBLEMS

1. An evacuated glass bulb weighs 27.9214 g. Filled with dry air at 1 atm and 25°C, it weighs 28.0529 g. Filled with a mixture of methane and ethane, it weighs 28.0140 g. Calculate the mol percent of methane in the gas mixture.

2. An oil bath maintained at 50.5°C loses heat to its surroundings at the rate of 4.53 kJ·min⁻¹. Its temperature is maintained by an electrically heated coil with a resistance of 60 Ω operated from a 110 V line. A thermoregulator switches the current on and off. What fraction of the time will the current be turned on? Suppose that in order to reduce temperature fluctuations the heater is operated through a variable voltage transformer. What voltage setting would you use to keep the heater on 95% of the time?

3. Calculate the work done on a 2000 kg car in accelerating it from rest to a speed of 100 km·hr⁻¹, neglecting friction.

4. A lead bullet is fired at a wooden plank. At what speed must it travel to melt on impact, if its initial temperature is 25°C and heating of the plank is neglected? The melting point of lead is 327°C and you can assume that it obeys the Dulong and Petit rule to estimate its specific heat. The heat of fusion of lead is 5.19 kJ·mol⁻¹. Does the neglect of heating of the plank introduce a large error in your estimate of speed?

5. What is the average power production in watts of a man who burns 2500 kcal of food in a day? Estimate the average additional power production of a 75 kg man who is climbing a mountain at the rate of 20 metres per minute.

6. Calculate the pressure exerted by 20 g of nitrogen in a closed 1-liter vessel at 25°C, using (a) the ideal-gas equation; (b) the van der Waals equation. Repeat calculation for a vessel of 100 cm³ volume.

7. Water has a surface tension of 72.75 dyne·cm⁻¹ at 20°C. Calculate the minimum work that would be required to convert 1 dm³ of water into droplets having a diameter of 1 μm.

8. Draw the van der Waals PV isotherm at 200 K for ethylene, based on values of a and b in Table 1.2. Mark the limits of the two-phase region on the isotherm, given that the vapor pressure of ethylene at 200 K is 4.379 atm. Calculate the areas between the loops of the van der Waals isotherm and the flat portion of the experimental PV curve.

9. The density of solid Al at 20°C is 2.70 g·cm⁻³, of the liquid at 660°C, 2.38 g·cm⁻³. Calculate the work done on surroundings when 100 kg of Al are heated under 1 atm pressure from 20° to 660°C.

*Thus $w = -\int_{V_A}^{V_B} P \, dV$ has the largest possible negative value for the reversible case.

10. What is the gravitational force in newtons between two masses of 1 kg that are 1 mm apart? What is the gravitational force between two neutrons 10^{-12} mm apart?

11. Calculate the work done on the surroundings when one mole of water (a) freezes at 0°C; (b) boils at 100°C. Compare these values with the corresponding latent heats.

12. At 318 K and 1 atm, N_2O_4 is dissociated into $2NO_2$ to the extent of 38%. Calculate the pressure developed if a 20-liter vessel containing 1 mol N_2O_4 is heated to 318 K. Give results in both $N \cdot m^{-2}$ and atm.

13. The coefficient of thermal expansion of ethanol is given by

$$\alpha = 1.0414 \times 10^{-3} + 1.5672 \times 10^{-6}\theta + 5.148 \times 10^{-8}\theta^2$$

where θ is the centigrade temperature. If 0° and 50° are taken as fixed points on a centigrade scale, what will be the reading of an ethanol thermometer when an ideal-gas thermometer reads 30°C?

14. At high pressures and temperatures a quite good equation of state is $P(V - nb) = nRT$, where b is a constant. Find $(\partial V/\partial T)_P$ and $(\partial V/\partial P)_T$ from this equation and hence dV from Eq. (1.23). Confirm (1.24) for this equation of state.

15. Calculate the compressibility factor z for N_2 at 80 atm and 100 K from the equations of van der Waals, Berthelot, Beattie–Bridgeman, and Benedict–Webb–Rubin. The necessary constants can be found in original papers or in Ch. 4 of *Molecular Theory of Gases and Liquids* by J. O. Hirschfelder and C. F. Curtiss (New York: John Wiley & Co., 1954).

16. Assume the ideal gas law $PV = nRT$, and derive an expression for the total fractional error in the pressure as calculated from measurements of the volume and the temperature of n moles of gas, if one assumes the error in T is independent of error in V.

17. An equation of state due to Dieterici is

$$P(V - nb') \exp \frac{na'}{RTV} = nRT$$

Evaluate the constants a' and b' in terms of the critical constants P_c, V_c, and T_c of a gas.

18. The gas density of SO_2 at 273.15 K is

P(atm)	0.25	0.50	1.00
ρ(g·dm^{-3})	0.71878	1.44614	2.92655

Compare these values with those calculated from the equation of D. Berthelot. Extrapolate these data to $P = 0$ and calculate the molar mass M of SO_2.

19. The volume of 1 kg of water between $\theta = 0$ and 40°C can be represented by

$$V = 999.87 - 6.426 \times 10^{-2}\theta + 8.5045 \times 10^{-3}\theta^2 - 6.79 \times 10^{-5}\theta^3$$

Calculate the temperature at which water has its maximum density.

20. One mole of ideal gas at 25°C is held in a cylinder by a piston at a pressure of 10^7 $N \cdot m^{-2}$. The piston pressure is released in three stages, first to 5×10^6, then to 10^6, then to 10^5 $N \cdot m^{-2}$. Calculate the work done by the gas during these irreversible isothermal expansions, and compare the total work with that done in an isothermal reversible expansion from 10^7 to 10^5 $N \cdot m^{-2}$ at 25°C.

21. A kilogram of ethylene is compressed from 10^{-1} m^3 to 10^{-2} m^3 at a constant temperature of 300 K. Calculate the minimum work that must be expended, assuming that the gas is (a) ideal (b) van der Waals.

22. Calculate the minimum work needed to remove a 10^3 kg mass from the gravitational field of the earth. Neglecting friction, estimate the initial velocity at which this mass would need to leave the earth's surface in order to escape if it were fired as a projectile. Suppose, however, it was a rocket that could accelerate at a constant rate for a period of 100s; in this case, what would be the force (thrust) required of the rocket engine in order to provide for escape of the 10^3 kg mass?

23. Calculate the reversible work required to separate a charge of $+1$ coulomb from a charge of -1 coulomb in vacuum from a distance of (a) 1 mm (b) 1 nm to infinite distance apart.

24. An elastic sphere of mass 100 g with a kinetic energy of 1 J strikes another such sphere of mass 1 g which is initially at rest. What is the maximum energy E_m that can be transferred from the moving to the stationary sphere? What is E_m if the mass of moving sphere is 1 g whereas that of the stationary sphere is 100 g?

25. It is sometimes necessary to correct weighings to allow for the buoyancy effect of air. Suppose you use brass weights (density $\rho = 8.40$ g·cm^{-3}) to weigh 10.0000 g aluminum in air at 25°C and 1 atm. What is the *in vacuo* weight?

26. Show that the maximum work that can be done by n moles of ideal gas in expanding at constant temperature T from V_1 to V_2 is $w(\text{max}) = nRT \ln (V_2/V_1)$.

27. A mad scientist has synthesized a most unusual gas called zapon which follows exactly the equation of state $(P + n^2a/V^2)V = nRT$, i.e. van der Waals' equation with $b = 0$, except that a is a function of temperature such that for $T \leq T_z, a = 0$, but for $T > T_z$, $a = a_0/T$ where a_0 is a constant. Plot α and β for this gas as functions of T.

28. The surface tensions of pure liquids fit quite well the empirical equation

$$\gamma = \gamma_0(1 - T_R)^{11/9}$$

where T_R is the reduced temperature and γ_0 is a constant. For liquid CS_2 we have

θ(°C)	-60	-50	-40	-30
γ(dyne·cm^{-1})	31.2	29.2	27.3	25.4

Calculate γ_0 and T_c, and compare T_c with the experimental value 546.2 K.

29. At 273.15 K the density of nitric oxide at various pressures was found to be:

P(atm)	1.0000	0.8000	0.5000	0.3000
ρ(g·dm^{-3})	1.3402	1.0719	0.66973	0.40174

Calculate (ρ/P) at each pressure and determine an exact value for the atomic weight of nitrogen from the limiting value of (ρ/P) as $P \to 0$. Take the atomic weight of oxygen as 15.9994.

30. Calculate the volume of a balloon with a lifting force of 200 kg at 25°C and 1 atm if the balloon is filled with (a) hydrogen, (b) helium. Calculate the volume of the helium balloon in the stratosphere at -60°C and 0.1 atm. Assume that air is 80% N_2, 20% O_2.

2

Energetics

It frequently happens, that in the ordinary affairs and occupations of life, opportunities present themselves of contemplating some of the most curious operations of Nature. . . . I have frequently had occasion to make this observation; and am persuaded that a habit of keeping the eyes open to every thing that is going on in the ordinary course of the business of life has oftener led, as it were by accident, or in the playful excursions of the imagination. . . to useful doubts, and sensible schemes for investigation and improvement, than all the more intense meditations of philosophers, in the hours expressly set aside for study.

Benjamin, Count of Rumford
1798*

The First Law of Thermodynamics is an extension of the principle of the conservation of mechanical energy. Such an extension became reasonable after it was shown that expenditure of work could cause the production of heat. Thus both work and heat were seen to be entities that described the transfer of energy from one system to another. If a temperature difference exists between two systems in thermal contact, energy can be transferred from one system to the other in the form of heat. Heat transfer between open systems can also occur by transport of matter from one system to the other. Work is the form in which energy is transferred to a system by displacement of parts of the system under the action of external forces.† We should not say that heat and work are "forms of energy" since they are entities that have significance only in terms of *energy transfers* between systems. We cannot speak of the "heat of a system" or "work of a system," although we can and do speak of the "energy of a system."‡

1. History of the First Law of Thermodynamics

The first quantitative experiments on conversion of work to heat were carried out by Benjamin Thompson, a native of Woburn, Massachusetts, who became

*"An Inquiry Concerning the Source of the Heat which is Excited by Friction" (read before the Royal Society, January 25, 1798).

†J. G. Kirkwood and I. Oppenheim, *Chemical Thermodynamics* (New York: McGraw-Hill Book Company, 1961) p. 18.

‡P. W. Bridgman, *The Nature of Thermodynamics* (Harvard University Press, 1941).

Count Rumford of The Holy Roman Empire. Commissioned by the Duke of Bavaria to supervise the boring of cannon at the Munich Arsenal, he was fascinated by the generation of heat during this operation. He suggested (1798) that the heat might arise from the mechanical energy expended, and was able to estimate the heat that would be produced by a horse working for an hour; in modern units his value for this *mechanical equivalent of heat* would be 0.183 cal·J^{-1}. Contemporary critics of these experiments claimed that the heat was evolved because metal in the form of fine turnings had a lower specific heat than bulk metal. Rumford then substituted a blunt borer, producing just as much heat with very few turnings. The advocates of the caloric hypothesis thereupon claimed that the heat must be due to the action of air on the metallic surfaces. In 1799, Humphry Davy provided further support for Thompson's theory by rubbing together two pieces of ice by clockwork in a vacuum and noting their rapid melting, showing that, even in the absence of air, this latent heat could be provided by mechanical work. Nevertheless, the times were not scientifically ready for a mechanical theory of heat until the work of Dalton and others provided an atomic theory of matter, and gradually an understanding of heat in terms of molecular motion.

By about 1840, the Law of Conservation of Energy was accepted in purely mechanical systems, the interconversion of heat and work was well established, and it was understood that heat was simply a form of motion of the smallest particles composing a substance. Yet the generalization of the conservation of energy to include heat changes had not yet been clearly made.

Thus we turn to the work of Julius Robert Mayer, one of the most curious figures in the history of science. He was born in 1814, the son of an apothecary in Heilbronn. He was always a mediocre student, but entered Tübingen University in 1832 to study medicine and obtained a good grounding in chemistry under Gmelin. He took his degree in 1838, presenting a short dissertation on the effect of santonin on worms in children. Nothing in his academic career suggested that he was about to make a great contribution to science.

Wishing to see the world, he signed as ship's doctor on the three-master *Java*, and sailed from Rotterdam in February, 1840. He spent the long voyage in idleness, lulled by the balmy off-shore breezes; according to Ostwald, in this manner he stored up the psychic energy which was to burst forth suddenly soon after he landed. According to Mayer's own story, his train of thought began abruptly on the dock at Surabaya, when several of the sailors needed to be bled. The venous blood was such a bright red that at first he thought he had opened an artery. The local physicians told him, however, that this color was typical of blood in the tropics, since the consumption of oxygen required to maintain the body temperature was less than in colder regions. Mayer began to think along these lines. Since animal heat was created by the oxidation of nutriments, the question arose of what happened if in addition to warmth the body also produced work. From an identical quantity of food, sometimes more and sometimes less heat could be obtained. If a fixed total yield of energy from food is obtainable, then one must conclude that work and heat are interchangeable quantities of the same kind. By burning the same amount of food the animal body can produce different proportions of heat and of work, but the sum of the two must be constant. Mayer spent his days on shipboard working feverishly on his theory. He became a man obsessed with one great idea, his whole life dedicated to it.

Actually Mayer was thoroughly confused about the distinctions between the concepts of force, momentum, work and energy, and the first paper that he wrote was not published by the editor of the journal to whom he submitted it. Poggendorf filed it away without even deigning to answer Mayer's letters. By the beginning of 1842, Mayer had straightened out his ideas and he could equate heat to kinetic energy and potential energy. In March, 1842, Liebig accepted his paper for the *Annalen der Chemie und Pharmazie.*

> From application of established theorems on the warmth and volume relations of gases, one finds . . . that the fall of a weight from a height of about 365 metres corresponds to the warming of an equal weight of water from 0 to 1°C.

This figure relates mechanical units of energy to thermal units. The conversion factor is called the mechanical equivalent of heat *J*. Hence,

$$w = Jq \tag{2.1}$$

In modern units, *J* is usually given as joules per calorie. To lift a weight of 1 g to a height of 365 m requires $365 \times 10^2 \times 981$ ergs of work, or 3.58 J. To raise the temperature of 1 g of water from 0 to 1°C requires 1.0087 cal. The value of *J* calculated by Mayer is therefore $3.56 \text{ J} \cdot \text{cal}^{-1}$. The accepted modern figure is 4.184. Mayer was able to state the principle of the conservation of energy, the First Law of Thermodynamics, in general terms, and to give one rather rough numerical example of its application. The exact evaluation of *J* and the proof that it is a constant independent of the method of measurement was accomplished by Joule.

2. The Work of Joule

Although Mayer was the philosophic father of the First Law, Joule's beautifully precise experiments firmly established the law on an experimental or inductive foundation. James Prescott Joule was born in 1818 near Manchester, the son of a wealthy brewer. He studied as a pupil of John Dalton. When he was 20, he began his independent researches in a laboratory provided by his father adjacent to the brewery. In later years he managed the business with good success, in addition to carrying on his extensive work in experimental chemistry and physics.

In 1840 he published his work on the heating effects of the electric current and established the following law:

> When a current of voltaic electricity is propagated along a metallic conductor, the heat evolved in a given time is proportional to the resistance of the conductor multiplied by the square of the electric intensity [current].

Thus,

$$q = I^2 R / J \tag{2.2}$$

The Joulean heat, as it is now called, can be considered as the frictional heat generated by the motion of the carriers of electric current.

In a long series of most careful experiments, Joule proceeded to measure the conversion of work into heat in various ways, by electrical heating, by compression of gases, by forcing liquids through fine tubes, and by the rotation of paddle wheels in water and mercury These studies culminated his great paper, "On the Mechanical Equivalent of Heat," read before the Royal Society in 1849. After all corrections, he obtained the final result that 772 foot pounds (ft lb) of work would produce the heat required to warm 1 lb of water 1°F. In our units, this corresponds to $J = 4.154$ J·cal^{-1}.

3. Formulation of the First Law

The philosophical argument of Mayer and the experimental work of Joule led to a definite acceptance of the conservation of energy. Hermann von Helmholtz placed the principle on a better mathematical basis in his work *Uber die Erhaltung der Kraft* (1847), which clearly stated the conservation of energy as a principle of universal validity and as one of the fundamental laws applicable to all natural phenomena.

We can use the principle to *define* a function U called the *internal energy*. A *closed system* is one for which there is no transfer of mass across the boundaries. Suppose any closed system undergoes a process by which it passes from state A to state B. If the only interaction with its surroundings is in the form of transfers of heat q to the system, or the performance of work w on the system, the change in U will be

$$\Delta U = U_B - U_A = q + w \tag{2.3}$$

Now, the First Law of Thermodynamics states that this energy difference ΔU depends only on the initial and final states and in no way on the path followed between them. Both q and w have many possible values, depending on exactly how the system passes from A to B, but their sum $q + w = \Delta U$ is invariable and independent of the path. If this were not true, it would be possible, by passing from A to B along one path and then returning from B to A along another, to obtain a net change in the energy of the closed system in contradiction to the principle of conservation of energy, the First Law of Thermodynamics. Therefore, we can say that (2.3) is a mathematical expression of the First Law. For a differential change, (2.3) becomes

$$dU = dq + dw \tag{2.4}$$

The energy function is undetermined to the extent of an arbitrary additive constant; it has been defined only in terms of the difference in energy between one state and another. Sometimes, as a matter of convenience, we may adopt a conventional standard state for a system and set its energy in this state equal to zero. For example, we might choose the state of the system at 0 K and 1 atm pressure as our standard. Then the energy U in any other state would be the change in energy in going from the standard state to the state in question.

The First Law has often been stated in terms of the universal human experience that is impossible to construct a perpetual motion machine, that is, a machine that produces useful work by a cyclic process, with no change in its surroundings. To see how this experience is embodied in the First Law, consider a cyclic process from state A to B and back to A again. If perpetual motion were ever possible, it would sometimes be possible to obtain a net increase in energy $\Delta U > 0$ by such a cycle. That this is impossible can be ascertained from (2.3), which indicates that for any such cycle, $\Delta U = (U_B - U_A) + (U_A - U_B) = 0$. A more general way of expressing this fact is to say that for any cyclic process the integral of dU vanishes:

$$\oint dU = 0 \qquad\qquad (2.5)$$

4.   The Nature of Internal Energy

In Section 1.7, we restricted the systems under consideration to those in a state of rest in the absence of gravitational or electromagnetic fields. With these restrictions, changes in the internal energy U include changes in the potential energy of the system, and in the energy transferred as heat. The potential energy changes may be considered to include also the energy changes caused by the rearrangements of molecular configurations that take place during changes in state of aggregation or in chemical reactions. If the system is moving, the kinetic energy is added to U. If the restriction on electromagnetic fields is removed, the definition of U is expanded to include the electromagnetic energy. Similarly, if gravitational effects are of interest, as in centrifugal operations, the energy of the gravitational field must be included in or added to U before applying the First Law.

In anticipation of future discussions, it may be mentioned that the interconversion of mass and energy can be readily measured in nuclear reactions. The First Law should, therefore, become a law of the conservation of mass-energy. The changes in mass theoretically associated with the energy changes in chemical reactions are so small they lie just outside the range of our present methods of measurement.* Thus, they need not be considered in ordinary chemical thermodynamics.

5.   A Mechanical Definition of Heat

Before continuing, we shall gives a better denfinition of heat. Figure 2.1 shows a system I separated from its surroundings II by an *adiabatic wall*. This is defined as a wall that separates two systems so that they are prevented from coming to thermal equilibrium with each other. Such a definition does not require the con-

---

*The change in energy corresponding to a change in mass of Δm is $\Delta U = c^2\,\Delta m$ where c is the speed of light. The most exothermic chemical reaction for a given mass of reactants is the recombination of two hydrogen atoms $(2H \rightarrow H_2)$, which has $\Delta U = -431$ kJ·mol^{-1} or $431 \times 10^3 \times 10^7 \times (\frac{1}{2}) = 2.16 \times 10^{12}$ erg·(gH)$^{-1}$. The decrease in mass on recombination of 2H would be $2.16 \times 10^{12}/(3 \times 10^{10})^2 = 2.4 \times 10^{-9}$ g·(gH)$^{-1}$. This is too small a change to be detected by present methods of weighing.

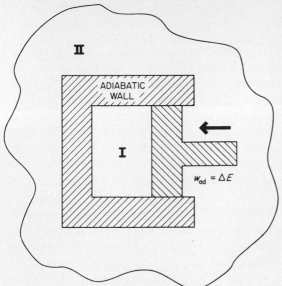

FIG. 2.1 A system I separated from surroundings II by an adiabatic wall.

cept of heat flow, and as shown in Section, 1.7, thermal equilibrium can be defined even without reference to temperature.

As the pressure on system I is increased, it is compressed from its initial state A to a new equilibrium state B; the adiabatic work done on the system is w_{ad}. The First Law of Thermodynamics can be stated in this form: When a system changes from state A to state B, the adiabatic work depends only on the initial and final states and can be set equal to the increase in a state function U, the internal energy. This statement follows at once from the impossibility of a perpetual motion machine. Thus, we can write

$$\Delta U = U_B - U_A = w_{ad} \tag{2.6}$$

Now suppose the system to be again in state A, and replace the adiabatic wall by a *diathermic wall*, defined as a wall that allows separated systems to come to thermal equilibrium. The system is brought from state A to state B by one of an infinite number of possible nonadiabatic paths. The work done on the system is w. The difference $w_{ad} - w$ is *defined* as the heat q transferred to the system in the change from A to B.

$$q = w_{ad} - w \tag{2.7}$$

Or, from (2.6),

$$q = \Delta U - w$$

Thus, the heat q transferred in a given process can be defined as the difference between the work done on the system along an adiabatic path from A to B and the work done along the given path from A to B. Although the concept of *heat* followed quite a different line in its historical development, this definition in terms of work gives a certain logical satisfaction.

6. Properties of Exact Differentials

We have seen in Section 1.23 that the work done by a system in going from one state to another is a function of the path between the states, and that $\oint dw$ is not in general equal to zero. The reason was readily apparent when the reversible process was considered. Then $\int_A^B dw = -\int_A^B P\,dV$. The differential expression $P\,dV$ cannot be integrated when only the initial and final states are known, since P is a function not only of the volume V but also of the temperature T, and this temperature may also change along the path of integration. On the other hand, $\int_A^B dU$ can always be carried out, giving $U_B - U_A$, since U is a function of the state of the system alone and is not dependent on the path by which that state is reached or on the previous history of the system.

There is nothing esoteric in the concept of a state function. Temperature T, volume V and pressure P are all state functions like energy U. Mathematically, therefore, we distinguish two classes of differential expressions. Those such as dU, dV, or dT are called *exact differentials*, since they are obtained by differentiation of some state function such as U, V, or T. Those such as dq or dw are *inexact differentials*, since they cannot be obtained by differentiation of a function of the state of the system alone. Conversely, dq or dw cannot be integrated to yield a q or w. The First Law states that although dq and dw are not exact differentials, their sum $dU = dq + dw$ is an exact differential.

The following statements are mathematically equivalent:

1. The function U is a function of the state of a system.
2. The differential dU is an exact differential.
3. The integral of dU about a closed path $\oint dU$ is equal to zero.

If $f(x_1, x_2, \ldots, x_i \cdots)$ is a function of a number n of independent variables $x_1, x_2 \ldots, x_i \cdots$ the total differential df is defined in terms of the partial derivatives $(\partial f/\partial x_i)_{x_1, x_2, \ldots}$, and the differentials of the independent variables, $dx_1, dx_2, \ldots, dx_i \cdots$, as

$$df = \sum \left(\frac{\partial f}{\partial x_i}\right)_{x_1, x_2 \cdots} dx_i \tag{2.8}$$

If there are only two independent variables, x and y, (2.8) becomes

$$df = \left(\frac{\partial f}{\partial x}\right)_y dx + \left(\frac{\partial f}{\partial y}\right)_x dy = M\,dx + N\,dy$$

where M and N are functions of x, y. Since the order of differentiation does not affect the result,

$$\frac{\partial}{\partial y}\left(\frac{\partial f}{\partial x}\right)_y = \frac{\partial}{\partial x}\left(\frac{\partial f}{\partial y}\right)_x$$

or

$$\left(\frac{\partial M}{\partial y}\right)_x = \left(\frac{\partial N}{\partial x}\right)_y \tag{2.9}$$

This is the *Euler reciprocity relation*, which will be useful in a number of derivations of thermodynamic equations. Conversely, if (2.9) holds, *df* is an exact differential.

7. Adiabatic and Isothermal Processes

Two kinds of process occur frequently both in laboratory experiments and in thermodynamic arguments. An *isothermal process* is one that occurs at constant temperature, $T = $ constant, $dT = 0$. To approach isothermal conditions, reactions are often carried out in thermostats. In an *adiabatic process*, heat is neither added to nor taken from the system; i.e., $q = 0$. For a differential adiabatic process, $dq = 0$; therefore, from (2.4), $dU = dw$. For an adiabatic reversible change in volume, $dU = -P \, dV$. For an adiabatic irreversible change, $dU = -P_{ex} \, dV$. Adiabatic conditions can be approached by careful thermal insulation of the system. High vacuum is the best insulator against heat conduction. Highly polished walls minimize heat loss by radiation. These principles are applied in Dewar vessels of various types.

8. Enthalpy

No mechanical work is done during a process carried out at constant volume; since $V = $ constant, $dV = 0$ and $w = 0$. It follows that the increase in energy equals the heat absorbed at constant volume,

$$\Delta U = q_V \tag{2.10}$$

If pressure is held constant, as for example in experiments carried out under atmospheric pressure, and no work is done except $P \, \Delta V$ work,

$$\Delta U = U_2 - U_1 = q + w = q - P(V_2 - V_1)$$

or

$$(U_2 + PV_2) - (U_1 + PV_1) = q_P \tag{2.11}$$

where q_P is the heat absorbed at constant pressure. We now define a new function, called the *enthalpy*, by

$$H = U + PV \tag{2.12}$$

Then, from (2.11),

$$\Delta H = H_2 - H_1 = q_P \tag{2.13}$$

The increase in enthalpy equals the heat absorbed at constant pressure when no work is done other than $P \, \Delta V$ work.

It will be noted that enthalpy H, like energy U or temperature T, is a function of the state of the system alone, and is independent of the path by which that state is reached. This fact follows from the definition in (2.12), since U, P, and V are all state functions.

9. Heat Capacities

Heat capacities are usually measured either at constant volume or at constant pressure. From the definitions in (1.42), (2.10), and (2.13), heat capacity at constant volume is

$$C_V = \frac{dq_V}{dT} = \left(\frac{\partial U}{\partial T}\right)_V \tag{2.14}$$

and heat capacity at constant pressure is

$$C_P = \frac{dq_P}{dT} = \left(\frac{\partial H}{\partial T}\right)_P \tag{2.15}$$

The heat capacity at constant pressure C_P cannot be smaller than that at constant volume C_V, because at constant pressure part of the heat added to a substance may be used in the work of expanding it, whereas at constant volume all the added heat produces a rise in temperature. An important equation for the difference $C_P - C_V$ can be obtained as follows:

$$C_P - C_V = \left(\frac{\partial H}{\partial T}\right)_P - \left(\frac{\partial U}{\partial T}\right)_V = \left(\frac{\partial U}{\partial T}\right)_P + P\left(\frac{\partial V}{\partial T}\right)_P - \left(\frac{\partial U}{\partial T}\right)_V \tag{2.16}$$

Since

$$dU = \left(\frac{\partial U}{\partial V}\right)_T dV + \left(\frac{\partial U}{\partial T}\right)_V dT \tag{2.17}$$

and

$$dV = \left(\frac{\partial V}{\partial T}\right)_P dT + \left(\frac{\partial V}{\partial P}\right)_T dP$$

by substituting this dV into the expression for dU and comparing coefficients, we find

$$\left(\frac{\partial U}{\partial T}\right)_P = \left(\frac{\partial U}{\partial V}\right)_T \left(\frac{\partial V}{\partial T}\right)_P + \left(\frac{\partial U}{\partial T}\right)_V$$

Substituting this expression in (2.16) gives

$$C_P - C_V = \left[P + \left(\frac{\partial U}{\partial V}\right)_T\right]\left(\frac{\partial V}{\partial T}\right)_P \tag{2.18}$$

The term $P(\partial V/\partial T)_P$ may be seen to represent the contribution to the heat capacity C_P caused by the change in volume of the system* against the *external pressure P*. The other term $(\partial U/\partial V)_T(\partial V/\partial T)_P$ is the contribution from the energy required for the change in volume against the internal cohesive or repulsive forces of the substance, represented by a change of the energy with volume at constant temperature. The term $(\partial U/\partial V)_T$ is called the *internal pressure*.† In the cases of liquids and solids, which have strong cohesive forces, this term is large. In the case of gases, on the other hand, the term $(\partial U/\partial V)_T$ is usually small compared to

---

*Substances usually expand with increase of temperature at constant pressure, but in exceptional cases there may be a contraction, for example, water between 1 and 4°C.

†Note that just as $\partial U/\partial r$, the derivative of the energy with respect to a displacement, is a force, the derivative with respect to volume, $\partial U/\partial V$, is a force per unit area or a pressure.

P. In fact, the first experiment to measure $(\partial U/\partial V)_T$ for gases, by Joule in 1843, failed to detect it at all.

10. The Joule Experiment

Joule's drawing of his apparatus is reproduced in Fig. 2.2. He described the experiment as follows:*

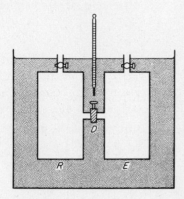

FIG. 2.2 The Joule experiment: free expansion of gas *R* into the evacuated container *E* follows opening of stopcock *D*. The temperature of the water surrounding the containers is measured.

I provided another copper receiver (*E*) which had a capacity of 134 cubic inches. . . . I had a piece *D* attached, in the center of which there was a bore 1/8 of an inch in diameter, which could be closed perfectly by means of a proper stopcock. . . . Having filled the receiver *R* with about 22 atmospheres of dry air and having exhausted the receiver *E* by means of an air pump, I screwed them together and put them into a tin can containing $16\frac{1}{2}$ lb. of water. The water was first thoroughly stirred, and its temperature taken by the same delicate thermometer which was made use of in the former experiments on mechanical equivalent of heat. The stopcock was then opened by means of a proper key, and the air allowed to pass from the full into the empty receiver until equilibrium was established between the two. Lastly, the water was again stirred and its temperature carefully noted.

Joule found no measurable temperature change and his conclusion was that "no change of temperature occurs when air is allowed to expand in such a manner as not to develop mechanical power" (i.e., as not to do external work).

The expansion in Joule's experiment, with the air rushing from *R* into the evacuated vessel *E*, is a typical irreversible process. Inequalities of temperature and pressure arise throughout the system, but eventually an equilibrium state is reached. There has been no change in the internal energy of the gas since no work was done by or on it, and it has exchanged no heat with the surrounding water (otherwise the temperature of the water would have changed). Therefore, $dU = 0$. Experimentally it is found that $dT = 0$. Joule concluded that the internal energy must depend only on the temperature and not on the volume. In mathematical terms, since

Phil. Mag., 26, 369 (1845).

$$dU = \left(\frac{\partial U}{\partial V}\right)_T dV + \left(\frac{\partial U}{\partial T}\right)_V dT = 0$$

$$\left(\frac{\partial U}{\partial V}\right)_T = -C_V \left(\frac{\partial T}{\partial V}\right)_U$$

Then,

$$\left(\frac{\partial U}{\partial V}\right)_T = 0 \quad \text{if} \quad \left(\frac{\partial T}{\partial V}\right)_U = 0$$

Joule's experiment, however, was not capable of detecting small effects since the heat capacity of his water calorimeter was extremely large compared to that of the gas used.

11. The Joule–Thomson Experiment

William Thomson (Kelvin) suggested a better procedure, and, working with Joule, carried out a series of experiments between 1852 and 1862. Their apparatus is shown schematically in Fig. 2.3. The idea was to throttle the gas flow from the

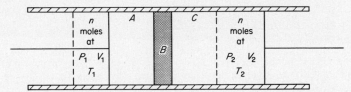

FIG. 2.3 Outline of the Joule–Thomson experiment: Gas at a higher pressure in A is expanded through the porous plug B to a lower pressure in C.

high pressure side A to the low pressure side C by interposing a porous plug B. In their first trials, this plug consisted of a silk handkerchief; in later work, porous meerschaum was used. In this way, by the time the gas emerges into C, it has already reached equilibrium and its temperature can be measured directly. The entire system is thermally insulated, so that the process is adiabatic and $q = 0$.

Suppose that the fore pressure in A is P_1, the back pressure in C is P_2, and the volumes of gas at these pressures are V_1 and V_2, respectively. The work done *on* the gas in forcing it through the plug is then $P_1 V_1$, and the work done *by* the gas in expanding on the other side is $P_2 V_2$. The net work done on the gas is therefore $w = P_1 V_1 - P_2 V_2$.

It follows that a Joule–Thomson expansion occurs at constant enthalpy:

$$\Delta U = U_2 - U_1 = q + w = 0 + w$$

$$U_2 - U_1 = P_1 V_1 - P_2 V_2$$

$$U_2 + P_2 V_2 = U_1 + P_1 V_1$$

$$H_2 = H_1$$

The Joule–Thomson coefficient, μ, is defined as the change of temperature with pressure at constant enthalpy,

$$\mu = \left(\frac{\partial T}{\partial P}\right)_H \tag{2.19}$$

This quantity is measured directly from the temperature change ΔT of the gas as it undergoes a pressure drop ΔP through the porous plug. Some experimental values of the Joule–Thomson coefficients, which are functions of temperature and pressure, are collected in Table 2.1 for a typical gas.

A positive μ corresponds to cooling on expansion, a negative μ, to warming. Most gases at room temperatures are cooled by a Joule–Thomson expansion. Hydrogen, however, is warmed if its initial temperature is above 193 K, but if it is first cooled below 193 K it can then be cooled further by a Joule–Thomson effect. The temperature 193 K at which $\mu = 0$ is called the *Joule–Thomson inversion temperature for hydrogen*. Inversion temperatures for other gases, except helium, lie considerably higher. The Joule–Thomson expansion provides one of the most important methods for liquefying gases.

TABLE 2.1
Joule–Thomson Coefficients for Carbon Dioxide[a]

μ (K·atm^{-1})

Temperature (K)	Pressure (atm)						
	0	1	10	40	60	80	100
220	2.2855	2.3035					
250	1.6885	1.6954	1.7570				
275	1.3455	1.3455	1.3470				
300	1.1070	1.1045	1.0840	1.0175	0.9675		
325	0.9425	0.9375	0.9075	0.8025	0.7230	0.6165	0.5220
350	0.8195	0.8150	0.7850	0.6780	0.6020	0.5210	0.4340
380	0.7080	0.7045	0.6780	0.5835	0.5165	0.4505	0.3855
400	0.6475	0.6440	0.6210	0.5375	0.4790	0.4225	0.3635

[a]From John H. Perry, *Chemical Engineers' Handbook* (New York: McGraw-Hill Book Company, 1941). Rearranged from *International Critical Tables*, Vol. 5, where further data may be found.

12. Application of the First Law to Ideal Gases

An analysis of the theory of the Joule–Thomson experiment must be postponed until the Second Law of Thermodynamics has been studied in the next chapter. It may be said, however, that the porous-plug experiments contradicted Joule's original conclusion that $(\partial U/\partial V)_T = 0$ for all gases. A real gas may have a considerable internal pressure, showing the existence of cohesive forces, and its energy depends on its volume as well as on its temperature.

An *ideal gas*, therefore, may be defined in thermodynamic terms as follows*:

1. The internal pressure $(\partial U/\partial V)_T = 0$.
2. The gas follows the equation of state, $PV = nRT$.

It follows from (2.17) that the energy of an ideal gas is a function of its temperature alone. For an ideal gas,

$$dU = \left(\frac{\partial U}{\partial T}\right)_V dT = C_V \, dT, \qquad C_V = \left(\frac{dU}{dT}\right)$$

The heat capacity of an ideal gas also depends only on its temperature. These conclusions greatly simplify the thermodynamics of ideal gases, so that many discussions are carried on in terms of the ideal gas model. Some examples follow.

Difference in heat capacities: When (2.18) is applied to an ideal gas, it becomes

$$C_P - C_V = P\left(\frac{\partial V}{\partial T}\right)_P$$

Then, since $PV = nRT$,

$$\left(\frac{\partial V}{\partial T}\right)_P = \frac{nR}{P}$$

and

$$C_P - C_V = nR \tag{2.20}$$

Temperature changes: Since $dU = C_V \, dT$ for an ideal gas,†

$$\Delta U = U_2 - U_1 = \int_{T_1}^{T_2} C_V \, dT \tag{2.21}$$

Likewise, for an ideal gas,†

$$dH = C_P \, dT$$

$$\Delta H = H_2 - H_1 = \int_{T_1}^{T_2} C_P \, dT \tag{2.22}$$

Isothermal reversible volume or pressure change: For an isothermal change in an ideal gas, the internal energy remains constant. Since $dT = 0$ and $(\partial U/\partial V)_T = 0$,

$$dU = dq - P \, dV = \left(\frac{\partial U}{\partial T}\right)_V dT + \left(\frac{\partial U}{\partial V}\right)_T dV = 0$$

Hence, from (2.4),

$$dq = -dw = P \, dV$$

---

*After a thermodynamic definition of temperature is obtained from the Second Law of Thermodynamics, we can derive 1 from 2, or derive 2 from 1 and Boyle's Law. Thus, 2 in itself completely defines an ideal gas.

†For any substance, at constant volume, $dU = C_V \, dT$, and at constant pressure, $dH = C_P \, dT$. For an ideal gas, U and H are functions only of T, and these relations hold even if V and P respectively are not constant.

Since

$$P = \frac{nRT}{V}$$

$$\int_1^2 dq = -\int_1^2 dw = \int_1^2 nRT \frac{dV}{V}$$

or,

$$q = -w = nRT \ln \frac{V_2}{V_1} = nRT \ln \frac{P_1}{P_2} \qquad (2.23)$$

Since the volume change is carried out reversibly, P always has its equilibrium value nRT/V, and the work $-w$ in (2.23) is the maximum work done in an expansion, or the minimum work needed to effect a compression. The equation tells us that the work required to compress a gas from 10 to 100 atm is just the same as that required to compress it from 1 to 10 atm.

Reversible adiabatic expansion: In this case, $dq = 0$, and $dU = dw = -P\,dV$. Since $dU = C_V\,dT$,

$$dw = C_V\,dT \qquad (2.24)$$

For a finite change,

$$w = \int_1^2 C_V\,dT \qquad (2.25)$$

We may write (2.24) as $C_V\,dT + P\,dV = 0$, whence,

$$C_V \frac{dT}{T} + nR \frac{dV}{V} = 0 \qquad (2.26)$$

Integrating between T_1 and T_2, and V_1 and V_2, the initial and final temperatures and volumes, we have

$$C_V \ln \frac{T_2}{T_1} + nR \ln \frac{V_2}{V_1} = 0 \qquad (2.27)$$

This integration assumes that C_V is a constant, not a function of T.

We may substitute for nR from (2.20), and using the conventional symbol γ for the heat capacity ratio C_P/C_V we find

$$(\gamma - 1) \ln \frac{V_2}{V_1} + \ln \frac{T_2}{T_1} = 0$$

Therefore,

$$\frac{T_1}{T_2} = \left(\frac{V_2}{V_1} \right)^{\gamma - 1} \qquad (2.28)$$

Since, for an ideal gas,

$$\frac{T_1}{T_2} = \frac{P_1 V_1}{P_2 V_2}$$

we have

$$P_1 V_1^\gamma = P_2 V_2^\gamma \qquad (2.29a)$$

It has been shown, therefore, that for a reversible adiabatic expansion of an ideal gas (with constant C_V)

$$PV^\gamma = \text{const.} \tag{2.29b}$$

We recall that for an isothermal expansion, $PV = \text{const.}$

These equations are plotted in Fig. 2.4. A given pressure fall produces a lesser volume increase in the adiabatic case, because the temperature also falls during the adiabatic expansion.

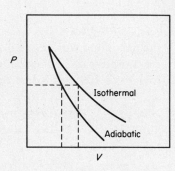

FIG. 2.4 Isothermal and adiabatic reversible expansions of an ideal gas from the same initial pressure and volume.

13.   Examples of Ideal-Gas Calculations

Let us take 1 m³ of gas at 273.2 K and 10 atm. We therefore have $10^4/22.414 = 446.1$ mol. We shall calculate the final volume and the work done in three different expansions to a final pressure of 1 atm. Let us assume that we have a monatomic gas such as neon. The molar heat capacity is then $C_{Vm} = \frac{3}{2}R$, independent of temperature.

Isothermal reversible expansion: In this case, the final volume

$$V_2 = \frac{P_1 V_1}{P_2} = \frac{(1)(10)}{(1)} = 10 \text{ m}^3$$

The work done by the gas in expanding equals the heat absorbed by the gas from its surroundings. From (2.23),

$$-w = q = nRT \ln \frac{V_2}{V_1}$$
$$= (446.1)(8.314)(273.2)(2.303)\log(10)$$
$$= 232.85 \text{ kJ}$$

Adiabatic reversible expansion: The final volume is calculated from (2.29a), with

$$\gamma = \frac{C_P}{C_V} = \frac{\frac{3}{2}R + R}{\frac{3}{2}R} = \frac{5}{3}$$

Thus,

$$V_2 = \left(\frac{P_1}{P_2}\right)^{1/\gamma} V_1, \qquad V_2 = (10)^{3/5}(1) = 3.981 \text{ m}^3$$

The final temperature is obtained from $P_2 V_2 = nRT_2$:

$$T_2 = \frac{P_2 V_2}{nR} = \frac{(1)\ 3.981}{(446.1)(8.206 \times 10^{-5})} = 108.8 \text{ K}$$

For an adiabatic process, $q = 0$, and $\Delta U = q + w = w$. Also, since C_V is constant, (2.21) gives

$$\Delta U = C_V \Delta T = n(\tfrac{3}{2}R)(T_2 - T_1) = -914.1 \text{ kJ}$$

The work *done by* the gas on expansion is therefore 914.1 kJ.

Irreversible adiabatic expansion: Suppose the pressure is suddenly released to 1 atm and the gas expands adiabatically against this constant pressure. Since this is not a reversible expansion, (2.29) cannot be applied. Since $q = 0$, $\Delta U = w$. The value of ΔU depends only on the initial and final states:

$$\Delta U = w = C_V(T_2 - T_1)$$

Also, for an expansion at constant pressure, we have from (1.44),

$$-w = P_2(V_2 - V_1) = P_2\left(\frac{nRT_2}{P_2} - \frac{nRT_1}{P_1}\right)$$

Equating the two expressions for w, we obtain

$$-C_V(T_2 - T_1) = P_2\left(\frac{nRT_2}{P_2} - \frac{nRT_1}{P_1}\right)$$

The only unknown is T_2.

$$-\tfrac{3}{2}nR(T_2 - 273.2) = 1\left(\frac{nRT_2}{1} - \frac{nR\ 273.2}{10}\right)$$

$$T_2 = 174.8 \text{ K}$$

Then,

$$\Delta U = w = \tfrac{3}{2}nR(174.8 - 273.2)$$
$$w = -547.4 \text{ kJ}$$

Note that there is considerably less cooling of the gas and less work done by the gas in the irreversible adiabatic expansion than in the reversible expansion.

14.   Thermochemistry—Heats of Reaction

Thermochemistry is the study of the heat effects that accompany chemical reactions, the formation of solutions, and changes in state of aggregation like melting or vaporization. Physicochemical changes are classified as *endothermic*, accompanied by the absorption of heat, or *exothermic*, accompanied by the evolution of heat. An example of an exothermic reaction is the burning of hydrogen.

$$H_2 + \tfrac{1}{2}O_2 \longrightarrow H_2O \text{ (gas)}, \qquad \Delta H = -241\ 750 \text{ J at 291.15 K}$$

The heat is given off *from the system* and therefore is written with a negative sign. A typical endothermic reaction would be the decomposition of water vapor,

$$H_2O \text{ (gas)} \longrightarrow H_2 + \tfrac{1}{2}O_2, \qquad \Delta H = 241\ 750 \text{ J at 291.15 K}$$

Like any other transfer of heat, the heat of a chemical reaction depends upon the conditions that hold during the process by which it is carried out. There are two particular conditions that are important because they lead to heats of reaction equal to changes in thermodynamic functions. The first such condition is that of *constant volume*. If the volume of a system is held constant, no work is done on the system,* and (2.3) for the First Law of Thermodynamics becomes

$$\Delta U = q_V \tag{2.30}$$

Thus the heat of reaction measured at constant volume is exactly equal to the change in internal energy ΔU of the reaction system. This condition is excellently approximated when the reaction is carried out in a bomb calorimeter. The other important special condition is that of *constant pressure*. During the course of an experiment under ordinary bench-top conditions, the pressure is effectively constant. Many calorimeters operate at this constant atmospheric pressure. From (2.13),

$$\Delta H = q_P \tag{2.31}$$

The heat of reaction measured at constant pressure is exactly equal to the change in enthalpy ΔH of the reaction system.

It is often necessary to use data obtained with a bomb calorimeter, which give ΔU, in order to calculate ΔH. From the definition of H in (2.12),

$$\Delta H = \Delta U + \Delta(PV) \tag{2.32}$$

By $\Delta(PV)$ we mean the change in PV for the entire system, or, in particular, the PV of the products minus the PV of the reactants for the indicated chemical reaction.

If all the reactants and products are liquids or solids, these PV values change only slightly during the reaction, provided the pressure is low [1 atm or so]. In such cases, $\Delta(PV)$ is usually so small compared with ΔH or ΔU that it may be neglected, and we set $q_P \cong q_V$.† For reactions at high pressures, however (e.g., at the bottom of the ocean), $\Delta(PV)$ may be considerable even for condensed phases.

For reactions in which gases occur in the reaction equation, the values of $\Delta(PV)$ depend on the change in the number of moles of gas as a result of reaction. From the ideal gas equation, we can write

$$\Delta(PV) = \Delta n(RT)$$

Therefore, from (2.32),

$$\Delta H = \Delta U + \Delta n(RT) \tag{2.33}$$

---

*It is assumed that no work except "PV work" would be possible in the experimental arrangement used.

†Note, however, that we cannot carry out a reaction at constant P, T and constant V, T and at the same time require that the initial and final P, V, T be the same in the two cases.

Thus, (2.32) becomes, in the general case,

$$q_P = \Delta U_P + P\,\Delta V$$

or

$$\Delta H_V = q_V + V\,\Delta P$$

depending on whether a condition of constant P or of constant V is chosen.

By Δn we mean the number of moles of gaseous products minus the number of moles of gaseous reactants.

Consider as an example the reaction

$$SO_2 + \tfrac{1}{2}O_2 \longrightarrow SO_3$$

The ΔU for this reaction as measured in a bomb calorimeter is $-97\,030$ J at 298 K. What is ΔH? The $\Delta n = 1 - 1 - \tfrac{1}{2} = -\tfrac{1}{2}$.
Therefore,

$$\Delta H = \Delta U - \tfrac{1}{2}RT$$
$$\Delta H = -97\,030 - \tfrac{1}{2}(8.314)\,(298) = -98\,270 \text{ J}$$

To specify the heat of reaction, it is necessary to write the exact chemical equation for the reaction and to specify the states of all reactants and products, noting particularly the constant temperature at which the measurement is made. Since most reactions are studied under conditions of essentially constant pressure, ΔH is usually the heat of reaction stated. Two examples follow:

$$CO_2(1 \text{ atm}) + H_2(1 \text{ atm}) \longrightarrow CO(1 \text{ atm}) + H_2O(g, 1 \text{ atm}), \quad \Delta H_{298} = \quad 41\,160 \text{ J}$$

$$AgBr + \tfrac{1}{2}Cl_2(1 \text{ atm}) \longrightarrow AgCl(c) + \tfrac{1}{2}Br_2, \qquad\qquad \Delta H_{298} = -28\,670 \text{ J}$$

As an immediate consequence of the First Law, ΔU or ΔH for any chemical reaction is independent of the path, that is, independent of any intermediate reactions that may occur. This principle, first established experimentally by G. H. Hess (1840), is called *The Law of Constant Heat Summation*. It is often possible, therefore, to calculate the heat of a reaction from measurements on quite different reactions. For example,

(1) $COCl_2 + H_2S \longrightarrow 2HCl + COS,$ $\qquad\qquad \Delta H_{298} = -78\,705 \text{ J}$
(2) $\quad COS + H_2S \longrightarrow H_2O(g) + CS_2(1),$ $\qquad \Delta H_{298} = \quad 3\,420 \text{ J}$
(3) $COCl_2 + 2H_2S \longrightarrow 2HCl + H_2O(g) + CS_2(1),$ $\quad \Delta H_{298} = -75\,285 \text{ J}$

15.   Enthalpies of Formation

A convenient standard state for a substance is the state in which it is stable at 298.15 K and 1 atm pressure—e.g., oxygen as $O_2(g)$, sulfur as S (rhombic crystal), mercury as Hg (l), etc. By convention, the enthalpies of the chemical elements in this particular standard state are set equal to zero. The *standard enthalpy* of formation $\Delta H_f^{\ominus}$ of any compound is the ΔH of the reaction by which it is formed from its elements, the reactants and products all being in a given standard state. For example, for a standard state at 298.15 K,

$$(1) \quad S + O_2 \longrightarrow SO_2, \qquad \Delta H_{298}^{\ominus} = -296.9 \text{ kJ}$$
$$(2)\ 2Al + \tfrac{3}{2}O_2 \longrightarrow Al_2O_3, \qquad \Delta H_{298}^{\ominus} = -1669.8 \text{ kJ}$$

The superscript $\ominus$ indicates we are writing a *standard* enthalpy of formation, so that the pressure for reactants and products is 1 atm; the absolute temperature is written as a subscript or in parentheses.

Thermochemical data are conveniently tabulated as standard enthalpies of formation $\Delta H_f^{\ominus}$. A few examples, selected from a recent compilation of the National Bureau of Standards,* are given in Table 2.2. The standard enthalpy of any reaction at 298.15 K is then readily found as the difference between the tabulated standard enthalpies of formation of the products and of the reactants.

TABLE 2.2
Standard Enthalpies of Formation at 298.15 K

Compound	State	$\Delta H_f^{\ominus}(298)$ (kJ·mol⁻¹)	Compound	State	$\Delta H_f^{\ominus}(298)$ (kJ·mol⁻¹)
H_2O	g	−241.826	H_2S	g	− 20.63
H_2O	l	−285.830	H_2SO_4	l	−814.00
H_2O_2	g	−133.2	SO_2	g	−296.8
HF	g	−271.1	SO_3	g	−395.7
HCl	g	− 92.312	CO	g	−110.523
HBr	g	− 36.40	CO_2	g	−393.513
HI	g	+ 26.48	$COCl_2$	l	−205.9
HIO_3	c	−238.6	S_2Cl_2	g	− 23.85
NO	g	+ 90.25	NH_3	g	− 46.11
N_2O	g	+ 82.05	HN_3	g	+294.1

Many of our thermochemical data have been obtained from measurements of heats of combustion. If the heats of formation of all its combustion products are known, the heat of formation of a compound can be calculated from its heat of combustion. For example,

(1) $\quad C_2H_6 + \frac{7}{2}O_2 \longrightarrow 2CO_2 + 3H_2O(l) \quad \Delta H_{298}^{\ominus} = -1560.1 \text{ kJ}$
(2) $C(\text{graphite}) + O_2 \longrightarrow CO_2 \quad\quad\quad\quad\quad \Delta H_{298}^{\ominus} = - 393.5 \text{ kJ}$
(3) $\quad\quad H_2 + \frac{1}{2}O_2 \longrightarrow H_2O(l) \quad\quad\quad\quad\quad \Delta H_{298}^{\ominus} = - 285.8 \text{ kJ}$
(4) $\quad\quad 2C + 3H_2 \longrightarrow C_2H_6 \quad\quad\quad\quad\quad\quad \Delta H_{298}^{\ominus} = - 84.3 \text{ kJ}$

The data in Table 2.3 were obtained from heats of combustion by F. D. Rossini and coworkers at the National Bureau of Standards. The standard state of carbon has been taken to be graphite.

When changes in state of aggregation occur, the appropriate latent heat must be added. For example,

$$S(\text{rhombic}) + O_2 \longrightarrow SO_2 \quad\quad \Delta H_{298}^{\ominus} = -296.90 \text{ kJ}$$
$$\underline{\quad\quad S(\text{rhombic}) \longrightarrow S \text{ (monoclinic)} \quad \Delta H_{298}^{\ominus} = \quad\quad 0.29 \text{ kJ}}$$
$$S(\text{monoclinic}) + O_2 \longrightarrow SO_2 \quad\quad \Delta H_{298}^{\ominus} = -297.19 \text{ kJ}$$

---

*The bureau publishes a comprehensive collection of thermodynamic data. The most recent compilations are in *Technical Notes* 270–3 (1968) and 270–4 (1969). [Superintendent of Documents, U. S. Government Printing Office, Washington, D. C. 20402]. When the tables are completed, they will be published in a single volume (about 1972) (*Selected Values of Chemical Thermodynamic Properties*).

TABLE 2.3
Enthalpies of Formation of Gaseous Hydrocarbons

Substance	Formula	$\Delta H_f^{\ominus}(298)$ $(kJ \cdot mol^{-1})$
Paraffins		
Methane	CH_4	$-$ 74.75 $\pm$ 0.30
Ethane	C_2H_6	$-$ 84.48 $\pm$ 0.45
Propane	C_3H_8	-103.6 $\pm$ 0.5
n-Butane	C_4H_{10}	-124.3 $\pm$ 0.6
Isobutane	C_4H_{10}	-131.2 $\pm$ 0.6
Olefines		
Ethylene	C_2H_4	52.58 $\pm$ 0.28
Propylene	C_3H_6	20.74 $\pm$ 0.46
1-Butene	C_4H_8	1.60 $\pm$ 0.75
cis-2-Butene	C_4H_8	$-$ 5.81 $\pm$ 0.75
trans-2-Butene	C_4H_8	$-$ 9.78 $\pm$ 0.75
2-Methylpropene	C_4H_8	$-$ 13.41 $\pm$ 1.25
Acetylenes		
Acetylene	C_2H_2	226.9 $\pm$ 1.0
Methylacetylene	C_3H_4	185.4 $\pm$ 1.0

16. Experimental Thermochemistry*

One of the landmarks in the development of thermochemistry was the publication in 1780 by Lavoisier and Laplace of their memoir *Sur la Chaleur*. They described the use of an ice calorimeter in which the heat liberated was measured by the mass of ice melted. They measured the heat of combustion of carbon, finding that "one ounce of carbon in burning melts six pounds and two ounces of ice." This result corresponds to a heat of combustion of -413.6 kJ per mole, as compared to the best modern value of -393.5. The calorimeter is shown in Fig. 2.5(a). It was filled with ice in the regions *b* and *a*, the outer layer of ice preventing heat transfer into the calorimeter. Lavoisier and Laplace also measured the heat evolved by a guinea pig placed in the calorimeter and compared it with the amount of "dephlogisticated air" (oxygen) consumed by the animal. They reached the following conclusion:

> ... respiration is thus a combustion, to be sure very slow, but otherwise perfectly similar to that of carbon; it takes place in the interior of the lungs, without emitting visible light, because the matter of the fire on becoming free is soon absorbed by the humidity of these organs. The heat developed in the combustion is transferred

*Descriptions of the experimental equipment and procedures can be found in the publications of the Thermodynamics Section at the National Bureau of Standards: *J. Res. Nat. Bur. Std.*, *6*, 1(1931); *27*, 289 (1941). An excellent general account of experimental calorimetry is the article by J. M. Sturtevant in *Physical Methods of Organic Chemistry*, Vol. 1, Pt. 1, 3rd ed., ed. by A. Weissberger (New York: Interscience Publishers, 1959), pp. 523–654. Special methods and applications are described in *Experimental Thermodynamics*, ed. by J. P. McCullough and D. W. Scott and published for I.U.P.A.C. by Butterworths, London, 1968.

FIG. 2.5(a) The ice calorimeter of Lavoisier and Laplace (scale is in inches). It was filled with ice in regions *a* and *b* and heat evolution occurred in chamber *f*. Water formed by melting of ice was collected and weighed in *d*.

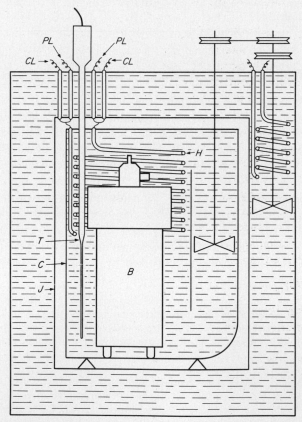

FIG. 2.5(b) Bomb calorimeter used at National Bureau of Standards. (*B*, bomb; *H*, heater; *C*, calorimeter vessel; *T*, resistance thermometer; *J*, jacket; *PL*, potential leads; *CL*, current leads.

to the blood which traverses the lungs, and from this is spread throughout all the animal system.

Calorimetry has always been one of the most exacting techniques of physical chemistry, and a prodigious amount of experimental ingenuity has been devoted to the design of calorimeters.* The usual measurement of the heat of a reaction consists essentially of (1) the careful determination of the amount of chemical reaction that produces a definite change in the calorimeter, and (2) the measurement of the electrical energy needed to produce exactly the same change in the calorimeter. The change in question is usually a temperature change. Electrical energy is almost always used because it can be measured with the highest precision. The type of calorimeter used in extensive measurements at the National Bureau of Standards is shown in Fig. 2.5b. If a potential difference E in volts (V) is impressed across a resistance R in ohms (Ω) for a time of t seconds, the energy dissipated is E^2t/JR or $E^2t/4.1840R$ defined calories.

A calorimeter can be used to determine the heat capacity of a substance by means of measurements of the input of electrical energy and the resultant rise in temperature.

For a measurement of the ΔH of an exactly specified chemical reaction, the reactants must be pure and the products must be precisely analyzed. The latter requirement can usually be satisfied with an ordinary combustion bomb, provided the compound burned contains only C, N, O, and H. For compounds of S, halogens, or metals, serious problems due to nonuniform combustion can arise. For example, suppose the products include water and H_2SO_4—unless the composition of the aqueous H_2SO_4 is uniform, the observed ΔH will not be constant, since the heat of solution of H_2SO_4 depends markedly on the composition. To circumvent such difficulties, the rotating bomb calorimeter, developed at the University of Lund, provides for stirring the contents of the bomb. An example of such an instrument as designed by Hubbard and coworkers at the Thermodynamics Laboratory of the U. S. Bureau of Mines is shown in Fig. 2.6. The rotating bomb has greatly extended the range of reliable thermochemical measurements. For example, the following reaction was studied by first burning the metal carbonyl in oxygen and then dissolving the products in nitric acid:

$$Mn_2(CO)_{10} + 4HNO_3 + 6O_2 \longrightarrow 2Mn(NO_3)_2 + 10CO_2 + 2H_2O$$

From the known heats of formation of the $Mn(NO_3)_2$, CO_2, H_2O, and HNO_3, the heat of formation of the carbonyl could be calculated.

The direct measurement of the heats of reactions other than combustion is called *reaction calorimetry*. An interesting example is the determination of the heat of formation of XeF_4 from

$$XeF_4 \text{ (c)} + 4\,KI \text{ (aq)} \longrightarrow Xe + 4\,KF \text{ (aq)} + 2\,I_2 \text{ (c)}$$

The result was $\Delta H_f^{\ominus}$ (298) $= -251$ kJ.

---

*G. T. Armstrong, *J. Chem. Educ.*, **41**, 297 (1964). "The Calorimeter and Its Influence on the Development of Chemistry."

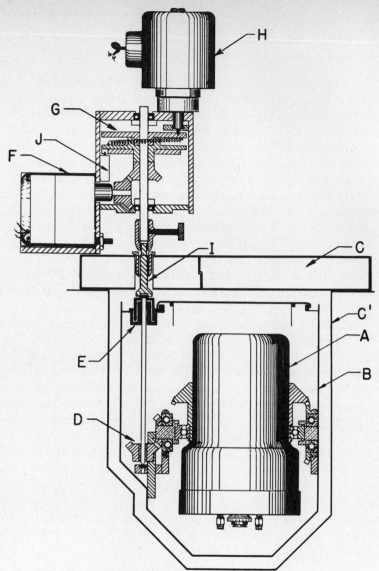

FIG. 2.6(a) A modern rotating-bomb calorimeter. The bomb *A* is shown in position in the calorimeter can *B*, which in turn is in position in the well *C'* of a constant-temperature jacket covered by constant-temperature lid *C*. The drive shaft *I*, actuated through a pair of miter gears by a 10-r.p.m. synchronous motor *F*, enters the calorimeter through an oil seal *E* and actuates the rotating mechanism through the miter gear *D*. When the bomb is rotating, a pin is held out of a hole in the wheel *G* by a solenoid *H*. A cam-activated switch *J* sends electrical impulses to a rotation counter, and the rotor is turned off automatically after a preset number of rotations. [W. D. Good, D. W. Scott, and G. Waddington, *J. Phys. Chem. 60*, 1080 (1950) give details of applications.]

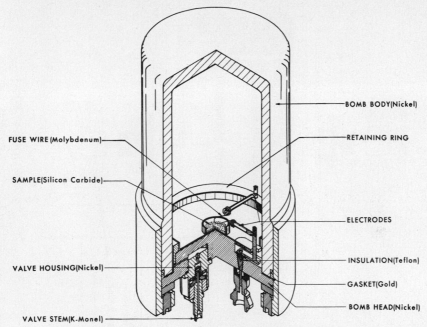

FUSE WIRE (Molybdenum)

SAMPLE (Silicon Carbide)

VALVE HOUSING (Nickel)

VALVE STEM (K-Monel)

BOMB BODY (Nickel)

RETAINING RING

ELECTRODES

INSULATION (Teflon)

GASKET (Gold)

BOMB HEAD (Nickel)

FIG. 2.6(b)   Details of the combustion bomb as used to determine ΔH (combustion) of SiC. [W. Hubbard et al., *J. Phys. Chem. 65*, 1168 (1961).]

17. Heat Conduction Calorimeters

The calorimeters so far described have been essentially adiabatic calorimeters, in that they were designed to minimize transfer of heat between the reaction vessel and its surroundings. If, however, the heat transfer to the surroundings takes place through a multijunction thermocouple designed to measure rapidly the temperature difference between the reactants and the surroundings, one can measure the heat transfer by simply integrating the thermocouple readings over time. A calorimeter of this type is called a *heat conduction calorimeter*. These calorimeters can be used either for batch operation or for flow operation. They are usually used as double calorimeters in which heat transfer from the reaction vessel is continuously balanced against transfer from a control vessel.

Figure 2.7(a) shows an instrument of this type designed by Benzinger and Kitzinger* for the study of biochemical reactions by the method of *heat-burst calorimetry*. The application of the heat-burst technique to measurement of ΔH for an antigen-antibody reaction is illustrated in Fig. 2.7(b). The total increase in temperature was only 10^{-5} K, and the heat measured was -1.21×10^{-2} J, equi-

---

*Methods Biochem. Analysis, **8**, 309 (1960).

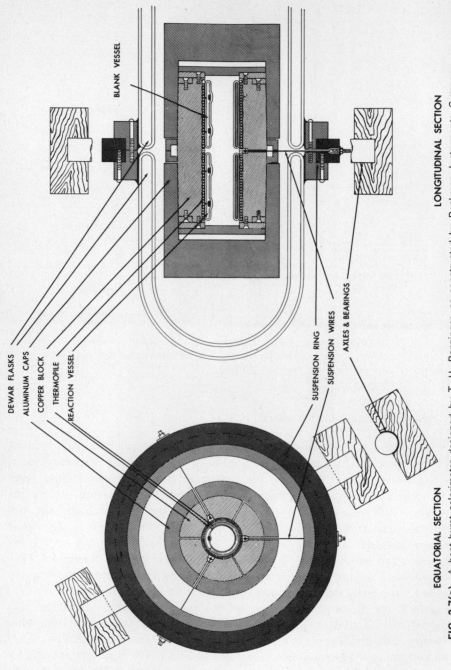

DEWAR FLASKS
ALUMINUM CAPS
COPPER BLOCK
THERMOPILE
REACTION VESSEL

BLANK VESSEL

SUSPENSION RING
SUSPENSION WIRES
AXLES & BEARINGS

EQUATORIAL SECTION

LONGITUDINAL SECTION

FIG. 2.7(a) A heat-burst calorimeter designed by T. H. Benzinger, as constructed by Beckman Instruments Company. *Left:* Equatorial section through instrument. *Right:* Longitudinal section through instrument.

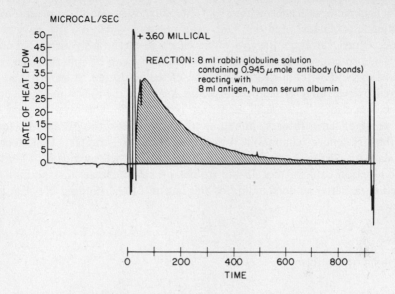

MICROCAL/SEC

+ 3.60 MILLICAL

REACTION: 8 ml rabbit globuline solution
containing 0.945 μmole antibody (bonds)
reacting with
8 ml antigen, human serum albumin

RATE OF HEAT FLOW

TIME

−0.69 MILLICAL

CONTROL: 8 ml rabbit globuline solution
pretreated with 0.016 ml antigen solution
reacting with
8 ml antigen, human serum albumin

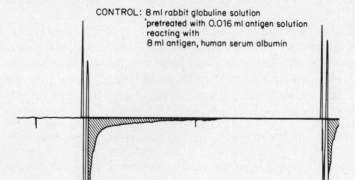

FIG. 2.7(b) Measurement of ΔH for an immunochemical reaction.
Top: specific antigen-antibody interaction. *Bottom:* After
removal of specific antibody by preinteraction with a
small amount of antigen. Time scale in seconds.

valent to $-30.5\ \text{kJ} \cdot \text{mol}^{-1}$ antibody. A much easier reaction to measure (since larger amounts of reactant were available) was the hydrolysis of ATP,

$$\text{ATP}^{4-} + \text{H}_2\text{O} \longrightarrow \text{ADP}^{3-} + \text{HPO}_4^{2-} + \text{H}^+$$

which yielded $\Delta H(298) = -22.2\ \text{kJ}$ at pH 7.0.

The heat conduction calorimeter is well adapted to flow calorimetry, in which a continuous stream of reactant and product are mixed inside the calorimeter.

Such an arrangement has been used by J. M. Sturtevant and his students for many interesting biochemical studies.* One example was work on the enzyme ribonuclease, which catalyzes the hydrolytic cleavage of the polynucleotide chain of RNA. The protein, which contains 124 amino acids, can be split between residues 20 and 21 to yield S-ribonuclease and the so-called S-peptide. Sturtevant and Hearn measured over a range of temperatures the ΔH when the S-ribonuclease combines with the S-peptide. As we shall see in the next chapter, the measurement of ΔH over a range of temperatures allows one to calculate also the ΔS and ΔG (entropy and Gibbs free energy changes) of a reaction. Therefore, complete basic thermodynamic data on chemical reactions can be readily obtained by means of convenient heat conduction calorimeters. In the studies on ribonuclease, the thermodynamic data led to a better understanding of the nature of the binding between protein molecules.

18. Heats of Solution

In many chemical reactions, one or more of the reactants are in solution, and the investigation of heats of solution is an important branch of thermochemistry. It is necessary to distinguish the *integral heat of solution* and the *differential heat of solution.*

Most of us can remember the first time we prepared a dilute solution of sulfuric acid. As we slowly poured the acid into the water with constant stirring, the solution became steadily hotter, but we noted that the rate of heating was less toward the end, when we were adding H_2SO_4 to an already fairly concentrated aqueous solution. We can write an equation for the change involved in adding 1 mole of liquid H_2SO_4 to n_1 moles of water:

$$H_2SO_4 \text{ (l)} + n_1 H_2O \longrightarrow H_2SO_4 \text{ (}n_1 H_2O\text{)}$$

The enthalpy change ΔH_s in this reaction per mole of H_2SO_4 is called the *integral heat of solution* per mole of H_2SO_4 to yield a solution of the final composition specified. In this case, in terms of mole fractions, the final composition would be $X_2(H_2SO_4) = 1/(n_1 + 1)$ and $X_1(H_2O) = n_1/(n_1 + 1)$.

In Table 2.4, we show the measured ΔH_s for a series of different values of n_1. We note that as n_1 increases and the final solution becomes more dilute, $-\Delta H_s$ per mole of H_2SO_4 steadily increases toward a limiting value of $\Delta H_s = -96.19$ kJ·mol^{-1}, which is called the *integral heat of solution to infinite dilution*. The data in Table 2.4 are also shown plotted in Fig. 2.8 in the form of ΔH_s vs. n_1/n_2, the ratio of amount of H_2O to amount of H_2SO_4.

The difference between the integral heats of solution to two different specified concentrations gives the *heat of dilution*. For example, from Table 2.4,

*J. M. Sturtevant, "Flow Calorimetry," *Fractions* (Beckman Instrument Co.) 1969, No.1, p. 1; J. M. Sturtevant and P. A. Lyons, *J. Chem. Thermodyn. 1*, 201 (1969). R. W. Menkins, G. D. Watt, and J. M. Sturtevant, *Biochemistry, 8*, 1874 (1969).

TABLE 2.4
Integral Heats of Solution

$$H_2SO_4(l) + n_1 H_2O \longrightarrow H_2SO_4(n_1 H_2O)$$

n_1/n_2 (mol H$_2$O/mol H$_2$SO$_4$)	$-\Delta H_s(298.15 \text{ K})$ (kJ·mol^{-1} H$_2$SO$_4$)
0.5	15.73
1.0	28.07
1.5	36.90
2.0	41.92
5.0	58.03
10.0	67.03
20.0	71.50
50.0	73.35
100.0	73.97
1000.	78.58
10000.	87.07
100000.	93.64
∞	96.19

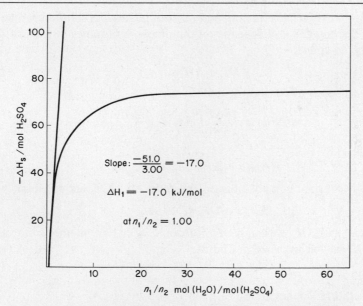

Slope: $\dfrac{-51.0}{3.00} = -17.0$

$\Delta H_1 = -17.0$ kJ/mol

at $n_1/n_2 = 1.00$

FIG. 2.8   The integral heat of solution of H$_2$SO$_4$ in H$_2$O. The slope of the curve at any composition, from Eq. (2-37), gives ΔH_1, the differential heat of solution of H$_2$O at that composition. The calculation shown at $n_1/n_2 = 1$ (equimolar solution) gives $\Delta H_1 = 17.00$ kJ/mol. From (2-37) and Table 2-4, $\Delta H_2 = -11.07$ kJ/mol.

$$H_2SO_4\left(\frac{n_1}{n_2} = 1.0\right) \longrightarrow H_2SO_4\left(\frac{n_1}{n_2} = 5.0\right), \qquad \Delta H_{\text{dil}} = -29.96 \text{ kJ·mol}^{-1} \text{ H}_2SO_4$$

or

$$H_2SO_4 (1.0 \text{ H}_2O) + 4 \text{ H}_2O \longrightarrow H_2SO_4 (5.0 \text{ H}_2O)$$

The ΔH_s shown in Table 2.4 and Fig. 2.8 are called integral heats of solution because they represent a summation of all the ΔH as H_2SO_4 is added to solutions that vary in composition from pure water to the final concentration of n_1 moles of water per mole of H_2SO_4.

Suppose we measured the enthalpy change per mole of H_2SO_4 caused by adding H_2SO_4 to a solution of H_2SO_4 and H_2O at some specified constant composition, say, n_1 moles H_2O and n_2 moles H_2SO_4. The enthalpy change in such a process depends on the composition specified, so that we can write it as a function of n_1 and n_2, namely, $\Delta H_2(n_1, n_2)$. This quantity is called the *differential heat of solution* of H_2SO_4 at the composition specified. Practically speaking, of course, we cannot dissolve H_2SO_4 in a solution of H_2SO_4 and H_2O without changing the composition of the solution, so that we must define the differential heat as the limit of $(\Delta H/\Delta n_2)$ at constant n_1 as $\Delta n_2 \to 0$. Thus,

$$\Delta H_2 = \lim_{\Delta n_2 \to 0} \left(\frac{\Delta H}{\Delta n_2}\right)_{n_1} = \left(\frac{\partial \Delta H}{\partial n_2}\right)_{n_1} \tag{2.34}$$

We see, therefore, that the slope of the curve in Fig. 2.8 in which the integral heat of solution is plotted against n_2/n_1 gives the differential heat of solution ΔH_2 at any value of the composition n_2/n_1.

The relation between the integral and differential heats of solution can be shown as follows. The integral ΔH depends on the numbers of moles of each of the two components, n_1 and n_2.

$$\Delta H_s = \Delta H_s(n_1, n_2)$$

Hence, for a change at constant T and P,

$$d(\Delta H_s) = \left(\frac{\partial \Delta H_s}{\partial n_1}\right)dn_1 + \left(\frac{\partial \Delta H_s}{\partial n_2}\right)dn_2 \tag{2.35}$$

From (2.34),

$$d(\Delta H_s) = \Delta H_1 \, dn_1 + \Delta H_2 \, dn_2 \tag{2.36}$$

Integrating at constant composition, so that ΔH_1 and ΔH_2 are constant, we have

$$\Delta H_s = \Delta H_1 \, n_1 + \Delta H_2 \, n_2 \tag{2.37}$$

Knowing ΔH_2 and ΔH_s, we can get ΔH_1 from (2.37). Other methods of evaluating ΔH_1 and ΔH_2 will be discussed in Chapter 7.

19. Temperature Dependence of Enthalpy of Reaction

Sometimes the ΔH of a reaction is measured at one temperature and we need to know its value at another. Such a situation is shown in schematic form as follows:

$$T_2: \qquad \text{REACTANTS} \xrightarrow{\Delta H_{T2}} \text{PRODUCTS}$$

$$C_P^{\text{re}}(T_2 - T_1) \quad \uparrow \qquad\qquad \uparrow \quad C_P^{\text{pr}}(T_2 - T_1)$$

$$T_1: \qquad \text{REACTANTS} \xrightarrow{\Delta H_{T1}} \text{PRODUCTS}$$

In this diagram, it is assumed that the heat capacities C_P are constant over the temperature range, The term C_P^{re} means the sum of the heat capacities for all the reactants in the stoichiometric equation for the reaction—similarly, for C_P^{pr}. From the First Law, it is evident that

$$\Delta H_{T1} + C_P^{\text{pr}}(T_2 - T_1) = C_P^{\text{re}}(T_2 - T_1) + \Delta H_{T2}$$

so that

$$\Delta H_{T2} - \Delta H_{T1} = (C_P^{\text{pr}} - C_P^{\text{re}})(T_2 - T_1) \tag{2.38}$$

If we write the difference $C_P^{\text{pr}} - C_P^{\text{re}}$ as ΔC_P, (2.38) becomes

$$\frac{\Delta H_{T2} - \Delta H_{T1}}{T_2 - T_1} = \Delta C_P \tag{2.39}$$

In the limit as $(T_2 - T_1) \longrightarrow 0$, the difference equation yields the differential form,

$$\frac{d(\Delta H)}{dT} = \Delta C_P \tag{2.40}$$

These equations were first obtained by G. R. Kirchhoff in 1858. They show that the rate of change of the enthalpy of reaction with temperature is equal to the difference in heat capacities C_P of products and reactants.

There is one oversimplification in the treatment given, since actually the heat capacities themselves vary with the temperature. Often, however, it is sufficiently accurate to use the average value of the heat capacity over the range of temperature considered.

As an example of the use of (2.39), consider the reaction

$$H_2O\,(g) \longrightarrow H_2 + \tfrac{1}{2}O_2, \qquad \Delta H^{\ominus} = 241\ 750 \text{ J at } 291.15 \text{ K}$$

What would be the $\Delta H^{\ominus}$ at 298.15 K? Over the small temperature range, the effectively constant C_P values per mole are

$$C_P(H_2O) = 33.56, \qquad C_P(H_2) = 28.83, \qquad C_P(O_2) = 29.12 \text{ J·K}^{-1}\text{·mol}^{-1}$$

Hence,

$$\Delta C_P = C_P(H_2) + \tfrac{1}{2}C_P(O_2) - C_P(H_2O) = 28.83 + \tfrac{1}{2}\,29.12 - 33.56 = 9.83 \text{ J·K}^{-1}$$

From (2.39),

$$\frac{\Delta H^{\ominus}_{298} - 241\ 750}{298 - 291} = 9.83$$

Thus,

$$\Delta H^{\ominus}_{298} = 241\ 820 \text{ J}$$

To integrate (2.40) more exactly, we require expressions for the heat capacities of reactants and products over the temperature range of interest.

The experimental heat capacity data can be represented by a power series:

$$C_P = a + bT + cT^2 + \cdots \tag{2.41}$$

Examples of such heat capacity equations are given in Table 2.5. These three-term equations fit the experimental data to within about 0.5% over a temperature

TABLE 2.5

Heat Capacity of Gases at 1 atm Pressure (273 to 1500 K)[a]

$C_P = a + bT + cT^2$ (C_P in $J \cdot K^{-1} \cdot mol^{-1}$)

Gas	a	$b \times 10^3$	$c \times 10^7$
H_2	29.07	−0.836	20.1
O_2	25.72	12.98	−38.6
Cl_2	31.70	10.14	−2.72
Br_2	35.24	4.075	−14.9
N_2	27.30	5.23	−0.04
CO	26.86	6.97	−8.20
HCl	28.17	1.82	15.5
HBr	27.52	4.00	6.61
H_2O	30.36	9.61	11.8
CO_2	26.00	43.5	−148.3
Benzene	−1.71	326.0	−1100.
n-Hexane	30.60	438.9	−1355.
CH_4	14.15	75.5	−180.

[a]H. M. Spencer, *J. Am. Chem. Soc.*, *67*, 1858 (1945); H. M. Spencer and J. L. Justice, *J. Am. Chem. Soc.*, *56*, 2311 (1934).

range from 273 to 1500 K. When the series expression for ΔC_P is substituted* into (2.40), the integration can be carried out analytically. Thus at a constant pressure, for the standard enthalpy change,

$$d(\Delta H^{\ominus}) = \Delta C_P \, dT = (A + BT + CT^2 + \cdots) \, dT$$

$$\Delta H_T^{\ominus} = \Delta H_0^{\ominus} + AT + \tfrac{1}{2}BT^2 + \tfrac{1}{3}CT^3 + \cdots \qquad (2.42)$$

where A, B, C, etc. are the sums of the individual a, b, c, ... in (2.41). Here, $\Delta H_0^{\ominus}$ is the constant of integration. Any one measurement of $\Delta H^{\ominus}$ at a known temperature T makes it possible to evaluate the constant $\Delta H_0^{\ominus}$ in (2.42). Then the $\Delta H^{\ominus}$ at any other temperature (within the range of validity of the heat capacity equations) can be calculated from the equation.

Rather extensive enthalpy tables are becoming available, which give $(H_T - H_0)$ as a function of T over a wide range of temperatures. The use of these tables makes direct reference to heat capacities unnecessary for calculations of $\Delta H_T^{\ominus}$.

20. Bond Enthalpies

Ever since the time of van't Hoff, chemists have sought to express the structures and properties of molecules in terms of bonds between the atoms. In many cases, to a good approximation, it is possible to express the heat of formation of a mole-

*For a typical reaction,

$$\tfrac{1}{2}N_2 + \tfrac{3}{2}H_2 \longrightarrow NH_3$$

$$\Delta C_P = C_P(NH_3) - \tfrac{1}{2}C_P(N_2) - \tfrac{3}{2}C_P(H_2)$$

cule as an additive property of the bonds forming the molecule. This formulation has led to the concepts of *bond energy* and *bond enthalpy*.

Consider a reaction in which a bond A—B is broken between atom A and atom B:

$$A—B\,(g) \longrightarrow A\,(g) + B\,(g) \qquad (2.43)$$

In terms of this reaction, the *bond energy* A—B has been variously defined by different authors as:

1. The energy change at absolute zero, $\Delta U_0^\ominus$;
2. The enthalpy change at absolute zero, $\Delta H_0^\ominus$;
3. The enthalpy change at 298.15 K, $\Delta H_{298}^\ominus$.

The first two definitions are useful in discussions of molecular structure, which often refer to spectroscopic data on dissociation energies of molecules. The last definition is more convenient for use with thermochemical data and calculations of heats of reaction. We shall adopt definition 3 and follow the notation suggested by Benson* in his excellent "resource paper" on bond energies. Thus, the bond energy $DH^\ominus$ (A—B) of the bond A—B is defined as the $\Delta H_{298}^\ominus$ of the reaction in (2.43). It is evident that this quantity would more correctly be called a *bond enthalpy*, and even though this name is not yet in common use, we shall adopt it for our discussion.

The entities A and B in (2.43) need not be atoms; they may be fragments of molecules. For example, the $DH^\ominus$ of the C—C bond in ethane would be the $\Delta H_{298}^\ominus$ of the reaction,

$$C_2H_6 \longrightarrow 2CH_3$$

It is important to realize that for a given type of bond, the $DH^\ominus$ depends on the particular molecule in which the bond occurs and on its particular situation in that molecule. Consider, for example, the molecule CH_4 and imagine that the H atoms are removed from it one at a time.

$$
\begin{aligned}
&(1)\ CH_4 \longrightarrow CH_3 + H \\
&(2)\ CH_3 \longrightarrow CH_2 + H \\
&(3)\ CH_2 \longrightarrow CH\ + H \\
&(4)\ CH\ \longrightarrow C\ \ + H
\end{aligned}
$$

If we could determine ΔH for each of these reactions, they would give the $DH^\ominus$ of the CH bond in these four different cases. Actually, it is difficult to obtain such data accurately but the approximate values are (1) 422, (2) 364, (3) 385, and (4) 335 kJ·mol⁻¹.

For many purposes, a much simpler kind of information would be adequate. Thus, the four C—H bonds in methane are certainly all equivalent, and if we could imagine the carbon atom reacting with four hydrogen atoms to form methane, we could set one-quarter of this overall enthalpy of reaction equal to the average $DH^\ominus$ of a C—H bond in CH_4. The reaction in question would be

$$CH_4 \longrightarrow C\,(g) + 4H$$

*S. Benson, *J. Chem. Educ.*, **42**, 502 (1965).

To calculate such average $DH^{\ominus}$ values, therefore, we need the enthalpies of formation of the molecules from the atoms. If we know the ΔH of atomization for all the elements, we can use these data to calculate the bond enthalpies from the ordinary standard enthalpies of formation. In most cases, it is not too difficult to obtain the ΔH for converting the elements to monatomic gases. In the case of the metals, this ΔH is simply the heat of sublimation to the monatomic form. For example,

$$Mg(c) \longrightarrow Mg(g), \qquad \Delta H^{\ominus}_{298} = 150.2 \text{ kJ}$$

$$Ag(c) \longrightarrow Ag(g), \qquad \Delta H^{\ominus}_{298} = 289.2 \text{ kJ}$$

In other cases, the heats of atomization can be obtained from the dissociation energies of diatomic gases. For example,

$$\tfrac{1}{2}Br_2(g) \longrightarrow Br(g), \qquad \Delta H^{\ominus}_{298} = 111.9 \text{ kJ}$$

$$\tfrac{1}{2}O_2(g) \longrightarrow O(g), \qquad \Delta H^{\ominus}_{298} = 249.2 \text{ kJ}$$

In a few cases, however, it has proved to be difficult to obtain the ΔH of atomization. The most notorious case is also the most important one, since all the bond enthalpies of organic molecules depend upon it—namely, the heat of sublimation of graphite:

$$C(\text{crystal, graphite}) \longrightarrow C \text{ (gas)}$$

Even today not all scientists are agreed on the correct value for the heat of sublimation of graphite, but the most reasonable value appears to be $\Delta H^{\ominus}_{298} = 716.68 \text{ kJ}$.

Some standard enthalpies for conversion of elements from their standard states to the form of monatomic gases (ΔH of atomization) are given in Table 2.6.

TABLE 2.6
Standard Enthalpies of Atomization of Elements[a]

Element	$\Delta H^{\ominus}_{298}$ (kJ)	Element	$\Delta H^{\ominus}_{298}$ (kJ)
H	217.97	N	472.70
O	249.17	P	314.6
F	78.99	C	716.68
Cl	121.68	Si	455.6
Br	111.88	Hg	60.84
I	106.84	Ni	425.14
S	278.81	Fe	404.5

[a]Data from NBS *Circular* 500 and NBS *Technical Notes* 270–1 and 270–2.

With these data, it is possible to calculate average bond enthalpies from standard enthalpies of formation. Consider, for example, the application of thermochemical data to determine the average $DH^{\ominus}$ of the two O—H bonds in water.

$$H_2 \longrightarrow 2H, \quad \Delta H^{\ominus}_{298} = \quad 436.0 \text{ kJ}$$

$$O_2 \longrightarrow 2O, \quad \Delta H^{\ominus}_{298} = \quad 498.3 \text{ kJ}$$

$$H_2 + \tfrac{1}{2}O_2 \longrightarrow H_2O, \; \Delta H^{\ominus}_{298} = -241.8 \text{ kJ}$$

Therefore,

$$2H + O \longrightarrow H_2O, \quad \Delta H^{\ominus}_{298} = -927.2 \text{ kJ}$$

This is the $\Delta H^{\ominus}_{298}$ for the formation of two O—H bonds, so that the average $DH^{\ominus}$ for the two O—H bonds in water can be taken as 927.2/2 = 463.6 kJ. This value is quite different from the dissociation enthalpy for HOH $\longrightarrow$ H + OH, which is 498 kJ.

The main sources of data for the determination of bond energies are molecular spectroscopy, thermochemistry, and electron-impact studies. The electron-impact method employs a mass spectrometer; the energy of the electrons in the ion source is gradually increased until the molecule is broken into fragments by the impact of an electron. The spectroscopic method is discussed in Section 17.10. The bond dissociation energy calculated from spectroscopic data is actually $\Delta U^{\ominus}_0$. It is usually convenient to calculate from this the $\Delta H^{\ominus}_{298}$. From (2.33), $\Delta H^{\ominus}_0 = \Delta U^{\ominus}_0$ and from (2.38),

$$\Delta H^{\ominus}_{298} = \Delta H^{\ominus}_0 + \Delta C_P \cdot \Delta T$$

For the reaction (2.43), we can assume that reactants and products behave as ideal gases, and that only translational and rotational degrees of freedom contribute to the heat capacity at 298 K (p.147). Hence, $\Delta C_P = 2(\frac{5}{2})R - \frac{7}{2}R = \frac{3}{2}R$, and

$$\Delta H^{\ominus}_{298} = \Delta U^{\ominus}_0 + \tfrac{3}{2}R \cdot 298 = \Delta U^{\ominus}_0 + 3\cdot75 \text{ kJ}$$

The $DH^{\ominus}$ of a bond A—B will be approximately constant in a series of similar compounds. This fact makes it possible to compile tables of average bond enthalpies, which can be used for estimating the $\Delta H^{\ominus}$ values for chemical reactions. The $\Delta H^{\ominus}$'s so obtained are often close enough to the experimental values to provide rapid estimates of reaction enthalpies and enthalpies of formation. Of course, different bond types must be distinguished, e.g., single, double, and triple bonds in the case of C—C bonds. Table 2.7 is a summary of average single bond enthalpies $DH^{\ominus}$ as given by Pauling. Table 2.8 gives a few individual $DH^{\ominus}$ values for specified molecules.

TABLE 2.7
Average Single Bond Enthalpies[a] (kJ·mol⁻¹)

	S	Si	I	Br	Cl	F	O	N	C	H
H	339	339	299	366	432	563	463	391	413	436
C	259	290	240	276	328	441	351	292	348	
N					200	270		161		
O		369			203	185	139			
F		541	258	237	254	153				
Cl	250	359	210	219	243					
Br		289	178	193						
I		213	151							
Si	227	177								
S	213									

[a]After L. Pauling, *Nature of the Chemical Bond*, 3rd ed. (Ithaca: Cornell University Press, 1960)

TABLE 2.8
Single and Multiple Bond Enthalpies (kJ·mol⁻¹)

Triple bonds	$DH^\ominus$	Double bonds	$DH^\ominus$	Single bonds	$DH^\ominus$	Single bonds	$DH^\ominus$
N≡N	946	CH_2=CH_2	682	CH_3—CH_3	368	CH_3—H	435
HC≡CH	962	CH_2=O	732	H_2N—NH_2	243	NH_2—H	431
HC≡N	937	O=O	498	HO—OH	213	OH—H	498
C≡O	1075	HN=O	481	F—F	159	F—H	569
		HN=NH	456	CH_3—Cl	349	CH_3—NH_2	331
		CH_2=NH	644	NH_2—Cl	251	CH_3—OH	381
				HO—Cl	251	CH_3—F	452
				F—Cl	255	CH_3—I	234
						F—I	243

It is interesting to compare the experimental $DH^\ominus$ of a given bond in a variety of different compounds. When appreciable variations are found, the reason is sought in terms of special factors in the molecular structure, such as partial double bond character, ionic character, and steric strains or repulsions.

As an example of the use of tabulated bond enthalpies in estimating a standard enthalpy of formation, consider the case of C_2H_5OH:

$$
\begin{array}{cc}
\begin{array}{c}
\text{H } \text{ H} \\
| \quad | \\
\text{H—C—C—O—H} \\
| \quad | \\
\text{H } \text{ H}
\end{array}
&
\begin{array}{ll}
\textit{Bonds} & DH^\ominus(\text{kJ}) \\
\text{1 C—C} & 348 \\
\text{5 C—H} & 5 \times 413 \\
\text{1 C—O} & 351 \\
\underline{\text{1 O—H}} & \underline{463}
\end{array}
\end{array}
$$

Hence

$$2C\,(g) + O\,(g) + 6H\,(g) \longrightarrow C_2H_5OH, \qquad \Delta H^\ominus_{298} = -3227 \text{ kJ}$$

From the enthalpies of atomization in Table 2.6,

$$
\begin{array}{lll}
2C\,(\text{graphite}) \longrightarrow 2C\,(g) & 2 \times 717 = & 1434 \\
\tfrac{1}{2}O_2 \qquad\qquad \longrightarrow O & & 249 \\
3H_2 \qquad\qquad \longrightarrow 6H & 6 \times 218 = & \underline{1308} \\
& & 2991 \text{ kJ}
\end{array}
$$

Therefore,

$$2C\,(\text{graphite}) + \tfrac{1}{2}O_2 + 3H_2 \longrightarrow C_2H_5OH\,(g), \qquad \Delta H^\ominus_{298} = -236 \text{ kJ}$$

The experimental value is $\Delta H^\ominus_{298} = -237$ kJ, so that the estimate is good to within about 0.5%, which is better than usual.

21.   Chemical Affinity

Many careful determinations of heats of reactions were made by Julius Thomsen and Marcellin Berthelot in the latter part of the nineteenth century. They were inspired to carry out a vast program of thermochemical measurements by the conviction that the heat of reaction was the quantitative measure of the *chemical affinity* of the reactants. In the words of Berthelot, in his *Essai de Mécanique chimique* (1878):

> Every chemical change accomplished without the intervention of an external energy tends toward the production of the body or the system of bodies that sets free the most heat.

Although, as Ostwald remarked in an unusually sarcastic vein, priority for this erroneous principle does not rest with Berthelot,

> what undoubtedly belongs to Berthelot are the numerous methods which he found to explain the cases in which the so-called principle is in contradiction with the facts. In particular, in the assumption of partial decomposition or dissociation of one or several of the reacting substances he discovered a never failing method for calculating an overall evolution of heat in cases where the experimental observation showed directly that there was an absorption of heat.

The principle of Thomsen and Berthelot is incorrect: It would imply that no endothermic reaction could occur spontaneously, and it fails to consider the reversibility of most chemical reactions. To understand the true nature of chemical affinity and the driving force in chemical reactions, we must go beyond the First Law of Thermodynamics and include the results of the Second Law. In the next chapter we shall see how this has been done.

PROBLEMS

1. Derive the expression $(\partial U/\partial T)_P = C_P - P(\partial V/\partial T)_P$.

2. Show that for an ideal gas $(\partial H/\partial V)_T = 0$ and $(\partial C_V/\partial V)_T = 0$.

3. An average man produces about 10^4 kJ of heat a day through metabolic activity. If a man were a closed system of mass 70 kg with the heat capacity of water, estimate his temperature rise in one day. Man is actually an open system and the main mechanism of heat loss is evaporation of water. How much water would need to be evaporated per day to maintain his temperature constant at 37°C? The ΔH (vaporization) of water at 37°C is 2405 J·g^{-1}.

4. Calculate ΔU and ΔH when 1 kg of (a) helium (b) neon are heated from 0 to 100°C in a closed container of 1 m^3 volume. Assume gases are ideal with $C_V = \frac{3}{2}R$ per mole. What information would you need to do this problem if the gases were not ideal?

5. One mole of ideal gas at 300 K is expanded adiabatically and reversibly from 20 to 1 atm. What is the final temperature of the gas, assuming $C_V = \frac{3}{2}R$ per mole?

6. Suppose that a piece of metal with a volume of 100 cm^3 at 1 atm is compressed adiabatically by a shock wave of 10^5 atm to a volume of 90 cm^3. Calculate the ΔU and ΔH of metal. (Assume the compression occurs at a constant pressure of 10^5 atm.)

7. Ammonia at 27.0°C and 1 atm is passed at the rate of 41 cm^3·s^{-1} into an insulated tube where it flows over an electrically heated wire of resistance 100 ohm(Ω) while the heating current is 0.050 ampere(A). The gas leaves the tube at 31.09°C. Calculate C_P and C_V per mole for NH_3.

8. When tungsten carbide WC was burned with excess oxygen in a bomb calorimeter,

it was found for the reaction

$$WC(c) + \tfrac{5}{2}O_2 \longrightarrow WO_3(c) + CO_2$$

that $\Delta U(300 \text{ K}) = -1192$ kJ. What is ΔH at 300 K? What is the ΔH_f of WC from it elements if the ΔH of combustion of pure C and pure W at 300 K are -393.5 kJ and -837.5 kJ, respectively?

9. From the data in Table 2.2, calculate $\Delta H^\ominus$ (298) for the following reactions:
   (a) $2HCl + CO_2 \longrightarrow COCl_2 + H_2O$ (g)
   (b) $2HN_3 + 2NO \longrightarrow H_2O_2 + 4N_2$

10. The $\Delta H_f^\ominus(298)$ for benzene, cyclohexene, and cyclohexane are 82.93, -7.11, and -123.1 kJ·mol^{-1}, respectively. Calculate $\Delta H^\ominus(298)$ for the reaction

$$C_6H_6 \text{ (g)} + 3H_2 \longrightarrow C_6H_{12} \text{ (g)}$$

Compare this value with three times the $\Delta H^\ominus(298)$ for

$$C_6H_{10} \text{ (g)} + H_2 \longrightarrow C_6H_{12} \text{ (g)}$$

Comment on the difference you find.

11. From the data for hydrogenation of benzene to cyclohexane given in Problem 10 and the heat capacity data in Table 2.5, calculate $\Delta H^\ominus$ for the reaction at 500 K. For cyclohexane gas. $C_P = 106.3$ J·K^{-1}·mol^{-1} at 298 K. Estimate the error caused in your calculation by assuming this C_P is constant from 298 to 500 K.

12. The restoring force F on a stretched elastic substance is a function of length l and temperature T; $F(l, T)$. If U is also $U(l, T)$, show that the heat capacity at constant F is

$$C_F = \left(\frac{\partial U}{\partial T}\right)_l + \left[\left(\frac{\partial U}{\partial l}\right)_T - F\right]\left(\frac{\partial l}{\partial T}\right)_F$$

13. The ΔH for oxidation of ferrocytochrome-c by ferricyanide ion,

$$\text{Fe}^{\text{II}}-\text{cyt-}c + \text{Fe}^{\text{III}} \longrightarrow \text{Fe}^{\text{III}}-\text{cyt-}c + \text{Fe}^{\text{II}}$$

was studied in a flow microcalorimeter at 25°C over a range of pH with following results:*

pH	6.00	7.00	8.00	9.00	9.75	10.00
$-\Delta H$(kJ·mol^{-1})	53.6	52.8	45.4	37.1	7.87	-2.93

Plot ΔH vs. pH. How would you interpret these results?

14. A gas follows the equation of state

$$PV = nRT$$

For this gas, $C_P = 29.4 + 8.40 \times 10^{-3}T$(J·K^{-1}·mol^{-1}).
   (a) Calculate C_V as a function of T.
   (b) Given 1 mol of this gas and the points $P_1 = 20.00$ atm, $V_1 = 2.00$ liter, and $P_2 = 5.00$ atm, $V_2 = 8.00$ liter, design an adiabatic process by which these points can be connected.
   (c) Calculate ΔU and ΔH for the gas for the process given in (b).

15. The following is a problem of interest in meteorology. Compute the specific energy changes in joule per kilogram for the following processes:

*G. D. Watt and J. M. Sturtevant, *Biochemistry*, **8**, 4567 (1969).

(a) Dry air is heated at constant volume from 0 to 10°C.

(b) Dry air is expanded isothermally from a volume of 0.9 to 1.0 $dm^3 \cdot g^{-1}$ at 10°C.

(c) The horizontal speed of air is increased from $10 \, m \cdot s^{-1}$ to $50 \, m \cdot s^{-1}$

16. The equation of state of a monatomic solid is

$$PV + nG = BU$$

where G is a function only of the molar volume (V/n), and B is a constant. Prove that

$$B = \frac{\alpha V}{\beta C_V}$$

where α is the thermal expansivity and β is the isothermal compressibility. This is the famous *Gruneisen equation* of interest in studies of the solid state.

17. Ten moles of nitrogen at 300 K are held by a piston under 40 atm pressure. The pressure is *suddenly* released to 10 atm and the gas expands adiabatically. If C_V for N_2 is assumed to be constant and equal to $20.8 \, J \cdot K^{-1} \cdot mol^{-1}$, calculate the final T of the gas. Assume the gas is ideal. Calculate ΔU and ΔH for the change in the gas.

18. The enthalpy of formation of the solid solution of NaCl and NaBr at 298 K as a function of the mole fraction X_2 of NaBr is given by

$$\Delta H_m(kJ \cdot mol^{-1}) = 5.996X_2 - 6.761X_2^2 + 0.765X_2^3$$

Calculate (a) ΔH when 0.5 mol NaCl and 0.5 mol NaBr form a solid solution, and (b) the differential heats of solution ΔH_1 and ΔH_2 of NaCl and NaBr in the 50 mol % solution.

19. Estimate the ΔH of combustion of glucose from the bond energies in Table 2.7. Compare your result with the experimental value of $14.7 \, kJ \cdot g^{-1}$.

20. It is estimated that the adult human brain consumes the equivalent of 10 g glucose per hour. Estimate the power output of the brain in watts (W).

21. The integral heat of solution of m moles of NaCl in 1000 g H_2O at 298 K is given as

$$\Delta H(kJ) = 3.861 \, m + 1.992 \, m^{3/2} - 3.038 \, m^2 + 1.019 \, m^{5/2}$$

Calculate (a) ΔH per mole of NaCl to form one molal solution, (b) ΔH per mole of NaCl to infinite dilution, (c) the ΔH of dilution per mole NaCl from 1 to 0.1 m, and (d) the differential ΔH per mole at 1.0 m NaCl.

22. Show that $(\partial U / \partial P)_V = \beta C_V / \alpha$.

23. If a compound is burned under adiabatic conditions so that all the heat evolved is used in heating the product gases, the maximum temperature reached is called the *adiabatic flame temperature*. Calculate this temperature for the burning of methane (a) in oxygen and (b) in air (80% N_2, 20% O_2), sufficient for complete combustion to CO_2 and H_2O (g). Use the heat capacities in Table 2.5, but neglect the terms in T^2.

24. Calculate ΔH for the combustion of CO to CO_2 (a) at 298 K and (b) at 1680 K, approximately the temperature of a blast furnace.

25. When 1 mol of H_2SO_4 is mixed with n moles of water at 25°C,

$$\Delta H = \frac{-75.6n}{n + 1.80}$$

where ΔH is in kilojoules and $n < 20$. Calculate the differential heat of solution of water, $\Delta H_w = (\partial \, \Delta H/\partial n_w)_{n_s}$ and that of sulfuric acid, $\Delta H_s = (\partial \, \Delta H/\partial n_s)_{n_w}$ when $n = 1$ and when $n = 10$.

26. The volume of a quartz crystal at 30°C is given by

$$V = V_0(1 - 2.658 \times 10^{-10}P + 24.4 \times 10^{-20}P^2)$$

where V_0 is the volume at $P = 0$ and P is given in kilograms per square metre.
(a) How large a pressure would be required to reduce the volume by 1%?
(b) Plot the compressibility at 30°C vs. P.
(c) Calculate the work done on 1 kg crystal in an isothermal reversible compression by 1%.

27. Show that $(\partial H/\partial P)_T = - \mu C_P$. From data in Tables 2.1 and 2.5, calculate the change in enthalpy of 1 mol of CO_2 which is isothermally compressed from 0 to 100 atm at 300 K. Comment briefly on the value found and how it compares with the ideal gas value.

28. The air in a room is heated from 0 to 20°C, the air pressure being maintained constant. What is the ΔU (change in internal energy) of the room? (You can ignore effects due to walls.)

29. The curved surface of a right cone is

$$A = \pi r\sqrt{r^2 + h^2}$$

where r is radius of the base and h is the altitude. Calculate $dA = Mdr + Ndh$ and apply Euler's rule to show that dA is a perfect differential.

3
Entropy
and Free Energy

Science owes more to the steam engine than the steam engine owes to Science.

L. J. Henderson
1917

The experiments of Joule showed that heat was not conserved in physical processes, since it could be generated by mechanical work. The reverse transformation, the conversion of heat into work, had been of interest to the practical engineer ever since the development of the steam engine by James Watt in 1769. Such an engine operates essentially as follows: A source of heat (e.g., a coal fire) is used to heat a *working substance* (e.g., steam), causing it to expand through a valve into a cylinder fitted with a piston. The expansion drives the piston forward, and by suitable coupling, mechanical work can be obtained from the engine. The working substance, which is cooled by the expansion, is withdrawn from the cylinder through a valve. A flywheel arrangement returns the piston to its original position, in readiness for another expansion stroke. In simplest terms, therefore, any such heat engine withdraws heat from a hot reservoir, converts some of this heat into work, and discards the remainder to a cold reservoir. In practice, frictional losses of work occur in the various moving components of the engine.

As the Industrial Revolution got underway in early nineteenth-century England, steady improvements were made in steam engines. Each engine had its own ratio of work output to coal burned; since this ratio increased with each basic advance in technology, no limit to the efficiency of the engines was forseen. There was no general theory to predict the efficiency ε, defined as

$$\varepsilon = \frac{-w}{q_2} \qquad (3.1)$$

where $-w$ was the work output and q_2 the heat input.

In 1824, the theory of this *English machine* was taken up by a young French engineer, Sadi Carnot, in a monograph *Réflexions sur la Puissance Motrice du Feu*. With remarkable insight, he made an abstract model of the essential features of the heat engine, and analyzed its operation with cool and faultless logic.

1.   The Carnot Cycle

Carnot devised a cycle to represent the operation of an idealized engine, in which heat is transferred from a hot reservoir at temperature θ_2, partly converted into work, and partly discarded to a colder reservoir at temperature θ_1 [Fig. 3.1(a)].

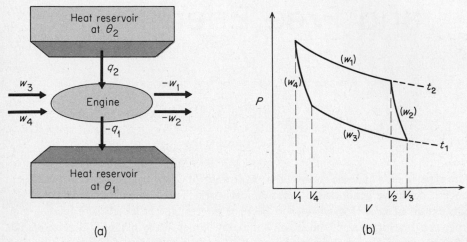

(a)   (b)

FIG. 3.1   Essential features of the Carnot heat engine (a) and the
Carnot cycle for its operation shown on the indicator
diagram (b).

The working substance through which these operations are carried out is returned at the end to the same state that it initially occupied, so that the entire process constitutes a complete cycle. We have written the temperatures as θ_1 and θ_2 to indicate that they are empirical temperatures, measured on any convenient scale.

The various steps in the cycle are carried out reversibly. To make the operation more definite, we may consider the working substance to be a gas (not necessarily ideal) and we may represent the cyclic process by the indicator diagram of Fig. 3.1(b). The steps in the working of the engine for one complete cycle are then [the sign convention is based on the engine (gas) as the system]:

1. The gas is placed in contact with the heat reservoir at θ_2, and expands reversibly and isothermally at θ_2 from V_1 to V_2, absorbs heat q_2 from the hot reservoir, and does work $-w_1$ on its surroundings.
2. The gas, insulated from any heat reservoir, expands reversibly and adiabatically ($q = 0$) from V_2 to V_3, and does work $-w_2$, as its temperature falls from θ_2 to θ_1.
3. The gas is placed in contact with the heat reservoir at θ_1, is compressed reversibly and isothermally at θ_1 from V_3 to V_4, as work w_3 is done upon it, and it gives up heat $-q_1$ to the heat reservoir.

4. The gas, insulated from any heat reservoir, is compressed reversibly and adiabatically ($q = 0$) from V_4 to the initial volume V_1, as work w_4 is done upon it, and its temperature rises from θ_1 to θ_2.

The First Law of Thermodynamics requires that for the cyclic process, $\Delta U = 0$. Now ΔU is the sum of all the heat added to the gas, $q = q_2 + q_1$, plus the sum of all the work done on the gas, $w = w_1 + w_2 + w_3 + w_4$.

$$\Delta U = q + w = q_1 + q_2 + w = 0$$

The work output of the engine is equal, therefore, to the heat taken from the hot reservoir minus the heat returned to the cold reservoir: $-w = q_2 - (-q_1) = q_2 + q_1$.

The efficiency of the engine is

$$\varepsilon = \frac{-w}{q_2} = \frac{q_2 + q_1}{q_2} \tag{3.2}$$

Since every step in this cycle is carried out reversibly, the maximum possible work is obtained for the particular working substance and temperatures considered.*

Consider now another engine operating with a different working substance. Let us assume that this second engine II working between the same two empirical temperatures θ_2 and θ_1 is more efficient than engine I; that is, it can deliver a greater amount of work from the same amount of heat q_2 taken from the hot reservoir (see Fig. 3.2). It could accomplish this result only by discarding less heat to the cold reservoir.

Let us now imagine that after the completion of a cycle by this supposedly more efficient engine, the original engine is run in reverse. It therefore acts as a

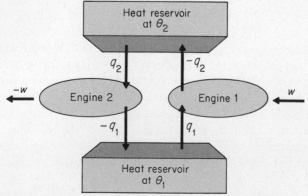

FIG. 3.2   Diagram to represent Engine II operating in forward direction and Engine I in reverse (as a heat pump).

*In the isothermal steps, the maximum work is obtained on expansion and the minimum work done in compression of the gas (cf. p. 51). In the adiabatic steps, $\Delta U = w$, and the work terms are determined only by the initial and final states.

heat pump. Since the original Carnot cycle is reversible, all the heat and work terms are changed in sign but not in magnitude. The heat pump takes in q_1 of heat from the cold reservoir; work w is provided to the pump by an external source; heat $-q_2$ is delivered from the pump to the hot reservoir.

For the forward process (engine II), $\qquad -w' = q_2 + q_1'$

For the reverse process (engine I), $\qquad \dfrac{w + q_1 = -q_2}{}$

Therefore, the net result is $\qquad -w' + w = -q_1 + q_1'$

Since $w' > w$, and $q_1' < q_1$, the net result of the combined operation of the engine and the heat pump is that heat $q'' = q_1' - q_1$ has been abstracted from a reservoir at constant temperature θ_1 and work $w'' = w - w'$ has been obtained without any other change taking place in the entire system.

In this result there is nothing contrary to the First Law of Thermodynamics, for energy has been neither created nor destroyed. The work done would be equivalent to the heat extracted from the reservoir. Nevertheless, an isothermal conversion of heat into work without any concomitant change in the system has never been observed. Think what it would imply: A ship would not need to carry any fuel; it could propel itself by taking heat from the ocean. Such a continuous extraction of work from the heat of the environment has been called *perpetual motion of the second kind*, whereas the production of work by a cyclic process with no change at all in the surroundings was called *perpetual motion of the first kind*. The impossibility of the latter is postulated by the First Law of Thermodynamics; the impossibility of the former is postulated by the Second Law.

If the supposedly more efficient Carnot engine delivered the same amount of work $-w$ as the original engine, it would need to withdraw less heat $q_2' < q_2$ from the hot reservoir. Then the result of running engine II forward and engine I in reverse, as a heat pump, would be

(II) $\qquad -w = q_2' + q_1'$

(I) $\qquad \dfrac{w + q_1 = -q_2}{}$

Net result $\qquad q_2 - q_2' = q_1' - q_1 = q$

This result would be the transfer of heat q from the cold reservoir at θ_1 to the hot reservoir at θ_2 without any other change in the system. In this case, also, there would be nothing contrary to the First Law, but it would be even more obviously contrary to experience than is perpetual motion of the second kind. We know that heat always flows from the hotter to the colder region. If we place a hot body and a cold body together, the hot one never becomes hotter, the cold one colder. In fact, considerable work must be expended to refrigerate something, to pump heat out of it. Heat never flows uphill, i.e., against a temperature gradient, of its own accord.

2. The Second Law of Thermodynamics

This Second Law may be expressed precisely in various equivalent forms.

The principle of Thomson: It is impossible to devise an engine which, *working in a cycle*, shall produce no effect other than the extraction of heat from a reservoir and the performance of an equal amount of work.

The principle of Clausius: It is impossible to devise an engine which, working in a cycle, shall produce no effect other than the transfer of heat from a colder to a hotter body.

In these statements of the Second Law, the phrase *working in a cycle* must be emphasized. The cycle allows us to specify that the working substance returns to exactly its initial state, so that the process can be carried out repeatedly. It is easy enough to convert heat into work isothermally if a cyclic process is not required: for example, simply expand a gas in contact with a heat reservoir.

We have now seen that to assume that any reversible cycle can exist that is more efficient than any other one operating between the same two temperatures leads to a direct contradiction of the Second Law of Thermodynamics. We therefore conclude that *all reversible Carnot cycles operating between the same initial and final temperatures must have the same efficiency.* Since the cycles are reversible, this efficiency is the maximum possible. It is completely independent of the working substance and is a function only of the working temperatures:

$$\varepsilon = g(\theta_1, \theta_2) \tag{3.3}$$

3. The Thermodynamic Temperature Scale

William Thomson (Kelvin) was the first to use the Second Law to define a *thermodynamic temperature scale*, which is completely independent of any thermometric substance. From (3.2) and (3.3) we can write for the efficiency ε of a reversible Carnot cycle, independent of the nature of the working substance,

$$\varepsilon = \frac{q_2 + q_1}{q_2} = g(\theta_1, \theta_2) \tag{3.4}$$

Since $g(\theta_1, \theta_2) - 1$ is also a universal function of the temperatures, say, $f(\theta_1, \theta_2)$, (3.4) becomes

$$\frac{q_1}{q_2} = f(\theta_1, \theta_2) \tag{3.5}$$

Consider now two Carnot cycles sharing an isotherm at θ_2, as shown in Fig. (3.3). Let the heat absorbed by the fluid in expansion along isotherms at $\theta_1, \theta_2, \theta_3$ be q_1, q_2, q_3, respectively. From (3.5),

$$\frac{q_1}{q_2} = f(\theta_1, \theta_2); \qquad \frac{q_2}{q_3} = f(\theta_2, \theta_3); \qquad \frac{q_1}{q_3} = f(\theta_1, \theta_3)$$

Hence,

$$f(\theta_1, \theta_3) = f(\theta_1, \theta_2) f(\theta_2, \theta_3) \tag{3.6}$$

Since θ_2 is an independent variable, the only way (3.6) can be satisfied for any choice of θ_2 is if the function $f(\theta_1, \theta_2)$ has the special form

$$f(\theta_1, \theta_2) = \frac{F(\theta_1)}{F(\theta_2)}$$

where $F(\theta)$ denotes an arbitrary function of the single variable θ. It follows from (3.5) that

$$\frac{q_1}{q_2} = \frac{F(\theta_1)}{F(\theta_2)} \tag{3.7}$$

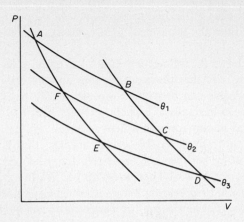

FIG. 3.3 Two Carnot cycles sharing the isotherm at θ_2.

Kelvin decided to use (3.7) as the basis of a *thermodynamic temperature scale*. He took the functions $F(\theta_1)$ and $F(\theta_2)$ to define the thermodynamic temperature function T. Thus, a temperature ratio on the Kelvin scale was defined as equal to the ratio of the heat absorbed to the heat rejected in the working of a reversible Carnot cycle.

$$\frac{q_2}{-q_1} = \frac{T_2}{T_1} \tag{3.8}$$

The efficiency of the cycle, from (3.2), then becomes

$$\varepsilon = \frac{q_2 + q_1}{q_2} = \frac{T_2 - T_1}{T_2} \tag{3.9}$$

The zero point of the thermodynamic scale is physically fixed as the temperature of the cold reservoir at which the efficiency becomes equal to unity, i.e., the heat engine becomes perfectly efficient. From (3.9), in the limit as $T_1 \rightarrow 0$, $\varepsilon \rightarrow 1$.

The efficiency calculated from (3.9) is the maximum thermal efficiency that can be approached by a heat engine. Since it is calculated for a reversible Carnot cycle, it represents an ideal, which actual irreversible cycles can never achieve. Thus, with a heat source at 393 K and a sink at 293 K, the maximum thermal efficiency is 100/393 = 25.4%. If the heat source is at 493 K and the sink still at 293 K, the efficiency is raised to 200/493 = 40.6%. It is easy to see why the trend in power plant design has been to higher temperatures for the heat source. In practice, the efficiency of steam engines seldom exceeds 80% of the theoretical value. Steam turbines generally can operate somewhat closer to their maximum thermal efficiencies, since they have fewer moving parts and consequently lower frictional losses.

In the case of a refrigeration cycle, the maximum efficiency would be that of a reversible Carnot cycle acting as a heat pump. In this case, the efficiency ε' would be

$$\varepsilon' = \frac{q_1}{w + q_1} = \frac{T_1}{T_2} \tag{3.10}$$

Suppose for example, we wish to maintain a system at 273 K in a room at 303 K. The maximum efficiency of the refrigerator would be $\varepsilon' = \frac{273}{303} = 90.1\%$.

4. Relation of Thermodynamic and Ideal-Gas Temperature Scales

Temperature on the Kelvin, or thermodynamic, scale has been denoted by the symbol T, which is the same as that used previously for the absolute temperature scale based on the thermal expansion of an ideal gas. It can be shown that these scales are indeed numerically the same by running a Carnot cycle with an ideal gas as the working substance.

Applying (2.23) and (2.24) to the four steps, we have

1. Isothermal expansion: $\qquad -w_1 = q_2 = RT_2 \ln (V_2/V_1)$

2. Adiabatic expansion: $\qquad -w_2 = \int_{T_1}^{T_2} C_V \, dT; \qquad q = 0$

3. Isothermal compression: $\qquad w_3 = -q_1 = -RT_1 \ln (V_4/V_3)$

4. Adiabatic compression: $\qquad w_4 = \int_{T_1}^{T_2} C_V \, dT; \qquad q = 0$

By summation of these terms, the total work obtained is

$$-w = -w_1 - w_2 - w_3 - w_4 = RT_2 \ln \frac{V_2}{V_1} + RT_1 \ln \frac{V_4}{V_3}$$

Since, from (2.23) and (2.26), $V_2/V_1 = V_3/V_4$,

$$-w = R(T_2 - T_1) \ln \frac{V_2}{V_1}$$

$$\varepsilon = \frac{-w}{q_2} = \frac{T_2 - T_1}{T_2}$$

Comparison with (3.9) completes the proof of the identity of the ideal-gas and thermodynamic temperature scales.

5. Entropy

Carnot's theorem, Equation (3.9), for a reversible Carnot cycle operating between T_2 and T_1 irrespective of the working substance may be written as

$$\frac{q_2}{T_2} + \frac{q_1}{T_1} = 0 \qquad\qquad (3.11)$$

We shall now extend this theorem to any reversible cycle and thereby demonstrate that the Second Law leads to a new state function, the entropy.

Figure 3.4 shows a general cyclic process ANA represented on a PV diagram.* We have superimposed on the cycle a system of adiabatics. These adiabatics can be drawn as closely together as we wish, so that they divide the general cycle into a set of cycles with infinitesimal parts such as AA' and BB', joined by pairs of adiabatics. We must envisage a large number of heat reservoirs at successive tempera-

---

*See P. S. Epstein, *Textbook of Thermodynamics* (New York: John Wiley & Sons, Inc., 1938), p. 57.

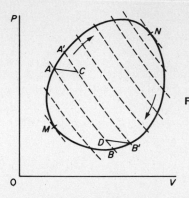

FIG. 3.4 A general cyclic process. *ANA* on a *PV* diagram is traversed by a set of adiabatics shown as dashed lines. This diagram is used in the proof that Carnot's theorem (3.11) can be extended to a cyclic process.

tures that differ by only infinitesimal amounts; heat is transferred from these reservoirs when the working substance is brought successively into contact with them as it traverses the cycle.

Now let us draw the infinitesimal *isotherm* at T_2 from A to C. Let us call dq_2 the actual heat transfer along AA' and dq'_2 the transfer along the isothermal segment AC. When we apply the First Law of Thermodynamics to the infinitesimal cycle $AA'CA$, we find

$$-dw = dq_2 - dq'_2$$

Since dw is given by the area of the little cycle, it is an infinitesimal of the second order, and must be neglected compared with $dq_2 \approx dq'_2$. That is, we can set the heat transferred to the working substance in each strip of the cycle [defined by a pair of adjacent adiabatics] equal to a corresponding transfer in an isothermal process. The same argument obviously applies to dq_1 and dq'_1 at the other end of the adiabatics.

Since $ACB'D$ is a Carnot cycle, we can apply (3.11) to it, to obtain

$$\frac{dq_2}{T_2} + \frac{dq_1}{T_1} = 0$$

We proceed in like manner with each of the strips that comprise the general cycle, and thus obtain for the entire cycle,

$$\oint \frac{dq}{T} = 0 \quad \text{(reversible)} \tag{3.12}$$

This equation holds true for *any reversible cyclic process* whatsoever.

Recall (p. 44) that the vanishing of the cyclic integral means that the integrand is a perfect differential of some function of the state of the system. We can thus define a new state function S by

$$dS = \frac{dq}{T} \quad \text{(for a reversible process)} \tag{3.13}$$

For a change from state A to state B,

$$\Delta S = \int_A^B \frac{dq}{T}$$

Hence,

$$\oint dS = \int_{A}^{B} dS + \int_{B}^{A} dS = S_B - S_A + S_A - S_B = 0$$

The function S was first introduced by Clausius in 1850; he called it the *entropy*, (Greek, $\tau\rho\acute{\epsilon}\pi\epsilon\iota\nu$, *to give a direction*). Equation (3.13) indicates that when the inexact differential expression dq is multiplied by $1/T$, it becomes an exact differential; $1/T$ is called an *integrating factor*. The integrand $\int_{A}^{B} dq_{rev}$ is dependent on the path, whereas $\int_{A}^{B} dq_{rev}/T$ is independent of the path. This is an alternative statement of the Second Law of Thermodynamics.

The TS diagram in Fig. 3.5 is analogous to the PV diagram of Fig. 3.1. In the

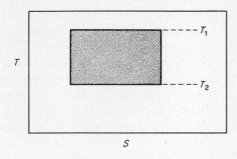

FIG. 3.5 Carnot cycle on a TS diagram. The shaded area is $\oint T dS$, the heat transferred reversibly to the system.

PV case, the area under the curve is a measure of the work done in traversing the indicated path. In the TS diagram, the area under the curve is a measure of the heat added to the system. Temperature and pressure are intensity factors; entropy and volume are capacity factors. The products $P\,dV$ and $T\,dS$ both have the dimensions of energy.

6. First and Second Laws Combined

From (3.13) and (2.4), we obtain an important relation sometimes called the *combined First and Second Laws:*

$$dU = T\,dS - P\,dV \tag{3.14}$$

(This relation applies to any system of constant composition in which only PV work is considered.) If we consider U as a function of S and V, $U(S, V)$,

$$dU = \left(\frac{\partial U}{\partial S}\right)_V dS + \left(\frac{\partial U}{\partial V}\right)_S dV$$

Comparison with (3.14) gives a new equation for the temperature,

$$\left(\frac{\partial U}{\partial S}\right)_V = T \tag{3.15}$$

and for the pressure,

$$\left(\frac{\partial U}{\partial V}\right)_S = -P \tag{3.16}$$

By means of these equations, the intensive variables P and T are given in terms of extensive variables of the system, U, V, and S.

7. The Inequality of Clausius

Equation (3.12) is valid for a *reversible* cycle. Clausius showed that for a cycle into which irreversibility enters at any stage, the integral of dq/T is always less than zero.

$$\oint \frac{dq}{T} < 0 \quad \text{(irreversible)} \tag{3.17}$$

We should note that the T in (3.17) is the temperature of the reservoir that supplies the heat, and not the temperature of the body to which the heat is supplied. In the case of a reversible process, this distinction is, of course, unnecessary, because no temperature gradient can exist under reversible (equilibrium) conditions.

The proof of (3.17) is based on the fact that the efficiency of an irreversible Carnot cycle is always less than that of a reversible cycle operating between the same two temperatures. In the reversible cycle, the isothermal expansion yields the maximum work and the isothermal compression requires the minimum work, so that the efficiency is highest for the reversible case. For the irreversible case, we therefore conclude from (3.9) that

$$\frac{q_2 + q_1}{q_2} < \frac{T_2 - T_1}{T_2}$$

Rearranging this inequality, we have

$$\frac{q_2}{T_2} + \frac{q_1}{T_1} < 0$$

This relation is extended to the general irreversible cycle, by following the argument based on Fig. (3.4). Instead of (3.12), which applies to the reversible case, we obtain the *inequality of Clausius*, given by (3.17).

8. Entropy Changes in an Ideal Gas

The calculation of entropy changes in an ideal gas is particularly simple because in this case $(\partial U/\partial V)_T = 0$, and energy terms due to cohesive forces never need to be considered. For a reversible process in an ideal gas, the First Law requires that

$$dq = dU + P\,dV = C_V\,dT + \frac{nRT\,dV}{V}$$

Therefore,

$$dS = \frac{dq}{T} = \frac{C_V\,dT}{T} + \frac{nR\,dV}{V} \tag{3.18}$$

On integration,

$$\Delta S = S_2 - S_1 = \int_1^2 C_V\,d\ln T + \int_1^2 nR\,d\ln V$$

If C_V is independent of temperature,

$$\Delta S = C_V \ln \frac{T_2}{T_1} + nR \ln \frac{V_2}{V_1} \tag{3.19}$$

For the special case of a temperature change at constant volume, the increase in entropy with increase in temperature is, therefore,

$$\Delta S = C_V \ln \frac{T_2}{T_1} \tag{3.20}$$

If the temperature of 1 mol of ideal gas with $C_V = 12.5 \, \text{J} \cdot \text{K}^{-1} \cdot \text{mol}^{-1}$ is doubled, the entropy is increased by $12.5 \ln 2 = 8.63 \, \text{J} \cdot \text{K}^{-1}$.

For the case of an isothermal expansion, the entropy increase becomes

$$\Delta S = nR \ln \frac{V_2}{V_1} = nR \ln \frac{P_1}{P_2} \tag{3.21}$$

If 1 mol of ideal gas is expanded to twice its original volume, its entropy is increased by $R \ln 2 = 5.74 \, \text{J} \cdot \text{K}^{-1}$.

9. Change of Entropy in Change of State of Aggregation

An example of a change in state of aggregation is the melting of a solid. At a fixed pressure, the melting point is a definite temperature T_m at which solid and liquid are in equilibrium. To change some of the solid to liquid, heat must be added to the system. As long as both solid and liquid are present, this added heat does not change the temperature of the system, but is absorbed by the system as the *latent heat of melting* ΔH_m of the solid. Since the change occurs at constant pressure, the latent heat, by (2.13), equals the difference in enthalpy between liquid and solid. Per mole of substance,

$$\Delta H_m = H(\text{liquid}) - H(\text{solid})$$

At the melting point, liquid and solid exist together in equilibrium. The addition of a little heat would melt some of the solid, the removal of a little heat would solidify some of the liquid, but the equilibrium between solid and liquid would be maintained. The latent heat at the melting point is necessarily a reversible heat, because the process of melting follows a path consisting of successive equilibrium states. We can therefore evaluate the entropy of melting ΔS_m at the melting point by a direct application of the relation $\Delta S = q_{\text{rev}}/T$, which applies to any reversible isothermal process:

$$S(\text{liquid}) - S(\text{solid}) = \Delta S_m = \frac{\Delta H_m}{T_m} \tag{3.22}$$

For example, ΔH_m for ice is $5980 \, \text{J} \cdot \text{mol}^{-1}$, so that $\Delta S_m = 5980/273.2 = 21.90$ $\text{J} \cdot \text{K}^{-1} \cdot \text{mol}^{-1}$.

By similar argument, the entropy of vaporization ΔS_v, the latent heat of vaporization ΔH_v, and the boiling point T_b, are related by

$$S(\text{vapor}) - S(\text{liquid}) = \Delta S_v = \frac{\Delta H_v}{T_b} \tag{3.23}$$

A similar equation holds for a change from one form of a polymorphic solid to another, if the change occurs at a T and P at which the two forms are in equilibrium, and if there is a latent heat associated with the transformation. For example, grey tin and white tin are in equilibrium at 286 K and 1 atm, and $\Delta H_t = 2090$ J·mol⁻¹. Then $\Delta S_t = 2090/286 = 7.31$ J·K⁻¹·mol⁻¹.

10.   Entropy Changes in Isolated Systems

The change in entropy in going from an equilibrium state A to an equilibrium state B is always the same, irrespective of the path between A and B, since the entropy is a function of the state of the system alone. It makes no difference whether the path is reversible or irreversible. Only if the path is reversible, however, is the entropy change given by $\int dq/T$:

$$\Delta S = S_B - S_A = \int_A^B \frac{dq}{T} \quad \text{(reversible)} \tag{3.24}$$

To evaluate the entropy change for an irreversible process, we must devise a reversible method for going from the same initial to the same final state, and then apply (3.24). In the kind of thermodynamics we have formulated in this chapter (sometimes called *thermostatics*), the entropy S is defined only for equilibrium states. Therefore, to evaluate a change in entropy, we must design a process that consists of a succession of equilibrium states—i.e., a reversible process.

In any completely isolated system, we are restricted to adiabatic processes because no heat can either enter or leave such a system.* For a *reversible* process in an isolated system, therefore, $dS = dq/T = 0/T = 0$, so that on integration, $S = $ constant. If one part of the system increases in entropy, the remaining part must decrease by an exactly equal amount.

A fundamental example of an irreversible process is the transfer of heat from a warmer to a colder body. We can make use of an ideal gas to carry out the transfer reversibly, and thereby calculate the entropy change. The process is summarized in Fig. 3.6. (1) The gas is placed in thermal contact with the warm reservoir at T_2 and expanded reversibly and isothermally until it takes up heat equal to q. To simplify the argument, we assume that the reservoirs have heat capacities so large that changes in their temperatures on adding or withdrawing heat q are negligible. (2) The gas is then removed from contact with the hot reservoir and allowed to expand reversibly and adiabatically until its temperature falls to T_1. (3) Next, it is placed in contact with the colder reservoir at T_1 and compressed isothermally until it gives up heat equal to q.

The hot reservoir has now lost entropy equal to q/T_2, whereas the cold reservoir has gained entropy equal to q/T_1. The net entropy change of the reservoirs has therefore been $\Delta S = q/T_1 - q/T_2$. Since $T_2 > T_1$, $\Delta S > 0$, and their entropy has

---

*The completely isolated system is, of course, a figment of imagination. Perhaps our whole universe might be considered an isolated system, but no small section of it can be rigorously isolated. As usual, the precision and sensitivity of our experiment must be allowed to determine how the system is to be defined.

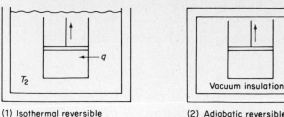 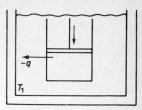

(1) Isothermal reversible expansion at T_2.

(2) Adiabatic reversible expansion. $T_2 \rightarrow T_1$ $q = 0$.

(3) Isothermal reversible compression at T_1.

FIG. 3.6(a)   Reversible transfer of heat from T_2 to T_1.

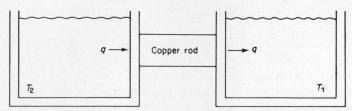

FIG. 3.6(b)   Irreversible transfer of heat from T_2 to T_1.

increased. The entropy of the ideal gas, however, has decreased by an exactly equal amount, so that for the entire isolated system of ideal gas plus heat reservoirs, $\Delta S = 0$ for the reversible process. If the heat transfer had been carried out irreversibly—e.g., by placing the two reservoirs in direct thermal contact and allowing heat q to flow along the finite temperature gradient thus established—there would have been no compensating entropy decrease. The entropy of the isolated system would have increased during the irreversible process, by the amount $\Delta S = q/T_1 - q/T_2$.

We shall now prove that *the entropy of an isolated system always increases during an irreversible process*. The proof of this theorem is based on the inequality of Clausius. Consider in Fig. 3.7 a perfectly general irreversible process in an iso-

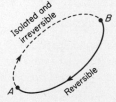

FIG. 3.7   A cyclic process in which the isolated system passes from A to B by an irreversible process and returns to its initial state A by a reversible process during which it is not isolated.

lated system, leading from state A to state B. It is represented by the dashed line. Next consider that the system is returned to its initial state A by a reversible path represented by the solid line from B to A. During this reversible process, the system need not be isolated, and can exchange heat and work with its environment. Since the entire cycle is in part irreversible, (3.17) applies, and

$$\oint \frac{dq}{T} < 0$$

Writing the cycle in terms of its two sections, we therefore obtain

$$\int_A^B \frac{dq_{\text{irrev}}}{T} + \int_B^A \frac{dq_{\text{rev}}}{T} < 0 \tag{3.25}$$

The first integral is equal to zero, since during the process $A \longrightarrow B$ the system is by hypothesis isolated and therefore no transfer of heat is possible. The second integral, from (3.24), is equal to $S_A - S_B$. Therefore, (3.25) becomes

$$S_A - S_B < 0 \quad \text{or} \quad S_B - S_A > 0$$

We have therefore proved that the entropy of the final state B is always greater than that of the initial state A, if A passes to B by an irreversible process in an isolated system.

Since all naturally occurring processes are irreversible, any change that actually occurs spontaneously in nature is accompanied by a net increase in entropy. This conclusion led Clausius to his famous concise statement of the laws of thermodynamics. "The energy of the universe is a constant; the entropy of the universe tends always toward a maximum."

11. Entropy and Equilibrium

Now that the entropy function has been defined and a method outlined for the evaluation of entropy changes, we have gained a powerful technique for analyzing the fundamental problem of physicochemical equilibrium. In our introductory chapter, the position of equilibrium in purely mechanical systems was shown to be the position of minimum potential energy. What is the criterion for equilibrium in a thermodynamic system?

We can combine (3.24) and (3.25) into one expression,

$$\int \frac{dq}{T} \leq \Delta S \tag{3.26}$$

in which the equality refers to reversible processes and the inequality to irreversible ones. Applied to an isolated system, for which $dq = 0$, (3.26) yields

$$\Delta S \geq 0 \quad \text{(isolated system)} \tag{3.27}$$

As previously shown, the entropy of an isolated system can never decrease. Since entropy is defined only for equilibrium states, the fact that it can increase in an isolated system implies that we can still produce certain changes in such a system, even though it remains isolated.

Let us therefore examine more closely what is implied by the condition that the system be isolated. Any system that we single out from the world may be subject to numerous *constraints*, which are imposed when we select or design the system. Let us first specify the constraints that are necessary and sufficient conditions for the system to be isolated in the thermodynamic sense, and then see what further kinds of constraints we are still at liberty to impose upon, or remove from, the system. From the First Law, the energy U of the isolated system must be constant, or $dU = 0$. Also, since $dU = dq - P\,dV$ and $dq = 0$ for the isolated system, we must have $dV = 0$ also. Thus, the necessary constraints on an isolated system are that U and V cannot change.

As an example of an isolated system subject to a constraint, consider two volumes, of H_2 gas and Br_2 gas, separated by a barrier held in place by a lock outside the isolated system. We can operate the lock from outside the system in such a way that neither U nor V of the system changes. As soon as the constraint of the barrier is removed, the H_2 and Br_2 will mix together and react, $H_2 + Br_2 \rightleftharpoons 2HBr$. By such an arrangement, we have therefore carried out a chemical reaction in an isolated system. The system has passed from one equilibrium state (separated $H_2 + Br_2$) to a quite different equilibrium state (equilibrium mixture of H_2, Br_2, and HBr). We know from (3.27) that $\Delta S > 0$ for this change. The original entropy of the system (with the constraint imposed) has increased to a higher value (with that particular constraint removed).

Now let us examine the new equilibrium state that has been reached. Subject to the now existing set of constraints, there is no possible variation of the state of the system that would allow it to move to an equilibrium state of lower entropy. Quite reasonably you may object that since the system is in equilibrium, there is no possible variation of the system at all unless we further relax the constraints. This is quite true, and to describe the equilibrium conditions, we introduce the idea of *virtual displacements* of the state variables of the system. A virtual displacement is not an actual physical displacement, but one of a class of mathematically conceivable displacements. Suppose δx_1, δx_2, δx_3, ..., etc., are the virtual displacements of the variables x_1, x_2, x_3, ..., etc. Then the virtual change in entropy would be

$$\delta S = \left(\frac{\partial S}{\partial x_1}\right)_{x_2, x_3, \cdots} \delta x_1 + \left(\frac{\partial S}{\partial x_2}\right)_{x_1, x_3, \cdots} \delta x_2 + \left(\frac{\partial S}{\partial x_3}\right)_{x_1, x_2, \cdots} \delta x_3 + \cdots$$

The equilibrium condition in the isolated system requires that for any such virtual displacement,

$$(\delta S)_{U,V} \leq 0 \quad \text{(for equilibrium)} \tag{3.28}$$

This criterion for thermodynamic equilibrium in an isolated system can be stated as follows: In a system at constant energy and volume, the entropy is a maximum. At constant U and V, the S is a maximum. As we have seen, the maximum may be subject to any additional constraints we have imposed on the system. If we remove all such constraints, the equilibrium condition is, of course, the absolute maximum of S at constant U and V.

If instead of a system at constant U and V, a system at constant S and V is considered, the equilibrium criterion takes the following form: At constant S and V, the U is a minimum. This is just the condition applicable in ordinary mechanics, in which thermal effects are excluded. In terms of a virtual change in U, it is written

$$(\delta U)_{V,S} \geq 0 \tag{3.29}$$

12.   Thermodynamics and Life

A question often raised is whether living organisms can escape in some way from the rigorous requirements of thermodynamics. Suppose, for example, we enclosed a baby elephant in an isolated system with plenty of air, water, and food. When it had grown to an elephant weighing about 6000 kg, would we find that the entropy of the system had decreased? Suppose we say that the answer is *no*. If we consider the elephant alone, the entropy of the matter comprising his body may have decreased, but there may have been an increase in the entropy of the matter in the rest of the system more than sufficient to cause a net increase in entropy. However, the problem is more tricky than this, since the system is not at equilibrium to begin with, nor is it at equilibrium when the elephant has achieved its full growth. Therefore, although we can define the energy U and volume V of the system, we cannot define its entropy S in the nonequilibrium state. Thus, we have no method for calculating the entropy change. The same difficulty is illustrated by the problem arising if the elephant dies and we ask what is ΔS for the change, live elephant $\longrightarrow$ dead elephant. Although the dead elephant could, in principle, be in a state of equilibrium (kept cold enough, like a mammoth in Siberian ice), the live elephant certainly is not. At best, it is in a steady state with respect to input and output of fuel, oxygen, combustion products, etc.

The same problem arises when we consider the spontaneous emergence of life from a primeval soup of prebiotic molecules. We are not able to define either the initial or the final state as an equilibrium state. A solution to the basic problem might be found if we could define the entropy also for nonequilibrium states. (We shall see to what extent this may be possible after we have examined the calculation of entropy by statistical mechanics.) Ordinary equilibrium thermodynamics can therefore tell us nothing directly about processes in living systems, although, of course, it can provide much useful information about the properties of various chemicals composing such systems. If we wish to apply thermodynamics to living systems under steady state conditions, we shall need to formulate a new science of *irreversible thermodynamics* (see Section 9.18).

13.   Equilibrium Conditions for Closed Systems

The drive, or perhaps better the drift, of physicochemical systems toward equilibrium is determined by two factors. One is the tendency toward minimum energy, the bottom of the potential energy curve. The other is the tendency toward maximum entropy. Only if U is held constant can S achieve its maximum; only if S is held constant can U achieve its minimum. What happens when U and S are forced to strike a compromise?

Chemical reactions are rarely studied under conditions of constant entropy or constant energy. Usually the chemist places his systems in thermostats and investigates them under conditions of approximately constant temperature and pressure. Sometimes changes at constant volume and temperature are followed, as in bomb calorimeters. It is most desirable, therefore, to obtain criteria for thermodynamic equilibrium that are applicable under these practical conditions. A system under these conditions is called a *closed system*, since no mass can be transferred across the boundary of the system, although transfer of energy is allowed.

Let us first consider a closed system under a condition of constant volume and temperature. Such a system would have perfectly rigid walls so that no $P\,dV$ work could be done on it. The rigid container would be surrounded by a heat bath of virtually infinite heat capacity at a constant temperature T, so that heat transfers could take place between the system and the heat bath without changing the temperature of the latter. In equilibrium, the temperature of the system would be held constant at T.

Helmholtz introduced a new state function especially suitable for discussions of the system at constant T and V, or, indeed, whenever we wish to specify its state in terms of the independent variables T, V, and appropriate composition variables. This function is called the *Helmholtz free energy* and is defined as

$$A = U - TS \tag{3.30}$$

Its complete differential would be

$$dA = dU - T\,dS - S\,dT \tag{3.31}$$

From the First Law, $dU = dq + dw$, so that (3.31) becomes

$$dA = dq - T\,dS + dw - S\,dT \tag{3.32}$$

If only $P\,dV$ work is considered,

$$dA = dq - T\,dS - P\,dV - S\,dT \tag{3.33}$$

At constant T and V,

$$dA = dq - T\,dS$$

From (3.26), $dq \leq T\,dS$ where the equality applies to reversible and the inequality to spontaneous changes. Therefore, at constant T and V, for any change in the independent variables of the system,

$$dA \leq 0$$

Hence, the equilibrium condition in terms of a virtual displacement from equilibrium δA becomes

$$\delta A \geq 0 \quad \text{(at constant } T \text{ and } V, \text{ no work)} \tag{3.34}$$

Thus, in a closed system at constant T and V, with no work done on system, the Helmholtz function A is a minimum at equilibrium.

14. The Gibbs Function—
Equilibrium at Constant *T* and *P*

Probably the most frequently encountered condition for a closed system is that of constant temperature and pressure. To a first approximation, this would be the common operating condition in a thermostat at atmospheric pressure, but a more precise definition would require the system to be placed in a temperature bath of virtually infinite heat capacity at fixed temperature T, and to be enclosed in a system designed to regulate the pressure at constant P.

The special function most suitable for such conditions, in which the state of the system is to be specified by T, P, and the necessary composition variables, was invented by J. Willard Gibbs. It has been called the *Gibbs free energy*, the *free enthalpy* or simply the *Gibbs function* and is denoted by G. The definition is

$$G = H - TS = U + PV - TS \qquad (3.35)$$

or

$$G = A + PV$$

Its complete differential becomes

$$dG = dU + P\,dV + V\,dP - T\,dS - S\,dT$$

At constant T and P,

$$dG = dq + dw + P\,dV - T\,dS$$

From (3.26), $dq \leq T\,dS$, and therefore, for any change in the independent variables of the system,

$$dG \leq 0 \quad \text{(at constant } T \text{ and } P, \text{ only } PV \text{ work)}$$

The equilibrium condition in terms of virtual changes in G at equilibrium becomes

$$\delta G \geq 0 \quad \text{(at constant } T \text{ and } P, \text{ only } PV \text{ work)} \qquad (3.36)$$

In a closed system at constant T and P, with only PV work allowed, the Gibbs function G is a minimum at equilibrium.

15.   Isothermal Changes in A and G

Consider in Fig. 3.8 a system that changes from state 1 to state 2 via different processes at constant temperature. There is an infinite number of isothermal paths that the system might traverse between states 1 and 2, but only one path is reversible. Since A is a state function, $\Delta A = A_2 - A_1$ does not depend on the path. From (3.30), for an isothermal change

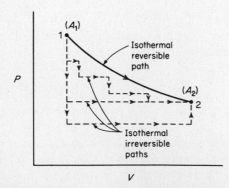

FIG. 3.8 Isothermal paths traversed by a system between state (1) and state (2).

$$A_2 - A_1 = U_2 - U_1 - TS_2 + TS_1$$

or

$$\Delta A = \Delta U - T\,\Delta S \quad \text{(constant } T) \tag{3.37}$$

From (3.32), for the *isothermal reversible path*, $dA = -PdV$ so that $\Delta A = w_{rev}$, the reversible work done on the system. The maximum work that can be *done by the system* on its surroundings in the change from 1 to 2 is $-\Delta A = -w_{rev}$.

Let us consider, for example, that the change in question is the oxidation of 1 mol of 2, 2, 4,-trimethylpentane ("iso-octane") at 298 K and 1 atm.

$$C_8H_{18}\,(g) + 12\tfrac{1}{2}O_2 \longrightarrow 8CO_2 + 9H_2O\,(g)$$

We can find ΔU for this reaction by measuring the heat of combustion of C_8H_{18} in a bomb calorimeter. It is $\Delta U_{298} = -5109$ kJ. The ΔS can be found by calorimetric methods to be described later. It is $\Delta S_{298} = 422$ J·K^{-1}. From (3.37),

$$\Delta A = \Delta U - T\Delta S$$
$$\Delta A_{298} = -5109 - 298(0.422) = -5235 \text{ kJ}$$

This is ΔA for the reaction at 298 K and 1 atm, irrespective of how it is carried out. It tells us that 5235 kJ is the maximum work that can possibly be obtained from the oxidation of 1 mol of iso-octane at 298 K and 1 atm. Note that this work is actually greater than $-\Delta U$ for the change, simply because the ΔS is positive. This result does not contradict the First Law because the system is not isolated.

If we burned the iso-octane in a calorimeter, we would obtain no work at all, $w = 0$. If we burned it in an internal combustion engine, we would obtain some work, perhaps as much as 1000 kJ. We might burn the octane in a fuel cell, and obtain considerably more useful work, perhaps as much as 3000 kJ. There would, however, be no practical way to obtain $-w_{rev} = 5235$ kJ, since to achieve the reversible process it would be necessary to eliminate all frictional losses and to carry out the process in the cell infinitely slowly, with an opposing emf always almost equal to the driving force. Nevertheless, it is useful to know that $-\Delta A$ for the chemical reaction gives the upper limit for the work that can be obtained.

The change in the Gibbs free energy of a system in an *isothermal* process leading from state 1 to state 2 is given from (3.35) as

$$G_2 - G_1 = H_2 - H_1 - T(S_2 - S_1)$$
$$\Delta G = \Delta H - T\,\Delta S \quad \text{(constant } T) \tag{3.38}$$

If we consider a process at constant pressure, from (3.35),

$$\Delta G = \Delta A + P\,\Delta V \quad \text{(constant } P)$$

Let us compute ΔG for the burning of iso-octane. From the reaction equation,

$$P\,\Delta V = \Delta n\,RT = (17 - 13.5)RT = 3.5RT$$
$$= (3.5)(8.314)(298) = 8680 \text{ J}$$

since $\Delta A = -5235$ kJ, we obtain $\Delta G = -5226$ kJ.

16.   Thermodynamic Potentials

In mechanics, the potential energy U serves as the potential function in terms of which equilibrium can be specified and from which forces acting on the system can be derived. If the potential U is given as a function of variables $r_1, r_2, \ldots, r_j \ldots$ defining the state of the system, a generalized force acting on the system is

$$F_j = -\left(\frac{\partial U}{\partial r_j}\right)_{r_1, r_2, \ldots}$$

The force is the gradient of the potential. For example, if the r's are ordinary Cartesian coordinates, x, y, z, the gradients

$$F_x = -\left(\frac{\partial U}{\partial x}\right)_{y,z}, \qquad F_y = -\left(\frac{\partial U}{\partial y}\right)_{x,z}, \qquad F_z = -\left(\frac{\partial U}{\partial z}\right)_{x,y}$$

give the components of the force acting on the system in each of three principal directions.

As a result of our study of the First and Second Laws of Thermodynamics, we have found two new functions that can be used as *thermodynamic potentials*. For a pure substance these functions can be written* $U(S, V)$ and $H(S, P)$. We specify the independent variables in each case since *in terms of its natural set of independent variables each function gives a simple equilibrium condition*. The two new functions A and G are also thermodynamic potentials, written in terms of their natural variables as $A(V, T)$ and $G(P, T)$.

We can consider the gradients of any of these thermodynamic potentials as generalized forces. For example, in systems at constant T and P, as we shall see later, it is convenient to think of gradients of G as driving forces for chemical and physical processes.

17.   Legendre Transformations

The total differentials of the thermodynamic functions can be related by means of *Legendre transformations*, which we shall first define in mathematical terms and then apply to the thermodynamic equations. Suppose we have a function $f(x_1, x_2, \ldots, x_n)$. Here f is the dependent variable and $x_1, x_2, \ldots, x_n$ are the independent variables. The complete differential of f is

$$df = f_1 \, dx_1 + f_2 \, dx_2 + \cdots + f_n \, dx_n \tag{3.39}$$

where

$$f_1 = \left(\frac{\partial f}{\partial x_1}\right)_{x_2 \cdots x_n}, \text{ etc.}$$

Now consider the function,

$$g = f - f_1 x_1 \tag{3.40}$$

---

*The use of the same symbol U for the mechanical *potential energy* and the thermodynamic *internal energy* may not be entirely justified. Why not?

On differentiation,

$$dg = df - d(f_1 x_1) = df - f_1\, dx_1 - x_1\, df_1$$

Or, from (3.39),

$$dg = -x_1\, df_1 + f_2\, dx_2 + \cdots + f_n\, dx_n$$

We have thus changed an independent variable from x_1 to f_1 and the dependent variable from f to g. Equation (3.40) is called a *Legendre transformation*.

Let us apply a Legendre transformation to the basic equation (3.14):

$$U = U(V, S)$$
$$dU = -P\, dV + T\, dS$$

In accord with (3.40),

$$H = U - \left(\frac{\partial U}{\partial V}\right)_S V = U + PV$$

so that,

$$dH = V\, dP + T\, dS$$

Similarly, since $(\partial U/\partial S)_V = T$,

$$A = U - \left(\frac{\partial U}{\partial S}\right)_V S = U - TS$$
$$dA = dU - T\, dS - S\, dT = -P\, dV - S\, dT$$

Finally, since $(\partial H/\partial S)_P = T$,

$$G = H - \left(\frac{\partial H}{\partial S}\right)_P S = H - TS$$
$$dG = dH - T\, dS - S\, dT = V\, dP - S\, dT$$

Mathematically, therefore, the introduction of the new thermodynamic potentials H, A, and G is achieved by performing Legendre transformations on the basic $U(S, V)$.

18. Maxwell's Relations

Let us summarize the four important relations for differentials of thermodynamic potentials:

$$
\begin{aligned}
dU &= -P\, dV + T\, dS \\
dH &= V\, dP + T\, dS \\
dA &= -P\, dV - S\, dT \\
dG &= V\, dP - S\, dT
\end{aligned}
\tag{3.41}
$$

From these we can at once derive relations between partial differential coefficients. Since,

$$dU = \left(\frac{\partial U}{\partial V}\right)_S dV + \left(\frac{\partial U}{\partial S}\right)_V dS$$

$$dH = \left(\frac{\partial H}{\partial P}\right)_S dP + \left(\frac{\partial H}{\partial S}\right)_P dS$$

$$dA = \left(\frac{\partial A}{\partial V}\right)_T dV + \left(\frac{\partial A}{\partial T}\right)_V dT \qquad (3.42)$$

$$dG = \left(\frac{\partial G}{\partial P}\right)_T dP + \left(\frac{\partial G}{\partial T}\right)_P dT$$

we can equate the coefficients of the differentials in (3.41) and (3.42) to obtain

$$\left(\frac{\partial U}{\partial V}\right)_S = -P \qquad \left(\frac{\partial U}{\partial S}\right)_V = T$$

$$\left(\frac{\partial H}{\partial P}\right)_S = V \qquad \left(\frac{\partial H}{\partial S}\right)_P = T$$

$$\left(\frac{\partial A}{\partial V}\right)_T = -P \qquad \left(\frac{\partial A}{\partial T}\right)_V = -S \qquad (3.43)$$

$$\left(\frac{\partial G}{\partial P}\right)_T = V \qquad \left(\frac{\partial G}{\partial T}\right)_P = -S$$

We have seen some of these equations before, but it is worthwhile to bring them all together.

If we apply Euler's relation (2.9) to the differentials in (3.41), we obtain valuable relations between first partial differential coefficients, which are known as *Maxwell's equations:*

$$\left(\frac{\partial T}{\partial V}\right)_S = -\left(\frac{\partial P}{\partial S}\right)_V$$

$$\left(\frac{\partial T}{\partial P}\right)_S = \left(\frac{\partial V}{\partial S}\right)_P$$

$$\left(\frac{\partial P}{\partial T}\right)_V = \left(\frac{\partial S}{\partial V}\right)_T \qquad (3.44)$$

$$\left(\frac{\partial V}{\partial T}\right)_P = -\left(\frac{\partial S}{\partial P}\right)_T$$

Finally, we derive two equations called *thermodynamic equations of state* because they give U and H in terms of P, V, and T. For U, from (3.30),

$$\left(\frac{\partial U}{\partial V}\right)_T = \left[\frac{\partial(A + TS)}{\partial V}\right]_T = \left(\frac{\partial A}{\partial V}\right)_T + T\left(\frac{\partial S}{\partial V}\right)_T$$

From (3.43) and (3.44),

$$\left(\frac{\partial U}{\partial V}\right)_T = -P + T\left(\frac{\partial P}{\partial T}\right)_V \qquad (3.45)$$

For H, from (3.35),

$$\left(\frac{\partial H}{\partial P}\right)_T = \left[\frac{\partial(G + TS)}{\partial P}\right]_T = \left(\frac{\partial G}{\partial P}\right)_T + T\left(\frac{\partial S}{\partial P}\right)_T$$

From (3.43) and (3.44),

$$\left(\frac{\partial H}{\partial P}\right)_T = V - T\left(\frac{\partial V}{\partial T}\right)_P \qquad (3.46)$$

We have now probably given enough examples to illustrate the ways in which thermodynamic functions can be derived and combined to yield a wonderful variety of forms providing both pleasure and utility (to borrow a phrase from Guy de Maupassant). The thermodynamic potentials and some of their most important relations are summarized in Table 3.1.

TABLE 3.1
The Thermodynamic Potentials

Name of function	Symbol and natural variables	Definition	Differential expression	Corresponding Maxwell relation
Internal energy	$U(S, V)$		$dU = T\,dS - P\,dV$	$\left(\dfrac{\partial T}{\partial V}\right)_S = -\left(\dfrac{\partial P}{\partial S}\right)_V$
Enthalpy	$H(S, P)$	$H = U + PV$	$dH = T\,dS + V\,dP$	$\left(\dfrac{\partial T}{\partial P}\right)_S = \left(\dfrac{\partial V}{\partial S}\right)_P$
Helmholtz free energy Helmholtz function	$A(T, V)$	$A = U - TS$	$dA = -SdT - P\,dV$	$\left(\dfrac{\partial S}{\partial V}\right)_T = \left(\dfrac{\partial P}{\partial T}\right)_V$
Gibbs free energy Gibbs function Free enthalpy	$G(T, P)$	$G = H - TS$	$dG = -S\,dT + V\,dP$	$\left(\dfrac{\partial S}{\partial P}\right)_T = -\left(\dfrac{\partial V}{\partial T}\right)_P$

19. Pressure and Temperature Dependence of Gibbs Function

From (3.41) we have

$$\left(\frac{\partial G}{\partial P}\right)_T = V \tag{3.47}$$

For an isothermal change from state (1) to state (2), therefore, $dG = V\,dP$ and

$$\Delta G = G_2 - G_1 = \int_1^2 V\,dP \quad \text{(constant } T) \tag{3.48}$$

To integrate this equation, we must know the variation of V with P for the substance being studied. Then, if G is known at one pressure and temperature, it can be calculated at any other pressure for that same temperature. If an equation of state is available, it can be solved for V as a function of P, and (3.48) can be integrated after substituting this $V(P)$ for V. In the simple case of an ideal gas, $V = nRT/P$, and

$$\Delta G = G_2 - G_1 = \int_1^2 nRT\frac{dP}{P} = nRT\ln\frac{P_2}{P_1} \tag{3.49}$$

For example, if 1 mol of an ideal gas is compressed isothermally at 300 K to twice its original pressure, the change in G is

$$\Delta G = (1)(8.314)(300)\ln 2 = 1730 \text{ J}$$

As another example, let us calculate the ΔG when 1 mol of mercury is compressed from 1 to 101 atm at 298 K. The molar volume of mercury is $M/\rho = 200.61/\rho$, where ρ is the density. Thus,

$$\Delta G = \int_1^2 V \, dP = 200.61 \int_1^2 \frac{dP}{\rho} \tag{3.50}$$

To compute this integral exactly, we would need to know $\rho = f(P)$. Let us assume, however, that ρ is almost constant over this pressure range, and equal to 13.5 g·cm^{-3}. Then

$$\Delta G = \frac{200.6}{13.5} \cdot 100 = 1486 \text{ cm}^3 \cdot \text{atm}$$

$$= \frac{1486}{9.866} \text{ J} = 150.6 \text{ J}$$

The temperature dependence of G at constant P is given from (3.43),

$$\left(\frac{\partial G}{\partial T}\right)_P = -S \tag{3.51}$$

Since $G = H - TS$, (3.51) can also be written

$$\left(\frac{\partial G}{\partial T}\right)_P = -S = \frac{G - H}{T} \tag{3.52}$$

The ways in which G and H behave as functions of T are shown in Fig. 3.9. Equa-

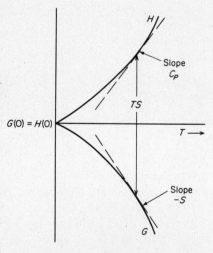

FIG. 3.9 Variation of free enthalpy G and enthalpy H of a pure substance with T at constant P. The limiting slopes as $T \longrightarrow 0$ are both zero, so that C_P and $S \longrightarrow 0$ as $T \longrightarrow 0$.

tion (3.52) is one form of the *Gibbs–Helmholtz equation*. It can also be written in alternative ways, as follows,

$$\left(\frac{\partial (G/T)}{\partial T}\right)_P = \frac{1}{T}\left(\frac{\partial G}{\partial T}\right)_P - \frac{G}{T^2} = \frac{-H}{T^2}$$

Also,

$$\left(\frac{\partial(G/T)}{\partial(1/T)}\right)_P = \left(\frac{\partial(G/T)}{\partial T}\right)_P \left(\frac{\partial T}{\partial(1/T)}\right)_P = \left(\frac{\partial(G/T)}{\partial T}\right)_P (-T^2) = H$$

If we have two states (1) and (2) of a system, with

$$\Delta G = G_2 - G_1, \qquad \Delta H = H_2 - H_1, \qquad \Delta S = S_2 - S_1$$

(3.52) becomes

$$\left(\frac{\partial \,\Delta G}{\partial T}\right)_P = -\Delta S = \frac{\Delta G - \Delta H}{T} \tag{3.53}$$

Equation (3.53) is especially useful when applied to chemical reactions. For example, the change in the Gibbs function ΔG for a chemical reaction [$(G$ of products) $- (G$ of reactants)] might be studied at a series of different constant temperatures, always under the same constant pressure. Equation (3.53) then describes how the observed ΔG depends on the temperature at which the reaction occurs.

As an example of the use of (3.52), let us calculate the change in G of 1 mol N_2 when we raise its temperature from 298 to 348 K at 2 atm pressure, given that the entropy of N_2 is $S = A + B \ln T$ with $A = 25.1$ and $B = 29.3 \text{ J} \cdot \text{K}^{-1}$. (This is actually a rough approximation.) We then have

$$\left(\frac{\partial G}{\partial T}\right)_P = -S = -(A + B \ln T)$$

$$\Delta G = G_2 - G_1 = -\int_{298}^{348} (A + B \ln T)\, dT$$

$$= -(A - B)\,\Delta T - B(T_2 \ln T_2 - T_1 \ln T_1)$$

$$= 210 - 9960 = -9750 \text{ J}$$

20. Pressure and Temperature Variation of Entropy

One of the Maxwell equations (3.44) provides us with the pressure dependence of the entropy,

$$\left(\frac{\partial S}{\partial P}\right)_T = -\left(\frac{\partial V}{\partial T}\right)_P \tag{3.54}$$

We recall that $(\partial V/\partial T)_P$ is related to the thermal expansivity α in (1.21),

$$\alpha = \frac{1}{V}\left(\frac{\partial V}{\partial T}\right)_P$$

Thus, we integrate (3.54) at constant temperature to obtain

$$\Delta S = S_2 - S_1 = -\int_{P_1}^{P_2} \alpha V \, dP \tag{3.55}$$

To evaluate this integral, we must know V and α as functions of pressure. These data can be calculated from an equation of state for the substance under consideration. For an ideal gas, a simple result is obtained:

$$PV = nRT \quad \text{and} \quad \left(\frac{\partial V}{\partial T}\right)_P = \alpha V = \frac{nR}{P}$$

Accordingly, (3.55) becomes

$$\Delta S = -\int_{P_1}^{P_2} nR \frac{dP}{P} = nR \ln \frac{P_1}{P_2} = nR \ln \frac{V_2}{V_1}$$

as already shown in Section 3.8.

The temperature variation of the entropy is readily calculated from (3.43), either at constant volume or at constant pressure, as we may prefer.

$$\left(\frac{\partial U}{\partial S}\right)_V = T \quad \text{and} \quad \left(\frac{\partial H}{\partial S}\right)_P = T$$

Hence,

$$\left(\frac{\partial S}{\partial U}\right)_V = T^{-1} \quad \text{and} \quad \left(\frac{\partial S}{\partial H}\right)_P = T^{-1}$$

Thus,

$$\left(\frac{\partial S}{\partial T}\right)_V = \left(\frac{\partial S}{\partial U}\right)_V\left(\frac{\partial U}{\partial T}\right)_V = T^{-1}C_V \quad \text{and} \quad \left(\frac{\partial S}{\partial T}\right)_P = \left(\frac{\partial S}{\partial H}\right)_P\left(\frac{\partial H}{\partial T}\right)_P = T^{-1}C_P \quad (3.56)$$

When we know the heat capacities as functions of temperature, we can integrate (3.56) to obtain the entropy changes with temperature:

At constant volume: At constant pressure:

$$dS = C_V \frac{dT}{T} \qquad\qquad dS = C_P \frac{dT}{T}$$

$$S = \int \frac{C_V}{T}\, dT \qquad\qquad S = \int \frac{C_P}{T}\, dT \qquad (3.57)$$

$$\Delta S = \int_{T_1}^{T_2} C_V\, d\ln T \qquad \Delta S = \int_{T_1}^{T_2} C_P\, d\ln T$$

These integrations are conveniently carried out graphically, as in Fig. 3.10,

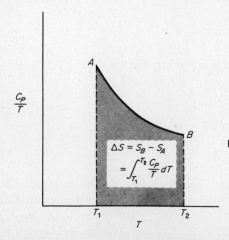

FIG. 3.10 Graphical evaluation of the entropy change with temperature. Data on the heat capacity C_P of substance as a function of T allows one to calculate ΔS by this integration procedure.

which shows C_P/T plotted against T. The area under the curve gives the ΔS between the initial and final states. If a phase transition occurs, the corresponding $\Delta S_t = \Delta H_t/T_t$ must be included.

For example, given the molar heat capacities in $J \cdot K^{-1}$, $C_P(\text{ice}) = 2.09 + 0.126T$, $C_P(\text{water}) = 75.3$, and $\Delta H_m = 6000\ J \cdot mol^{-1}$, calculate the ΔS when 1 mol of water is heated from 263 to 283 K. From (3.57), we can write at once

$$\Delta S = \int_{263}^{273} (2.09 + 0.126T)\frac{dT}{T} + \frac{6000}{273} + \int_{273}^{283} 75.3\frac{dT}{T}$$

$$\Delta S = 2.09 \ln\frac{273}{263} + 1.26 + 22.0 + 75.3 \ln\frac{283}{273}$$

$$\Delta S = 26.1\ J \cdot K^{-1}$$

21. Applications of Thermodynamic Equations of State

With the help of our thermodynamic equations of state (3.45), we can easily prove the statement in Section 2.12 that a gas must have a zero internal pressure, $(\partial U/\partial V)_T = 0$, if it obeys the equation of state $PV = nRT$. For such a gas,

$$\left(\frac{\partial P}{\partial T}\right)_V = \frac{nR}{V}$$

Hence, from (3.45),

$$\left(\frac{\partial U}{\partial V}\right)_T = -P + T\frac{nR}{V} = -P + P = 0$$

Also with the help of (3.45), we can derive a most useful expression for $C_P - C_V$. Equation (2.18) becomes

$$C_P - C_V = \left[P + \left(\frac{\partial U}{\partial V}\right)_T\right]\left(\frac{\partial V}{\partial T}\right)_P = T\left(\frac{\partial P}{\partial T}\right)_V\left(\frac{\partial V}{\partial T}\right)_P$$

Then, from (1.24),

$$C_P - C_V = \frac{\alpha^2 V T}{\beta} \tag{3.58}$$

An important application of (3.46) is the theoretical analysis of the Joule–Thomson coefficient. From (2.19),

$$\mu = \left(\frac{\partial T}{\partial P}\right)_H = -\frac{1}{C_P}\left(\frac{\partial H}{\partial P}\right)_T$$

From (3.46),

$$\mu = \frac{T(\partial V/\partial T)_P - V}{C_P} \tag{3.59}$$

It is apparent that the Joule–Thomson effect can be either a warming or a cooling of the substance, depending on the relative magnitudes of the two terms in the numerator of (3.59). In general, a gas will have one or more inversion points at which the sign of the coefficient changes as it passes through zero. From (3.59), the condition for an inversion point is that

$$T\left(\frac{\partial V}{\partial T}\right)_P = \alpha VT = V \quad \text{or} \quad \alpha = T^{-1}$$

For an ideal gas, this is always true (Law of Gay–Lussac), so that μ is always zero in this case. For other equations of state, it is possible to derive μ from (3.59) without direct measurement, if C_P data are available. This theory is of basic importance in the design of equipment for the liquefaction of gases.

22.   The Approach to Absolute Zero

To evaluate the difference between the entropy S_0 of a substance at 0 K and its entropy S at some temperature T, we can rewrite (3.57) as follows:

$$S = \int_0^T \left(\frac{C_P}{T}\right) dT + S_0 \tag{3.60}$$

If any changes of state occur between the temperature limits, the corresponding entropy changes must be included in S. For a gas at temperature T, the general expression for the entropy therefore becomes

$$S = \int_0^{T_m} \frac{C_P(\text{crystal})}{T} dT + \frac{\Delta H_m}{T_m} + \int_{T_m}^{T_b} \frac{C_P(\text{liquid})}{T} dT$$

$$+ \frac{\Delta H_v}{T_b} + \int_{T_b}^{T} \frac{C_P(\text{gas})}{T} dT + S_0 \tag{3.61}$$

This equation tells us how to calculate the entropy of a substance from calorimetric measurements of (1) its heat capacity over a range of temperature starting near 0 K, and (2) the latent heats of all changes of state between 0 K and the final T. All the terms in (3.61) can be measured in a calorimeter, except for S_0. The evaluation of S_0 becomes possible by virtue of the third fundamental law of thermodynamics. The limiting value of the entropy of a substance as T approaches 0 K is S_0.

We shall first consider methods used to attain very low temperatures, and then see how the entropy behaves in this region. The science of the production and use of low temperatures is called *cryogenics*. Some remarkable properties of matter become evident only at temperatures within a few degrees of absolute zero—e.g., superconductivity of metals and the superfluid state of helium.

A gas will be cooled in a Joule–Thomson expansion provided the coefficient $\mu > 0$. In 1860, William Siemens devised a countercurrent heat exchanger, which greatly enhanced the utility of the Joule–Thomson method for producing low temperatures. It was applied in the Hampson and Linde process for the production of liquid air.* The cycle used is shown in Fig. 3.11. Chilled compressed gas is cooled further by passage through a throttling valve. The expanded gas passes back over

---

*Oxygen and nitrogen were first liquefied by Cailletet in 1877 by means of rapid expansion of cold compressed gas. The first large-scale production of liquid air was achieved by Claude in 1902, using expansion against a piston.

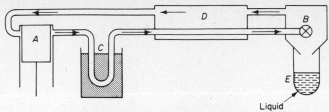

FIG. 3.11 The Hampson-Linde Process for liquefying gases. The gas is compressed in *A*, cooled by refrigerant in *C* and then passes through an expansion valve *B*, where further cooling occurs through the Joule–Thomson effect. The cooled gas passes back through a counter-current heat exchanger *D*. After repeated cycles, the cooling at *B* suffices to liquefy some of the gas, and the liquid is collected in *E*.

the inlet tube, cooling the unexpanded gas. When the cooling is sufficient to cause condensation, the liquid air can be drawn off at the bottom of the apparatus. Liquid nitrogen boils at 77 K, liquid oxygen at 90 K, and they are easily separated by fractional distillation.

To liquefy hydrogen, it must first be chilled below its Joule–Thomson inversion temperature at 193 K; the Siemens process can then be used to bring it below its critical temperature at 33 K. The production of liquid hydrogen was first achieved in this way by James Dewar in 1898.

The boiling point of hydrogen is 20 K. In 1908, Kammerlingh–Onnes, founder of the famous cryogenic laboratory at Leiden, used liquid hydrogen to cool helium below its inversion point at 100 K, and then liquefied it by an adaptation of the Joule–Thomson principle. Temperatures as low as 0.84 K have been obtained with liquid helium boiling under reduced pressures. This temperature is about the limit of such a method, since gigantic pumps become necessary to carry off the gaseous helium.

In 1926, William Giauque and Peter Debye independently proposed a new refrigeration technique, called *adiabatic demagnetization*.* Giauque brought the method to experimental realization in 1933. Certain salts, notably of the rare earths, have high paramagnetic susceptibilities.† The cations act as little magnets, which line up in the direction of an applied external magnetic field. The salt is then *magnetized*. When the field is removed, the alinement of the little magnets disappears, and the salt is *demagnetized*.

An apparatus used for adiabatic demagnetization is shown in Fig. 3.12(a). The salt, gadolinium sulfate, for example, is placed within the inner chamber of a double Dewar vessel. It is magnetized while being cooled with liquid helium, which is introduced around the inner chamber. The liquid helium is then pumped away, leaving the chilled magnetized salt thermally isolated from its environment by the adiabatic barrier of the evacuated space. The magnetic field is then reduced

*J. Am. Chem. Soc., **49**, 1870 (1927).
†Compare Section 17.26.

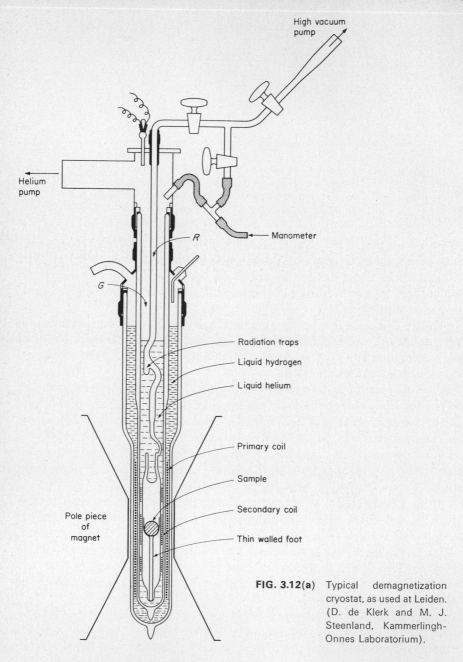

High vacuum
pump

Helium
pump

R

— Manometer

G

———— Radiation traps

———— Liquid hydrogen

———— Liquid helium

———— Primary coil

———— Sample

———— Secondary coil

Pole piece
of
magnet

———— Thin walled foot

FIG. 3.12(a) Typical demagnetization cryostat, as used at Leiden. (D. de Klerk and M. J. Steenland, Kammerlingh-Onnes Laboratorium).

to zero; this process effects the *adiabatic demagnetization* of the salt; since there is no heat transfer, $q = 0$.

This demagnetization is not strictly reversible, but no error in the argument is caused if we consider it to be so. Why not? For an adiabatic reversible demagnetization, $\Delta S = 0$. Figure 3.13 therefore represents the experiment of Giauque on a

FIG. 3.12(b) Demagnetization apparatus mounted in the great electromagnet of C.N.R.S. Laboratory, Bellevue, France. The diameter of the coils is about 200 cm. (N. Kurti, University of Oxford).

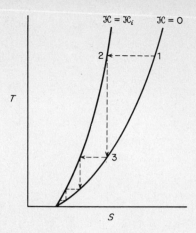

FIG. 3.13 Successive stages of cooling by adiabatic demagnetization represented on a T-S diagram.

TS diagram. Two curves are shown for the salt, one in the demagnetized state in the absence of a field ($\mathcal{H} = 0$) and one in the magnetized state in the presence of a field ($\mathcal{H} = \mathcal{H}_i$). In a particular experiment, $\mathcal{H}_i$ was 0.800 tesla (T) and the initial isothermal magnetization was done at 1.5 K.

At any constant temperature, the magnetized salt is in a state of lower entropy than the demagnetized salt. At constant T, the transition

demagnetized salt $\longrightarrow$ magnetized salt

goes from a state of higher entropy and energy to a state of lower entropy and energy. (It is analogous in this respect to the transition, liquid $\longrightarrow$ solid.) The transition is shown as $1 \longrightarrow 2$ on the TS diagram. When the field is reduced to zero, there is an isentropic ($\Delta S = 0$) change in the system back to the $\mathcal{H} = 0$ curve. This change is shown as $2 \longrightarrow 3$ on the TS diagram. It is evident that the temperature must fall. In the experiment cited, the temperature fell from 1.5 to 0.25 K.

In 1950, workers at Leiden reached 0.0014 K by adiabatic demagnetization of a paramagnetic salt. Still lower temperatures have been reached by applying the same principle to nuclear magnetic moments,[*] about 2×10^{-5} K being the current record.[†] The measurement of such low temperatures presents special problems. A vapor-pressure thermometer using helium is satisfactory down to 1 K. Below this, the magnetic properties themselves can provide a temperature scale. For example, the Curie–Weiss Law for the paramagnetic susceptibility can be used: $\chi = \text{constant}/T$.

23.   The Third Law of Thermodynamics

The approach to within 2×10^{-5} K of absolute zero does not mean that there remains only a small step, which soon will be taken. On the contrary, the detailed analysis of these experiments near 0 K definitely indicates that the absolute zero is absolutely unattainable.

---

[*]Compare Section 14.13.
[†]D. de Klerk, M. J. Steenland, and C. J. Gorter, *Physica*, *16*, 571 (1950).

The situation we face is shown in Fig. 3.13. In the successive stages of isothermal magnetization and adiabatic demagnetization, the fractional cooling obtained in each stage steadily decreases. Thus, even if perfect reversibility could be achieved, we would attain absolute zero only as the limit of an infinite series of steps. All possible cooling processes are subject to this same limitation. As we did for the First and Second Laws, we therefore postulate the Third Law of Thermodynamics as an inductive generalization:

> It is impossible by any procedure no matter how idealized to reduce the temperature of any system to the absolute zero in a finite number of operations.*

If we again refer to Fig. 3.13, we can see that the unattainability of absolute zero is connected with the fact that in the limit as $T \rightarrow 0$, the entropies of the magnetized and demagnetized states approach each other. For the isothermal magnetization, therefore, in the limit as $T \rightarrow 0$, ΔS must go to zero. There is nothing special about the magnetic case; any cooling procedure can be reduced to a TS diagram of this sort. Thus, for any isothermal reversible process $a \rightarrow b$, the Third Law requires that in the limit as $T \rightarrow 0$, $S_0^a \rightarrow S_0^b$, or

$$S_0^a - S_0^b = \Delta S_0 = 0 \tag{3.62}$$

This statement of the Third Law is similar to the famous *heat theorem* proposed by Walther Nernst in 1906.

Actually, the first satisfactory statement of the Third Law was given in 1923 in the first edition of the book *Thermodynamics and the Free Energy of Chemical Substances*, by G. N. Lewis and M. Randall:

> If the entropy of each element in some crystalline state be taken as zero at the absolute zero of temperature, every substance has a finite positive entropy; but at the absolute zero of temperature the entropy may become zero, and does so become in the case of perfect crystalline substances.

24.   An Illustration of the Third Law

Only changes or differences in entropy have any physical meaning in thermodynamics. When we speak of the entropy of a substance at a certain temperature, we mean the difference between its entropy at that temperature and its entropy at some other temperature, usually 0 K. Since the chemical elements are unchanged in any physicochemical process, we can assign any arbitrary values to their entropies S_0 at 0 K without affecting in any way the values of ΔS for any chemical change. It is most convenient, therefore, to take the value of S_0 for all the *chemical elements* as equal to zero. This is a convention first proposed by Max Planck in 1912, and incorporated in the statement of Lewis and Randall. It then follows that the entropies S_0 of all pure chemical *compounds* in their stable states at 0 K are also zero, since from (3.62) $\Delta S_0 = 0$ when they are formed from their elements.

As an example, consider the application of the Third Law to data on elemental sulfur. Let us set $S_0 = 0$ for rhombic sulfur and determine experimentally S_0 for

---

*R. H. Fowler and E. A. Guggenheim, *Statistical Thermodynamics* (London: Cambridge University Press, 1940), p. 224.

monoclinic sulfur to see how the Third Law is followed. The transition temperature for S (rhombic) $\longrightarrow$ S (monoclinic) is 368.5 K, and the latent heat of transition is 401.7 J·mol^{-1}. From (3.61),

$$S^{rh}_{368.5} = S^{rh}_0 + \int_0^{368.5} \frac{C_P}{T}\, dT$$

$$S^{mono}_{368.5} = S^{mono}_0 + \int_0^{368.5} \frac{C_P}{T}\, dT$$

To evaluate S^{mono}_0, it is necessary to have heat capacities for supercooled monoclinic sulfur from 0 to 368.5 K. This measurement causes no difficulties, since the rate of change of monoclinic to rhombic sulfur is extremely slow at low temperatures. Excellent heat capacities are therefore available for both monoclinic* and rhombic* sulfurs. The integrations of the C_P/T vs. T curves yield

$$S^{rh}_{368.5} = S^{rh}_0 + 36.86(\pm 0.20) \quad J\cdot K^{-1}\cdot mol^{-1}$$

$$S^{mono}_{368.5} = S^{mono}_0 + 37.82(\pm 0.40) \quad J\cdot K^{-1}\cdot mol^{-1}$$

Thus,

$$S^{rh}_{368.5} - S^{mono}_{368.5} = S^{rh}_0 - S^{mono}_0 - 0.96 \pm 0.65 \quad J\cdot K^{-1}\cdot mol^{-1}$$

But

$$S^{rh}_{368.5} - S^{mono}_{368.5} = \frac{-401.7}{368.5} = -1.09 \pm 0.01 \quad J\cdot K^{-1}\cdot mol^{-1}.$$

Hence,

$$S^{rh}_0 - S^{mono}_0 = -0.15 \pm 0.65 \quad J\cdot K^{-1}\cdot mol^{-1}$$

which is zero within experimental error. Thus, if we set $S^{rh}_0 = 0$, we shall have $S^{mono}_0 = 0$ also.

Many checks of this kind have been made for both elements and crystalline compounds. We must not forget, however, that $S_0 = 0$ is restricted to *perfect crystalline substances*.† Thus, glasses, solid solutions, and crystals retaining structural disorder even near absolute zero, are excluded from the rule $S_0 = 0$. The discussion of such exceptions follows so naturally from the statistical interpretation of entropy that we shall not pursue it further until this subject is introduced in Chapter 5.

25. Third-Law Entropies

It is now possible to use heat capacity data extrapolated to 0 K to determine *Third Law entropies*. As an example, the determination of the standard entropy, $S^{\ominus}_{298}$, for hydrogen chloride gas is shown in Table 3.2. The value $S^{\ominus}_{298} = 185.8 \pm 0.4$ J·K^{-1}·mol^{-1} is for HCl at 298.15 K and 1 atm pressure. A small correction due to

*E. D. Eastman and W. C. McGavock, *J. Am. Chem. Soc.*, *59*, 145 (1937); E. D. West, *J. Am. Chem. Soc.*, *81*, 29 (1959).

†An exception is liquid helium, for which $S \longrightarrow 0$ as $T \longrightarrow 0$. As $T \longrightarrow 0$, helium becomes a perfect superfluid.

TABLE 3.2
Evaluation of the Entropy of HCl from Measurements of its Heat Capacity

Contribution	$J \cdot K^{-1} \cdot mol^{-1}$
1. Extrapolation from 0–16 K (Debye Theory, Section 18.34)	1.3
2. $\int C_P \, d \ln T$ for solid I from 16–98.36 K	29.5
3. Transition, solid I $\longrightarrow$ solid II, 1190/98.36	12.1
4. $\int C_P \, d \ln T$ for solid II from 98.36–158.91 K	21.1
5. Fusion, 1992/158.91	12.6
6. $\int C_P \, d \ln T$ for liquid from 158.91–188.07 K	9.9
7. Vaporization, 16 150/188.07	85.9
8. $\int C_P \, d \ln T$ for gas from 188.07–298.15 K	13.5
	$S^{\ominus}_{298.15} = 185.9$

nonideality of the gas raises the figure to 186.6. A number of Third Law entropies are collected in Table 3.3.

TABLE 3.3
Third Law Entropies
(Substances in the Standard State at 298.15 K)

Substance	$(S^{\ominus}_{298}/J \cdot K^{-1} \cdot mol^{-1})$	Substance	$(S^{\ominus}_{298}/J \cdot K^{-1} \cdot mol^{-1})$
		Gases	
H_2	130.59	CO_2	213.7
D_2	144.77	H_2O	188.72
HD	143.7	NH_3	192.5
N_2	191.5	SO_2	248.5
O_2	205.1	CH_4	186.2
Cl_2	223.0	C_2H_2	200.8
HCl	186.6	C_2H_4	219.5
CO	197.5	C_2H_6	229.5
		Liquids	
Mercury	76.02	Benzene	173
Bromine	152	Toluene	220
Water	70.00	Bromobenzene	208
Methanol	127	*n*-Hexane	296
Ethanol	161	Cyclohexane	205
		Solids	
C(diamond)	2.44	I_2	116.1
C(graphite)	5.694	NaCl	72.38
S(rhombic)	31.9	LiF	37.1
S(monoclinic)	32.6	LiH	247
Ag	42.72	$CuSO_4 \cdot 5H_2O$	305
Cu	33.3	$CuSO_4$	113
Fe	27.2	AgCl	96.23
Na	51.0	AgBr	107.1

The standard entropy change $\Delta S^\ominus$ in a chemical reaction can be calculated immediately, if the standard entropies of products and reactants are known,

$$\Delta S^\ominus = \sum v_i S_i^\ominus$$

Here, v_i is the stoichiometric mole number in the reaction equation, taken as positive for products, negative for reactants. One of the most satisfactory experimental checks of the Third Law is provided by the comparison of $\Delta S^\ominus$ values obtained in this way from low temperature heat capacity measurements, with $\Delta S^\ominus$ values derived either from measured equilibrium constants K and reaction heats or from the temperature coefficients of the electromotive forces of cells (Section 12.7). Examples of such comparisons are shown in Table 3.4. Except for the formation

TABLE 3.4
Checks of the Third Law of Thermodynamics

Reaction	Temperature (K)	Third Law $\Delta S^\ominus$ ($J \cdot K^{-1} \cdot mol^{-1}$)	Experimental $\Delta S^\ominus$	Method
$Ag\ (c) + \frac{1}{2}Br_2\ (l) \longrightarrow AgBr\ (c)$	265.9	-12.6 ± 0.7	-12.6 ± 0.4	emf
$Ag\ (c) + \frac{1}{2}Cl_2\ (g) \longrightarrow AgCl\ (c)$	298.15	-57.9 ± 1.0	-57.4 ± 0.2	emf
$Zn\ (c) + \frac{1}{2}O_2\ (g) \longrightarrow ZnO\ (c)$	298.15	-100.7 ± 1.0	-101.4 ± 0.2	K and ΔH
$C + \frac{1}{2}O_2\ (g) \longrightarrow CO\ (g)$	298.15	84.8 ± 0.8	89.45 ± 0.2	K and ΔH
$CaCO_3\ (c) \longrightarrow CaO\ (c) + CO_2\ (g)$	298.15	160.7 ± 0.8	159.1 ± 0.8	K and ΔH

of CO, the agreements are within experimental errors. The Third Law entropy of CO is too low by about $4.7\ J \cdot K^{-1} \cdot mol^{-1}$. The reason for this discrepancy will become evident when the statistical interpretation of the Third Law is considered in Chapter 5. In brief, CO does not revert to a perfect crystal as $T \longrightarrow 0$, as a consequence of the two possible orientations of CO molecules, head-to-tail ($\overrightarrow{CO}\ \overrightarrow{CO}$) and head-to-head ($\overrightarrow{CO}\ \overleftarrow{OC}$).

The great utility of Third Law measurements in the calculation of chemical equilibria has led to an intensive development of techniques for measuring heat capacities at low temperatures, with the use of liquid hydrogen and helium as refrigerants. The experimental procedure consists essentially in a careful measurement of the temperature rise caused in an insulated sample by a carefully measured energy input.

PROBLEMS

1. A steam engine operates between 138 and 38°C. What is the minimum amount of heat that must be withdrawn from the hot reservoir to obtain 1000 J work?

2. Compare the maximum thermal efficiencies of heat engines operating with (a) steam between 130 and 40°C, and (b) mercury vapor between 380 and 50°C.

3. In regions with mild winters, heat pumps can be used for space heating in winter and cooling in summer. Assuming ideal thermodynamic efficiency for the pump, compare the cost of keeping a room at 25°C in winter and in summer when outside temperatures are 12 and 38°C, respectively. What assumptions did you make in this calculation? Suppose the heat pump were running 50% of the time when the outside temperature was 12°C. If the outside temperature fell to 0°C, could this heat pump maintain the inside at 25°C?

4. A cooling system is designed to maintain a refrigerator at −25°C in a room at ambient temperature +25°C. The heat transfer into the refrigerator is estimated as 10^4 J·min^{-1}. If the unit is assumed to operate at 50% of its maximum thermodynamic efficiency, estimate its power requirement.

5. An imaginary ideal-gas heat engine operates on the following cycle: (1) increase in pressure of gas at constant volume V_2 from P_2 to P_1; (2) adiabatic expansion from (P_1, V_2) to (P_2, V_1); (3) decrease in volume of gas at constant pressure P_2 from V_1 to V_2. Draw the cycle on a PV diagram and show that the thermal efficiency is

$$\epsilon = 1 - \gamma \frac{(V_1/V_2) - 1}{(P_1/P_2) - 1}$$

where $\gamma = C_P/C_V$ and heat capacities are assumed to be independent of temperature.

6. Prove that it is impossible for two reversible adiabatics on a PV diagram to intersect.

7. Consider an arbitrary initial state P_0, V_0 for a gas on a PV diagram. Prove that in the neighborhood of this state there exist other states P_i, V_i which are not accessible from P_0, V_0 along reversible adiabatic paths. This is an example in two dimensions of the Principle of Caratheodory, which, in its general form, can be used as a statement of the Second Law.* Show that the proof of inaccessibility implies that $dq(\text{reversible})/T = dS$, where dS is a perfect differential.

8. What is the exit velocity of helium gas that is expanding from 100 atm and 1000 K through a jet nozzle to 1 atm? Assume ideal gas and reversible adiabatic flow?

9. Consider the system shown in Fig. 3.6b, in which heat q is conducted down the metal bar between reservoirs at T_2 and T_1. What can you say about the ΔS of the conducting bar?

10. One mole of ideal gas is heated at constant pressure from 273 to 373 K. Calculate ΔS of the gas if $C_V = \frac{3}{2}R$.

11. Calculate ΔS when 0.5 mol liquid water at 273 K is mixed with 0.5 mol liquid water at 373 K, and the system is allowed to reach equilibrium in an adiabatic enclosure. Assume $C_P = 77$ J·K^{-1}·mol^{-1} from 273 to 373 K.

12. An electric current of 10 A flows through a resistor of 10 Ω, which is kept at a constant temperature of 10°C by running water. In 10 s, what is the ΔS (a) of the resistor and (b) of the water?

13. An electric current of 10 A flows through a thermally insulated resistor of 10 Ω, initially at a temperature of 10°C for 10 s. If the resistor has a mass of 10 g and its $C_P = 1.00$ J·g^{-1}·K^{-1}, what is ΔS of the resistor and ΔS of its surroundings?

14. One mole of an ideal gas is expanded adiabatically, but irreversibly, from V_1 to V_2,

*See, e.g., H. Reiss, *Methods of Thermodynamics*, (New York: Blaisdell Publ. Co., 1965), pp. 22, 71.

and no work is done, $w = 0$. Does the temperature of the gas change? (a) What is ΔS of the gas and ΔS of its surroundings? (b) If the expansion is performed reversibly and isothermally, what would be ΔS of the gas and of its surroundings?

15. Consider the TS diagram for a given ideal gas. Show that the ratio of the slopes of any isobaric line and any isovolumic line at the same temperature is C_P/C_V.

16. One mole of ideal gas, initially at 400 K and 10 atm, is adiabatically expanded against a constant pressure of 5 atm until equilibrium is reattained. If $C_V = 18.8 + 0.021T$ $J \cdot K^{-1} \cdot mol^{-1}$, calculate ΔU, ΔH, and ΔS for the change in the gas.

17. Calculate the ΔS per mole of mixture at 300 K and 1 atm when 1 mol each of N_2, O_2, and H_2, each at 300 K and 1 atm, are mixed. You can assume that the ΔS for each gas is that corresponding to its expansion from 1 atm to its partial pressure in an ideal mixture of ideal gases.

18. Calculate ΔS (vaporization) of the following liquids at their normal boiling points: benzene, nitrogen, *n*-pentane, mercury, diethyl ether, cyclohexane, water. (You will need data on T_b and ΔH_v from handbooks.) Comment briefly on ΔS_v values obtained.

19. Calculate ΔU, ΔH, ΔS, ΔA, and ΔG in expanding 1 mol of ideal gas at 25°C from 10 to 100 dm³.

20. At $-5°C$, the vapor pressure of ice is 3.012 mm and that of supercooled liquid water is 3.163 mm. The latent heat of fusion of ice is 5.85 kJ·mol⁻¹ at $-5°C$. Calculate ΔG and ΔS per mole for the transition *water* $\longrightarrow$ *ice* at $-5°C$.

21. For each of the following processes, state which of the quantities, ΔU, ΔH, ΔS, ΔG, and ΔA, are equal to zero for system specified.
    (a) A nonideal gas is taken around a Carnot cycle.
    (b) A nonideal gas is adiabatically expanded through a throttling value.
    (c) An ideal gas is adiabatically expanded through a throttling valve.
    (d) Liquid water is vaporized at 100°C and 1 atm.
    (e) H_2 and O_2 react to form H_2O in a thermally isolated bomb.
    (f) HCl and NaOH react to form H_2O and NaCl in an aqueous solution at constant T and P.

22. Derive the following relations:
    (a) $(\partial H/\partial P)_T = T(\partial S/\partial P)_T + V$
    (b) $(\partial C_P/\partial P)_T = -T(\partial^2 V/\partial T^2)_P$
    (c) $C_P = -T(\partial^2 G/\partial T^2)_P$

23. At $P = 0$, the molar volume of mercury at 273 K is 14.72 cm³·mol⁻¹, and the compressibility is $\beta = 3.88 \times 10^{-11}$ m²·N⁻¹. Assuming β is constant over the pressure range, calculate ΔG_m for compression of mercury from 0 to 3000 kg·cm⁻².

24. In the transition $CaCO_3$ (aragonite) $\longrightarrow$ $CaCO_3$ (calcite), $\Delta G_m^{\ominus}$ (298) $= -800$ J and $\Delta V_m = 2.75$ cm³. At what pressure would aragonite become the stable form at 298 K?

25. Prove that a gas that obeys Boyle's Law and has $(\partial U/\partial V)_T = 0$ follows the equation of state $PV = nRT$.

26. Prove that $C_P - C_V = \alpha^2 VT/\beta$. Calculate $\gamma = C_P/C_V$ for solid copper at 500 K, given $V_m = 7.115$ cm³·mol⁻¹, $\alpha = 5.42 \times 10^{-5}$ K⁻¹, and $\beta = 8.37 \times 10^{-12}$ m²·N⁻¹.

27. Derive an equation for the Joule–Thomson coefficient μ of a gas obeying the equation of state $P(V - nb) = nRT$.

28. Show that for a van der Waals gas, $(\partial U/\partial V)_T = a/V^2$.

29. A thermodynamic study was made of the propeller-shaped molecule triptycene* $(C_{20}H_{14}$; 9, 10-benzeno–9, 10-dihydroanthracene) by measurements of the heat capacity C_P from 10 to 550 K. The compound melts at 527.18 K with $\Delta H_m = 7236$ cal·mol^{-1}. The molar heat capacities were as follows.

			Solid					
T.K	10	20	30	40	50	60	70	80
cal·K^{-1}·mol^{-1}	0.863	4.303	7.731	10.649	13.17	15.40	17.43	19.33
	90	100	120	140	160	180	200	
	21.16	22.98	26.67	30.55	34.63	38.91	43.37	
	220	240	260	280	298.15	320		
	48.01	52.83	57.79	62.88	67.56	73.16		
	350	400	450	500	527.18			
	80.67	92.53	103.85	113.98	119.38			

		Liquid	
	527.18	530	550
	130.86	130.90	133.45

1 cal = 4.1840 J. Calculate the third-law entropy $S^{\ominus}$ for triptycene at 298.15 K and for the liquid at 550 K in units of J·K^{-1}mol^{-1}.

30. The reversible work done by a magnetic induction B in increasing the magnetization M of a paramagnetic solid is $dw = B\, dM$.
    (a) By analogy with C_V and C_P, we can define two heat capacities C_M and C_B for magnetic systems. Prove that

$$C_B - C_M = -T\left(\frac{\partial M}{\partial T}\right)_B\left(\frac{\partial B}{\partial T}\right)_M$$

   (b) At moderate temperatures and fields, paramagnetic solids follow the Curie equation, $M = \zeta B/T$, where ζ is a constant. Show that in this case, $C_B - C_M = M^2/\zeta$.
   (c) The Curie constant for gadolinium sulfate is tabulated as 15.7 cm^3·K·mol^{-1}. Calculate the work done on 1 mol of this salt when it is reversibly magnetized by raising the field from 0 to 1.0 tesla (T).

31. The efficiency of a heat engine is $\epsilon = (T_2 - T_1)/T_2$, so that as $T_1 \longrightarrow 0$, $\epsilon \longrightarrow 1$. But if $\epsilon = 1$, all the heat taken from the hot reservoir would be converted into work, contrary to the Second Law. Therefore, T_1 cannot be zero. It seems from this argument that the Third Law can be deduced from the Second Law. Can you find any flaws in this purported deduction?

*J.T.S. Andrews and E. F. Westrum, *J. Chem. Thermodynamics 2*, 245 (1970).

4

Kinetic Theory

Thermodynamics deals with pressures, volumes, masses, temperatures, energies, and the relations between them, without seeking to elucidate further the nature of these properties.

Thermodynamics allows us to derive relationships between large scale (*macroscopic*) properties of systems, but in no case can it explain why a system has a certain numerical value for a given property. For example, we can derive from thermodynamics an equation to relate the melting point of a solid to the external pressure [Eq. (6.20)], but we cannot derive from thermodynamics the fact that the melting point of silver at 1 atm pressure is 1234 K. To understand at all why the macroscopic properties of matter have their actual values, we must have a theory that explores matter on a finer scale (a *microscopic* theory) in terms of elementary particles, fields of force, and other principles of structure and interaction.

1. Atomic Theory

The word atom is derived from the Greek $\alpha\tau o\mu o\varsigma$, meaning indivisible; atoms were believed to be the ultimate and eternal particles of which all material things were made. Our knowledge of Greek atomism comes mainly from the long poem of the Roman, Lucretius, *De Rerum Natura*, "Concerning the Nature of Things," written in the first century before Christ. Lucretius expounded the theories of Epicurus and of Democritus:

> . . . this world of ours was made
> By natural process, as the atoms came
> Together, willy nilly, quite by chance,
> Quite casually and quite intentionless
> Knocking against each other, massed, or spaced
> So as to colander others through, and cause
> Such combinations and conglomerates
> As form the origin of mighty things,
> Earth, sea and sky, and animals and men.
> Face up to this, acknowledge it. I tell you
> Over and over—out beyond our world
> There are, elsewhere, other assemblages
> Of matter, making other worlds, Oh, ours
> Is not the only one in air's embrace.*

The properties of substances were determined by the forms of their atoms. Atoms of iron were hard and strong with spines that locked them together into a solid; atoms of water were smooth and slippery like poppy seeds; atoms of salt were sharp and pointed and pricked the tongue; whirling atoms of air pervaded all matter.

Later philosophers were inclined to discredit the atomic theory. They found it hard to explain in terms of atoms the many qualities of materials. This problem was foreshadowed in one of the fragments from Democritus that have survived from about 420 B.C.

> The Intellect: "Apparently there is color, apparently sweetness, apparently bitterness, actually there are only atoms and the void." The Senses: "Poor Intellect, do you hope to defeat us, while from us you borrow your very evidence. Your victory is in fact your defeat."

The four elements of Heraclitus and Artistotle, earth, air, fire, and water, provided at least a symbolic representation of the textures of a sensual world. Atoms were almost forgotten till the seventeenth century, as the alchemists sought the philosopher's stone by which the Aristotelian elements could be blended to make gold.

The writings of Descartes (1596–1650) helped to restore the idea of a corpuscular structure of matter. Gassendi (1592–1655) introduced many of the concepts of the present atomic theory; his atoms were rigid, moved at random in a void, and collided with one another. These ideas were extended by Hooke, who first proposed (1678) that the *elasticity* of a gas was the result of collisions of its atoms with the retaining walls. In 1738, Daniel Bernoulli provided a mathematical treatment of this model and correctly derived Boyle's Law by considering the collisions of atoms with the container wall. This work was overlooked for about 120 years, until it was "rediscovered" in 1859.†

---

*From *The Way Things Are*, the *De Rerum Natura* of Titus Lucretius Carus, transl. by Rolfe Humphries (Bloomington, Ind.: Indiana University Press, 1968).

†E. Mendoza, *Physics Today*, *14*, 36 (1961).

2. Molecules

Boyle had discarded the alchemical notion of elements and defined them as substances that had not been decomposed in the laboratory, but until the work of Antoine Lavoisier from 1772 to 1783 chemical thought was dominated by the phlogiston theory of Georg Stahl, which was actually a survival of alchemical conceptions. With Lavoisier, the elements took on their modern meaning and chemistry became a quantitative science. The Law of Definite Proportions and the Law of Multiple Proportions had become fairly well established by 1808, when John Dalton published his *New System of Chemical Philosophy*.

Dalton proposed that the atoms of each element had a characteristic atomic mass, and that these atoms were the combining units in chemical reactions. This hypothesis provided a clear explanation for the Laws of Definite and Multiple Proportions. Dalton had no unequivocal way of assigning atomic masses, and he made the unfounded assumption that in the most common compound between two elements, one atom of each was combined. According to this system, water would be HO, and ammonia NH. If the mass of the combining unit of hydrogen was set equal to unity, the analytical data would then give O = 8, N = 4.5, in Dalton's system.

About this time, Gay–Lussac was studying the chemical combinations of gases, and he found that the ratios of the *volumes* of reacting gases were small whole numbers. This discovery provided a more logical method for assigning atomic masses. Gay–Lussac, Berzelius, and others thought that the volume occupied by the atoms of a gas must be very small compared to the total gas volume, so that equal volumes of gas should contain equal numbers of atoms. The masses of such equal volumes would therefore be proportional to the atomic masses. This idea was received coldly by Dalton and many of his contemporaries, who pointed to reactions such as that which they wrote as $N + O \rightarrow NO$. Experimentally, the nitric oxide was found to occupy the same volume as the nitrogen and oxygen from which it was formed, although it evidently contained only half as many "atoms."[*]

Not till 1860 was the solution to this problem understood by most chemists, although half a century earlier it had been given by Amadeo Avogadro. In 1811, he published in the *Journal de Physique* an article that clearly drew the distinction between the *molecule* and the *atom*. The "atoms" of hydrogen, oxygen, and nitrogen are in reality *molecules* containing two atoms each. Equal volumes of gases should contain the same number of molecules (*Avogadro's Principle*).

Since one mole of any substance by definition contains the same number of molecules, according to Avogadro's Principle the molar volumes of all gases should be the same. The extent to which real gases conform to this rule may be seen from the molar volumes in Table 4.1, calculated from the measured gas densities. For an ideal gas at 0°C and 1 atm, the molar volume would be 22414 cm³. The number of molecules in 1 mol is now called *Avogadro's Number, L*.

---

[*]The elementary corpuscles of compounds were then called "atoms" of the compound.

TABLE 4.1
Molar Volumes of Gases in cm³ at 0°C and 1 atm Pressure

Hydrogen	22 432	Argon	22 390
Helium	22 396	Chlorine	22 063
Methane	22 377	Carbon dioxide	22 263
Nitrogen	22 403	Ethane	22 172
Oxygen	22 392	Ethylene	22 246
Ammonia	22 094	Acetylene	22 085

The work of Avogadro was ignored until it was forcefully presented by Cannizzaro at the Karlsruhe Conference in 1860. The reason for this neglect was probably the deeply rooted belief that chemical combination occurred by virtue of an affinity between unlike elements. After the electrical discoveries of Galvani and Volta, this affinity was generally ascribed to the attractions between unlike charges. The idea that two identical atoms of hydrogen might combine into the compound molecule H_2 was abhorrent to the chemical philosophy of the early nineteenth century.

3. The Kinetic Theory of Heat

Through frictional phenomena, even primitive peoples knew the connection between heat and motion. As the kinetic theory became accepted during the seventeenth century, heat became identified with the mechanical motion of the atoms or corpuscles. In the words of Francis Bacon:

> When I say of motion that it is the genus of which heat is a species I would be understood to mean, not that heat generates motion or that motion generates heat (though both are true in certain cases) but that heat itself, its essence and quiddity, is motion and nothing else. . . . Heat is a motion of expansion, not uniformly of the whole body together, but in the small parts of it . . . the body acquires a motion alternative, perpetually quivering, striving, and struggling, and initiated by repercussion, whence springs the fury of fire and heat.*

Although such ideas were widely discussed during the intervening years, the caloric theory, considering heat as a weightless fluid, was the working hypothesis of most natural philosophers until the quantitative work of Rumford and Joule brought about the general adoption of the mechanical theory. This was rapidly developed by Boltzmann, Maxwell, Clausius, and others, from 1860 to 1890.

According to the tenets of the kinetic theory, both temperature and pressure are thus manifestations of molecular motion. Temperature is a measure of the average translational kinetic energy of the molecules, and pressure arises from the average force resulting from repeated impacts of molecules with containing walls.

The Novum Organon or a True Guide to the Interpretation of Nature, 1st (Latin) ed. London, 1620.

4. The Pressure of a Gas

We shall now give a modern version of Bernoulli's derivation of the pressure of a gas in terms of the properties of its constituent molecules.*

Consider in Fig. 4.1 a planar wall of area A and a collection of N molecules

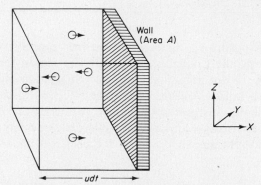

FIG. 4.1 Collision with an area A normal to the X axis by molecules of gas with component of velocity between u and $u + du$. The pressure of the gas on the wall is calculated from the rate of transfer of momentum to the wall (force) per unit area.

moving with random velocities in a volume V adjacent to the wall. The simplest kinetic-theory model of a gas assumes that the volume occupied by the molecules can be neglected compared to the total volume. It assumes also that the molecules behave as rigid spheres, with no forces of attraction or repulsion between one another except during the contacts of actual collisions. A gas that follows such a kinetic-theory model is called a *perfect gas*.

Let us focus attention on those molecules with a component of velocity in the X direction that falls between u and $u + du$. Note that the positive X axis is directed normal to the wall. The fraction of molecules $dN(u)/N$ that have a velocity component between u and $u + du$ will be specified by a *density function* $f(u)$, such that

$$\frac{dN(u)}{N} = f(u)\,du \tag{4.1}$$

We can also interpret $f(u)\,du$ as the *probability* that the molecular velocity component lies between u and $u + du$. Since u can range from $-\infty$ to $+\infty$, the probability is unity that its value lies somewhere in this range:

$$\int_{-\infty}^{+\infty} f(u)\,du = 1 \tag{4.2}$$

*W. Kauzmann, *Kinetic Theory of Gases* (New York: W. A. Benjamin, Inc., 1966) pp. 50–59.

Later in this chapter, we shall derive an explicit expression for $f(u)$, but we do not need this for the present analysis.

In a time interval dt, every molecule with a positive velocity component ($u > 0$) between u and $u + du$ will collide with the wall, provided it is initially within a distance $u\,dt$ from the wall. The number of such molecules striking the area A of the wall is accordingly the number of molecules in the specified velocity range that at $t = 0$ lie within a volume of base A and length $u\,dt$, or volume $Au\,dt$. From (4.1), the number of molecules in the specified velocity range per unit volume is $(N/V)f(u)\,du$, and hence the number of these special collisions in time dt becomes $(N/V)f(u)Au\,du\,dt$. In each such collision, a molecule undergoes a change in momentum, from $+mu$ to $-mu$, or of magnitude $2mu$. Hence, the momentum change dp due to these collisions in time dt becomes

$$dp = 2mu\frac{N}{V}f(u)Au\,du\,dt = 2mu^2\frac{N}{V}A\,f(u)\,du\,dt$$

The contribution of these particular collisions to the pressure is the force (rate of change of momentum) per unit area,

$$dP = \frac{dp/dt}{A}$$

or

$$dP = 2mu^2\frac{N}{V}f(u)\,du \tag{4.3}$$

Since only positive velocities ($0 < u < \infty$) contribute to the pressure, the total pressure becomes

$$P = 2m\frac{N}{V}\int_0^\infty u^2 f(u)\,du$$

The *mean square velocity component* is defined by

$$\overline{u^2} = \int_{-\infty}^\infty u^2 f(u)\,du$$

The density function for positive velocities must be the same as that for negative ones, so that

$$\overline{u^2} = 2\int_0^\infty u^2 f(u)\,du$$

Hence we obtain the kinetic theory expression for the pressure of a perfect gas:

$$P = Nm\frac{\overline{u^2}}{V} \tag{4.4}$$

The magnitude of the velocity of a gas molecule is related to the magnitudes of its components, u, v, and w along the three rectangular axes by

$$c^2 = u^2 + v^2 + w^2$$

Since no particular component is preferred to any other (the gas is isotropic so far as velocities and other molecular properties are concerned),

$$\overline{u^2} = \overline{v^2} = \overline{w^2} = \frac{\overline{c^2}}{3}$$

Hence, (4.4) becomes

$$P = Nm\frac{\overline{c^2}}{3V} \tag{4.5}$$

The quantity $\overline{c^2}$ is the *mean square speed* of the gas molecules. We recall that *velocity* is a vector quantity whereas *speed* is a scalar.

The total translational kinetic energy of the molecules is

$$E_k = N(\tfrac{1}{2}m\overline{c^2})$$

Hence, from (4.5), $P = 2E_k/3V$, or

$$PV = \tfrac{2}{3}E_k \tag{4.6}$$

Since the kinetic energy is a constant, unchanged by elastic collisions, (4.6) is equivalent to Boyle's Law. A perfect gas might also be defined as one in which all the energy is kinetic energy.

5.   Gas Mixtures and Partial Pressures

If we use the perfect-gas model of Section 4.4 to calculate the pressure of a gas mixture, we obtain a sum of terms like (4.6), one for each gas,

$$P_1 = \frac{2}{3}\frac{E_{k1}}{V}$$

$$P_2 = \frac{2}{3}\frac{E_{k2}}{V}, \quad \text{etc.}$$

The P_1 is the pressure that gas (1) would exert if it occupied the total volume all by itself. It is called the *partial pressure* of gas (1).

According to our model, the gas molecules can interact only by elastic collisions, so that the kinetic energy of the mixture must equal the sum of the individual kinetic energies.

$$E_k = E_{k1} + E_{k2} + \cdots + E_{kc}$$

From (4.6), the total pressure of the mixture is

$$P = \frac{2}{3}\frac{E_k}{V}$$

Therefore,

$$P = P_1 + P_2 + \cdots + P_c \tag{4.7}$$

This is *Dalton's Law of Partial Pressures*, which is valid for ideal gas mixtures. The extent of the deviation for nonideal gases may be considerable, as is shown in the rather typical example of a mixture of 50.06% argon and 49.94% ethylene:

Calculated from Dalton's Law	30.00	70.00	110.00
Actual pressure, atm, 25°C	29.15	64.55	101.85

6. Kinetic Energy and Temperature

The concept of temperature was first introduced in connection with the study of thermal equilibrium. When two bodies are placed in contact, energy flows from one to the other until a state of equilibrium is reached. The two bodies are then at the same temperature. We have found that the temperature can be measured conveniently by means of an ideal-gas thermometer, this empirical scale being identical with the thermodynamic scale derived from the Second Law.

A distinction was drawn in thermodynamics between mechanical work and heat. According to the kinetic theory, the transformation of mechanical work into heat is simply a degradation of large-scale motion into motion on the molecular scale. An increase in the temperature of a body is equivalent to an increase in the average translational kinetic energy of its constituent molecules. We may express this equivalence mathematically by saying that temperature is a function of E_k alone, $T = f(E_K)$. We know that this function must have the special form $T = \frac{2}{3}(E_k/nR)$, or

$$E_k = \tfrac{3}{2}nRT \tag{4.8}$$

so that (4.6) may be consistent with the ideal-gas relation, $PV = nRT$.

Temperature is thus not merely a function of, but in fact proportional to, the average translational kinetic energy of the molecules. The interpretation of absolute zero by the kinetic theory is thus the complete cessation of all molecular motion— the zero point of kinetic energy.*

If the total internal energy of a gas is given by this translational kinetic energy,

$$U = E_k = \tfrac{3}{2}nRT$$

The heat capacity of such a gas would be

$$C_V = \left(\frac{\partial U}{\partial T}\right)_V = \tfrac{3}{2}nR \tag{4.9}$$

If the heat capacity of a gas exceeds this value, we can conclude that this gas is taking up some form of energy other than translational kinetic energy.

The average translational kinetic energy may be resolved into components in the three *degrees of freedom* corresponding to velocities parallel to the three rectangular coordinates. Thus, for 1 mol of gas, where L is the Avogadro Number,

$$E_k = \tfrac{1}{2}Lm\overline{c^2} = \tfrac{1}{2}Lm\overline{u^2} + \tfrac{1}{2}Lm\overline{v^2} + \tfrac{1}{2}Lm\overline{w^2}$$

For each translational degree of freedom, therefore, from (4.8),

$$E_{1_k} = \tfrac{1}{2}Lm(\overline{u^2}) = \tfrac{1}{2}RT \text{ (per mole)} \tag{4.10}$$

This is a special case of a more general theorem known as the *Principle of Equipartition of Energy.*

---

*It will be seen later that this picture has been somewhat changed by quantum theory, which requires a small residual energy even at the absolute zero.

7. Molecular Speeds

Equation (4.5) may be written

$$\overline{c^2} = \frac{3P}{\rho} \tag{4.11}$$

where $\rho = Nm/V$ is the density of the gas. From (1.19) and (4.11), we obtain for the *mean square speed* $\overline{c^2}$, if M is the molar mass,

$$\overline{c^2} = \frac{3RT}{Lm} = \frac{3RT}{M}$$

The root mean square (rms) speed is

$$c_{rms} = (\overline{c^2})^{1/2} = \left(\frac{3RT}{M}\right)^{1/2} \tag{4.12}$$

The *average speed* $\bar{c}$, as we shall see later, differs only slightly from the rms speed, being

$$\bar{c} = \left(\frac{8RT}{\pi M}\right)^{1/2} \tag{4.13}$$

From (4.12) or (4.13), we can readily calculate average or rms speeds of the molecules of a gas at any temperature. Some results are shown in Table 4.2.

TABLE 4.2
Average Speeds of Gas Molecules at 273.15 K

Gas	m·s^{-1}	Gas	m·s^{-1}
Ammonia	582.7	Hydrogen	1692.0
Argon	380.8	Deuterium	1196.0
Benzene	272.2	Mercury	170.0
Carbon dioxide	362.5	Methane	600.6
Carbon monoxide	454.5	Nitrogen	454.2
Chlorine	285.6	Oxygen	425.1
Helium	1204.0	Water	566.5

In accordance with the Principle of Equipartition of Energy, we note that at any constant temperature, the lighter molecules have the higher average speeds. The average molecular speed of hydrogen at 298 K is 1768 m·s^{-1} or 6365 km·h^{-1}, about the speed of a rifle bullet. The average speed of a mercury vapor atom would be only about 638 km·h^{-1}.

8. Molecular Effusion

A direct experimental illustration of the different average speeds of molecules of different gases can be obtained from the phenomenon called *molecular effusion*. Consider the arrangement shown in Fig. 4.2. Molecules from a vessel of gas under

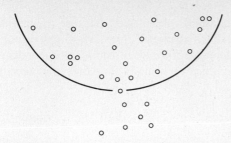

FIG. 4.2 Effusion of gases. In molecular effusion, each molecule moves independently through the orifice.

pressure escape through a tiny orifice, so small that the distribution of the velocities of the gas molecules remaining in the vessel is not affected in any way; i.e., no appreciable mass flow occurs in the direction of the orifice. The number of molecules escaping in unit time is then equal to the number that, in their random motion, happen to hit the orifice, and this number is proportional to the average molecular speed. From (4.13), the relative rate of effusion of two different gases would be

$$\frac{V_{E1}}{V_{E2}} = \frac{\bar{c}_1}{\bar{c}_2} = \left(\frac{M_2}{M_1}\right)^{1/2} \tag{4.14}$$

Thus, at constant temperature the rate of effusion varies inversely as the square root of the molar mass. Thomas Graham (1848) was the first to obtain experimental evidence for this law. Some of his data are shown in Table 4.3.

TABLE 4.3
The Effusion of Gases[a]

Gas	Relative velocity of effusion	
	Observed	Calculated from (4.14)
Air	(1)	(1)
Nitrogen	1.0160	1.0146
Oxygen	0.9503	0.9510
Hydrogen	3.6070	3.7994
Carbon dioxide	0.8354	0.8087

[a]*Source:* Graham, "On the Motion of Gases," *Phil. Trans. Roy. Soc.* (London), **36**, 573 (1846).

From the work of Graham and later experimenters, it appears that (4.14) is not perfectly obeyed. The equation fails for higher pressures and larger orifices. Under these conditions, the molecules can collide many times with one another in passing through the orifice, and a hydrodynamic flow toward the orifice is set up throughout the container, leading to the formation of a jet of escaping gas.*

Graham's Law suggests that effusive flow may provide a good method for separating gases of different molecular masses. Permeable barriers with fine pores

*For a discussion of jet flow, see H. W. Liepmann and A. E. Puckett, *Introduction to Aerodynamics of a Compressible Fluid* (New York: John Wiley & Sons, Inc., 1947), pp. 32ff.

are used in the separation of volatile compounds of isotopes. Because the lengths of the pores are considerably greater than their diameters, the flow of gas through such barriers does not follow the simple equation of orifice effusion. The dependence on molecular mass is the same, however, since each molecule passes through the barrier independently of any others.*

9. Imperfect Gases—
The van der Waals Equation

The calculated properties of the *perfect gas* of kinetic theory are the same as the experimental properties of the *ideal gas* of thermodynamics. Extension of the model of the perfect gas may therefore provide an explanation for observed deviations from ideal-gas behavior.

The first improvement of the model is to abandon the assumption that the volume of the molecules themselves can be neglected compared to the total gas volume. The finite volume of the molecules decreases the available void space in which the molecules are free to move. Instead of the V in the perfect gas equation, we must write $V - nb$, where b is called the *excluded volume per mole*. This is not just equal to the volume occupied by L molecules but actually to four times that volume. This result can be seen in a qualitative way by considering the two molecules of Fig. 4.3(a), regarded as impenetrable spheres each with a *diameter d*. The

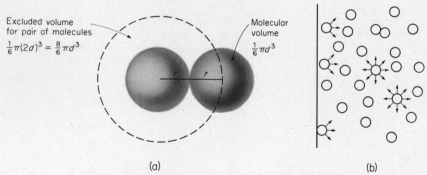

FIG. 4.3   Corrections to perfect-gas law. (a) Excluded volume.
(b) Intermolecular forces.

centers of these two molecules cannot approach each other more closely than the distance d; the excluded volume for the pair is therefore a sphere of *radius d*. This volume is $(\frac{4}{3})\pi d^3$ per pair, or $(\frac{2}{3})\pi d^3$ per molecule, which is four times the volume of the molecule. Consideration of the finite molecular volumes leads, therefore, to a gas equation of the form

$$P(V - nb) = nRT$$

*Martin Knudsen, *Kinetic Theory of Gases* [London: Methuen, 1950].

A second correction to the equation of state of a perfect gas comes from the forces of cohesion between molecules. We recall that the thermodynamic definition of the ideal gas includes the requirement that $(\partial U/\partial V)_T = 0$. If this condition is not fulfilled, the energy of a gas will depend on its volume. The molecular model explains such a dependence by the potential energy of interactions between molecules. The way in which such interactions affect the gas equation may be seen by considering Fig. 4.3(b). Molecules in the interior of the gas volume are in a uniform field of force, whereas molecules near to or colliding with the container walls experience a net attraction toward the body of the gas. This attraction decreases the pressure compared to that which would be exerted in the absence of such attractive forces.

The total inward force at the wall is proportional to the number of molecules (v) in the surface layer and to the number of molecules (v) in the adjacent inner layers of the gas. Note that these numbers v are assumed to be the same. The attractive force is therefore proportional to v^2. At any given temperature, v is inversely proportional to V/n, the volume per mole of the gas. Therefore, the correction to the pressure due to these cohesive forces is proportional to $(V/n)^{-2}$, or may be set equal to $a(V/n)^{-2}$, where a is the constant of proportionality. We must therefore add n^2a/V^2 to the experimental pressure P to compensate for the attractive forces.

By this line of reasoning, van der Waals in 1873 obtained his famous equation of state,

$$\left(P + \frac{n^2a}{V^2}\right)(V - nb) = nRT$$

This equation provides a good representation of the behavior of gases at moderate densities, but deviations become large at higher densities. The values of the constants a and b are obtained from the experimental PVT data at moderate densities, or from the critical constants of the gas. Some of these values were collected in Table 1.1.

10. Intermolecular Forces and the Equation of State

The van der Waals model gives a pictorial representation of factors causing deviations from perfect gas behavior, and, considering its simplicity, the model is remarkably effective. A closer analysis of the problem soon indicates, however, the need for a more realistic treatment of the forces between molecules. The van der Waals b measures a sort of repulsive force that comes suddenly into operation when two molecules, considered as rigid spheres, come into contact. The van der Waals a measures the effect of attractive forces, but says nothing about the nature of these forces beyond the requirement that they cause a decrease in the perfect gas pressure proportional to the square of the concentration of molecules in the gas. As a consequence of the interactions between the electric fields of the negative electrons and positive nuclei from which molecules are constructed, there are forces of interaction between any pair of molecules, which depend on the nature of the molecules and the distance separating them. For many purposes, it suffices to distinguish two principal forces: (1) a repulsive force due primarily to the electrostatic repulsions between outer electron clouds of the molecular structure; (2) an attractive force due to correlations of the positions of electrons in one molecule with those in the other, which occur in such a way as to cause a net electrical attraction. The quantitative theory of these forces will be discussed later, after necessary background in quantum mechanics and molecular structure has been provided. For the present, we shall simply outline the final conclusions of the theory.

The attractive forces* decrease with separation in accord with an inverse seventh-power law, and if the molecules can be approximated by spheres (so that no special directions occur in the field of force), the attractive force can be represented simply by

$$F_L = -k_L \ r^{-7}$$

where k_L is a positive constant. The repulsive force is considerably *shorter in range*, i.e., it falls off more rapidly with separation r. It can be represented conveniently by an inverse thirteenth-power law,

$$F_R = k_R \ r^{-13}$$

where k_R is a positive constant.

Instead of considering the forces themselves, it is usually more convenient for us to consider the potential energy functions $U(r)$ from which they can be derived by $F = -dU/dr$.

If the sum of attractive and repulsive forces is $F = k_R \ r^{-13} - k_L \ r^{-7}$,

$$U = \frac{k_R}{12} r^{-12} - \frac{k_L}{6} r^{-6} + \text{const.}$$

At $r = \infty$, $U = 0$, so that the constant of integration is 0. The intermolecular potential energy is usually written in the standard form,

$$U(r) = 4\epsilon \left\{ \left(\frac{\sigma}{r} \right)^{12} - \left(\frac{\sigma}{r} \right)^{6} \right\} \tag{4.15}$$

in which ϵ is the maximum energy of attraction (depth of potential well) and σ is one of the values of r for which $U(r) = 0$ (the other is $r = \infty$). This potential is called the *Lennard–Jones 6–12 potential* in honor of the English physicist who first made extensive applications of it in the theory of gas imperfections.

In Fig. 4.4, we have plotted each of the terms in (4.15) as well as their sum, the total intermolecular potential energy, for the general case. The Lennard–Jones potential curves for several actual gases are also shown. The constants ϵ and σ are obtained by fitting experimental PVT data to a theoretical equation of state based on (4.15). The theoretical equations that allow us to calculate the equation of state from the intermolecular potential are obtained through an application of statistical mechanics. Once again, we promise to derive them later, and simply give the result in this chapter.

---

*The principal contribution to the attractive forces are the London or dispersion forces discussed in Section 19.7.

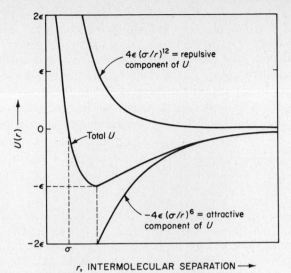

FIG. 4.4(a)   The Lennard-Jones 6–12 potential.

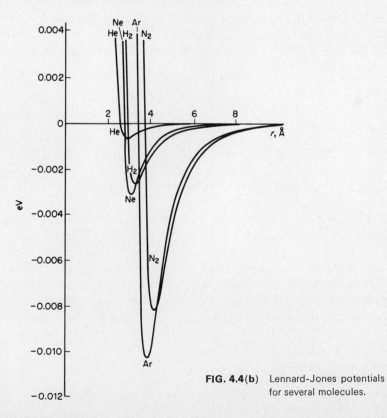

FIG. 4.4(b)   Lennard-Jones potentials for several molecules.

The virial equation of state was

$$\frac{PV}{nRT} = 1 + B(T)\frac{n}{V} + \cdots$$

where $B(T)$ is the second virial coefficient. The second virial coefficient is obtained theoretically by considering only interactions between pairs of molecules. The third virial coefficient would require the introduction of interactions between sets of three molecules, etc. The result of a statistical mechanical calculation (Section 19.8) is that the second virial coefficient for the intermolecular pair potential $U(r)$ is given by

$$B(T) = 2\pi L \int_0^\infty [1 - e^{-U(r)/kT}] \, r^2 \, dr \tag{4.16}$$

11. Molecular Velocities—Directions

The kinetic theory will show us how to derive many measurable properties if we can specify the behavior of the molecules that compose the gas. The data that we need are the masses and velocities of the gas molecules and the laws of force between them. Of course, we cannot expect to know the individual velocities of all the molecules in a gas, and a statistical kind of knowledge—what fractions of the molecules have velocities within given ranges—suffices for our derivations.

Since velocity is a vector quantity, it has direction as well as magnitude. In Fig. 4.5, we show a representation of a velocity vector as an arrow from the origin

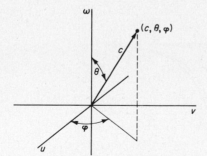

FIG. 4.5 Spherical polar coordinates in velocity space.

to a point in a three-dimensional space. This space, however, is not our usual three-dimensional space with coordinates x, y, z, but a *velocity space* in which the distances along the axes specify u, v, w, the three components of the velocity along the X, Y, Z directions.* Thus, any point in this space specifies the magnitude c of the

---

*In analytical mechanics, momentum p is a more fundamental variable than velocity. Instead of velocity space, therefore, a momentum space is often used, with $p_j = mv_j$. The momentum of a particle is represented by a point in momentum space, with components, for example, p_x, p_y, p_z. We can now combine three coordinates x, y, z with the three momenta p_x, p_y, p_z to define a six-dimensional Euclidean space, called *phase space*. A point in this space specifies both the coordinates and the momentum components of the particle.

velocity, from

$$c^2 = u^2 + v^2 + w^2, \quad c = \sqrt{u^2 + v^2 + w^2}$$

and its direction, as the vector from the origin to the point (u, v, w). The direction can be specified in terms of two angles, φ, a sort of longitude, and θ, a sort of colatitude (latitude measured from the North Pole). Thus c, θ, and φ specify both the magnitude and the direction of the velocity vector in polar coordinates in velocity space.

If we are dealing with a collection of actual gas molecules, we can hardly ask how many have an exact direction specified by θ and φ. Since the angles are continuously variable, we must allow a certain range of angular variables to phrase the question meaningfully. Suppose we circumscribe about the origin a sphere of radius c. We can now consider a direction to be specified by an element of solid angle between ω and $\omega + d\omega$. Just as an ordinary angle is defined as the ratio of a length of circular arc to its radius, so a solid angle is specified as the ratio of an area of spherical surface to the square of its radius, $d\omega = dA/c^2$. The total surface of the sphere is $4\pi c^2$, and hence the total solid angle subtended by a sphere is $\omega = 4\pi$. The fraction of the surface covered by $d\omega$ is accordingly $d\omega/4\pi$.

From Fig. 4.6, we see that

$$dA = c \sin \theta \, d\varphi \cdot c \, d\theta = c^2 \sin \theta \, d\theta \, d\varphi$$

Hence, the element of solid angle is

$$d\omega = \sin \theta \, d\theta \, d\varphi \tag{4.17}$$

We shall make use of this expression in calculating the average number of molecules approaching a surface from a given direction.

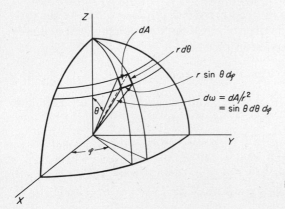

FIG. 4.6 Element of solid angle in spherical polar coordinates.

The specification of the direction of motion of a molecule is easy enough, because in a volume of gas *at equilibrium*, all directions of molecular motion are equally probable. If the gas is in the form of a jet, or beam, or if it is flowing through a tube, all directions are no longer equally probable.

12.   Collisions of Molecules with a Wall

Consider in Fig. 4.7 an area of wall in contact with a volume V of gas containing N molecules. Our problem is to calculate the frequency of molecular collisions with the wall, i.e., the number of molecules that strike the wall in unit time. We

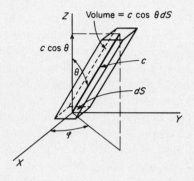

FIG. 4.7 Element of volume from which molecules coming from a specified direction (θ, φ) with a speed c strike the surface in unit time.

first obtain an expression for the number of molecules striking the wall from a given direction and then integrate over all possible directions. Consider a particular velocity of magnitude between c and $c + dc$, the direction of which lies within a solid angle $d\omega$ at an angle θ to the wall. Since all directions of molecular motion are equally probable, a fraction $d\omega/4\pi$ of molecular velocities will have this direction, and a fraction $f(c)\,dc$ will lie in the specified range of speed, where $f(c)$ is the density function. In a time dt, molecules with velocity c will hit the wall provided they are within a distance $c \cos \theta\,dt$ from the wall, where $c \cos \theta$ is the component of velocity normal to the wall. If there are N/V molecules per unit volume, the number of such molecules striking unit area is $(N/V)cf(c) \cos \theta\,dc\,dt$. The product of the various factors just outlined gives the number of molecules dN colliding with unit area of the wall from the specified direction in time dt,

$$dN = \frac{N}{V} c \cos \theta \frac{d\omega}{4\pi} f(c)\,dc\,dt$$

Introducing $d\omega = \sin \theta\,d\theta\,d\varphi$ from (4.17), and integrating over all directions from the wall, we obtain

$$\frac{dN}{dt} = Z = \frac{N}{4\pi V} \int_0^\infty \int_0^{2\pi} \int_0^{\pi/2} \sin \theta \cos \theta\,d\theta\,d\varphi\ cf(c)\,dc$$

Since

$$\int_0^{\pi/2} \sin \theta \cos \theta\,d\theta = \frac{1}{2}$$

integration over the angular variables gives

$$Z = \frac{N}{4V} \int_0^\infty cf(c)\,dc$$

But

$$\int_0^\infty cf(c)\,dc = \bar{c}$$

the average speed of the molecules, so that

$$Z_{wall} = \frac{1}{4}\frac{N}{V}\bar{c} \qquad (4.18)$$

This is an important formula in the theory of surface reactions. It also applies directly to the problem of gaseous effusion, for if we imagine that an area of the solid wall is opened to form an orifice, the rate of effusion through that area is exactly equal to the rate at which molecules would strike that area. Thus, (4.18) is also a quantitative expression for the rate of effusion per unit area.

Let us calculate from (4.18) the frequency of collisions with the wall for nitrogen gas at 300 K and 1 atm. Since $PV = nRT$ and $n = N/L$,

$$\frac{N}{V} = \frac{PL}{RT} = \frac{1\,(atm) \times 6.02 \times 10^{23}}{82.05\,(cm^3 \cdot atm \cdot K^{-1}) \times 300\,K} = 2.45 \times 10^{19} cm^{-3}$$

$$\bar{c} = \left(\frac{8RT}{\pi M}\right)^{1/2} = \left(\frac{8 \times 8.314 \times 10^7 \times 300}{3.142 \times 28}\right)^{1/2} = 4.76 \times 10^4\ cm \cdot s^{-1}$$

Hence,

$$Z_{wall} = \tfrac{1}{4}(2.45 \times 10^{19})(4.76 \times 10^4) = 2.92 \times 10^{23}\ s^{-1} \cdot cm^{-2}$$

13. Distribution of Molecular Velocities

In their constant motion, the molecules of a gas collide many times with one another, and these collisions provide the mechanism through which the velocities of individual molecules are continually changing. As a result, there exists a distribution of velocities among the molecules; most have velocities with magnitudes quite close to the average, and relatively few have magnitudes far above or far below the average.

The distribution of velocities among molecules in a gas was partly guessed and partly calculated by the Scotch theoretical physicist, James Clerk Maxwell, long before it was measured experimentally or derived exactly. From 1860 to 1865, Maxwell was professor of Natural Philosophy at Kings College, London. During these years, he published his derivation of the distribution law,[*] and also the great paper that laid the foundations of electromagnetic theory.

Let us consider a collection of N molecules in a volume of gas at equilibrium at some temperature T. The velocities of the molecules can be specified in terms of components u, v, w, parallel to the three Cartesian axes X, Y, Z. As shown in Fig. 4.5, we can represent the velocities as points in a velocity space, such that the magnitude and direction of each velocity is represented by a vector from the origin to a point with coordinates (u, v, w). The magnitude c of the velocity vector (the molecular speed) is given by $c^2 = u^2 + v^2 + w^2$.

We now wish to describe the probability that one velocity component for a molecule—say, u—lies in a range from u to $u + du$. The probability that any

*J. C. Maxwell, *Phil. Mag.*, *19*, 31 (1860).

given molecule has the component between u and $u + du$ is thus $f(u)\, du$. Similarly, $f(v)$ and $f(w)$ can be defined for the other two components.

Next we ask the question, what is the number of molecules having *simultaneously* velocity components between u and $u + du$, v and $v + dv$, w and $w + dw$? We can define a density function $F(u, v, w)$ to express the result as

$$NF(u, v, w)\, du\, dv\, dw$$

At this point, Maxwell introduces an apparently innocent assumption, which, however, leads to an exact specification of the function $F(u, v, w)$. The assumption is that the probability of having one velocity component in a given range (say, u to $u + du$) is completely independent of the probability of having another component in a given range (say, v to $v + dv$). If the probabilities are independent, the combined probability is simply the product of the independent probabilities. (As an analogy, the chance of drawing a spade from a deck of cards is $\frac{1}{4}$ and the chance of drawing a queen is $\frac{1}{13}$, so that the chance of drawing the queen of spades is $\frac{1}{4} \times \frac{1}{13} = \frac{1}{52}$.) Thus, the assumption of independent probabilities allows us to write

$$F(u, v, w) = f(u)f(v)f(w) \qquad (4.19)$$

Can we justify the assumption of independent probabilities for the velocity components? Let us say first of all that we can easily derive the assumption from quantum mechanics; when we obtain the solution for the problem of a *particle in a box* (Section 13.20), we shall see that the Schrödinger equation for the wave function ψ of the system separates in terms of Cartesian coordinates, x, y, z, so that

$$\psi(x, y, z) = X(x)Y(y)Z(z)$$

and consequently, it can be shown that the velocity components of the particle do have independent probabilities. This result, of course, was not available to Maxwell in 1860. A satisfactory proof of the distribution law can, however, be obtained by means of classical mechanics alone, and it was first accomplished by Boltzmann in 1896. This rigorous proof is quite difficult,* and we shall be satisfied here with the Maxwell derivation, which is easy, once the assumption of independent probabilities for the velocity components has been accepted.

We need one further statement about $F(u, v, w)$. Since the gas is isotropic and at equilibrium, the function $F(u, v, w)$ must be a function of c only, $F(c)$; that is to say, the particular direction of the velocity cannot influence its probability. If this were not true, there would be a tendency for the gas to flow in a certain direction, contradicting the assumption of equilibrium. Hence, we can write

$$F(u, v, w) = F(c) = f(u)f(v)f(w) \qquad (4.20)$$

Taking logarithms of both sides of (4.20), we obtain

$$\ln F(c) = \ln f(u) + \ln f(v) + \ln f(w)$$

On differentiation,

$$\left(\frac{\partial \ln F}{\partial u}\right)_{v,w} = \frac{d \ln F}{dc}\left(\frac{\partial c}{\partial u}\right) = \frac{u}{c}\frac{d \ln F}{dc} = \frac{d \ln f}{du}$$

*A good discussion is given by E. H. Kennard, *The Kinetic Theory of Gases* (New York: McGraw-Hill Book Company, 1938), pp. 32–48.

Rearrangement gives

$$\frac{d \ln F}{c \, dc} = \frac{d \ln f}{u \, du} \tag{4.21}$$

Since exactly the same result would be obtained in terms of components v and w,

$$\frac{d \ln F}{c \, dc} = \frac{d \ln f}{u \, du} = \frac{d \ln f}{v \, dv} = \frac{d \ln f}{w \, dw}$$

or

$$\frac{d \ln f}{u \, du} = \text{constant} \equiv -\gamma$$

On integration, we have

$$f(u) = a e^{-\gamma u^2} \tag{4.22}$$

We evaluate the constant a from the condition that

$$\int_{-\infty}^{+\infty} f(u) \, du = 1 = a \int_{-\infty}^{+\infty} e^{-\gamma u^2} \, du \tag{4.23}$$

This condition simply states that the integration of the probabilities over all possible velocity components must equal unity. Since the integral in (4.23) equals $(\pi/\gamma)^{1/2}$, $a = (\gamma/\pi)^{1/2}$ and our density function becomes

$$f(u) = \left(\frac{\gamma}{\pi}\right)^{1/2} e^{-\gamma u^2}$$

We evaluate the constant γ by calculating, in terms of $f(u)$, the mean kinetic energy for one degree of freedom, which we already know equals $\frac{1}{2}kT$.

The *mean value theorem** is as follows: If $p(x)$ is the density function for any variable x, so that $p(x) \, dx$ is the probability that the variable lies between x and $x + dx$, the mean value of any function of x, $g(x)$ is given by

$$\overline{g(x)} = \int_{-\infty}^{+\infty} p(x) g(x) \, dx \tag{4.24}$$

We apply (4.24) to calculate the mean value of the kinetic energy in one degree of freedom, $\frac{1}{2}mu^2$,

$$\overline{\tfrac{1}{2}mu^2} = \tfrac{1}{2}kT = \int_{-\infty}^{+\infty} \left(\frac{\gamma}{\pi}\right)^{1/2} e^{-\gamma u^2} (\tfrac{1}{2}mu^2) \, du$$

or

$$\frac{kT}{m} \pi^{1/2} = \gamma^{1/2} \int_{-\infty}^{+\infty} e^{-\gamma u^2} u^2 \, du$$

---

*We can derive this theorem from the ordinary definition of the mean value in a discrete distribution. Suppose that in a set of trials or experiments to find the value of $g(x)$, the value for x_1 occurs n_1 times; for x_2, n_2 times, etc. Then the mean value of $g(x)$ is

$$\overline{g(x)} = \frac{\sum n_j g(x_j)}{\sum n_j}$$

But $n_j / \sum n_j$ is simply the probability p_j for the value x_j, so that

$$\overline{g(x)} = \sum p_j(x_j) g(x_j)$$

where the sum is over all the discrete values of x_j. If we now let the separation between the discrete values pass in the limit to zero, the corresponding limit of the sum becomes an integral over dx, or

$$\overline{g(x)} = \int_{-\infty}^{+\infty} p(x) g(x) \, dx \tag{4.24}$$

The integral equals $\pi^{1/2}/2\gamma^{3/2}$, so that

$$\gamma = \frac{m}{2kT}$$

We can now write the density function as

$$p(u) = \left(\frac{m}{2\pi kT}\right)^{1/2} e^{-mu^2/2kT} \tag{4.25}$$

The density function for a component of molecular velocity is closely related to the *normal density function* $\varphi(x)$ of probability theory, often called the *Gaussian density function*. This is defined by

$$\varphi(x) = \frac{1}{(2\pi)^{1/2}} e^{-x^2/2} \tag{4.26}$$

Hence, setting $x^2 = mu^2/kT$, we would have

$$p(u) = \left(\frac{m}{kT}\right)^{1/2} \varphi(x) \tag{4.27}$$

The integral

$$\Phi(x) = \frac{1}{(2\pi)^{1/2}} \int_{-\infty}^{x} e^{-y^2/2} \, dy = \int_{-\infty}^{x} \varphi(y) \, dy \tag{4.28}$$

is called the *normal distribution function* or the *Gaussian distribution*. In the language of statistics or probability theory, the *density function* gives the fraction of a population that lies between x and $x + dx$, whereas the *distribution function* gives the cumulative fraction between $-\infty$ and an upper limit x. The functions $\varphi(x)$ and $\Phi(x)$ have been extensively tabulated and excellent tables should be available in all libraries.*

The functions $\varphi(x)$ and $\Phi(x)$ are plotted in Fig. 4.8. The domain bounded by the graph of $\varphi(x)$ and the x axis has unit area,

$$\Phi(\infty) = \int_{-\infty}^{+\infty} \varphi(x) \, dx = 1$$

Also,

$$\Phi(0) = \int_{-\infty}^{0} \varphi(x) \, dx = \tfrac{1}{2}$$

Instead of the standardized variable x, we sometimes use a new variable $z = x/h$, where z is called the *deviation*. In the case of the velocity distribution, for exam-

---

Tables of the Error Function and Its Derivative, National Bureau of Standards Applied Mathematics Series 41 (Washington, D.C., 1954). The error function is

$$\text{erf}\,(x) = \frac{2}{\sqrt{\pi}} \int_{0}^{x} e^{-t^2} \, dt = \frac{1}{\sqrt{\pi}} \int_{-x}^{+x} e^{-t^2} \, dt$$

so that

$$\text{erf}\,(x) = 2\Phi(\sqrt{2}\,x) - 1$$

Note that $\text{erf}\,(0) = 0$ and $\text{erf}\,(\infty) = 1$.

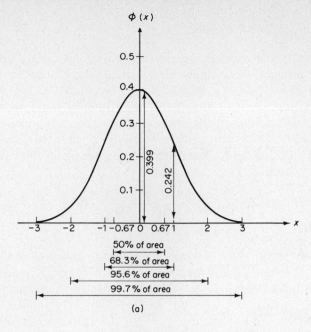

(a)

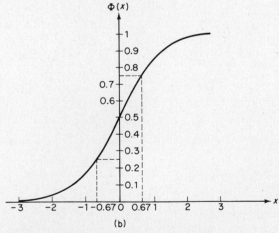

(b)

FIG. 4.8 (a) The normal density function. (b) The normal distribution function. [From W. Feller, *An Introduction to Probability Theory and Its Applications* (New York: John Wiley,
1957).]

ple, $z \equiv u = (kT/m)^{1/2}x$, or $h = (m/kT)^{1/2}$. We can easily show that the *mean
deviation* is

$$\bar{z} = \frac{1}{h\sqrt{\pi}} \longrightarrow \left(\frac{kT}{\pi m}\right)^{1/2}$$

14.   Velocity in One Dimension

It is worthwhile to examine in more detail the one-dimensional density function of (4.25). For instance, let us apply it to calculate the probability that the velocity component of a N_2 molecule lies between 999.5 and 1000.5 $m \cdot s^{-1}$ at 300 K. We can then set $du \simeq \Delta u = 1\ m \cdot s^{-1}$, the small velocity range required. In a large number N_0 of molecules, the fraction dN/N_0 with a velocity component between u and $u + du$ is simply $p(u)\ du$ from (4.25); in other words, the probability that any molecule out of the N_0 has a velocity component in this range is $p(u)\ du$. In the example chosen,

$$p(u)\ du \simeq p(u)\ \Delta u = \left(\frac{28 \times 10^{-3}}{2\pi \times 8.317 \times 300} \right)^{1/2} \exp \left(\frac{-28 \times 10^{-3} \times 10^{6}}{2 \times 8.317 \times 300} \right)$$

$$= 4.84 \times 10^{-6}$$

Note that instead of m/kT, we have used M/RT, where M is the mass of 1 mol of N_2 and R is the gas constant per mole. We compute, therefore, that about 5 molecules out of each million will have a velocity component in the specified range.

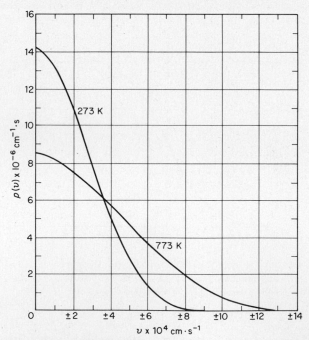

FIG. 4.9 The density functions of one component of molecular velocity for nitrogen at 273 K and 773 K. These curves are examples of the normal density function shown in Fig. 4.8. They are symmetrical about $v = 0$ and only one half of the complete function is plotted here.

The one-dimensional velocity distribution is plotted for nitrogen at 273 and 773 K in Fig. 4.9. The most probable value for the velocity component is $u = 0$. The fraction of the molecules with a velocity component in a given range declines at first slowly and then rapidly as the velocity is increased. From the curve and from a consideration of (4.25), it is evident that as long as $\frac{1}{2}mu^2 < kT$, the fraction of molecules having a velocity u falls slowly with increasing u. When $\frac{1}{2}mu^2 = 10kT$, the fraction has decreased to e^{-10}, or 5×10^{-5} times its value at $\frac{1}{2}mu^2 = kT$. Thus, only a small fraction of any lot of molecules can have kinetic energies much greater than kT per degree of freedom.

15. Velocity in Two Dimensions

Instead of a one-dimensional gas (one degree of freedom of translation), consider now a two-dimensional gas. The probability that a molecule has a given u in no way depends on the value of its Y component v. The fraction of the molecules having simultaneously velocity components between u and $u + du$, and v and $v + dv$, is simply the product of the two individual probabilities.

$$p(u)p(v) \, du \, dv = \frac{m}{2\pi kT} \exp\left[\frac{-m(u^2 + v^2)}{2kT}\right] du \, dv \qquad (4.29)$$

This sort of distribution may be graphically represented as in Fig. 4.10, where

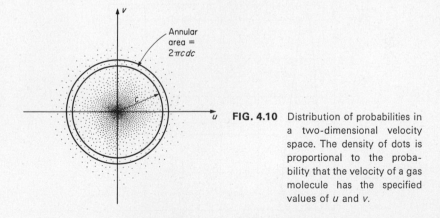

FIG. 4.10 Distribution of probabilities in a two-dimensional velocity space. The density of dots is proportional to the probability that the velocity of a gas molecule has the specified values of u and v.

a coordinate system with u and v axes has been drawn. Any point in the (u, v) plane represents a simultaneous value of u and v; the plane is a two-dimensional velocity space. The dots have been drawn so as to represent schematically the density of points in this space, i.e., the relative frequency of occurrence of sets of simultaneous values of u and v.

The graph looks like a target that has been peppered with shots by a marksman aiming at the bull's-eye. In the molecular case, each individual component of molecular velocity, u or v, aims at the value zero. The resulting distribution represents the statistical summary of the results. The more skillful the marksman, the more closely will his results cluster about the center of the target. For the molecules, the skill of the marksman has its analog in the coldness of the gas. The lower the temperature, the better the chance that a component of molecular velocity will come close to zero.

If, instead of the individual components, u and v, the resultant speed c is considered, where $c^2 = u^2 + v^2$, it is evident that its most probable value is not zero. The reason is that the number of ways in which c can be made up from u and v increases in direct proportion with c, whereas at first the probability of any value of u or v declines rather slowly with increasing velocity. From Fig. 4.10, it appears that we can obtain the distribution of c, regardless of direction, by integrating over the annular area between c and $c + dc$, which is $2\pi c\, dc$. The required fraction is then

$$\frac{dN}{N_0} = p(c)\, dc = \frac{m}{kT}\, \exp\!\left(\frac{-mc^2}{2kT}\right) c\, dc \tag{4.30}$$

16. Velocity in Three Dimensions

We can now obtain the three-dimensional velocity density function. The fraction of molecules having simultaneously a velocity component between u and $u + du$,

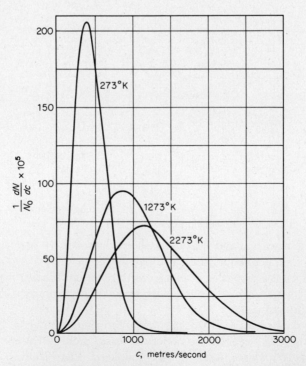

FIG. 4.11 Relative probabilities of molecular speeds in nitrogen at three different temperatures. The density function $4\pi(m/2\pi\, kT)^{3/2} \exp(-mc^2/2\, kT)\, c^2$ of (4.32) is plotted against c.

v and $v + dv$, and w and $w + dw$, is

$$\frac{dN}{N_0} = \left(\frac{m}{2\pi kT}\right)^{3/2} \exp\left[\frac{-m(u^2 + v^2 + w^2)}{2kT}\right] du\, dv\, dw \quad (4.31)$$

We wish an expression for the fraction of molecules with a speed between c and $c + dc$, regardless of direction, where $c^2 = u^2 + v^2 + w^2$. These are the molecules whose velocity points lie within a spherical shell of thickness dc at a distance c from the origin. The volume of this shell is $4\pi c^2\, dc$. Therefore, the desired density function is

$$\frac{dN}{N_0} = 4\pi\left(\frac{m}{2\pi kT}\right)^{3/2} \exp\left(\frac{-mc^2}{2kT}\right) c^2\, dc \quad (4.32)$$

This is the expression derived by Maxwell in 1860.

The equation (4.32) is plotted in Fig. 4.11 at several different temperatures. The curve becomes broader and less peaked at the higher temperatures, as the average speed increases and the distribution about the average becomes wider.

We can now calculate the average molecular speed $\bar{c}$. Using (4.32) and (4.24), we have

$$\bar{c} = \int_0^\infty f(c)c\, dc = 4\pi\left(\frac{m}{2\pi kT}\right)^{3/2} \int_0^\infty e^{-mc^2/2kT}\, c^3\, dc$$

The evaluation of this integral can be obtained* from

$$\int_0^\infty e^{-ax^2} x^3\, dx = \frac{1}{2a^2}$$

Making the appropriate substitutions, we find

$$\bar{c} = \left(\frac{8kT}{\pi m}\right)^{1/2} \quad (4.33)$$

Similarly, the average kinetic energy is

$$\tfrac{1}{2}mc^2 = \frac{m}{2} \int_0^\infty f(c)c^2\, dc = 2\pi m\left(\frac{m}{2\pi kT}\right)^{3/2} \int_0^\infty e^{-mc^2/2kT} c^4\, dc$$

Since the integral,

$$\int_0^\infty e^{-ax^2} x^4\, dx = \frac{3\pi^{1/2}}{8a^{5/2}}$$

we obtain,

$$\tfrac{1}{2}(mc^2) = \tfrac{3}{2}kT \quad (4.34)$$

*Letting $x^2 = z$, we have

$$\int_0^\infty e^{-ax^2} x\, dx = \frac{1}{2}\int_0^\infty e^{-az}\, dz = \frac{1}{2}\left(\frac{e^{-az}}{-a}\right)_0^\infty = \frac{1}{2a}$$

Then,

$$\int_0^\infty e^{-ax^2} x^3\, dx = -\frac{d}{da}\int_0^\infty e^{-ax^2} x\, dx = \frac{1}{2a^2}$$

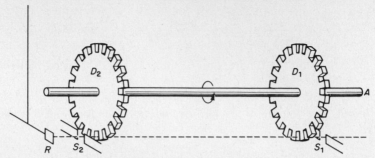

FIG. 4.12 Schematic device for separating beam of gas molecules according to ranges of velocity.

17.   Experimental Velocity Analysis

Several ingenious experiments have been devised to check the Maxwell equation for molecular speeds. A basic type of apparatus is shown in Fig. 4.12. A beam of gas molecules is defined by slits S_1 and S_2, and intercepted by two notched disks D_1 and D_2, which can be rapidly rotated on a common shaft. If a molecule enters through a notch in D_1, it can exit through D_2 only if its transit time for the distance d from D_1 to D_2 equals some integral multiple of the time required to turn one notch in D_2 to the next one. If the molecule has a velocity v in the direction $S_1 S_2$, and if the angular velocity of the disks is ω, their radius is r, and they contain b notches,

$$\frac{d}{v} = n\frac{2\pi r}{b} \cdot \frac{1}{r\omega}$$

or

$$v = db\omega/2\pi n$$

With a receiver and detector to measure the intensity of the transmitted beam, this device acts as a *velocity analyzer*. Velocity distributions measured with such devices have agreed with the Maxwell equation within experimental error. Similar velocity analyzers have been used in kinetic studies with *molecular beams*, which will be described in Chapter 9.

18.   The Equipartition of Energy

Equation (4.34) gives the average translational kinetic energy of a molecule in a gas. Note that the average energy is independent of the mass of the molecule. Per mole of gas,

$$E_k(\text{translational}) = \tfrac{3}{2}LkT = \tfrac{3}{2}RT$$

For a monatomic gas, like helium, argon, or mercury vapor, this translational kinetic energy is the total kinetic energy of the gas. For diatomic gases, like N_2 or

Cl_2, and polyatomic gases, like CH_4 or N_2O, there is also energy associated with rotational and vibrational motions.

A useful model for a molecule is obtained by supposing that the masses of the constituent atoms are concentrated at points. Almost all the atomic mass is in fact concentrated in a tiny nucleus, the radius of which is about 10^{-13} cm. Since the overall dimensions of molecules are of the order of 10^{-8} cm, a model based on point masses is reasonable. Consider a molecule composed of N atoms. To represent the instantaneous locations in space of N mass points, we require $3N$ coordinates. The number of coordinates required to locate all the mass points (atoms) in a molecule is called the number of its *degrees of freedom*. Thus, a molecule of N atoms has $3N$ degrees of freedom.

The atoms comprising each molecule move through space as a connected entity, and we can represent the translational motion of the molecule as a whole by the motion of the *center of mass* of its constituent atoms. Three coordinates (degrees of freedom) are required to represent the instantaneous position of the center of mass. The remaining $(3N - 3)$ coordinates represent the so-called *internal degrees of freedom*.

The internal degrees of freedom may be further subdivided into *rotations* and *vibrations*. Since the molecule has moments of inertia I about suitably chosen axes, it can be set into rotation about these axes. If its angular velocity about an axis is ω, the rotational kinetic energy is $\frac{1}{2}I\omega^2$. The vibrational motion, in which the atoms in a molecule oscillate about their equilibrium positions, is associated with both kinetic and potential energies, being in this respect exactly like the vibration of an ordinary spring. The vibrational kinetic energy is also represented by a quadratic expression, $\frac{1}{2}\mu v^2$. The vibrational potential energy can *in some cases* be represented also by a quadratic expression, but in the coordinates q rather than in the velocities, for example, $\frac{1}{2}\kappa q^2$. Each vibrational degree of freedom would then contribute two quadratic terms to the total energy of the molecule.*

By an extension of the derivation leading to (4.34), it can be shown that each of these quadratic terms that comprise the total energy of the molecule has an average value of $\frac{1}{2}kT$. This conclusion, a direct consequence of the Maxwell–Boltzmann distribution law, is the most general expression of the Principle of the Equipartition of Energy.

19. Rotation and Vibration of Diatomic Molecules

We may visualize the rotation of a diatomic molecule by reference to the rigid rotor model in Fig. 4.13 which might represent a molecule such as H_2, N_2, HCl, or CO. The masses of the atoms, m_1 and m_2, are concentrated at points, distant r_1 and r_2, respectively, from the center of mass. The molecule, therefore, has moments of inertia about the X and Z axes, but not about the Y axis on which the mass points lie. The distance between the mass points is fixed at r, so that such a molecule is one type of *rigid rotor*.

*The quantities μ and κ will be defined later.

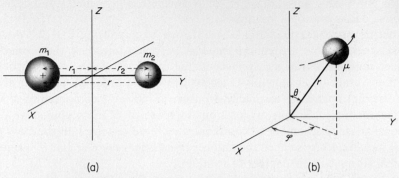

FIG. 4.13 Model of diatomic molecule as a rigid rotor. The energy of the system in (a) is identical with that in (b), in which a single mass μ rotates at a fixed distance r from the origin.

The energy of rotation of a solid body is given by

$$E_{\text{rot}} = \tfrac{1}{2}I_1\omega_1^2 + \tfrac{1}{2}I_2\omega_2^2 + \tfrac{1}{2}I_3\omega_3^2 \tag{4.35}$$

where the ω's are angular velocities of rotation, and the I's are moments of inertia referred to the principal axes of rotation. For the rigid diatomic rotor $\omega_1 = \omega_2$, $I_3 = 0$, and $I_1 = I_2 = I$ with

$$I = m_1 r_1^2 + m_2 r_2^2$$

The distances r_1 and r_2 from the center of mass are

$$r_1 = \frac{m_2}{m_1 + m_2} r, \qquad r_2 = \frac{m_1}{m_1 + m_2} r$$

Thus,

$$I = \frac{m_1 m_2}{m_1 + m_2} r^2 = \mu r^2 \tag{4.36}$$

The quantity

$$\mu = \frac{m_1 m_2}{m_1 + m_2} \tag{4.37}$$

is called the *reduced mass* of the molecule. The rotational motion is equivalent to that of a mass μ at a distance r from the intersection of the axes.

Only two coordinates are required to describe such a rotation completely; for example, two angles θ and φ suffice to fix the orientation of the rotor in space. There are thus two degrees of freedom for the rotation of a diatomic rotor. According to the Principle of the Equipartition of Energy, the average molar rotational energy should therefore be $\bar{E}_{\text{rot}} = 2L(\tfrac{1}{2}kT) = RT$.

The simplest model for a vibrating diatomic molecule (Fig. 4.14) is the harmonic oscillator. From mechanics, we know that simple harmonic motion occurs when a particle is acted on by a restoring force F directly proportional to its distance from the equilibrium position $x = r - r_e$. Thus,

$$F = -\kappa x = m \frac{d^2 x}{dt^2} \tag{4.38}$$

The same equation applies to the vibrating diatomic molecule when we set $m = \mu$,

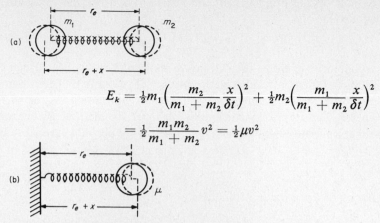

$$E_k = \tfrac{1}{2}m_1\left(\frac{m_2}{m_1+m_2}\frac{x}{\delta t}\right)^2 + \tfrac{1}{2}m_2\left(\frac{m_1}{m_1+m_2}\frac{x}{\delta t}\right)^2$$

$$= \tfrac{1}{2}\frac{m_1 m_2}{m_1+m_2}v^2 = \tfrac{1}{2}\mu v^2$$

FIG. 4.14 Model of a diatomic vibrator. An atom of mass m_1 is separated from an atom of mass m_2 by the distance r_e at equilibrium, and $r_e + x$ during the vibration, where x is a function of time t. The kinetic energy of the system in (a) is equal to that of the system in (b), in which a single mass μ is held by an identical spring to a rigid support.

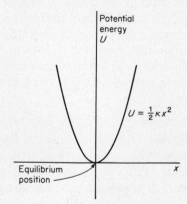

FIG. 4.15 Potential-energy curve of harmonic oscillator.

the reduced mass. The constant κ is called the *force constant*. The motion of a particle under the influence of such a restoring force may be represented by a potential energy function $U(x)$, such that

$$F = -\left(\frac{\partial U}{\partial x}\right) = -\kappa x \quad \text{and} \quad U(x) = \tfrac{1}{2}\kappa x^2 \tag{4.39}$$

The potential energy curve is thus a parabola. An example is drawn in Fig. 4.15. The motion of the system, as has been pointed out in previous cases, is analogous to that of a puck moving on such a surface. Starting from rest at any position x, it has only potential energy, $U = \tfrac{1}{2}\kappa x^2$. As it slides down the surface, it gains kinetic energy up to a maximum at position $x = 0$, where $r = r_e$ the equilibrium interatomic distance. The kinetic energy is then reconverted to potential energy as the puck slides up the other side of the incline. The total energy at any time is always a constant,

$$E_{\mathrm{vib}} = \tfrac{1}{2}\mu \left(\frac{dx}{dt}\right)^2 + \tfrac{1}{2}\kappa x^2$$

It is apparent, therefore, that vibrating molecules when heated can take up energy as both potential and kinetic energy of vibration. The Equipartition Principle states that the average energy for each vibrational degree of freedom is therefore kT, $\tfrac{1}{2}kT$ for the kinetic energy plus $\tfrac{1}{2}kT$ for the potential energy.

For a diatomic molecule, the total average energy per mole therefore becomes

$$\bar{E} = \bar{E}_{\mathrm{trans}} + \bar{E}_{\mathrm{rot}} + \bar{E}_{\mathrm{vib}} = \tfrac{3}{2}RT + RT + RT = \tfrac{7}{2}RT$$

20. Motions of Polyatomic Molecules

The motions of polyatomic molecules can also be represented by the simple mechanical models of the rigid rotator and the harmonic oscillator. If the molecule contains N atoms, there are $(3N - 3)$ internal degrees of freedom. In the case of the diatomic molecule, $3N - 3 = 3$. Two of the three internal coordinates are required to represent the rotation, leaving one vibrational coordinate. In the case of a triatomic molecule, $3N - 3 = 6$. To divide these six internal degrees of freedom into rotations and vibrations, we must first consider whether the molecule is linear or bent. If it is linear, all the atomic mass points lie on one axis, and there is therefore no moment of inertia about this axis. A linear molecule behaves like a diatomic molecule in regard to rotation, and there are only two rotational degrees of freedom. For a linear triatomic molecule, there are thus $3N - 3 - 2 = 4$ vibrational degrees of freedom. The average energy of such molecules, according to the Equipartition Principle, would therefore be

$$\bar{E} = \bar{E}_{\mathrm{trans}} + \bar{E}_{\mathrm{rot}} + \bar{E}_{\mathrm{vib}} = 3(\tfrac{1}{2}RT) + 2(\tfrac{1}{2}RT) + 4(RT) = 6\tfrac{1}{2}RT \text{ per mole}$$

A nonlinear (bent) triatomic molecule has three principal moments of inertia, and therefore three rotational degrees of freedom. Any nonlinear polyatomic molecule has $3N - 6$ vibrational degrees of freedom. For the triatomic case, there are therefore three vibrational degrees of freedom. The average energy according to the Equipartition Principle would be

$$\bar{E} = 3(\tfrac{1}{2}RT) + 3(\tfrac{1}{2}RT) + 3(RT) = 6RT \text{ per mole}$$

Examples of linear triatomic molecules are HCN, CO_2, and CS_2. Bent triatomic molecules include H_2O and SO_2.

The vibratory motion of a collection of mass points bound together by linear restoring forces [i.e., a polyatomic molecule in which the individual atomic displacements obey (4.38)] may be quite complicated. It is always possible, however, to represent the complex vibratory motion by means of a number of simple motions, the so-called *normal modes of vibration*. In a normal mode of vibration, each atom in the molecule is oscillating with the same frequency. Examples of the normal modes for linear and bent triatomic molecules are shown in Fig. 4.16. The bent molecule has three distinct normal modes, each with a characteristic frequency. The frequencies, of course, have different numerical values in different compounds. In the case of the linear molecule, there are four normal modes: two correspond to

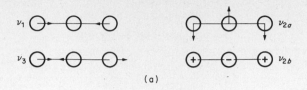

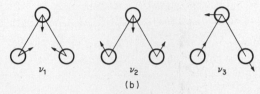

FIG. 4.16 Normal modes of vibration of triatomic molecules. (a) Linear, ex.: CO_2; (b) Bent, ex.: H_2O. The arrows indicate displacements in the plane of the paper whereas + and − indicate displacements normal to that plane.

stretching of the molecule (ν_1, ν_3) and two correspond to bending (ν_{2a}, ν_{2b}). The two bending vibrations differ only in that one is in the plane of the paper and one (denoted by + and −) is normal to the plane. These vibrations have the same frequency, and are called *degenerate vibrations*.

When we described the translational motions of molecules and their consequences for the kinetic theory of gases, it was desirable at first to employ a simplified model. The same procedure has been followed in this discussion of the internal molecular motions. Thus, diatomic molecules do not really behave as rigid rotors, since, at rapid rotation speeds, centrifugal force tends to separate the atoms by stretching the bond between them. Likewise, a more detailed theory shows that the vibrations of the atoms are not strictly harmonic.

21. The Equipartition Principle and Heat Capacities

According to the Equipartition Principle, a gas on warming should take up energy in all its degrees of freedom, $\frac{1}{2}RT$ per mole for each translational or rotational coordinate, and RT per mole for each vibration. The heat capacity at constant volume, $C_V = (\partial U/\partial T)_V$, could then be readily calculated from the average energy.

From (4.8), the translational contribution to C_{Vm} is $\frac{3}{2}R$. Since $R = 8.314 \text{ J} \cdot \text{K}^{-1}$, the molar heat capacity is 12.48 J·K⁻¹. When this figure is compared with the experimental values in Table 4.4, it is found to be confirmed only for the monatomic gases, He, Ne, Ar, Hg, which have no internal degrees of freedom. The observed heat capacities of diatomic and polyatomic gases are always higher, and increase with temperature, so we may surmise that rotational and vibrational contributions are occurring.

For a diatomic gas, the Equipartition Principle predicts an average energy of $\frac{7}{2}RT$, or $C_{Vm} = \frac{7}{2}R = 29.10 \text{ J} \cdot \text{K}^{-1} \cdot \text{mol}^{-1}$. For the diatomic gases in the table, this value is reached only at high temperatures. For polyatomic gases, the discrepancy

TABLE 4.4
Molar Heat Capacity of Gases

C_V in $J \cdot K^{-1} \cdot mol^{-1}$ at Temperature K

Gas	298.15	400	600	800	1000	1500	2000
He	12.48	12.48	12.48	12.48	12.48	12.48	12.48
H_2	20.52	20.87	21.01	21.30	21.89	23.96	25.89
O_2	21.05	21.79	23.78	25.43	26.56	28.25	29.47
Cl_2	25.53	26.99	28.29	28.89	29.19	29.69	29.99
N_2	20.81	20.94	21.80	23.12	24.39	26.54	27.68
H_2O	25.25	25.93	27.98	30.36	32.89	38.67	42.77
CO_2	28.81	33.00	39.00	43.11	45.98	40.05	52.02

with the simple classical theory is even more marked. The Equipartition Principle cannot explain why the observed C_V is less than predicted, why C_V increases with temperature, or why the C_V values differ for the various diatomic gases. The theory is thus satisfactory for translational motion, but it does not apply to rotations and vibrations. Since the Equipartition Principle is a direct consequence of the kinetic theory, and in particular of the Maxwell–Boltzmann distribution law, it is evident that an entirely new basic theory will be required to cope with the problem of heat capacities. Such a development is found in the quantum theory introduced in Chapter 13.

22. Collisions Between Molecules

The most interesting events in the life of a molecule occur when it makes a collision with another molecule. Chemical reactions between molecules depend upon such collisions. Important *transport processes* in gases, through which energy (by heat conduction), mass (by diffusion), and momentum (by viscosity) are transferred from one point to another involve collisions between gas molecules. We shall give first a brief oversimplified account of molecular collisions, neglecting the distribution of molecular velocities, and then give a more exact account, including the velocity distribution.

Suppose there are two kinds of molecules, A and B, which interact as rigid spheres with diameters d_A and d_B. The contact between two such molecules is shown

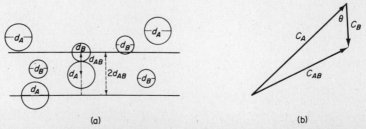

(a)   (b)

FIG. 4.17 (a) Molecular collisions. (b) Relative velocity.

in Fig. 4.17. A collision occurs whenever the distance between centers becomes as small as $d_{AB} = (d_A + d_B)/2$. Let us imagine a sphere of *radius* d_{AB} to be circumscribed about the center of A. Whenever the center of a B molecule comes within this sphere, B can be said to make a *rigid-sphere collision* with A.

Let us now suppose that all the B molecules are at rest and the A molecule is rushing through the volume of stationary B molecules with an average speed $\bar{c}_A$. We can imagine that the moving A molecule in unit time sweeps out a cylindrical volume $\pi d_{AB}^2 \bar{c}_A$. If N_B/V is the number of B molecules per unit volume, there will be $\pi d_{AB}^2 \bar{c}_A N_B/V$ centers of B molecules encountered per unit time in the sweep by the A molecule. We can thus estimate the collision frequency for one A molecule as

$$z_{AB} = \frac{\pi d_{AB}^2 \bar{c}_A N_B}{V}$$

If (N_A/V) molecules of A per unit volume, the total collision frequency between A and B molecules would be

$$Z_{AB} = \frac{\pi d_{AB}^2 \bar{c}_A N_A N_B}{V^2}$$

This estimate neglects the effect of the excluded volumes of the molecules (similar to the van der Waals b correction to a perfect gas model), but at low pressures this correction would be small.

A much more important error in the derivation comes from the assumption that the B molecules are stationary while the A sweep through the volume. Actually, of course, it is the *speed of A relative to B* that determines the frequency of collisions. This relative speed c_{AB} is the magnitude of the *vector difference* between the velocities of A and B. As shown in Fig. 4.17, the magnitude c_{AB} depends on the angle between c_A and c_B.

$$c_{AB} = (c_A^2 + c_B^2 - 2c_A c_B \cos \theta)^{1/2} \tag{4.40}$$

Thus, our expressions for the collision frequencies should be

$$z_{AB} = \frac{\pi d_{AB}^2 \bar{c}_{AB} N_B}{V} \tag{4.41}$$

$$Z_{AB} = \frac{\pi d_{AB}^2 \bar{c}_{AB} N_A N_B}{V^2} \tag{4.42}$$

Even if the molecules cannot be considered as rigid spheres, we can express the experimental results of various collisional processes in terms of an *effective collision cross section* σ_{AB}. For rigid spheres $\sigma_{AB} = \pi d_{AB}^2$. For other cases, σ_{AB} can be calculated from expressions for the forces between molecules.

In the next section, we shall derive rigorously the expression for collision frequency. Then we shall find that

$$\bar{c}_{AB} = \sqrt{\frac{8kT}{\pi \mu}}$$

where μ is the reduced mass of (4.37).

If we consider a volume of a single kind of gas, $m_1 = m_2$ in (4.37), and the relative speed becomes simply

$$\bar{c}_{AA} = \sqrt{2}\,\sqrt{\frac{8kT}{\pi m}} = \sqrt{2}\,\bar{c}$$

The number of collisions experienced by one molecule in unit time is then

$$z_{AA} = \frac{\sqrt{2}\,\pi d^2 \bar{c} N_A}{V} \tag{4.43}$$

The total number of collisions in unit volume and unit time is

$$Z_{AA} = \frac{\tfrac{1}{2}\sqrt{2}\,\pi d^2 \bar{c} N_A^2}{V^2} \tag{4.44}$$

(The factor of $\tfrac{1}{2}$ is necessary so as not to count every collision twice.)

An important quantity in kinetic theory is the average distance a molecule travels between collisions. This is called the *mean free path*. The average number of collisions experienced by one molecule in unit time is z_{AA} in (4.43). In this time, the molecule has traveled a distance $\bar{c}$. The mean free path λ is therefore $\bar{c}/z_{AA}$, or

$$\lambda = \frac{1}{\sqrt{2}\,\pi(N/V)d^2} \tag{4.45}$$

Consider the example of an ideal gas at STP for which $N/V = (6.02 \times 10^{23})/22\,414$ molecules$\cdot$cm^{-3}. For a typical $d = 4 \times 10^{-8}$ cm, $\lambda = 427 \times 10^{-8}$ cm.

23. Derivation of Collision Frequency

Let us consider again a volume of gas containing N_A, N_B molecules of A, B. The number of A molecules with velocity components between u and $u + du$, v and $v + dv$, w and $w + dw$, from (4.31), is

$$dN_A = N_A\left(\frac{m_A}{2\pi kT}\right)^{3/2} \exp\left[\frac{-m_A(u^2 + v^2 + w^2)}{2kT}\right] du\,dv\,dw \tag{4.46a}$$

Similarly, the number of B molecules with velocity components between u' and $u' + du'$, v' and $v' + dv'$, w' and $w' + dw'$, is

$$dN_B = N_B\left(\frac{m_B}{2\pi kT}\right)^{3/2} \exp\left[\frac{-m_B(u'^2 + v'^2 + w'^2)}{2kT}\right] du'\,dv'\,dw' \tag{4.46b}$$

The number of collisions in unit volume in unit time between A and B molecules with velocity components in these respective ranges [from (4.42)] is

$$dZ_{AB} = \frac{dN_A\,dN_B\,\pi d_{AB}^2\,c_{AB}}{V^2} \tag{4.47}$$

Here, c_{AB} is the relative speed of the A and B molecules in the specified individual velocity range, i.e.,

$$c_{AB} = [(u - u')^2 + (v - v')^2 + (w - w')^2]^{1/2} \tag{4.48}$$

To calculate the total number of collisions in unit time between all the A and B molecules, we substitute (4.46a), (4.46b), and (4.48) into (4.47), and integrate over all values of u, v, w, u', v', w'. This result is written as

$$Z_{AB} = \tfrac{1}{8} N_A N_B \pi d_{AB}^2 \frac{(m_A m_B)^{3/2}}{(\pi k T)^3} \int\int\int\int\int\int_{-\infty}^{+\infty} [(u - u')^2 + (v - v')^2 + (w - w')^2]^{1/2}$$

$$\exp\left[\frac{-m_A(u^2 + v^2 + w^2) - m_B(u'^2 + v'^2 + w'^2)}{2kT}\right] du\, dv\, dw\, du'\, dv'\, dw' \qquad (4.49)$$

To perform this integration, we must transform the variables from the six ordinary velocity components (in so-called laboratory coordinates) to a set of six new velocity variables. Three of these specify the components of the relative speed c_{AB},

$$u_{AB} = u - u', \quad v_{AB} = v - v', \quad w_{AB} = w - w' \qquad (4.50)$$

The other three variables give the velocity components of the center of mass of the system of two particles

$$U = \frac{m_A u + m_B u'}{m_A + m_B}, \qquad V = \frac{m_A v + m_B v'}{m_A + m_B}, \qquad W = \frac{m_A w + m_B w'}{m_A + m_B} \qquad (4.51)$$

The use of the center of mass coordinates here is similar to the separation of translational from internal degrees of freedom in Section 4.18. It allows us to separate out the kinetic energy terms due to motion of the center of mass and thus, as we shall see, allows us to integrate (4.49), which becomes:

$$Z_{AB} = \tfrac{1}{8} N_A N_B \pi d_{AB}^2 \frac{(m_A m_B)^{3/2}}{(\pi k T)^3} \int\int\int_{-\infty}^{+\infty} \exp\left[\frac{-(m_A + m_B)(U^2 + V^2 + W^2)}{2kT}\right] dU\, dV\, dW$$

$$\int\int\int_{-\infty}^{+\infty} (u_{AB}^2 + v_{AB}^2 + w_{AB}^2)^{1/2} \exp\left[\frac{-\mu(u_{AB}^2 + v_{AB}^2 + w_{AB}^2)}{2kT}\right] du_{AB}\, dv_{AB}\, dw_{AB} \qquad (4.52)$$

In the preceding expression, we have introduced the reduced mass from (4.37). Also, in going from (4.49) to (4.52) we have used the fact that

$$du\, dv\, dw\, du'\, dv'\, dw' = du_{AB}\, dv_{AB}\, dw_{AB}\, dU\, dV\, dW$$

We prove this relation for one set of components (since the argument holds for the other two as well). When we change from the variables u, u' to the new variables $u_{AB}(u, u')$ and $U(u, u')$, the products of the differentials are related by

$$du_{AB}\, dU = \frac{\partial(u_{AB}, U)}{\partial(u, u')} du\, du'$$

where $\partial(u_{AB}, U)/\partial(u, u')$ is the *Jacobian* of the transformation, defined as the determinant,

$$\frac{\partial(u_{AB}, U)}{\partial(u, u')} = \begin{vmatrix} \dfrac{\partial u_{AB}}{\partial u} & \dfrac{\partial U}{\partial u} \\[2mm] \dfrac{\partial u_{AB}}{\partial u'} & \dfrac{\partial U}{\partial u'} \end{vmatrix}$$

which, in this case, from (4.50) and (4.51), is equal to

$$\begin{vmatrix} 1 & \dfrac{m_A}{m_A + m_B} \\[2mm] -1 & \dfrac{m_B}{m_A + m_B} \end{vmatrix} = 1$$

The first three integrals in (4.52) are of the form

$$\int_{-\infty}^{+\infty} \exp\left[\frac{-(m_A + m_B)U^2}{2kT}\right] dU = \left(\frac{2\pi kT}{m_A + m_B}\right)^{1/2} \tag{4.53}$$

from the evaluation already given of (4.23), if we let $\gamma = (m_A + m_B)/2kT$.

The second triple integral can be evaluated after transformation to polar coordinates,

$$u_{AB}^2 + v_{AB}^2 + w_{AB}^2 = c_{AB}^2$$

$$du_{AB}\, dv_{AB}\, dw_{AB} = c_{AB}^2 \sin\theta\, dc_{AB}\, d\theta\, d\phi$$

so that the triple integral becomes (from p. 141),

$$\int_0^\infty \int_0^\pi \int_0^{2\pi} \sin\theta\, c_{AB}^3 \exp\left(\frac{-\mu c_{AB}^2}{2kT}\right) d\phi\, d\theta dc_{AB}$$

$$= 4\pi \int_0^\infty c_{AB}^3 \exp\left(\frac{-\mu c_{AB}^2}{2kT}\right) dc_{AB} = 8\pi\left(\frac{kT}{\mu}\right)^2 \tag{4.54}$$

Collecting the results of (4.53), (4.54), and substituting into (4.52), we finally obtain

$$Z_{AB} = \tfrac{1}{8} N_A N_B\, \pi d_{AB}^2 \frac{(m_A m_B)^{3/2}}{(\pi kT)^3}\left(\frac{2\pi kT}{m_A + m_B}\right)^{3/2} 8\pi\left(\frac{kT}{\mu}\right)^2$$

$$Z_{AB} = N_A N_B\, \pi d_{AB}^2\left(\frac{8kT}{\pi\mu}\right)^{1/2} \tag{4.55}$$

For the special case in which all the molecules are the same, so that $m_A = m_B = m$, and $N_A = N_B = N/2$, μ becomes $m_A/2$, and

$$Z_{AA} = \frac{N^2}{2}\pi d^2 \sqrt{2}\left(\frac{8kT}{\pi m}\right)^{1/2} = \tfrac{1}{2}\sqrt{2}\,\pi d^2 N^2 \bar{c} \tag{4.56}$$

24. The Viscosity of a Gas

The concept of viscosity is first met in problems of fluid flow, treated by hydrodynamics and aerodynamics, as a measure of the frictional resistance that a fluid in motion offers to an applied shearing force. The nature of this resistance may be seen from Fig. 4.18(a). If a fluid is flowing past a stationary plane surface, the layer of fluid adjacent to the plane boundary is stagnant; successive layers have increasingly higher velocities. The frictional force F, resisting the relative motion of any two adjacent layers, is proportional to S, the area of the interface between them, and to dv/dr, the velocity gradient between them. This is Newton's Law of Viscous Flow,

$$F = \eta S \frac{dv}{dr} \tag{4.57}$$

The proportionality constant η is called the *coefficient of viscosity*. It is evident that the dimensions of η are (mass) (length)$^{-1}$ (time)$^{-1}$. The SI unit is kg·m^{-1}·s^{-1}. The cgs unit, called the *poise* (P), is equal to one-tenth the SI unit.

The kind of flow governed by this relationship is called *laminar* or *streamline*

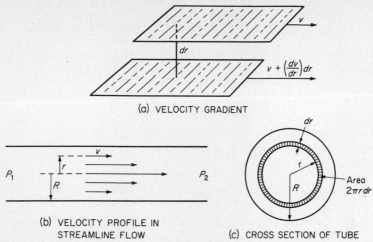

(a) VELOCITY GRADIENT

(b) VELOCITY PROFILE IN
    STREAMLINE FLOW

(c) CROSS SECTION OF TUBE

FIG. 4.18   Aspects of streamline flow considered in derivation of the
Poiseuille equation.

flow. It is evidently quite different in character from the effusive (or diffusive) flow previously discussed, since it is a massive flow of fluid, in which there is super-imposed on all the random molecular velocities a component of velocity in the direction of flow. An important case of streamline flow is the flow through pipes or tubes when the diameter of the tube is large compared to the mean free path in the fluid. The study of flow through tubes has been the basis for many experimental determinations of the viscosity coefficient.

The theory of the process was first worked out by J. L. Poiseuille, in 1844. Consider an incompressible fluid flowing through a tube of circular cross section with radius R and length l. The fluid at the walls of the tube is assumed to be stagnant, and the rate of flow increases to a maximum at the center of the tube [see Fig. 4.18(b)]. Let v be the linear velocity at any distance r from the axis of the tube. A cylinder of fluid of radius r experiences a viscous drag given by (4.57) as

$$F_r = -\eta \frac{dv}{dr} \cdot 2\pi r l$$

For steady flow, this force must be exactly balanced by the force driving the fluid in this cylinder through the tube. Since pressure is the force per unit area, the driving force is

$$F_r = \pi r^2 (P_1 - P_2)$$

where P_1 is the fore pressure and P_2 the back pressure.

Thus for steady flow,

$$-\eta \frac{dv}{dr} \cdot 2\pi r l = \pi r^2 (P_1 - P_2)$$

$$dv = -\frac{r}{2\eta l}(P_1 - P_2) \, dr$$

On integration,

$$v = -\frac{(P_1 - P_2)}{4\eta l} r^2 + \text{const.}$$

According to our hypothesis, $v = 0$ when $r = R$; this boundary condition determines the integration constant, so that we obtain finally

$$v = \frac{(P_1 - P_2)}{4\eta l} (R^2 - r^2) \tag{4.58}$$

The total volume of fluid flowing through the tube per second is calculated by integrating over each element of cross-sectional area, given by $2\pi r\, dr$ [see Fig. 4.18(c)]. Thus,

$$\frac{dV}{dt} = \int_0^R 2\pi r v\, dr = \frac{\pi(P_1 - P_2)R^4}{8l\eta} \tag{4.59}$$

This is the *Poiseuille equation*. It was derived for an incompressible fluid and therefore may be satisfactorily applied to liquids but not to gases. For gases, the volume is a strong function of the pressure. The average pressure along the tube is $(P_1 + P_2)/2$. If P_0 is the pressure at which the volume is measured, the equation becomes

$$\frac{dV}{dt} = \frac{\pi(P_1 - P_2)R^4}{8l\eta} \cdot \frac{P_1 + P_2}{2P_0} = \frac{\pi(P_1^2 - P_2^2)R^4}{16l\eta P_0} \tag{4.60}$$

By measuring the volume rate of flow through a tube of known dimensions, we can determine the viscosity η of the gas. Some results of such measurements are included in Table 4.5.

TABLE 4.5
Transport Properties of Gases

(at 273 K and 1 atm)

Gas	Mean free path λ m ($\times 10^9$)	Viscosity η kg·m^{-1}·s^{-1} ($\times 10^6$)	Thermal conductivity κ J·K^{-1}·m^{-1}·s^{-1} ($\times 10^3$)	Specific heat capacity c_V J·K^{-1}·kg^{-1} ($\times 10^{-3}$)	$\dfrac{\eta c_v}{\kappa}$
NH_3	44.1	9.76	21.5	1.67	0.76
Ar	63.5	21.0	16.2	0.314	0.41
CO_2	39.7	13.8	14.4	0.640	0.61
CO	58.4	16.8	23.6	0.741	0.43
Cl_2	28.7	12.3	7.65	0.342	0.55
C_2H_4	34.5	9.33	17.0	1.20	0.65
He	179.8	18.6	140.5	3.11	0.41
H_2	112.3	8.42	169.9	10.04	0.50
N_2	60.0	16.7	24.3	0.736	0.51
O_2	64.7	18.09	24.6	0.649	0.50

25. Kinetic Theory of the Viscosity of Gases

The kinetic picture of gas viscosity has been represented by the following analogy: Two railroad trains are moving in the same direction, but at different speeds, on parallel tracks. The passengers on these trains amuse themselves by jumping back and forth from one to the other. When a passenger jumps from the more rapidly moving train to the slower one he transports momentum $m \, \Delta v$, where m is his mass and Δv the excess velocity of his train. He tends to speed up the more slowly moving train when he lands upon it. A passenger who jumps from the slower to the faster train, on the other hand, tends to slow it down. The net result of the jumping game is thus a tendency to equalize the velocities of the two trains. An observer from afar who could not see the jumpers might simply note this result as a *frictional drag between the trains.*

The mechanism by which one layer of flowing gas exerts a viscous drag on an adjacent layer is similar, the gas molecules taking the role of the playful passengers. Consider in Fig. 4.19 a gas in a state of laminar flow parallel to the Y axis. Its velocity

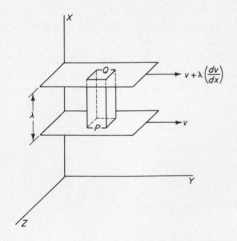

FIG. 4.19 Kinetic theory of gas viscosity. The diagram represents two layers of gas in streamline flow separated by a mean free path λ. Momentum transfer between the layers is calculated to obtain the viscous force.

increases from zero at the plane $x = 0$ with increasing x. If a molecule at P crosses to Q, in one of its free paths between collisions, it will bring to Q, on the average, an amount of momentum less than that common to molecules at the position Q by virtue of their distance along the X axis. Conversely, if a molecule travels from Q to P, it will transport to the lower, more slowly moving layer, momentum in excess of that of a molecule belonging to that layer. The net result of the random thermal motions of the molecules is to decrease the average velocities of the molecules in the layer at Q and to increase those in the layer at P. This transport of momentum tends to counteract the velocity gradient set up by the shear forces acting on the gas.

It is evident from the picture of Fig. 4.19 that a gas in a state of viscous flow is certainly not in the equilibrium state described by the Maxwell–Boltzmann equation. The occurrence of transport processes implies that the gas is nonuniform.

The basic theoretical study of this subject is entitled *The Mathematical Theory of Non-uniform Gases.** In the case of viscous flow, the nonuniformity consists in the gradient of velocity (in the *x* direction in our example). There is a directed mass motion of the gas molecules in the *y* direction. As momentum is transferred from one layer to the next, this directed motion is degraded into the random thermal motion of the molecules. Thus, the viscous flow is accompanied by degradation of energy of mass motion into energy of molecular motion. The temperature of the gas will rise as a consequence of this viscous or frictional dissipation of mechanical energy into heat. Although the flowing gas is certainly not in an equilibrium state, we can nevertheless use the average equilibrium values of properties such as velocity and mean free path to make an approximate kinetic theory of viscosity. In doing this, we implicitly assume that the Maxwell distribution is not appreciably disturbed by the mass flow, and we simply add the velocity of mass flow onto the Maxwellian velocities.

In view of the formidable difficulties in any reasonably exact theory, we shall give here an oversimplified derivation, which does serve to bring out some of the basic factors that govern the viscosity of a gas. The length of the mean free path λ may be taken as the average distance over which momentum is transferred. If the velocity gradient is dv/dx, the difference in velocity between the two ends of the free path is $\lambda \, dv/dx$. A molecule of mass m, passing from the upper to the lower layer, thus transports momentum equal to $m\lambda \, dv/dx$. From Fig. 4.19, the number crossing unit area *up and down* per unit time is $\frac{1}{2}N\bar{c}/V$. The momentum transport per unit time is then $\frac{1}{2}(N\bar{c}\cdot m\lambda/V)(dv/dx)$. This momentum change with time is equivalent to the frictional force of (4.57), which was $F = \eta \, (dv/dx)$ per unit area. Hence,

$$\eta \frac{dv}{dx} = \frac{1}{2}\frac{Nm\bar{c}\lambda}{V}\frac{dv}{dx}$$

$$\eta = \frac{1}{2}\frac{Nm\bar{c}\lambda}{V} = \frac{1}{2}\rho\bar{c}\lambda \tag{4.61}$$

By eliminating λ between (4.45) and (4.61), one obtains

$$\eta = \frac{m\bar{c}}{2(2)^{1/2}\pi d^2} \tag{4.62}$$

This equation indicates that the viscosity of a gas is independent of its density. This seemingly improbable result was predicted by Maxwell, and its subsequent experimental verification was one of the triumphs of the kinetic theory. The physical reason for the result is clear from the preceding derivation: at lower densities, fewer molecules jump from layer to layer in the flowing gas, but because of the longer free paths, each jump carries proportionately greater momentum. For imperfect gases, the equation fails and the viscosity increases with density.

The second important conclusion from (4.62) is that the viscosity of a gas increases with increasing temperature, linearly with $\bar{c} \propto T^{1/2}$. That η increases with T has been well confirmed, but the dependence is somewhat stronger than the pre-

*Sydney Chapman and T. G. Cowling (London: Cambridge University Press, 1952).

dicted $T^{1/2}$. The reason is that the molecules are not actually hard spheres but must be regarded as somewhat soft, or surrounded by fields of force. The higher the temperature, the faster the molecules are moving, and hence the deeper one molecule can penetrate into the field of force of another before it is repelled away. This correction to (4.62) was formulated by Sutherland (1893) as a correction to the hard-sphere molecular diameter d:

$$d^2 = d_\infty^2\left(1 + \frac{A}{T}\right) \qquad (4.63)$$

Here, d_∞ and A are constants, d_∞ being interpreted as the value of d as T approaches infinity. More recent work has sought to express the temperature coefficient of the viscosity in terms of the laws of force between the molecules.

26. Molecular Diameters and Intermolecular Force Constants

Data on gas viscosity and other transport properties of gases have provided some of the best sources of information about intermolecular forces. If we use the simple hard-sphere model for molecules, the transport data yield values for the effective molecular diameters. Any measurement that yields a value for the mean free path λ immediately gives a molecular diameter d through (4.45). Table 4.6 summarizes values for molecular diameters obtained in this way from gas viscosities.

Values of d estimated by other methods are also included in the table. The estimation from van der Waals b is based on the fact that $b = 4Lv_m$, where $v_m = \pi d^3/6$ is the volume of a molecule considered as a rigid sphere. The value of

TABLE 4.6
Molecular Diameters

[Nanometre (nm)]

Molecule	From gas viscosity	From van der Waals b	From molecular refraction*	From closest packing
Ar	0.286	0.286	0.296	0.383
CO	0.380	0.316	–	0.430
CO_2	0.460	0.324	0.286	–
Cl_2	0.370	0.330	0.330	0.465
He	0.200	0.248	0.148	–
H_2	0.218	0.276	0.186	–
Kr	0.318	0.314	0.334	0.402
Hg	0.360	0.238	–	–
Ne	0.234	0.266	–	0.320
N_2	0.316	0.314	0.240	0.400
O_2	0.296	0.290	0.234	0.375
H_2O	0.272	0.288	0.226	–

*The theory of this method is discussed in *Atoms and Molecules*, by M. Karplus and R. N. Porter (New York: W. A. Benjamin, 1970), p. 255.

d from the closest packing of molecules in crystal structures is based on the fact (Chapter 18) that such packing leaves a void volume of 26%, so that $\pi d^3/6 = 0.74 \, m/\rho$, where ρ is the density of the closest packed crystal structure. The diversity of values obtained when "molecular diameters" are calculated by different methods indicates that the rigid-sphere model is only a first approximation even for simple molecules.

When one moves beyond the rigid-sphere approximation, the data on gas viscosity can be interpreted in terms of models of intermolecular forces, such as those summarized in the Lennard-Jones potentials. When using the Lennard-Jones potential to calculate experimental properties, it is advisable to use force constants obtained from viscosity data when you wish to calculate transport properties, and force constants obtained from second virial coefficients when you wish to calculate thermodynamic equilibrium properties. Table 4.7 contains a few exam-

TABLE 4.7
Force Constants for the Lennard-Jones 6–12 Potential (Eq. 4.15)

Gas	Force constants from viscosity		Force constants from second virial coefficients	
	ϵ/k (K)	σ (nm)	ϵ/k (K)	σ (nm)
He	10.22	0.2576	10.22	0.2556
Ne	35.7	0.2789	35.6	0.2749
Ar	124	0.3418	119.8	0.3405
Kr	190	0.361	171	0.360
Xe	229	0.4055	221	0.4100
H_2	38.0	0.2915	37.00	0.2928
N_2	91.5	0.3681	95.05	0.3698
O_2	113	0.3433	117.5	0.358
CO_2	190	0.3996	189	0.4486
CH_4	137	0.3882	148.2	0.3817

ples of Lennard-Jones force constants obtained from these sources. The necessary theoretical derivations and complete tables can be found in the standard reference book on this subject.[*]

27. Thermal Conductivity

Gas viscosity depends on the transport of momentum across a momentum (velocity) gradient. A similar theoretical treatment is applicable to thermal conductivity and to diffusion. The thermal conductivity of a gas is a consequence of the transport of kinetic energy across a temperature (i.e., kinetic energy) gradient. Diffusion in a gas is the transport of mass across a concentration gradient.

The thermal conductivity coefficient κ is defined as the heat flow per unit time

[*]J. O. Hirschfelder, C. F. Curtiss, and R. B. Bird, *Molecular Theory of Gases and Liquids* (New York: John Wiley & Sons, Inc., 1954).

$\dot{q}$, per unit temperature gradient across unit cross-sectional area, i.e., by

$$\dot{q} = \kappa \cdot S \cdot \frac{dT}{dx}$$

By comparison with (4.61),

$$\kappa \frac{dT}{dx} = \tfrac{1}{2} N V^{-1} \bar{c} \lambda \frac{d\epsilon}{dx}$$

where $d\epsilon/dx$ is the gradient of ϵ, the average kinetic energy per molecule. Now

$$\frac{d\epsilon}{dx} = \frac{dT}{dx}\frac{d\epsilon}{dT} \quad \text{and} \quad \frac{d\epsilon}{dT} = mc_V$$

where m is the molecular mass and c_V is the specific heat capacity (heat capacity per unit mass). It follows that

$$\kappa = \tfrac{1}{2} N V^{-1} mc_V \bar{c} \lambda = \tfrac{1}{2} \rho c_V \bar{c} \lambda = \eta c_V \qquad (4.64)$$

Some thermal conductivity coefficients are included in Table 4.5. It should be emphasized that, even for an ideal gas, the simple theory is approximate, since it assumes that all the molecules are moving with the same speed $\bar{c}$ and that energy is exchanged completely at each collision.*

28. Diffusion

Consider in Fig. 4.20 two different gases A and B at constant T and P. The gas A is confined between $x = -l$ and $x = +l$, while the gas B fills the remaining space in the region from $-\infty$ to $-l$ and from $+l$ to $+\infty$. This choice of geometry gives

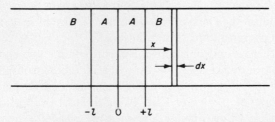

FIG. 4.20 A diffusion system in one dimension showing the initial planes of separation of gas A from gas B and a layer of gas between x and $x + dx$.

but one of a large number of problems that have been solved. Suppose that the barrier planes between the two gases are instantaneously removed. The gases will then begin to mix by the process of *diffusion*. The random thermal motions and collisions of the gas molecules lead to a continuous mixing until the total volume has a uniform composition. In the situation in Fig. 4.20, mixing occurs through

*A readable introduction to more exact theories of the transport processes is given by James Jeans, *Introduction to the Kinetic Theory of Gases* (London: Cambridge University Press, 1959).

diffusion in the $+x$ and $-x$ directions only, and the problem is simplified by this restriction to one dimension.

Consider a thin layer of gas between x and $x + dx$. The number of molecules of A or B per unit volume at any time t is a function only of x, i.e., $C_A(x, t)$ and $C_B(x, t)$ denote these respective concentrations. The diffusive flux J_A of molecules of A across a plane at x is the net number of molecules of A passing through unit area of the plane in the positive x direction in unit time. It is proportional to the concentration gradient of A at x, $\partial C_A/\partial x$,

$$J_A = -D_{AB}\frac{\partial C_A}{\partial x} \tag{4.65}$$

The proportionality constant D_{AB} is called the *diffusion coefficient*. Expression (4.65) is called *Fick's First Law of Diffusion*.

At constant T and P throughout the system, the total number of molecules per unit volume is independent of x, so that

$$\frac{\partial(C_A + C_B)}{\partial x} = \frac{\partial C_A}{\partial x} + \frac{\partial C_B}{\partial x} = 0 \tag{4.66}$$

From (4.65) and (4.66), the total flux of A and B molecules across any plane must also be zero,

$$J_A + J_B = 0 \tag{4.67}$$

If we write

$$J_B = -D_{BA}\frac{\partial C_B}{\partial x}$$

in accord with (4.65), it follows from (4.66) and (4.67) that $D_{AB} = D_{BA}$, which might as well be written simply as D. Hence, in a two-component solution, there is only one diffusion coefficient to consider; it is often called the *interdiffusion coefficient* of A and B. Of course, the value of this diffusion coefficient may, and usually does, depend upon the composition of the solution.

The fact that there is only one diffusion coefficient in a two-component solution is sometimes perplexing to students who have learned about isotopic tracer methods, which enable us to label one species and to measure its diffusion in a solution by following the course of the tracer label. With a small proportion of radioactive A^* or B^* molecules, we can define and measure *tracer diffusion coefficients*,

$$J_A^* = -D_A^*\frac{\partial C_A^*}{\partial x}$$

$$J_B^* = -D_B^*\frac{\partial C_B^*}{\partial x}$$

The D_A^* and D_B^* measure independently the diffusion of A^* and B^* through the solution of A and B. (Since there are now more than the two original components A and B, we are not contradicting our previous statement about the existence of only a single interdiffusion coefficient D.) In the ideal case, which is generally applicable to diffusion in gases at moderate pressures, we find that D is a weighted average of D_A^* and D_B^*,

$$D = \frac{C_B D_A^* + C_A D_B^*}{C_A + C_B} = X_B D_A^* + X_A D_B^* \tag{4.68}$$

where X is a mole fraction.

The fundamental differential equation for diffusion (in one dimension) can be derived as follows. We consider a volume, having unit cross section, of the region between x and $x + dx$ in Fig. 4.20 and write an expression for the increase in concentration of A with time, $\partial C_A / \partial t$. This increase will equal the excess of A molecules diffusing into the region over those diffusing out, divided by the volume (dx). Thus,

$$\frac{\partial C_A}{\partial t} = \frac{1}{dx}[J_A(x) - J_A(x + dx)]$$

But

$$J_A(x + dx) = J_A(x) + \left(\frac{\partial J_A}{\partial x}\right) dx$$

Hence, from (4.65),

$$\frac{\partial C_A}{\partial t} = -\left(\frac{\partial J_A}{\partial x}\right) = \frac{\partial}{\partial x}\left(D \frac{\partial C_A}{\partial x}\right) \tag{4.69}$$

In case D is independent of x, (4.69) becomes

$$\frac{\partial C_A}{\partial t} = D\left(\frac{\partial^2 C_A}{\partial x^2}\right) \tag{4.70}$$

This equation is called *Fick's Second Law of Diffusion*.

It has the same form as the partial differential equation for heat conduction,

$$\frac{\partial T}{\partial t} = \beta\left(\frac{\partial^2 T}{\partial x^2}\right)$$

where β is the thermal diffusivity, equal to the thermal conductivity divided by the heat capacity per unit volume. Thus, all the solutions for heat conduction problems, which have been obtained for a great variety of boundary conditions, can be readily applied to diffusion problems.*

In the derivation of (4.61), we calculated the viscosity coefficient from the transport of momentum across a velocity gradient. The diffusion coefficient measures the transport of molecules across a concentration gradient. Hence, the simple mean free path treatment yields

$$D = \frac{\eta}{\rho} = \tfrac{1}{2}\lambda \bar{c} \tag{4.71}$$

This expression would hold for the self-diffusion coefficient, as measured in a pure gas by means of tracer molecules. For a mixture of two different gases, from (4.68),

$$D = \tfrac{1}{2}\lambda_1 \bar{C}_1 X_1 + \tfrac{1}{2}\lambda_2 \bar{C}_2 X_1$$

where X_1 and X_2 are the mole fractions.

The results of the theoretical treatments of the transport processes are sum-

---

*J. Crank, *The Mathematics of Diffusion* (Oxford: The Clarendon Press, 1956); H. S. Carslaw and J. C. Jaeger, *Conduction of Heat in Solids* (Oxford: The Clarendon Press, 1959).

marized in Table 4.8. Instead of the approximate formulas derived in the text, the table gives the exact theoretical results for a rigid-sphere model.

TABLE 4.8
Transport Processes in Gases

Process	Entity transported	Exact theoretical expression for rigid spheres	SI units of coefficient
Viscosity	Momentum, mv	$\eta = 0.499\rho\bar{c}\lambda$	$\text{kg} \cdot \text{m}^{-1} \cdot \text{s}^{-1}$
Thermal conductivity	Kinetic energy, $\frac{1}{2}mv^2$	$\kappa = 1.261\rho\bar{c}\lambda c_V$	$\text{J} \cdot \text{m}^{-1} \cdot \text{s}^{-1} \cdot \text{K}^{-1}$
Diffusion	Mass, m	$D = 0.599\lambda\bar{c}$	$\text{m}^2 \cdot \text{s}^{-1}$

29. Solutions of Diffusion Equation

The partial differential equation (4.70) is second-order, linear, and homogeneous. It is of the parabolic type and the solution must be made to fit a boundary condition and an initial condition. The simplest boundary condition would specify the values of $C(t)$ for all times at the boundaries of the domain through which diffusion is taking place. The initial condition would specify the value of $C(x)$ at some time which may be taken as $t = 0$.

For example, we can readily verify by substitution into (4.70) that a solution of the equation is

$$C = \alpha t^{-1/2} \exp\left(\frac{-x^2}{4Dt}\right) \tag{4.72}$$

where α is a constant. To what boundary and initial conditions does this solution correspond? As $t \to 0$, it gives $C = 0$ everywhere except at the origin $x = 0$, where $C \to \infty$. This situation is called an *instantaneous plane source* at the origin. It would correspond to the diagram in Fig. 4.20 if the width from $-l$ to $+l$ were compressed into a plane of zero thickness at $x = 0$. The constant α is related to the strength of the source, i.e., the number N of molecules B initially present at $x = 0$. Since mass is conserved, we can write for any time t,

$$N = \int_{-\infty}^{+\infty} C \, dx = \alpha \int_{-\infty}^{+\infty} t^{-1/2} \exp\left(\frac{-x^2}{4Dt}\right) dx$$

$$N = 2\alpha(\pi D)^{1/2}$$

Hence, (4.72) becomes

$$C = \frac{N}{2(\pi Dt)^{1/2}} \exp\left(\frac{-x^2}{4Dt}\right) \tag{4.73}$$

Note that for this one-dimensional problem, the concentration C is given as number of molecules per unit distance. If we consider diffusion through a cross section of unit area, the concentration would become the usual *number per unit volume*.

When the solution in (4.73) is plotted in Fig. 4.21 for three different values of

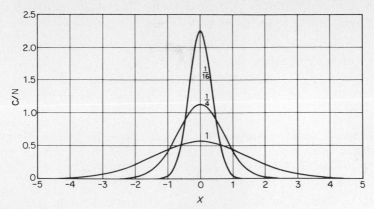

FIG. 4.21 Concentration–distance curves for diffusion from an instantaneous plane source. Numbers on curves are values of Dt.

Dt, one can visualize how the molecules of B spread outward by diffusion from the planar source. [We shall leave the extension of the solution to the source of finite width (Fig. 4.20) as a problem (4.24).]

An interesting and important way to look at the diffusion process in Fig. 4.21 is to focus attention on an individual molecule of B and to ask, what is the probability that it will have diffused a distance x in a time t? We must, of course, allow a certain spread of distance, and so we call $p(x)\,dx$ the probability that the molecule has diffused to a region between x and $x + dx$. This probability is simply the number of B molecules between x and $x + dx$ divided by the total number in the original source. Hence,

$$p(x)\,dx = \frac{C(x)\,dx}{N} = \frac{1}{2(\pi Dt)^{1/2}} \exp\left(\frac{-x^2}{4Dt}\right) dx$$

With this interpretation in mind, we can now ask what is the mean square distance $\overline{x^2}$ diffused by a molecule in time t. (We cannot use simply $\bar{x}$, since diffusion in $+$ or $-$ directions is equally probable and $\bar{x} = 0$.) Thus,

$$\overline{x^2} = \int_{-\infty}^{+\infty} x^2 p(x)\,dx \tag{4.74}$$

Substituting (4.73) into (4.74) and evaluating the integral, we find

$$\overline{x^2} = 2Dt \tag{4.75}$$

This simple relation is used time and again to provide rapid estimates of mean diffusion distances, not only in gases but in liquid and solid systems as well.

For example, geologists are concerned with the question of whether diffusion in the solid state can be a mechanism for transport of constituents of minerals. A typical diffusion coefficient in a solid at ordinary temperature would be that for helium in tourmaline, 10^{-8} cm$^2 \cdot$s^{-1}. How far would a helium atom be expected to diffuse in a tourmaline crystal in a million years? From (4.75),

$$\overline{x^2} = 2 \times 10^{-8} \times 10^6 \times 3.156 \times 10^7 = 6.212 \times 10^5$$
$$(\overline{x^2})^{1/2} = 7.88 \times 10^2 \text{ cm}$$

Such figures are of interest when we try to answer questions relative to effects of cosmic rays on mineral deposits.

PROBLEMS

1. At what speeds would molecules of (a) He and (b) N_2 have to leave the surface of (α) Earth and (β) moon in order to escape into space? At what temperatures would the average speeds $\bar{c}$ of these molecules equal these "speeds of escape"? The mass of the moon can be taken as $\frac{1}{80}$ that of Earth.

2. Consider a column of air assumed to be at a uniform temperature T in the gravitational field of the earth. Show that the pressure at an altitude x is given by $P = P_0 \exp(-mgx/kT)$, where P_0 is the pressure at $x = 0$, m is the average molecular mass of the molecules in air and g is the gravitational acceleration. Calculate P at an altitude of 8 km if $P_0 = 760$ mmHg and $T = 273$ K.

3. Suppose that T decreases with altitude x according to the formula $T = T_0 \exp(-x/h)$, where T_0 is the temperature at $x = 0$ and h is the thickness of an atmosphere assumed to be homogeneous with density equal to that at $x = 0$ at all altitudes. Show that the barometric formula then becomes $P = P_0 \exp(1 - e^{x/h})$. Calculate the pressure at an altitude of 8 km for $T_0 = 273$ K and compare the result with that found in Problem 2.

4. The quantity $\gamma = -dT/dx$, where x is the height in the atmosphere, is called the *temperature lapse rate*. Show that if γ is constant, $P = P_0(T/T_0)^{g/R\gamma}$, where g is the gravitational acceleration.

5. In the method of Knudsen,* the vapor pressure is determined by the rate at which the substance, under its equilibrium pressure, diffuses through an orifice. In one experiment, beryllium powder was placed inside a molybdenum box having an effusion hole 3.18 mm in diameter. At 1537 K, it was found that 8.88 mg of Be effused in 15.2 min. Calculate the vapor pressure of Be at 1537 K.

6. In the derivation of the van der Waals a term, it is stated that the pressure of the perfect gas is reduced as a result of cohesive forces between the gas molecules. Thus, a molecule colliding with the wall does not carry as much momentum owing to the "pull" from the molecules in the gas. But one might expect the molecule leaving the wall to carry extra momentum, due to the attractive forces in the gas. Thus, there should be no van der Waals correction to the pressure. Criticize this argument.

7. Consider a Dewar vessel in which the space between the walls has been evacuated to a pressure of 10^{-8} atm. If the vessel contains liquid nitrogen at 77 K and the outside temperature is 300 K, estimate the heat conduction into the vessel per square metre of surface. (Assume hot and cold surfaces are parallel.) By obtaining suitable information from the literature, estimate the heat transfer into the vessel due to radiation. Comment briefly on factors important in the design of Dewar vessels.

8. Perrin studied the distribution of uniform spherical [0.212 micrometre (μm) radius]

*M. Knudsen, *Ann. Physik*, **29**, 179 (1909).

grains of gamboge (density $1.206 \text{ g} \cdot \text{cm}^{-3}$) suspended in water at 15°C, by taking counts on four equidistant horizonal planes across a cell 100 μm deep. The relative numbers of grains at the four levels were:

Level	5 μm	35 μm	65 μm	95 μm
Numbers	100	47	22.6	12

Estimate the Avogadro Number L from these data.

9. The permeability at 20°C of Pyrex glass to helium is given as $4.9 \times 10^{-13} \text{ cm}^3 \cdot \text{s}^{-1}$ per millimetre thickness and per millibar (mbar) partial pressure difference. The He content of the atmosphere at sea level is about 5×10^{-4} mol %. Suppose a 100-cm³ round Pyrex flask (0.7 mm wall) was evacuated to 10^{-12} atm and sealed. What would be the pressure in the flask at the end of one month due to inward diffusion of helium?

10. Suppose that the oxygen pressure in an ultrahigh-vacuum system has been reduced to 10^{-10} torr at 30°C. A clean surface of a germanium crystal is formed by cleavage within the vacuum system. If every oxygen molecule striking this surface is converted to GeO, estimate the time required to cover half the Ge sites with oxygen. (You will need at least one additional numerical quantity. State any assumptions made.)

11. What is the *chance* that an oxygen molecule in air at 273 K and 1 atm travels (a) 10^{-5} mm, (b) 1 mm without experiencing a collision?

12. Find the probability that a random variable x with a normal Gaussian density function [Eq. (4.26)] will fall within the following intervals: (a) $-\sigma$ to $+\sigma$; (b) -2σ to $+2\sigma$; (c) -3σ to $+3\sigma$, where σ is the standard deviation from the mean ($x = 0$).

13. This is the problem of the crocked sailor in one dimension. A sailor leaves a waterfront bar still able to walk but unable to navigate. He is just as likely to take a step east as a step west, i.e., each step has a probability $\frac{1}{2}$ of being in either direction. After he has taken N steps, what is the probability that he is n steps away from his starting point? Calculate n for $N = 100$; 1000. (This is the problem called a *one-dimensional random walk*. A good solution and discussion are given by Chandrasekhar in his classic review article.*)

14. Derive the distribution law for velocities of gas molecules in one dimension from the random walk result in Problem 13.

15. What fraction of the molecules in (a) hydrogen, and (b) mercury vapor have a kinetic energy within $0.9 \, kT$ and $1.1 \, kT$ at (a′) 300 K, and (b′) 1000 K?

16. What fractions of the molecules in (a) H_2 and (b) N_2 have kinetic energies greater than 1.0 eV at (a) 1000 K (b) 3000 K?

17. A pinhole 0.2 μm in diameter is punctured in a 1-dm³ vessel containing Cl_2 at 300 K and at a pressure of 1 mmHg (torr). If the gas effuses into a vacuum, how long would it take for the pressure to fall to 0.5 mmHg (torr)? What assumption did you make in this calculation? Can you justify it? Suppose the inside pressure were initially 700 mmHg. How would you solve the problem in this case?

18. Consider a gas in two dimensions in a rectangular container. Calculate the rate at which molecules strike unit length of the container.

*S. Chandrasekhar, "Stochastic Problems in Physics and Astronomy," *Rev. Mod. Phys.*, *15*, 1–89 (1943).

19. A glass cylinder closed at both ends has a sintered glass filter in the middle dividing it into two compartments of equal volume. One side is initially filled with H_2 at pressure P and the other side with N_2 at pressure $2P$. The temperature is constant. The pores of the filter have a diameter much less than the mean free paths of the gases. Describe what happens starting at $t = 0$.

20. A cylinder with a porous disk similar to that in Problem 19 is filled on both sides with H_2 at pressure P, and the temperature on one side T_2 is greater than the temperature T_1 on other side. Describe what happens at $t > 0$. Suppose a wide tube also connects the two sides. What happens? Comment briefly on application of the First and Second Laws of Thermodynamics to these happenings.

21. Calculate the second virial coefficient for a gas of rigid spheres of diameter d without attractive forces, i.e., $U(r) = \infty$ for $0 \leq r \leq d$, and $U(r) = 0$ for $r > d$.

22. A tube 1 m long is mounted on a vertical axis midway between its ends and spun at 2000 revolutions per second (rps) at 315 K. If the tube were filled with UF_6 vapor at 1 atm pressure, calculate the isotopic enrichment in $^{235}UF_6$ compared to its normal concentration of 1 part to 140 parts $^{238}UF_6$.

23. Show that the fraction of molecules in a gas having kinetic energies between E and $E + dE$ is

$$f(E) = \frac{2\pi}{(\pi k T)^{3/2}} e^{-E/kT} E^{1/2} \, dE$$

24. Suppose that pairs of molecules in a gas have a repulsive potential energy $U(r) = A'r^{-12}$, where A' is a positive constant and r is the distance between centers. Show that the second virial coefficient is

$$B = \frac{2\pi L}{3} \left(\frac{A'}{kT} \right)^{1/4} \Gamma(\tfrac{3}{4})$$

where Γ is the gamma function. Considering the virial coefficients plotted in Fig. 1.11, has this model any range of practical usefulness?

5
Statistical Mechanics

If anyone has ever maintained that the universe is a pure throw of the dice, the theologians have abundantly refuted him. "How often," says Archbishop Tillotson, "might a man, after he had jumbled a set of letters in a bag, fling them out upon the ground before they would fall into an exact poem, yea, or so much as make a good discourse in prose! And may not a little book be as easily made by chance as this great volume of the world?"

Charles S. Peirce
1878*

The central problem in physical chemistry is how to calculate the macroscopic properties of a system from data about the structures and properties of the atoms and molecules of which it is composed.† The macroscopic properties of systems in equilibrium are those described by the usual variables of equilibrium thermodynamics. The macroscopic properties of systems not in equilibrium are described by the variables pertaining to various rate processes and transport phenomena. In principle, we should be able to calculate both equilibrium and nonequilibrium properties of systems from the properties of their constituent molecules. If this task were entirely accomplished, we could regard physical chemistry as a completed chapter in the book of science. The status of the subject today is a fair way from any such finality. As we shall see, considerable progress has been made with the theory of equilibrium properties, but the more difficult nonequilibrium theory is still in an early stage of development.

1. The Statistical Method

As a typical system, let us consider 1 g of oxygen gas in a volume of one liter at 300 K. Suppose we could obtain complete and quantitative information about the properties of all the individual molecules and their interactions with one another, as functions of time. These properties would include all those by which ordinary mechanics would describe the molecules: positions, momenta, kinetic and potential

*"The Order of Nature" in *Philosophical Writings of Peirce* (New York: Dover Publications, 1955), p. 223.

†In subsequent discussions, we shall use the term *molecules* to include atoms and molecules, both charged and uncharged. In some books, these elementary units are called *systems*, and what we call a *system* is called an *assembly*.

energies. The thermodynamic properties to be calculated would be those that describe the state of the system at equilibrium under the specified conditions: pressure P, entropy S, internal energy U, Gibbs energy G, etc.

Any large-scale system contains an enormous number of molecules—about 1.82×10^{22} in the liter of oxygen. We could not possibly keep track of the variables describing so many individual molecules. Any theory that seeks to interpret the behavior of macroscopic systems in terms of molecular properties must therefore rely on statistical methods. Such methods are designed to consider the typical or average behavior to be expected from large collections of objects. In fact, the larger the collection the more reliable are the results obtained by statistical methods. For example, nobody can determine in advance whether a given radium atom will disintegrate within the next 10 min, the next 10 days, or the next 10 centuries. If a milligram of radium is studied, however, we know that close to 2.23×10^{10} atoms will disintegrate in any 10-min period.

The discipline that allows us to make the theoretical connection between microscopic mechanical properties and macroscopic thermodynamic properties is, therefore, rightly called *statistical mechanics*. We can symbolize the connection as follows:

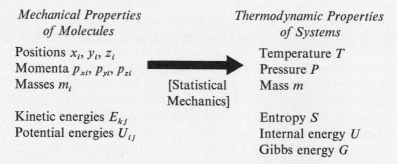

Mechanical Properties of Molecules		*Thermodynamic Properties of Systems*
Positions x_i, y_i, z_i		Temperature T
Momenta p_{xi}, p_{yi}, p_{zi}		Pressure P
Masses m_i	[Statistical Mechanics]	Mass m
Kinetic energies E_{kj}		Entropy S
Potential energies U_{ij}		Internal energy U
		Gibbs energy G

In addition to the preceding variables there exist *external variables*, which are common to both the mechanical and the thermodynamic descriptions. In most cases that we shall consider, the only external variable is the volume V. In problems dealing with surface films, we would have also the surface area A^σ. Other specifications might include an external electric field $\mathbf{E}$, and so on.

The problem is *how to make the arrow work* in the preceding program. A curious thing about the program is that we do not know, and indeed cannot know, the values of the mechanical variables for all the molecules in the system. What we do know, however, is a good deal about the *possible values* that these mechanical variables may take for any single molecule. Another curious and important point is a fundamental difference between mechanics and thermodynamics. Thermodynamic systems change with time always in the same direction—toward the equilibrium state. There is nothing in the mechanical properties of an individual molecule to indicate why this should be so. We can see this distinction clearly in the process of diffusion. Figure 5.1 represents two gases A and B that diffuse into each other to form a mixture. Once they are mixed, they will not become unmixed by a reversal of the diffusional process. But if we focus attention on any single indi-

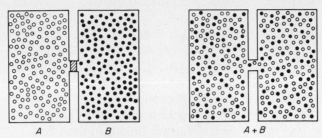

FIG. 5.1 Mixture of two different gases by diffusion. When the plug between the two vessels is removed, the gases mix spontaneously but the mixture will not spontaneously revert to its initial state.

vidual *A* molecule, and follow its trajectory, there is no reason why all its velocity components should not be reversed, so that it would retrace its path back to any given starting point. Whereas thermodynamic processes are inherently irreversible, mechanical processes are inherently reversible.

If the two gases are ideal, there is no energy change when they mix by interdiffusion. The driving force of the diffusion is entirely due to the increase in entropy in the process. What is the explanation of this increase in entropy in terms of the properties of the molecules? It cannot be due to any properties of the molecules as individuals. It must be related to some property of an entire collection of molecules.

2. Entropy and Disorder

We can gain an insight into the increase in entropy when two gases mix by interdiffusion by considering first a mixing process that involves a much smaller number of individuals. Suppose we have a new pack of cards in which all the suits are stacked separately in rank order, from the ace of spades to the deuce of clubs. There is only one way in which the cards can be stacked so as to give this completely ordered arrangement. If a single card is displaced, the perfect order is destroyed. There are 51 different places where one could insert a designated displaced card. If we simply require that any one card be displaced, there would be $W = (51)^2 = 2601$ different arrangements that satisfy this requirement. In the nomenclature of the physical systems to be discussed later, we would say that there are 2601 *distinguishable states* of the deck in which one card is displaced.

Note that each separate one of the displaced arrangements is just as precisely defined as the perfectly ordered arrangement of the new deck we started with. There are $W = 2601$ different arrangements, which are all described by saying "one card is displaced," without specifying which card it is and exactly where it has been replaced in the deck. If we are willing to sacrifice this precise *information*, we can say that the arrangement with one card displaced is 2601 times more *probable* than an arrangement where the position of every card is specified, or in particular, the original completely ordered arrangement. The probability is proportional to the number of states W,* so that the relative probability of any two arrangements would be W_1/W_2.

Suppose now that we give the cards a thorough shuffle. The result is to destroy completely the ordered arrangement. The final result will indeed be a particular one out of the 52! possible arrangements, but we do not know which one. We have *lost all the information* we originally possessed about the arrangement of the cards in the deck. We could look at the cards and thus regain information as extensive as that which we had lost, but we could not expect to get back to our original situation in any reasonable time by further shuffling of the deck.

Why is the disorder produced by shuffling essentially irreversible? It is not because any specified mixed arrangement is more probable than any specified unmixed one, but rather because the number of disordered arrangements is so much greater than the number of ordered ones. Each particular arrangement has the same probability, 1/52!, but there are so many disordered arrangements of various kinds that the chances are overwhelming that any random mixing process (shuffle) will terminate in one of these many possibilities.

Therefore we can summarize the results of the mixing as follows:

> Decrease in order
> Increase in disorder
> Loss of information

and we add: *Increase in entropy*

The analogy between the shuffling of cards and the mixing of gases by inter-diffusion is evident. The initial state of the system of two gases with all A molecules in one compartment and all B molecules in the other is like an ordered deck with all red cards together and all black cards together. The interdiffusion occurs by virtue of the thermal motions of the molecules, whereas the cards are mixed mechanically. The final mixture of A and B molecules is less ordered than the initial system of pure A and pure B. The mixed state is more probable than the unmixed state, because there are more ways of distributing the molecules of A and B so as to yield mixed states than there are ways to yield pure states. We have lost information in the sense that we can no longer say that a given A molecule is in a certain part of the container. The mixed state has a higher entropy than the unmixed state.

To define a quantitative relationship between entropy S and the number W of distinguishable states of a system, we recall that entropy is an additive function, whereas W is multiplicative. If we consider a system divided into two parts, the entropy of the whole is the sum of the entropies of the parts, $S = S_1 + S_2$. On the other hand, W of the combined system is the product of the W's of the two parts; $W = W_1 W_2$, since any of the W_1 states of part I can be combined with

---

*The probability is a fraction, whereas W itself may be a large number. In the case of the displaced card, the probability would be $p = W/N!$ Since the total number of arrangements of the deck is 52!, $p = 2601/52!$

any of the W_2 states of part II. Therefore, the relation between S and W must be a logarithmic one, the most general form of which is

$$S = a \ln W + b \tag{5.1}$$

The value of the constant a may be derived by analyzing, from the viewpoint of probability, a simple change for which ΔS is known from thermodynamics. Consider the expansion of 1 mol of ideal gas, originally at pressure P_1 in a container of volume V_1, into an evacuated container of volume V_2. The final pressure is P_2 and the final volume is $V_1 + V_2$. For this change, since $R = Lk$,

$$\Delta S = S_2 - S_1 = R \ln \frac{V_1 + V_2}{V_1} = k \ln \left(\frac{V_1}{V_1 + V_2} \right)^{-L} \tag{5.2}$$

When the containers are connected, the probability of finding one given molecule in the first container is simply the ratio of the volume V_1 to the total volume $V_1 + V_2$. Since probabilities are multiplicative, the chance of finding all L molecules in the first container—i.e., the probability p of the original state of the system—is $p_1 = [V_1/(V_1 + V_2)]^L$. In the final state, all the molecules must be in one or the other of the containers, so that the probability $p_2 = 1^L = 1$. Therefore, $p_2/p_1 = W_2/W_1 = [V_1/(V_1 + V_2)]^{-L}$.

Thus, from (5.1),

$$\Delta S = S_2 - S_1 = a \ln \frac{W_2}{W_1} = a \ln \left(\frac{V_1}{V_1 + V_2} \right)^{-L}$$

Comparison with (5.2) shows that a is equal to k, the Boltzmann constant, and (5.1) becomes

$$S = k \ln W + b \tag{5.3}$$

For a change for state 1 to state 2,

$$\Delta S = S_2 - S_1 = k \ln \frac{W_2}{W_1}$$

This relation was first given by Boltzmann in 1896.

The relative probability of observing a decrease in entropy of ΔS below the equilibrium value may be obtained from (5.3) as

$$\frac{W}{W_{eq}} = e^{-\Delta S/k}$$

For 1 mol of helium, S/k at 273 K $= 9 \times 10^{24}$. The chance of observing an entropy decrease one-millionth of this amount is about $e^{-10^{19}}$. Such a fluctuation on a macroscopic scale is so improbable that it is "never" observed. No one watching a book lying on a desk would expect to see it spontaneously fly up to the ceiling as it experienced a sudden chill. Yet it is not impossible to imagine a situation in which all the molecules in the book moved spontaneously in a given direction. Such a situation is only extremely improbable, since there are so many molecules in any macroscopic portion of matter. Anyone who sees a book flying spontaneously into the air is dealing with a poltergeist and not an entropy fluctuation (probably!). Only when the system is very small is there an experimental chance of observing an appreciable *relative decrease* in entropy.

3. Entropy and Information

We have seen that an increase in the information specified for a system corresponds to a decrease in the entropy of the system. Is it possible to obtain a quantitative relation between entropy and information?

The first step is to consider the quantitative measure of information provided by the *information theory* of Weaver and Shannon.* Suppose that in a given situation N different possible things might happen each having the same a priori probability. As we gain information about the situation, suppose we find that one particular outcome actually occurs. The larger the initial N, the more information we gain by narrowing the outcomes from N to one. We define the information gained by

$$I = K \ln N \tag{5.4}$$

where K is a constant yet to be specified. The logarithmic relation in (5.4) is required by the fact that information is an additive property. If we have two independent systems, with N_1 equally probable outcomes for the first and N_2 for the second, the total number of possible outcomes would be $N = N_1 N_2$ (since any one outcome of the first set can be combined with any of the N_2 outcomes of the second). Hence,

$$I = K \ln N_1 N_2 = K \ln N_1 + K \ln N_2 = I_1 + I_2$$

The choice of K is determined by the unit of information. It is often convenient to transfer information by means of a binary code (e.g., in a computer, a switching point that can be either on [1] or off [0]). If a message contains n such symbols, there would be $N = 2^n$ possibilities for their arrangements. Then $I = K \ln N = Kn \ln 2$. If the constant K is now chosen so that $K \ln 2 = 1$, then

$$I = n = \log_2 N$$

The unit of information thus defined is the *binary digit* or *bit*. As an example, suppose we wish to specify one of a set of 32 different cards, $I = \log_2 32 = 5$ (since $2^5 = 32$), so that the specification requires five bits of information. (Five successive divisions by two of the 32 cards will locate a particular card.)

We can also measure information in thermodynamic units by choosing the constant in (5.4) so that $K = k$, the Boltzmann constant. Then, if we initially have information $I_0 = 0$ corresponding to N_0 possibilities and we finally have $I_1 > 0$ corresponding to N possibilities,

$$I_1 = k \ln \left(\frac{N_0}{N_1} \right)$$

If the possibilities are equivalent to the distinguishable states in Boltzmann's model of a thermodynamic system,

$$\Delta S = S_1 - S_0 = k \ln \left(\frac{N_1}{N_0} \right)$$

*See L. Brillouin, *Science and Information Theory* (New York: Academic Press, Inc., 1956).

Therefore,

$$-I_1 = S_1 - S_0 = \Delta S \tag{5.5}$$

We can thus interpret entropy as negative information (*neginformation*) or information as negative entropy (*negentropy*).

4. Stirling Formula for N!

In computations of the number of different possible arrangements of the elementary microscopic units (molecules, oscillators, ions, etc.) that make up macroscopic systems, it is often necessary to evaluate the factorials of large numbers. A useful formula due to Stirling (1730) can be obtained as follows. Since

$$N! = N(N-1)(N-2)\cdots(2)(1)$$

$$\ln N! = \sum_{m=1}^{N} \ln m$$

As m becomes large, this sum can be approximated ever more closely by an integral, so that

$$\ln N! \approx \int_1^N \ln m \, dm = |m \ln m - m|_1^N$$

Since when $N \gg 1$, the lower limit is negligible,

$$\ln N! \approx N \ln N - N \tag{5.6}$$

The approximation in (5.6) will be adequate for our needs, but there is also a more accurate formula,*

$$N! = \sqrt{2\pi N}\left(\frac{N}{e}\right)^N\left(1 + \frac{1}{12N} + \frac{1}{288N^2} + \cdots\right) \tag{5.7}$$

5. Boltzmann

During the last years of the nineteenth century, a great intellectual battle was waged in the field of scientific theory, between those who believed in atoms as real fundamental particles and those who considered them simply as useful models for mathematical discussions in a world based entirely on transformations of energy. The leader of the forces of atomism was Ludwig Boltzmann of the University of Vienna, whose name is linked with that of Maxwell as cofounder of the kinetic theory of gases, and with that of Gibbs in the development of statistical mechanics.

In 1895, a conference was summoned at Lübeck to discuss the conflicting views of world structure. The report in favor of energetics was made by Helm of Dresden. Behind him stood Wilhelm Ostwald, and behind both was ranged the powerful positivist philosophic system of the absent Ernst Mach. The leading

*Harold Jeffreys and Bertha S. Jeffreys, *Methods of Mathematical Physics*, 2nd ed. (London: Cambridge University Press, 1950), p. 464.

opponent of energetics was Boltzmann, seconded by Felix Klein. Arnold Sommerfeld reported that the struggle between Boltzmann and Ostwald equaled, outwardly and inwardly,

> the struggle of the bull with the supple matador. But this time the bull conquered the matador despite all his finesse. The arguments of Boltzmann drove through. All the young mathematicians stood on his side.

Boltzmann was subject to sudden changes from happiness to affliction, which he ascribed to the fact that he was born during the dying hours of a gay Mardi Gras ball. Others have suggested that his depressions were caused by concern for the atomic theory. The enemies of traditional atomism, under the leadership of Mach, called Boltzmann the last pillar of that bold edifice of thought. Boltzmann personally felt each tremor of what was then believed to be a "tottering edifice." In a sudden intensification of depression, he committed suicide by drowning at Duino near Trieste during a summer outing in 1906.

From our present vantage point, it might seem that the attacks of Mach and Ostwald on atomism were only minor setbacks in a broad advance of the basic theory, which was to achieve some of its greatest successes in the first half of the twentieth century. It may be unduly romantic to call Boltzmann, as some have done, a martyr for the atomic theory. His memorial is a white marble bust by Ambrosi, under which is engraved a short formula:

$$S = k \ln W \tag{5.8}$$

6. How the State of a System Is Defined

In the discussion of statistical mechanics that follows, we shall often refer to the *state* or *states* of a system. In thermodynamics, we found that the state of a system was determined by the specification of the numerical values of a number of properties called *state functions*. In statistical mechanics, we also define the state of a system by the specification of the values of certain quantities, but, because we are now dealing with the individual molecules that comprise the system, we need to specify many more values. The specification of the state is quite different, depending upon whether we use classical mechanics or quantum mechanics to describe the system.

In classical mechanics, the state of a system of individual particles is determined when we specify for each particle three coordinates giving its position in space, and three components of velocity (or of momentum) at that point. For a system of N atoms, therefore, values for $6N$ variables must be specified. If the atoms are combined into molecules, the number of coordinates and velocities is not altered, although we may divide them into internal degrees of freedom and translations of the center of mass, as described previously. The coordinates might be ordinary Cartesian coordinates x, y, z, or some other type, such as spherical polar coordinates, r, θ, ϕ. In short, we can use a triplet of generalized coordinates q_1, q_2, q_3. Similarly, the velocities can be considered as components $\dot{x}$, $\dot{y}$, $\dot{z}$, or as a generalized triplet $\dot{q}_1, \dot{q}_2, \dot{q}_3$.

The classical definition of the state of the system implies that we can somehow keep track of the individual molecules. Classical mechanics does not require that the molecules themselves be intrinsically distinguishable in any way, but only that some perspicacious observer may exist who in principle* can track a given molecule throughout its various collisions and trajectories. In the classical picture, when a molecule A collides with a molecule B, we can always calculate the final positions and momenta of A and B, provided we know their initial values before collision.

Quantum mechanics† describes the state of a system in quite a different way from that of classical mechanics, and we must say a few words here about the quantum mechanical description, even though we plan to discuss it in more detail later, in Chapter 14. Quantum mechanics specifies the state of a system by giving the value of a function Ψ. For a single particle, Ψ is a function of the coordinates q_1, q_2, q_3, and of the time, $\Psi(q_1, q_2, q_3, t)$. Actually, the time-dependent part can be separated to give

$$\Psi = \psi(q_1, q_2, q_3)\rho(t)$$

where $\psi(q_1, q_2, q_3)$ specifies a *stationary state* of the system. In this chapter, we shall be concerned only with these stationary states. Note that the function defining the stationary state $\psi(q_1, q_2, q_3)$ does not depend upon the velocities or momenta. We could, however, obtain another state function $\Phi(p_1, p_2, p_3)$, in which momenta instead of coordinates were the independent variables.‡ Either ψ or Φ is sufficient to define the state, and to give all possible information about both coordinates and momenta.

The function ψ is a probability amplitude—i.e., it does not tell us exactly what the coordinates or momenta are for the given state, but allows us to calculate the probability that a coordinate be within a certain range q_1 to $q_1 + dq_1$, or a momentum component in a range p_1 to $p_1 + dp_1$.

In general, ψ is a complex quantity having real and imaginary parts, $\psi = \alpha + \beta i$. The probability is then given by the product of ψ by its complex conjugate, $\psi^* = \alpha - \beta i$. Thus, $\psi\psi^* \, dq_1$ gives the probability that the coordinate lies between q_1 and $q_1 + dq_1$. Note that $\psi\psi^*$ is always real. Although one might think that the quantum mechanical definition of state is more economical than the classical one, because we do not need to specify both positions and momenta, the fact that we need to specify both a real and an imaginary part of the ψ function means that just as many quantities are needed in either case. We should emphasize, however, that in the quantum mechanical case, $\psi(q_1, q_2, q_3)$ specifies the state of the particle, and

---

*The cynic remarks that anything said to be possible "in principle" is apt to be impossible in practice. It must be confessed that a device to track simultaneously some 10^{12} particles would be one of the marvels of the world.

†Is any apology due the reader for the introduction of quantum mechanical ideas at this point? In this Aquarian age, it would be an unduly sheltered mind that came to physical chemistry in complete innocence of quantum mechanics. High school physics texts introduce the basic concepts of quantized energy levels, and freshman chemistry courses have long since forsaken the realism of fertilizers and blast furnaces for abstract displays of rainbow colored orbitals.

‡The function Φ is a Fourier transform of ψ.

$$\Phi(p_1, p_2, p_3) = h^{-3/2} \int\int\int \psi(q_1, q_2, q_3) \, e^{-(2\pi i/h)(p_1 q_1 + p_2 q_2 + p_3 q_3)} \, dq_1 dq_2 dq_3$$

the probability of finding coordinates or momenta in a given range is calculated from the ψ. In the classical case, the values of $q_1, q_2, q_3, p_1, p_2, p_3$ are themselves exactly stated so as to specify the state.

Furthermore, we may mention that quantum mechanics does not permit a simultaneous fixing of both coordinates and momenta to any arbitrary accuracy. They are subject to the famous Heisenberg uncertainty relation,

$$\Delta q_1 \cdot \Delta p_1 \geq \frac{h}{4\pi}$$

where Δq_1 and Δp_1 are the uncertainties in q_1 and its conjugate momentum p_1. Thus, an exact specification of a coordinate q_1 ($\Delta q_1 \to 0$) implies a complete loss of information about its conjugate momentum p_1 ($\Delta p_1 \to \infty$), and vice versa. Because of these uncertainties in the values of position and momentum, one cannot, even *in principle*, keep track of individual molecules according to the quantum mechanical description. If a collision between two identical molecules A and B is followed, *in some cases* the range of uncertainties will be such that we shall not be able, after the collision, to say which molecule was originally A and which was originally B.

For a system containing two or more identical particles, the quantum mechanical state is still specified by a function ψ. For example, for two particles, ψ would be a function of the six coordinates,

$$\psi(x_1, y_1, z_1, x_2, y_2, z_2)$$

The stationary states are characterized by definite values for their energies, so that a state ψ_j has a definite energy E_j. Sometimes more than one state ψ may have the same value for the energy E. In such a case, we say that the energy level has a *degeneracy g* equal to the number of ψ functions with the same energy. In the subsequent discussion and enumeration of states, it will often be convenient to think of a set of definite quantized energy levels, each specified by the energy E_j and a degeneracy g_j.

7. Ensembles

In the formulation of statistical mechanics given originally by Boltzmann, the determination of the number of states of a system was made by considering directly the molecules in the system and their distribution among the allowed energy levels or energy ranges. Each distinct way of distributing the N molecules of the system among the molecular energy states gave a *distinguishable state* of the system.

J. Willard Gibbs (1900) introduced the idea of an *ensemble* of systems. As we shall see, this concept leads to certain advantages in computing average properties of systems. In particular, it obviates the *impossible task of time averaging* for systems containing of the order of L molecules. An ensemble is a mental construction made for the purpose of mathematical analysis of the statistical mechanical problem. An ensemble consists of a large number of replicas of the system under discussion, each of which is subject to the same thermodynamic restrictions as the original system.*

If the original system is an isolated mass of gas, the number of molecules it contains N and its volume V are fixed, and its energy E is fixed within an extremely narrow range. We can construct an ensemble of $\mathscr{N}$ such systems, each member having the same N, V, E. This ensemble is shown schematically in Fig. 5.2. It is called a *microcanonical ensemble*.

There are two basic postulates required to relate the concept of the ensemble to practical calculations of average properties:

1. *First postulate.* The average of any mechanical variable M over a long time in the actual system is equal to the ensemble average of M, provided that the systems of the ensemble replicate the thermodynamic state and surroundings of the actual system of interest. Strictly speaking, this postulate holds only in the limit as $\mathscr{N} \rightarrow \infty$.

2. *Second postulate.* In an ensemble representative of an isolated thermodynamic system (microcanonical ensemble), the members are distributed with equal probability over the possible *quantum states* consistent with the specified values of N, V, E. This is the *principle of equal a priori probabilities*.

We can have any number of members of the ensemble to represent each state

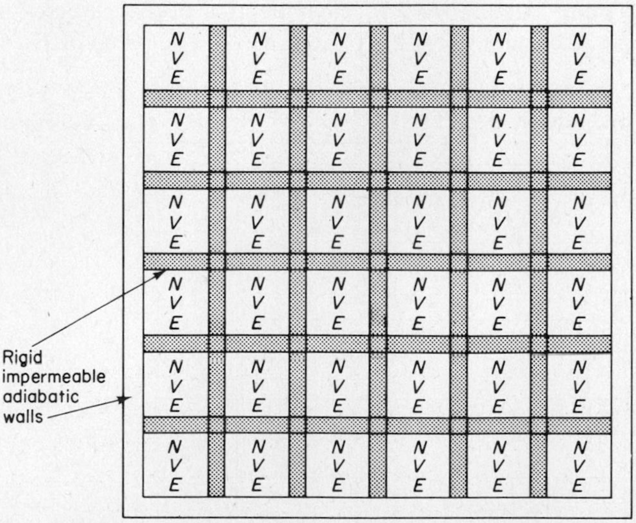

FIG. 5.2 Schematic representation of a microcanonical ensemble of $\mathscr{N}$ systems each with same N, V, E.

*In classical mechanics, the state of the system, specified by giving the coordinates q_i and p_i of every molecule in the system, would be represented by one point in phase space. The ensemble would be represented by a collection of points in phase space, corresponding to all the possible distributions of coordinates and momenta among the molecules of the actual system.

of the system (so long as it is the same number for each one). The smallest number $\mathscr{N}$ of systems in the ensemble would include one system for each state of the original system that is the basis for the ensemble. The possibility of allowing $\mathscr{N} \longrightarrow \infty$ is one important advantage of the ensemble method, for it means that we never need to worry about the use of the Stirling approximation for $N!$ when we use this method.

When we have discussed quantum mechanics later in the book, we shall see that for a system of definite volume, the allowed energy states can have only discrete quantized values. We can then think of the distinguishable states as being the discrete allowed energy states of quantum theory. In classical mechanics, we can imagine phase space to be divided into cells of volume $\delta\tau$; we can then imagine a microscopic state to be specified by giving the particular cell in phase space in which the system is situated.

The postulate of equal a priori probabilities is absolutely fundamental for statistical mechanics. It is this postulate that allows us to make the transition between mechanics and thermodynamics (the arrow on p. 168). The postulate corresponds to the interpretation of probability which allowed us to say that the probability of drawing any card at random from the deck of 52 cards is 1/52.

A second type of ensemble introduced by Gibbs is the *canonical ensemble*, shown schematically in Fig. 5.3. This ensemble consists of a large number of systems, each having the same value of N, V, and T. We can consider the systems to be separated by diathermic walls, which permit the passage of energy but not of material particles. At equilibrium, therefore, each member of the canonical ensemble will have the same T, but not necessarily the same E. The value of E will fluc-

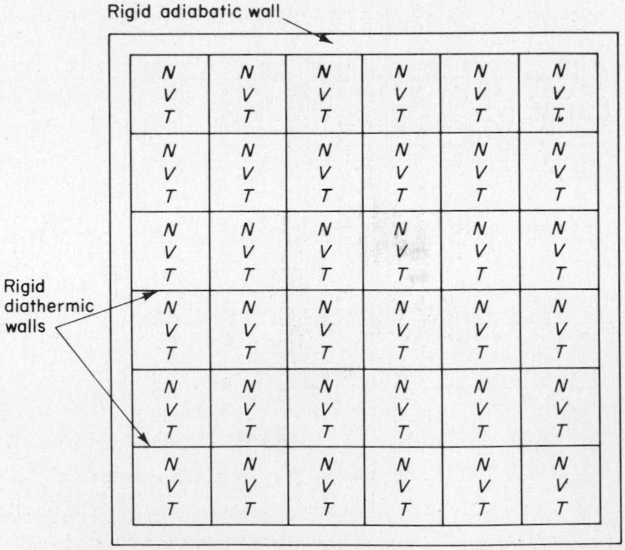

Rigid adiabatic wall

Rigid diathermic walls

N V T	N V T	N V T	N V T	N V T	N V T_e
N V T	N V T	N V T	N V T	N V T	N V T
N V T	N V T	N V T	N V T	N V T	N V T
N V T	N V T	N V T	N V T	N V T	N V T
N V T	N V T	N V T	N V T	N V T	N V T
N V T	N V T	N V T	N V T	N V T	N V T

FIG. 5.3 Schematic representation of a canonical ensemble of $\mathscr{N}$ systems each with same N, V, T.

tuate about some average value, the ensemble average $\bar{E}$. This average value $\bar{E}$ determines T for the entire ensemble.

8. Lagrange Method for Constrained Maximum

We must at this point introduce a mathematical procedure that will be required in the subsequent discussion. The Lagrange method is designed to secure the maximum (or minimum) of some function $f(x_1, x_2, \ldots, x_n)$ subject to some additional condition or conditions on the independent variables $(x_1 \cdots x_n)$, such as $g(x_1, x_2, \ldots, x_n) = 0$, etc. In the absence of any such constraint, we would have the ordinary conditions for an extremum,

$$\delta f = \sum_{j=1}^{n} \frac{\partial f}{\partial x_j} \, \delta x_j = 0 \tag{5.9}$$

where the δx_j are the variations of the x's. Since the δx_j would be independent, they could be given any arbitrary values, so that the only way to satisfy (5.9) would be to require that each coefficient of the δx_j vanish,

$$\frac{\partial f}{\partial x_j} = 0 \quad \text{for all } j$$

In case there is a constraint on the x's, we would have a relation such as

$$g(x_i \cdots x_n) = 0$$

or, a condition,

$$\delta g = 0 = \sum_{j=1}^{n} \frac{\partial g}{\partial x_j} \, \delta x_j \tag{5.10}$$

Now the δx_j are no longer all independent, because if we knew $n-1$ of them, we could calculate the nth from (5.10).

Lagrange's method tells us to multiply (5.10) by an undetermined parameter λ and to add the product to (5.9). This operation gives

$$\sum_{j=1}^{n} \left(\frac{\partial f}{\partial x_j} + \lambda \frac{\partial g}{\partial x_j} \right) \delta x_j = 0 \tag{5.11}$$

Because of (5.10), the δx_j are not independent; we can think of (5.11) as an equation that gives one of the δx_j in terms of the other $n - 1$ of them. Suppose we designate a particular one δx_n and consider it to be the dependent one. Since we have not so far imposed any condition at all on λ, the trick is to require now that

$$\frac{\partial f}{\partial x_n} + \lambda \frac{\partial g}{\partial x_n} = 0 \tag{5.12}$$

or

$$\lambda = - \frac{\partial f / \partial x_n}{\partial g / \partial x_n}$$

The result is that the term in the sum in (5.11) that contains the δx_n vanishes, and we are left with a sum of terms including only independent variations δx_j. By the same argument that was applied to (5.9), the coefficients of the $n - 1$ independent δx_j must vanish:

$$\frac{\partial f}{\partial x_j} + \lambda \frac{\partial g}{\partial x_j} = 0, \qquad j = 1, 2, \ldots, n-1 \quad \text{(i.e., all except } n\text{)} \tag{5.13}$$

If we combine (5.13) with our choice of λ from (5.12), we can write

$$\frac{\partial f}{\partial x_j} + \lambda \frac{\partial g}{\partial x_j} = 0 \quad \text{for all } j \tag{5.14}$$

In case there are several constraints, we can easily extend the Lagrange method by introducing a separate undetermined multiplier, $\lambda_1, \lambda_2, \ldots$, etc., for each such constraint.

9. Boltzmann Distribution Law

We now come to one of the most important problems studied by Boltzmann, the distribution of a large number of particles among the different energy states accessible to them. We have already seen a special case of this problem in the Maxwell distribution law for kinetic energies. Boltzmann was concerned with the more general problem of the distribution of particles among a set of energy states of any kind. In the context of quantum mechanics (which came later), these energy states would be the discrete quantized stationary states of the particles under study.

Let us consider a system consisting of N noninteracting individual particles (atoms, molecules, or elementary units of some other kind). The assumption of *noninteracting particles* must, in fact, be relaxed somewhat to allow a weak interaction sufficient to preserve an equilibrium condition. To make the system more definite, suppose it is a liter of gas, and that the particles are the gas molecules, considered to obey the requirements for the perfect gas, i.e., the potential energy negligible in comparison to the kinetic energy. The gas molecules can still interact with one another directly on collisions, or indirectly through collisions with the container walls. Such interactions allow equilibrium conditions to be established and maintained. Nevertheless, we can specify the energy state of each individual molecule at any time *without reference to interaction energies between molecules.* Thus, we can still call these molecules *noninteracting particles.*

This problem was actually treated by Boltzmann without employing the concept of the ensemble. In our example, however, the liter of gas can be considered to be a canonical ensemble, and the individual weakly interacting molecules are the members that comprise the ensemble. Each molecule can be considered to be in a heat bath consisting of the $N - 1$ other molecules, so that its T is specified.* The volume V (the total gas volume), and the number of particles (one per member) are all constant. Each molecule has the same set of allowed energy states, ϵ_1, $\epsilon_2, \epsilon_3, \ldots$, etc. We can specify the occupation numbers N_j as the number of

---

*Note that the energy of the molecule may fluctuate widely, but since the molecule is in equilibrium with the heat bath, its T is fixed. The T is a measure of the *average energy* of all the molecules.

molecules found in each of the allowed states. These energy states can be either the discrete energy levels given by quantum theory or the cells in phase space specified by a small range of energies ϵ_j to $\epsilon_j + d\epsilon_j$ for each particle.

Since its N, V, and E are fixed, the liter of gas is *one member of a microcanonical ensemble*. We can now apply the two basic postulates to a microcanonical ensemble consisting of $\mathcal{N}$ liters of perfect gas. We thereby can calculate the ensemble average of any property of the liter of gas molecules and set this equal to the time average value of the property. In taking the average, we assign equal a priori probabilities to each member of the ensemble. The Boltzmann problem is to calculate the ensemble average value of the occupation number N_j of the noninteracting particles (molecules) in a state with particle energy ϵ_j, subject to the condition of constant total energy and constant number of particles,

$$\sum N_j \, \epsilon_j = E \tag{5.15}$$

$$\sum N_j = N \tag{5.16}$$

Let us suppose that the N particles are assigned to the energy levels in such a way that there are N_1 in level ϵ_1, N_2 in ϵ_2, or, in general, N_j in level ϵ_j. Since permuting the particles within a given energy level does not produce a new distribution, the number of ways of realizing a distribution is the total number of permutations $N!$, divided by the number of permutations of the particles within each level, $N_1! \, N_2! \cdots N_j! \cdots$. The required number is

$$W_n = \frac{N!}{N_1! \, N_2! \cdots N_j! \cdots} = \frac{N!}{\Pi_j N_j!} \tag{5.17}$$

The subscript n denotes that this is simply one of a large number of possible distributions. *The total number of distinguishable states would be obtained by summing W_n over all the possible distributions.*

$$W = \sum W_n = \sum_{\substack{\text{(over all} \\ \text{distributions)}}} \frac{N!}{\Pi_j N_j!} \tag{5.18}$$

As an example of this formula, let us consider in Table 5.1 two different distributions of four particles, a, b, c, d, among four energy states $\epsilon_1, \epsilon_2, \epsilon_3, \epsilon_4$. Note that interchanges of the particles *within level ϵ_1* are not significant, but the interchange of particles between levels ϵ_1, ϵ_3, and ϵ_4 leads to a new distribution.

There are a number of ways to solve the mathematical problem of computing the ensemble average of the N_j. For an ensemble containing a large number of particles, the *average values* of the set of N_j at equilibrium can be set equal to the *most probable values* of the N_j for the system.* The most probable set of values for the N_j will be that giving the greatest number of distinguishable states W_n. The procedure is therefore to write down W_n for a set of population numbers N_j and to maximize this W_n with respect to all possible variations of the N_j's subject to the constraints (5.15) and (5.16).†

---

*T. L. Hill, *Introduction to Statistical Thermodynamics* (Reading, Mass.: Addison-Wesley Publishing Co., Inc., 1960), p. 478.

†Note that W_n is only one term in the sum for W in (5.18). We maximize W_n so that it corresponds to the largest term in the sum, i.e., the most probable set of distribution numbers.

TABLE 5.1

Examples of Two Possible Distributions of Four Molecules among Four Energy States

Distribution (1) $N_1 = 2, N_2 = 0, N_3 = 1, N_4 = 1$	Distribution (2) $N_1 = 3, N_2 = 0, N_3 = 1, N_4 = 0$
States	*States*

<table>
<tr><td>ϵ_1</td><td>ϵ_2</td><td>ϵ_3</td><td>ϵ_4</td><td></td><td>ϵ_1</td><td>ϵ_2</td><td>ϵ_3</td><td>ϵ_4</td></tr>
<tr><td>ab</td><td></td><td>c</td><td>d</td><td></td><td>abc</td><td></td><td>d</td><td></td></tr>
<tr><td>ab</td><td></td><td>d</td><td>c</td><td></td><td>abd</td><td></td><td>c</td><td></td></tr>
<tr><td>ac</td><td></td><td>b</td><td>d</td><td></td><td>acd</td><td></td><td>b</td><td></td></tr>
<tr><td>ac</td><td></td><td>d</td><td>b</td><td></td><td>bcd</td><td></td><td>a</td><td></td></tr>
<tr><td>ad</td><td></td><td>b</td><td>c</td><td></td><td></td><td></td><td></td><td></td></tr>
<tr><td>ad</td><td></td><td>c</td><td>b</td><td></td><td></td><td></td><td></td><td></td></tr>
<tr><td>bc</td><td></td><td>a</td><td>d</td><td></td><td></td><td></td><td></td><td></td></tr>
<tr><td>bc</td><td></td><td>d</td><td>a</td><td></td><td></td><td></td><td></td><td></td></tr>
<tr><td>bd</td><td></td><td>a</td><td>c</td><td></td><td></td><td></td><td></td><td></td></tr>
<tr><td>bd</td><td></td><td>c</td><td>a</td><td></td><td></td><td></td><td></td><td></td></tr>
<tr><td>cd</td><td></td><td>a</td><td>b</td><td></td><td></td><td></td><td></td><td></td></tr>
<tr><td>cd</td><td></td><td>b</td><td>a</td><td></td><td></td><td></td><td></td><td></td></tr>
</table>

$$W_{(1)} = \frac{4!}{2!\,0!\,1!\,1!} = \frac{4 \cdot 3 \cdot 2 \cdot 1}{2 \cdot 1 \cdot 1 \cdot 1} = 12 \qquad W_{(2)} = \frac{4!}{3!\,0!\,1!\,0!} = \frac{4 \cdot 3 \cdot 2 \cdot 1}{3 \cdot 2 \cdot 1 \cdot 1 \cdot 1} = 4$$

By taking the logarithm of both sides of (5.17), the continued product is reduced to a summation.

$$\ln W_n = \ln N! - \sum_j \ln N_j! \tag{5.19}$$

The condition for a maximum* in W_n is that the variation† of W_n, and hence of $\ln W_n$, be zero. Since $\ln N!$ is a constant,

$$\delta \ln W_n = 0 = \sum \delta \ln N_j! \tag{5.20}$$

From the Stirling formula (5.6), (5.20) becomes

$$\delta \sum N_j \ln N_j - \delta \sum N_j = 0$$

$$\sum \ln N_j \, \delta N_j = 0 \tag{5.21}$$

The two restraints in (5.15) and (5.16), since N and E are constants, can be written

$$\delta N = \sum \delta N_j = 0$$

$$\delta E = \sum \epsilon_j \, \delta N_j = 0$$

These two equations are multiplied by two arbitrary constants, α and β, and added

---

*This condition is for a maximum or a minimum. The minimum of (5.18) would place all the particles in one level, a solution of no practical interest. The maximum of $\ln x$ is at same value of x as the maximum of x, since $\ln x$ is a monotonic $f(x)$.

†A useful reference is Chapter 6, "Calculus of Variations," H. Margenau and G. M. Murphy, *The Mathematics of Physics and Chemistry* (Princeton, N. J.: D. Van Nostrand Co., 1956).

to (5.21), yielding

$$\sum \alpha \, \delta N_j + \sum \beta \epsilon_j \, \delta N_j + \sum \ln N_j \, \delta N_j = 0 \qquad (5.22)$$

The variations δN_j may now be considered to be perfectly arbitrary (the restraining conditions have been removed), so that for (5.22) to hold, each term in the summation must vanish. As a result,

$$\ln N_j + \alpha + \beta \epsilon_j = 0$$
$$N_j = \exp(-\alpha - \beta \epsilon_j) \qquad (5.23)$$

The constant α is evaluated from the condition (5.16), $\sum N_j = N$, whence

$$e^{-\alpha} \sum e^{-\beta \epsilon_j} = N$$

Therefore, Eq. (5.23) becomes

$$\frac{N_j}{N} = \frac{e^{-\beta \epsilon_j}}{\sum e^{-\beta \epsilon_j}} \qquad (5.24)$$

The constant β is determined by calculating a known average property of a molecule in a perfect gas, using (5.24) to compute the average with our mean value formula from p. 135. The property chosen will be the average kinetic energy in one degree of freedom. $\bar{\epsilon} = \tfrac{1}{2}kT$.

$$\bar{\epsilon} = \frac{\sum \epsilon_j N_j}{\sum N_j} = \frac{\sum \epsilon_j e^{-\beta \epsilon_j}}{\sum e^{-\beta \epsilon_j}} \qquad (5.25)$$

In terms of any component of momentum, say p_{xj},

$$\epsilon_j = \frac{1}{2m} p_{xj}^2$$

From (5.25), therefore,

$$\bar{\epsilon} = \frac{\sum\limits_j (p_{xj}^2/2m) \exp(-\beta p_{xj}^2/2m)}{\sum\limits_j \exp(-\beta p_{xj}^2/2m)}$$

The summations are taken over all the momentum components available to the gas molecule. In the classical case, the momenta are continuously variable and the sums can be replaced by integrals,

$$\bar{\epsilon} = \frac{\dfrac{1}{2m} \displaystyle\int_{-\infty}^{+\infty} p_x^2 \exp(-\beta p_x^2/2m) \, dp_x}{\displaystyle\int_{-\infty}^{+\infty} \exp(-\beta p_x^2/2m) \, dp_x} \qquad (5.26)$$

Since

$$\int_{-\infty}^{+\infty} e^{-ax^2} \, dx = \left(\frac{\pi}{a} \right)^{1/2}$$

and

$$\int_{-\infty}^{+\infty} x^2 e^{-ax^2} \, dx = \frac{1}{2} \left(\frac{\pi}{a^3} \right)^{1/2}$$

the integration of (5.26) yields

$$\bar{\epsilon} = \frac{1}{2\beta}$$

Since we know that $\bar{\epsilon} = \frac{1}{2}kT$, $\beta = 1/kT$. Equation (5.24) then gives us the *Boltzmann distribution law*,

$$\frac{N_j}{N} = \frac{e^{-\epsilon_j/kT}}{\sum e^{-\epsilon_j/kT}} \tag{5.27}$$

The denominator of this expression,

$$z = \sum e^{-\epsilon_j/kT} \tag{5.28}$$

is called the *particle partition function*, or, in case the particles in question are molecules, the *molecular partition function*.

Often, the Boltzmann distribution law is applied to compute the ratio of the numbers of particles in two different discrete energy states—e.g., N_0 in state with energy ϵ_0, and N_1 in state with energy ϵ_1. From (5.27),

$$\frac{N_1}{N_0} = e^{-(\epsilon_1-\epsilon_0)/kT} \tag{5.29}$$

We must distinguish this form of the distribution law from that in (5.27), which gives the number in a given state divided by the total number, e.g., N_1/N.

An example of a Boltzmann distribution is shown in Table 5.2, for a set of energy levels equally spaced a distance kT apart.

It is convenient at this point to make one extension of the distribution law in (5.27). It is possible that there may be more than one state corresponding with the energy level ϵ_j. If this is so, the level is said to be *degenerate* and is assigned a *statistical weight* g_j, equal to the number of superimposed levels. Then (5.27) becomes

$$\frac{N_j}{N} = \frac{g_j e^{-\epsilon_j/kT}}{\sum\limits_j g_j e^{-\epsilon_j/kT}} \tag{5.30}$$

This is the Boltzmann distribution law in its most general form.

TABLE 5.2
An Example of a Boltzmann Distribution

Number of level	ϵ_j/kT	$e^{-\epsilon_j/kT}$	N_j for $N = 1000$
0	0	1.000	633
1	1	0.368	233
2	2	0.135	85
3	3	0.050	32
4	4	0.018	11
5	5	0.007	4
6	6	0.002	1
7	7	0.001	1
8	8	0.0003	0
9	9	0.0001	0
10	10	0.0000	0

$$z = \sum e^{-\epsilon_j/kT} = 1.582$$

The average energy $\bar{\epsilon}$ is given by [see (5.25)]

$$\bar{\epsilon} = \frac{\sum N_j \epsilon_j}{\sum N_j} = \frac{\sum g_j \epsilon_j e^{-\epsilon_j/kT}}{\sum g_j e^{-\epsilon_j/kT}} = kT^2 \left(\frac{\partial \ln z}{\partial T}\right)_V \tag{5.31}$$

The molecular partition function is really useful only when the system of interest can be considered to be made up of noninteracting elements—e.g., a perfect gas of molecules with no appreciable intermolecular forces. The reason for this restriction is that only in this case can we define and enumerate the states of the system in terms of quantum mechanical energy states of individual molecules or classical positions and momenta of individual molecules. As soon as interactions between molecules occur, the description of the state of the system must include potential energy terms, such as $U(r_{ij})$, which are functions of the intermolecular distances.

10. Statistical Thermodynamics

Up to now, we have been concerned with noninteracting particles. We must find a more general treatment to deal with models of actual gases, liquids, or solids.

Thermodynamics does not deal with individual particles, but with systems containing very large numbers of particles. The usual thermodynamic measure, the mole, contains 6.02×10^{23} molecules. To apply statistical mechanics to the calculation of thermodynamic functions, we find it most convenient to use the Gibbs canonical ensemble. We can mentally construct such an ensemble, as was shown in Fig. 5.3, to consist of a large number $\mathcal{N}$ of systems, each containing 1 mol (L molecules) of the substance under consideration. These systems—members of the ensemble—are separated from one another by diathermic walls that permit heat conduction but no passage of matter. Each system is kept at the same constant volume V. The entire ensemble of systems is surrounded by a rigid adiabatic wall that isolates it completely from the rest of the world, so that it has a fixed constant energy E_t.

Now we note that the canonical ensemble is itself a system of fixed N, V, and $E(= E_t)$ and hence is itself *one member* of a microcanonical ensemble. Therefore, every possible state of the canonical ensemble has an equal a priori probability in accord with postulate 1 on p. 177. We thus can calculate average properties for the canonical ensemble by giving each *state of the ensemble* the same weight in accord with postulate 2. Suppose our original system is 1 mol of O_2 gas. Our canonical ensemble will consist of $\mathcal{N}$ molar volumes of O_2, each at the same temperature T, but each having its own energy E_j.* We can use the canonical ensemble to obtain statistical mechanical formulas for properties of 1 mol of O_2 without making any assumptions about the behavior of the molecules in each member of the ensemble. Whether we can evaluate numerically the formulas so obtained is another problem.

*This energy would be obtained from the solution to a many-body quantum mechanical problem.

If we can compute the average value of a property over all the systems in the ensemble, the average should give us the time average behavior of the actual thermodynamic system. It is well to recall that we are speaking of systems at equilibrium, so that the properties calculated are the ordinary molar thermodynamic equilibrium quantities.

The allowed energy states of any particular system in the ensemble will be specified by $E_1, E_2, \ldots, E_j$, etc. Since the systems all have the same V, T, N, they all have the same set of allowed values of E_j, although they may occupy different levels of the set of E_j, i.e., have different distributions over the E_j. If n_j is the number of systems of the ensemble in a state with energy E_j, the total energy

$$E_t = \sum n_j E_j \tag{5.32}$$

and the total number of systems

$$\mathcal{N} = \sum n_j \tag{5.33}$$

Since each system in the ensemble has the same volume V and number of molecules N, the volume of the ensemble would be $\mathcal{N} V$ and its total number of molecules $\mathcal{N} N$. If we could at one instant of time determine the energy states E_j to which each system in the ensemble belonged, we might find n_1 systems in state E_1, n_2 in E_2, n_3 in E_3, and in general, n_j in E_j. The set of numbers $n_1, n_2, n_3, \cdots, n_j$ $\cdots$ is called a *distribution*. There are many possible distributions of the systems among the energy states, all subject, of course, to the constraints (5.32) and (5.33).

For any given distribution **n** (where **n** stands for all the distribution numbers $n_1, n_2, n_3, \cdots, n_j \cdots$), there will be a large number of ways of assigning the systems of the ensemble to the energy states E_j. In fact, this number will be given by the now familiar formula as

$$W_t(\mathbf{n}) = \frac{\mathcal{N}!}{n_1! n_2! \cdots} = \frac{\mathcal{N}!}{\prod_j n_j!}$$

If we select a system at random out of the canonical ensemble, the probability p_j that it will be in the state E_j is simply the *average value of n_j* divided by the total number of systems $\mathcal{N}$,

$$p_j = \frac{\bar{n}_j}{\mathcal{N}} = \frac{1}{\mathcal{N}} \frac{\sum_n W_t(\mathbf{n}) n_j(\mathbf{n})}{\sum_n W_t(\mathbf{n})} \tag{5.34}$$

Put into words, (5.34) states that the value of n_j averaged over all possible distributions is taken by weighting the $n_j(\mathbf{n})$ for a given distribution by the number of different states $W_t(\mathbf{n})$ that assign the systems of the ensemble to that particular distribution. In accord with basic postulate 2, each of these states has the same a priori probability.

The canonical ensemble average of a mechanical property, such as the energy, for example, is simply

$$\bar{E} = \sum_j p_j E_j$$

If we allow $\mathcal{N}$ to become very large (in the limit $\mathcal{N} \to \infty$), the most probable distribution and those distributions virtually indistinguishable from it completely dominate the set of distributions in (5.34). In fact, we can carry out the calculation of (5.34) by including only the most probable distribution. This result is equivalent to setting the sum over all the distributions equal to the largest term in the sum— a result that at first seems rather fantastic, but which can be shown to be valid as $\mathcal{N} \to \infty$.* Thus, in this limit, we include only the weight $W_t(\mathbf{n}^*)$ where $\mathbf{n}^*$ denotes the most probable distribution. Then

$$p_j = \frac{\bar{n}_j}{\mathcal{N}} = \frac{n_j^*}{\mathcal{N}}$$

The problem now becomes similar mathematically to that treated in the derivation of the Boltzmann distribution—namely, to determine the most probable n_j^* subject to the restraints (5.31) and (5.32). The result is

$$\frac{n_j^*}{\mathcal{N}} = \frac{e^{-\beta E_j}}{\sum e^{-\beta E_j}}$$

We could show again that in this case also $\beta = 1/kT$, but we shall omit the rather lengthy proof.

Thus,

$$\frac{n_j^*}{\mathcal{N}} = \frac{e^{-E_j/kT}}{\sum e^{-E_j/kT}} \tag{5.35}$$

We could define $Z = \sum e^{-E_j/kT}$, but for future use it is worthwhile to introduce explicitly the factor g_j, the statistical weight of the energy state E_j. It may happen that several states have the same energy (or are within such a narrow range of a given energy as to have practically the same energy). The factor g_j gives the number of such states with practically the same energy. We then define

$$Z(V, T, N) = \sum g_j e^{-E_j/kT} \tag{5.36}$$

where Z is the *canonical ensemble partition function*. The summation in (5.36) is now taken over the different values of the energy levels, rather than over all the distinct states, because the introduction of g_j has lumped together states with the same energy. In terms of Z, we find

$$\bar{E} = kT^2 \left(\frac{\partial \ln Z}{\partial T}\right)_{V,N} \tag{5.37}$$

This $\bar{E}$, the canonical ensemble average of the energy, can be identified with the thermodynamic U of the system upon which the emsemble was based.

From (5.37), the molar heat capacity at constant volume then becomes

$$C_V = \left(\frac{\partial U}{\partial T}\right)_{V,N} = \frac{\partial}{\partial T}\left(kT^2 \frac{\partial \ln Z}{\partial T}\right)_{V,N}$$

$$C_V = \frac{k}{T^2}\left(\frac{\partial^2 \ln Z}{\partial(1/T)^2}\right)_{V,N} \tag{5.38}$$

*We omit the derivation and refer to T. L. Hill, *Introduction to Statistical Thermodynamics* (Addison-Wesley Publishing Co., Inc., 1960), p. 478. (We are also indebted to Hill for the treatment of the canonical-ensemble problem given in the text.)

11. Entropy in Statistical Mechanics

To calculate the entropy by statistical mechanics we return to the fundamental theorem of (3.13),

$$dS = \frac{dq \text{ (reversible)}}{T}$$

in which we found that T^{-1} is an integrating factor for the differential of reversible heat. We now seek to obtain an expression for dq_{rev} in terms of the canonical ensemble partition function Z of (5.36), and hence to calculate dS. To facilitate the mathematical operations, we again write $\beta = (kT)^{-1}$ and define a function*

$$B = \ln Z = \ln \sum g_j e^{-E_j/kT} = \ln \sum g_j e^{-\beta E_j}$$

For the canonical ensemble (see Fig. 5.3), Z and hence B are functions of T, V, and N. (We can imagine that the ensemble is immersed in heat baths at various values of T to study the variation of Z with T, and we can imagine that each member of the ensemble is connected with an identical device for carrying out a change in volume so as to do work.)

From $B(V, T)$, therefore,

$$dB = \left(\frac{\partial B}{\partial \beta}\right)_V d\beta + \sum_j \left(\frac{\partial B}{\partial E_j}\right)_T dE_j \tag{5.39}$$

From (5.37),

$$\left(\frac{\partial B}{\partial \beta}\right)_V = -U$$

and from (5.35) and (5.36),

$$\left(\frac{\partial B}{\partial E_j}\right)_T = -\frac{\beta}{\mathcal{N}} n_j$$

Therefore (5.39) becomes

$$dB = -U d\beta - \frac{\beta}{\mathcal{N}} \sum n_j \, dE_j \tag{5.40}$$

or

$$d(B + \beta U) = \beta \left(dU - \frac{1}{\mathcal{N}} \sum n_j \, dE_j \right) \tag{5.41a}$$

We have thus obtained the most interesting result that β is an integrating factor of $[dU - (1/\mathcal{N}) \sum n_j dE_j]$, converting it into the perfect differential of a function $(B + \beta U)$. A suspicion at this point that $d(B + \beta U)$ is something closely related to dS will soon be justified.

Consider

$$dS = \frac{dq_{\text{rev}}}{T} = \frac{dU - dw_{\text{rev}}}{T} \tag{5.41b}$$

---

*Actually the derivation that will be given demonstrates that $\beta = (kT)^{-1}$, and thus supplements the discussion in Section 5.10. Our derivation of the entropy is based upon that given by Erwin Schrödinger in his wonderful little book, *Statistical Thermodynamics* (Cambridge University Press, 1946).

It is evident that $k\, d(B + \beta U)$ will be identified as dS if we can show that $(1/\mathcal{N}) \sum n_j\, dE_j$ is the ensemble average of the work done on a system in the canonical ensemble.

Actually the mathematical analysis would already force us to this conclusion, since if we wish to relate the thermodynamic expression (5.41b) to the statistical mechanical expression (5.41a), there are no state functions that will fit except dS and $(kT)^{-1} = \beta$. We can also, however, see the result by interpreting the mathematical expression in (5.41a) in terms of variations in the canonical ensemble.

Let us require that all the systems in the ensemble are coupled to identical mechanisms of "screws, pistons or what-not" (to use Schrödinger's phrase), which we can manipulate and thereby change the states of the systems. When we do this, we change the energy levels E_j. All the E_j's for all the $\mathcal{N}$ identical systems of the ensemble are thus changed in exactly the same way, so that we still have a canonical ensemble of systems. It is evident, therefore, that $\sum n_j\, dE_j$ represents the work done on all systems in the canonical ensemble and $(1/\mathcal{N}) \sum n_j\, dE_j$ is the ensemble average of the work done on one member of the ensemble. By our basic postulate, therefore, $(1/\mathcal{N}) \sum n_j\, dE_j$ corresponds to the thermodynamic term dw_{rev}. Thus the term in parenthesis on the right hand side of (5.41a) is the reversible heat and β is its integrating factor. We have proved that

$$dS = k\, d\!\left(B + \frac{U}{kT}\right)$$

Upon integration and substitution of $B = \ln Z$,

$$S = k \ln Z + \frac{U}{T} + \text{const.} \qquad (5.42)$$

As we should expect from our discussion in Section 3.23, an absolute value for the entropy S is not determined since the expression (5.42) contains an arbitrary constant. The constant, however, is independent of $(B + U/kT)$ and hence independent of N, V, and T, the variables upon which that function depends. Thus the constant will always drop out of any calculations of entropy changes ΔS for chemical and/or physical changes in a system.

12.   The Third Law in Statistical Thermodynamics

There are two parts to the discussion of the Third Law on the basis of statistical mechanics. The first problem is to consider the constant in Eq. (5.42). The entropy S has no physically fixed zero level, but to set const. $= 0$ in (5.42) would be equivalent to adopting such a definite zero level for the entropy in all cases. As was shown in Section 5.11, for any system under study the constant is indeed independent of the parameters of the system, so that the difference in entropy ΔS between any two states of the system differing in values of the defining parameters (volume, magnetic field, pressure, etc.) will approach zero at $T = 0$. Furthermore, this statement that $\Delta S \longrightarrow 0$ at $T = 0$ applies also to any possible chemical changes in the system.

Schrödinger* has discussed this point in the following way:

> A typical case would be a system consisting of L iron atoms and L sulphur atoms. In one of the two thermodynamical states they form a compact body, 1 gram-molecule† of FeS; in the other, 1 gram-atom of Fe and 1 gram-atom† of S, separated by a diaphragm, so that they can under no circumstances unite; the much lower energy levels of the chemical compound are made inaccessible.
>
> Now in all such cases it is only a question of believing in the possibility of transforming one state into the other by small reversible steps, so that the system never quits the state of thermodynamical equilibrium, to which all our considerations apply. All the small, slow steps of this process can then be regarded as small, slow changes of certain parameters, changing the values of the ϵ_j's. Then the 'const.' will not change in all these processes—and the statement applies.
>
> For instance, in the example mentioned, you would gradually heat the gram-molecule of FeS till it evaporates; then go on heating till it dissociates as completely as desired; then separate the gases with the help of a semi-permeable diaphragm; then condense them separately by lowering the temperature (of course with an impermeable diaphragm between them) and cool them down to zero. Having once or twice gone through such considerations, you no longer bother to think them out in detail, but just declare them as 'thinkable'—and the statement applies.
>
> After this has been thoroughly turned over in the mind the simplest way of codifying it once and for all is, of course, to decide to put 'const.' = zero in all cases. It is possibly the only way to avoid confusion—no alternative suggests itself. But to regard this 'putting equal to zero' as the essential thing is certainly apt to create confusion and to detract attention from the point really at issue.

The second part of the statistical treatment of the Third Law is the derivation of an expression for S_0, the value of the entropy predicted by (5.42) for $T = 0$. From (5.36), we find‡

$$S_0 = k \ln g_0 \qquad (5.43)$$

where g_0 is the statistical weight (degeneracy) of the lowest possible energy state of the system.

As an example, consider a perfect crystal at the absolute zero. There will usually be one and only one equilibrium arrangement of its constituent atoms, ions, or molecules. In other words, the statistical weight of the lowest energy state is unity, and from (5.43) the entropy at 0 K becomes zero.

Statistical Thermodynamics, p. 17 (by kind permission of the Cambridge University Press).
†We should now say *1 mol*.
‡As $T \longrightarrow 0$ we can certainly neglect all terms in Z except for the first two, so that (5.42) becomes

$$S = \frac{1}{T} \frac{g_0 E_0 e^{-E_0/kT} + g_1 E_1 e^{-E_1/kT}}{g_0 e^{-E_0/kT} + g_1 e^{-E_1/kT}} + k \ln (g_0 e^{-E_0/kT} + g_1 e^{-E_1/kT})$$

Near the limit

$$e^{-E_1/kT} \ll e^{-E_0/kT}$$

so that

$$S = \frac{E_0}{T} + \frac{1}{T} \frac{g_1}{g_0} e^{-(E_1 - E_0)/kT} + k \ln g_0 - \frac{E_0}{T} + k \frac{g_1}{g_0} e^{-(E_1 - E_0)/kT}$$

In the limit as $T = 0$, the remaining exponential terms decrease much more rapidly than T, so that we are left with only

$$S_0 = k \ln g_0$$

In certain cases, however, the particles in a crystal may persist in more than one geometrical arrangement even at absolute zero. An example is nitrous oxide. Two adjacent molecules of N_2O can be oriented either as (ONN NNO) or as (NNO NNO). The energy difference ΔU between these alternative configurations is so slight that their relative probability, $\exp(\Delta U/RT)$, is practically unity even at low temperatures. When the crystal has cooled to the *extremely low* temperature at which even a tiny ΔU might produce considerable reorientation at equilibrium, the *rate* of rotation of the molecules within the crystal has become vanishingly slow. Thus, the random orientations are effectively frozen. As a result, heat capacity measurements will not include a residual entropy S_0 equal to the entropy of mixing of the two arrangements. From (5.3), this amounts to

$$S_0 = -R \sum X_i \ln X_i = -R(\tfrac{1}{2} \ln \tfrac{1}{2} + \tfrac{1}{2} \ln \tfrac{1}{2}) = R \ln 2 = 5.77 \text{ J} \cdot \text{K}^{-1} \cdot \text{mol}^{-1}$$

It is found that the entropy calculated from statistics is actually $4.77 \text{ J} \cdot \text{K}^{-1} \cdot \text{mol}^{-1}$ higher than the Third Law value; this difference checks the calculated 5.77 within the experimental uncertainty of ± 1.1 in S_0. A number of examples of this type have been carefully studied.*

Another source of residual entropy of mixing at 0 K arises from the isotopic constitution of the elements. This effect can usually be ignored, since in most chemical reactions the isotopic ratios change very slightly.

When we have found how to evaluate Z from spectroscopic data on the properties of molecules (Sections 5.14 and 5.15), we shall be able to use (5.42) to compute the standard molar entropies $S^\ominus$ of ideal gases. Some examples of the results of such calculations are compared in Table 5.3 with the best values from the Third Law (measurements of heat capacities upwards from very low temperatures).

TABLE 5.3
Comparison of Statistical (Spectroscopic) and Third Law (Heat Capacity) Entropies

Gas	Entropy as ideal gas at 1 atm, 298.15 K $(\text{J} \cdot \text{K}^{-1} \cdot \text{mol}^{-1})$	
	Statistical	Third Law
N_2	191.5	192.0
O_2	205.1	205.4
Cl_2	223.0	223.1
HCl	186.8	186.2
HBr	198.7	199.2
HI	206.7	207.1
H_2O	188.7	185.3
N_2O	220.0	215.2
NH_3	192.2	192.1
CH_4	185.6	185.4
C_2H_4	219.5	219.6

*For the interesting case of ice, see L. Pauling, *J. Am. Chem. Soc.*, **57**, 2680 (1935).

13. Evaluation of Z for Noninteracting Particles

The Helmholtz function, $A = U - TS$, is important in statistical thermodynamics because it is related so simply to the partition function Z and so directly to the pressure (and hence the equation of state) of a substance. From (5.42),

$$A = -kT \ln Z \tag{5.44}$$

and

$$P = -\left(\frac{\partial A}{\partial V}\right)_T = kT\left(\frac{\partial \ln Z}{\partial V}\right)_T \tag{5.45}$$

The preceding equations promise a way to calculate theoretically the values of all thermodynamic properties of all substances, *if* we can only determine $Z(V, T)$ (it is a big *if*). Just as Archimedes, given a fulcrum, would use his lever to tilt the world, so any physical chemist, given the partition functions Z, would calculate all the equilibrium properties of matter. We can anticipate that in most cases it will not be easy to obtain the desired Z.

The only case in which evaluation of Z can be accomplished without mathematical difficulties is for a system of noninteracting particles. In this case, the energy of the system can be written as the sum of the energies of the individual particles,

$$E = \epsilon_a + \epsilon_b + \epsilon_c + \cdots$$

Let us first suppose that the individual particles are distinguishable from one another, so that the subscripts a, b, c, etc. imply that a system with particle a in energy state 1 and particle b in energy state 2 can be physically distinguished from a system with b in 1 and a in 2.

Let us recall the definition of the single particle partition functions,

$$z_a = \sum e^{-\epsilon_{aj}/kT}, \qquad z_b = \sum e^{-\epsilon_{bj}/kT}$$

We can now see that the product of the z's, one for each particle in the system, generates all the possible values of the total energy,

$$Z = \sum e^{-E_j/kT} = \left(\sum e^{-\epsilon_{aj}/kT}\right)\left(\sum e^{-\epsilon_{bj}/kT}\right) \cdots = z_A z_B \cdots$$

Suppose, for example, there are two particles, one with three energy states, $\epsilon_{a1}, \epsilon_{a2}, \epsilon_{a3}$, and one with two states, ϵ_{b1} and ϵ_{b2}. Then,

$$
\begin{aligned}
z_a z_b &= (e^{-\epsilon_{a1}/kT} + e^{-\epsilon_{a2}/kT} + e^{-\epsilon_{a3}/kT})(e^{-\epsilon_{b1}/kT} + e^{-\epsilon_{b2}/kT}) \\
&= e^{-(\epsilon_{a1}+\epsilon_{b1})/kT} + e^{-(\epsilon_{a1}+\epsilon_{b2})/kT} + e^{-(\epsilon_{a2}+\epsilon_{b1})/kT} \\
&\quad + e^{-(\epsilon_{a2}+\epsilon_{b2})/kT} + e^{-(\epsilon_{a3}+\epsilon_{b1})/kT} + e^{-(\epsilon_{a3}+\epsilon_{b2})/kT}
\end{aligned}
$$

We see that all possible energy states for the system of two particles are included in the sum.

In this case, the molecules were all different, as denoted by a, b, c, etc. If they are all molecules of the same species, they cannot be physically or chemically distinguished from one another. Suppose, for example, we consider again the liter of O_2 gas. If two oxygen molecules in the gas volume could be interchanged, the state of the gas would be exactly the same after the interchange as it was before. Interchange of the spacial coordinates between a pair of atoms does not lead to a different state for the volume of gas. An energy state $\epsilon_{a1} + \epsilon_{b2}$ is in no way distinguishable from a state $\epsilon_{a2} + \epsilon_{b1}$. If the N molecules were chemically of the same species, but distinguishable units, $Z = z_A z_B z_C \cdots$ would become, simply, $Z = z^N$.

We must, however, correct this expression so as not to count certain states too many times. When we form the partition function

$$Z = \sum e^{-E_j/kT}$$

a state like $\epsilon_{a1} + \epsilon_{b2}$ should be counted only once, not twice. In general, terms of the following kind, in Z

$$e^{-(\epsilon_{ai} + \epsilon_{bj} + \epsilon_{ck} + \cdots)/kT}$$

where $i \neq j \neq k$, will occur $N!$ times in the summation (since N molecules can be permuted among the N states in $N!$ ways). If this were the only kind of extra term in the sum-over-states, the problem of correcting for indistinguishability of particles would be easy—simply divide z^N by $N!$

Unfortunately, there are also terms of the type,

$$e^{-(\epsilon_{ai} + \epsilon_{bi} + \epsilon_{ci} + \cdots)/kT}$$

which place two or more molecules in the same energy state i. At ordinary temperatures and gas densities, many more states are available than there are molecules to fill them. Thus, the chance that more than one molecule occupies the same state is very low. When the number of states is much larger than the number of particles, the number of multiply occupied states becomes negligibly small compared to the number of singly occupied states.*

We shall see later (Chapter 13), from the quantum mechanical specification of allowed states, that at ordinary temperatures and densities there are so many available states for gas molecules that we are justified in neglecting multiple occu-

*For instance, suppose there are 10 states available to two particles. We can assign the particles to the states as follows:

States of a

	1	2	3	4	5	6	7	8	9	10
1	$^1/_1$									
2		$^1/_1$								
3			$^1/_1$							
4				$^1/_1$						
5					$^1/_1$					
6						$^1/_1$				
7							$^1/_1$			
8								$^1/_1$		
9									$^1/_1$	
10										$^1/_1$

States of b

Only the pairs of states along the diagonal correspond to doubly occupied states. There are only 10 of these out of 100 possible arrangements, or 10%. If there are 100 states for two particles, there will be only 10^2 double states out of 10^4 total states, or 1%.

pancies. Therefore, we can write

$$Z = \frac{z^N}{N!} \tag{5.46}$$

As a consequence of (5.46), we can evaluate Z for a system of noninteracting, indistinguishable particles (perfect gas), provided we know the molecular partition function z. To calculate z, we need know only the allowed energy states of the molecules. These energy states can be determined experimentally from sufficiently detailed experimental spectroscopic data (Chapter 17). For simple gas molecules, however, we can use theoretical expressions for the energy states, derived from quantum mechanics.

As discussed in Section 4.18, the energy of a molecule can be divided into the translational kinetic energy ϵ_t of the center of mass of the molecule and energy terms ϵ_I (both kinetic and potential) associated with internal degrees of freedom. Hence, we can write

$$\epsilon = \epsilon_t + \epsilon_I \tag{5.47}$$

From (5.28), it follows that

$$z = z_t z_I \tag{5.48}$$

so that the translational part of the molecular partition function ($z_t = \sum e^{-\epsilon_{tt}/kT}$) can be separated from the internal part.

14.   Translational Partition Function

In Section 13.20, we shall solve the Schrödinger equation of quantum mechanics to derive the following expression for the allowed translational energy levels of a particle of mass m confined within a parallelepiped with sides of length a, b, c:

$$E = \frac{h^2}{8m}\left(\frac{n_1^2}{a^2} + \frac{n_2^2}{b^2} + \frac{n_3^2}{c^2}\right)$$

Here, h is the Planck constant, 6.62×10^{-34} J·s, and n_1, n_2, n_3 are integers called *quantum numbers*. The quantum numbers specify the allowed energy levels.

The molecular partition function is

$$z = \sum\sum\sum \exp\left[-\frac{h^2}{8mkT}\left(\frac{n_1^2}{a^2} + \frac{n_2^2}{b^2} + \frac{n_3^2}{c^2}\right)\right]$$

where n_1, n_2, n_3 are each summed from 0 to ∞. The energy levels are so closely packed together (owing to the smallness of h^2) that the sums can be replaced by integrals,

$$z = \int_0^\infty\int_0^\infty\int_0^\infty \exp\left[\frac{-h^2}{8mkT}\left(\frac{n_1^2}{a^2} + \frac{n_2^2}{b^2} + \frac{n_3^2}{c^2}\right)\right] dn_1\, dn_2\, dn_3$$

We then have a product of three integrals, each of form

$$\int_0^\infty e^{-n^2 h^2/8mkTa^2}\, dn$$

Letting

$$x^2 = \frac{n^2 h^2}{8ma^2 kT}$$

we have

$$z = \frac{a}{h}(8mkT)^{1/2} \int_0^\infty e^{-x^2}\, dx = \frac{(2\pi mkT)^{1/2} a}{h}$$

For each of the three degrees of translational freedom, a similar expression is found, and since $abc = V$, we therefore obtain

$$z = \frac{(2\pi mkT)^{3/2} V}{h^3} \tag{5.49}$$

The partition function Z per mole is

$$Z = \frac{1}{L!} z^L = \frac{1}{L!}\left[\frac{(2\pi mkT)^{3/2} V}{h^3}\right]^L \tag{5.50}$$

The molar energy is, therefore, from (5.37),

$$U_m = \bar{E} = LkT^2 \frac{\partial \ln z}{\partial T} = RT^2 \frac{\partial \ln z}{\partial T} = RT^2 \frac{3}{2}\cdot\frac{1}{T} = \frac{3}{2}RT$$

This is the simple result expected from the Principle of Equipartition.

We evaluate the entropy from (5.42) with the help of the Stirling formula, $L! \approx (L/e)^L$. Thus,

$$Z = \left[\frac{(2\pi mkT)^{3/2} eV}{Lh^3}\right]^L$$

$$\ln Z = L \ln\left[\frac{eV}{Lh^3}(2\pi mkT)^{3/2}\right]$$

The molar entropy is, therefore,

$$S_m = \frac{3}{2}R + R \ln \frac{eV}{Lh^3}(2\pi mkT)^{3/2}$$
$$S_m = R \ln \frac{e^{5/2} V}{Lh^3}(2\pi mkT)^{3/2} \tag{5.51}$$

Sackur and Tetrode first obtained this famous equation by somewhat unsatisfactory arguments in 1913. As an example, we shall use it to calculate the entropy of argon at 298.2 K and 1 atm.

$$R = 8.314\ \text{J}\cdot\text{K}^{-1} \qquad\qquad \pi = 3.1416$$
$$\text{e} = 2.718 \qquad\qquad m = 6.63 \times 10^{-26}\ \text{kg}$$
$$V = 22.465 \times 10^{-3}\ \text{m}^3 \qquad k = 1.38 \times 10^{-23}\ \text{J}\cdot\text{K}^{-1}$$
$$L = 6.02 \times 10^{23} \qquad\qquad T = 298.2\ \text{K}$$
$$h = 6.62 \times 10^{-34}\ \text{J}\cdot\text{s}$$

Substituting these quantities into (5.51), we calculate the entropy to be 154.7 $\pm$ 0.1 J·K^{-1}·mol^{-1}. What is the principal source of the probable error of ± 0.1 J·K^{-1}·mol^{-1}? The Third Law value is 154.6 $\pm$ 0.2 J·K^{-1}·mol^{-1}.

15. Partition Functions for Internal Molecular Motions

If we knew the internal energy states of a molecular species from spectroscopic data, we would be able to calculate a partition function for internal molecular motions, and hence the contribution of the internal degrees of freedom to the thermodynamic properties of the substance. Thus,

$$z_I = \sum g_j e^{-\epsilon_j/kT} \tag{5.52}$$

To a quite good approximation, it is possible to take the internal energy as a sum of independent terms, one each for rotational, vibrational, and electronic energies,

$$\epsilon_I = \epsilon_r + \epsilon_v + \epsilon_e \tag{5.53}$$

Quantum mechanics provides theoretical expressions for these separate energy terms for diatomic and polyatomic molecules. If we had these quantum mechanical energy formulas, we could substitute them into (5.52) to obtain useful expressions for the different contributions to the molecular partition functions,

$$z_I = z_r \, z_v \, z_e \tag{5.54}$$

Instead of proceeding further with such a development at this point, we are simply going to list the final formulas in Table 5.4 and postpone the derivations until the quantum mechanical energy formulas are derived in Chapter 14.

The formulas of Table 5.4 are quite useful, but they are not to be regarded as the final answer to the calculation of thermodynamic quantities from data on molecular structure. They are based on a complete separation of vibrational and rotational motions, which is only an approximation, as we shall see in Chapter 17. The fundamental and rigorous solution to the problem (for a gas of noninteracting molecules) is obtained through (5.52) and the actual experimental energy levels. The formulas in Table 5.4, based on (5.53) and (5.54), are good approximations for simple molecules under most conditions.

TABLE 5.4
Molecular Partition Functions

Motion	Degrees of freedom	Partition function[a]	Order of magnitude
Translational	3	$\dfrac{(2\pi mkT)^{3/2}}{h^3} V$	$10^{24} - 10^{25} V$
Rotational (linear molecule)	2	$\dfrac{8\pi^2 IkT}{\sigma h^2}$	$10 - 10^2$
Rotational (nonlinear molecule)	3	$\dfrac{8\pi^2 (8\pi^3 ABC)^{1/2}(kT)^{3/2}}{\sigma h^3}$	$10^2 - 10^3$
Vibrational (per normal mode)	1	$\dfrac{1}{1 - e^{-h\nu/kT}}$	$1 - 10$

[a]The term σ is a symmetry number equal to the number of indistinguishable positions into which a molecule can be turned by rigid rotations; A, B, and C are moments of inertia.

As an example of the application of these equations, consider the calculation of the molar entropy of F_2 at 298.15 K, assuming translational and rotational contributions only. From (5.51), the translational entropy is calculated to be 154.7 J·K⁻¹. Then the rotational entropy per mole is

$$S_{rm} = RT\frac{\partial \ln z_r}{\partial T} + k \ln z_r^L = R + R \ln z_r = R + R \ln \frac{8\pi^2 IkT}{2h^2}$$

Note that the rotational energy is simply RT in accord with the Equipartition Principle. Substituting $I = 32.5 \times 10^{-40}$ g·cm², we obtain $S_{rm} = 48.1$ J·K⁻¹. Adding the translational term, we have

$$S_m = S_{rm} + S_{tm} = 48.1 + 154.7 = 202.8 \text{ J·K}^{-1}\text{·mol}^{-1}$$

Let us now calculate the vibrational contribution to the entropy of F_2 at 298.15 K. The fundamental vibration frequency is $v = 2.676 \times 10^{13}$ s⁻¹. Hence,

$$x = \frac{hv}{kT} = \frac{(6.62 \times 10^{-27})(2.676 \times 10^{13})}{(1.38 \times 10^{-16})(298.15)} = 4.305$$

The vibrational entropy per mole is

$$S_{vm} = RT\left(\frac{\partial \ln z_v}{\partial T}\right) + R \ln z_v$$

$$= R\left[\frac{x}{e^x - 1} - \ln(1 - e^{-x})\right]$$

With $x = 4.30$,

$$S_{vm} = R(0.0590 + 0.0136) = R(0.0726)$$
$$= 0.605 \text{ J·K}^{-1}\text{·mol}^{-1}$$

Although the vibrational contribution is small at 298 K, it can, of course, become much greater at higher temperatures. The statistical entropy of F_2 per mole at 298.15 K is thus

$$S_m = S_{tm} + S_{rm} + S_{vm} = 154.7 + 48.1 + 0.6 = 203.4 \text{ J·K}^{-1}\text{·mol}^{-1}$$

which compares with the experimental value of $S_m = 203.2$ J·K⁻¹·mol⁻¹, obtained from heat capacity measurements and the Third Law.

We shall postpone the discussion of other statistical calculations of thermodynamic functions—in particular, the historic and important problem of heat capacity—until we have discussed the quantum theory of internal energy levels in Chapter 14.

16. Classical Partition Function

In Section 4.11, we introduced the idea of the phase space of a particle. Instead of an individual particle, suppose we have a macroscopic system with s degrees of freedom—e.g., a gas containing $s/3$ atoms. We can define the state of the system at any time by specifying s values of the coordinates and s values of the momentum components. The phase space pertaining to the system would have $2s$ dimensions, and any point in this space would specify the state of the system.

The concept of phase space can be extended to systems of any number of mass points. For example, the phase space of a mole of monatomic gas would have $6L$ dimensions, corresponding to $3L$ coordinates q_i and $3L$ momenta p_i. A differential volume element in this phase space would be defined by

$$d\tau = dq_1\, dq_2\, dq_3 \cdots dq_{3L-2}\, dq_{3L-1}\, dq_{3L}\, dp_1\, dp_2\, dp_3 \cdots dp_{3L-2}\, dp_{3L-1}\, dp_{3L}$$

The state of a system with s degrees of freedom can be represented by a point in a $2s$-dimensional phase space. A canonical ensemble of systems can then be represented by a collection of points in phase space, one for each member of the ensemble. The summation over discrete energy states in (5.36) is replaced by an integration over the total volume of phase space, to yield a classical partition function:

$$Z = \frac{1}{N!h^s} \int \cdots \int_{\substack{\text{phase}\\\text{space}}} e^{-\mathscr{H}(q_1 \cdots p_s)/kT}\, dq_1 \cdots dp_s \qquad (5.54)$$

where $\mathscr{H}$ (the classical Hamiltonian) is the sum of the kinetic and potential energies for the system. Note especially the factor h^s preceding the integral. It is the volume of a cell in phase space. Since phase space is a combined momentum and coordinate space, an element of volume $dp\, dq$ has the dimensions of ml^2t^{-1}, a quantity known in mechanics as *action*. Since Z in (5.36) is dimensionless, it is obviously necessary to introduce a factor with the dimensions of $(\text{action})^{-s}$ into the classical expression (5.54) to preserve this dimensionless character, since the integral in (5.54) itself has the dimensions $(pq)^s$ or $(\text{action})^s$.

The equivalence of the factor having dimensions of action with the Planck constant h can be established by using (5.54) to calculate the molecular partition function z for a particle in a box. Since there is only kinetic energy, for each degree of freedom we have, as on p. 194,

$$z = \frac{1}{h} \int_0^a \int_{-\infty}^{+\infty} e^{-p^2/2mkT}\, dp\, dq$$

$$z = \frac{(2\pi mkT)^{1/2}\, a}{h}$$

Thus, the classical partition function is identical with that calculated from quantum mechanics, as it must be, provided the volume of the cell in phase space is h^s.

The classical partition function of (5.54) has many important applications. In Chapter 7, we shall apply it to the theory of solutions, and in Chapter 19, to the theory of imperfect gases and liquids.

PROBLEMS

1. Each of the seven letters *timsech* is written on a card. The cards are then shuffled and laid out in a row. What is the probability of obtaining the word *chemist*?

2. The isotopic composition of lead in atoms percent is 204, 1.5%; 206, 23.6%; 207, 22.6%; 208, 52.3%. Calculate the entropy of mixing per mole of Pb at 0 K.

3. The isotopic composition of chlorine is 75.4 atom % ^{35}Cl and 24.6 atom % ^{37}Cl.

Suppose that in the limit of 0 K, crystals of Cl_2 contain a random mixture of isotopic species. Calculate the ΔS of mixing per mole of Cl_2. If Cl_2 were completely dissociated into atoms, what would be the ΔS of mixing?

4. In the English language, there are 27 letters (including blanks). Supposing all letters are equally probable, calculate the number of bits of information per letter.

5. Suppose that the different letters have different probabilities P_j. Show that the information content in bits per letter is then

$$I = -K \sum_{j=1}^{27} P_j \ln P_j$$

where $K = 1/(\ln 2)$.* Using probability data on the English language,† calculate I.

6. This problem requires two players. One player writes a sentence with 15 or more letters. The other player guesses each letter starting with the first (including blanks). After each guess, he is given an answer *yes* or *no*, and proceeds. Each guess represents one bit of information. Calculate the number of bits per letter in the sentence used. Compare the figure with that obtained in Problem 5, and discuss the result briefly.

7. It has been estimated that the amount of information contained in a bacterial cell like *E. coli* is 10^{11} to 10^{13} bits.‡ The mass of the cell is about 10^{-13} g. Calculate the negative entropy per gram of cells associated with their information content. How would you estimate the information per cell?

8. Write the formula for the ΔS of mixing of N_1 molecules of gas (1), N_2 of gas (2), and N_3 of gas (3), in an ideal mixture with a total pressure of 1 atm. Compare this formula with the Shannon formula for information given in Problem 5. Discuss the comparison.

9. From Eqs. (5.45) and (5.49), show that the equation of state of a perfect gas is $PV = nLkT$.

10. The number of ways of distributing N indistinguishable particles among the various states of a system is given by the product $W = \prod_j g_j^{N_j}/N_j!$ where g_j is the statistical weight of the state j, and N_j is the number of particles in state j. Derive an expression for the average number of particles in state j subject to the condition $\sum g_j N_j = N$ and $\sum g_j N_j \epsilon_j = E$.

11. Calculate the molecular translational partition functions z for (a) H_2, (b) CH_4, (c) C_8H_{18}, in a volume of 1 cm³ at $T = 298$ K.

12. Calculate the molar rotational partition function Z_r at 298 K for $^{14}N_2$ and $^{14}N^{15}N$, given that the internuclear distance is 0.1095 nanometre (nm) for both molecules.

13. Derive general formulas for U, S, A, and C_V for a substance that can exist in only two states separated by an energy ϵ.

14. The ground electronic state of Cl_2 is a doublet with a separation of 881 cm^{-1}. From the formulas derived in Problem 13, calculate the electronic parts only of U, S, A, and C_V for Cl_2 and plot them vs. T from 0 to 1000 K.

---

*The formula is the work of C. E. Shannon. See *Bell System Tech. J.*, *30*, 50 (1951).

†For example, L. Brillouin, *Science and Information Theory*, p. 5, or any book on codes and ciphers.

‡H. J. Morowitz, *Bull. Math. Biophys.*, *85*, 17 (1955).

15. Calculate the standard molar entropies $S^{\ominus}$ (298 K) of the group–0 gases, helium to radon, and plot them against the molar mass M and against $M^{3/2}$.

16. Calculate the heat capacity C_V per mole of CO_2 at intervals of 100 K from 0 to 1000 K. Assume CO_2 is an ideal gas. It is a linear molecule with moment of intertia $I = 71.67 \times 10^{-40}$ g·cm². There are four degrees of vibrational freedom in CO_2, corresponding to wave numbers of $\sigma_1 = 2349$ cm⁻¹, $\sigma_2 = 1320$ cm⁻¹, and $\sigma_3 = 667$ cm⁻¹ (doubly degenerate) (see Fig. 17.11). Compare the calculated C_V values with experimental values found in the literature.*

17. Calculate the equilibrium constant K_P for the reaction $I_2 \longrightarrow 2I$ at 1000 K. The fundamental vibration of I_2 is at $\sigma = 214.4$ cm⁻¹, and the internuclear distance is 0.2667 nm. The ground state of I is a doublet $^2P_{3/2,\,1/2}$ with a separation of 7603 cm⁻¹. What would be the calculated K_P if the ground state was assumed to be a singlet?

18. The ionization potential of potassium is 4.33 electron volts (eV) (i.e., this is ΔU for $K \longrightarrow K^+ + e$). Calculate the degree of dissociation of K at 5×10^3 K and 10^{-3} atm.

19. Calculate the translational partition function Z for helium with $V = 1$ cm³ and $T = 100$ K from Eq. (5.50). Considering that the allowed states of the helium atom are given by the discrete quantum mechanical energy levels of a particle in a cubical box of side 1 cm [Eq. (13.58)], estimate the number of states that contribute appreciably to the Z found from (5.50). What is the significance of this result for the validity of the derivation of Boltzmann statistics given in Section 5.9?

20. Compute the thermodynamic probability W for N molecules when (a) all the molecules have the same velocity $+c$ in direction and magnitude; (b) one-half have velocities $+c$ and one-half $-c$; (c) one-sixth each have velocities $\pm c_1$, $\pm c_2$, $\pm c_3$, respectively. Show that for $N \longrightarrow \infty$, each distribution is infinitely more probable than the preceding one. Use Stirling's approximation $N! \approx (2\pi N)^{1/2}(N/e)^N$.

21. Consider a set of three equally spaced energy levels with a spacing ϵ. Assign particles to these levels in a geometrical progression—e.g., 2000, 200, 20. Does this assignment correspond to a Boltzmann distribution? Calculate $W = N!/N_1!N_2!N_3!$ for this assignment. Show that any other assignment with the same total energy will have a lower W (e.g., shift 10 particles from middle level, 5 into lowest level, and 5 into highest level).

22. Show that the average total kinetic energy of the molecules in a gas that cross a given plane of unit area in unit time is $2kT$. Comment on the fact that this value is larger than $\frac{3}{2}kT$, the average kinetic energy of all the molecules in the gas.

23. In 1871, J. C. Maxwell created the *sorting demon*, "a being whose faculties are so sharpened that he can follow every molecule in its course, and would be able to do what is at present impossible to us. . . . Let us suppose that a vessel is divided into two portions, A and B by a division in which there is a small hole, and that a being who can see the individual molecules opens and closes this hole, so as to allow only the swifter molecules to pass from A to B, and only the slower ones to pass from B to A. He will, thus, without expenditure of work raise the temperature of B and lower that of A, in contradiction to the second law of thermodynamics." Discuss and criticize this passage from Maxwell. Can you save the Second Law from the demon?

24. Consider 1 mol of krypton at 300 K and volume V, and 1 mol of helium at the same

*For example, *Thermodynamic Functions of Gases*, ed. F. Din (London: Butterworth, 1962).

volume. What must be the temperature of the helium if both gases are to have the same entropy? How would you interpret this result in terms of the relation of entropy to probability?

25. Molecular N_2 is heated in an electric arc, and spectroscopic observations indicate that the relative numbers of molecules in excited vibrational states with energies given by $\epsilon = (v + \frac{1}{2})h\nu$ are

v	0	1	2	3
N_v/N_0	1.00	0.26	0.07	0.00

(a) Show that the gas is in thermodynamic equilibrium with respect to the distribution of vibrational energy.

(b) What is the temperature of the gas?

(c) What fraction of the total energy of the gas is vibrational energy?

6
Changes of State

Chemistry, by means of visible operations, analyses bodies into certain gross and tangible principles, salts, sulfurs, and the like. But physics, by means of delicate speculations, acts upon these principles just as chemistry has acted upon the bodies themselves; it resolves them into other principles, yet more simple, into little particles designed and moved in an infinity of fashions: here we have the basic difference between physics and chemistry. The spirit of chemistry is more complex, more involved; it resembles those mixtures where the principles are intimately entwined one with another. The spirit of physics is neater, more simple and free: finally it ascends even to the primal origins. The other spirit does not go to the very end of things.

Fontenelle
1733*

Changes such as the melting of ice, the solution of sugar in water, the vaporization of benzene, or the transformation of graphite to diamond are called *changes in state of aggregation* or *phase changes*. They are characterized by discontinuous changes in certain properties of the system at some definite temperature and pressure. The word *phase* is derived from the Greek $\phi\alpha\sigma\iota\varsigma$, meaning *appearance*. We need to distinguish phase changes from chemical changes, which involve chemical reactions, and physical changes, such as expansion or compression that occur continuously with variations in temperature or pressure. In the solid state especially, the distinction between a chemical change and a phase change is not always possible to maintain, since certain solid phases exist over a range of compositions within which the structures may exhibit various degrees of disorder.†

1. Phases

If a system is, in the words of J. Willard Gibbs, "uniform throughout, not only in chemical composition, but also in physical state," it is said to be *homogeneous*, or to consist of only *one phase*. Examples are a volume of air, a noggin of rum, or a cake of ice. Mere difference in shape or in degree of subdivision is not enough to determine a new phase. Thus a mass of cracked ice is still only one phase. We are thus assuming, at this stage in our analysis, that a variable surface area has no appreciable effect on the properties of a substance.

A system consisting of more than one phase is called *heterogeneous*. Each

---

*Bernard le Bovier de Fontenelle, *Histoire de l'Académie Royale des Sciences*, 1733.
†J. S. Anderson, *Advan. Chem. Ser.*, *39*, 1 (1963).

physically or chemically different, homogeneous, and mechanically separable part of a system constitutes a distinct phase. Thus, water with cracked ice in it is a two-phase system. The contents of a flask of liquid benzene in contact with benzene vapor and air is a two-phase system; if we add a measure of sucrose (practically insoluble in benzene), we obtain a three-phase system: a solid, a liquid, and a vapor phase.

In systems consisting entirely of gases, only one phase can exist at equilibrium, since all gases are miscible in all proportions (unless, of course, a chemical reaction intervenes, e.g., $NH_3 + HCl$). With liquids, depending on their mutual miscibility, one or more phases can arise. Hildebrand has shown a curious picture of a test tube containing 10 stable liquid layers.* Solids usually enjoy limited intersolubility, and many different solid phases can coexist in a system at equilibrium.

2. Components

The composition of a system can be completely described in terms of the *components* that are present in it. The ordinary meaning of the word *component* is somewhat restricted in this technical usage. We wish to impose a requirement of economy on our description of the system. This is done by using the *minimum* number of chemically distinct constituents necessary to describe the composition of each phase in the system. The constituents so chosen are the *components*. If the concentrations of the components are stated for each phase, then the concentrations in each phase of any and all substances present in the system are uniquely fixed. This definition may be expressed more elegantly by saying that the components are those constituents the concentrations of which may be *independently varied* in the various phases.

A more practical way of defining the *number of components* is to set it equal to the total number of independent chemical constituents in the system minus the number of distinct chemical reactions that can occur in the system between these constituents. By *number of independent constituents* we mean total number minus the number of any restrictive conditions, such as material balance or charge neutrality. By a *distinct chemical reaction* we mean one that cannot be written simply as a sequence of other reactions in the system.

Consider, for example, the system consisting of calcium carbonate, calcium oxide, and carbon dioxide. There are three distinct chemical constituents, $CaCO_3$, CaO, and CO_2. One reaction occurs between them: $CaCO_3 \rightarrow CaO + CO_2$. Hence, the number of components $c = 3 - 1 = 2$.

A more complex example is the system formed by mixing NaCl, KBr, and H_2O. Suppose that we can isolate from this system also the constituents KCl, NaBr, $NaBr \cdot H_2O$, $KBr \cdot H_2O$ and $NaCl \cdot H_2O$. The possible distinct chemical reactions between these constituents are

$$NaCl + KBr \longrightarrow NaBr + KCl; \quad NaCl + H_2O \longrightarrow NaCl \cdot H_2O$$
$$KBr + H_2O \longrightarrow KBr \cdot H_2O; \quad NaBr + H_2O \longrightarrow NaBr \cdot H_2O$$

There are eight constituents but there is a condition of material balance since the

---

*Joel H. Hildebrand and Robert L. Scott, *Regular Solutions* (Englewood Cliffs, N. J.: Prentice-Hall, Inc., 1962).

moles of KCl must always equal the sum of the moles of NaBr and $NaBr \cdot H_2O$. Therefore, the number of components, $c = (8 - 1) - 4 = 3$. If we remove the condition that all the NaBr and KCl are formed from the original reactants, and allow these salts to be added separately to the system, we would have $c = 4$. An easy way to see this is to consider that the composition of any phase can be specified in terms of four ions (Na^+, K^+, Cl^-, and Br^-) plus H_2O, but that the condition of electroneutrality requires that $Na^+ + K^+ = Cl^- + Br^-$.

A chemical reaction included in calculating the number of components must be one that actually occurs in the system, and not simply a *possible* reaction that does not occur because of the absence of a suitable catalyst or other condition necessary to give it a measurable rate. Thus, a mixture of water vapor, hydrogen, and oxygen would be a three-component system if conditions were such that the reaction $H_2 + \frac{1}{2}O_2 \longrightarrow H_2O$ did not actually occur. If, however, a suitable catalyst was present, or if the temperature was high enough for reaction, the system would become one with $c = 3 - 1 = 2$ components. If we imposed the condition that all the H_2 and O_2 come from dissociation of H_2O, we should have a system of only one component.

Careful examination of each individual system is necessary to decide the best choice of components. It is generally wise to choose as components those constituents that cannot be converted into one another by reactions occurring within the system. Thus, $CaCO_3$ and CaO would be a possible choice for the $CaCO_3 \rightleftharpoons CaO + CO_2$ system, but a poor choice because the concentration of CO_2 would have to be expressed by negative quantities. While the *identity* of the components is subject to some degree of choice, the *number* of components is always definitely fixed for any given case.

3. Degrees of Freedom

For the complete description of a system, the numerical values of certain variables must be reported. These variables are chosen from among the state functions of the system, such as pressure, temperature, volume, energy, entropy, and the concentrations of the various components in the different phases. Values for all the possible variables need not be explicitly stated, for a knowledge of some of them definitely determines the values of the others. For any complete description, however, at least one capacity factor is required, since otherwise the mass of the system is undetermined.

An important feature of equilibria between phases is that they are independent of the actual amounts of the phases that may be present.* Thus, the vapor pressure of water above liquid water in no way depends on the volume of the vessel or on whether a few milliliters or many liters of water are in equilibrium with the vapor phase. Similarly, the concentration of a saturated solution of salt in water is a fixed and definite quantity, regardless of whether a large or a small excess of undissolved salt is present.

---

*This statement is proved in the next section. It is true as long as variations in surface area are not considered (see Chapter 11).

In discussing phase equilibria, we therefore need not consider the capacity factors, which express the masses of the phases. We consider only the intensity factors, such as temperature, pressure, and concentrations. Of these variables, a certain number may be independently varied, but the rest are fixed by the values chosen for the independent variables and by the thermodynamic requirements for equilibrium. The number of the intensive state variables that can be independently varied without changing the number of phases is called the *number of degrees of freedom* of the system, or, sometimes, the *variance*.

For example, the state of a certain amount of a pure gas may be specified completely by any two of the variables, pressure, temperature, and density. If any two of these are known, the third can be calculated. Therefore, this system has two degrees of freedom; it is a *bivariant* system.

In the system water – water vapor, only one variable need be specified to determine the state. At any given temperature, the pressure of vapor in equilibrium with liquid water is fixed in value. This system has one degree of freedom, or is said to be *univariant*.

4. General Equilibrium Theory: The Chemical Potential

Under conditions of constant temperature and pressure, any change in the system proceeds from a state of higher Gibbs free energy G_1 to a state of lower Gibbs free energy G_2. For this reason, it became natural to think of the Gibbs function G as a thermodynamic potential, and to think of any change in a system as a passage from a state of higher to a state of lower potential. Of course, the choice of G as the potential function is due to the choice of the condition of constant T and P. At constant T and V, the suitable potential function would be A; at constant P and S, it would be H; etc.

If a system contains more than one component in a given phase, we cannot specify its state without some specification of the *composition* of that phase. In addition to P, V, and T, we need to introduce new variables to measure the amounts of the different chemical constituents in the system. As usual, the mole will be chosen as the chemical measure, with the symbols $n_1, n_2, n_3, \ldots, n_i$ representing the *number of moles* of component $1, 2, 3, \ldots, i$ in the particular phase we are considering.

It then follows that each thermodynamic function depends on these n_i's as well as on P, V, T. Thus, $U = U(V, T, n_i)$; $G = G(P, T, n_i)$; etc. Consequently, a complete differential—e.g., of the Gibbs function, becomes

$$dG = \left(\frac{\partial G}{\partial T}\right)_{P, n_i} dT + \left(\frac{\partial G}{\partial P}\right)_{T, n_i} dP + \sum_i \left(\frac{\partial G}{\partial n_i}\right)_{T, P, n_j} dn_i \qquad (6.1)$$

From (3.43), $dG = -S\,dT + V\,dP$ for any system of constant composition, i.e., when all the $dn_i = 0$. Therefore, Eq. (6.1) becomes

$$dG = -S\,dT + V\,dP + \sum \left(\frac{\partial G}{\partial n_i}\right)_{T, P, n_j} dn_i \qquad (6.2)$$

The coefficient $(\partial G/\partial n_i)_{T, P, n_j}$ was introduced by Gibbs, who called it the *chem-*

ical potential, and gave it the special symbol μ_i. Hence,

$$\mu_i = \left(\frac{\partial G}{\partial n_i}\right)_{T, P, n_j} \tag{6.3}$$

It is the change in the Gibbs free energy of the phase with a change in the number of moles of component i, the temperature, the pressure, and the number of moles of all other components being kept constant. The chemical potentials therefore measure how the Gibbs energy of a phase depends on any changes in its composition.

Equation (6.2) can now be written

$$dG = -S\,dT + V\,dP + \sum \mu_i\,dn_i \tag{6.4}$$

An equation like (6.4), which includes the variation of a thermodynamic function with the number of moles of the different components, is said to apply to an *open system*. We can change the amount of any component i in an open system by adding or removing dn_i of this component. At constant temperature and pressure, (6.4) becomes

$$dG = \sum \mu_i\,dn_i \quad \text{(constant } T, P) \tag{6.5}$$

An equation like this would apply to each phase of a system of several phases, and the transfers of mass dn_i might occur from one phase to another. If we consider the phase to be *closed*, so that no transfer of mass across its boundaries is allowed, (3.36) applies, and we have

$$\sum \mu_i\,dn_i = 0 \quad \text{(constant } T, P, \text{ closed phase)} \tag{6.6}$$

We might, however, consider the entire system of several phases to be closed. We then have the relation

$$\sum_i \mu_i^\alpha\,dn_i^\alpha + \sum_i \mu_i^\beta\,dn_i^\beta + \sum_i \mu_i^\gamma\,dn_i^\gamma + \cdots = 0 \tag{6.7}$$

where $\alpha, \beta, \gamma, \ldots$ represent the several phases. We can still transfer components across the phase boundaries in this system, but no mass can enter or leave the system as a whole.

We shall explore other aspects and applications of the chemical potential in the next chapter on solutions. Now we shall use the new function μ to follow the derivation given by Gibbs of the *phase rule*, the fundamental equation governing equilibria between phases.

5. Conditions for Equilibrium
Between Phases

In a system containing several phases, certain thermodynamic requirements for the existence of equilibrium may be derived.

For thermal equilibrium, it is necessary that the temperatures of all the phases be the same. Otherwise, heat would flow from one phase to another. This intuitively recognized condition may be proved by considering two phases α and β at temperatures T^α and T^β. The condition for equilibrium at constant volume and composition given in (3.28) is $\delta S = 0$. Let S^α and S^β be the entropies of the two phases, and suppose there is a virtual transfer of heat δq from α to β at equilibrium. Then,

$$\delta S = \delta S^\alpha + \delta S^\beta = 0 \quad \text{or} \quad -\frac{\delta q}{T^\alpha} + \frac{\delta q}{T^\beta} = 0$$

whence,

$$T^\alpha = T^\beta \tag{6.8}$$

For mechanical equilibrium it is necessary that the pressures of all the phases be the same. Otherwise, one phase would increase in volume at the expense of another. This condition may be derived from the equilibrium condition at constant overall volume and temperature, $\delta A = 0$. Suppose one phase expanded into another by δV. Then,

$$\delta A = P^\alpha \delta V - P^\beta \delta V = 0$$
$$P^\alpha = P^\beta \tag{6.9}$$

In addition to the conditions given by (6.8) and (6.9), a condition is needed that expresses the requirements of chemical equilibrium. Let us consider the system with phases α and β maintained at constant temperature and pressure, and denote by n_i^α, n_i^β, the amounts of some particular component i in the two phases. The equilibrium condition $\delta G = 0$ of (3.36) becomes

$$\delta G = \delta G^\alpha + \delta G^\beta = 0 \tag{6.10}$$

Suppose that a virtual process occurred by which δn_i moles of component i were taken from phase α and added to phase β. (This process might be a chemical reaction or a change in aggregation state.) Then, by virtue of (6.7), (6.10) becomes

$$\delta G = -\mu_i^\alpha \, \delta n_i + \mu_i^\beta \, \delta n_i = 0$$
$$\mu_i^\alpha = \mu_i^\beta \tag{6.11}$$

This is the general condition for equilibrium with respect to transport of matter between phases in a closed system, including chemical equilibrium between phases. For any component i in the system, the value of the chemical potential μ_i must be the same in every phase, when the system is in equilibrium at constant T and P.

The various equilibrium conditions are given in the following summary:

Capacity factor	Intensity factor	Equilibrium condition
S	T	$T^\alpha = T^\beta$
V	P	$P^\alpha = P^\beta$
n_i	μ_i	$\mu_i^\alpha = \mu_i^\beta$

6. The Phase Rule

Between 1875 and 1876, Josiah Willard Gibbs, Professor of Mathematical Physics at Yale University, published in the *Transactions of the Connecticut Academy of Sciences* a series of papers entitled "On the Equilibrium of Heterogeneous Substances." With brilliant beauty and precision, Gibbs disclosed in these papers the basic science of heterogeneous equilibrium.

The Gibbs phase rule provides a general relationship among the degrees of freedom of a system f, the number of phases p, and the number of components c. This relationship always is

$$f = c - p + 2 \tag{6.12}$$

The derivation proceeds as follows.

The number of degrees of freedom is equal to the number of intensive variables required to describe a system, minus the number that cannot be independently varied. The state of a system containing p phases and c components is specified at equilibrium if we specify the temperature, the pressure, and the amounts of each component in each phase. The number of variables required is therefore $pc + 2$.

Let n_i^α denote the number of moles of a component i in a phase α. Since the size of the system, or the actual amount of material in any phase, does not affect the equilibrium, we are really interested in the relative amounts of the components in the different phases and not in their absolute amounts. Therefore, instead of the mole numbers n_i^α, the mole fractions X_i^α should be used. These are given by

$$X_i^\alpha = \frac{n_i^\alpha}{\sum_i n_i^\alpha} \tag{6.13}$$

For each phase, the sum of the mole fractions equals unity.

$$X_1 + X_2 + X_3 + \cdots + X_c = 1$$

or

$$\sum X_i = 1 \tag{6.14}$$

If all but one mole fraction are specified, that one can be calculated from (6.14). If there are p phases, there are p equations similar to (6.14) and therefore p mole fractions that need not be specified since they can be calculated. The total number of independent variables to be specified is thus $pc + 2 - p$ or $p(c - 1) + 2$.

At equilibrium, the conditions given by (6.11) impose a set of further restraints on the system by requiring that the chemical potentials of each component be the same in every phase. These conditions are expressed by a set of equations, such as

$$\mu_1^\alpha = \mu_1^\beta = \mu_1^\gamma = \cdots$$
$$\mu_2^\alpha = \mu_2^\beta = \mu_2^\gamma = \cdots$$
$$\begin{array}{ccc} \cdot & \cdot & \cdot \\ \cdot & \cdot & \cdot \\ \cdot & \cdot & \cdot \end{array} \tag{6.15}$$
$$\mu_c^\alpha = \mu_c^\beta = \mu_c^\gamma = \cdots$$

Each equality sign in this set of equations signifies a condition imposed on the system, decreasing its variance by one. Inspection shows that there are therefore $c(p - 1)$ of these conditions.

The degrees of freedom equal the total required variables minus the restraining conditions. Therefore,

$$f = p(c - 1) + 2 - c(p - 1)$$
$$f = c - p + 2 \tag{6.16}$$

7. Phase Diagram for One Component

From the phase rule, when $c = 1, f = 3 - p$, and three different cases are possible:

$$p = 1, \quad f = 2, \quad \text{bivariant system}$$
$$p = 2, \quad f = 1, \quad \text{univariant system}$$
$$p = 3, \quad f = 0, \quad \text{invariant system}$$

Since the maximum number of degrees of freedom is two, the equilibrium conditions for a one-component system can be represented by a phase diagram in two dimensions, the most convenient choice of variables being P and T. If we wish, however, also to display the volume changes in the system, we can construct a model in three dimensions of the complete PVT surface. Every point on this surface denotes a set of equilibrium values P, V, T for the substance. We usually plot the volume per mole (V_m) or per gram (v).

Figure 6.1 shows such a PVT surface for carbon dioxide, a substance that contracts on freezing. In the case of a substance that expands on freezing, like water, the solid–liquid surface would slope in the opposite direction.

Let us follow an isotherm on the surface, by increasing P at constant T. We begin at point a, which corresponds to CO_2 gas at $P = 1.00$ atm, $T = 293.15$ K, and $V_m = 24570$ cm^3. As we increase pressure, volume decreases along line ab until at point b liquid CO_2 begins to form. The pressure at this point is 56.3 atm and the molar volume of the vapor in equilibrium with liquid is $V_m = 230.4$ cm^3. The volume of the liquid is given by the point c as $V_m = 56.5$ cm^3. The line bc is called a *tie-line*, since it connects points representing phases in equilibrium with each other. On our 293.15 K isotherm, the pressure remains constant at 56.3 atm until the vapor is completely converted to liquid at c. Any point between b and c represents a two-phase region of coexistence of liquid and vapor; depending upon the relative amounts of liquid and vapor, the volume can have any value between that for pure vapor at b and that for pure liquid at c. Increase in the pressure beyond c is applied to the pure liquid phase, which has a low compressibility, so that the isotherm rises steeply until it intercepts the melting point curve at d, which is very close to 4950 atm at 293.15 K. The densities of liquid and solid CO_2 at this pressure have not been directly measured, although Bridgman determined the decrease in volume on freezing, ΔV_m from d to e on diagram, to be 3.94 cm$^3 \cdot$ mol^{-1}. From a rather wild extrapolation of the data of Holser and Kennedy,* we can estimate V_m of liquid CO_2 at d to be about 35 cm^3, so that the V_m of the solid CO_2 at e would be about 31 cm$^3 \cdot$ mol^{-1}. The isotherm continues with further compression of solid CO_2 along the line ef and beyond.

Figure 6.1 also shows the projections of the PVT surface on the PT and PV planes. The PT projection is the one usually used as a "phase rule diagram." On the PT diagram, states with two phases in equilibrium are represented by lines. On

*S. P. Clark, ed., *Handbook of Physical Constants*, Memoir 97 (Geological Society of America, 1966), p. 371.

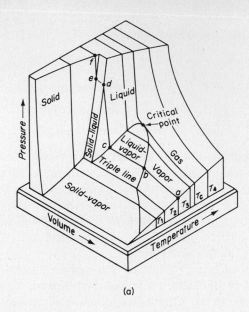

(a)

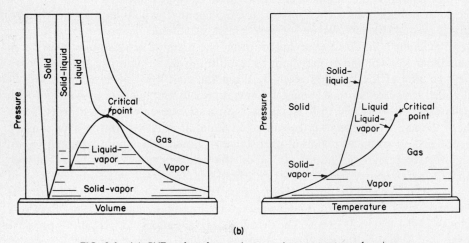

(b)

FIG. 6.1   (a) *PVT* surface for a substance that contracts on freezing.
(b) Projections of the surface on the *PT* and *PV* planes.
[After F. W. Sears, *An Introduction to Thermodynamics,
The Kinetic Theory of Gases and Statistical Mechanics*
(Reading, Mass.: Addison-Wesley), 1950.]

VT or *VP* diagrams, however, such states are represented by areas of the plane, because *V* (unlike *P*) is not the same for the two phases in equilibrium.

At sufficiently high pressures, solid CO_2 can exist well above the critical temperature of the liquid $\rightleftharpoons$ vapor transition. There has been a lively debate as to whether a critical point ever exists for a solid–liquid transition. At present, the "nays" seem to have the better of it, on the basis of the argument that the solid-liquid transition requires a change in the symmetry of the structure of matter, so that continuity of states [Section 1.18] between a symmetrical crystal structure and an isotropic liquid would not be possible.

8. Thermodynamic Analysis of *PT* Diagram

To investigate the thermodynamic conditions for a change in phase, we take as the starting point the relation

$$\mu^\alpha = \mu^\beta$$

Suppose we have $\mu^\alpha(P_0, T_0)$ at a particular point on the *PT* curve (it may be for liquid–gas, liquid–solid, solid–gas, or solid–solid equilibrium; the thermodynamic theory is the same in all cases). The problem is to find $\mu^\alpha(P, T)$ at a closely neighboring point.

Let us therefore expand $\mu^\alpha(T, P)$ about $\mu^\alpha(P_0, T_0)$ by Taylor's theorem:

$$\mu^\alpha(T, P) = \mu^\alpha(T_0, P_0) + dP\left(\frac{\partial \mu^\alpha}{\partial P}\right)_T + dT\left(\frac{\partial \mu^\alpha}{\partial T}\right)_P + \cdots \tag{6.17}$$

Similarly, we expand $\mu^\beta(T, P)$. We know that $\mu^\alpha(T_0, P_0) = \mu^\beta(T_0, P_0)$. We must determine the ratio of dP and dT so that

$$\mu^\alpha(P_0 + dP, T_0 + dT) = \mu^\beta(T_0 + dT, P_0 + dP)$$

From the first two terms in the expansion (6.17),

$$\frac{dP}{dT} = \frac{-\left(\dfrac{\partial \mu^\alpha}{\partial T} - \dfrac{\partial \mu^\beta}{\partial T}\right)}{\left(\dfrac{\partial \mu^\alpha}{\partial P} - \dfrac{\partial \mu^\beta}{\partial P}\right)} \tag{6.18}$$

But $\partial\mu^\alpha/\partial T = -S^\alpha$, $\partial\mu^\alpha/\partial P = V^\alpha$, etc. Therefore, (6.18) becomes

$$\frac{dP}{dT} = \frac{S^\alpha - S^\beta}{V^\alpha - V^\beta} = \frac{\Delta S}{\Delta V} \tag{6.19}$$

where ΔS is the the entropy change and ΔV the volume change for the phase transition. This is the famous Clapeyron–Clausius Equation. First proposed by the French engineer Clapeyron in 1834, it was placed on a firm thermodynamic basis by Clausius, some 30 years later.

If the latent heat of the phase transformation is ΔH_t, ΔS_t is simply $\Delta H_t/T$ where T is the temperature at which the phase change is occurring. Equation (6.19) thus becomes

$$\frac{dP}{dT} = \frac{\Delta H_t}{T \, \Delta V_t} \tag{6.20}$$

This equation applies to any change of state (fusion, vaporization, sublimation, and changes between crystalline forms), provided the appropriate latent heat is employed.

To integrate Eq. (6.20) exactly, it would be necessary to know both ΔH_t and ΔV_t as functions of temperature and pressure.* The variations of ΔV_t would be

---

*A good discussion of the temperature variation of ΔH_v is given by E.A. Guggenheim in *Modern Thermodynamics* (London: Methuen & Co., Ltd., 1933), p. 57. ΔH_v varies considerably with *T*, but its variations with *P* can be neglected for moderate pressures.

equivalent to data on the densities of the two phases over the desired ranges of T and P. In most calculations over short temperature ranges, however, both ΔH_t and ΔV_t may be taken as constant.

As an example, let us estimate the melting point of ice under a pressure of 400 atm. The densities of ice and water at 273.15 K and 1 atm pressure are $\rho_I = 0.917$ g$\cdot$cm^{-3} and $\rho_W = 0.9998$ g$\cdot$cm^{-3}, respectively. The latent heat of fusion, $\Delta H_f/M = 333.5$ J$\cdot$g$^{-1} = 3291$ cm$^3\cdot$atm$^{-1}\cdot$g^{-1}. If we assume that ΔH_f and the ρ's are practically constant over the pressure range, with $V = M/\rho$, (6.20) gives*

$$\frac{\Delta T}{\Delta P} = \frac{MT(\rho_W^{-1} - \rho_I^{-1})}{\Delta H_f}$$

$$\frac{\Delta T}{T} = (400)\frac{(1.0002 - 1.0905)}{3291}$$

$$\frac{\Delta T}{T} = -1.128 \times 10^{-3}$$

with $T = 273$, $\Delta T = -3.08°$, so that the melting point would be 270.07 K.

For the change liquid $\longrightarrow$ vapor, (6.20) becomes

$$\frac{dP}{dT} = \frac{\Delta H_v}{T_b(V_g - V_l)} \tag{6.21}$$

Several good approximations can be made in this equation. If we neglect the volume of the liquid compared to that of the vapor, and assume ideal-gas behavior for the latter, $V_g = nRT/P$, and (6.21) becomes

$$\frac{d \ln P}{dT} = \frac{\Delta H_v}{nRT^2} \tag{6.22}$$

A similar equation would be a good approximation for the sublimation curve.

As we showed for (3.54), we may also write (6.22) as

$$\frac{d \ln P}{d(1/T)} = -\frac{\Delta H_v}{nR} \tag{6.23}$$

If the logarithm of the vapor pressure is plotted against $1/T$, the slope of the curve at any point multiplied by $-R$ yields a value for the heat of vaporization per mole. In many cases, since ΔH_v is effectively constant over short temperature ranges, the curve is a straight line. This fact is useful to remember in extrapolating data on vapor pressures.

Over any extended range of temperature, the latent heat of vaporization varies considerably. It must decrease with increasing temperature and approach zero at the critical point. Figure 6.2 shows how the latent heat of vaporization of water varies with temperature.

When ΔH_v can be taken as constant, however, the integrated form of (6.23) is

$$\ln \frac{P_2}{P_1} = \frac{\Delta H_v}{nR}\left(\frac{1}{T_1} - \frac{1}{T_2}\right) \tag{6.24}$$

*We have approximated $\ln [(T - \Delta T)/T]$ by $-\Delta T/T$

$$\ln \left(1 - \frac{\Delta T}{T}\right) = -\frac{\Delta T}{T} - \frac{1}{2}\left(\frac{\Delta T}{T^2}\right) - \cdots$$

so that the error in approximation is about 0.5%.

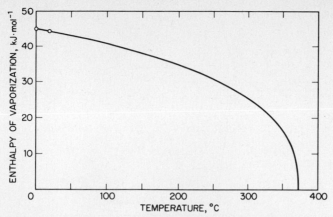

FIG. 6.2   The heat of vaporization of water as a function of temperature.

An approximate value for ΔH_v can often be obtained from *Trouton's rule* (1884):

$$\frac{\Delta H_v}{nT_b} \approx 92 \text{ J} \cdot \text{K}^{-1} \cdot \text{mol}^{-1}$$

The rule is followed fairly well by many nonpolar liquids. It is equivalent to the statement that the entropy of vaporization per mole is approximately the same for all such liquids.

The usual change of state (solid to liquid, liquid to vapor, etc.) is called a *first-order transition*. At the transition temperature T_t at constant pressure, the Gibbs energies G of the two forms are equal, but there is a discontinuous change in the slope of the G vs. T curve for the substance. Since $(\partial G/\partial T)_P = -S$, there is therefore a break in the S vs. T curve, the value of ΔS at T_t being related to the observed latent heat for the transition by $\Delta S = \Delta H_t/T_t$. There is also a discontinuous change ΔV in volume, since the densities of the two forms are not the same.

A number of transitions have been studied in which no latent heat or density change can be detected. Examples are the transformations of certain metals from ferromagnetic to paramagnetic solids at their Curie points, and the transitions of some metals at low temperatures to a condition of electric superconductivity. In these cases, there is an abrupt change in slope, but no discontinuity, in the S vs. T curve at T_t. As a result, there is a discontinuity ΔC_P in the heat capacity curve, since $C_P = T(\partial S/\partial T)_P$. A change of state of this kind is called a *second-order transition*.

9.   The Helium System

The properties of helium at low temperatures are sometimes startling. All other substances become solids at sufficiently low temperatures under their own vapor pressures. In the case of helium even in the limit as $T \longrightarrow 0$, the solid phase does not form unless an elevated pressure is applied.

Helium exists in two stable isotopes, ^{4}He and ^{3}He, the latter having an abundance of only about 1 part in 10^6 in atmospheric helium. The phase diagram of

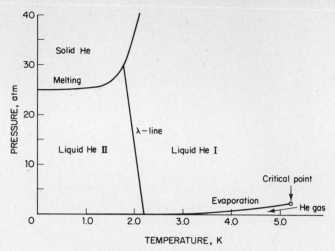

FIG. 6.3   Phase diagram of Helium-4.

⁴He is shown in Fig. 6.3. As the temperature is reduced along an isobar at 1 atm, a transition occurs somewhat above 2 K from ordinary liquid He-I to a second liquid phase, liquid He-II. This is the only known system in which two liquid phases coexist for the same substance.* The transition curve between the two liquids is called the λ curve. Liquid He-II behaves *as if* it were composed of two fluids that mix freely with each other without any viscosity at all between them. One of the two fluids is called the *normal fluid*, with density ρ_n, and the other, the *superfluid*, with density ρ_s. The density of liquid He-II is thus $\rho = \rho_n + \rho_s$. The value of ρ_n increases from 0 at 0 K to ρ at the λ curve, whereas ρ_s increases from 0 at the λ curve to ρ at 0 K. The superfluid has a viscosity $\eta = 0$. Since liquid He-II is composed entirely of ordinary atoms of ⁴He, we cannot believe that it contains two physically different fluids, but its properties are represented mathematically by such a model.

The transition liquid He-II $\rightleftharpoons$ liquid He-I does not behave like an ordinary first-order phase transition, with a latent heat ΔH_t and change in volume ΔV_t. If we plot the heat capacity C_V vs. T on both sides of the transition, we find, as shown in Fig. 6.4, that there is a singularity at the λ point, at which $C_V \rightarrow \infty$. The shape of the resultant curve resembles a Greek λ, and this is the origin of the name *lambda transition*.

---

*We should mention the mysterious *ortho-water* of Deryagin and others, reported to have a density about 1.5 times that of ordinary water. [See "Polywater" by E. R. Lippincott, *et al.*, *Science, 164*, 1482 (1969).] At present, such a form of water seems like one of the *hronir* of Borges' story "Tlon, Uqbar, Orbis Tertius." In 1963, Kurt Vonnegut published a noteworthy science fiction novel *Cats Cradle* based on a new form of water, unfortunately stable with respect to ordinary water. The destruction of all life on earth was the consequence of this fictional phase change. "Ice-nine was the last gift Felix Hoenikker created for mankind before going to his just reward. ... He had made a chip of ice-nine. It was blue white. It had a melting point of one-hundred-fourteen-point-four-degrees Fahrenheit."

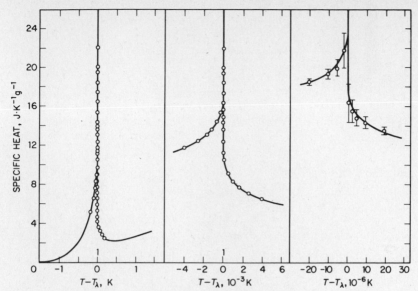

FIG. 6.4 The specific heat of liquid ^{4}He under the saturated vapor
pressure as a function of $T - T_\lambda$. The width of the small
vertical line just above the origin indicates the portion of the
diagram shown expanded (in width) in the curve directly
to the right. [After M. J. Buckingham and W. M. Fairbank,
Prog. Low Temp. Phys., 3, 80 (1961).]

According to the classification given by Ehrenfest,* a second-order transition
would have a finite break in the C_V-T curve. In view of the logarithmic singularity
at the λ point, the transition in liquid helium cannot be classified as a second-order
transition, and indeed seems to escape the Ehrenfest classification altogether.

10. Vapor Pressure and External Pressure

If the external pressure on a liquid is increased, its vapor pressure is raised. In
crude terms, molecules are squeezed out of the liquid into the vapor phase. An
idealized system is shown in Fig. 6.5 in which external pressure can be applied to
a liquid through a piston that is permeable to vapor but not to liquid.

The treatment of this situation was originally given by Gibbs. We refer again
to (6.17) and the equilibrium condition $\mu^g = \mu^l$. At constant temperature, only the
second term in (6.17) need be considered, to give

$$dP^g \left(\frac{\partial \mu^g}{\partial P}\right)_T = dP^l \left(\frac{\partial \mu^l}{\partial P}\right)_T \tag{6.25}$$

From (3.49), for a pure substance, $(\partial \mu / \partial P)_T = V_m$, the molar volume. Hence,

*A good discussion is given by M. Zemansky, *Heat and Thermodynamics,* 5th ed. (New York:
McGraw-Hill Book Company, 1968), p. 377.

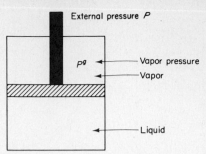

External pressure P

p^g ← Vapor pressure

← Vapor

Liquid

FIG. 6.5 An idealized arrangement to apply pressure to a liquid through a piston permeable to the vapor.

(6.25) yields the *Gibbs equation,*

$$V_m^g \, dP^g = V_m^l \, dP^l$$

or

$$\frac{dP^g}{dP^l} = \frac{V_m^l}{V_m^g} \tag{6.26}$$

If the vapor behaves as an ideal gas, $V_m^g = RT/P^g$, and (6.26) becomes

$$\frac{d \ln P^g}{dP^l} = \frac{V_m^l}{RT} \tag{6.27}$$

Since the molar volume of the liquid does not vary greatly with pressure, we may assume a constant V_m^l in integrating (6.27) to obtain

$$\ln \frac{P_1^g}{P_2^g} = \frac{V_m^l \, (P_1^l - P_2^l)}{RT} \tag{6.28}$$

In theory, we can measure the vapor pressure of a liquid under an applied hydrostatic pressure in two ways: (1) with an atmosphere of "inert" gas; (2) with an ideal membrane semipermeable to the vapor. In practice, the inert gas may dissolve in the liquid, in which case the application of the Gibbs equation to the problem is of doubtful validity.

As an example of the use of (6.28), let us estimate the vapor pressure of mercury under an external pressure of 1000 atm at 373.2 K. The density is 13.352 g·cm⁻³. Hence,

$$V_m^l = \frac{M}{\rho} = \frac{200.61}{13.352} = 15.025 \text{ cm}^3$$

and

$$\ln \frac{P_1^g}{P_2^g} = \frac{15.025(1000 - 1)}{82.05 \times 373.2} = 0.4902$$

Therefore, $P_1^g/P_2^g = 1.633$. The vapor pressure at 1 atm is 0.273 torr, so that the calculated vapor pressure at 1000 atm is 0.455 torr.

11. Statistical Theory of Phase Changes

One of the most difficult yet most fascinating problems facing theoretical chemistry today is how to make a quantitative theory for changes of state. The thermodynamic relations applicable to phase changes are perfectly clear, but they provide no explanation of the numerical values of melting points, boiling points, critical constants, latent heats, and other such quantities, in terms of interactions between molecules and the statistical properties of their ensembles. In the most general terms, in a phase change at constant pressure, a system passes from a low temperature state characterized by low enthalpy H and entropy S to a high temperature state of higher H and S. Thus,

$$\text{low } H \longrightarrow \text{high } H, \qquad \Delta H_t$$

$$\text{low } S \longrightarrow \text{high } S, \qquad \Delta S_t = \frac{\Delta H_t}{T}$$

At equilibrium at constant T and P,

$$\Delta G = 0 = \Delta H - T\,\Delta S$$

so that

$$\Delta S_t = \frac{\Delta H_t}{T_t}$$

The system "pays for" the transition into a state of higher energy by means of the greater randomness and disorder that correspond to the higher entropy. At equilibrium, the two driving forces are compensated exactly, so that, for phases α and β, $G_m^\alpha = G_m^\beta$, where G_m is the Gibbs free energy per mole.

The behavior of the free energy functions at a typical first-order phase transition, the melting of a crystal, is shown in Fig. 6.6. The sharp transition point is marked by a sharp discontinuity in the slope of the G vs. P curve. In mathematical terms, there is a *singularity* in the function $G(P)$ at the point of the phase transition.

The problem of treating phase changes by statistical thermodynamics can be divided into two steps. The first step is to establish the potential energy function U for the interaction between the molecules, atoms, or ions that comprise the system. In principle, these potential energies can be calculated by quantum mechanics, but such calculation is usually too difficult, and the potential functions are often based on empirical data. Examples are the Lennard–Jones potentials mentioned in connection with nonideal gases. The Lennard–Jones type of potential

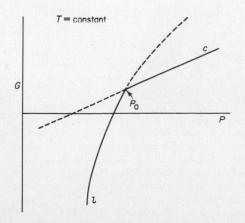

FIG. 6.6 The Gibbs energy G of a pure substance as a function of pressure (constant T), showing a crystal $\rightleftharpoons$ liquid transition. The solid lines represent the G of the phase that is stable, and the dashed lines represent the extrapolated values of G for the phase that is metastable. The point P_0 is the pressure of the melting point of the crystal at specified temperature T.

gives a simple interaction between pairs of particles, so that $U = U(r_{ij})$ where r_{ij} is the distance between the ith and the jth particles.*

The second step in the theoretical study of phase changes is to calculate a partition function for the system of interacting particles and to see whether the function correctly predicts a phase change. An example of this approach was the work of Joseph Mayer on the theory of condensation. He was able to show that if the expression for the intermolecular potential energy contained both an attractive term and a repulsive term, then the Gibbs free energy for the system, $G(P, T)$ would necessarily have a singularity below a certain temperature, which could be identified with the critical temperature T_c.

Even the simple van der Waals equation can be used to indicate the change of state *vapor* ⇌ *liquid*. We recall that the equation displays a maximum and a minimum in the two-phase region, as shown in Fig. 6.7. We might look upon these loops in the van der Waals curve as its desperate effort to represent a phase transition, but the sections of the curve where $(\partial P/\partial V)_T > 0$ (pressure increasing on expansion) correspond to no physical reality. The van der Waals model is far too simple to yield the sharp discontinuities of a phase transition.

Nevertheless, we can find the predicted end points of the two-phase region from the condition that the shaded area above the tie-line (AB) must equal that below

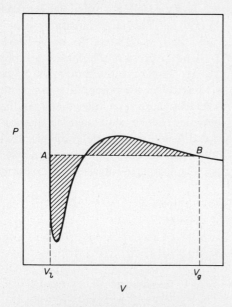

FIG. 6.7 Phase transition as calculated from van der Waals equation. The isotherm shown is actually that for CO_2 at 253 K. The points A and B are fixed by the condition that the areas of the two shaded loops must be equal.

*Such pairwise interactions are only a first approximation, however, to the interactions that occur between particles in dense media. The interaction energy between i and j is then modified by the presence of other particles k, l, etc. so that the potentials become complicated functions of various intermolecular distances. Essentially one has a *many-body problem,* with all the mathematical difficulties that this implies. A difficulty that becomes especially important in solid state theory is that the energy of the system cannot always be expressed in terms of interaction energies between definite particles. In many cases, *collective interactions* must be considered, which pertain to the system as a whole. In the theory of metals, e.g., the electrons act collectively, as a sort of negatively charged plasma, and not as individual particles.

it. This condition arises simply from the equality of the Gibbs free energies for liquid and vapor states in equilibrium, $G^g = G^l$, so that

$$\Delta G = G^g - G^l = 0$$

But

$$\Delta G = \int_l^g V \, dP$$

From the rule for integration by parts, if P_e is the liquid–vapor equilibrium pressure,

$$\int_l^g V \, dP = P_e(V^g - V^l) - \int_l^g P \, dV = 0$$

Inspection of Fig. 6.7 shows that this expression is just the difference between the two shaded areas under the loops in the van der Waals curve.

Since $A = -kT \ln Z$ and $G = A + PV$, the singularity in $G(T)$ at the transition point corresponds to a singularity in the partition function Z of (5.36). The mathematical difficulty of the phase transition reverts to the question as to how an apparently well behaved function like Z can suddenly lead to a singularity. It can, in fact, be shown that if Z were simply the sum over a finite number $\mathcal{N}$ of exponential terms, $\sum e^{-E_i/kT}$, it could never lead to such a singularity. This important theorem was proved by van Hove.* It is only when we go to the limit, $\mathcal{N} \rightarrow \infty$, in our statistical mechanical formulation of thermodynamic functions, that the mathematical possibility of a phase transition arises. Thus, from (5.42) the Helmholtz free energy is obtained from

$$A = \lim_{\mathcal{N} \to \infty} [-kT \ln \sum_{\substack{\text{over} \\ \mathcal{N}}} e^{-E_i/kT}]$$

A phase transition is a *cooperative phenomenon*. If a small region of a crystal melts, the molten region of disorder spreads like wildfire through the entire crystal, so that even the part of the system farthest away from any given small region contributes to the thermodynamic state of the system in that region. The disorder (or order in the reverse change) changes suddenly and sharply since it is a cooperative property of the system as a whole.

The Mayer approach to condensation was by way of the theory of imperfect gases. Another way to attack the problem of phase changes has been to start with a model of the solid state, in which the molecules are arranged on a regular lattice (see Section 18.4). Such lattice models have been applied to liquids also and even to gases. Some statistical mechanicians, resolved to find a mathematical model, have even devoted considerable attention to the model of the "one-dimensional lattice gas." The basic model for all these lattice theories was proposed originally by E. Ising in 1925† to treat ferromagnetism. In two dimensions it is shown in Fig. 6.8. The arrows in (b) represent electron spins, which can have either one of two orientations. Given a law of interaction between the little magnets associated with the spins, the Ising problem was to derive the net magnetization of the system as

---

*See G. E. Uhlenbeck, "Remarks on the Condensation Problem," in *Lectures in Statistical Mechanics* (Providence, R. I.: American Mathematical Society, 1963).

†*Z. Physik, 31*, 253 (1925).

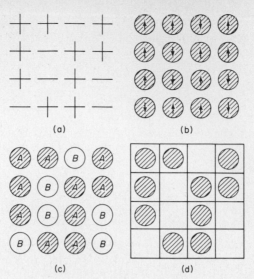

FIG. 6.8 A configuration of the Ising model, (a), may represent: (b) An arrangement of spins; (c) An arrangement of atoms in a binary alloy; (d) A configuration of a 'lattice gas.' [After J. M. Ziman, *Principles of the Theory of Solids* (London: Cambridge University Press, 1964).]

a function of T. Here, again, we see that there is a combinatorial problem in calculating the number of configurations associated with each set of spin states. The entropy of disordering the spins competes with the attractive energy of ordered spins. At some temperature T, there should be a transition from ordered (magnetized) to disordered (demagnetized) state.

We shall give in Chapter 11 a solution to the one-dimensional Ising problem, where it is applied to find the adsorption isotherm of a gas on lattice sites on a solid surface. The one-dimensional problem, however, does not lead to a phase change, since no long-range ordering is possible. The phenomenon of a phase change first appears in the two-dimensional problem. The mathematical solution to this problem was given by Lars Onsager in 1944, for the case $X_A = X_B = \frac{1}{2}$ (equal numbers of $+$ and $-$ spins). Despite intensive efforts by theoretical chemists and mathematicians, the three-dimensional problem has not yet been solved. If and when a solution is obtained, it will very likely mark a major breakthrough toward a mathematical theory of phase changes.

12. Solid–Solid Transformations—
The Sulfur System

Solid sulfur occurs in two different crystalline forms, a low temperature orthorhombic form and a high temperature monoclinic form. The phase diagram is shown in Fig. 6.9. The pressure scale has been made logarithmic to bring the interesting low pressure regions into prominence.

Monoclinic sulfur melts under its own vapor pressure of 0.025 torr at 393.2 K, the point E on the diagram. From E to the critical point F there extends the vapor pressure curve of liquid sulfur EF. Also from E, there extends the curve ED, the melting point curve of monoclinic sulfur. The density of liquid sulfur is less than that of the monoclinic solid, the usual situation in a solid–liquid transformation, and hence ED slopes to the right as shown. The point E is a triple point, at which three phases coexist, $S_m - S_{liq} - S_{vap}$.

The curve AB is the vapor pressure curve of solid orthorhombic sulfur. At point B, it intersects the vapor pressure curve of monoclinic sulfur BE, and also the transformation curve for orthorhombic $\rightleftharpoons$ monoclinic sulfur, BD. This intersection determines the triple point B, at which orthorhombic sulfur, monoclinic sulfur, and sulfur vapor coexist. Since there are three phases and one component, $f = c - p + 2 = 3 - 3 = 0$, and point B is an invariant point. It occurs at 0.01 torr pressure and 368.7 K.

The density of monoclinic sulfur is less than that of rhombic sulfur, and therefore the transition temperature $(S_o \rightarrow S_m)$ increases with increasing pressure.

The slope of ED is greater than that of BD, so that these curves intersect at D, forming a third triple point on the diagram, $S_o - S_m - S_{liq}$. This occurs at 428

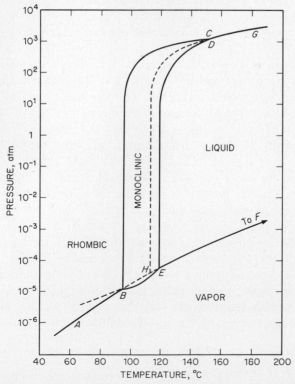

FIG. 6.9 The sulfur system on a PT diagram. Note the logarithmic pressure scale.

K and 1290 atm. At pressures higher than this, S_o is again the stable solid form, and DG is the melting point curve of S_o in this region of high pressure. The range of stable existence of S_m is confined to the totally enclosed area BED.

In addition to the stable equilibria represented by solid lines, a number of metastable equilibria are easily observed. If orthorhombic sulfur is heated quite rapidly, it will pass the transition point B without change and finally melt to liquid sulfur at 387 K (point H). The curve EH is the metastable vapor pressure curve of supercooled liquid sulfur. Extending from H to D is the metastable melting point curve of orthorhombic sulfur. Point H is a metastable triple point, $S_o - S_{liq} - S_{vap}$. All these metastable equilibria are quite easily studied because of the sluggishness that characterizes the rate of attainment of equilibrium between the solid phases.

13. Measurements at High Pressures

It is only a truism that our attitude toward the physical world is conditioned by the scale of magnitudes provided in our terrestrial environment. We tend to classify pressures and temperatures as high or low by comparing them with the 10^5 $N \cdot m^{-2}$ and 293 K of a spring day in the laboratory, despite the fact that almost all matter in the universe exists under very different conditions. Even at the center of Earth, by no means a large astronomical body, the pressure is about 4000 kbar,* and substances would have properties quite unlike those with which we are familiar. At the center of a comparatively small star, like our sun, the pressure would be about 10^7 kbar.

The pioneer work of Gustav Tammann on high pressure measurements was extended by P. W. Bridgman and his coworkers at Harvard. Pressures up to 400 kbar were achieved, and methods were developed for measuring the properties of substances at 100 kbar.† The attainment of such pressures was made possible by the construction of pressure vessels of alloys such as Carboloy (tungsten carbide cemented with cobalt). In the multiple-chamber technique, the container for the substance to be studied is enclosed in another vessel, and pressure is applied both inside and outside the inner container, usually by means of hydraulic presses. Thus, although the absolute pressure in the inner vessel may be 100 kbar, the pressure differential that its walls must sustain is only 50 kbar.

The practical limit of any multiple-stage apparatus is soon reached, since, although the maximum pressure may increase linearly with the number of stages, the sheer bulk of the apparatus increases exponentially. Thus, most modern apparatus for ultrahigh pressures is based on the idea of using some of the mechanical force that produces the pressure also to support the apparatus. A major advantage of such support is based on the fact that carbide pistons fail under high pressures not through flow but by fracture. Pyrophyllite (*wonderstone*), a hydrated

---

*Workers in the field of high pressures now prefer to use the pressure unit *kilobar* (kbar) $= 10^8$ $N \cdot m^{-2} = 0.9869 \times 10^3$ atm.

†For details, see P. W. Bridgman, *The Physics of High Pressures* (London: Bell and Co., 1949), and his review article, *Rev. Mod. Phys., 18*, 1 (1946). Recent developments are surveyed in *High Pressure Physics and Chemistry*, ed. by R. S. Bradley (New York: Academic Press, Inc., 1963).

aluminum silicate, has been used as a gasketing material, because it transmits the applied pressure without too great frictional loss and is stable under most operating conditions.

A high pressure apparatus that provides considerable support to the inner piston is the *tetrahedral anvil*, shown in Fig. 6.10. As the pistons advance, some pyrophyllite is extruded between the anvils, and this effect helps to support the carbide faces at the region of greatest pressures. This apparatus, designed by Tracy Hall, was used for the first commercial production of synthetic diamonds.

The carbon system is shown in Fig. 6.11. Diamonds can be made by compressing graphite at an elevated temperature.* Without catalytic action, it was estimated that a pressure of 200 kbar at about 4000 K would be required to transform graphite into diamond. No available containers could withstand these conditions. By the use of metallic catalysts, such as tantalum and cobalt, a rapid transformation can be achieved at about 70 kbar and 2300 K.

FIG. 6.10 The tetrahedral anvil, used to achieve high pressures and temperatures.

*F. P. Bundy, *J. Chem. Phys.*, *38*, 631 (1963).

The highest laboratory pressures (up to about 2000 kbar) have been achieved by dynamic methods in which a shock wave produced by compressed gas or explosives is caused to travel through the specimen. High pressure, high velocity gases accelerate the material of the specimen, and a high pressure is produced in the specimen owing to the inertia of the material that has just been accelerated. Within a few microseconds (μs), the entire specimen reaches a high velocity and high pressure, and as the shock front passes through the material, the rarefaction wave spreads back, reducing the pressure again. The various motions are followed by high speed photography, and the data so obtained allow one to calculate the maximum pressure achieved. In 1961, B. J. Alder and R. M. Christian* found evidence for diamond formation in graphite that was shock loaded at high temperatures. Alder remarked, "We were millionaires for a microsecond."

Measurements on water at high pressures have yielded the results shown in the phase diagram of Fig. 6.12. The melting point of ordinary ice (ice I) falls on compression, until a value of $-22.0°C$ is reached at 2040 atm. Further increase in pressure results in the transformation of ice I into a new modification, ice III, whose melting point increases with pressure. Altogether six different polymorphic forms of ice have been found. There are five triple points shown on the water diagram. At a pressure of about 20 000 atm, liquid water freezes to ice VII at about 100°C. Ice IV is not shown. Its existence was indicated by the work of Tammann, but it was not confirmed by Bridgman.

Answers to major geochemical problems will require further data on the properties of minerals at high pressures. As icebergs float in the oceans, mountains float in a sort of sea of plastic rock that flows readily under pressure. The discontinuity between the lighter minerals of the crust and the underlying denser minerals is the famous Mohorovičić or M discontinuity. Under the continents, this is about 40 km below the surface, but under the deep ocean floor, it is only 7 to 10 km down.

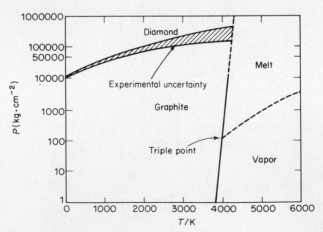

FIG. 6.11(a)   The phase diagram of carbon with logarithmic pressure scale.

*B. J. Alder and R. M. Christian, *Phys. Rev. Letters*, *4*, 450 (1961).

FIG. 6.11(b) Large diamond crystals can be grown from a molten
catalyst metal, such as nickel, at pressures around 60
kbar in the range from 1700–1800 K. The photograph
shows a freshly grown single crystal of diamond, half-
buried in the center of its catalyst metal growth medium,
now frozen. Surrounding the metal is its refractory
insulation. The carbon tube furnace surrounding the
insulation has been removed. The crystal is about 5 mm
in diameter (one carat). [R. H. Wentorf, General Electric
Research and Development Center, Schenectady, New
York. Details of the process were reported at the
Symposium on Rate Processes at High Pressure,
Toronto, May 1970.]

The M discontinuity was detected originally by a sudden increase in the velocity
of seismic waves. Early theories postulated a difference in chemical composition at
the M discontinuity, but the current hypothesis is that the discontinuity marks the
locus of phase transformations from low density crystalline forms of silicate minerals
like albite and calcium feldspars to high density forms like jadeite and garnet.
The density changes from about 2.95 to 3.50 $g \cdot cm^{-3}$, and the high density material
is, of course, the form stable at the higher pressures. The transformation pressure
is at 15 to 20 kbar, depending on temperature. According to this theory, the M
discontinuity is simply a natural expression of the PT transformation curve of these

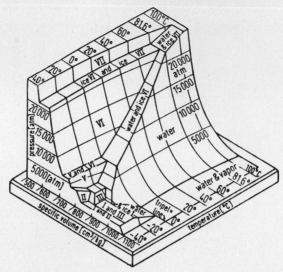

FIG. 6.12 *PVT* surface for H_2O. Adapted with changes from Zemansky (1968) by D. Eisenberg and W. Kauzmann, *The Structure and Properties of Water* (Oxford: Oxford University Press, 1969).

two classes of minerals, a sort of large-scale plot of a phase diagram. The process of mountain building would then occur whenever there was a temperature fluctuation beneath the surface causing a sudden fall in the stable level of the phase transformation and the formation of a large amount of new material of lower density. A temperature fluctuation of 10% at the *M* discontinuity would displace the equilibrium sufficiently to elevate peaks to the heights of the Rocky Mountains. But what could cause a fluctuation of the order of 10%? To be sure, the fluctuation is not sudden—it may take several million years. According to current theories, upward convection of hot rock in the mantle could cause the requisite fluctuations of temperature.

PROBLEMS

1. Naphthalene melts at 80°C. Its heat of fusion is 19.29 kJ·mol⁻¹ at 80°C and 19.20 kJ·mol⁻¹ at 70°C. The heat capacity of the liquid is 223 J·K⁻¹·mol⁻¹, and that of the solid is 214 J·K⁻¹·mol⁻¹. Calculate the entropy change of the naphthalene and of the surroundings when 1 mol of liquid naphthalene supercooled to 70°C freezes to solid at 80°C.

2. The normal melting point of mercury is −38.87°C. At this temperature, the specific volume of the liquid is 0.07324 cm³·g⁻¹ and that of the solid is 0.07014 cm³·g⁻¹. The heat of fusion is 11.63 J·g⁻¹. Assume these quantities are all independent of *T* and *P* and calculate (a) Δ*S* and Δ*G* when 1 g liquid mercury freezes at −38.87°C, and (b) the melting point of mercury under a pressure of 200 atm.

3. An equation for the temperature variation of the latent heat λ of a phase change along the equilibrium PT curve was derived by M. Planck* as

$$\frac{d\lambda}{dT} = \Delta C_P + \frac{\lambda}{T} - \lambda \left(\frac{\partial \ln \Delta V}{\partial T} \right)_P$$

   Derive the Planck equation, starting from

$$d\lambda = \left(\frac{\partial \lambda}{\partial T} \right)_P dT + \left(\frac{\partial \lambda}{\partial P} \right)_T dP$$

4. From the Planck equation in Problem 3, calculate $d\lambda/dT$ for water, given the coefficients of thermal expansion α of water and ice at 273 K as -6.0×10^{-5} and 11.0×10^{-5}, respectively. The latent heat of fusion at 273 K is $5.983 \text{ kJ} \cdot \text{mol}^{-1}$, and the specific heat capacities are $4.184 \text{ J} \cdot \text{K}^{-1} \cdot \text{g}^{-1}$ and $2.092 \text{ J} \cdot \text{K}^{-1} \cdot \text{g}^{-1}$, respectively. The specific volume of ice is $1.093 \text{ cm}^3 \cdot \text{g}^{-1}$.

5. Show that for a solid $\longrightarrow$ vapor or liquid $\longrightarrow$ vapor transition, the Planck equation of Problem 3 becomes, to a good approximation, $(d\lambda/dT) \approx \Delta C_P$. Would this approximation also be useful for solid $\longrightarrow$ liquid? Problems 3, 4, and 5 are discussed by K. Denbigh.†

6. Sketch roughly the phase diagram of acetic acid from these data: (a) the low pressure α form melts at 16.6°C under its own vapor pressure of 9.1 torr; (b) there is a high pressure β form that is denser than α, but both α and β are denser than the liquid; (c) the normal boiling point of liquid is 118°C; (d) phases α, β, and liquid are in equilibrium at 55°C and 2000 atm.

7. Pure wolfram melts at 3370°C. The liquid has a vapor pressure of 5.00 torr at 4337°C and of 60 torr at 5007°C. Calculate the normal boiling point of wolfram.

8. In an experimental determination of the latent heat of vaporization ΔH_v of a pure liquid from vapor pressure measurements over a range of temperatures and the Clapeyron–Clausius equation, a systematic error of 0.2°C is made in the temperature measurements. What error will this cause in the ΔH_v if its value is $40.0 \text{ kJ} \cdot \text{mol}^{-1}$?

9. When a metal is heated, it emits electrons (thermionic emission). If the difference in energy between a motionless electron in the gas and in the metal is ϕ, show that the partition function for the electron vapor is $Z_e = 2e^{-\phi/kT}(2\pi m_e kT/h^2)^{3/2} V$. What would be the number of electrons per unit volume, in equilibrium with metallic wolfram at 4000 K if $\phi = 4.25 \text{ eV}$?

10. The vapor pressure of solid iodine is 0.25 torr and its density is $4.93 \text{ g} \cdot \text{cm}^{-3}$ at 293 K. Assuming the Gibbs equation to hold, estimate the vapor pressure of iodine under a 1000-atm pressure of argon.

11. From the following data, sketch the phase diagram of nitrogen at low temperatures. There are three crystal forms, α, β, γ, which coexist at 4650 atm and 44.5 K. At this triple point, the volume changes ΔV in $\text{cm}^3 \cdot \text{mol}^{-1}$ are $\alpha \longrightarrow \gamma$, 0.165; $\beta \longrightarrow \gamma$, 0.208; $\beta \longrightarrow \alpha$, 0.043. At 1 atm and 36 K, $\beta \longrightarrow \alpha$ with $\Delta V = 0.22$. The ΔS values for the transitions cited are 1.25, 5.88, 4.59, and $6.52 \text{ J} \cdot \text{K}^{-1} \cdot \text{mol}^{-1}$, respectively.‡

Treatise on Thermodynamics, 3rd ed. (New York: Dover Publications, Inc., 1927), p. 154.

†*The Principles of Chemical Equilibrium* (London: Cambridge University Press, 1964). Highly recommended.

‡J. Swenson, *Chem. Phys.*, **23**, 1963 (1955).

12. Show that the vapor pressure of a solid is given by

$$\ln P = \ln \frac{kT}{V} + \ln \left(\frac{Z_g}{Z_s}e^{-\lambda/kT}\right)$$

where Z_g, Z_s are partition functions for gas and solid, and λ is the latent heat of vaporization per molecule. The partition function for solid copper at 1000 K is exp (5.20) (as derived in Chapter 18). Calculate the vapor pressure of copper at 1000 K.

13. The vapor pressures of liquid gallium are

T(K)	1302	1427	1623
P(torr)	0.01	0.10	1.00

Calculate $\Delta H^\ominus$, $\Delta G^\ominus$, and $\Delta S^\ominus$, for the vaporization of gallium at 1427 K.

14. Sketch graphs of G, S, V, and C_P, against T at constant P for typical first- and second-order phase transitions. Derive the Ehrenfest equation for a second-order transition,

$$\frac{dP}{dT} = \frac{\Delta\alpha}{\Delta\beta}$$

where α is the expansivity and β the compressibility.

15. The vapor pressure of water at 298 K is 23.76 torr. What is $\Delta G^\ominus_{298}$ for the change $H_2O(g) \longrightarrow H_2O(l)$?

16. At 298 K, the standard heat of combustion of diamond is 395.3 kJ·mol⁻¹, and that of graphite is 393.4 kJ·mol⁻¹. The molar entropies are 2.439 and 5.694 J·K⁻¹·mol⁻¹, respectively. Find $\Delta G^\ominus$ for the transition graphite $\longrightarrow$ diamond at 298 K and 1 atm. The densities are 3.513 g·cm⁻³ for diamond and 2.260 g·cm⁻³ for graphite. Assuming densities and ΔH_s are independent of pressure, calculate the pressures at which diamond and graphite would be in equilibrium at 298 and 1300 K.

17. The heat capacity of liquid zinc from 419.5 to 907°C can be represented by C_P (Zn, l) $= 7.09 + 1.15 \times 10^{-3}\, T$ in units of cal. K⁻¹·mol⁻¹. Zinc forms a monatomic gas with C_P(Zn, g) $= \frac{5}{2}R$. The normal boiling point of Zn is 907°C. Calculate the vapor pressure of Zn at 500°C, assuming it to give an ideal gas. What would be the result of the calculation if you neglected the variation of the heat of vaporization with T?

18. From the diagram in Fig. 6.12, estimate the triple point for ice VI–ice VII–liquid. What is the approximate density of liquid water at this triple point?

19. N_2 and H_2 are reacted over a catalyst at high pressure to yield a gas containing 13 mol % NH_3. The product circulates through a cooler and passes out at 30°C and 300 atm. The density of liquid NH_3 at 30°C is 0.595 g·cm⁻³, and its vapor pressure is 11.5 atm. What fraction of the NH_3 entering the cooler is condensed? Assume NH_3 behaves as ideal gas.

20. The vapor pressure of water is given by

$$\log_{10} P = A - \frac{2121}{T}$$

where A is a constant the value of which depends on units chosen for P. The ΔH_{vap} of water at 373 K is 2550 J·g⁻¹. (a) Ten grams of water are introduced into an evacuated vessel of constant volume, 10 dm³. At a temperature of 323 K, what will be the mass of liquid water? (b) The temperature is raised gradually. At what temperature will all the water be vaporized?

7

Solutions

In the winter of the year 1729, I exposed beer, wine, vinegar, and brine in large open vessels, to the frost; which, thus, congeal'd nearly all the water of these liquors into a soft, spongy kind of ice, and united the strong spirits of the fermented liquors; so that by piercing the ice, they might be pour'd off and separated from the water, which diluted them before freezing; and the more intense the frost is, the more perfect this separation: whence we see that cold unfits water to dissolve alcohol, and the salt of vinegar: and it is probable, that the utmost possible cold in nature would deprive water of all its dissolving power.

Hermann Boerhaave*
1732

A solution is any phase containing more than one component. The solution may be gas, liquid, or solid. Below the critical point gases that do not react chemically are miscible in all proportions. Fluids above the critical point may have a limited range of intersolubility. Liquids often dissolve a wide variety of gases, solids, or other liquids, and the composition of these liquid solutions can be varied over a wide or narrow range depending on the solubility relationships in the particular system. Solid solutions are formed when a gas, a liquid, or another solid dissolves in a solid. They are often characterized by very limited concentration ranges, although many pairs of solids are known (e.g., copper and nickel) that are mutually soluble in all proportions.

1. Measures of Composition

The *mole fraction* is the most convenient way to describe the composition of a solution in theoretical discussions. Suppose a solution contains n_A moles of component A, n_B moles of component B, n_C moles of component C, etc. Then the mole fraction X_A of component A is

$$X_A = \frac{n_A}{n_A + n_B + n_C + \cdots} \tag{7.1}$$

If there are only two components,

$$X_A = \frac{n_A}{n_A + n_B}$$

*Elements of Chemistry, 1732.

The molality m_B of a component B in a solution is defined as the amount of component B per unit mass of some other component chosen as the solvent. The S.I. unit of molality is mole per kilogram ($\text{mol} \cdot \text{kg}^{-1}$). One advantage of molality is that it is easy to prepare a solution of a given molality by accurate weighing procedures. The relation between molality m_B (in $\text{mol} \cdot \text{kg}^{-1}$) and mole fraction X_B for a solution of two components, in which the solvent A has molecular weight M_A, is

$$X_B = \frac{m_B}{(1000/M_A) + m_B}$$

or

$$X_B = \frac{m_B M_A}{1000 + m_B M_A} \tag{7.2}$$

Notice that in dilute solution, as $m_B M_A$ becomes much less than 1000, the mole fraction becomes proportional to the molality, $X_B \propto m_B M_A / 1000$.

The concentration c of a component B in a solution is the amount of the component in a unit volume of the solution, $c = n/V$. For some purposes, the number concentration C is convenient, the number of particles (atoms, ions, or molecules) in a unit volume of solution, $C = N/V$. For a solution of two components, in which M_A, M_B are the molecular weights of solvent and solute, respectively, and ρ is the density of the solution and c is in units of $\text{mol} \cdot \text{dm}^{-3}$,

$$X_B = \frac{c_B}{[(1000\rho - c_B M_B)/M_A] + c_B}$$

or

$$X_B = \frac{c_B M_A}{1000\rho + c_B(M_A - M_B)} \tag{7.3}$$

In dilute solutions, ρ approaches ρ_A, the density of pure solvent, and the term $c_B(M_A - M_B)$ becomes much less than $1000 \rho_A$, so that

$$X_B \approx \frac{c_B M_A}{1000 \rho_A}$$

The mole fraction then becomes proportional to concentration. In dilute aqueous solutions, $\rho \approx \rho_A \approx 1$, and concentration becomes approximately equal numerically to molality.

Since the density of a solution varies with the temperature, we see from (7.3) and (7.2) that the concentration c_B must also vary with temperature, whereas the molality m_B and mole fraction X_B are independent of temperature. The methods of describing the composition of solutions are summarized in Table 7.1.

2. Partial Molar Quantities:
Partial Molar Volume

The equilibrium properties of solutions are described in terms of state functions, such as P, T, V, U, S, G, H. The basic problem in the thermodynamic theory of solutions is how these functions depend on the composition of the solution.

TABLE 7.1
The Composition of Solutions

Name	Symbol	Definition	Usual SI unit
Molality	m	Amount of solute in unit mass of solvent	$\text{mol} \cdot \text{kg}^{-1}$
Concentration	c	Amount of solute in unit volume of solution	$\text{mol} \cdot \text{dm}^{-3}$
Volume molality	m'	Amount of solute in unit volume of solvent	$\text{mol} \cdot \text{dm}^{-3}$
[Mass] percent	$\%$	Mass of solute in 100 unit masses of solution	Dimensionless
Mole fraction	X_A	Amount of component A divided by total amount of all components	Dimensionless

Consider a solution containing n_A moles of A and n_B moles of B. Let us assume that the volume V of solution is so large that the addition of one extra mole of A or of B does not change the concentration to an appreciable extent. Now add 1 mol of A to this large amount of solution and measure the resulting increase in volume of solution at constant temperature and pressure. This increase in volume per mole of A is called the *partial molar volume* of A in the solution at the specified pressure, temperature, and composition. It is denoted by the symbol V_A.* It is the change of volume with moles of A, at constant temperature, pressure, and moles of B, and is therefore written as

$$V_A = \left(\frac{\partial V}{\partial n_A} \right)_{T, P, n_B} \tag{7.4}$$

One reason for introducing such a function is that the volume of a solution is not, in general, simply the sum of the volumes of the individual components. For example, if 100 cm³ of alcohol are mixed at 25°C with 100 cm³ of water, the volume is not 200 cm³ but about 190 cm³.

If dn_A moles of A and dn_B of B are added to a solution, the increase in volume at constant temperature and pressure, since $V = V(n_A, n_B)$, is given by

$$dV = \left(\frac{\partial V}{\partial n_A} \right)_{n_B} dn_A + \left(\frac{\partial V}{\partial n_B} \right)_{n_A} dn_B \tag{7.5}$$

or, from (7.4),

$$dV = V_A \, dn_A + V_B \, dn_B \tag{7.6}$$

This expression can be integrated, which corresponds physically to increasing the volume of the solution without changing its composition, V_A and V_B hence being constant.† The result is

$$V = n_A V_A + n_B V_B \tag{7.7}$$

---

*We have used the same symbol for the molar volume of pure A and the partial molar volume of A in solution. In fact, the partial molar volume becomes the molar volume in the case of a pure component. It is only rarely that any confusion can arise between these two quantities. If necessary, we shall denote the molar volume of pure A as $V_A^{\bullet}$.

This equation tells us that the volume of the solution equals the number of moles of A times the partial molar volume of A, plus the number of moles of B times the partial molar volume of B.

Differentiation of (7.7) gives

$$dV = V_A \, dn_A + n_A \, dV_A + V_B \, dn_B + n_B \, dV_B$$

By comparison with (7.6), we find

$$n_A \, dV_A + n_B \, dV_B = 0$$

or

$$dV_A = -\frac{n_B}{n_A} \, dV_B = \frac{X_B}{X_B - 1} \, dV_B \qquad (7.10)$$

Equation (7.10) is one example of the *Gibbs–Duhem equation*. This particular application is in terms of the partial molar volumes, but any other partial molar quantity can be substituted for the volume.

We can define partial molar quantities for any extensive state function. For example,

$$S_A = \left(\frac{\partial S}{\partial n_A}\right)_{T, P, n_B}, \qquad H_A = \left(\frac{\partial H}{\partial n_A}\right)_{T, P, n_B}, \qquad G_A = \left(\frac{\partial G}{\partial n_A}\right)_{T, P, n_B} \qquad (7.11)$$

The partial molar quantities are themselves intensity factors, since they are capacity factors per mole. The partial molar Gibbs free energy G_A is equivalent to the chemical potential μ_A.

All the thermodynamic relations derived in earlier chapters can be applied to the partial molar quantities. For example,

$$\left(\frac{\partial G_A}{\partial P}\right)_T = \left(\frac{\partial \mu_A}{\partial P}\right)_T = V_A; \qquad \left(\frac{\partial \mu_A}{\partial T}\right)_P = -S_A; \qquad \left(\frac{\partial H_A}{\partial T}\right)_P = C_{PA} \qquad (7.12)$$

---

†Mathematically, the integration is equivalent to the application of Euler's theorem to the homogeneous function $V(n_A, n_B)$.

A function $f(x, y, z)$ is called homogeneous of degree n if, when x, y, z are multiplied by any positive number k,

$$f(kx, ky, kz) = k^n f(x, y, z)$$

If we differentiate this equation with respect to k, we obtain

$$\frac{df}{dk} = k^{-1}\left(x\frac{\partial f}{\partial x} + y\frac{\partial f}{\partial y} + z\frac{\partial f}{\partial z}\right) = nk^{n-1} f$$

Note that

$$\frac{df}{dk} = \frac{\partial f}{\partial(kx)}\frac{d(kx)}{dk} + \frac{\partial f}{\partial(ky)}\frac{d(ky)}{dk} + \frac{\partial f}{\partial(kz)}\frac{d(kz)}{dk}$$

If we set $k = 1$,

$$x\frac{\partial f}{\partial x} + y\frac{\partial f}{\partial y} + z\frac{\partial f}{\partial z} = nf \qquad (7.8)$$

This is Euler's theorem on homogeneous functions. Now $V(n_A, n_B)$ is a homogeneous function of degree 1 of the variables n_A and n_B. (If all the n's are multiplied by k, V is multiplied by k also.) Hence, (7.8) gives

$$n_A\left(\frac{\partial V}{\partial n_A}\right) + n_B\left(\frac{\partial V}{\partial n_B}\right) = V \qquad (7.9)$$

The thermodynamic theory of solutions is expressed in terms of these partial molar functions just as the theory for pure substances is based on the ordinary thermodynamic functions.

Consider the formation of a binary solution from n_A moles of component A and n_B moles of component B:

$$n_A A + n_B B \longrightarrow \text{solution}$$

At given T and P, the ΔG for the solution process is

$$\Delta G = G \,(\text{solution}) - n_A G_A^\bullet - n_B G_B^\bullet$$

where $G_A^\bullet$, $G_B^\bullet$ denote the molar free energies of the pure components. By analogy with (7.7),

$$G \,(\text{solution}) = n_A G_A + n_B G_B$$

Thus,

$$\Delta G = n_A (G_A - G_A^\bullet) + n_B (G_B - G_B^\bullet) \tag{7.13}$$

(Note that $G_A \equiv \mu_A$, $G_B \equiv \mu_B$.)

Exactly similar equations can be written for the change of any extensive thermodynamic state variable, U, H, S, V, A, C_V, C_P, etc. We can thus define the *enthalpy of solution, entropy of solution*, etc. It may be convenient also to write the changes in the partial molar quantities on solution as ΔG_A, ΔG_B so that (7.13) becomes

$$\Delta G = n_A \,\Delta G_A + n_B \,\Delta G_B \tag{7.14}$$

[We have already seen such an equation for the enthalpy of solution in Eq. (2.37)].

3. Activities and Activity Coefficients

Instead of the chemical potential μ_A, it is often convenient to use a related function, the *absolute activity** λ_A, defined by

$$\mu_A = RT \ln \lambda_A \tag{7.15}$$

Relations written in terms of μ are easily expressed in terms of λ. For example, the condition in (6.11) for equilibrium of the component A between gas and liquid phases, would become

$$\lambda_A^g = \lambda_A^l \tag{7.16}$$

In dealing with solutions, we are often concerned with the difference between the value of μ_A in the solution and its value in some reference state. This difference can be written

$$\mu_A - \mu_A^\ominus = RT \ln \frac{\lambda_A}{\lambda_A^\ominus} = RT \ln a_A \tag{7.17}$$

The ratio of the absolute activity to the activity in some reference state defines a relative activity a_A. In dealing with nonelectrolyte solutions of *liquids*, a convenient reference state is the pure liquid at $P = 1$ atm and the temperature specified

---

*The word *absolute* does not imply any absolute choice of zero levels for energy or entropy. It is used simply to distinguish λ from the relative activity a.

for the solution. If we call the values in this reference state $\mu_A^\bullet$ and $\lambda_A^\bullet$, Eq. (7.17) becomes

$$\mu_A - \mu_A^\bullet = RT \ln \frac{\lambda_A}{\lambda_A^\bullet} = RT \ln a_A$$

with

$$a_A = \frac{\lambda_A}{\lambda_A^\bullet} \tag{7.18}$$

The relative activity defined in this way is usually called, simply, the *activity*.

The ratio of the activity a_A to the mole fraction X_A is called the *activity coefficient* γ_A:

$$a_A = \gamma_A X_A \tag{7.19}$$

4. Determination of Partial Molar Quantities

The evaluation of the partial quantities will now be described, using the partial molar volume as an example. The methods for H_A, S_A, G_A, etc., are exactly similar.

The partial molar volume V_A, defined by (7.4), is equal to the slope of the curve obtained when the volume of the solution is plotted against the molality m_A of A. This follows since m_A is the number of moles of A in a constant quantity, namely 1 kg, of component B.

The determination of partial molar volumes by this *slope method* is rather inaccurate; the *method of intercepts* is therefore usually preferred. To employ this method, a quantity is defined, called the *mean molar volume of the solution* V_m, which is the volume of the solution divided by the total number of moles of the various constituents. For a two-component solution,

$$V_m = \frac{V}{n_A + n_B}$$

Then,

$$V = V_m (n_A + n_B)$$

and

$$V_A = \left(\frac{\partial V}{\partial n_A}\right)_{n_B} = V_m + (n_A + n_B)\left(\frac{\partial V_m}{\partial n_A}\right)_{n_B} \tag{7.20}$$

Now, the derivative with respect to mole number n_A of A is transformed into a derivative with respect to mole fraction X_B of B,

$$\left(\frac{\partial V_m}{\partial n_A}\right)_{n_B} = \frac{dV_m}{dX_B}\left(\frac{\partial X_B}{\partial n_A}\right)_{n_B}$$

since

$$X_B = \frac{n_B}{n_A + n_B}, \qquad \left(\frac{\partial X_B}{\partial n_A}\right)_{n_B} = -\frac{n_B}{(n_A + n_B)^2}$$

Thus, (7.20) becomes

$$V_A = V_m - \frac{n_B}{n_A + n_B}\frac{dV_m}{dX_B}$$

$$V_m = X_B \frac{dV_m}{dX_B} + V_A \tag{7.21}*$$

The application of this equation is illustrated in Fig. 7.1, where the mean molar volume V_m of a solution is plotted against the mole fraction. The line $S_1 S_2$ is drawn tangent to the curve at point P, corresponding to a definite mole fraction X_B'. The intercept $O_1 S_1$ at $X_B = 0$ is V_A, the partial molar volume of A at the particular composition X_B'. It can readily be shown that the intercept on the other axis, $O_2 S_2$, is the partial molar volume of B, V_B.

This convenient method of intercepts is usually used to determine partial molar quantities in binary solutions. It is not restricted to volumes, but can be applied to any extensive state function, S, H, U, G, etc., given the necessary data. It can also be applied to heats of solution, and the partial molar heats of solution so obtained are the same as the differential heats described in Chapter 2.

If the variation with concentration of a partial molar quantity is known for one component in a binary solution, the Gibbs–Duhem equation (7.10) permits the calculation of the variation for the other component. This calculation can be accomplished by integration of (7.10). For example,

$$\int dV_A = -\int \frac{n_B}{n_A}dV_B = \int \frac{X_B}{X_B - 1}dV_B$$

where X is the mole fraction. If $X_B/(X_B - 1)$ is plotted against V_B, the area under the curve gives the change in V_A between the upper and lower limits of integration. The V_A of pure A is simply the molar volume $V_A^\bullet$ of pure A, which can be used as the starting point for the evaluation of V_A at any other concentration.

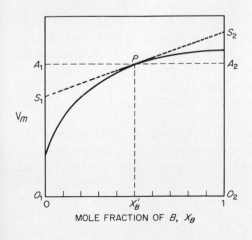

FIG. 7.1 Determination of partial molar volumes by intercept method. The dashed line is the tangent to the curve of V_m as a function of X_B at the particular mole fraction X_B'. The intercept $O_1 S_1$ gives V_A, the partial molar volume of A at X_B', and the intercept $O_2 S_2$ gives V_B, the partial molar volume of B at X_B'.

*Recall the standard slope-intercept form of the straight line: $y = mx + b$. Equation (7.21) thus fits the straight line $S_1 S_2$ in Fig. 7.1.

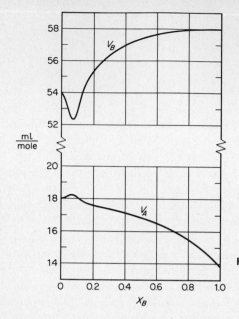

FIG. 7.2 Partial molar volumes in solutions of water and ethanol at 20°C. V_A (water), V_B (ethanol), X_B (mole fraction ethanol).

Figure 7.2 shows the partial molar volumes of both components in solutions of water and ethanol.

5. The Ideal Solution—Raoult's Law

The concept of the ideal gas has played an important role in discussions of the thermodynamics of gases and vapors. Many cases of practical interest are treated adequately by means of the ideal-gas approximations, and even systems deviating largely from ideality are conveniently referred to the norm of behavior set by the ideal case. It would be most helpful to find some similar concept to act as a guide in the theory of solutions, and fortunately this is indeed possible. Ideality in a gas implies a complete absence of cohesive forces; the internal pressure $(\partial U/\partial V)_T = 0$. Ideality in a solution is defined by complete uniformity of cohesive forces. If there are two components A and B, the intermolecular forces between A and A, B and B, and A and B are all the same.

An important property in the theory of solutions is the vapor pressure of a component above the solution. This partial vapor pressure is a good measure of the tendency of the given species to escape from the solution into the vapor phase. The tendency of a component to escape from solution is a direct reflection of the physical state of affairs within the solution, so that by studying the escaping tendencies, or partial vapor pressures, as functions of temperature, pressure, and concentration, we obtain a description of the properties of the solution. We may think of an analogy in which a nation represents a solution and its citizens the molecules. If life in the nation is good, the tendency to emigrate will be low. This presupposes, of course, the absence of artificial barriers.

For any component A of a solution in equilibrium with its vapor,

$$\mu_A^{sol} = \mu_A^{vap}$$

If the vapor behaves as an ideal gas, μ_A^{vap} can be related to the partial vapor pressure P_A of A above the solution. For an ideal gas mixture, $dP/P = dP_A/P_A$. From (7.12), at constant T,

$$d\mu_A^{vap} = V_A\,dP = RT\frac{dP}{P} = RT\frac{dP_A}{P_A}$$

On integration, writing $\mu_A^{\ominus,v}$ for the value of μ_A^{vap} when $P_A = 1$ atm, we obtain

$$\mu_A^{vap}(T,P) = \mu_A^{\ominus,v}(T) + RT\ln P_A$$

Hence also, at equilibrium,

$$\mu_A^{sol}(T,P) = \mu_A^{\ominus,v}(T) + RT\ln P_A \tag{7.22}$$

For the pure liquid in equilibrium with its vapor,

$$\mu_A^{\bullet}(T,P) = \mu_A^{\ominus,v}(T) + RT\ln P_A^{\bullet}$$

It follows from the last two equations and (7.17) that

$$\mu_A^{\bullet} - \mu_A^{sol} = RT\ln\frac{P_A^{\bullet}}{P_A} = RT\ln\frac{\lambda_A^{\bullet}}{\lambda_A}$$

Thus,

$$a_A = \frac{\lambda_A}{\lambda_A^{\bullet}} = \frac{P_A}{P_A^{\bullet}} \tag{7.23}$$

As long as the vapor behaves as an ideal gas, the relative activity of a component in solution can be directly measured by $P_A/P_A^{\bullet}$, the ratio of the partial pressure of A above the solution to the vapor pressure of pure A.

An ideal solution is defined as one for which

$$a_A = X_A \tag{7.24}$$

or, from (7.19),

$$\gamma_A = \frac{a_A}{X_A} = 1$$

It follows from (7.23) that for such an ideal solution,

$$\frac{P_A}{P_A^{\bullet}} = X_A \tag{7.25}$$

In 1886, François Marie Raoult first reported extensive vapor pressure data on solutions that closely approximated the relation (7.25), which is therefore called *Raoult's Law*.

The following is an example of the determination of activities from vapor pressures. At 298.15 K, the vapor pressure of pure water is $P_A^{\bullet} = 23.76$ torr, and that of pure propanol–1, is $P_B^{\bullet} = 21.80$ torr. Above a solution in which the mole fraction of water is $X_A = 0.400$, the partial vapor pressure of water is 19.86 torr, and that of propanol is 15.52 torr. Hence, the activities are

$$a_A = \frac{19.86}{23.76} = 0.836; \qquad a_B = \frac{15.52}{21.80} = 0.712$$

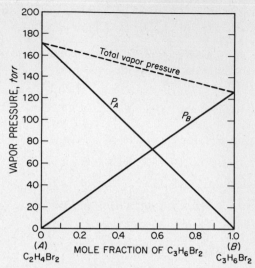

FIG. 7.3 Pressures of vapors above solutions of ethylene bromide and propylene bromide at 85°C. The solutions follow Raoult's Law.

The activity coefficients are

$$\gamma_A = \frac{0.836}{0.400} = 2.09; \qquad \gamma_B = \frac{0.712}{0.600} = 1.19$$

If the component B added to pure A lowers the vapor pressure, (7.25) can be written in terms of a relative vapor pressure lowering,

$$\frac{P_A^{\bullet} - P_A}{P_A^{\bullet}} = (1 - X_A) = X_B \tag{7.26}$$

This form of the equation is especially useful for solutions of a relatively involatile solute in a volatile solvent.

The vapor pressures of the system ethylene bromide + propylene bromide are plotted in Fig. 7.3. The experimental results almost coincide with the theoretical curves predicted by (7.25). In this instance, the agreement with Raoult's Law is excellent.

We seldom find solutions that follow Raoult's Law closely over an extended range of concentrations, because ideality in solutions implies a complete similarity of interaction between the components, which can rarely be achieved. Solutions of isotopes, however, provide good examples of ideal solutions, even in the solid state.

6. Thermodynamics of Ideal Solutions

When we put Raoult's Law, (7.25), into (7.22), we obtain

$$\mu_A(T, P) = \mu_A^{\ominus}(T) + RT \ln P_A^{\bullet} + RT \ln X_A \tag{7.27}$$

The first two terms are independent of composition, and we can combine them as $\mu_A^{\bullet}$ (T, P). Hence, for any component in an ideal solution,

$$\mu_A(T, P) = \mu_A^{\bullet}(T, P) + RT \ln X_A \tag{7.28}$$

From (7.27), we can calculate the partial molar volume of A in the solution, since

$$V_A = \left(\frac{\partial \mu_A}{\partial P}\right)_T$$

The first and third terms on the right in (7.27) are independent of pressure. Thus, from (6.27),

$$V_A = RT \left(\frac{\partial \ln P_A^{\bullet}}{\partial P}\right)_T = V_A^{\bullet} \tag{7.29}$$

which proves that the partial molar volume of a component in an ideal solution equals the molar volume of the pure component. There is no change in volume $(\Delta V = 0)$ when components are mixed to form an ideal solution.

In a similar way, from $\partial(\mu_A/T)/\partial T = -H_A/T^2$, (6.22) and (7.28), we can show that

$$H_A = H_A^{\bullet} \tag{7.30}$$

Thus, there is no heat of solution $(\Delta H = 0)$ when components are mixed to form an ideal solution.

The entropy change on mixing components to form an ideal solution is calculated as follows. From (7.12) and (7.28),

$$S_A = -\left(\frac{\partial \mu_A}{\partial T}\right)_P = S_A^{\bullet} - R \ln X_A$$

For mixing n_A moles of A and n_B moles of B,

$$\Delta S = n_A (S_A - S_A^{\bullet}) + n_B (S_B - S_B^{\bullet})$$

$$\Delta S = -n_A R \ln X_A - n_B R \ln X_B$$

Per mole,

$$\frac{\Delta S}{(n_A + n_B)} = -R (X_A \ln X_A + X_B \ln X_B)$$

This result can be extended to any number of components in a solution, to give for the entropy of mixing per mole

$$\Delta S_m = -R \sum_i X_i \ln X_i \tag{7.31}$$

Let us calculate the entropy of mixing the elements in air, taking the composition by volume to be 79% N_2, 20% O_2, and 1% argon.

$$\Delta S_m = -R (0.79 \ln 0.79 + 0.20 \ln 0.20 + 0.01 \ln 0.01)$$

$$\Delta S_m = 4.60 \text{ J} \cdot \text{K}^{-1} \cdot \text{mol}^{-1} \text{ of mixture}$$

7. Solubility of Gases in Liquids—Henry's Law

Consider a solution of component B, which may be called the solute, in A, the solvent. If the solution is sufficiently diluted, a condition ultimately is attained in which each molecule of B is effectively completely surrounded by component A.

The solute B is then in a uniform environment irrespective of the fact that A and B may form solutions that are far from ideal at higher concentrations. In such a very dilute solution, the escaping tendency of B from its uniform environment is proportional to its mole fraction, but the proportionality constant k no longer is $P_B^\bullet$ as it was for the ideal solution. We may write

$$P_B = kX_B \tag{7.32}$$

This equation was established and extensively tested by William Henry in 1803 in a series of measurements of the dependence on pressure of the solubility of gases in liquids. Henry's Law is not restricted to gas–liquid systems, however, but is followed by a wide variety of fairly dilute solutions and by *all* solutions in the limit of extreme dilution.*

Some data on the solubility of gases in water over a wide range of pressures are shown in Fig. 7.4. If Henry's Law were obeyed exactly, these solubility curves would all be straight lines. Actually, the curves for H_2, He, and N_2 are quite linear up to about 100 atm, but for O_2 we can see definite deviations even in this range. The high solubility of H_2 in water is interesting. We should expect the solubility at moderate pressures to depend on the attractive energy between solute and solvent molecules, and on this basis H_2 and He might be expected to have about the same solubility. In nonpolar solvents, such as CCl_4 or benzene, hydrogen does not display a particularly high solubility (e.g., never so high as nitrogen). It is hard to resist the conclusion, therefore, that some form of specific interaction is occurring between the dissolved H_2 molecules and the liquid water structure.†

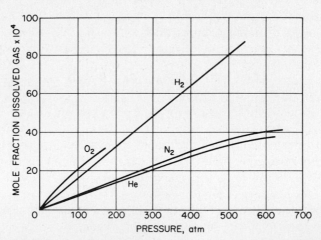

FIG. 7.4 Solubility of gases in water at 298.15 K as a function of pressure.

*The form of the law for dissociated solutes, such as electrolytes, is discussed in Section 10.15.

†Would it be reasonable to suppose that the H_2 molecule may act as a weak bridge between two H_2O molecules?

$$\begin{array}{ccc} H & & H \\ \diagdown & & \diagup \\ O \;\ldots\; H{-}H \;\ldots\; O & \\ \diagup & & \diagdown \\ H & & H \end{array}$$

8. Mechanism of Anesthesia

One of the most fascinating unsolved problems in medical physiology is the mechanism by which gases produce anesthesia and narcosis. Many anesthetics, such as krypton and xenon, are apparently inert chemically; in fact, it would appear that all gases produce an anesthetic effect at high enough pressures. Cousteau, in *The Silent World*, gives a memorable account of the nitrogen narcosis experienced at great depths, *l'ivresse des grandes profondeurs*, which has claimed the life of more than one diver.*

Table 7.2 summarizes some data on mice, which were tested by a criterion based on their righting reflex. Animals in a test chamber were rocked off their feet; if a mouse did not replace all four feet on the floor within 10 s, the anesthetic was awarded a "knockout." The results indicate that even helium produces narcosis at high pressures.

Early attempts to understand the causes of anesthesia were made by Meyer (1899) and Overton (1901), who found a good correlation between the solubility of a gas in a lipid (olive oil) and its narcotic efficacy. Since nerve cell membranes are composed mainly of lipids and proteins, it was suggested that the anesthetic molecules dissolve in the membranes and block the process of nerve conduction in some way as yet unknown. The activity of dissolved anesthetic necessary to produce anesthesia is in the range 0.02 to 0.05, so that the anesthetic can certainly alter the properties of the membrane to some considerable extent.

In 1961, Linus Pauling† suggested a novel theory of anesthesia, based on the idea that gaseous anesthetics react with water near the surfaces of nerve membranes to form deposits of microcrystalline hydrates. At body temperatures, the partial pressures of anesthetics in nervous tissues are considerably below the dissociation pressures of the known hydrates. Pauling therefore suggested that formation of hydrate crystals might be enhanced close to membrane surfaces by a sort of cooperative ordering of the water molecules due to the large intermolecular forces acting there.

Actually, the correlation of anesthetic potency with hydrate formation was as good as that with lipid solubility, since both phenomena are in turn related to intermolecular attractive forces. To make a test between the two theories, it would be necessary to find anesthetics that dissolved in lipids readily but formed hydrates with difficulty. The fluorinated compounds CF_4 and SF_6 seemed to provide such a possibility, and when the test was made,‡ the results, as shown in Figs. 7.5(a) and (b), appeared to favor a lipid environment rather than an aqueous one as the site of anesthetic action.§ Additional experiments will be needed to settle this

---

*Inert gas narcosis should not be confused with *decompression sickness* ("the bends"), which is caused by rapid release of gas dissolved in the tissues with consequent bubble formation and damage to cells.

†L. Pauling, *Science*, *134*, 15 (1961). See also S. L. Miller, *Proc. Natl. Acad. Sci.*, *47*, 1515 (1961).

‡K. W. Miller, W. D. M. Paton, and E. B. Smith, *Brit. J. Anaesthesia*, *39*, 910 (1962).

§Newer data on the dissociation pressure of the CF_4 hydrate have brought it close to the curve in Fig. 7.5(b) (point 13), but SF_6 still seems anomalous. [S. L. Miller, E. I. Eger, and C. Lundgren, *Nature*, *221*, 469 (1969).]

TABLE 7.2
Best Estimates of Anesthetic Pressures for Mice (Righting Reflex)

Key no.	Gas	Pressure (atm)	Key no.	Gas	Pressure (atm)
1	He	190	10	C_2H_4	1.1
2	Ne	>110	11	C_2H_2	0.85
3	Ar	24	12	Cyclo–C_3H_6	0.11
4	Kr	3.9	13	CF_4	19
5	Xe	1.1	14	SF_6	6.9
6	H_2	85	15	CF_2Cl_2	0.4
7	N_2	35	16	$CHCl_3$	0.008
8	N_2O	1.5	17	Halothane	0.017
9	CH_4	5.9	18	Ether	0.032

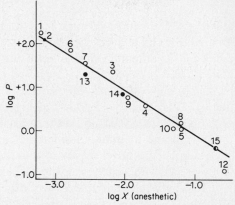

FIG. 7.5(a) Graph of log anesthetic pressure against log solubility in olive oil at 37°C. For key to numbers in this and following figures, see Table 7.2 ; solid circles are fully fluorinated compounds, the half-solid circle a partially fluorinated compound. The line is of unit slope.

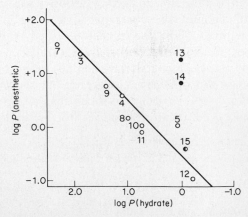

FIG. 7.5(b) Graph of log anesthetic pressure against log hydrate dissociation pressure at 0°C. The line is of unit slope.

problem and to provide finally an understanding of anesthesia at the molecular level.

9. Two-Component Systems

For systems of two components, the phase rule, $f = c - p + 2$, becomes $f = 4 - p$. The following cases are possible:

$$p = 1, \quad f = 3 \qquad \text{trivariant system}$$
$$p = 2, \quad f = 2 \qquad \text{bivariant system}$$
$$p = 3, \quad f = 1 \qquad \text{univariant system}$$
$$p = 4, \quad f = 0 \qquad \text{invariant system}$$

The maximum number of degrees of freedom is three. A complete graphical representation of a two-component system therefore requires a three-dimensional diagram, with coordinates corresponding to pressure, temperature, and composition. Since a three-dimensional representation is usually inconvenient, we may hold one variable constant while we plot the behavior of the other two. In this way, we obtain planar graphs showing pressure vs. composition at constant temperature, or pressure vs. temperature at constant composition.

10. Pressure–Composition Diagrams

The example of a PX diagram in Fig. 7.6 shows the system 2-methylpropanol-1 + propanol-2, which obeys Raoult's Law quite closely over the entire range of compositions. The straight upper line (*liquidus curve*) represents the dependence of the total vapor pressure above the solution on the mole fraction in the liquid. The

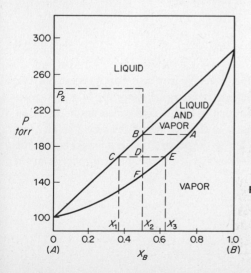

FIG. 7.6 Pressure–composition (mole fraction) diagram at 60°C for the system 2-methylpropanol-1 (*A*) + propanol-2 (*B*) which form practically ideal solutions.

curved lower line represents the dependence of the total vapor pressure on the composition of the vapor.

Consider a liquid of composition X_2 at a pressure P_2. This point lies in a one-phase region, in which there would be three degrees of freedom. One of these is used by the requirement of constant temperature for the diagram. Thus, for any arbitrary composition X_2, the liquid solution at constant T can exist over a range of different pressures.

As pressure is decreased along the dashed line of constant composition, nothing happens until the liquidus curve is reached at B. At this point, liquid begins to vaporize. The vapor that is formed is richer than the liquid in the more volatile component, propanol-2. The composition of the first vapor to appear is given by point A on the vapor curve.

As pressure is further reduced below B, a two-phase region on the diagram is entered. This represents the region of stable coexistence of liquid and vapor. The dashed line passing horizontally through a typical point D in the two-phase region is called a *tie-line*; it connects liquid and vapor compositions that are in equilibrium.

In the two-phase region, the system is bivariant. One of the degrees of freedom is used by the requirement of constant temperature, and only one remains. When the pressure is fixed in this region, therefore, the compositions of both the liquid and the vapor phases are also definitely fixed. They are given, as we have seen, by the end points of the tie-line.

The overall composition of the system at point D in the two-phase region is X_2. This is made up of liquid having a composition X_1, and vapor having a composition X_3. We can calculate the relative amounts of liquid and vapor required to yield the overall composition. Let n_l and n_v be the sum of the numbers of moles of both components A and B in liquid and in vapor, respectively. From a material balance applied to component B,

$$X_2(n_l + n_v) = X_1 n_l + X_3 n_v$$

or

$$\frac{n_l}{n_v} = \frac{X_3 - X_2}{X_2 - X_1} = \frac{DE}{DC} \tag{7.33}$$

This expression is called the *lever rule*. It applies to any two compositions of two phases in equilibrium connected by a tie-line in a phase diagram of a two-component system. If the diagram is plotted in mass fractions instead of mole fractions, the ratio of the line segments gives the ratio of the masses of the two phases.

As pressure is still further decreased along BF, more liquid is vaporized, until finally, at F, no liquid remains. Further decrease in pressure then proceeds in the one-phase, all-vapor region.

11.   Temperature–Composition Diagrams

The temperature–composition diagram of the liquid–vapor equilibrium is the boiling point diagram of the solutions at the constant pressure chosen. If the pressure is 1 atm, the boiling points are the normal ones. The diagram for the 2-methylpropanol-1 + propanol-2 system is shown in Fig. 7.7.

The boiling point diagram for an ideal solution can be calculated if the vapor pressures of the pure components are known as functions of temperature. The two end points of the boiling point diagram shown in Fig. 7.7 are the temperatures at which the pure components have vapor pressures of 760 torr, viz., 82.3 and 108.5°C. The composition of the solution that boils anywhere between these two temperatures—say, at 100°C—is found as follows:

If X_A is the mole fraction of C_4H_9OH, from Raoult's Law we have

$$760 = P_A^{\bullet} X_A + P_B^{\bullet}(1 - X_A)$$

At 100°C, the vapor pressure of C_3H_7OH is 1440 torr; of C_4H_9OH, it is 570 torr. Thus, $760 = 570 X_A + 1440(1 - X_A)$, or $X_A = 0.781$, $X_B = 0.219$. This gives one intermediate point on the liquidus curve; the others are calculated in the same way.

The composition of the vapor is given by Dalton's Law:

$$X_A^{\text{vap}} = \frac{P_A}{760} = \frac{X_A^{\text{liq}} P_A^{\bullet}}{760} = 0.781 \times \frac{570}{760} = 0.585$$

$$X_B^{\text{vap}} = \frac{P_B}{760} = \frac{X_B^{\text{liq}} P_B^{\bullet}}{760} = 0.219 \times \frac{1440}{760} = 0.415$$

The curve for vapor-composition vs. pressure is therefore readily constructed from the liquidus curve.

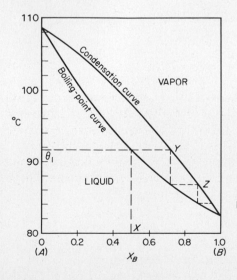

FIG. 7.7 Boiling point vs. composition diagram for the system 2-methylpropanol-1 (*A*) + propanol-2 (*B*), which form practically ideal solutions.

12.   Fractional Distillation

The application of the boiling point diagram to a simplified representation of distillation is shown in Fig. 7.7. The solution of composition X begins to boil at temperature θ_1. The first vapor that is formed has a composition Y, richer in the more volatile component. If this is condensed and reboiled, vapor of composition Z is obtained. This process is repeated until the distillate is composed of pure component B. In practical cases, the successive fractions will each cover a range of compositions, but the vertical lines in Fig. 7.7 may be considered to represent average compositions within these ranges.

A fractionating column is a device that carries out automatically the successive condensations and vaporizations required for fractional distillation. An especially clear example is the bubble-cap type of column in Fig. 7.8. As vapor ascends from the boiler, it bubbles through a film of liquid on the first plate. This liquid is somewhat cooler than that in the boiler, so that a partial condensation takes place. The vapor that leaves the first plate is therefore enriched in the more volatile component compared to the vapor from the boiler. A similar enrichment takes place on each successive plate. Each attainment of equilibrium between liquid and vapor corresponds to one of the steps in Fig. 7.7.

The efficiency of a distilling column is measured by the number of such equi-

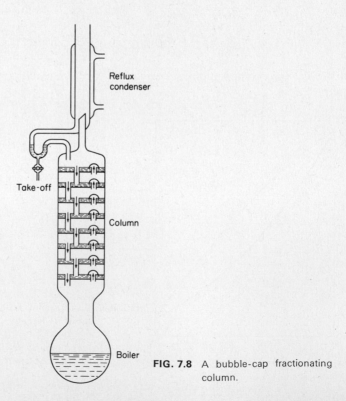

FIG. 7.8   A bubble-cap fractionating column.

librium stages that it achieves. Each stage is called a *theoretical plate*. In a well designed bubble-cap column, each unit acts very nearly as one theoretical plate. We also describe the performance of various types of packed columns in terms of theoretical plates. The separation of liquids whose boiling points lie close together requires a column with a considerable number of plates. The number actually required depends on the *cut* that is taken from the head of the column, i.e., the ratio of distillate taken off to that returned to the column.*

Let us suppose, for example, that we start with a solution with mole fraction $X_A = 0.500$ of butanol (A) in propanol (B), and distill it in a column with two theoretical plates. The first distillate that is taken will have a composition, as read from Fig. 7.7, of $X_B = 0.952$ propanol.

13. Solutions of Solids in Liquids

Solubility curve and *freezing point depression curve* are two different names for the same thing—that is, a temperature vs. composition curve for a solid–liquid equilibrium at some constant pressure, usually chosen as one atmosphere. Such a diagram is shown in Fig. 7.9(a) for the system benzene + naphthalene. The curve

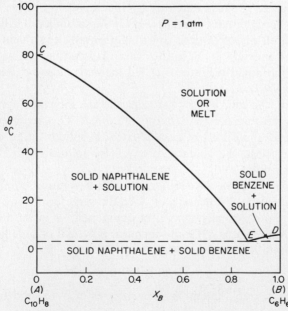

FIG. 7.9(a) Temperature–composition diagram for the system naphthalene (A) and benzene (B). The solids are mutually insoluble and the liquid solution is practically ideal.

*For details of methods for determining the number of theoretical plates in a column, see C. S. Robinson and E. R. Gilliland, *Fractional Distillation* (New York: McGraw-Hill Book Company, 1950).

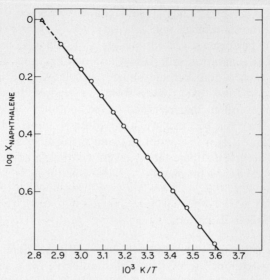

FIG. 7.9(b)   Solubility of naphthalene in benzene. Data of 7.9(a) plotted as log X_A vs T^{-1}.

CE may be considered to illustrate either (1) the depression of the freezing point of naphthalene by the addition of benzene, or (2) the solubility of solid naphthalene in the solution. Both interpretations are fundamentally equivalent: in one case, we consider T as a function of X; in the other, X as a function of T. The lowest point E on the solid-liquid diagram is called the *eutectic point* (from the Greek, $\epsilon\upsilon\tau\eta\kappa\tau\acute{o}\varsigma$, *easily melted*).

In this diagram, the solid phases that separate are shown as pure naphthalene (A) on one side and pure benzene (B) on the other. This representation is not exactly correct, since there must be a small extent of solid solution of B in A and of A in B. Nevertheless, the absence of any solid solution is in many cases an excellent approximation.

For a pure solid A to be in equilibrium with a solution containing A, it is necessary that the chemical potentials of A be the same in the two phases, $\mu_A^s = \mu_A^l$. From (7.28), the chemical potential of component A in an ideal solution is $\mu_A^l = \mu_A^{\bullet l} + RT \ln X_A$, where $\mu_A^{\bullet l}$ is the chemical potential of pure liquid A. Thus, the equilibrium condition can be written

$$\mu_A^s = \mu_A^{\bullet l} + RT \ln X_A$$

Now μ_A^s and $\mu_A^{\bullet l}$ are simply the molar free energies of pure solid and pure liquid. Hence,

$$\frac{G_A^{\bullet s} - G_A^{\bullet l}}{RT} = \ln X_A \qquad (7.34)$$

Since we have $\partial(G/T)/\partial T = -H/T^2$ from (3.52), differentiation of (7.34) with respect to T yields (with ΔH_f the latent heat of fusion),

$$\frac{H_A^{\bullet l} - H_A^{\bullet s}}{RT^2} = \frac{\Delta H_f}{RT^2} = \frac{d \ln X_A}{dT} \tag{7.35}$$

It is a good approximation to take ΔH_f independent of T over moderate ranges of temperature. Integrating (7.34) from $T^\bullet$, the freezing point of pure A, mole fraction unity, to T, the temperature at which pure solid A is in equilibrium with solution of mole fraction X_A, we obtain

$$\frac{\Delta H_f}{R}\left(\frac{1}{T^\bullet} - \frac{1}{T}\right) = \ln X_A \tag{7.36}$$

This is the equation for the temperature variation of the solubility X_A of a pure solid in an ideal solution. From Fig. 7.9(b), we see that the linear relation between $\ln X$ and T^{-1} is followed closely for solubility of naphthalene in benzene.

As an example of (7.36), let us calculate the solubility of naphthalene in an ideal solution at 25°C. Naphthalene melts at 80°C, and its heat of fusion at the melting point is 19.29 kJ·mol^{-1}. Thus, from (7.36),

$$\frac{19290}{8.314}(353.2^{-1} - 298.2^{-1}) = 2.303 \log X_A$$

$$X_A = 0.298$$

This is the mole fraction of naphthalene in any ideal solution, whatever the solvent may be. Actually, the solution will approach ideality only if the solvent is rather similar in chemical and physical properties to the solute. Typical experimental values for the solubility X_A of naphthalene in various solvents at 25°C are as follows: chlorobenzene, 0.317; benzene, 0.296; toluene, 0.286; acetone, 0.224; hexane, 0.125.

We can rewrite (7.36) in terms of $X_B = 1 - X_A$, the mole fraction of solute:

$$\frac{\Delta H_f}{R}\left(\frac{T - T^\bullet}{TT^\bullet}\right) = \ln(1 - X_B)$$

For depressions of the freezing point $(T^\bullet - T) = \Delta T_f$ small compared to $T^\bullet$, we can set $TT^\bullet \approx T^{\bullet 2}$, and thus, on expanding the log in a power series,

$$\frac{\Delta H_f(\Delta T_f)}{RT^{\bullet 2}} = -\ln(1 - X_B) = X_B + \frac{1}{2}X_B^2 + \frac{1}{3}X_B^3 + \cdots$$

For dilute solutions, $X_B \ll 1$, and

$$\Delta T_f \simeq \frac{RT_f^{\bullet 2}}{\Delta H_f} \cdot X_B = K_F m_B \tag{7.37}$$

where m_B is the molality of B, and the molal *freezing-point-depression constant* is

$$K_F = \frac{RT_f^{\bullet 2}M_A}{(\Delta H_f)} \tag{7.38}$$

For water, $K_F = 1.855$; benzene, 5.12; camphor, 40.0; etc. Because of its exceptionally large K_F, camphor is used in a micromethod for molecular weight determination by freezing point depression.

14.   Osmotic Pressure

Properties of dilute solutions that depend upon the number of solute molecules rather than on their kind have been called *colligative properties*.* These properties can all be related to the lowering of the vapor pressure when a nonvolatile solute is dissolved in a solvent. They include boiling point elevation,† freezing point depression, and osmotic pressure.

In 1748, J. A. Nollet described an experiment in which a solution of "spirits of wine" was placed in a cylinder, the mouth of which was closed with an animal bladder and immersed in pure water. The bladder was observed to swell greatly and sometimes even to burst. The animal membrane is semipermeable; water can pass through it, but alcohol cannot. The increased pressure in the tube, caused by diffusion of water into the solution, was called the *osmotic pressure* (from the Greek, ωσμός, *impulse*).

The first detailed quantitative study of osmotic pressure is found in a series of researches by W. Pfeffer, published in 1887. Ten years earlier, Moritz Traube had observed that colloidal films of cupric ferrocyanide acted as semipermeable membranes. Pfeffer deposited this colloidal precipitate within the pores of earthenware pots, by soaking them first in copper sulfate and then in potassium ferrocyanide solution.

Extensive measurements of osmotic pressure were carried out by H. N. Morse, J. C. W. Frazer, and their colleagues at Johns Hopkins, and R. T. Rawdon (Berkeley) and E. G. J. Hartley at Oxford.‡

The method used by the Hopkins group is shown in Fig. 7.10(a). The porous cell impregnated with cupric ferrocyanide is filled with water and immersed in a vessel containing the aqueous solution. The pressure is measured by means of an attached manometer. The system is allowed to stand until there is no further increase in pressure. Then the osmotic pressure is just balanced by the hydrostatic pressure in the column of solution. Pressures extending up to several hundred atmospheres were measured by a variety of ingenious methods. These included the calculation of the pressure from the change in the refractive index of water on compression, and the application of piezoelectric gauges.

The English workers used the apparatus shown in Fig. 7.10(b). Instead of waiting for equilibrium to be established and then reading the pressure, they applied an external pressure to the solution just sufficient to balance the osmotic

---

*From the Latin *colligatus*, bound or fastened together. Evidently the word was intended to connote that the properties depended on the *collection* of solute particles.

†A derivation of the equation for boiling point elevation is now included in most general chemistry textbooks and can be left as an exercise (Problem 14).

‡An excellent detailed discussion of this work is to be found in J. C. W. Frazer's article, "The Laws of Dilute Solutions," in *A Treatise on Physical Chemistry*, 2nd ed., ed. by H. S. Taylor (New York: D. Van Nostrand Co., Inc., 1931), pp. 353–414. A good survey of experimental methods and applications is "Determination of Osmotic Pressure," by R. H. Wagner and L. D. Moore, in *Physical Methods of Organic Chemistry*, Part 1, 3rd ed., by A. Weissberger (New York: Interscience Publishers, 1959), pp. 815–894.

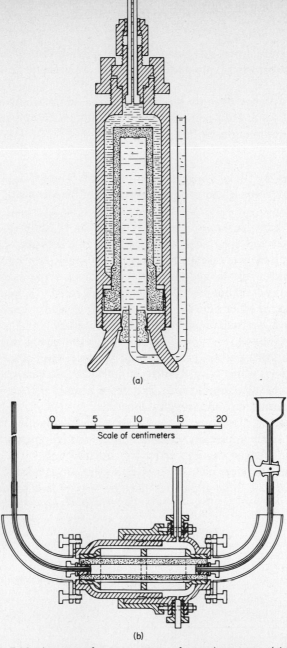

(a)

Scale of centimeters
0 5 10 15 20

(b)

FIG. 7.10 Apparatus for measurement of osmotic pressure: (a) The static method of Morse and Frazer. The semipermeable membrane is supported in the porous walls of a cylindrical cell. The interior of the cell is filled with pure solvent and the solution to be measured is placed in the volume surrounding the cell. (b) The dynamic method of Berkeley and Hartley. The internal cell contains pure solvent and the solution is outside the cell. Through the small-bore tubing that leads to the external chamber, hydrostatic pressure is applied until osmotic flow stops.

pressure. They could detect this balance precisely by observing the level of liquid in the capillary tube, which would fall rapidly if there was any flow of solvent into the solution.

Some typical precise osmotic pressure data are summarized in Table 7.3.

In 1885, J. H. van't Hoff pointed out that in dilute solutions the osmotic pressure Π obeyed the relationship $\Pi V = nRT$, or

$$\Pi = cRT \tag{7.39}$$

where $c = n/V$ is the molar concentration of solute. The validity of the equation can be judged by comparison of the calculated and experimental values of Π in Table 7.3.

An osmotic pressure arises when two solutions of different concentrations (or a pure solvent and a solution) are separated by a semipermeable membrane. A simple illustration is a gaseous solution of hydrogen and nitrogen. Thin palladium foil is permeable to hydrogen, but practically impermeable to nitrogen. If pure nitrogen is put on one side of a palladium barrier and a solution of nitrogen and hydrogen on the other side, the requirements for osmosis are satisfied. Hydrogen flows through the palladium from the hydrogen-rich to the hydrogen-poor side of the membrane. This flow continues until the chemical potential of the H_2, $\mu(H_2)$ is the same on both sides. In this example, the nature of the semipermeable membrane is rather clear. Hydrogen molecules are catalytically dissociated into atoms at the palladium surface, and these atoms, perhaps in the form of protons and electrons, diffuse through the barrier.

A mechanism based on some kind of distinction in solubility is responsible for many cases of semipermeability. For example, protein membranes, like those employed by Nollet, can dissolve water but not alcohol.

In other cases, the membrane may act as a sieve, or as a bundle of capillaries. The cross sections of these capillaries may be very small, so that they can be

TABLE 7.3
Osmotic Pressures of Solutions of Sucrose in Water at 20°C

Molality (m)	Molar concentration (c) $(mol \cdot dm^{-3})$	Observed osmotic pressure (atm)	Calculated osmotic pressure		
			Eq. (7.39)	Eq. (7.43)	Eq. (7.41)
0.1	0.098	2.59	2.36	2.40	2.44
0.2	0.192	5.06	4.63	4.81	5.46
0.3	0.282	7.61	6.80	7.21	7.82
0.4	0.370	10.14	8.90	9.62	10.22
0.5	0.453	12.75	10.9	12.0	12.62
0.6	0.533	15.39	12.8	14.4	15.00
0.7	0.610	18.13	14.7	16.8	17.40
0.8	0.685	20.91	16.5	19.2	19.77
0.9	0.757	23.72	18.2	21.6	22.15
1.0	0.825	26.64	19.8	24.0	24.48

permeated by small molecules, such as water, but not by large molecules, such as carbohydrates or proteins.

Irrespective of the mechanism by which the semipermeable membrane operates, the final result is the same. Osmotic flow continues until the chemical potential of the diffusing component has the same value on both sides of the barrier. If the flow takes place into a closed volume, the pressure inside necessarily increases. The final equilibrium osmotic pressure can be calculated by thermodynamic methods.

15. Osmotic Pressure and Vapor Pressure

Consider a pure solvent A, which is separated from a solution of B in A by a membrane permeable to A alone. At equilibrium, an osmotic pressure Π has developed. The condition for equilibrium is that the chemical potential of A is the same on both sides of the membrane, $\mu_A^\alpha = \mu_A^\beta$. Thus, at equilibrium, the μ_A in the solution must equal that of the pure A. There are two factors tending to cause the value of μ_A in the solution to depart from that in pure A. These factors must therefore have exactly equal and opposite effects on μ_A. The first is the change in μ_A produced by dilution of A in the solution. This change causes a lowering of μ_A equal to $\Delta\mu = RT \ln P_A/P_A^\bullet$. Exactly counteracting this is the increase in μ_A in the solution due to the imposed pressure Π. From (7.12), $d\mu_A = V_A \, dP$, so that $\Delta\mu_A = \int_0^\Pi V_A \, dP$.

At equilibrium, therefore, in order that μ_A in solution should equal $\mu_A^\bullet$ in the pure liquid,

$$\int_0^\Pi V_A \, dP = -RT \ln \frac{P_A}{P_A^\bullet}$$

If it is assumed that the partial molar volume V_A is independent of pressure, i.e., the solution is practically incompressible,

$$V_A \Pi = RT \ln \frac{P_A^\bullet}{P_A} \tag{7.40}$$

The significance of this equation can be stated as follows: *The osmotic pressure is the external pressure that must be applied to the solution to raise the vapor pressure of the solvent A to that of pure A.*

In most cases, the partial molar volume V_A of solvent in solution can be well approximated by the molar volume of the pure liquid $V_A^\bullet$. In the special case of an ideal solution, (7.40) then becomes

$$\Pi V_A^\bullet = -RT \ln X_A \tag{7.41}$$

By replacing X_A by $(1 - X_B)$ and expanding as in Section 7.13, we obtain the formula for a dilute solution,

$$\Pi V_A^\bullet = RTX_B \tag{7.42}$$

Since the solution is dilute,

$$\Pi = \frac{RT}{V_A^\bullet} \cdot \frac{n_B}{n_A} \approx RTm' \tag{7.43}$$

This is the equation used by Frazer and Morse as a better approximation than the van't Hoff equation. As the solution becomes very dilute, m', the volume molality, approaches c, the molar concentration, and we find as the end product of the series of approximations

$$\Pi = RTc$$

The adequacy with which Eqs. (7.41), (7.43), and (7.39) represent the experimental data can be judged from the comparisons in Table 7.3.*

16. Deviations of Solutions from Ideality

Only a few of the many liquid solutions that have been investigated follow Raoult's Law over the complete range of concentrations. For this reason, most practical applications of the ideal equations are made in the treatment of dilute solutions. As a solution becomes more dilute, the behavior of the solute B approaches more closely to that given by Henry's Law. *Henry's Law is thus a limiting law that is followed by all solutes in the limit of extreme dilutions*, as $X_B \longrightarrow 0$. The behavior of the solvent, as the solution becomes more dilute, approaches more closely that given by Raoult's Law. In the limit of extreme dilution, as $X_A \longrightarrow 1$, all solvents obey Raoult's Law as a limiting law.

One of the most instructive ways of discussing the properties of nonideal solutions is in terms of their deviations from ideality. The first extensive measurements of vapor pressure, permitting such comparisons, were made by Jan von Zawidski around 1900.

Two types of deviation from ideality can be distinguished: cases in which $a_A > X_A$ or $\gamma_A > 1$ are *positive deviations;* those in which $a_A < X_A$ or $\gamma_A < 1$ are *negative deviations*. In some cases, a solution may have positive deviation in one range of concentrations, negative in another.

A system exhibiting a positive deviation from Raoult's Law is water + dioxane, whose vapor pressure vs. composition diagram is shown in Fig. 7.11(a). An ideal solution would follow the dashed lines. The positive deviation is characterized by vapor pressures higher than those calculated for ideal solution. The escaping tendencies of the components in the solution are accordingly higher than the escaping tendencies in the individual pure liquids. The effect has been ascribed to cohesive forces between unlike components smaller than those within the pure liquids, resulting in a trend away from complete miscibility. To put it naively, the components are happier by themselves than when they are mixed together; they are unsociable. A scientific translation is obtained by equating a happy component to one in a state of low free energy. We should expect this incipient immiscibility to be reflected in an increase in volume on mixing and also in an absorption of heat on mixing.

The other general type of departure from Raoult's Law is the negative devia-

*The osmotic pressures of solutions of high polymers and proteins provide some of the best data on the thermodynamic properties of these macromolecules. A typical investigation is that of M. J. Schick, P. Doty, and B. H. Zimm, *J. Am. Chem. Soc.*, *72*, 530 (1950).

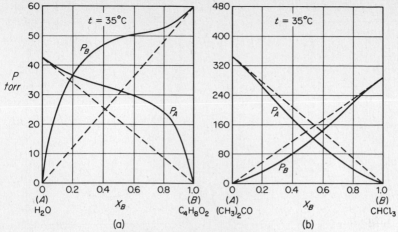

FIG. 7.11   (a) Positive deviation from ideality. Partial vapor pressures in water + dioxane system at 35°C. (b) Negative deviation from ideality. Partial vapor pressures in the acetone + chloroform system at 35°C. (The Raoult's Law values are shown as dashed straight lines.)

tion, illustrated by the chloroform + acetone system in Fig. 7.11(b). In this case, the escaping tendency of a component from solution is less than it would be from the pure liquid. This fact may be the result of attractive forces between the unlike molecules in solution greater than those between the like molecules in the pure liquids. In some cases, actual association or compound formation may occur in the solution. As a result, in cases of negative deviation, we should expect a contraction in volume and an evolution of heat on mixing.

In some cases of deviation from ideality, the simple picture based on differences in cohesive forces may not be adequate. For example, positive deviations are often observed in aqueous solutions. Pure water is itself strongly associated, and addition of a second component may depolymerize the water to some extent, causing an increased partial vapor pressure.

A sufficiently great positive deviation from ideality may lead to a maximum in the PX diagram, and a sufficiently great negative deviation, to a minimum. An illustration of such behavior is shown in Fig. 7.12(a). At a maximum or minimum in the vapor pressure curve, the vapor and the liquid must have the same composition.

If we have measured one of the partial vapor pressure curves in a binary system like those in Figs. 7.11 and 7.12(a), we can always calculate the other one from a Gibbs–Duhem type of equation. By analogy with (7.10),

$$d \ln P_A = \left(\frac{X_B}{X_B - 1} \right) d \ln P_B$$

A related expression includes explicitly the slopes of the P vs. X curves,

$$(1 - X_B)\left(\frac{\partial \ln P_A}{\partial X_B} \right)_T + X_B \left(\frac{\partial \ln P_B}{\partial X_B} \right)_T = 0 \tag{7.44}$$

This form is called the *Duhem–Margules equation*.

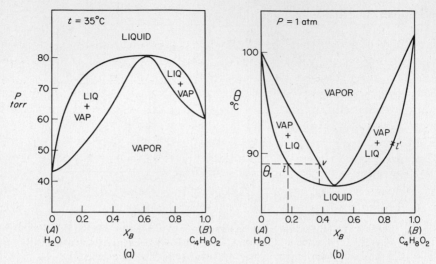

FIG. 7.12 The dioxane + water system illustrates positive deviation from Raoult's Law. (a) *PX* diagram at 35°C ; (b) *TX* diagram at 1 atm (normal boiling-point diagram).

17.   Boiling Point Diagrams

The *PX* diagram in Fig. 7.12(a) has its counterpart in the boiling point (*TX*) diagram in Fig. 7.12(b). A maximum in the *PX* curve corresponds to a minimum in the *TX* curve.

A solution with the composition corresponding to a maximum or minimum point on the boiling point diagram is called an *azeotropic solution* (from the Greek ζηιν, *to boil*, and α-τρόπος, *unchanging*), since there is no change in composition on boiling. Such solutions cannot be separated by distillation at constant pressure. In fact, at one time it was thought that they were real chemical compounds; however, changing the pressure changes the composition of an azeotropic solution.

The distillation of a system with a maximum or minimum boiling point can be discussed by reference to Fig. 7.12(b). If the temperature of a solution having the composition *l* is raised, it begins to boil at the temperature θ_1. The first vapor that distills has the composition *v*, richer than the original liquid in component *B*. The residual solution therefore becomes richer in *A*; and if the vapor is continuously removed, the boiling point of the residue rises, as its composition moves along the liquidus curve from *l* toward pure *A*. If a fractional distillation is carried out, a final separation into pure *A* and the azeotropic solution is achieved. Similarly, a solution of original composition *l'* can be separated into pure *B* and azeotrope.

18. Solubility of Liquids in Liquids

If the positive deviation from Raoult's Law becomes sufficiently large, the components may no longer form a continuous series of solutions. As successive portions of one component are added to the other, a limiting solubility is finally reached, beyond which two distinct liquid phases are formed. Usually, but not always, increasing temperature tends to promote solubility, as the thermal kinetic energy overcomes the reluctance of the components to mix freely. In other words, the $T \, \Delta S$ term in $\Delta G = \Delta H - T \, \Delta S$ becomes more important. A solution that displays a large positive deviation from ideality at elevated temperatures therefore frequently splits into two phases when it is cooled.

An example of such behavior is the n-hexane + nitrobenzene system shown in the TX diagram of Fig. 7.13. At the temperature and composition indicated by the point x, two phases coexist, the conjugate solutions represented by y and z. The relative amounts of the two phases are proportional, as usual, to the segments of the tie-line. As the temperature is increased along the isopleth XX', the amount of the hexane-rich phase decreases and the amount of nitrobenzene-rich phase increases. Finally, at Y, the hexane-rich phase disappears completely, and at temperatures above Y, there is only one solution.

This gradual disappearance of one solution is characteristic of systems having all compositions except one. The exception is the composition corresponding to the maximum in the TX curve. This composition is called the *critical composition* and the temperature at the maximum is the *critical solution temperature* or *upper consolute temperature*. As a two-phase system having the critical composition is gradually heated (line cc' in Fig. 7.13), there is no gradual disappearance of one phase. Even in the immediate neighborhood of the maximum d, the ratio of the segments of the tie-line remains practically constant. The compositions of the two

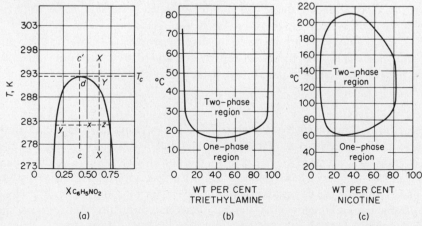

FIG. 7.13 Partial miscibility of two liquids. (a) *n*-hexane + nitrobenzene; (b) triethylamine + water; (c) nicotine + water.

conjugate solutions gradually approach each other until, at the point d, the boundary line between the two phases suddenly disappears and a single phase remains.

As the critical temperature is slowly approached from above, a curious phenomenon is observed. Just before the single homogeneous phase passes over into two distinct phases, the solution is suffused by a pearly opalescence. This *critical opalescence* is believed to be caused by the scattering of light from small regions of slightly differing density, which are formed in the liquid in the incipient separation of the two phases. X-ray studies have revealed that such regions may persist even several degrees above the critical point.*

Strangely enough, some systems exhibit a lower consolute temperature. At high temperatures, two partially miscible solutions are present, which become completely intersoluble when sufficiently cooled. An example is the triethylamine + water system in Fig. 7.13(b), with a lower consolute temperature of 18.5°C, at 1 atm pressure. Note the great increase in solubility as the temperature approaches this point. This strange behavior suggests that large negative deviations from Raoult's Law (e.g., compound formation) become sufficient at the lower temperatures to counteract the positive deviations responsible for the immiscibility.

Finally, systems have been found with both upper and lower consolute temperatures. These are more common at elevated pressures, and we might expect all systems with a lower consolute temperature to display an upper one at sufficiently high temperature and pressure. An example at atmospheric pressure is the nicotine + water system of Fig. 7.13(c).

19. Thermodynamic Condition for Phase Separation

The interpretation of the *solubility gap* in systems with limited intersolubility is based upon the free energy of the mixture. Figure 7.14 shows the Gibbs energy of mixing G_M [$= \Delta G$ in Eq. (7.14)] as a function of the composition of a binary solution for three different cases. We might find such results at three different temperatures for a system like one of those in Fig. 7.13.

In (a), the components are miscible in all proportions. The criterion for this condition is that the G_M vs. X curve is convex downward over the entire range of X. The criterion for complete miscibility is, therefore, that for all X,

$$\frac{\partial^2 G_M}{\partial X^2} > 0$$

In case (b), we have drawn a dashed curve between points X'_A and X''_A to show the course of G_M in this region computed for a single liquid phase. There are two points of inflection where $(\partial^2 G_M / \partial X^2) = 0$. We note that anywhere in the region between X'_A and X''_A the G_M can be decreased if the system breaks into two distinct liquid phases, one of the composition X'_A and the other of composition X''_A. These compositions represent ends of a tie-line between conjugate solutions on the usual TX diagram.

*G. Brady, *J. Chem. Phys.*, *32*, 45 (1960).

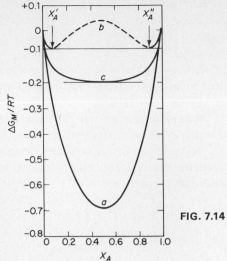

FIG. 7.14 Gibbs energy of mixing: (a) one liquid phase; (b) two liquid phases; (c) at critical point.

In (c), we see the limiting case of a system showing a critical point or consolute temperature; the points of inflection marking the limits of solubility have moved together until they coincide at the critical point. The condition for the critical point is that both $(\partial^2 G_M/\partial X^2)$ and $(\partial^3 G_M/\partial X^3)$ equal 0.

From a theoretical point of view, we can expect these higher derivatives of G_M to be exquisitely sensitive to small variations in the format and parameters of the laws of intermolecular force upon which one seeks to build an analysis of the properties of solutions.

20. Thermodynamics of Nonideal Solutions

Thermodynamic properties of nonideal solutions can often be displayed most clearly by calculating the differences between their values in the actual solution and the values they would have in an ideal solution of the same composition. Such differences are called *excess functions*. For example, consider the Gibbs free energy for any component A:

Actual solution $G_A - G_A^{\bullet} = RT \ln a_A = RT \ln X_A + RT \ln \gamma_A$

Ideal solution $G_A^{\text{id}} - G_A^{\bullet} = RT \ln X_A$

Excess function $G_A^{\text{ex}} = G_A - G_A^{\text{id}} = RT \ln \gamma_A$

For a binary solution of A and B,

$$\Delta G_M^{\text{ex}} = RT(X_A \ln \gamma_A + X_B \ln \gamma_B) \tag{7.45}$$

Since for an ideal solution ΔH and ΔV of mixing are zero, the excess functions

ΔH^{ex} and ΔV^{ex} are simply the mixing functions. We calculate the excess entropy from

$$\Delta S_M^{\mathrm{ex}} = -\left(\frac{\partial\, \Delta G_M^{\mathrm{ex}}}{\partial T}\right)_{P,X}$$

Hildebrand introduced (1929) the concept of a *regular solution*, in which the entropy of mixing is virtually ideal, whereas the ΔH_M may depart markedly from zero.

If there is a change in volume on mixing, this ΔV will itself cause some change in the entropy. Therefore, the usual ΔS_P^{ex}, as measured at constant pressure, should be corrected to ΔS_V^{ex}, measured at constant volume, before one makes any comparison with theoretical models for the ΔS (e.g., calculations from statistical mechanics). The correction,* due to Scatchard, is

$$\Delta S_P - \Delta S_V = \frac{\alpha}{\beta}\, \Delta V \tag{7.46}$$

where α is thermal expansivity and β is compressibility.

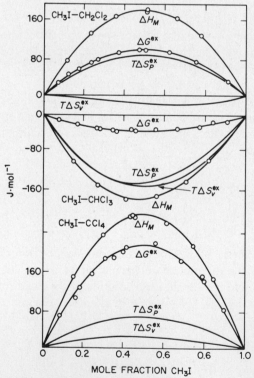

FIG. 7.15 Excess thermodynamic functions: CH_3I + chloromethanes at 298 K.

$$^*S_P = S_V + \int_V^{V+\Delta V} \left(\frac{\partial S}{\partial V}\right)_T dV'$$

From the Maxwell equation (3.46), $(\partial S/\partial V)_T = (\partial P/\partial T)_V = \alpha/\beta$. If we assume α/β is constant, the integral is $(\alpha/\beta)\,\Delta V$.

In Fig. 7.15, the excess functions are shown for solutions of CH_3I in three different chloromethanes.[*] It is interesting to note that the ΔS_V^{ex} is quite small for the CH_2Cl_2 and CCl_4 systems, which thus nearly fulfill the Hildebrand criteria for regularity.

Table 7.4 shows a few excess functions for solution of liquids in liquids.[†] As we would expect for liquid solutions, $\Delta G_P \approx \Delta A_V$.

TABLE 7.4
Thermodynamic Excess Functions of Mixing at Constant Pressure
and at Constant Volume at $X = 0.5$

System	T K	ΔV_P^{ex} cm$^3\cdot$mol^{-1}	ΔG_P^{ex} J$\cdot$mol^{-1}	ΔA_V^{ex} J$\cdot$mol^{-1}	ΔH_P^{ex} J$\cdot$mol^{-1}	ΔU_V^{ex} J$\cdot$mol^{-1}
Ethylene chloride +Benzene	298	0.24	25.9	26.8	60.7	−32.6
Carbon tetrachloride +Benzene	308	0.01	81.6	81.6	109	106
Carbon bisulfide +Acetone	308	1.06	1050	1040	1460	1120
Carbon tetrachloride +Neopentane	273	−0.5	318	318	314	427
n-Perfluorohexane +n-Hexane	298	4.84	1350	1320	2160	1230

21. Solid–Liquid Equilibria: Simple Eutectic Diagrams

Two-component solid–liquid equilibria, in which the liquids are intersoluble in all proportions and in which there is no appreciable solid–solid solubility, give the simple diagram of Fig. 7.16. Examples of systems of this type are collected in Table 7.5.

Consider the behavior of a solution of composition X on cooling along the isopleth XX'. When point P is reached, pure solid A begins to separate from the solution. As a result, the residual solution becomes richer in the other component B, its composition falling along the line PE. At any point Q in the two-phase region, the relative amounts of pure A and residual solution are given as usual by the ratio of the tie-line segments. When point R is reached, the residual solution has the eutectic composition E. Further cooling now results in the simultaneous precipitation of a mixture of A and B in relative amounts corresponding to E.

The eutectic point is an invariant point on a constant pressure diagram; since three phases are in equilibrium $f = c - p + 2 = 2 - p + 2 = 4 - 3 = 1$, and the single degree of freedom is used by the choice of a condition of constant pressure.

[*]E. A. Moelwyn–Hughes and R. W. Missen, *Trans. Faraday Soc.*, **53**, 607 (1957).
[†]R. L. Scott, *J. Phys. Chem.*, **64**, 1241 (1963).

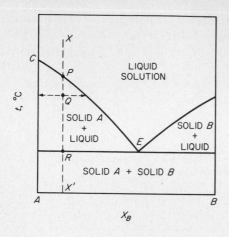

FIG. 7.16 Simple eutectic diagram for two components, *A* and *B*, completely intersoluble as liquids but with negligible solid–solid solubility.

TABLE 7.5
Systems with Simple Eutectic Diagrams, such as Fig. 7.16

Component *A*	Melting point of *A* (K)	Component *B*	Melting point of *B* (K)	Eutectic	
				(K)	(Mol % *B*)
CHBr$_3$	280.5	C$_6$H$_6$	278.5	247	50
CHCl$_3$	210	C$_6$H$_5$NH$_2$	267	202	24
Picric acid	395	TNT	353	333	64
Sb	903	Pb	599	519	81
Cd	594	Bi	444	417	55
KCl	1063	AgCl	724	579	69
Si	1685	Al	930	851	89
Be	1555	Si	1685	1363	32

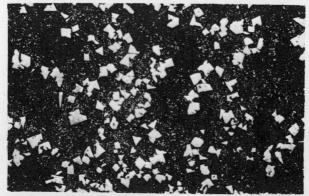

FIG. 7.17 Photomicrograph at 50 X of 80% Pb + 20% Sb, showing crystals of Sb in a eutectic matrix. [Arthur Phillips, Yale University.]

Microscopic examination of alloys often reveals a structure indicating that they have been formed from a melt by a cooling process similar to that considered along the isopleth XX' of Fig. 7.16. Crystallites of pure metal are found dispersed in a matrix of finely divided eutectic mixture. An example is shown in Fig. 7.17.

22. Formation of Compounds

If aniline and phenol are melted together in equimolar proportions, a definite compound crystallizes on cooling, $C_6H_5OH \cdot C_6H_5NH_2$. Pure phenol melts at 313 K, pure aniline at 267 K, and the compound melts at 304 K. The complete TX diagram for this system, in Fig. 7.18, is typical of many instances in which stable compounds occur as solid phases. A convenient way to look at such a diagram is to imagine that it is made up of two diagrams of the simple eutectic type placed side-by-side. In this case, one such diagram would be the phenol + compound diagram, and the other the aniline + compound diagram. The phases corresponding with the various regions of the diagram are labeled.

A maximum, such as the point C, indicates the formation of a compound with a *congruent* melting point, since if a solid having the composition $C_6H_5OH \cdot C_6H_5NH_2$ is heated to 304 K, it melts to a liquid of identical composition.

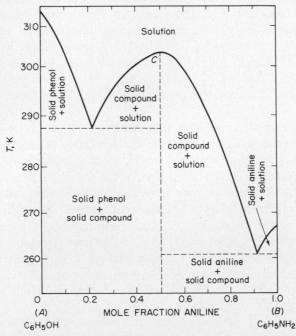

FIG. 7.18 The system phenol + aniline, illustrating the formation of an intermolecular compound.

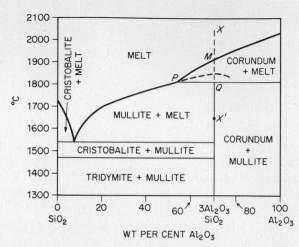

FIG. 7.19 The system silica + alumina displays a peritectic point at *P* above which the compound mullite $3Al_2O_3 \cdot SiO_2$ melts incongruently to yield solid corundum Al_2O_3 + a liquid phase.

In some systems, solid compounds are formed that do not melt to a liquid having the same composition, but instead decompose before such a melting point is reached. An example is the silica + alumina system (Fig. 7.19), which includes a compound, $3Al_2O_3 \cdot SiO_2$, called *mullite*. If a melt containing 40% Al_2O_3 is prepared and cooled slowly, solid mullite begins to separate at about 2053 K. If some of this solid compound is removed and reheated along the line $X'X$, it decomposes at 2573 K into solid corundum and a liquid solution (melt) having the composition *P*. Thus, $3Al_2O_3 \cdot SiO_2 \rightarrow Al_2O_3$ + solution. Such a change is called *incongruent melting*, since the composition of the liquid differs from that of the solid.

The point *P* is the incongruent melting point or *peritectic point* (from the Greek, $\tau\eta\kappa\tau o\varsigma$, *melting*, and $\pi\epsilon\rho\iota$, *around*). The suitability of this name becomes evident if one follows the course of events as a solution with composition $3Al_2O_3 \cdot SiO_2$ is gradually cooled along XX'. When the point *M* is reached, solid corundum (Al_2O_3) begins to separate from the melt, the composition of which therefore becomes richer in SiO_2, falling along the line *MP*. When the temperature falls below that of the peritectic at *P*, the following change occurs: liquid + corundum $\rightarrow$ mullite. The solid Al_2O_3 that has separated reacts with the surrounding melt to form the compound mullite. If a specimen taken at a point such as *Q* is examined, the solid material is found to consist of two phases, a core of corundum surrounded by a coating of mullite. It was from this characteristic appearance that the term *peritectic* originated.

The microstructures of solid phases can often be controlled by the rate at which the melt is cooled or quenched. Figure 7.20 shows an unusual structure obtained by extremely rapid quenching from the iron + aluminum system.

FIG. 7.20 A molten alloy of 10% Fe in Al was quenched from 1200°C at a rate in excess of 500°C per minute. In a matrix of a stable compound Al_6Fe, a metastable phase crystallized as ten-point dendritic stars. An electron-probe micro-analysis indicated that the metastable compound was considerably richer in iron than Al_6Fe. (C. Adam and L. M. Hogan, Department of Mining and Metallurgical Engineering, University of Queensland.)

23. Solid Solutions

In the theory of phase equilibria, solid solutions are no different from other kinds of solution: They are simply solid phases containing more than one component. The phase rule makes no distinction between the kind of phase (gas, liquid, or solid) that occurs, being concerned only with how many phases are present. Therefore, most of the diagrams typical of liquid–vapor and liquid–liquid systems are found to have counterparts among solid–liquid and solid–solid systems.

Two general classes of solid solution can be distinguished on structural grounds. A *substitutional solid solution* is one in which solute atoms, or groups of atoms,

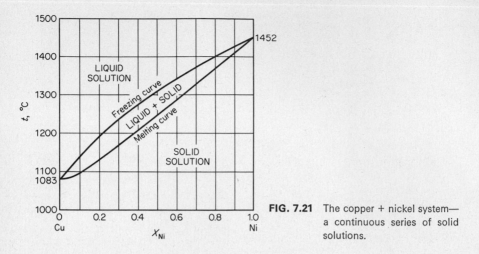

FIG. 7.21 The copper + nickel system— a continuous series of solid solutions.

are substituted for solvent atoms or groups in the crystal structure. For example, nickel has a face-centered cubic structure; if some of the nickel atoms are replaced at random by copper atoms, a solid solution is obtained. This substitution of one group for another is possible only when the substituents do not differ greatly in size. An *interstitial solid solution* is one in which the solute atoms or groups occupy interstices in the crystal structure of the solvent. For example, carbon atoms may occupy some of the interstices in the nickel structure. Interstitial solid solution can occur to an appreciable extent only when the solute atoms are small compared with the solvent atoms.

An example of a system with a continuous series of solid solutions is copper + nickel, Fig. 7.21. Important industrial alloys, such as Constantan (60Cu, 40Ni) and Monel (60Cu, 35Ni, 5Fe) are solid solutions of this kind.

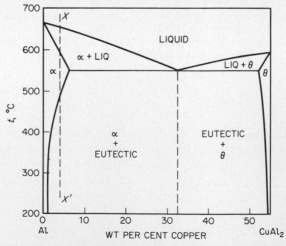

FIG. 7.22 A section of the aluminum + copper system. Alloys containing up to 6 percent copper exhibit age hardening.

For intermetallic systems, the simple eutectic diagram of Fig. 7.16 is usually an oversimplification. In many cases, however, the solubility gap extends across *almost* the entire diagram. Usually, too, the gap increases considerably with decreasing temperature. An interesting case is shown in the aluminum + copper diagram of Fig. 7.22. Only the portion of the system extending from pure Al to the intermetallic compound $CuAl_2$ is covered. The solid solution of copper in aluminum is called the α phase, and the solid solution of aluminum in the compound $CuAl_2$ is called the θ phase.

The phenomenon of age hardening of alloys is interpreted in terms of the effect of temperature on the solubility gap in solid solutions. If a melt containing

FIG. 7.23 Direct transmission electron micrograph of 33% Cu + 67% Al, the eutectic composition in Fig. 7.22. The light ribbons are α-phase, the dark are θ-phase. [N. Takahashi and K. Ashinuma, University of Yamanashi.]

about 4% Cu and 96% Al is cooled along *XX'*, it first solidifies to a solid solution α. This solid solution is soft and ductile. If it is quenched rapidly to room temperature, it becomes metastable. Changes in the solid state are usually sluggish, so that the metastable solution can persist for some time. It changes slowly, however, to the stable form, which is a mixture of two phases—solid solution α and solid solution θ. This two-phase alloy is much less plastic than the homogeneous solid solution α. The exact mechanism of the hardening is still not completely elucidated, but it is always associated with the change from single-phase to two-phase alloy.

Figure 7.23 shows a remarkable lamellar structure in an alloy of 33% Cu and 67% Al examined under the electron microscope. The dark ribbons are the θ phase, the lighter ones, the α phase.

24. The Iron–Carbon Diagram

No discussion of phase diagrams should omit the iron–carbon system, which is the theoretical basis for ferrous metallurgy. The part of the diagram of greatest interest extends from pure iron to the compound iron carbide, or *cementite*, Fe₃C. This section is reproduced in Fig. 7.24.

Pure iron exists in two different modifications. The stable crystalline form up to 910°C, called α iron, has a body-centered cubic structure. At 910°C, transition occurs to a face-centered cubic structure, γ iron; but at 1401°C, γ iron transforms

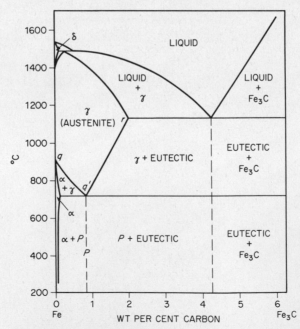

FIG. 7.24 A portion of the iron–carbon phase diagram. [After J. B. Austin, *Metals Handbook* (Cleveland: American Society for Metals, 1948), p. 1181.]

back to a body-centered cubic structure, now called δ iron. This is an interesting, but not unique, example of an allotrope that is stable, at constant pressure, both below and above a certain temperature range. The solid solutions of carbon in the iron structures are called *ferrite*.

Apart from the small section concerned with δ ferrite, the upper portion of the diagram is a typical example of limited solid–solid solubility.

The curve qq' shows how the transformation temperature of α to γ ferrite is lowered by interstitial solution of carbon in the iron. The region labeled α represents the range of solid solutions of C in α iron. The region labeled γ represents the range of solid solutions of C in γ iron, which are given a special name, *austenite*. The decrease in the transition temperature $\alpha \longrightarrow \gamma$ is terminated at q', where the curve intersects the solid solubility curve rq' of carbon in γ iron. A point such as q', which has the properties of a eutectic but occurs in a completely solid region, is called a *eutectoid*.

The two phases formed by the eutectoid decomposition of austenite are α ferrite and cementite. These phases form a lamellar structure of alternate bands called *pearlite*. If the composition is close to the eutectoid, the steel is composed entirely of pearlite. If the composition is richer in carbon, or *hypereutectoid*, it

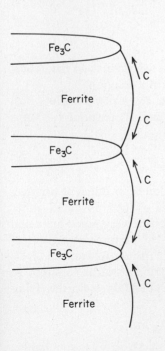

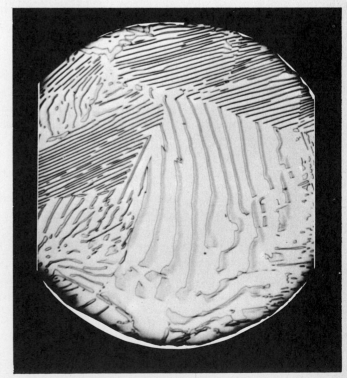

FIG. 7.25 Formation and appearance of pearlite. The photomicrograph is at 1250 X. [U. S. Steel Corporation Research Center.]

may contain other grains of cementite in addition to those occurring in the pearlite. If the composition is poorer in carbon, or *hypoeutectoid*, and the steel is cooled slowly, it may contain additional grains of ferrite. Figure 7.25 shows the formation and appearance of pearlite. The first stage in the formation seems to be the nucleation of a crystallite of cementite. As this grows, it removes carbon from the surrounding austenite. Nucleation of ferrite then occurs at the surface of the cementite, because low carbon favors the transformation of γ to α.

The diagram Fig. 7.24 explains the distinction between the steels and the cast irons. Any composition below 2% carbon can be heated until a homogeneous solid solution (austenite) is obtained. In this condition, the alloy is readily hot rolled or submitted to other forming operations. On cooling, segregation of two phases occurs. Cementite is a hard, brittle material, and its occurrence in the pearlitic steels is responsible for their high strength. The way in which the cooling is carried out determines the rate of segregation of the two phases and their grain sizes, and provides a great many possibilities for obtaining different mechanical properties by annealing and tempering.

Compositions above 2% in carbon belong to the general class of cast irons. They cannot be brought into a homogeneous solid solution by heating, and therefore cannot conveniently be mechanically worked. They are formed by casting from the molten state, and used where hardness and corrosion resistance are desirable and where brittleness due to high content of cementite is not deleterious.

25. Statistical Mechanics of Solutions

If we know the properties of a pair of pure components, we can make a theoretical treatment of what happens when they are mixed to form a solution.

First let us look at the statistical mechanical model for a perfect solution of N_A molecules of A and N_B molecules of B. If Z_A and Z_B are the partition functions for pure A and B, the partition function for the solution is

$$Z_{AB} = \frac{(N_A + N_B)!}{N_A! N_B!} Z_A Z_B \tag{7.47}$$

The combinatorial factor $(N_A + N_B)!/N_A!N_B!$ is the number of different arrangements of A and B molecules arising from interchanges of their positions within the solution. In this model, the A and B molecules must be very much alike and each molecule must be in the same intermolecular field of force regardless of the identities of its neighbors. This situation and the partition function in (7.47) correspond exactly to the criteria described previously for an ideal solution. In particular, since $A = -kT \ln Z$, the partition function of (7.47) gives for the Helmholtz free energy of mixing,

$$\Delta A_M = (N_A + N_B) kT(X_A \ln X_A + X_B \ln X_B)$$

Since $\Delta A = \Delta U - T \Delta S$ for the isothermal mixing, and $\Delta U = 0$ for this model,

$$\Delta S_M = -(N_A + N_B) k(X_A \ln X_A + X_B \ln X_B)$$

as in (7.31).

If the solution is not ideal, ΔU_M will differ from zero, and we shall have the problem of evaluating the partition function Z_{AB} of the mixture for a model that includes energy terms for the different interactions between A and A, B and B, A and B.

The first approximation that is made is the separation of translational from internal degrees of freedom, so that, as in the case of the gas (Eq. 5.48) we can write for the solution

$$Z = Z_t \cdot Z_I \tag{7.48}$$

This will be an excellent approximation for mixtures of nearly spherical molecules, such as CCl_4 and $SiCl_4$. For nonspherical but nonpolar molecules, the approximation will still be fairly good, since the forces between such molecules will not vary strongly with direction. For polar molecules, however, (7.48) becomes a poor approximation, since the rotation of a polar molecule will depend markedly upon the position and orientation of its neighbors. We shall restrict ourselves to cases where (7.48) is satisfactory. In such cases, the excess free energy (ΔG^{ex} or ΔA^{ex}) depends entirely upon the translational partition function, since the contributions of the internal degrees of freedom cancel.

It is most convenient to employ the classical form of Z_t as in (5.54),

$$Z_t = \frac{1}{N!}\frac{1}{h^{3N}} \int \cdots \int \exp\left(\frac{-\mathscr{H}}{kT}\right) dp_1 \cdot dp_{3N} \, dq_1 \cdots dq_{3N} \tag{7.49}$$

The integration of the momenta extends from $-\infty$ to $+\infty$, and of the coordinates, over the volume of the system. The Hamiltonian is

$$\mathscr{H} = \frac{1}{2m} \sum_{i=1}^{3N} p_i^2 + U(q_1 \cdots q_{3N})$$

When this is substituted into (7.49), the integration over momenta yields (as on p. 198)

$$\left(\frac{2\pi mkT}{h^2}\right)^{3N/2}$$

This factor cannot contribute to the free energy of mixing, and it is usually absorbed into Z_I to give Z_I'.

The remaining factor is

$$Q = \frac{1}{N!} \int \cdots \int \exp\left(\frac{-U}{kT}\right) dq_1 \cdots dq_{3N} \tag{7.50}$$

It is called the *configuration integral* or the *configurational partition function*. If we could evaluate this function Q from a knowledge of the properties of individual molecules, we would have gone a long way toward a complete statistical mechanical theory of liquids and imperfect gases. The extension of Q to a solution of two components A and B is, clearly,

$$Q = \frac{1}{N_A! N_B!} \int \cdots \int \exp\left(\frac{-U}{kT}\right) dq_1 \cdots dq_{3N} \tag{7.51}$$

We shall postpone any discussion of attempts to evaluate Q itself until Chapter

19 (Liquids), and comment only briefly on how (7.51) has been used in the theory of solutions.*

One approach to the problem is to evaluate Q on the basis of a *lattice model*. Each molecule of A and B is assumed to occupy a definite site on a rigid lattice. Thus, the volume is fixed by $N_A + N_B$, and $\Delta V^{ex} = 0$ for this model. It is then assumed that the potential energy U can be split into two terms: (1) the interaction between the molecules at rest at their equilibrium positions on the lattice points; (2) the energy due to vibration of the molecules about the lattice points. Therefore,

$$Q = Q(lattice) \times Q(vibration)$$

In the simplest formulation, it is supposed that only $Q(lattice)$ is changed by mixing the components. Thus, the simple model attempts to calculate the free energy of mixing by calculating $Q(lattice)$.

We further simplify the model by assuming that only interactions between *nearest neighbors* need be considered. If each molecule has z nearest neighbors, there are altogether $(N_A + N_B)z/2$ *pairs* of nearest neighbors, N_{AA} of type AA, N_{BB} of type BB, and N_{AB} of type AB. We see that

$$zN_A = 2N_{AA} + N_{AB}$$
$$zN_B = 2N_{BB} + N_{AB}$$

Let u_{AA}, u_{BB}, and u_{AB} be the pairwise interaction energies. Then the lattice energy becomes

$$E(lattice) = N_{AA}u_{AA} + N_{AB}u_{AB} + N_{BB}u_{BB}$$
$$= \tfrac{1}{2}zN_A u_{AA} + \tfrac{1}{2}zN_B u_{BB} + N_{AB}(u_{AB} - \tfrac{1}{2}u_{AA} - \tfrac{1}{2}u_{BB})$$

Hence, $w = u_{AB} - \tfrac{1}{2}u_{AA} - \tfrac{1}{2}u_{BB}$ is the energy gained on mixing by the creation of one AB nearest neighbor pair. We can also write $E_A = \tfrac{1}{2}zN_A u_{AA}$ and $E_B = \tfrac{1}{2}zN_B u_{BB}$ as the lattice energies of pure A and B, respectively. Therefore,

$$Q(lattice) = N \sum_{AB} g(N_A, N_B, N_{AB}) \exp\left[\frac{-(E_A + E_B + N_{AB}w)}{kT}\right]$$
$$= \exp\left[\frac{-(E_A + E_B)}{kT}\right] \sum_{AB} g(N_A, N_B, N_{AB}) \exp\left(\frac{-N_{AB}w}{kT}\right)$$

Here $g(N_A, N_B, N_{AB})$ is the number of different arrangements of the N_A molecules of A and N_B molecules of B that give N_{AB} pairs of nearest neighbors of type AB. The Helmholtz free energy of mixing becomes

$$\Delta A_M = -kT \ln\left[\sum g(N_A, N_B, N_{AB}) \exp\left(\frac{-N_{AB}w}{kT}\right)\right] \tag{7.52}$$

The evaluation of the sum in (7.52) is equivalent to the *Ising problem* discussed in the previous chapter.

If we set $w = 0$ in (7.52) we see that

$$\sum_{AB} g(N_A, N_B, N_{AB}) = \frac{(N_A + N_B)!}{N_A!N_B!}$$

---

*The standard reference is I. Prigogine, *The Molecular Theory of Solutions* (Amsterdam: North Holland Publishing Co., 1957).

and, since $N_A + N_B = N$,

$$\Delta A_M = -kT \ln \frac{N!}{N_A! N_B!}$$

as given by (7.47) for perfect solutions.

26.   The Bragg–Williams Model

The most simple assumption about AB pairs is to assume complete randomness in the mixing. This model was first introduced, by Bragg and Williams, in the theory of metallic solid solutions. A completely random distribution would give the maximum term in the summation of $g(N_A, N_B N_{AB})$ in (7.52), and the maximum term is, of course, $N!/N_A! N_B!$ To a good approximation we can replace the summation of g by this maximum term. We also insert the average value $\overline{N}_{AB}$ to obtain

$$\Delta A_M = -kT \ln \frac{N!}{N_A! N_B!} + \overline{N}_{AB} w$$

But the average value of N_{AB} is simply

$$\overline{N}_{AB} = z \frac{N_A N_B}{N_A + N_B}$$

Thus, for the Bragg–Williams model,

$$\Delta A_M = kT[N_A \ln X_A + N_B \ln X_B] + zw \frac{N_A N_B}{N_A + N_B} \qquad (7.53)$$

The first term is the value for ideal solutions, and the second term is the excess ΔA^{ex}. On this approximation, the excess entropy $\Delta S^{\text{ex}} = 0$, and the entire departure from ideality is due to the excess energy ΔU^{ex}. As we saw in Fig. 7.15 and Table 7.4, however, the excess free energy is found experimentally to be quite evenly divided between the entropy and energy terms, except for *regular solutions*. The Bragg–Williams model, therefore, may be reasonable for regular solutions but it cannot cope with other types of deviation from ideality.

The Bragg–Williams model does predict a phase separation at lower temperatures for all instances where $w > 0$. In Fig. 7.26(a), we have plotted $\Delta A_M/kT$ in accord with the Bragg–Williams formula (7.53). We see that when the interaction energy zw becomes considerably greater than kT, the free energy curve will display exactly the behavior described in Fig. 7.14 as leading to separation of phases.

We can also calculate vapor pressure curves from the Bragg–Williams model. From (7.53), the chemical potential (per molecule) can be obtained by differentiation with respect to N_A.

$$\mu_A - \mu_A^\bullet = kT \ln X_A + (1 - X_A)^2 zw$$

Hence, from (7.23),

$$a_A = \frac{\lambda_A}{\lambda_A^\bullet} = \frac{P_A}{P_A^\bullet} = X_A \exp\left[\frac{(1 - X_A)^2 zw}{kT}\right]$$

In Fig. 7.26(b), we have plotted $P_A/P_A^\bullet$ for $zw/kT = 1, 0$, and -2. The case $w = 0$,

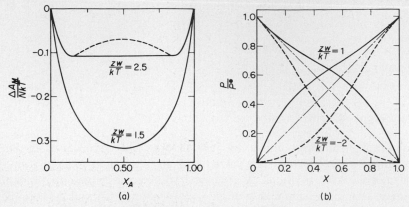

FIG. 7.26   (a) Free energy of mixing according to the Bragg–Williams approximation. (b) Partial vapor pressures according to the Bragg–Williams approximation for $zw/kT = 1$, 0, and -2.

of course, is Raoult's Law. For $w > 0$, there is a positive deviation, and for $w < 0$, a negative deviation. Thus, the Bragg–Williams theory, based on the lattice model, gives a quite good general interpretation of some properties of solutions. The model should not, however, be taken too seriously for liquid solutions, since no ordered lattice structure actually exists in liquids. For solid solutions, however, the model should be quite realistic.

It would indeed be interesting to follow further developments in the statistical theory of nonideal solutions, but the subject becomes increasingly complex and perhaps enough has been said to give some feeling for the nature of the problems. It is important to go beyond the model of random mixing and to consider cases in which the intermolecular forces are sufficiently specific to cause deviations from complete randomness that will actually be reflected in an excess entropy of mixing (either $+$ or $-$). The standard treatment by Prigogine would be an excellent reference to these fascinating theoretical developments.

The theory of solutions is of great importance to any chemist who must choose a suitable solvent in which to carry out a synthetic reaction. So far, the chemist has been guided not so much by the theory of solutions as by familiarity with an immense collection of empirical data on solvent effects. As our understanding of solutions improves, it should become possible to fit the proper solvent to each reaction by means of soundly based theoretical considerations.

PROBLEMS

1. Solutions are prepared at 25°C containing n mol NaCl in 1.000 kg water. The volumes in cubic centimetres are found to vary with n as $V = 1001.38 + 16.6253n + 1.7738n^{3/2} + 0.1194n^2$. Draw graphs showing the partial molar volumes of H_2O and NaCl in the solutions as a function of molality from 0 to 2 molal. Show that Eq. (7.21) applies to some particular point on your graphs where V_A is the partial molar volume of water.

2. Suppose you have a mixture of n-propane ($X_A = 0.4$) and n-butane ($X_B = 0.6$) in an ideal liquid solution at 77 K. Devise an isothermal reversible process to separate the solution into its pure components. What limitations would there be to the use of your process in practice? Calculate the minimum work necessary to separate 1 mol of the solution into its pure components.

3. The boiling point of pure toluene is 110.60°C. A solution containing 5.00 g of diphenyl, $C_{12}H_{10}$ in 100 g toluene boils at 111.68°C. A solution containing 6.00 g of an unknown nonvolatile substance in 200 g toluene boils at 112.00°C. Calculate the mass per mole of the unknown.

4. A solution containing 25.3 mol % benzene and 74.7 mol % toluene boils at 100°C and 1 atm. The liquid obtained by condensing the vapor boils at 94.8°C. Calculate the composition of this liquid. The vapor pressure of pure benzene is 1357 torr at 100°C and 1108 torr at 94.8°C. Assume ideal behavior of solutions and vapor.

5. An unknown compound is immiscible with water. It is steam distilled at 98.0°C and $P = 737$ torr. The vapor pressure of water is 707 torr at 98.0°C, the distillate was 75.0% by weight water. Calculate the molecular weight of the unknown.

6. A dilute solution contains m moles of solute A in 1 kg of a solvent with a boiling point elevation constant K_b. The solute dimerizes in solution in accord with the reaction $2 A \rightleftharpoons A_2$, with equilibrium constant K. Show that

$$K = \frac{K_b(K_b m - \Delta T_b)}{(2 \Delta T_b - K_b m)^2}$$

where ΔT_b is the elevation of the boiling point in a solution of molality m in A.

7. Phenol melts at 40°C; α-naplthylamine melts at 50°C. In the binary system, there are eutectics at 75 mol % phenol and 17°C, and 36 mol % phenol and 23°C. A compound is formed at 50 mol % phenol with a melting point of 28°C. All these data are at 1 atm. (a) Sketch the phase diagram $T—X$ at 1 atm for this system. (b) Describe clearly what happens if a mixture containing 40 mol % phenol is cooled from 50 to 10°C. (c) Describe clearly what happens if a mixture containing 85 mol % phenol is cooled from 50 to 10°C.

8. The solubility of boric acid in water is 50.4 g per kg water at 20°C and 116 g per kg water at 50°C. Calculate the average ΔH (solution) over this temperature range. Is this an integral heat of solution or a differential heat of solution? Explain.

9. The partial molar volume V_2 of K_2SO_4 in water solutions at 298 K is given by

$$V_2(\text{cm}^3) = 32.280 + 18.216m^{1/2} + 0.0222m$$

Obtain an equation for V_1, the partial molar volume of H_2O. Take $V_1^{\bullet} = 17.963$ cm$^3 \cdot$mol^{-1} for H_2O.

10. The heat of fusion of ice is 6008 J$\cdot$mol^{-1} at 0°C and $C_P = 37.20$ J$\cdot$K$^{-1}\cdot$mol^{-1} for ice, 75.42 J$\cdot$K$^{-1}\cdot$mol^{-1} for liquid water, and 32.97 J$\cdot$K$^{-1}\cdot$mol^{-1} for water vapor. (Give a brief qualitative explanation of why C_P for liquid water is so much higher than that for ice or vapor.) The vapor pressure of water at the triple point of 273.16 K is 4.58 torr. The *standard* enthalpy change for vaporization of liquid water is $\Delta H_{298}^{\ominus} = 44.85$ kJ$\cdot$mol^{-1}. Calculate the vapor pressures of (a) liquid water and (b) ice at -10°C.

11. Pure water is saturated with an equimolar mixture of H_2, N_2, and O_2 at a total pressure of 5 atm. The water is then boiled and the gases removed. Calculate the composition of the gas mixture obtained in mole percent (after drying). Assume

Henry's Law is obeyed, $P_B = kX_B$, where P_B is in atmospheres and $k \times 10^{-4} =$ 7.80, 8.45, and 4.68, for H_2, N_2, and O_2, respectively.

12. Calculate the mass of (a) methanol and (b) ethylene glycol, which when dissolved in 5 kg water would just prevent the formation of ice at $-10°C$. Discuss briefly the relative advantages of these two compounds as antifreeze additives to cooling systems for motor cars.

13. For regular solutions in which the components have molecules of the same size, the Gibbs energy of mixing is

$$\Delta G_M = RT(n_1 \ln X_1 + n_2 \ln X_2) + (n_1 + n_2)X_1 X_2 w$$

where n denotes the amount of component, X the mole fraction, and w is a parameter that measures the deviation from Raoult's Law. Show that for such solutions, the chemical potentials are

$$\mu_1 = \mu_1^\bullet + RT \ln X_1 + wX_2^2$$
$$\mu_2 = \mu_2^\bullet + RT \ln X_2 + wX_1^2$$

where $\mu_1^\bullet$, $\mu_2^\bullet$ refer to the pure liquids. Show that the activity coefficients are given by

$$\ln \gamma_1 = X_2^2\left(\frac{w}{RT}\right)$$

$$\ln \gamma_2 = X_1^2\left(\frac{w}{RT}\right)$$

Benzene and CCl_4 form regular solutions with $w = 324 \, J \cdot mol^{-1}$ at 298 K. For an equimolar solution, calculate mixing functions, ΔH_M and ΔS_M^{ex}. Calculate γ_1 and γ_2 over a range of mole fractions from $X_1 = 0$ to 1, and plot the results.*

14. By a procedure similar to that used to derive Eq. (7.37), derive the expression for the elevation of the boiling point ΔT_b of a liquid A by addition of a nonvolatile solute at low molality m_B,

$$\Delta T_b = K_B m_B = \frac{RT_b^{\bullet 2} M_A}{(\Delta H_v) 1000} m_B$$

where M_A is the molecular weight of solvent A, ΔH_v is its enthalpy of vaporization per mole, and $T_b^\bullet$ is its boiling point.

15. Calculate the minimum work necessary to separate 1 mol of pure $C_2H_4Br_2$ from (a) a large volume of an equimolar solution of $C_2H_4Br_2$ and $C_3H_6Br_2$, and (b) a solution of exactly 2 mol of each component. The solutions are ideal.

16. When cells of the skeletal vacuole of a frog were placed in a series of NaCl solutions of different concentrations at 25°C, it was observed microscopically that they remained unchanged in 0.7% NaCl solution, shrank in more concentrated solutions, and swelled in more dilute solutions. Water freezes from the 0.7% salt solution at $-0.406°C$. What is the osmotic pressure of the cell cytoplasm at 25°C?

17. The human body contains about 150 g of potassium in ionized form distributed so that the concentration inside cells is 0.155 $mol \cdot dm^{-3}$ and that outside cells is 0.005 $mol \cdot dm^{-3}$. What is the total Gibbs energy ΔG associated with this unequal distribution compared to that for a uniform concentration?

*K.S. Pitzer and L. Brewer, *Thermodynamics* [2nd ed. of a book by G. N. Lewis and M. Randall] (New York: McGraw-Hill, 1961), Ch. 19.

18. The vapor pressure of a liquid that obeys Trouton's rule increases by 20 torr·K^{-1} at temperatures around its normal boiling point. Estimate the ΔH (vaporization) and T_b for this liquid.

19. At 1 atm pressure of CO_2, 1.7 g CO_2 will dissolve in 1 kg water at 20°C, and 1.0 g CO_2 will dissolve in 1 kg water at 40°C. If a bottle is unsafe with a pressure of more than 2 atm of gas inside, what is the maximum pressure of CO_2 at 20°C that is safe for a bottled beverage that might be exposed to 40°C? Assume solutions follow Henry's Law.

20. A new polypeptide antibiotic has been isolated and only 2 mg are available. By an ultracentrifugal method, the molecular weight is found to be $M = 12500$. It is desired to check this by another method. Calculate the freezing point depression, boiling point elevation, lowering of vapor pressure, and osmotic pressure at 20°C for the substance dissolved in water. Which method would you advise on the basis of your calculations? Why? What special precautions would you take in the measurements? What would you estimate the probable error in your determination of M to be if best modern techniques were used?

21. For the ideal solutions in Fig. 7.7, draw a curve showing how the mole fraction X_1 of propanol-2 in the vapor varies with that in the liquid. Use this curve to estimate the number of theoretical plates required in a distilling column to obtain a distillate with mole fraction $X_1 = 0.9$ from a solution with $X_1 = 0.1$. Assume total reflux.

22. The melting points and heats of fusion of o-, p-, and m-dinitrobenzene are 390.1, 446.7, and 363.0 K, and 16.33, 14.00, and 17.90 kJ·mol^{-1}, respectively. Assuming ideal solubility behavior, calculate the ternary eutectic temperature and composition for mixtures of o, m, and p compounds.

23. At 387.5°C, the vapor pressure of K is 3.25 and that of Hg is 1280 torr. Over a solution 50 mol % of K in Hg, the vapor pressure of K is 1.07 and that of Hg. 13.0 torr. Calculate the activities and activity coefficients of K and Hg in the solution. Calculate the ΔG_M of 0.5 mol of K and 0.5 mol of Hg at 387.5°C. If the ΔS_M is ideal, calculate the ΔH_M for the equimolar solution.

24. Review the latest literature on the mechanism of anesthesia. Can you as yet reach a decision on the relative validity of the Pauling theory and the theory of lipid solubility? Try to outline some experiments that might provide critical tests of the two theories.

25. Figure 7.20 shows the dendritic growth of the aluminum-rich phase to give star-shaped inclusions that usually display 10 spokes. Can you suggest a mechanism for such a pattern? In Chapter 18, we shall see that 10-fold rotational symmetry is forbidden for a crystal structure. How can these inclusions display what might seem to be a forbidden symmetry?

26. From the data in Problem 23, calculate w/kT from the Bragg–Williams model that yielded Eq. (7.53). Assume $z = 12$. (Why?)

27. Draw the phase diagram for the magnesium + nickel system in the condensed region. Magnesium melts at 651°C and Ni at 1452°C. They form a compound $MgNi_2$ that melts at 1145°C, and a compound Mg_2Ni that decomposes at 770°C into a liquid containing 50% by weight Ni and 50% of $MgNi_2$. The eutectics are at 23% Ni and 510°C and 89% Ni and 1080°C.

28. The following thermodynamic data are available at 1100 K:

	$H_T^{\ominus} - H_0^{\ominus}$ $(kJ \cdot mol^{-1})$	$-(G_T^{\ominus} - H_0^{\ominus})/T$ $(kJ \cdot K^{-1} \cdot mol^{-1})$
Cementite	112.71	154.5
α-Fe	34.71	40.59
Graphite	15.05	12.86

The solubility of cementite in α-Fe was measured as

$$\log (\text{wt} \% \, C) = \frac{-9700}{4.575T} + 0.41$$

Calculate the solubility of graphite in α-Fe at 1100 K.

29. J. J. Van Laar* gave a useful semiempirical equation for the excess free enthalpy of solutions,

$$\Delta G^{\text{ex}} = \frac{b_{12} X_1 X_2}{b_1 X_1 + b_2 X_2}$$

where b_{12}, b_1, and b_2 are characteristic constants. Show that the Van Laar relation implies that

$$\sqrt{\frac{1}{\ln \gamma_1}} = \frac{\sqrt{A_{12}}}{B_{12}}\left(\frac{X_1}{X_2}\right) + \frac{1}{\sqrt{A_{12}}}$$

$$\frac{1}{\sqrt{\ln \gamma_2}} = \frac{\sqrt{B_{12}}}{A_{12}}\left(\frac{X_2}{X_1}\right) + \frac{1}{\sqrt{B_{12}}}$$

where $A_{12} = (b_{12}/b_2 RT)$, $B_{12} = (b_{12}/b_1 \, RT)$.

30. The following data are available for solutions of acetone and chloroform at 50°C.

Mole fraction of acetone		Total pressure
Liquid	*Vapor*	*torr*
0	0	521
0.10	0.071	495
0.20	0.165	474
0.30	0.279	463
0.38	0.380	458
0.40	0.408	460
0.50	0.550	469
0.60	0.684	489
0.70	0.789	511
0.80	0.890	540
0.90	0.955	576
1.00	1.000	612

Calculate the activity coefficients of both components and plot them in accord with the Van Laar expressions in Problem 29, thereby determining A_{12} and B_{12}. Do the Van Laar coefficients have any interpretation in terms of interactions between components in the solutions?

*J. J. Van Laar, *Z. Physik. Chem.*, 72, 723 (1910), and *Z. Physik. Chem.*, 83, 599 (1913).

8
Chemical Affinity

"For instance," said the Captain, "what we call limestone is a more or less pure calcareous earth in close combination with a volatile acid known to us in gaseous form. If we put a piece of this limestone in a weak solution of sulphuric acid, the latter will take possession of the lime and appear with it in the form of gypsum; but the delicate gaseous acid will escape. Here we see a case of separation, and of a new combination, so that we think we are correct in using here the term 'elective affinity,' as it really looks as though one relation had been deliberately chosen in preference to another."

"Excuse me as I excuse the scientist," said Charlotte. "But in this case I should never think of a choice but of a compelling force—and not even that. After all, it may be merely a matter of opportunity. Opportunity makes connections as it makes thieves. As to your chemical substances, the choice seems to be exclusively in the hands of the chemist who brings these elements together. But once together, and they are united, God have mercy on them! In the present case I am only sorry for the poor gaseous acid which must again roam about in infinite space."

"The acid has only to combine with water to refresh the healthy and sick as a mineral spring," the Captain retorted.

"That is easy for the lime to say," said Charlotte. "The lime is taken care of; it is a substance; but that other displaced element may have much trouble until it finds a home again."

Johann Wolfgang von Goethe
(1809)*

The alchemists endowed their chemicals with almost human natures, and believed that reactions occurred when the reactants loved each other. Robert Boyle, in *The Sceptical Chymist* (1661), took a dim view of such theories:

> I look upon amity and enmity as affections of intelligent beings, and I have not yet found it explained by any, how those appetites can be placed in bodies inanimate and devoid of knowledge or of so much as sense.

In that same year, Isaac Newton entered Trinity College, Cambridge, at the age of 19. He always had a great interest in chemical experimentation and spent long hours in a laboratory in the garden behind his rooms in Cambridge.

> He very rarely went to bed before two or three of the clock, sometimes not till five or six . . . expecially at spring and fall of the leaf, at which time he used to employ about six weeks in his laboratory, the fire scarcely going out night or day, he sitting up one night and I another, till he finished his chemical experiments.

Thus reported Humphrey Newton, his cousin and assistant.

*Elective Affinities, trans. by Elizabeth Mayer and Louise Bogan (Chicago: Henry Regnery Co., 1963), p. 41.

Although he never published a book on chemistry, Newton formulated a number of important chemical questions in the *Queries* at the end of his *Opticks*. Probably as a result of his work on the gravitational attractions between bodies, he considered whether the affinity between different chemical substances could be due to attractive forces between their atoms or corpuscles. We must remember that at this time there was no clear understanding of the distinction between mixtures, solutions, and chemical compounds. In fact, it was well into the nineteenth century before such distinctions could be made at all precisely. In Query 31, Newton asked:

> Have not the small Particles of Bodies certain Powers, Virtues, or Forces by which they act at a distance, not only upon the Rays of Light for reflecting, refracting and inflecting them, but also upon one another for producing a great part of the Phenomena of Nature? For it's well known that Bodies act one upon another by the Attractions of Gravity, Magnetism, and Electricity; and these instances shew the tenor and course of Nature, and make it not improbable but that there may be more attractive Powers than these. For Nature is very constant and conformable to Herself. How these Attractions may be performed, I do not here consider. What I call attraction may be performed by impulse, or by some other means unknown to me. . . .

The origin of the affinity between different chemical substances is, of course, one of the great problems of chemical science. We now think gravitational attraction has no connection with chemical affinity, and the new kinds of force suggested by Newton are not required. We shall see later how the question Newton published in 1714 was finally to receive an answer in 1926, after the application of quantum theory to chemical problems: in essence, the answer would be that chemical attraction is electrical in its origin.

As usual, however, quite apart from the detailed microscopic mechanism of chemical affinities, the powerful methods of thermodynamics provide a mathematical analysis of the phenomena themselves, and describe exactly how chemical affinity is influenced by factors such as temperature, pressure, and concentration.

Experimental data on chemical reactivities were summarized in the early *Tables of Affinity*, such as that of Etienne Geoffroy in 1718, which recorded the order in which acids would expel weaker acids from combination with bases.

Claude Louis Berthollet, in 1801, in his book, *Essai de Statique Chimique*, contended that these tables were wrong in principle, since the quantity of reagent present plays an important role and a reaction can be reversed by adding a sufficient excess of one of the products. While serving as scientific adviser to Napoleon with the expedition to Egypt in 1799, he noted that sodium carbonate was being deposited along the shores of the salt lakes. The reaction $Na_2CO_3 + CaCl_2 \rightarrow CaCO_3 + 2NaCl$ as carried out in the laboratory was known to proceed to completion as $CaCO_3$ precipitated. Berthollet recognized that the large excess of sodium chloride present in evaporating brines could cause the reaction to be reversed, converting limestone into sodium carbonate. He went too far, however, and finally maintained that the actual *composition* of chemical compounds could be changed by varying the proportions of the reaction mixture. In the ensuing controversy with Louis Proust, the Law of Definite Proportions was well established, but Berthollet's ideas on chemical equilibrium, the good with the bad, were discredited, and consequently neglected for some 50 years. We now recognize many examples of definite departures from stoichiometric composition in various solid inorganic compounds, such as metallic oxides and sulfides, which are appropriately called *berthollides*, as distinguished from *daltonides*, in which definite proportions are rigorously maintained.

1. Dynamic Equilibrium

The correct form of what we now call the *Law of Chemical Equilibrium* was obtained as the result of a series of studies, not of equilibria but of chemical reaction rates. In 1850, Ludwig Wilhelmy investigated the hydrolysis of sugar with acid catalysts and found that the rate was proportional to the concentration of sugar remaining undecomposed. In 1862, Marcellin Berthelot and Péan de St. Gilles reported similar results in their famous paper* on the hydrolysis of esters, data from which are shown in Table 8.1. The effect on products of varying concentrations of reactants is readily apparent.

In 1863, the Norwegian chemists C. M. Guldberg and P. Waage expressed these relations in a very general form and applied the results to the problem of chemical equilibrium. They recognized that chemical equilibrium is a dynamic and not a static condition. It is characterized not by cessation of all interaction but by the fact that the rates of forward and reverse reactions have become the same.

Consider the general reaction, $A + B \rightleftharpoons C + D$. According to the *Law of Mass Action*, the rate of the forward reaction is proportional to the concentrations of A and B. If these are written as $[A]$ and $[B]$, $v_{\text{forward}} = k_f[A][B]$. Similarly, v_{back}

TABLE 8.1
Data of Berthelot and St. Gilles on the Reaction
$$C_2H_5OH + CH_3COOH \rightleftharpoons CH_3COOC_2H_5 + H_2O$$
(1 mol of acetic acid is mixed with varying amounts of alcohol,
and the amount of ester present at equilibrium is found)

Moles of alcohol	Moles of ester produced	Equilibrium constant $K = \dfrac{[EtAc][H_2O]}{[EtOH][HAc]}$
0.05	0.049	2.62
0.18	0.171	3.92
0.50	0.414	3.40
1.00	0.667	4.00
2.00	0.858	4.52
8.00	0.966	3.75

*Ann. Chim. Phys. (3), *65*, 385 (1862).

$= k_b[C][D]$. At equilibrium, therefore, $v_{forward} = v_{back}$, so that

$$k_f[A][B] = k_b[C][D]$$

Thus,

$$\frac{[C][D]}{[A][B]} = \frac{k_f}{k_b} = K$$

More generally, if the reaction is $aA + bB \rightleftharpoons cC + dD$, at equilibrium,

$$\frac{[C]^c [D]^d}{[A]^a [B]^b} = K \tag{8.1}$$

Equation (8.1) is a statement of Guldberg and Waage's *Law of Chemical Equilibrium*. The constant K is called the *equilibrium constant* of the reaction. It provides a quantitative expression for the dependence of chemical affinity on concentrations of reactants and products. By convention, the concentration terms for the reaction *products* are always placed in the *numerator* of the expression for the equilibrium constant.

Actually, this work of Guldberg and Waage does not constitute a general proof of the equilibrium law, since it is based on a special type of rate equation, which is certainly not always obeyed, as we shall see when we take up the study of chemical kinetics. Their work was important because they recognized that chemical affinity is influenced by two factors, a *concentration effect* and what might be called a *specific affinity*, which depends on the chemical natures of the reacting species, their temperature, and pressure. We shall later derive the equilibrium law from thermodynamic principles.

2. Free Enthalpy and Chemical Affinity

The Gibbs free energy function G (free enthalpy) described in Chapter 3 provides the true measure of chemical affinity under conditions of constant temperature and pressure. The free enthalpy change in a chemical reaction can be defined as $\Delta G = G(\text{products}) - G(\text{reactants})$. When the free enthalpy change is zero, there is no net work obtainable by any change or reaction at constant temperature and pressure. The system is in a state of equilibrium. When the free enthalpy change is positive for a proposed reaction, net work must be put into the system to effect the reaction; otherwise it cannot take place. When the free enthalpy change is negative, the reaction can proceed spontaneously with accomplishment of net work. The larger the amount of this work that can be accomplished, the farther removed is the reaction from equilibrium. For this reason, $-\Delta G$ has often been called the *driving force* of the reaction. From the statement of the equilibrium law, it is evident that this driving force depends on the concentrations of reactants and products. It also depends on their specific chemical constitutions, and on the temperature and pressure, which determine the molar free energies of reactants and products.

If we consider a reaction at constant temperature, e.g., one conducted in a thermostat, $-\Delta G = -\Delta H + T \Delta S$. The driving force is made up of two parts, a $-\Delta H$ term and a $T \Delta S$ term. The $-\Delta H$ term is the reaction heat at constant pressure, and the $T\,\Delta S$ term is the heat when the process is carried out reversibly. The difference is the amount of the reaction heat at constant pressure that can be converted into net work, i.e., total heat minus unavailable heat.

If a reaction at constant volume and temperature is considered, the decrease in the Helmholtz function, $-\Delta A = -\Delta U + T\,\Delta S$, would be used as the proper measure of the affinity of the reactants, or the driving force of the reaction. The condition of constant volume is met less frequently in laboratory practice.

We now can see why the principle of Berthelot and Thomsen (Section 2.21) was wrong. They considered only one of the two factors that make up the driving force of a chemical reaction, namely, the heat of reaction. They neglected the $T\,\Delta S$ term. The reason for the apparent validity of their principle was that, for many reactions, the ΔH term far outweighs the $T\,\Delta S$ term. This is especially so at low temperatures; at higher temperatures, the $T\,\Delta S$ term naturally increases.

The fact that the driving force for a reaction is large (ΔG is a large negative quantity) does not mean that the reaction will necessarily occur under any given conditions. An example is a bulb of hydrogen and oxygen on the laboratory shelf. For the reaction, $H_2 + \frac{1}{2}O_2 \longrightarrow H_2O(g)$, when $T = 298$ K and each of reactants and products are at $P = 1$ atm, $\Delta G_{298} = -228.6$ kJ. Despite the large negative ΔG, the reaction mixture can be kept for years without any detectable formation of water vapor; but if, at any time, a pinch of platinum-sponge catalyst is added, the reaction takes place with explosive violence. The necessary affinity certainly existed, but the *rate* of attainment of equilibrium depended on entirely different factors.

Another example is the resistance to oxidation of reactive metals, such as aluminum and magnesium: $2Mg + O_2$ (1 atm) $\longrightarrow 2MgO(c)$; $\Delta G_{298} = -570.6$ kJ. In this case, after the metal is exposed to air, it becomes covered with a very thin layer of oxide, and further reaction occurs at an extremely slow rate because reactants must diffuse through the oxide film. Thus, the equilibrium condition is not attained. The incendiary bomb and the thermite reaction, on the other hand, remind us that the large $-\Delta G$ for this reaction is a valid measure of a great affinity between the reactants.

3. Condition for Chemical Equilibrium

We shall now give a more exact mathematical derivation of the condition for equilibrium. Consider a chemical reaction,

$$v_1 A_1 + v_2 A_2 + \cdots \longrightarrow v_n A_n + v_{n+1} A_{n+1} + \cdots \tag{8.2}$$

It can be written briefly as

$$\sum v_i A_i = 0 \tag{8.3}$$

if we recall the convention that stoichiometric mole numbers v_i are positive for products and negative for reactants.

We can denote the *extent of reaction** by the symbol ξ. A change from ξ to $\xi + d\xi$ means that $v_1\,d\xi$ moles of A_1, $v_2\,d\xi$ moles of A_2, etc., have reacted to form

*Formerly, ξ was called the *degree of advancement* of the reaction.

$v_n \, d\xi$ moles of A_n, etc. Thus, ξ is a convenient measure of *extent of reaction*. The number of moles reacted of any component i is

$$dn_i = v_i \, d\xi \tag{8.4}$$

Let us consider a system containing the reactants and products of (8.2) in equilibrium at constant T and P. To derive the equilibrium condition, we follow the procedure previously used in discussing phase equilibria (Sect. 6.5). We suppose that there could be a reaction of extent $\delta\xi$. The change in the Gibbs energy of the system would be given from Eq. (6.5) as

$$\delta G = \sum \mu_i \, \delta n_i$$

From (8.4), therefore,

$$\delta G = \sum v_i \, \mu_i \, \delta\xi$$

Hence,

$$\frac{\delta G}{\delta\xi} = \sum v_i \, \mu_i \tag{8.5}$$

At equilibrium, however, G must be a minimum with respect to any virtual displacement of the reaction. Thus,

$$\left(\frac{\delta G}{\delta\xi}\right)_{T,P} = 0 \tag{8.6}$$

We thus derive from (8.5) and (8.6) the equilibrium condition,

$$\left(\sum v_i \, \mu_i\right)_{\text{eq}} = 0 = \Delta G_{\text{eq}} \tag{8.7}$$

In 1922, the Belgian thermodynamicist de Donder introduced a new function called the *Affinity*, defined by

$$\mathscr{A} = -\left(\frac{\partial G}{\partial\xi}\right)_{P,T} \tag{8.8}$$

At equilibrium, $\mathscr{A} = 0$.

4. Standard Free Enthalpies

In Chapter 2, we introduced the definition of standard states in order to simplify calculations with energies and enthalpies. Similar conventions are helpful for use with free enthalpy, and various choices of standard state have been made.

A standard state frequently used is the *state of the substance under 1 atm pressure*. This is a useful definition for gas reactions; for reactions in solution, other choices of standard state may be more convenient and will be introduced as needed. The superscript "$\ominus$" will be used to indicate a standard state. The absolute temperature will be written as a subscript. We shall often write 298, when 298.15 K (25°C) is understood.

The most stable form of an element in the standard state (1 atm pressure) and at a temperature of 298.15 K will by convention be assigned a free enthalpy of zero. The *standard free enthalpy of formation of a compound* is the free enthalpy of the reaction by which it is formed from its elements, when all reactants and products are in standard states. For example,

$$H_2 \text{ (1 atm)} + \tfrac{1}{2}O_2 \text{ (1 atm)} \longrightarrow H_2O \text{ (g; 1 atm)}, \qquad \Delta G^{\ominus}_{298} = -228.61 \text{ kJ}$$

$$S(\text{rhombic crystal}) + 3F_2 \text{ (1 atm)} \longrightarrow SF_6 \text{ (g; 1 atm)}, \qquad \Delta G^{\ominus}_{298} = -983.2 \text{ kJ}$$

In this way, it is possible to make tabulations of standard free enthalpies such as those given by the National Bureau of Standards. Examples are collected in Table 8.2. The methods used to determine these values will be described later.

Free enthalpy equations can be added and subtracted just as thermochemical equations, so that the free enthalpy of any reaction can be calculated from the sum of the free enthalpies of products minus the sum of the free enthalpies of reactants:

$$\Delta G^{\ominus} = G^{\ominus}(\text{products}) - G^{\ominus}(\text{reactants})$$

If we adopt the convention that the stoichiometric number v_i of moles of a reactant is negative in the summation, this equation can be written concisely as

$$\Delta G^{\ominus} = \sum v_i \, G_i^{\ominus} \tag{8.9}$$

For example,

$$Cu_2O(c) + NO(g) \longrightarrow 2CuO(c) + \tfrac{1}{2}N_2(g)$$

From Table 8.2,

$$\Delta G^{\ominus}_{298} = 2(-127.2) + \tfrac{1}{2}(0) - (-146.4) - 86.57 = -194.6 \text{ kJ}$$

Since $\Delta G^{\ominus}$ often varies rapidly with the temperature, it is not a suitable function for tables of thermodynamic data from which interpolation is usually necessary. Thus, $-(G_T^{\ominus} - H_{298}^{\ominus})/T$ or $-(G_T^{\ominus} - H_0^{\ominus})/T$ is usually tabulated.* In these functions, the free enthalpy is expressed with reference to the enthalpy at either 298 or 0 K. An example of such a tabulation of free enthalpy data is shown in Table 8.3.

5. Free Enthalpy and Equilibrium in Reactions of Ideal Gases

Many important applications of equilibrium theory are in the field of homogeneous gas reactions—i.e., reactions taking place entirely between gaseous products and reactants. To a good approximation, in many such cases, the gases may be considered to obey the ideal gas laws.

At constant temperature, the differential of free enthalpy for an ideal gas is given from (3.49) as

$$dG = V \, dP = nRT \, d \ln P$$

When we integrate from $G^{\ominus}$ and $P^{\ominus}$, the free enthalpy and pressure in the chosen standard state, to G and P, the values in any other state,

$$G - G^{\ominus} = nRT \ln \left(\frac{P}{P^{\ominus}} \right) \tag{8.10}$$

---

*See K. S. Pitzer and L. Brewer, *Thermodynamics*, revision of classic text of G. N. Lewis and M. Randall (New York: McGraw-Hill Book Company, 1961), pp. 166, 669.

TABLE 8.2
Standard Free Enthalpies of Formation at 298.15 K

Compound	State	$\Delta G_f^{\ominus}(298.15)$ $(kJ \cdot mol^{-1})$	Compound	State	$\Delta G_f^{\ominus}(298.15)$ $(kJ \cdot mol^{-1})$
$AgCl$	c	-109.70	HCN	aq	120.0
$AgBr$	c	-95.94	(ionized)	aq	172.0
AgI	c	-66.32	HDO	l	-241.9
Al_2O_3	Corundum	-1582.0		g	-233.13
As_2O_5	c	-782.4	HF	g	-273.0
B_2O_3	c	-493.7	HN_3	g	328.0
$CaCO_3$	c	-1128.8	HNO_3	aq	-111.3
$CaSO_4$	c	-1320.3	H_2O	l	-237.18
CCl_4	l	-65.27		g	-228.59
	g	-60.63	H_2O_2	l	-120.4
CF_4	g	-879.0		g	-105.6
CH_3OH	l	-166.4	H_2S	g	-33.6
CH_4	g	-40.75	H_2SO_4	aq	-744.63
$CHCl_3$	l	-73.72	H_3PO_4	aq	-1142.7
	g	-70.37	(ionized)	aq	-1019.0
CH_3COOH	l	-390.0	$NaCl$	c	-384.03
(ionized)	aq	-369.4	NH_3	g	-16.5
(un-ionized)	aq	-396.6	NH_4Cl	c	-203.00
C_2H_2	g	209.2	NH_4CNO	aq	-177.0
C_2H_4	g	68.12	NH_4N_3	c	274.0
C_2H_6	g	-32.9	NH_4NO_3	c	-184.0
C_2H_5OH	l	-174.9	NH_4OH	aq	-263.8
	g	-168.6	(ionized)	aq	-236.6
C_6H_6	l	124.50	NO	g	86.57
	g	129.66	NO_2	g	51.30
CO	g	-137.15	N_2O	g	104.2
CO_2	g	-394.36	N_2O_4	g	102.0
	aq	-386.0	N_2O_5	g	115.0
CO_3^{2-}	aq	-527.9		c	114.0
$CO(NH_2)_2$	c	-196.8	O	g	231.75
COS	g	-169.3	O_3	g	163.0
CS_2	l	-65.27	OH	g	34.2
	g	-67.15	P_4	g	24.5
CuO	c	-127.2	PCl_3	l	-272.0
Cu_2O	c	-146.4		g	-268.0
$CuBr_2$	c	-127.0	PF_3	g	-897.5
D	g	206.5	PH_3	g	13.4
D_2O	l	-243.49	PH_4I	c	0.8
	g	-234.55	S_8	g	49.66
Fe_2O_3	c	-741.0	SO_2	g	-300.19
Fe_2S_2	c	-166.7	SO_3	g	-371.1
H	g	203.26	SiF_4	g	-1572.7
H^+	aq	$[0.0]$	SiO_2	c (α-quartz)	-856.67
Hg	g	1.72	$ZnCl_2$	c	-369.43
HBr	g	-53.43	ZnO	c	-318.3
HCl	g	-95.300	$ZnSO_4$	c	-874.5
	aq	-131.26	$ZnSO_4 \cdot H_2O$	c	-1132.1
HCO_3^-	aq	-586.85	$ZnSO_4 \cdot 6H_2O$	c	-2324.8
HCN	g	125.0	$ZnSO_4 \cdot 7H_2O$	c	-2563.1

*National Bureau of Standards Technical Note 270–3, 270–4, *Selected Values of Chemical Thermodynamic Properties* (Washington: U.S. Government Printing Office, 1968, 1969). The standard states for the aqueous acids in the table correspond to unit activity on the molality scale.

TABLE 8.3
Free Enthalpy Functions $(G^\ominus - H_0^\ominus)/T$, from 0 to 4000 K
$(J \cdot K^{-1} \cdot mol^{-1})$

Compound	Formula	State	Temperature (K)										
			0	298.15	400	600	800	1000	1500	2000	2500	3000	4000
Oxygen	O_2	g	0	−175.98	−184.56	−196.51	−205.20	−212.12	−225.13	−235.73	−242.38	−248.81	−259.23
Hydrogen	H_2	g	0	−102.19	−110.55	−122.19	−130.48	−136.98	−148.91	−157.61	−164.55	−170.37	−179.86
Hydroxyl	OH	g	0	−154.07	−162.77	−174.77	−183.28	−189.89	−204.70	−210.94	−218.00	−223.93	−233.56
Water	H_2O	g	0	−155.53	−165.30	−178.94	−188.89	−196.72	−211.8	−223.34	−232.84	−240.96	
Nitrogen	N_2	g	0	−162.41	−170.96	−182.79	−191.25	−197.93	−210.39	−219.57	−226.89	−232.99	−242.85
Nitric oxide	NO	g	0	−179.83	−188.84	−201.21	−210.05	−217.00	−229.97	−239.86	−247.04	−253.34	−263.45
Carbon	C	graphite	0	−2.164	−3.450	−6.180	−8.945	−11.59	−17.49				
Carbon monoxide	CO	g	0	−168.82	−177.37	−189.21	−197.71	−204.43	−217.00	−226.26	−233.64	−239.80	−249.73
Carbon dioxide	CO_2	g	0	−182.234	−191.74	−206.02	−217.13	−226.39	−244.68	−258.78	−270.29	−280.79	

287

Since $P^\ominus = 1$ atm, this becomes

$$G - G^\ominus = nRT \ln P \tag{8.11}$$

Equation (8.11) gives the free enthalpy of an ideal gas at pressure P (atm) and temperature T, minus its free enthalpy in a standard state at $P = 1$ atm and temperature T.

If an ideal mixture of ideal gases is considered, Dalton's Law of Partial Pressures holds, and the total pressure is the sum of the pressures that the gases would exert if each one occupied the entire volume by itself. These pressures are called the *partial pressures* of the gases in the mixture, $P_1, P_2, \ldots, P_n$. Thus, if v_i is the number of moles of gas i in the mixture,

$$P_i V = v_i RT \tag{8.12}$$

For each individual gas i in the ideal mixture, (8.11) can be written

$$G_i - G_i^\ominus = RT \ln P_i \tag{8.13}$$

where G_i is the molar free enthalpy at P_i and $G_i^\ominus$ is the molar free enthalpy at $P_i^\ominus = 1$ atm. For a chemical reaction, therefore, from (8.9),

$$\Delta G - \Delta G^\ominus = RT \sum v_i \ln P_i \tag{8.14}$$

If we now consider the pressures P_i to be the equilibrium pressures in the gas mixture, ΔG must be equal to zero for the reaction at equilibrium [Eq. (8.7)]. Thus, we obtain the important relation,

$$-\Delta G^\ominus = RT \sum v_i \ln P_i^{eq}$$

or

$$\sum v_i \ln P_i^{eq} = \frac{-\Delta G^\ominus}{RT} \tag{8.15}$$

Since $\Delta G^\ominus$ is a function of the temperature alone, the left side of this expression is equal to a constant at constant temperature. For a typical reaction $aA + bB \rightleftharpoons cC + dD$, the summation can be written out as

$$\sum v_i \ln P_i^{eq} = \ln \frac{(P_C^{eq})^c (P_D^{eq})^d}{(P_A^{eq})^a (P_B^{eq})^b}$$

This expression is simply the logarithm of the equilibrium constant in terms of partial pressures, which we denote as $K_P(T)$. Equation (8.15) therefore becomes

$$-\Delta G^\ominus = RT \ln K_P \tag{8.16}$$

The analysis in this section has now established two important results. We have given a rigorous thermodynamic proof that for a reaction between ideal gases there exists an equilibrium constant K_P, defined by

$$K_P = \frac{P_C^c P_D^d}{P_A^a P_B^b} \quad \text{(at equilibrium)} \tag{8.17}$$

This constitutes a thermodynamic proof of the Law of Chemical Equilibrium. Secondly, an explicit expression has been derived, (8.16), which *relates the equilibrium constant to the standard free enthalpy change* in the chemical reaction. We are now able, from thermodynamic data, to calculate the equilibrium constant, and thus the concentration of products from any given concentration of reactants, which was one of the fundamental problems that chemical thermodynamics aimed to answer.*

From (8.16), K_P is a function of temperature, $K_P(T)$, and $\Delta G^\ominus$ is itself a function of T. However, K_P is independent of the total pressure and independent of variations in the individual partial pressures. These partial pressures are varied by changing the proportions of reactants and products in the initial reaction mixture. After the mixture comes to equilibrium, the partial pressures must conform to (8.17). It should not be forgotten, however, that our equilibrium theory has so far been restricted to ideal gas mixtures.

6. Equilibrium Constant in Concentration Units

Sometimes the equilibrium constant is expressed in terms of concentrations c_i. For an ideal gas, $P_i = n_i RT/V = c_i RT$. Substituting this into (8.17), we find

$$K_P = \frac{c_C^c c_D^d}{c_A^a c_B^b}(RT)^{c+d-a-b} = K_c\,(RT)^{\Delta v} \tag{8.18}$$

Here, K_c is the equilibrium constant in terms of concentrations (e.g., in units of $\text{mol}\cdot\text{dm}^{-3}$) and Δv is the number of moles of products *less* the number of moles of reactants in the stoichiometric equation for the reaction.

Another way of expressing the composition of an equilibrium mixture is in terms of mole fractions X_i. From (1.38),

$$P_i = X_i P \quad \text{and} \quad X_i = \frac{P_i}{P}$$

Therefore, the equilibrium constant in mole fractions is

$$K_X = \frac{X_C^c X_D^d}{X_A^a X_B^b} = K_P\,P^{-\Delta v} \tag{8.19}$$

Since K_P for ideal gases is independent of pressure, it is evident that K_X is a function of pressure except when $\Delta v = 0$. It is thus constant only with respect to variations of the X's at constant T and P.

7. Measurement of Homogeneous Gas Equilibria

Experimental methods for measuring gaseous equilibria can be classified as either static or dynamic. In the static method, measured amounts of reactants are introduced into suitable reaction vessels, which are closed and kept in a thermostat

---

*The dimensions of K_P sometimes cause difficulty. From (8.16) it is evident that K_P is dimensionless, but (8.17) might seem to imply that it has the dimensions of $P^{\Delta v}$. The apparent paradox is resolved when we consider that (8.11) was obtained from (8.10) by setting $P^\ominus = 1$ atm. Thus, the "pressures" that appear in (8.17) are really ratios of pressures to a standard pressure of 1 atm, and, hence, dimensionless. It is therefore necessary always to use the atmosphere as the pressure unit in K_P expressions, because we have chosen $P^\ominus = 1$ atm as our standard state.

until equilibrium has been attained. The contents of the vessels are then analyzed to determine the equilibrium concentrations. If the reaction proceeds very slowly at temperatures below those chosen for the experiment, it is sometimes possible to "freeze the equilibrium" by chilling the reaction vessel rapidly. The vessel may then be opened and its contents analyzed chemically. This was the procedure used by Max Bodenstein* in his classic investigation of the hydrogen + iodine equilibrium: $H_2 + I_2 \rightleftharpoons 2HI$. The reaction products were treated with an excess of standard alkali; iodide and iodine were determined by titration, and the hydrogen gas was collected and its volume measured. For the formation of hydrogen iodide, $\Delta v = 0$; there is no change in the number of moles during the reaction. Therefore, $K_P = K_c = K_X$.

If the initial numbers of moles of H_2 and I_2 are a and b, respectively, they will be reduced to $a - x$ and $b - x$ with the formation of $2x$ moles of HI. The total number of moles at equilibrium is, therefore, $a + b + c$, where c is the initial number of moles of HI. Accordingly, the equilibrium constant can be written,

$$K_P = K_X = \frac{(c + 2x)^2}{(a - x)(b - x)}$$

The $(a + b + c)$ terms required to convert numbers of *moles* into *mole fractions* have canceled between numerator and denominator. In a run at 721 K, Bodenstein mixed 22.13 cm³ of H_2 with 16.18 cm³ of I_2, and found 28.98 cm³ of HI at equilibrium. Hence,

$$K = \frac{(28.98)^2}{(22.13 - 14.49)(16.18 - 14.49)} = 65.0$$

In the dynamic method, the reactant gases are passed through a reactor at elevated temperature at a rate slow enough to allow complete attainment of equilibrium. This condition can be tested by making runs at successively lower flow rates, until there is no longer any change in the observed extent of reaction. The effluent gases are rapidly chilled and analyzed. Sometimes a catalyst is included in the hot zone to speed attainment of equilibrium. This is a safer method if a suitable catalyst is available, since it minimizes the possibility of any back reaction occurring after the gases leave the reaction chamber. A catalyst changes the rate of the reaction, not the position of final equilibrium.

These flow methods were extensively used by W. Nernst and F. Haber (*ca.* 1900) in their pioneer work on technically important gas reactions. An example is the *water-gas equilibrium*, which was studied both with and without an iron catalyst.† The reaction is

$$H_2 + CO_2 \rightleftharpoons H_2O + CO$$

with

$$K_P = \frac{P_{H_2O} P_{CO}}{P_{H_2} P_{CO_2}}$$

*Z. Physik Chem., 22, 1 (1897); 29, 295 (1899).
†Z. Anorg. Chem., 38, 5 (1904).

If the original mixture contains a moles of H_2, b moles of CO_2, c moles of H_2O, and d moles of CO, analysis of the data is as follows:

Constituent	Original moles	At Equilibrium		
		Moles	Mole fraction	Partial pressure
H_2	a	$a - x$	$(a - x)/(a + b + c + d)$	$[(a - x)/n]P$
CO_2	b	$b - x$	$(b - x)/(a + b + c + d)$	$[(b - x)/n]P$
H_2O	c	$c + x$	$(c + x)/(a + b + c + d)$	$[(c + x)/n]P$
CO	d	$d + x$	$(d + x)/(a + b + c + d)$	$[(d + x)/n]P$

Total moles at equilibrium: $a + b + c + d = n$

Substituting the partial pressures, we obtain

$$K_P = \frac{(c + x)(d + x)}{(a - x)(b - x)}$$

Values for the equilibrium composition, obtained by analysis of the product gases, have been used to calculate the constants in Table 8.4.

TABLE 8.4
The Water Gas Equilibrium, $H_2 + CO_2 \rightleftharpoons H_2O + CO$ (1259 K and 1 atm)

Initial composition (mol %)		Equilibrium composition (mol %)			K_P
CO_2	H_2	CO_2	H_2	$CO = H_2O$	
10.1	89.9	0.69	80.52	9.40	1.59
30.1	69.9	7.15	46.93	22.96	1.57
49.1	51.9	21.44	22.85	27.86	1.58
60.9	39.1	34.43	12.68	26.43	1.61
70.3	29.7	47.51	6.86	22.82	1.60

8. Principle of Le Chatelier and Braun

If a system of chemical reactants and products in stable equilibrium is perturbed by subjecting it to a small variation in one of the variables that define the equilibrium state, the system will tend to return to an equilibrium state, which usually is somewhat different from the initial state. Henry Le Chatelier (1888) and F. Braun* (1887) considered this problem theoretically and arrived at the general principle

*H. Le Chatelier, *Recherches sur les Équilibres Chimiques* (Paris: Vve. Ch. Dunod, 1888); *Ann. Mines*, *13*, 200 (1888). F. Braun, *Z. Physik Chem.*, *1*, 259 (1887).

that a thermodynamic system tends to balance or counteract the effects of any imposed stress. In the words of Le Chatelier,

> Tout systéme en équilibre chimique éprouve, du fait de la variation d'un seul des facteurs de l'équilibre, une transformation dans un sens tel que, si elle produisait seul, elle amènerait une variation de signe contraire du facteur considéré.*

The principle indicates, for example, that if heat is evolved in a chemical reaction, increasing the temperature tends to reverse the reaction; if the volume decreases in a reaction, increasing the pressure shifts the equilibrium position toward the product side.

A proof of the Le Chatelier–Braun Principle as applied to chemical reactions follows.† From (8.5) and the Gibbs equation (6.4),

$$dG = -S \, dT + V \, dP + \left(\sum_{i=1}^{c} \nu_i \mu_i\right) d\xi \tag{8.20}$$

At equilibrium, from (8.7) and (8.8), the affinity $\mathscr{A}$ vanishes, and

$$-\mathscr{A} = 0 = \left(\frac{\partial G}{\partial \xi}\right)_{T,P} = \sum \nu_i \mu_i \tag{8.21}$$

From (8.20), the total differential of $-\mathscr{A} = (\partial G/\partial \xi)$ is

$$-d\mathscr{A} = d\left(\frac{\partial G}{\partial \xi}\right)_{T,P} = -\left(\frac{\partial S}{\partial \xi}\right)_{T,P} dT + \left(\frac{\partial V}{\partial \xi}\right)_{T,P} dP + \left(\frac{\partial^2 G}{\partial \xi^2}\right)_{T,P} d\xi \tag{8.22}$$

For all equilibrium states,

$$-d\mathscr{A} = d\left(\frac{\partial G}{\partial \xi}\right)_{T,P} = 0$$

so that (8.22) yields

$$\left(\frac{\partial \xi_e}{\partial T}\right)_P = \frac{(\partial S/\partial \xi)_{T,P}}{(\partial^2 G/\partial \xi^2)_{T,P}} = \frac{(dq/d\xi)_{T,P}}{T(\partial^2 G/\partial \xi^2)_{T,P}} \tag{8.23}$$

and

$$\left(\frac{\partial \xi_e}{\partial P}\right)_T = -\frac{(\partial V/\partial \xi)_{T,P}}{(\partial^2 G/\partial \xi^2)_{T,P}} \tag{8.24}$$

where ξ_e is the equilibrium value of the extent of reaction, and dq is the reversible heat added to the system. Now, for a stable equilibrium, $(\partial^2 G/\partial \xi^2)_{T,P} > 0$ always (condition of minimum in Gibbs free energy). Therefore, (8.23) shows that if the T of an equilibrium reaction system is increased at constant P, the extent of reaction increases in that direction in which heat is absorbed by the system at constant T and P. Similarly, (8.24) shows that an increase in P at constant T results in a change in reaction in that direction in which the volume of the system is decreased at constant T and P.

---

*Any system in chemical equilibrium, as a result of the variation in one of the factors determining the equilibrium, undergoes a change such that, if this change had occurred by itself, it would have introduced a variation of the factor considered in the opposite direction.

†After J. G. Kirkwood and I. Oppenheim, *Chemical Thermodynamics* (New York: McGraw-Hill Book Company, 1961).

9. Pressure Dependence of Equilibrium Constant

The equilibrium constants K_P and K_c are independent of the pressure for ideal gases; the constant K_X is pressure dependent. Since $K_X = K_P P^{-\Delta v}$,

$$\ln K_X = \ln K_P - \Delta v \ln P$$

$$\frac{d \ln K_X}{dP} = \frac{-\Delta v}{P} = \frac{-\Delta V}{RT} \tag{8.25}$$

When a reaction occurs without any change in the total number of moles of gas in the system, $\Delta v = 0$. An example is the previously considered water-gas reaction. In such instances, the constant K_P is the same as K_X or K_c, and for ideal gases, the position of equilibrium does not depend on the total pressure. When Δv is not equal to zero, the pressure dependence of K_X is given by (8.25). When there is a decrease in the mole numbers ($\Delta v < 0$), and thus a decrease in the volume, K_X increases with increasing pressure. If there is an increase in v and V ($\Delta v > 0$), K_X decreases with increasing pressure.

An important class of reactions for which $\Delta v \neq 0$ is that of molecular dissociations. An example is the dissociation of dinitrogen tetroxide into dioxide, $N_2O_4 \longrightarrow 2NO_2$. In this case,

$$K_P = \frac{P^2_{NO_2}}{P_{N_2O_4}}$$

If 1 mol of N_2O_4 is dissociated at equilibrium to a fractional extent a, $2a$ mol of NO_2 are produced. The total number of moles at equilibrium then becomes $(1 - a) + 2a = 1 + a$. It follows that

$$K_X = \frac{[2a/(1 + a)]^2}{(1 - a)/(1 + a)} = \frac{4a^2}{1 - a^2}$$

Since for this reaction $\Delta v = +1$,

$$K_P = K_X P = \frac{4a^2}{1 - a^2} P$$

When a is small compared to unity, this expression predicts that the degree of dissociation a varies inversely as the square root of the pressure.

Even at room temperatures, N_2O_4 is appreciably dissociated. As a result, its pressure is greater than that predicted by the ideal gas law, since each mole yields $(1 + a)$ moles of gas after dissociation. Thus, $P(\text{ideal}) = nRT/V$, whereas $P(\text{observed}) = (1 + a) nRT/V$. Hence,

$$a = \left(\frac{V}{nRT}\right)(P_{ob} - P_{id})$$

This behavior provides a simple means for measuring a. For example, in an experiment at 318 K and 1 atm, a is found to be 0.38. Therefore,

$$K_X = 4(0.38)^2/(1 - 0.38^2) = 0.67$$

At 10 atm, $K_X = 0.067$ and $a = 0.128$.

Dissociations of elementary gases are important in high temperature processes and in research on the upper atmosphere. Constants for a few of these equilibria are collected in Table 8.5.

An inert gas added to a mixture of reacting ideal gases produces no effect if $\Delta v = 0$ for the reaction. If, however, $\Delta v \neq 0$, the addition of inert gas will influence the extent of reaction at equilibrium (see Problem 6).

TABLE 8.5.
Equilibrium Constants of Dissociation Reactions

	K_P (atm)				
$T(K)$	$O_2 \rightleftharpoons 2O$	$H_2 \rightleftharpoons 2H$	$N_2 \rightleftharpoons 2N$	$Cl_2 \rightleftharpoons 2Cl$	$Br_2 \rightleftharpoons 2Br$
600	1.4×10^{-37}	3.6×10^{-33}	1.3×10^{-56}	4.8×10^{-16}	6.18×10^{-12}
800	9.2×10^{-27}	1.2×10^{-23}	5.1×10^{-41}	1.04×10^{-10}	1.02×10^{-7}
1000	3.3×10^{-20}	7.0×10^{-18}	1.3×10^{-31}	2.45×10^{-7}	3.58×10^{-5}
1200	8.0×10^{-16}	5.05×10^{-14}	2.4×10^{-25}	2.48×10^{-5}	1.81×10^{-3}
1400	1.1×10^{-12}	2.96×10^{-11}	7.5×10^{-21}	8.80×10^{-4}	3.03×10^{-2}
1600	2.5×10^{-10}	3.59×10^{-9}	1.8×10^{-17}	1.29×10^{-2}	2.55×10^{-1}
1800	1.7×10^{-8}	1.52×10^{-7}	7.6×10^{-15}	0.106	
2000	5.2×10^{-7}	3.10×10^{-6}	9.8×10^{-13}	0.570	

10. Temperature Dependence of Equilibrium Constant

An expression for the variation of K_P with temperature is derived by combining (8.16) and (3.53). Since

$$-\Delta G^{\ominus} = RT \ln K_P \tag{8.26}$$

and

$$\left[\frac{\partial}{\partial T} \left(\frac{\Delta G^{\ominus}}{T} \right) \right]_P = \frac{-\Delta H^{\ominus}}{T^2} \tag{8.27}$$

therefore, since K_P is function only of T,

$$\left(\frac{\partial \ln K_P}{\partial T} \right)_P = \frac{d \ln K_P}{dT} = \frac{\Delta H^{\ominus}}{RT^2} \tag{8.28}$$

If the reaction is endothermic ($\Delta H^{\ominus} > 0$), the equilibrium constant increases with temperature; if the reaction is exothermic ($\Delta H^{\ominus} < 0$), the equilibrium constant decreases as temperature is raised. Equation (8.28) can also be written

$$\frac{d \ln K_P}{d(1/T)} = \frac{-\Delta H^{\ominus}}{R} \tag{8.29}$$

Thus, if $\ln K_P$ is plotted against $1/T$, the slope of the curve at any point is equal to $-\Delta H^{\ominus}/R$. As an example of this treatment, data* for variation with tempera-

*A. H. Taylor and R. H. Crist, *J. Am. Chem. Soc.*, 63, 1377 (1941).

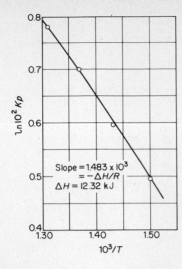

FIG. 8.1 Variation with temperature of $K_p = P_{H_2}P_{I_2}/P^2HI$, illustrating the van't Hoff equation (8.29).

ture of the $2HI \rightleftharpoons H_2 + I_2$ equilibrium are plotted in Fig. 8.1. The curve is practically a straight line, indicating $\Delta H^\ominus$ is constant for the reaction over the experimental temperature range. The value calculated from the slope is $\Delta H^\ominus = 12.32$ kJ.

It is also possible to measure the equilibrium constant at one temperature and with a value of $\Delta H^\ominus$ obtained from thermochemical data to calculate the constant at other temperatures. Equation (8.28) can be integrated to give

$$\ln \frac{K_P(T_2)}{K_P(T_1)} = \int_{T_1}^{T_2} \frac{\Delta H^\ominus}{RT^2} dT$$

Since, over a short temperature range $\Delta H^\ominus$ may often be nearly constant, we obtain

$$\ln \frac{K_P(T_2)}{K_P(T_1)} = \frac{-\Delta H^\ominus}{R}\left(\frac{1}{T_2} - \frac{1}{T_1}\right) \tag{8.30}$$

If the heat capacities of reactants and products are known as functions of temperature, an explicit expression for the temperature dependence of $\Delta H^\ominus$ can be derived from (2.42). This expression for $\Delta H^\ominus$ as a function of temperature can then be substituted into (8.28), whereupon integration yields an explicit equation for K_P as a function of temperature. This has the form

$$\ln K_P = \frac{-\Delta H_0^\ominus}{RT} + A \ln T + BT + CT^2 + \cdots + I \tag{8.31}$$

The value of the integration constant I can be determined if the value of K_P is known at any one temperature, either experimentally or by calculation from $\Delta G^\ominus$. Recall that one value of $\Delta H^\ominus$ is needed to determine $\Delta H_0^\ominus$, the integration constant of the Kirchhoff equation.

Therefore, from a knowledge of the heat capacities of reactants and products, and of one value each for $\Delta H^\ominus$ and K_P, we may calculate the equilibrium constant at any temperature. For example,

$$CO + H_2O(g) \rightleftharpoons H_2 + CO_2, \qquad K_P = \frac{P_{H_2}P_{CO_2}}{P_{CO}P_{H_2O}}$$

From Table 8.2, the standard free enthalpy change at 298 K is

$$\Delta G^{\ominus}_{298} = -394.36 - (-228.59 - 137.15) = -28.62 \text{ kJ}$$

Thus,

$$\ln K_P(298) = \frac{28.62}{298R} = 11.48 \quad \text{or} \quad K_P(298) = 9.55 \times 10^4$$

From the enthalpies of formation in Table 2.2,

$$\Delta H^{\ominus}_{298} = -393.50 - (-242.21 - 110.54) = -41.15 \text{ kJ}$$

The heat capacity Table 2.5 yields, for this reaction,

$$\Delta C_P = C_P(CO_2) + C_P(H_2) - C_P(CO) - C_P(H_2O)$$
$$= -2.155 + 26.1 \times 10^{-3}T - 12.5 \times 10^{-6}T^2 \text{ J·K}^{-1}$$

From (2.42),

$$\Delta H^{\ominus} = \Delta H^{\ominus}_0 - 2.155T + 13.1 \times 10^{-3}T^2 - 4.17 \times 10^{-6}T^3$$

Substituting $\Delta H^{\ominus} = -41.15$, $T = 298$ K, and solving for $\Delta H^{\ominus}_0$, we get $\Delta H^{\ominus}_0 = -41.51$ kJ. Then the temperature dependence of the equilibrium constant, (8.31), becomes

$$\ln K_P = \frac{41.51}{RT} - \frac{2.155}{R}\ln T + \frac{13.1 \times 10^{-3}}{R}T - \frac{4.17 \times 10^{-6}}{2R}T^2 + I$$

By inserting the value of $\ln K_P$ at 298 K, and $R = 8.314 \times 10^{-3}$kJ·K^{-1}·mol^{-1}, we can evaluate the integration constant as $I = -3.97$. Now K_P can be readily calculated at any temperature. For example, at 800 K, $\ln K_P = 1.63$, $K_P = 5.10$.

11. Equilibrium Constants from Heat Capacities and the Third Law

We have now seen how a knowledge of the heat of reaction and of the temperature variation of the heat capacities of reactants and products allows us to calculate the equilibrium constant at any temperature, provided there is a single experimental measurement of either K_P or $\Delta G^{\ominus}$ at some one temperature. If we have an independent method for finding the integration constant I in (8.31), we can calculate K_P without any recourse to experimental measurements of the equilibrium or of the free enthalpy change. This calculation would be equivalent to the evaluation of the entropy change $\Delta S^{\ominus}$ from thermal data alone, i.e., from heats of reaction and heat capacities. If we know $\Delta S^{\ominus}$ and $\Delta H^{\ominus}$, K_P can be found from

$$\Delta G^{\ominus} = \Delta H^{\ominus} - T\Delta S^{\ominus}$$

From (3.57), the entropy of a substance at temperature T is given by

$$S = \int_0^T \frac{C_P}{T} dT + S_0$$

where S_0 is the entropy at 0 K. The Third Law of Thermodynamics allows us to set $S_0 = 0$ for the perfect crystal at 0 K. Thus, it becomes possible to evaluate

$\Delta G^{\ominus}$, and hence K_P, entirely from calorimetric data. The historic problem of the relation of chemical affinity to the thermal properties of matter is thus solved.

12. Statistical Thermodynamics of Equilibrium Constants

In Chapter 5, we boasted that given the partition function Z we would venture to calculate all the equilibrium properties of a substance. Thus, given the Z's for products and reactants, we can calculate the equilibrium constant for the reaction. For the perfect gas, we can calculate Z, and hence K_P, from spectroscopic data on the energy levels of individual noninteracting molecules.

The equilibrium constant K_P of a chemical reaction between ideal gases is obtained from the partition functions of the reactants and products by means of the relation,

$$\Delta G^{\ominus} = -RT \ln K_P$$

From (5.44) and (5.46), the Helmholtz free energy per mole is

$$A_m = -kT \ln Z = -kT \ln \left(\frac{z^L}{L!} \right)$$

Since $G_m = A_m + PV_m = A_m + RT$, and $L! = (L/e)^L$ (Stirling formula),

$$G_m = -RT \ln \left(\frac{z}{L} \right) \tag{8.32}$$

Let us write

$$z = z_I \frac{(2\pi mkT)^{3/2} V}{h^3} \tag{8.33}$$

where z_I designates $z_r z_v z_e$, the product of the nontranslational contributions to z. If we consider 1 mol of ideal gas in its standard state of $P = 1$ atm,

$$V_m = \frac{RT}{P} = RT$$

and (8.33) becomes, for the partition function of an ideal gas in the standard state,

$$z^{\ominus} = z_I \frac{(2\pi mkT)^{3/2}}{h^3} (RT) \tag{8.34}$$

Then, from (8.32),

$$G_m^{\ominus} = -RT \ln \left(\frac{z^{\ominus}}{L} \right) \tag{8.35}$$

Let us consider the application of this theory to a simple reaction,

$$A \rightleftarrows B$$

Now A and B might represent two isomers—e.g., butane and isobutane. Figure 8.2 shows two sets of energy levels, $\epsilon_i(A)$ belonging to A and $\epsilon_j(B)$ belonging to B. In each case, the zero level of energy in this diagram corresponds to complete dissociation of the molecule into atoms in the ground state. The difference in the lowest energy levels, specified by $j = 0$, of the two compounds is

$$\Delta \epsilon_0 = \epsilon_0(B) - \epsilon_0(A) \tag{8.36}$$

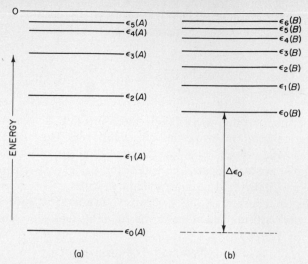

FIG. 8.2 Sets of energy levels for two isomeric molecules, A and B.

When the reaction $A \rightleftharpoons B$ has come to equilibrium, the molecules of A will be distributed in accord with a Boltzmann distribution in the energy levels $\epsilon_j(A)$, and the molecules of B similarly distributed in their levels $\epsilon_j(B)$.

If we take the zero level of energy for the entire system as the lowest energy level of the A molecules, the partition function for A is simply

$$z_A = \sum_{j=0}^{\infty} \exp\left[\frac{-\epsilon_j(A)}{kT}\right]$$

To discuss equilibrium between A and B, we must reckon the energy levels of B from the same zero level as that used for A. Thus,

$$z_B = \sum_{j=0}^{\infty} \exp\left(\frac{-[\epsilon_j(B) + \Delta\epsilon_0]}{kT}\right) = \exp\left(\frac{-\Delta\epsilon_0}{kT}\right) \sum_{j=0}^{\infty} \exp\left(\frac{-\epsilon_j(B)}{kT}\right)$$

Having made such an assignment of a common zero level for the energies, we can obtain the statistical expression for the equilibrium constant from (8.35). For the reaction $A \rightleftharpoons B$,

$$-RT \ln K_P = \Delta G^\ominus = -RT \ln\left[\frac{(z_B^\ominus/L)}{(z_A^\ominus/L)} \exp\left(\frac{-\Delta\epsilon_0}{kT}\right)\right]$$

or

$$K_P = \frac{z_B^\ominus}{z_A^\ominus} \exp\left(\frac{-\Delta\epsilon_0}{kT}\right)$$

We can thus give the equilibrium constant a simple statistical interpretation. It is the sum of the probabilities that the system be found at equilibrium in one of the energy levels of B divided by the sum of the probabilities that it be found in one of the levels of A.

The reaction $A \rightleftharpoons B$ is a special case, in that $\Delta\nu$, the change in the number of moles on reaction, is zero. Consider now the general reaction,

$$aA + bB \rightleftharpoons cC + dD$$

From (8.35),

$$\Delta G^{\ominus} = -RT \ln\frac{(z_C^{\ominus}/L)^c(z_D^{\ominus}/L)^d}{(z_A^{\ominus}/L)^a(z_B^{\ominus}/L)^b} \exp\left(\frac{-\Delta\epsilon_0}{kT}\right)$$

and

$$K_P = \frac{(z_C^{\ominus}/L)^c(z_D^{\ominus}/L)^d}{(z_A^{\ominus}/L)^a(z_B^{\ominus}/L)^b} \exp\left(\frac{-\Delta\epsilon_0}{kT}\right) \tag{8.37}$$

The value of $L\,\Delta\epsilon_0 = \Delta U_0$ gives the ΔU of the reaction at 0 K. This quantity is often tabulated, but it can be calculated from ΔU_{298} or ΔH_{298} by means of the Kirchhoff equation (2.40) and heat capacity data.

13. Example of a Statistical Calculation of K_P

We shall calculate K_P at 1000 K for the dissociation reaction,

$$Na_2 \rightleftharpoons 2Na$$

The energy of dissociation of Na_2 was measured spectroscopically as $\Delta\epsilon_0^{\ominus} = 0.73$ eV. The fundamental vibration frequency of Na_2 is at $\lambda^{-1} = 159.23 \text{ cm}^{-1}$, and the internuclear distance is 0.3078 nm. From (8.37),

$$K_P = \frac{[z^{\ominus}(Na)]^2}{z^{\ominus}(Na_2)} \cdot \frac{1}{L} \exp\left(\frac{-\Delta\epsilon_0^{\ominus}}{kT}\right)$$

From (8.34) and Table 5.4,

$$K_P = \frac{(2\pi m_{Na}kT)^3/h^6}{(2\pi m_{Na_2}kT)^{3/2}/h^3} \cdot \frac{RT}{L} \cdot \frac{\sigma h^2[g^2(Na)/g(Na_2)]}{8\pi^2 IkT}(1 - e^{-h\nu_0/kT}) \exp\left(\frac{-\Delta\epsilon_0^{\ominus}}{kT}\right)$$

$$m_{Na} = \tfrac{1}{2}m_{Na_2} = 23/(6.02 \times 10^{23}) = 3.82 \times 10^{-23} \text{ g}$$

$$I = \mu r^2 = (1.91 \times 10^{-23})(3.078 \times 10^{-8})^2 = 1.81 \times 10^{-38} \text{ cm}^2\cdot\text{g}$$

$$\frac{h\nu_0}{kT} = \frac{6.62 \times 10^{-27} \times 3 \times 10^{10} \times 159.23}{1.38 \times 10^{-16} \times 10^3} = 0.229 \tag{8.38}$$

$$1 - e^{-h\nu_0/kT} = 1 - 0.795 = 0.205$$

$$\exp\left(\frac{-\Delta\epsilon_0^{\ominus}}{kT}\right) = \exp\frac{-0.73 \times 1.602 \times 10^{-12}}{1.38 \times 10^{-16} \times 10^3}$$

$$= \exp(-8.47) = 2.09 \times 10^{-4}$$

Notice that the R in (8.38) has the dimensions $\text{cm}^3\cdot\text{atm}\cdot\text{K}^{-1}$ since it was introduced in connection with the definition of the standard state of 1 atm.

The ground state of Na_2 is a singlet, whereas the ground state of Na is a doublet (^2S). Two almost superimposed states make up the ground state of Na, and hence its statistical weight $g = 2$. Making the substitutions, we find

$$K_P = 2.428$$

14. Equilibria in Nonideal Systems—Fugacity and Activity

The development of the theory of the equilibrium constant for ideal gases started with the equation

$$dG = V\,dP - S\,dT \tag{8.39}$$

We introduced $V = nRT/P$ and obtained at constant temperature,

$$dG = nRT\,d\ln P \tag{8.40}$$

Upon integration, we found for the free enthalpy *per mole*,

$$G_m = G_m^\ominus + RT\ln P \tag{8.41}$$

For the general case of a component A in a nonideal solution, instead of (8.39), we have

$$d\mu_A = V_A\,dP - S_A\,dT \tag{8.42}$$

Or, at constant temperature,

$$d\mu_A = V_A\,dP \tag{8.43}$$

Equation (8.41) led to results in such convenient form for equilibrium calculations, that we should like to keep as close to it as possible. With this end in view, G. N. Lewis introduced a new function, called the *fugacity f*. He defined it by an equation analogous to (8.40),

$$d\mu_A = dG_A = RT\,d\ln f_A = V_A\,dP \tag{8.44}$$

Integrating between the given state and some freely chosen standard state, we obtain

$$\mu_A = \mu_A^\ominus + RT\ln \frac{f_A}{f_A^\ominus} \tag{8.45}$$

The fugacity is the true measure of the escaping tendency of a component in solution. We can think of it as a sort of *idealized partial pressure* or partial vapor pressure. It becomes equal to the partial pressure only when the vapor behaves as an ideal gas.

Comparison of (8.45) and (7.17) indicates that the fugacity is proportional to the absolute activity λ. Therefore,

$$\frac{f}{f_A^\ominus} = \frac{\lambda_A}{\lambda_A^\ominus} = a_A \tag{8.46}$$

The ratio of the fugacity of A to its fugacity in a standard state equals the corresponding ratio of absolute activities. As mentioned in Section 7.3, this ratio is called the *activity a_A*. The activity a is a dimensionless quantity. Whenever we talk about activity, we must know what standard state has been chosen. In terms of the activity, Eq. (8.45) becomes

$$\mu_A = \mu_A^\ominus + RT\ln a_A \tag{8.47}$$

When the treatment of equilibrium in Section 8.5 is carried through in terms of chemical potentials and activities, we obtain an expression K_a for the equilibrium constant, which is always valid, not just for ideal gases but also for nonideal gases and solutions.

$$K_a = \frac{a_C^c a_D^d}{a_A^a a_B^b} \tag{8.48}$$

$$\Delta\mu^{\ominus} = -RT \ln K_a \tag{8.49}$$

To translate the equilibrium calculations back into measurable terms, we must have some way of computing the actual concentrations in the reaction mixture from the calculated activities.

15. Nonideal Gases—Fugacity and Standard State

We define the standard state of a real gas A as that state in which the gas has unit fugacity, $f_A^{\ominus} = 1$, and in which, furthermore, the gas behaves as if it were ideal. Note that for a gas, therefore, the activity equals the fugacity

$$a_A = f_A \quad \text{(for a gas)} \tag{8.50}$$

The definition of the standard state may seem curious, since it is not a real state of the gas, but a *hypothetical state*. We want to make the gas behave ideally in its standard state, so that we can compare properties of different gases in ideal standard states with one another and with theoretical calculations.

The definition of the standard state of unit fugacity should be clear from Fig. 8.3. At a sufficiently low pressure, every gas behaves ideally and its fugacity then becomes equal to its pressure. To get the property of a gas in its standard state, we must move along the experimental curve (as a function of pressure) until it joins the ideal curve, and then move back along the ideal curve until we reach the point of unit fugacity. There is no problem in calculating the change in a property along the ideal curve, since we have simple equations for the properties of an ideal gas.

The fugacity of a pure gas or of a gas in a mixture can be evaluated if sufficiently detailed *PVT* data are available. In the case of a pure gas,

$$dG = n\, d\mu = V\, dP \tag{8.51}$$

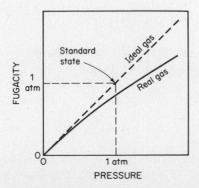

FIG. 8.3 Definition of standard state of unit fugacity.

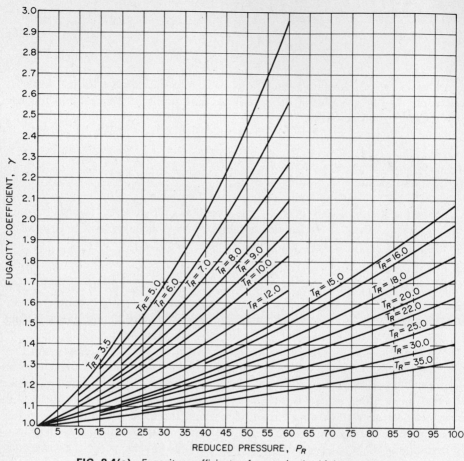

FIG. 8.4(a) Fugacity coefficients of gases in the high temperature range. Each curve corresponds to a particular value of the reduced temperature T_R.

If the gas is ideal, $V = nRT/P$. For a nonideal gas, we write

$$n\alpha = V\,(\text{ideal}) - V\,(\text{real}) = \left(\frac{nRT}{P}\right) - V$$

whence $V = n(RT/P) - n\alpha$. Substituting this expression into (8.44), we find that

$$RT\,d\ln f = d\mu = RT\,d\ln P - \alpha\,dP$$

The equation is integrated from $P' = 0$ to P.

$$RT \int_{f,\,P=0}^{f} d\ln f' = RT \int_{P=0}^{P} d\ln P' - \int_{0}^{P} \alpha\,dP'$$

As its pressure approaches zero, a gas approaches ideality, and for an ideal gas, the fugacity equals the pressure, $f = P$ [cf. (8.10) and (8.45)]. The lower limits of the first two integrals must therefore be equal, so that we obtain

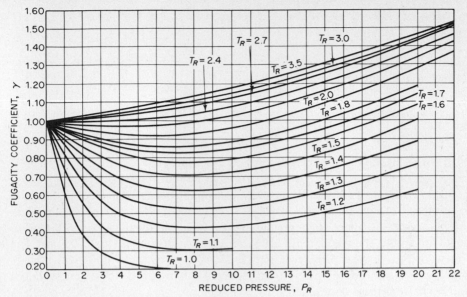

FIG. 8.4(b) Fugacity coefficients of gases in the intermediate temperature range.

$$RT \ln f = RT \ln P - \int_0^P \alpha \, dP' \tag{8.52}$$

This equation enables us to evaluate the fugacity at any pressure and temperature, provided *PVT* data for the gas are available. If the deviation from ideality of the gas volume is plotted against *P*, the integral in (8.52) can be evaluated graphically. Alternatively, an equation of state can be used to calculate an expression for α as a function of *P*, making it possible to evaluate the integral by analytic methods.

The ratio of fugacity to pressure defines the *fugacity coefficient*:

$$\gamma = \frac{f}{P} \tag{8.53}$$

For an ideal gas, $\gamma = 1$.

In Chapter 1, we saw that different gases displayed approximately the same deviations from ideality if they were in corresponding states. This rule is illustrated by the fact that different gases have almost the same fugacity coefficient γ, when they are at the same reduced temperature and pressure, $T_R = T/T_c$ and $P_R = P/P_c$.

Figure 8.4 shows a family of curves* relating the fugacity coefficient of a gas to P_R at various values of the reduced temperature T_R. To the approximation that the Law of Corresponding States is valid, all gases will fit this single set of curves. We thus can estimate the fugacity of a gas solely from a knowledge of its critical constants, T_c and P_c.

*R. H. Newton, *Ind. Eng. Chem.*, 27, 302 (1935).

16.   Use of Fugacity in Equilibrium Calculations

From (8.48), (8.50), and (8.53), we find for the equilibrium constant of a reaction between nonideal gases,

$$K_f = \frac{f_C^c f_D^d}{f_A^a f_B^b} = \frac{\gamma_C^c \gamma_D^d}{\gamma_A^a \gamma_B^b} \cdot \frac{P_C^c P_D^d}{P_A^a P_B^b} \qquad (8.54)$$

or

$$K_f = K_\gamma K_P$$

Of course, K_γ is not an equilibrium constant but simply the ratio of fugacity coefficients needed to convert the partial pressures in K_P into the fugacities in K_f.

As an example of the use of fugacities in equilibrium problems, let us consider the synthesis of ammonia, $\frac{1}{2}N_2 + \frac{3}{2}H_2 \rightarrow NH_3$. The industrially important reaction is carried out under high pressures, at which the ideal gas approximation would fail badly. The reaction has been carefully investigated up to 3500 atm.[*] The percent of NH_3 in equilibrium with a 3:1 H_2—N_2 mixture at 723 K and various total pressure is shown in Table 8.6. In the third column of the table are the values calculated from these data of

$$K_P = \frac{P_{NH_3}}{P_{N_2}^{1/2} P_{H_2}^{3/2}}$$

Since K_P for ideal gases should be independent of the pressure, these results show the large departure from ideality at the higher pressures.

In this case, the yield of ammonia is increased at high pressures by the increase in K_P as well as by the direct effect of pressure on the position of equilibrium (Le Chatelier effect).

Let us calculate the equilibrium constant K_f by the use of fugacity coefficients from Newton's graphs. We thereby adopt the approximation that the activity coefficient of a gas in a mixture is determined only by the temperature and by the *total pressure*. This approximation ignores specific interactions between components in the mixture of gases. Consider the calculation of the fugacity coefficients γ at 723 K and 600 atm.

	P_c	T_c	P_R	T_R	$\gamma = f/P$
N_2	33.5	126	17.9	5.74	1.35
H_2	12.8	33.3	46.8	21.7	1.19
NH_3	111.5	405.6	5.38	1.78	0.85

The values of γ are read from the graphs, at the proper values of reduced pressure P_R and reduced temperature T_R.

In this case, $K_\gamma = \gamma_{NH_3}/\gamma_{N_2}^{1/2} \gamma_{H_2}^{2/3}$. The values of K_γ and K_f are shown in Table 8.6. There is a marked improvement in the constancy of K_f as compared with K_P. Only at 1000 atm and above does the approximate treatment of the fugacities

[*]L. J. Winchester and B. F. Dodge, *Am. Inst. Chem. Eng. J.*, 2, 431 (1956).

TABLE 8.6

Equilibrium in the Ammonia Synthesis at 723 K with 3:1 Ratio of H_2 to N_2

Total pressure (atm)	% NH_3 at equilibrium	K_P	K_γ	K_f (approximate)
10	2.04	0.00659	0.995	0.00655
30	5.80	0.00676	0.975	0.00659
50	9.17	0.00690	0.945	0.00650
100	16.36	0.00725	0.880	0.00636
300	35.5	0.00884	0.688	0.00608
600	53.6	0.01294	0.497	0.00642
1000	69.4	0.02496	0.434	0.01010
2000	89.8	0.1337	0.342	0.0458
3500	97.2	1.0751		

appear to fail. If we had the correct K_γ, the K_f would of course remain constant. To carry out an exact thermodynamic treatment, it would be necessary to calculate the fugacity of each gas in the particular mixture under study. Extensive *PVT* data on the mixture would be needed for such a calculation.

Often, knowing $\Delta G^\ominus$ for the reaction, we wish to calculate the equilibrium concentrations in a reaction mixture. The procedure is to obtain K_f from $-\Delta G^\ominus = RT \ln K_f$, to estimate K_γ from the graphs, and then to calculate the partial pressures from $K_P = K_f/K_\gamma$.

17. Standard States for Components in Solution

The expression obtained in (8.48) for the equilibrium constant K_a in terms of activities represents a perfectly general solution to the problem of chemical equilibrium in solutions. Before we can apply it to practical cases, we must choose and define the standard states for components in a solution.

There are two different standard states in common use. With increasing dilution, the solvent always approaches the ideal behavior specified by Raoult's Law, and the solute always approaches the behavior specified by Henry's Law. One standard state (I) is therefore based on Raoult's Law as a limiting law, and the other standard state (II) is based on Henry's Law. We may choose whichever definition seems more convenient for any component in a particular solution.

Case I. Standard state for a component considered as a solvent. In this case, the standard state of a component A in a solution is taken to be the pure liquid or pure solid at 1 atm pressure and at the temperature in question. This choice of standard state was used in the discussion of liquid solutions in the previous chapter.

In this case, the activity

$$a_A = \frac{f_A}{f_A^\bullet} \approx \frac{P_A}{P_A^\bullet} \tag{8.55}$$

where $P_A^\bullet$ is the vapor pressure of pure A under 1 atm total pressure (see Section 7.5). It is almost always sufficiently accurate to take the activity equal to the ratio of the partial pressure P_A of A above the solution to the vapor pressure of pure A (at 1 atm pressure). It is always possible, however, to convert these vapor pressures to fugacities, should the vapors depart appreciably from ideal gas behavior.

With this choice of standard state, Raoult's Law becomes

$$a_A = \frac{P_A}{P_A^\bullet} = X_A$$

Thus, for the ideal solution, or for any solution in the limit as $X_A \to 1$, we have $X_A = a_A$.

We define an activity coefficient γ_A by

$$a_A = \gamma_A X_A \tag{8.56}$$

Thus, as $X_A \to 1$, $\gamma_A \to 1$.

Case II.   Standard state for a component considered as a solute. In this case, we choose the standard state so that in the limit of extreme dilution, as $X_B \to 0$, $a_B \to X_B$. As long as Henry's Law is obeyed, as shown in Fig. 8.5,

$$f_B = kX_B \tag{8.57}$$

The standard state is obtained by extrapolating the Henry's Law line to $X_B = 1$. Thus, we see that the fugacity in the standard state, $f_B^\ominus$, is simply equal to k, the constant of Henry's Law:

$$f_B^\ominus = k \tag{8.58}$$

As in the case of the nonideal gas, the standard state is a *hypothetical state*. We can think of it in physical terms as a state in which the pure solute B ($X_B = 1$) has the properties it would have in an infinitely dilute solution in the solvent A.

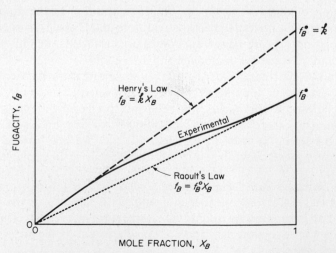

FIG. 8.5 Definition of standard state for a solute B, based on Henry's Law in dilute solution.

For all theoretical discussions, the use of mole fractions to express the composition of solutions eventually may supercede the other common composition variables—m, the molality, and c, the molar concentration. At present, however, there is a large body of data expressed in these other variables. We therefore need to define activities and activity coefficients based on m and c, which we may denote as $^m a$, $^m \gamma$, $^c a$, $^c \gamma$.

We begin with the fact that in a sufficiently dilute solution, the activity coefficients will approach unity, so that a standard state based on Henry's Law can be selected, just as depicted in Fig. 8.5, except that now the fugacity is plotted against m or c instead of X:

$$f_B = {}^m k m_B \quad \text{or} \quad f_B = {}^c k c_B \tag{8.59}$$

The experimental straight lines at high dilutions are extrapolated to $m_B = 1$ or $c_B = 1$ to define the corresponding standard states. We also define activity coefficients $^m \gamma$ and $^c \gamma$ by

$$^m a = {}^m \gamma\, m \quad \text{and} \quad {}^c a = {}^c \gamma\, c \tag{8.60}$$

Relations between the three activity coefficients are readily derived.*

18. Activities of Solvent and Nonvolatile Solute from Vapor Pressure of Solution

As an example of this important method, let us consider how the activities of water A and of sucrose B are determined from data on the vapor pressures of the solution. The same method has been applied to obtain activities of amino acids, peptides, and other solutes of biochemical interest.

Sucrose is nonvolatile, so that the total vapor pressure above the solution in this case equals the partial vapor pressure of the water, P_A. If we neglect the small correction for nonideality of the water vapor, we can readily tabulate the activities a_A of the water from

$$a_A = \frac{P_A}{P_A^{\bullet}}$$

The results are shown in Table 8.7 for the particular temperature of 323.2 K. For example, at 323.2 K, $P_A^{\bullet}$, the vapor pressure of pure water, is 92.51 torr. The vapor pressure of a sucrose solution in which the mole fraction of water is $X_A = 0.9665$ is $P_A = 88.97$ torr. Hence, $a_A = 88.97/92.51 = 0.9617$. The activity coefficient $\gamma_A = 0.9617/0.9665 = 0.9949$ at this concentration. Note that the standard state for the solvent, water, is chosen as the pure liquid at 1 atm pressure. As $X_A \to 1$, $a_A \to X_A$ and $\gamma_A \to 1$.

The activity of the sucrose obviously cannot be determined from its partial vapor pressure because this is immeasurably low. If we had a volatile solute—e.g.,

*S. Glasstone, *An Introduction to Electrochemistry* (Princeton, N.J.: D. Van Nostrand Co., Inc., 1942), p. 134.

alcohol—we would doubtless take pure alcohol to be its standard state and compute its activity from (8.55). In the case of sucrose, on the other hand, we clearly should choose the second definition of standard state, that which is based on Henry's Law for the solute.

The activity of the sucrose can be calculated provided we know the vapor pressure of the solvent water over the whole range of solute concentration from $X_B = 0$ up to the highest concentration of interest. From the Gibbs–Duhem equation (7.10),

$$n_A \, d\mu_A + n_B \, d\mu_B = 0$$

From (8.47), this becomes

$$n_A \, d\ln a_A + n_B \, d\ln a_B = 0$$

Dividing by $n_A + n_B$, we obtain

$$X_A \, d\ln a_A + X_B \, d\ln a_B = 0$$

Hence,

$$\int d\ln a_B = -\int \left(\frac{X_A}{X_B}\right) d\ln a_A = -\int \left(\frac{X_A}{1 - X_A}\right) d\ln a_A \qquad (8.61)$$

We wish to calculate a_B from the measurements giving a_A as a function of X_A. There appears to be some difficulty in using the preceding integration, since as $X_B \longrightarrow 0$, $X_A \longrightarrow 1$, and the integrals approach ∞. This difficulty is easily avoided by starting the integration, not at $X_A = 1$, but at a value of X_A at which the solvent begins to follow Raoult's Law, i.e., at which $X_A = a_A$. At this value, $X_B = a_B$, where a_B is defined on the basis of Henry's Law. Therefore, the integrals in (8.61) have lower limits corresponding to extremely dilute solutions. The results of such a calculation of the activities of sucrose in aqueous solution are shown in Table 8.7.

In view of the importance of this method of determining activity coefficients,

TABLE 8.7
**Activities of Water and Sucrose in Their Solutions at 323.2 K Obtained
from Vapor Pressure Lowering and the Gibbs–Duhem Equation**

Mole fraction of water X_A	Activity of water a_A	Mole fraction of sucrose X_B	Activity of sucrose a_B
0.9940	0.9939	0.0060	0.0060
0.9864	0.9934	0.0136	0.0136
0.9826	0.9799	0.0174	0.0197
0.9762	0.9697	0.0238	0.0302
0.9665	0.9617	0.0335	0.0481
0.9559	0.9477	0.0441	0.0716
0.9439	0.9299	0.0561	0.1037
0.9323	0.9043	0.0677	0.1390
0.9098	0.8758	0.0902	0.2190
0.8911	0.8140	0.1089	0.3045

we shall also show explicitly how it is used when the composition of the solution and the activity coefficients are referred to the molality m_B. The composition of the solution is then based on 1 kg ($1000/M_A$ mol) of solvent. If the solvent is water, $m_A = 55.51$. The activity of the solute is $^m\gamma_B m_B = {^m}a_B$. The Gibbs–Duhem equation becomes

$$m_B \, d \ln a_B + m_A \, d \ln a_A = 0$$

giving

$$m_B \, d \ln m_B + m_B \, d \ln \gamma_B = -55.51 \, d \ln a_A = -55.51 \, d \ln \left(\frac{P_A}{P_A^{\bullet}} \right) \qquad (8.62)$$

We define the *molal osmotic coefficient* ϕ *of solute* by

$$\phi = \frac{-m_A \ln a_A}{m_B} = \frac{-55.51 \ln a_A}{m_B} \qquad (8.63)$$

Thus, ϕ is determined by the vapor pressure of the solvent and the molality of the solute. From (8.63),

$$d(\phi m_B) = \phi \, dm_B + m_B \, d\phi = -55.51 \, d \ln {^m}a_B$$

Hence, by equating the left sides of (8.62) and (8.63), we obtain

$$d \ln \gamma_B = (\phi - 1) \, d \ln m_B + d\phi$$

On integration from pure solvent to final solution of molality m_B,

$$\int_0^{m_B} d \ln \gamma_B = \int_0^{m_B} (\phi - 1) \, d \ln m_B' + \int_0^{m_B} d\phi$$

The activity of pure solvent is unity, so that as $m_B \to 0$, $\phi \to 1$.* The integration thus gives

$$\ln {^m}\gamma_B = \int_0^{m_B} \frac{(\phi - 1) \, dm_B'}{m_B'} + (\phi - 1) \qquad (8.64)$$

A convenient way to determine the activity of water for use in computations based on (8.64) is the *isopiestic* method. We take a set of reference standards for the activity of water. Sucrose solutions are convenient for this purpose. We place the reference solution and the solution of unknown activity in open vessels in an evacuated chamber such as a vacuum desiccator. Water will evaporate from the solution of higher vapor pressure and condense into the solution of lower vapor pressure, until equilibrium is reached when the water vapor pressure is the same for the two solutions. At this *isopiestic point*, the activity of water is the same in the two solutions. We measure the composition of the two solutions. We then know the activity of water at the measured composition. By repeating the method for different compositions, we determine the values of ϕ over the range of molalities required to calculate the activity coefficient of the solute $^m\gamma_B$ from (8.64). Some values for compounds of biochemical interest measured in this way are summarized in Table 8.8.

*We see that $\ln a_A \to \ln X_A \to \ln (1 - X_B) \to -X_B$ so that $\phi = \frac{n_A X_B}{n_A} \approx 1$.

TABLE 8.8

Molal Activity Coefficients $^m\gamma$ of Some Amino Acids and Peptides
in Aqueous Solution at 298.15 K

Compound \ Molality m	0.2	0.3	0.5	1.0	1.5	2.0
Glycine	0.961	0. 44	0.913	0.854	0.812	0.782
Alanine	1.005	1.007	1.012	1.024	1.027	
Threonine	0.989	0.984	0.975	0.959	0.951	0.944
Proline	1.019	1.028	1.048	1.097	1.149	1.205
ϵ-Aminocaproic acid	0.971		0.951	0.942	1.002	1.072
Glycylglycine	0.912	0.879	0.828	0.745	0.697	
Glycylalanine	0.935	0.912	0.883	0.855		

19. Equilibrium Constants in Solution

The relation,

$$\Delta G^{\ominus} = -RT \ln K_a$$

is universally valid, but, in fact, it simply summarizes the mathematical analysis of the equilibrium problem. We might paraphrase the content of this equation as follows:

1. Define a standard state for each of the reactants and products in a chemical equilibrium, $aA + bB \rightleftharpoons cC + dD$.
2. Compute $\Delta G^{\ominus}$ for the reaction in which all the components are in this standard state.
3. Then there will always be a function $K_a(T, P)$, which will be related to the activities of the components in the equilibrium mixture, as

$$K_a = \frac{a_C^c a_D^d}{a_A^a a_B^b}$$

To obtain any information on the actual composition of this equilibrium mixture, or, conversely, to calculate K_a, and hence $\Delta G^{\ominus}$, from the equilibrium composition, we must be able to relate the activities to some composition variables. The most logical choice is to define an activity coefficient γ such that $a = \gamma X$, where X is the mole fraction. Hence,

$$K_a = \left(\frac{\gamma_C^c \gamma_D^d}{\gamma_A^a \gamma_B^b}\right)\left(\frac{X_C^c X_D^d}{X_A^a X_B^b}\right) = K_\gamma K_X$$

Note that K_γ is not an equilibrium constant, but simply the indicated product of activity coefficients. In general, K_X will not be an equilibrium constant either, since it will not be constant at constant T and P as we vary the composition of the equilibrium mixture. In some cases, however, it may happen that K_γ does not change much as we vary the composition. In particular, in dilute solutions, in which the solutes approximately follow Henry's Law and the solvent approximately follows Raoult's Law, we can choose the standard states (as shown in the previous section) so that all the γ's approach unity. In this case, $K_a \rightarrow K_X$, and

$$\Delta G^{\ominus} \longrightarrow -RT \ln K_X$$

To the extent that these approximations are satisfactory, we shall be able to use an equilibrium constant K_X in terms of the mole fractions in the solution. Let us not disdain such an approximate equilibrium constant, because often the experimental data do not justify a more elaborate thermodynamic treatment of the equilibrium anyway. We must, however, keep clearly in mind the choice of standard states for the $\Delta G^{\ominus}$. For all the reactants, the standard states would be those at $X = 1$. In the case of the solvent, this would be the pure liquid. In the case of the solutes, these would be the hypothetical states in which $X = 1$, but the environment would be that in an extremely dilute solution.

An example of the calculation of K_X was shown in Table 8.1, for an esterification equilibrium. Actually, there are not many careful studies of equilibrium in solution which do not involve electrolytes, and hence effects due to ionization. (Such systems are discussed in Chapter 10.) One example often cited is the early work (1895) of Cundall on the dissociation $N_2O_4 \rightleftharpoons 2NO_2$ in chloroform solution. Some of his data are shown in Table 8.9, with the calculated K_X.

We have also included the computed values of an equilibrium constant in terms of concentrations c in moles per liter (or per dm^3).

$$K_c = \frac{c_{NO_2}^2}{c_{N_2O_4}}$$

Note carefully that the use of K_c implies a derivation from $\Delta G^{\ominus} = -RT \ln K_a$ based on a new and distinctive choice of standard states. We then must have a new set of activity coefficients $^c\gamma$, so that $a = {^c\gamma}\, c$, and as $c \rightarrow 0$, $^c\gamma \rightarrow 1$, and $K_a \rightarrow K_c$. The corresponding standard state is the hypothetical state of the solute at a concentration of 1 mol·dm^{-3} ($c = 1$), but with an environment the same as that in an extremely dilute solution.

For the choice of standard states consistent with K_X, we find for the reaction $N_2O_4 \rightleftharpoons 2NO_2$

$$\Delta G^{\ominus}_{(X)} = -RT \ln K_X = 53.97 \text{ kJ}$$

TABLE 8.9
Dissociation of N_2O_4 in Chloroform Solution at 8.2°C

$X(N_2O_4)$	$X(NO_2)$	K_X	$c(N_2O_4)$	$c(NO_2)$	K_c
1.03×10^{-2}	0.93×10^{-6}	8.37×10^{-11}	0.129	1.17×10^{-3}	1.07×10^{-5}
1.81	1.28	9.05	0.227	1.61	1.14
2.48	1.47	8.70	0.324	1.85	1.05
3.20	1.70	9.04	0.405	2.13	1.13
6.10	2.26	8.35	0.778	2.84	1.04
		Mean 8.70×10^{-11}			Mean 1.09×10^{-5} mol·dm^{-3}

For the choice of standard states consistent with K_c, we find

$$\Delta G^{\ominus}_{(c)} = -RT \ln K_c = 26.74 \text{ kJ}$$

Caveat lector: It is folly to use $\Delta G^{\ominus}$ values for reactions in solution unless you are sure you understand the exact standard state upon which they are based.

20. Thermodynamics of Biochemical Reactions

Our sun is the ultimate source of energy for all life on earth. Through the intermediacy of photosynthetic reactions in green plants, some of this energy is stored in the form of chemical free energy in carbohydrates,

$$nCO_2 + nH_2O \xrightarrow{hv} (CH_2O)_n + nO_2$$

The carbohydrate fuel taken into the animal body is converted to the following principal end uses:

1. maintenance of temperature (in idiotherms);
2. movement—muscular work;
3. synthesis of structural materials, such as proteins;
4. electrical activity of nervous systems;
5. pumping ions and molecules against concentration gradients.

Since the processes in the animal body occur at approximately constant temperature and pressure, the Gibbs free energy G is the correct thermodynamic potential to use in discussing driving forces, affinities, and equilibrium compositions in various processes of physiological chemistry. The change ΔG represents the available energy for the reactions, or the energy that is free to be used to work muscles, pump ions, or produce electrical signals.

Chemical reactions in biological systems are usually characterized by a rather precise fitting of the driving force of the reaction to the demands of the process that is to be driven. In this way, little free energy is wasted. Although the physiological processes are not reversible (even a snail does not take forever to traverse a path), the delicate balance of each driving force against a considerable opposing force ensures that the utilization of free energy in the living organism will be remarkably efficient. In other words, the burning of the carbohydrate fuel does not occur in a single flash (as in an inefficient internal combustion engine), but rather in a series of stages, each one of which lowers the thermodynamic potential G by a relatively small step.

Man's brain contains 6×10^9 neurons (nerve cells), and the fraction of total power that is used to operate the electrical systems of this computing center is surprisingly large. In a newborn infant, 50% of the total oxygen consumption is used by the brain; in the adult, this demand drops to from 20 to 25%, but a 1.50 kg brain constitutes only about 2.5% of the mass of an adult human. Such a brain consumes 5.5 g of glucose per hour (thus operating at an average power of 25 W).

The electrical activity of the brain cells uses free energy that has been stored

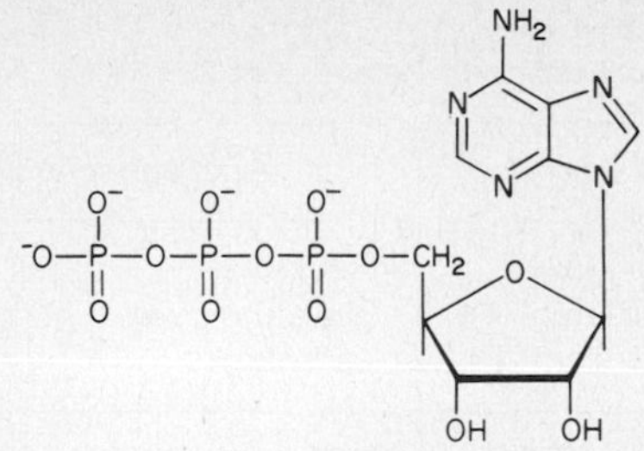

NH$_2$

FIG. 8.6 Adenosine triphosphate (ATP).

in the form of gradients of ionic concentration across cell membranes. The ionic composition within the cells is 150 mmol·dm^{-3} K$^+$ and 15 mmol·dm^{-3} Na$^+$, and in the interstitial fluid outside the cells it is 5 mmol·dm^{-3} K$^+$ and 150 mmol·dm^{-3} Na$^+$. The electrical impulses along a nerve cell membrane (*action potentials*) are due primarily to an inrush of Na$^+$ ions through temporarily activated areas of membrane, which in the resting state would have very low permeability to ions such as Na$^+$. The sodium ions that enter the cells must be pumped out *against* the high concentration gradient. This process is accomplished by "sodium pumps" in the membranes, the mechanism of which is not yet really understood. It is known, however, that the source of free energy for the sodium pump is ATP (adenosine triphosphate) (Fig. 8.6). ATP is also the source of free energy for muscular contraction and protein synthesis. More than 90% of the free energy used by the brain is required to operate the sodium pumps, and it is this great demand that causes the brain to require such a large supply of oxygen.

21. Free Enthalpy of Formation
of Biochemicals in Aqueous Solution

Biochemical reactions proceed in an aqueous medium at rather closely controlled pH and ionic composition. These conditions are quite different from the usual standard states for reactions in gases or nonpolar liquids, and there is a problem of translating the thermodynamic data from the standard condition that is usual in calorimetric work to conditions of physiological interest. For example, we can obtain standard free enthalpies of formation of biochemicals in the crystalline state at 298.15 K by means of data on enthalpies of formation and Third Law entropies from heat capacity measurements on the crystals. Examples of such thermodynamic data are given in Table 8.10.

The affinity in biochemical reactions will be determined by the ΔG of the reaction in an aqueous physiological medium. Thus, instead of the $\Delta G_f^{\ominus}(c)$ of the crystalline compounds, we wish to know the $\Delta G_f^{\ominus}(w)$ of the compounds as solutes in aqueous solution. The appropriate standard state will usually be at unit activity on the molality scale—i.e., at $^m a = 1$. We can readily calculate $\Delta G_f^{\ominus}(w)$ from $\Delta G_f^{\ominus}(c)$ by a two-step process:

1. Dissolve the crystals in water to form a *saturated solution*. Since crystals and solute are in equilibrium in a saturated solution, for this step $\Delta G = 0$.

TABLE 8.10

Thermodynamic Data for Amino Acids and Peptides at 298.15 K[a]

Compound	Crystalline state			Aqueous solution		
	$\Delta H_f^{\ominus}$ (kJ·mol⁻¹)	$\Delta S_f^{\ominus}$ (J·K⁻¹ ·mol⁻¹)	$\Delta G_f^{\ominus}$ (kJ·mol⁻¹)	Solubility (molality sat. soln.)	m_γ	$\Delta G_f^{\ominus \, b}$ (kJ·mol⁻¹)
DL-Alanine	−563.6	−644	−372.0	1.9	1.046	−373.6
DL-Alanyglycine	−777.8	−967	−489.5	3.161	0.73	−491.6
L-Aspartic acid	−973.6	−812	−731.8	0.0377	0.78	−723.0
Glycine	−528.4	−431	−370.7	3.33	0.729	−372.8
Glycylglycine	−745.2	−854	−490.4	1.7	0.685	−490.8
DL-Leucine	−640.6	−975	−349.4	0.0756	1.0	−343.1
DL-Leucylglycine	−860.2	−1310	−469.9	0.126	1.0	−468.4

[a] From F. H. Carpenter, *J. Am. Chem. Soc.*, *82*, 1120 (1960), where references to original sources are given.
[b] For the dipolar-ionic form at standard state $^m a = 1$ and pH of isoelectric point (at this pH, the concentration of dipolar ions is at a maximum).

2. Calculate ΔG for changing the solute activity from its value at saturation a_{sat} to its unit activity in the standard state.

$$\Delta G = RT \ln \frac{1}{a^{sat}} = -RT \ln (^m \gamma \, m)^{sat}$$

To carry out this computation, we need only know, besides the molality of the saturated solution, the activity coefficient at that molality. Hence,

$$\Delta G_f^{\ominus}(w) = \Delta G_f^{\ominus}(c) - RT \ln(^m \gamma \, m)^{sat} \qquad (8.65)$$

For example, the solubility of glycine in water at 298.15 K is 3.30 molal. From Table 8.10, $^m \gamma = 0.729$, and

$$\Delta G_f^{\ominus}(c) = -370.7 \text{ kJ·mol}^{-1}$$

Hence,

$$\Delta G_f^{\ominus}(w) = -370.7 - (8.314)(298.15)(10^{-3})(2.303) \log (3.30 \times 0.729)$$

$$\Delta G_f^{\ominus}(w) = -370.7 - 2.2 = -372.9 \text{ kJ·mol}^{-1}$$

We can use such $\Delta G_f^{\ominus}(w)$ values to calculate equilibrium constants K_a for reactions of interest. To be sure, the physiological solvent medium will not be pure water, and for precise calculations we should like to know the various activity coefficients in particular solutions containing inorganic salts as well as organic solutes. Furthermore, a temperature of 310.7 K is more typical than 298.2 K for biochemical reactions *in vivo*, and often, by choice, *in vitro* as well.

Many important biochemicals are acids or bases, and the pH of the medium can have a considerable effect on affinities and equilibrium constants. Fortunately, the physiological medium is well buffered close to pH 7.0, so that it is usually safe to assume a hydrogen ion concentration of 10^{-7} molal. If, however, in a given reaction, H⁺ ions are used up or set free, the driving force ΔG can be extremely sensitive to pH. The ΔG for dilution of H⁺ from unit activity to 10^{-7} molal at

300 K would be about

$$\Delta G = RT \ln \frac{10^{-7}}{1} = -40.2 \text{ kJ} \cdot \text{mol}^{-1}$$

As an example of the importance of the effects of pH and ionic concentrations on biochemical equilibria, let us consider the thermodynamics of the hydrolysis of ATP as analyzed by Alberty.* The hydrolytic reaction is

$$\text{ATP}^{4-} + \text{H}_2\text{O} \longrightarrow \text{ADP}^{3-} + \text{HPO}_4^{2-} + \text{H}^+ \tag{8.66}$$

but the species actually measured in the usual experiments are the total ATP and total adenosine diphosphate (ADP) in ionized and un-ionized forms, so that the empirical equilibrium constant is usually written

$$K = \frac{(\text{ADP})(\text{P}_i)}{(\text{ATP})} \tag{8.67}$$

where P_i is the phosphate ion.

Both ATP and ADP form complexes with univalent ions, such as Na^+ and K^+, and with bivalent ions, such as Mg^{2+} and Ca^{2+}, which occur in physiological media. To simplify the analysis, however, Alberty considered only the effects of Mg^{2+} in a medium in which the ionic strength (see Section 10.18) was maintained constant in 0.2 M tetra-n-propylammonium chloride, a salt with a large cation, which would complex to only a minimal extent with the ATP and ADP anions in the presence of H^+ and Mg^{2+}.

By a computer analysis of all the simultaneous equilibria, the thermodynamic parameters, K, $\Delta G^\ominus$, $\Delta H^\ominus$, and $T \Delta S^\ominus$ were calculated for reaction (8.66) as functions of pH and pMg $[-\log c(\text{Mg}^{2+})]$, and the results were plotted on contour diagrams as shown in Fig. 8.7. In view of the key role of ATP as the source of free energy for most physiological processes, these diagrams represent one of the thermodynamic foundations for living systems (subject always to the restrictions discussed on p.92).

The hydrolysis of ATP at pH 9 produces 1 mol of H^+ per mole of ATP hydrolyzed. The dilution of H^+ from the standard state to 10^{-9} molar makes a large contribution to $\Delta S^\ominus$ $[2.303(R)(\text{pH}) = 172 \text{ J} \cdot \text{K}^{-1} \cdot \text{mol}^{-1}$ at pH 9], and hence to the overall $\Delta G^\ominus$ for the hydrolysis reaction.

Table 8.10 summarizes some typical thermodynamic data of biochemical interest. Let us use data from the table to calculate the $\Delta G^\ominus$ and K_a for the synthesis of a simple dipeptide at 298.2 K by the reaction,

$$\text{alanine} + \text{glycine} \rightleftharpoons \text{alanylglycine} + \text{H}_2\text{O}$$

We shall ignore the effect of pH and assume that all reactants and products are completely in the form of dipolar ions.† The $\Delta G_f^\ominus$ for liquid water is $-237.2 \text{ kJ} \cdot \text{mol}^{-1}$ from Table 8.2. Hence,

$$\Delta G^\ominus = (-491.6 - 237.2) - (-373.6 - 372.8) = 17.6 \text{ kJ}$$

$$K_a = \exp(-\Delta G^\ominus/RT) = 8.13 \times 10^{-4} \quad \text{at 298.15 K}$$

---

*R. A. Alberty, *J. Biol. Chem.*, **244**, 3290 (1969).

†An amino acid $\text{R} \cdot \text{CH}_2 \cdot \text{NH}_2 \cdot \text{COOH}$ exists in neutral solutions principally in the form $\text{R} \cdot \text{CH}_2 \cdot \text{NH}_3^+ \cdot \text{COO}^-$, when R indicates the side chain of the amino acid.

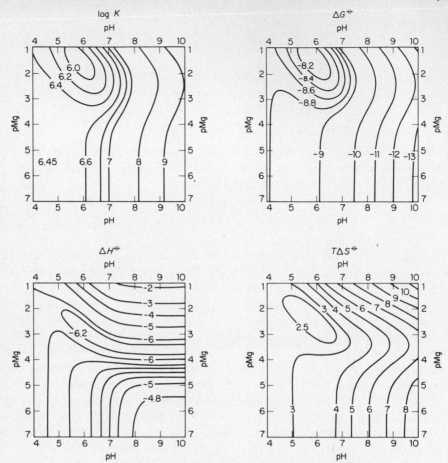

FIG. 8.7 Thermodynamic parameters for ATP + $H_2O \rightleftharpoons$ ADP + Pi
as a function of pH and pMg in 0.2 ionic strength tetra-*n*-
propylammonium chloride. The numerical values on the
contours are in units of kilocalories (kcal) where 1 kcal =
4.1840 kJ by definition. [R. A. Alberty, *J. Biol. Chem.*, *244*,
3290 (1969).]

We can conclude that the synthesis of peptide bonds will not occur spontaneously
to an appreciable extent from amino acids in aqueous solution. Some external
source of free energy will be required. Students of biochemistry will be familiar
with the mechanisms whereby free energy from the sun, which has been stored in
the form of ATP, can be released to drive the processes of peptide bond formation
in protein synthesis.

In Chapter 3, it was stated rather categorically that thermodynamics was
competent to deal only with equilibrium systems and hence could tell us nothing
about what goes on in living cells. Yet we are now blithely using thermodynamic
arguments to derive conclusions about biochemical reactions. The thermodynamic
calculations indicate which reactions are possible and which are impossible in a
specified *in vitro* environment that simulates the physiological milieu in a living

cell. To apply these results to a living cell, we need some additional postulate or assumption. It would seem to be sufficient to require that the process we wish to consider in the cell (e.g., synthesis of a peptide bond) is not coupled in any way with any other physical or chemical processes occurring in the cell. All we can say is that if such coupling does not exist, then we can apply our equilibrium thermodynamics to the reaction within the cell. The demonstration that the coupling is absent is an empirical problem. For example, living cells deprived of their supply of ATP do not synthesize peptide bonds. To make a more general theoretical argument, we shall still be forced to extend the formulation of thermodynamics in such a way as to allow us to define state functions such as G and S in systems that are not at equilibrium.

22. Pressure Effects on Equilibrium Constants

The equilibrium constant K_X is most convenient for describing pressure effects on equilibrium in ideal solutions. From (7.12) and (8.49), we have

$$\left(\frac{\partial \ln K_X}{\partial P}\right)_T = \frac{-\Delta V^\bullet}{RT}$$

Here, $\Delta V^\bullet$ is the volume of pure liquid products minus the volume of pure liquid reactants. For nonideal solutions, the activities and activity coefficients based on mole fractions can be used, so that

$$\left(\frac{\partial \ln K_a}{\partial P}\right)_T = \frac{-\Delta V^\bullet}{RT} \tag{8.68}$$

If the reaction proceeds with an appreciable ΔV, there can be a large effect of pressure on the equilibrium constant. For example, with $\Delta V = -20$ cm³, a pressure of 1000 atm would increase K about 2.3 times, and a pressure of 10 000 atm, about 3500 times.

For reactions of nonpolar molecules in nonpolar solvents, ΔV is usually not more than a few cubic centimetres, unless there is a net change in the number of covalent bonds. When there is an increase in the number of bonds, ΔV decreases, and *vice versa*. Such volume changes are due to the relative shortness of the covalent bond distance between atoms as compared with nonbonded distances. Polymerization reactions may be expected to have a considerable negative ΔV, and thus to be markedly favored by elevated pressures. The dimerization of NO_2 was studied in CCl_4 solution,* $2NO_2 \rightleftharpoons N_2O_4$. Some of the results are shown in Table 8.11. At 324.7 K, K_m increases fourfold between 1 and 1500 atm, corresponding to an average $\Delta V = -23$ cm³. To integrate (8.68), we would need to know ΔV as a function of pressure, but it is a decent approximation to use a constant ΔV equal to its value at 1 atm.

The volume change ΔV in a chemical reaction can be measured directly with a *dilatometer*, which usually consists of a reaction bulb to which is attached a cali-

---

*A. H. Ewald, *Discussions Faraday Soc.*, **22**, 138 (1956).

TABLE 8.11
Effect of Pressure on the Equilibrium $2NO_2 \rightleftharpoons N_2O_4$ in CCl_4 Solution

Temperature (K)	Pressure (atm)	$K_m(P \text{ atm})/ K_m(1 \text{ atm})$
295.2	750	2.08
	1500	3.77
324.7	750	2.30
	1500	4.06

brated capillary tube. Linderstrøm–Lang and his coworkers at the Carlsberg Laboratories* have used this simple equipment in an important series of researches on the ΔV's in reactions of biochemical interest, such as denaturation of proteins and inactivation or inhibition of enzymes. Changes in the conformation of proteins often cause appreciable volume changes in the aqueous medium. When a peptide bond is hydrolyzed, there is almost always a decrease in volume, because new electrically charged groups are formed (an NH_3^+ and a COO^-), and these produce a contraction of the surrounding water, an effect known as *electrostriction*. The effect is seen also in a simple reaction like

$$CH_3COOH + NH_3 \longrightarrow CH_3COO^- + NH_4^+$$

which has $\Delta V = -17.4 \, cm^3$ at 25°C.

23. Effect of Pressure on Activity

We saw in Section 8.16 how the activity (fugacity) of a gaseous component is related to its pressure. In contrast with the strong dependence of activity on pressure for a gas, the activity of a component in a liquid or solid phase depends only weakly on pressure. At moderate pressures, we may even neglect entirely the effect of pressure on activities in condensed phases, without serious error. At high pressures, however, the total effect of pressure can become appreciable, and in geochemical, oceanographic, and astrochemical studies, high pressure conditions will often be of major interest. We certainly should know, therefore, how to include in our thermodynamic theory of equilibrium the effect of pressure on condensed phases.

At any constant temperature, we can calculate the dependence of the fugacity of a component A on the total pressure from (8.44),

$$d\mu_A = RT \, d\ln f_A = V_A \, dP$$

The ratio Γ of the fugacity at any pressure P_2 to the fugacity at 1 atm is therefore given by

$$\ln \Gamma = \ln \frac{f^{(P_2)}}{f^{(1 \text{ atm})}} = \frac{1}{RT} \int_1^{P_2} V_A \, dP \qquad (8.69)$$

where we have written

$$\Gamma = \frac{f_A^{(P_2)}}{f_A^{(1 \text{ atm})}}$$

*See *Comptes Rendus* of Carlsberg Laboratories from 1940 to date.

Hence,

$$\frac{f_A^{(P_2)}}{f_A^{\ominus}} = \Gamma \frac{f_A^{(1\ atm)}}{f_A^{\ominus}}$$

Or, since $f_A^{\ominus}$ is the fugacity of A in its standard state,

$$a \text{ (at } P_2) = \Gamma a \text{ (at 1 atm)} \tag{8.70}$$

If we use the standard state based on mole fractions, (8.70) becomes

$$a = \Gamma \gamma X \tag{8.71}$$

For a pure liquid or solid, the activity *at 1 atm* equals unity if the standard state is defined as a pure liquid or solid at 1 atm.

As an example of the use of (8.69), let us calculate the activity of pure liquid water at 323.3 K and 10^4 atm from the data of Bridgman.* In this case, $V_A = V_A^{\bullet}$, and the integration in (8.69) can be carried out graphically by plotting the molar volume of water against the pressure and taking the area under the curve. The calculation is shown in Fig. 8.8. The integration yields $\Gamma = 439$, so that the activity of pure liquid water at 323.2 K and 10^4 atm is $a = 439$. At 100 atm, the activity would be about 1.07.

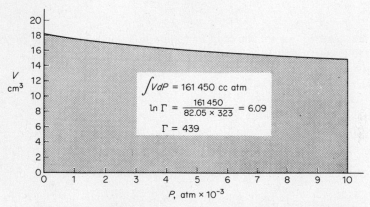

FIG. 8.8 Calculation of the activity of water at 323.2 K and 10^4 atm by graphical evaluation of the integral in Eq. (8.69). V is the molar volume of water.

24. Chemical Equilibria Involving Condensed Phases

The simplest examples of such equilibria include pure liquid or solid reactants. Let us consider a typical case:

$$\text{NiO(c)} + \text{CO} \rightleftharpoons \text{Ni(c)} + \text{CO}_2$$

The general expression for the equilibrium constant is

$$K_a = \frac{a_{\text{Ni}}\, a_{\text{CO}_2}}{a_{\text{NiO}}\, a_{\text{CO}}}$$

*P. W. Bridgman, *The Physics of High Pressure* (London: G. Bell & Sons, Ltd., 1949), p. 130.

This expression may be rewritten as

$$K_a = \frac{f_{CO_2}}{f_{CO}} \cdot \frac{\Gamma_{Ni}}{\Gamma_{NiO}}$$

At moderate pressures, Γ_{Ni} and $\Gamma_{NiO} = 1$ (the activities of the pure solids $= 1$). If we consider that the gases behave ideally, K_a becomes, simply,

$$K_P = \frac{P_{CO_2}}{P_{CO}}$$

The practical rule is that for reactions at moderate pressures, the activities of pure solid and liquid substances are simply set equal to unity in the equilibrium constant, so that no terms are included for such reactants. For the reaction cited at 1500 K, $\Delta G^{\ominus} = -81.09$ kJ, so that

$$K_P = \frac{P_{CO_2}}{P_{CO}} = 6.68 \times 10^2$$

Any ratio of P_{CO_2} to P_{CO} less than 6.68×10^2 will reduce NiO to Ni. Any ratio greater than this will oxidize Ni to NiO.

PROBLEMS

1. From the standard free enthalpies of formation in Table 8.2, calculate $\Delta G^{\ominus}$ and K_P for the following reactions at 298 K:
   (a) $D_2O(g) + H_2O(g) \longrightarrow 2HDO(g)$
   (b) $NO_2 + CO \longrightarrow NO + CO_2$
   (c) $PH_3(g) + \frac{3}{2}F_2 \longrightarrow PF_3(g) + 3HF(g)$
   (d) $SiO_2(c) + 2H_2 \longrightarrow Si(c) + 2H_2O(l)$

2. The equilibrium constant of the reaction

   $$I_2 + cyclopentene \longrightarrow 2HI + cyclopentadiene$$

   was measured spectrophotometrically in the gas phase between 175 and 415°C.[*]

   $$\log_{10} K_P(atm) = 7.55 - \frac{22160}{4.575T}$$

   Calculate $\Delta G^{\ominus}$, $\Delta H^{\ominus}$, and $\Delta S^{\ominus}$ for the reaction at 300°C. If equal initial amounts of I_2 and cyclopentene are mixed at a total pressure of 1 atm and 300°C, what will be the equilibrium partial pressure of I_2? What at 10 atm?

3. The equilibrium constant of $COCl_2 \rightleftharpoons CO + Cl_2$ was measured as follows:[†]

T (K)	635.7	670.4	686.0	722.2	760.2
K_P (atm)	0.01950	0.04414	0.07575	0.1971	0.5183

   Plot these results and extrapolate to 298 K. Calculate $\Delta H^{\ominus}$, $\Delta S^{\ominus}$ and $\Delta G^{\ominus}$ for the reaction at 298 K.

4. For the reaction $\frac{1}{2}N_2 + \frac{3}{2}H_2 \longrightarrow NH_3$ at 298 K, plot the ΔG vs. the extent of reaction ξ from $\xi = 0$ to $\xi = 1$. Also plot the affinity $\mathscr{A}$ of the reaction vs. ξ.

[*]S. Furuyama, D. M. Golden, and S. W. Benson, *J. Chem. Thermodynamics*, 2, 161 (1970).
[†]A. Lord and H. O. Pritchard, *J. Chem. Thermodynamics*, 2, 187 (1970).

5. For the reaction $N_2O_4 \rightleftharpoons 2NO_2$, calculate K_P, K_X, K_c at 298 K and 1 atm from the $\Delta G_f^{\ominus}$ data in Table 8.2. At what pressure would N_2O_4 be 50% dissociated at 298 K? For two choices of standard state, $P = 1$ atm and $c = 1$ mol·cm^{-3}, calculate $\Delta G^{\ominus}$ for the reaction at 298 K.

6. The vapor of PCl_5 decomposes as $PCl_5 \longrightarrow PCl_3 + Cl_2$. The density of a partially dissociated sample of PCl_5 vapor at 1 atm and 403 K was 4.800 g·dm^{-3}. Calculate the degree of dissociation α and $\Delta G^{\ominus}$ (403 K) for the reaction. Calculate α at 403 K if the total pressure is still 1 atm, but 0.5 atm is a partial pressure of argon.

7. If the gases in the ammonia synthesis reaction $N_2 + 3H_2 \longrightarrow 2NH_3$ are all ideal, prove that the maximum concentration of ammonia at equilibrium is attained when the ratio of H_2 to N_2 is 3:1.

8. Calculate the equilibrium constant K_P for $Cl_2 \rightleftharpoons 2Cl$ at 1200 K by statistical thermodynamics. The fundamental vibration frequency of Cl_2 is at $\sigma = 565$ cm^{-1}, and the equilibrium $Cl-Cl$ distance is 0.199 nm. The ΔU_o of dissociation of Cl_2 is 2.48 eV.

9. Calculate the equilibrum constant of the reaction $H_2 + D_2 \rightleftharpoons 2HD$ at 300 K, given:

	H_2	HD	D_2
Fundamental vibration σ (cm^{-1})	4371	3786	3092
Moment of inertia I (g·cm^2 $\times 10^{40}$)	0.458	0.613	0.919

Use the formulas in Table 5.4 for the partition functions.

10. The following reagents were mixed and allowed to come to equilibrium at 25°C: ethyl acetate 1.800 g; ethanol, 1.570 g; glacial acetic acid, 1.052 g. Five cubic centimeters of exactly 3N HCl (5.23 g) was used as a catalyst. The entire sample at equilibrium was titrated to the neutral point with 34.50 cm^3 exactly 1N NaOH. Calculate the equilibrium constant K_X for the reaction,

$$C_2H_5OH + CH_3COOH \rightleftharpoons CH_3COOC_2H_5 + H_2O$$

11. The following thermodynamic data are available at 298 K for the reagents in the reaction, $ZnS(c) + H_2 \longrightarrow Zn(c) + H_2S$

	$\Delta H^{\ominus}$ (kJ·mol^{-1})	$S^{\ominus}$ (J·K^{-1}·mol^{-1})	C_P (J·K^{-1}·mol^{-1})
$H_2S(g)$	-22.18	205.6	$36.0 + 13.0 \times 10^{-3}T$
$Zn(c)$		41.6	$22.0 + 11.3 \times 10^{-3}T$
$ZnS(c)$	-184.10	57.7	$53.6 + 4.2 \times 10^{-3}T$
$H_2(g)$		130.5	$29.5 - 0.8 \times 10^{-3}T$

Calculate $\Delta G^{\ominus}$ at 1000 K for the reaction, and hence the partial pressure of H_2S when H_2 at 1 atm is passed over ZnS heated to 1000 K.

12. For the reaction, $6CH_4 \longrightarrow C_6H_6(g) + 9H_2$, $\Delta H^{\ominus}(298$ K$) = 127.0$ kcal·mol^{-1}, $\Delta S^{\ominus}(298$ K$) = 77.7$ cal·K^{-1}·mol^{-1}, and

$$\Delta C_P = 42.0 - 32.1 \times 10^{-3}T + 3.83 \times 10^{-6}T^2 \text{ cal·K}^{-1}\cdot\text{mol}^{-1}$$

If other reactions could be ignored, would this reaction provide a suitable way for production of benzene from natural gas? [1 cal = 4.1840 J]

13. At 1000°C, K_P for $CO_2(g) + C(c) \rightleftharpoons 2CO(g)$ is 121.5 atm. (a) Calculate K_c. (b) If a vessel initially contained carbon and CO_2 at 10 atm and 1000°C, what would be the total pressure at equilibrium?

14. Derive a formula for the fugacity of a gas that obeys the equation of state

$$PV_m = RT + AP + BP^2$$

where V_m is the molar volume.

15. Derive the relations between the activity coefficients $^c\gamma$, $^m\gamma$, and γ (based on molar concentration, molality, and mole fraction, respectively).

16. A gas follows an equation of state

$$P = \frac{nRT}{V} + \frac{n^2 B}{V^2}$$

Derive an expression for its fugacity.

17. The synthesis of methanol, $CO + 2H_2 \rightarrow CH_3OH$, is carried out at high pressures. At 1000 K, for CH_3OH, $(G - H_0^\ominus)/T = 257.7 \text{ J·K}^{-1}\text{·mol}^{-1}$ and $\Delta H_0^\ominus = -190.6$ kJ·mol^{-1}; for CO, $\Delta H_0^\ominus = -113.8$. The free enthalpy data are in Table 8.3 for CO and H_2. For CH_3OH, $T_c = 513.2$ K and $P_c = 78.5$ atm. Critical constants for H_2 and CO are in Table 1.2. By means of Newton's graphs, calculate K_f and K_P for the methanol synthesis at 1, 100, 500, and 1000 atm.

18. For the reaction, $NiO(c) + CO(g) \rightarrow Ni(c) + CO_2(g)$,

T (K)	936	1027	1125
K_P	4.54×10^3	2.55×10^3	1.58×10^3

Calculate $\Delta G^\ominus$, $\Delta H^\ominus$, $\Delta S^\ominus$ for the reaction at 1000 K. Is ΔC_P for the reaction > 0 or < 0? Would an atmosphere of 20% CO_2, 5% CO, and 75% N_2 oxidize Ni at 1000 K?

19. Deoxygenated nitrogen is often prepared in the laboratory by passing tank nitrogen over hot copper turnings. The reaction is $2Cu(c) + \frac{1}{2}O_2(g) \rightarrow Cu_2O(c)$, for which $\Delta G^\ominus = -39\,850 + 15.06T$ (cal). What would be the residual concentration of oxygen in the nitrogen if equilibrium were achieved at 600°C? How could you measure this concentration?

20. From data in Tables 8.2 and 8.10, calculate $\Delta G^\ominus(298 \text{ K})$ in aqueous solution for

$$\text{glycine} + \text{DL-leucine} \rightarrow \text{DL-leucylglycine} + H_2O$$

What proportion of the overall $\Delta G^\ominus$ is made up of the $T\Delta S^\ominus$ for forming 1 mol of liquid water?

21. H. J. Morowitz* has given an interesting estimate of the $\Delta S^\ominus$ of forming a protein from amino acids in solution. He finds

$$\Delta S = -kn\left[\ln\left(\frac{V}{\sum n_i V_i}\right)\left(\frac{x}{y}\right)e - \sum z_i \ln z_i\right]$$

where n_i is the number of amino acid molecules of species i, and volume V_i, and $n = \sum n_i$, $z_i = n_i/n$, and x/y is the ratio of the number of possible orientations of the free monomers compared with the number of orientations of the protein structure. The contribution of the x/y to the ΔS is approximated as the loss of one degree of rotational freedom. For a typical protein of $M = 30\,000$, calculate ΔS from the formula. From this ΔS, one must subtract the ΔS of n molecules of liquid water. Make this correction. Discuss the final result and possible errors in the estimations.

*H. J. Morowitz, *Energy Flow in Biology* (New York: Academic Press, Inc., 1968), p. 92.

22. Amagat measured the following molar volumes for CO_2 at 333 K:

P (atm)	13.01	35.42	53.65	74.68	85.35
V_m (cm$^3 \cdot$ mol^{-1})	2000	666.7	400	250	200

Calculate the fugacity f and the fugacity coefficient $\gamma = f/P$ for CO_2 at 333 K and $P = 10, 20, 40, 80$ atm.

23. From the data in Table 8.7, plot the log of ratio of the activity coefficients of water and sucrose $\ln (\gamma_A/\gamma_B)$ vs. X_A, the mole fraction of water from $X_A = 0.9$ to $X_A = 1$. Show that

$$\int_0^1 \ln \left(\frac{\gamma_A}{\gamma_B}\right) dX_A = 0$$

24. For n-pentane (g) and isopentane (g) at 298 K, $\Delta G_f^{\ominus} = -194.4$ and -200.8 kJ$\cdot$mol^{-1}, respectively. The vapor pressures of the liquids are given by

$$n\text{-pentane} \qquad \log_{10} P \text{ (atm)} = 3.9714 - \frac{1065}{T - 41}$$

$$\text{isopentane} \qquad \log_{10} P \text{ (atm)} = 3.9089 - \frac{1020}{T - 40}$$

Calculate K_P for the isomerization, n-pentane $\rightleftharpoons$ isopentane in the gas phase at 298 K, and K_X for isomerization in the liquid phase, assuming ideal solutions.

25. Suppose that a large molecule P, such as an enzyme, contains n reactive sites at which a small molecule A could be bound. Assume that the n combining sites on P are all equivalent and independent, i.e., the reactions $P + A \rightleftharpoons PA$, $PA + A \rightleftharpoons PA_2$, etc., all have the same equilibrium constant for association K_{as}. Show that the average number of occupied sites per molecule is

$$\bar{\nu} = \frac{nK_{as}[A]}{1 + K_{as}[A]}$$

where $[A]$ is the concentration of A.*

*See J. T. Edsall and J. Wyman, *Biophysical Chemistry* (New York: Academic Press, Inc., 1958), p. 610, for a discussion and extension of this model, which includes many important biochemical equilibrium problems.

9
Chemical Reaction Rates

Now forasmuch as the Elements, unless they be altered, cannot constitute mixt
bodies; nor can they be altered unless they act and suffer one from another; nor
can they act and suffer unless they touch one another, we must first speak
a little concerning contact or mutual touching, Action, passion and Reaction.

Daniel Sennert
(1660)*

There are two basic questions in physical chemistry: "Where are chemical reactions
going?" and "How fast are they getting there?" The first is the problem of equilibrium, or chemical statics; the second is the problem of the rate of attainment of
equilibrium, or chemical kinetics.

Chemical kinetics is subdivided into the study of *homogeneous reactions*, which
occur entirely within one phase, and *heterogeneous reactions*, which occur at an
interface between phases. Some reactions consisting of a number of steps may
begin in a surface, continue in a homogeneous phase, and sometimes terminate
on a surface.

The study of the kinetics of any reaction can be divided into two parts:
(1) the formulation of the reaction rate in terms of concentrations of interacting
species and rate constants; (2) the explanation of the values of the rate constants
in terms of the structures and dynamics of the interacting species.

1. The Rate of Chemical Change

The *rate of reaction* $\mathbf{v}$ is defined as

$$\mathbf{v} \equiv \frac{d\xi}{dt} \tag{9.1}$$

where ξ is the extent of reaction described in Section 8.3. For a general stoichiometric equation,

$$v_1 A_1 + v_2 A_2 + \cdots \longrightarrow v_1' A_1' + v_2' A_2' + \cdots, \text{ etc.}$$

Thirteen Books of Natural Philosophy (London: Peter Cole, 1660).

$$\mathbf{v} \equiv \frac{d\xi}{dt} = -\frac{1}{v_1}\frac{dn_1}{dt} = -\frac{1}{v_2}\frac{dn_2}{dt} = \frac{1}{v_1'}\frac{dn_1'}{dt} = \frac{1}{v_2'}\frac{dn_2'}{dt} = \cdots, \text{ etc.} \qquad (9.2)$$

where n_1 is the amount of reactant A_1, etc. If the volume of the system in which the reaction occurs is independent of the time, we can conveniently express the reaction rate in terms of concentrations $c_j = n_j/V$. Thus,

$$\mathbf{v} \equiv \frac{d\xi}{dt} = -\frac{V}{v_1}\frac{dc_1}{dt} = -\frac{V}{v_2}\frac{dc_2}{dt} = \frac{V}{v_1'}\frac{dc_1'}{dt} = \frac{V}{v_2'}\frac{dc_2'}{dt} = \cdots, \text{ etc.} \qquad (9.3)$$

In systems of constant volume, the rate of reaction per unit volume, $\mathbf{v}/V$, is often referred to as simply the *rate of reaction*.

Qualitative observations of the rates of chemical reactions were recorded by early writers on metallurgy, brewing, and alchemy, but the first significant quantitative investigation was by L. Wilhelmy, in 1850. He studied the inversion of sucrose in aqueous solutions of acids, following the change with a polarimeter:

$$H_2O + C_{12}H_{22}O_{11} \longrightarrow C_6H_{12}O_6 + C_6H_{12}O_6$$
$$\text{sucrose} \qquad\qquad \text{glucose} \qquad \text{fructose}$$

The rate of decrease in the concentration of sugar c with time t was proportional to the concentration of sugar remaining unconverted,

$$-\frac{dc}{dt} = k_1 c$$

The constant k_1 is called the *rate constant* or *specific rate* of the reaction.* Its value was found to be proportional to the concentration of acid. Since the acid does not appear in the stoichiometric equation for the reaction, it is acting as a catalyst, increasing the reaction rate without being consumed.

Wilhelmy integrated the differential equation for the rate, obtaining

$$\ln c = -k_1 t + C$$

At $t = 0$, the concentration has its initial value c_0, so that $C = \ln c_0$. Therefore, $\ln c = -k_1 t + \ln c_0$, or

$$c = c_0\, e^{-k_1 t}$$

The experimental concentrations of sucrose closely followed this exponential decrease with time. On the basis of this work, Wilhelmy deserves to be called the founder of chemical kinetics.

The important paper of Guldberg and Waage, which appeared in 1863 (Section 8.1), emphasized the dynamic nature of chemical equilibrium. Van't Hoff later set the equilibrium constant equal to the ratio of rate constants for forward and reverse reactions: $K = k_f/k_b$.

In 1865 and 1867, Harcourt and Esson[†] studied the reaction between potassium permanganate and oxalic acid. They showed how to calculate rate constants for a reaction in which the rate was proportional to the product of concentrations of two reactants. They also discussed the theory of consecutive reactions.

---

*The words *rate*, *speed*, and *velocity* are all synonymous in chemical kinetics, although not so in physical mechanics; however, the recommended term is *reaction rate*.

†A. V. Harcourt and W. Esson, *Proc. Roy. Soc.*, *14*, 470 (1865); *Phil. Trans. Roy. Soc. London*, *Ser. A*, *156*, 193 (1866); *157*, 117 (1867).

2.   Experimental Methods in Kinetics

An experimental determination of reaction rate usually requires a thermostat to maintain the system at constant temperature and a good clock to measure the passage of time. These two requisites are not hard to obtain. The pursuit of the third variable, the concentration of reactant or product, is the source of most difficulties. A reaction cannot be turned on and off like a stopcock, although a reaction occurring at an elevated temperature can often be virtually stopped by chilling the system. It is difficult to determine the concentration c at a definite time t by any sampling technique.

The best method of analysis is, therefore, one that is practically continuous, being based on some physical property that does not require removal of successive samples from the reaction mixture. Wilhelmy's use of optical rotation is a case in point. Other physical methods include the following:

1. absorption spectra and colorimetric analysis;
2. measurement of dielectric constant;*
3. measurement of refractive index;†
4. dilatometric methods, based on the change in volume due to reaction;
5. change in pressure in some gas reactions.

To follow the pressure without concurrent analysis of the reaction products can lead to deceptive results. For example, the decomposition of ethane according to $C_2H_6 \rightarrow C_2H_4 + H_2$ changes the pressure, but actually some methane, CH_4, is included among the products.

Flow systems are often used for the study of rapid reactions, as discussed in Section 9.16. When one wishes to begin measurement of the reaction close to a given set of initial conditions, the flow method is often not feasible. In such cases, the *stopped-flow method* can be used. It has been extensively applied to reactions in solution, especially enzymatic reactions, for which it has the added advantage of being economical of materials. An example of a stopped-flow apparatus for a gas phase reaction is shown in Fig. 9.1 (a). It was designed to study $2NO_2 + O_3 \rightarrow N_2O_5 + O_2$ under conditions in which reaction is complete within 0.1 s. A stream of $O_2 + NO_2$ was mixed with a stream of $O_3 + O_2$ in a chamber with tangential jets. After mixing, which was complete within 0.01s, a magnetically operated steel gate trapped a portion of the gas mixture. The disappearance of NO_2, which is brown, was followed by the change of intensity of a beam of transmitted light. The beam was chopped with a rotating sector wheel 300 times per second, and the pulses were allowed to fall on a photomultiplier tube, the output of which was connected to an oscilloscope. The pulsations on the screen were photographed, the height of each peak giving the NO_2 concentration at intervals of $\frac{1}{300}$ s.

*T. G. Majury and H. W. Melville, *Proc. Roy. Soc. London, Ser. A*, *205*, 496 (1951).
†N. Grassie and H. W. Melville, *Proc. Roy. Soc. London, Ser. A*, *207*, 285 (1951).

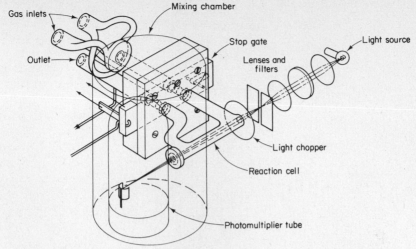

FIG. 9.1(a) Apparatus of Johnston and Yost for study of a rapid gas reaction: isometric projection of mixing chamber, stop gate, and 2 mm diameter reaction cell, with schematic drawing of the light source, filters, and lenses. [*J. Chem. Phys. 17*, 386 (1949).]

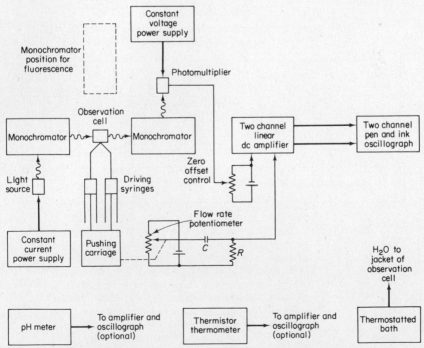

FIG. 9.1(b) Block diagram of stopped-flow apparatus of type used to study rapid reactions in solution.

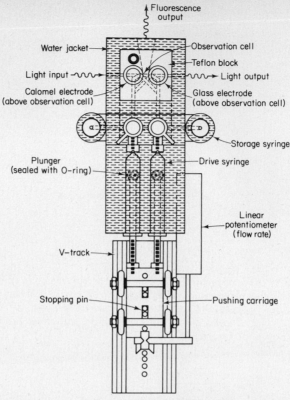

FIG. 9.1(c) Schematic plan view of fluid handling assembly of stopped-flow apparatus shown in Fig. 9.1(b).

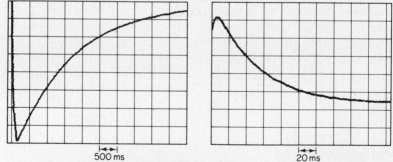

500 ms   20 ms

FIG. 9.1(d) An oscilloscope recording of a transient reaction obtained by the stopped-flow procedure. Transient changes in the absorption of a substrate chromophore (furyl-acrylamido) in the reaction of N[β (2-furyl) acryloyl]-L-tryptophan methyl ester with a molar excess of α-chymotrypsin; pH-5.30, 330 nm. (a) At a scan speed of 500 msec per major division. (b) The same reaction at a scan speed of 20 msec per major division. The extent of time between mixing and observation is of the order of a few milliseconds. The time resolution thereafter is of the order of 0.1 msec.

A stopped-flow cell designed for reactions in solution is shown in Figs. 9.1(b and c). Application of such equipment to very fast reactions is limited by the time necessary to fill the observation cell with mixed reactants, and by the "stopping disturbance" when the gates are shut. These two times cannot both be minimized, since the faster are the inlet velocities of reactants, the greater is the disturbance when the cell is isolated. Figure 9.1(d) shows typical data obtained from a cell like the one shown in (c).

For reactions much faster than the millisecond range, various *relaxation methods* are available, which will be discussed in Section 9.17. These methods avoid the problem of initial mixing entirely by starting the observations after a sharp initial disturbance (e.g., in temperature or electric field) of the existent steady state or equilibrium condition.

3. Order of a Reaction

The experimental data of chemical kinetics are records of concentrations of reactants and products at various times, temperature usually being held constant throughout any one run. On the other hand, theoretical expressions for reaction rates as functions of concentrations of reactants, and sometimes of products, are differential equations of the general form,

$$\frac{dc_1}{dt} = f(c_1, c_2, \cdots, c_n)$$

Here, c_1 is concentration of the particular product or reactant which is being followed to measure the rate of reaction. Before comparison of theory with experiment, it is necessary either to integrate the theoretical rate law or to differentiate the experimental concentration vs. time curve.

The rate laws are of practical importance since they provide concise expressions for the course of the reaction and can be applied in calculating reaction times, yields, and optimum economic conditions. Also, the laws often afford an insight into the *mechanism*[*] by which the reaction proceeds. On the molecular scale, the course of a reaction may be complex, but sometimes the form of the empirical rate law will suggest the particular path via which the reaction takes place.

In many instances, the rate, which is proportional to the decrease in concentration of reactant A, $-dc_A/dt$, is found to depend on the product of concentration terms. For example,

$$-\frac{dc_A}{dt} = k'c_A^a c_B^b \cdots c_N^n$$

The *order* of the reaction is defined as the sum of the exponents of the concentration terms in this rate law. For example, the decomposition of nitrogen pentoxide, $2N_2O_5 \longrightarrow 4NO_2 + O_2$, is found to follow the law,

$$\frac{-d[N_2O_5]}{dt} = k_1[N_2O_5]$$

[*]The word *chemism* seems preferable, but it has not been popular in English. In German, *Chemismus* is often used.

where we use square brackets to indicate concentrations. This is therefore a *first-order reaction*. The decomposition of nitrogen dioxide, $2NO_2 \rightarrow 2NO + O_2$, follows the law,

$$\frac{-d[NO_2]}{dt} = k_2[NO_2]^2$$

This is a *second-order reaction*. The reaction rate in benzene solution of triethyl-amine and ethyl bromide,

$$(C_2H_5)_3N + C_2H_5Br \longrightarrow (C_2H_5)_4NBr$$

follows the equation,

$$\frac{-d[C_2H_5Br]}{dt} = k_2[C_2H_5Br][(C_2H_5)_3N]$$

This reaction is also second order. It is said to be *first order with respect to* C_2H_5Br, *first order with respect to* $(C_2H_5)_3N$, and *second order overall*. The decomposition of acetaldehyde, $CH_3CHO \rightarrow CH_4 + CO$, in the gas phase at 720 K fits the rate expression,

$$\frac{-d[CH_3CHO]}{dt} = k'[CH_3CHO]^{3/2}$$

This is a reaction of the *three-halves order*.

The order of a reaction need not be a whole number but may be zero or fractional. It is determined solely by the best fit of a rate equation with the empirical data. It is important to realize that there is no *necessary* connection between the form of the stoichiometric equation for the reaction and the kinetic order. Thus, the decompositions of N_2O_5 and of NO_2 have equations of identical form, yet one is a first-order and the other a second-order reaction.

The units of the rate constant depend on the order of the reaction. For first order, $-dc/dt = k_1 c$; hence the usual units of k_1 $(mol \cdot dm^{-3} \cdot s^{-1})/(mol \cdot dm^{-3}) = s^{-1}$. For second order, $-dc/dt = k_2 c^2$; the units of k_2 $(mol \cdot dm^{-3} \cdot s^{-1})/(mol \cdot dm^{-3})^2 = mol^{-1} \cdot dm^3 \cdot s^{-1}$. In general, for a reaction of the nth order, the dimensions of the constant k_n are $(time)^{-1}$ $(concentration)^{1-n}$.

4.   Molecularity of a Reaction

Many chemical reactions are not kinetically simple; they proceed through a number of steps or stages between initial reactants and final products. Each of the individual steps is called an *elementary reaction*. Complex reactions are made up of a sequence of elementary reactions, each of which proceeds in a single step.

In the earlier literature, the terms *unimolecular, bimolecular,* and *trimolecular* were used to denote reactions of the first, second, and third orders. We now reserve the concept of the *molecularity* of a reaction to indicate the molecular mechanism by which it proceeds. For example, careful studies have been made of the reaction

$$NO + O_3 \longrightarrow NO_2 + O_2$$

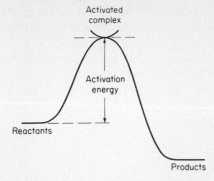

FIG. 9.2 Energy barrier surmounted by a system in chemical reaction.

When an NO molecule strikes an O_3 with sufficient kinetic energy, it can capture an O atom, thus completing the reaction. This elementary reaction involves two molecules and it is therefore called a *bimolecular reaction*.

It will be shown later that before the usual chemical reaction can take place, the molecule or molecules involved must be raised to a state of higher potential energy. They are then said to be *activated* or to form an *activated complex*. This process is shown in schematic form in Fig. 9.2. Reactants and products are both at stable potential-energy minima; the activated complex is the state at the top of the potential-energy barrier.

The molecularity of a reaction can be defined as the number of molecules* of reactants that are used to form the activated complex. In the case of the preceding example, the complex is formed from two molecules, $NO + O_3$, and the reaction is bimolecular. Clearly, the molecularity of a reaction must be a whole number, and, in fact, is found to be one, two, or, rarely, three.

Experimental measurements show that the rate of the reaction of NO with O_3 is

$$\frac{-d[\mathrm{NO}]}{dt} = k_2[\mathrm{NO}][\mathrm{O}_3]$$

This reaction is therefore second order. All bimolecular reactions are second order, but the converse is not true; some second-order reactions are not bimolecular.

A good example of a unimolecular reaction is a radioactive decay, e.g., Ra $\longrightarrow$ Rn $+ \alpha$. Only one atom is involved in each disintegration, and the reaction is unimolecular. It also follows a first-order law, $-dC_{\mathrm{Ra}}/dt = k_1 C_{\mathrm{Ra}}$, where C_{Ra} is the number concentration of radium atoms present at any time.

This example is a nuclear reaction and not a chemical reaction. Chemical reactions that are unimolecular are either isomerizations or decompositions. The isomerization of cyclopropane to propene has become one of the most studied prototypes of unimolecular reaction:

$$\begin{array}{c} \mathrm{CH_2} \\ \diagup \diagdown \\ \mathrm{CH_2 - CH_2} \end{array} \longrightarrow \mathrm{CH_3 - CH = CH_2}$$

*The word *molecule* is used in its general sense to include atomic reactants also.

Many decomposition reactions have complex mechanisms, but it is possible to isolate certain steps that appear to be unimolecular. An example is the dissociation of ethane into two methyl radicals, $C_2H_6 \longrightarrow 2CH_3$.

The concept of molecularity should be applied only to individual elementary reactions. If a reaction proceeds through several stages, we do not talk about its molecularity, since one step may involve two molecules, another three, and so on. At the risk of repetition, *reaction order* applies to the experimental rate equation; *molecularity* applies to the theoretical mechanism.

5.   Reaction Mechanisms

Two meanings for the term *reaction mechanism* are in common use. In one sense, *reaction mechanism* means the particular sequence of elementary reactions leading to the overall chemical change whose kinetics are under study. In the second sense, *reaction mechanism* means the detailed analysis of how the chemical bonds (or the nuclei and electrons) in the reactants rearrange to form the activated complex. For the present, we shall understand the mechanism of a reaction to be established when we have found a sequence of elementary reactions that explain the observed kinetic behavior. We admit that each of these elementary reactions itself has a definite *mechanism*, but theory is not yet able to elucidate this finer detail.

Consider, for example, the gas reaction,

$$2O_3 \longrightarrow 3O_2$$

We cannot predict the kinetic law this reaction follows simply by looking at its stoichiometric equation. If it were a bimolecular elementary reaction, it would follow the rate law

$$\frac{-d[O_3]}{dt} = k_2[O_3]^2$$

(Such a rate law is a necessary but not a sufficient condition for the reaction to be bimolecular.) As a matter of fact, experiment shows that the rate law is

$$\frac{-d[O_3]}{dt} = \frac{k_a[O_3]^2}{[O_2]}$$

With this information, we can suggest a reasonable mechanism:

$$O_3 \underset{k_{-1}}{\overset{k_1}{\rightleftarrows}} O_2 + O$$

$$O + O_3 \xrightarrow{k_2} 2O_2$$

The reversible dissociation is assumed to be rapid, leading to an equilibrium concentration of oxygen atoms,

$$[O] = \frac{K[O_3]}{[O_2]}$$

where $K = k_1/k_{-1}$. Then the slower second step gives the net rate of decomposition of O_3,

$$\frac{-d[O_3]}{dt} = k_2[O][O_3] = \frac{k_2K[O_3]^2}{[O_2]}$$

Thus, the suggested mechanism leads to the observed rate law. This agreement does not prove that the mechanism is correct. It is a necessary but not a sufficient condition for its correctness.

A skeptic might say that a sufficient condition can never be found in an experimental subject like chemical kinetics, but we shall be content with a reasonably sufficient condition based on the weight of the evidence. (A kinetic problem, like a crime, calls for a proof beyond "reasonable doubt" and not a mathematical proof.) Once a mechanism has been found that gives the observed kinetics, it can be tested in various ways. One might, for example, measure independently the individual reaction rates and equilibrium constants involved, to see whether the predicted relation is, in fact, confirmed. In the ozone decomposition, for instance,

$$k_a = k_2 K$$

One could measure or calculate K and measure k_2 by introducing oxygen atoms of known concentration into ozone.

The proof of a mechanism is not easy. Thus, there are many reactions whose kinetics are well known; there are some for which reasonable mechanisms have been proposed; but there are only a few for which the mechanisms have been proved beyond reasonable doubt.

6. First-Order Rate Equations

Consider a reaction, $A \longrightarrow B + C$. Let the initial concentration of A be a mol·dm^{-3}. If after a time t, x mol·dm^{-3} of A have decomposed, the remaining concentration of A is $a - x$, and x mol·dm^{-3} of B or C have been formed. The rate of formation of B or C is thus dx/dt. For a first-order reaction this rate is proportional to the instantaneous concentration of A, so that

$$\frac{dx}{dt} = k_1(a - x) \tag{9.4}$$

Separating the variables and integrating, we obtain

$$-\ln(a - x) = k_1 t + C$$

where C is the constant of integration. The usual initial condition is that $x = 0$ at $t = 0$, whence $C = -\ln a$, and the integrated equation becomes

$$\ln \frac{a}{a - x} = k_1 t \tag{9.5}$$

or

$$x = a(1 - e^{-k_1 t})$$

If $\ln[a/(a - x)]$ is plotted against t, a straight line passing through the origin is obtained, the slope of which is the first-order rate constant k_1.

If (9.4) is integrated between limits x_1 to x_2 and t_1 to t_2, the result is

$$\ln \frac{a - x_1}{a - x_2} = k_1(t_2 - t_1)$$

This interval formula can be used to calculate the rate constant from any pair of concentration measurements.

Applications of these equations to the first-order decomposition of gaseous N_2O_5 are shown in Table 9.1 and Fig. 9.3.*

Another test of a first-order reaction is found in its *half-life*, τ, the time required to reduce the concentration of A to half its initial value. In (9.5), when $x = a/2$, $t = \tau$, and

$$\tau = \frac{\ln 2}{k_1} \tag{9.6}$$

Thus, the half-life of a first-order reaction is independent of the initial concentration of reactant. In a first-order reaction, it would take just as long to reduce the reactant concentration from 0.1 $mol \cdot dm^{-3}$ to 0.05 as it would to reduce it from 10 to 5.

TABLE 9.1
Decomposition of Nitrogen Pentoxide (T = 318.2 K)

Time, t (s)	$P_{N_2O_5}$ (Torr)	k_1 (s^{-1})	Time, t (s)	$P_{N_2O_5}$ (Torr)	k_1 (s^{-1})
0	348.4		4200	44	0.000478
600	247		4800	33	0.000475
1200	185	0.000481	5400	24	0.000501
1800	140	0.000462	6000	18	0.000451
2400	105	0.000478	7200	10	0.000515
3000	78	0.000493	8400	5	0.000590
3600	58	0.000484	9600	3	0.000467
			∞	0	

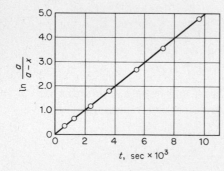

FIG. 9.3 A first-order reaction. The thermal decomposition of nitrogen pentoxide, plotted according to Eq. (9.5).

7. Second-Order Rate Equations

Consider a reaction written as $A + B \rightarrow C + D$. Let the initial concentrations at $t = 0$ be a $mol \cdot dm^{-3}$ of A and b $mol \cdot dm^{-3}$ of B. After a time t, x $mol \cdot dm^{-3}$ of A, and of B, will have reacted, forming x $mol \cdot dm^{-3}$ of C and of D. If a second-

*F. Daniels, *Chemical Kinetics* (Ithaca: Cornell University Press, 1938), p. 9.

order rate law is followed,

$$\frac{dx}{dt} = k_2(a - x)(b - x) \tag{9.7}$$

Separating the variables, we have

$$\frac{dx}{(a - x)(b - x)} = k_2 \, dt$$

The expression on the left is integrated by breaking it into partial fractions.* The integration yields

$$\frac{\ln(a - x) - \ln(b - x)}{a - b} = k_2 t + C$$

When $t = 0$, $x = 0$, and hence $C = \ln(a/b)/(a - b)$. Therefore, the integrated second-order rate law is

$$\frac{1}{a - b} \ln \frac{b(a - x)}{a(b - x)} = k_2 t \tag{9.8}$$

A reaction found to be second order is that between ethylene bromide and potassium iodide in 99% methanol,

$$C_2H_4Br_2 + 3KI \longrightarrow C_2H_4 + 2KBr + KI_3$$

Note that the stoichiometric equation in this case bears no simple relation to the order of reaction. Sealed bulbs containing the reaction mixture were kept in a thermostat. At intervals of 2 or 3 min, a bulb was withdrawn, and its contents were analyzed for I_2 ($= KI_3$) by means of the thiosulfate titration. The second-order rate law is

$$\frac{d[I_2]}{dt} = \frac{dx}{dt} = k_2[C_2H_4Br_2][KI] = k_2(a - x)(b - 3x)$$

The integrated equation is†

$$\frac{1}{3a - b} \ln \frac{b(a - x)}{a(b - 3x)} = k_2 t$$

Figure 9.4 is a plot of the left side of this equation against time. The excellent linearity confirms the second-order law. The slope of the line is the rate constant, $k_2 = 0.299$ dm$^3 \cdot$mol^{-1} min^{-1}.

*Let

$$\frac{1}{(a - x)(b - x)} = \frac{A}{(a - x)} + \frac{B}{(b - x)} = \frac{A(b - x) + B(a - x)}{(a - x)(b - x)}$$

Then,

$$(bA + aB) - (A + B)x = 1$$

and

$$\begin{array}{ll} bA + aB = 1 & A = \dfrac{-1}{(a - b)} \\[2mm] A + B = 0 & B = \dfrac{1}{(a - b)} \end{array}$$

†Note that we can express all concentrations in equivalents per unit volume and always use Eq. (9.8).

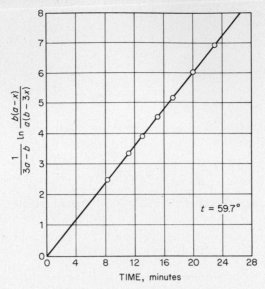

FIG. 9.4 The second-order reaction, $C_2H_4Br_2 + 3 KI \longrightarrow C_2H_4 +$ $2KBr + KI_3$. [From R. T. Dillon, *J. Am. Chem. Soc., 54,* 952 (1932).]

A special case of the general second-order equation (9.7) arises when the initial concentrations of both reactants are the same, $a = b$. This condition can be purposely arranged in any case, but it will be necessarily true whenever only one reactant is involved in a second-order reaction. An example is the decomposition of gaseous hydrogen iodide, $2HI \longrightarrow H_2 + I_2$, which follows the rate law,

$$-d[HI]/dt = k_2[HI]^2$$

In these cases, the integrated equation (9.8) cannot be applied, since when $a = b$ it reduces to $k_2t = 0/0$, which is indeterminate. It is best to return to the differential equation, which becomes

$$\frac{dx}{dt} = k_2(a - x)^2$$

Integration yields

$$\frac{1}{(a - x)} = k_2t + C$$

When $t = 0$, $x = 0$, so that $C = a^{-1}$. The integrated rate law is, therefore,

$$\frac{x}{a(a - x)} = k_2t \tag{9.9}$$

The half-life of a second-order decomposition is found from (9.9) by setting $x = a/2$ when $t = \tau$, so that

$$\tau = \frac{1}{k_2a} \tag{9.10}$$

The half-life varies inversely as the initial concentration.

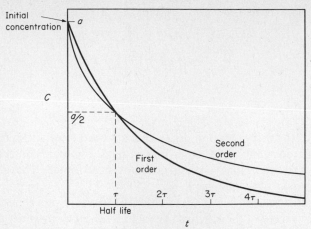

FIG. 9.5 Concentration vs. time curves for first- and second-order reactions with same initial concentrations *a* and same half-lives *τ*.

Figure 9.5 shows the concentration vs. time curve of a first-order and a second-order reaction having the same half-lives. The second-order reaction is of the type $2A \longrightarrow$ products.

8. Third-Order Rate Equations

In the gas phase, third-order reactions are quite rare, and all that have been studied fall into the class,

$$
\begin{array}{ccc}
2A & + & B & \longrightarrow \text{products} \\
a - 2x & & b - x & & x
\end{array}
$$

The differential rate equation is, accordingly,

$$\frac{dx}{dt} = k_3(a - 2x)^2(b - x).$$

This equation can be integrated by breaking it into partial fractions. The result, after we apply the initial condition, $x = 0$ at $t = 0$, is

$$\frac{1}{(2b - a)^2}\left[\frac{(2b - a)2x}{a(a - 2x)} + \ln\frac{b(a - 2x)}{a(b - x)}\right] = k_3 t$$

Examples of gas reactions of the third order, following this rate law, are

$$2NO + O_2 \longrightarrow 2NO_2$$
$$2NO + Br_2 \longrightarrow 2NOBr$$
$$2NO + Cl_2 \longrightarrow 2NOCl$$

In every case, $-d[NO]/dt = k_3 [NO]^2 [X_2]$.

The recombination of atoms in the gas phase usually requires the presence of a third body M to take up the excess energy of the exothermic reaction. Thus, these reactions are often third order. For example,

$$Cl + Cl + M \longrightarrow Cl_2 + M, \quad etc.$$

$$\frac{-d[Cl]}{dt} = k_3 \, [Cl]^2 \, [M]$$

9.   Determination of Reaction Order

In simple reactions of the first or second order, it is not hard to establish the order and evaluate the rate constants. The experimental data are simply inserted into different integrated rate equations until a constant k is found. The graphical methods leading to linear plots are useful.

In more complicated reactions, it is often desirable to adopt other methods for at least a preliminary survey of the kinetics. The *initial reaction rate* often provides helpful information, for in a sufficiently slow reaction, the rate dx/dt can be found with some precision before there has been any extensive chemical change. It is then possible to assume that all the reactant concentrations are still effectively constant at their initial values. If $A + B + C \longrightarrow$ products, and the initial con- centrations are a, b, c, the rate can be written quite generally as

$$\frac{dx}{dt} = k \, (a - x)^{n_1} \, (b - x)^{n_2} \, (c - x)^{n_3}$$

If x is very small, the initial rate will be

$$\frac{dx}{dt} = k \, a^{n_1} \, b^{n_2} \, c^{n_3}$$

While we keep b and c constant, the initial concentration a can be varied, and the resultant change of the initial rate measured. In this way, the value of n_1 is esti- mated. Similarly, by keeping a and c constant while we vary b, a value of n_2 is found; and with a and b constant, variation of c yields n_3. The initial-rate method is especially useful in those reactions that cannot be trusted to progress to any appreciable extent without becoming involved in labyrinthine complications. If the order of reaction found by using initial rates differs from that found by using the integrated rate equation, it is probable that the products are interacting with the initial reactants.

A systematic way to find the reaction order with respect to each reactant is the *isolation method*, devised by W. Ostwald. The reaction mixture is made up so that one particular reactant A has an initial concentration that is low compared with the concentrations of all the other reactants. As the reaction proceeds, the frac- tional change in the concentration of A will be much greater than that of any other reactant. In fact, at least during the initial stages of the reaction, the concen- trations of reactants B, C, D, etc., may be taken as effectively constant at their initial values. To a good approximation, the rate equation may then take the form,

$$\frac{dx}{dt} = k \, (a - x)^{n_1} \, b^{n_2} \, c^{n_3} \cdots = k' \, (a - x)^{n_1}$$

By comparing the data with integrated forms of this equation for various choices of n_1, it is possible to determine the order of the reaction with respect to component A. The orders with respect to B, C, etc., are found in like manner.

A situation like that in the isolation method often occurs for reactions in solution when one reactant is the solvent. For example, in the hydrolysis of ethyl acetate,

$$CH_3COOC_2H_5 + H_2O \longrightarrow CH_3COOH + C_2H_5OH$$

the ester concentration is much lower than that of the solvent, water. The reaction follows a rate law that is first order with respect to the ester, and the effectively constant water concentration does not appear in the rate equation:

$$\frac{-d[CH_3COOC_2H_5]}{dt} = k_2 [CH_3COOC_2H_5][H_2O] = k_1 [CH_3COOC_2H_5]$$

Such a reaction is sometimes called a *pseudo first-order* reaction. Another way of holding constant the concentration of a reactant is to use a saturated solution with an excess of pure solute phase always present.

10. Opposing Reactions

In many reactions, the position of equilibrium is so far on the product side, at the temperature and pressure chosen for the experiment, that for all practical purposes one can say that the reaction *goes to completion*. This is the case in the N_2O_5 decomposition and the oxidation of iodide ion that have been described. There are other cases in which a considerable concentration of reactants remains when equilibrium is reached. A well-known example is the hydrolysis of ethyl acetate in aqueous solution,

$$CH_3COOC_2H_5 + H_2O \rightleftharpoons CH_3COOH + C_2H_5OH$$

In such instances, as the product concentrations are gradually increased, the velocity of the reverse reaction becomes appreciable. The measured net rate of change is thereby decreased, and to deduce a rate equation to fit the empirical data, we must take into consideration the opposing reaction.

For *opposing first-order reactions*, $A \rightleftharpoons B$, let the first-order rate constant in the forward direction be k_1, in the reverse, k_{-1}. Initially, at $t = 0$, the concentration of A is a, and that of B is b. If after a time t, a concentration x of A has been transformed into B, the concentration of A is $a - x$, and that of B is $b + x$. The differential rate equation is, therefore,

$$\frac{dx}{dt} = k_1(a - x) - k_{-1}(b + x)$$

or

$$\frac{dx}{dt} = (k_1 + k_{-1})(m - x)$$

where $m = (k_1a - k_{-1}b)/(k_1 + k_{-1})$. Integration yields

$$-\ln(m - x) = (k_1 + k_{-1})t + C$$

When $t = 0$, $x = 0$, so that $C = -\ln m$. Thus,

$$\ln \frac{m}{m - x} = (k_1 + k_{-1})t \tag{9.11}$$

By Guldberg and Waage's principle, the equilibrium constant $K = k_1/k_{-1}$. Thus, equilibrium measurements can be combined with rate data to separate the forward and reverse constants in (9.11).

Such reversible first-order reactions are found in some intramolecular rearrangements and isomerizations.* The cis–trans isomerization of styryl cyanide vapor was followed by the change in the refractive index of the solution obtained on condensation.

$$\begin{matrix} C_6H_5\text{—CH} & & C_6H_5\text{—CH} \\ \| & \rightleftharpoons & \| \\ NC\text{—CH} & & CH\text{—CN} \end{matrix}$$

Equilibrium at 573 K is at about 80% trans isomer.

The case of *opposing second-order reactions* was first treated by Max Bodenstein in his classic study of the combination of hydrogen and iodine.† Between 523 and 773 K, the reaction $H_2 + I_2 \rightarrow 2HI$ can be conveniently studied, but at higher temperatures, the equilibrium lies too far on the reactant side. Even in the cited temperature range, we must consider the reverse reaction to obtain satisfactory rate constants. The concentrations at time t will be denoted as follows:

$$H_2 \quad + \quad I_2 \quad \underset{k_{-2}}{\overset{k_2}{\rightleftharpoons}} \quad 2HI$$

$$a - \left(\frac{x}{2}\right) \quad b - \left(\frac{x}{2}\right) \quad\quad x$$

The net rate of formation of HI is

$$\frac{d[HI]}{dt} = \frac{dx}{dt} = k_2\left(a - \frac{x}{2}\right)\left(b - \frac{x}{2}\right) - k_{-2}\,x^2 \tag{9.12}$$

When the equilibrium constant $K = k_2/k_{-2}$ is introduced into (9.12) and the equation is integrated, the result is

$$k_2 = \frac{2}{mt}\left[\ln\left(\frac{\dfrac{a + b + m}{1 - 4K^{-1}} - x}{\dfrac{a + b - m}{1 - 4K^{-1}} - x}\right) + \ln\left(\frac{a + b - m}{a + b + m}\right)\right](1 - 4K)^{-1}$$

with

$$m = \sqrt{(a + b)^2 - 4ab(1 - 4K)^{-1}}$$

Good constants obtained from this rather formidable expression are shown

*G. B. Kistiakowsky, et al., *J. Am. Chem. Soc.*, **54**, 2208 (1932); **56**, 638 (1934); **57**, 269 (1935); **58**, 2428 (1936).

† *Z. Physik Chem.*, **13**, 56 (1894); **22**, 1 (1897); **29**, 295 (1898). The mechanism of these apparently simple reactions is still not established, and the current state of our understanding will be discussed in Section 9.34.

TABLE 9.2
Rate Constants for the Reaction $H_2 + I_2 \rightleftharpoons 2HI$

T (K)	$cm^3 \cdot mol^{-1} \cdot s^{-1}$		$K = k_2/k_{-2}$
	k_2	k_{-2}	
300	2.04×10^{-16}	2.24×10^{-19}	912
400	6.61×10^{-9}	2.46×10^{-11}	371
500	2.14×10^{-4}	1.66×10^{-6}	129
600	2.14×10^{-1}	2.75×10^{-3}	77.8
700	3.02×10^{1}	5.50×10^{-1}	54.9

in Table 9.2 for a number of temperatures, together with values of K and k_{-2} from separate experiments.*

11. Principle of Detailed Balancing

When a system reaches equilibrium, the forward rate of any chemical reaction equals the reverse rate. For example, if interconversion of A and C occurs by a reversible first-order process,

$$A \underset{k_{-1}}{\overset{k_1}{\rightleftharpoons}} C$$

At equilibrium,

$$\frac{d[A]}{dt} = 0 = -k_1[A] + k_{-1}[C]$$

Suppose, however, there is an alternative reaction pathway (or mechanism) from A to C via an intermediate B:

$$\begin{array}{ccc} & B & \\ {}^{k_2}\nearrow & & \searrow{}^{k_3} \\ A & \underset{k_{-1}}{\longleftarrow} & C \end{array}$$

Is it possible to maintain equilibrium between A and C by allowing A to go to C via intermediate B but C to return to A directly? We can certainly maintain a constant concentration of A by such a cyclic process:

$$\frac{d[A]}{dt} = 0 = -k_2[A] + k_{-1}[C]$$

Nevertheless, such a cyclic balancing of reaction rates is rigorously prohibited by

*A. F. Trotman-Dickenson and G. S. Milne, *Tables of Bimolecular Gas Reactions* (NSRDS–NBS 9) (Washington: U.S. Government Printing Office, 1967).

a general principle from statistical mechanics, called the *Principle of Detailed Balancing:* When equilibrium is reached in a reaction system, any particular chemical reaction and the exact reverse of that reaction must, on the average, be occurring at the same rate.

The Principle of Detailed Balancing is a consequence for large-scale systems of the *Principle of Microscopic Reversibility,** which applies on the scale of individual molecular processes. Consider a perfectly general collision process between two molecules. The states of the molecules will be defined by the coordinates and momenta of all their atoms. Now suppose that the directions of all the momenta are reversed to yield a second collision process that can be called the *reverse* of the first. At equilibrium, the probability of the existence of any particular configuration of molecules depends only on the energy of the configuration, and the energy is not altered by simply reversing the directions of the momentum vectors. Hence, the probability of the reverse collision is identical with the probability of the original forward collision. This result is an instance of the Principle of Microscopic Reversibility. If we extend this argument to all the possible forward and reverse collisions in a reacting system, we can thereby establish the Principle of Detailed Balancing.

For example, in the reaction

$$H_2 + I_2 \rightleftharpoons 2HI$$

at 663 K, about 10% of the forward reaction rate at equilibrium has been shown to proceed via a chain mechanism:

$$I_2 \rightleftharpoons 2I$$
$$I + H_2 \longrightarrow HI + H$$
$$H + I_2 \longrightarrow HI + I$$
$$H + I \longrightarrow HI$$

From the Principle of Detailed Balancing, we can conclude that 10% of the reverse reaction at equilibrium must proceed through the reverse of this chain mechanism.

The application of the Principle of Detailed Balancing to reactions at equilibrium is rigorously valid. One must be careful, however, in applying it to reactions not at equilibrium.[†] Each case must then be analyzed in some detail to decide whether it is likely or possible that no changes in mechanism occur as the reaction conditions depart from those in the equilibrium mixture. Under such circumstances, the Principle of Detailed Balancing can be a qualitative guide to the kinds of reactions that need be considered in the mechanism of the reverse process, once the forward mechanism has been elucidated.

---

[*] R. C. Tolman, *Principles of Statistical Mechanics* (Oxford: Oxford University Press, 1938), p. 163. The Principle of Microscopic Reversibility can be derived from the necessary mathematical form of quantum mechanical expressions for transition probabilities.

[†] R. M. Krupka, H. Kaplan, and K. J. Laidler, *Trans. Faraday Soc., 62*, 2755 (1966).

12. Rate Constants and Equilibrium Constants

Let us consider a general reaction at some constant temperature,

$$aA + bB \rightleftharpoons cC + dD$$

Apart from effects due to nonideality in the solution, we can always write an equilibrium constant for such a reaction in the form

$$K_c = \frac{C^c D^d}{A^a B^b}$$

or

$$\frac{A^a B^b}{C^c D^d} K_c = 1 \qquad (9.13)$$

where for simplicity in subsequent notation we have written C, D, A, B for the *equilibrium concentrations* of the reactants. At equilibrium, the rates of forward and reverse reactions must be equal, $\mathbf{v}_f = \mathbf{v}_r$. These rates will be some functions of the concentrations, so that

$$\frac{\mathbf{v}_f(A, B, C, D)}{\mathbf{v}_r(A, B, C, D)} = 1 \qquad (9.14)$$

In order for both (9.13) and (9.14) to hold, it is a sufficient condition* that

$$\frac{\mathbf{v}_f(A, B, C, D)}{\mathbf{v}_r(A, B, C, D)} = \left[\frac{A^a B^b}{C^c D^d} K_c\right]^s \qquad (9.15)$$

where s is any real constant.

In many cases, it will turn out that s is $+1$ or some small integer, so that a simple relation between the rates and the equilibrium expression can be obtained. When the rates are proportional to the products of concentration terms raised to small integral powers, Eq. (9.15) becomes

$$\frac{k_f A^{n_1} B^{n_2} C^{n_3} D^{n_4}}{k_b A^{n'_1} B^{n'_2} C^{n'_3} D^{n'_4}} = \left[\frac{A^a B^b}{C^c D^d} K_c\right]^s \qquad (9.16)$$

so that

$$n_1 - n'_1 = as, \qquad n_2 - n'_2 = bs, \qquad \text{etc.}$$

and

$$\frac{k_f}{k_b} = K_c^s \qquad (9.17)$$

These rules have often been used to determine the rate constant k_b of a reverse reaction from the values of k_f and K_c. For example,

$$2NO(g) + O_2(g) \rightleftharpoons 2NO_2(g)$$

$$K_c = \frac{[NO_2]^2}{[NO]^2[O_2]}$$

*Equation (9.14) is not a necessary condition, and more complicated solutions may occur. See A. Hollingsworth, *J. Chem. Phys.*, **20**, 921 (1952).

The rate* of the reverse reaction has been found to be

$$\frac{-d[NO_2]}{dt} = k_r\, [NO_2]^2$$

Thus, from (9.16),

$$\left[\frac{[NO]^2[O_2]}{[NO_2]^2}K_c\right]^s = \frac{k_f\,[NO]^{n_1}\,[O_2]^{n_2}\,[NO_2]^{n_3}}{k_r\,[NO_2]^2}$$

$$2s = n_1$$
$$s = n_2 \tag{9.18}$$
$$0 = n_3, n'_1, n'_2$$
$$2s = n'_3 = 2$$

Thus, $s = 1$, and $n_1 = 2$, $n_2 = 1$, so that the rate of the forward reaction at equilibrium must be

$$\mathbf{v}_f = k_f[NO]^2\,[O_2] \quad \text{and} \quad K = \frac{k_f}{k_r},$$

results which have been verified by experiments.

We must reemphasize, however, that these relations have been derived for reaction rates at equilibrium, and they must be used with caution for reactions not at equilibrium. Consider, for example, the reaction sequence†

$$A \underset{k_{-1}}{\overset{k_1}{\rightleftharpoons}} B \underset{k_{-2}}{\overset{k_2}{\rightleftharpoons}} C$$

At equilibrium,

$$\frac{[C]}{[A]} = \frac{k_1 k_2}{k_{-1}k_{-2}} = K$$

If we measured the disappearance of A in the initial stages of reaction, it would be, simply,

$$\frac{-d[A]}{dt} = k_1[A]$$

Similarly, the initial rate of disappearance of C would be $k_{-2}[C]$. It is clearly erroneous to deduce, from (9.17) and k_1, that $k_{-2} = K/k_1$. We can show that at equilibrium,

$$\mathbf{v}_f = \frac{k_1 k_2}{k_{-1} + k_2}[A]$$

$$\mathbf{v}_r = \frac{k_{-1} k_{-2}}{k_{-1} + k_2}[C]$$

The ratio of these two composite rate constants is $k_1 k_2 / k_{-1} k_{-2} = K$, as it must be at equilibrium.

---

*The constant volume V has not been explicitly included in these "rates of reaction" (Section 9.1).

†This example is discussed by K. J. Laidler in *Chemical Kinetics* (New York: McGraw-Hill Book Company, 1965), p. 329.

13. Consecutive Reactions

It often happens that the product of one reaction becomes itself the reactant of a following reaction. There may be a series of consecutive steps. Only in the simplest cases has it been possible to obtain analytic solutions to the differential equations of these reaction systems. They are especially important in polymerization and depolymerization processes. With modern computers, any of these sequential reaction schemes can be integrated numerically for the parameters and times of interest.

A simple consecutive reaction scheme that can be treated exactly is one involving only irreversible first-order steps. The general case of n steps has been solved,* but only the example of two steps will be discussed. This can be written

$$A \xrightarrow{k_1} B \xrightarrow{k'_1} C$$
$$\quad x \qquad y \qquad z$$

The simultaneous differential equations are

$$-\frac{dx}{dt} = k_1 x, \qquad -\frac{dy}{dt} = -k_1 x + k'_1 y, \qquad \frac{dz}{dt} = k'_1 y$$

The first equation can be integrated directly, giving $-\ln x = k_1 t + C$. When $t = 0$, let $x = a$, the initial concentration of A. Then $C = -\ln a$, and $x = a\,e^{-k_1 t}$. The concentration of A declines exponentially with the time, as in any first-order reaction.

Substitution of the value found for x into the second equation gives

$$\frac{dy}{dt} = -k'_1 y + k_1 a\,e^{-k_1 t}$$

This is a linear differential equation of the first order, with the solution†

$$y = e^{-k'_1 t}\left[\frac{k_1 a\,e^{(k'_1 - k_1)t}}{k'_1 - k_1} + C\right]$$

When $t = 0$, $y = 0$, so that $C = -k_1 a/(k'_1 - k_1)$.

We now have expressions for x and y. In the reaction sequence, there is no change in the total number of molecules, since every time an A disappears a B appears, and every time a B disappears a C appears. Thus, $x + y + z = a$, and z is calculated to be

$$z = a\left(1 - \frac{k'_1\,e^{-k_1 t}}{k'_1 - k_1} + \frac{k_1\,e^{-k'_1 t}}{k'_1 - k_1}\right) \qquad (9.19)$$

In Fig. 9.6, the concentrations x, y, z are plotted as functions of the time, for the case $k_1 = 2k'_1$. The intermediate concentration y rises to a maximum and

---

*H. Dostal, *Monatshefte für Chemie*, **70**, 324 (1937). For second-order steps, see P. J. Flory, *J. Am. Chem. Soc.*, **62**; 1057, 1561, 2255 (1940).

†W. A. Granville, P. F. Smith, and W. R. Longley, *Elements of Calculus* (Boston: Ginn & Co., 1957), p. 380.

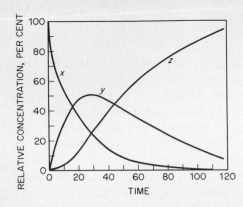

FIG. 9.6 Concentration changes in consecutive first-order reactions.

then falls asymptotically to zero, while the final product rises gradually to the value of a.

Such a reaction sequence was found in the thermal decomposition (pyrolysis) of acetone.[*]

$$(CH_3)_2CO \longrightarrow CH_2{=}CO + CH_4$$
$$CH_2{=}CO \longrightarrow \tfrac{1}{2}C_2H_4 + CO$$

The concentration of the intermediate, ketene, rises to a maximum and then declines during the course of the reaction. Actually, however, the decomposition is more complex than the simple equations would imply.

In dealing with consecutive first-order steps, we can apply the *bottle-neck principle*. If one of the stages has a *specific rate* much lower than any of the others, the overall rate will be controlled by the specific rate (rate constant) of this particular reaction. For instance, in the preceding example, if $k_1 \ll k'_1$, (9.19) reduces to

$$z = a(1 - e^{-k_1 t})$$

which is identical to (9.5) and includes only the lowest rate constant.

14. Parallel Reactions

Sometimes a given substance can react or decompose in more than one way. Then the alternative parallel reactions must be included in analyzing the kinetic data. Consider a schematic pair of parallel first-order reactions,

$$A \begin{cases} \xrightarrow{k_1} B \\ \xrightarrow{k_2} C \end{cases}$$

The rate equations for formation of B and C are

$$\frac{db}{dt} = k_1(a_0 - b - c)$$

$$\frac{dc}{dt} = k_2(a_0 - b - c)$$

[*]C. A. Winkler and C. N. Hinshelwood, *Proc. Roy. Soc. London, Ser. A, 149,* 340 (1935).

where b, c denote the concentrations of B, C, respectively, and a_0 is the initial concentration of a. The integrated rate expression for $b = c = 0$ at $t = 0$ becomes

$$b = \frac{k_1 a_0}{k_1 + k_2} [1 - e^{-(k_1 + k_2)t}]$$

$$c = \frac{k_2 a_0}{k_1 + k_2} [1 - e^{-(k_1 + k_2)t}]$$

(9.20)

In the case of such parallel processes, the most rapid rate determines the predominant path of the overall reaction. If $k_1 \gg k_2$, the decomposition of A will yield mostly B. For example, alcohols can be either dehydrated to olefins or dehydrogenated to aldehydes,

$$C_2H_5OH \begin{cases} \xrightarrow{k_1} C_2H_4 + H_2O \\ \xrightarrow{k_2} CH_3CHO + H_2 \end{cases}$$

By suitable choice of catalyst and temperature, one rate can be made much faster than the other. In such a case, the composition of the mixture of reaction products depends upon the relative rates and not upon the equilibrium constants for the two reactions.

15. Chemical Relaxation

The application of relaxation methods to the study of rapid chemical reactions in solutions was initiated by Manfred Eigen about 1950, and many of the subsequent developments have been made in his laboratory at the Max Planck Institute for Physical Chemistry, Göttingen.* The basic idea is to subject a reaction system that is already at equilibrium to a sharp variation in some physical parameter upon which the value of the equilibrium constant K is dependent. The system then must change to a new state of chemical equilibrium, and the rate of this change can be followed. The principal parameters that can be adapted to relaxation studies are temperature, pressure, and electric field. Ultrasonic absorption is also sometimes useful, as it occurs in a system subjected to a compression wave of high frequency and appreciable amplitude. An apparatus for the temperature (T) jump method is sketched in Fig. 9.7. The sample volume is usually about 1 cm³, and the T jump is produced by discharge of a capacitor through the reaction mixture, with an input of about 50 kJ in 1 μs (a power of 5×10^7 W).

Provided the displacement of the chemical equilibrium is small enough, the rate of restoration of equilibrium will always follow first-order kinetics, irrespective of the kinetics of the forward and reverse reactions. Thus, if Δx_0 is the initial displacement of some concentration x that describes the composition of the mixture, and Δx is the value of this displacement at any time t after the initial disturbance,

$$\Delta x = \Delta x_0 \, e^{-t/\tau}$$

(9.21)

where τ is the *chemical relaxation time* for the reaction system. The relation of

*M. Eigen, *Discussions Faraday Soc.*, **17**, 194 (1954).

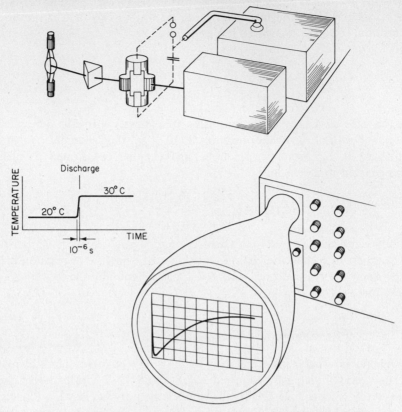

FIG. 9.7 Principle of the temperature-jump and electric field pulse relaxation technique. The field pulse may be generated either by condenser discharge (exponential, or critically damped pulse) or by a matched cable discharge (square pulse). T-jumps also are produced by high-power microwave pulses. [M. Eigen, *Nobel Symposium, 5,* 333 (1967).]

τ to the rate constants of the system has been worked out for most cases of interest.* We shall give two examples to show how it goes.

Consider a reversible first-order reaction,

$$A \underset{k_{-1}}{\overset{k_1}{\rightleftharpoons}} B$$

Let a be the concentration of $A + B$, and x the concentration of B at any time. Then,

$$\frac{dx}{dt} = k_1(a - x) - k_{-1}x$$

The rate of change of $\Delta x = x - x_e$ is, accordingly,

*G. H. Czerlinski, *Chemical Relaxation* (New York: Marcel Dekker, Inc., 1966).

$$\frac{d(\Delta x)}{dt} = k_1(a - \Delta x - x_e) - k_{-1}(\Delta x + x_e) \tag{9.22}$$

At equilibrium, however,

$$\frac{dx}{dt} = 0 = k_1(a - x_e) - k_{-1}x_e$$

Therefore, (9.22) becomes

$$\frac{d(\Delta x)}{dt} = -(k_1 + k_{-1})\,\Delta x$$

so that

$$\tau = (k_1 + k_{-1})^{-1} \tag{9.23}$$

A somewhat more complicated case would be a first-order forward reaction with a second-order reverse,

$$A \underset{k_2}{\overset{k_1}{\rightleftharpoons}} B + C$$

This type would include the ionization of a weak acid in a large excess of solvent, for instance,

$$CH_3COOH + H_2O \rightleftharpoons CH_3COO^- + H_3O^+$$

Let a be the concentration of $CH_3COOH + CH_3COO^-$, and x the concentration of H_3O^+ (equal to that of CH_3COO^-). Including the constant water concentration in k_1, we can write

$$\frac{dx}{dt} = k_1(a - x) - k_2\,x^2$$

Therefore, with $\Delta x = x - x_e$,

$$\frac{d(\Delta x)}{dt} = k_1(a - x_e - \Delta x) - k_2(x_e + \Delta x)^2 \tag{9.24}$$

At equilibrium,

$$\frac{dx}{dt} = 0 = k_1(a - x_e) - k_2\,x_e^2$$

Where departure from equilibrium Δx is very small, the term in $(\Delta x)^2$ can be neglected compared with those in Δx, and (9.24) becomes

$$\frac{d(\Delta x)}{dt} = -(k_1 + 2k_2 x_e)\,\Delta x$$

so that

$$\tau = (k_1 + 2k_2 x_e)^{-1} \tag{9.25}$$

If we combine (9.25) with $K = k_1/k_2$, we can obtain from τ and the equilibrium constant K the forward and reverse rate constants for the ionization reaction.

Table 9.3 lists a few of many interesting results that have been obtained by various relaxation methods. Experimental details can be found in the original papers.

TABLE 9.3
Experimental Rate Constants for Rapid Reactions in Aqueous Solutions[a]

Reaction	T (K)	Method	Reference	k_2 $dm^3 \cdot mol^{-1} \cdot s^{-1}$
$H^+ + HS^- \longrightarrow H_2S$	298	Wien effect	(1)	7.5×10^{10}
$H^+ + N(CH_3)_3 \longrightarrow N(CH_3)_3H^+$	298	NMR	(2)	2.5×10^{10}
$H^+ + (NH_3)_5CoOH^{2+} \longrightarrow$	285	T jump	(3)	1.4×10^9
$\quad H_2O + (NH_3)_5Co^{3+}$				
$H^+ + AlOH^{2+} \longrightarrow H_2O + Al^{3+}$	298	Wien effect	(4)	3.8×10^9
$OH^- +$ diethylmalonic acid $\longrightarrow$	298	T jump	(5)	2.4×10^8
$\quad H_2O +$ diethylmalonate				
$H^+ + OH^- \longrightarrow H_2O$	298	T jump	(6)	1.5×10^{11}

[a]References for Table 9.3 are as follows:
1. M. Eigen and K. Kustin, *J. Am. Chem. Soc.*, *82*, 5952 (1960):
2. E. Grunwald, *et al.*, *J. Chem. Phys.*, *33*, 556 (1960).
3. M. Eigen and W. Kruse, *Z. Naturforsch.*, *186*, 857 (1963).
4. L. P. Holmes, D. L. Cole, and E. M. Eyring, *J. Phys. Chem.* 72(1), 301 (1968).
5. M. H. Miles, *et al.*, *J. Phys. Chem.*, *70*, 3490 (1966).
6. M. Eigen, *Discussions Faraday Soc.*, *17*, 194 (1954).

16.   Reactions in Flow Systems*

All the rate equations discussed so far apply to *static systems*, in which the reaction mixture is enclosed in a vessel at constant volume and temperature. We must now consider *flow systems*, in which reactants enter continuously at the inlet of a reaction vessel, while the product mixture is withdrawn at the outlet. We shall describe two examples of flow systems: (a) a reactor in which there is no stirring; (b) a reactor in which complete mixing is effected at all times by vigorous stirring.

Figure 9.8 shows a tubular reactor through which the reaction mixture passes at a volume rate of flow u (e.g., in liters per second). Let us consider an element of volume dV sliced out of this tube, and focus attention on one particular component K, which enters this volume element at a concentration c_K and leaves at $c_K + dc_K$. If there is no longitudinal mixing, the net change with time of the amount of K within dV, (dn_K/dt), will be the sum of two terms, one due to chemical reaction within dV and the other equal to the excess of K entering dV over that leaving.

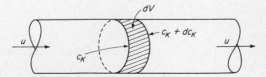

FIG. 9.8   Element of volume in a flow reactor.

*An excellent general reference is K. G. Denbigh, *Chemical Reactor Theory* (Cambridge University Press, 1965).

Thus,

$$\frac{dn_K}{dt} = R_K \, dV - u \, dc_K \tag{9.26}$$

The chemical reaction rate *per unit volume* is denoted by R_K. The explicit form of R_K is determined by the rate law for the reaction: for a reaction of first order with respect to K, $R_K = -k_1 c_K$; for second order, $R_K = -k_2 c_K^2$, etc.

After reaction in the flow system has continued for some time, a *steady state* is attained, in which the number of moles of each component in any volume element no longer changes with time, the net flow into the element exactly balancing the reaction within it. Then $dn_K/dt = 0$, and (9.26) becomes

$$R_K \, dV - u \, dc_K = 0 \tag{9.27}$$

After R_K is introduced as a function of c_K, the equation can be integrated. For example, with $R_K = -k_1 c_K$,

$$-k_1 \frac{dV}{u} = \frac{dc_K}{c_K}$$

The integration is carried out between the inlet and the outlet of the reactor.

$$\frac{-k_1}{u} \int_0^{V_0} dV = \int_{c_{K1}}^{c_{K2}} \frac{dc_K}{c_K}$$
$$-k_1 \frac{V_0}{u} = \ln \frac{c_{K2}}{c_{K1}} \tag{9.28}$$

The total volume of the reactor is V_0, and c_{K2} and c_{K1} are the concentrations of K at the outlet and inlet, respectively.

Equation (9.28) reduces to the integrated rate law for a first-order reaction in a static system if the time t is substituted for V_0/u. The quantity V_0/u is called the *contact time* for the reaction; it is the average time that a molecule would take to pass through the reactor. Thus (9.28) allows us to evaluate the rate constant k_1 from a knowledge of the contact time and of the concentrations of any reactant species at the inlet and outlet of the reactor. For other reaction orders also, the correct equation for a flow reactor is obtained by substituting V_0/u for t in the equation for the static system. Many reactions that are too swift for convenient study in a static system can be followed readily in a flow system, in which the contact time is reduced by use of a high flow rate and a small volume.*

The derivation of (9.27) tacitly assumed that there was no volume change ΔV as a result of the reaction. If $\Delta V \neq 0$, the flow rate at constant pressure would not be constant. In liquid-flow systems, effects due to ΔV are generally negligible, but for gaseous systems, the form of the rate equations is considerably modified when $\Delta V \neq 0$. A convenient collection of integrated rate laws including such cases is given by Hougen and Watson.†

An example of a *stirred-flow reactor*‡ is shown in Fig. 9.9. The reactants enter

---

*Experimental methods are described in *Techniques of Organic Chemistry* (New York: Interscience Publishers, 1953), pp. 669–738.

†O. A. Hougen and K. M. Watson, *Chemical Process Principles*, Pt. 3 (New York: John Wiley & Sons, Inc., 1947), p. 834.

‡K. G. Denbigh, *Trans. Faraday Soc.*, 40, 352 (1944); *Discussions Faraday Soc.*, 2, 263 (1947).

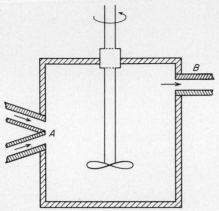

FIG. 9.9 A stirred-flow reactor. Reactants enter at A, and reaction mixture with the steady-state composition is withdrawn continuously at B.

the vessel at A, and stirring at high speed (about 3000 rpm) effects mixing within about 1s. The outflow of product mixture at B exactly balances the feed. After a steady state is attained, the composition of the mixture in the reactor remains unchanged as long as the composition and rate of supply of reactants is unchanged. An equation like (9.27) still applies, but in this case dV becomes V_0, the total reactor volume, and dc_K becomes $c_{K2} - c_{K1}$, where c_{K1} and c_{K2} are the initial and final concentrations of reactant K. Thus,

$$R_K = \frac{u}{V_0}(c_{K2} - c_{K1}) = \frac{dc_K}{dt}$$

With this method, there is no need to integrate the rate equation. One point on the rate curve is obtained from each steady state measurement, and a number of runs with different feed rates and initial concentrations is required to determine the order of the reaction.

An important application of the stirred-flow reactor is the study of transient intermediates, the concentration of which in a static system might quickly reach a maximum value and then fall to zero (as was shown in Fig. 9.6). For example, when Fe^{3+} is added to $Na_2S_2O_3$, a violet color appears, which fades within 1 or 2 min. In a stirred flow reactor, the conditions can be adjusted so that the color is maintained, and the intermediate responsible, which appears to be $FeS_2O_3^+$, can be studied by absorption spectrometry.

17.   Steady States and Dissipative Processes

In a static (or batch) reaction vessel, the concentrations of the reactants and products change with time until an equilibrium condition is attained. At constant T and P, this equilibrium represents the minimum in Gibbs free energy G with respect to fractional conversion of reactants to products.

In a static system, the change in composition occurs in the time coordinate, whereas in a flow system the change in composition is shifted to a space coordinate, either continuously along the axis of a tubular flow reactor, or discontinuously from tank to tank in a series of continuously stirred reactors. If the rate of feed to such a flow system is constant, the composition as a function of the appropriate space coordinate will generally (but not always or necessarily) approach a steady state value. This steady state conversion will not be at the minimum of G but at some other value determined by the flow rates and rate constants of the system. These flow systems are *open systems* (p. 206) in the thermodynamic sense, and hence their time-invariant states are not *equilibrium states* but *steady states*.

There is a basic analogy between a living cell and a continuous stirred flow reactor, and the same theoretical analysis can often be applied to chemical kinetics in both. In the case of the cell, there is no obvious internal stirrer, but distances from one part to any other part are usually small* so that mixing by diffusion should be adequate to maintain the "well stirred" condition. For a cell diameter of 10^{-3} cm, the mean diffusion time across the cell of a molecule with diffusion coefficient 10^{-5} cm$^2\cdot$s^{-1} would be about $\tau = (10^{-3})^2/10^{-5} = 10^{-1}$ s. The cell does not have a definite inlet and outlet like a reactor, but the entire cell wall or outer membrane serves these functions, and substances can enter or leave the cell either by transmembrane diffusion or by processes of active transport.

Essentially the same mathematical method as used in chemical kinetics can also be applied to the rates of transport of substances from one region to another within a living organism. For example, radioactive iodine injected parenterally becomes distributed between blood and thyroid gland. We may wish to calculate the rate of rise and fall of radio-iodine concentration in the blood, and its rate of uptake by the thyroid. The problem is quite analogous to a system of coupled chemical reactions. The mechanism by which radio-iodine is transported from blood to thyroid is not specified, but as long as the rate is proportional to the concentration of radio-iodine in the blood, it can be represented formally as a first-order process.†

When the relations between the compartments of such systems are linear, i.e., involve only first-order processes, the mathematical treatment can be based on the convenient methods available for solving systems of linear differential equations.‡ If the equations become nonlinear, however, the mathematical difficulties become much greater, and numerical integration with the digital computer is usually required.

Steady state chemical kinetic systems can sometimes give rise to a remarkable phenomenon, which may be the key to what we call "life." In the usual system of coupled linear processes, both equilibrium states and steady states are stable

---

*In the case of nerve cells (such as a motoneuron to a muscle in a giraffe), special mechanisms are required to transfer substances from one part of the cell to another.

†R. Aris, "Compartmental Analysis and the Theory of Residence Time Distributions," in *Intracellular Transport*, ed. K. B. Warren (New York: Academic Press, 1966).

‡These problems are readily handled by methods employing the Laplace transform, which converts the differential equations to simultaneous algebraic equations. See A. Rescigno and G. Segre, *Drug and Tracer Kinetics* (Waltham, Mass.: Blaisdell Publishing Company, 1966).

with respect to small disturbances in the parameters that characterize the system. Thus, for example, if we inject an extra pulse of a given reactant A into the inlet stream of a series of stirred reactors, there will be a transient wave of changed concentrations through the reactors, but the previous steady state is soon restored. In the case of certain systems of nonlinear reactions, however, it can be shown that some steady states are unstable with respect to disturbances. It is possible, therefore, to set up persistent oscillations in the concentrations of reactants and products, or even to cause the entire system to shift from one steady state to another one farther removed from equilibrium. We can see an analogy of this phenomenon in the Bénard effect in the hydrodynamics of a heated fluid. If a fluid is heated from below, under certain conditions it will shift from a regime that follows the ordinary equation of heat conduction, to a different regime in which convection currents are maintained. Convection is a mass motion, imparting a kinetic energy to the fluid as a whole. Thus, random thermal energy is being converted into large-scale mechanical energy. Since the system is not operating in a cycle, no contradiction of the Second Law of Thermodynamics is implied. The considerable flow of heat along the temperature gradient between the boundaries of the system provides the possibility for a *dissipative process* to occur, which maintains the interior of the system in a state of mass flow that is far removed from equilibrium or even from the steady state of heat conduction.

The first recognition that a similar dissipative process might take place in chemical reaction systems was apparently in a paper by Turing,[*] which dealt with chemical reactions coupled with diffusion of reactants. Prigogine[†] has initiated a general theoretical study of such processes. If there is a sequence of reactions

$$A \rightleftharpoons X_1 \rightleftharpoons X_2 \cdots \rightleftharpoons B$$

which includes nonlinear steps (e.g., second- or third-order steps), there may be more than one steady state solution for the concentrations of intermediates X_1, X_2, etc., corresponding to a given fixed nonequilibrium ratio A/B. In particular, one of these solutions may lead to large gradients of concentration between the values of $[X_1]$, $[X_2]$, etc. In other words, the system assumes a steady state in which the entropy of mixing is unusually low, or the negentropy is unusually high. The system therefore takes on an ordered state, and maintains itself in this steady state of high order at the expense of a dissipation of free energy, which is provided by the fixed nonequilibrium ratio of A/B for the reaction sequence. But is this not exactly what we mean by a living system—a localized region of order that maintains itself by feeding on the free energy stores of its environment?

An example of an ordered structure in a dissipative system of the kind predicted by the theory of Prigogine and Glansdorff[‡] has been discovered by Marcelle

---

[*]A. M. Turing, *Phil. Trans. Roy. Soc. London, Ser. B, 237,* 37 (1952). Some change in nomenclature would seem to be desirable so that we would not be forced to say that a life of "dissipation" is the only way to live.

[†]I. Prigogine, "Dissipative Structures in Chemical Systems," *Fifth Nobel Symp.* (New York: John Wiley & Sons, Inc., 1967), p. 371; "Structure, Dissipation and Life," in *Theoretical Physics in Biology* (Versailles: Institut de la Vie, 1969).

[‡]P. Glansdorff and I. Prigogine, *Thermodynamic Theory of Structure, Stability, and Fluctuations* (New York: John Wiley & Sons, Inc., 1971).

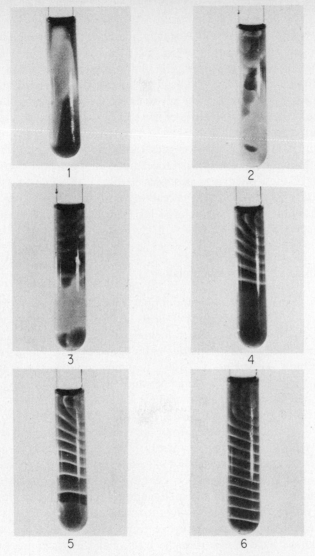

FIG. 9.10 Development of a dissipative structure in a homogeneous chemical reaction far from equilibrium. The reaction mixture consists of equal volumes of 1.2 M malonic acid, 0.35 M $KBrO_3$, 4×10^{-3} M $Ce_2(SO_4)_3$, and 1.5 M H_2SO_4, with a few drops of the indicator *ferroin*. The time course of the evolution of the dissipative structure in a 2.0 cm³ test tube maintained at 21.0°C is shown in photographs 1 to 6. The final structure is usually stable for 15 to 30 min and dissappears as the reaction approaches equilibrium. The lighter bands are blue regions with excess Ce^{4+} and the darker bands are red regions with excess Ce^{3+}. The reaction is complex, but the principal steps appear to be:

(1) $CH_2(COOH)_2 + 6Ce^{4+} + 2H_2O \rightarrow 2CO_2 + HCOOH + 6Ce^{3+} + 6H^+$

(2) $10Ce^{3+} + 2HBrO_3 + 10H^+ \rightarrow 10Ce^{4+} + Br_2 + 6H_2O$

(3) $CH_2(COOH)_2 + Br_2 \rightarrow CHBr(COOH)_2 + HBr$

Herschkowitz* in a homogeneous chemical reaction in which malonic acid reacts in an oxidizing solution made up from Ce^{3+} and BrO_3^- ions. The results of a typical experiment are shown in Fig. 9.10. The reacting solution first displays temporal oscillations, passing periodically from a red color, indicating an excess of Ce^{3+}, to a blue color, indicating an excess of Ce^{4+}. These oscillations do not occur at the same instant in all parts of the solution; they begin at one point and propagate in all directions at various rates. After a variable number of oscillations, a small region of inhomogeneous concentration appears, from which layers of coloration, alternately red and blue, proceed one by one until they fill the tube. The evolution of these structures is shown in Fig. 9.10. These dissipative structures appear only when the system is reacting far from equilibrium. As the reaction approaches equilibrium the colored layers disappear and the solution again becomes homogeneous.

18.   Nonequilibrium Thermodynamics

In all our discussions of thermodynamics up to this point, two basic facts about the subject have been emphasized: (1) it deals with relations between measurable properties of materials; (2) it has been formulated only for systems in equilibrium. There are many measurable properties of nonequilibrium systems (thermal conductivity, diffusion coefficient, viscosity, to name a few) which are similar to thermodynamic properties, such as temperature, density, or entropy, in that their definitions are not based upon any model for the structure of matter. Such properties are often called *phenomenological*. It is natural to ask, therefore, whether any theory exists that can provide relations between such *nonequilibrium* phenomelogical properties. The answer is found in the subject called *nonequilibrium thermodynamics* or the *thermodynamics of irreversible processes*.†

Thermodynamics is based on a few definitions of state functions (P, V, T) and three general laws. The laws of thermodynamics are postulates that are stated to be universally valid; they are certainly not restricted only to reversible processes or systems in equilibrium. Is it not curious, therefore, that all the calculations and deductions that we have made from the laws of thermodynamics (in Chapters 3 and 8 especially) have been concerned exclusively with systems in equilibrium? The reason is that some of the state functions, which are essential for descriptions of the systems of interest, were defined only for equilibrium states.

If we examine the variables of state, we find that they fall into two classes. Some of them can be used to describe nonequilibrium states with no difficulty of any kind. Examples of this class are volume V, mass m, concentration c, amount of substance n, and energy U. These functions are perfectly well defined for any

*M. Herschkowitz-Kaufman, *C. R. Acad. Sci.* (Paris), *270*, 1049 (1970).

†An introductory treatment of the subject has been written by I. Prigogine, *Introduction to Thermodynamics of Irreversible Processes* (New York: John Wiley & Sons, Inc., 1967). An excellent discussion of the basic theory is to be found in *Treatise on Irreversible and Statistical Thermophysics*, by W. Yourgrau, A. van der Merwe, and G. Raw (New York: The Macmillan Company, 1966).

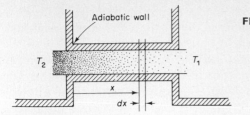

FIG. 9.11 Heat conduction in a metal bar between reservoirs at constant temperatures T_2 and T_1. The bar can be divided into cells of thickness dx and the temperature defined at any point x along the bar.

system or any part of a system, whether or not it is in equilibrium. On the other hand, serious difficulties arise when we attempt to use such functions as P, T, and S in the description of nonequilibrium systems. The impossibility of defining the pressure of a gas during an irreversible expansion was discussed in Section 1.7. The same sort of problem occurs if we wish to define the temperature T during an irreversible process. Consider, for example, in Fig. 9.11 two heat reservoirs at T_1 and $T_2 > T_1$ connected by a metal bar along which heat is being conducted from the hotter to the colder reservoir, a typical irreversible process. We do not have any way to define the temperature at any point along the bar, since T has been defined only for a system in equilibrium.

To overcome this roadblock so as to proceed with thermodynamic calculations on nonequilibrium systems, we must introduce a new postulate that will allow us to define P and T at any point in a system in which an irreversible process is occurring. We therefore introduce the *Postulate of Local Equilibrium:* A system can be divided (conceptually) into cells so small that each cell effectively corresponds to a given point in the system, but so large that each contains thousands of molecules. At a time t, the matter in a given cell is isolated from its surroundings and allowed to come to equilibrium, so that at $t + \delta t$, the P and T of the cell can be specified. The Postulate of Local Equilibrium is that we can take the P and T at any point in the original nonequilibrium system at time t to be equal to the P and T in the corresponding cell when equilibrium is reached at $t + \delta t$.

Before we can apply thermodynamics to a nonequilibrium system, we need one additional postulate, to wit: The relations between P, V, T, etc., as defined by the Postulate of Local Equilibrium, are identical with the relations between ordinary equilibrium functions, so that all the relations between state functions that were derived for equilibrium states also hold good for the functions we have now defined for nonequilibrium states.

The Postulate of Local Equilibrium is not unreasonable or unusual. In classical theoretical physics, the temperature function was treated in exactly this way, and problems in heat conduction were analyzed in terms of a temperature T defined at each point along a temperature gradient. We do not contend, however, that the postulate can be applied to *all* irreversible systems, but there should be a range of validity, comprising systems in which the properties are not varying too rapidly with time. In such cases the δt required to reach local equilibrium is short compared with the times required for measurable changes in the total system.* As is the

---

*Thus, we would feel confident in applying the postulates to heat conduction across a thermocouple junction but would feel uneasy about their application to a nuclear explosion.

case for any theoretical analysis based on stated postulates, the ultimate justification of the postulates will be found in the validity of the relations between experimental quantities that can be derived from them.

The definition of the entropy S might seem to present a special problem in irreversible thermodynamics, since the function is introduced into ordinary thermodynamics by means of a definition $dS = dq_{rev}/T$, explicitly in terms of a reversible transfer of heat. In practice, however, the entropies of various substances are calculated by means of other equations, which were derived from the initial definition—e.g., especially Eq. (3.61). We can therefore use our two postulates to calculate the entropy S of each cell of a system undergoing an irreversible process, and hence the entropy per unit mass $S/m = s$ at each point in the system. For example, in Fig. 9.11 the bar is divided into thin slices normal to the temperature gradient, and in each slice T and P are defined and constant. Then we can calculate S for any slice by using the equation that relates the entropy of the metal to its heat capacity C_P: $S = \int_0^T C_P \, d \ln T$.

One consequence of the new postulates is that we can apply the Gibbs equation (6.4) to each cell in the system,

$$T \, dS = dU + P \, dV - \Sigma \mu_i \, dn_i \tag{9.29}$$

As another example of a nonequilibrium system, consider a mixture of H_2 and O_2 gases in a container at T and P in the presence of a suitable catalyst for the reaction $H_2 + \frac{1}{2}O_2 \rightarrow H_2O$. At any time t, the container would contain definite amounts of H_2, O_2, and H_2O. Since we know the molar entropies of each of these substances, we could calculate the entropy of the system by adding up the entropies for the gases present at any instant and including the calculated entropy of mixing on the assumption that the composition was uniform throughout the container. Since the reaction does not proceed in the absence of catalyst, we could even experimentally stop the reaction at any instant and make any measurements necessary to determine deviation from ideal gas behavior in the partially reacted mixture of gases. Since the example chosen does not differ in any fundamental way from the generality of chemical reactions, there seems to be no reason why we could not determine S as a function of time t, and hence dS/dt, the rate of production of entropy during the reaction process. To maintain the temperature of the system constant, it might be necessary to have heat flows into or out of the container during reaction, so that we could write for the differential change in entropy

$$dS = d_i S + d_e S \tag{9.30}$$

where $d_i S$ is due to changes within the system and $d_e S$ is due to flow of entropy into the system from the exterior.

19. The Onsager Method

The extension of thermodynamics to irreversible processes began in the theoretical investigation by William Thomson (Kelvin) of the properties of thermocouples (1854–1857). He noted that two irreversible thermal effects, Joulean heat evolution

(I^2R) and heat conduction, were occuring simultaneously with two reversible effects, the Peltier heat transfer at the thermocouple junction, and the [Thomson] heat associated with current flow. The latter two effects reversed their sign with reversal of the direction of current flow. Thomson treated the reversible effects by simply ignoring the simultaneous irreversible effects. Despite the apparently unjustifiable theoretical basis for this analysis, the thermocouple equations* obtained were confirmed by experiments. An approach similar to that of Thomson was applied to thermal diffusion (diffusion in the presence of a temperature gradient) by Eastman and Wagner. In subsequent work, however, it became more difficult to decide which phenomena were to be called "reversible" and which "irreversible," and the theories became more arbitrary and less convincing. An adequate formulation of thermodynamics for irreversible processes was first provided by Onsager in 1931. This theory was refined by Casimir and forms the basis of most current approaches to the subject.

The Onsager formulation can be summarized in the following statements.

1. The theory is based on the *Principle of Microscopic Reversibility*: Under equilibrium conditions any process and the reverse of that process will be taking place on the average at the same rate.
2. One can write *thermodynamic equations of motion* for various transport processes, in which the rates of flow or *fluxes* are equal to a sum of terms, each of which is proportional to a *thermodynamic force*.† This linear proportionality between the components of the fluxes and the forces is an important restriction on the range of validity of the theory. In some cases (e.g., heat flow), a flux term is proportional to its corresponding force (the temperature gradient) over a wide range. In other cases, such as chemical reactions, the linear proportionality holds only for small departures from equilibrium.

The equations of motion for the case of two fluxes J_1 and J_2 take the general form

$$J_1 = L_{11}X_1 + L_{12}X_2$$
$$J_2 = L_{21}X_1 + L_{22}X_2$$

(9.31)

For example, J_1 might be the flow of heat and J_2 the flow of matter; X_1 would be the temperature gradient and X_2 would be the force appropriate for diffusion— i.e., the chemical potential gradient in the absence of external forces. The L_{ij} are called the *phenomenological coefficients*. The L_{ii} are the ordinary *direct coefficients*, which give the flux of a quantity in terms of a force that is related to a gradient of an intensity factor corresponding‡ to that quantity. The cross coefficients,

---

*See M. Zemansky, *Heat and Thermodynamics* (New York: McGraw-Hill Book Company, 1969), p. 409.

†The forces of irreversible thermodynamics are not Newtonian forces because they are not usually related to accelerations.

‡The product of a flux and its conjugate force must have the dimensions of rate of production of entropy.

L_{ij} (called *drag coefficients* by Eckart), give the flux of a quantity caused by a gradient in an intensity factor not directly related. For example, in (9.31), the flux of matter caused by a temperature gradient is determined by L_{21}, and the flux of heat caused by a chemical potential gradient is determined by L_{12}.

Onsager showed that with a proper choice of fluxes and forces, one always has $L_{21} = L_{12}$, or, in general,

$$L_{ij} = L_{ji} \tag{9.32}$$

A "proper choice" means that $J = dF/dt$, where F is a *state function*. Equation (9.32) is called the *Onsager Reciprocal Relation*. The derivation of the Onsager Relation is based upon the Principle of Microscopic Reversibility and the application of statistical mechanics to conditions *close to equilibrium*. The Onsager Relation, however, has been confirmed experimentally in a wide variety of situations departing considerably from equilibrium, and it can be used with confidence even if not with unquestioning faith.*

Some examples of generalized forces and fluxes for important irreversible processes are summarized in Table 9.4.

Some of the most valuable applications of nonequilibrium thermodynamics are in systems in which two or more forces are acting simultaneously. Whenever $L_{ij} \neq 0$, there will be *cross effects* in the phenomena observed. An important general principle gives a sufficient condition that $L_{ij} = 0$, i.e., the absence of coupling between the processes so that the force X_i cannot produce a flux J_i. The principle was originally stated by Pierre Curie in 1908 in the following form: *Macroscopic causes cannot have more elements of symmetry than the effects they produce.*

TABLE 9.4
Examples of Generalized Forces and Fluxes in Irreversible Processes

Process	Generalized force, X	Generalized flux, J	Relation between conventional and phenomenological coefficients
Chemical reaction $A \underset{k_{-1}}{\overset{k_1}{\rightleftharpoons}} B$	$\mathscr{A} = -\sum v_i \mu_i$	$\dfrac{1}{V}\dfrac{d\xi}{dt}$	$L = k_1 c_A^{eq}/RT$ [a]
Heat conduction	$-T^{-2}\,\mathrm{grad}\ T$	w (energy flow rate/ unit area)	$L_{11} = \lambda T^2$
Binary diffusion	$-\mathrm{grad}\ \mu_i$	J_i (material flow rate/ unit area)	$D = L(\partial\mu/\partial c)$ [b]
Electrical conductance of binary electrolyte	$-\mathrm{grad}\ \Phi$	i (current density)	$\dfrac{\kappa}{F^2} = z_1^2 L_{11}$ $+\, 2z_1 z_2 L_{12} + z_2^2 L_{22}$

[a]Expression holds only close to equilibrium.
[b]Note that there are no cross terms in a binary diffusion system (see p. 160).

*D. G. Miller, *Chem. Rev.*, *60*, 15 (1960).

An instance of Curie's Principle is that chemical affinity $\mathscr{A}$, which has no directional properties and is thus a scalar quantity, cannot produce a directed flow of heat, electricity, of matter, since these flows all have vectorial properties and are thus less symmetrical that the affinity $\mathscr{A}$. Since in this case $L_{ij} = 0$, it follows from Onsager's Law that $L_{ji} = 0$ also, and a vector force, such as a temperature gradient, cannot produce a chemical reaction rate.*

20. Entropy Production

The concept of entropy production in a system which is the site of irreversible processes plays an important role in nonequilibrium thermodynamics. Consider, for example, a system at constant T and P in which a chemical reaction is taking place. If we combine (9.29) and (9.30) and note that $Td_eS = dq = dU + PdV$, since heat can be transferred reversibly to the isothermal system, we obtain

$$d_iS = -\frac{1}{T} \sum \mu_i \, dn_i \qquad (9.33)$$

With the extent of reaction ξ from (8.4), (9.33) gives for the rate of entropy production,

$$\dot{S} = \frac{d_iS}{dt} = -\frac{1}{T} \frac{d\xi}{dt} \sum v_i \, \mu_i \qquad (9.34)$$

where the v_i are the stoichiometric coefficients of the reaction equation. The affinity $\mathscr{A} = -\sum v_i \mu_i$, so that

$$\frac{T}{V} \frac{d_iS}{dt} = \frac{d\xi}{dt} \frac{\mathscr{A}}{V} = v\mathscr{A} \qquad (9.35)$$

In this case, the generalized flux J is the reaction rate $v = d\xi/dt$ per unit volume, and the generalized force X is the affinity $\mathscr{A}$. We should not conclude, however, that the reaction rate v is always a linear function of $\mathscr{A}$, although such a relation is valid in the neighborhood of equilibrium.

21. Stationary States

We have already discussed several examples of stationary, nonequilibrium states in systems in which dissipative processes are occurring. In an equilibrium state, the rate of entropy production is zero, $\dot{S} = 0$. In a stationary state, on the other hand, it has been shown that the rate of entropy production has the minimum value consistent with the external restraints imposed upon the system. This theorem appears to have been first derived by Prigogine in 1947. Its validity is subject to the conditions that the phenomenological equations are linear, the coefficients

---

*In case the space in which the fluxes occur is not isotropic, the affinity $\mathscr{A}$ would not necessarily be a scalar. For example, in a cell membrane, coupling between reaction rate and transport of components may occur, as in the active transport of Na^+ ions that is coupled with hydrolysis of ATP.

L_{ij} are constant, and the Onsager Reciprocal Relations hold. Such nonequilibrium stationary states are stable with respect to small perturbations of the variables defining the system.

To obtain a proof (which can be made general) of the theorem, consider a system in which matter and energy are being transferred between two phases at different temperatures. The constraint on the system is that the temperatures of the two phases are maintained constant, and no flow of matter into or out of the system is allowed. The phenomenological laws are

$$J_e = L_{11}X_e + L_{12}X_m$$
$$J_m = L_{21}X_e + L_{22}X_m \tag{9.36}$$

where J_e and J_m are fluxes of energy and of matter. The rate of entropy production is

$$\frac{d_iS}{dt} = J_eX_e + J_mX_m > 0 \tag{9.37}$$

From (9.36),

$$\frac{d_iS}{dt} = L_{11}X_e^2 + 2L_{21}X_eX_m + L_{22}X_m^2 > 0 \tag{9.38}$$

The derivative of (9.38) with respect to X_m at constant X_e is

$$\frac{\partial}{\partial X_m}\left(\frac{d_iS}{dt}\right) = 2(L_{21}X_e + L_{22}X_m) = 2J_m$$

Therefore, if

$$\frac{\partial}{\partial X_m}\left(\frac{d_iS}{dt}\right) = 0$$

(i.e., if the rate of entropy production is a minimum*), then $J_m = 0$. But this is just the condition for the stationary state,

$$J_m = 0 = L_{21}X_e + L_{22}X_m$$

In such a system, heat flows from one phase to another but there is no net transfer of matter between the phases.

We shall not give here the proof that the stationary state is stable with respect to small perturbations. A straightforward demonstration is given by Prigogine. He shows that as the result of any irreversible process within a system, the rate of entropy production S can only decrease. Hence, once the system reaches a stationary state, it cannot be disturbed from it by any spontaneous irreversible perturbation.

22. Effect of Temperature on Reaction Rate

At this point, we shall leave the subject of classical chemical kinetics and turn to what has been called the *theory of absolute reaction rates,* by which is meant the theory of the rate constants of chemical reactions. Our ultimate aim would be to

---

*Since d_iS/dt in (9.38) is a positive definite quadratic expression, the extremal condition must be a minimum and not a maximum or inflection point.

calculate the rate constant of any elementary reaction from the structures of the reactive molecules and the properties of the medium in which they are reacting. This task will prove to be arduous indeed—in fact, beyond the reach of present-day theory except in the case of that most simple of reactions,

$$H + H_2 \longrightarrow H_2 + H$$

It was once said in another context that "it is better to travel hopefully than to arrive." In this field of physical chemistry, the hope is certainly lively, the landscapes are entrancing, and the destination is well beyond the horizon.

The effect of varying the temperature has been the most important key to the theory of rate processes. In 1889, Arrhenius pointed out that since the van't Hoff equation for the temperature coefficient of the equilibrium constant K_c was $d \ln K_c/dT = \Delta U/RT^2$, whereas the mass action law related equilibrium constant to a ratio of rate constants, $K_c = k_f/k_b$, a reasonable equation for the variation of rate constant k' with temperature might be

$$\frac{d \ln k'}{dT} = \frac{E_a}{RT^2} \qquad (9.39)$$

The quantity E_a is called the *activation energy* of the reaction.

If E_a is not temperature dependent, (9.39) yields, on integration,

$$\ln k' = \frac{-E_a}{RT} + \ln A \qquad (9.40)$$

where $\ln A$ is the constant of integration. Hence,

$$k' = A \exp \left(\frac{-E_a}{RT} \right) \qquad (9.41)$$

Here, A is called the *frequency factor* or *pre-exponential factor*. This is the famous Arrhenius equation for the rate constant.

From (9.40), it follows that a plot of logarithm of rate constant against reciprocal of absolute temperature should be a straight line. The validity of the Arrhenius equation has been excellently confirmed in this way for a large number of experimental rate constants. An example from the data of Bodenstein on the $H_2 + I_2 \rightarrow 2HI$ reaction is shown in Fig. 9.12. We shall see later that the Arrhenius equation is only a good approximate representation of the temperature dependence of k'.

According to Arrhenius, Eq. (9.41) indicated that molecules must acquire a certain critical energy E_a before they can react, the Boltzmann factor $e^{-E_a/RT}$ being the fraction of molecules that managed to obtain the necessary energy. This interpretation is still held to be essentially correct.

By referring to Fig. 9.2, we can obtain a picture of the activation energy as the potential-energy hill that must be climbed to reach the activated state. It is evident also that the heat of reaction at constant volume, ΔU_V, is the difference between the activation energies of forward and backward reactions,

$$\Delta U_V = E_f - E_b \qquad (9.42)$$

Variation of temperature allows one to vary the average energy of the reactants and, in accord with a Maxwell–Boltzmann distribution function, the pro-

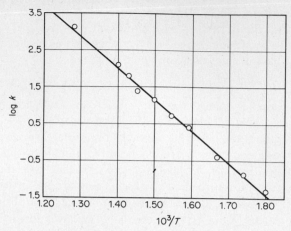

FIG. 9.12 Temperature dependence of rate constant for formation of hydrogen iodide, illustrating applicability of the Arrhenius equation.

portions of specially energetic molecules. In recent years, experimental methods have been developed for producing beams of molecules with closely defined energies. As we shall see later, these methods have overcome the limitations that averaging over velocities and directions has imposed upon the theory of reaction rates.

23.   Collision Theory of Gas Reactions

Reaction velocities have been studied in gaseous, liquid, and solid solutions, and at interfaces between phases. Homogeneous reactions in liquid solutions have been investigated most extensively, because they are of great practical importance and usually require relatively simple experimental methods. From a theoretical viewpoint, however, they have the disadvantage that the statistical mechanics of liquid solutions is still in a rather primitive stage of development, at least at the depth required to provide any quantitative calculations of rate constants. Homogeneous gas reactions, although more difficult to study experimentally, are more amenable to our available theoretical techniques. The statistical mechanics and kinetic theory of gases are already developed to the point at which calculations of rate constants from molecular properties are sometimes possible.

The first theories of gas reactions were based on the kinetic theory of gases.* They postulated that during collisions between gas molecules, a rearrangement of chemical bonds would sometimes occur, forming new molecules from the old ones. The rate of reaction was set equal to the number of collisions in unit time (frequency factor) multiplied by the fraction of collisions that result in the chemical changes.

In Chapter 4, on the kinetic theory of gases, frequency of collisions between gas molecules was calculated on the basis of a model that treated the molecules

*W. C. McC. Lewis, *J. Chem. Soc.*, *113*, 471 (1918); M. Trautz, *Z. Anorg. Chem.*, *96*, 1 (1916).

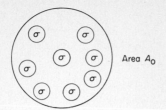

FIG. 9.13 The area A_o contains N circular targets each of area σ, the collision cross section.

as rigid spheres. In this case, there is no interaction between the molecules until the centers of the spheres reach a separation of $(d_1 + d_2)/2$, at which point an elastic collision occurs and the molecules rebound with velocities that can be computed from the laws of conservation of translational kinetic energy and of momentum. Such an idealized process is far removed from what must happen when chemical bonds are broken, formed, or rearranged during a *reactive collision* between two gas molecules, and we shall hardly be surprised if the rigid sphere model of an elastic collision provides only a sort of "zero-order approximation" for the complex processes that occur during such reactive collisions.

No matter what model is chosen for the collision, a convenient and concise way to describe the results of the process is in terms of a *collision cross section σ*. A cross section is the measure of the *probability* of a given collision process. Consider in Fig. 9.13 a planar area A_o covered by a collection of circular targets. If we fire a projectile normal to the area, the probability of hitting one of the targets is the ratio of target area to total area.

$$p = \frac{A}{A_o}$$

If the number of targets per unit area is Γ, then $A = \Gamma \sigma A_o$, and

$$p = \Gamma \sigma \tag{9.43}$$

We see that σ has the dimension of an area, which when multiplied by Γ gives for the probability p a number between 0 and 1.

In Fig. 9.14, we show a beam of molecules incident upon a scattering center at C. We can think of the center as a molecule assumed to be stationary while a molecule in the beam approaches at a relative velocity v. The scattering process depends on a quantity b called the *impact parameter*, which is the closest distance of approach to the scattering center by the trajectory of the incident particle extrapolated along its initial direction without any deflection. The lines in the figure therefore depict a section of the incident beam defined by impact parameters between b and $b + db$. Let us assume that the scattering center acts on the particles in the beam only through central forces [i.e., the potential energy of interaction $U(r)$ depends only on the separation of an incident particle from the center]. In this case, the set of trajectories for a given b will have a cylindrical symmetry about the line $b = 0$. The deflected segment of the beam can thus be characterized by two parameters, a distance of closest approach r_0, and a deflection angle θ. Incident molecules with impact parameters between b and $b + db$ will be deflected at angles between θ and $\theta + d\theta$.

Every collision for which θ differs from zero will correspond to a deflection of

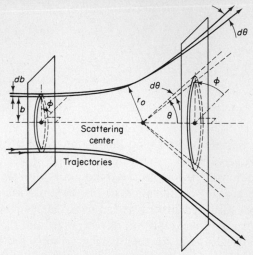

FIG. 9.14 Deflection of particle trajectories by a scattering center. The figure illustrates the relationship between θ and b and the cylindrical symmetry of the set of trajectories of equal b around the line for $b = 0$. The closest approach of these trajectories to the scattering center is r_0. This figure is adapted from an article by E. F. Greene and A. Kuppermann, *J. Chem. Educ.*, **45**, 361 (1968), which provides an excellent introduction to the use of reaction cross sections in chemical kinetics.

the incident beam. We therefore define a differential cross section as

$$d\sigma = 2\pi b(\theta)\, db \qquad (9.44)$$

The total collision cross section is, then,

$$\sigma = \int_{b_{\min}}^{b_{\max}} 2\pi b(\theta)\, db \qquad (9.45)$$

where $b_{\max}$ corresponds to $\theta = 0$ and $b_{\min}$ to $\theta = \pi$.

Figure 9.15 summarizes three models for collisions. We show the intermolecular potential energy function and a schematic representation of the collision for each case. In the collision diagrams, one molecule is assumed to be stationary, while the second molecule approaches it with the initial relative velocity.

Case (a) is the hard-sphere model previously discussed in Section 4.22. Here, $b_{\min} = 0$ and $b_{\max}$ is d_{12}, the hard-sphere collision diameter. Hence, from (9.45), $\sigma = \pi d_{12}^2$, as given on p. 149.

In Case (b) (hard spheres with some attractive potential), the result of the attractive potential is to divert the incoming molecule toward the target molecule. The collision cross section is somewhat increased [$U(d_{12})$ being negative], since the hard-sphere molecules can make contact even when the impact parameter is greater than d_{12}.

Case (c) (Lennard–Jones type of attractive and repulsive potentials) does not lead to any simple formula for σ, which can be either greater or less than the hard-

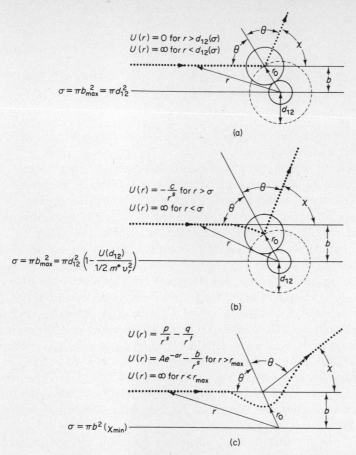

$$\sigma = \pi b_{max}^2 = \pi d_{12}^2$$

$U(r) = 0$ for $r > d_{12}(\sigma)$
$U(r) = \infty$ for $r < d_{12}(\sigma)$

(a)

$$\sigma = \pi b_{max}^2 = \pi d_{12}^2 \left(1 - \frac{U(d_{12})}{1/2\, m^*\, v_r^2} \right)$$

$U(r) = -\dfrac{c}{r^s}$ for $r > \sigma$

$U(r) = \infty$ for $r < \sigma$

(b)

$$\sigma = \pi b^2 (\chi_{min})$$

$U(r) = \dfrac{p}{r^s} - \dfrac{q}{r^t}$

$U(r) = Ae^{-ar} - \dfrac{b}{r^s}$ for $r > r_{max}$

$U(r) = \infty$ for $r < r_{max}$

(c)

FIG. 9.15 Trajectories for three different models for molecular collisions: (a) elastic hard spheres; (b) elastic hard spheres with
superposed weak central attractions; (c) molecules with
finite central repulsive and attractive forces. [After I. Amdur
and G. G. Hammes, *Chemical Kinetics, Principles and
Selected Topics* (New York: McGraw-Hill Book Company,
1966).]

sphere case, depending on relative values of attractive and repulsive terms in the
potential. It is possible, however, to make a computer calculation of the actual
course of the collision for any particular choice of parameters.

24. Reaction Rates and Cross Sections

We can extend the concept of a collision cross section to define a reaction cross
section, which is a measure of the probability that a chemical reaction occurs in a
given collisional process. Consider again the reaction $A + B \rightarrow C + D$ for the
case in which molecule A is approaching molecule B with a relative velocity v.

The reaction rate per unit volume can be expressed as

$$-\frac{dC_A}{dt} = \sigma_r(v)\, v C_A C_B \tag{9.46}$$

where C represents the number of molecules per unit volume and $\sigma_r(v)$ is the *reaction cross section*. Since $-dC_A/dt = k_2 C_A C_B$, we can write

$$k_2(v) = v\sigma_r(v) \tag{9.47}$$

where $k_2(v)$ is the bimolecular rate constant for molecules with a particular value v for the relative velocity.

Such reaction cross sections have been used for a long time in nuclear reactions, but their application to chemical reactions is a more recent development. In a typical nuclear reaction, a beam of particles of defined energy is incident upon a target, and we are interested in the probability that a particular reaction occurs in a particular kind of collision in which the energy is specified. For chemical reactions, however, we have had to be satisfied with a reaction rate averaged over all the different kinds of collisions in a volume of gas molecules, which are moving in random directions with relative velocities distributed in accord with the Maxwell–Boltzmann law, $f(v)$. In recent years, techniques have been devised to produce *monoenergetic molecular beams*; with these, we can study chemical reactions between pairs of molecules having exactly defined energies and directions of approach. We can obviously get a lot more information about reactions in this way (at the expense of a lot more experimental work). We should then be able to calculate the ordinary reaction rate in a gas mixture by taking an average over all the collisions, each weighted by its particular cross section. Such a treatment should also allow us to diagnose the causes for the failures of various simplified models for the kinetics of the reaction.

In accord with (9.47), therefore, we formulate the ordinary rate constant in a mixture of gas molecules as

$$k_2 = \int_0^\infty f(v)\, \sigma_r(v)\, v\, dv \tag{9.48}$$

This expression is valid for any distribution of relative speeds $f(v)$, whether or not it is Maxwellian. Thus, it would hold good for molecules produced by photochemical reactions or for molecules in a beam with a certain spread of speeds. In a mixture of reacting gases maintained at constant temperature, most collisions do not lead to chemical reaction. These nonreactive collisions provide the mechanism whereby the Maxwellian distribution of molecular speeds is maintained. Reactive collisions continually skim off the most energetic molecules from the end of the distribution, but the normal distribution is quickly restored by the nonreactive collisions of molecules with the container wall and with one another.

If we replace the relative speed v in (9.48) by the relative kinetic energy $E = \frac{1}{2}\mu v^2$, we obtain

$$k_2 = \frac{1}{\mu} \int_0^\infty f(E)\, \sigma_r(E)\, dE \tag{9.49}$$

This equation allows us to calculate k_2 for a bimolecular gas reaction, provided we know the distribution law $f(E)$ and the reaction cross section $\sigma_r(E)$. The subsequent development of the collision theory consists in calculating the results of various specific models for each of these factors.

25. Calculation of Rate Constants from Collision Theory

As yet, we do not have much quantitative information on the variation of σ_r with E, but we do know in a general way how the function must behave. Below a certain threshold value E_0, σ_r must be 0. It should rise to a maximum and then fall away again. The decline in σ_r at high energies is reasonable when we consider that at high speeds the time spent by any pair of molecules in the process of collision becomes small, and thus the probability of rearrangement of bonds to cause a chemical reaction becomes quite low. A typical dependence of σ_r upon E is shown in Fig. 9.16, for the reaction $T + H_2 \longrightarrow HT + H$ where T is tritium [^{3}H]. This curve was calculated by Karplus and coworkers* and is doubtless a correct general picture of $\sigma_r(E)$ for this reaction. We note that even the maximum value of σ_r is less (by a factor of about 10) than the cross section suggested by the size of the molecule as determined from transport phenomena, such as viscosity.

For most reactions of interest, the critical energy E_0 is much greater than kT. For example, 80 kJ·mol^{-1} is a moderate activation energy, but even at 500 K, it would correspond to about $E_0/RT = 80\,000/4157 = 19.25$. Only a small fraction of molecules at thermal equilibrium will have an energy that is much greater than kT. We can see this by considering the Maxwell–Boltzmann distribution in two degrees of freedom, (4.30), in the form†

$$\frac{\Delta N}{N_\circ} = \frac{1}{RT} e^{-E/RT} \Delta E$$

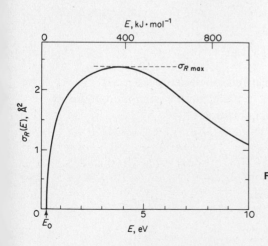

FIG. 9.16 Theoretical calculation of the variation with relative initial kinetic energy E of the reaction cross section for $T + H_2 \longrightarrow TH + H$.

*M. Karplus, R. N. Porter, and R. D. Sharma, *J. Chem. Phys.*, **43**, 3529 (1965).
†The result here is not exact, since we replaced dN by ΔN instead of taking

$$\int_{E_\circ}^{E_\circ + \Delta E} f(E)\, dE$$

For $\Delta E = 4$ kJ at $E = 80$ kJ, and RT $= 4$ kJ, $\Delta N/N_o = e^{-20}$. The fraction comprising all the molecules with energies greater than 80 kJ would be

$$\frac{\Delta N}{N_o} = \int_E^\infty \frac{1}{RT} e^{-E'/RT}\, dE' = -e^{-E/RT}\Big|_E^\infty = e^{-E/RT} = e^{-20}$$

Thus, at 500 K, virtually all molecules with energies greater than 80 kJ have energies very close to 80 kJ.

The exponential decline in the number of molecules with energies much greater than kT is so rapid that, practically speaking, almost all the energetic molecules that do have energies greater than the critical value E_0 are clustered in the immediate neighborhood of E_0 itself. To obtain experimental information about the σ_r vs. E curve, therefore, we cannot rely on ordinary kinetic studies at various constant temperatures. We must devise a method to study reactions of molecules that are selected so as to have a desired energy range. This sort of experiment is indeed possible by the method of molecular beams to be described in Section 9.32.

To calculate a rate constant from (9.49), let us first suppose that $f(E)$ is the ordinary Maxwell expression for the kinetic energy in three translational degrees of freedom, given from Eq. (4.54) as

$$f_3(E) = \left(\frac{1}{\pi\mu}\right)^{1/2}\left(\frac{2}{kT}\right)^{3/2} E\, e^{-E/kT} \tag{9.50}$$

A somewhat different model of collision theory postulates that only two degrees of freedom are utilized, which can be considered to be the components of translation of each molecule along the line of centers at the time of collision. In other words, only the velocity components in the direction of a "head-on" collision are effective. It is impossible to justify this restriction in any rigorous way, but it is interesting nevertheless to include the calculation for such a model. The distribution law $f(E)$ is the one based on velocities in two dimensions, Eq. (4.30), which yields,

$$f_2(E) = \left(\frac{1}{kT}\right) e^{-E/kT} \tag{9.51}$$

We can now use either of these distribution functions to calculate k_2 from (9.49), provided we know $\sigma_r(E)$ or care to make some assumption about it.

The simple hard-sphere model would correspond to

$$\begin{aligned} \sigma_r &= 0 && \text{for } E < E_a \\ \sigma_r &= \pi d_{12}^2 && \text{for } E > E_a \end{aligned} \tag{9.52}$$

From (9.50), this would yield

$$k_2 = \left(\frac{1}{\pi\mu}\right)^{1/2}\left(\frac{2}{kT}\right)^{3/2} (\pi d_{12}^2) \int_{E_a}^\infty E\, e^{-E/kT}\, dE$$

On integrating, we find

$$k_2 = \left(\frac{8kT}{\pi\mu}\right)^{1/2} (\pi d_{12}^2)\left(1 + \frac{E_a}{kT}\right) e^{-E_a/kT}. \tag{9.53}$$

In a similar fashion, the use of (9.51) and (9.52) leads to

$$k_2 = \left(\frac{8kT}{\pi\mu}\right)^{1/2} (\pi d_{12}^2)\, e^{-E_a/kT} \tag{9.54}$$

The rate constant calculated from (9.54) is seen to be somewhat smaller than that obtained from (9.53), as would be expected since one less degree of freedom is contributing to the energy of activation.

The rate constant as calculated in (9.49) through (9.54) was based on concentrations expressed as molecules per unit volume, so that, if the units in (9.49) were in the centimetre-gram-second (cgs) system, the units of k_2 would be (molecule/cm^3)$^{-1}$s^{-1}. To convert this into dm$^3\cdot$mol$^{-1}\cdot$s^{-1}, a more usual unit for a second-order rate constant, we need to multiply k_2 from (9.49) by $L/10^3$, where L is the Avogadro Number. For gas reactions, a common unit is cm$^3\cdot$mol$^{-1}\cdot$s^{-1} and if this is used, k_2 in (9.49) is multiplied by L.

26. Tests of Hard-Sphere Collision Theory

The simple collision theory might be compared with experiments in a number of ways, but in view of the various experimental uncertainties, the most clean-cut comparison is between the calculated and experimental pre-exponential factors A. The Arrhenius equation can be taken as an empirical expression for the rate constant,

$$k_2 = A\, e^{-E_a/RT} \quad \text{(experimental)} \tag{9.55}$$

Any theoretical equation like (9.54) can be written,

$$k_2 = B(T)\, e^{-U_a/RT} \quad \text{(theoretical)} \tag{9.56}$$

Because of the temperature-dependent function in the theoretical pre-exponential factor, some correction terms are needed before we can get a theoretical A from (9.56). Thus,

$$E_a = RT^2 \frac{d\ln k_2}{dT} = U_a + \theta RT$$

where

$$A = B\, e^{\theta} \quad \text{and} \quad \theta = T\left(\frac{d\ln B}{dT}\right) \tag{9.57}$$

The correlations in (9.57) can be used to compare any theoretical k_2 of the form (9.56) with experimental data expressed in the form (9.55).

The most satisfactory critical comparison by these methods was made for a series of 12 bimolecular reactions.* The molecular diameters d_1 and d_2 used to calculate the hard-sphere collision diameter $d_{12} = (d_1 + d_2)/2$ were derived from data on the dimensions of the molecules obtained from electron diffraction or spectroscopy. The volume of the molecule was calculated and the diameter d taken to be that of a sphere of equivalent volume. The final results of the calcula-

---

*D. R. Herschbach, H. S. Johnston, K. S. Pitzer, and R. E. Powell, *J. Chem. Phys.*, **25**, 736 (1956).

TABLE 9.5
Kinetic Parameters for Some Bimolecular Reactions

Reaction	Activation energy (kJ·mol^{-1})	Observed	Calculated transition-state theory [Eq. (9.64)]	Calculated simple collision theory	Reference[a]
		Logarithm of frequency factor (cm^3·mol^{-1}·s^{-1})			
1. $NO + O_3 \longrightarrow NO_2 + O_2$	10.5	11.9	11.6	13.7	a
2. $NO_2 + O_3 \longrightarrow NO_3 + O_2$	29.3	12.8	11.1	13.8	b
3. $NO_2 + F_2 \longrightarrow NO_2F + F$	43.5	12.2	11.1	13.8	c
4. $NO_2 + CO \longrightarrow NO + CO_2$	132.0	13.1	12.8	13.6	d
5. $2NO_2 \longrightarrow 2NO + O_2$	111.0	12.3	12.7	13.6	e
6. $NO + NO_2Cl \longrightarrow NOCl + NO_2$	28.9	11.9	11.9	13.9	f
7. $2NOCl \longrightarrow 2NO + Cl_2$	103.0	13.0	11.6	13.8	g
8. $NOCl + Cl \longrightarrow NO + Cl_2$	4.6	13.1	12.6	13.8	h
9. $NO + Cl_2 \longrightarrow NOCl + Cl$	84.9	12.6	12.1	14.0	i
10. $F_2 + ClO_2 \longrightarrow FClO_2 + F$	35.6	10.5	10.9	13.7	j
11. $2ClO \longrightarrow Cl_2 + O_2$	0.0	10.8	10.0	13.4	k
12. $COCl + Cl \longrightarrow CO + Cl_2$	3.5	14.6	12.3	13.8	h

[a]Table references are as follows:
  (a) H. S. Johnston and H. J. Crosby, *J. Chem. Phys.*, *22*, 689 (1954).
  (b) H. S. Johnston and D. M. Yost, *J. Chem. Phys.*, *17*, 386 (1949).
  (c) R. L. Perrine and H. S. Johnston, *J. Chem. Phys.*, *21*, 2200 (1953).
  (d) H. S. Johnston, W. A. Bonner, and D. J. Wilson, *J. Chem. Phys.*, *26*, 1002 (1957).
  (e) M. Bodenstein and H. Ramstetter, *Z. Physik Chem.*, *100*, 106 (1922).
  (f) E. C. Freiling, H. S. Johnston, and R. A. Ogg, *J. Chem. Phys.*, *20*, 327 (1952).
  (g) G. Waddington and R. C. Tolman, *J. Am. Chem. Soc.*, *57*, 689 (1935).
  (h) W. G. Burns and F. S. Dainton, *Trans. Faraday Soc.*, *48*, 39, 52 (1952).
  (i) P. G. Ashmore and J. Chanmugan, *Trans. Faraday Soc.*, *49*, 270 (1953).
  (j) P. J. Aynoneno, J. E. Sicre, and H. J. Schumacher, *J. Chem. Phys.*, *22*, 756 (1954).
  (k) G. Porter and F. J. Wright, *Discussions Faraday Soc.*, *14*, 23 (1953).

tions and comparisons with experiments are shown in Table 9.5. Also included in the table are pre-exponential factors calculated from *activated-complex theory* [*transition-state theory*], which will be discussed later.

The pre-exponential factors calculated from collision theory are all considerably higher than the experimental values. This failure of the hard-sphere model was discovered in the early days of collision theory, and an attempt was made to explain it by introducing a *steric factor p* into the theoretical expression, which would thus become

$$k_2 = pB(T)\, e^{-U_a/RT} \tag{9.58}$$

The *steric factor* was based on the idea that some collisions would be more effective than others, depending on certain directional factors. For example, in a reaction such as

$$CH_3 \cdot CH_2 \cdot CH_2 \cdot Br + Na \longrightarrow NaBr + CH_3 \cdot CH_2 \cdot CH_2$$

a sodium atom striking the methyl end of the molecule would be unlikely to capture the Br atom. Reasonable though it may be in qualitative terms, the concept of a steric factor has been of little use in quantitative calculations of rate constants. Nobody has proposed any satisfactory way to calculate p from the properties of molecules. The steric factor p is simply a numerical measure of the failure of the hard-sphere collision theory due to the fact that such a crude model has been taken for the variation of reaction cross section with energy and direction.

We have noted that the rate constants calculated by this collision theory are usually too high. What is even worse, however, is that the calculated pre-exponential A factors are all more or less the same. The theory does not show in any way the considerable variations in A due to changes in the structures of the reactants. A deeper theoretical approach will be needed to do this, and we shall see that the transition-state theory not only gives better quantitative agreement for the A factors, but also helps to explain how and why they differ from one molecule to another.

27. Reactions of Hydrogen Atoms and Molecules

When we consider the rate of a reaction in terms of simple hard-sphere collision theory, we are using a model that is clearly a pretty rude approximation. A red billiard ball hits a green billiard ball, they disappear instantaneously, two yellow balls career away. To accept this picture is to give up any hope of following the intricate gradual changes that take place in an actual reaction process.

For example, consider the reaction,

$$H_2 + D_2 \longrightarrow 2HD$$

The H_2 and D_2 molecules do not interact as two hard spheres. When we follow the motions of the two H nuclei and the two D nuclei, we see a gradual rearrangement as the reaction progresses. As H_2 approaches D_2, the H atoms begin to form tenuous bonds with the D atoms while the molecules are still quite far apart. At the same time, the H—H and D—D bonds are somewhat weakened and begin to stretch. The closer H_2 approaches D_2, the greater are these effects. In most cases, the molecules do not have enough kinetic energy to overcome their mutual repulsion and to approach each other closely enough to complete the processes of reaction. Sometimes, however, their kinetic energy is great enough, and they achieve a critical configuration from which they can proceed to the formation of the product, 2HD. The critical configuration may be drawn as a square complex (Fig. 9.17), in which the H—H and D—D bonds are considerably lengthened and weakened and definite H—D bonds have begun to form. This intermediate configuration, which is formed when the molecules have enough energy to react

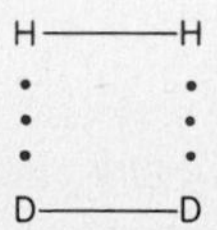

FIG. 9.17 Activated complex for a bi-molecular four-center reaction between H_2 and D_2.

($E > E_a$, the activation energy), is called the *activated complex* or *transition state*. If we consider the process of reaction to require the surmounting of a hill of potential energy, we can identify the activated complex as the configuration of the system at the maximum of the potential-energy barrier. Note that the same activated complex is formed for both the forward and reverse reactions, $H_2 + D_2 \rightleftharpoons 2HD$.

At the present time, a highly exact calculation of the potential energy at all stages of a reaction has been made only for the four-center exchange reactions,

$$H_2 + D_2 \longrightarrow 2HD, \qquad \text{etc.}$$

and for the three-center reactions,

$$H + H_2 \longrightarrow H_2 + H$$

This type of reaction can be studied as a thermal process by means of para H_2 or ortho D_2.*

$$\text{(1) } H + p\text{-}H_2 \longrightarrow o\text{-}H_2 + H$$
$$\text{(2) } D + o\text{-}D_2 \longrightarrow p\text{-}D_2 + D$$

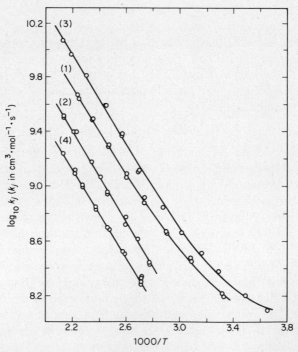

FIG. 9.18 Rate constants for the reactions $H + H_2 = H_2 + H$ (1), $D + D_2 = D_2 + D$ (2), $D + H_2 = DH + H$ (3), and $H + D_2$ = HD + D (4).

*In homonuclear diatomic molecules with nuclear spins, the spins can be either parallel or opposed to each other, giving rise to two nuclear-spin isomers.

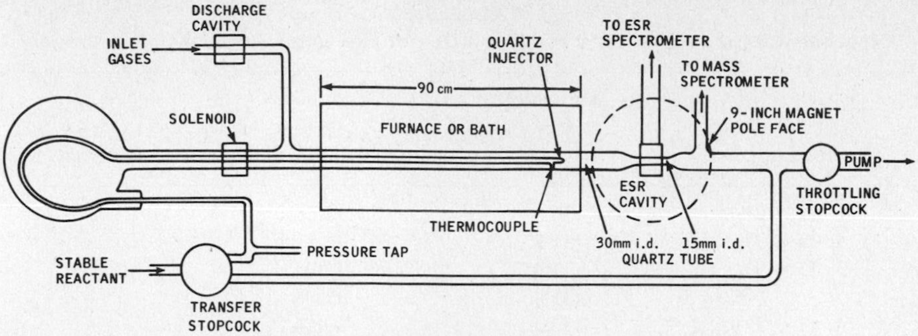

FIG. 9.19 Apparatus to study kinetics of atomic reactions by measurement of atomic concentrations by electron-spin resonance. Either the furnace shown or a cooling chamber can be used, so that a wide range of operating temperatures is possible.

The other experimental method is to follow isotopic exchange, as in

$$(3) \ D + H_2 \longrightarrow HD + D$$
$$(4) \ H + D_2 \longrightarrow HD + H$$

and similar reactions with tritium. Figure 9.18 shows the Arrhenius plots obtained in studies of these reactions at the University of Toronto.[*]

Important work on the reactions has also been done at the Applied Physics Laboratory (Johns Hopkins University).[†] This work used an interesting new type of apparatus, shown in Fig. 9.19, in which atoms of H or D were generated in an electric discharge and then mixed with the molecular species to study either reaction (3) or (4). The D atom concentration was measured by direct observation of its electron spin resonance (ESR) spectrum (Section 17.36). This technique is applicable to many atomic reactions, because atoms usually have unpaired electron spins and hence yield simple and characteristic ESR spectra. In the apparatus shown in Fig. 9.16, the reaction furnace can be replaced by a cooling chamber, so that it can be used to study a range of rate constants from 10^7 to 5×10^{11} cm³ ·mol⁻¹·s⁻¹.

The results of these studies of the three-center hydrogen reactions are summarized in the Arrhenius parameters in Table 9.6. In obtaining these parameters, we have ignored the curvature of the Arrhenius plots at low temperatures, which can be seen clearly in Fig. 9.18. This curvature is believed by some to be due to a quantum mechanical *tunnel effect*, by which a hydrogen atom can go *through* the top of an activation energy barrier rather than *over* it. This effect will be discussed in Section 13.21.

We have stressed the experimental work on these hydrogen reactions, because it provides the best material for rigorous testing of various theories of absolute reaction rates.

[*]Reference (a), Table 9.6.
[†]Reference (b), Table 9.6.

Experimental Rate Parameters for Three-Center Exchange Reactions of Hydrogen

Reaction	Activation energy E_a (kJ·mol⁻¹)	$\log_{10} A$ (cm³·mol⁻¹·s⁻¹)	Reference[a]
1. $H + H_2$	36.8	13.68	c
2. $D + D_2$	31.9	13.21	a
3. $D + H_2$	37.9	14.08	c
	31.8	13.64	b
4. $H + D_2$	30.5	12.64	a
	39.3	13.69	b

[a]Table references are as follows:
  (a) D. J. LeRoy, B. A. Ridley, and K. A. Quickert, *Discussions Faraday Soc.*, *44*, 92 (1967).
  (b) A. A. Westenberg and N. deHaas, *J. Chem. Phys.*, *47*, 1393 (1967).
  (c) B. A. Ridley, W. R. Schulz, and D. J. LeRoy, *J. Chem. Phys.*, *44*, 3344 (1966).

28. Potential Energy Surface for $H + H_2$

We shall describe the reaction in terms of $D + H_2$ in order to facilitate the distinction between the target molecule and the approaching atom. To describe the configuration of this reacting system at any instant, we need to have three spacial coordinates, which might be taken to be the atomic distance between H and H, the distance of D from the midpoint of the H—H bond, and the angle between the H—H bond and the vector from the midpoint of the bond to D. If we wished to represent the potential energy of this system as a function of the coordinates, we should need to tabulate the representation in a four-dimensional space consisting of the three coordinates and the energy *E*. The only way we could represent such a function graphically would be to maintain one of the coordinates constant and plot the others in three dimensions. The computational problems in such a complete representation of the system are obviously formidable, and, in fact, it has never yet been accomplished.

Fortunately, a great simplification of the problem results from the fact that one particular direction of approach of D to H—H is energetically more favorable than any other. This is the case in which D approaches H—H along the line of centers—in other words, the angle θ is always taken to be 180°. We shall give a quantitative justification for this statement later, after results of calculations of the potential-energy surface have been considered.

The potential energy when the reacting system maintains this linear configuration will therefore give the most favorable path along which the reaction can proceed. In this case, *E* is a function of only two coordinates, namely, the distance D—H and the distance H—H. We plot one distance along the abscissa and the other distance along the ordinate in a plane, and plot the energy on a vertical axis. We can thus represent the potential energy of the system as a three-dimensional surface.

What makes such a calculation at all feasible is the fact that one can separate the motions of the nuclei of the atoms from the motions of their electrons. This independence of the electronic and nuclear motions is called the *Born–Oppenheimer approximation*. It underlies most of the quantitative calculations that have been made of the energies of molecular systems. Essentially one considers the nuclei of the atoms fixed in some definite positions and then calculates the interactions of the electrons with the fixed nuclei and with one another. The total electrostatic energy of such an interaction is computed by the methods of molecular quantum mechanics. We call this total energy the *potential energy* of the system of atoms fixed at the positions chosen for the nuclei. It includes the potential energy of interaction of the nuclear and electronic charges and both the potential *and* kinetic energies of the electrons. Nevertheless, it is quite conventional to refer to it as the *potential energy* of the system. We then shift the nuclei to some new positions and repeat the calculation. In this way, we calculate the *potential energy* of the system as a function of the positions of the nuclei. We can then consider that the nuclei move in a potential field given by this computed energy function.

The idea that a chemical reaction can be represented by such a potential energy surface was suggested by Marcelin* in 1915, but the first surface was actually computed in 1931 in a paper by Eyring and Polanyi† that is one of the great landmarks in the progress of chemical kinetics. With the techniques and computers available at that time, they were not able to perform a purely theoretical calculation, but relied on a semiempirical approach that used spectroscopic data. In any event, the surface they constructed by a combination of calculation and intuition has turned out to be essentially correct. It is shown in Fig. 9.20 in the form of drawings of a three-dimensional model.

We can trace the *path of the reaction* over this surface, as the pathway from reactant side to product side that follows the contour of minimum potential energy.‡ The path traverses a deep valley $(D + H_2)$, rises over a mountain pass to a saddle point or col at the linear configuration of the activated complex (D—H—H), and then passes down the other side of the pass into another deep valley (DH + H).

We shall follow the *reaction path* in greater detail by means of a contour map, based not on the historic Eyring–Polanyi surface, but on the exact theoretical calculations of Shavitt and coworkers,§ as shown in Fig. 9.21.

Consider a cut taken through the map at $r_2 = 0.38$ nm, i.e., at a H—H separation sufficiently large to leave the D—H molecule practically undistorted. The cross section, shown in Fig. 9.21(b), is then simply the potential-energy curve for the HD molecule (as shown in Fig. 15.4).

If one travels along the valley floor, following the dashed line on the map, the

*_Ann. Phys._, *3*, 158 (1915).

†H. Eyring and M. Polanyi, *Z. Physik Chem.*, B, *12*, 279 (1931).

‡If interconversion of translational and vibrational energy is considered, the path of reaction is more like that of a bobsled sliding on its run.

§I. Shavitt, R. M. Stevens, F. L. Minn, and M. Karplus, *J. Chem. Phys.*, *48*, 2700 (1968).

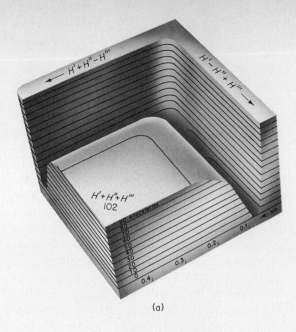

(a)

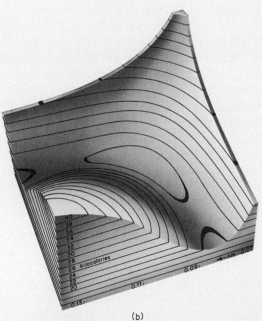

(b)

FIG. 9.20 (a) Drawing made from a photograph of a model of the potential-energy surface for the reaction $H + H_2 \longrightarrow H_2 + H$, constructed by Goodeve (1934) according to the contour lines calculated by Eyring and Polyani (1931). In this model, only the resonance forces are considered. (b) Close-up of the saddle region of a model in which coulombic as well as resonance forces have been taken into account. [After F. H. Johnson, H. Eyring, and M. J. Polissar, *The Kinetic Basis of Molecular Biology* (New York: John Wiley & Sons, Inc., 1954), p. 16.]

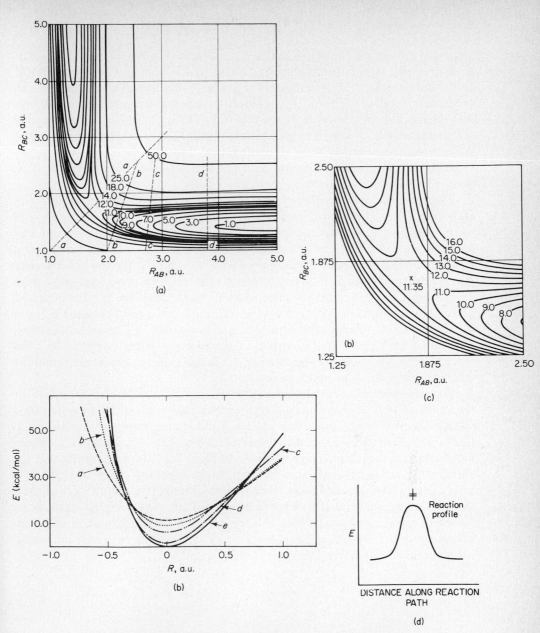

FIG. 9.21 (a) Contour map of the potential-energy surface for linear H_3. The dashed lines indicate the positions of the sections plotted in (b). Energy in kilocalories relative to H_2, H limit. (b) Cross sections through the linear H_3 potential surface perpendicular to the minimum-energy path. The positions of sections *a–d* are indicated by the dashed lines in (a); section *e* is for an isolated hydrogen molecule. (c) Contour map of potential-energy surface in the saddle point region for linear H_3: energy in kilocalories relative to H_2 + H limit. (d) Potential-energy profile along reaction path. [1 Kcal = 4.184 KJ, 1 a.u. = 0.1 nm]

view to right and left looks like the cross section in (b). The elevation, however, gradually rises as one traverses the mountain pass, reaching a height of 47.5 kJ at the saddle point.* This is the configuration of the activated complex, which occurs at $r_1 = r_2 = 0.093$ nm. This distance is considerably greater than the normal internuclear separation in H_2, which is 0.074 nm. When the system reaches this configuration, it can either decompose into $DH + H$ by moving down into the other valley, or return along its original path into $D + H_2$.

If the potential energy is drawn as a function of distance along the reaction path, Fig. 9.21(d) is obtained.

The potential-energy surface gives a picture of a chemical reaction from beginning to end. In any reaction, there is always a certain configuration at the top of the mountain pass. In many respects this configuration of atoms is like an ordinary molecule, except that it has no stable equilibrium state. It is called the *activated complex*, and its properties are designated with the symbol‡. We can now consider that any reaction takes place in two stages: (1) the reactants come together to form an activated complex; (2) the activated complex decomposes into the products. These stages are not sharply separated in any way, and from a dynamic point of view, the reaction process is smooth and continuous. We can, however, designate a *transition state* as the highest region of the potential energy surface along the reaction path, the maximum elevation of the saddle point or col leading from the valley of reactants to the valley of products.†

The same method of calculation used to obtain the potential-energy surface for the approach of D to H_2 along the line of nuclear centers ($\theta = 180°$) was also used to calculate the approach at other angles. Although the results of these calculations confirm our surmise that these other directions are less favorable (i.e., require the reactant to surmount a higher mountain pass), they do not indicate that the various angular approaches can be completely neglected. Even though noncollinear trajectories make harder reaction paths, there are so many more of them that the "average transition state" is bent to about 20°. Any complete quantitative theory of reaction rates will need to take such nonlinear encounters into consideration, thus making even more difficult a calculation that is already impossible with current computers and techniques.

---

*This value is with reference to the value of $H + H_2$, calculated on the basis of the same wave functions. Since the calculated H_2 value is known to be somewhat too high (relative to either the experiment or to an exact calculation), a corrected value for the calculated activation energy would be 38 kJ·mol⁻¹.

†In this discussion, we are always making the assumption that the reaction path is a continuous path over a single potential energy surface. In the theory of rate processes, a reaction that obeys this criterion is called an *adiabatic reaction*. It is likely that in some rather unusual instances the reaction path changes from one potential energy surface to another. This type is called a *nonadiabatic* reaction. The term *adiabatic* as used here is not the same as the *adiabatic* used in thermodynamics.

29. Activated-Complex Theory*

In terms of the potential-energy surface, we can think of the course of a chemical reaction as the trajectory of a representative point in configuration space from the reactant side to the product side of the surface. It is possible, however, to formulate the reaction rate entirely in terms of properties of the reactants and of the transition state at which the interacting molecules have formed the *activated complex* for the particular reaction. The rate of reaction is the number of activated complexes passing per second over the top of the potential-energy barrier. This rate is equal to the concentration of activated complexes times the average frequency with which a complex moves across to the product side.

The calculation of the concentration of activated complexes is greatly simplified if we assume that they are in equilibrium with the reactants. This equilibrium can then be treated formally by means of thermodynamics or statistical mechanics. The activated complex is not a state of stable equilibrium, since it lies at a saddle point and not a minimum of potential energy. Yet, more detailed calculations have shown that there is probably little error in treating the equilibrium by ordinary thermodynamic or statistical methods, except in the case of extremely rapid reactions.† The fundamental difficulty is therefore not the assumption of equilibrium between activated complexes and reactants, but the impossibility at the present time of calculating the structure and properties of the activated complex.

We shall consider the transition-state formulation of the specific rate of a simple bimolecular reaction:

$$A + B \longrightarrow [AB^\ddagger] \longrightarrow \text{products}$$

We shall follow a derivation‡ of the activated-complex theory for the bimolecular rate constant k_2 that may be somewhat more transparent than the original. The curve RXP in Fig. 9.21(d) shows a section of the path of reaction along a potential energy surface such as those in Figs. 9.20 and 9.21. A narrow region of the col, of arbitrary length δ, defines the region of existence of the activated complex; this region is the transition state for the chemical reaction.

If the activated complexes $[AB^\ddagger]$ are in equilibrium with reactants, the equi-

---

*The quantitative formulation of absolute reaction rates in terms of activated complexes was first extensively used in the work of H. Eyring [*J. Chem. Phys.*, *3*, 107 (1935); *Chem. Rev.*, *17*, 65 (1935)]. This theory has been applied to a wide variety of *rate processes* besides chemical reactions, such as the flow of liquids, diffusion, dielectric loss, and internal friction in high polymers. Other noteworthy contributions to the basic theory were made by M. G. Evans and M. Polanyi [*Trans. Faraday Soc.*, *31*, 875 (1935)], and H. Pelzer and E. Wigner [*Z. Physik Chem.*, *B*, *15*, 445 (1932)].

†R. D. Present, *J. Chem. Phys.*, *31*, 747 (1959). Even when the activation energy is as low as $5RT$, the reaction rate is only about 8% less than that predicted by equilibrium theory. It is possible, however, to derive a transition-state expression for the rate constant of a bimolecular reaction without recourse to the untenable hypothesis of equilibrium [John Ross and Peter Mazur, *J. Chem. Phys.*, *35*, 19 (1961)]. This derivation also leads to a deeper understanding of the parameters that enter the theory.

‡K. J. Laidler and J. C. Polanyi, *Prog. Reaction Kinetics*, *3*, 1 (1965).

librium constant for the formation of complexes is

$$\frac{[AB^\ddagger]}{[A][B]} = K^\ddagger$$

The concentration of complexes then would become

$$[AB^\ddagger] = K^\ddagger[A][B]$$

When $K^\ddagger$ is written in terms of the molecular partition functions per unit volume,*

$$[AB^\ddagger] = [A][B]\frac{z_\ddagger'}{z_A'z_B'}\ e^{-E_0/kT} \tag{9.59}$$

where $z_\ddagger'$ is the partition function per unit volume of the activated complex and E_0 is the height of the lowest energy level of the complex above the sum of the lowest energy levels of the reactants $A + B$.

According to the activated-complex theory, the rate of the reaction is

$$\frac{-d[A]}{dt} = k_2[A][B] = [AB^\ddagger] \times \text{(frequency of passage of complex over barrier)}$$

$$\tag{9.61}$$

The frequency $\nu^\ddagger$ of passage of $AB^\ddagger$ over the barrier is equal to the frequency with which a complex flies apart into products. The complex flies apart when one of its vibrations becomes a translation, and what was formerly one of the bonds holding the complex together becomes the direction of translation of the fragments of the separated complex.

From (9.59) and (9.61), we can therefore write for the rate constant,

$$k_2 = \nu^\ddagger\frac{z_\ddagger'}{z_A'\,z_B'}\ e^{-E_0/kT} \tag{9.62}$$

If we examine $z_\ddagger'$, we may conclude that it is just like a partition function for a normal molecule, except that one of its vibrational degrees of freedom is in the act of passing over to the translation along the reaction coordinate. From Table 5.4, the ordinary expression for a partition function in one vibrational degree of freedom would be

$$z_v^\ddagger = (1 - e^{-h\nu^\ddagger/kT})^{-1} \tag{9.63}$$

For this particular anomalous vibration along the reaction coordinate, we can

---

*It must be noted that the equilibrium constant used here is in terms of concentrations, K_c. We can see directly that for any reaction $aA + bB \rightleftharpoons cC + dD$,

$$K_c = \frac{[C]^c[D]^d}{[A]^a[B]^b} = \frac{(z_C/V)^c(z_D/V)^d}{(z_A/V)^a(z_B/V)^b}\ e^{-E_0/kT}$$

If we write z' for the partition functions per unit volume.

$$K_c = z_C'^c z_D'^d / z_A'^a z_B'^b e^{-E_0/kT} \tag{9.60}$$

When we use the translational partition function of Eq. (5.49), therefore, we divide out the volume. If the other factors in (8.27) are in the usual cgs units, the volume is in cubic centimetres, and hence the concentrations in K_c will be in molecules per cubic centimetre. The units of the bimolecular rate constant k_2 calculated in this section will accordingly be $(\text{molecule/cm}^3)^{-1}\cdot\text{s}^{-1}$. To convert to $(\text{mol/cm}^3)^{-1}\cdot\text{s}^{-1}$, the calculated k_2 [Eq. (9.65)] must be multiplied by L.

be sure that $h\nu/kT \ll 1$, since, at any temperature at which reaction is detectable, this "decomposition vibration" of the complex must, by hypothesis, be completely excited. Hence, when we expand

$$e^{-h\nu\ddagger/kT} = 1 - \frac{h\nu^{\ddagger}}{kT} + \frac{1}{2}\left(\frac{h\nu^{\ddagger}}{kT}\right)^2 - \cdots$$

we can discard terms beyond the first power in $(h\nu^{\ddagger}/kT)$. Expression (9.63) becomes

$$z_v^{\ddagger} = \left(\frac{h\nu^{\ddagger}}{kT}\right)^{-1} = \frac{kT}{h\nu^{\ddagger}}$$

Our next step is to factor this particular $z_v^{\ddagger}$ out of the complete $z_{\ddagger}'$, so that

$$z_{\ddagger}' = z_v^{\ddagger} z^{\ddagger\prime} = \left(\frac{kT}{h\nu^{\ddagger}}\right) z^{\ddagger\prime}$$

When we substitute this into (9.62), we obtain the famous *Eyring equation* for the rate constant,

$$k_2 = \frac{kT}{h} \frac{z^{\ddagger\prime}}{z_A' z_B'} \ e^{-E_0/kT} \tag{9.64}$$

This expression for k_2 should be multiplied by a factor κ, the *transmission coefficient*, which is the probability that the complex will dissociate into products instead of back into reactants. For most reactions, κ is between 0.5 and 1.0. Thus,

$$k_2 = \kappa \frac{kT}{h} \frac{z^{\ddagger}}{z_A z_B} \ e^{-E_0/kT} \tag{9.65}$$

This is the theoretical expression given by activated-complex theory for the bimolecular rate constant. One can see at once that it includes explicitly terms that depend upon the properties of the reactant molecules and the activated complex. Thus, it is likely to be much better than the hard-sphere collision theory equation (9.53), which really provided no way to understand variations in the observed pre-exponential factors.

In the same paper that gave the comparison of experimental bimolecular rate constants of gas reactions with calculations on hard-sphere collision theory (our Table 9.5), the authors included theoretical rate constants calculated from activated-complex theory in accord with (9.64). They calculated the partition functions for the reactant molecules from the known molecular structures, and they assumed structures for the activated complexes, which were based upon known bond distances and angles. Figure 9.22 shows the models so constructed for the activated complexes in some of these reactions. The resultant calculated pre-exponential factors are given in Table 9.5. In almost all these examples, the activated-complex equation has given better results than the hard-sphere collision theory. Also, the activated-complex theory has brought the properties of the molecules into the picture in a realistic way. Instead of an unsatisfactory correction by an arbitrary steric factor, we can now estimate quantitatively the dependence of the pre-exponential factor on the molecular structure.

Let us take a numerical example of the calculation of the pre-exponential

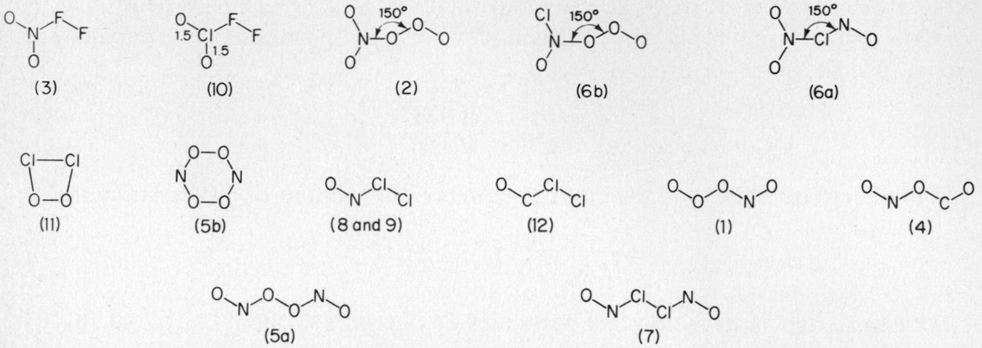

(3)   (10)   (2)   (6b)   (6a)

(11)   (5b)   (8 and 9)   (12)   (1)   (4)

(5a)   (7)

FIG. 9.22 Models of activated complexes for the reactions in Table 9.5.

factor in k_2 from the transition-state equation (9.64), the reaction $2ClO \rightarrow Cl_2 + O_2$. The activated complex is shown in Fig. 9.22. The following molecular parameters are available, obtained from spectroscopic data for the reactants, and estimated for the complex from data on known structures with similar geometries and bonds.

^{35}ClO: $g = 2$, $\sigma = 1$, $I = 4.3 \times 10^{-39}\,\text{g}\cdot\text{cm}^2$, $\lambda^{-1} = 800\,\text{cm}^{-1}$

$(^{35}ClO)^{\ddagger}_2$: $g = 1$, $\sigma = 2$, $\mathscr{A}BC = 2.20 \times 10^{-114}\,(\text{g}\cdot\text{cm}^2)^3$, $\lambda^{-1}_i = 1500, 700,$
$$800, 600, 200\,\text{cm}^{-1}$$

Note that the diatomic molecule ClO, has two rotational degrees of freedom with a single moment of inertia I, and one vibrational degree of freedom. The complex has three rotational degrees of freedom and the product of the three moments of inertia $\mathscr{A}BC$ has been computed from the dimensions of the molecule and the atomic masses. There would be $3n - 6 = 6$ vibrational degrees of freedom in a normal molecule, but the complex has only 5, having lost one into the reaction coordinate. Although vibration frequencies of the complex are only estimated values, it turns out that the expression for the rate constant is very insensitive to the values of v, so that good calculated values can be expected even from rather inaccurate frequencies v_i.

We write out explicitly the pre-exponential factor in (9.64), using the formulas for partition functions given in Table 5.4:

$$\frac{kT}{h}\frac{z'^{\ddagger}}{(z_{ClO})^2} = \frac{kT}{h} \cdot \frac{(2\pi m^{\ddagger}kT)^{3/2}/h^3}{(2\pi mkT)^3/h^6}\frac{g^{\ddagger}}{g^2}\frac{8\pi^2(8\pi^3\mathscr{A}BC)^{1/2}(kT)^{3/2}/\sigma^{\ddagger}h^3}{(8\pi^2IkT)^2/\sigma^2h^4}$$
$$\times \frac{\prod_{i=1}^{5}(1 - e^{-hv_i/kT})^{-1}}{(1 - e^{-hv/kT})^{-2}}$$

We first insert the numerical values of all the constants π, k, h, to obtain

$$A = 2.649 \times 10^{-65}T^{-1}\frac{m^{\ddagger\,3/2}}{m^3}\frac{g^{\ddagger}\sigma^2}{g^2\sigma^{\ddagger}}\frac{(\mathscr{A}BC)^{1/2}}{I^2}\frac{\prod_{i=1}^{5}(1 - e^{-v^{\ddagger}_i/\theta T})^{-1}}{(1 - e^{-v/\theta T})^{-2}} \qquad (9.66)$$

where $\theta = 2.083 \times 10^{10}\,\text{K}^{-1}\cdot\text{s}^{-1}$. We now introduce the molecular parameters and calculate A at 400 K.

$$A = (2.469 \times 10^{-65})(2.5 \times 10^{-3})(3.64 \times 10^{33})(1)(5.90 \times 10^{19})(2.405)$$
$$A = 3.20 \times 10^{-14} \text{ cm}^3 \cdot \text{s}^{-1}$$

The terms in parentheses correspond to the terms in (9.66). In units of $(\text{mol}/\text{cm}^3)^{-1} \cdot \text{s}^{-1}$,

$$A = (3.20 \times 10^{-14})(6.02 \times 10^{23}) = 1.93 \times 10^{10}$$

For this reaction, since $E_0 = 0$, $A = k_2$ from (9.41)

30. Transition-State Theory in Thermodynamic Terms

The formalism of transition-state theory is often expressed in terms of thermodynamic functions instead of partition functions. If we consider again the reaction

$$A + B \longrightarrow (AB^{\ddagger}) \longrightarrow \text{products}$$

with

$$K^{\ddagger} = \frac{[AB^{\ddagger}]}{[A][B]}$$

we can write the rate constant from (9.64) as

$$k_2 = \frac{kT}{h} K^{\ddagger} \tag{9.67}$$

Since

$$\Delta G^{\ominus\ddagger} = -RT \ln K^{\ddagger} \tag{9.68}$$

and

$$\Delta G^{\ominus\ddagger} = \Delta H^{\ominus\ddagger} - T \Delta S^{\ominus\ddagger}$$

Eq. (9.67) can be written as*

$$k_2 = \frac{kT}{h} e^{-\Delta G^{\ominus\ddagger}/RT} = \frac{kT}{h} e^{\Delta S^{\ominus\ddagger}/R} e^{-\Delta H^{\ominus\ddagger}/RT} \tag{9.69}$$

The quantities $\Delta G^{\ominus\ddagger}$, $\Delta H^{\ominus\ddagger}$, and $\Delta S^{\ominus\ddagger}$ are called the *Gibbs free energy of activation*, the *enthalpy of activation*, and the *entropy of activation*.†

---

*In this formulation, we have ignored the abnormal vibration in the activated complex along the reaction coordinate and treated the complex as if it were a normal molecule. This procedure has no appreciable effect on the practical applications of the equation.

†The temperature coefficient of the rate constant is conveniently derived from (9.67) by taking logarithms and differentiating,

$$\frac{d \ln k_2}{dT} = \frac{1}{T} + \frac{d \ln K^{\ddagger}}{dT}$$

Since $K^{\ddagger}$ is an equilibrium constant in terms of concentrations,

$$\frac{d \ln K^{\ddagger}}{dT} = \frac{\Delta U^{\ddagger}}{RT^2}$$

Therefore,

$$\frac{d \ln k_2}{dT} = \frac{RT + \Delta U^{\ddagger}}{RT^2}$$

Thus, the Arrhenius activation energy of (9.39) is

$$E_a = RT + \Delta U^{\ddagger}$$

The standard state to which $K^{\ddagger}$ and $\Delta G^{\ominus\ddagger}$ are referred is usually taken to be 1 mol·cm^{-3} for gas reactions, in which case the corresponding units of k_2 in (9.67) are (mol/cm^3)$^{-1}$·s^{-1}. If we prefer to base K_c on molar concentrations, however, the units of k_2 would be (mol/dm^3)$^{-1}$·s^{-1}.

The experimental entropy of activation* can be calculated from the rate constant at a given temperature and the experimental activation energy. As an example, consider the dimerization of butadiene:

$$2C_4H_6 \longrightarrow C_8H_{12} \text{ (3-vinylcyclohexene)}$$

From 440 to 660 K, the experimental rate constant is

$$k_2 = 9.2 \times 10^9 \exp(-99.12 \text{ kJ/RT}) \text{cm}^3 \cdot \text{mol}^{-1} \cdot \text{s}^{-1}$$

From

$$E_a = \Delta H^{\ominus\ddagger} + 2RT$$

at 600 K,

$$\Delta H^{\ominus\ddagger} = 99.12 - 9.96 = 89.16 \text{ kJ mol}^{-1}$$

From

$$k_2 = e^2 \left(\frac{kT}{h}\right) \exp\left(\frac{\Delta S^{\ominus\ddagger}}{R}\right) \exp\left(\frac{-E_a}{RT}\right)$$

at 600 K,

$$9.2 \times 10^9 = (7.360)(1.25 \times 10^{13}) \exp\left(\frac{\Delta S^{\ominus\ddagger}}{R}\right)$$

$$\Delta S^{\ominus\ddagger} = -76.6 \text{ J} \cdot \text{K}^{-1}$$

We should note that the standard state is at a unit concentration of 1 mol·cm^{-3}.

The concept of *activation entropy* is a definite improvement over the less precise concept of the *steric factor*, which was used in collision theory The experimental $\Delta S^{\ominus\ddagger}$ provides one of the best indications of the nature of the transition state. A positive activation entropy $\Delta S^{\ddagger}$ means that the entropy of the complex is greater than the entropy of the reactants. A loosely bound complex has a higher entropy than a tightly bound one. More often there is a decrease in entropy in passing to the activated state. In bimolecular reactions, the complex is formed by association of two individual molecules, and there is a loss of translational and rotational free-

---

From (2.12), $\Delta U^{\ddagger} = \Delta H^{\ddagger} - \Delta(PV)^{\ddagger}$. In liquid and solid systems, $\Delta(PV)^{\ddagger}$ would be very small at ordinary pressures, and one can take

$$E_a \approx \Delta H^{\ddagger} + RT \tag{9.70}$$

For reactions of ideal gases, from (2.33),

$$\Delta H^{\ddagger} = \Delta U^{\ddagger} + \Delta n^{\ddagger}RT$$

where $\Delta n^{\ddagger}$ is the number of moles of complex always equal to 1, minus the number of moles of reactants. In a unimolecular reaction, therefore, $\Delta n^{\ddagger} = 0$, and (9.70) is valid. For a bimolecular reaction, $\Delta n^{\ddagger} = -1$, and

$$E_a = \Delta H^{\ddagger} + 2RT$$

*The idea of an entropy of activation was developed by Rodebush, La Mer, and others, before the advent of transition-state theory, which gave it a precise formulation. W. H. Rodebush, *J. Am. Chem. Soc.*, 45, 606 (1923); *J. Chem. Phys.*, 4, 744 (1936). V. K. La Mer, *J. Chem. Phys.*, 1, 289 (1933).

dom, so that $\Delta S^{\ddagger}$ is usually negative. In fact, sometimes $\Delta S^{\ddagger}$ is not notably different from ΔS for the complete reaction. When this situation occurs in reactions of the type $A + B \longrightarrow AB$, it indicates that the activated complex $[AB]^{\ddagger}$ is similar in structure to the product molecule AB. Formerly, such reactions were considered to be "abnormal," since they have unusually low steric factors. With the advent of transition-state theory, it became clear that the low steric factor is the result of the increase in order, and consequent decrease in entropy, when the complex is formed.

31. Chemical Dynamics—
Monte Carlo Methods

We have discussed two important lines of attack on the theoretical calculation of rate constants of chemical reactions, collision theory and activated-complex theory. There is another way to attack the problem, more direct and more exciting to many contemporary theorists, which has been made possible by the development of digital computers with high speeds and capacious memories. This method is based on the idea of actually doing a "computer experiment" in which thousands of collisions between reactants are simulated on the computer and the resulting reaction probabilities averaged to derive the rate constant. The calculation involves integration of the simultaneous (usually classical) equations of motion of the bodies involved in the collisions.

Calculations of this type require some mathematical technique for making a random choice of initial conditions (coordinates, energies, velocities, impact parameters), subject to the limitation that the set of molecules conforms to a Maxwell–Boltzmann equilibrium distribution. The element of chance in the selection of initial parameters can be introduced by generation of a sequence of random numbers. Such a process evoked memories of the clicking of the boule in a spinning roulette wheel, and this kind of computer experiment came to be called a *Monte Carlo calculation*.

The calculations for the $H + H_2$ reaction are certainly of interest in view of all the information available from other methods, but we must realize that (1) the computer experiments may not yet provide a reliable model of the actual collisions, and (2) even if they do, the extension of these results to more complex reactions may not be justified. With these cautions, we show in Fig. 9.23 a series of computed collisions for this reaction. The collisions are almost always very simple, with an interaction time just about equal to that required for an atom to pass unimpeded by a molecule. There is no collisional complex with a lifetime long enough to allow us to apply statistical mechanics to the energy distribution among its degrees of freedom. This result does not necessarily contradict the formalism of activated-complex theory, although it suggests that Eq. (9.65) must be regarded as a sort of first approximation to some more exact result. The computed reactive collisions also showed that vibrational energy can be used to provide some of the energy needed to surmount the activation energy barrier, but rotational energy is not used.

Interesting Monte Carlo calculations of more complex reactions have already been made* and the method seems capable of further important developments. The investigation of chemical reactions in molecular beams provides experimental data that are directly comparable with the results of the dynamical calculations of individual reaction trajectories, especially in regard to angular distributions of reaction products.

32. Reactions in Molecular Beams

An experimental apparatus for the study of chemical reactions in molecular beams is shown in Fig. 9.24. It includes two sources, one for each reactant species, and a detector of products, all mounted in a large high-vacuum chamber. This apparatus would be used to study reaction in crossed beams. In many cases, however, only one reactant has been in the form of a beam, which is allowed to intercept a region containing the second reactant in more diffuse form. The most exacting work

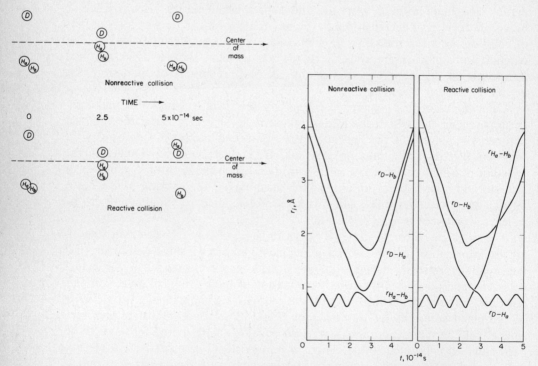

FIG. 9.23 Illustration of nonreactive and reactive collisions between D and H_2. (a) (*left*), relative locations of the three particles D,H, H at three times. (b) (*right*), the variation of the three internuclear distances with time for $E = 2$ eV. Note the rotation of the product molecule H_aD after the reactive collision.

*See especially the series of papers by D. C. Bunker and coworkers, for example, *J. Chem. Phys., 41*, 2377 (1964); *Sci. Am., 211*, 100 (1964).

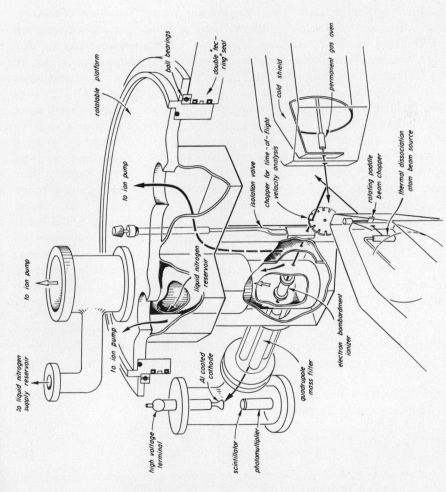

FIG. 9.24 Molecular beam reactive scattering apparatus with electron-bombardment detector. [Y. T. Lee, J. D. McDonald, P. R. LeBreton, and D. R. Herschbach, *Rev. Sci. Inst., 40,* 1402 (1969).]

rotatable platform

ball bearings

double "tec-ring" seal

to ion pump

liquid nitrogen reservoir

to ion pump

to liquid nitrogen supply reservoir

to ion pump

high voltage terminal

Al coated cathode

scintillator

photomultiplier

quadrupole mass filter

electron bombardment ionizer

isolation valve

chopper for time-of-flight velocity analysis

cold shield

permanent gas oven

rotating paddle beam chopper

thermal dissociation atom beam source

requires that each beam be subjected to *velocity analysis* before it enters the reactive zone. Ideally, therefore, the velocities of the reactant molecules would be independently controlled, and the directions of intersection of the crossed beams could be selectively determined.

The first experiment with crossed beams was performed by Taylor and Datz* on the reaction

$$K + HBr \longrightarrow H + KBr$$

They used *thermal beams*, i.e., no velocity analysis was used. The choice of reaction was due to the fact that a sensitive detector was available that could distinguish between K atoms and KBr molecules. This was the Kingdon–Langmuir surface-ionization detector. Whenever a K atom strikes a hot tungsten wire, it loses an electron and leaves the wire as a K^+ ion. The positive ion current can be monitored to measure the K-atom concentration.

If you build an apparatus and set it up in your laboratory, then you will naturally describe any event occurring in your apparatus in terms of a frame of reference that gives a set of laboratory [L] coordinates for every point of interest. The event to be studied may be a collision between molecules A and B. The molecules couldn't care less where you have placed your coordinate axes, and the kinematics of their encounter is under no obligation to fit conveniently into a description in terms of L coordinates. A different frame of reference that provides a more concise representation of the collision has its origin at the center of mass of molecules A and B. Coordinates based on this moving origin are called *center-of-mass* [C] coordinates. In L coordinates, the center of mass is moving, but in C coordinates, the center of mass is stationary. The velocity v in L coordinates is that observed with the apparatus. If c is the velocity of the center of mass, the velocity u in C coordinates is

$$u = v + c$$

The relations between the velocities are summarized† in Fig. 9.25, which is obtained by applying the laws of conservation of mass and linear momentum to the encounter.

Figure 9.26 shows the angular distribution in C coordinates of the alkali-halide products from the reactions,‡

$$K + Br_2 \longrightarrow KBr + Br \tag{A}$$
$$K + CH_3I \longrightarrow KI + CH_3 \tag{B}$$

The transformation from L to C coordinates involves some approximations, but the result is believed to be essentially correct. The completely different angular dependences of the product distributions in these reactions are indeed striking.

*E. H. Taylor and S. Datz, *J. Chem. Phys.*, **23**, 1711 (1955).

†From R. D. Evans, *The Atomic Nucleus* (New York: McGraw-Hill Book Company, 1955). Appendix B contains a clear account of the kinematic problem.

‡D. R. Herschbach, *Adv. Chem. Phys.*, **10**, 319 (1966).

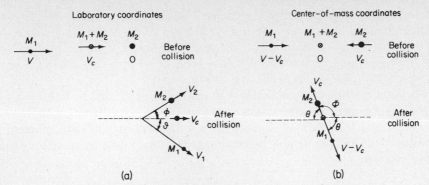

Laboratory coordinates   Center-of-mass coordinates

FIG. 9.25 (a) In L coordinates, the center mass, marked $\otimes$, moves with constant velocity V_c before, during, and after the collision. (b) In C coordinates, the center of mass, marked $\otimes$, is stationary.

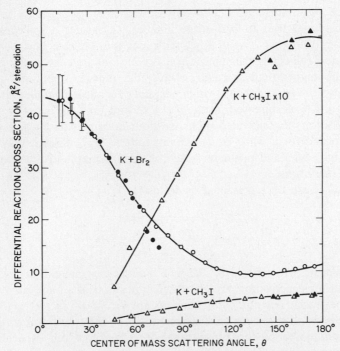

FIG. 9.26 Comparison of angular distributions (in the center-of-mass system) of alkali halide products from the $K + CH_3I$ and $K + Br_2$ reactions, as derived in the fixed-velocity approximation.

Reaction (A) is a typical *stripping reaction*. The maximum product intensity is at 0°, or in the same direction as the relative velocity of the two reactant molecules. On the other hand, in reaction (B), the maximum intensity of the product is at 180°. This is a typical *rebound reaction*. The product KI bounces back in the opposite direction to that of the incident K atom.

The total reaction cross section σ_r can readily be calculated by integrating the differential $d\sigma$ over all angles. For reaction (A), $\sigma_r = 2.10 \text{ nm}^2$, and for (B), $\sigma_r = 0.30 \text{ nm}^2$. The value for (B) is close to σ for a collision of hard spheres, but the value for (A) is considerably higher, suggesting that forces are quite long range in this reaction.

33. Theory of Unimolecular Reactions

From 1918 to 1935, a number of gas reactions were found to be kinetically of the first order and apparently simple unimolecular decompositions. These reactions presented a paradox: the necessary activation energy must evidently come from kinetic energy transferred during collisions, yet the reaction velocity did not depend on collision frequency.

In 1922, F. A. Lindemann (Cherwell) showed how a collisional mechanism for activation could lead to first-order kinetics.* Consider a molecule A that decomposes according to $A \rightarrow B + C$, with a first-order rate law, $-d[A]/dt = k_{ex}[A]$. In a vessel full of $[A]$, the intermolecular collisions are continually producing molecules with higher than average energy, and, indeed, sometimes molecules with an energy above some critical value necessary for the *activation* that precedes decomposition. Let us suppose that there is a certain time lag between activation and decomposition; the activated molecule does not immediately fall to pieces, but moves around for a while in its activated state. Sometimes it may meet an energy-poor molecule, and in the ensuing collision, it may be robbed of enough energy to be *deactivated*.

The situation can be represented as follows:

$$A + A \underset{k_{-2}}{\overset{k_2}{\rightleftharpoons}} A + A^\star$$

$$A^\star \overset{k_1}{\longrightarrow} B + C$$

Activated molecules are denoted by $A^\star$. The bimolecular rate constant for activation is k_2, and for deactivation, k_{-2}. The decomposition of an activated molecule is a true unimolecular reaction with rate constant k_1.

The process called *activation* consists essentially in the transfer of translational kinetic energy into energy stored in internal degrees of freedom, especially vibrational degrees of freedom. The mere fact that a molecule is moving rapidly, i.e., has a high translational kinetic energy, does not make it unstable. To cause reaction, the energy must get into the chemical bonds, where vibrations of high ampli-

*Trans. Faraday Soc., 17, 598 (1922). An essentially correct interpretation had previously been given by I. Langmuir, *J. Am. Chem. Soc.*, 42, 2190 (1920).

tude will lead to ruptures and rearrangements. The transfer of energy from translation to vibration can occur only in collisions with other molecules or with the wall. The situation is like that of two rapidly moving cars; their kinetic energies will not wreck them unless they happen to collide and the kinetic energy of the whole is transformed into internal energy of the parts.

The point of the Lindemann theory is that there is a lag between the activation of the internal degrees of freedom and the subsequent decomposition. The reason is that a polyatomic molecule can take up collisional energy into a number of its $3N - 6$ vibrational degrees of freedom, and then some time may elapse before enough energy flows into the one bond that breaks. The differential equations for the Lindemann mechanism are

$$\frac{d[A^\star]}{dt} = k_2[A]^2 - k_{-2}[A^\star][A] - k_1[A^\star]$$

$$\frac{-d[A]}{dt} = k_2[A]^2 - k_{-2}[A^\star][A]$$

$$\frac{d[B]}{dt} = k_1[A^\star]$$

This set of equations is not soluble in closed form and one has recourse to the *steady state approximation*. After the reaction has been under way for a short time, the rate of formation of activated molecules can be set equal to their rate of disappearance, so that the net rate of change in $[A^\star]$ is zero, $d[A^\star]/dt = 0$. In justification of this assumption, it can be said that there are not many activated molecules present, so that the value of $[A^\star]$ is necessarily small, and its rate of change will also be small, and can usually be set equal to zero without serious error.

With $d[A^\star]/dt = 0$, the first of the preceding equations gives for the steady state concentration of $A^\star$

$$[A^\star] = \frac{k_2[A]^2}{k_{-2}[A] + k_1}$$

The reaction velocity is the rate at which $A^\star$ decomposes into B and C, or

$$\frac{d[B]}{dt} = k_1[A^\star] = \frac{k_1 k_2[A]^2}{k_{-2}[A] + k_1}$$

If the rate of decomposition of $A^\star$ is much greater than its rate of deactivation, $k_1 \gg k_{-2}[A]$, and the net rate reduces to

$$\frac{d[B]}{dt} = k_2[A]^2$$

This expression is the ordinary second-order law.

On the other hand, if the rate of deactivation of $A^\star$ is much greater than its rate of decomposition, $k_{-2}[A] \gg k_1$, and the overall rate becomes

$$\frac{d[B]}{dt} = \frac{k_1 k_2}{k_{-2}}[A] = k[A]$$

It is evident that first-order kinetics can be obtained with a collisional mechanism for activation. This will be the result whenever the activated molecule has so long a lifetime that it is usually deactivated by collision before it has a chance to break into fragments.

As the pressure in the reacting system is decreased, the rate of deactivation, $k_{-2}[A^*][A]$, must likewise decrease, and at low enough pressures the condition for first-order kinetics must always fail when $k_{-2}[A]$ is no longer much greater than k_1. The observed first-order rate constant should therefore decline at low pressures.

If the data are expressed in terms of an experimental rate constant k_{ex}, from

$$-\frac{d[A]}{dt} = k_{ex}[A]$$

it is evident that

$$k_{ex} = \frac{k_1 k_2 [A]}{k_{-2}[A] + k_1} \tag{9.71}$$

The limiting first-order rate constant at high pressures is

$$k_\infty = \frac{k_1 k_2}{k_{-2}}$$

It is convenient to rewrite (9.71) as

$$\frac{1}{k_{ex}} = \frac{1}{k_\infty} + \frac{1}{k_2[A]} \tag{9.72}$$

The theory would thus predict a linear relation between $1/k_{ex}$ and $1/[A]$ ([A] being proportional the pressure of A).

In Fig. 9.27 are plotted the rate constants obtained by Trotman-Dickenson and his coworkers for the first-order thermal isomerization of cyclopropane at various pressures.

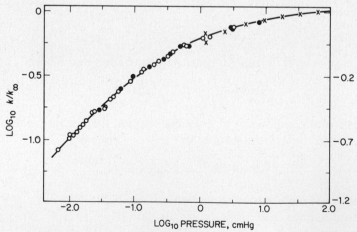

FIG. 9.27 The dependence of the rate of isomerization of cyclo-propane on pressure. Upper curve: ○, runs to about 30% conversion; ●, runs to about 70% conversion; x, results of Chambers & Kistiakowsky, [H. O. Pritchard, R. G. Sowden, and A. F. Trotman-Dickenson, *Proc. Roy. Soc., A., 217,* 563 (1953).]

$$
\begin{array}{c}
CH_2 \\
\diagup \diagdown \\
CH_2\!-\!CH_2
\end{array}
\longrightarrow CH_3\!-\!CH\!=\!CH_2
$$

The marked fall-off in k_{ex} as the pressure is lowered confirms the theoretical prediction, in a qualitative way, but the quantitative prediction of (9.72) has failed. This
example is typical of all the cases that have been studied.

If the decrease of k_{ex} at low pressures is merely the result of a lowered probability of deactivation of A^*, it should be possible to restore the initial rate by adding a sufficient pressure of a completely inert gas. This effect of inert gas has been
confirmed in a number of cases. Table 9.7 summarizes the relative efficiencies of
various gases in restoring the rate constant of the cyclopropane isomerization to
its high pressure value.

The Lindemann theory of unimolecular reactions is plausible and provides
the best explanation for many experiments. Table 9.8 lists some reactions now

TABLE 9.7
Relative Efficiencies of Added Gases in Maintaining the Rate of Isomerization of Cyclopropane

Molecule	Efficiency (pressure/pressure)	Collision diameter (nm)	Efficiency (collision/collision)
Cyclopropane	1.000	0.50	1.000
Helium	0.060 ± 0.011	0.22	0.048
Argon	0.053 ± 0.007	0.36	0.070
Hydrogen	0.24 ± 0.03	0.27	0.12
Nitrogen	0.060 ± 0.003	0.38	0.070
Carbon monoxide	0.072 ± 0.009	0.38	0.084
Methane	0.27 ± 0.03	0.41	0.24
Water	0.79 ± 0.11	0.40	0.74
Propylene	~ 1.0	0.50	~ 1.0
Benzotrifluoride	1.09 ± 0.13	0.85	0.75
Toluene	1.59 ± 0.13	0.80	1.10
Mesitylene	1.43 ± 0.26	0.90	0.89

TABLE 9.8
Unimolecular Gas Phase Decompositions[a]

Reactant	Products	$\log A$ (s^{-1})	E_a (kJ·mol^{-1})
$CH_3 \cdot CH_2Cl$	$C_2H_4 + HCl$	14.6	254
$CCl_3 \cdot CH_3$	$CCl_2\!=\!CH_2 + HCl$	12.5	200
t-Butyl bromide	Isobutene + HBr	14.0	177
t-Butyl alcohol	Isobutene + H_2O	11.5	228
$ClCOOC_2H_5$	$C_2H_5Cl + CO_2$	10.7	123
$ClCOOCCl_3$	$COCl_2$	13.15	174
Cyclobutane	C_2H_4	15.6	262
Perfluorocyclobutane	C_2F_4	15.95	310
N_2O_4	NO_2	16	54

[a]From S. W. Benson, *The Foundations of Chemical Kinetics* (New York: McGraw-Hill Book Company, 1960).

believed to be true unimolecular processes. Many reactions once thought to be simple unimolecular decompositions have been shown to proceed via complex chain mechanisms, which often yield deceptively simple rate laws. This aspect of gas kinetics will be considered a little later in the chapter.

When the first-order rate constant begins to fall at lower pressures, the rate of formation of activated molecules is no longer much greater than their rate of decomposition, and, in fact, the overall rate is beginning to be determined by the rate of supply of activated molecules. According to simple collision theory, therefore, the rate at this point should be about $Z_{11}e^{-E/RT}$. When this prediction was compared with an experiment in a typical case, such as the cyclopropane isomerization, it was found that the reaction was going about 5×10^5 times faster than was permissible by simple collision theory.

A solution to this contradiction was given by Hinshelwood.* The $e^{-E/RT}$ term used to calculate the fraction of activated molecules is based on the condition that the critical energy is acquired in two translational degrees of freedom only. If energy in vibrational degrees of freedom also can be transferred in collisions, the probability that a molecule gets the necessary E is much enhanced. Instead of a simple $e^{-E/RT}$ term, the chance is now†

$$P_E = \frac{e^{-E/RT}(E/RT)^{s-1}}{(s-1)!} = f_s\, e^{-E/RT} \tag{9.73}$$

Here, s is the number of vibrations in which the energy E can be acquired.

The rate of activation may now be increased by a large factor f_s. For the cyclopropane case, with $E = 61.8$ kJ, $T = 764$ K, the factor $f_s = 5 \times 10^5$ when $s = 7$. Since the molecule contains nine atoms, there are altogether $3N - 6 = 21$ vibrations. The theory of Hinshelwood would include a third of these in the activation process. It has always been possible to find a value of s less than $3N - 6$ that fits the observed activation rate‡ by means of (9.73).

Since the Lindemann–Hinshelwood theory failed to explain quantitatively the fall-off in k_{ex} with pressure, the underlying model evidently required some

---

*C. N. Hinshelwood, *Proc. Roy. Soc. London, Ser. A, 113*, 230 (1926). See also G. N. Lewis and D. F. Smith, *J. Am. Chem. Soc., 47*, 1508 (1925).

†This is a good approximate formula when $E \gg RT$. A derivation is given by E. A. Moelwyn–Hughes, *Physical Chemistry*, 2nd ed. (New York: Pergamon Press, 1961), p. 42.

‡R. A. Ogg, *J. Chem. Phys., 21*, 2079 (1953), has proposed an interesting mechanism for the N_2O_5 case, which previously appeared exceptional.

$$N_2O_5 \underset{2}{\overset{1}{\rightleftharpoons}} NO_2 + NO_3$$

$$NO_2 + NO_3 \overset{3}{\longrightarrow} NO + O_2 + NO_2$$

$$NO + NO_3 \overset{4}{\longrightarrow} 2NO_2$$

The steady-state treatment applied to $[NO_3]$ and $[NO]$ gives

$$\frac{-d[N_2O_5]}{dt} = k_0[N_2O_5] = \frac{2k_3k_1}{k_2 + 2k_3}[N_2O_5]$$

so that the first-order rate constant k_0 is composite.

modification. About 1928, Kassel, Rice, and Ramsperger pinpointed the basic defect in the model. It had been assumed that the lifetimes of all the energized molecules $A^{\star}$ would be the same, independent of the amount of internal energy that they had acquired by collisional activation. In the Rice-Ramsperger-Kassel model, the more excess energy $A^{\star}$ obtains, the greater is its probability of decomposing in a given time interval, and hence the less its chance of being deactivated before it can decompose. The probability of decomposition was shown to be proportional to $[1 - (E_a/E)]^{s-1}$, where E_a is the critical minimum energy for any decomposition. The theoretical equations obtained from this model, and from subsequent refinements of it, are in good agreement with experiments, as shown in Fig. 9.27, for example.

In the high pressure limit, the transition-state formulation can be applied directly to unimolecular reactions, since there will be a Maxwell–Boltzmann equilibrium between reactant molecules A and activated complexes $A^{\ddagger}$. In the formulation due to Marcus,

$$A + A \rightleftharpoons A^{\star} + A$$
$$A^{\star} \rightleftharpoons A^{\ddagger} \longrightarrow \text{products}$$

The energized molecule $A^{\star}$ must undergo a considerable change in passing to the transition state $A^{\ddagger}$ from which it decomposes. From the point of view of transition-state theory, however, such intermediate processes are irrelevant, and one can write

$$\frac{[A^{\ddagger}]}{[A]} = K^{\ddagger}$$

From (9.64), this yields

$$k_1 = \frac{kT}{h} K^{\ddagger} = \frac{kT}{h} \cdot \frac{z^{\ddagger}}{z} \tag{9.74}$$

The problem now is the always difficult one of how to obtain information that will permit a reasonable calculation of $z^{\ddagger}$.

34. Chain Reactions:
Formation of Hydrogen Bromide

After Bodenstein had completed his study of the hydrogen–iodine reaction, he turned to $H_2 + Br_2 \longrightarrow 2HBr$, probably expecting to find another example of second-order kinetics. The results* were surprisingly different, for the reaction rate was found to fit the rather complicated expression,

$$\frac{d[HBr]}{dt} = \frac{k_a[H_2][Br_2]^{1/2}}{k_b + [HBr]/[Br_2]}$$

where k_a and k_b are constants. Thus, the rate is inhibited by the product, HBr. In the initial stages of the combination, $[HBr]/[Br_2] \ll k_b$, so that $d[HBr]/dt = k'[H_2][Br_2]^{1/2}$, with an overall order of $\frac{3}{2}$.

There was no interpretation of this curious rate law for 13 years. Then the problem was solved independently and almost simultaneously by Christiansen,

*M. Bodenstein and S. C. Lind, *Z. Physik Chem.*, 57, 168 (1906).

Herzfeld, and Polanyi. They proposed a chain of reactions with the following steps:

$$\text{Chain initiation} \quad (1)\ \text{Br}_2 \xrightarrow{k_1} 2\text{Br}$$

$$\text{Chain propagation} \quad (2)\ \text{Br} + \text{H}_2 \xrightarrow{k_2} \text{HBr} + \text{H}$$

$$(3)\ \text{H} + \text{Br}_2 \xrightarrow{k_3} \text{HBr} + \text{Br}$$

$$\text{Chain inhibition} \quad (4)\ \text{H} + \text{HBr} \xrightarrow{k_4} \text{H}_2 + \text{Br}$$

$$\text{Chain breaking} \quad (5)\ 2\text{Br} \xrightarrow{k_5} \text{Br}_2$$

The reaction is initiated by bromine atoms from the thermal dissociation $\text{Br}_2 \rightarrow 2\text{Br}$. The chain propagating steps (2) and (3) form two molecules of HBr and regenerate the bromine atom, ready for another cycle. Thus, very few bromine atoms are needed to cause an extensive reaction. Step (4) is introduced to account for the observed inhibition by HBr; since this inhibition is proportional to the ratio $[\text{HBr}]/[\text{Br}_2]$, it is evident that HBr and Br_2 compete, so that the atom being removed must be H rather than Br.

To derive the kinetic law from the chain mechanism, the stationary state treatment is applied to the reactive atoms, which must be present in low concentrations.

$$\frac{d[\text{Br}]}{dt} = 0 = 2k_1[\text{Br}_2] - k_2[\text{Br}][\text{H}_2] + k_3[\text{H}][\text{Br}_2] + k_4[\text{H}][\text{HBr}] - 2k_5[\text{Br}]^2$$

$$\frac{d[\text{H}]}{dt} = 0 = k_2[\text{Br}][\text{H}_2] - k_3[\text{H}][\text{Br}_2] - k_4[\text{H}][\text{HBr}]$$

These two simultaneous equations are solved for the steady state concentrations of the atoms, giving

$$[\text{Br}] = \left[\frac{k_1}{k_5}[\text{Br}_2]\right]^{1/2}, \qquad [\text{H}] = k_2 \frac{(k_1/k_5)^{1/2}[\text{H}_2][\text{Br}_2]^{1/2}}{k_3[\text{Br}_2] + k_4[\text{HBr}]}$$

The rate of formation of the product, HBr, is

$$\frac{d[\text{HBr}]}{dt} = k_2[\text{Br}][\text{H}_2] + k_3[\text{H}][\text{Br}_2] - k_4[\text{H}][\text{HBr}]$$

Introducing the values for [H] and [Br] and rearranging, we find

$$\frac{d[\text{HBr}]}{dt} = 2\frac{k_3 k_2 k_4^{-1} k_1^{1/2} k_5^{-1/2}[\text{H}_2][\text{Br}_2]^{1/2}}{k_3 k_4^{-1} + [\text{HBr}][\text{Br}_2]^{-1}}$$

This agrees exactly with the empirical expression, but now the constants k_a and k_b are interpreted as composites of constants for step reactions in the chain. Note that $k_1/k_5 = K$ is the equilibrium constant for the dissociation $\text{Br}_2 \rightleftharpoons 2\text{Br}$.

The $\text{H}_2 + \text{Cl}_2 \rightarrow 2\text{HCl}$ reaction is more difficult to study. It is exceedingly sensitive to light, which starts a chain reaction by photodissociation of chlorine, $\text{Cl}_2 + h\nu \rightarrow 2\text{Cl}$. The subsequent reaction steps are similar to those with Br_2. The thermal reaction proceeds similarly, but it is complicated by wall effects and traces of moisture and oxygen.

For many years after the pioneer kinetic studies of Bodenstein, the reaction $\text{H}_2 + \text{I}_2 \rightarrow 2\text{HI}$ and the reverse reaction $2\text{HI} \rightarrow \text{H}_2 + \text{I}_2$ were considered to be

purely bimolecular processes. In 1934, however, a curious anomaly was noticed.*
If the reaction was allowed to go to equilibrium and pure para-H_2 was added to
the equilibrium mixture, the rate of the interconversion, p-$H_2 \rightleftharpoons$ o-H_2, was much
faster than could be explained by the molecular reactions

$$H_2 + I_2 \rightleftharpoons 2HI \tag{A}$$

The suggestion was made that atomic iodine was being formed to some extent and
acting to catalyze the para–ortho H_2 conversion. This observation was not sufficiently dramatic to influence the textbook citations of $H_2 + I_2$ as a purely bimolecular process. But in 1955, Benson and Srinivasan† discovered that the activation
energies of both the forward and reverse reactions in (A) increased appreciably
with temperature. It looked as if some new mechanism with higher overall activation energy was occurring at higher temperatures. They suggested a chain mechanism like that in $H_2 + Br_2$:

$$I_2 \rightleftharpoons 2I$$
$$I + H_2 \longrightarrow HI + H$$
$$H + I_2 \longrightarrow HI + I$$
$$H + HI \longrightarrow H_2 + I$$
$$I + HI \longrightarrow I_2 + H$$

In 1967, J. H. Sullivan‡ suggested that the nonchain part of the $H_2 + I_2$
reaction also occurred with iodine atoms as intermediates

$$I_2 \overset{K}{\rightleftharpoons} 2I \tag{A}$$
$$I + H_2 \longrightarrow (IH_2) \tag{B}$$
$$I + (IH_2) \longrightarrow 2HI \tag{C}$$

The last two steps might be combined to give a trimolecular reaction

$$H_2 + 2I \overset{k_3}{\longrightarrow} 2HI \tag{D}$$

From reaction (A), $K = [I]^2/[I_2]$, so that

$$[I] = K^{1/2} [I_2]^{1/2}$$

From (D),

$$\frac{d[HI]}{dt} = 2k_3[H_2] [I]^2$$

$$= 2k_3 K[H_2] [I_2] = k_{ex}[H_2] [I_2]$$

Thus, the experimental second-order rate constant would be a composite

$$k_{ex} = 2k_3 K$$

There is some question now as to whether any simple four-center reactions
proceed by a bimolecular mechanism. Noyes has obtained evidence§ that certain

*E. J. Rosenbaum and T. R. Hogness, *J. Chem. Phys.*, 2, 267 (1934).
†S. W. Benson and R. Srinivasan, *J. Chem. Phys.*, 23, 200 (1955).
‡J. H. Sullivan, *J. Chem. Phys.*, 46, 73 (1967).
§P. R. Walton and R. M. Noyes, *J. Am. Chem. Soc.*, 88, 4324 (1966).

interhalogen reactions in solution are bimolecular. For example,

$$2IBr \rightleftharpoons I_2 + Br_2$$

In the gas phase,

$$HCl + Br_2 \longrightarrow HBr + BrCl$$

may be bimolecular.*

35. Free-Radical Chains

In 1900, Moses Gomberg discovered that hexaphenylethane dissociates in solution into two triphenylmethyl radicals,

$$(C_6H_5)_3C\text{—}C(C_6H_5)_3 \longrightarrow 2(C_6H_5)_3C$$

Such compounds with trivalent carbon atoms were at first believed to be chemical anomalies capable of occurring only in complex molecules.

One of the first suggestions that simple radicals might act as chain carriers in chemical reactions was made in 1925 by Hugh S. Taylor.† If a mixture of hydrogen and mercury vapor is irradiated with ultraviolet light of $\lambda = 253.7$ nm, the mercury atoms are raised to a higher electronic state. They then react with hydrogen molecules, producing hydrogen atoms:

$$Hg(^1S_0) + h\nu(253.7 \text{ nm}) \longrightarrow Hg(^3P_1)$$
$$Hg(^3P_1) + H_2 \longrightarrow HgH + H$$

If ethylene is added to the reaction mixture, there is a rapid reaction to form ethane, butane, and some higher polymeric hydrocarbons. Taylor suggested that the hydrogen atom combined with ethylene, forming a free ethyl radical, C_2H_5, which then started a chain reaction.

$$H + C_2H_4 \longrightarrow C_2H_5$$
$$C_2H_5 + H_2 \longrightarrow C_2H_6 + H, \quad \text{etc.}$$

In 1929, F. Paneth and W. Hofeditz obtained good evidence that aliphatic free radicals occur in the decomposition of molecules of the metallic alkyls, such as mercury dimethyl and lead tetraethyl. The experiment of Paneth‡ is represented in Fig. 9.28. A current of pure nitrogen at 2-mm pressure was saturated with lead-tetramethyl vapor by passing over the liquid in A. The vapors were next passed through a tube heated at B to about 450°C. Decomposition of $Pb(CH_3)_4$ deposited a lead mirror in the tube at the heated section. When the vapors from the decomposition, after flowing down the tube a distance of 10 to 30 cm, passed over a previously deposited mirror of lead at 100°C, this mirror was gradually removed. It appears, therefore, that the metal alkyl first breaks into free methyl radicals, $Pb(CH_3)_4 \rightarrow Pb + 4CH_3$. These are carried along in the stream of nitrogen for a considerable distance before they recombine to stable hydrocarbons.

*R. M. Noyes, *J. Am. Chem. Soc.*, **88**, 4318 (1966).
†*Trans. Faraday Soc.*, **21**, 560 (1925).
‡*Berichte*, **62**, 1335 (1929).

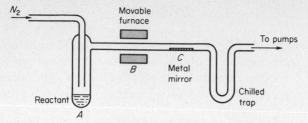

FIG. 9.28 Paneth experiment. The lifetime of the radical can be calculated from position of mirror and rate of its removal.

They remove metallic mirrors by reacting with the metal to form volatile alkyls. Thus, if the mirror is zinc, $Zn(CH_3)_2$ can be recovered; if it is antimony, $Sb(CH_3)_3$ is recovered as the mirror is removed. From 1932 to 1934, F. O. Rice* and co-workers showed that the thermal decomposition by the Paneth technique of many organic compounds such as $(CH_3)_2CO$, C_2H_6, and other hydrocarbons, gave products that would remove metal mirrors. They therefore concluded that free radicals were formed in the primary steps of the decomposition of all these molecules.

In 1935, Rice and Herzfeld† showed how free-radical chain mechanisms could be devised that would lead to a simple overall kinetics. The products from the decompositions were in good agreement with the proposed radical mechanisms. A typical example is the following possible mechanism for the decomposition of acetaldehyde, $CH_3CHO \rightarrow CH_4 + CO$:

$$(1)\ CH_3CHO \xrightarrow{\ k_1\ } CH_3 + CHO$$

$$(2)\ CH_3CHO + CH_3 \xrightarrow{\ k_2\ } CH_4 + CO + CH_3$$

$$(3)\ 2CH_3 \xrightarrow{\ k_3\ } C_2H_6$$

One primary split into methyl radicals can result in the decomposition of many CH_3CHO molecules, since the chain carrier, CH_3, is regenerated in step (2). The steady state treatment of the CH_3 concentration yields

$$\frac{d[CH_3]}{dt} = 0 = k_1[CH_3CHO] - 2k_3[CH_3]^2$$

so that

$$[CH_3] = \left(\frac{k_1}{2k_3}\right)^{1/2}[CH_3CHO]^{1/2}$$

The reaction rate based on methane formation is then

$$\frac{d[CH_4]}{dt} = k_2[CH_3][CH_3CHO] = k_2\left(\frac{k_1}{2k_3}\right)^{1/2}[CH_3CHO]^{3/2}$$

---

*F. O. Rice, *J. Am. Chem. Soc.*, *53*, 1959 (1931); F. O. Rice, W. R. Johnston, and B. L. Evering, *J. Am. Chem. Soc.*, *54*, 3529 (1932); F. O. Rice and A. L. Glasebrook, *J. Am. Chem. Soc.*, *56*, 2381 (1934).

†*J. Am. Chem. Soc.*, *56*, 284 (1934).

The free-radical scheme predicts an order of $\frac{3}{2}$, in reasonable accord with experiments.

A primary split into free radicals usually requires a high activation energy, whereas E for an elementary decomposition into the final products may be considerably lower. A rapid reaction is possible despite the high initial E because of the long chain of steps of low activation energy following the formation of the radicals. Sometimes the scales may be delicately balanced between the two mechanisms, and in certain temperature ranges the radical mechanism and the intramolecular-decomposition mechanism simultaneously occur to appreciable extents. Free-radical chains play an important role in the pyrolyses of hydrocarbons, aldehydes, ethers, ketones, metal alkyls, and many other organic compounds.

Sometimes a good test for the radical mechanism can be made by studying a mixture of isotopically substituted species. Suppose, for example, we heat a mixture of CH_3CHO and CD_3CDO. If the *intramolecular* mechanism is followed,

$$CH_3CHO \longrightarrow CH_4 + CO$$
$$CD_3CDO \longrightarrow CD_4 + CO$$

We should obtain a mixture of CH_4 and CD_4 in the products. If the chain mechanism is followed, we should obtain also CH_3D and CD_3H from the steps,

$$CH_3 + CD_3CDO \longrightarrow CH_3D + CO + CD_3$$
$$CD_3 + CH_3CHO \longrightarrow CD_3H + CO + CH_3$$

Actually all the isotopically mixed methanes are found, so that the radical mechanism is indicated.*

36. Branching Chains—Explosive Reactions

The theory of chain reactions gives a good interpretation of many of the peculiar features of explosions.

The formation of H_2O from H_2 and O_2 when the mixture is heated or reaction is otherwise initiated has been the subject of hundreds of papers, and is still a problem for active research. This reaction displays the upper and lower pressure limits characteristic of many explosions, as shown in Fig. 9.29. If the pressure of a 2:1 mixture of H_2 and O_2 is kept below the lower line on the diagram, the thermal reaction proceeds slowly. At a temperature of 500°C, this lower pressure limit is shown at 1.5 mm, but its value depends on the size of the reaction vessel. If the pressure is raised above this value, the mixture explodes. As the pressure is raised still further, there is a limit of 50 mm at 500°C above which there is no longer an explosion, but once again a comparatively slow reaction. This second explosion limit is strongly temperature dependent, but it does not vary with vessel size.

If an exothermic reaction is carried out in a confined space, the heat evolved often cannot be dissipated. The temperature therefore increases, so that the rate of reaction increases and there is a corresponding rise in the rate of heat produc-

*L. A. Wall and W. J. Moore, *J. Phys. Chem.*, **55**, 965 (1951).

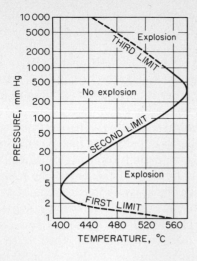

FIG. 9.29 Explosion limits of a stoichio-metric hydrogen–oxygen mix-ture in a spherical KCl-coated vessel of 7.4 cm diameter. [After B. Lewis and G. v. Elbe, *Combustion, Flames and Explosions of Gases* (New York: Academic Press, 1953), p. 29.]

tion. The reaction velocity increases practically without bound and the result is called a *thermal explosion*. The third explosion limit in Fig. 9.29 arises in this way.

In other systems, the thermal effects are less decisive, and the explosion is due to a different cause, namely, the occurrence of *branched chains* in the reaction mechanism. In the chain reactions discussed so far, each propagating sequence led to the formation of a molecule of product and the regeneration of *one* chain carrier. In a branched chain, more than one carrier is produced in each reaction sequence. In the following schematic chain reaction, R represents the reactive chain carrier:

$$A \xrightarrow{k_1} R$$

$$R + A \xrightarrow{k_2} P + \alpha R$$

$$R \xrightarrow{k_3} \text{destruction}$$

In this scheme, P is the final product and α is the number of chain carriers formed from one initial R in the chain propagating step. The destruction of chain carriers can occur in two ways. They may diffuse to the walls of the reaction vessel where they become adsorbed and combine in a surface reaction, or they may be destroyed in the gas phase. If the preceding scheme is to yield a steady reaction rate, $d[R]/dt$ must be zero.

$$\frac{d[R]}{dt} = 0 = k_1[A]^n - k_2[R][A] + \alpha k_2[R][A] - k_3[R]$$

or

$$[R] = \frac{k_1[A]^n}{k_2[A](1 - \alpha) + k_3}$$

The probability of destruction, proportional to k_3, can be written as the sum of two terms, k_g for the gas phase reaction and k_w for the wall reaction, so that,

$$[R] = \frac{k_1[A]^n}{k_2[A](1 - \alpha) + k_g + k_w} \tag{9.75}$$

In all the cases previously treated, α has been unity, so that $(1 - \alpha) = 0$, leaving a radical concentration proportional to the rate of formation divided by the rate of destruction.

If α is greater than unity, chain branching occurs. In particular, a critical situation arises when α becomes so large that $k_2[A](\alpha - 1) = k_g + k_w$, for then the denominator becomes zero and the carrier concentration $[R]$ goes to infinity. The reaction rate is proportional to the concentration of the carrier, so that it also increases without bound at this critical condition. The steady-state treatment fails completely, and the reaction goes so rapidly that there is an explosion.

It is now clear why there can be both upper and lower explosion limits. The destruction rate at the wall k_w depends on diffusion of carriers to the wall, which is more rapid at low pressures. Thus, when the pressure falls to a point at which chain carriers are being destroyed at the wall as rapidly as they are being produced, an explosive reaction is no longer possible. This lower pressure limit therefore depends on the size and material of the reaction vessel: in a larger vessel, fewer radicals reach the wall. The upper explosion limit is reached when destructive collisions in the gas phase outweigh the chain branching.

For the hydrogen–oxygen reaction, a chain scheme somewhat like the following appears to be reasonable:

$$(1) \quad H_2 + O_2 \longrightarrow HO_2 + H$$
$$(2) \quad H_2 + HO_2 \longrightarrow OH + H_2O$$
$$(3) \quad OH + H_2 \longrightarrow H_2O + H$$
$$(4) \quad O_2 + H \longrightarrow OH + O$$
$$(5) \quad H_2 + O \longrightarrow OH + H$$
$$(6) \quad HO_2 + wall \longrightarrow removal$$
$$(7) \quad H + wall \longrightarrow removal$$
$$(8) \quad OH + wall \longrightarrow removal$$

The hydroxyl radical, OH, has been spectroscopically detected in the reaction mixture. Chain branching occurs in steps (4) and (5), since OH, O, and H are all active chain propagators.*

37. Trimolecular Reactions

The necessity of a third body to carry off the excess energy in atom recombinations is well established. Studies have shown that such reactions as $M + H + H \longrightarrow H_2 + M$ and $M + Cl + Cl \longrightarrow Cl_2 + M$ are of the third order. The factors that determine the relative efficiencies of different third bodies M in promoting recombination are of great interest in connection with the problem of energy transfer between molecules. In a study of the recombination of iodine and bromine atoms produced by thermal decomposition of the molecules, Rabinowitsch†

*R. C. Anderson, "Combustion and Flame," *J. Chem. Ed.*, **44**, 248 (1967) gives further references to this reaction in a general review of the fascinating kinetic problems of reactions in flames.

†*Trans. Faraday Soc.*, **33**, 283 (1937).

measured the rate constants for the reaction,

$$X + X + M \longrightarrow X_2 + M: \frac{-d[X]}{dt} = k_3[X]^2[M]$$

He found the following relative values of k_3:

$M =$	He	Ar	H_2	N_2	O_2	CH_4	CO_2	C_6H_6
$X = Br$	0.76	1.3	2.2	2.5	3.2	3.6	5.4	
$X = I$	1.8	3.8	4.0	6.6	10.5	12	18	100

It is difficult to calculate the number of *triple collisions* that occur in a gas, but a fairly good estimate should be obtained by setting the ratio of binary collisions Z_{12} to triple collisions Z_{121} equal to the ratio of mean free path to molecular diameter, λ/d. Since d is of the order of 10^{-8} cm, and λ at 1 atm pressure is about 10^{-5} cm for most gases, the ratio is about 1000. Rabinowitsch found that this ratio, Z_{12}/Z_{121}, closely paralleled the rate constants of the recombination reactions of halogen atoms. In this case at least, the efficiency of the third body seems to depend mainly on the number of triple collisions it undergoes.

Besides three-body recombinations, the only known gas reactions that may be trimolecular are the third-order reactions of nitric oxide mentioned on p. 337. Trautz showed that these may actually consist of two bimolecular reactions; for example,

$$(1) \qquad 2NO \rightleftharpoons N_2O_2$$

$$(2) \qquad N_2O_2 + O_2 \xrightarrow{k_2} 2NO_2$$

If equilibrium is set up in (1), $K = [N_2O_2]/[NO]^2$. Then, from (2)

$$\frac{d[NO_2]}{dt} = k_2[N_2O_2][O_2] = k_2K[NO]^2[O_2]$$

The observed third-order constant is $k_3 = k_2K$. How could you tell whether the reaction proceeded through formation of an N_2O_2 intermediate or through three-body collision?

38. Reactions in Solution

We cannot make a complete theoretical analysis of the rates of reactions in liquid solutions, although many special aspects of such reactions are quite well understood. It might seem that collision theory should hardly be applicable at all, since there is no unequivocal way of calculating collision frequencies. It turns out, however, that even the gas-kinetic expressions sometimes give reasonable values for the frequency factors.

First-order reactions, such as the decomposition of N_2O_5, Cl_2O, or CH_2I_2, and the isomerization of pinene, proceed at about the same rate in gas phase and in solution. It appears, therefore, that the rate is the same whether a molecule becomes activated by collision with solvent molecules or by collisions in the gas phase with others of its own kind. It is more remarkable that many second-order, presumably bimolecular, reactions have rates close to those predicted from the

TABLE 9.9
Examples of Reactions in Solution

Reaction	Solvent	E_a (kJ·mol^{-1})	A (Eq. 9. 41) (dm^3·mol^{-1}·s^{-1})	A_{calc}/A_{obs}
$C_2H_5ONa + CH_3I$	C_2H_5OH	81.6	2.42×10^{11}	0.8
$C_2H_5ONa + C_6H_5CH_2I$	C_2H_5OH	83.3	0.15×10^{11}	14.5
$NH_4CNO \longrightarrow (NH_2)_2CO$	H_2O	97.1	42.7×10^{11}	0.1
$CH_2ClCOOH + OH^-$	H_2O	108.4	4.55×10^{11}	0.6
$C_2H_5Br + OH^-$	C_2H_5OH	89.5	4.30×10^{11}	0.9
$(C_2H_5)_3N + C_2H_5Br$	C_6H_6	46.9	2.68×10^2	1.9×10^9
$CS(NH_2)_2 + CH_3I$	$(CH_3)_2CO$	56.9	3.04×10^6	1.2×10^5
$C_{12}H_{22}O_{11} + H_2O \longrightarrow 2C_6H_{12}O_6$ (sucrose)	$H_2O(H^+)$	107.9	1.5×10^{15}	1.9×10^{-4}

gas-kinetic collision theory. Some examples are shown in the last column of Table 9.9.

The explanation of such an agreement seems to be the following. Any given reactant solute molecule will have to diffuse for some distance through the solution before it meets another reactant molecule. Thus, the number of such encounters will be lower than in the gas phase. Having once met, however, the two reactant molecules will remain close to each other for a considerable time, being surrounded by a *cage* of solvent molecules. Thus repeated collisions may occur between the same pair of reactant molecules. The net result is that the effective collision number is not much different from that in the gas phase.

There are other cases in which the calculated constant deviates by factors ranging from 10^9 to 10^{-9}. A high frequency factor corresponds to a large positive $\Delta S^{\ddagger}$, and a low frequency factor to a negative $\Delta S^{\ddagger}$. The remarks on the significance of $\Delta S^{\ddagger}$ in gas reactions apply equally well here. We expect association reactions to have low frequency factors owing to the decrease in entropy when the activated complex is formed. An example is the Menschutkin reaction, combination of an alkyl halide with a tertiary amine:

$$(C_2H_5)_3N + C_2H_5Br \longrightarrow (C_2H_5)_4NBr$$

Such reactions have values of $\Delta S^{\ddagger}$ from -140 to -200 J·K^{-1}·mol^{-1}, usually nearly equal to the $\Delta S^{\ominus}$ for the complete reaction.

In a gas, an upper limit for the rate of a bimolecular reaction is set by the collision frequency. In a liquid, an upper limit would be set by the frequency of *first encounters* between reactant molecules moving in random Brownian motion through the solution. In 1917, Smoluchowski* treated a similar problem in his theory of the growth of a colloidal particle by the accretion of particles that diffused toward it and became incorporated at its surface. Debye† applied this theory to reactions in solution and extended it to the situation in which there was a definite intermolecular potential energy $U(r)$ between the molecules.

*M. v. Smoluchowski, *Z. Physik Chem.*, *92*, 129 (1917).
†P. Debye, *Trans. Electrochem. Soc.*, *82*, 265 (1942).

Consider molecules A that are moving through the solution in random Brownian fashion and encountering stationary molecules B, with which they react on every encounter. The flux of A through unit area is

$$-J_A = -D_A\left[\frac{\partial C_A}{\partial r} + \frac{C_A}{kT}\frac{\partial U}{\partial r}\right] \tag{9.76}$$

This is an extension of Ficks First Law, Eq. (4.65), in which we have added to the diffusive flux a term caused by the force $\partial U/\partial r$ of the potential $U(r)$. The generalized mobility v_A (velocity per unit force) is related to the diffusion coefficient by the Einstein relation, Eq. (10.19),

$$\frac{D_A}{kT} = v_A$$

The flux of A molecules across the surface of a sphere of radius r is

$$I_A = 4\pi r^2 J_A = -4\pi r^2 D_A\left[\frac{\partial C_A}{\partial r} + \frac{C_A}{kT}\frac{\partial U}{\partial r}\right] \tag{9.77}$$

The boundary conditions for (9.77) are that when $r = \infty$, $C_A = C_A^\circ$ and $U = 0$; and when $r = d_{12}$, the collision diameter, $C_A = 0$ and $U = U(d_{12})$. We can thus write* the integral of (9.77) as

$$I_A\int_{d_{12}}^{\infty}\frac{e^{U/kT}}{r^2}\,dr = 4\pi D_A\int_0^{C_A^\circ} d(C_A\,e^{U/kT})$$

$$I_A = -\frac{4\pi D_A\,C_A^\circ}{\displaystyle\int_{d_{12}}^{\infty} e^{U/kT}(dr/r^2)}$$

Since both A and B molecules are mobile, we must replace D_A by $D_A + D_B$. If concentration of B molecules is C_B, the limiting reaction rate becomes $I_A C_B^\circ$, and the second-order rate constant becomes

$$k_2 = \frac{4\pi(D_A + D_B)}{\displaystyle\int_{d_{12}}^{\infty} e^{U/kT}\,(dr/r^2)} \tag{9.78}$$

In the special case for which $U = 0$, $\int_{d_{12}}^{\infty} dr/r^2 = 1/d_{12}$ and (9.78) yields

$$k_2 = 4\pi d_{12}(D_A + D_B)$$

This rate constant would be in units based on molecular concentration, and conversion to units based on molar concentrations would give

$$k_2 = 4\pi d_{12}(D_A + D_B)L \tag{9.79}$$

With typical values of $d_{12} = 5 \times 10^{-8}$ cm, $D = 10^{-5}$ cm$^2\cdot$s^{-1}, (9.79) gives $k_2 = 4 \times 10^9$ mol$^{-1}\cdot$dm$^3\cdot$s^{-1}. For a gas reaction, the collision frequency would correspond to a maximum k_2 of about 10^{11} mol$^{-1}\cdot$dm$^3\cdot$s^{-1}. As already mentioned, owing to the cage effect, the collision frequency in liquid solutions may be about

*Note that

$$d[C_A\,e^{U/kT}] = e^{U/kT}\left[C_A\frac{dU}{kT} + dC_A\right]$$

the same as in gases, but the rate constant in liquids cannot reach this limiting value, since diffusion control will take over from collision control at about $k_2 = 10^9 \text{ mol}^{-1} \cdot \text{dm}^3 \cdot \text{s}^{-1}$.

39. Catalysis

The word catalysis (*Katalyse*) was coined by Berzelius in 1835: "Catalysts are substances which by their mere presence evoke chemical reactions that would not otherwise take place." The Chinese *Tsoo Mei* is more picturesque; it also means the "marriage broker," and so implies a theory of catalytic action. The idea of catalysis extends far back into chemical history. In a fourteenth-century Arabian manuscript, Al Alfani describes the "Xerion, aliksir, noble stone, magisterium, that heals the sick, and turns base metals into gold, without in itself undergoing the least change." Early workers were fascinated by the idea that a mere trace of catalyst sometimes sufficed to produce great changes.

Catalytic action has been likened to that of a coin inserted in a slot machine that yields valuable products and also returns the coin. In a chemical reaction the catalyst enters at one stage and leaves at another. The essence of catalysis is not the entering but the falling out.

Wilhelm Ostwald was the first to emphasize that a catalyst influences the rate of a chemical reaction but has no effect on the position of equilibrium. His famous definition was, "A catalyst is a substance that changes the velocity of a chemical reaction without itself appearing in the end products." Ostwald's proof was based on the First Law of Thermodynamics. Consider a gas reaction that proceeds with a change in volume. The gas is confined in a cylinder fitted with a piston; the catalyst is in a small receptacle within the cylinder, and can be alternately exposed and covered. If the equilibrium position were altered by exposing the catalyst, the pressure would change, the piston would move up and down, and a perpetual-motion machine would be available.

Since a catalyst changes rate but not equilibrium, it *must accelerate the forward and reverse reactions in the same proportion*. Thus, catalysts that accelerate hydrolysis of esters must also accelerate esterification of alcohols; dehydrogenation catalysts such as nickel and platinum are also good hydrogenation catalysts; enzymes, such as pepsin and papain that catalyze splitting of the peptide bond must also catalyze its formation.

A distinction is made between *homogeneous catalysis*, in which the entire reaction occurs in a single phase, and *heterogeneous catalysis*, in which reaction occurs at interfaces between phases. The latter is also called *contact* or *surface catalysis*, and will be discussed in Chapter 11. Most examples of homogeneous catalysis have been studied in liquid solutions. In fact, catalysis in solution is the rule rather than the exception, and it can even be maintained that most reactions in liquid solutions would not proceed at an appreciable rate if catalysts were rigorously excluded. Examples of catalysis by acids and bases are discussed in Section 10.29.

40. Homogeneous Catalysis

An example of homogeneous catalysis in the gas phase is the effect of iodine vapor on the decomposition of aldehydes and ethers. The addition of a few percent of iodine often increases the rate of pyrolysis several hundredfold. The reaction rate follows the equation,

$$\frac{-d[\text{ether}]}{dt} = k_2\, [\text{I}_2]\, [\text{ether}]$$

Dependence of the rate on catalyst concentration is characteristic of homogeneous catalysis.

The catalyst acts by providing a mechanism for the decomposition that has a considerably lower activation energy than the uncatalyzed mechanism. In this instance, the uncatalyzed pyrolysis has $E_a = 210$ kJ, whereas with added iodine, the E_a drops to 140 kJ. The most likely mechanism is $\text{I}_2 \longrightarrow 2\text{I}$, followed by an attack of I atoms on the ether to yield radicals.

Transition-state theory provides the more general statement that catalysis is based on the existence of some mechanism for lowering the *free energy of activation*, $\Delta G^{\ddagger}$. A new reaction sequence is introduced, with a new transition state, which is at a lower free energy than the one in the uncatalyzed reaction. Although the activation energy may be lowered in many catalyzed reactions, it is not unusual to find instances in which the enhanced rate of a catalyzed reaction is due mainly to an increase in the entropy of activation.

41. Enzyme Reactions

Catalysts devised by man have been remarkable accelerators of chemical reaction rates. Yet their successes appear insignificant compared with the catalytic activity of enzymes synthesized by living cells. Consider one example among many—the formation of proteins. The synthesis of a protein in the laboratory requires elaborate equipment, yet it is carried out rapidly and continuously by living cells. The isotopic tracer experiments of R. Schoenheimer* showed that proteins in rat liver have an average lifetime of only 10 days. In addition to this continuous self-replacement, the liver synthesizes glycogen or animal starch from glucose; it manufactures urea which is excreted as the end product of nitrogen metabolism; and it also undertakes to detoxicate any number of unwanted substances, rendering them harmless to the animal organism.

H. Büchner was the first to establish, in 1897, that the intact cell is not necessary for these catalytic actions, since cell-free filtrates could be prepared containing the *enzymes* in solution. Since all known enzymes are proteins, they fall in the range of particle diameter from 10 to 100 nm. Enzyme catalysis is therefore midway between

---

*R. Schoenheimer, *The Dynamic State of Body Constituents* (Cambridge, Mass.: Harvard University Press, 1946).

homogeneous and heterogeneous catalysis, and a theoretical discussion can be based either on intermediate compound formation between enzyme and substrate molecules in solution, or on adsorption of substrate at the surface of the enzyme.

Enzymes are specific in their catalytic actions. *Urease* will catalyze the hydrolysis of urea, $(NH_2)_2CO$, in dilutions as high as one part of enzyme in 10^7 of solution, yet it has no detectable effect on the hydrolysis rate of substituted ureas, e.g., methyl urea, $(NH_2)(CH_3NH)CO$. *Pepsin* will catalyze the hydrolysis of the peptide glycyl-L-glutamyl-L-tyrosine, but it is completely ineffective if one of the amino acids has the opposite optical configuration of the D form, or if the peptide is slightly different, e.g., L-glutamyl-L-tyrosine. Such specificity is not absolute, however, and many enzymes are effective to some extent with any substrate having a structure quite similar to that of the natural one.

Almost all enzymes fall into one of two large classes, the hydrolytic enzymes and the oxidation-reduction enzymes. The enzymes of the first class appear to be complex acid-base catalysts, accelerating ionic reactions, principally the transfer of hydrogen ions. The oxidation-reduction enzymes catalyze electron transfers, perhaps through intermediate radical formation.

42. Kinetics of Enzyme Reactions

Mechanisms for enzyme kinetics usually begin with a first step in which enzyme E combines with substrate S to form a complex. This formulation was given originally by V. Henri* in 1903 and extended by Michaelis and Menten† in 1913, and Briggs and Haldane‡ in 1925. Henri assumed that the complex was in equilibrium with reactants, and that the rate determining step in the catalyzed reaction was the decomposition of the complex to yield the products P.

The simplest mechanism of this type would be

$$E + S \underset{k_{-1}}{\overset{k_1}{\rightleftharpoons}} ES \underset{k_{-2}}{\overset{k_2}{\rightleftharpoons}} E + P \tag{9.80}$$

The reverse reaction between E and P to reform ES is often slow enough to be neglected, and will always become negligible in the early stages of reaction when P is very low. With this approximation, the rate of change of ES, after an initial transient, reaches the steady state,

$$\frac{d[ES]}{dt} = k_1[E][S] - k_{-1}[ES] - k_2[ES] = 0$$

The steady state concentration $[ES]$ is, therefore,

$$[ES] = \frac{k_1[E][S]}{k_{-1} + k_2} = \frac{[E][S]}{K_m} \tag{9.81}$$

where $K_m = (k_{-1} + k_2)/k_1$ is often called the *Michaelis constant*.

From (9.81), the rate of the enzyme catalyzed reaction (per unit volume) would

---

*V. Henri, *Compt. Rend. Acad. Sci., Paris*, *135*, 916 (1902).

†L. Michaelis and M. L. Menten, *Biochem. Z.*, *49*, 333 (1913).

‡G. E. Briggs and J. B. S. Haldane, *Biochem. J.*, *19*, 338 (1925).

become

$$v = -\frac{d[S]}{dt} = k_2[ES] = \frac{k_2[E][S]}{K_m}$$

In this form, the equation is not so useful for practical purposes, since it includes the concentration of free enzyme $[E]$, whereas the experimentally known quantity is $[E_0]$ the total concentration of enzyme, both free and combined, in the reaction mixture. Since $[E_0] = [E] + [ES]$, we obtain from (9.81),

$$[ES] = \frac{[E_0][S]}{K_m + [S]}$$

and the rate is given by the *Michaelis–Menten equation,*

$$v = \frac{-d[S]}{dt} = k_2[ES] = \frac{k_2[E_0][S]}{K_m + [S]} \qquad (9.82)$$

It is often convenient to rewrite this equation in terms of the maximum rate V which is reached when $[S]$ becomes so large that $[S] \gg K_m$, and hence, from (9.82), $V = k_2[E_0]$, and

$$\frac{v}{V} = \frac{[S]}{[S] + K_m} \qquad (9.83)$$

The variation of v/V with $[S]$ for an example with $K_m = 0.2$ is shown in Fig. 9.30. When $[S] = K_m$, $v = V/2$.

In practical application, it is best to transform (9.83) into a linear equation, in one of the following ways:

$$\text{(Lineweaver–Burk):} \qquad \frac{1}{v} = \frac{K_m}{V}\frac{1}{[S]} + \frac{1}{V} \qquad (9.84)$$

$$\text{(Eadie):} \qquad v = -\frac{v}{[S]}K_m + V \qquad (9.85)$$

$$\text{(Dixon):} \qquad \frac{[S]}{v} = \frac{K_m}{V} + \frac{1}{V}[S] \qquad (9.86)$$

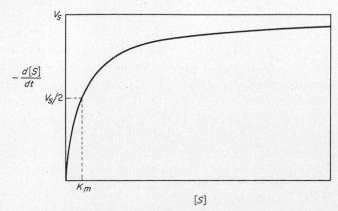

FIG. 9.30 Plot of the Michaelis-Menten equation (9.82). Note that when the rate is half the maximum, the substrate concentration $[S] = K_m$.

The value of v as a function of $[S]$ can usually be obtained from the initial slopes of rate curves. Then, through one of the linear plots, one can obtain K_m and V. Since $[E_0]$ is known, V gives k_2. From k_2 and K_m, one can obtain the equilibrium constant for formation of $[ES]$.

43. Enzyme Inhibition

Studies of the inhibition of enzyme catalyzed reactions have contributed to our understanding of enzyme mechanisms, and to the elucidation of the molecular basis for many biological mechanisms, including control processes in the cell and the mode of action of various drugs.

Several kinds of inhibition have been characterized. In *competitive inhibition*, the inhibitor molecule attaches itself to the enzyme at the site normally occupied by the substrate molecule (or, in some cases, an inhibitor combines with the substrate at the site normally used to attach the enzyme). Several different mechanisms of *noncompetitive inhibition* can be envisaged in which the inhibitor may alter the activity of the enzyme without actually blocking the active site.

Let us consider a case of competitive inhibition in accord with the following kinetics:

$$E + S \underset{k_{-1}}{\overset{k_1}{\rightleftharpoons}} ES$$

$$E + I \underset{k_{-3}}{\overset{k_3}{\rightleftharpoons}} EI, \qquad K_I = \frac{[E][I]}{[EI]}$$

$$ES \overset{k_2}{\longrightarrow} E + P$$

Note that it is conventional in enzyme literature to write equilibrium constants as *dissociation constants*. The concentration of free enzyme $[E]$ is now related to the concentration $[E_0]$ added to the reaction system by

$$[E] = [E_0] - [ES] - [EI]$$

The reaction rate becomes

$$v = \frac{V[S]}{[S] + K_m[1 + ([I]/K_I)]} \tag{9.87}$$

The maximum rate V_m is the same as in absence of inhibition, but the rate at any given substrate concentration is reduced. When (9.87) is thrown into a Lineweaver–Burk form,

$$\frac{1}{v} = \frac{K_m}{V}\left[1 + \frac{[I]}{K_I}\right]\cdot\frac{1}{[S]} + \frac{1}{V},$$

a straight line is still obtained, but the slope is a linear function of the concentration of inhibitor $[I]$. This result is a good diagnostic test for competitive inhibition.

An example of noncompetitive inhibition is the case in which inhibitor I combines with ES and renders it inactive.

$$E + S \rightleftharpoons ES \longrightarrow P + E$$
$$ES + I \rightleftharpoons ESI \qquad K_{SI}$$
$$E + I \rightleftharpoons EI \qquad K_I$$

If we suppose that $K_{SI} = K_I$, i.e., the affinity of the inhibitor site for inhibitor does not depend on whether E is bound to S, the rate becomes

$$v = \frac{V[S]}{([S] + K_m)(1 + [I]/K_I)}$$

In the Lineweaver–Burk form,

$$\frac{1}{v} = \left\{1 + \frac{[I]}{K_I}\right\}\left\{\frac{K_m}{V} \cdot \frac{1}{[S]} + \frac{1}{V}\right\}$$

In this case, both the slope and the intercept are multiplied by the same factor.

44. An Exemplary Enzyme–Acetylcholinesterase

For many years, acetylcholinesterase (ACE) has been studied intensively because of its physiological importance, and also because inhibitors of this enzyme include effective insecticides and nerve gases. The reaction catalyzed is the hydrolysis of acetylcholine (AC):

$$CH_3)_3N^+ \cdot CH_2CH_2O\overset{\overset{O}{\|}}{C} \cdot CH_3 + H_2O \rightarrow H^+ + CH_3\overset{\overset{O}{\|}}{C}-O^- + CH_3)_3N^+ \cdot CH_2 \cdot CH_2OH$$

Acetylcholine is the chemical transmitter at synapses between nerve endings and muscles, and probably also between many neurons in the central nervous system. The AC is released from the nerve ending upon the arrival of an electrical impulse along the nerve; it diffuses across a synaptic cleft about 20 nm wide, and depolarizes the postsynaptic membrane by abruptly increasing its conductance of Na^+ and K^+ ions. (How AC does this is a mystery.) The enzyme ACE catalyzes the removal of AC from the junction within a fraction of a millisecond, leaving the channel clear for another transmission pulse. Acetylcholinesterase is one the most active enzymes known. If we measure activity by a *turnover number* (T. N.) equal to the number of substrate molecules reacting in unit time for a single enzyme molecule, the T. N. for ACE is about 10^6 min^{-1} at 37°C.

The Michaelis constant for dissociation of the *ES* complex is $K_m = 4.5 \times 10^{-4}$ at 25°C. Among the competitive inhibitors of ACE are alkylated ammonium ions—for example, $(CH_3)_4N^+$ with $K_I = 1.62 \times 10^{-3}$, and $(C_2H_5)_4N^+$ with $K_I = 4.5 \times 10^{-4}$. A much more powerful competitive inhibitor is eserine with $K_I = 6.1 \times 10^{-8}$.

An example of a noncompetitive inhibitor that reacts with the enzyme-substrate complex is Tensilon with $K_I = 3.3 \times 10^{-7}$.

$$(CH_3)_2(C_2H_5)N^+ \hspace{-0.5em}\left\langle \bigcirc \right\rangle \hspace{2em} \text{Tensilon}$$
$$\text{OH}$$

Acetylcholinesterase is also subject to *irreversible inhibition* in which the inhibitor reacts with the active site, but cannot be recovered unchanged. Various organophosphorus compounds fall into this class. For example, one of the earliest nerve gases was Sarin.

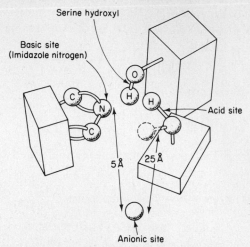

FIG. 9.31(a) Proposed arrangement of functional groups at the active center of acetylcholinesterase.

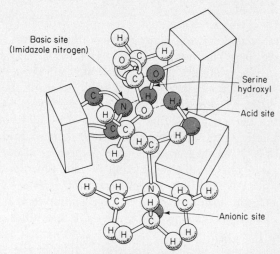

FIG. 9.31(b) A suggested structure for the Michaelis complex between acetylcholine and the enzyme; the dotted lines indicate electrostatic attractions and also bonds that are formed during acetylation. There is also electrostatic attraction between the imidazole nitrogen atom and the carbonyl carbon atom. The three bonds broken are indicated on the diagram. [R. M. Krupka and K. J. Laidler, *J. Am. Soc.*, *83*, 1458 (1961).]

$$
\begin{array}{c}
CH_3 \quad\quad O \\
CH_3 \quad\quad P \\
\backslash\;\;\;\; / \quad \backslash \\
C-O \quad\; F \\
/ \\
CH_3
\end{array}
\qquad \text{Sarin}
$$

Such irreversible inhibitors can be used to fix the active site of the enzyme, by experiments that show where the inhibitor is bound after partial hydrolysis of the enzyme to small peptides. As shown in Fig. 9.31(a), ACE has two binding sites, one called the *anionic site* and the other the *esteratic site*. The organophosphorus inhibitors act by phosphorylating the —OH group of a serine residue that forms part of the esteratic site.

The actual structure of pure crystalline ACE has not yet been determined, so that we cannot depict the exact conformation of the active site. Indirect evidence supports the schematic conformation in Fig. 9.31(b). The mechanism of the reaction suggested by Krupka* is as follows. The double-bonded nitrogen of the imidazole ring abstracts a proton from the hydroxyl of serine, and the resulting alkoxide anion then makes a nucleophilic attack on the acetyl group of the substrate. A conformation change in the enzyme brings the acetylated serine close to a second imidazole group, which transfers a proton from water to the acetyl group, thus effecting the hydrolysis.

PROBLEMS

1. The following data were obtained for the reaction between sodium thiosulfate and *n*-propyl bromide at 37.50°C. The unreacted $S_2O_3^{2-}$ was measured by titration with I_2. The I_2 titer is given as cubic centimetres of 0.02572 N iodine per 10.02 cm³ sample of reaction mixture.

t (s)	0	1110	2010	3192	5052	7380	11232	78840
I_2 (titer)	37.63	35.20	33.63	31.90	29.80	28.04	26.01	22.24

Write a balanced equation for reaction. Determine the order of reaction and the rate constant in standard units. What is the extent of reaction ξ at 4000 s?

2. The first-order gas reaction $SO_2Cl_2 \longrightarrow SO_2 + Cl_2$ has $k_1 = 2.20 \times 10^{-5}$ s^{-1} at 593 K. What percent of a sample of SO_2Cl_2 would be decomposed by heating at 593 K for 2.00 h?

3. The reaction $A + B \longrightarrow C$ takes place in two steps by the mechanism $2A \rightleftharpoons D$ followed by $B + D \xrightarrow{k_2} A + C$. The first step comes to a rapid equilibrium (constant K_1). Derive an expression for the rate of formation of C in terms of K_1, k_2, [A], [B].

4. The reaction $CH_3CH_2NO_2 + OH^- \longrightarrow CH_3CHNO_2^- + H_2O$ was carried out at 273 K with an initial concentration of each reactant of 5.00×10^{-3} mol·dm^{-3}. The OH$^-$ concentration fell to 2.60×10^{-3} after 5 min, 1.70×10^{-3} after 10 min,

*R. M. Krupka, *Biochemistry*, 5, 1988 (1966).

and 1.30×10^{-3} after 15 min. Show graphically that the reaction was second order and determine k_2 from the slope of the graph.

5. Assuming that the water concentration is so large that the inversion of sucrose is first order, derive a formula for the rate constant k_1 in terms of the angle of rotation of polarized light at the beginning of reaction, at time t, and at the end of the reaction.

6. If all reactants have initial concentration a and the reaction has an order n, show that the half-life is given by

$$\tau = \frac{(2^{n-1} - 1)}{a^{n-1} k_s (n - 1)}$$

where k_s is the rate constant.

7. Consider the exchange reaction,

$$CH_3S^*H + C_2H_5SH \rightleftharpoons CH_3SH + C_2H_5S^*H$$

where S* is a radioactive isotope present in tracer amounts, initially all in the CH_3SH molecules. Show that the rate of exchange of S* between the two molecular species will be first order regardless of the orders of the individual forward or reverse reactions.*

8. The reaction $2NO + 2H_2 \rightarrow N_2 + 2H_2O$ was studied with equimolar quantities of NO and H_2 at various initial pressures.

Initial P (torr)	354	340.5	375	288	251	243	202
Half-life, τ (min)	81	102	95	140	180	176	224

Calculate the overall order of the reaction (see Problem 6). Can you devise a mechanism consistent with this order?

9. A possible mechanism for the hydrogenation of ethylene, $C_2H_4 + H_2 \rightarrow C_2H_6$, in the presence of mercury vapor is

$$Hg + H_2 \xrightarrow{k_1} Hg + 2H$$

$$H + C_2H_4 \xrightarrow{k_2} C_2H_5$$

$$C_2H_5 + H_2 \xrightarrow{k_3} C_2H_6 + H$$

$$H + H \xrightarrow{k_4} H_2$$

Determine the rate of formation of C_2H_6 in terms of the rate constants and the concentrations [Hg], [H_2], and [C_2H_4]. Assume that H and C_2H_5 reach steady-state concentrations.

10. In what proportion of binary collisions does the kinetic energy along the line of centers exceed 100 kJ at 300 K, 600 K, 1200 K?

11. A certain reaction is 20% complete in 12.6 min at 300 K and in 3.20 min at 340 K. Estimate its activation energy E_a.

12. Given the reaction,

$$A(g) \underset{k_2}{\overset{k_1}{\rightleftharpoons}} B(g) + C(g)$$

where $k_1 = 0.20 \text{ s}^{-1}$ and $k_2 = 5.0 \times 10^{-4} \text{ s}^{-1} \cdot \text{atm}^{-1}$ at 298 K, and each value is doubled on going to 310 K, calculate (a) the equilibrium constant at 298 K, (b) the

---

*See A. A. Frost and R. G. Pearson, *Kinetics and Mechanism* (New York: John Wiley & Sons, Inc., 1961), p. 192.

activation energies for the forward and reverse reactions, (c) $\Delta H^{\ominus}$ for the overall reaction, and (d) the time for P (total) to reach 1.5 atm if one starts with only A at $P = 1$ atm and 298 K.

13. The thermal isomerization of bicyclo [2, 1]-pent-2-ene is a unimolecular reaction with $\log_{10} k_1(s^{-1}) = 14.21 - 112.4\theta^{-1}$ where $\theta = 2.303RT \, kJ \cdot mol^{-1}$. Calculate the Arrhenius activation energy E_a and $\Delta S^{\ddagger}$. Comment on the $\Delta S^{\ddagger}$ found.*

14. At temperatures below 800 K, the dimerization of tetrafluoroethylene, $2C_2F_4 \longrightarrow$ cyclo-C_4F_8, follows second-order kinetics with $k_2 = 10^{11.07} \exp(-107 \, kJ/RT)$ $cm^3 \cdot mol^{-1} \cdot s^{-1}$. The molecular diameter obtained from electron diffraction is 5.12 $\times 10^{-10}$ m. Calculate k_2 by simple hard-sphere collision theory, and hence estimate the steric factor p at 725 K. Discuss the result briefly.†

15. The reaction between CO and O was studied in a flow system from 409 to 503 K at 1 atm pressure.‡ The reaction was third order

$$CO + O + M \longrightarrow CO_2 + M$$

where M represents CO or O_2 and $k_3 = 10^{11.83} \exp(12.5 \, kJ/RT) \, cm^6 \cdot mol^{-2} \cdot s^{-1}$. How would you interpret the negative activation energy? Suppose that O atoms at a concentration of $10^{-4} \, mol \cdot dm^{-3}$ are introduced at the inlet of a tube with a flowing stream of CO at 500 K and 1 atm. If the flow rate is 10 $cm^3 \cdot s^{-1}$, what is the concentration of oxygen atoms at the exit of a tube 50 cm long and 1 cm in diameter?

16. The compound $CH_3 - O - N = O$ undergoes a cis-trans isomerization by internal rotation about the O—N bond. The half-life of the first-order conversion of the cis form was measured by NMR techniques as 10^{-6} s at 298 K. Assuming $\Delta S^{\ddagger} = 0$ for this reaction, calculate the height of the barrier to rotation. How good an assumption do you think $\Delta S^{\ddagger} = 0$ would be? Explain.

17. For the decomposition of N_2O_5,

θ (°C)	25	35	45	55	65
$10^5 \, k_1(s^{-1})$	1.72	6.65	24.95	75	240

Calculate A and E_a for the reaction in the equation $k_1 = Ae^{-E_a/RT}$. Calculate $\Delta G^{\ddagger}$. $\Delta S^{\ddagger}$, $\Delta U^{\ddagger}$, $\Delta H^{\ddagger}$ for the reaction at 50°C.

18. The rate constant for the second-order gas reaction, $H_2 + I_2 \longrightarrow 2HI$, is 0.0234 $mol^{-1} \cdot dm^3 \cdot s^{-1}$ at 400°C, and the activation energy $E_a = 150 \, kJ \cdot mol^{-1}$. Calculate $\Delta H^{\ddagger}$ and $\Delta S^{\ddagger}$.

19. In the reversible reaction

$$D - R_1R_2R_3CBr \rightleftharpoons L \cdot R_1R_2R_3CBr$$

both the forward and reverse reactions are first order with $\tau = 10$ min. If you start with 1.000 mol of the D-bromide, how many moles of L-bromide will be present after 10 min?

20. A possible mechanism for the reaction $C_2H_6 + H_2 \longrightarrow 2CH_4$ is the sequence,

$$C_2H_6 \rightleftharpoons 2CH_3 \qquad K$$
$$CH_3 + H_2 \longrightarrow CH_4 + H \qquad k_1$$
$$H + C_2H_6 \longrightarrow CH_4 + CH_3 \qquad k_2$$

*D. M. Golden and J. I. Branman, *Trans. Faraday Soc.*, **65**, 464 (1969).
†G. A. Drennan and R. A. Matula, *J. Phys. Chem.*, **72**, 3462 (1968).
‡V. N. Kondratiev and E. I. Intezarova, *Int. J. Chem. Kin.*, **1**, 105 (1969).

If the first reaction is at equilibrium and H is in a steady state, derive the rate law for the formation of CH_4, $d[CH_4]/dt = 2k_1 K^{1/2}[C_2H_6]^{1/2}[H_2]$. How would you test this rate law against experimental or theoretical data?

21. A possible mechanism for the reaction $2NO + O_2 \longrightarrow 2NO_2$ is

$$
\begin{array}{llll}
(1) & NO + NO & \longrightarrow N_2O_2 & k_1 \\
(2) & N_2O_2 & \longrightarrow 2NO & k_2 \\
(3) & N_2O_2 + O_2 & \longrightarrow 2NO_2 & k_3
\end{array}
$$

(a) Apply the steady state approximation to $[N_2O_2]$ to obtain the rate law

$$\frac{d[NO_2]}{dt} = \frac{2k_1 k_3 [NO]^2 [O_2]}{k_2 + k_3 [O_2]}$$

(b) If only a very small fraction of the N_2O_2 formed in (1) goes to form products in (3), while most of the N_2O_2 reverts to NO in reaction (2), and if the activation energies are $E_1 = 82$ kJ, $E_2 = 205$ kJ, and $E_3 = 82$ kJ, what is the overall activation energy?

22. Formulate an equation based on transition-state theory for the effect of pressure on the rate constant of a reaction, by defining and introducing the concept of the volume of activation $\Delta V^{\ddagger}$. What would be the $\Delta V^{\ddagger}$ per mole of a reaction for which the rate constant was doubled by an increase in pressure from 1 to 3000 atm at 298 K?

23. The rate constants for solvolysis of benzyl chloride in an acetone-water solution at 298 K were

| P (kbar) | 0.001 | 0.345 | 0.689 | 1.033 |
| $k \times 10^6$ (s^{-1}) | 7.18 | 9.58 | 12.2 | 15.8 |

Calculate $\Delta V^{\ddagger}$, the volume of activation. How would you interpret the result in terms of a reaction mechanism?*

24. Show that for a first-order reaction, the time required for 99.9% completion of the reaction is 10 times the time for 50.0% completion.

25. In the polymerization reactions, $2A \rightarrow A_2$; $A_2 + A \rightarrow A_3$; $A_3 + A \rightarrow A_4 \cdots$, etc., if all the rate constants are the same k_2, show that the integrated rate expression is $y = ak_2 t[(4 + k_2 t)/(2 + k_2 t)^2]$, where y is the total amount of polymer $(A_2 + A_3 + A_4 + \cdots + A_n)$ and a is the initial amount of reactant A. What is dy/dt, the polymerization rate? What is the apparent order of reaction in a single run? What is the order based on the initial rates if the concentration a is varied?

26. The reaction $N + C_2H_4 \rightarrow HCN + CH_3$ was studied in a stirred flow reactor of 10 cm^3 volume by introducing atomic N at a flow rate of 10^{-6} mol·s^{-1} in a stream of N_2 at 3.6×10^{-5} mol·s^{-1}. The flow rate of C_2H_4 was 6.0×10^{-6} mol·s^{-1}. The rate constant for reaction at 313 K was 1.6×10^{-13} (molecules·cm^{-3})$^{-1}$·s^{-1}. What was the concentration of CH_3 radicals within the reactor?†

27. The hydrolysis of sucrose by the enzyme invertase was followed by measuring the initial rate of change in polarimeter readings at various initial concentrations of sucrose.

*K. J. Laidler and R. Martin, *Int. J. Chem. Kin.*, *1*, 113 (1969).
†E. Milton and H. Dunford, *J. Chem. Phys.*, *34*, 51 (1961).

Sucrose (mol·dm⁻³)	0.0292	0.0584	0.0876	0.117	0.146	0.175	0.234
Initial rate	0.182	0.265	0.311	0.330	0.349	0.372	0.371
Initial rate ($2M$ urea)	0.083	0.111	0.154	0.182	0.186	0.192	0.188

By means of a Lineweaver–Burk plot, calculate the Michaelis constant K_m for the enzyme-substrate complex.*

28. The effect of an enzyme inhibitor I on the simple kinetic scheme of Michaelis and Menten will depend on whether I is competitive or noncompetitive, i.e., on whether I binds to the same active site as the substrate or to some different site on the enzyme.

(1) Competitive
$$I + E \underset{k_{-3}}{\overset{k_3}{\rightleftharpoons}} EI$$

(2) Noncompetitive
$$EI + S \underset{k_{-y}}{\overset{k_y}{\rightleftharpoons}} ESI \overset{k_s}{\longrightarrow} EI + P$$

Derive the equations corresponding to (9. 83) for the cases (1) and (2). The reaction in Problem 27 was also studied in the presence of $2\,\mathrm{M}$ urea, which acted as a reversible inhibitor. Apply your theoretical analysis to determine whether urea is a competitive or noncompetitive I for this reaction.†

29. The luminescent reaction of luciferin catalyzed by the enzyme luciferase obtained from the crustacean *Cypridina* was measured at an enzyme concentration of $10^{-9}\,\mathrm{M}$. The initial luciferin concentration is given as cm^3 of $8.0 \times 10^{-5}\,\mathrm{M}$ luciferin in 20 cm^3 of reaction mixture.

Luciferin conc.	0.04	0.06	0.08	0.10	0.20	0.40	0.90
Initial rate (15°C)	15	19	23	26	44	56	76
Millivolts per minute (22°C)	18	22	33	37	62	91	123

The conversion factor from millivolt reading of the photoelectric integrator to moles of luciferin reacted was $8.6 \times 10^{-12}\,\mathrm{mol·mV^{-1}}$ at 22°C and 7.7×10^{-12} at 15°C. For this reaction, estimate $\Delta H^{\ddagger}$, $\Delta S^{\ddagger}$ and $\Delta H^{\ominus}$, $\Delta S^{\ominus}$ for the formation of the enzyme-substrate complex. What standard state is used?‡

---

*A. M. Chase, H. C. V. Meier, and V. J. Menna, *J. Cellular Comp. Physiol.*, **59**, 1 (1962).
†M. Dixon, *Biochem. J.*, **55**, 170 (1953).
‡W. J. Kauzmann, A. M. Chase, and E. H. Brigham, *Arch. Biochem.*, **24**, 281 (1949).

10

Electrochemistry: Ionics

The electrical matter consists of particles extremely subtile, since it can permeate common matter, even the densest metals, with such ease and freedom as not to receive any perceptible resistance.
If any one should doubt whether the electrical matter passes thro' the substance of bodies, or only over and along their surfaces, a shock from an electrified large glass jar, taken through his own body, will probably convince him.

Benjamin Franklin
(1749)*

All chemical interactions are electrical at the atomic level so that in a sense all chemistry is electrochemistry. In a more restricted sense, electrochemistry has come to mean the study of solutions of electrolytes and of phenomena occurring at electrodes immersed in these solutions. The electrochemistry of solutions may claim our special interest because physical chemistry first emerged as a distinct science in this field. Its first journal, *Zeitschrift für physikalische Chemie*, was founded in 1887 by Wilhelm Ostwald, and the early volumes are devoted mainly to researches in electrochemistry by Ostwald, van't Hoff, Nernst, Kohlrausch, Arrhenius, and others of their school.

1. Electricity

William Gilbert, Queen Elizabeth's physician, coined the word *electric* in 1600 (from the Greek, ἤλεκτρον, *amber*). He applied it to bodies that when rubbed with fur attracted small bits of paper or pith. Gilbert was unwilling to admit the possibility of "action at a distance," and in his treatise *De Magnete* he advanced an ingenious theory for the electrical attraction.

> An effluvium is exhaled by the amber and is sent forth by friction. Pearls, carnelian, agate, jasper, chalcedony, coral, metals, and the like, when rubbed are inactive; but is there nought emitted from them also by heat and friction? There is indeed, but what is emitted from the dense bodies is thick and vaporous [and thus not

*Benjamin Franklin, "Opinions and Conjectures concerning the Properties and Effects of the electrical Matter, arising from Experiments and Observations, made at Philadelphia, 1749."

mobile enough to cause attractions]. A breath, then . . . reaches the body that is to be attracted and as soon as it is reached it is united to the attracting electric. For as no action can be performed by matter save by contact, these electric bodies do not appear to touch, but of necessity something is given out from the one to the other to come into close contact therewith, and to be a cause of incitation to it.

Further investigation revealed that materials such as glass, after being rubbed with silk, exerted forces opposed to those from amber. Two varieties of electric fluid were thus distinguished, vitreous and resinous. Frictional machines were devised for generating high electrostatic potentials, and used to charge condensers in the form of Leiden jars.

Benjamin Franklin (1747) simplified matters by proposing a one-fluid theory, according to which bodies rubbed together acquire a surplus or deficit of electric fluid, depending on their relative attractions for it. The resultant difference in charge causes the observed forces. Franklin established the convention that the vitreous type of electricity is positive (fluid in excess), and the resinous type is negative (fluid in defect).

In 1791, Luigi Galvani accidentally brought the bare nerve of a partially dissected frog's leg into contact with a discharging electrical machine. The sharp convulsion of the leg muscles led to the discovery of galvanic electricity,* for it was soon found that the electric machine was unnecessary and that twitching could be produced simply by bringing nerve ending and end of leg into contact through a metal strip. The action was enhanced when two dissimilar metals completed the circuit. Galvani, a physician, named the new phenomenon "animal electricity" and believed it to be characteristic of living tissues only.

Alessandro Volta, a physicist, Professor of Natural Philosophy at Pavia, soon discovered that electricity could have an inanimate origin. From a stack of dissimilar metals in contact with moist paper, he was able to charge an electroscope. In 1800, he constructed his famous *pile*, consisting of many consecutive plates of silver, zinc, and cloth soaked in salt solution. From the terminals of the pile he drew the shocks and sparks previously observed only with electrostatic devices.

The news of Volta's pile aroused an enthusiasm and amazement like those caused by the uranium pile in 1945. In May of 1800, Nicholson and Carlisle decomposed water by means of electric current, oxygen appearing at one pole of the pile and hydrogen at the other. Solutions of various salts were soon decomposed, and in 1806 and 1807, Humphry Davy used a pile to isolate sodium and potassium from their hydroxides. The theory that atoms in a compound were held together by attraction between unlike charges immediately gained a wide acceptance.

2. Faraday's Laws and Electrochemical Equivalents

In 1813 Michael Faraday, then 22 years old and a bookbinder's apprentice, went to the Royal Institution as Davy's laboratory assistant. In the following years, he

---

*Vans Gravesande and Adanson independently discovered the intense discharges of electric fish in 1750.

carried out the series of researches that were the foundations of electrochemistry and electromagnetism. He studied intensively the decomposition of solutions of salts, acids, and bases by the electric current. With the assistance of William Whewell, Faraday devised the elegant nomenclature used in these studies: *electrode, electrolysis, electrolyte, ion, anion, cation*. The electrode *toward* which the cations move in a cell is called the *cathode*. The electrode *toward* which the anions move is called the *anode*.

Faraday proceeded to study quantitatively the relation between the amount of electrolysis, or chemical action produced by the current, and the quantity of electricity. The unit of electric quantity is now the coulomb (C) or ampere second (A·s). His results were summarized as follows:*

> The chemical power of a current of electricity is in direct proportion to the absolute quantity of electricity which passes. . . . The substances into which these [electrolytes] divide, under the influence of the electric current, form an exceedingly important general class. They are combining bodies, are directly associated with the fundamental parts of the doctrine of chemical affinity; and have each a definite proportion, in which they are always evolved during electrolytic action. I have proposed to call . . . the numbers representing the proportions in which they are evolved *electrochemical equivalents*. Thus hydrogen, oxygen, chlorine, iodine, lead, tin, are ions; the three former are anions, and the two metals are cations, and 1, 8, 36, 125, 104, 58 are their electrochemical equivalents nearly.

> Electrochemical equivalents coincide, and are the same, with ordinary chemical equivalents. I think I cannot deceive myself in considering the doctrine of definite electrochemical action as of the utmost importance. It touches by its facts more directly and closely than any former fact, or set of facts, has done, upon the beautiful idea that ordinary chemical affinity is a mere consequence of the electrical attractions of different kinds of matter.

We now recognize that ions in solution may bear more than one elementary charge, and that the electrochemical equivalent mass is atomic mass M divided by number of charges on the ion $|z|$. The constant amount of electricity always associated with one equivalent of electrochemical reaction is called the faraday (F). It is equal to 96 487 C. Thus, Faraday's Laws of electrolysis can be summarized in the equation

$$\frac{m}{M} = \frac{It}{|z|F} = \frac{Q}{|z|F} \tag{10.1}$$

Here m is the mass of an element of atomic mass M liberated at an electrode, by the passage of current I through a solution for a time t.

The fact that a definite quantity of electric charge, or a small integral multiple thereof, is always associated with each charged atom in solution, strongly suggested that electricity itself is atomic in nature. Hence, in 1874, G. Johnstone Stoney addressed the British Association as follows:

> Nature presents us with a single definite quantity of electricity which is independent of the particular bodies acted on. To make this clear, I shall express Faraday's Law in the following terms. . . . For each chemical bond which is ruptured within an electrolyte a certain quantity of electricity traverses the electrolyte which is the same in all cases.

*Phil. Trans. Roy. Soc. London, Ser. A, 124, 77 (1834).

In 1891, Stoney proposed that this natural unit of electricity should be given a special name, the *electron*.* Hence, 1 mol of electrons would equal 1 F of electric charge.

$$F = Le \qquad (10.2)$$

3. Coulometers

A careful measurement of the amount of chemical reaction caused by the passage of a certain amount of electric charge through an electrolytic cell gives a precise measure of the amount of electric charge that passed. Such a device for measuring charge passed is called a *coulometer*.

An example is the *silver coulometer*, which uses platinum electrodes in aqueous silver nitrate. The gain in mass of the cathode is measured after a current is passed through a solution of $AgNO_3$. The reaction at the cathode can be written

$$Ag^+ + e^- \longrightarrow Ag$$

One mole of silver, 107.870 g, is deposited on the cathode for each faraday passed through the coulometer. Thus, 1 C is equivalent to

$$m(Ag) = \frac{M(Ag)}{|z|\, F} = \frac{107.870 \text{ g mol}^{-1}}{96487 \text{ C mol}^{-1}}.\, 1\, C = 1.118 \times 10^{-3} \text{ g of silver}$$

4. Conductivity Measurements

One of the earliest theoretical problems in electrochemistry was how solutions of electrolytes conducted an electric current.

Metallic conductors were known to obey Ohm's Law,

$$I = \frac{\Delta\Phi}{R} \qquad (10.3)$$

where I is the current (amperes), $\Delta\Phi$ is the difference in electric potential between the terminals of the conductor (volts), and the proportionality constant R is the *resistance* (ohms). The resistance depends on the dimensions of the conductor. For a conductor of uniform cross section,

$$R = \frac{\rho l}{A} \qquad (10.4)$$

Here, l is the length and A the cross-sectional area, and the specific resistance ρ ($\Omega \cdot m$) is called *resistivity*. The reciprocal of resistance is *conductance* (Ω^{-1}) and the reciprocal of resistivity is *specific conductance* or *conductivity* κ ($\Omega^{-1} \cdot m^{-1}$).

The first studies of the conductivity of solutions were made with rather large direct currents. The resulting electrochemical action was so great that erratic

*Later, an elementary particle was discovered with a charge of $-e$, and this particle was given the name *electron*. The unit of charge e is 1.6021×10^{-19} C, the charge of the proton.

results were obtained, and it appeared that Ohm's Law was not obeyed—i.e., the conductivity seemed to depend on the $\Delta\Phi$. This result was largely due to *polarization* at the electrodes of the conductivity cell—i.e., a departure from equilibrium conditions in the surrounding electrolyte.

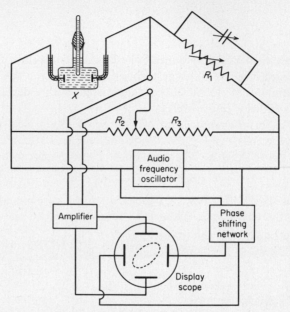

FIG. 10.1 AC Wheatstone bridge for measurement of conductance of electrolytes.

These difficulties were overcome by the use of an alternating-current (a-c) bridge, such as that shown in Fig. 10.1. With a-c frequencies in the audio range [1000 to 4000 hertz (Hz)], the direction of the current changes so rapidly that polarization effects are eliminated. One difficulty with the a-c bridge is that the cell acts as a capacitance in series with a resistance, so that even when the resistance arms are balanced there is a residual unbalance due to the capacitances. This effect can be partially overcome by inserting a variable capacitance in the other arm of the bridge, but for very precise work further refinements are necessary.*

The balance point of the bridge is indicated on the cathode-ray oscilloscope. The voltage from the bridge midpoint is filtered, amplified, and fed to the vertical plates of the oscilloscope. A small portion of the bridge input signal is fed to the horizontal plates through a suitable phase-shifting network. When the two signals are properly phased, the balance of capacitance is indicated by the closing of the

*T. Shedlovsky, *J. Am. Chem. Soc.*, *54*, 1411 (1932); W. F. Luder, *J. Am. Chem. Soc.*, *62*, 89 (1940); J. Braunstein and G. D. Robbins, *J. Chem. Ed.*, *48*, 52 (1971). The last authors analyze the sources of capacitance in a-c bridge measurements of electrolytic solutions and show that the principal capacitance is in series with the electrolyte and arises from the charging and discharging of the double layer at the surface of the electrodes. (See Section 11.19.)

loop on the oscilloscope screen, and the balance of resistance is indicated by the tilt of the loop from horizontal.

A typical conductivity cell is also shown in Fig. 10.1. Instead of measuring their dimensions, we now usually calibrate these cells before use with a solution of known conductivity, such as one-molar potassium chloride. The cell must be well thermostatted, since conductivity increases with temperature.

As soon as reliable conductivity data were available, it became apparent that solutions of electrolytes followed Ohm's Law. Resistance was independent of potential difference,* and the smallest applied voltage sufficed to produce a current of electricity. Any conductivity theory would have to explain this fact: the electrolyte is always ready to conduct electricity, and this capability is not something produced by the applied electric field.

On this score, the ingenious theory proposed in 1805 by C. J. von Grotthuss must be judged inadequate. He supposed the molecules of electrolyte to be polar, with positive and negative ends. An applied field lined them up in a chain. Then the field caused the molecules at the end of the chain to dissociate, the free ions thus formed being discharged at the electrodes. Thereupon, there was an exchange of partners along the chain. Before further conduction could occur, each molecule had to rotate under the influence of the field to reform the original oriented chain. Despite its shortcomings, the Grotthuss theory was valuable in emphasizing the necessity of having free ions in the solution to explain the observed conductivity. We shall see later that a mechanism similar to that of Grotthuss actually occurs in some cases.

In 1857, Clausius proposed that especially energetic collisions between undissociated molecules in electrolytes maintained at equilibrium a small number of charged particles. These particles were believed to be responsible for the observed conductivity.

5. Molar Conductances

From 1869 to 1880, Friedrich Kohlrausch and his coworkers published a long series of careful conductivity investigations. The measurements were made over a range of temperatures, pressures, and concentrations.

Typical of this painstaking work was the extensive purification of the water used as a solvent. After 42 successive distillations *in vacuo*, they obtained a *conductivity water* with $\kappa = 4.3 \times 10^{-6} \ \Omega^{-1} \cdot m^{-1}$ at 18°C. Ordinary distilled water in equilibrium with the carbon dioxide of the air has a conductivity of about $70 \times 10^{-6} \ \Omega^{-1} \cdot m^{-1}$.

To reduce conductivities to a common concentration basis, a function called the *molar conductance* is defined by

$$\Lambda = \frac{\kappa}{c} \tag{10.5}$$

---

*At high electric field strengths, however, departures from Ohm's Law will be observed.

For this equation, the usual unit of the concentration c is mole per cubic centimetre. To specify Λ, we must specify the formula unit with concentration c. Thus, $\Lambda(MgSO_4) = 2\Lambda(\frac{1}{2}MgSO_4)$.

Some values for Λ are plotted in Fig. 10.2. We can distinguish two classes of

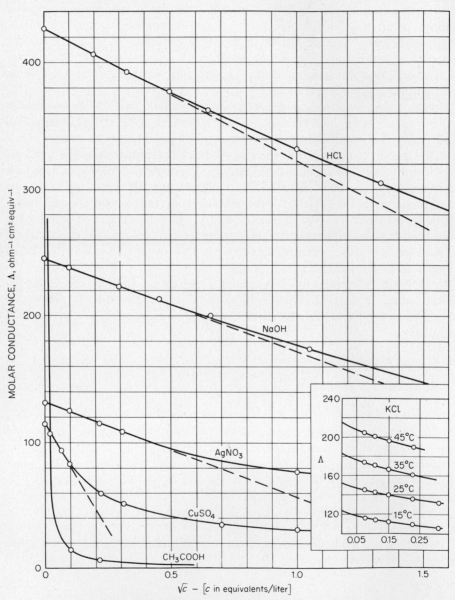

FIG. 10.2   Molar conductances at 298.15 K of electrolytes in aqueous solution vs. square roots of concentrations. Insert shows variation with temperature of Λ for KCl.

electrolytes on the basis of their conductivities. Strong electrolytes—such as most salts and acids like hydrochloric, nitric, and sulfuric—have high molar conductances that increase only moderately with increasing dilution. Weak electrolytes, such as acetic and other organic acids and aqueous ammonia, have much lower molar conductances at high concentrations, but the values increase greatly with increasing dilution.

The value of Λ extrapolated to zero concentration is called *molar conductance at infinite dilution*, Λ_0. The extrapolation is made readily for strong electrolytes but is impossible to make accurately for weak electrolytes because of their steep increase in Λ at high dilutions, where the experimental measurements become very uncertain. It was found that data for strong electrolytes were fairly well represented by an empirical equation,

$$\Lambda = \Lambda_0 - k_c c^{1/2} \tag{10.6}$$

where k_c is an experimental constant.

Kohlrausch discovered certain interesting relations between the values of Λ_0 for different electrolytes: the difference in Λ_0 for pairs of salts having a common ion was always approximately constant. For example (at 298.15 K) in units of $\Omega^{-1} \cdot cm^2 \cdot mol^{-1}$,

	Λ_0		Λ_0		Λ_0
NaCl	128.1	$NaNO_3$	123.0	NaOH	246.5
KCl	149.8	KNO_3	145.5	KOH	271.0
	21.7		22.5		24.5

Thus, no matter what the anion may be, there is an approximately constant difference between the Λ_0 values of potassium and sodium salts. This behavior can be readily explained if Λ_0 is the sum of two independent terms, one characteristic of the anion and one of the cation. Thus,

$$\Lambda_0 = \Lambda_0^+ + \Lambda_0^- \tag{10.7}$$

where Λ_0^+ and Λ_0^- are the *molar ionic conductances* at infinite dilution. This is Kohlrausch's *Law of the Independent Migration of Ions*.

This rule makes it possible to calculate the Λ_0 for weak electrolytes such as organic acids from values for their salts, which are strong electrolytes. For example (at 298.15 K),

$$\Lambda_0(HAc) = \Lambda_0(NaAc) + \Lambda_0(HCl) - \Lambda_0(NaCl)$$
$$= 91.0 + 425.0 - 128.1 = 387.9 \ \Omega^{-1} \cdot cm^2 \cdot mol^{-1}$$

6.   The Arrhenius Ionization Theory

From 1882 to 1886, Julius Thomsen published data on the heats of neutralization of acids and bases. He found that the heat of neutralization of a strong acid by a strong base in dilute solution was always nearly constant, being about 57.7 kJ per equivalent at 298.15 K. The neutralization heats of weak acids and bases were lower, and indeed the "strength" of an acid appeared to be proportional to its heat of neutralization by a strong base, such as NaOH.

These results and available conductivity data led Svante Arrhenius in 1887 to propose a new theory for the behavior of electrolyte solutions. He suggested that an equilibrium exists in the solution between undissociated solute molecules and ions that arise from these by electrolytic dissociation. Strong acids and bases being almost completely dissociated, their interaction was in every case equivalent to $H^+ + OH^- \rightarrow H_2O$, thus explaining the constant heat of neutralization.

While Arrhenius was working on this theory, osmotic-pressure studies by van't Hoff appeared, which provided a striking confirmation of the new ideas. We recall (Section 7.14) that van't Hoff found that the osmotic pressures of dilute solutions of nonelectrolytes often followed the equation $\Pi = cRT$. The osmotic pressures of electrolytes were always higher than predicted from this equation, often by a factor of two, three, or more, so that a modified equation was written as

$$\Pi = icRT \tag{10.8}$$

The van't Hoff *i factor* for strong electrolytes was close to the number of ions that would be formed if a solute molecule dissociated according to the Arrhenius theory. Thus, for NaCl, KCl, and other uniunivalent electrolytes, $i = 2$; for $BaCl_2$, K_2SO_4, and other unibivalent species, $i = 3$; for $LaCl_3$, $i = 4$.

On April 13, 1887, Arrhenius wrote to van't Hoff as follows:

> It is true that Clausius had assumed that only a minute quantity of dissolved electrolyte is dissociated, and that all other physicists and chemists had followed him, but the only reason for this assumption, so far as I can understand, is a strong feeling of aversion to a dissociation at so low a temperature, without any actual facts against it being brought forward. . . . At extreme dilution all salt molecules are completely dissociated. The degree of dissociation can be simply found on this assumption by taking the ratio of the equivalent conductivity of the solution in question to the equivalent conductivity at the most extreme dilution.

Thus, Arrhenius would write the degree of dissociation α as

$$\alpha = \frac{\Lambda}{\Lambda_0} \tag{10.9}$$

The van't Hoff *i* factor can also be related to α. If one molecule of solute capable of dissociating into v ions per molecule is dissolved, the total number of particles present will be $i = 1 - \alpha + v\alpha$. Therefore,

$$\alpha = \frac{i-1}{v-1} \tag{10.10}$$

Values of α for weak electrolytes calculated from (10.9) and (10.10) were in good agreement.

By calculating an equilibrium constant for ionization, Ostwald obtained a *dilution law*, governing the variation of molar conductance Λ with concentration. For a binary electrolyte AB with degree of dissociation α, whose concentration is c,

$$AB \rightleftharpoons A^+ + B^-$$

$$c(1 - \alpha) \quad \alpha c \quad \alpha c$$

$$K = \frac{\alpha^2 c}{(1 - \alpha)}$$

From (10.9), therefore,

$$K = \frac{\Lambda^2 c}{\Lambda_0 (\Lambda_0 - \Lambda)} \tag{10.11}$$

This equation was closely followed by weak electrolytes in dilute solutions. An example is shown in Table 10.1. In this case, the "dilution law" is obeyed at concentrations below about 0.1 molar, but discrepancies appear at higher concentrations, because the calculated K no longer remains constant.

TABLE 10.1
Test of Ostwald's Dilution Law—Acetic Acid at 25°C, $\Lambda_0 = 387.9$[a]

c (mol·dm^{-3})	Λ (Ω^{-1}·cm^2·mol^{-1})	% Dissociation $[100\alpha = 100(\Lambda/\Lambda_0)]$	Eq. (10.11) $K \times 10^5$ (mol·dm^{-3})
1.011	1.443	0.372	1.405
0.2529	3.221	0.838	1.759
0.06323	6.561	1.694	1.841
0.03162	9.260	2.389	1.846
0.01581	13.03	3.360	1.846
0.003952	25.60	6.605	1.843
0.001976	35.67	9.20	1.841
0.000988	49.50	12.77	1.844
0.000494	68.22	17.60	1.853

[a]D. A. MacInnes and T. Shedlovsky, *J. Am. Chem. Soc.*, 54, 1429 (1932).

The accumulated evidence gradually won acceptance for the Arrhenius theory, although chemists at the time still found it hard to believe that a stable molecule placed in water could spontaneously dissociate into ions. This criticism was, in fact, justified, and it soon became evident that the solvent must play more than a passive role in the formation of an ionic solution.

7. Solvation of Ions

We now know that the crystalline salts are themselves formed of ions in regular array, so that there is no question of "ionic dissociation" when they dissolve. The process of solution simply allows the ions to separate from one another. The separation is particularly easy in aqueous solutions owing to the high dielectric constant of water, $\epsilon = 78.5$ at 298.15 K. If we compare, for water and a vacuum, the energy necessary to separate two ions, say Na^+ and Cl^-, from a distance of

0.2 nm to infinity, we find

Vacuum	*Water*

$$\Delta E = \int_{0.2}^{\infty} F dr = \int_{0.2}^{\infty} \frac{e_1 e_2}{4\pi\epsilon_0 r^2} dr \qquad\qquad \Delta E = \int_{0.2}^{\infty} \frac{e_1 e_2}{4\pi\epsilon_0 \epsilon r^2} dr = \frac{\Delta E \,(\text{vacuum})}{\epsilon}$$

$$= \frac{(1.60 \times 10^{-19})^2}{4\pi(8.854 \times 10^{-12})(2 \times 10^{-10})} \qquad\qquad = \frac{1.15 \times 10^{-18}}{78.5}$$

$$= 1.15 \times 10^{-18} \text{ J} \qquad\qquad\qquad\qquad = 1.47 \times 10^{-20} \text{ J}$$

$$= 7.19 \text{ eV} \qquad\qquad\qquad\qquad\qquad = 0.0915 \text{ eV}$$

A similar argument was used by Born* to estimate the free energy of solvation of an ion of radius a. If the ion is transferred from a medium with dielectric constant ϵ_1 to one with ϵ_2, the electrical free energy change is

$$\Delta G_e = \frac{-z^2 e^2}{8\pi\epsilon_0 a}\left[\frac{1}{\epsilon_1} - \frac{1}{\epsilon_2}\right] \tag{10.12}$$

With $\epsilon_1 = 1$ for an ion in vacuum. (10.12) gives a value for the solvation free energy of the ion. The equation is not accurate, however, because bulk dielectric constants are not valid in the immediate neighborhood of an ion. Latimer† and coworkers tried to correct for this effect by using an effective radius of the ion that was larger than the crystal radius, thus excluding a volume of solvent around each ion from the bulk solvent with dielectric constant ϵ_2. For univalent cases, they arbitrarily added 0.085 nm to a for cations, 0.010 nm for anions. Table 10.2 gives calculated ΔG_e for a number of ions.

TABLE 10.2
Calculated Free Energies of Hydration of Ions (kJ·mol⁻¹ at 293 K)

	Li⁺	Na⁺	K⁺	Rb⁺	Cs⁺	F⁻	Cl⁻	Br⁻	I⁻
Born equation	−1004	−699	−515	−460	−418	−515	−377	−347	−310
Latimer equation	−481	−377	−305	−280	−255	−477	−351	−326	−293
Mean hydration number[a]	4	3	2	1	—	3	2	—	0.7

[a]J. O'M. Bockris, *Quart. Rev. London*, 3, 173 (1949).

The hydration number N_w of an ion is defined as the number of water molecules that have lost their translational degrees of freedom because of their association with the ion. Different methods give roughly concordant values for N_w. Smaller ions bind more water than larger ions, and cations somewhat more water than anions (since the positive charge is more effective in polarizing the negative electronic clouds of the solvent molecules).

The ionic radius of Na⁺ in crystals is about 0.095 nm, and of K⁺, about 0.133

*M. Born, *Z. Physik*, 1, 45 (1920).

†W. M. Latimer, K. S. Pitzer, C. M. Slansky, *J. Chem. Phys.*, 7, 108 (1939).

nm. In aqueous solution, this order is reversed and the effective radii of the hydrated ions are 0.24 nm for Na^+ and 0.17 nm for K^+. As a direct consequence of this difference in size of the hydrated ions, the membranes of living cells are generally much more permeable to K^+ ions than to Na^+ ions. Typically, the interior of the cell will have a higher concentration of K^+ ions than the exterior, and the reverse will be true for Na^+ ions. Such ionic concentration gradients are associated with differences in electric potential across the cell membranes. Many important physiological mechanisms are thus based on the hydration of ions and the consequent effects on ionic mobilities.

8. Transport Numbers and Mobilities

The fraction of the current carried by a given ion in solution is called the *transport number* or *transference number* of that ion.

From the Kohlrausch equation (10.7), the transference numbers t_0^+ and t_0^- of cation and anion at infinite dilution may be written

$$t_0^+ = \frac{\Lambda_0^+}{\Lambda_0}, \qquad t_0^- = \frac{\Lambda_0^-}{\Lambda_0} \tag{10.13}$$

The *mobility* u of an ion is defined as its velocity in the direction of an electric field of unit strength. The SI units are $m \cdot s^{-1}/V \cdot m^{-1}$ ($m^2 \cdot s^{-1} \cdot V^{-1}$). The conductivity κ can be defined by $i = \kappa E$, where i is the current through unit area and E is the electric field. It follows that

$$\kappa = Cu|ze| \tag{10.14}$$

where C is the number of charge carriers in unit volume and $|ze|$ is the absolute value of their charge. If there are several different carriers, we add their contributions to give $\kappa = \sum C_i |z_i e| u_i$. We see that two factors always determine the value of a conductivity: the concentration of mobile charges, and the mobility of the charge carriers.

The conductivity calculated for 1 F of charge in unit volume is the molar conductance Λ_i. Hence, when $N|ze| = F$ in (10.14), $\kappa = \Lambda_i$. Thus,

$$\Lambda_i = Fu_i = t_i \Lambda \tag{10.15}$$

This relation applies to each ion in a solution. If we know the transference number t_i of an ion, we can therefore calculate its mobility from the molar conductance Λ of the solution.

9. Measurement of Transport Numbers—Hittorf Method

The method of Hittorf is based on concentration changes in the neighborhood of the electrodes, caused by the passage of current through the electrolyte. Figure 10.3 illustrates the principle of the method by means of a cell divided into three compartments. The situation of the ions before the passage of any current is

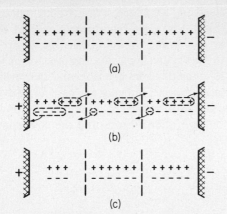

FIG. 10.3 Transport numbers (Hittorf method).

represented schematically as (a), each $+$ or $-$ sign indicating one equivalent of the corresponding ion.

Now let us assume that the mobility of the positive ion is three times that of the negative ion, $u_+ = 3u_-$. Let 4 F of electric charge be passed through the cell. At the anode, therefore, four equivalents of negative ions are discharged, and at the cathode, four equivalents of positive ions. Four faradays must pass across any boundary plane drawn through the electrolyte parallel to the electrodes. Since the positive ions travel three times faster than the negative ions, 3 F are carried across the plane from left to right by the positive ions while 1 F is being carried from right to left by the negative ions. This transfer is depicted in part (b) of the figure. The final situation is shown in (c). The change in number of equivalents around the anode is $\Delta n_a = 6 - 3 = 3$; around the cathode it is $\Delta n_c = 6 - 5 = 1$. The ratio of these concentration changes is necessarily identical with the ratio of the ionic mobilities: $\Delta n_a / \Delta n_c = u_+/u_- = 3$.

Suppose the amount of electricity Q passed through the cell was measured by a coulometer in series. Provided the electrodes are inert, Q/F equivalents of cations have therefore been discharged at the cathode, and Q/F equivalents of anions at the anode. The net loss of solute from the cathode compartment is

$$\Delta n_c = \frac{Q}{F} - t_+ \frac{Q}{F} = \frac{Q}{F}(1 - t_+) = \frac{Q t_-}{F}$$

Thus,

$$t_- = \frac{\Delta n_c}{Q/F}, \qquad t_+ = \frac{\Delta n_a}{Q/F} \tag{10.16}$$

where Δn_a is net loss of solute from the anode compartment. Since $t_+ + t_- = 1$, both transport numbers can be determined from measurements on either compartment, but it is useful to have both analyses as a check.

10. Transport Numbers—
Moving Boundary Method

This method is based on the early work of Oliver Lodge (1886), who used an indicator to follow the migration of ions in a conducting gel. More recent applications discard the gel and indicator and use an apparatus such as that in Fig. 10.4, to follow the moving boundary between two liquid solutions. The electrolyte to be studied, CA, is introduced into the apparatus in a layer above a solution of a salt with a common anion, $C'A$, and a cation with a mobility considerably less than that of the ion C^+. As an example, a layer of KCl solution could be introduced above a layer of $CdCl_2$ solution. The mobility of Cd^{2+} is considerably lower than

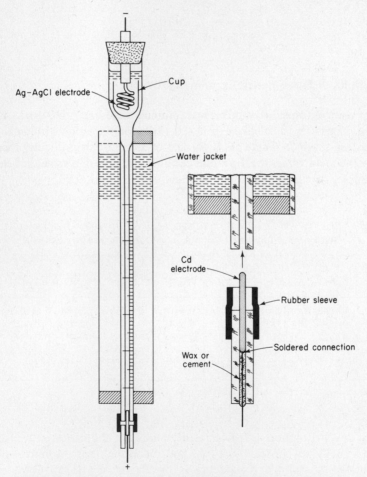

FIG. 10.4 Cell for measurement of transference number by the moving boundary method. [D. P. Shoemaker and C. W. Garland, *Experiments in Physical Chemistry* (New York: McGraw-Hill Book Company, 1967), p. 185.]

that of K^+. When a current is passed through the cell, A^- ions move downward toward the anode, while C^+ and C'^+ ions move upward toward the cathode. A sharp boundary is preserved between the two solutions, because the more slowly moving C'^+ ions never overtake the C^+ ions; nor do the following ions, C'^+, fall far behind, because if they began to lag, the solution behind the boundary would become more dilute, and its higher resistance and therefore steeper potential drop would increase the ionic velocity. Even with colorless solutions, a sharp boundary is visible owing to the different refractive indices.

Suppose the boundary moves a distance x for the passage of Q coulombs. The number of equivalents transported is then Q/F, of which $t_+ Q/F$ are carried by the positive ion. The volume of solution swept out by the boundary during the passage of Q coulombs is $t_+ Q/Fcz_+$. If a is the cross-sectional area of the tube, $xa = t_+ Q/Fcz_+$, or

$$t_+ = \frac{Fxacz_+}{Q} \tag{10.17}$$

11.   Results of Transference Experiments

Some measured transport numbers are summarized in Table 10.3. With these values, it is possible to calculate from (10.15) the molar ionic conductances, some of which are given in Table 10.4. By use of the Kohlrausch rule, they may be combined to yield values for the molar conductances Λ_0 of a variety of electrolytes. For example, Λ_0 ($\frac{1}{2}$ $BaCl_2$) would be $63.64 + 76.34 = 139.98$.

TABLE 10.3
Transport Numbers of Cations in Water Solutions at 298.15 K[a]

Solution normality	$AgNO_3$	$BaCl_2$	LiCl	NaCl	KCl	KNO_3	$LaCl_3$	HCl
0.01	0.4648	0.440	0.3289	0.3918	0.4902	0.5084	0.4625	0.8251
0.05	0.4664	0.4317	0.3211	0.3876	0.4899	0.5093	0.4482	0.8292
0.10	0.4682	0.4253	0.3168	0.3854	0.4898	0.5103	0.4375	0.8314
0.50		0.3986	0.300		0.4888		0.3958	
1.00		0.3792	0.287		0.4882			

[a]L. G. Longsworth, *J. Am. Chem. Soc.*, 57, 1185 (1935); 60, 3070 (1938).

Since ions in solution are undoubtedly hydrated, the observed transport numbers are actually not those of bare ions but of solvated ions. Mobilities of ions can be calculated from molar conductances by means of (10.15). Some results are given in Table 10.5. The effect of hydration is shown in the set of values for Li^+, Na^+, K^+. Although Li^+ is undoubtedly the smallest bare ion, it has the lowest mobility, i.e., the resistance to its motion through the solution is highest. This resistance must be partly due to a tightly held sheath of water molecules, bound by the intense electric field of the small ion.

TABLE 10.4
Molar Ionic Conductances at Infinite Dilution, Λ_0, 298.15 K[a]

Cation	$10^4\Lambda_0^+$ $(\Omega^{-1}\cdot m^2\cdot mol^{-1})$	Anion	$10^4\Lambda_0^-$ $(\Omega^{-1}\cdot m^2\cdot mol^{-1})$
H^+	349.82	OH^-	198.0
Li^+	38.69	Cl^-	76.34
Na^+	50.11	Br^-	78.4
K^+	73.52	I^-	76.8
NH_4^+	73.4	NO_3^-	71.44
Ag^+	61.92	CH_3COO^-	40.9
$\frac{1}{2}Ca^{2+}$	59.50	ClO_4^-	68.0
$\frac{1}{2}Ba^{2+}$	63.64	$\frac{1}{2}SO_4^{2-}$	79.8
$\frac{1}{2}Sr^{2+}$	59.46		
$\frac{1}{2}Mg^{2+}$	53.06		
$\frac{1}{3}La^{3+}$	69.6		

[a]D. MacInnes, *Principles of Electrochemistry* (New York: Reinhold Publishing Corp., 1939).

TABLE 10.5
Mobilities of Ions in Water Solutions at 298.15 K

Cations	Mobility $(m^2\cdot s^{-1}\cdot V^{-1})$	Anions	Mobility $(m^2\cdot s^{-1}\cdot V^{-1})$
H^+	36.30×10^{-8}	OH^-	20.52×10^{-8}
K^+	7.62×10^{-8}	SO_4^{2-}	8.27×10^{-8}
Ba^{2+}	6.59×10^{-8}	Cl^-	7.91×10^{-8}
Na^+	5.19×10^{-8}	NO_3^-	7.40×10^{-8}
Li^+	4.01×10^{-8}	HCO_3^-	4.61×10^{-8}

12. Mobilities of Hydrogen and Hydroxyl Ions

Table 10.5 reveals that, with two exceptions, the ionic mobilities in aqueous solutions do not differ in order of magnitude, all being about 6×10^{-8} $m^2\cdot s^{-1}\cdot V^{-1}$. The exceptions are the hydrogen and hydroxyl ions with the abnormally high mobilities of 36.3×10^{-8} and 20.5×10^{-8} $m^2\cdot s^{-1}\cdot V^{-1}$ respectively.

The high mobility of the hydrogen ion is observed only in hydroxylic solvents, such as water and the alcohols, in which it is strongly solvated—e.g., in water to the hydronium ion, OH_3^+. It is believed to be an example of a Grotthuss type of conductivity, superimposed on the normal transport process. Thus, the OH_3^+ ion is able to transfer a proton to a neighboring water molecule,

$$
\underset{+}{H-\overset{\overset{\displaystyle H}{|}}{O}-H} + \overset{\overset{\displaystyle H}{|}}{O}-H \longrightarrow H-\overset{\overset{\displaystyle H}{|}}{O} + \underset{+}{H-\overset{\overset{\displaystyle H}{|}}{O}-H}
$$

This process may be followed by the rotation of the donor molecule so that it is again in a position to accept a proton.

$$
\begin{array}{ccc}
\text{H} & & \text{H} \\
| & \longrightarrow & | \\
\text{H—O} & & \text{O—H}
\end{array}
$$

The high mobility of the hydroxyl ion in water is also believed to be caused by a proton transfer, between hydroxyl ions and water molecules,

$$
\begin{array}{cccc}
\text{H} & \text{H} & \text{H} & \text{H} \\
| & | & | & | \\
\text{O} + \text{H—O} & \longrightarrow & \text{O—H} + \text{O} \\
{\scriptstyle -} & & & {\scriptstyle -}
\end{array}
$$

Protons play an important part in electrical phenomena in living systems, and the exact mechanisms by which they move are being actively investigated. Eigen* has suggested that the predominant form of the proton in water is the ion $H_9O_4^+$, consisting of a hydronium ion OH_3^+ that holds three water molecules by hydrogen bonds (Section 15.29).

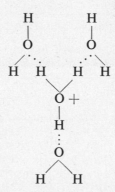

Other water molecules might be more loosely attached to this complex, but the structure shown has exceptional stability. When a proton passes along the hydrogen bond, a new hydronium ion is formed, which at first is not coordinated with its three water molecules. There will therefore be a certain time needed for rotation of the water molecules before this coordination sheath is established about the newly formed OH_3^+. The reformation of the complex $H_9O_4^+$ is believed to be the slow step that determines the overall mobility of protons in water.† It is a sort of "structural diffusion" of the water bound to the OH_3^+. The mobility of the proton in ice appears to be about 50 times higher than that in water at 273 K. In the structure of ice, a proton can simply pass from one site to the next in the rigid structure, and there is no migration of water associated with its motion. Hence, the higher mobility in ice reflects the true rate of the proton transfer along the hydrogen bond.‡ It is so fast as to suggest a quantum mechanical tunneling of the H^+ from one site to the next (Section 13.22).

*M. Eigen, *Proc. Roy. Soc. London, Ser. A, 247*, 505 (1958).
†B. E. Conway, J. O'M. Bockris, and H. Linton, *J. Chem. Phys., 24*, 834 (1956).
‡B. E. Conway and J. O'M. Bockris, *J. Chem. Phys., 28*, 354 (1958).

13.   Diffusion and Ionic Mobility

The speed v of an ion of charge Q in an electric field $\mathbf{E}$ is related to its mobility u by $v = Eu$. The driving force in such ionic migration is the negative gradient of the *electric potential* Φ: $E = -\partial\Phi/\partial x$. Even in the absence of an external electric field, however, ions can migrate if there is a difference in *chemical potential* μ between different parts of the system. The migration of a substance under the action of a difference in chemical potential is called *diffusion*. Just as the electric force (per unit charge) on each particle equals the negative gradient of electric potential, so the diffusive force equals the negative gradient of chemical potential. In one dimension, therefore, the force on a particle of the ith kind is

$$F_i = -\frac{1}{L}\left(\frac{\partial\mu_i}{\partial x}\right)_T$$

Since μ_i refers to 1 mol of particles, it has been divided by Avogadro's Number L. The velocity under the action of unit force is u/Q, so that $v_i = (-u_i/LQ_i)(\partial\mu_i/\partial x)$. The net flow of material through unit cross section in unit time is, therefore,

$$S_{ix} = -\frac{u_i}{Q_i}\frac{c_i}{L}\cdot\left(\frac{\partial\mu_i}{\partial x}\right)_T$$

where c_i is the molar concentration. For a sufficiently dilute solute,

$$\mu_i = RT\ln c_i + \mu_i^{\ominus}, \quad\text{and}\quad \left(\frac{\partial\mu_i}{\partial x}\right)_T = \frac{RT}{c_i}\left(\frac{\partial c_i}{\partial x}\right)_T$$

Hence,

$$S_{ix} = -kT\frac{u_i}{Q_i}\left(\frac{\partial c_i}{\partial x}\right)_T$$

In 1855, Fick stated in his empirical First Law of Diffusion that the flow S_{ix} is proportional to the gradient of concentration:

$$S_{ix} = -D_i\frac{\partial c_i}{\partial x} \tag{10.18}$$

Thus, the *diffusion coefficient*,

$$D_i = \frac{kT}{Q_i}u_i \tag{10.19}$$

Or, from (10.15),

$$D_i = \frac{RT}{\mathbf{F}^2}\frac{\Lambda_i}{|z_i|}$$

This equation, derived by Nernst[*] in 1888, indicates that diffusion experiments can yield information about ionic mobilities similar to that obtained from conductivity data.[†] Equation (10.19) obviously applies to the diffusion of a single

---

[*]*Z. Physik Chem.*, **2**, 613 (1888).

[†]Experimental methods for measuring diffusion coefficients in solutions are discussed in detail by A. L. Geddes and R. B. Pontius, "Determination of Diffusivity," in *Technique of Organic Chemistry*, Vol. 1, Pt. 2, 3rd ed., ed. by A. Weissberger (New York: Interscience Publishers, 1960), pp. 895–1006.

ionic species only. An experimental example would be the diffusion of a small amount of HCl dissolved in a solution of KCl. The Cl⁻ concentrations would be constant throughout the system, and the experiment would measure diffusion of the H⁺ ions alone. In other cases, such as diffusion of salts from concentrated to dilute solution, it is necessary to use a suitable average value of the diffusion coefficients of the ions to represent the overall D. For instance, Nernst showed that for electrolytes of type CA, the proper average is

$$D = \frac{2D_C D_A}{D_C + D_A}$$

14.   Defects of the Arrhenius Theory

After the controversy over ionic dissociation, it began to be realized that the Arrhenius theory was unsatisfactory on a number of points, none of which was among those urged against it by its fierce original opponents.

The behavior of strong electrolytes presented many anomalies. The Ostwald Dilution Law was not closely followed by moderately strong electrolytes such as dichloroacetic acid, although it agreed well with the data for weak acids, such as acetic acid. Also, values for the degree of dissociation α of strong electrolytes obtained from conductance ratios were not in agreement with those from van't Hoff i factors, and the "dissociation constants" calculated by the mass action law were far from constant. The absorption spectra of dilute solutions of strong electrolytes revealed no evidence for undissociated molecules.

Another discrepancy was in the heats of neutralization of strong acids and bases. Although one of the first supports for the ionization theory was the constancy of these ΔH values for different acid–base pairs, more critical examination indicated that the ΔH values were actually too concordant to satisfy the theory. According to Arrhenius, there should have been small differences in the extents of ionization of acids such as HCl, H_2SO_4, and HNO_3, at any given concentration, and these differences should have been reflected in corresponding differences in the ΔH values, but such distinctions were not, in fact, observed.

As early as 1902, a possible explanation of many of the deficiencies of the simple dissociation theory was suggested by van Laar, who called attention to the strong electrostatic forces that must be present in an ionic solution and their influence on the behavior of the dissolved ions. In 1912, S. R. Milner gave a detailed discussion of this problem, but his excellent results were not widely understood.

In 1923, P. Debye and E. Hückel devised a theory which is the basis for the modern treatment of strong electrolytes. It starts with the assumption that in strong electrolytes the solute is completely dissociated into ions. The observed deviations from ideal behavior, e.g., apparent degrees of dissociation of less than 100%, are ascribed entirely to the electrical interactions between ions in solution. These deviations are therefore greater with more highly charged ions and in more concentrated solutions.

The electrical-interaction theory can be applied to equilibrium problems, and also to the important transport problems in the theory of electrical conductivity. Before describing these applications, we shall discuss the nomenclature and conventions employed for the thermodynamic properties of electrolytic solutions.

15. Activities and Standard States

As shown in Section 8.17., the standard state for a component considered as a solute B is based on Henry's Law. In the limit as $X_B \longrightarrow 0$, $a_B \longrightarrow X_B$, and the activity coefficient $\gamma_B \longrightarrow 1$. The departure of γ_B from unity is a measure of the departure of the behavior of the solute from that prescribed by Henry's Law. Henry's Law implies the absence of interaction between molecules of solute (component B sees only solvent A surrounding it). Therefore, the deviations of the activity coefficients from unity measure the effects of interactions between solute species in solution.

We need to make two changes in the definition of standard state given in Section 8.17 before we use it for solutions of electrolytes. The composition of electrolytic solutions is almost always expressed in molalities instead of mole fractions. Also, we need to consider the effect of dissociation of the electrolyte, by which a molecule of added solute yields two or more molecules or ions in the solution. Thus, for NaCl in water, the limiting form of Henry's Law would be*

$$f_B = km_B^2 \tag{10.20}$$

where f_B and m_B are the fugacity and molality of the NaCl. For an electrolyte yielding ν particles on dissociation,

$$f_B = km_B^\nu$$

The usual standard state for an electrolyte in solution is illustrated in Fig. 10.5, for the case of NaCl. It is a hypothetical state in which the solute would exist at unit molality and 1 atm pressure but would still have the environment typical of an extremely dilute solution that followed Henry's Law [Eq. (7.32)].

The activity of B is, therefore,

$$a_B = \frac{f_B}{f_B^\ominus} \quad \text{where} \quad f_B^\ominus = k$$

and

$$a_B = \gamma_B m_B$$

We have not used a special symbol to distinguish these activities and standard states from those based on the choice of mole fraction as the variable of composition. It is clear that they are not the same, but no confusion can arise because we shall use only molalities throughout this chapter.

---

*Consider a molecule that dissociates in solution as $B \longrightarrow 2A$. Then the equilibrium constant $K_a = a_A^2/a_B$. In very dilute solution, dissociation is practically complete, and $a_A \longrightarrow m_A$. Hence, $a_B = K_a^{-1}m_A^2$. But when dissociation is complete, m_A is twice the molality of B initially added, $2m_B^0$, so that $a_B = K_a^{-1}(2m_B^0)^2$.

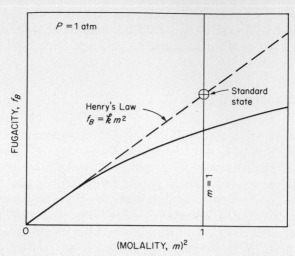

FIG. 10.5 Definition of standard state for a 1 : 1 electrolyte like NaCl in solution, based on Henry's Law in dilute solutions.

16. Ion Activities

In dealing with electrolytic solutions, it would apparently be most convenient to use the activities of the different ionic species, but there are serious difficulties in the way of such a procedure. The requirement of overall electrical neutrality in the solution prevents any increase in the charge due to positive ions without an equal increase in the charge due to negative ions. For example, we can change the concentration of a solution of sodium chloride by adding equal numbers of sodium and chloride ions. If we could possibly add sodium ions alone or chloride ions alone, the solution would acquire a net electric charge. The properties of ions in such a charged solution would differ considerably from their properties in the normal uncharged solution. There is no way to separate effects due to positive ions from those due to the accompanying negative ions in an uncharged solution. Thus there is no way to *measure* individual ion activities.

Nevertheless, it is convenient to define an expression for the activity of an electrolyte in terms of the ions into which it dissociates. Consider, for example, a solute like NaCl, dissociated in solution according to $NaCl \rightarrow Na^+ + Cl^-$. The activity a of the NaCl is readily measurable from osmotic pressure, freezing point depression, or in other ways. If we denote the activity of the cation as a_+ and that of the anion as a_-, we can write *as definitions*,

$$a = a_+ a_- = a_\pm^2 \tag{10.21}$$

If we recall that an activity is always a ratio of a fugacity to the fugacity in a standard state which we are free to choose, we may regard* (10.21) as defining the

*Another way of looking at (10.21) is that it amounts to setting $\Delta G^\ominus = 0$ for the reaction $NaCl \rightarrow Na^+ + Cl^-$, since then $K = a_+ a_-/a = e^{-\Delta G^\ominus/RT} = 1$. In other words, we choose the standard states of Na^+ and Cl^- so that $\Delta G^\ominus = 0$.

standard states for the conventional individual ionic activities a_+ and a_-. The quantity $a_\pm$, the *geometric mean* of a_+ and a_-, is called the *mean activity* of the ions.

For more complex types of electrolytes, these definitions can be generalized. Consider an electrolyte that dissociates as

$$C_{v_+}A_{v_-} \longrightarrow v_+C^+ + v_-A^-$$

The total number of ions is $v = v_+ + v_-$. We then write

$$a = a_+^{v_+} a_-^{v_-} = a_\pm^v \tag{10.22}$$

For example,

$$La_2(SO_4)_3 \longrightarrow 2La^{+3} + 3SO_4^{-2}$$

$$a = a_{La}^2 a_{SO_4}^3 = a_\pm^5$$

We can also define individual ionic activity coefficients γ_+ and γ_- by

$$a_+ = \gamma_+ m_+ \quad \text{and} \quad a_- = \gamma_- m_- \tag{10.23}$$

The experimentally measured activity coefficient is $\gamma_\pm$, the geometric mean of the individual ionic coefficients, where

$$\gamma_\pm^v = \gamma_+^{v_+} \gamma_-^{v_-} \tag{10.24}$$

Equation (10.22) can then be written

$$a = m_+^{v_+} m_-^{v_-} \gamma_+^{v_+} \gamma_-^{v_-}$$

or

$$a_\pm = a^{1/v} = (m_+^{v_+} m_-^{v_-} \gamma_+^{v_+} \gamma_-^{v_-})^{1/v} \tag{10.25}$$

Substituting (10.24) into (10.25), we obtain

$$\gamma_\pm = \frac{a_\pm}{(m_+^{v_+} m_-^{v_-})^{1/v}} \tag{10.26}$$

This equation applies in any solution, whether the ions are added together as a single salt or added separately as a mixture of salts.

For a solution of a single salt, of molality m,

$$m_+ = v_+ m \quad \text{and} \quad m_- = v_- m$$

In this case, (10.26) becomes

$$\gamma_\pm = \frac{a_\pm}{m(v_+^{v_+} v_-^{v_-})^{1/v}} = \frac{a_\pm}{m_\pm} \tag{10.27}$$

In the case of $La_2(SO_4)_3$, for example, $v_+ = 2$ and $v_- = 3$, so that

$$\gamma_\pm = \frac{a_\pm}{m(2^2 \cdot 3^3)^{1/5}} = \frac{a_\pm}{108^{1/5} m} \tag{10.28}$$

The activity coefficient as defined in (10.27) becomes unity at infinite dilution.

The activities can be determined by several different methods. Among the most important are measurements of the colligative properties of solutions, such as freezing point depression and osmotic pressure, measurements of the solubilities of sparingly soluble salts, and methods based on the emf of electrochemical cells. We shall describe the emf method in the next chapter.

17. Activity Coefficients from Freezing Points

For a two-component solution with solute (1) and solvent (0), the Gibbs–Duhem equation (7.10) for chemical potentials in a binary solution may be written

$$n_1 \, d\mu_1 + n_0 \, d\mu_0 = 0$$

Combination with (8.47) yields

$$n_1 \, d\ln a_1 + n_0 \, d\ln a_0 = 0$$

Equation (7.35) applies to an ideal solution. For a dilute, nonideal solution, we have, instead,

$$\frac{d\ln a_0}{dT} = \frac{\Delta H_f}{RT^2}$$

The freezing point depression $\Delta T = T_0 - T$, and since $T^2 \approx T_0^2$,

$$-d\ln a_0 = \frac{\Delta H_f}{RT_0^2} \, d(\Delta T)$$

Therefore,

$$d\ln a_1 = -\left(\frac{n_0}{n_1}\right) d\ln a_0 = \left(\frac{n_0}{n_1}\right)\left(\frac{\Delta H_f}{RT_0^2}\right) d(\Delta T) = \frac{1}{mK_F} d(\Delta T)$$

Here, K_F is the molal freezing point depression constant (p. 249), and m is molality of solute.

From (10.27),

$$a_1 = a_\pm^\nu = \gamma_\pm^\nu m^\nu (\nu_+^{\nu_+} \nu_-^{\nu_-})$$

so that,

$$d\ln a_\pm = d\ln \gamma_\pm m = d\ln \gamma_\pm + d\ln m = \frac{d(\Delta T)}{\nu m K_F} \tag{10.29}$$

Let $j = 1 - (\Delta T/\nu m K_F)$, whereupon,

$$dj = \frac{-d(\Delta T)}{\nu m K_F} + \left(\frac{\Delta T}{\nu m^2 K_F}\right) dm$$

or,

$$\frac{d(\Delta T)}{\nu m K_F} = -dj + (1 - j)\frac{dm}{m}$$

By comparing this with (10.29), we have

$$d\ln \gamma_\pm = -dj - j \, d\ln m \tag{10.30}$$

As m approaches 0, the solution approaches ideality, $\gamma_\pm \to 1$, and $j \to 0$ (since for an ideal solution $\Delta T/\nu m K_F = 1$.) Therefore, on integration of (10.30), we obtain

$$\int_1^{\gamma_\pm} d\ln \gamma_\pm' = \int_0^m -j \, d\ln m' - \int_0^j dj'$$

$$\ln \gamma_\pm = -j - \int_0^m \left(\frac{j}{m'}\right) dm' \tag{10.31}$$

The integration in this expression can be carried out graphically from a series of measurements of the freezing point depression in solutions of known low concentrations. We plot j/m vs. m, extrapolate to zero concentration, and measure the area under the curve. A similar treatment is applicable to osmotic-pressure data.

18. The Ionic Strength

Many properties of ionic solutions depend on electrostatic interactions between ionic charges. The electrostatic force between a pair of doubly charged ions is four times the force between a pair carrying unit charges. A useful function of ionic concentration, devised to include such effects of ionic charge, is the ionic strength I, defined by

$$I = \tfrac{1}{2} \sum m_i z_i^2 \tag{10.32}$$

The summation is taken over all the different ions in a solution, multiplying the molality of each by the square of its charge.

For example, a 1.00 molal solution of NaCl would have an ionic strength $I = \tfrac{1}{2}(1.00) + \tfrac{1}{2}(1.00) = 1.00$. A 1.00 molal solution of $La_2(SO_4)_3$ would have

$$I = \tfrac{1}{2}[2(3)^2 + 3(2^2)] = 15.0$$

In dilute solutions, the activity coefficients of electrolytes, the solubilities of sparingly soluble salts, rates of ionic reactions, and other related properties become functions of ionic strength.

If the molar concentration c is used instead of the molality m,

$$c = \frac{m\rho}{1 + mM}$$

where ρ is the density of the solution and M is the molar mass of the solute. In dilute solution, this relation approaches $c = \rho_0 m$, where ρ_0 is the density of the solvent. Therefore,

$$I = \tfrac{1}{2} \sum m_i z_i^2 \approx \frac{1}{2\rho_0} \sum c_i z_i^2 \tag{10.33}$$

19. Experimental Activity Coefficients

Mean activity coefficients obtained by various methods* are summarized in Table 10.6 and plotted in Fig. 10.6. For comparison, the activity coefficient of a typical nonelectrolyte, sucrose, is also shown. Quite typically the coefficients for electrolytes decline markedly with increasing concentration in dilute solution, but then pass through minima and rise again in more concentrated solutions. The interpretation of this behavior constitutes one of the principal problems in the theory of strong electrolytes.

*An extensive tabulation was given by H. S. Harned and B. B. Owen, *The Physical Chemistry of Electrolytic Solutions* (New York: Reinhold Publishing Corp., 1950).

TABLE 10.6
Mean Molal Activity Coefficients of Electrolytes

m	0.001	0.002	0.005	0.01	0.02	0.05	0.1	0.2	0.5	1.0	2.0	4.0
HCl	0.966	0.952	0.928	0.904	0.875	0.830	0.796	0.767	0.758	0.809	1.01	1.76
HNO_3	0.965	0.951	0.927	0.902	0.871	0.823	0.785	0.748	0.715	0.720	0.783	0.982
H_2SO_4	0.830	0.757	0.639	0.544	0.453	0.340	0.265	0.209	0.154	0.130	0.124	0.171
NaOH						0.82		0.73	0.69	0.68	0.70	0.89
$AgNO_3$			0.92	0.90	0.86	0.79	0.72	0.64	0.51	0.40	0.28	
$CaCl_2$	0.89	0.85	0.785	0.725	0.66	0.57	0.515	0.48	0.52	0.71		
$CuSO_4$	0.74		0.53	0.41	0.31	0.21	0.16	0.11	0.068	0.047		
KCl	0.965	0.952	0.927	0.901		0.815	0.769	0.719	0.651	0.606	0.576	0.579
KBr	0.965	0.952	0.927	0.903	0.872	0.822	0.777	0.728	0.665	0.625	0.602	0.622
KI	0.965	0.951	0.927	0.905	0.88	0.84	0.80	0.76	0.71	0.68	0.69	0.75
LiCl	0.963	0.948	0.921	0.89	0.86	0.82	0.78	0.75	0.73	0.76	0.91	1.46
NaCl	0.966	0.953	0.929	0.904	0.875	0.823	0.780	0.730	0.68	0.66	0.67	0.78

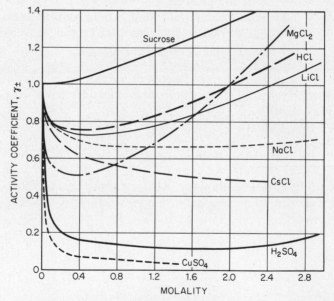

FIG. 10.6   Mean molal activity coefficients of electrolytes, with values
for sucrose shown for comparison.

We can see in a general way why the activity coefficient as a function of concentration passes through a minimum. Consider the chemical potential of the
ionic solute,

$$\mu = \mu^\ominus + RT \ln \gamma m$$

The interionic attractions lower the free energy of the ions, and hence γ tends to
decrease. On the other hand, the ions also exert attractive forces on the water

molecules, so that the free energy of the water is lowered,* and, as shown in the Gibbs–Duhem relation (Section 8.18), this effect raises the activity coefficient of the solute. The two opposite effects outlined may lead to a minimum in γ vs. m. Actually, however, the effects in concentrated solutions are too complicated for any simple model.

20. A Review of Electrostatics

Newton's Law of Gravitation was stated on p. 4. An equation of exactly similar form governs the electrostatic force between two charges Q_1 and Q_2 separated by a distance r in a vacuum. The magnitude of the force is

$$F = \frac{Q_1 Q_2}{K r^2} \tag{10.34}$$

This is Coulomb's Law. Two choices are available with regard to the proportionality constant K. If K is set equal to unity, Eq. (10.34) defines a unit of charge (electrostatic units, esu). The force between two charges of 1 esu each at a distance of 1 cm is 1 dyne (dyn). In the SI system, the unit of charge is the coulomb, and the constant K is written as $(4\pi\epsilon_0)$ so that (10.34) becomes

$$F = \frac{Q_1 Q_2}{4\pi\epsilon_0 r^2} \tag{10.35}$$

The ϵ_0 is called the *permittivity of vacuum* and has the value 8.854×10^{-12} $C^2 \cdot N^{-1} \cdot m^{-2}$ (or $C^2 \cdot J^{-1} \cdot m^{-1}$). At the slight inconvenience of including the extra factor in Coulomb's Law, we have obtained consistency between electrical and mechanical units in the SI system. The force between two charges of 1 C separated by a distance of 1 m is $(4\pi\epsilon_0)^{-1}$N. Also, the joule = newton metre = volt coulomb.

The electric field $\mathbf{E}$ at any point is the force exerted on a unit test charge (1 C) placed at that point. The field at a distance r from a charge Q at the origin of coordinates is, therefore,

$$\mathbf{E} = \frac{Q\,\mathbf{r}}{4\pi\epsilon_0 r^3} \tag{10.36}$$

The equation indicates that the direction of the vector field due to a positive charge is the same as the direction of the vector from the origin to the point of measurement [Fig. 10.7(a)]. The absolute magnitude of the field would be written

$$|E| = \frac{Q}{4\pi\epsilon_0 r^2}$$

The field due to any collection of charges may be obtained by simply adding the fields due to each individual charge. Since the force (and the field) are vectorial quantities, the addition must be made of the vectors. A unit electric field exerts a force of 1 N on a charge of 1 C. The unit of field strength is therefore newton per

*For example, if the hydration number of NaCl is about 6, in a 1.0 molal solution of NaCl only about $\frac{49}{55}$ of the water is "free solvent."

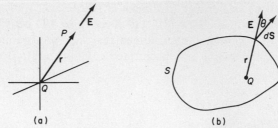

FIG. 10.7   (a) The electric field at P due to a positive charge Q at the origin is equal in magnitude to the force **F** on a unit test charge placed at P. The direction of the field is that of the vector from Q to P and the magnitude of the field is $|E| = Q/4\pi\epsilon_0 r^2$. (b) To illustrate Gauss's theorem we show an arbitrary surface S surrounding an electric charge Q. Associated with each little element of area of the surface $d\mathbf{S}$ is a vector normal to that area $d\mathbf{S}$. The flux of electric field through $d\mathbf{S}$ is then $\mathbf{E} \cdot d\mathbf{S} = E \cos \theta \, dS$, where θ is the angle between the field direction and the normal to $d\mathbf{S}$.

coulomb, or, since joule = volt coulomb = newton metre, the unit is usually cited as volt per metre, $V \cdot m^{-1}$.

At any point in space, the electric field will be specified by a single vector. If the test charge is moved about, we can draw vectors showing the direction of the field at each point. If these vectors are joined, they yield a representation of the *lines of force* characteristic of the vector field. (The field vectors give the tangents of the lines of force at any point). The number of lines of force passing through a small unit area normal to the direction of **E** is made proportional to the magnitude of the field strength, E. Thus, in regions where lines of force are dense, the field is higher than where lines of force are sparse. Figure 10.8 shows this graphic representation of the field in the neighborhood of various charge distributions.

Just as Eq. (1.11) represented the mechanical force as the gradient of a potential U, so the electric field can be represented as the gradient of an electric potential Φ. In terms of the three components of the vector field,

$$E_x = -\frac{\partial \Phi}{\partial x}, \qquad E_y = -\frac{\partial \Phi}{\partial y}, \qquad E_z = -\frac{\partial \Phi}{\partial z} \qquad (10.37)$$

or, in terms of vector notation,

$$\mathbf{E} = -\text{grad } \Phi = -\nabla\Phi \qquad (10.38)$$

The negative sign indicates that a positive charge will move from higher to lower potential, and work must be done on it to move it in the opposite direction. It is usually easier to carry out mathematical calculations with the scalar potential function $\Phi(x, y, z)$, and it is a simple matter to calculate from Φ the field at any point by means of (10.37) or (10.38). From (10.37), it is evident that the unit of electric potential is the *volt*.

We shall next demonstrate Gauss's theorem by reference to Fig. 10.7(b), which shows a closed surface S surrounding a charge Q. The number of lines of

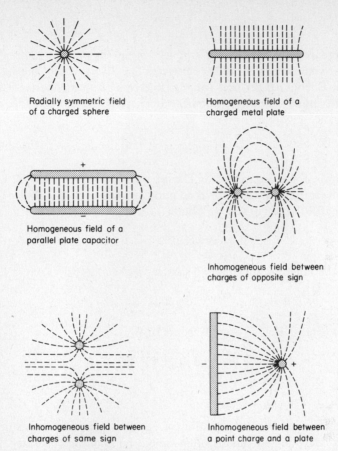

FIG. 10.8 Lines of force in different electric fields.

force passing through an element of area dS is $\mathbf{E} \cdot d\mathbf{S} = E \cos \theta \, dS$, where θ is the angle between $\mathbf{E}$ and the normal to $d\mathbf{S}$. The element of solid angle $d\omega$ subtended by dS at the charge Q is $dS \cos \theta / r^2$, and the total solid angle subtended by the closed surface S at any point within the surface is 4π. Hence, we can write, from (10.36),

$$\int \mathbf{E} \cdot d\mathbf{S} = \int E \cos \theta \, dS = \frac{Q}{4\pi\epsilon_0} \int d\omega = \frac{Q}{\epsilon_0} \qquad (10.39)$$

which is Gauss's theorem.

If the surface S encloses not an individual charge but a distribution of charge of varying density ρ, the theorem can be written

$$\int \mathbf{E} \cdot d\mathbf{S} = \frac{1}{\epsilon_0} \int \rho \, dV \qquad (10.40)$$

where the charge density is integrated over the volume V enclosed by the surface.

We now apply to (10.40) the divergence theorem of vector analysis. The divergence of any vector $\mathbf{E}$ is

$$\text{div } \mathbf{E} = \nabla \cdot \mathbf{E} = \frac{\partial E_x}{\partial x} + \frac{\partial E_y}{\partial y} + \frac{\partial E_z}{\partial z} \tag{10.41}$$

The divergence of a vector is a simple concept in physical terms.* Consider a differential volume element $dV = dx\, dy\, dz$. The divergence of the vector $\mathbf{E}$ is then the flux of $\mathbf{E}$ from dV per unit volume. In terms of lines of force, it would be the number leaving dV minus the number entering dV. We can see at once, therefore, the divergence theorem

$$\int \text{div } \mathbf{E} \, dV = \int \mathbf{E} \cdot d\mathbf{S} \tag{10.42}$$

The integral of the divergence of $\mathbf{E}$ over the volume enclosed by a surface S is equal to the integral of the normal component of $\mathbf{E}$ over the surface S.

From (10.40) and (10.42), we obtain

$$\int \text{div } \mathbf{E} \, dV = \frac{1}{\epsilon_0} \int \rho \, dV$$

Or, since the integrals are equal no matter what volume is taken,

$$\text{div } \mathbf{E} = \frac{\rho}{\epsilon_0}$$

If we now express $\mathbf{E}$ as $-$ grad Φ from (10.40),

$$\text{div grad } \Phi = \nabla\nabla \cdot \Phi = \frac{-\rho}{\epsilon_0}$$

$$\nabla^2 \Phi = \frac{-\rho}{\epsilon_0} \tag{10.43}$$

This is the famous Poisson equation of electrostatics. The operator div grad $= \nabla^2$ is read "del squared" and is often called the *Laplacian operator*. If a region is free of charge, $\rho = 0$, and (10.43) becomes

$$\nabla^2 \Phi = 0 \tag{10.44}$$

which is the Laplace equation.

The Laplacian operator ∇^2 can be written in ordinary Cartesian coordinates as

$$\nabla^2 \equiv \frac{\partial^2}{\partial x^2} + \frac{\partial^2}{\partial y^2} + \frac{\partial^2}{\partial z^2} \tag{10.45}$$

It is an interesting exercise† to transform this expression into spherical polar coordinates to obtain

$$\nabla^2 \equiv \frac{\partial^2}{\partial r^2} + \frac{2}{r}\frac{\partial}{\partial r} + \frac{1}{r^2}\frac{\partial^2}{\partial \theta^2} + \frac{\cos\theta}{r^2 \sin\theta}\frac{\partial}{\partial \theta} + \frac{1}{r^2 \sin^2\theta}\frac{\partial^2}{\partial \phi^2}. \tag{10.46}$$

---

*More detailed mathematical derivations of (10.42) can be found in any book that includes vector analysis, e.g., M. L. Boas, *Mathematical Methods in the Physical Sciences* (New York: John Wiley & Sons, Inc., 1966), p. 242.

†The transformation is worked out in detail in H. Hameka, *Introduction to Quantum Theory* (New York: Harper & Row, Publishers, 1967), Appendix A.

21. The Debye–Hückel Theory

The theory of Debye and Hückel is based on the assumption that strong electrolytes are completely dissociated into ions. Observed deviations from ideal behavior are then ascribed to electrical interactions between ions. To obtain theoretically the equilibrium properties of ionic solutions, it is necessary to calculate the extra free energy arising from these electrostatic interactions.

If the ions were distributed completely at random, the chances of finding either a positive or a negative ion in the neighborhood of a given ion would be identical. Such a random distribution would have no electrostatic energy, since, on the average, attractive configurations would be exactly balanced by repulsive ones. It is evident that this cannot be the physical situation, since in the immediate neighborhood of a positive ion, a negative ion is more likely to be found than another positive ion. Indeed, were it not for the fact that ions are continually being batted about by molecular collisions, an ionic solution might acquire a well ordered structure similar to that of an ionic crystal. The thermal motions effectively prevent any complete ordering, but the final situation is a dynamic compromise between the electrostatic interactions tending to produce ordered configurations and the kinetic collisions tending to destroy them. Our problem is to calculate the average electric potential Φ of a given ion in the solution due to all the other ions. Knowing Φ, we can calculate the work that must be expended to charge the ions reversibly to this potential, and this work will be the free energy due to electrostatic interactions. The extra electric free energy is directly related to the ionic activity coefficient, since both are a measure of the deviation from ideality.

Since the Debye–Hückel theory is an *electrostatic theory*, there is nothing to prevent our using it to calculate activity coefficients for single ions without reference to the difficulties mentioned in the thermodynamic definition of these quantities. Thermodynamics is concerned with relations between measurable quantities. Electrostatics is based on an abstract model for natural phenomena (point charges, electrostatic fields, etc.). As with computations based on any other model, it will be necessary finally to combine the calculated quantities in such a way as to provide a predicted value of some *measurable quantity*.

22. The Poisson–Boltzmann Equation

On the average, a given ion will be surrounded by a spherically symmetrical distribution of other ions, forming the *ionic atmosphere*. Figure 10.9 depicts a central ion with a section of a spherical shell at a distance r. The closest approach of any other ion to the center is designated by a. We wish to calculate the average electrostatic potential $\Phi(r)$ due to the central ion and its surrounding atmosphere. The value of $\Phi(r)$ is determined by the average density of electric charge ρ, which can vary from point to point in the solution, being equal to the number of charges in any small region of the solution divided by the volume of that region, Q/dV.

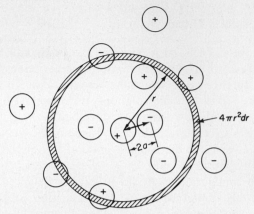

FIG. 10.9 A central positive ion of radius *a* surrounded by an ionic atmosphere of positive and negative ions of radius *a*. The problem is to calculate from electrostatic theory the electric potential at the central ion due to the surrounding ionic atmosphere.

The potential Φ is related to the charge density ρ by the Poisson differential equation. In the case of a spherically symmetric distribution of charge about a central ion, the potential Φ is a function only of r. In this case, therefore, the Poisson equation (10.43) becomes

$$\frac{1}{r^2}\frac{d}{dr}\left(r^2\frac{d\Phi}{dr}\right) = \frac{-\rho}{\epsilon_0\epsilon} \tag{10.47}$$

The use of the ordinary (macroscopic) dielectric constant ϵ is a basic flaw in the Debye–Hückel model, since dipolar solvent molecules in the immediate neighborhood of an ion will not be able to orient themselves freely in an external electric field, and, consequently, the effective (microscopic) dielectric constant may be much lower than the value for bulk solvent.*

To obtain solutions of (10.47) for $\Phi(r)$, we must calculate the charge density ρ as a function of Φ and substitute it into the equation. Debye and Hückel used the Boltzmann theorem (Section 5.9) to calculate ρ. Equation (5.29) indicates that if C_i is the average number of ions of kind i in unit volume of solution, the number C_i' in unit volume [of ions] which have an energy E greater than the average, is given by

$$C_i' = C_i\,e^{-E/kT} \tag{10.48}$$

The work required to bring an ion of charge Q_i into a region of potential Φ is $Q_i\Phi$. Since over the whole solution of positive and negative ions this work averages to zero, we can set the extra energy E above the average in (10.48) equal to $Q_i\Phi$, so that

$$C_i' = C_i\,e^{-Q_i\Phi/kT} \tag{10.49}$$

*E. Hückel, *Physik. Z.*, *26*, 93 (1925).

The charge density ρ in a region at potential Φ is simply the summation of the number of ions per unit volume given by (10.49), over all the different kinds of ions present in the solution, each multiplied by its own charge Q_i. Thus,

$$\rho = \sum C_i' Q_i = \sum C_i Q_i \, e^{-Q_i \Phi / kT} \tag{10.50}$$

Combination of (10.47) with (10.50) then yields

$$\frac{1}{r^2} \frac{d}{dr} r^2 \left(\frac{d\Phi}{dr} \right) = -\frac{1}{\epsilon_0 \epsilon} \sum C_i Q_i \, e^{-Q_i \Phi / kT} \tag{10.51}$$

The Debye–Hückel treatment now considers a solution so dilute that ions will rarely be close together. In this case, the interionic potential energy is usually small, and indeed much less than the average thermal energy, so that $Q_i \Phi \ll kT$. Then the exponential factor in (10.50) may be expanded as follows:

$$e^{-Q_i \Phi / kT} = 1 - \frac{Q_i \Phi}{kT} + \frac{1}{2!} \left(\frac{Q_i \Phi}{kT} \right)^2 - \cdots$$

If terms higher than the second are negligible, (10.50) becomes

$$\rho = \sum C_i Q_i - \frac{\Phi}{kT} \sum C_i Q_i^2$$

The first term vanishes by virtue of the requirement of overall electrical neutrality, and with $Q_i = z_i e$,

$$\rho = -\frac{e^2 \Phi}{kT} \sum C_i z_i^2 \tag{10.52}$$

Note that $\sum C_i z_i^2$ is closely related to the *ionic strength*, which was defined as $I = \frac{1}{2} \sum m_i z_i^2$.

This linearization of the Boltzmann distribution is a second serious problem in the Debye–Hückel model, since the assumption that the electrostatic energy $z_1 e \Phi$ is much less than the thermal energy kT only becomes true at a distance from the central ion about equal to the diameter of one H_2O molecule. In the neighborhood of a central ion at higher concentrations ($> 10^{-2}$ to 10^{-3} mol·dm^{-3}), we can expect $z_1 e \Phi \geq kT$. In such cases, the linear distribution function

$$\frac{C_i'}{C_i} = 1 - \frac{z_i e \Phi}{kT}$$

would give absurd values for the numbers of ions in the ionic atmosphere.

The defect of the Debye–Hückel model at other than low C's is even deeper than this. To discard the linear approximation, and use instead the exact Boltzmann equation (10.49), is to jump from the frying pan into the fire. The application of the Boltzmann equation assumed that we could set the potential energy that appears in (10.48) equal to the electrostatic energy $z_i e \Phi$ where Φ is given by the Poisson equation.* The Poisson equation, however, is based on the idea of a continuous or time-averaged distribution of charge density ρ. If we actually com-

---

*This assumption has been discussed by many authors, including J. G. Kirkwood, *J. Chem. Phys.*, **2**, 767 (1934); J. C. Poirier in *Chemical Physics of Ionic Solutions*, ed. by B. G. Conway and R. G. Barradas (New York: John Wiley & Sons, Inc., 1934), p. 767.

pute from the Boltzmann factor the contributions to Φ due to ions in different spherical shells about the central ion, we find in a typical case that for a strong 1–1 electrolyte at 10^{-2} mol·dm^{-3}, 0.8 ions in a first shell contribute 50% of the total electrostatic potential of the ion cloud (in other words, *one* counter-ion present 80% of the time). Such a distribution of a few localized ions would give large fluctuations in a time-averaged charge density ρ. Thus, the Debye–Hückel model lacks mathematical consistency, at least until concentrations fall below about 3×10^{-3} mol·dm^{-3} for 1–1 electrolytes.

As we proceed with the derivation of the Debye–Hückel equations, we should keep in mind the facts that not only is the linearization procedure unacceptable except for large ions and low concentrations ($< 10^{-2}$ to 10^{-3} mol·dm^{-3}), but also abandonment of the linearization leads to no improvement.

By substituting (10.51) into (10.47), one obtains the linearized *Poisson–Boltzmann equation*,

$$\frac{1}{r^2} \cdot \frac{d}{dr}\left(r^2 \frac{d\Phi}{dr}\right) = \frac{e^2 \Phi}{\epsilon_0 \epsilon kT} \sum C_i z_i^2$$

$$\frac{d}{dr}\left(r^2 \frac{d\Phi}{dr}\right) = b^2 r^2 \Phi \tag{10.53}$$

where,

$$b^2 = \frac{e^2}{\epsilon_0 \epsilon kT} \sum C_i z_i^2 \tag{10.54}$$

The quantity b^{-1} has the dimensions of length, and is called the *Debye length*. It is an approximate measure of the *thickness of the ionic atmosphere*, i.e., the distance over which the electrostatic field of an ion extends with appreciable strength. Values of b^{-1} at various concentrations for different types of electrolyte are summarized in Table 10.7.

Equation (10.53) can be readily solved by making the substitution $u = r\Phi$, whence $d^2u/dr^2 = b^2 u$, so that

$$u = Ae^{-br} + Be^{br} \tag{10.55}$$

$$\Phi = \frac{A}{r}e^{-br} + \frac{B}{r}e^{br}$$

A and B are constants of integration, to be determined from the boundary conditions. In the first place, Φ must vanish as r goes to infinity, so that

$$0 = \frac{Ae^{-\infty}}{\infty} + \frac{Be^{\infty}}{\infty}$$

This can be true only if $B = 0$, since the limit of e^r/r as r goes to infinity is not zero. We then have left

$$\Phi = \frac{A}{r}e^{-br} \tag{10.56}$$

This potential consists of an ordinary coulomb potential A/r multiplied by a screening factor, e^{-br}. The *screened Coulomb potential* has been used in a variety of other interesting applications, such as the work of Niels Bohr on nuclear colli-

TABLE 10.7
**Debye Length (nm) (Effective Radius of Ionic Atmosphere)
in Aqueous Solutions at 298 K**

Concentration mol·dm^{-3} / Salt type	1:1	1:2	2:2	1:3
10^{-1}	0.96	0.55	0.48	0.39
10^{-2}	3.04	1.76	1.52	1.24
10^{-3}	9.6	5.55	4.81	3.93
10^{-4}	30.4	17.6	15.2	12.4

sions in the solid state* and the theory of conductivity of alloys of Mott and Friedel.†
The value of A can be determined by substituting (10.56) back into (10.47),

$$\rho = \frac{-Ab^2\epsilon_0\epsilon}{r}e^{-br}$$

which gives the charge density in the ion cloud as a function of r. But the total charge of the ion cloud must be just equal to the charge of the central ion, but opposite in sign. Hence,

$$\int_a^\infty 4\pi r^2 \rho(r)\, dr = -z_i e$$

$$Ab^2\epsilon_0\epsilon \int_a^\infty 4\pi r\, e^{-br}\, dr = z_i e$$

which gives

$$A = \frac{z_i e}{4\pi\epsilon_0\epsilon}\frac{e^{ba}}{1 + ba} \tag{10.57}$$

The final result for the potential Φ is, accordingly,

$$\Phi = \frac{z_i e}{4\pi\epsilon_0\epsilon}\frac{e^{ba}}{1 + ba}\frac{e^{-br}}{r} \tag{10.58}$$

The potential in (10.58) is made up of two parts: that contributed by the central ion itself, which is $z_i e/4\pi\epsilon_0\epsilon r$, and that due to the ionic atmosphere. The ionic atmosphere contribution is

$$\Phi' = \frac{z_i e}{4\pi\epsilon_0\epsilon r}\left(\frac{e^{ba}}{1 + ba}e^{-br} - 1\right) \tag{10.59}$$

Since no ions from the atmosphere can approach the central ion more closely than $r = a$, the potential that exists at the site of the central ion due to the ionic atmosphere is obtained from (10.59) by setting $r = a$,

*N. Bohr, *The Penetration of Atomic Particles through Matter* (Copenhagen: Munksgaard, 1953).
†J. Friedel, *Advan. Phys.*, **3**, 446 (1954).

$$\Phi'_{r=a} = \frac{-z_i e}{4\pi\epsilon_0 \epsilon}\left(\frac{b}{1 + ba}\right) \tag{10.60}$$

In extremely dilute solutions, as shown in Table 10.7, $b \approx 10^6 \text{ cm}^{-1}$, so that with $a \approx 10^{-8} \text{ cm}$, $ba \ll 1$ and (10.60) becomes

$$\Phi'_{r=a} = \frac{-z_i e b}{4\pi\epsilon_0 \epsilon} \tag{10.61}$$

The extra potential due to the ionic atmosphere is related to the extra free enthalpy of the ionic solution, from which one can calculate the value of the activity coefficient.

23. The Debye–Hückel Limiting Law

Let us imagine a given ion to be introduced into the solution in an uncharged state (this process requires negligible electrical energy). Then let us increase the charge Q gradually to its final value ze. For the case of extreme dilution, the constant potential Φ is given by (10.61),* so that the electrical energy required *per ion* is

$$\Delta G = \int_0^{ze} \Phi \, dQ = \int_0^{ze} \frac{-bQ}{4\pi\epsilon_0 \epsilon} \, dQ = -\frac{bz^2 e^2}{8\pi\epsilon_0 \epsilon} \tag{10.62}$$

On the assumption that deviations of the dilute ionic solution from ideality are caused entirely by the electrical interactions, it can now be shown that this extra electric free enthalpy per ion is simply $kT \ln \gamma_i$, where γ_i is the conventional ion activity coefficient. Let us write for the chemical potential of an ionic species i,

$$\mu_i = RT \ln a_i + \mu_i^\ominus$$

$$\mu_i = \mu_i (\text{ideal}) + \mu_i (\text{electric})$$

Since

$$\mu_i (\text{ideal}) = RT \ln m_i + \mu_i^\ominus$$

and

$$a_i = \gamma_i m_i$$

$$\mu_i (\text{electric}) = RT \ln \gamma_i$$

This is the extra electric free enthalpy per mole. The extra free enthalpy per ion is therefore equal to $kT \ln \gamma_i$, but this is equal to the expression in (10.62). Therefore,

$$\ln \gamma_i = -\frac{z_i^2 e_i^2 b}{8\pi\epsilon_0 \epsilon k T} \tag{10.63}$$

*If we take Φ as given by (10.60), the extra electric free enthalpy would be

$$\Delta G = -\frac{z^2 e^2}{8\pi\epsilon_0 \epsilon}\frac{b}{1 + ba} \tag{10.61a}$$

and the expression finally obtained for the activity coefficient in (10.66) would be

$$\ln \gamma_\pm = -|z_+ z_-|\left(\frac{e^2}{8\pi\epsilon_0 \epsilon k T}\right)\left(\frac{b}{1 + ba}\right) \tag{10.66a}$$

We may substitute for b from (10.54)

$$b = \left(\frac{e^2}{\epsilon_0 \epsilon k T} \sum C_i z_i^2\right)^{1/2}$$

Since C_i and c_i are related by $C_i = c_i L$,

$$b = \left(\frac{L^2 e^2}{\epsilon_0 \epsilon R T} \sum c_i z_i^2\right)^{1/2}$$

In the dilute solutions being considered, $c_i = m_i \rho_0$, where ρ_0 is the solvent density, so that the ionic strength, (10.32), may be introduced, giving

$$b = \left(\frac{2 L^2 e^2 \rho_0}{\epsilon_0 \epsilon R T}\right)^{1/2} I^{1/2} = B I^{1/2} \tag{10.64}$$

Since the individual ion activity coefficients cannot be measured, the mean activity coefficient is calculated to obtain an expression that can be compared with experimental data. From (10.24),

$$(v_+ + v_-) \ln \gamma_\pm = v_+ \ln \gamma_+ + v_- \ln \gamma_-$$

Therefore, from (10.63),

$$\ln \gamma_\pm = -\left(\frac{v_+ z_+^2 + v_- z_-^2}{v_+ + v_-}\right) \frac{e^2 b}{8 \pi \epsilon_0 \epsilon k T}$$

Since $|v_+ z_+| = |v_- z_-|$,

$$\ln \gamma_\pm = -|z_+ z_-|\left(\frac{e^2 b}{8 \pi \epsilon_0 \epsilon k T}\right) \tag{10.65}$$

The valence factor can be evaluated as follows for the different electrolyte types:

Type	Example	Ionic charges	Valence factor $\|z_+ z_-\|$
uniunivalent	$NaCl$	$z_+ = 1,\ z_- = -1$	1
unibivalent	$MgCl_2$	$z_+ = 2,\ z_- = -1$	2
unitrivalent	$LaCl_3$	$z_+ = 3,\ z_- = -1$	3
bibivalent	$MgSO_4$	$z_+ = 2,\ z_- = -2$	4
bitrivalent	$Fe_2(SO_4)_3$	$z_+ = 3,\ z_- = -2$	6

Let us now transform (10.65) into base-10 logarithms and introduce the values of the universal constants. If e is taken as 1.602×10^{-19} C, R must be 8.314 J·K^{-1}· mol^{-1}. The result is the Debye–Hückel *limiting law* for the activity coefficient,

$$\log \gamma_\pm = -1.825 \times 10^6 |z_+ z_-|\left(\frac{I \rho_0}{\epsilon^3 T^3}\right)^{1/2} = -A |z_+ z_-| I^{1/2} \tag{10.66}$$

For water at 298 K, $\epsilon = 78.54$, $\rho_0 = 0.997$ kg·dm^{-3}, and the equation becomes

$$\log \gamma_\pm = -0.509 |z_+ z_-| I^{1/2} \tag{10.67}$$

In the derivation of the limiting law we consistently assumed that the analysis applied only to dilute solutions. It is not to be expected, therefore, that the equation should hold for concentrated solutions. As solutions become more and more dilute, however, the equation should represent the experimental data more and more closely. This expectation has been fulfilled by numerous measurements, so that the Debye–Hückel theory for very dilute solutions may be considered to be well substantiated. For example, in Fig. 10.10, some experimental activity co-

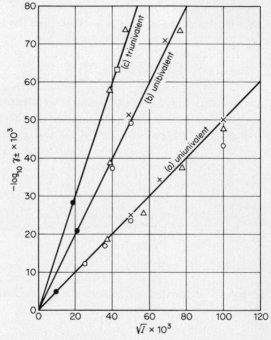

FIG. 10.10   Activity coefficients of sparingly soluble salts in salt solutions. [After Brønsted and LaMer, *J. Am. Chem. Soc.,* *46,* 555 (1924).]

efficients are plotted against the square roots of the ionic strengths. These data were obtained from the solubilities* of sparingly soluble complex salts in the presence of added salts, such as $NaCl$, $BaCl_2$, and KNO_3. The straight lines indicate the theoretical curves predicted by the limiting law, and it is evident that these limiting slopes are followed at low ionic strengths.

Another successful experimental test has been the measurement of activity coefficients for the same electrolyte in solvents with various dielectric constants.†

The Debye–Hückel theory is interesting in many ways, yet it is of little quantitative use in calculating properties of solutions except in the limit of extreme dilution. It is greatly improved and clarified by the theory of ionic association

---

*The method is described by S. Glasstone, *An Introduction to Electrochemistry* (Princeton, N.J.: D. Van Nostrand Co., Inc., 1942), p. 175.

†H. S. Harned, *et al., J. Am. Chem. Soc.,* **61,** 49 (1939).

(Section 10.25), which removes inconsistencies and some of the inaccuracy. The general statistical mechanical theory of concentrated ionic solutions, like the general theory of the liquid state, remains as one of the major challenges to future generations of theoretical chemists.

24. Theory of Conductivity

The interionic attraction theory was also applied by Debye and Hückel to the electric conductivity of solutions. An improved theory for point charges was given by Lars Onsager in 1928, and extended by Fuoss and Onsager in 1955 to charged spheres. The calculation of the conductivity is a difficult problem, and we shall content ourselves with a qualitative discussion.*

Under the influence of an electric field, an ion moves through a solution not in a straight line, but in a series of zigzag steps similar to those of Brownian motion. The persistent effect of the potential difference ensures an average drift of the ions in the field direction. Opposing the electric force on the ion is first of all the frictional drag of the solvent. Although the solvent is not a continuous medium, Stoke's Law is frequently used to estimate this effect, $F = 6\pi\eta av$, where η is the viscosity of the medium, a is the ionic radius, and v is the ionic drift velocity. Since the molecules of solvent and the ions are about the same size, it is more likely that the ion moves by jumping from one "hole" to another in the liquid.

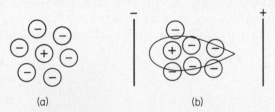

FIG. 10.11   (a) Ionic atmosphere of ion at rest; (b) asymmetric cloud around moving ion.

In addition to this *viscous effect*, two important electrical effects must be considered. As shown in Fig. 10.11(a), an ion in any static position is surrounded by an ionic atmosphere of opposite charge. If the ion jumps to a new position, it will tend to drag with it this oppositely charged aura. The ionic atmosphere, however, has a certain inertia, and cannot instantaneously readjust itself to the new position of its central ion. Thus, around a moving ion, the atmosphere becomes asymmetric, as in Fig. 10.11(b). Behind the ion there is a net accumulation of opposite charge, which exerts an electrostatic drag, decreasing the ionic velocity in the field direction. This retardation is called the *asymmetry effect*. It will obviously be greater at higher ionic concentrations. A second electrical action that lowers the mobility

---

*A definitive account of the subject is the book *Electrolytic Conductance*, by R. M. Fuoss and F. Accascina (New York: Interscience Publishers, 1959).

of the ions is called the *electrophoretic effect*. The ions comprising the atmosphere around a given central ion are themselves moving, on the average in the opposite direction, under the influence of the applied field. Since they are solvated, they tend to carry along with them their associated solvent molecules, so that there is a net flow of solvent in a direction opposite to the motion of any given (solvated) central ion, which is thus forced to "swim upstream" against this current.

The steady state of motion of an ion can be found by equating the electric driving force to the sum of frictional, asymmetric, and electrophoretic retardations. Onsager calculated each of the terms, and thereby obtained a theoretical equation for the limiting molar conductance in dilute solutions,

$$\Lambda = \Lambda_0 - [A\Lambda_0 + B]c^{1/2} = \Lambda_0 - Sc^{1/2} \tag{10.68}$$

The equation contains one empirical parameter $\Lambda_0(T)$, and A and B involve the viscosity η and dielectric constant ϵ of the solvent, the charge type of the electrolyte, the temperature, and universal constants. For a 1:1 electrolyte, (10.68) becomes

$$\Lambda = \Lambda_0 - \left(\frac{8.204 \times 10^5}{(\epsilon T)^{3/2}}\Lambda_0 + \frac{82.50}{(\epsilon T)^{1/2}\eta}\right)c^{1/2} \tag{10.69}$$

In these equations, c is the concentration of *ionized* solute in moles per cubic decimetre. If dissociation is not complete, c must be calculated from the degree of dissociation, α. Strictly speaking, the Onsager theory predicts only the slope S of the conductance curve in the limit as $c \rightarrow 0$. It does this very well for solutes of various salt types in solvents of various dielectric constants and temperatures. As an equation for calculating conductances, however, it is not much use above 10^{-3} mol·dm^{-3} for 1:1 electrolytes, or even lower concentrations for higher charge types.

25. Ionic Association

A real difficulty in many of these electrochemical theories is to obtain a definite answer to the question, "What is the degree of dissociation of an electrolyte in solution?"

There is no such difficulty in a gas such as $N_2O_4 \rightleftharpoons 2NO_2$. When a N_2O_4 molecule dissociates into two NO_2 molecules, the NO_2 on the average will exist for a considerable time (order of seconds) before recombining with another NO_2. During this time, the NO_2 travels freely through the gas for 1 km or so in a zigzag path with collisions. Most of the time it is either definitely dissociated or definitely associated. All measurements made on the gas (density, absorption spectra, heat capacity, etc.) give the same value for the degree of dissociation.

In the case of a molecule such as HNO_3 in water, the situation is different in two respects. The dissociation of the molecule to give H^+ (hydrated) and NO_3^- requires the separation of the oppositely charged ions. The electrostatic attraction between these two ions decreases relatively slowly as they separate, so that some kind of association still exists even when they are several molecular diameters apart. Also, the rates of dissociation and reformation of molecules or complexes from ions in solution are extraordinarily high. Thus, the mean lifetime of a complex or of a dissociated ion may be only of the order 10^{-10} s instead of 1 s, as in a gas. In this short time, few ions can become really free, and the most likely step after they separate is an almost immediate reassociation. Therefore, one method that gives a value for the degree of dissociation of HNO_3 in solution may yield a result quite different from that of a second method. For example, in more concentrated solutions, Raman spectra reveal bands for both HNO_3 and NO_3^-. We could calculate a degree of dissociation from the intensities of these bands,* but we would get a different value from a measurement of osmotic pressure or conductance. From X-ray studies of concentrated ionic solutions we can sometimes get direct information about ionic association.† Figure 10.12 shows some results obtained from a concentrated solution of erbium chloride. There is a firmly held octahedral arrangement of H_2O molecules about the Er^{3+} ion, and definite Er^{3+}–$(Cl^-)_2$ ion pairs.

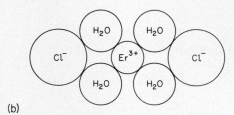

(b)

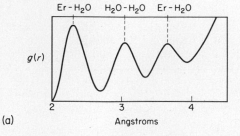

(a)   Angstroms

FIG. 10.12 (a) X-ray radial distribution curves (cf. Section 19.2). Results of X-ray diffraction of concentrated $ErCl_3$ solutions. (b) Planar model of solution around Er^{3+} ion. There are also H_2O molecules above and below Er^{3+} to give octahedral complex.

The *ionic association theory* for more concentrated solutions has been developed independently by N. Bjerrum and by R. M. Fuoss and C. Kraus. They consider that definite, although transient, *ion pairs* are brought together by electrostatic attraction. The formation of pairs will be greater the lower is the dielectric constant of the solvent and the smaller are the ionic radii, both of these factors tending to increase the electrostatic attractions.

The degree of association may become appreciable even in a solvent of high dielectric constant ϵ, such as water. Bjerrum has calculated that in one-molar

---

*The dissociation constant for HNO_3 at 298 K, $K_a = [a(H^+)a(NO_3^-)]/a(HNO_3)$, is calculated as $K_a = 21.4$ from the spectral data. See O. Redlich, *Chem. Rev.*, **39**, 333 (1946).

†G. W. Brady, *J. Chem. Phys.*, **33**, 1079 (1960).

aqueous solution, uniunivalent ions having a diameter of 0.282 nm are 13.8% associated; of 0.176 nm 28.6% associated. Such association into ion pairs would drastically lower the value of the ionic activity coefficients.

The failure of the Onsager conductance equation for dilute solutions of salts *MA* of multivalent ions is evidence for appreciable ionic association, and quantitative association constants can be obtained from the conductance data. If the ion pair is assumed to make no contribution to the conductance, (10.68) can be written

$$\Lambda = \alpha[\Lambda_0 - S(\alpha\,|z|\,c)^{1/2}] \tag{10.70}$$

where α is the fraction of free ions of charge z at concentration c. The dissociation constant for a 2:2 salt, $MA \rightleftharpoons M^{2+} + A^{2-}$, is

$$K = \frac{\gamma_{\pm}^2 \alpha^2 c}{(1 - \alpha)} \tag{10.71}$$

We shall take as an example an analysis of some data on $MgSO_4$ given by Davies.* Table 10.8 gives the measured Λ over a range of c. The degree of disso-

TABLE 10.8
Ionic Association in Aqueous $MgSO_4$ at 298.15 K from Conductance Data

$c \times 10^4$ (mol·dm⁻³)	Λ (Ω^{-1}·cm²·equiv⁻¹)	α	pK'
1.6196	127.31	0.9891	2.141
3.2672	124.27	0.9791	2.125
5.3847	121.34	0.9689	2.090
8.5946	117.85	0.9563	2.046
12.011	114.92	0.9459	2.003
16.759	111.61	0.9339	1.956

ciation obtained from (10.70) for the 2:2 salt is

$$\alpha = \frac{\Lambda}{133.06 - 343.1(\alpha c)^{1/2}}$$

The fourth column gives the square root of the ionic strength and the last column, pK', when K' is the equilibrium constant in (10.71) without the activity coefficient correction ($pK = -\log_{10} K$). The association of the ions is seen to be appreciable even in such dilute solutions. We shall leave as an exercise (Problem 18) the test of the K' data against the Debye–Hückel limiting law for the activity coefficient $\gamma_\pm$.

Ionic association is a basic factor that must be considered in any studies of aqueous ionic solutions. In solvents of lower dielectric constant, it will be even more important. Many specific factors, such as ionic size, polarizability, and effect on water structure, must be considered in the interpretation of pK values for ionic

*C. W. Davies, *Ionic Associations* (London: Butterworth & Co., Ltd., 1962).

association, and one should not, therefore, fall into the error of believing that the ionic strength can serve as a general coverall for these specific ion effects.

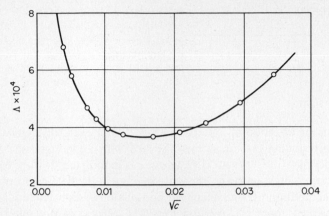

FIG. 10.13 Molar conductance of tetraisoamyl-ammonium nitrate in dioxane–water. [Fuoss and Kraus, *J. Am. Chem. Soc., 55,* 2387 (1933).]

In more concentrated solutions, we may obtain molar conductance vs. concentration curves like that in Fig. 10.13. The minima in such curves are explained by the ionic association theory.* The ion pairs $(+ \; -)$ are electrically neutral and do not contribute to the molar conductance, which therefore falls as more pairs are formed. As the solution becomes still more concentrated, triple ions, either $(+ \; - \; +)$ or $(- \; + \; -)$, begin to be formed from some of the pairs. Because these triplets bear a net charge, they contribute to the molar conductance, which therefore increases from its minimum value.

26.   Effects of High Fields

Ohm's Law can be written

$$i = \kappa E \qquad\qquad (10.72)$$

where i is the current density (SI unit: $A \cdot m^{-2}$). The validity of this law implies that the conductivity κ is a constant independent of the magnitude of the electric field E. The measurements that demonstrated the validity of Ohm's Law for electrolytes were actually restricted to rather low values of E, and in 1927 Wien showed that deviations from (10.72) occurred at field strengths in the range of 10^7 $V \cdot m^{-1}$. Some results for $MgSO_4$ are shown in Fig. 10.14(a).

Such effects of field strength can be explained easily on the basis of the ionic atmosphere model in Fig. 10.11. The motion of an ion through solution must follow a zigzag path similar to a Brownian motion, upon which is superimposed a

*R. Fuoss and C. Kraus, *J. Am. Chem. Soc., 55,* 21 (1933).

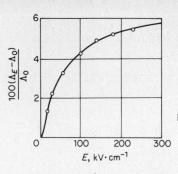

FIG. 10.14(a) First Wien effect: Effect of electric field on conductance of $MgSO_4$ solutions.

drift velocity in the direction of the applied field. After each little jump, the ionic atmosphere surrounding the ion is reestablished with a characteristic relaxation time of the order of $(10^{-10}/cz)$ s at 25°C, where c is in $mol \cdot dm^{-3}$. If the electric field becomes sufficiently high, however, the drift velocity in the field direction becomes higher than the random thermal velocities, and the ionic atmosphere is left behind and cannot reform about the moving ion. The relaxation and electrophoretic effects on ionic mobility then disappear and the observed conductivity κ rises above its value for low fields.

A second effect of high fields, also discovered by Wien, is called the *field dissociation effect*. It is the increase in degree of dissociation α of a weak electrolyte in a strong electric field. An example of this effect for acetic acid is shown in Fig. 10.14(b). A theoretical analysis made by Onsager* in 1934 was extended and simplified by Bass† in 1968. The dissociation process is assumed to take place in two stages: a breaking of the covalent bond to form a Bjerrum ion pair, followed by separation of the ion pair. For acetic acid,

$$CH_3COOH \rightleftarrows CH_3COO^-H^+ \rightleftarrows CH_3COO^- + H^+$$

The high field acts on the second step. Onsager found that for a strong field acting on a weak 1:1 electrolyte, the ratio of dissociation constant with the field to that in absence of the field was

$$\frac{K(E)}{K(0)} = \left(\frac{2}{\pi}\right)^{1/2} (8b)^{-3/4} \exp(8b)^{1/2} \tag{10.73}$$

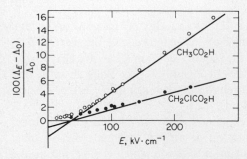

FIG. 10.14(b) Second Wien Effect: Effect of electric field on dissociation of weak electrolytes.

*L. Onsager, *J. Chem. Phys.*, **2**, 599 (1934).
†L. Bass, *Trans. Faraday Soc.*, **64**, 2153 (1968).

where

$$b = \frac{e^3 |E|}{8\pi\epsilon_0 \epsilon k^2 T^2}$$

This is a particularly interesting result because it predicts a definitely nonlinear field effect, and hence may be the basis of important physiological mechanisms, such as the "firing" of volleys of electrical impulses along nerve fibers. Although the potential differences across resting nerve membranes are only about 7×10^{-2} V, the thickness of the membranes is only about 7×10^{-7} cm, so that the effective electric field may reach the range of 10^5 V·cm^{-1} required for appreciable Wien dissociation effects on the membrane constituents. It would be quite a surprise to find that this rather esoteric electrochemical phenomenon was the basis for nerve conduction.*

27. Kinetics of Ionic Reactions

Electrostatic forces between ions, which are so important in determining properties such as activity coefficients and conductances in ionic solutions, will also have important special effects on the rate constants of reactions between ions. We can expect that reactions between ions of the same sign will be much slower, and those between ions of opposite sign will be much faster, as compared with similar reactions in which one or both of the reactants are uncharged. The dielectric constant ϵ of the medium will also be important in ionic reactions, since the lower is ϵ, the greater will be the electrostatic interaction.

A simple theoretical model, shown in Fig. 10.15, allows us to estimate the magnitudes of some of these electrostatic effects. When the two ions with charges $z_A e$ and $z_B e$ are a long distance apart, the electrostatic interaction is zero, but at a distance r, the force between them has the magnitude

$$F = \frac{z_A z_B e^2}{4\pi\epsilon_0 \epsilon r^2}$$

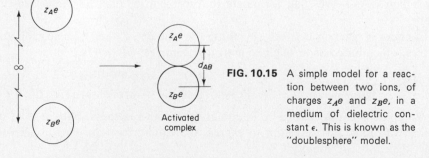

FIG. 10.15   A simple model for a reaction between two ions, of charges $z_A e$ and $z_B e$, in a medium of dielectric constant ϵ. This is known as the "doublesphere" model.

*L. Bass and W. J. Moore, in *Structural Chemistry and Molecular Biology*, ed. by A. Rich and N. Davidson (San Francisco: W. H. Freeman, 1968).

To decrease this separation by a distance dr requires work

$$dw = \frac{-z_A z_B e^2}{4\pi\epsilon_0 \epsilon r^2}\, dr$$

The total electrostatic work needed to bring the ions from infinite separation to their collision diameter d_{AB} is

$$w = -\int_{\infty}^{d_{AB}} \frac{z_A z_B e^2\, dr}{4\pi\epsilon_0 \epsilon r^2} = \frac{z_A z_B e^2}{4\pi\epsilon_0 \epsilon\, d_{AB}} \tag{10.74}$$

To calculate the effect of this term on the rate constant, we use expression (9.67) from transition-state theory,

$$k_2 = \frac{kT}{h}\mathrm{e}^{-\Delta G\ddagger/RT}$$

The free energy of activation $\Delta G^{\ddagger}$ is the sum of a nonelectrostatic part and an electrostatic part given by (10.74)

$$\Delta G^{\ddagger} = \Delta G_0^{\ddagger} + \Delta G_E^{\ddagger} = \Delta G_0^{\ddagger} + \frac{z_A z_B e^2 L}{4\pi\epsilon_0 \epsilon\, d_{AB}}$$

Substituting this $\Delta G^{\ddagger}$ into (9.67) and taking logs, we obtain

$$\ln k_2 = \ln k_0 - \frac{z_A z_B e^2}{4\pi\epsilon_0 \epsilon kT d_{AB}} \tag{10.75}$$

This equation predicts that $\ln k_2$ should be a linear function of $1/\epsilon$, when the rates of the same reaction are measured in a series of solvents of different ϵ. Experimental data are in good agreement with this prediction, until ϵ is so low that ion association becomes a complicating factor.

28. Salt Effects on Kinetics of Ionic Reactions

Pioneering work on salt effects on ionic reactions was done by J. N. Brønsted before the Debye–Hückel theory was available. This work was one of the earliest applications of the idea of an activated complex to the quantitative interpretation of reaction rates.* We shall formulate the problem in terms of transition-state theory.

Consider a reaction between two ions, A^{z_A} and B^{z_B}, z_A and z_B being the ionic charges, which proceeds through an activated complex, $(AB)^{z_A+z_B}$.

$$A^{z_A} + B^{z_B} \longrightarrow (AB)^{z_A+z_B} \longrightarrow \text{products}$$

$$\textit{Example:} \quad Fe^{3+} + I^- \longrightarrow (Fe\text{—}I)^{2+} \longrightarrow Fe^{2+} + \tfrac{1}{2}I_2$$

The complex is considered to be in equilibrium with reactants, but since we are dealing with ions, it is necessary to express the equilibrium constant in activities rather than concentrations:

*Z. Physik Chem., 102, 169 (1922). The early work was reviewed by V. K. LaMer, Chem. Rev., 10, 185 (1932).

$$K^{\ddagger} = \frac{a^{\ddagger}}{a_A a_B} = \frac{c^{\ddagger}}{c_A c_B} \cdot \frac{\gamma^{\ddagger}}{\gamma_A \gamma_B}$$

The a's and γ's are the activities and activity coefficients. The concentration of activated complexes is $c^{\ddagger} = c_A c_B K^{\ddagger}(\gamma_A \gamma_B)/\gamma^{\ddagger}$. From (9.61), the reaction rate is $-(dc_A/dt) = k_2 c_A c_B = (kT/h)c^{\ddagger}$. The rate constant is

$$k_2 = \frac{kT}{h} K^{\ddagger} \frac{\gamma_A \gamma_B}{\gamma^{\ddagger}} = \frac{kT}{h} \cdot \frac{\gamma_A \gamma_B}{\gamma_{\ddagger}} e^{\Delta S^{\ddagger}/R} e^{-\Delta H^{\ddagger}/RT} \tag{10.76}$$

In dilute solution, the activity coefficient terms can be estimated from the Debye–Hückel theory. From (10.67), at 298 K in an aqueous solution, $\log_{10} \gamma_i = -0.509 \, z_i^2 I^{1/2}$. Taking the $\log_{10}$ of (10.76) and substituting the Debye–Hückel expression, we get

$$\log_{10} k_2 = \log_{10} \frac{kT}{h} K^{\ddagger} + \log_{10} \frac{\gamma_A \gamma_B}{\gamma^{\ddagger}}$$
$$= B + [-0.509 z_A^2 - 0.509 z_B^2 + 0.509(z_A + z_B)^2] I^{1/2}$$
$$\log_{10} k_2 = B + 1.018 z_A z_B I^{1/2} \tag{10.77}$$

The constant, $\log_{10}(kT/h)K^{\ddagger}$, has been written as B.

The Brønsted equation (10.77) predicts that the plot of $\log_{10} k_2$ vs. the square root of the ionic strength should be a straight line. For a water solution at 298 K, the slope is nearly equal to $z_A z_B$, the product of the ionic charges. Three special cases can occur:

1. If z_A and z_B have the same sign, $z_A z_B$ is positive and the rate constant increases with the ionic strength.
2. If z_A and z_B have different signs, $z_A z_B$ is negative and the rate constant decreases with the ionic strength.
3. If one of the reactants is uncharged, $z_A z_B$ is zero and the rate constant is independent of the ionic strength.

These theoretical conclusions have been verified in a number of experimental studies. A few examples are illustrated in Fig. 10.16. This change of k_2 with I is called the *primary kinetic salt effect*. The ionic strength I is calculated from $\Sigma \frac{1}{2} m_i z_i^2$, and the summation is extended over all the ionic species present in solution, not merely the reactant ions.

Much of the early work on ionic reactions is comparatively useless because the salt effect was not understood. It is now often the practice in following the rate of an ionic reaction to add a considerable excess of inert salt, e.g., NaCl, to the solution, so that the ionic strength is effectively constant throughout the reaction. If pure water is used, the change in ionic strength as the reaction proceeds may lead to erratic rate constants.

We should understand that the Brønsted equation, which seems to be so beautifully confirmed by the data in Fig. 10.16, cannot be expected to hold at salt concentrations much beyond the range of validity of the Debye–Hückel theory. In more concentrated solutions, it is not possible to summarize all the salt effects in a simple ionic-strength factor. Specific interactions between the ions in solution

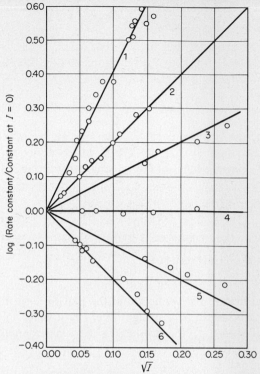

FIG. 10.16 Variations of rates of ionic reactions with the ionic strength. The circles are experimental values; the lines are theoretical, from Eq. (10.77).

(1) $2[Co(NH_3)_3Br]^{2+} + Hg^{2+} + 2H_2O \longrightarrow$
     $2[Co(NH_3)_5H_2O]^{3+} + HgBr_2$

(2) $S_2O_8^= + 2I^- \rightarrow I_2 + 2SO_4^=$

(3) $[NO_2NCOOC_2H_5]^- + OH^- \longrightarrow$
     $N_2O + CO_3^= + C_2H_5OH$

(4) Inversion of cane sugar

(5) $H_2O_2 + 2H^+ + 2Br^- \longrightarrow 2H_2O + Br_2$

(6) $[Co(NH_3)_5OH]^{2+} + OH^- \longrightarrow$
     $[Co(NH_3)_5OH]^{2+} + Br^-$

will influence the reaction rates. For example, in the reaction

$$S_2O_8^{2-} + 2I^- \longrightarrow 2SO_4^{2-} + I_2$$

the salt effect follows the Brønsted equation in general, but is strongly dependent on the identity of the cation, with the magnitude of the effect decreasing in the order, $Cs > Rb > K > Na > Li$. For reactions between ions of the same charge sign, the salt effect often appears to be governed predominantly by the concentrations and charges of those added ions with a sign opposite to that of the reactant ions.* Ionic association is also an important factor in reaction rates especially when multivalent ions are involved either as reactants or as added ions.†

*A. R. Olson and T. R. Simonson, *J. Chem. Phys.*, *17*, 1167 (1949).

†C. W. Davies, *Prog. Reaction Kinetics*, *1*, 161 (1961), has surveyed a number of interesting examples.

29.   Acid–Base Catalysis

Among the most interesting cases of homogeneous catalysis are reactions catalyzed by acids and bases. This acid–base catalysis is of the utmost importance, governing the rates of a great number of organic reactions, and many of the processes of physiological chemistry since it is likely that many enzymes act as acid–base catalysts.

The earliest studies in this field were those by Kirchhoff in 1812 on the conversion of starch to glucose by the action of dilute acids, and by Thénard in 1818 on the decomposition of hydrogen peroxide in alkaline solutions. The classic investigation of Wilhelmy in 1850 dealt with the rate of inversion of cane sugar by acid catalysts. The hydrolysis of esters, catalyzed by both acids and bases, was extensively studied in the latter half of the nineteenth century. The catalytic activity of an acid in these reactions became one of the accepted measures of acid strength, often used by Arrhenius and Ostwald in the early days of the ionization theory.

TABLE 10.9
Data from Ostwald on the Catalytic Constants of Different Acids

Acid	Relative conductivity	k'' (ester)	k'' (sugar)
HCl	100	100	100
HBr	101	98	111
HNO_3	99.6	92	100
H_2SO_4	65.1	73.9	73.2
CCl_3COOH	62.3	68.2	75.4
$CHCl_2COOH$	25.3	23.0	27.1
HCOOH	1.67	1.31	1.53
CH_3COOH	0.424	0.345	0.400

Table 10.9 gives some of Ostwald's results on inversion of sucrose and hydrolysis of methyl acetate. If we write the acid as HA, these reactions are

$$C_{12}H_{22}O_{11} + H_2O + HA \longrightarrow C_6H_{12}O_6 + C_6H_{12}O_6 + HA$$

$$CH_3COOCH_3 + H_2O + HA \longrightarrow CH_3COOH + CH_3OH + HA$$

The reaction rate may be written $dx/dt = k'$ $[CH_3COOCH_3]$ $[H_2O]$ $[HA]$. Since water is present in large excess, its concentration is effectively constant. The rate therefore reduces to $dx/dt = k''$ $[HA]$ $[CH_3COOCH_3]$. Now k'' is called the *catalytic constant*. The values in Table 10.9 are all relative to 100 for k'' with HCl.

Ostwald and Arrhenius showed that the catalytic constant of an acid is proportional to its molar conductance. They concluded that the nature of the anion was unimportant, the only active catalyst being the hydrogen ion H^+. In other reactions, however, it was necessary to consider the effect of the OH^- ion and also the rate of the uncatalyzed reaction. The result was a three-term equation for the observed rate constant, $k_2 = k_0 + k_{H^+}[H^+] + k_{OH^-}[OH^-]$. Since in aqueous solution $K_w = [H^+][OH^-]$,

$$k_2 = k_0 + k_{H^+}[H^+] + \frac{k_{OH^-}K_w}{[H^+]} \tag{10.78}$$

Since K_w is about 10^{-14}, in 0.1 N acid $[OH^-]$ is 10^{-13}, and in 0.1 N base $[OH^-]$ is 10^{-1}. There is a 10^{12}-fold change in $[OH^-]$ and $[H^+]$ in passing from dilute acid to dilute base. Therefore, the OH^- catalysis will be negligible in dilute acid and the H^+ catalysis negligible in dilute base, except in the unusual event that the catalytic constants for H^+ and OH^- differ by as much as 10^{10}. By measurements in acid and basic solutions, it is therefore generally possible to evaluate k_{H^+} and k_{OH^-} separately.

If $k_{H^+} = k_{OH^-}$, a minimum in the overall rate constant occurs at the neutral point. If either k_{H^+} or k_{OH^-} is very low, there is no rise in k_2 on the corresponding side of the neutral point. These and other varieties of rate constant vs. pH curves, arising from different relative values of k_0, k_{H^+}, and k_{OH^-}, are shown in Fig. 10.17.

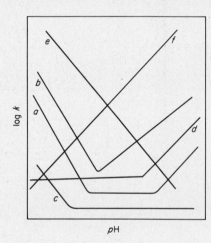

FIG. 10.17 Acid–base catalysis: the influence of pH on rate constants. The reactions are identified in the text.

Examples of each of the different types have been studied experimentally.* They include the following:

(a) the mutation of glucose;

(b) hydrolysis of amides, γ-lactones, and esters, and halogenation of ketones;

(c) hydrolysis of alkyl orthoacetates;

(d) hydrolysis of β-lactones, decomposition of nitramide, halogenation of nitroparaffins;

(e) inversion of sugars, hydrolysis of diazoacetic ester, acetals;

(f) depolymerization of diacetone alcohol, and decomposition of nitrosoacetonamine.

*See A. Skrabal, *Z. Elektrochem.*, *33*, 322 (1927); R. P. Bell, *Acid-Base Catalysis* (New York: Oxford University Press, 1941) and *The Proton in Chemistry* (Ithaca: Cornell University Press, 1959).

30. General Acid–Base Catalysis

The influence of added salts in the primary kinetic salt effect has already been noted. In addition to this direct dependence of reaction rate on ionic strength, there is an indirect influence which is important in catalyzed reactions. In solutions of weak acids and bases, added salts, even if they do not possess a common ion, may change the H^+ or OH^- ion concentration through their effect on activity coefficients. For an acid $HA \rightarrow H^+ + A^-$,

$$K_a = \frac{a_H \cdot a_{A^-}}{a_{HA}} = \frac{c_H \cdot c_{A^-}}{c_{HA}} \cdot \frac{\gamma_H \cdot \gamma_{A^-}}{\gamma_{HA}}$$

Any change in the ionic strength of the solution affects the γ terms and hence the concentration of H^+. Consequently, if the reaction is catalyzed by H^+ or OH^- ions, the rate is dependent on the ionic strength, by virtue of this *secondary kinetic salt effect*. Unlike the primary effect, it does not alter the *rate constant* provided this is calculated from the true H^+ or OH^- concentration.

The broader picture of the nature of acids and bases given by the work of Brønsted and Lowry implies that not only H^+ and OH^- but also the undissociated acids and bases may be effective catalysts. The essential feature of catalysis by an acid is the transfer of a proton from acid to substrate,* and catalysis by a base involves the acceptance of a proton by the base. Thus, in Brønsted–Lowry nomenclature, the substrate acts as a base in acid catalysis, or as an acid in basic catalysis. In the case of hydrogen-ion catalysis in aqueous solution, the acid is really the hydronium ion, OH_3^+.

For example, the hydrolysis of nitramide is susceptible to basic but not to acid catalysis, the mechanism being,

$$NH_2NO_2 + OH^- \longrightarrow H_2O + NHNO_2^-$$

$$NHNO_2^- \longrightarrow N_2O + OH^-$$

Not only the OH^- ion but also other bases can act as catalysts—e.g., the acetate ion,

$$NH_2NO_2 + CH_3COO^- \longrightarrow CH_3COOH + NHNO_2^-$$

$$NHNO_2^- \longrightarrow N_2O + OH^-$$

$$OH^- + CH_3COOH \longrightarrow H_2O + CH_3COO^-$$

The reaction rate with different bases B is always $v = k_B(B)\,(NH_2NO_2)$. Brønsted found that there was a relation between the catalytic constant k_B and the dissociation constant K_b of the base, namely,

$$k_B = CK_b^\beta \tag{10.79}$$

or

$$\log k_B = \log C + \beta \log K_b$$

Here, C and β are constants for bases of a given charge type. Thus, the stronger

---

*Substance the reaction of which is being catalyzed.

the base, the higher the catalytic constant.* Figure 10.18 shows data on the nitramide decomposition that are in good accord with the Brønsted relation (10.79).

The nitramide hydrolysis displays *general basic catalysis*. Other reactions provide examples of *general acid catalysis*, with a relation like (10.79) between k_A and K_a. Some reactions also occur with both general acid and general basic catalysis.

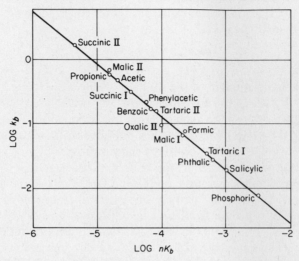

FIG. 10.18 General basic catalysis of the nitramide decomposition.

Since a solvent like water can act as either an acid or a base, it is often itself a catalyst. What was formerly believed to be the uncatalyzed reaction, represented by k_0 in (10.78), is in most cases a reaction catalyzed by the solvent acting as acid or base.

The Brønsted equation is a special case of a more general rule known as a *linear free-energy relationship*, which indicates a linear dependence of a free energy of activation $\Delta G^\ddagger$ on a free energy of reaction $\Delta G^\ominus$. In a homologous series of reactions, we may expect that the greater the affinity, measured by $\Delta G^\ominus$ for the reaction, the greater will be the reaction rate, measured by $\Delta G^\ddagger$. The Brønsted equation is equivalent to

$$\Delta G^\ddagger = \beta \, \Delta G^\ominus + C$$

Any acid–base catalysis involves the intermediacy of both an acid and a base, but if one step has a rate constant much lower than the others, the kinetics can appear to be catalyzed by either the acid or the base, depending on which one reacts in the slow step of the mechanism. An example is the base-catalyzed aldol condensation of acetaldehyde, which has the following mechanism:†

*For polybasic bases, a correction must be made. See R. P. Bell, *Acid-Base Catalysis* (New York: Oxford University Press, 1941), p. 83.

†R. P. Bell, *The Proton in Chemistry* (Ithaca: Cornell University Press, 1959).

$$CH_3CHO + B \underset{k_{-1}}{\overset{k_1}{\rightleftarrows}} CH_2 \cdot CHO^- + A$$

$$CH_3CHO + CH_2CHO^- \overset{k_2}{\rightleftarrows} CH_3CHO^- \cdot CH_2CHO$$

$$A + CH_3CHO^- \cdot CH_2CHO \rightleftarrows CH_3CHOHCH_2CHO + B$$

The first step is much slower than the other two, so that the reaction displays general catalysis by base B.

PROBLEMS

1. After passage of an electric current for 40 min, 8.95 mg of silver are deposited in a coulometer. Calculate the average current I. If the measurement of the mass is accurate to ± 0.01 mg and of the time to ± 0.1 s, what is accuracy of measurement of I?

2. A conductivity cell filled with 0.1 molar KCl at 25°C has measured resistance of 24.36 Ω. The conductivity κ of 0.1 molar KCl is $1.1639 \, \Omega^{-1} \cdot m^{-1}$, and water with $\kappa = 7.5 \times 10^{-6} \, \Omega^{-1} \cdot m^{-1}$ was used to make up solutions. Filled with 0.01 molar acetic acid, the cell resistance is $1982 \, \Omega$ at 25°C. Calculate the molar conductance Λ of 0.01 molar acetic acid.

3. A solution of LiCl was electrolyzed in a Hittorf cell. On passing 5000C of electricity, the mass of LiCl in the anode compartment decreased by 0.6720 g. Compute the transference number t^+ of the Li$^+$ ion.

4. A 4.000 molal solution of $FeCl_3$ was electrolyzed between platinum electrodes. After electrolysis, the cathode portion in a Hittorf apparatus was 3.15 molal in $FeCl_3$ and 1.00 molal in $FeCl_2$. What was the transference number of Fe^{3+}?

5. A current of 10.00 mA was passed through a $AgNO_3$ solution in a Hittorf cell with silver electrodes for 80.0 min. After electrolysis, the cathode solution weighed 40.28 g and was titrated with 86.00 cm^3 of 0.0200 molar KSCN. The anode solution weighed 40.40 g and required 112.00 cm^3 of 0.0200 molar KSCN. Calculate the transference number of Ag$^+$.

6. The solubility of sodium propionate NaOPr in water at 25°C is $9.80 \, \text{mol} \cdot \text{dm}^{-3}$, and $\log \gamma_\pm = -0.2454 + 0.103 \, c$ near this concentration c in $\text{mol} \cdot \text{dm}^{-3}$. Calculate $\Delta G^\ominus(298)$ for NaOPr(c) $\rightarrow$ NaOPr (aq solution). What would be the approximate molality of NaOPr in its standard state in solution?

7. The following conductivities were measured for chloroacetic acid in aqueous solution at 25°C.

c^{-1} (dm$^3 \cdot$mol^{-1})	16	32	64	128	256	512	1024
Molar conductance, Λ	53.1	72.4	96.8	127.7	164	205.8	249.2

   If $\Lambda_0 = 362 \, \Omega^{-1} \cdot cm^2 \cdot mol^{-1}$, are these data in accord with the Ostwald Dilution Law, Eq. (10.11)? Do they fit the Onsager conductance equation (10.68)? Compare the experimental value of the limiting slope S with the theoretical value from (10.69).

8. The conductivity of a saturated solution of AgCl in pure water at 20°C is 1.26×10^{-4} $\Omega \cdot m^{-1}$ higher than that of the water itself. Calculate the solubility of AgCl in water at 20°C.

9. From the mean molal activity coefficients $\gamma_{\pm}$ in Table 10.6, calculate the mean ionic activity $a_{\pm}$ and the activity a_2 in 0.100 molal solutions of $AgNO_3$, $CuSO_4$, H_2SO_4, $CaCl_2$.

10. From the Born equation (10.12), estimate the free enthalpy ΔG_e in the transfer of Na^+ and K^+ ions from aqueous solution in cytoplasm into the lipid environment of a membrane ($\epsilon = 5$) if the ionic radii are 0.095 and 0.133 nm, respectively. Can you derive from the Born equation expressions for the ΔH_e and ΔS_e of the transfer? (*Hint*: Assume only the ϵ's are functions of T.)

11. Some molar ionic conductances Λ_0 in $\Omega^{-1} \cdot cm^2 \cdot mol^{-1}$ at three temperatures are as follows:

°C	H⁺	Li⁺	K⁺	½Ca²⁺	Cl⁻	I⁻
0	225	19.4	40.7	31.2	41.0	41.4
25	350	38.7	73.5	59.5	76.35	76.8
45	441	58.0	103.5	88.2	108.9	108.6

The viscosity of water in centipoises is 1.792 at 0°C, 0.894 at 25°C, and 0.599 at 45°C. Can the variations with T of the Λ_0's be explained satisfactorily in terms of motions of spherical ions through a solution of viscosity η? If not, what other factors would you suggest may need to be taken into consideration?

12. The diffusion coefficient of KCl in water at 25°C in the limit of extreme dilution is $1.962 \times 10^{-5}\ cm^2 \cdot s^{-1}$. How does this value compare with that calculated from the molar ionic conductances in Table 10.4, through the Nernst–Einstein relation, Eq. (10.19)?

13. Calculate the "thickness of the ionic atmosphere" according to Debye–Hückel theory for 0.1 and 0.01 molal solutions of 1:1 electrolyte in 70% ethanol–water at 25°C ($\epsilon = 38.5$). Compare these values with those in water solution as given in Table 10.7.

14. The solubility of AgCl in water at 25°C is $10^{-4.895}\ mol \cdot dm^{-3}$. By means of the Debye–Hückel theory, calculate $\Delta G^{\ominus}$ for the change: $AgCl(c) \longrightarrow Ag^+ + Cl^-$ (aq). Calculate the solubility of AgCl in a solution of KNO_3 in which the ionic strength is $I = 0.010\ mol \cdot dm^{-3}$.

15. If pure water has a conductivity $\kappa = 6.2 \times 10^{-6}\ \Omega^{-1} \cdot m^{-1}$ at 298 K, calculate κ of a saturated solution of CO_2 in H_2O at 298 K, if the CO_2 pressure is 20 torr and the equilibrium constant for $H_2O(l) + CO_2(aq) \longrightarrow HCO_3^- + H^+$ is $K_c = 4.16 \times 10^{-7}$. The solubility of CO_2 in H_2O follows Henry's Law with $k = 0.0290$ $mol \cdot dm^{-3} \cdot atm^{-1}$.*

16. From the following freezing point depressions for aqueous solutions of NaCl, calculate $\gamma_{\pm}$ on molal scale for NaCl in 0.05 molal solution.

Molality	0.01	0.02	0.05	0.10	0.20	0.50
Freezing point depression, K	0.0361	0.0714	0.1758	0.3470	0.6850	1.677

17. The potential difference across the nerve membrane in a giant axon of the squid is $\Delta\Phi = 70\ mV$, and the thickness of the membrane is 7.0 nm. If the dielectric constant of the membrane is $\epsilon = 5$, calculate the change in pH of the membrane when it is

*See D. A. MacInnes and D. Belcher, *J. Am. Chem. Soc.*, 55, 2630 (1933).

suddenly depolarized to $\Delta\Phi = 50$ mV by an electric impulse. Base your analysis on Eq. (10.73) for the Wien effect.

18. Consider the values in Table 10.8 of pK' for the dissociation $MgSO_4 \rightarrow Mg^{2+} + SO_4^{2-}$ in dilute aqueous solution at 298 K. These values have not been corrected for $\gamma_\pm$ as in Eq. (10.71). Calculate $\gamma_\pm$ from the Debye–Hückel limiting law, and hence calculate K from K'.

19. According to Bjerrum, a pair of oppositely charged ions will form an electrostatic complex when the distance between them becomes less than a critical value q. From the Boltzmann distribution function, the number of ions of type i in a spherical shell of thickness dr at a distance r from a central ion j is

$$dN_i = N_i \exp\left(\frac{-z_i e\Phi}{kT}\right) 4\pi r^2 \, dr$$

where the electrostatic potential due to the central ion is $\Phi = z_j e / 4\pi\epsilon_0\epsilon r$. Show that the Bjerrum $q = |z_i z_j| e^2 / 8\pi\epsilon_0\epsilon kT$. Calculate the q for $1:1$ and $2:2$ ion pairs in media of dielectric constants $\epsilon = 80$ and 30. Comment briefly on the results.

20. Bjerrum calculated the degree of association $1 - \alpha$ of ion pairs by integrating the distribution law (as obtained in Problem 19) between the limits a, the closest approach of a pair of ions considered as rigid spheres, and q, the maximum distance for electrostatic association. Repeat this calculation to show that

$$1 - \alpha = 4\pi Lc \int_a^q \exp\left(\frac{-z_i z_j e^2}{4\pi\epsilon_0\epsilon kTr}\right) r^2 \, dr$$

where c is in moles per cubic centimetre.* Evaluate the integral and hence calculate $1 - \alpha$ for the four cases cited in Problem 19, for $a = 0.20$ nm and $c = 10^{-5}$ mol·cm^{-3}.

21. The molar conductances Λ of molten nitrates were measured over a range of temperatures and represented as $\Lambda = A[\exp(-B/T)]$, with following results:

	LiNO$_3$	NaNO$_3$	KNO$_3$
$A\ (\Omega^{-1}\cdot\text{mol}^{-1}\cdot\text{cm}^2)$	968	706	657
$B\ (K)$	1795	1608	1784

A. E. Stearn and H. Eyring† derived the following equation from transition-state theory:

$$\Lambda_i = \frac{|z_i| e F \bar{l}^2}{6h} \exp\left(\frac{-\Delta G^\ddagger}{RT}\right)$$

where $\bar{l}^2$ is the mean square jump distance in the melt, and Λ_i is the molar ionic conductance of an ion of charge z_i. Discuss the fit of this equation with the experimental data for the nitrates.

22. The temperature dependence of Λ at constant volume and at constant pressure, and the pressure dependence at constant temperature, yielded the following coefficients for molten LiNO$_3$ at 400°C:

$$RT^2 \left(\frac{\partial \ln \Lambda}{\partial T}\right)_P = 14.2 \text{ kJ·mol}^{-1}$$

*H. Bjerrum, *Kgl. Danske Videnskab. Selskab, Mat. Fys. Medd.* (7), 9 (1926).
†*J. Phys. Chem.*, 44, 955 (1940).

$$RT^2\left(\frac{\partial \ln \Lambda}{\partial T}\right)_V = 13.7 \text{ kJ} \cdot \text{mol}^{-1}$$

$$-RT\left(\frac{\partial \ln \Lambda}{\partial P}\right)_T = 0.5 \text{ cm}^3 \cdot \text{mol}^{-1}$$

Calculate $\Delta H^\ddagger$, $\Delta U^\ddagger$, $\Delta S^\ddagger$, and $\Delta V^\ddagger$ for the migration rate process at 400°C. Use data from Problem 21, as necessary.

23. A metal cylinder 1 m long and 0.3 m in diameter has a coaxial cylindrical metal rod 0.04 m in diameter. The space between the rod and the cylinder is filled with 0.5 molal KCl solution and a current of 2 A is passed from inner to outer cylinder. Derive an expression for the current density as a function $f(r)$ where r is the distance from the center of the cylinder. Derive $E(r)$, where E is the electric field. What is the potential difference $\Delta\Phi$ between the two cylinders? What is the resistance R between the two cylinders?

24. The rate constant k_2 of the reaction $S_2O_8^{2-} + 2I^- \longrightarrow 2SO_4^{2-} + I_2$ was measured at various ionic strengths I at 25°C:

I (mol $\cdot$ dm^{-3})	0.00245	0.00365	0.00645	0.00845	0.01245
k_2 (dm$^3 \cdot$ mol$^{-1} \cdot$ s^{-1})	1.05	1.12	1.18	1.26	1.39

Do these results follow the Brønsted equation (10.77)?

25. The catalytic rate constant k_c for exchange of ^{18}O between acetone and water in solution of an acid HA at 298 K would be expected to include terms for specific H^+ catalysis and general acid–base catalysis:

$$k_c = k_H[H^+] + k_{HA}[HA] + k_A[A^-]$$

The k_c was measured at various ratios $[HA]/[A^-]$:

$[HA]/[A^-]$	0.2	0.2	1.0	1.0
k_c (s^{-1})	1.5×10^4	2.7×10^4	4.3×10^4	4.8×10^4
$[HA]$ (mol $\cdot$ dm^{-3})	0.0225	0.100	0.0135	0.050

Calculate each of the rate constants k_H, k_{HA}, and k_A, and discuss briefly a likely mechanism for the reaction.*

26. As a general rule, the dissociation constant K_H of an acid HA measured in H_2O is greater than K_D for the acid DA measured in heavy water. For example, for acetic acid HAc, $K_H/K_D = 3.33$ at 25°C. How would you interpret this difference? If the pH of a solution of HAc in H_2O was 6.0, what would be the pD of a solution of the same molar concentration of DAc in D_2O?

27. The rate constant of the reaction $H^+ + OH^- \longrightarrow H_2O$ at 25°C was measured by an ultrasonic relaxation method as $k_r = 1.3 \times 10^{11}$ dm$^3 \cdot$ mol$^{-1} \cdot$ s^{-1}. If this rate is to be explained as a diffusion-controlled process in accord with Eq. (9.79), what would be the value of d_{12}, the collision diameter? How would you interpret this calculated value of d_{12}?

*W. C. Gardiner, *Rates and Mechanisms of Chemical Reactions* (New York: W. A. Benjamin, Inc., 1969).

11

Interfaces

The body of an unicorn, which is entirely free from poison, repels every poisonous thing. Place a living spider inside a circle formed by a strip of the skin of an unicorn, and the spider will not be able to pass. But if the circle be composed of some envenomed substance, the spider with no difficulty will cross the line, which is homogeneous to its own nature.

Basil Valentine
(1660)*

If the optic nerve of a salamander is cut, new nerve fibers will sprout from the stump and find their way back to the brain in such a way as to reestablish the original connections and restore normal vision to the animal.† Thousands of definite contacts are formed as the result of a specific recognition between the ends of nerve fibers and certain cell surfaces in the brain. A similar renewal of nerve contacts has been observed in many cold-blooded animals but, for reasons as yet unknown, it does not occur in mammals. This example is but one of the many interfacial phenomena important in living systems, so that an understanding of the physical chemistry of surfaces is essential for research in molecular biology.

In the interpretation of surface phenomena, as in the theory of solutions, the general theoretical problem is to calculate the properties of the system in terms of the electronic structures and resultant interactions of molecules. A difficult problem in homogeneous solutions, it becomes even more so in surface regions and at interfaces between two different phases, where the system is not homogeneous. In Chapter 6, in fact, a phase was defined as a part of a system that was "homogeneous throughout." Yet it is clearly a contradiction to say that matter exists deep in the interior of a sample under the same conditions as it does near the surface. For example, at the surface of a liquid in contact with its vapor, the molecules are subject to a net attraction toward the bulk liquid. Thus, all liquid surfaces in the absence of external forces tend to contract to a minimum area; for example, freely suspended volumes of liquid assume a spherical shape to achieve a minimal ratio of surface to volume. It will be necessary to introduce a number of new variables of state to develop the thermodynamic and statistical theory of interfaces.

The Triumphal Chariot of Antimony (London: Thomas Bruster, 1660).
†R. W. Sperry, *Proc. Natl. Acad. Sci.*, **50**, 703 (1963); *J. Neurophysiol.*, **8**, 15 (1945).

Thomas Graham in 1861 introduced the word *colloid* to describe suspensions of one material in another that did not separate on long standing. Colloids thus consist of a *dispersed phase* and a *dispersion medium*. Dispersed materials with a particle size of less than about 0.2 μm would generally be classed in the colloidal state. The limit of resolution of a microscope using ordinary light is about 0.2 μm, so that direct observation of colloidal particles usually requires the use of an electron microscope, although larger ones can be visualized from scattered light by dark-field observation.

Colloidal suspensions are called *sols*. If the bulk dispersed phase spontaneously enters the dispersion medium, they are *lyophilic sols*. Examples are high polymer solutions, such as proteins in water or rubber in benzene. These solutions display many of the physical properties of colloidal suspensions owing to the high molar mass of the solute. Colloidal suspensions of essentially insoluble materials are called *lyophobic sols*. They may be prepared by condensation or dispersion methods. In the simplest terms, they owe their stability to the fact that the particles bear electrical charges of same sign and thus cannot approach one another closely enough to coalesce.

Since the small particle size of colloidal dispersions leads to high surface-to-volume ratios, surface properties are of primary importance in the study of colloids.

1. Surface Tension

The system shown in Fig. 11.1(a) is a liquid in contact with its vapor. To extend the area of the interface, it is necessary to bring molecules from the interior into the surface, so that work must be done against cohesive forces in the liquid. It follows that the molar free energy of the surface region of the liquid is higher than that of the bulk liquid.

In 1805, Thomas Young showed that the mechanical properties of the surface could be related to those of a hypothetical membrane stretched over the surface. This membrane was supposed to be in a state of tension. A *tension* is a negative pressure and pressure is force per unit area, so that surface tension is force per unit length. Surface tension acts parallel to the surface and pulls inward to oppose any attempt to extend the surface area. The unit of surface tension in the SI system is newton per meter ($N \cdot m^{-1}$).

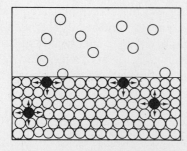

FIG. 11.1(a) Liquid–vapor interface. Work must be done on the system to extend its surface by bringing molecules from the interior to the surface.

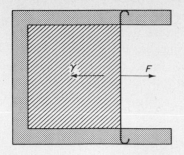

FIG. 11.1(b) Liquid–vapor interface viewed from above. Work must be done against the surface tension to extend the surface area.

As a matter of fact, an interface separating two phases, α and β, must be a region of some finite thickness, in which there is a gradual change from properties of α to those of β. The great contribution of Young was to show that so far as its mechanical properties are concerned, such an interfacial region can be replaced by the conceptual model of a stretched membrane of infinitesimal thickness. The location of this dividing plane between the two regions is called the *surface of tension*. It can be proved rigorously that the properties of the surface layer suffice to establish completely (1) the position of this surface of tension, and (2) the value of the surface tension acting in it.*

2. Equation of Young and Laplace

The idea of a surface tension was one of those great simplifying concepts that open up the future development of a scientific field. By means of the concept, Young, and later independently Laplace, were able to derive explicitly the conditions for mechanical equilibrium at a general curved surface between two phases. Figure 11.2 depicts a spherical surface with radius of curvature r. Consider an element $\delta\mathscr{A}$ of the surface and the forces acting upon it. The pressure (force per unit area) across the element $\delta\mathscr{A}$ is $P'' - P'$, so that the force acting on it is $(P'' - P')\,\delta\mathscr{A}$. The component of force in the z direction is $(P'' - P')\,\delta\mathscr{A}\cos\alpha = (P'' - P')\,\delta\mathscr{A}'$. This z component of force summed over the entire area of the spherical cap is, therefore, $(P'' - P')\,\pi a^2$. The surface tension exerts a force $\gamma\,\delta l$

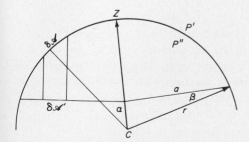

FIG. 11.2 Derivation of the equation of Young and Laplace.

*A proof is given by R. Defay, I. Prigogine, A. Belleman, and D. M. Everett, *Surface Tension and Adsorption* (London: Longmans, Green & Company, Ltd., 1966), p. 4.

on each element δl of the circumference of the base of the cap, and the component of this force along CZ is $- \gamma \cos \beta \, \delta l$. The total force F in direction CZ due to surface tension becomes $-\gamma \cos \beta (2\pi a)$, or, since $\cos \beta = a/r$, $F = -2\pi \gamma a^2/r$. Because the system is symmetrical about CZ, all forces normal to CZ must add to zero. The condition for equilibrium is thus that the z components also cancel, or

$$(P'' - P')\pi a^2 - \frac{2\pi \gamma a^2}{r} = 0$$

This gives the equation of Young and Laplace for a spherical surface,

$$P'' - P' = \frac{2\gamma}{r} \tag{11.1}$$

This equation says that as a consequence of the existence of surface tension in a spherical surface of curvature r, mechanical equilibrium is maintained between two fluids at different pressures P'' and P'. (Note that the fluid on the concave side of a surface is at a pressure P'', which is greater than P' on the convex side.) In the case of a plane surface, of course, the radius of curvature goes to infinity and the condition of equilibrium is simply $P'' = P'$. A general surface can be characterized by two radii of curvature at any point, r_1 and r_2, and for this case, the equation of Young and Laplace becomes

$$P'' - P' = \gamma \left(\frac{1}{r_1} + \frac{1}{r_2} \right) \tag{11.2}$$

As an example of (11.1), consider a droplet of mercury with $r = 10^{-4}$ m. For the mercury–air interface, $\gamma = 476 \times 10^{-3}$ N·m^{-1} at 293 K. Hence,

$$P'' - P' = \frac{2(476 \times 10^{-3})}{10^{-4}} = 9.52 \times 10^3 \text{ N·m}^{-2}$$

or

$$\frac{9.52 \times 10^3}{1.01325 \times 10^5} = 9.51 \times 10^{-2} \text{ atm}$$

(Note that this is the difference in hydrostatic pressure across the interface and, of course, does not refer to the vapor pressure of the mercury.)

3. Mechanical Work on a Capillary System

Consider in Fig. 11.3 an arrangement of two connected cylinders each with separate pistons. The small piston in the cylinder at the left allows us to change both the volume and the surface area of the system at the right. The system at the right contains a liquid at volume V'' and pressure P'' separated by the surface SS' from its vapor of volume V' at pressure P'. Let the left piston be moved so as to cause the right piston to withdraw slightly. If $d\mathscr{A}$ is the change in the area of the interface SS', the work done on the system is

$$dw = -P' \, dV' - P'' \, dV'' + \gamma \, d\mathscr{A} \tag{11.3}$$

This equation is a general expression for the mechanical work done on a capillary

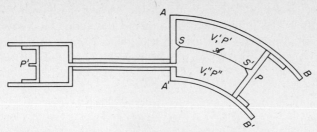

FIG. 11.3 Mechanical work done on a capillary system. Piston on the right is subjected to surface tension and to hydrostatic pressure.

system. If the surface-tension model is to be mechanically equivalent to the real system, the surface $\dot{S}S'$ cannot be arbitrary but must coincide with Young's *surface of tension*. The work done on a *plane* surface, however, is independent of its location, so that (11.3) would be true for any arbitrarily defined plane surface SS'.

From (11.3), the First Law of Thermodynamics applied to a system with surface SS' would be

$$dU = dq + dw = dq - P' \, dV' - P'' \, dV'' + \gamma \, d\mathscr{A} \qquad (11.4)$$

Since $dA = dU - T \, dS - S \, dT$, (11.4) yields

$$dA = -S \, dT - P' \, dV' - P'' \, dV'' + \gamma \, d\mathscr{A} \qquad (11.5)$$

From $dG = dA + V \, dP + P \, dV$, (11.5) gives

$$dG = -S \, dT + V' \, dP' + V'' \, dP'' + \gamma \, d\mathscr{A} \qquad (11.6)$$

From (11.6), we see that

$$\left(\frac{\partial G}{\partial \mathscr{A}} \right)_{T,P} = \gamma \qquad (11.7)$$

Equation (11.7) states that the surface tension equals the partial derivative of the Gibbs free energy with respect to surface area at constant T and P.*

4. Capillarity

Awareness of capillary phenomena begins when babies in bathtubs discover that some things get soaked and others do not. Children are taught that water rises in small tubes as a result of "capillary action," often presented as a consequence of attraction of a solid surface for the molecules of a liquid.

The rise or fall of liquids in capillary tubes and its applications to the measurement of surface tension may be treated quantitatively as a consequence of the fundamental equation (11.2). Whether a liquid rises in a glass capillary, like water,

---

*Note that this is not the same as the free energy per unit area, as has sometimes been erroneously stated.

or is depressed, like mercury, depends on the relative magnitude of the forces of cohesion between the liquid molecules themselves, and the forces of adhesion between the liquid and the walls of the tube. These forces determine the contact angle θ, which the liquid makes with the tube walls (Fig. 11.4). If this angle is less

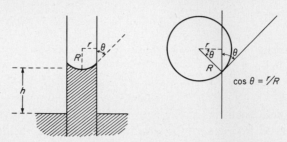

FIG. 11.4 Capillary rise of a liquid that wets the walls of a tube.

than 90°, the liquid is said to wet the surface and a concave meniscus is formed; a contact angle greater than 90° corresponds to a convex meniscus. The occurrence of a concave meniscus leads to a capillary rise, as shown in Fig. 11.4. As soon as the meniscus is formed, the pressure P' under the curved surface is less than that P'' above it, but this P'' is the same as the pressure at the plane surface. The liquid thus rises in the tube until the weight of the liquid column just balances the pressure difference $P'' - P'$ and restores the hydrostatic equilibrium. *The liquid column acts as a manometer to register the pressure difference across the curved meniscus.*

Consider in Fig. 11.4 a cylindrical tube whose radius r is sufficiently small that the surface of the meniscus can be taken as a section of a sphere with radius R. Then, since $\cos \theta = r/R$, from (11.1) $P = (2\gamma \cos \theta)/r$. If the capillary rise is h, and if ρ and ρ_0 are the densities of liquid and gas, the weight of the cylindrical liquid column is $\pi r^2 gh(\rho - \rho_0)$, where g is the gravitational acceleration. The force per unit area balancing the pressure difference is $gh(\rho - \rho_0)$, so that $(2\gamma \cos \theta/r) = gh(\rho - \rho_0)$, and

$$\gamma = \frac{1}{2} gh(\rho - \rho_0) \frac{r}{\cos \theta} \tag{11.8}$$

Equation (11.8) should be corrected for the weight of the liquid above the bottom of the meniscus. To a first approximation, the meniscus is a hemisphere of radius r, or volume $\frac{2}{3}\pi r^3$, so that the volume of this liquid is $\pi r^3 - \frac{2}{3}\pi r^3 = \frac{1}{3}\pi r^3$, and (11.8) becomes

$$\gamma = \frac{g(\rho - \rho_0)r}{2 \cos \theta}\left(h + \frac{r}{3}\right) \tag{11.9}$$

For large capillaries, it is no longer a good approximation to assume a spherical shape for the meniscus, and graphical or tabular correction factors have been computed. When all necessary corrections have been made, the capillary-rise method is probably the most accurate way to measure surface tension, good to about 2 parts in 10^4. Other methods, such as maximum bubble pressure, weight of drops, and the ring method of duNoüy, are discussed in reference works* and in problems at the end of this chapter.

To provide some idea of the range of values of surface tensions, we present a small selection of data in Table 11.1. Liquids with exceptionally high surface tensions are those in which cohesive forces are large. The theoretical problem is to calculate the surface tension directly from the basic theory of intermolecular forces. Surface tensions of liquid inert gases are the most amenable to theoretical calculation.†

TABLE 11.1
Surface Tensions of Liquids

A. *Surface Tensions of Pure Substances at 293* K $(\text{N·m}^{-1} \times 10^4)$

Isopentane	137.2	Ethyl iodide	299
Nickel carbonyl	146	Benzene	288.6
Diethyl ether	171.0	Carbon tetrachloride	266.6
n-Hexane	184.3	Methylene iodide	507.6
Ethyl mercaptan	218.2	Carbon bisulfide	323.3
Ethyl bromide	241.6	Water	727.5

B. *Surface Tensions of Liquid Metals and Molten Salts* $(\text{N·m}^{-1} \times 10)$

	K	γ		K	γ
Ag	1243	8	AgCl	725	1.26
Au	1343	10	NaF	1283	2.60
Cu	1403	11	NaCl	1273	0.98
Hg	273	4.7	NaBr	1273	0.88

5. Enhanced Vapor Pressure of Small Droplets—Kelvin Equation

One of the most interesting consequences of surface tension is the fact that the vapor pressure of a liquid is greater when it is in the form of small droplets than when it has a plane surface. This result was first deduced by William Thomson (Kelvin).‡

Consider a spherical drop of pure liquid with curvature $1/r$ in equilibrium with its vapor. From (11.1), the condition for mechanical equilibrium is

---

*A. W. Adamson, *Physical Chemistry of Surfaces* (New York: Interscience Publishers, 1968), Chap. 1.

†Namely, J. G. Kirkwood and F. P. Buff, *J. Chem. Phys.*, *17*, 338 (1949). See also S. Ono and S. Kondo, "Molecular Theory of Surface Tension of Liquids," in *Handbuch der Physik*, ed. by Flügge, Vol. X (Berlin: Springer Verlag, 1960), p. 134.

‡*Phil. Mag.* (4), *42*, 448 (1871).

$$dP'' - dP' = d\left(\frac{2\gamma}{r}\right)$$ (11.10)

For physicochemical equilibrium between two phases,

$$\mu_i' = \mu_i'' \quad \text{or} \quad d\mu_i' = d\mu_i''$$

From

$$d\mu_i = -S_i\, dT + V_i\, dP$$

at constant temperature, $V_i'\, dP' = V_i''\, dP''$. On combining this with (11.10), we have

$$d\left(\frac{2\gamma}{r}\right) = \left(\frac{V_i' - V_i''}{V_i''}\right) dP'$$ (11.11)

If we neglect the molar volume of the liquid V_i'' in comparison with V_i', and treat the latter as an ideal gas volume, $V_i' = RT/P'$, (11.11) becomes

$$d\left(\frac{2\gamma}{r}\right) = \frac{RT}{V_i''}\frac{dP'}{P'}$$

This equation is integrated (taking V_i'' as constant) from zero curvature ($1/r = 0$), at which $P' = P_0$, the normal vapor pressure, to a curvature $1/r$, at which the vapor pressure is P, thus giving the *Kelvin equation*.

$$\ln\frac{P}{P_0} = \frac{2\gamma}{r}\frac{V_i''}{RT} = \frac{2M\gamma}{RT\rho r}$$ (11.12)

where $V_i'' = M/\rho$, M being molar mass and ρ the liquid density. A similar equation can be derived for the solubility of small particles, by means of the relation between vapor pressure and solubility developed in Section 7.5.

Application of (11.12) to water droplets gives the following calculated pressure ratios (water at 293 K, $P_0 = 17.5$ torr):

r (mm)	10^{-3}	10^{-4}	10^{-5}	10^{-6}
P/P_0	1.001	1.011	1.114	2.95

The conclusions of the Kelvin equation have been experimentally verified. Thus there can be no doubt that small droplets of liquid have a higher vapor pressure than bulk liquid, and that small particles of solid have a greater solubility than bulk solid.

These results lead to the rather curious problem of how new phases can ever arise from old ones. For example, if a container filled with water vapor at slightly below the saturation pressure is chilled suddenly, perhaps by adiabatic expansion as in the Wilson cloud chamber, the vapor may become supersaturated with respect to liquid water. It is then in a metastable state, and we may expect condensation to take place. A reasonable molecular model of condensation would seem to be that several molecules of water come together to form a tiny droplet, and that this embryo of condensation grows by accretion as additional molecules from the vapor happen to hit it. The Kelvin equation, however, indicates that this tiny droplet, being less than 10^{-6} mm in diameter, would have a vapor pressure many times that of bulk liquid. With respect to the embryos, the vapor would not be supersaturated at all. Such embryos should immediately reevaporate, and the emergence of a new phase at the equilibrium pressure, or even moderately above it, should be impossible.

There are two ways of escaping this dilemma. In the first place, we know the statistical basis of the Second Law of Thermodynamics. In any system at equilibrium there are always fluctuations about the average condition, and if the system contains few molecules, these fluctuations may be relatively large (cf. Chapter 5). There is always a chance that an appropriate fluctuation may lead to the formation of an embryo of a new phase, even though it would have only a short existence. The chance of such a fluctuation is $e^{-\Delta S/k}$, where ΔS is the deviation of the entropy from the equilibrium value. This fluctuation mechanism is called *spontaneous nucleation*. In many cases, however, the chance $e^{-\Delta S/k}$ is very small, and it is then more likely that dust particles act a nuclei for condensation in supersaturated vapors or solutions.

6. Surface Tensions of Solutions

Let us consider aqueous solutions as examples. Substances that markedly lower the surface tension of water, such as the fatty acids, contain both a polar hydrophilic group and a nonpolar hydrophobic group. The hydrophilic group, e.g., —COOH in fatty acids, makes the molecule reasonably soluble if the nonpolar residue is not too large. Hydrocarbon residues of fatty acids would be extremely

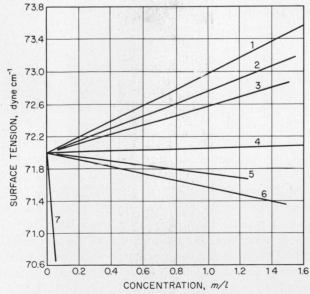

FIG. 11.5 Surface tensions at 298 K of aqueous amino acid solutions. (1) glycine; (2) β-alanine; (3) α-alanine; (4) β-aminobutyric acid; (5) ϵ-aminocaproic acid; (6) 1-aminobutyric acid; (7) 1-aminocaproic acid. [J. R. Pappenheimer, M. P. Lepie, and J. Wyman, *J. Am. Chem. Soc.*, *58*, 1851 (1936).]

"uncomfortable" in the interior of an aqueous solution (in a state of high free energy), and little work is required to bring them from the interior to the surface. This qualitative picture leads to the conclusion that whenever a solute lowers the surface tension, the surface layers of the solution are enriched in that solute. The solute is said to be *positively adsorbed* at the interface.

On the other hand, solutes such as ionic salts usually increase the surface tension of aqueous solutions above the value for pure water, although these increases are smaller than the decreases produced by fatty acids and similar compounds. The reason for the observed increase is that the dissolved ions, by virtue of ion–dipole attractions, tend to pull water molecules into the interior of the solution. To create new surface, additional work must be done against electrostatic forces. It follows that in such solutions the surface layers are poorer in solute. The solute is said to be *negatively adsorbed* at the interface.

Examples of surface tension vs. concentration curves are shown in Fig. 11.5. These data demonstrate the effect of increasing the size of the hydrophobic side chain in a series of amino acids.

7. Gibbs Formulation of Surface Thermodynamics

To provide a quantitative discussion of the surface properties of solutions, we shall follow the elegant analysis of the thermodynamics of surfaces given by Willard Gibbs.

Figure 11.6 represents two phases, α and β, separated by an interfacial region.

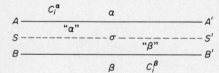

FIG. 11.6 Diagram for definition of a surface phase.

The exact position of this region depends on how we choose to draw the boundary planes AA' and BB'. Let us place these planes so that the following condition is satisfied: there is no appreciable inhomogeneity in the properties of bulk phase α up to the surface AA', nor in those of bulk phase β up to the surface BB'. In particular, the concentration of any component i up to AA' is uniformly c_i^α, and of any component i up to BB', uniformly c_i^β. Within the interfacial region, properties of the system vary continuously from purely α- at AA' to purely β- at BB'. Because of the rather short range of intermolecular forces, the thickness of the interfacial region will usually be not more than about ten molecular diameters.

The phases α and β can be arbitrarily separated by any surface SS' drawn within the interfacial region parallel to the surfaces AA' and BB'. Such a surface SS' is called a *surface phase*, and is designated by a superscript σ. It is a strictly two-dimensional phase. Its area will be denoted by $\mathscr{A}$.

From the conservation of mass, we can write the total amount of component *i* in the system as

$$n_i = n_i^\alpha + n_i^\beta + n_i^\sigma \tag{11.13}$$

In terms of the concentrations *c*,

$$n_i^\alpha = c_i^\alpha V^\alpha \quad \text{and} \quad n_i^\beta = c_i^\beta V^\beta$$

In other words, we define n_i^α and n_i^β by the convention that the composition of the bulk phases continues to be constant right up to the interface *SS'*. We should emphasize that n_i^σ may then be positive or negative, as needed to conserve the mass of *i* in the expression (11.13) for n_i.

Gibbs defined the *adsorption* of *i* at the interface as

$$\Gamma_i = \frac{n_i^\sigma}{\mathscr{A}} \tag{11.14}$$

This quantity has the dimensions of a surface concentration (amount of substance per unit area), and it may, of course, be positive or negative in accord with n_i^σ.

We can extend the treatment of (11.13) and (11.14) to other extensive variables of the system. For example,

surface energy $U^\sigma = U - U^\alpha - U^\beta$
surface entropy $S^\sigma = S - S^\alpha - S^\beta$,   etc.

The choice of the dividing surface has so far been left arbitrary, but to make the mechanical properties of the model system conform to those of the real system, the division must be made at the surface of tension defined in Section 11.1. In the case of a plane surface, however, the work done in extending the surface does not depend on the location of the surface of tension, and we can accordingly place the Gibbs surface *SS'* wherever we please.

From (11.7), (6.4), and (11.13), we can write for the surface Gibbs function,

$$dG^\sigma = -S^\sigma \, dT + \gamma \, d\mathscr{A} + \sum \mu_i \, dn_i^\sigma \tag{11.15}$$

Note that at equilibrium $\mu_i^\sigma = \mu_i^\alpha = \mu_i^\beta$. At constant *T*, (11.15) becomes

$$dG^\sigma = \gamma \, d\mathscr{A} + \sum \mu_i \, dn_i^\sigma \tag{11.16}$$

If we integrate this expression at constant γ and μ_i (or apply Euler's theorem, p. 232),

$$G^\sigma = \gamma \mathscr{A} + \sum \mu_i \, n_i^\sigma$$

When this is differentiated,

$$dG^\sigma = \gamma \, d\mathscr{A} + \mathscr{A} \, d\gamma + \sum \mu_i \, dn_i^\sigma + \sum n_i^\sigma \, d\mu_i \tag{11.17}$$

On combining (11.17) with (11.16),

$$\sum n_i^\sigma \, d\mu_i + \mathscr{A} \, d\gamma = 0 \tag{11.18}$$

(The derivation just given is similar to that which led to the Gibbs–Duhem equation on p. 232.) Division by $\mathscr{A}$ and introduction of Γ_i from (11.14) give the important *Gibbs surface tension equation.*

$$d\gamma = -\sum \Gamma_i \, d\mu_i \tag{11.19}$$

The explicit form of this equation for a solution of two components is

$$dy = -\Gamma_1\, d\mu_1 - \Gamma_2\, d\mu_2 \tag{11.20}$$

At first sight it would appear that we could use the Gibbs equation to determine the *adsorption* of each component from the variation of surface tension with composition, as, for example, from (11.20)

$$\Gamma_1 = -\left(\frac{\partial \gamma}{\partial \mu_1}\right)_{T,\mu_2}$$

We cannot do this, however, because we cannot independently vary μ_1 and μ_2. If the surface were not planar, we should indeed have another degree of freedom, the curvature of the surface, but with planar surfaces we cannot determine the absolute adsorptions Γ_i from the Gibbs equation.

8. Relative Adsorptions

Gibbs met this problem by introducing the idea of relative adsorptions. Let us place the dividing surface in Fig. 11.6 in such a way that the adsorption of one component (say, 1) is zero. The adsorption of all other components $i \neq 1$ at this surface are their adsorptions relative to that of component 1, and are written $\Gamma_{i,1}$. Thus, since $\Gamma_{1,1} = 0$, (11.20) becomes

$$dy = -\Gamma_{2,1}\, d\mu_2$$

or

$$\Gamma_{2,1} = -\left(\frac{\partial \gamma}{\partial \mu_2}\right)_T \tag{11.21}$$

This equation is called the *Gibbs (relative) adsorption isotherm*. For an ideal solution, defined by $\mu_2 = \mu_2^\ominus + RT \ln c_2$, it becomes

$$\Gamma_{2,1} = -\frac{1}{RT}\left(\frac{\partial \gamma}{\partial \ln c_2}\right)_T \tag{11.22}$$

An example* of the application of (11.22) is the treatment of the data on surface tensions of ethanol (2) and water (1) solutions at 25°C. The data are summarized in Table 11.2. If the vapor is taken as an ideal gas mixture with P_2 the partial pressure of ethanol, $\mu_2 = \mu_2^\ominus(T) + RT \ln P_2$, and (11.22) becomes

$$\Gamma_{2,1} = -\frac{1}{RT}\left(\frac{\partial \gamma}{\partial \ln P_2}\right)$$

By plotting γ vs. $\ln P_2$, the slopes given in the table were obtained. The computed values of $\Gamma_{2,1}$ are listed in the last two columns of Table 11.2, in units of mol·cm^{-2} and molecules·nm^{-2}.

Several experimental methods have been devised to test the Gibbs adsorption equation. One of the first was a direct approach by J. W. McBain,† who set up a

*E. A. Guggenheim and N. K. Adam, *Proc. Roy. Soc. London, Ser. A, 139*, 231 (1933).
†J. W. McBain and L. A. Wood, *Proc. Roy. Soc. London, Ser. A, 174*, 286 (1940).

TABLE 11.2

**Data on the Vapor Pressures and Surface Tensions of Ethanol–Water
Solutions and Their Application to the Calculation of the Gibbs
Surface Adsorption of Ethanol at the Surface of the Solutions**

Mole fraction ethanol X_2	Partial Pressures (torr)		Surface tension γ dyne·cm⁻¹	$-d\gamma/d \ln P_2$	Surface adsorption $\Gamma_{2,1}$	
	water P_1	ethanol P_2			mol·cm⁻² $\times 10^6$	molecules nm⁻²
0	23.75	0.0	72.2	0.0	0.0	0.0
0.1	21.7	17.8	36.4	15.6	6.3	3.8
0.2	20.4	26.8	29.7	16.0	6.45	3.9
0.3	19.4	31.2	27.6	14.6	5.9	3.6
0.4	18.35	34.2	26.35	12.6	5.1	3.1
0.5	17.3	36.9	25.4	10.5	4.25	2.6
0.6	15.8	40.1	24.6	8.45	3.4	2.06
0.7	13.3	43.9	23.85	7.15	2.9	1.75
0.8	10.0	48.3	23.2	6.2	2.5	1.50
0.9	5.5	53.3	22.6	5.45	2.2	1.33
1.0	0.0	59.0	22.0	5.2	2.1	1.27

swiftly moving microtome to slice off the top layer (about 0.05 mm) from a solution and scoop it into a sample tube for analysis. He could calculate $\Gamma_{2,1}$ from the relation

$$\Gamma_{2,1} = \frac{m_1}{\mathscr{A}}\left[\left(\frac{m_2}{m_1}\right) - \left(\frac{c_2'}{c_1'}\right)\right]$$

where m_2/m_1 is the ratio of masses of solute to solvent in the surface sample and c_2'/c_1' is the ratio of their concentrations (in masses per unit volume) in the bulk of the solution. Another method is based on the use of solutes containing radioactive tracers that are β emitters of low penetrating power.* The different methods give adsorptions in quite good agreement with those calculated from the Gibbs equation.

9.　Insoluble Surface Films

In 1765, Benjamin Franklin, after making some observations on the spreading of oils over the surface of the pond at Clapham Common, estimated that the thinnest layers of oil that could be formed were "2.5 nm" thick. Much later, Pockels† and Rayleigh‡ discovered that some sparingly soluble substances would spread over the surface of a liquid to form films exactly one molecule thick, i.e., a *unimolecular film* or *monolayer*.

*G. Nilsson, *J. Phys. Chem., 61*, 1135 (1957).

†A. Pockels, *Nature, 43*, 437 (1891).

‡Rayleigh, *Proc. Roy. Soc. London, Ser. A, 47*, 364 (1890); *Phil. Mag., 48*, 337 (1899).

In 1917, Irving Langmuir devised a method for measuring directly the *surface pressure* exerted by surface films on liquids. The essential features of his *film balance* are shown in Fig. 11.7. The *fixed barrier*, which may be a strip of mica, floats on the surface of the water and is suspended from a torsion wire. At the ends of the floating barrier are attached strips of platinum foil or waxed threads,

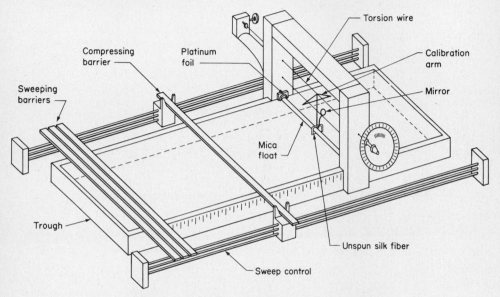

FIG. 11.7 A modern form of a Langmuir film balance.

which lie upon the water surface and connect the ends of the barrier to the sides of the trough, thus preventing leakage of surface film past the float. A *movable barrier* rests upon the sides of the trough and is in contact with the water surface. A number of movable barriers are provided for sweeping the surface clean.

In a typical experiment, a tiny amount of insoluble spreading substance is introduced onto a clean water surface. For example, a dilute solution of stearic acid in benzene might be used; the benzene evaporates rapidly, leaving a film of stearic acid. Then the moving barrier is advanced toward the floating barrier. The surface film exerts a surface pressure on the float, pushing it backward. The torsion wire, attached to a calibrated circular scale, is twisted until the float is returned to its original position. The required force divided by the length of the float is the force per unit length or surface pressure.

Surface pressure is simply another way of expressing the lowering of surface tension caused by a surface film. On one side of the float is a clean water surface with tension γ_0, and on the other side a water surface covered to a certain extent with stearic acid molecules, with lowered surface tension γ. The surface pressure f is simply the negative of the change in surface tension.

$$f = -\Delta\gamma = \gamma_0 - \gamma \tag{11.23}$$

Different substances in unimolecular films display a great variety of surface area vs. surface pressure $(f - \mathscr{A})$ isotherms. Sometimes, the film behaves like a two-dimensional gas, sometimes like a two-dimensional liquid or solid. In addition, there are other types of monolayers having no exact analogs in the three-dimensional world. They can be recognized, however, to be definite surface phases by the discontinuities in the $f - \mathscr{A}$ diagram that signal their occurrence. If a surface film behaves as an ideal two-dimensional gas, we have an equation* similar to the three-dimensional $PV = nRT$:

$$f\mathscr{A} = n^\sigma RT \tag{11.24}$$

If we introduce an excluded area b^σ, a two-dimensional analog of the van der Waals excluded volume correction b, the equation of state becomes

$$f(\mathscr{A} - n^\sigma b^\sigma) = n^\sigma RT$$

10. Structure of Surface Films

The type of $f - \mathscr{A}$ isotherm observed when an organic compound is spread over water depends on the structure of the compound. It is not easy, however, to deduce the exact conformation and packing of the molecules in the surface layer. In the case of molecules with polar end groups and long hydrocarbon chains, the hydrophilic end group is in the water and the hydrophobic chain is in the air. This conclusion was reached by Langmuir as a result of his early observation that the surface area per molecule was the same, about 0.20 nm^2, for close packed surface films of all the normal fatty acids from C_{14} to C_{18}.

The $f - \mathscr{A}$ isotherms for stearic acid $C_{17}H_{35}COOH$ and n-hexatriacontanoic acid $C_{35}H_{71}COOH$ are shown in Fig. 11.8(a), as well as that for isostearic acid. For comparison, the measurement of tri-paracresylphosphate is also shown. If the steep linear portion of the $f - \mathscr{A}$ curve, believed to correspond to compression of a closely packed surface layer, is extrapolated to $f = 0$, the areas are the same for the two straight-chain acids, but the area of the acid with a single branch at the end of the hydrocarbon chain is considerably greater. In Fig. 11.8(b), the structures of these molecules are shown in the surface orientation that occurs in closest packing.

These monolayer structures are closely related to the structure of soap or detergent films. When a soap film thins, it ultimately reaches a "black film" stage—its thickness is now below that required to give interference colors. The limiting thickness for sodium stearate black films is about 5.0 nm. This is about twice the length of the fully extended molecule, and the structure of the film is a bimolecular leaflet in which the polar end groups are in contact and the nonpolar chains are exposed to the exterior, as shown in Fig. 11.9(a).

---

*This can be proved by working through the derivation in Section 4.4 for a two-dimensional case. We find $f\mathscr{A} = \frac{1}{2}NmC^2 = E_K$ and $E_K = RT$.

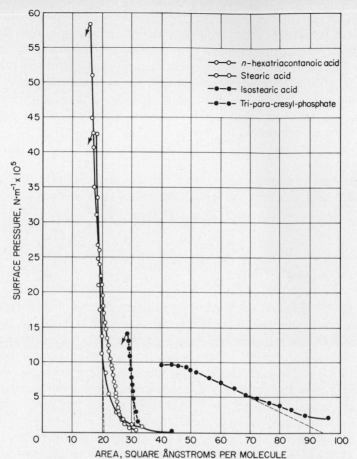

FIG. 11.8(a) Pressure–area curves of four films show collapse pressures (*small arrows*) and cross section of molecules (*lowest point on broken lines*). Film compressibility is related to slope of curves. Tri-*para*-cresyl phosphate film collapses slowly. [H. E. Ries, *Sci. Am., 244*, No. 3, 152 (1961); H. E. Ries and W. A. Kimball, *Proc. 2nd Int. Conf. Surface Activity* (London: Butterworth, 1957), p. 75.]

One of the most interesting applications of surface film studies was made in 1925 by Gorter and Grendel.* They found that the lipids extracted from red-cell membranes would spread on water to a thickness just about half that of the membrane itself. Assuming in accord with Langmuir that the lipid layer was one molecule thick, they concluded that the membrane was essentially a double layer of lipid molecules, probably with some protein on its surfaces. In 1943, Davson and

*J. Expl. Med., 41, 439 (1925).

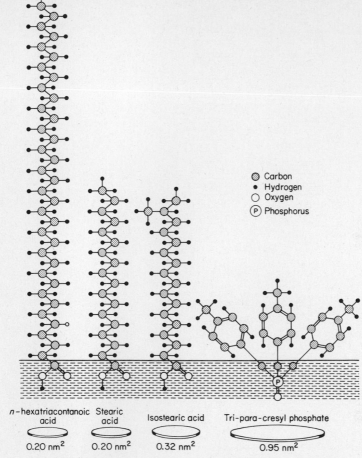

Carbon
Hydrogen
Oxygen
P Phosphorus

n-hexatriacontanoic acid	Stearic acid	Isostearic acid	Tri-para-cresyl phosphate
0.20 nm²	0.20 nm²	0.32 nm²	0.95 nm²

FIG. 11.8(b)   Molecules of film-forming substances at a water–air interface are oriented with their polar groups in the water (*broken lines*) and their nonpolar portions in the air. Cross-sectional areas of molecules appear at bottom. [H. E. Ries, *Sci. Am., 244,* No. 3, 152 (1961).]

Danielli,† on the basis of this work and data on the low surface tensions of membranes, advanced the model shown in Fig. 11.9(b) for the outer membranes of animal cells. Their model was widely accepted by biologists, but it is now under severe attack, primarily on the basis of evidence from electron microscopy, which indicates that the microstructure of the membrane persists even after most of the lipid has been extracted. Such studies suggest that membrane proteins form a structural framework in the interstices of which the lipid constituents are held.

†H. Davson and J. F. Danielli, *Permeability of Natural Membranes* (New York: The Macmillan Company, 1943).

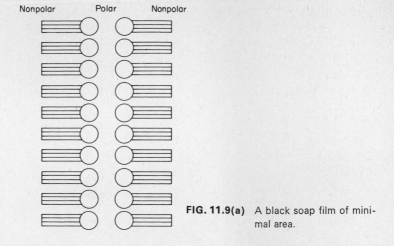

FIG. 11.9(a) A black soap film of minimal area.

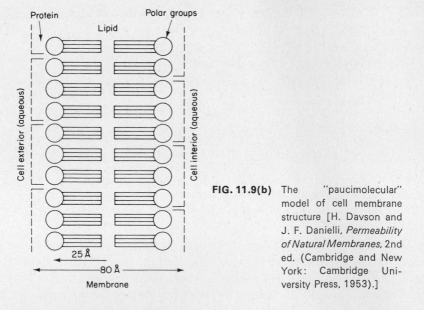

FIG. 11.9(b) The "paucimolecular" model of cell membrane structure [H. Davson and J. F. Danielli, *Permeability of Natural Membranes,* 2nd ed. (Cambridge and New York: Cambridge University Press, 1953).]

11. Dynamic Properties of Surfaces

If we could measure the surface tension γ of a solution just momentarily after its top layer had been sliced off by a McBain microtome, we should find a value different from the equilibrium or static value. The reason for this effect is evident: there has not been enough time to reestablish adsorption equilibrium by diffusion of solute from bulk solution to surface layer. The purely *dynamic surface tension* would be the value of γ when all adsorptions $\Gamma_{i,1} = 0$, i.e., when the composition of the surface is the same as that of the bulk. Dynamic surface tension is an important factor in the theory of rate processes in surface layers. We can only refer to the existence of a fascinating literature in this field, dealing with such topics as oscillating jets, surface waves, and impingement of jets on surfaces.[*]

Another important group of dynamic properties includes *surface viscosity* and related phenomena. In 1869, the blind Belgian physicist Plateau first noted a difference between surface and bulk viscosities in his experiments on the damping of oscillations of compass needles suspended in and on various liquids. If an element of surface $dx\,dy$ flows in its plane xy with a velocity $u(y)$ in the x direction, it experiences a frictional resistance F (force) from the adjacent monolayer elements, given by

$$F = \eta^{\sigma}\frac{du}{dy}\,dx\,dy$$

where η^{σ} is the coefficient of surface viscosity (cf. Section 4.24). What are the dimensions of surface viscosity, and what is its SI unit?

Measurements of surface viscosity are useful in the study of monolayers, since changes in η^{σ} are often sensitive indicators of surface phase transitions. Molecular interactions in films of proteins and proteolipids are also reflected in η^{σ} and may provide clues to factors important in formation of natural membranes. So far, however, no detailed molecular theory of η^{σ} exists. The problem is clearly even more difficult than the calculation of an equilibrium property like surface tension, and we can hardly expect a satisfactory solution in less than a generation.

In pure liquids, a surface viscosity η^{σ} different from the bulk η has not yet been demonstrated. The effects observed by Plateau were actually due to adsorption of dissolved impurities at the surface. The Italian physicist Marangoni[†] was the first to point out that a compass needle moving on a surface would leave a cleanly swept area behind it (with surface tension γ_2) and would tend to concentrate adsorbed surfactant substances in front (with surface tension γ_1). See Fig. 11.10. Thus, $\gamma_1 < \gamma_2$ and the motion of the needle would be damped by the resultant surface pressure. This *Marangoni effect* tends also to stabilize films and surfaces against dilatational deformations (changes in area) and is responsible for most of the dilatational viscosity of surface films (as distinct from the shear viscosity previously discussed).

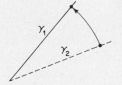

FIG. 11.10 Marangoni effect on moving needle in surface. As needle moves counterclockwise it leaves behind a clean surface with $\gamma_2 > \gamma_1$.

---

[*]A detailed bibliography is given by R. Defay and I. Prigogine, *Surface Tension and Adsorption* (London: Longmans, Green & Company, Ltd., 1966), p. 68.

[†]See L. E. Scriven and C. V. Sternling, "The Marangoni Effects," *Nature*, *187*, 186 (1960).

Many fascinating dynamic surface phenomena are related to similar local differences in surface tensions. Classic examples are the *camphor dance*, in which a small piece of camphor scoots over a water surface, and the *tears of strong wine*, in which alcohol-rich droplets form on the sides of wine glasses.* Similar motions due to surface tension are likely to be important in biological phenomena, such as transport of bacteria or subcellular entities in tissues, pinocytosis, discharge of secretory granules and vesicles, and the behavior of mucous secretions in the lung. In such systems, the interfaces between proteolipid cell membranes and cytoplasm provide the milieu for dynamic effects of surface tensions.

12.   Adsorption of Gases on Solids

Experimental methods for studying gas–solid systems are so different from those used in gas–liquid systems that the adsorption data obtained seem to pertain to two quite different disciplines, until one sees the unifying thermodynamic theory. An experimental apparatus for a volumetric method is shown in Fig. 11.11. The

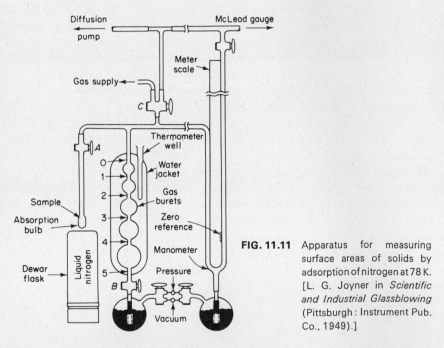

FIG. 11.11 Apparatus for measuring surface areas of solids by adsorption of nitrogen at 78 K. [L. G. Joyner in *Scientific and Industrial Glassblowing* (Pittsburgh: Instrument Pub. Co., 1949).]

vapor of the adsorbate is contained in a calibrated gas buret, and its pressure is measured with the manometer. The adsorbent is contained in a thermostatted sample tube, separated from the adsorbate by a stopcock or cutoff. All volumes

*H. J. Tress, *Trans. Soc. Glass Technol.*, **38**, 89 (1954).

in the apparatus are calibrated. When vapor is admitted to the adsorbent sample, the amount adsorbed can be calculated from the pressure reading after equilibrium is attained. A series of measurements at different pressures determines the adsorption isotherm.

Two typical isotherms are shown in Fig. 11.12. Instead of pressure, *relative pressure* $P/P^{\bullet}$ is used as a coordinate, where $P^{\bullet}$ is the vapor pressure of the adsorbate at the temperature of the isotherm. These isotherms illustrate two kinds of

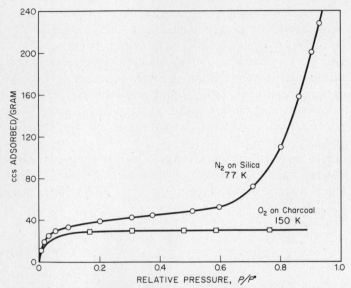

FIG. 11.12 Adsorption isotherms exhibiting physisorption (N_2 on silica at 77 K) and chemisorption (O_2 on charcoal at 150 K). The pressure scale for O_2 is expanded 10 ×, so that it runs from 0 to 0.1.

adsorption behavior that can usually be distinguished. The case of nitrogen on silica gel at 78 K is an example of *physical adsorption* or *physisorption*. The case of oxygen on charcoal at 150 K is a typical *chemical adsorption* or *chemisorption*.

Physisorption is due to the operation of forces between the solid surface and the adsorbate molecules that are similar to the van der Waals forces between molecules. These forces are undirected and relatively nonspecific. They ultimately lead to the condensation of vapor to liquid, when P becomes equal to $P^{\bullet}$. The energies of adsorption involved are of the order of 300 to 3000 $J \cdot mol^{-1}$. The adsorption increases rapidly at high $P/P^{\bullet}$, finally leading to condensation on the surface. At relative pressures around 0.8, even before condensation occurs, there may be several superimposed layers of adsorbate on the surface. Physical adsorption is generally reversible, i.e., on decreasing the pressure the adsorbed gas is desorbed along the same isotherm curve. An exception to this rule is observed when the adsorbent contains many fine pores or capillaries.*

In contrast with physisorption, chemisorption is the result of much stronger binding forces, comparable with those leading to the formation of chemical compounds. Such adsorption may be regarded as the formation of a sort of surface compound. The energies of adsorption range from about 40 to 400 $kJ \cdot mol^{-1}$. At low temperatures, chemisorption is seldom reversible. Generally, the solid must be heated to higher temperature and pumped at high vacuum to remove chemisorbed gas. Sometimes the gas that is desorbed is not the same as that adsorbed; for example, after oxygen is adsorbed on charcoal at 150 K, heating and pumping will cause desorption of carbon monoxide. On the other hand, hydrogen chemisorbed on nickel, presumably with the formation of surface Ni—H bonds, can be recovered as H_2. Chemisorption is completed when a surface is covered by an adsorbed monolayer. Sometimes a physically adsorbed layer may form on top of an underlying chemisorbed layer. The same system may display physisorption at one temperature and chemisorption at some higher temperature. Thus, nitrogen is physisorbed on iron at 78 K and chemisorbed with the formation of surface iron nitride at 800 K.

The most direct way to distinguish physical from chemical adsorption is by examination of the infrared (IR) spectra of the adsorbed molecules, as shown in Fig. 11.13(a). Figure 11.13(b) shows IR spectra of acetylene in solution and adsorbed

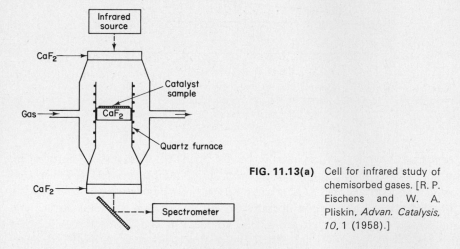

FIG. 11.13(a) Cell for infrared study of chemisorbed gases. [R. P. Eischens and W. A. Pliskin, *Advan. Catalysis*, *10*, 1 (1958).]

on silica and on palladium-coated silica. The spectrum on silica is like that in solution except for a small shift to lower frequencies, but the spectrum on palladium is completely different. We can conclude that the latter case is a chemisorp-

---

*Just as the vapor pressure of liquids with convex surfaces (e.g., droplets) is greater than that for plane surfaces, so the vapor pressure of liquids with concave surfaces is less. The change is given by the Kelvin equation. Therefore, condensation in capillaries is facilitated, whereas evaporation from capillaries is inhibited. When capillary condensation occurs, the adsorption isotherm exhibits hysteresis on desorption.

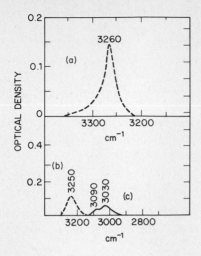

FIG. 11.13(b)   Infrared spectra: (a) acetylene in liquid solution;
(b) acetylene on porous
silica glass; (c) acetylene
on porous silica glass
coated with palladium.

tion with formation of new bonds, whereas the former case is a typical physical adsorption.*

13.   The Langmuir Adsorption Isotherm

The first quantitative theory of the adsorption of gases was given in 1916 by Irving Langmuir, who based his model on the following assumptions:

1. The solid surface contains a fixed number of adsorption sites. At equilibrium at any temperature and gas pressure, a fraction θ of the sites is occupied by adsorbed molecules, and a fraction $1 - \theta$ is not occupied.
2. Each site can hold one adsorbed molecule.
3. The heat of adsorption is the same for all the sites and does not depend on the fraction covered θ.
4. There is no interaction between molecules on different sites. The chance that a molecule condenses at an unoccupied site or leaves an occupied site does not depend on whether or not neighboring sites are occupied.

We can derive the Langmuir adsorption isotherm from a kinetic discussion of the condensation and evaporation of gas molecules at the surface. If θ is the fraction of the surface area covered by adsorbed molecules at any time, the rate of evaporation of molecules from the surface is proportional to θ or equal to $k_d \theta$, where k_d is a constant at constant T. The rate of condensation of molecules on the surface is proportional to the fraction of surface that is bare, $1 - \theta$, and to the rate at which molecules strike the surface, which, at a given temperature, varies directly with the gas pressure. The rate of condensation is therefore set

*L. H. Little, H. Sheppard, and D. J. C. Yates, *Proc. Roy. Soc. London, Ser. A, 259*, 242 (1960).

equal to $k_a P(1 - \theta)$. At equilibrium, the rate of condensation equals the rate of evaporation,

$$k_d\theta = k_a P(1 - \theta)$$

Solving for θ, we obtain

$$\theta = \frac{k_a P}{k_d + k_a P} = \frac{bP}{1 + bP} \tag{11.25}$$

where b is the ratio of rate constants, k_a/k_d, called the *adsorption coefficient*.

The Langmuir isotherm of Eq. (11.25) is plotted in Fig. 11.14(a). Sometimes it is more convenient to plot it in the form of a straight line,*

$$\frac{1}{\theta} = 1 + \frac{1}{bP} \tag{11.26}$$

Figure 11.14(b) shows some data for the adsorption of gases on silica plotted in this form. The good straight lines indicate that the adsorptions conform to the Langmuir isotherm.

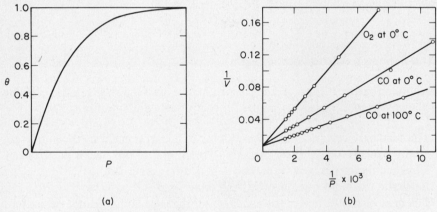

(a)   (b)

FIG. 11.14   (a) Langmuir isotherm. (b) Adsorption of gases on silica plotted in accord with Eq. (11.26). The volume adsorbed (proportional to θ) is in cm³ at STP per gram and the pressure is in torr. [E. C. Markham and A. F. Benton, *J. Am. Chem. Soc., 53,* 497 (1931).]

Two limiting cases of the Langmuir isotherm are often of interest. Where $bP \ll 1$, for instance, when the pressure is low or the adsorption coefficient is very small,

$$\theta = bP$$

Such a linear dependence of θ on P is always found in the low pressure region of the adsorption curve. When $bP \gg 1$, at high pressures or, with particularly strong adsorption, at lower pressures, the isotherm reduces to

*Similar reciprocal plots were later introduced into enzyme kinetics by H. Lineweaver and D. Burk, *J. Am. Chem. Soc., 56,* 658 (1934). See Section 9.42.

$$1 - \theta = \frac{1}{bP}$$

This expression holds in the flat upper region of the isotherm: the fraction of bare surface becomes inversely proportional to the pressure.

14. Adsorption on Nonuniform Sites

Even the smoothest solid surface is rough on a 1-nm scale. Examination of the cleavage faces of crystals by the most refined optical techniques* reveals that they have terrace-like surfaces. Experiments on photoelectric or thermionic emission from metals indicate that the surfaces are a patchwork of areas with different work functions. F. C. Frank† has elucidated a mechanism by which crystals often grow from vapor or solution. New atoms or molecules are not deposited on the planar surfaces, but at jogs in the surface associated with *dislocations* in the crystal structure. The resultant surface structure is a miniature replica of the spiral growth pattern of the Babylonian ziggurat (Section 18.23).

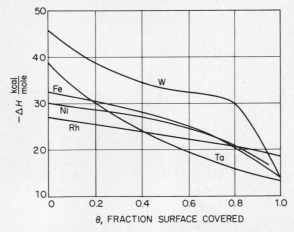

FIG. 11.15 Heat of adsorption of hydrogen on clean metal surfaces.
[O. Beeck, *Discussions Faraday Soc., 8*, 118 (1950).]

The heat of adsorption often declines markedly with increasing surface coverage. Typical results are shown in Fig. 11.15. This effect obviously indicates a nonuniform surface. The lack of uniformity, however, may either pre-exist in the different adsorption sites, or be caused by the repulsive forces between adsorbed atoms or molecules. Especially if the surface-to-adsorbate bond is partially ionic, as much recent evidence suggests, the repulsions may become large, markedly lowering the heat of adsorption at higher coverages.

*S. Tolansky, *Multiple Beam Interferometry* (London: Methuen & Co., Ltd., 1948).
†F. C. Frank, *Advan. Phys., 1* (1952).

Since the model for the Langmuir isotherm is a set of uniform adsorption sites, it is hardly surprising, in view of the nonuniformity of actual surfaces, that many cases of strong adsorption do not fit this isotherm. In some instances, an empirical isotherm due to Freundlich is more successful,

$$\theta = KP^{1/m} \tag{11.27}$$

where K is a constant and m is a number greater than one. It can be shown* that (11.27) corresponds to a nonuniform surface in which the heat of adsorption q^a decreases with log θ.

A linear variation of q^a with θ is usually more in accord with experiment.

$$q^a = q_0^a(1 - \alpha\theta)$$

This behavior of q^a corresponds to the Temkin isotherm,

$$\theta = \frac{RT}{q_0^a\alpha} \ln(A_0 P) \tag{11.28}$$

where α and A_0 are constants for the given system at constant T. The chemisorption of N_2 and H_2 on iron should follow the Temkin isotherm.

The variations of q^a with θ for the three isotherms mentioned are shown in Fig. 11.16.

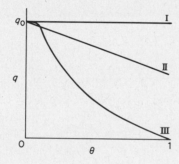

FIG. 11.16 Variation of heat of adsorption with surface coverage: q^a–θ curves for various isotherms. I. Langmuir. II. Temkin. III. Freundlich.

15. Surface Catalysis

Many reactions that are slow in a homogeneous gas or liquid phase proceed swiftly if a suitable solid surface is available. The earliest instance of this *contact action* or *contact catalysis* was the dehydrogenation of alcohols by metals studied by van Marum in 1796. In 1817, Davy and Döbereiner investigated the glowing of certain metals in a mixture of air and combustible gases; and in 1825, Faraday worked on the catalytic combination of hydrogen and oxygen. These studies laid the experimental foundations of heterogeneous kinetics.

An interesting example was found† in the bromination of ethylene: $C_2H_4 + Br_2 \rightarrow C_2H_4Br_2$. This reaction goes readily in a glass vessel at 200°C; it was at

*G. Halsey, *Advan. Catalysis*, **4**, 259 (1952).
†R. G. W. Norrish, *J. Chem. Soc.*, 3006 (1923).

first thought to be an ordinary homogeneous combination, but the rate seemed to be higher in smaller reaction vessels. When a vessel was packed with glass tubing or beads, the rate was considerably enhanced. This method is frequently used for detecting wall reactions. An increased rate in a packed vessel indicates that a considerable share of the observed reaction is heterogeneous, on the packing and wall, rather than homogeneous, in the gas phase only.

The decomposition of formic acid illustrates the specificity often displayed by surface reactions. If the acid vapor is passed through a heated glass tube, the reaction is about one-half dehydration and one-half dehydrogenation.

(1) $$HCOOH \longrightarrow H_2O + CO$$
(2) $$HCOOH \longrightarrow H_2 + CO_2$$

If the tube is packed with Al_2O_3, only reaction (1) occurs; but if it is packed with ZnO, (2) is the exclusive result. Thus, different surfaces can accelerate different parallel paths, and so, in effect, determine the nature of the products.

A surface reaction can usually be divided into the following elementary steps:

1. diffusion of reactants to surface;
2. adsorption of reactants at surface;
3. chemical reaction on the surface;
4. desorption of products from surface;
5. diffusion of products away from the surface.

These are consecutive steps, and if any one has a much slower rate constant than all the others, it will become rate determining. Steps 1 and 5 are usually fast. Only with extremely active catalysts might they determine the overall rate. Diffusion has a $T^{1/2}$ and chemical reaction has an $e^{-E/RT}$ temperature dependence. Therefore, if a catalytic reaction rate increases only slightly with temperature, it may be diffusion controlled. Steps 2 and 4 generally have higher specific rates than step 3, but reactions are known in which they may be the rate determining stages. Usually, however, the actual reaction on the surface, step 3, is rate determining. In some cases, instead of reaction entirely on the surface, a molecule from the fluid phase may react with an adsorbed species.

The kinetics of many surface reactions can be treated successfully on the basis of the following assumptions:

1. The rate determining step is a reaction of adsorbed molecules.
2. The reaction rate per unit surface area is proportional to θ, the fraction of surface covered.
3. The value of θ is given by a Langmuir isotherm.

Examples are given in Problem 15.

16. Activated Adsorption

Often, the potential-energy barrier that must be surmounted before adsorption can occur is small or negligible, and the adsorption rate is governed by the rate of supply of gas. Sometimes, however, a considerable activation energy, E_a, may be required for adsorption, and the rate constant for adsorption, $k_a = A \exp(-E_a/RT)$, may become slow enough to determine the overall rate of a surface reaction. Adsorption that requires an appreciable activation energy has been called *activated adsorption*.

The chemisorption of gases on metals usually does not require any appreciable activation energy. The work of J. K. Roberts* showed that the adsorption of hydrogen on carefully cleaned metal filaments proceeds rapidly even at 25 K, to form a tightly held monolayer of adsorbed hydrogen atoms. The heat of adsorption is close to that expected for the formation of covalent metal-hydride bonds.

One important exception to this type of behavior has been found in the adsorption of nitrogen on an iron catalyst at about 373 K.† This adsorption is a slow activated adsorption and it seems to be the rate determining step in the synthesis of ammonia on these catalysts. If S is the catalyst surface, the reaction can be represented as follows:

$$N_2 \longrightarrow \left.\begin{matrix} N \\ N \end{matrix}\right\} S$$

$$\cdots\cdots \longrightarrow NH\} S \longrightarrow NH_2\} S \longrightarrow NH_3\} S \longrightarrow NH_3$$
$$\text{surface reaction} \qquad\qquad\qquad \text{desorption}$$

$$H_2 \longrightarrow \left.\begin{matrix} H \\ H \end{matrix}\right\} S$$

activated adsorption

The adsorption and activation of the hydrogen was ruled out as the slow step because the exchange reaction $H_2 + D_2 \rightleftharpoons 2HD$ occurs on the catalyst even at liquid-air temperatures, presumably via the dissociation of H_2 and D_2 into adsorbed atoms. Then it was found that the hydrogens in NH_3 are readily exchanged with deuterium from D_2 on the catalyst at room temperature. This indicates that processes involving N—H bonds are not likely to be rate determining. The only possible slow step seemed to be the activated adsorption of N_2 itself, which, therefore, appeared to govern the speed of the important synthetic ammonia reaction. The idea that slow activated adsorption of N_2 was the rate determining step was strengthened by measurements of the adsorption rate by vacuum microbalance techniques.‡ At this point, however, some difficulties in the activated adsorption model appeared, when it was found that the chemisorption of N_2 was accelerated by simultaneous chemisorption of H_2.§ During the actual synthesis of NH_3, the rate of adsorption of N_2 is about 10 times the synthesis rate.

*J. K. Roberts, *Some Problems in Adsorption* (London: Cambridge University Press, 1939).
†P. H. Emmett and S. Brunauer, *J. Am. Chem. Soc.*, 62, 1732 (1940).
‡J. J. F. Scholten and P. Zweitering, *Trans. Faraday Soc.*, 56, 262 (1960).
§K. Tamaru, *Trans. Faraday Soc.*, 59, 979 (1963).

A new analysis of the mechanism, undertaken by Taylor and coworkers,* indicated that most of the nitrogen on the surface was in the form of adsorbed —NH (imine radicals). Thus, all the following reactions need to be considered:

$$N_2 \longrightarrow 2N \text{ (ads)}$$
$$H_2 \longrightarrow 2H \text{ (ads)}$$
$$N \text{ (ads)} + H \text{ (ads)} \longrightarrow NH \text{ (ads)}$$
$$NH \text{ (ads)} + H \text{ (ads)} \longrightarrow NH_2 \text{ (ads)}$$
$$NH_2 \text{ (ads)} + H \text{ (ads)} \longrightarrow NH_3 \text{ (ads)}$$

17. Statistical Mechanics of Adsorption

The statistical mechanical theory of the adsorption isotherm not only is fascinating in itself but also illustrates a type of theoretical treatment that can be applied to other fields, such as high polymer solutions, phase transitions, and ferromagnetism. A one-dimensional model will be discussed, which is shown in Fig. 11.17 together with the symbolism used. It consists of a linear set of lattice sites, some occupied, some empty, with an energy of interaction $U = w$ between each pair of neighboring occupied sites and $U = 0$ between occupied sites farther apart. This model is called a *one-dimensional lattice gas*, since atoms adsorbed on the sites can exchange freely with one another. This statistical problem, first discussed by Ising in 1925 in connection with the theory of ferromagnetism, was described in Section 6.11.

The one-dimensional problem is an excellent illustration of the power of the method of the *canonical ensemble* in statistical mechanics. The problem could hardly be treated at all by means of the microcanonical ensemble because it involves an *interaction between particles*. We shall follow the derivation given by Hill.†

A configuration of N molecules on M sites with Q pairs of type *ox* will have a total interaction energy of

$$N_{xx}w = \left(N - \frac{Q}{2}\right)w$$

Let the statistical weight of this energy state be $g(N, M, Q)$, i.e., there are g ways of arranging N molecules on the M sites so that there are Q pairs of type *ox*.

$$\underline{x} \quad \underline{x} \mid \underline{o} \quad \underline{o} \mid \underline{x} \mid \underline{o} \mid \underline{x} \quad \underline{x} \quad \underline{x} \mid \underline{o}$$

$$M = 10 \qquad N = 6 \qquad M - N = 4 \qquad Q = 5 \qquad N_{xx} = 3$$

FIG. 11.17 Model of section of a one-dimensional lattice with M sites and N adsorbed molecules. Occupied sites are designated by x, and empty sites by o. (The two end sites are considered to make an ox pair.)

*A. Ozaki, H. S. Taylor, M. Boudart, *Proc. Roy. Soc. London, Ser. A*, *258*, 47 (1960).

†T. Hill, *An Introduction to Statistical Thermodynamics* (Reading, Mass.: Addison-Wesley Publishing Co., Inc., 1960), Chap. 14.

Then, from (5.36), the partition function Z is

$$Z(N, M, T) = z^N \sum_Q g(M, N, Q) \exp\{-[N - (Q/2)]w/kT\}$$
$$= (ze^{-w/kT})^N \sum_Q g(M, N, Q)(e^{w/2kT})^Q \qquad (11.29)$$

Here z is the partition function for a single adsorbed molecule, and the sum is taken over all possible values of Q for the given values of M and N.

Let us first of all consider what happens to (11.29) when $w = 0$. This case would correspond to the *Langmuir model of adsorption* in which there is no interaction between adsorbed molecules. Equation (11.29) would then become

$$Z(N, M, T) = z^N \sum_Q g(M, N, Q) = z^N \frac{M!}{N!(M - N)!} \qquad (11.30)$$

since $M!/N!\,(M - N)!$ is just the total number of configurations of given M and N (the number of permutations of M sites divided by number of permutations of occupied sites times number of permutations of unoccupied sites).

Since the surface area $\mathscr{A}$ is proportional to the number of sites M, we can write $d\mathscr{A} = \alpha\, dM$ where α is a proportionality factor. If we use a chemical potential μ *per molecule*, (11.15) becomes

$$dG^\sigma = -S^\sigma\, dT + \gamma\, d\mathscr{A} + \sum \mu\, dN$$

For the surface phase, the Gibbs and Helmholtz free energies are identical, so that from (5.44)

$$G^\sigma = -kT \ln Z^\sigma$$

and

$$\left(\frac{\partial G^\sigma}{\partial N}\right)_T = \mu = -kT\left(\frac{\partial \ln Z^\sigma}{\partial N}\right)_T$$

From (11.30) and Stirling's formula,

$$\frac{\mu}{kT} = -\left(\frac{\partial \ln Z}{\partial N}\right)_{M,T} = \ln \frac{N}{z(M - N)} = \ln \frac{\theta}{z(1 - \theta)}$$

where $\theta = N/M$ is the fraction of sites covered.

If the adsorbed phase is in equilibrium with gas at pressure P,

$$\mu = \mu^0 + kT \ln P$$

or

$$\frac{\mu}{kT} = \frac{\mu^0}{kT} + \ln P = \ln \frac{\theta}{z(1 - \theta)}$$

or

$$\theta(P, T) = \frac{b(T)P}{1 + b(T)P}$$

which is the Langmuir adsorption isotherm of Eq. (11.25) with

$$b(T) = z(T)\, e^{\mu^0/kT} \qquad (11.31)$$

Let us now return to the case with an interaction energy between neighboring sites, $w > 0$. When we are dealing with approximately 1 mol of gas molecules and adsorption sites, in the summation of (11.29) M, N, and Q are enormously large numbers. We shall evaluate $g(M, N, Q)$ in (11.29) by breaking the linear array of sites into groups of x's and o's, as shown in Fig. 11.17. In the example shown in the figure, we have $(Q + 1)/2$ groups of x's and $(Q + 1)/2$ groups of o's, since an ox pair occurs at each boundary between an x group and an o group. How many ways can we arrange the N x's into the $(Q + 1)/2$ groups? Each x group must have at least one x in it; thus, the number of arrangements is the number of ways of distributing the remaining $N - (Q + 1)/2$ x's among the $(Q + 1)/2$ groups, or, with neglect of unity compared with the large numbers,

$$\frac{N!}{(Q/2)![N - (Q/2)]!}$$

The corresponding number of $(M - N)$ o's is given by this same expression with N replaced by $(M - N)$. Then g is *twice* the product of these two distribution numbers, since each particular linear array could be written backwards as well as forwards as shown. Hence,

$$g(N, M, Q) = \frac{2N!(M - N)!}{[N - (Q/2)]![M - N - (Q/2)]![(Q/2)!]^2} \tag{11.32}$$

We now substitute this expression for g into (11.29) and evaluate the maximum term in the summation

$$Z(N, M, T) = (z\,e^{-w/kT})^N \sum_Q t(N, M, Q) \tag{11.33}$$

with

$$t(N, M, Q) = g(e^{w/2kT})^Q \tag{11.34}$$

From (11.34), the condition that t have its maximum value t^* is that

$$\frac{\partial \ln t}{\partial Q} = 0 = \frac{\partial \ln g}{\partial Q} + \frac{w}{2kT} \tag{11.35}$$

The Stirling formula applied to (11.32) gives for the nonconstant terms (those containing Q):

$$\ln g = -\left(N - \frac{Q}{2}\right)\ln\left(N - \frac{Q}{2}\right) - \left(M - N - \frac{Q}{2}\right)\ln\left(M - N - \frac{Q}{2}\right) - Q\ln\left(\frac{Q}{2}\right)$$

Thus (11.35) yields

$$\frac{\partial \ln g}{\partial Q} = -\frac{w}{2kT} = \frac{1}{2}\ln\left(N - \frac{Q^*}{2}\right) + \frac{1}{2}\ln\left(M - N - \frac{Q^*}{2}\right) - \ln\left(\frac{Q^*}{2}\right) \tag{11.36}$$

where Q^* is the value of Q that corresponds to the maximum term t^* in the summation.

If we denote the fraction of sites occupied as $\theta = N/M$ and let $y = Q^*/2M$, (11.36) becomes

$$\frac{(\theta - y)(1 - \theta - y)}{y^2} = e^{-w/kT} \tag{11.37}$$

Solution of the quadratic equation (11.37) gives

$$y = \frac{2\theta(1 - \theta)}{\beta + 1} \tag{11.38}$$

where

$$\beta = [1 - 4\theta(1 - \theta)(1 - e^{-w/kT})]^{1/2}$$

We now have in terms of θ and w the value of Q that makes t a maximum. As we did in Section 5.9 in deriving the Boltzmann distribution, we replace the sum in (11.33) by its maximum term to evaluate the log of the partition function Z,

$$\ln Z = N \ln(z \, e^{-w/kT}) + \ln t^*$$

The chemical potential of the adsorbed species becomes

$$-\frac{\mu}{kT} = \left(\frac{\partial \ln Z}{\partial N}\right)_{M,T} = \ln(z \, e^{-w/kT}) + \left(\frac{\partial \ln t^*}{\partial N}\right)_{M,T}$$

or

$$-\frac{\mu}{kT} = \ln(z \, e^{-w/kT}) + \left(\frac{\partial \ln g^*}{\partial N}\right)_{M,T}$$

where g^* is the value of g corresponding to the maximum term t^*. From (11.32) and (11.38),

$$\frac{\partial \ln g^*}{\partial N} = \frac{N[M - N - (Q^*/2)]}{(M - N)[N - (Q^*/2)]} = \frac{\theta(1 - \theta - y)}{(1 - \theta)(\theta - y)}$$

Thus,

$$-\frac{\mu}{kT} = \ln(z \, e^{-w/kT}) + \frac{\theta(1 - \theta - y)}{(1 - \theta)(\theta - y)}$$

or

$$\lambda z \, e^{-w/kT} = \frac{1 - \theta}{\theta}\left(\frac{\theta - y}{1 - \theta - y}\right)$$

where $\lambda = e^{\mu/kT}$.

Introducing y from (11.38), we obtain

$$x \equiv \lambda z \, e^{-w/kT} = \frac{\beta - 1 + 2\theta}{\beta + 1 - 2\theta} \tag{11.39}$$

For an ideal gas,

$$\mu = \mu^0 + kT \ln P$$

or

$$\lambda = e^{\mu/kT} = e^{\mu^0/kT} P$$

Thus, λ is proportional to the pressure of a gas in an adsorption system, so that (11.39) is the equation of the adsorption isotherm corresponding to the one-dimensional Ising model. Figure 11.18 shows a plot of (11.39) for the case $w = 0$ (Langmuir adsorption isotherm) and a case with an attractive interaction energy between the adsorbed particles. A physical situation corresponding more directly with the one-dimensional model would be adsorption on the units of a long polymer chain.

The process of adsorption of a molecule from a gas onto the surface of a solid is always accompanied by a decrease in entropy, since some of the freedom of the gaseous state is inevitably lost by adsorption. Yet the decrease in entropy may be less than that for condensation to liquid. Usually $\Delta U < 0$ as the molecules are adsorbed. We can consider the ΔU to measure the decrease in potential energy U of the molecules as they pass to the adsorbed state. There will be a potential

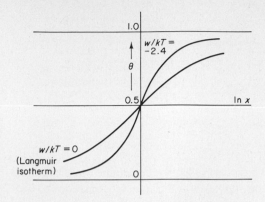

FIG. 11.18 Computed adsorption isotherms based on the one-dimensional Ising model.

energy curve for a molecule approaching a solid surface much like the curves shown in Fig. 4.4 for interaction of a pair of molecules, but the distance of effective interaction will be more extended in the case of the surface. The principal terms in the attractive energy will correspond to London (or van der Waals) dispersion forces (see Section 19.7). Instead of the r^{-6} dependence of the London potential between a pair of molecules, there will be an r^{-3} dependence for the London interaction between a molecule and a solid, as shown in Fig. 11.19. This longer range interaction is the basic reason why gases are adsorbed on solid surfaces at pressures well below those at which they condense to liquids or solids. In the case of dipolar molecules, there will also be electrical interaction between the adsorbed dipoles and the electric field at the surface.

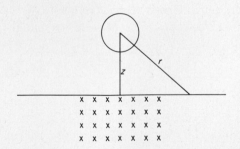

FIG. 11.19 Calculation of interaction of a molecule in gas phase with an extended solid surface. Integration of a potential energy $U = Ar^{-6}$ over all the interactions between the gas molecule and the extended surface at a distance z leads to an interaction energy of form Bz^{-3} (see Problem 17).

18.   Electrocapillarity

The presence of a net electric charge on a surface lowers the surface tension, because repulsion between like charges decreases the work done in extending the surface area. In 1875, G. Lippmann* made the first quantitative measurements of this effect by means of his *capillary electrometer*, shown in Fig. 11.20. This device

*Ann. Chem. Phys., 5, 494 (1875).

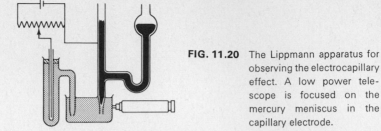

FIG. 11.20 The Lippmann apparatus for observing the electrocapillary effect. A low power telescope is focused on the mercury meniscus in the capillary electrode.

consists of an electrochemical cell with a mercury electrode contained in a capillary tube and a nonpolarizable reference electrode (such as the calomel electrode shown). An external voltage source permits adjustment of the electric potential between the mercury capillary electrode and the mercury lead of the calomel electrode.

The mercury electrode in contact with the nonreactive salt solution can be considered to behave as an *ideal polarizable electrode*. No transfer of electrons or ions takes place between the mercury and the solution and hence the only effect of an applied electric potential difference $\Delta\Phi$ is to change the density of charge $Q/\mathscr{A}$ at the mercury surface. As $Q/\mathscr{A}$ changes, the surface tension γ of the mercury also changes, and the position of the mercury meniscus shifts in the capillary (observed with a low power microscope). The change in γ is measured by the change in height of the mercury reservoir necessary to restore the original position of the meniscus. One thus obtains a curve of γ vs. Φ called the *electrocapillary curve*. Some examples are shown in Fig. 11.21.

The ideal nonpolarizable electrode* and the ideal polarizable electrode are limiting cases that can be treated exactly by thermodynamics. The work of David Grahame† and others has provided an exact thermodynamic treatment of the electrical double layer at ideal polarizable electrodes. One basic relation is the Lippmann equation,

$$\left(\frac{\partial\gamma}{\partial\Phi}\right)_{T,P,\mu} = -\frac{Q}{\mathscr{A}} \tag{11.40}$$

According to this, the slope of the electrocapillary curve yields the surface density of charge on the electrode. We shall refer to the paper of Grahame for the derivation of (11.40), which is based on the Gibbs equation (11.18). However, we can see the significance of (11.40) by rewriting it as an equilibrium condition at constant T, P, and composition:

$$\mathscr{A}\,d\gamma + Q\,d\Phi = 0$$

This equation indicates that the variation in Gibbs free energy due to the change in γ just balances that due to the change in the electric potential of the charge Q at the surface.

---

*Discussed in Chapter 12.
†D. C. Grahame and R. W. Whitney, *J. Am. Chem. Soc.*, **64**, 1548 (1942).

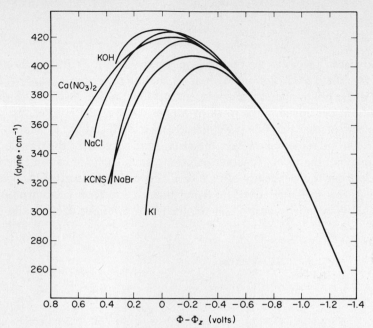

FIG. 11.21 Electrocapillary curves for mercury and different electrolytes at 18°C. Potentials referred to Φ for an electrolyte (sodium fluoride) without specific adsorption. [D. C. Grahame, *Chem. Rev., 41*, 441 (1947).]

From (11.40) we note that the maximum of the electrocapillary curve corresponds to a surface charge density of zero. For many electrolytes, the potential of zero charge at the mercury surface is about 0.5 V (referred to a normal calomel electrode in KCl), but for some electrolytes it deviates considerably from that value. In such cases, *specific adsorption of ions* may occur at the mercury surface. Mercury electrodes have been used in most of these studies because they are uniform, pure, and free of strains. The basic theory is equally applicable to other electrodes and to membrane surfaces, but the experimental problems are more difficult.

The *capacitance of the double layer* is defined by

$$C = \frac{1}{\mathscr{A}}\left(\frac{\partial Q}{\partial \Phi}\right)_{T,P,\mu} = \left(\frac{\partial^2 \gamma}{\partial \Phi^2}\right)_{T,P,\mu} \tag{11.41}$$

Measurements of capacitance by a-c bridge techniques are often used to obtain experimental data about double layers.

Polarizable electrodes have been widely used in physiological studies—e.g., to measure resting potential differences across the membranes of nerve and muscle cells and the action potentials responsible for nerve conduction and muscle contraction.

19.   Structure of the Double Layer

The thermodynamic theory of the double layer can be used to provide information concerning the relative adsorptions of ions at the interface, but it cannot reveal the statistical distribution of these ionic charges. As in the case of the theory of ionic solutions developed in the work of Milner, Debye, and Hückel, the introduction of electrostatic equations is required to develop a statistical theory of the double layer. In one respect, however, the electrostatic theory of the double layer is more flexible that the theory for ionic solutions, since each ion can interact with the planar electrode independently of its interaction with other ions. Thus, we can separate effects due to the electric field from those due to ionic concentrations, a separation that is not really possible in the theory of solutions of mutually inter-acting ions.

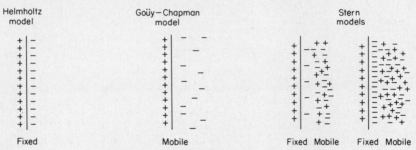

FIG. 11.22   Models for electrical double layers.

The earliest model of the double layer was given in 1879 by Helmholtz, who suggested the picture shown in Fig. 11.22, a layer of ions at the solid surface and a rigidly held layer of oppositely charged ions in the solution. The electric potential corresponding to such a charge distribution is also shown. The Helmholtz double layer is equivalent to a simple parallel-plate capacitor. If λ is the distance separating the oppositely charged plates and ε the dielectric constant of the medium, the capacitance per unit area of interface is $\varepsilon\varepsilon_0/\lambda$. If $Q/\mathscr{A}$ is the surface charge density, the potential difference $\Delta\Phi$ across the double layer is

$$\Delta\Phi = \frac{\lambda Q}{\varepsilon_0 \varepsilon \mathscr{A}}$$

The Helmholtz model of the double layer is basically inadequate, because the thermal motions of liquid molecules could scarcely permit such a rigid array of charges at the interface.

The theory of a diffuse double layer with a statistical distribution of ions in the electric field was given by Goüy in 1910 and Chapman in 1913. This work was contemporary with that of Milner on ionic solutions (1912) and well before that of Debye and Hückel in 1923.

The Poisson–Boltzmann equation in this case (compare Section 10.22) is one-dimensional. With x the distance from the solid surface and n_i^0 the concentration of ions of species i in the bulk of solution (where $\Phi \longrightarrow 0$ as $x \longrightarrow \infty$),

$$\frac{d^2\Phi}{dx^2} = -\frac{1}{\varepsilon_0\varepsilon} \sum z_i e n_i^0 \exp -\frac{z_i e\Phi}{kT} \tag{11.42}$$

This equation is subject to the usual unsatisfactory assumption that there exists a constant effective dielectric constant ε even in the neighborhood of the charged electrode.

For the case of a symmetrical binary electrolyte with $z_- = -z_+$ (11.42) becomes

$$\frac{d^2\Phi}{dx^2} = -\frac{1}{\varepsilon_0\varepsilon} zen^0 \left[\exp\left(-\frac{ze\Phi}{kT}\right) - \exp\left(\frac{ze\Phi}{kT}\right) \right]$$

$$\frac{d^2\Phi}{dx^2} = \frac{2zen^0}{\varepsilon_0\varepsilon} \sinh\left(\frac{ze\Phi}{kT}\right) \tag{11.43}$$

Following the method of Verwey and Overbeek,* we write

$$y = \frac{ze\Phi}{kT}$$

$$w = \frac{ze\Phi_0}{kT} \tag{11.44}$$

$$\kappa^2 = \frac{2n^0 e^2 z^2}{\varepsilon_0\varepsilon kT}$$

$$\xi = \kappa x$$

where Φ_0 is the potential at the interface ($x = 0$). Equation (11.43) thus becomes

$$\frac{d^2 y}{d\xi^2} = \sinh y$$

On integrating once,

$$\frac{dy}{d\xi} = -2 \sinh\left(\frac{y}{2}\right) + C_1$$

At $\xi = \infty$, $dy/d\xi = 0$ and $y = 0$, yielding $C_1 = 0$. On integrating again,

$$\ln \frac{e^{y/2} - 1}{e^{y/2} + 1} = -\xi + C_2$$

We find C_2 from the other boundary condition: $y = w$ at $\xi = 0$,

$$C_2 = \ln \frac{e^{w/2} - 1}{e^{w/2} + 1}$$

Hence, the final solution is

$$e^{y/2} = \frac{e^{w/2} + 1 + (e^{w/2} - 1)e^{-\xi}}{e^{w/2} + 1 - (e^{w/2} - 1)e^{-\xi}} \tag{11.44}$$

Although this equation looks a bit formidable, we find on plotting that it gives an approximately exponential decrease in Φ through the double layer, as shown in

---

Theory of the Stability of Lyophobic Colloids (Amsterdam: Elsevier, 1948). Also O. F. Devereux and P. L. De Bruyn, *Interaction of Plane Parallel Double Layers* (Cambridge, Mass.: M.I.T. Press, 1963).

Fig. 11.23(b). The quantity κ^{-1} (analogous to Debye's thickness of the ionic atmosphere) is a measure of the thickness of the double layer. The values computed at 25°C for different ionic concentrations and valencies are shown in Table 11.3.

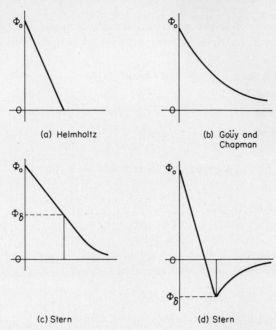

FIG. 11.23 Potential in different types of double layer, showing variation with distance from surface.

TABLE 11.3
Thickness of the Double Layer Calculated
from Goüy–Chapman Theory

c (mol·dm^{-3})	$z = 1$ (κ^{-1} cm)	$z = 2$ (κ^{-1} cm)
10^{-5}	10^{-5}	0.5×10^{-5}
10^{-3}	10^{-6}	0.5×10^{-6}
10^{-1}	10^{-7}	0.5×10^{-7}

If we know the potential function $\Phi(x)$ we can calculate the charge per unit area of the surface (or in the diffuse layer).

$$\frac{Q}{\mathscr{A}} = -\int_0^\infty \rho \, dx = -\varepsilon_0 \varepsilon \int_\theta^\infty \frac{d^2\Phi}{dx^2} \, dx = -\varepsilon_0 \varepsilon \left(\frac{d\Phi}{dx}\right)_{x=0}$$

Thus, the surface charge denisty can be obtained from the initial slope of the potential function.

A serious defect of the Goüy–Chapman theory is that it treats the ions as point charges. It thus leads to absurdly high values for the charge concentration in the immediate neighborhood of the interface. In 1924, Stern* provided a suitable correction in the form of an adsorbed layer of ions with a thickness δ about equal to the ionic diameters. This layer was assumed to be held fixed at the surface. The potential function $\Phi(x)$ for the Stern modification of the Goüy–Chapman model is shown in Fig. 11.23. There is a linear fall of potential in the rigid Stern layer, followed by a Goüy–Chapman type of fall in the diffuse layer. The total potential fall in the diffuse layer is called Φ_δ.

The properties of double layers are fundamental for the theoretical analysis of colloidal behavior and the repulsion between similarly charged double layers is responsible for the stability of lyophobic sols.

20. Colloidal Sols

Preparations of lyophobic sols may be divided into (a) dispersion methods and (b) condensation methods. Dispersion methods include simple grinding in a ball mill, electrical dispersion by striking an arc through the dispersion medium from electrodes of the material to be dispersed,† passage of ultrasonic vibrations through a coarse suspension. Condensation methods are based on changes in solubility with temperature of solvent or on formation of colloidal particles by chemical precipitation reactions.

Whether or not condensation of a second phase within a dispersion medium will yield a stable colloidal suspension is a complex problem, which involves the kinetics of nucleation from a homogeneous supersaturated phase and subsequent growth of the nuclei by diffusion of reactants to their surfaces. The stability of the dispersion depends on the rate of accretion of the particles to form aggregates that precipitate from the sol. An important factor is the extent of initial supersaturation, since more nuclei are formed at higher supersaturation and generally this effect leads to a smaller final particle size. Sol formation in precipitation reactions is promoted by a low concentration of electrolytes, since ions in solution tend to neutralize the electric charge on the sol particles, thus facilitating their coagulation. For example, the reaction

$$2H_2S + SO_2 \longrightarrow 3S + 2H_2O$$

readily leads to formation of stable sulfur sols because no ionic products are formed. On the other hand,

$$AgNO_3 + KBr \longrightarrow AgBr + KNO_3$$

leads to a precipitate of AgBr, since the K^+ and NO_3^- ions decrease electrical repulsions between the AgBr particles.

Usually the particles in a sol have a fairly wide distribution of size, but methods

---

*O. Stern, *Z. Elektrochem.*, *30*, 508 (1924).
†V. M. Bredig, *Z. Elektrochem.*, *4*, 514, 547 (1898).

have been developed to produce *monodisperse sols* of uniform particle size. Such sols are convenient for theoretical studies. The basic idea in preparing a mono-disperse sol is to adjust the condition of condensation or precipitation so that self-nucleation occurs only in one burst during a short period of time, after which the embryos grow from a solution of lower supersaturation. Good examples of this procedure are found in the work of LaMer.* Electron microscopy is the most direct way to examine the form of sol particles. An example is shown in Fig. 11.24.

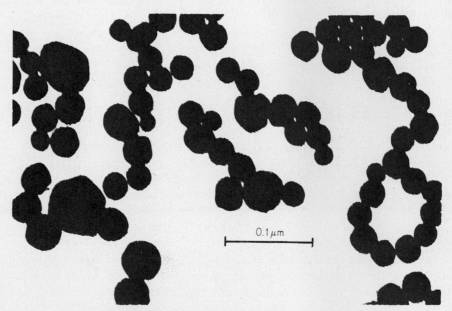

0.1 μm

FIG. 11.24 A gold sol protected by cations shown at the onset of flocculation. (Professor Heinrich Thiele, University of Kiel.)

The reason for the stability of sols is that the particles are surrounded by electrical double layers that repel each other. Thus, the typical sol is extremely sensitive to the concentrations and charges of ions in the dispersion medium. For example, if a 10^{-3} molar KI solution is titrated with 10^{-1} molar $AgNO_3$, an AgI sol is formed in which the particles have a negative diffuse double layer, due to the preferential adsorption of I^- ions. When the concentration of I^- in the solution is reduced to about 10^{-10} molar, the sol rapidly coagulates. On the other hand, if 10^{-3} molar $AgNO_3$ is titrated with 10^{-1} molar KI, a positively charged sol is obtained due to excess adsorption of Ag^+ ions. This sol coagulates when the I^- concentration reaches 10^{-6} molar.

*V. K. LaMer and R. H. Dinegan, *J. Am. Chem. Soc.*, **72**, 4847 (1950); and V. K. LaMer, *Ind. Eng. Chem.*, **44**, 1270 (1952).

21.   Electrokinetic Effects

If two ionic solutions are separated by a membrane or porous barrier across which an electric field or pressure difference is impressed, a number of interesting electrokinetic effects can be observed. A schematic experimental arrangement is shown in Fig. 11.25. Temperature and composition are uniform throughout the solutions, and the two regions differ only in hydrostatic pressure ΔP and electric potential $\Delta \Phi$. This situation calls for the methods of nonequilibrium thermodynamics.

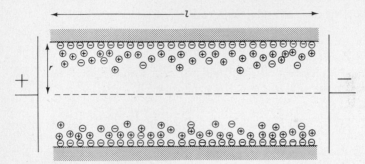

FIG. 11.25(a)   A model for electro-osmotic flow and electro-osmotic pressure across a single capillary tube. [G. Kortüm, *Lehrbuch der Elektrochemie* (Weinheim: Verlag Chemie, 1966), p. 403.]

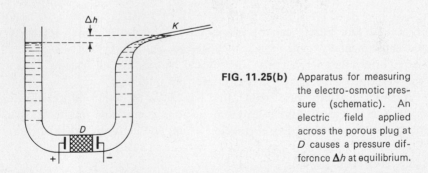

FIG. 11.25(b)   Apparatus for measuring the electro-osmotic pressure (schematic). An electric field applied across the porous plug at D causes a pressure difference Δh at equilibrium.

If the flux of charge is the electric current I and the flux of matter is the volume flow of liquid J, the phenomenological equations can be written

$$I = L_{11}\, \Delta \Phi + L_{12}\, \Delta P$$

$$J = L_{21}\, \Delta \Phi + L_{22}\, \Delta P$$

with the Onsager relation,

$$L_{12} = L_{21}$$

The electrokinetic effects are now defined as follows:

1. *Streaming potential*, the potential difference per unit pressure difference at zero electric current.

$$\left(\frac{\Delta\Phi}{\Delta P}\right)_{I=0} = -\frac{L_{12}}{L_{11}}$$

2. *Electro-osmosis*, the flow of matter per unit electric current, when the pressure is uniform.

$$\left(\frac{J}{I}\right)_{\Delta P=0} = \frac{L_{21}}{L_{11}}$$

3. *Electro-osmotic pressure*, the pressure difference per unit potential difference when the flow of matter is zero.

$$\left(\frac{\Delta P}{\Delta\Phi}\right)_{J=0} = -\frac{L_{21}}{L_{22}}$$

4. *Streaming current*, the current flow per unit matter flow at zero potential difference.

$$\left(\frac{I}{J}\right)_{\Delta\Phi=0} = \frac{L_{12}}{L_{22}}$$

As a consequence of the Onsager relation, each osmotic effect is related to a streaming effect:

$$\left(\frac{\Delta\Phi}{\Delta P}\right)_{I=0} = -\left(\frac{J}{I}\right)_{\Delta P=0} \tag{11.45}$$

$$\left(\frac{\Delta P}{\Delta\Phi}\right)_{J=0} = -\left(\frac{I}{J}\right)_{\Delta\Phi=0} \tag{11.46}$$

Since all four effects can be measured independently, these relations were tested experimentally[*] and found to hold very well. Since they are derived from nonequilibrium thermodynamics, they should be valid for all systems, irrespective of the structure of the barrier that separates the solutions.

Electrokinetic phenomena have often been used to obtain information on the properties of electrical double layers at solution–membrane interfaces. This calculation requires introduction of a model of the double layer, either of Helmholtz or Stern type, as shown in Fig. 11.23. A calculation of this kind[†] for the electro-osmotic pressure yielded

$$\left(\frac{\Delta P}{\Delta\Phi}\right)_{J=0} = \frac{8\varepsilon_0\varepsilon\zeta l}{r^2}$$

for a capillary tube of length l and radius r, where ζ is the effective potential across the double layer. The dielectric constant ε in the double layer is probably not the same as the bulk ε, and the *zeta potential* ζ is the result of a rather brutal averaging process applied to a double layer in a nonequilibrium (flowing) state. We can take ζ as approximately equal to Φ_δ in Fig. 11.23.

[*]D. G. Miller, *Chem. Rev.*, 60, 15 (1960).
[†]A. Sheludko, *Colloid Chemistry* (Amsterdam: Elsevier, 1966), p. 141.

PROBLEMS

1. The surface tension of liquid mercury is 470 dyn·cm^{-1} at 273 K. Calculate from Eq. (11.9) the capillary depression in a tube of 1 mm diameter if the contact angle is 140°.

2. Show that for a one-component system, the surface enthalpy is $H^\sigma = G^\sigma + TS^\sigma = \gamma - T(d\gamma/dT)$. The γ of liquid mercury at 288 K is 487 dyn·cm^{-1}; at 273 K it is 470 dyn·cm^{-1}. Calculate G^σ, H^σ, and S^σ for mercury at 280 K.

3. An empirical equation for the surface tensions of pure liquids is $\gamma = \gamma_0(1 - T_R)^{11/9}$, where T_R is the reduced temperature T/T_c and γ_0 is a constant characteristic of the liquid. Show that when this equation holds, $S^\sigma = \frac{11}{9}(\gamma_0/T_c)(1 - T_R)^{2/9}$. Calculate γ_0 and S^σ for several of the liquids listed in Table 11.1, using handbook data for T_c.

4. For a cubic crystal containing 10^{21} atoms, the cohesive energy is 10 eV per surface atom at 300 K. If the cohesive energy per surface atom remains constant, at what size of crystal will the cohesive energy just equal the thermal energy $3kT$ per atom in the crystal? Is it reasonable to expect the cohesive energy per surface atom to be a constant?

5. What pressure would be necessary to blow capillary water at 20°C out of the pores of a sintered glass filter with a uniform pore diameter of 0.20 μm? Give a schematic design of an apparatus to measure the distribution of pore sizes in a filter.

6. If bubbles of air 10^{-3} mm in diameter but no other nuclei are present in water just below the boiling point, how much could the water be superheated before boiling began?

7. Because of the high surface tension of water, macromolecules such as proteins in water solution may exhibit the phenomenon of *hydrophobic interaction*, whereby hydrophobic groups contact each other so as to decrease the total extent of the interface between water and the nonpolar groups. If the effective interfacial tension is 70 dyn·cm^{-1}, calculate the ΔG per mole of the hydrophobic interaction between protein molecules of $M = 60\,000$, assuming 25% of the nonpolar area of each molecule can be involved in a *hydrophobic interaction*. Can you cite any factors that you have neglected in estimating the ΔG?

8. Can surface tension alone explain the rise of sap in plants? Explain, with use of appropriate numerical estimates. If you conclude the answer is "no," what other factors are needed to explain the rise?

9. Derive an equation for the solubility of small particles from the Kelvin equation for vapor pressure of small drops. If the bulk solubility of dinitrobenzene in water is 10^{-3} mol·dm^{-3} at 25°C, estimate the solubility of crystallites of 0.01 μm diameter. Take the interfacial tension as 25.7 dyn·cm^{-1}.

10. The surface tensions of dilute solutions of phenol in water at 303 K were

wt % phenol	0.024	0.047	0.118	0.471
γ, dyn·cm^{-1}	72.6	72.2	71.3	66.5

Calculate Γ_2 from the Gibbs adsorption isotherm for a 0.100% solution.

11. The mean activity coefficient of KCl in aqueous solution at 0.002 molal is $\gamma_\pm = 0.9648$ at 25°C. What size droplets of 0.002 M KCl solution at 25°C would be in equilibrium with a plane surface of pure H_2O?

12. In 1923, D. B. Macleod* suggested an empirical equation for the surface tension

$$M\gamma^{1/4} = \mathscr{P}(\rho_l - \rho_v)$$

where ρ_l and ρ_v are the densities of liquid and vapor and the constant $\mathscr{P}$ was called the *parachor* by S. Sugden.† The parachor has been shown to be an additive function of atoms and groups in a molecule, and is approximately independent of T. Use the values of the parachor as given by O. R. Quale‡ and appropriate data on the densities to calculate γ at 20°C for ethyl acetate, ethyl bromide, and ethyl iodide; compare with experimental values.

13. An empirical equation due to Szyszkowski gives the surface tension of dilute aqueous solutions of organic compounds as

$$\frac{\gamma}{\gamma_w} = 1 - 0.411 \log_{10}\left(1 + \frac{X}{a}\right)$$

where γ_w is the surface tension of water, X is the mole fraction of solute, and a is a constant characteristic of the organic compound. For *n*-valeric acid, $a = 1.7 \times 10^{-4}$. Calculate the average area occupied by a molecule of *n*-valeric acid adsorbed from aqueous solution at 25°C onto the surface when $X = 0.01$.

14. Show how the Gibbs adsorption isotherm can be used in the thermodynamic treatment of adsorption of gases on solid surfaces to obtain the equation

$$df = \frac{RTv}{sV_m} d\ln P$$

where f is the surface pressure $(-\Delta\gamma)$, v is the volume of gas adsorbed per unit mass of solid, s is the area of the solid per unit mass, and V_m is the molar volume of the gas. Suppose that the adsorbed gas follows an equation of state $f = b - a\mathscr{A}$, where a and b are constants and $\mathscr{A}$ is the area per molecule. Calculate the resulting P–v adsorption isotherm.§

15. Explain the following observations from the standpoint of the adsorption properties of reactant and product molecules, as based on Langmuir isotherms. (a) The decomposition of NH_3 on W is zero order. (b) The decomposition of N_2O on Au is first order. (c) The recombination of H atoms on Au is second order. (d) The decomposition rate of NH_3 on Pt depends on $P(NH_3)/P(H_2)$. (e) The decomposition rate of NH_3 on Mo is strongly retarded by N_2 but does not approach zero as the surface becomes saturated with nitrogen.

16. Let us consider the statistical mechanical expression (11.31) for the adsorption coefficient in a Langmuir adsorption isotherm. Consider the adsorption of argon from the gas phase onto a uniform surface of a crystal on which the adsorbed atoms are freely mobile. Calculate $b(T)$ (treating the adsorbed phase as a two-dimensional gas) and calculate $\theta(P, T)$ for $P = 1$ atm, $\theta = 200$ K, and $\Delta\varepsilon_0$ (adsorption) $= 0.35$ eV.

*Trans. Faraday Soc., *19*, 38 (1923).
†*J. Chem. Soc.*, 22 (1924).
‡*Chem. Rev.*, *53*, 439 (1953).
§W. D. Harkins and G. Jura, *J. Am. Chem. Soc.*, **68**, 1941 (1946).

17. With reference to Fig. 11.19, calculate the interaction of a molecule in the gas phase with an extended solid surface, under the assumption that the attractive potential energy between the gas molecule and each atom of the surface has the form $U = -Ar^{-6}$. Calculate the interaction between krypton gas and krypton solid assuming that A is taken from the Lennard–Jones expression in (4.15) with $\varepsilon/k = 171$ K and $\sigma = 0.36$ nm. At what temperature would the calculated attractive potential energy equal $\frac{3}{2}kT$?

18. The ΔH of adsorption when the amount of substance adsorbed is held constant is called the *isosteric heat of adsorption* ΔH_V. The adsorption of N_2 on charcoal amounted to 0.894 cm³ (STP)·g⁻¹ at $P = 4.60$ atm and 194 K and at $P = 35.4$ atm and 273 K. Calculate ΔH_V, and hence $\Delta S_V^{\ominus}$ and $\Delta G_V^{\ominus}$, for the adsorption at 273 K. Discuss briefly the factors that determine $\Delta S_V^{\ominus}$. Would $\Delta S_V^{\ominus}$ be greater if the specific surface area of the charcoal were greater?

19. From the data in Fig. 11.21, calculate the surface density of charge when $\Phi - \Phi_z = -1.0$ V. Express the charge density in coulomb per square metre and also in elementary charges per square centimetre.

20. Using the treatment of viscosity in Section 4.24 as a model and the simple Hemholtz expression on p. 510 for the potential difference across the double layer, show that the electro-osmotic flow through a capillary tube is

$$J = \frac{AE\varepsilon\varepsilon_0\,\Delta\Phi}{\eta}$$

where η is the viscosity and A the cross-sectional area of the tube. If $\Delta\Phi$ for a water vs. glass interface is -0.050 V, calculate the electro-osmotic flow of water through a glass capillary 1 cm long and 1 mm diameter when $E = 100$ V·cm⁻¹.

21. Calculate the average molecular weight and the molecular area of egg albumin from the following data [H. Bull, *J. Amer. Chem. Soc.*, 67, 4 (1945)]. The data refer to a monolayer of egg albumin spread on a film balance at 25°C.

Π (mN·m⁻¹)	0.07	0.11	0.18	0.20	0.26	0.33	0.38
$\mathscr{A}$ (m²·mg⁻¹)	2.00	1.64	1.50	1.45	1.38	1.36	1.32

22. The following data were obtained by McBain and Britton (1930) for the adsorption of ethylene on activated charcoal at 273 K. Show by means of a suitable graph that these results fit the Langmuir isotherm. From the graph determine the values of the two constants in the Langmuir isotherm and calculate the surface area of one gram of charcoal. (One molecule of ethylene can be taken to occupy 0.21 nm² of surface.)

Pressure C_2H_4 (atm)	4.0	9.7	13.4		19.0	27.1
C_2H_4 ads. g/g charcoal	0.163	0.189	0.198		0.206	0.206

12

Electrochemistry— Electrodics

If a piece of zinc and a piece of copper be brought in contact with each other, they will form a weak electrical combination, of which the zinc will be positive, the copper negatve; this may be learnt by the use of a delicate condensing electrometer; or by pouring zinc filings through holes, in a plate of copper, upon a common electrometer; but the power of the combination may be most distinctly exhibited in the experiments, called Galvanic experiments, *by connecting the two metals, which must be in contact with each other, with a nerve and muscle in the limb of an animal recently deprived of life, a frog for instance; at the moment the contact is completed, or the circuit made, one metal touching the muscle, the other the nerve, violent contractions of the limb will be occasioned.*

Humphry Davy*

In Chapter 10 we considered the physical chemistry of solutions of electrolytes— the subject called *ionics*. We shall now consider the thermodynamics and kinetics of the processes that occur at electrodes immersed in these solutions and connected via an external conductor. The reaction at the surface of an electrode is a transfer of charge, usually in the form of electrons to or from neutral molecules or ions. An electrode acting as a source of electrons is a *cathode;* an electrode acting as a sink for electrons is an *anode*.

A pair of electrodes dipping into an ionic solution and connected by an external metallic conductor constitutes a typical *electrochemical cell*. If the cell is used to supply electrical energy—that is, if it converts the free energy of a physical or chemical change into electrical free energy—it is called a *galvanic cell*. A cell in which an external supply of electrical energy is used to bring about a physical or chemical change is called an *electrolytic cell*.

1. Definitions of Potentials

To discuss electrode phenomena, it is necessary to start with precise definitions for the various potential differences that can occur in the rather complex systems con-

*Elements of Chemical Philosophy (London: J. Johnson & Co., 1812). This interesting work has been called the "first textbook of physical chemistry."

sisting of several phases and interfaces. We shall use a set of definitions and symbols due principally to Erich Lange, which have been widely adopted by writers in this field.

Let us first consider purely electrostatic effects. We must emphasize that it is always possible to define and measure a difference in electrostatic potential between two points in the same phase, or between two pieces of the same chemical substance. We cannot, however, measure a difference in electrostatic potential between two points in different phases, or between two pieces of different chemical substances. A difference in potential is measured by the work required to move a test charge from one point to the other. In electrostatic theory, the work is determined by the distribution of electric point charges in the medium through which the test charge is moved. If, however, the test charge moves across an interface between two different phases, there will be contributions to the work due to the difference in chemical potential caused by the local interaction of the test charge with the chemically different surrounding media. There is no way to separate this "chemical work term" from the "electrostatic work term." Hence arises the impossibility of measuring a purely electrostatic potential difference between different phases, a restriction first pointed out by Gibbs* and reemphasized by Guggenheim.†

With this restriction in mind, let us consider in Fig. 12.1(a), a spherical mass of a homogeneous material substance (phase α) situated in a vacuum. The sphere carries a net charge Q, so that the electrostatic potential $\mathcal{V}$ at any point outside the sphere at a distance R from its center is $Q/4\pi\varepsilon_0 R$. (The potential at $R = \infty$ is taken to be $\mathcal{V} = 0$.) Let us bring a unit test charge from infinity ($\mathcal{V} = 0$) toward the surface of phase α, stopping at a point about 10^{-6} cm away, so that image forces

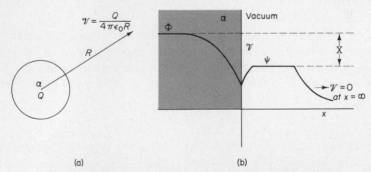

(a)   (b)

FIG. 12.1   (a) Electrostatic potential in vacuum at a distance R from the center of a uniformly charged sphere of phase α. (b) Electrostatic potentials measured by work done on positive test charge brought from $x = \infty$ in vacuum into interior of phase α.

Collected Works of J. Willard Gibbs, Vol. 1. Thermodynamics (New Haven: Yale University Press, 1906), p. 429.

†*J. Phys. Chem.*, **33**, 842 (1929). See also *A Commentary on the Scientific Writings of J. Willard Gibbs, Vol. 1. Thermodynamics*, ed. by F. G. Donnan and A. Haas (New Haven: Yale University Press, 1936), pp. 181–211, 331–349.

and any chemical changes in α due to the approaching charge are negligible. The variation of the potential $\mathscr{V}$ with distance is usually as shown in Fig. 12.1(b), $\mathscr{V} \propto 1/R$ to within 10^{-5} to 10^{-6} cm from the surface, and then $\mathscr{V}$ approximately constant owing to short-range interactions. The *outer potential* or *Volta potential* ψ is defined as the work necessary to bring the test charge from ∞ to, say, 10^{-6} cm from surface. The ψ is a measurable quantity because it is an electric potential difference between two points in the same medium (in this case, vacuum).

The surface of phase α will usually be the locus of a double layer of charge which may arise from various causes. In the case of water, for instance, it has been shown that the molecular dipoles are oriented with their positive ends out of the surface and negative ends inward. In metals, energetic electrons can partially escape to give a negative electron film at the surface with compensating positive metal ions just below, thus forming a dipolar layer with its negative side outward. These dipolar layers contribute no net charge, but, nevertheless, work must be done to carry a test charge through such a layer. Thus, to take the test charge from just outside the surface into the medium α will require further work. The electric potential change resulting from the surface dipole layer is called the *surface potential* χ.

The *inner potential* is then defined as

$$\Phi = \psi + \chi$$

where ψ is due primarily to the charge Q, and χ to the surface dipoles. The *inner potential* Φ is also called the *Galvani potential*. Neither Φ nor χ is experimentally measurable, since the test charge in the medium α will cause rearrangements of electronic structures in the real substance α of the type we call *chemical*, and the chemical work terms cannot be distinguished from those due to the purely electrostatic potentials χ and Φ. We can express the results of the chemical interactions by the chemical potential μ.

The work done in taking the test charge from $R = \infty$ into the interior of phase α is, of course, a measurable quantity, even though we cannot measure separately its electrostatic and its chemical parts. This measurable work allows us to define the *electrochemical potential* $\bar{\mu}_i$. If we compute $\bar{\mu}_i$ for 1 mol of a component i in the phase α, we can write formally

$$\bar{\mu}_i = \mu_i + z_i F \Phi \tag{12.1}$$

where z_i is the charge number (including sign) on the ions of component i, and F is the Faraday constant. We should note that the subscript "i" assigns a chemical identity to our previously abstract electrostatic test charge. For example, it might be a Na^+ ion and phase α might be NaCl solution, or it might be Cu^+ ion with α a piece of copper. Frequently, we shall take component "i" to be an electron. In solid state physics, the electrochemical potential of an electron is called the *Fermi energy*.

The equilibrium condition of Section 6.4 for uncharged components can now be generalized for charged components. At equilibrium for component i between phases α and β,

$$\bar{\mu}_i^\alpha = \bar{\mu}_i^\beta \tag{12.2}$$

Since $\bar{\mu}_i$ and ψ are measurable quantities, there is another measurable potential of some importance,

$$\Upsilon_i = \bar{\mu}_i - z_i F \psi \tag{12.3}$$

This Υ_i is called the *real potential* or the *work function*. It is measured by the work required to move a charged component "i" from the interior of a phase to a point just outside the range of any surface effects, i.e., about 10^{-6} cm away. The work function can be determined by thermionic or photoelectric methods.

If we consider the phase α in contact with another phase β (instead of a vacuum), we can use the preceding potentials to define various potential differences between the phases. If α has a certain surface concentration σ of fixed charges, we may expect in general to find that opposite charges in β will be distributed close to the interface so as to achieve local electric neutrality as nearly as possible. Thus, an electric *double layer* will be produced, the properties of which are of major importance in many surface and colloidal phenomena. The potential across the double layer will be $\Delta\Phi$, the Galvani potential difference, which cannot be directly measured.

The conceptual difficulty in this subject arises from the application of differential equations derived from an electrostatic theory based on point charges in homogeneous structureless fluid media (characterized by dielectric constants ε) to actual chemical systems composed of molecules and ions which themselves possess electric "microstructures" that are not inert with respect to the test charges. If one becomes familiar with the definitions and nomenclature that have been outlined, many traditional difficulties will become less formidable. In particular, one will find a resolution of the historic controversy as to whether the Galvani or the Volta potential determines the emf of an electrochemical cell, and it should be possible to interpret the paradoxical epigraph that H. Davy contributed to this chapter.

2. Electric Potential Difference for a Galvanic Cell

Figure 12.2 shows a typical cell. A zinc electrode dips into a solution of $1.0\ m$ $ZnSO_4$ and a copper electrode dips into a solution of $1.0\ m$ $CuSO_4$. The two solutions are separated by a porous barrier, which allows electrical contact but prevents excessive mixing of the solutions by interdiffusion. This cell is represented by the diagram,

$$Zn\,|\,Zn^{2+}\,(1.0\ m)\,|\,Cu^{2+}\,(1.0\ m)\,|\,Cu \tag{A}$$

The vertical lines denote phase boundaries.

It is always possible to measure the difference in electric potential between two pieces of the same kind of metal. We therefore attach to each electrode a length of copper wire and connect these copper leads to a voltmeter or some other device for measuring this difference in potential. In the cell in Fig. 12.2, the copper electrode is the observed $+$ terminal.

The Galvani potential difference $\Delta\Phi$ of the cell is equal in sign and magnitude

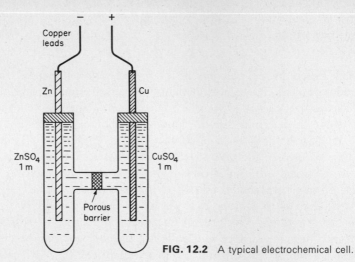

FIG. 12.2 A typical electrochemical cell.

to the electric potential of a metallic conducting lead on the right minus that of an identical lead on the left.

$$\Delta\Phi = \Phi_R - \Phi_L \tag{12.4}$$

Note that we cannot measure Φ_R or Φ_L separately, but their difference is a directly measurable quantity. The meaning of "left" and "right" refers to the cell *as written* in (A); clearly it has nothing to do with how the actual cell is arranged on the laboratory bench.

We need next to establish a convention that relates the written form of the cell to the chemical reaction occurring in the cell. If the equation for a typical chemical reaction is written as

$$\tfrac{1}{2}Zn + \tfrac{1}{2}Cu^{2+} \longrightarrow \tfrac{1}{2}Zn^{2+} + \tfrac{1}{2}Cu \tag{B}$$

the cell diagram (A) signifies that this reaction (B) takes place when positive electricity flows through the cell from left to right. Through the external circuit, the flow of electrons is from left to right.* If the electrodes of the cell written in (A) are connected through a resistance, and positive electricity does flow through the electrolyte of the cell from left to right, the electric potential difference $\Delta\Phi$ will be positive.

On the other hand, if the cell reaction is written

$$\tfrac{1}{2}Cu + \tfrac{1}{2}Zn^{2+} \longrightarrow \tfrac{1}{2}Cu^{2+} + \tfrac{1}{2}Zn \tag{C}$$

the corresponding cell diagram would be

$$Cu \,|\, Cu^{2+} \,|\, Zn^{2+} \,|\, Zn \tag{D}$$

The process occurring in the experimental cell obviously must be the same regard-

---

*We can also express the rule as follows: The reaction taking place at the left-hand electrode of the cell is written to indicate that electrons are supplied to the external circuit (i.e., an oxidation process), and the reaction occurring at the right-hand electrode is written to indicate an acceptance of electrons from the external circuit (i.e., a reduction process).

less of how the cell is specified on paper. In the cell shown in Fig. 12.2, unless the ratio $[Zn^{2+}]/[Cu^{2+}]$ becomes extremely small, oxidation occurs at the zinc electrode.

3. Electromotive Force (emf) of a Cell

The electromotive force (emf) E of the cell is defined as the limiting value of the electric potential difference $\Delta\Phi$ as the current through the cell goes to zero.

$$E = \Delta\Phi_{(I\to 0)} \tag{12.5}$$

The definition of emf states that the potential difference is measured while no current is being drawn from the external leads. In practice, it is possible to measure E under conditions in which the current drawn from the cell is so small as to be negligible. The method, devised by Poggendorf, uses a circuit known as the *potentiometer*.

A basic potentiometer circuit is shown in Fig. 12.3. The slide wire is calibrated

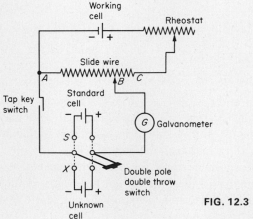

FIG. 12.3   Basic circuit of direct reading potentiometer.

with a scale so that any setting of the contact corresponds to a certain voltage. With the double-pole, double-throw switch in the standard cell position S, we set the slide wire to the voltage reading of the standard cell, and adjust the rheostat until no current flows through the galvanometer G. At this point, the potential difference between A and B, the IR along the section AB of the slide wire, just balances the emf of the standard cell. We then set the switch in the unknown cell position X, and readjust the slide wire contact until no current flows through G. From the new setting, we can read directly from the scale of the slide wire the emf of the unknown cell.

The most widely used standard is the Weston cell, shown in Fig. 12.4, which is written

$$Cd(Hg) \mid CdSO_4 \cdot \tfrac{8}{3}H_2O \text{ (s)}, Hg_2SO_4 \text{ (s)} \mid Hg$$

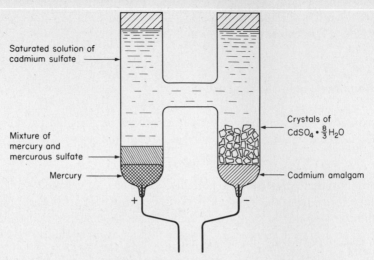

Saturated solution of cadmium sulfate

Crystals of CdSO$_4 \cdot \frac{8}{3}$H$_2$O

Mixture of mercury and mercurous sulfate

Mercury

Cadmium amalgam

FIG. 12.4 Weston standard cell.

The cell reaction is

$$\text{Cd(Hg)} + \text{Hg}_2\text{SO}_4 \text{ (s)} + \tfrac{8}{3}\text{H}_2\text{O (l)} \longrightarrow \text{CdSO}_4 \cdot \tfrac{8}{3}\text{H}_2\text{O (s)} + 2\text{Hg (l)}$$

Its emf in volts at $\theta°$C is given by

$$E = 1.01845 - 4.05 \times 10^{-5}(\theta - 20) - 9.5 \times 10^{-7}(\theta - 20)^2$$

Thus, at 20°C, $E = 1.01845$ V; at 25°C, $E = 1.01832$ V. The small temperature coefficient of E is one advantage of this cell. Since individual cells may differ slightly in emf, those used in the laboratory should be calibrated by the National Bureau of Standards, or checked against such calibrated cells.

The accuracy of the compensation method for measuring an emf is limited only by the accuracy of the standard E and of the various resistances in the circuit. The precision of the method will be determined mainly by the sensitivity of the galvanometer used to detect the balance between unknown and standard (working) emf and by the closeness of the control of temperature. Since it is not difficult to balance such a circuit so that less than 10^{-12} A is drawn from the cell, we can satisfy for all practical purposes the condition specified by the definition of emf, i.e., measurement of $\Delta\Phi$ as $I \to 0$.

We can formally express the emf in terms of the Galvani potential differences between the different phases in the cell. For the example of the cell shown in Fig. 12.2, as $I \to 0$,

$$E = (\Phi_{\text{Cu}} - \Phi_{\text{Cu}^{2+}}) + (\Phi_{\text{Cu}^{2+}} - \Phi_{\text{Zn}^{2+}}) + (\Phi_{\text{Zn}^{2+}} - \Phi_{\text{Zn}}) + (\Phi_{\text{Zn}} - \Phi'_{\text{Cu}})$$

The Φ'_{Cu} refers to the copper lead from the zinc electrode. Therefore,

$$E = \Phi_{\text{Cu}} - \Phi'_{\text{Cu}} = \Delta\Phi_1 + \Delta\Phi_2 + \Delta\Phi_3 + \Delta\Phi_4$$

Thus, each $\Delta\Phi$ contributes to the emf, and the old controversy as to the origin of the emf reduces to a question of relative magnitudes of $\Delta\Phi$ terms. Since they cannot be measured, the only way to resolve the question would be to calculate the $\Delta\Phi$ from some kind of model.

4. The Polarity of an Electrode

When you connect a cell to a potentiometer and balance the potentials, you must, of course, connect the $+$ and $-$ terminals of the cell to the $+$ and $-$ terminals of the potentiometer, respectively. The polarity of the potentiometer terminals is determined by how the $+$ and $-$ terminals of the standard cell are connected. To achieve a balance point, the $+$ electrode of the standard cell must be connected through the external circuit to the $+$ electrode of the unknown.

You need not worry how to decide which electrode of a cell is $+$ and which is $-$. The $+$ electrode is the one connected to the $+$ terminal of the potentiometer when balance is achieved. If you have connected the $-$ of cell to the $+$ of potentiometer, you will not be able to obtain a balance, and you will therefore not have any measurement to worry about. Thus, a few minutes of experimentation tells which electrode is $+$ and which is $-$.

You might ask, however, how Edward Weston knew that the cadmium electrode was negative when he made his first standard cell in 1892. The answer goes back to the choice between the vitreous and the resinous electric fluids. When Franklin proposed the one-fluid theory, he picked the vitreous fluid as *the* electrical fluid, and this kind of electricity became *positive*. The resinous kind of electricity was then considered to result from a deficiency of electrical fluid, and was called, therefore, *negative electricity*. Franklin might have made the opposite choice, since the assignment of names was quite arbitrary.

As it turned out, we live in a world which is not symmetrical where $+$ and $-$ electricity is concerned. The ordinary carriers of positive electricity are the massive positive ions, but the ordinary carriers of negative electricity are the much lighter electrons. A positive ion is an atom that has lost one or more electrons. If we had Franklin's choice to make over again today, we would probably reverse his signs to avoid the semantic difficulty of having negative carriers of a positive fluid.

In any case, the choice of $+$ and $-$ is consistent throughout the science of electricity. When we say that the copper wire attached to the cadmium electrode in the Weston cell is more negative than the copper wire attached to the mercury electrode, we mean that it contains an excess of the same kind of electricity found in amber rubbed with cat's fur.

5. Reversible Cells

An electrode dipping into a solution is said to constitute a *half-cell*. Thus $Zn\,|\,Zn^{2+}$ $(0.1\ m)$ is a half-cell. The typical cell is the combination of two half-cells.

We shall be interested primarily in the class of cells called *reversible cells*. These may be recognized by the following criterion: The cell is connected with a potentiometer arrangement for emf measurement by the compensation method. The emf of the cell is measured (a) with a small current flowing through the cell in one direction, (b) then with an imperceptible flow of current, and (c) finally with a small flow in the opposite direction. If a cell is reversible, its emf changes only slightly during this sequence, and there is no discontinuity in the value of the emf at the point of balance (b). Reversibility implies that any chemical reaction occurring in the cell can proceed in either direction, depending on the flow of current, and at the null point the driving force of the reaction is just balanced by the compensating emf of the potentiometer. If a cell is reversible, it follows that the half-cells comprising it are both reversible.

One source of irreversibility in cells is the *liquid junction*, like that in the Daniell cell of Fig. 12.2. At the junction between $ZnSO_4$ and $CuSO_4$, we have the following situation:

$$\begin{array}{c|c}
Zn^{2+} & Cu^{2+} \\
(1.0\ m) & (1.0\ m) \\
SO_4^{2-} & SO_4^{2-}
\end{array}$$

If we pass a small current through the cell from left to right, it is carried across the junction by Zn^{2+} ions and by SO_4^{2-} ions. But, if we go through the balance point, and pass a small current in the opposite direction, it is carried across the junction from right to left by Cu^{2+} and SO_4^{2-} ions. Thus, the cell with such a liquid junction is inherently not reversible.

Before we can apply reversible thermodynamics to such cells, we must eliminate the liquid junction. We can do this with considerable success by means of a *salt bridge*. This device consists of a connecting tube filled with a concentrated solution of a salt, usually KCl. The solution can be made up in a gel (for instance, agar) to decrease mixing with the solutions in the two half-cells. Now most of the current will be carried across the junction by K^+ and Cl^- ions. There will still be some irreversible effects where the bridge enters the two solutions, but these are regarded as minimal. When the liquid junction has been "eliminated" by such a device, the cell is written with a double bar in the middle:

$$Zn\,|\,Zn^{2+}\,||\,Cu^{2+}\,|\,Cu$$

A better way to avoid irreversible effects is to avoid liquid junctions, by using a single electrolyte. The Weston cell does this with a solution of $CdSO_4$ which is also saturated with the sparingly soluble Hg_2SO_4. We shall discuss later other examples of cells without liquid junctions. Even in such cells, however, changes in electrolyte concentration around the electrodes, as a consequence of the cell reaction, may introduce small irreversible effects.

6. Free Energy and Reversible emf

The electric work done on a charge Q when its potential is changed by an amount $\Delta\Phi$ is $Q\,\Delta\Phi$. Consider a cell in which $|z|$ equivalents of reactant are converted to products. Then $|z|\,F$ coulombs of electric charge pass through the cell. If $E > 0$, from (12.4) and (12.5), $\Phi_R > \Phi_L$. Thus, $|z|\,F$ coulombs of charge are transferred through the cell from Φ_L to Φ_R. Therefore, the electrical work done on the cell is $-|z|\,FE$.

Under reversible conditions, the work done on the cell is $w = \Delta A$ where A is the Helmholtz free energy (Section 3.13). Considering the mechanical (PV) work done at constant pressure on all the different phases in the cell, we can write

$$\Delta A = -\sum_\alpha P^\alpha\,\Delta V^\alpha - |z|\,FE$$

But, from (3.35),

$$\Delta G = \Delta A + \sum_\alpha P^\alpha\,\Delta V^\alpha$$

Therefore,

$$\Delta G = -|z|\,FE \tag{12.6}$$

Thus, *the reversible emf is equal to the decrease in free enthalpy for the cell reaction* per unit charge (e.g., per coulomb, if the free enthalpy is in joules.) In the case of the Daniell cell, $E = 1.100$ V at 25°C. Hence, for the reaction

$$Zn + CuSO_4\ (1\ m) \longrightarrow Cu + ZnSO_4\ (1\ m)$$
$$\Delta G = -2(96\,487)(1.100) = -212\,300\ \text{V·C (J)*}$$

A reaction can proceed spontaneously only if $\Delta G < 0$. Thus, only if $E > 0$ can a cell reaction proceed spontaneously, so that the cell can serve as a source of electrical energy. When $E > 0$, the cell reaction therefore can proceed as written. In this case, oxidation occurs at the left-hand electrode, and the resulting positive ions migrate through the cell from left to right. The electrons flow through the external circuit from left to right. For the Daniell cell, therefore, when it acts as a galvanic cell,

$$-\ \underline{\quad\quad\quad\quad \overset{-e\ \longrightarrow}{\quad\quad\quad\quad}\quad\quad\quad\quad}\ +$$

$$Zn\left|Zn^{2+}\ \overset{\longleftarrow\ SO_4^{2-}}{\underset{\longrightarrow}{}}\right|\ \overset{\longleftarrow\ SO_4^{2-}}{Cu^{2+}\ \longrightarrow}\ \bigg|Cu$$

7. Entropy and Enthalpy of Cell Reactions

The application of the Gibbs–Helmholtz equation (3.52) to the relation $\Delta G = -|z|\,FE$ allows us to calculate the ΔH and ΔS of a cell reaction from the temperature coefficient of the reversible emf. (Note that the heat absorbed in the reaction in the *reversible cell* is $T\,\Delta S$, and not ΔH.)

$$\Delta S = -\left(\frac{\partial\,\Delta G}{\partial T}\right)_P = zF\left(\frac{\partial E}{\partial T}\right)_P \tag{12.7}$$

---

*Another useful unit is the *volt Faraday* (V·F); 1 V·F = 96 487 J. Note especially that 1 V·F·mol^{-1} = 1 eV per molecule. For the Daniell cell, $\Delta G = -2(1.100) = -2.200$ V·F at 25°C.

Since, at constant T,

$$\Delta H = \Delta G + T \Delta S$$

$$\Delta H = -|z|\, FE + |z|\, FT \left(\frac{\partial E}{\partial T} \right)_P \tag{12.8}$$

Let us apply these relations to the Weston standard cell. At 25°C,

$$E = 1.01832 \text{ V}$$

and

$$\frac{dE}{dT} = -5.00 \times 10^{-5} \text{ V} \cdot \text{K}^{-1}$$

Hence,

$$\Delta G = -2(96\,487)(1.01832) = -196\,509 \text{ J}$$
$$\Delta S = 2(96\,487)(-5.00 \times 10^{-5}) = -9.65 \text{ J} \cdot \text{K}^{-1}$$

and,

$$\Delta H = -196\,509 - 2876 = -199\,385 \text{ J}$$

In this example, $T \Delta S = (298.5 \times -9.649) \text{ J} = 2876 \text{ J}$.

The study of the temperature dependence of the emf of cells led Nernst to his discovery of The Third Law of Thermodynamics.*

8.   Types of Half-Cells (Electrodes)

One of the simplest half-cells consists of a *metal electrode* in contact with a solution containing ions of the metal, e.g., silver in silver nitrate solution. Such a half-cell is represented as $\text{Ag}\,|\,\text{Ag}^+(c)$, where c is the silver ion concentration, and the vertical bar denotes the phase boundary. The reaction occurring at this electrode is the solution or deposition of the metal, according to $\text{Ag} \rightleftharpoons \text{Ag}^+ + e$.

It is sometimes convenient to form a metal electrode by using an amalgam instead of a pure metal. A liquid amalgam has the advantage of eliminating nonreproducible effects due to strains in the solid metals or polarization at the electrode surface.† In some instances, a dilute amalgam electrode can be successfully employed whereas the pure metal would react violently with the solution, for example, in the sodium amalgam half-cell, $\text{NaHg}(c_1)\,|\,\text{Na}^+(c_2)$. If the amalgam is saturated with the solute metal, the electrode is equivalent to a solid metal electrode, since the chemical potential of a component in its saturated solution equals the chemical potential of pure solute.‡ If the amalgam is not saturated, methods are available for calculating the emf of a pure metal electrode from a series of measurements at different amalgam concentrations.

---

*See A. Sommerfeld, *Thermodynamics and Statistical Mechanics*, Lectures on Theoretical Physics, Vol. 5 (New York: Academic Press, Inc., 1956), pp. 71-75, 113-121.

†The high hydrogen overpotential on mercury (Section 12.27) helps to eliminate polarization.

‡The solid phase may itself be an alloy of mercury with the metal, in which case the activity of metal in the liquid amalgam is definitely below that of the pure metal. The cadmium—mercury system is of this kind.

Gas electrodes can be constructed if one places a strip of nonreactive metal, usually platinum or gold, in contact with both the solution and a gas stream. The hydrogen electrode consists of a platinum strip exposed to a current of hydrogen, and partly immersed in an acid solution. The hydrogen is probably dissociated into atoms at the catalytic surface of the platinum, the electrode reactions being

$$\tfrac{1}{2}H_2 \longrightarrow H$$

$$H \longrightarrow H^+ + e$$

$$\text{overall:} \quad \tfrac{1}{2}H_2 \longrightarrow H^+ + e$$

The chlorine electrode operates similarly, negative chloride ions being formed in the solution: $e + \tfrac{1}{2}Cl_2 \longrightarrow Cl^-$. The chlorine passes into the solution over an inert metal electrode.

In *nonmetal–nongas electrodes*, the inert metal is in contact with a liquid or solid phase. An example is in the bromine–bromide half-cell: $Pt\,|\,Br_2\,|\,Br^-$.

In an *oxidation–reduction electrode*, an inert metal dips into a solution containing ions in two different oxidation states, e.g., ferric and ferrous ions in the half-cell $Pt\,|\,Fe^{2+}, Fe^{3+}$. When electrons are supplied to the electrode, the reaction is $Fe^{3+} + e \rightarrow Fe^{2+}$. Since it is the function of electrodes either to accept electrons from, or to donate electrons to, ions in the solution, they are all in a sense oxidation–reduction electrodes. The difference between the silver electrode and the ferric–ferrous electrode is that in the former the concentration of the lower oxidation state, metallic silver, cannot be varied.

Metal–insoluble-salt electrodes consist of a metal in contact with one of its slightly soluble salts; in the half-cell, this salt is in turn in contact with a solution containing a common anion. An example is the silver–silver-chloride half-cell: $Ag\,|\,AgCl\,|\,Cl^-(c_1)$. The electrode reaction can be considered in two steps:

$$AgCl(s) \rightleftharpoons Ag^+ + Cl^-$$

$$Ag^+ + e \rightleftharpoons Ag(s)$$

Or, overall,

$$AgCl(s) + e \rightleftharpoons Ag(s) + Cl^-$$

Such an electrode is thermodynamically equivalent to a chlorine electrode ($Cl_2\,|\,Cl^-$) in which the gas is at a pressure equal to the dissociation pressure of AgCl according to $AgCl \rightleftharpoons Ag + \tfrac{1}{2}Cl_2$. This is a useful fact in view of the experimental difficulties involved in the use of reactive-gas electrodes. The metal–insoluble-salt electrode is reversible with respect to its common anion.

Metal–insoluble-oxide electrodes are similar to the metal–insoluble-salt type. An example is the antimony–antimony-trioxide electrode, $Sb\,|\,Sb_2O_3\,|\,OH^-$. An antimony rod is covered with a thin layer of oxide and dipped into a solution containing OH^- ions. The electrode reaction is

$$Sb(s) + 3OH^- \rightleftharpoons \tfrac{1}{2}Sb_2O_3 + \tfrac{3}{2}H_2O(l) + 3e$$

The electrode is reversible with respect to OH^- ions. Since OH^- and H^+ ions can establish rapid equilibrium, the electrode is also reversible with respect to H^+ ions.

9. Classification of Cells

When two suitable half-cells are connected, we have an *electrochemical cell*. The connection is made by bringing the solutions in the half-cells into contact, so that ions can pass between them. If these two solutions are the same, there is no liquid junction, and we have a *cell without transference*. If the solutions are different, the transport of ions across the junction will cause irreversible changes in the two electrolytes, and we have a *cell with transference*.

The decrease in free enthalpy $-\Delta G$ that provides the driving force in a cell may come from a chemical reaction or from a physical change. Cells in which the driving force is a change in concentration (almost always a dilution process) are called *concentration cells*. The change in concentration can occur either in the electrolyte or in the electrodes. Examples of changes in concentration in electrodes are found in amalgams or alloy electrodes with different concentrations of the solute metal and in gas electrodes with different pressures of the gas.

The varieties of electrochemical cells can be classified as follows:

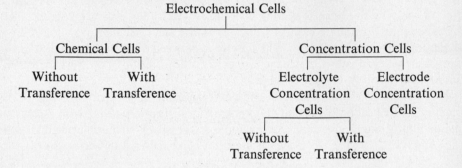

10. The Standard emf of Cells

Let us consider the generalized cell reaction: $aA + bB \rightleftharpoons cC + dD$. By comparison with (8.14), the free enthalpy change in terms of activities of the reactants is

$$\Delta G = \Delta G^\ominus + RT \ln \frac{a_C^c\, a_D^d}{a_A^a\, a_B^b}$$

Since $\Delta G = -|z|\,FE$,

$$E = E^\ominus - \frac{RT}{|z|F} \ln \frac{a_C^c\, a_D^d}{a_B^b\, a_A^a} \tag{12.9}$$

We shall call (12.9) the *Nernst equation.* *

When the activities of all the products and reactants are unity, the value of the emf is $E^\ominus = -\Delta G^\ominus/|z|\,F$. This $E^\ominus$ is called the *standard emf* of the cell. It is related to the equilibrium constant of the cell reaction, since

---

*A similar expression was given by Nernst in terms of concentrations instead of activities.

$$E^{\ominus} = -\frac{\Delta G^{\ominus}}{|z|F} = \frac{RT}{|z|F}\ln K_a \tag{12.10}$$

The determination of the standard emf of a cell is therefore one of the most important procedures in electrochemistry. We shall illustrate a useful method by means of a typical example.

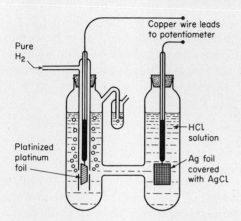

FIG. 12.5 Hydrogen electrode and silver–silver-chloride electrode in arrangement for determination of standard emf.

Consider the cell shown in Fig. 12.5 consisting of a hydrogen electrode and a silver–silver-chloride electrode immersed in a solution of hydrochloric acid:

$$Pt(H_2)\,|\,HCl(m)\,|\,AgCl\,|\,Ag$$

This is a chemical cell without a liquid junction. The electrode reactions are

$$\tfrac{1}{2}H_2 \rightleftarrows H^+ + e$$

$$AgCl + e \rightleftarrows Ag + Cl^-$$

The overall reaction is, accordingly,

$$AgCl + \tfrac{1}{2}H_2 \rightleftarrows H^+ + Cl^- + Ag$$

From (12.9), the emf of the cell is

$$E = E^{\ominus} - \frac{RT}{F}\ln\frac{a_{Ag}\,a_{H^+}\,a_{Cl^-}}{a_{AgCl}\,a_{H_2}^{1/2}}$$

Setting the activities of the solid phases equal to unity, and choosing the hydrogen pressure so that $a_{H_2} = 1$ (for ideal gas, $P = 1$ atm), we obtain the equation,

$$E = E^{\ominus} - \frac{RT}{F}\ln a_{H^+}\,a_{Cl^-}$$

Introducing the mean activity of the ions defined by (10.21), we have

$$E = E^{\ominus} - \frac{2RT}{F}\ln a_{\pm} = E^{\ominus} - \frac{2RT}{F}\ln \gamma_{\pm}m \tag{12.11}$$

On rearrangement,

$$E + \frac{2RT}{F}\ln m = E^{\ominus} - \frac{2RT}{F}\ln \gamma_{\pm}$$

According to the Debye–Hückel theory, in dilute solutions, $\ln \gamma_{\pm} = Am^{1/2}$, where A is a constant. Hence, for dilute solution our equation becomes

$$E + \frac{2RT}{F} \ln m = E^{\ominus} - \frac{2RTA}{F} m^{1/2}$$

If the quantity on the left is plotted against $m^{1/2}$, and extrapolated back to $m = 0$, the intercept at $m = 0$ gives the value of $E^{\ominus}$.* For this cell, one obtains $E^{\ominus} = 0.2225$ V at 25°C.

Once the standard emf has been determined in this way, (12.11) can be used to calculate mean activity coefficients for HCl from the measured emf's E in solutions of different molalities m. This method has been the most important source of precise data on ionic activity coefficients.

11. Standard Electrode Potentials

Rather than tabulate data for all the numerous cells that have been measured, it would be much more convenient to make a list of *single-electrode potentials* of the various half-cells. Cell emf's could then be obtained simply by taking differences between these electrode potentials. The status of single-electrode potentials is similar to that of single-ion activities. In 1899, Gibbs pointed out that it is not possible to devise any experimental procedure that will measure a difference in *electric potential* between two points in media of different chemical composition, for instance, a metal electrode and the surrounding electrolyte. What we always, in fact, measure is a difference in potential between two points of the same chemical composition, such as two brass terminals of a potentiometer.

Consider an ion of copper: (a) in metallic copper, (b) in a solution of copper sulfate. Its state is determined by its chemical environment, usually expressed by its chemical potential μ, and by its electrical environment, expressed by its electrical potential Φ. But there is no way of experimentally separating these two factors, since there is no way of separating electricity from matter and the phenomena we call "chemical" are all "electrical" in origin. Thus, we can measure only the *electrochemical potential* of an ion, $\bar{\mu} = \mu + zF\Phi$. It may sometimes be convenient to make an arbitrary separation of this quantity into two parts, but there is no way to give the separation an experimental meaning.

Although we cannot measure absolute single-electrode potentials, we can solve the problem of reducing the cell emf's to a common basis by expressing all the values relative to the same reference electrode. The choice of a conventional reference state does not affect the values of differences between the electrode potentials, i.e., the cell emf's. The reference electrode is taken to be the *standard hydrogen electrode*, which is assigned by convention the value $E^{\ominus} = 0$. It is the hydrogen electrode in which (a) the pressure of hydrogen is 1 atm (strictly, unit

---

*A. S. Brown and D. A. MacInnes, *J. Am. Chem. Soc.*, **57**, 1356 (1935). In practice, an extended form of the Debye–Hückel theory is often used to give a somewhat better extrapolation function.

fugacity, but the gas may be taken to be ideal), (b) the solution contains hydrogen ion at a mean ionic activity ($a_\pm$) of 1. Thus,

$$Pt\,|\,H_2\ (1\ atm)\,|\,H^+(a_\pm = 1)$$

If a cell is formed by combining any electrode X with the standard hydrogen electrode, the measured potential of the electrode X relative to that of the standard hydrogen electrode taken as zero is called the *relative electrode potential*, or in short, the *electrode potential*, of X. Thus if the electrode X is positive with respect to the *standard hydrogen electrode*, the electrode potential of X is positive. The sign of the electrode potential always is the observed sign of its polarity when it is coupled with a standard hydrogen electrode.*

For example,

$$Pt\,|\,H_2(1\ atm)\,|\,H^+(a_\pm = 1)\,||\,X^+\,|\,X$$

Then, the emf is

$$E = \Phi_R - \Phi_L$$
$$E(X^+/X) - E^\ominus(H_2/H^+) = E(X^+/X)$$

If the emf is the standard emf $E^\ominus$, the electrode potential is the *standard electrode potential*. When we speak of an *electrode potential*, we usually mean this standard potential, which is the one given in tables. The electrode potential for any other choice of activities can be calculated from (12.9).

For example, the zinc electrode would be written

$$Zn^{2+}\,|\,Zn$$

This notation enables us to recall that the half-cell (electrode) is written at the right, so that the complete cell is

$$Pt\,|\,H_2\,|\,H^+\,||\,Zn^{2+}\,|\,Zn$$

The reaction that takes place at the zinc electrode is

$$Zn^{2+} + 2e^- \longrightarrow Zn \qquad\qquad (A)$$

a reduction of Zn^{2+} ion to metallic zinc. The reaction at the hydrogen electrode is

$$H_2 \longrightarrow 2H^+ + 2e \qquad\qquad (B)$$

Thus, the cell reaction is

$$Zn^{2+} + H_2 \longrightarrow Zn + 2H^+ \qquad\qquad (C)$$

---

*The electrode potential is defined here in accord with the 1953 Stockholm Convention. It is also similar to the definition given by Gibbs. The electrode potential so defined is a *sign invariant quantity*. By this we mean that it has a definite sign, which in no way depends on how the electrode is written on paper. The sign is the *experimental polarity* of the electrode when coupled with a standard hydrogen electrode.

The electrode potential as defined is the *half-cell emf* of the electrode

$$X^+\,|\,X$$

This is the emf of the cell,

$$Pt\,|\,H_2\,|\,H^+\,||\,X^+\,|\,X$$

in which the cell reaction is

$$\tfrac{1}{2}H_2 + X^+ \longrightarrow H^+ + X$$

TABLE 12.1
Standard Electrode Potentials at 25°C

Electrode	Electrode reaction (acid solutions)	$E^\ominus$ (V)
Li^+/Li	$Li^+ + e \rightleftharpoons Li$	-3.045
K^+/K	$K^+ + e \rightleftharpoons K$	-2.925
Cs^+/Cs	$Cs^+ + e \rightleftharpoons Cs$	-2.923
Ba^{2+}/Ba	$Ba^{2+} + 2e \rightleftharpoons Ba$	-2.906
Ca^{2+}/Ca	$Ca^{2+} + 2e \rightleftharpoons Ca$	-2.866
Na^+/Na	$Na^+ + e \rightleftharpoons Na$	-2.714
Mg^{2+}/Mg	$Mg^{2+} + 2e \rightleftharpoons Mg$	-2.363
Al^{3+}/Al	$Al^{3+} + 3e \rightleftharpoons Al$	-1.662
Zn^{2+}/Zn	$Zn^{2+} + 2e \rightleftharpoons Zn$	-0.7628
Fe^{2+}/Fe	$Fe^{2+} + 2e \rightleftharpoons Fe$	-0.4402
Cd^{2+}/Cd	$Cd^{2+} + 2e \rightleftharpoons Cd$	-0.4029
Sn^{2+}/Sn	$Sn^{2+} + 2e \rightleftharpoons Sn$	-0.136
Pb^{2+}/Pb	$Pb^{2+} + 2e \rightleftharpoons Pb$	-0.126
Fe^{3+}/Fe	$Fe^{3+} + 3e \rightleftharpoons Fe$	-0.036
$D^+/D_2/Pt$	$2D^+ + 2e \rightleftharpoons D_2$	-0.0034
$H^+/H_2/Pt$	$2H^+ + 2e \rightleftharpoons H_2$	0
$Sn^{+4}, Sn^{+2}/Pt$	$Sn^{4+} + 2e \rightleftharpoons Sn^{2+}$	$+0.15$
$Cu^{2+}, Cu^+/Pt$	$Cu^{2+} + e \rightleftharpoons Cu^+$	$+0.153$
$S_2O_3^{2-}, S_4O_6^{2-}/Pt$	$S_4O_6^{2-} + 2e \rightleftharpoons 2S_2O_3^{2-}$	$+0.17$
Cu^{2+}/Cu	$Cu^{2+} + 2e \rightleftharpoons Cu$	$+0.337$
$I^-/I_2/Pt$	$I_2 + 2e \rightleftharpoons 2I^-$	$+0.5355$
$Fe(CN)_6^{4-}, Fe(CN)_6^{3-}/Pt$	$Fe(CN)_6^{3-} + e \rightleftharpoons Fe(CN)_6^{4-}$	$+0.69$
$Fe^{2+}, Fe^{3+}/Pt$	$Fe^{3+} + e \rightleftharpoons Fe^{2+}$	$+0.771$
Ag^+/Ag	$Ag^+ + e \rightleftharpoons Ag$	$+0.7991$
Hg^{2+}/Hg	$Hg^{2+} + 2e \rightleftharpoons Hg$	$+0.854$
$Hg_2^{2+}, Hg^{2+}/Pt$	$2Hg^{2+} + 2e \rightleftharpoons Hg_2^{2+}$	$+0.92$
$Br^-/Br_2/Pt$	$Br_2 + 2e \rightleftharpoons 2Br^-$	$+1.0652$
$Mn^{2+}, H^+/MnO_2/Pt$	$MnO_2 + 4H^+ + 2e \rightleftharpoons Mn^{2+} + 2H_2O$	$+1.23$
$Cr^{3+}, Cr_2O_7^{2-}, H^+/Pt$	$Cr_2O_7^{2-} + 14H^+ + 6e \rightleftharpoons 2Cr^{3+} + 7H_2O$	$+1.33$
$Cl^-/Cl_2/Pt$	$Cl_2 + 2e \rightleftharpoons 2Cl^-$	$+1.3595$
$Ce^{3+}, Ce^{4+}/Pt$	$Ce^{4+} + e \rightleftharpoons Ce^{3+}$	$+1.61$
$Co^{2+}, Co^{3+}/Pt$	$Co^{3+} + e \rightleftharpoons Co^{2+}$	$+1.808$
$SO_4^{2-}, S_2O_8^{2-}/Pt$	$S_2O_8^{2-} + 2e \rightleftharpoons 2SO_4^{2-}$	$+2.01$

(basic solutions)		
$OH^-/Ca(OH)_2/Ca/Pt$	$Ca(OH)_2 + 2e \rightleftharpoons 2OH^- + Ca$	-3.02
$H_2PO_2^-, HPO_3^{2-}, OH^-/Pt$	$HPO_3^{2-} + 2e \rightleftharpoons H_2PO_2^- + 3OH^-$	-1.565
$ZnO_2^{2-}, OH^-/Zn$	$ZnO_2^{2-} + 2H_2O + 2e \rightleftharpoons Zn + 4OH^-$	-1.215
$SO_3^{2-}, SO_4^{2-}, OH^-/Pt$	$SO_4^{2-} + H_2O + 2e \rightleftharpoons SO_3^{2-} + 2OH^-$	-0.93
$OH^-/H_2/Pt$	$2H_2O + 2e \rightleftharpoons H_2 + 2OH^-$	-0.82806
$OH^-/Ni(OH)_2/Ni$	$Ni(OH)_2 + 2e \rightleftharpoons Ni + 2OH^-$	-0.72
$CO_3^{2-}/PbCO_3/Pb$	$PbCO_3 + 2e \rightleftharpoons Pb + CO_3^{2-}$	-0.509
$OH^-, HO_2^-/Pt$	$HO_2^- + H_2O + 2e \rightleftharpoons 3OH^-$	$+0.878$

When all components in the cell are in their standard states, the emf is -0.763 V at 25°C. The standard electrode potential of the zinc electrode is therefore -0.763 V at 25°C. The negative value indicates that the cell reaction as written cannot spontaneously proceed when the reactants are in their standard states, i.e., H_2 gas cannot reduce Zn^{2+} under these conditions.

Table 12.1 is a collection of standard electrode potentials.*

12. Calculation of the emf of a Cell

As a typical example, let us calculate the emf at 25°C of the cell,

$$Zn\,|\,ZnSO_4\ (1.0\ m)\,\|\,CuSO_4\ (0.1\ m)\,|\,Cu$$

The cell reaction is

$$Zn + CuSO_4 \longrightarrow ZnSO_4 + Cu$$

From (12.4), the standard emf $E^\ominus$, when we take the electrode potentials from Table 12.1, is

$$E^\ominus = E_R^\ominus - E_L^\ominus = +0.337 - (-0.763) = +1.100\ \text{V}$$

The Nernst equation becomes

$$E = E^\ominus - \frac{RT}{2F} \ln \frac{a(ZnSO_4)\,a(Cu)}{a(CuSO_4)\,a(Zn)}$$

or, since $a(Cu) = a(Zn) = 1$ (p. 320),

$$E = 1.100 - 0.0295 \log \frac{a_\pm^2(ZnSO_4)}{a_\pm^2(CuSO_4)}$$

From (10.25)

$$a_\pm = \gamma_\pm m$$

From Table 10.6, for $CuSO_4$ at $m = 0.10\ \text{mol·kg}^{-1}$, $\gamma_\pm = 0.16$; $ZnSO_4$ at $m = 1.00\ \text{mol·kg}^{-1}$, $\gamma_\pm = 0.045$.

Thus,

$$E = 1.100 - 0.059 \log \frac{0.045}{0.016} = 1.073\ \text{V}$$

This example shows that, provided activity coefficients are available for the electrolytes used, we can calculate the cell emf from the tabulated standard electrode potentials and the Nernst equation. In many cases, the use of molalities or concentrations instead of activities (i.e., the assumption that all activity coefficients are unity) will provide an adequate estimated value of E.

---

*The most comprehensive survey of the data on electrode reactions is *Standard Aqueous Electrode Potentials and Temperature Coefficients at 25°C* by A. J. de Bethune and N. A. S. Loud (Skokie, Ill.: C. A. Hampel, 1964). This booklet also contains a theoretical discussion and many problems (with solutions) on chemical applications of electrode potentials.

13.   Calculation of Solubility Products

Standard electrode potentials can be combined to yield the $E^\ominus$ and thus the $\Delta G^\ominus$ and equilibrium constant for the solution of salts. In this way, we can calculate the solubility of a salt even when an extremely low value makes direct measurement difficult.

As an example, consider silver iodide, which dissolves according to: $AgI \longrightarrow Ag^+ + I^-$. The *solubility product constant* is $K_{sp} = a(Ag^+)a(I^-)$. A cell whose net reaction corresponds to the solution of silver iodide can be formed by combining a $Ag|AgI$ electrode with a Ag electrode,

$$Ag|Ag^+|I^-, AgI(s)|Ag$$

The electrode reactions are

Electrode potentials (V)

$$AgI(s) + e \longrightarrow Ag + I^- \qquad E^\ominus = -0.1518$$
$$Ag \longrightarrow Ag^+ + e \qquad E^\ominus = +0.7991$$

Overall reaction: $AgI(s) \longrightarrow Ag^+ + I^- \qquad E^\ominus = -0.9509\ [E_R - E_L]$

Then, from $\Delta G^\ominus = -|z|\,FE^\ominus = -RT \ln K_{sp}$,

$$\log_{10} K_{sp} = \frac{(-0.9509 \times 96\,487)}{(2.303 \times 8.314 \times 298.2)} = -16.07$$

The activity coefficients will be unity in the very dilute solution of AgI, so that K_{sp} corresponds to a solubility of 2.17×10^{-6} g·dm^{-3} at 25°C.

14.   Standard Free Enthalpies and Entropies of Aqueous Ions

The theoretically interesting quantity $\Delta G_i^\ominus$, the standard free enthalpy of formation of an ion in solution, cannot be measured in absolute value for an individual ion, but, as in the analogous case of the single electrode potentials, it is possible to obtain values relative to a reference ion. The reference standard is the hydrogen ion at $a_\pm = 1$, which is taken to have a standard $\Delta G^\ominus(H^+) = 0$ in aqueous solution at 25°C. As an example, consider the reaction

$$Cd + 2H^+ \longrightarrow Cd^{2+} + H_2, \qquad E^\ominus = 0.403 \text{ V} \quad (25°C)$$

If all the reactants are in their standard states,

$$-|z|\,FE^\ominus = \Delta G^\ominus = \bar{\mu}^\ominus(Cd^{2+}) + \bar{\mu}^\ominus(H_2) - \bar{\mu}^\ominus(Cd) - 2\,\bar{\mu}^\ominus(H^+)$$

Now $\bar{\mu}^\ominus(H_2)$ and $\bar{\mu}^\ominus(Cd)$ are zero because the free enthalpies of the elements are taken as zero in their standard states at 25°C, and $\bar{\mu}^\ominus(H^+)$ is zero by our convention. It follows that

$$\bar{\mu}^\ominus(Cd^{2+}) = \Delta G^\ominus = -|z|\,FE^\ominus = -2 \times 0.403 \times 96\,487 = -77.74 \text{ kJ·mol}^{-1}$$

In addition to the standard ionic free enthalpies, it is useful to obtain also the standard ionic entropies, $S_i^\ominus$. These are the partial molar entropies of the ions in solution relative to the conventionally chosen standard that sets the entropy of the hydrogen ion at unit activity equal to zero, $S^{\ominus}(H^+) = 0$.

One method for calculating an ionic entropy may be illustrated in terms of our example of the Cd^{2+} ion. Consider again the reaction: $Cd + 2H^+ \rightarrow Cd^{2+} + H_2$. The standard entropy change is

$$\Delta S^{\ominus} = S^{\ominus}(Cd^{2+}) + S^{\ominus}(H_2) - 2S^{\ominus}(H^+) - S^{\ominus}(Cd)$$

Now $S^{\ominus}(Cd)$ and $S^{\ominus}(H_2)$ have been evaluated from Third Law measurements and statistical calculations, being 51.5 and 130.7 $J \cdot K^{-1} \cdot mol^{-1}$ at 25°C. By our convention, $S^{\ominus}(H^+)$ is zero. Therefore, $S^{\ominus}(Cd^{2+}) = \Delta S^{\ominus} - 79.2$. The value of $\Delta S^{\ominus}$ can be obtained from $\Delta S^{\ominus} = (\Delta H^{\ominus} - \Delta G^{\ominus})/T$. If cadmium is dissolved in a large excess of dilute acid, the heat of solution per mole of cadmium is the standard enthalpy change $\Delta H^{\ominus}$, since in the extremely dilute solution all the activity coefficients approach unity. This experiment yields the value $\Delta H^{\ominus} = -69.87$ kJ. The $\Delta G^{\ominus}$ from the cell emf was found to be -77.74 kJ. Therefore, $\Delta S^{\ominus} = 7870/298.2 = 26.4 \, J \cdot K^{-1}$. It follows that $S^{\ominus}(Cd^{2+}) = -52.8 \, J \cdot K^{-1} \cdot mol^{-1}$.

Table 12.2 gives some standard ionic entropies determined by such methods. It must be emphasized that these values always refer to a neutral combination of the given ions with the H_3O^+ ion. For example, the value for K^+ is $S^{\ominus}(K^+) - S^{\ominus}(H^+)$, the value for Mg^{2+} is $S^{\ominus}(Mg^{2+}) - 2S^{\ominus}(H^+)$, the value for Cl^- is $S^{\ominus}(Cl^-) + S^{\ominus}(H^+)$, etc. It is possible to obtain a close estimate of the absolute entropy of the H^+ ion in aqueous solution.* Various calculations converge to a value of about $-21 \, J \cdot K^{-1}$.

TABLE 12.2
Standard Entropies of Ions in Gas State and in Aqueous Solution at 25°C ($J \cdot K^{-1} \cdot mol^{-1}$)

Cation	$S^{\ominus}$ (gas)	$S^{\ominus}$ (water)	Anion	$S^{\ominus}$ (gas)	$S^{\ominus}$ (water)
H_2O^+	108.8	0[a]	F^-	145.6	-9.6
Li^+	133.1	19.7	Cl^-	153.6	56.5
Na^+	147.7	58.6	Br^-	159.4	82.4
K^+	154.4	101.3	I^-	169.0[b]	105.9
Rb^+	164.4	120.1	OH^-		10.4
Cs^+	169.9	133.1	HSO_4^-		128.0
Ag^+	166.9	73.2	SO_4^{2-}		18.4
Mg^{2+}	148.5	-132.2	NO_3^-		125.1
Ca^{2+}	154.8	-53.1	PO_4^{3-}		-217.6
Cu^{2+}	161.1	-110.9	HCO_3^-		92.9
Zn^{2+}	161.1	-107.5	CO_3^{2-}		-54.4
Fe^{2+}	159.0	-108.4			
Fe^{3+}	159.0	-255.2			
Al^{3+}	149.8	-318.0			

[a]Values of $S^{\ominus}$ (water) are with reference to $S^{\ominus}$ (water) = 0 for the H_3O^+ ion.
[b]Values for compound ions in gas phase are not given, since they would include rotational and vibrational contributions and necessary data are not available.

*J. O'M. Bockris and B. E. Conway, *Modern Aspects of Electrochemistry*, Vol. 1 (London: Butterworth & Co., Ltd., 1954).

The entropies of the gaseous ions are easily calculated from the Sackur–Tetrode equation (5.51), which yields, at 298 K,

$$S_i \text{ (gas)} = 108.8 + 28.7 \log_{10} M$$

where M is molecular weight. The difference between $S^{\ominus}$ (water) and $S^{\ominus}$ (gas) in Table 12.2 gives the entropy of hydration of the ion. The hydration entropies are always negative. There is always a loss of translational entropy as the gaseous ion passes into solution. In addition, there is the effect of the electric field on the water molecules surrounding the ion. In the case of multiply charged ions, this effect is especially large and suggests a virtual immobilization or "freezing" of the water structure around a central ion.

15. Electrode-Concentration Cells

Consider a cell consisting of two hydrogen electrodes at different pressures placed in the same solution of hydrochloric acid:

$$Pt\,|\,H_2(P_1)\,|\,HCl(a)\,|\,H_2(P_2)\,|\,Pt$$

At the left electrode: $\quad \frac{1}{2}H_2(P_1) \longrightarrow H^+(a_{\pm}) + e$

At the right: $\quad H^+(a_{\pm}) + e \longrightarrow \frac{1}{2}H_2(P_2)$

The overall change is accordingly $\frac{1}{2}H_2(P_1) \longrightarrow \frac{1}{2}H_2(P_2)$, the transfer of one equivalent of hydrogen from pressure P_1 to P_2. The emf of the cell is

$$E = \frac{-RT}{2F} \ln \frac{P_2}{P_1}$$

Another type of electrode-concentration cell consists of two amalgam electrodes of different concentrations in contact with a solution containing ions of the metal that is dissolved in the amalgam. For example,

$$Cd\text{—}Hg(a_1)\,|\,CdSO_4\,|\,Cd\text{—}Hg(a_2)$$

The emf of this cell arises from the free energy of transferring cadmium from an amalgam in which its activity is a_1 to an amalgam in which its activity is a_2. The

TABLE 12.3
Cadmium Amalgam Electrode-Concentration Cells[a]

Grams cadmium per 100 g mercury		Observed emf	Calculated emf
Electrode (1)	Electrode (2)	V	V
1.000	0.1000	0.02966	0.02950
0.1000	0.01000	0.02960	0.02950
0.01000	0.001000	0.02956	0.02950
0.001000	0.0001000	0.02950	0.02950

[a] G. Hulett, *J. Am. Chem. Soc.*, **30**, 1805 (1908).

emf is, therefore,

$$E = \frac{-RT}{2F} \ln \frac{a_2}{a_1}$$

If the amalgams are considered to be ideal solutions, we may replace the activities by mole fractions. In Table 12.3 are some experimental data for these cells, together with the calculated emf's based on ideal solutions. Note how the observed values approach the theoretical ones with increasing dilution of the amalgam.

16. Electrolyte-Concentration Cells

In the following cell,

$$Pt\,|\,H_2\,|\,HCl(c)\,|\,AgCl\,|\,Ag$$

the cell reaction is

$$\tfrac{1}{2}H_2 + AgCl \longrightarrow Ag + HCl(c)$$

Measurements with two different concentrations c of hydrochloric acid yielded the following results at 25°C:

c (mol·dm⁻³)	E (V)	$-\Delta G$ (J)
0.0010	0.5795	55 920
0.0539	0.3822	36 880

If two such cells oppose each other, the combination constitutes a cell that can be written

$$Ag\,|\,AgCl\,|\,HCl(c_2)\,|\,H_2\,|\,HCl(c_1)\,|\,AgCl\,|\,Ag$$

The overall change in this cell is simply the difference between the changes in the two separate cells: for the passage of each faraday, the transfer of 1 mol of HCl from concentration c_2 to c_1, $HCl(c_2) \rightarrow HCl(c_1)$. Note, however, that there can be no direct transference of electrolyte from one side to the other. The HCl is removed from the left side by the reaction $HCl + Ag \rightarrow AgCl + \tfrac{1}{2}H_2$. It is added to the right side by the reverse of this reaction. From the preceding data, if $c_1 = 0.001$ and $c_2 = 0.0539$ mol·dm⁻³, the Gibbs free energy of dilution is $\Delta G = -19\,040$ J and $E = (19\,040/96\,487) = 0.1973$ V. This cell is an example of a concentration cell without transference.

If the two HCl solutions of different concentrations contact each other directly, as through a sintered glass plug, we have a concentration cell with transference.

$$Ag\,|\,AgCl\,|\,HCl(c_2) :: HCl(c_1)\,|\,AgCl\,|\,Ag$$

Ions can now pass across the liquid junction between the two solutions. As 1 F passes through the cell, it is carried across the liquid junction partly by the H⁺ ion $(t_+\,F)$ and partly by the Cl⁻ ion $(t_-\,F)$, where t_+ and t_- are transference numbers. In the case of the cell shown, the electrodes are reversible to the Cl⁻ ion, so that 1 F of Cl⁻ ions enters the electrolyte at the right and leaves at the left. There is, therefore, a net transfer of t_+ mole of HCl from left to right, and the emf of the concentration cell would be

$$E = t_+ \frac{RT}{F} \ln \frac{a_2}{a_1} \qquad (12.12)$$

or just t_+ times the emf for the cell without transference. For a case in which the electrodes are reversible to the H^+ ion (hydrogen electrodes), the transference number t_- would appear in (12.12).

Actually, the argument given here is not rigorous, since the cell with transference is not a reversible cell. There is always a diffusion process taking place at the liquid junction, and consequently a *liquid-junction potential* arises that cannot be treated by equilibrium thermodynamics. More exact treatments show, however, that (12.12) is a very good approximation to the cell emf.[*]

17. Nonosmotic Membrane Equilibrium

Figure 12.6 depicts two solutions α and β separated by a membrane. We shall denote quantities in α by unprimed symbols and those in β by primed ones. In this section, we shall assume that the membrane is not permeable to solvent, so that

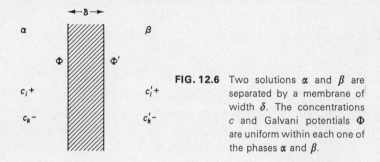

FIG. 12.6 Two solutions α and β are separated by a membrane of width δ. The concentrations c and Galvani potentials Φ are uniform within each one of the phases α and β.

effects due to osmotic pressure can be ignored. The concentrations of positive ions are c_{i^+}, c'_{i^+} and of negative ions, c_{k^-}, c'_{k^-}. If a condition of equilibrium is reached, the current of each ion across the membrane must vanish,

$$I_{i^+} = 0, \quad I_{k^-} = 0 \qquad \text{for all } i, k$$

If the membrane is permeable to all the ions, equilibrium will be reached only when all the ionic concentrations have become equalized across the membrane, and when there is no potential difference across the membrane, $\Delta\Phi = 0$.

To have a nonvanishing potential difference, $\Delta\Phi \neq 0$, at equilibrium, the membrane must be nonpermeable to one or more of the ions on either side. The simplest example is that described by Nernst in 1888, in which only one ion, say, j, can pass through the membrane. The equilibrium condition is then

$$\bar{\mu}_j = \bar{\mu}'_j$$

[*]D. G. Miller, *J. Phys. Chem.*, **70**, 2639 (1966).

From (12.1), therefore,

$$\mu_j + z_j F\Phi = \mu'_j + z_j F\Phi' \tag{12.13}$$

or, since $\mu_j = \mu_j^0 + RT \ln a_j$,

$$\Delta\Phi = \frac{RT}{z_j F} \ln \frac{a_j}{a'_j} \tag{12.14}$$

In dilute solution,

$$\frac{a_j}{a'_j} = \frac{c_j}{c'_j} \quad \text{and} \quad \Delta\Phi = \frac{RT}{z_j F} \ln \frac{c_j}{c'_j} \tag{12.15}$$

We can look upon this difference as just the electric potential necessary to prevent equalization of the ionic concentrations of j by diffusion across the membrane. It is possible to fix the ratio of ionic concentrations, c_j/c'_j, whereupon an equilibrium potential $\Delta\Phi$ will be established that corresponds to the ratio chosen. It is important to note that this situation is possible because the membrane is *permeable to only one ion*. If several ions could pass through the membrane, it would no longer be possible to obtain an equilibrium $\Delta\Phi$ corresponding to any chosen ratio of ionic concentrations.

Membranes of typical mammalian nerve cells are permeable to K^+ but, in their resting states, relatively impermeable to Na^+, Cl^-, and other ions. The concentration $[K^+]$ inside the cell is about 20 times the $[K^+]$ outside. From (12.15), the Nernst equilibrium potential across the membrane at 25°C would therefore be about

$$\Delta\Phi = \frac{(8.314)(298.2)}{96\,487} \ln\left(\frac{1}{20}\right) = 25.7 \ln\left(\frac{1}{20}\right) = -77.5 \text{ mV}$$

(the inside being negative relative to the outside). This $\Delta\Phi$ is close to the experimentally observed value, and, to a first approximation, the nerve membrane potential can be regarded as an equilibrium K^+ potential. Closer consideration indicates, however, that since the membrane is permeable to some extent to ions other than K^+, the observed potential must be interpreted as a steady state potential for a system in which the K^+ permeability is considerably greater than that of any other ion. It is not possible to satisfy the equilibrium condition (12.15) simultaneously for all the ions.

The individual ionic activities in (12.14) are not measurable quantities, and we cannot measure the electrical potential difference between two different phases α and β. If the solution is sufficiently dilute to justify replacing the activity ratio by a concentration ratio, we can calculate $\Delta\Phi$ from (12.15). If we place inert electrodes, such as platinum wires, in the two solutions, we can measure $\Delta\Phi$ between the two platinum leads. It is only an approximation to equate this $\Delta\Phi$ to that between the two solutions α and β, but the approximation is valid to about the same degree as that used in calculating $\Delta\Phi$ from the concentration ratio. Most of the experimental data in electrophysiological studies rely on approximations of this kind.

It should be pointed out that if the electrodes used are reversible with respect to the permeant ion (K^+ in the preceding example), the $\Delta\Phi$ between the leads from the two sides of the membrane would be zero, since each electrode would reach equilibrium with the solution in which it is immersed, and $\Delta G = 0$ for the cell at equilibrium.

18. Osmotic Membrane Equilibrium

In this case, we assume that the membrane is permeable to solvent and to some, but not all, ions. A theory for such membrane equilibria was first given by F. G. Donnan,* and hence one often refers to the *Donnan equilibrium* and the *Donnan potential* in discussions of this subject. A simple example is shown in Fig. 12.7, in which the solution on one side contains a cation P^{z+} which cannot pass through the membrane; e.g., it might be a substance of high molecular weight, such as a protein. The presence of the nonpermeating ions causes an unequal distribution of the other ions on the two sides of the membrane.

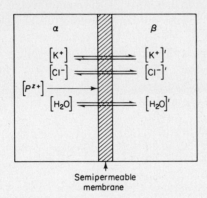

FIG. 12.7 An example of a Donnan membrane equilibrium with high polymer cations P^{z+} and KCl as the neutral salt.

The equilibrium conditions are

$$\bar{\mu}_{K^+} = \bar{\mu}'_{K^+}, \qquad \bar{\mu}_{Cl^-} = \bar{\mu}'_{Cl^-}, \qquad \mu_{H_2O} = \mu'_{H_2O}$$

where quantities for phase β are primed and those for α unprimed. If the activities of water molecules are different on the two sides of the membrane, from (7.40) there will be at equilibrium an osmotic pressure

$$\Pi = \frac{RT}{V(H_2O)} \ln \frac{a'(H_2O)}{a(H_2O)} \tag{12.16}$$

For the ions, from (12.1), (8.44), and (8.46),

$$\Pi V(K^+) - RT \ln \frac{a'(K^+)}{a(K^+)} = F \Delta \Phi \tag{12.17}$$

$$\Pi V(Cl^-) - RT \ln \frac{a'(Cl^-)}{a(Cl^-)} = -F \Delta \Phi \tag{12.18}$$

When Π is eliminated between (12.16) and either (12.17) or (12.18),

*F. G. Donnan, *Z. Elektrochem.*, *17*, 572 (1911).

$$\Delta\Phi = \frac{RT}{F} \ln \frac{a'(K^+)\,[a(H_2O)]^{r^+}}{a(K^+)\,[a'(H_2O)]^{r^+}} = \frac{RT}{F} \ln \frac{a(Cl^-)\,[a'(H_2O)]^{r^-}}{a'(Cl^-)\,[a(H_2O)]^{r^-}} \qquad (12.19)$$

where

$$r^+ = \frac{V(K^+)}{V(H_2O)} \quad \text{and} \quad r^- = \frac{V(Cl^-)}{V(H_2O)}$$

It is usually a satisfactory approximation to take the activity of water to be the same in the two solutions, so that (12.19) is simplified to

$$\Delta\Phi = \frac{RT}{F} \ln \frac{a'(K^+)}{a(K^+)} = \frac{RT}{F} \ln \frac{a(Cl^-)}{a'(Cl^-)}$$

When the nonpermeating ion P^{z+} is in low concentration, the activity ratios for the small ions can be replaced by concentration ratios, so that

$$\Delta\Phi = \frac{RT}{F} \ln \frac{a'(K^+)}{a(K^+)} \longrightarrow \frac{RT}{F} \ln \frac{c'(K^+)}{c(K^+)}$$

and

$$c(K^+)\,c(Cl^-) = c'(K^+)\,c'(Cl^-) = c'^2$$

Then the conditions of electrical neutrality give

$$c(K^+) + zc(P^{z+}) = c(Cl^-)$$
$$c'(K^+) = c'(Cl^-) = c'$$

Thus,

$$c'^2 = c(K^+)\,[c(K^+) + zc(P^{z+})]$$

Taking the square root of both sides, but neglecting terms quadratic and higher in $c(P^+)$, we obtain

$$\frac{c'}{c(K^+)} = 1 + \frac{zc(P^{z+})}{2c(K^+)} \qquad (12.20)$$

Table 12.4 gives calculated Donnan effects for several different ionic concentrations. We note that large differences in concentrations of diffusible ions may

TABLE 12.4
Examples of Donnan Membrane Equilibria at 25°C[a]

	(Concentrations in mol·dm^{-3})			$\dfrac{c'_+}{c_+} = \dfrac{c_-}{c'_-}$	$\Delta\Phi$ (mV)
$zc(P^{z+})$	$c'_+ = c'_-$	c_+	c_-		
0.002	0.0010	0.00041	0.00241	2.44	22.90
	0.0100	0.00905	0.01105	1.10	2.56
	0.100	0.0990	0.1010	1.01	2.58
0.02	0.0010	0.00005	0.02005	20.05	76.96
	0.0100	0.00414	0.02414	2.41	22.65
	0.100	0.0905	0.1105	1.10	2.56

[a]Adapted from *Physical Chemistry of Macromolecules*, by Charles Tanford (New York: John Wiley & Sons, Inc., 1961), p. 226.

arise across membranes as the consequence of the presence of nondiffusible ions on one side. Appreciable Donnan potential differences may also occur.*

19. Steady State Membrane Potentials

If two different ionic solutions are separated by a permeable barrier, a potential difference can arise between them. Such potentials occur in galvanic cells where the electrode compartments are separated by salt bridges or porous barriers. Similar potential differences across the outer membranes of living cells are closely related to essential physiological processes, such as propagation of electric impulses by nerve cells and the so-called *active transport* of ions and metabolites, i.e., transport against gradients of electrochemical potential across cell walls. Since these living systems are not in thermodynamic equilibrium, these potentials are not equilibrium potentials. In many instances, however, they can be treated as steady state potentials.

The motion of ions through a membrane under the combined influences of electric fields and concentration gradients leads to a typical *electrodiffusion problem*. The analysis given in 1890 by Max Planck† has been the starting point of all subsequent theoretical work in this field. The system studied by Planck is shown in Fig. 12.8. The concentrations of positive ions are denoted by c_i with $i = 1 \cdots n$, and the concentrations of negative ions by $\bar{c}_k$ with $k = 1 \cdots m$. In Planck's treatment, all the ions are assumed to have the same charge number $|z|$. The theory has been extended to ions of different $|z|$ by Schlögl.‡

The membrane in Fig. 12.8 is regarded as an infinite plane region of dielectric

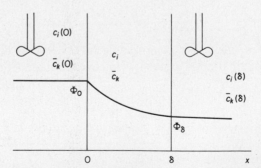

FIG. 12.8 Steady-state system leading to membrane potentials [Planck model]. The membrane extends from $x = 0$ to $x = \delta$. The solutions are well stirred so that the ionic concentrations are uniform throughout each solution.

*Although the Donnan equilibrium has been discussed in the context of osmotic equilibrium, a similar effect would be observed even if the membrane were not permeable to the solvent, but in such a case, of course, there would be no equilibrium osmotic pressure.

†M. Planck, *Ann. Physik Chem.*, *39*, 161 (1890).

‡R. Schlögl, *Z. Physik Chem.*, N. F. *1*, 305 (1954).

constant ϵ between $x = 0$ and $x = \delta$. The ionic concentrations on either side are maintained constant and uniform. In the case of a galvanic cell with a liquid junction, this condition could be achieved in some kind of flowing contact between the two electrode compartments. In the case of living cells, active transport across the membrane maintains effectively constant concentrations. Since there are steady fluxes J_i, J_k of individual ions through the membrane, the system is not in an equilibrium state but rather in a *steady state*, characterized by the fact that the *total electric current* through the membrane is zero. The steady state flux of each ionic species can be written

$$J_i = -D_i \frac{dc_i}{dx} + u_i c_i E = \text{constant} \tag{12.21}$$

where u_i is the electric mobility and E is the strength of the electric field. Since, from (10.19),

$$\frac{u_i}{D_i} = \frac{z_i e}{kT}$$

(12.21) becomes

$$J_i = -D_i \frac{dc_i}{dx} + \frac{D_i z_i e c_i E}{kT}$$

or, on dividing through by D_i,

$$\frac{-dc_i}{dx} + \frac{z_i e}{kT} E c_i = A_i, \qquad i = 1 \cdots n \tag{12.22}$$

where A_i is a constant. Similarly,

$$\frac{-d\bar{c}_k}{dx} + \frac{z_k e}{kT} E \bar{c}_k = B_k, \qquad k = 1 \cdots m \tag{12.23}$$

In addition to the $m + n$ equations for the fluxes, the Poisson equation should hold at any point in the membrane:

$$\frac{dE}{dx} = \frac{-d^2\Phi}{dx^2} = \frac{e}{\epsilon_0 \epsilon} \sum (z_i c_i + z_k \bar{c}_k) \tag{12.24}$$

There are thus $m + n + 1$ equations to be solved, with $m + n + 1$ integration constants.

In the absence of an external electric field, the net charge transfer vanishes in the steady state,

$$\sum (I_i + I_k) = \sum (z_i e J_i + z_k e J_k) = 0$$

and the ionic concentrations at $x = 0$ and $x = \delta$ are all fixed, so that $c_i(0)$, $c_i(\delta)$, $\bar{c}_k(0)$, $\bar{c}_k(\delta)$ are prescribed. Hence, there are $2(m + n) + 1$ conditions, which suffice to fix the $m + n + 1$ integration constants and the $m + n$ fluxes (or constants A_i, B_k). The problem is thus well defined mathematically, although the general integration of the set of equations presents major difficulties.

We shall consider the simplified case treated by Planck, in which $z_i = 1$, $z_k = -1$. We introduce dimensionless variables,

$$\zeta = \frac{x}{\delta} \qquad \phi = \frac{e\Phi}{kT}$$

$$p = \frac{\sum c_i}{\bar{C}} \qquad n = \frac{\sum c_k}{\bar{C}}$$

where $\bar{C} = \langle \sum c_i + \sum \bar{c}_k \rangle = $ const. is the space average.
The Poisson equation then becomes

$$\frac{\bar{\lambda}^2}{\delta^2} \frac{d^2\phi}{d\zeta^2} = -(p - n) \tag{12.25}$$

where $\bar{\lambda}^2 = \varepsilon_0 \epsilon kT/e^2\bar{C}$. We note that $(p - n)$ is the excess of positive over negative charges in the membrane, and λ is the thickness of the ionic atmosphere as introduced in the Debye–Hückel theory. The various approximate solutions of the Planck equations arise from different assumptions about the ratio $(\bar{\lambda}^2/\delta^2)$ in (12.25).

Planck (1890) made an approximation equivalent to $\bar{\lambda}^2/\delta^2 \ll 1$, which would become more satisfactory as the membrane became thicker or the ionic concentration higher. From (12.25), it would then follow that $p \simeq n$ (electrical neutrality within the membrane), despite the fact that $d^2\phi/d\zeta^2 \neq 0$ (electric field within the membrane not constant).

In 1943, D. Goldman* made the opposite approximation, viz., $\bar{\lambda}^2/\delta^2 \gg 1$. From (12.25), this assumption corresponds to

$$\frac{d^2\phi}{d\zeta^2} = -(p - n)\frac{\delta^2}{\lambda^2} \simeq 0$$

In other words, even if $p \neq n$ (so that the membrane is not electrically neutral), $E = -d\Phi/dx$ is a constant throughout the membrane. Thus, the Goldman approximation is a *constant-field model*. The Goldman treatment has a practical advantage in that it yields an explicit expression for the membrane potential $\Delta\Phi$, which is given by Planck only implicitly in a pair of transcendental equations.

As will be seen from (12.26), the results of the Planck and Goldman treatments coincide for the special case in which $\bar{C}(0) = \bar{C}(\delta)$, i.e., where the total ionic concentrations are the same on both sides of the membrane. In this case, the Planck model also yields a constant field across the membrane.

Some examples of concentrations of ions across cell membranes are shown in Table 12.5. In the case of nerve cells under physiological conditions, the total ionic concentrations are, in fact, approximately the same on both sides of the membrane, being high in K^+ inside the cell and high in Na^+ outside. The resting potentials across natural membranes under these conditions can be calculated to a fair degree of accuracy from either the Goldman or the Planck approximations, and, therefore, they shed little light on the problem of which model Nature herself may prefer. The problem essentially reduces to the question of what is the unknown concentration of ions within the 10-nm membranes.

The integration of the Planck equations (with the approximation of electroneutrality) gave

*J. Gen. Physiol., 27, 37 (1943).

TABLE 12.5
Concentrations of Ions Across Cell Membranes

Preparation	Ionic concentrations ($mmol \cdot kg^{-1}$)					
	[K$^+$]		[Na$^+$]		[Cl$^-$]	
	In	Out	In	Out	In	Out
Animal cells						
Squid nerve axon	410	10	49	460	40	540
Frog nerve	110	2.5	37	120		120
Frog muscle	125	2.5	15	120	1.2	120
Rat muscle	140	2.7	13	150		140
Plant Cells						
Nitella clavata	54	0.005	10	0.02	91	1
Chara ceratophylla	88	1.2	142	60	225	75

$$\frac{\xi V_\delta - V_0}{U_\delta - \xi U_0} = \frac{\ln (C_\delta/C_0) - \ln \xi}{\ln (C_\delta/C_0) + \ln \xi} \cdot \frac{\xi C_\delta - C_0}{C_\delta - \xi C_0} \tag{12.26}$$

where

$$U = \sum D_i C_i \quad \text{and} \quad V = \sum D_k \bar{C}_k$$

The potential difference across the membrane is given by

$$\Delta\Phi = \Phi_\delta - \Phi_0 = \frac{kT}{e} \ln \xi \tag{12.27}$$

To use the Planck theory, one must solve (12.26) for ξ and then calculate the potential difference $\Delta\Phi$ from (12.27). A simple computer program can be written to solve (12.26) by Newton's method.* These Planck equations have been widely used to compute liquid-junction potentials.†

In the case of the Goldman approximation, we cannot add up Eqs. (12.22) and (12.23) to obtain a single equation in $C(x)$, since the electroneutrality condition $\sum c_i = \sum \bar{c}_k$ does not hold. On the other hand, because E is constant, each equation can be integrated immediately, and the conditions $c_i(0)$, $c_i(\delta)$, etc., determine the $(n + m)$ integration constants and $(n + m)$ ionic fluxes. Then the steady state condition of zero net charge transfer is applied to calculate $\Delta\Phi$ as

$$\Delta\Phi = \frac{kT}{e} \ln \frac{\sum D_i C_i(\delta) + \sum D_k C_k(0)}{\sum D_i C_i(0) + \sum D_k C_k(\delta)} \tag{12.28}$$

The Goldman equation is not restricted to AB electrolytes, and can be applied to mixtures of ions of various charges.

*E. L. Stiefel, *An Introduction to Numerical Mathematics* (New York: Academic Press, Inc., 1963), p. 79.

†A good description is given by D. MacInnes, *The Principles of Electrochemistry* (New York: Dover Publications, Inc., 1961), Chap. 13.

20. Nerve Conduction

In 1850, Helmholtz successfully performed an experiment then generally regarded as impossible: he measured the conduction velocity in a frog nerve, finding it to be about 30 m·s⁻¹. Only 16 years later, Bernstein began his researches on the origin of the nerve impulse. This work was before the advent of the ionization theory of Arrhenius in 1883 and the brilliant researches in electrochemistry that marked the later years of the nineteenth century. Nernst's paper on electrochemical cells appeared in 1888 and Planck's analysis of the electrodiffusion problem in 1890. In 1902, Bernstein published a definitive statement of his membrane theory of the nerve impulse, which was based on the depolarization of a nerve cell membrane selectively permeable to K^+ ions. In that same year, however, Overton showed that Na^+ ions played an essential role in the excitation of an *action potential*, as the impulse propagated along a nerve was called.

Experimental investigation of potentials across nerve membranes and ionic currents through them was delayed by the small sizes of nerve cells, but major advances became feasible after J. Z. Young in 1936 called attention to giant nerve axons in certain squid, which had diameters as large as 10^3 μm, compared to the usual 0.1 to 20 μm. With these giant axons it was possible to introduce electrodes into the internal medium and even to squeeze out this cellular medium (*cytoplasm*) and replace it with ionic solutions of any desired composition. As a result of bril-

FIG. 12.9 Electron micrograph at 126 000X of a thin section through the optic nerve of an adult rat. The transverse section of one myelinated axon is shown with portions of four others. The nerve would contain several thousand of such axons. The spiral layers of the myelin are formed from the outer membranes of special cells called *oligodendrocytes,* which wrap around the axons of the neurons to form an insulating sheath. At regular intervals of about 100 times the axonal diameter along the length of the neuron, there are openings in the sheath called *nodes of Ranvier.* The depolarization pulse of the action potential jumps from one node to the next, in a process called *saltatory conduction.* Dispersed through the *axoplasm* are dark circular sections which are transverse cuts through *neurotubules* that run through the axon between the body of the neuron and the synapses at the ends of the axon. The neurotubules provide a means of transport of materials from the cell body to the synapses. [Photograph by Alan Peters, Department of Anatomy, Boston University School of Medicine from *The Fine Structure of the Nervous System,* by A. Peters, S. L. Palay, and H. de F. Webster (New York: Harper and Row, 1970).]

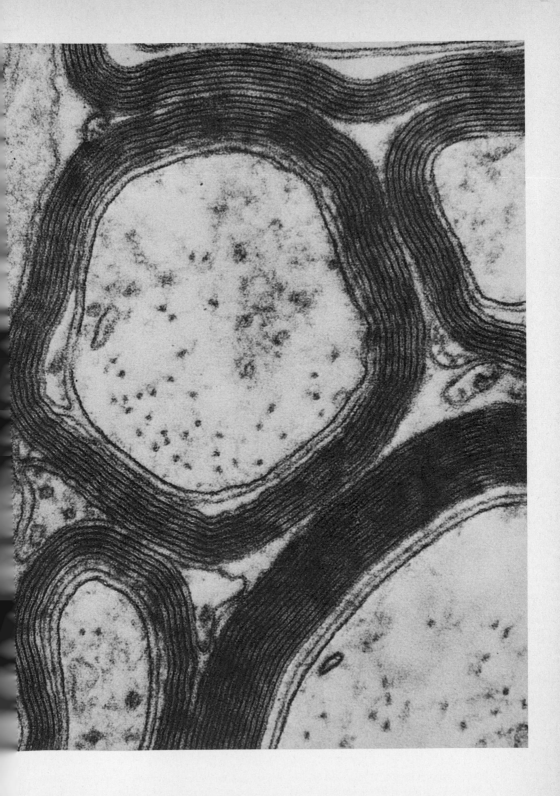

551

liant work by Hodgkin and Huxley,* and by Cole,† the main ionic features of the nerve impulse were delineated. At the present time, however, a great unsolved problem still remains: the explanation of the electrical properties of nerve membranes in terms of their structures of constituent proteins and lipids. Figure 12.9 shows an electron micrograph of a cross section of a nerve-cell axon at high magnification.

Figure 12.10 depicts the electrochemical events during the course of an action

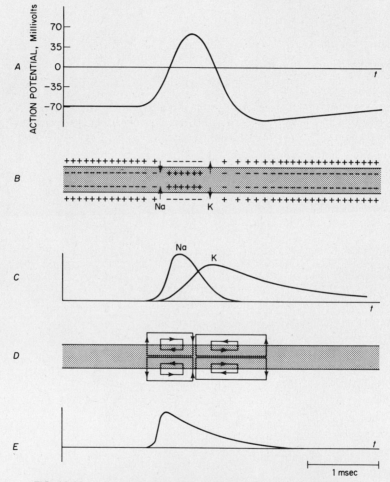

FIG. 12.10 Summary of the events occurring during propagation of an impulse along a squid giant axon. A, action potential; B, polarity of potential difference across the membrane; C, changes in sodium and potassium permeabilities; D, local circuit current flows; E, variation in total membrane conductance.

*A. L. Hodgkin, *The Conduction of the Nervous Impulse* (Liverpool: University Press, 1964).
†K. S. Cole, *Membranes, Ions and Impulses* (Berkeley: University of California Press, 1968).

potential along a typical axon. In the resting state, the potential difference is about -70 mV at 25°C, which is close to the value calculated for a Nernst equilibrium potential,

$$\Delta\Phi = \frac{-RT}{F} \ln \frac{[K^+]_{in}}{[K^+]_{out}}$$

We should not conclude from this result that the system is in equilibrium, however, because the steady state treatment of Planck or Goldman would give almost the same result, provided the mobility of K^+ in the membrane is much greater than that of other ions, such as Na^+ and Cl^-.

When a depolarization pulse of 20 mV or more is applied suddenly across the nerve membrane, so that $\Delta\Phi$ drops from -70 to about -50 mV, there is a rapid increase in its permeability to Na^+ and also a slower increase in its permeability to K^+, as shown in Fig. 12.10. As a consequence of this further depolarization, areas of membrane adjacent to the site of the initial depolarization pulse also become depolarized, as a result of IR drop along the axon, by more than the critical amount needed to initiate an action potential. The result is a rapidly propagated wave of depolarization constituting the nerve impulse.*

21. Electrode Kinetics

We turn now to a consideration of the rates of processes at electrode surfaces, the basic subject matter of *electrochemical kinetics*.† In many respects, a chemical reaction at an electrode resembles one occurring on the surface of a solid catalyst. We may, in fact, consider a metallic electrode to be a catalytic surface that facilitates the transfer of electrons to and from chemical reactant molecules and ions. Thus, an electrode reaction can be viewed as a succession of steps similar to those in heterogeneous catalysis, as listed in Section 11.15:

1. diffusion of reactants to electrode;
2. (reaction in the layer of solution adjacent to electrode);
3. adsorption of reactants on electrode;
4. transfer of electrons to or from adsorbed reactant species;
5. desorption of products from electrode;
6. (reaction in layer of solution adjacent to electrode);
7. diffusion of products away from electrode.

Since the reaction sequence does not necessarily include steps 2 and 6, they have been put into parentheses. (An example of such a step would be the decomposition

---

*A remarkable feature of the electrochemical behavior of the nerve membrane is that, when depolarization is held fixed ("voltage-clamp method") the sodium permeability reverts to normal after about a millisecond but the potassium permeability does not.

†An excellent review is "The Kinetics of Electrode Processes" by K. J. Laidler, *J. Chem. Educ.*, **47**, 600 (1970).

or formation of a complex ion in the solution phase before or after an electron-transfer step.)

The general theory of reaction rates, as given in Section 9.28, showed how the system in a chemical reaction moves along a potential-energy surface between two minima of free energy. Some source of energy is necessary to provide the free energy that allows the system to surmount the activation barrier. There are three principal ways in which molecules can acquire the necessary activation energy. In Chapter 9, we studied reactions in which the energy of activation came from thermal energy associated with translations of molecules and their internal vibrations and rotations. In Chapter 17, we shall study activation by photochemical processes, the absorption of radiation. In this chapter, we study the third principal way of activation, in which the energy of an electric field is used to activate charged species, such as ions and electrons, so as to help them to surmount an activation-energy barrier. Since electrochemical reactions are always studied at some temperature $T > 0$, there will also be a thermal contribution to the energy of activation; consequently, electrochemical kinetics is based upon a combination of thermal and electrical activation. In fact, the principal theoretical problem in electrochemical kinetics is to discover exactly how the thermal and electrical activations combine to determine the actual reaction rate; this problem leads to the theory of the *transfer coefficient* α, which will be discussed later. We shall now consider the various steps in electrode reactions as listed previously.

22.   Polarization

At equilibrium, the rate of electron transfer across an electrode in one direction is exactly balanced by an equal rate of electron transfer in the opposite direction, so that

$$\overrightarrow{i} = \overleftarrow{i}$$

The equilibrium difference in electric potential $\Delta\Phi_e$ is determined by this condition. As in any chemical reaction, the condition of equilibrium is not the cessation of all exchange, but the equality of the forward and reverse reaction rates. The exchange-current density at equilibrium is designated by i_0. Since there is no net forward or reverse current, we cannot measure i_0 directly, but we can obtain it from data on rates of exchange of radioactive tracers between electrode and solution. Values of i_0 for some electrode reactions are given in Table 12.6. We see that these exchange reaction rates vary over many orders of magnitude.

When an electrochemical cell is operating under nonequilibrium conditions, $\overrightarrow{i} \neq \overleftarrow{i}$ and there is a net current density $i = \overrightarrow{i} - \overleftarrow{i}$. The electric potential difference between the terminals of the cell departs from the equilibrium value $\Delta\Phi_e = E$, the emf. If the cell is converting chemical free energy into electrical energy, $\Delta\Phi < E$. If the cell is using an external source of electrical energy to carry out a chemical reaction, $\Delta\Phi > E$. The actual value of $\Delta\Phi$ depends on the current density i at the electrodes. The difference

TABLE 12.6
Exchange-Current Densities i_0 at 25°C for Some
Electrode Reactions[a]

Metal	System	Medium	Log i_0 (A·cm^{-2})
Mercury	Cr^{3+}/Cr^{2+}	KCl	−6.0
Platinum	Ce^{4+}/Ce^{3+}	H_2SO_4	−4.4
Platinum	Fe^{3+}/Fe^{2+}	H_2SO_4	−2.6
Palladium	Fe^{3+}/Fe^{2+}	H_2SO_4	−2.2
Gold	H^+/H_2	H_2SO_4	−3.6
Platinum	H^+/H_2	H_2SO_4	−3.1
Mercury	H^+/H_2	H_2SO_4	−12.1
Nickel	H^+/H_2	H_2SO_4	−5.2

[a]After J. O'M. Bockris, *Modern Electrochemistry* (New York: Plenum Press, 1970).

$$\Delta\Phi(i) - \Delta\Phi(0) = \eta \tag{12.29}$$

is called the *polarization* of the cell. The value of η is determined in part by the potential (IR) necessary to overcome the resistance R in the electrolyte and leads. The corresponding electrical energy I^2R is dissipated as heat, being analogous to frictional losses in irreversible mechanical processes. The remaining part of η, which is the part of theoretical interest, is due to rate-limiting processes at the electrodes;

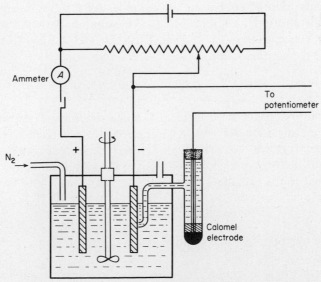

FIG. 12.11 Apparatus for measuring the potential of an electrode relative to that of a standard reference electrode. The potential is measured as a function of the current density at the electrode under study.

the corresponding electrical energy is being used to provide the free energy of activation in one or more of the steps (previously outlined) in the electrode reaction.

In electrochemical kinetics, we usually wish to study reactions at a particular electrode, i.e., in a particular half-cell. This result is accomplished by introducing into the cell an auxiliary reference electrode with a lead into the electrolyte very close to the experimental electrode. Such an arrangement is depicted in Fig. 12.11. There are a number of technical problems* in obtaining true electrode properties with this system, but we shall assume that they have been overcome and proceed with the theoretical discussion.

23.   Diffusion Overpotential

If the various electrochemical reaction steps (steps 2 to 6) are sufficiently rapid, the overall rate of the electrode reaction may be controlled by a diffusion process. We recall that rapid reactions in solution and at catalytic surfaces can also become subject to diffusion control (p. 501). The detailed theory of the diffusion process depends on the form of the electrode, whether it is moving or stationary, and on the extent of stirring of the electrolyte solution.

One of the earliest discussions was given by Nernst in 1904 for the case of steady state diffusion toward a stationary planar electrode in a solution so vigorously stirred that only a thin layer adjacent to the electrode, about 10^{-2} to 10^{-3} cm thick, functioned as a diffusion barrier. (This layer can be observed by the *Schlieren* technique.) For concreteness, let us consider the reaction in question to be the discharge of Cu^{2+} and deposition of copper on the cathode. As Cu^{2+} is removed from the solution at the electrode, a layer of electrolyte solution of thickness δ is formed, in which the concentration $[Cu^{2+}]$ is depleted. By Fick's First Law, the rate of deposition of copper (amount n) will be

$$-\frac{dn}{dt} = D\mathscr{A}\frac{dc}{dx} = D\mathscr{A}\frac{c_0 - c_1}{\delta}$$

where $\mathscr{A}$ is the area of the electrode, and c_0 and c_1 are the concentrations of Cu^{2+} in the bulk electrolyte solution and at the electrode surface, respectively. The diffusion coefficient D of Cu^{2+} ion is assumed to be independent of its concentration. In steady state of diffusion, the concentration gradient is linear across the layer δ.

The current to the cathode is

$$I_\delta = -zF\frac{dn}{dt} = \frac{zFD\mathscr{A}(c_0 - c_1)}{\delta} \tag{12.30}$$

where z is the number of faradays transferred in the half-reaction, in this case 2 for the reduction of Cu^{2+}.

The difference in activity of Cu^{2+} across the diffusion layer will lead to a potential difference of the form calculated for a concentration cell, namely

*G. Kortüm, *Lehrbuch der Elektrochemie*, 4th ed. (Weinheim: Verlag Chemie, 1966), pp. 416–420.

$$\eta_D = \frac{RT}{zF} \ln \frac{a_1}{a_0} \tag{12.31}$$

This η_D is called the *diffusion overpotential*.* From (12.31),

$$c_1 = c_0 \left(\frac{\gamma_0}{\gamma_1} \right) \exp\left(\frac{zF\eta_D}{RT} \right) \tag{12.32}$$

where γ_0/γ_1 is the ratio of activity coefficients corresponding to the ratio of concentrations c_0/c_1.

From (12.30) and (12.32),

$$\eta_D = \frac{RT}{zF} \ln \left[\frac{\gamma_1}{\gamma_0} \left(1 - \frac{\delta I_\delta}{\mathscr{A} c_0 D |z| F} \right) \right]$$

The limiting value of I_δ would correspond to the discharge of every ion striking the electrode, so that $c_1 = 0$ and

$$I_{max} = \frac{zFD\mathscr{A} c_0}{\delta} \tag{12.33}$$

As $I \to I_{max}$, $\eta \to -\infty$, but before this happens, some other ion will of course begin to be discharged. Since a typical value of δ is about 10^{-2} cm and $D \simeq 10^{-5}$ cm$^2 \cdot$s^{-1}, from (12.33) $I_{max}/\mathscr{A}$ is usually about $10^2\, c_0$ A$\cdot$cm^{-2} when c_0 is in mol$\cdot$cm^{-3}.

24. Diffusion in Absence of a Steady State—Polarography

There are many interesting situations in which rates of electrochemical reactions involve diffusional processes that depend upon the time (i.e., nonsteady states). Examples include initial transients at the beginning of a reaction, systems in which the electrolyte solution is not stirred, and systems in which the electrode surface is continuously renewed. These problems are of interest and often of practical importance in applied electrochemistry, but they do not introduce any new principles or concepts. Thus, we shall not discuss them here, except for a brief discussion of polarography, a technique that illustrates many of the perils of nonsteady states.

Let us consider the electrolysis of a solution containing several different cations, Cu^{2+}, Tl^+, Zn^{2+}, etc. There is a certain reversible potential at which each ion is discharged at the cathode. This potential depends on the standard potential of the electrode M^{z+}/M, and on the concentration of M^{z+} in the solution. At 25°C, the Nernst equation corresponding to the reduction of M^{z+} to M takes the form

$$E = E^0 + \frac{0.0592}{|z|} \log a(M^{z+})$$

Thus, a tenfold change in the ionic activity changes the discharge potential of the

---

*Since it leads to a polarization of the electrode associated with differences in the concentration of the electroactive species (e.g., Cu^{2+}) at the electrode surface and in the bulk solution, it was previously called the *concentration polarization*. Most electroanalytical chemists now call η_D the *concentration overpotential*.

ions by $0.0592/|z|$ V. A factor of 10^2 in activity corresponds to $0.1184/|z|$ V.

If we gradually increase the potential applied to the cell, the cation that is most easily reducible deposits first. As we continue to increase the applied potential, the current also increases. As the current rises, the concentration of the ion being discharged becomes more and more depleted in the neighborhood of the cathode, particularly if the solution is not stirred. This is the phenomenon of concentration polarization. Eventually, the limiting value of I_δ is reached, given by (12.33) for the case of a stationary electrode, and the I_δ vs. voltage curve becomes flat and remains so until the applied potential rises to a value at which the next most easily reducible cation can be discharged at the cathode. When this happens, the second ion begins to be discharged, even though there may still be an appreciable concentration of the first ion in the bulk of the solution. At the same time, the current increases again until a value of applied potential is reached at which a new limiting I_δ is attained (governed by the sum of the c_0 values for both reducible cations). With increasing E, this process may be repeated with a third ion, etc.

Is there any way to use such a sequence of discharges of ions to identify the ions and to measure their concentrations in the solution? In 1922, Jaroslav Heyrovsky of Prague devised an elegant method. In 1924, Heyrovsky and Shikata perfected an automatic instrument called the *polarograph* based on this method.

If we use the concentration polarization to differentiate reducible substances in a solution, we must have a cathode of tiny area, since otherwise the current through the cell would become impossibly high. We must prevent stirring of the solution and eliminate electrical migration on the part of the electroactive ion so that the current is governed solely by a diffusion-controlled process. Also, the cathode surface should be clean, reproducible, and, preferably, readily renewable. These conditions are met by the *dropping mercury electrode*, which provides a continual flow of droplets of mercury, about 0.5 mm in diameter. A pool of mercury located at the bottom of the cell can serve as a reference electrode and anode, which because of its large area is negligibly polarized. Alternatively, standard reference electrodes of large area can be used as reference anodes.

A schematic diagram of the polarograph is shown in Fig. 12.12. Since oxygen is more easily reduced than many other electroactive species of interest, it is desirable to remove dissolved oxygen from the electrolyte solution by bubbling an inert gas through it. A typical current vs. voltage curve is shown in Fig. 12.13 for a solution containing 10^{-4} M Cu^{2+}, Tl^+, and Zn^{2+} and 0.1 M KNO_3. The KNO_3 is added in large excess to carry essentially all the cell current. Although the current will be carried by the K^+ and the NO_3^- ions, the K^+ ion will not be discharged* at the cathode even though its concentration greatly exceeds that of the Cu^{2+}, Tl^+, or Zn^{2+} ion. The high overvoltage for the discharge of hydrogen ion on mercury is a great advantage in *polarography*, for it allows us to study many ions that are normally reduced with greater difficulty than H^+. Even Na^+ and K^+ may be discharged before H^+ at the dropping mercury electrode.

As each mercury drop grows and falls, the current oscillates between a maxi-

---

*K^+ *is* reduced at about -2.0 V *vs.* a standard calomel electrode.

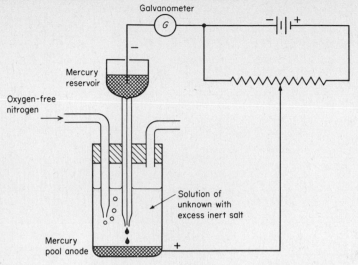

FIG. 12.12   Essential features of polarograph.

mum and a minimum. The overall rise from one flat portion of the curve to the next is called the *polarographic wave*. The *half-wave potential* (shown in Fig. 12.13) serves to identify the reducible ion. The value of the diffusion-limited current for each ion is proportional to the concentration of the ion.

The theoretical calculation of the diffusion current presents a difficult (if not impossible) problem, because it requires a solution of the diffusion equation with an unusual boundary condition, determined by the change in area of the growing drop. In 1938, Ilkovic gave an approximate solution*,

$$I = |z|\, F\mathscr{A} D \frac{c - c_0}{(\tfrac{3}{7}\pi D t)^{1/2}} \tag{12.34}$$

where $\mathscr{A}$ is the area of the mercury drop at time t after the beginning of its formation, D is the diffusion coefficient of the ion, and c, c_0 are the concentrations of that ion in the bulk of solution and at the surface of the mercury drop, respectively. If the surface of the growing mercury drop is assumed to be spherical, the area would be

$$\mathscr{A} = 4\pi r^2 = 4\pi \left(\frac{3 i_m t}{4\pi \rho} \right)^{2/3} \tag{12.35}$$

where i_m is the flow rate of mercury through the capillary tip (mass/time) and ρ is the density of mercury.

Whereas the Ilkovic equation describes the factors that control the height of the polarographic wave, i.e., the diffusion current, another equation due to Heyrovsky and Ilkovic describes the shape of the wave and its position along the potential axis. We can derive this equation from the Nernst equation for the reac-

---

*D. Ilkovic, *J. Chim. Phys.*, **35**, 129 (1938).

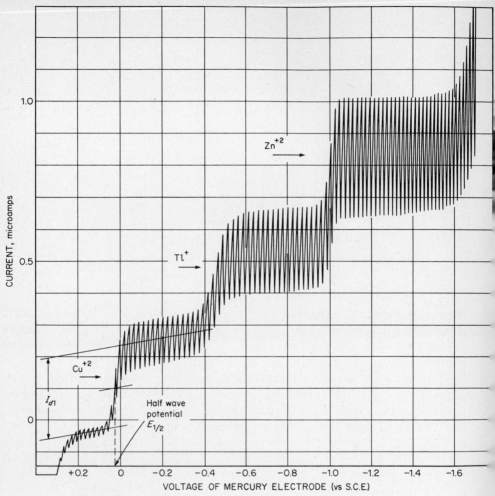

FIG. 12.13 Polarogram of an aqueous solution containing 10^{-4} M Cu^{2+}, Tl^+, and Zn^{2+} and 0.1 M KNO_3 as supporting electrolyte. Voltages are referred to the saturated calomel electrode [W. B. Schaap, Indiana University].

tion at the electrode surface and the law of diffusion.* For cathodic reduction of an ion at the dropping mercury electrode (d.m.e.),

$$E_{\text{d.m.e.}} \cong E^0 - \frac{RT}{|z|F} \ln \frac{I}{(I_\delta - I)} \qquad (12.36)$$

where I_δ is the mean limiting cathodic diffusion current (see Fig. 12.12), and $E^0 = E_{1/2}$ is the half-wave potential.

*J. Heyrovsky and J. Kuta, *Principles of Polarography* (Prague: Czechoslovak Academy of Science, 1965), p. 122ff.

25. Activation Overpotential

As the current density becomes high, the transport of ions to the electrode will become the rate-determining step in an electrode reaction. At lower current densities, some other one of the steps cited in Section 21 will become rate determining. These processes at lower current densities are especially interesting because they include the various types of electrode kinetics.

The way in which an applied electric field can activate reactant molecules is shown in Fig. 12.14, where the Gibbs free energy (free enthalpy) along a reaction coordi-

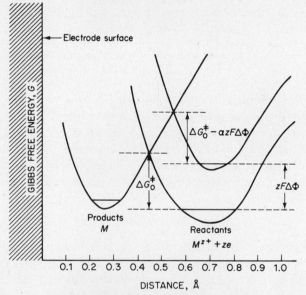

FIG. 12.14 Schematic Gibbs free-energy curves for the electrode reaction $M^{z+} + ze \longrightarrow M$ which show how the electric potential $\Delta\Phi$ lowers the free energy of activation $\Delta G_0^{\ddagger}$ by an amount $\alpha z F \Delta\Phi$.

nate is plotted schematically for an electrode reaction. The lower curve on the right depicts the free-enthalpy barrier for a purely thermal chemical reaction, (i.e., one in which any effect of electric field on the ions has been excluded). For an electrochemical reaction, with ionic reactants, the Galvani potential difference $\Delta\Phi$ across the double layer at the electrode surface will assist the transfer of an ion through the double layer in one direction but will inhibit its transfer in the opposite direction. We do not know, however, exactly how $\Delta\Phi$ varies through the double layer, so that when the activated complex is reached, at the maximum in the free enthalpy barrier, only some fraction α of the electrical energy difference $zF \Delta\Phi$ has been used. The fraction α is called the *transfer coefficient*. Generally, α will be in the neighborhood of 0.5, but its exact calculation would obviously be difficult; in fact, at present, this calculation is a major unsolved problem in electrochemical kinetics. Some experimental values of α are summarized in Table 12.7.

TABLE 12.7
Experimental Values of Transfer Coefficient α^a

Electrode metal	Reaction	α
Platinum	$Fe^{3+} + e \longrightarrow Fe^{2+}$	0.58
Platinum	$Ce^{4+} + e \longrightarrow Ce^{3+}$	0.75
Mercury	$Ti^{4+} + e \longrightarrow Ti^{3+}$	0.42
Mercury	$2H^+ + 2e \longrightarrow H_2$	0.50
Nickel	$2H^+ + 2e \longrightarrow H_2$	0.58
Silver	$Ag^+ + e \longrightarrow Ag$	0.55

[a]From J. O'M. Bockris and A. K. N. Reddy, *Modern Electrochemistry* (New York: Plenum Press, 1970).

At equilibrium, the currents of each ionic species to and from the electrode must be equal. Suppose, for example, that a copper electrode is dipping into a $CuSO_4$ solution. When the equilibrium electrode potential is established, the cathodic current due to the reduction of Cu^{2+} ions moving across the double layer toward the metal electrode is just equal to the anodic current arising from the oxidation of copper metal to Cu^{2+} ions, which leave the electrode and move across the double layer into the solution,

$$I(\text{cathodic}) = I(\text{anodic}) = I_0$$

where I_0 is the *exchange current*. There is no net current to or from the electrode at equilibrium. However, as the potential difference $\Delta\Phi$ departs from its equilibrium value, there will be either a net anodic or cathodic current I, which is, in turn, a measure of the rate of the electrode reaction. This net current I is the difference between the currents in cathodic and anodic directions,

$$I = I(\text{cathodic}) - I(\text{anodic})$$

This net current is related to an extra electric potential difference η_t, over and above the equilibrium value $\Delta\Phi$, which is called the *activation overpotential*,

$$\eta_t = \Delta\Phi - \Delta\Phi_{\text{rev}}$$

As Fig. 12.14 shows, a change from the equilibrium potential increases the current in one direction and decreases it in the other. By convention, a net reduction current is taken as positive and a net oxidation current as negative. Thus, α corresponds to the cathodic (reduction) process and $(1 - \alpha)$ to the anodic (oxidation) process.

From the general equations for rate processes given by the transition-state theory (Section 9.29), we can write the anodic and cathodic current densities as the rate of transfer of ions over the free-enthalpy barriers at the surface of the electrode.

Anodic $\quad i_a = zFk_a c_{0R} \exp\left[\dfrac{-\Delta G_a^{\ddagger} - (1-\alpha)zF\,\Delta\Phi}{RT}\right]$

Cathodic $\; i_c = zFk_c c_{00} \exp\left(\dfrac{-\Delta G_c^{\ddagger} + \alpha zF\,\Delta\Phi}{RT}\right)$

In these expressions, k_a and k_c are the pre-exponential parts of the rate constants for the forward (anodic) and reverse (cathodic) electron transfers, and c_{0R} and c_{00} represent the surface concentrations of reduced reactant and oxidized products of the electrochemical reaction, $O + ze^- \longrightarrow R$. The $\Delta G_a^{\ddagger}$ and $\Delta G_c^{\ddagger}$ are the thermal free-enthalpy barriers for the anodic and cathodic electrode reactions, respectively. In a regime in which diffusion overpotential is negligible, the ionic concentrations adjacent to the electrode can be taken as constant, independent of i and also of time (when the extent of reaction is small).

At equilibrium, we can thus write for the exchange current per unit area,

$$i_0 = zFk_a c_{0R} \exp -\left[\dfrac{\Delta G_a^{\ddagger} + (1-\alpha)zF\,\Delta\Phi_{\text{rev}}}{RT}\right]$$

$$= zFk_c c_{00} \exp -\left(\dfrac{\Delta G_c^{\ddagger} - \alpha zF\,\Delta\Phi_{\text{rev}}}{RT}\right)$$

In terms of i_0 and $\eta = \Delta\Phi - \Delta\Phi_{\text{rev}}$,

$$i = i_c - i_a = i_0\left[\exp\dfrac{\alpha zF\eta_t}{RT} - \exp\dfrac{(1-\alpha)zF\eta_t}{RT}\right] \qquad (12.37)$$

This is the important *Butler–Volmer equation*, which yields the dependence of current density on overpotential shown in Fig. 12.15.

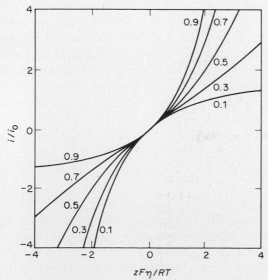

FIG. 12.15 Plot of i/i_0 against $zF\eta/RT$ for different values of transfer coefficient [H. Gerischer and W. Vielstich, Z. *Physik Chem.*, Frankfurt, *3*, 16 (1955)].

The slope of the curve at $\eta_t = 0$ gives the exchange current,

$$\left(\frac{\partial i}{\partial \eta}\right)_{\eta \to 0} = \frac{zF}{RT} i_0 \qquad (12.38)$$

where the i_0 does not depend on α.

If the overpotential has large positive or negative values, $|\eta| \gg (RT/zF)$, one of the partial currents becomes much greater than the other, which is then negligible. In this case, either

$$\ln i_a = \ln i_0 - \frac{(1 - \alpha)zF}{RT} \eta \qquad (12.39)$$

or

$$\ln i_c = \ln i_0 + \frac{\alpha zF}{RT} \eta \qquad (12.40)$$

This type of logarithmic dependence of i upon η was found empirically in 1905 by Tafel, and we can refer to (12.39) and (12.40) as *Tafel equations*. From the slope of the linear log i vs. η plots, the transfer coefficient α can be determined. However, a linear log i vs. η plot does not prove that η is an activation overpotential, since a similar form can occur with a diffusion overpotential.

We shall not give any more details of electrode kinetics here, but refer the reader to specialized treatises. We may mention, however, that overpotentials can be caused by slow reactions in the solution adjacent to the electrode (reaction overpotential η_R) and in the process of deposition of a solid product on the electrode (crystallization overpotential η_C). In many cases, a diffusion overpotential (concentration polarization) will occur together with an activation overpotential, and methods are available for separating these factors. Thus, a great variety of interesting chemical kinetics of both homogeneous and heterogeneous reactions can be studied by electrodic techniques. We can anticipate a major increase in the practical importance of this branch of kinetics as a result of developments in such fields as fuel cells, storage batteries for vehicle propulsion, and electrochemical synthesis (where great selectivity can be achieved through choice of the electrode and a carefully controlled potential).

26. Kinetics of Discharge of Hydrogen Ions

We shall consider one example of electrode kinetics in some detail, the discharge of a hydrogen ion and the resultant hydrogen overpotential. Ever since the pioneering work of Tafel, much effort has been devoted to the study of the discharge of H_3O^+ ions on metal electrodes. No comprehensive theoretical model yet exists which can explain the observed phenomena in all the different metals that have been studied. Some of the experimental overpotential vs. current density data are shown in Fig. 12.16.

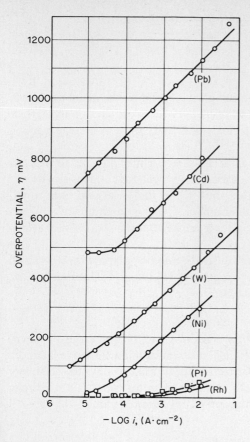

FIG. 12.16 Variation of overpotential for discharge of H⁺ on various metals with current density. The linear parts of the curves are in accord with the Tafel equation (12.40) [J. O'M. Bockris].

The overall reaction can be written

$$2H_3O^+ + 2e^- \longrightarrow H_2 + 2H_2O$$

In accord with our general analysis, the reaction can be broken down into the following successive stages:

1. transport of H_3O^+ ions to the phase boundary;
2. discharge of the H_3O^+ ion by one of the following mechanisms:
    (a) By formation of adsorbed H atoms on electrode surface sites not already covered (Volmer reaction)

    $$H_3O^+ + Me + e^- \rightleftarrows Me\text{-}H + H_2O$$

    where Me represents an electrode metal;
    (b) By reaction with H atoms already adsorbed on the electrode surface (Heyrovsky reaction)

    $$H_3O^+ + MeH + e^- \rightleftarrows Me + H_2 + H_2O$$

3. recombination of two adsorbed H atoms to yield H_2 (Tafel reaction)

    $$Me\text{-}H + Me\text{-}H \rightleftarrows 2Me + H_2$$

4. desorption of H_2 from the surface into the solution;
5. transport of H_2 molecules away from the electrode:
    (a) By diffusion;
    (b) By gas evolution.

Which one of these steps is rate determining will depend on the metal used, and, to a considerable extent, on the condition of its surface. Some sites on the surface may provide active centers having a high catalytic activity for a particular reaction step.

With platinized platinum electrodes in acid solutions, it is possible to detect a diffusion control of the H_3O^+ discharge at extremely high current densities (about $100 \text{ A} \cdot \text{cm}^{-2}$ at a concentration of $1.0 \text{ mol} \cdot \text{dm}^{-3}$), but under less extreme conditions we need not consider step 1 as rate determining. The demonstration of diffusion control at all is, nevertheless, important, since it proves that the species discharging is H_3O^+ and not H_2O ($2H_2O + 2e^- \rightarrow H_2 + 2OH^-$).

The other transport step (step 5) occurs at strong anodic polarization with metal electrodes that are catalytically active for splitting of H_2 into $2H$. On these metals, diffusion control takes over for the reaction,

$$H_2 + 2H_2O \longrightarrow 2H_3O^+ + 2e^-$$

The discharge of H_3O^+ ions provides a good example of the interesting problems that arise when we try to elucidate the detailed mechanism of an electrode reaction. Viewed from the standpoint of the most effective application of physical chemistry to technological advances, it is hard to escape the conclusion that electrode kinetics has been relatively underplayed as a field for basic research.

PROBLEMS

1. Devise a cell in which the reaction is $2AgBr + H_2 \rightarrow 2Ag + 2HBr$ (aq). Calculate $E^\ominus$ at 298 K for the cell from the standard electrode potentials. The solubility of AgBr is 2.10×10^{-6} M at 298 K.

2. The standard electrode potentials of $I^-|AgI(c)|Ag(c)$ and $I^-|I_2(c)$ are -0.152 and $+0.536$ V at 298 K, respectively. (a) Write the total reaction for a cell consisting of these two electrodes in the same solution of iodide ions. (b) What is the emf of the cell when the substances are all in their standard states? (c) The emf increases with T by $1.00 \times 10^{-4} \text{ V} \cdot \text{K}^{-1}$. Calculate $\Delta G^\ominus$, $\Delta S^\ominus$, and $\Delta H^\ominus$ for the cell reaction at 298 K.

3. Write the balanced cell reaction and calculate the emf at 298 K of the cell

$$\text{Pt}|\text{Sn}^{2+}(a = 0.100), \text{Sn}^{4+}(a = 0.0100)\,\|\,\text{Fe}^{3+}(a = 0.200)|\text{Fe}$$

The standard electrode potentials are in Table 12.1.

4. Calculate the emf of the following cell at 25°C:

$$\text{Ag}|\text{AgCl(c)}|\text{NaCl}(a_2 = 0.0100)|\text{NaCl}(a_1 = 0.0250)|\text{AgCl(c)}|\text{Ag}$$

The transference number of Na^+ is 0.390. The liquid junction potential (diffusion potential) in such a cell can be estimated from the equation

$$E_D = -\frac{RT}{F}\frac{\Lambda_+ - \Lambda_-}{\Lambda_+ + \Lambda_-}\ln\frac{a_\pm(2)}{a_\pm(1)}$$

where Λ_i are the ionic conductances (Table 10.4). Estimate E_D for the cell.

5. Precise data by Harned and Nims* for the cell $Ag|AgCl|NaCl(4\ m)|NaHg|NaCl$ $(0.1\ m)|AgCl|Ag$ are

θ (°C)	15	20	25	30	35
E (V)	0.16265	0.18663	0.19044	0.19407	0.19755

Calculate ΔH for the cell reaction at 298.15 K. If E can be measured to ± 0.0001 V, and θ to $\pm 0.010°C$, how precise is the value of ΔH?

6. The emf at 298 K of the concentration cell with transference,

$$H_2(1\ atm)|HCl(a = 0.0100)|HCl(a = 0.100)|H_2(1\ atm)$$

is 0.0190 V. What is the average transference number of the H^+ ion?

7. The emf of the cell, $H_2(1\ atm)|HCl(0.01\ m)|AgCl(c)|Ag$ is given by E(in volts) $= -0.096 + 1.90 \times 10^{-3}\ T - 3.041 \times 10^{-6}\ T^2$. Calculate ΔG, ΔS, ΔH, and ΔC_p for the cell reaction (state the reaction) at 298 K.

8. The standard electrode potentials for Fe^{2+}, $Fe^{3+}|Pt$ and Ce^{3+}, $Ce^{4+}|Pt$ are 0.771 and 1.61 V, respectively. Calculate the potential at the end point in the titration of 0.050 mol Fe^{2+} with Ce^{4+} solution. Assume total volume of solution at end point is 1 dm³ and that all activity coefficients are unity.

9. The measured voltage of the concentration cell $Ag|AgNO_3(0.0100\ m)\|AgNO_3(y)|Ag$ was 0.0650 V at 25°C. Calculate the molality y of the right-hand half-cell (a) assuming $\gamma_\pm = 1$ (b) estimating $\gamma_\pm$ from the Debye–Hückel equation.

10. For the cell $H_2(1\ atm)|HCl|AgCl|Ag$, $E^\ominus = 0.2220$ V at 298 K. If the measured $E = 0.396$ V, what is the pH of the HCl solution? Cite any approximations made.

11. From the values of $E^\ominus$ in Table 12.1 and $S_i^\ominus$ in Table 12.2, calculate the $E^\ominus$ for the Fe^{2+}, $Fe^{3+}|Pt$ electrode at 273 K.

12. The standard electrode potentials $E^\ominus$(298 K) for $Cu^{2+}|Cu$ and $Cu^+|Cu$ are 0.337 and 0.530 V, respectively. In general, is it easier to oxidize Cu to the +2 or +1 state? Can you suggest any reason for your answer based on the electronic structures of the ions and related factors? What is the equilibrium constant for the reaction $2Cu^+ \longrightarrow Cu^{2+} + Cu$ at 298 K? Describe what happens when a piece of cuprous oxide Cu_2O is dissolved in dilute H_2SO_4.

13. The emf's of the cell $H_2(1\ atm)|HCl(m)|AgCl|Ag$ at 25°C were measured by Harned and Ehlers.†

m (mol·kg⁻¹)	E (V)	m	E
0.01002	0.46376	0.05005	0.38568
0.01010	0.46331	0.09642	0.35393
0.01031	0.46228	0.09834	0.35316
0.04986	0.38582	0.20300	0.31774

*H. S. Harned and L. F. Nims, *J. Am. Chem. Soc. 54*, 423 (1932).
†*J. Am. Chem. Soc.*, *54*, 1350 (1932).

Calculate $E^\ominus$ by graphical extrapolation, based on $E = E^\ominus - A \log a(\text{HCl})$ where $A = 2.303RT/F$, and the Debye–Hückel expression $\log \gamma_\pm = -B\sqrt{m}$. Thus, $E + 2A \log m = E^\ominus + 2AB\sqrt{m}$.

14. From the values of $E^\ominus$ obtained in Problem 13 and the measured E's, calculate the mean activity coefficients $\gamma_\pm$ of HCl at $m = 0.010$, 0.050, 0.100, and 0.200 mol·kg^{-1}.

15. The emf of the cell $H_2(P)|0.1\ m\ \text{HCl}|\text{HgCl}|\text{Hg}$ was measured as a function of P at 298 K.

P (atm)	1.0	37.9	51.6	110.2	286.6	731.8	1035.2
E (mV)	399.0	445.6	449.6	459.6	473.4	489.3	497.5

Calculate the fugacity coefficients ($\gamma = f/P$) and plot them as a function of P over the range 1 to 1000 atm. Compare these experimental values with those estimated from Newton's graphs in Section 8.15.

16. Consider a membrane of a plant cell, which is permeable to Na$^+$, Cl$^-$, and H$_2$O but not to proteins. Suppose that initially there is a NaCl solution 0.050 mol·dm^{-3} on each side of the membrane, and a protein P at 0.001 mol·dm^{-3} concentration inside the cell, which ionizes to give P^{z+} with $z = 10$ plus 10 Cl$^-$ ions. Calculate the Donnan potential across the cell wall at equilibrium. (Approximate activities by concentrations.)

17. The circular tip of a microelectrode used for intracellular recording has an inside diameter of 0.36 μm and is 1 cm long. It is filled with 0.100 M KCl solution at 35°C. What is the maximum value of the capacitance in the circuit in parallel with this electrode if we wish the time constant of its response to intracellular potential changes to be less than 100 μs?

18. According to one theory of permeability of cell membranes,* the permeability of a membrane to an ion depends on the existence of pores which have a diameter closely matched to that of the ions. Suppose the diameters must match to within ± 0.01 nm. Calculate the probability that a Na$^+$ ion with a diameter of 0.72 nm can pass through a randomly chosen pore in the membrane of a frog muscle if the diameters of the pores have a normal Gaussian distribution with a mean of 0.78 nm and a standard deviation of 0.02 nm.

19. When a piece of metal is cold worked, it can store a considerable amount of energy as a consequence of strains and defects introduced by the mechanical stresses. Devise an electrochemical method to measure the Gibbs free energy per unit mass of copper that is stored in a copper rod after cold working. Give numerical estimates of the range of values of quantities measured.

20. Set up the solution of Eq. (12.26) for ξ by Newton's method, and hence calculate the Planck membrane potential from (12.27) for the following cases. On one side of the membrane we have 0.100 M NaCl solution and on the other side (a) 0.100 M KCl, (b) 0.200 M KCl. Assume that the diffusion coefficients of the ions are proportional to the mobilities in Table 10.5.

21. For the situations in Problem 20, calculate the membrane potentials from the Goldman equation (12.28). If the membrane is 0.10 nm thick and has an effective dielectric constant $\epsilon = 10$, would you prefer the Planck or the Goldman equation?

*L. J. Mullins, *J. Gen. Physiol.*, *43*(s), 105 (1960).

22. A small efficient battery for hearing aids is composed of zinc, potassium hydroxide, water, mercuric oxide, and mercury. (a) Write an overall equation for the cell reaction in which Zn and KOH are consumed, Hg is deposited, and $K_2Zn(OH)_4$ is formed. (b) Write the reactions occurring at the electrodes. (c) From literature data, compute $E^\ominus$ for the battery and estimate E when the KOH solution is 1.0 M. (d) If the battery is designed to operate continuously for 1000 h at a power of 5 mW, estimate the minimum mass of the unit exclusive of encapsulation.

23. The adult human brain operates at a power of about 25 W. Most of the power is used to operate "sodium pumps" in the membranes of the nerve cells, which maintain the internal ionic composition at about 15 mM Na^+, while the external composition is about 150 mM Na^+. Assuming that *all* the power is used for such pumps, and that they have an overall efficiency of 50%, calculate the total flux of Na^+ ions out of the nerve cells in the brain per second. Taking the figure of 10^{10} nerve cells in the brain, and a Na^+ uptake per nerve impulse of 10^{-11} mol Na^+ per cell, estimate the average firing rate of a brain cell. Correct for a steady Na^+ leakage of 10^{-11} mol·cm^{-2}·s^{-1}. (Why do we spend one-third of our lives unconscious? Your answer must at present be speculative.)

24. The solubility of iodine in water at 25°C is 1.33×10^{-3} mol·dm^{-3}. The electrode $I^-|I_2|Pt$ has $E^\ominus = 0.5355$ V and the electrode $I_3^-, I^-|Pt$, $E^\ominus = 0.5365$ V. What is the concentration of I_3^- in a saturated solution of iodine in water at 25°C when $[I^-]$ is 0.500 M?

25. A dropping mercury electrode is set up as in Fig. 12.11 with a capillary 0.15 mm in diameter and the drop rate adjusted to $i_m = 2.00$ mg·s^{-1}. The solution contains $TlNO_3$ at $c = 10^{-4}$ M and a 100-fold excess of KNO_3. Calculate the variation of diffusion current with time from the beginning of the drop until its fall. For Tl^+, $D = 2.0 \times 10^{-5}$ cm^2·s^{-1} at 25°C. Assume $c_0 = 0$.

26. One liter of a 1.0 molar $CdSO_4$ solution is electrolyzed between platinum electrodes of 50 cm^2 area at a constant current of 0.050 A. What fraction of the cadmium will be deposited before hydrogen evolution begins at the cathode? Take the overvoltage from Fig. 12.15.

27. From the data in Problem 25, calculate the average diffusion current from $t = 0$, the beginning of the mercury drop, to $t = t_1$, when the drop leaves the capillary.

28. The experimentally measured exchange current density for the electrode $Pt|Fe^{2+}$, Fe^{3+} is $i_0 = 0.50$ A·cm^{-2} at 298 K. The $\Delta H^\ddagger$ for i_0 is 36.5 kJ·mol^{-1}, and $\alpha = 0.58$. Calculate at (a) 298 K and (b) 323 K the relative current density i/i_0 as a function of overpotential η from -1.0 to $+1.0$ V.

13

Particles and Waves

*If therefore angels are not composed of matter and form, as was said above,
it follows that it would be impossible to have two angels of the same species
... The motion of an angel can be continuous or discontinuous as it wishes ...
And thus an angel can be at one instant in one place, and at another instant in
another place, not existing at any intermediate time.*

Thomas Aquinas
(1268)*

According to Karl Marx, "conflict is the engine of progress." The historical development of the physical sciences can be viewed as a dialectic between two opposing concepts of the ultimate basis of physical reality. On the one hand, there was the long tradition of atomism. Atomism sought to understand matter and its interactions in terms of fundamental particles, which were allowed to have only a few intrinsic properties—in particular, position, mass, velocity, charge, and spin. On the other hand, there were the powerful continuum or field theories. Field theories found their expression in partial differential equations that described the behavior of functions continuous in space and time, free of any particulate character. Examples were electric, magnetic, and gravitational field strengths.

Because their logical premises were not compatible, conflict between particle theory and field theory was inevitable. The mathematical concept of continuous space is not compatible with the concept of an elementary particle, for what is to prevent division of the particle into halves, division again, and so on, *ad infinitum*? This dilemma could be resolved by a decision that space itself is not infinitely divisible, that there exists a fundamental quantum of distance, and that time is also quantized into smallest elementary units. Such a resolution would represent a victory for the particle theory. Conversely, an advance in field theory could demonstrate that elementary particles are simply kinks, vortices, or other singularities in the continuous fields. Such an outcome would be a victory for field theory.

The contemporary situation is that a powerful mathematical technique has been developed which provides, at least temporarily and for a limited range of problems, a satisfactory synthesis of particle and field concepts. This technique is called *quantum mechanics*. A wide range of important physical phenomena, espe-

Summa Theologica, I. 50, 4.

cially those comprised by the theory of relativity and gravitation, stand rather desolately outside the realm of this synthesis, testimony to the incompleteness of quantum mechanics as a physical theory. Whatever may be the faults of quantum mechanics as a general theory of the universe, however, its success in the analysis of atomic and molecular phenomena has been impressive. In principle,* all of chemistry can be derived from quantum mechanics plus a few basic empirical properties of electrons, protons, and neutrons. In practice, such derivations are restricted to simple systems: as of now, atoms about as complex as oxygen or molecules as complex as water.

The success of quantum mechanics in resolving some of the computational problems of field–particle duality has given birth to a wide-ranging philosophical method called *complementarity*. Encouraged by the quantum synthesis of incompatible concepts, the philosophy of complementarity urges that such a synthesis of opposites provides a necessary and permanent tension in our world. The reconciliation of mercy with justice is seen to reflect the wave–particle duality of the electron. In the light of our short history as rational animals, it would seem unlikely that the permanent resolution of even the dialectic of field and particle has already been found.

1. Simple Harmonic Motion

Before we outline the development of quantum theory, we shall briefly review some elementary aspects of oscillatory and wave motions.

The vibration of a simple harmonic oscillator, discussed in Section 4.19, is an example of a motion periodic in time. The simplest model of such a system, shown in Fig. 13.1(a), is a mass m attached to a rigid support by a spring of force constant κ. The spring is assumed to have no mass itself, and to be perfectly elastic (i.e., no viscous or damping forces are present in the spring that would cause dissipation of its stored energy into heat).

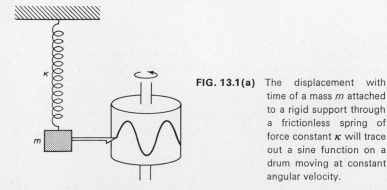

FIG. 13.1(a) The displacement with time of a mass m attached to a rigid support through a frictionless spring of force constant κ will trace out a sine function on a drum moving at constant angular velocity.

*From *en principe, oui,* a French expression meaning *non.*

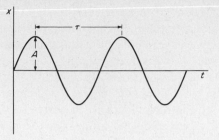

FIG. 13.1(b) Simple harmonic vibration, displacement x as a
function of time t.

The equation of motion, $F = ma$, becomes

$$m\frac{d^2x}{dt^2} = -\kappa x \tag{13.1}$$

This is a simple linear differential equation.* It can be solved by first making the substitution $v = dx/dt$. Then $d^2x/dt^2 = dv/dt = (dv/dx)(dx/dt) = v(dv/dx)$, and the equation becomes

$$v\left(\frac{dv}{dx}\right) + \left(\frac{\kappa}{m}\right)x = 0$$

Integration gives

$$v^2 + \left(\frac{\kappa}{m}\right)x^2 = C$$

The constant C can be evaluated from the fact that when the oscillator is at the extreme limit of its vibration, $x = A$, the kinetic energy is zero, and hence $v = 0$. Thus $C = (\kappa/m)A^2$, and

$$v^2 = \left(\frac{dx}{dt}\right)^2 = \frac{\kappa}{m}(A^2 - x^2)$$

$$\frac{dx}{dt} = \left[\frac{\kappa}{m}(A^2 - x^2)\right]^{1/2}$$

$$\frac{dx}{(A^2 - x^2)^{1/2}} = \left(\frac{\kappa}{m}\right)^{1/2} dt$$

$$\sin^{-1}\frac{x}{A} = \left(\frac{\kappa}{m}\right)^{1/2} t + C'$$

If the initial condition is $x = 0$ at $t = 0$, the integration constant $C' = 0$.

The solution of the equation of motion of the simple harmonic oscillator is, accordingly,

$$x = A \sin \left(\frac{\kappa}{m}\right)^{1/2} t \tag{13.2}$$

If we set

$$\left(\frac{\kappa}{m}\right)^{1/2} = 2\pi\nu \tag{13.3}$$

*See, for example, W. A. Granville, P. F. Smith, and W. R. Longley, *Elements of Calculus* (Boston: Ginn and Company, 1957), p. 379.

(13.2) becomes

$$x = A \sin 2\pi vt \tag{13.4}$$

The simple harmonic vibration can be represented graphically by this sine function, as shown in Fig. 13.1(b). The constant v is the *frequency* of motion, the number of vibrations in unit time. The reciprocal of frequency, $\tau = 1/v$, is the *period*, the time required for a single vibration. Whenever $t = n(\tau/2)$, where n is an integer, the displacement x passes through zero. Note that for constant κ the frequency depends inversely on the square root of the mass,

$$v = \frac{1}{2\pi}\sqrt{\frac{\kappa}{m}} \tag{13.5}$$

The quantity A, the maximum value of the displacement, is the *amplitude* of the vibration. At the position $x = A$, the oscillator reverses its direction of motion. At this point, therefore, the kinetic energy is zero, and all the energy is potential energy E_p. At position $x = 0$, all the energy is kinetic energy E_k. Since the total energy, $E = E_p + E_k$, is always constant, it must equal the potential energy at $x = A$. In Section 4.19, the potential energy of the oscillator was shown to be equal to $\frac{1}{2}\kappa x^2$, so that the total energy is

$$E = \frac{1}{2}\kappa A^2 \tag{13.6}$$

The total energy is proportional to the square of the amplitude. This relation holds true for all periodic motions.

2. Wave Motion

The motion described in Fig. 13.1 is an oscillatory motion, but not a wave motion. There is no propagation of energy along the spring.

A simple example of wave motion in one dimension can be provided by a displacement moving along a string, as illustrated in Fig. 13.2(a). The transverse displacement of the string at point x and time t can be represented by

$$u = f(x, t)$$

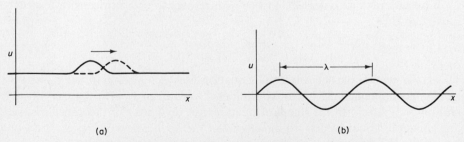

(a)   (b)

FIG. 13.2   (a) A wave consisting of a single pulse moving in the $+x$ direction. The solid line shows the pulse at a given time t and the dashed line indicates the pulse at some later time $t + \Delta t$. (b) Profile of a sine wave of wavelength λ.

Suppose we focus attention on some particular state of displacement, for example, the maximum displacement $u = A$. If this state travels along the positive x direction with a velocity v, at any time t

$$u = f(x, t) = f(x - vt) \tag{13.7}$$

Equation (13.7) is the mathematical definition of a *wave*.

Suppose that at time $t = 0$, the wave has the particular form

$$u = A \sin 2\pi\sigma x \tag{13.8}$$

as shown in Fig. 13.2(b). Such an instantaneous "snapshot" of the waveform is called the *profile* of the wave. If the sine wave is displaced a distance λ along the x axis, it will be exactly superimposed upon its original profile. The quantity λ is the *wavelength*. It is a measure of the periodicity of the wave in space. The profile of the sine wave has the form

$$u = A \sin 2\pi \frac{x}{\lambda} \tag{13.9}$$

and thus from (13.8) $\sigma = 1/\lambda$, which is called the *wave number*. The moving sine wave, from (13.7) and (13.9), becomes

$$u = A \sin \frac{2\pi}{\lambda}(x - vt) \tag{13.10}$$

The velocity v is called the *phase velocity* because it gives the velocity of propagation of a given phase of the wave. Now it is evident that $\lambda/v = \tau$, the period of the wave, or the time between successive crests at any given point x. Hence, the frequency

$$v = \frac{1}{\tau} = \frac{v}{\lambda}$$

and (13.10) can be written

$$u = A \sin 2\pi(\sigma x - vt) \tag{13.11}$$

The wave motion described by (13.11) is called a *plane wave*, since the displacement is constant in any plane normal to the direction of propagation. In other words, u is a function only of one spacial coordinate, x.

If we differentiate u in (13.11) twice with respect to t and twice with respect to x, we find that

$$\frac{\partial^2 u}{\partial x^2} = \frac{1}{v^2}\frac{\partial^2 u}{\partial t^2} \tag{13.12}$$

This is the general partial differential equation for wave motion in one dimension.

We have obtained (13.12) by working backward from a particular case of wave motion on a string. In the standard treatment in textbooks of theoretical physics, this equation would be derived directly from the application of Newton's Second Law to an element of string with mass μ per unit length under a tension γ.*

---

*For example, J. C. Slater and N. H. Frank, *Mechanics* (New York: McGraw-Hill Book Company, 1947), p. 146.

It can be shown by direct substitution that Eq. (13.12) is satisfied by any function of the form (13.7). It is also satisfied, however, by any function of the form

$$u = g(x + vt) \tag{13.13}$$

Just as (13.7) represents a wave traveling in the positive x direction, so (13.13) represents a wave traveling in the negative x direction. The general solution of (13.12) is, therefore,

$$u = f(x - vt) + g(x + vt) \tag{13.14}$$

where f and g are arbitrary functions.*

3. Standing Waves

So far we have tacitly assumed that the length of the string is infinite, so that waves can travel freely in either the positive or negative direction. Let us suppose now that the string has a definite length L, extending from $x = 0$ to $x = L$. What can we say about wave motion on this finite length of string?

It is evident that neither (13.7) nor (13.13) will be satisfactory solutions to the wave equation, since they represent traveling waves that cannot satisfy the boundary conditions, which would require the displacement to halt abruptly at $x = 0, L$. We thus have what is known as a *boundary value problem*. It is quite a common situation in mathematical physics to obtain a solution to a differential equation, and then to be required to make this solution fit a specified set of boundary conditions. In many cases the solution will contain some parameter, the values of which must be selected in order to meet the boundary conditions. These selected values are called the *eigenvalues* or *characteristic values* of the problem. The solutions that correspond to the eigenvalues are called the *eigenfunctions* or *characteristic functions*.

Instead of the solution in terms of arbitrary functions, we can solve the wave equation (13.12) in another way, which will be especially useful when we consider quantum mechanical problems. Since (13.12) is a linear differential equation with constant coefficients, we can *separate the variables* so that the solution can be written

$$u(x, t) = X(x)T(t) \tag{13.15}$$

That is, the displacement u is a product of a function of x alone and a function of t alone. From (13.15),

$$\frac{\partial^2 u}{\partial x^2} = T(t)\frac{\partial^2 X}{\partial x^2}$$

$$\frac{\partial^2 u}{\partial t^2} = X(x)\frac{\partial^2 T}{\partial t^2}$$

---

*The general solution of a second-order partial differential equation will always contain two arbitrary functions, just as the general solution of a second-order ordinary differential equation will always contain two arbitrary constants.

so that from (13.12),

$$\frac{1}{X}\frac{\partial^2 X}{\partial x^2} = \frac{1}{v^2 T}\frac{\partial^2 T}{\partial t^2}$$

The only way in which the left side of this equation, which depends only on x, can always equal the right side, which depends only on t, is if both sides are equal to the same constant, which we shall call $-\omega^2/v^2$. This *separation constant* is so far undetermined. Thus, we have

$$\frac{1}{X}\frac{d^2 X}{dx^2} = \frac{1}{v^2 T}\frac{d^2 T}{dt^2} = -\frac{\omega^2}{v^2} \tag{13.16}$$

Equations (13.16) are equivalent to two ordinary [not partial] differential equations, so that we have indeed "separated the variables" by the substitution (13.15), to obtain:

$$\frac{d^2 T}{dt^2} = -\omega^2 T, \qquad \frac{d^2 X}{dx^2} = -\frac{\omega^2}{v^2} X \tag{13.17}$$

Solutions of (13.17), by inspection or by reference to Section 13.1 where a similar equation was solved,* are

$$T = e^{\pm i\omega t}, \qquad X = e^{\pm i\omega x/v} \tag{13.18}$$

A solution to (13.12) is thus

$$u = T(t)X(x) = e^{\pm i\omega t}e^{\pm i\omega x/v} \tag{13.19}$$

We can use any of the four possible combinations of signs, and can multiply (13.19) by any arbitrary complex constant $Ae^{i\delta}$ to secure a solution of any chosen amplitude and phase. We recognize, of course, that (13.19) represents a wave motion, with

$$v = \frac{\omega}{2\pi}\lambda$$

$$\omega = \frac{2\pi}{\lambda}v = 2\pi\nu$$

so that ω is the angular frequency, usually measured in radians per second. We can also use the solution in the form of real sine and cosine functions, writing (13.19) as

$$u = \frac{\sin}{\cos}\,\omega\!\left(t \pm \frac{x}{v}\right) \quad \text{or} \quad \frac{\sin}{\cos}\,\omega t\,\frac{\sin\omega x}{\cos v} \tag{13.20}$$

We can readily fit this solution to the boundary condition

$$u = 0 \quad \text{at} \quad x = 0, \qquad x = L \quad \text{for all} \quad t \geq 0$$

For u to vanish at $x = 0$, the cosine function of x must be deleted, giving

$$u = \sin\frac{\omega x}{v}\frac{\sin}{\cos}\,\omega t$$

*The reader should verify the solutions by substitution into (13.17).

In order that $u = 0$ at $x = L$, we must have

$$\sin \frac{\omega L}{v} = 0 \quad \text{or} \quad \frac{\omega L}{v} = n\pi$$

where n is an integer. This condition restricts the allowed values of ω. Hence, (13.20) becomes

$$u_n = \frac{\sin}{\cos} \omega_n t \sin \frac{n\pi x}{L} \tag{13.21}$$

The expression in (13.21) means that we can use either the sine or the cosine as the function of t, or any desired combination of the form

$$u_n = (A_n \sin \omega_n t + B_n \cos \omega_n t) \sin \frac{n\pi x}{L} \tag{13.22}$$

The solution in (13.22) clearly represents what is called a *standing wave* or *stationary wave*. The function of x always has the same form, no matter what the value of t may be. Thus, where $\sin (n\pi x/L) = 0$, there is a node in the wave at all times, and where $\sin (n\pi x/L) = 1$, the wave always has its maximum amplitude, although the value of this maximum amplitude oscillates with time in accord with the function indicated in (13.22). Strictly speaking, a standing wave is an oscillatory motion, but not a wave motion since there is no propagation of energy along the string, simply an interchange between kinetic and potential energy similar to that occurring in the harmonic oscillator shown in Fig. 13.1.

Some standing waves on the string of length L are shown in Fig. 13.3. All the standing waves must satisfy the condition

$$n\frac{\lambda}{2} = L \tag{13.23}$$

This is the simplest form of the *eigenvalue condition*, which results from the boundary value problem for a vibrating string. The functions u_n in (13.22) are then the

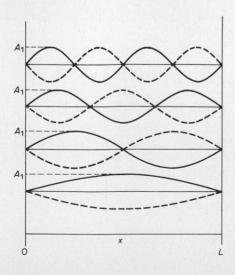

FIG. 13.3 Examples of allowed standing waves on a string of length L. The extremes of the oscillations are shown by solid and dashed lines. All the allowed waves conform to the condition $n\lambda/2 = L$ where n is an integer.

eigenfunctions for this problem. As the wavelength decreases, the allowed standing waves on a string of length L become more numerous. Thus, the possible high-frequency oscillations greatly outnumber the low-frequency ones. From (13.23), standing waves can occur on a string of length L only for certain values of the wavelength given by

$$\lambda = \frac{2L}{n}$$

If we consider a range in which L is much greater than λ, $(L \gg \lambda)$, we can approximate the set of integers n by a continuous function $n(\lambda)$. Thus, the number of standing waves that can occur in any range of wavelengths from λ to $\lambda + d\lambda$ will be

$$dn = -\frac{2L}{\lambda^2}\, d\lambda \tag{13.24}$$

The negative sign indicates that the number in a given range decreases as λ increases.

The corresponding problem in three dimensions is concerned with the number of standing waves in an enclosure of volume V. The result* is

$$dn = -\frac{4\pi V}{\lambda^4}\, d\lambda \tag{13.25}$$

We may note that (13.22) is not the most general solution to the equation (13.12), since any linear combination of the solutions in (13.22) would also be a solution. This result is an example of the *principle of superposition*. The general solution can thus be written

$$u = \sum_{n=1}^{\infty} (A_n \cos \omega_n t + B_n \sin \omega_n t) \sin \frac{n\pi x}{L} \tag{13.26}$$

The arbitrary constants of this general solution can be determined by fitting to (13.26) the initial conditions of the problem, $u(x, 0)$ and $\dot{u}(x, 0)$.

Since any arbitrary function of period $2L$ can be expressed as a Fourier series of the form (13.26), the solution in (13.14), given in terms of arbitrary functions, is equivalent to the one in (13.26), obtained by separating the partial differential equation into two ordinary differential equations.

4. Interference and Diffraction

The familiar construction of Huygens depicts the interference of light waves. Consider, for example, in Fig. 13.4, an effectively plane wave front from a single source, incident upon a set of slits, a prototype of the diffraction grating. Each slit can now be regarded as a new light source from which spreads a semicircular wave (or hemispherical, in the three-dimensional case). If the wavelength of the radiation is λ, a series of concentric semicircles of radii λ, 2λ, 3λ may be drawn with these sources as centers. Points on these circles represent the consecutive

---

*A complete derivation can be found in *Heat and Thermodynamics*, 5th ed., by J. K. Roberts and A. R. Miller (New York: Interscience Publishers, 1960), p. 526.

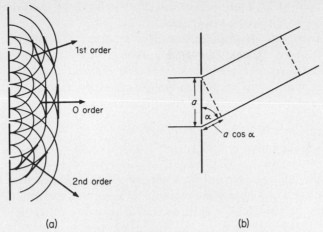

(a)   (b)

FIG. 13.4   Diffraction: (a) Huygens construction; (b) path difference.

maxima in amplitude of the new wavelets. Now, following Huygens, the new resultant wave fronts are the curves or surfaces that are simultaneously tangent to the secondary wavelets. These are called the *envelopes* of the wavelet curves and are shown in the illustration.

An important result of this construction is that there are a number of possible envelopes. The one that moves straight ahead in the same direction as the original incident light is called the *zero-order beam*. On either side of this are *first-*, *second-*, *third-*, *etc.*, *order diffracted beams*. The angles by which the diffracted beams deviate from the original direction evidently depend on the wavelength of the incident radiation. The longer the wavelength, the greater is the diffraction.

We can derive the condition for formation of a diffracted beam from Fig. 13.4, where attention is focused on two adjacent slits. If the two diffracted rays are to reinforce each other, they must be in phase, otherwise the resultant amplitude is cut down by interference. The condition for reinforcement is therefore that the difference in path for the two rays must be an integral number of wavelengths. If α is the angle of diffraction and a the separation of the slits, this path difference is $a \cos \alpha$ and the condition becomes

$$a \cos \alpha = h\lambda \tag{13.27}$$

where h is an integer.

This equation applies to a linear set of slits. For a two-dimensional plane grating, there are two similar equations to be satisfied. For the case of light incident normal to the grating,

$$a \cos \alpha = h\lambda$$
$$b \cos \beta = k\lambda$$

It will be noted that appreciable energy occurs in higher-order diffracted beams only when the spacings of the grating are not much larger than the wavelength of the incident light. To obtain diffraction effects with X rays, for example, the spacings should be of the order of 0.1 nm.*

Max von Laue, in 1912, realized that the interatomic spacings in crystals were probably of the order of magnitude of the wavelengths of X rays. Crystal structures should therefore serve as three-dimensional diffraction gratings for X rays. This prediction was immediately verified in the critical experiment of Friedrich, Knipping, and Laue. An example of an X-ray diffraction picture is shown in Fig. 18.16.

5. Black-Body Radiation

The first definite failure of the old wave theory of light was found in the study of *black-body radiation*. All objects are continually absorbing and emitting radiation. Their properties as absorbers or emitters may be extremely diverse. Thus, a pane of window glass does not absorb much visible light but absorbs most of the ultraviolet. A sheet of metal absorbs both the visible and ultraviolet but may be reasonably transparent to X rays.

For a body to be in equilibrium with its environment, the radiation it is emitting must be equivalent (in wavelength and energy) to the radiation it is absorbing. It is possible to conceive of objects that are perfect absorbers of radiation, the so-called *ideal black-bodies*. Actually, no substances approach very closely this ideal over an extended range of wavelengths. The best laboratory black-body is not a substance at all, but a cavity. This cavity is constructed with insulating walls, in one of which a small orifice is made. When the cavity is heated, the radiation from the orifice is a good sample of the equilibrium radiation within the heated enclosure, which is practically ideal black-body radiation.

There is an analogy between the behavior of radiation within such a cavity and that of gas molecules in a box. Both molecules and radiation are characterized by a density and both exert pressure on the confining walls. One difference is that gas density is a function of volume and temperature, whereas radiation density is a function of temperature alone. Analogous to the velocities distributed among the gas molecules are the frequencies distributed among the oscillations that comprise the radiation. At any given temperature, there is a characteristic distribution of the gas velocities given by Maxwell's equation. The corresponding problem of the spectral distribution of black-body radiation—i.e., the fraction of the total energy radiated within each range of wavelength—was first explored experimentally (1877–1900) by O. Lummer and E. Pringsheim. Some of their results are shown in Fig. 13.5. At high temperatures, the position of the maximum is shifted to shorter wavelengths—an iron rod glows first dull red, then orange, then white as its temperature is raised and higher frequencies become appreciable in the radiation. These curves have a resemblance to those of the Maxwell distribution law. When these data of Lummer and Pringsheim appeared, attempts were made to explain

---

*It is also possible to use larger spacings and work with extremely small angles of incidence. The complete equation, corresponding to Eq. (13.27) for incidence at an angle α_0, is $a(\cos \alpha - \cos \alpha_0) = h\lambda$.

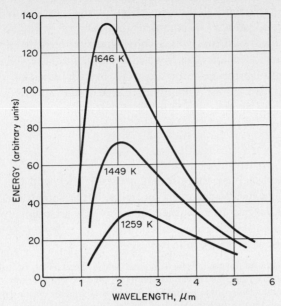

FIG. 13.5 Experimental measurements of Lummer and Pringsheim on spectral distribution of radiation from a black body at three different temperatures.

them theoretically by arguments based on the wave theory of light and the Principle of Equipartition of Energy.

Without going into the details of these efforts, which were all unsuccessful, it is possible to see why they were bound to fail. According to the Principle of Equipartition of Energy, an oscillator in thermal equilibrium with its environment should have an average energy equal to kT, $\frac{1}{2}kT$ for its kinetic energy and $\frac{1}{2}kT$ for its potential energy, where k is the Boltzmann constant. This classical theory states that the average energy depends in no way on the *frequency* of the oscillator. In a system containing 100 oscillators, 20 with a frequency v_1 of 10^{10} Hz and 80 with $v_2 = 10^{14}$ Hz, the Equipartition Principle predicts that 20% of the energy shall be in the low-frequency oscillators and 80% in the high-frequency oscillators. The radiation within a cavity can be considered to consist of standing waves of various frequencies. According to classical theory, the problem of the energy distribution over the various frequencies (intensity I vs. v) reduces to determination of the number of allowed vibrations in any range of frequencies.

From (13.25), since for electromagnetic waves $v = c/\lambda$, where c is the phase velocity,

$$dn = \frac{4\pi V}{c^3} v^2 \, dv \tag{13.28}$$

Since there are many more permissible high frequencies than low frequencies, and since by the Equipartition Principle all frequencies have the same average energy, it follows that the intensity I of black-body radiation should rise continuously with increasing frequency. This conclusion follows inescapably from classical Newtonian mechanics, yet it is in complete disagreement with the experimental data of Lummer and Pringsheim, which show that the intensity of the radiation rises to a maximum and then falls steeply with increasing frequency. This breakdown of classical mechanical principles when applied to radiation was regarded with dismay by the physicists of the time. They called it the *ultraviolet catastrophe*.

6. The Quantum of Energy

Max Planck was born in 1858, the son of a professor of law at the University of Kiel. At first he was inclined to become a musician, but he soon turned to the study of physics, although he had been warned that this was a closed subject in which new discoveries of any importance could scarcely be expected. In his *Scientific Autobiography*,* Planck wrote as follows:

> My original decision to devote myself to science was a direct result of a discovery which since my early youth has never ceased to fill me with enthusiasm—the comprehension of the far-from-obvious fact that the laws of human reasoning coincide with the laws that govern the sequences of impressions we receive from the world about us, and that therefore pure reasoning can enable man to gain an insight into the mechanism of the outside world. In this connection it is important to realize that the outside world is something independent from man, something absolute. The quest for the laws that apply to this absolute appeared to me to be the most sublime scientific pursuit in life.

In 1892 Planck became professor of physics at Berlin and soon took up the problem of the interpretation of the measurements of Lummer and Pringsheim. On October 19, 1900, he was able to announce to the Berlin Physical Society the mathematical form of the law that governs the energy distribution.† To obtain this law, he needed to introduce a new physical constant h, whose meaning, however, he could not deduce from his thermodynamic theories. Therefore, he went back to the atomic theory to discover how to interpret this constant and to obtain a physical picture that would lead to his energy law.

Classical mechanics was founded on the ancient maxim, *natura non facit saltum* (nature does not make a jump). Thus an oscillator was expected to take up energy *continuously* in arbitrarily small increments. Although matter was believed to be atomic, energy was assumed to be strictly continuous. Planck discarded this precept and suggested that an oscillator could acquire energy only in discrete units, called *quanta*. The quantum theory began therefore as an *atomic theory of energy*. Planck introduced the postulate that the magnitude of the quantum of energy ϵ was not fixed, but depended on the frequency v of the oscillator:

$$\epsilon = hv \tag{13.29}$$

Planck's constant h has the dimensions of energy times time, a quantity called

---

*M. Planck, *A Scientific Autobiography* (London: Williams and Norgate, Ltd., 1950).
†*Verhandl. Deut. Ges. Phys.*, 2, 237 (1900).

action. (Angular momentum also has the same dimensions.) In the SI system,
$h = (6.6262 \pm 0.0001) \times 10^{-34}$ J·s.

7. The Planck Distribution Law

Consider a collection of N oscillators having a fundamental vibration frequency
v. If these can take up energy only in increments hv, the allowed energies are 0,
hv, $2hv$, $3hv$, etc. Now, according to the Boltzmann formula, if N_0 is the number of
systems in the lowest energy state, the number having an energy ϵ_i above this
state is

$$N_i = N_0 \, e^{-\epsilon_i/kT} \tag{13.30}$$

In the collection of oscillators, for example,

$$N_1 = N_0 \, e^{-hv/kT}$$
$$N_2 = N_0 \, e^{-2hv/kT}$$
$$N_3 = N_0 \, e^{-3hv/kT}$$

The total number of oscillators in all energy states is, therefore,

$$N = N_0 + N_0 \, e^{-hv/kT} + N_0 \, e^{-2hv/kT} + \cdots = N_0 \sum_{i=0}^{\infty} e^{-ihv/kT}$$

The total energy E of all the oscillators equals the energy of each state times the
number in that state:

$$E = 0N_0 + hvN_0 \, e^{-hv/kT} + 2hvN_0 \, e^{-2hv/kT} + \cdots = N_0 \sum_{i=0}^{\infty} ihv \, e^{-ihv/kT}$$

The average energy of an oscillator is, therefore,

$$\bar{\epsilon} = \frac{E}{N} = \frac{hv \sum i \, e^{-ihv/kT}}{\sum e^{-ihv/kT}}$$

Or, if $x = hv/kT$,

$$\bar{\epsilon} = hv \frac{\sum i \, e^{-ix}}{\sum e^{-ix}} = \frac{hv}{e^x - 1} = \frac{hv}{e^{hv/kT} - 1} \tag{13.31}*$$

According to this expression, the mean energy of an oscillator with fundamental
frequency v approaches the classical value kT when hv becomes much less than
kT.† Using this equation in place of the classical Equipartition of Energy, Planck
derived an energy-distribution formula in excellent agreement with the experimental
data for black-body radiation. The energy density $E(v) \, dv$ is simply the number of
oscillators per unit volume between v and $v + dv$ [from Eq. (13.28)‡] times the

---

*In Eq. (13.31) let $e^{-x} = y$, then the denominator $\sum y^i = 1 + y + y^2 + \cdots = 1/(1 - y)$,
$(y < 1)$. The numerator, $\sum iy^i = y(1 + 2y + 3y^2 + \cdots) = y/(1 - y)^2$, $(y < 1)$ so that Eq.
(13.31) becomes $hvy/(1 - y) = hv/(e^{hv/kT} - 1)$.
†When $hv \ll kT$, $e^{hv/kT} \approx 1 + (hv/kT)$.
‡For electromagnetic radiation, there is an extra factor of two in (13.28), because both electric
and magnetic fields are oscillating.

average energy of an oscillator. Hence, Planck's Law is

$$E(v)\,dv = \frac{8\pi h v^3}{c^3}\frac{dv}{e^{hv/kT} - 1} \tag{13.32}$$

This equation gave agreement with the experimental data of Fig. 13.5. Quantum theory had achieved its first great success.

8. Photoelectric Effect

In 1887, Heinrich Hertz observed that a spark would jump a gap more readily when the gap electrodes were illuminated by light from another spark gap than when they were kept in the dark. Work during the next 15 years established that this phenomenon was due to the emission of electrons from the surfaces of solids upon incidence of light having suitable wavelengths. An apparatus for study of the photoelectric effect is shown in Fig. 13.6. By variation of the potential Φ applied

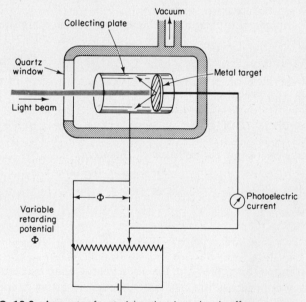

FIG. 13.6 Apparatus for studying the photoelectric effect.

to the collecting plate, it is possible to repel photoelectrons having a kinetic energy less than $\frac{1}{2}mv^2 = e\Phi$, and thus to determine the maximum kinetic energy of the photoelectrons. Lenard (1902) found that the maximum kinetic energy depended on the *frequency* v of the incident light, and below a certain frequency, v_0, no electrons were ejected.

In 1905 Albert Einstein, then working as an examiner in the Patent Office at Bern, in his spare time devised a theory for the photoelectric effect. Planck had assumed that the oscillators comprising a black-body radiator were quantized, but Einstein extended this concept to suggest that radiation itself was quantized. Thus, the light incident on the metal probe in Fig. 13.6 was composed of quanta of energy hv. When they struck the surface of the metal, some of the energy was used to overcome the attractive potential Υ which held the electron in the metal, and the remaining energy was converted into the kinetic energy of the ejected electrons. Thus, according to Einstein,

$$hv = \tfrac{1}{2}mv^2 + e\Upsilon \tag{13.33}$$

or

$$\tfrac{1}{2}mv^2 = hv - e\Upsilon = h(v - v_0)$$

The Einstein equation for the photoelectric effect was confirmed by many careful measurements, which provide a convenient way to evaluate the Planck constant h. This work was a major support of the quantum theory in its early days.

9. Spectroscopy

In 1663, Isaac Newton, at the age of 20, began his experimental work in optics by grinding his own lenses for the construction of a telescope. He was soon concerned with the problem of reducing chromatic aberration. In 1666, he bought a glass prism, "to try therewith the phenomena of the colours."

> Having darkened my chamber, and made a small hole in the window shuts, to let in a convenient quantity of the sun's light, I placed my prism at his entrance, that it might thereby be refracted to the opposite wall.

He found that the length of the colored area was much greater than its breadth, and from further experiments he deduced that white light was split by the prism into what he called a *spectrum* of the colors, as a result of the different *refrangibility* (refractive index) of the glass for different colors of light. The "phenomenon of the colors" had been known for a long time, but Newton was the first to interpret it correctly.

These experiments were the beginnings of spectroscopy, but little progress was made for almost a century after Newton. In 1777, Carl Scheele studied the darkening of silver chloride by light of different colors and found that the effect was greatest toward the violet end of the spectrum. In 1803, Inglefield suggested that there might be invisible rays beyond the violet, which could also darken the salt; the existence of such *ultraviolet rays* was demonstrated by Ritter and Wollaston. In 1800, William Herschel had discovered *infrared rays* beyond the red by their heating effect on a thermometer bulb.

A great advance in the experimental methods for studying spectra was made by Josef Fraunhofer. Fraunhofer was born in 1787 in Straubing, a village near Munich. His father was a glazier. The family was large and poor and Josef received practically no formal education, being apprenticed at the age of 11 to a mirror-maker in Munich. It was soon evident that young Fraunhofer was an experimental genius. At 20, he was optical foreman, at 22 a director of the company, and at 24 he had complete charge of the glass making. In 1814, he was faced with the problem of defining more accurately the colors of the light used in measuring the properties of various glasses. He was thus led to make a detailed examination of the spectrum of sunlight. The spectroscope as used by Newton consisted merely of a slit, a prism, and a lens which was placed between the slit and the prism to focus the slit image on a screen. Fraunhofer had the excellent idea of viewing the slit with a theodolite telescope placed behind the prism, and this instrument enabled him to make exact measurements of the angles. When he viewed the spectrum from sunlight, he found that it was crossed with *an almost countless number of strong and weak vertical lines.* He proved that these dark lines were really in the sunlight and measured about 700 of them. These lines provided the first precise standards for measuring the dispersion of optical glasses, and this work of Fraunhofer may be said to mark the beginning of spectroscopy as an exact science.

Fraunhofer also discovered the transmission grating, and in 1823 he constructed the first glass grating by ruling lines with a diamond point. Thus he was able to measure the exact wavelength of the lines, whereas previously only the angles had been used.

Of course, Fraunhofer had no idea of the origin of the black lines in the solar spectrum, and their explanation was not worked out until about 35 years later. Meanwhile, many workers had studied the spectra from flames and had noticed the characteristic bright lines produced by addition of various salts. In 1848, Foucault observed that a sodium flame that emitted a D line would also absorb this same line from the light of a sodium arc placed behind it. It remained for Kirchhoff to state the general laws connecting the emission and absorption of light. He was at that time professor of physics at Heidelberg, and in 1859 he announced his famous law, which states, "The relation between the power of emission and the power of absorption for rays of the same wavelength is constant for all bodies at the same temperature." If a body absorbs light of a certain wavelength, it also emits this same wavelength. Thus, the Fraunhofer lines are due to the absorption of certain wavelengths by the sun's atmosphere. One can discover what elements exist in the sun if one can find what elements on earth give bright lines of emission at the same wavelengths as the dark Fraunhofer absorption lines in the solar spectrum.

At that time, the professor of chemistry at Heidelberg was Robert Bunsen, who joined Kirchhoff in an extensive series of researches on the spectra of the elements. In 1861, while studying the alkali metals, Li, Na, K, they observed certain new lines which they traced to two new alkali metals, Cs and Rb. Ever since, spectroscopic identification has been the best proof for the existence of a new element.

In 1868, a monumental work on the solar spectrum was published by A. J. Ångstrom of Uppsala. He gave the wavelengths of about 1000 Fraunhofer lines to six significant figures in units of 10^{-10}m. This unit was subsequently called the *Ångstrom unit* in his honor.

In 1885, J. J. Balmer discovered a regular relationship between the frequencies of the atomic hydrogen lines in the visible region of the spectrum. The wave numbers σ are given by

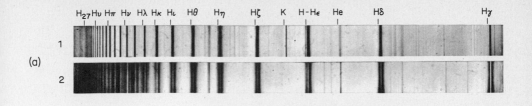

(a)

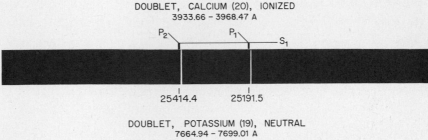

DOUBLET, CALCIUM (20), IONIZED
3933.66 – 3968.47 A

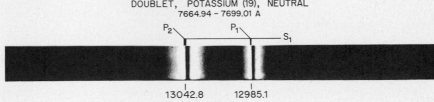

(b)

DOUBLET, POTASSIUM (19), NEUTRAL
7664.94 – 7699.01 A

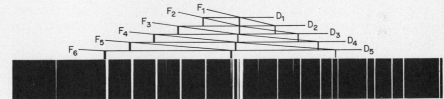

SEXTET, VANADIUM (23), NEUTRAL
4379.24 – 4429.80 A

(c)

FIG. 13.7 Examples of atomic spectra. [Charlotte Sitterly, National Bureau of Standards.] (a) Balmer series of hydrogen as observed in spectra of two stars. No. 1 is the spectrum from *Zeta Tauri*. Note successive hydrogen lines converging to a series limit. No. 2 is the spectrum from *11 Camelopardalis*. Note the marked self absorption of the hydrogen lines by atomic hydrogen in the stellar atmosphere. [From observations made at the University of Michigan Observatory.] (b) The doublets of potassium (K) and singly ionized calcium (Ca+) at high resolution. These lines arise from transitions between the ground state $^2S_{1/2}$ and the two states $^2P_{3/2}$ and $^2P_{1/2}$ as explained in Section 14.20. [From the collection of W. F. Meggers.] (c) A section of the atomic spectrum of vanadium, showing lines arising from combinations between D and F states. [From the collection of W. F. Meggers.]

$$\sigma = \mathscr{R}\left(\frac{1}{2^2} - \frac{1}{n_1^2}\right)$$

with $n_1 = 3, 4, 5, \ldots$, etc. The bright red H_α line at $\lambda = 656.28$ nm corresponds to $n_1 = 3$, the blue H_β line at 486.13 nm, to $n_1 = 4$, etc. The constant $\mathscr{R}$ is called the *Rydberg constant* and has the value 109 677.581 cm^{-1}. It is one of the most accurately known physical constants.

Other hydrogen series were discovered later which obeyed the more general formula

$$\sigma = \mathscr{R}\left(\frac{1}{n_2^2} - \frac{1}{n_1^2}\right) \tag{13.34}$$

Lyman found the series with $n_2 = 1$ in the far ultraviolet, and others were found in the infrared. Many similar series have been observed in the atomic spectra of other elements. Some examples of atomic spectra are shown in Fig. 13.7.

10.   The Interpretation of Spectra*

Although spectroscopy developed rapidly after the work of Kirchhoff and Bunsen, there was only a slow growth in understanding the origin of spectra. Throughout the nineteenth century it was believed that the line spectra of atoms were produced simultaneously by each individual atom behaving as an oscillator with a large number of different periods of vibration. In 1907, Arthur Conway proposed that a single atom produces a single spectral line at a time. He suggested that a single electron in an atom is in an "abnormal state," which can produce vibrations of a specific frequency. In 1908, Walther Ritz showed that the observed frequencies are the differences between certain spectral terms taken in pairs. In 1911, John Nicholson at Cambridge applied the Rutherford model of the atom to spectra and suggested that quantum jumps take place between definite states corresponding to the term values of Ritz, but he failed to grasp Conway's idea that only a single electron is involved.

Despite this intense scientific activity and steady progress, the first correct application of quantum theory to the interpretation of spectra was not made in the field of atomic spectra but in connection with the absorption spectra of molecules. This advance was made by the Danish chemist, Niels Bjerrum, in a paper "On the Infrared Spectra of Gases" published in 1912. He showed that infrared absorption by molecules is due to uptake of rotational and vibrational energy in definite quanta. We shall discuss such molecular spectra in some detail in Chapter 17, since they provide the most complete information about the internal structure and motions of molecules.

*The historical development given here follows closely that outlined by Edmund Whittaker in his *History of Theories of Aether and Electricity*, Vol. 2 (London: Th. Nelson and Sons, Ltd., 1953).

11. The Work of Bohr on Atomic Spectra

The problem of the interpretation of atomic spectra was finally solved in a work of genius by a young Dane who at the time was one of Rutherford's research students at Manchester, and who was destined to become the most influential physicist of our times, discovering a new way of looking at the world and founding a philosophical school which has been compared (by physicists) to that of Plato.

Niels Bohr was born in 1885; his father was Professor of Physiology at the University of Copenhagen. Niels spent many hours as a boy in his father's laboratory, but both he and his brother Harald, who became a noted mathematician, found time to become famous throughout Scandinavia as football players. "I grew up," said Bohr, "in a house with a rich intellectual life where scientific discussions were the order of the day. Indeed, for my father, there was scarcely a sharp distinction between his own scientific work and his lively interest in all the problems of human life."

Niels Bohr was destined to bring together two main streams of physics—the German school of theoretical physics exemplified by Planck and Einstein, and the English school of experimental physics of Thomson and Rutherford. The model of the nuclear atom proposed by Rutherford in 1911 had been based firmly on the facts of the experiment, with little reference to the theoretical ideas then generally accepted. According to electromagnetic theory, the atom of Rutherford had no right to exist. The electrons revolving about the nucleus are accelerated charged particles; therefore, they should continuously emit radiation, lose energy, and execute descending spirals until they fall into the positive center. But the electrons were unaware of what was expected of them, and the facts of chemistry and physics pointed clearly to the Rutherford model. For Bohr there was only one conclusion: the old principles of theoretical physics must be false.

Bohr solved the problem of atomic spectra by taking what was correct in the old ideas and throwing away what was incorrect, and then adding exactly the necessary new ideas. Thus he accepted Conway's principles:

(1) Spectral lines are produced by atoms one at a time.

(2) A single electron is responsible for each line.

He retained Nicholson's principles:

(3) The Rutherford nuclear atom is the correct model.

(4) The quantum laws apply to jumps between different states characterized by discrete values of angular momentum and, added Bohr, energy.

He applied Ehrenfest's rule to the angular momentum of the electron:

(5) The angular momentum $L = n(h/2\pi)$, where n is an integer.

Then he advanced the distinctly new principles:

(6) Two different states of the electron in the atom are involved. These are called *allowed stationary states*, and the spectral terms of Ritz correspond to these states.

(7) The Planck–Einstein equation $\epsilon = hv$ holds for the emission and absorption. Thus, if the electron makes a transition between two states with energies E_1 and E_2, the frequency of the spectral line is given by

$$hv = E_1 - E_2 \qquad (13.35)$$

Finally Bohr proposed a revolutionary concept, which aroused a storm of controversy among scientists and philosophers:

(8) We must renounce all attempts to visualize or to explain classically the behavior of the active electron during a transition of the atom from one stationary state to another.

12. Bohr Orbits and Ionization Potentials

To specify which orbits of the electrons around the nucleus are allowed, Bohr used condition (5). The angular momentum of a particle of mass m moving with a velocity v in a circular path of radius r is $L = mvr$, and hence the condition becomes

$$mvr = \frac{nh}{2\pi} \qquad (13.36)$$

The integer n is called the *principal quantum number*.

The electron is held in its orbit by the electrostatic force that attracts it to the nucleus. If the nucleus has a charge Ze, this force is $Ze^2/4\pi\epsilon_0 r^2$ from Coulomb's Law. For a stationary state, it must exactly balance the centrifugal force mv^2/r. Hence,

$$\frac{Ze^2}{4\pi\epsilon_0 r^2} = \frac{mv^2}{r}$$

$$r = \frac{Ze^2}{4\pi\epsilon_0 mv^2} \qquad (13.37)$$

Then from (13.36),

$$r = \frac{\epsilon_0 n^2 h^2}{\pi me^2 Z} \qquad (13.38)$$

In the case of hydrogen, $Z = 1$, and the smallest orbit would be that with $n = 1$, which would have a radius

$$a_0 = \frac{\epsilon_0 h^2}{\pi me^2} = 5.292 \times 10^{-11} \text{ m} = 0.05292 \text{ nm} \qquad (13.39)$$

This a_0 is called *the radius of the first Bohr orbit.** Values of the same magnitude for the radius of a hydrogen atom had been estimated from the kinetic theory of gases, and the theoretical calculation indicated to Bohr that he was on the right track in his theory.

He could now demonstrate that the Balmer series arises from transitions be-

---

*The Bohr theory was extended to deal also with elliptical orbits. See L. Pauling and E. B. Wilson, *Introduction to Quantum Mechanics* (New York: McGraw-Hill Book Company, 1935), Chap. 2.

tween an orbit with $n = 2$ and outer orbits; in the Lyman series, the lower term corresponds to the orbit with $n = 1$; the other series are explained similarly. Bohr could calculate the energy of the electron in each allowed orbit, and then calculate the spectral line frequencies from (13.35). The energy levels so obtained are plotted in Fig. 13.8, and the transitions responsible for absorption and emission of a quantum of radiation are shown as vertical lines.

The energy levels are calculated as follows. The total energy E in any state is the sum of the kinetic and potential energies,

$$E = E_k + E_p = \frac{mv^2}{2} - \frac{Ze^2}{4\pi\epsilon_0 r}$$

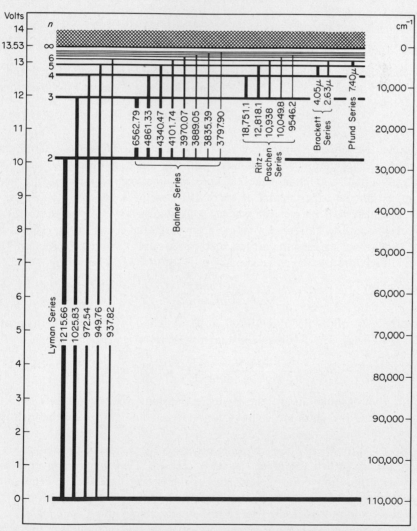

FIG. 13.8   Energy levels of the hydrogen atom. The wavelengths of spectral lines corresponding to the transitions are given in Ångstrom units (1Å = 0.1 nm).

From (13.37),

$$E = \frac{-Ze^2}{4\pi\epsilon_0 r} + \frac{Ze^2}{8\pi\epsilon_0 r} = \frac{-Ze^2}{8\pi\epsilon_0 r}$$

Notice that the potential energy is twice the magnitude of the kinetic energy and opposite in sign. This result will be true at equilibrium for any system in which the forces are central, i.e., dependent only on the distance between two centers. Therefore, from (13.38),

$$E = -\left(\frac{me^4}{8\epsilon_0^2 h^2}\right)\left(\frac{Z^2}{n^2}\right)$$

The frequency of a spectral line due to the transitions between levels with quantum numbers n_1 and n_2 is

$$\nu = \left(\frac{1}{h}\right)(E_{n_1} - E_{n_2}) = \left(\frac{me^4 Z^2}{8\epsilon_0^2 h^3}\right)\left(\frac{1}{n_2^2} - \frac{1}{n_1^2}\right) \tag{13.40}$$

This theoretical expression has exactly the form of the experimental law found by Rydberg, and hence a theoretical value for the Rydberg constant can be obtained,

$$\mathscr{R} = \frac{me^4}{8\epsilon_0^2 ch^3} = 109\,737\,\text{cm}^{-1} \tag{13.41}$$

The excellent agreement with the experimental value of $109\,677.576 \pm 0.012$ cm^{-1} may without exaggeration be called a triumph for the Bohr theory.

Actually a small correction makes the agreement between experiment and theory exact to within the experimental error of the constants and measurements. In (13.41), the mass m of the electron was used. Actually, however, the electron does not revolve about the stationary mass point given by the center of the proton, but about the center of mass of the proton–electron system. As shown in Section 4.19, we should therefore use the reduced mass μ of the two particles, where

$$\mu = \frac{mm_p}{m + m_p}$$

m being the mass of the electron (9.1090×10^{-28} g) and m_p the mass of the proton (1.6725×10^{-24} g). The Rydberg constant for the H atom thus becomes

$$\mathscr{R} = \frac{me^4}{8\epsilon_0^2\, ch^3[1 + (m/m_p)]} = 109\,678\,\text{cm}^{-1}$$

For other hydrogenlike atoms, the Rydberg constant will vary slightly with the nuclear mass, as a result of the slight variation in μ from its value for the hydrogen atom.

In Fig. 13.8, the energy levels become more closely packed as the height above the lowest state (called the *ground state*) increases. They finally converge to a limit whose height above the ground level is the energy necessary to remove the electron completely from the field of the nucleus. In the observed spectrum, lines become more and more densely packed and finally merge into a *continuum*, i.e., a region of continuous absorption or emission of radiation without any line structure. The reason for the continuum is that once an electron is completely free from the nucleus, it is no longer restricted to quantized discrete energy states, but may take up continuously the ordinary kinetic energy of translation, corresponding to its speed in free space, $\frac{1}{2}mv^2$.

The difference in energy between the series limit and the ground level is called the *ionization potential*. If an atom contains more than one electron, there will be first, second, third, etc., ionization potentials. For example, the first ionization potential of a lithium atom is the energy required for Li $\rightarrow$ Li$^+$ + e; the second ionization potential is the energy for Li$^+$ $\rightarrow$ Li^{2+} + e.

The potential energy of an electron in the first Bohr orbit in the hydrogen atom is

$$-\frac{e^2}{4\pi\epsilon_0 a_0} = -4.359 \times 10^{-18}\ \text{J}$$

This value would correspond to

$$\frac{-4.359 \times 10^{-18}}{1.6022 \times 10^{-19}} = -27.21\ \text{eV}$$

This energy defines an *atomic unit of energy*, sometimes called the *hartree*.

TABLE 13.1
Ionization Potentials Relative to That of Hydrogen
(In Atomic Units of Energy, Equal to 27.21 eV)

Element	Z, atomic number	Configuration of outer electrons	I_1	I_2	I_3	I_4	I_5	I_6	I_7	I_8
H	1	s	0.50							
He	2	s^2	0.92	2.00						
Li	3	s	0.20	3.00	4.5					
Be	4	s^2	0.35	0.67	5.65	8.0				
B	5	s^2p	0.31	0.93	1.4	9.65	12.5			
C	6	s^2p^2	0.42	0.90	1.76	2.37	12.0	18.0		
N	7	s^2p^3	0.54	1.09	1.75	2.72	3.60	20.3	24.5	
O	8	s^2p^4	0.50	1.29	2.02	2.85	4.04	5.07	27.3	32
F	9	s^2p^5	(0.67)	1.29	2.31	3.20	3.78	5.50	6.8	35
Ne	10	s^2p^6	0.79	1.51	2.34					
Na	11	s	0.19	1.75	2.62					
Mg	12	s^2	0.28	0.55	2.95					
Al	13	s^2p	0.22	0.69	1.05	4.5				
Si	14	s^2p^2	0.30	0.60	1.23	1.66	6.24			
P	15	s^2p^3	0.40	0.73	1.11	1.77	2.39			
S	16	s^2p^4	0.38	0.86	1.29	1.74	2.47	3.24		
Cl	17	s^2p^5	0.48	0.87	1.47	1.75	2.50	(3.4)	4.0	
Ar	18	s^2p^6	0.58	1.03	1.51	6.3				
K	19	s	0.16	1.17	1.74					
Ca	20	s^2	0.23	0.44	1.88					
Sc	21	s^2d	0.25	0.48	0.91	2.67				
Ti	22	s^2d^2	0.25	0.50	1.02	1.59	3.54			
V	23	s^2d^3	0.25	0.52	0.98	1.80	2.53	4.53		
Cr	24	sd^5	0.25	0.62	(1.0)	(1.85)	(2.7)			
Mn	25	s^2d^5	0.25	0.58	(1.2)	(1.86)	(2.8)			

In the same way, $a_0 = 0.052917$ nm, the Bohr radius of the hydrogen atom, defines an *atomic unit of length*, sometimes called the *bohr*.

Examples of ionization potentials are given in Table 13.1 in atomic units. The alkali metals have low first ionization potentials, which are consistent with their tendency to form singly positive ions and their strong electropositive character. The inert gases, however, have high ionization potentials in accord with their reluctance to enter chemical reactions.

13. Particles and Waves

In intensive work on the Bohr theory from 1913 to 1926, some important successes were achieved, yet there were also many troublesome failures. The theory could not explain, for example, the spectra of helium and more complex atoms. Also, it was increasingly evident that the logical foundations of the theory were incomplete. In any theory certain postulates must be made without proof, but in the Bohr theory there were too many of these underived postulates. It seemed that there should be some way of proving some of them from more fundamental principles. This hope was to be realized by revolutionary work in which Bohr himself played an important part. The basic discoveries, however, were made by three younger men—de Broglie, Heisenberg, and Schrödinger.

There is one branch of physics in which, as we have seen, integral numbers occur naturally, namely in the stationary state solutions of the equation for wave motion. This fact suggested the next great advance in physical theory: the idea that electrons, and in fact all material particles, possess wavelike properties. It was known that radiation displayed both particle and wave aspects. Now it was to be shown, first theoretically and soon afterward experimentally, that the same is true of matter.

This new way of thinking was first proposed in 1923 by Louis de Broglie. He described his approach later, in his Nobel Prize Address.*

> ...When I began to consider these difficulties [of contemporary physics] I was chiefly struck by two facts. On the one hand the quantum theory of light cannot be considered satisfactory, since it defines the energy of a light corpuscle by the equation $\epsilon = h\nu$, containing the frequency ν. Now a purely corpuscular theory contains nothing that enables us to define a frequency; for this reason alone, therefore, we are compelled, in the case of light, to introduce the idea of a corpuscle and that of periodicity simultaneously.

> On the other hand, determination of the stable motion of electrons in the atom introduces integers; and up to this point the only phenomena involving integers in Physics were those of interference and of normal modes of vibration. This fact suggested to me the idea that electrons too could not be regarded simply as corpuscles, but that periodicity must be assigned to them also.

A simple two-dimensional illustration of this idea may be seen in Fig. 13.9, which shows two possible electron waves of different wavelengths, for the case

*L. de Broglie, *Matter and Light* [New York: Dover Publications (1st ed., W. W. Norton Co.), 1946].

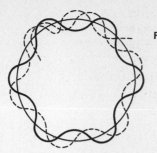

FIG. 13.9 Schematic drawing of an electron wave constrained to move around the nucleus. The solid line represents a possible stationary wave. The dashed line shows how a wave of somewhat different wavelength would be destroyed by interference.

of an electron revolving around an atomic nucleus. In one case, the circumference of the electron orbit is an integral multiple of the wavelength of the electron wave. In the other case, this condition is not fulfilled, and as a result the wave is destroyed by interference, and the supposed state is nonexistent. The introduction of integers associated with the permissible states of electronic motion therefore occurs quite naturally once the electron is given wave properties. The situation is analogous to the occurrence of stationary waves on a vibrating string. The necessary condition for a stable orbit of radius r_e is that

$$2\pi r_e = n\lambda \tag{13.42}$$

A free electron is associated with a progressive wave so that any energy is allowable. A bound electron is represented by a standing wave, which can have only certain definite frequencies or energies.

In the case of a photon, there are two fundamental equations to be obeyed: $\epsilon = h\nu$ and $\epsilon = mc^2$. When these are combined, we obtain $h\nu = mc^2$ or $\lambda = c/\nu = h/mc = h/p$, where p is the momentum of the photon. De Broglie considered that a similar equation governed the wavelength of the electron wave. Thus,

$$\lambda = \frac{h}{mv} = \frac{h}{p} \tag{13.43}$$

If we eliminate λ between (13.42) and (13.43), we obtain $mvr_e = nh/2\pi$, which is simply the original Bohr condition (13.36) for a stable orbit. Thus, the idea that electrons have wavelike properties suffices to give the rather mysterious Bohr condition directly.

The de Broglie equation (13.43) is a fundamental relation between the momentum of the electron considered as a particle and the wavelength of the electron considered as a wave. Consider, for example, an electron that has been accelerated through a potential difference $\Delta\Phi$ of 10 kV. Then $e\Delta\Phi = \frac{1}{2}mv^2$, and its velocity would be 5.9×10^7 m·s^{-1}, about one-fifth that of light. The wavelength of such an electron is

$$\lambda = \frac{h}{mv} = \frac{6.62 \times 10^{-34}}{(9.11 \times 10^{-31})(5.9 \times 10^7)} = 1.20 \times 10^{-11} \text{ m}$$

which is about the same wavelength as that of a rather hard X ray. Table 13.2 lists the theoretical wavelengths associated with various particles and other objects.

TABLE 13.2
Wavelengths of Various Objects

Particle	Mass (kg)	Speed (m·s⁻¹)	Wavelength (μm)
1-V electron	9.1×10^{-31}	5.9×10^{5}	1.2×10^{-3}
100-V electron	9.1×10^{-31}	5.9×10^{6}	1.2×10^{-4}
10 000-V electron	9.1×10^{-31}	5.9×10^{7}	1.2×10^{-5}
100-V proton	1.67×10^{-7}	1.38×10^{5}	2.9×10^{-6}
100-V α particle	6.6×10^{-7}	6.9×10^{4}	1.5×10^{-6}
H_2 molecule at 200°C	3.3×10^{-7}	2.4×10^{3}	8.2×10^{-6}
α Particle from radium	6.6×10^{-7}	1.51×10^{7}	6.6×10^{-9}
Rifle bullet (22)	1.9×10^{-3}	3.2×10^{2}	1.1×10^{-27}
Golf ball	4.5×10^{-2}	3×10	4.9×10^{-28}
Snail	1.0×10^{-2}	1×10^{-3}	6.6×10^{-23}

14. Electron Diffraction

If electrons have wave properties, a 10^{-2} nm electron wave should be diffracted by a crystal structure in much the same way as X-ray waves. Experiments along this line were first carried out by two groups of workers, who shared a Nobel prize for their discoveries. C. Davisson and L. H. Germer worked at the Bell Telephone Laboratories in New York, and G. P. Thomson, the son of J. J. Thomson, and A. Reid were at the University of Aberdeen. One of the first diffraction diagrams obtained by Thomson by passing beams of electrons through thin gold foils is shown in Fig. 13.10. The wave nature of the electron was unequivocally demonstrated by these researches. Figure 13.11 shows a striking diffraction diagram recently obtained from a thin film of chromium, part of which was a single crystal

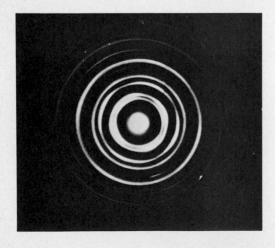

FIG. 13.10 One of the first electron diffraction pictures, which showed the wave nature of electrons, obtained by G. P. Thomson from gold foil.

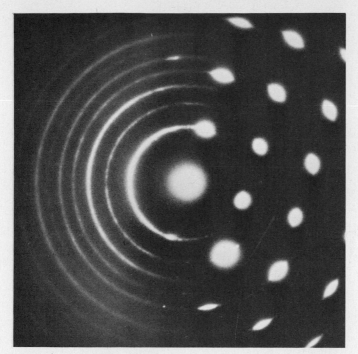

FIG. 13.11   Electron diffraction diagram of a thin chromium film,
which was partly polycrystalline (rings) and partly a single
crystal (spots) [I. B. M. Laboratories, Kingston, N.Y.].

and part a compact of tiny crystals. Diffraction patterns have also been obtained
from crystals placed in beams of neutrons or hydrogen atoms, so that these more
massive particles also display wave properties.

Electron beams, owing to their negative charge, have one advantage over X
rays as a means of investigating the fine structure of matter, in that appropriate
arrangements of electric and magnetic fields can be designed to act as lenses for
electrons. These arrangements have been applied in the development of electron
microscopes capable of resolving images as small as 5×10^{-10} m in diameter.
Figure 13.12 shows an electron microscope similar to one designed by E. Ruska
and B. v. Borries. Several electron microscope pictures are reproduced in this book.

15.   Waves and the Uncertainty Principle

The de Broglie wavelengths of ordinary objects are vanishingly small, and a
baseball batter need not consider diffraction phenomena when he swings at an
inside curve.* But in the subatomic world, h/mv is no longer so small as to be negli-

*And, by our British and Australian sportsmen, even less regard need be paid to the wave
properties of a cricket ball.

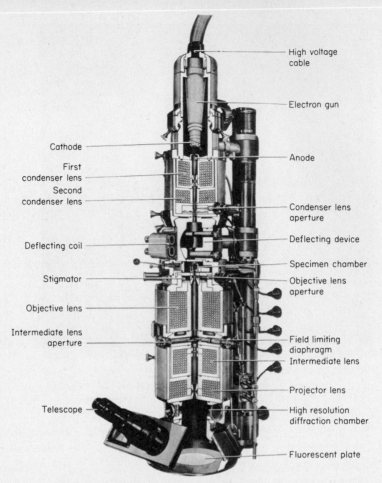

FIG. 13.12   Cross-sectional diagram of an electron microscope [Japan Electron Microscope Company].

gible. The de Broglie wavelengths of electrons are of such a magnitude that diffraction effects occur in molecules and crystals.

A fundamental tenet of classical mechanics is that it is possible to measure simultaneously the position and the momentum of any body. The strict determinism of mechanics rested upon this principle. Knowing the position and velocity of a particle at any instant and the forces on it at all times, Newtonian mechanics ventured to predict its position and velocity at any other time, past or future. Systems were completely reversible in time, past configurations being obtained simply by substituting $-t$ for t in the dynamical equations. But, if a particle has some of the properties of a wave, is it really possible to measure simultaneously its position and its velocity? The possible methods of measurement must be analyzed before an answer can be given.

In Fig. 13.13, suppose a particle of mass m moves along the positive x direction

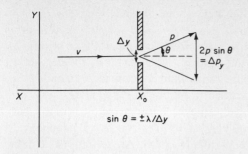

FIG. 13.13 Illustration of the uncertainty principle by the diffraction of a beam of particles at a slit. The momentum of the incoming beam is all in the *x* direction. But as a result of diffraction at the slit, the diffracted beam has momentum *p* with components in both *x* and *y* directions.

with a speed v. Its y component of momentum p_y is known to be zero, but we do not know anything at all about its coordinate y. At some point x_0 we try to measure y by placing in the path of the particle a slit of width Δy. Since the particle has wave properties, the wavelength of its de Broglie wave will be $\lambda = h/mv$ and thus diffraction will occur at the slit. The form of the diffracted intensity is depicted on the screen placed beyond x_0. From the discussion of Fig. 13.4, we know that the distance between the first two minima in the diffraction pattern would correspond to a difference of just one wavelength between the lengths of paths traversed by the waves diffracted from the two edges of the slit. Thus, from Fig. 13.13,

$$\Delta y \cdot \sin \theta = \lambda$$

As a result of the diffraction, the new direction of the momentum cannot be defined more closely than to within an angular spread $\pm\theta$, so that, from the figure,

$$\Delta p_y = 2p \sin \theta = \frac{2p\lambda}{\Delta y}$$

Hence, the product

$$\Delta y \cdot \Delta p_y \simeq 2p\lambda = 2h$$

The product of the uncertainty in the coordinate Δy times the uncertainty in its conjugate momentum Δp_y is therefore of the order of h. More precisely,

$$\Delta y \cdot \Delta p_y \geq \frac{h}{4\pi} = \frac{1}{2}\hbar \tag{13.44}$$

where $\hbar = h/2\pi$.

This is the famous *Uncertainty Principle* or *Principle of Indeterminacy** first stated by Werner Heisenberg in 1926. If we try to define precisely the position of a particle, we must sacrifice information about its momentum (or velocity). If we have exact data on its velocity, we cannot at the same time expect to know exactly where it is located in space.

As an example of (13.44), consider a microscopic particle 10^3 nm in diameter, with a mass of 6×10^{-13} g. Then

$$\Delta y \cdot \Delta v_y \simeq \frac{h}{m} \simeq \frac{6 \times 10^{-27}}{6 \times 10^{-13}} \simeq 10^{-14} \text{ cm}^2 \cdot \text{s}^{-1}$$

*W. Kauzmann, *Quantum Chemistry* (New York: Academic Press, Inc., 1957), Chap. 7.

If we measured the position to within 1.0 nm, about the resolving power of an electron microscope, $\Delta y = 10^{-7}$ cm and hence $\Delta v = 10^{-7}$ cm·s^{-1}. With this indeterminacy in velocity, the position 1 s later would be uncertain to within 2.0 nm or about 0.2% of the diameter of the particle. Thus, even in the case of an ordinary microscopic particle, indeterminacy may limit exact measurements. With particles of atomic or subatomic size, the effect would be much greater.

When waves are associated with particles, an uncertainty principle is a necessary consequence. If the wavelength or frequency of an electron wave is to have a definitely fixed value, the wave must have an infinite extent. Any attempt to confine a wave within boundaries requires destructive interference at the boundaries to reduce the amplitudes there to zero. It follows that an electron wave of perfectly fixed frequency, or momentum, must be infinitely extended and therefore must have a completely indeterminate position. To fix the position, we require superimposed waves of different frequencies, and thus as the position becomes more closely defined, the frequency and hence the momentum become less precisely specified.*

The uncertainty relation of (13.44) can be expressed also in terms of energy and time. Thus,

$$\Delta E \cdot \Delta t \geq \frac{h}{4\pi} \qquad (13.45)$$

To measure the energy of a system with an accuracy ΔE, the measurement must be extended over a period of time of order of magnitude $h/\Delta E$. This equation is used to estimate the sharpness of spectral lines. In general, lines arising from transitions from the ground state of an atom are sharp, because the optical electron spends a long time Δt in the ground state and ΔE, the uncertainty in the energy level, is correspondingly small. The line breadth is related to ΔE by $\Delta v = \Delta E/h$. The lifetime of excited states may sometimes be very short and transitions between such states give rise to diffuse lines as a result of the indeterminacy of the energy levels.†

16. Zero-Point Energy

According to the old quantum theory, the energy levels of a harmonic oscillator were given by $E_v = vhv$. If this were true, the lowest energy level, with $v = 0$, would have zero energy. This would be a state of complete rest, represented by the minimum in the potential-energy curve in Fig. 4.15.

The uncertainty principle does not allow such a state of completely defined

---

*The superimposed electron waves of different frequencies form a *wave packet*. Thus, a localized electron moving through space is represented not by a wave of definite frequency but by such a wave packet. The wave packet plays an important role in wave mechanics. See, for example, C. W. Sherwin, *Introduction to Quantum Mechanics* (New York: Holt, Rinehart & Winston, Inc., 1959), p. 130.

†This is not the only cause of broadening of spectral lines. In addition, there is a *pressure broadening* due to interaction with the electric fields of neighboring atoms or molecules, and a *Doppler broadening*, due to motion of the radiating atom or molecule with respect to the observer.

position and completely defined (in this case, zero) momentum. Consequently, the wave treatment shows that the energy levels of the oscillator are given by

$$E_v = (v + \tfrac{1}{2})hv \tag{13.46}$$

Now, when $v = 0$, the ground state, there is a residual *zero-point energy* amounting to

$$E_0 = \tfrac{1}{2}hv \tag{13.47}$$

This must be added to the Planck expression for the mean energy of an oscillator, which was derived in (13.31).

17. Wave Mechanics—The Schrödinger Equation

In 1926, Erwin Schrödinger and Werner Heisenberg independently discovered the basic principles for a new kind of mechanics, which provided mathematical techniques competent to deal with the wave–particle duality of matter and energy. The formalism of Schrödinger was called *wave mechanics* and that of Heisenberg, *matrix mechanics*. Despite their quite different mathematical formalisms, the two methods are essentially equivalent* at the deeper level of their basic physical concepts. They represent two different forms of the fundamental theory called *quantum mechanics*.

The mathematics of the Schrödinger method is more familiar to the chemist, and it is usual, therefore, to use the Schrödinger wave equation as the basis for chemical applications of quantum mechanics. Strictly speaking, we cannot *derive* the wave equation from any more fundamental postulates. It occupies in quantum mechanics a position analogous to that of Newton's $F = m(d^2x/dt^2)$ in classical mechanics. Nevertheless, the theory did not spring fully grown from the mind of its creator one sunny day on the beach near Kiel.

A more likely development is the following. The general differential equation of wave motion in one dimension was given by (13.12) as

$$\frac{\partial^2 u}{\partial x^2} = \frac{1}{v^2}\frac{\partial^2 u}{\partial t^2}$$

where $u(x, t)$ is the displacement and v the velocity. To separate the variables, we let $u(x, t) = w(x)e^{2\pi i v t}$, as shown in Eq. (13.19). By substituting this into the partial differential equation, we obtain an ordinary differential equation for the time-independent function $w(x)$,

$$\frac{d^2 w}{dx^2} + \frac{4\pi^2 v^2}{v^2}w = 0 \tag{13.48}$$

This is the wave equation with the time dependence removed.

To apply this equation to a *matter wave*, we introduce the de Broglie relation as follows: The total energy E is the sum of the potential energy U and the kinetic

*P. A. M. Dirac, *Nature*, *203*, 115, 771 (1964).

energy $p^2/2m$,

$$E = \frac{p^2}{2m} + U$$

$$p = [2m(E - U)]^{1/2}$$

Therefore,

$$\lambda = \frac{h}{p} = h[2m(E - U)]^{-1/2}$$

Or, since $\nu = v/\lambda$,

$$\nu^2 = \frac{v^2}{\lambda^2} = \frac{2mv^2(E - U)}{h^2}$$

Substituting this into (13.48) and letting $w = \psi$, the amplitude of the matter wave, one obtains

$$\frac{d^2\psi}{dx^2} + \frac{8\pi^2 m}{h^2}(E - U)\psi = 0 \tag{13.49}$$

This is the famous Schrödinger equation in one dimension. For three dimensions, it takes the form

$$\nabla^2\psi + \frac{8\pi^2 m}{h^2}(E - U)\psi = 0 \tag{13.50}$$

As defined in Eq. (10.45), the operator ∇^2 in Cartesian coordinates is

$$\nabla^2 \equiv \frac{\partial^2}{\partial x^2} + \frac{\partial^2}{\partial y^2} + \frac{\partial^2}{\partial z^2}$$

The *Hamiltonian operator* $\hat{H}$ is defined as

$$\hat{H} = -\frac{h^2}{8\pi^2 m}\nabla^2 + U$$

In terms of $\hat{H}$, the Schrödinger equation can be written simply as

$$\hat{H}\psi = E\psi \tag{13.51}$$

The solutions of (13.51) must satisfy the particular boundary conditions imposed upon the system. Just as the simple wave equation for a vibrating string yields a discrete set of stationary-state solutions when the eigenvalue condition (13.23) is satisfied, so appropriate solutions are obtained for the Schrödinger equation for certain energy values E.

In the case of the Schrödinger equation for an electron, discrete energy values E are found whenever the electron is constrained to move in a defined space, whereas a continuous range of values for E is found for the electron moving freely through space, not confined in any way. The allowed energy values are called the *characteristic values* or *eigenvalues* for the system. The corresponding *wave functions* are called the *characteristic functions* or *eigenfunctions*.

18.   Interpretation of the ψ Functions

The wave function ψ is a sort of amplitude function. In the case of a light wave, the intensity of light or energy of the electromagnetic field at any point is proportional to the square of the amplitude of the wave at that point. In terms of light quanta or photons $h\nu$, the more intense the light at any place, the more photons are falling on that place. This fact can be expressed in another way by saying that the greater the value of the amplitude of a light wave in any region, the greater is the probability of a photon being within that region.

A similar interpretation is the most useful one for the eigenfunctions ψ of Schrödinger's equation. They are therefore sometimes called *probability amplitude functions*. If $\psi(x)$ is a solution of the wave equation for an electron, then the relative probability of finding the electron within the range from x to $x + dx$ is given by $\psi^2(x)\,dx$.*

The physical interpretation of the wave function as a probability amplitude implies that it must obey certain mathematical conditions. We require that $\psi(x)$ be single valued, finite, and continuous for all physically possible values of x. It must be single valued because the probability of finding the electron at any point x must have only one value. It cannot be infinite at any point, for then the electron would be fixed at exactly that point, which would be inconsistent with its wave properties. The requirement of continuity is helpful in the selection of physically reasonable solutions for the wave equation.

19.   Solution of the Schrödinger Equation—The Free Particle

The simplest application of the Schrödinger equation is the case of a free particle, i.e., one moving in the absence of any potential field. In this case, we may set $U = 0$ in (13.49), and the one-dimensional equation becomes

$$\frac{d^2\psi}{dx^2} + \frac{8\pi^2 m}{h^2} E\psi = 0 \qquad (13.52)$$

This equation has the same form as Eq. (13.1), and its solution is, therefore,

$$\psi = A \sin kx + B \cos kx \qquad (13.53)$$

---

*Since the function ψ may be a complex quantity, the probability is written more generally as $\psi^*\psi$, where ψ^* is the complex conjugate of ψ. Thus, e.g., if $\psi = e^{-ix}$, $\psi^* = e^{ix}$. The interpretation of ψ^2 as a probability is due to Max Born. The actual probability of finding the electron between x and $x + dx$ would be

$$\frac{\psi^2(x)\,dx}{\int_{-\infty}^{+\infty} \psi^2(x)\,dx}$$

For any case except a free particle, the wave function must be quadratically integrable so that this probability can be evaluated.

where

$$k = \frac{2\pi}{h}(2mE)^{1/2} \tag{13.54}$$

From the relation of sines and cosines to complex exponentials, we can rewrite (13.53) as

$$\psi = Ce^{ikx} + De^{-ikx} \tag{13.55}$$

where C and D are arbitrary constants. If $D = 0$, the solution $\psi = Ce^{ikx}$ corresponds to a beam of particles moving in the positive x direction. If $C = 0$, the solution $\psi = De^{-ikx}$ corresponds to a beam moving in the negative x direction. The de Broglie wavelength associated with the particles is $\lambda = 2\pi/k$, and hence the momentum of the particles is

$$p_x = \frac{kh}{2\pi} \quad \text{or} \quad \frac{-kh}{2\pi}$$

The sum of the two functions in (13.53) or *either* the sine *or* the cosine function represents the superposition of two beams traveling in opposite directions.

We may note that there are no restrictions on the values of k for the free particle. The wave function ψ in (13.55) always meets the conditions that it be single valued, finite, and continuous. Thus, the kinetic energy E of a free particle can have any positive value. This result corresponds to the observation that the complete dissociation of an electron from an atom is marked by the onset of a *continuum* of light absorption. As long as the electron is bound to the rest of the atom, its energy levels are discrete and quantized. As soon as the electron is completely free, its energy is continuous and nonquantized.

20.   Solution of Wave Equation— Particle in Box

What is the effect of imposing a constraint upon the free particle by requiring that its motion be confined within fixed boundaries? In three dimensions, this is the problem of a particle enclosed in a box. The one-dimensional problem is that of a particle required to move between set points on a straight line. The potential function that corresponds to such a condition is shown in Fig. 13.14. For values of x between 0 and a, the particle is completely free, and $U = 0$. At the boundaries, however, the particle is constrained by an infinite potential wall over which there is no escape; thus, $U = \infty$ when $x = 0$, $x = a$. Outside the domain $0 \leq x \leq a$, the wave function $\psi = 0$.

The situation is now similar to that of the vibrating string considered at the beginning of the chapter. Restricting the electron wave within fixed boundaries corresponds to fixing the ends of the string so that $\psi = 0$ at $x = 0, a$. To obtain standing waves, it is necessary to restrict the allowed wavelengths so that there is an integral number of half-wavelengths between 0 and a; i.e., $n(\lambda/2) = a$. Some of the allowed electron waves are shown in Fig. 13.14, superimposed upon the potential-energy diagram.

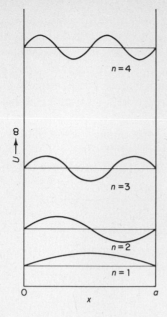

FIG. 13.14 Electron in a one-dimensional box. Allowed electron waves and energy levels. Note that E and ψ are plotted on the same graph, but the zero of ψ is different for each E.

The allowed energy values, the *eigenvalues* of the solution to the Schrödinger equation, can be obtained from (13.53). To have $\psi = 0$ at $x = 0$, the cosine function must vanish, a condition that requires $B = 0$. Thus,

$$\psi = A \sin \frac{2\pi}{h}(2mE)^{1/2}x \tag{13.56}$$

To have $\psi = 0$ at $x = a$, we must have

$$\sin \frac{2\pi}{h}(2mE)^{1/2}a = 0$$

$$\frac{2\pi}{h}(2mE)^{1/2}a = n\pi \tag{13.57}$$

This condition, therefore, restricts the allowed values of E to certain discrete eigenvalues, which from (13.57) are

$$E_n = \frac{n^2h^2}{8ma^2} \tag{13.58}$$

The first four of these energy levels are shown in Fig. 13.14.

From (13.58), two important consequences can be deduced which hold true for the energy of electrons, not only in this special case but quite generally. First, it is apparent that as the value of a increases, the kinetic energy E_n decreases. Other factors being the same, the more room the electron has in which to move, the lower will be its kinetic energy. *The more localized is its motion, the higher will be its kinetic energy.* Remember that the lower the energy, the greater the stability of a system. Such a *delocalization* of the motion of an electron can occur in molecules with certain kinds of structures, notably conjugated and aromatic carbon compounds, and it always leads to enhanced stability for the compound.

Secondly, the integer n is a typical *quantum number*, which now appears quite naturally and without ad hoc hypothesis. Its function is to specify the *number of nodes* in the electron wave. When $n = 1$, there are no nodes. When $n = 2$, there is a node in the center of the box; when $n = 3$, there are two nodes, etc. The value of the energy depends directly on n^2, and therefore rises rapidly as the number of nodes increases. This result will be true also for solutions of the Schrödinger equation for other systems.

We can easily extend the results for one dimension to the case of a three-dimensional box in the form of a parallelpiped of sides a, b, c. The potential U equals zero everywhere within the box, and the Schrödinger equation (13.50) is

$$\nabla^2\psi \equiv \left(\frac{\partial^2\psi}{\partial x^2} + \frac{\partial^2\psi}{\partial y^2} + \frac{\partial^2\psi}{\partial z^2}\right) = \frac{-8\pi^2 m}{h^2}E\psi \tag{13.59}$$

The equation can be separated by the substitution

$$\psi(x, y, z) = X(x)Y(y)Z(z) \tag{13.60}$$

which gives

$$\frac{1}{X}\frac{\partial^2 X}{\partial x^2} + \frac{1}{Y}\frac{\partial^2 Y}{\partial y^2} + \frac{1}{Z}\frac{\partial^2 Z}{\partial z^2} = -\frac{8\pi^2 mE}{h^2} \tag{13.61}$$

Since this equation must be true for all values of the independent variables x, y, z, we can conclude that each term on the left must equal some constant. Thus we can write

$$\frac{1}{X}\frac{d^2 X}{dx^2} = -k_x^2$$

$$\frac{1}{Y}\frac{d^2 Y}{dy^2} = -k_y^2 \tag{13.62}$$

$$\frac{1}{Z}\frac{d^2 Z}{dz^2} = -k_z^2$$

with $k_x^2 + k_y^2 + k_z^2 = +8\pi^2 mE/h^2 = k^2$. The equations in (13.62) are similar to (13.52) previously solved for the one-dimensional case, so that

$$X(x) = A_x \sin k_x x + B_x \cos k_x x$$
$$Y(y) = A_y \sin k_y y + B_y \cos k_y y \tag{13.63}$$
$$Z(z) = A_z \sin k_z z + B_z \cos k_z z$$

The boundary conditions $\psi(x, y, z) = 0$ at $x = a$, $y = b$, or $z = c$ require that $X(x) = 0$ at $x = a$, $Y(y) = 0$ at $y = b$, $Z(z) = 0$ at $z = c$, so that the cosine terms must vanish, and $B_x = B_y = B_z = 0$. The boundary condition, $\psi(x, y, z) = 0$ whenever $x = 0$ or a, $y = 0$ or b, or $z = 0$ or c, requires that

$$k_x = \frac{n_1\pi}{a}, \qquad k_y = \frac{n_2\pi}{b}, \qquad k_z = \frac{n_3\pi}{c} \tag{13.64}$$

Thus the solutions (13.60) are the eigenfunctions,

$$\psi_{n_1 n_2 n_3}(x, y, z) = A \sin\frac{n_1\pi x}{a} \sin\frac{n_2\pi y}{b} \sin\frac{n_3\pi z}{c} \tag{13.65}$$

specified by a set of three quantum numbers n_1, n_2, n_3. The allowed energy levels, from (13.62) and (13.64), are

$$E = \frac{h^2 k^2}{8\pi^2 m} = \frac{h^2}{8m}\left(\frac{n_1^2}{a^2} + \frac{n_2^2}{b^2} + \frac{n_3^2}{c^2}\right) \tag{13.66}$$

The eigenvalues E for this three-dimensional problem depend upon three distinct integral quantum numbers, n_1, n_2, n_3.

The amplitude A in (13.65) is fixed by the normalization condition, which is the requirement that the probability equal unity that the electron is to be found somewhere within the box.

$$\int_0^c \int_0^b \int_0^a \psi^2(x, y, z)\, dx\, dy\, dz = 1$$

From (13.65), therefore,

$$A = \left(\frac{8}{abc}\right)^{1/2}$$

If the box is cubical with side a, Eq. (13.66) becomes

$$E = \frac{h^2}{8ma^2}(n_1^2 + n_2^2 + n_3^2) \tag{13.67}$$

An important new feature now appears, namely, the occurrence of more than one distinct eigenfunction corresponding to the same eigenvalue for the energy.

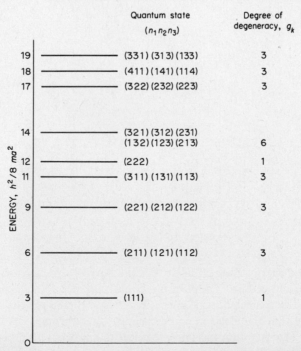

	Quantum state $(n_1 n_2 n_3)$	Degree of degeneracy, g_k
19	(331) (313) (133)	3
18	(411) (141) (114)	3
17	(322) (232) (223)	3
14	(321) (312) (231) (132) (123) (213)	6
12	(222)	1
11	(311) (131) (113)	3
9	(221) (212) (122)	3
6	(211) (121) (112)	3
3	(111)	1

ENERGY, $h^2/8\, ma^2$

FIG. 13.15 Allowed energy levels for the particle in the cubic box.

For example, the three eigenfunctions

$$\psi_{1,2,1}, \qquad \psi_{2,1,1}, \qquad \text{and} \quad \psi_{1,1,2}$$

correspond to different distributions in space, but they all have the same energy,

$$E = \frac{h^2}{8ma^2} \cdot 6$$

The energy level E_{211} is said to have a threefold *degeneracy*. In any statistical treatment of energy levels of the system, this level would have a *statistical weight* of $g_k = 3$.

Figure 13.15 shows some of the energy levels for a cubical box of side a, with the specification of their quantum states in terms of n_1, n_2, n_3 and the degree of degeneracy g_k.

21.   Penetration of a Potential Barrier

In the problem of the particle in a box (both in one and in three dimensions), there was no possibility that the particle could escape through the walls of the box. The box was a perfect trap for the particle because at the walls the potential U went to ∞. We state this reason here, but will prove it later in this section.

Let us consider now a one-dimensional problem in which the potential barrier is not infinite, but has a certain finite height U and width a. This situation is shown in Fig. 13.16. We assume that a particle, e.g., an electron, comes from the left with

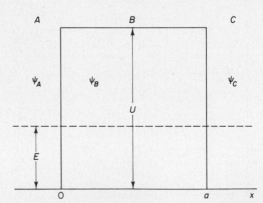

FIG. 13.16   An electron wave of kinetic energy E is incident upon a square potential barrier of height $U > E$ and width a.

a certain kinetic energy $E < U$ and strikes the barrier. In classical mechanics, the result is simple and certain. The electron would experience an elastic collision with the barrier and be reflected back in the negative x direction. So long as $E < U$, the probability that the electron could escape either through or over the barrier would be nil. The quantum mechanical result is amazingly different. It indicates that so long as the barrier is neither infinitely high nor infinitely wide, there is always a finite probability that the electron will penetrate the barrier and continue on its path in the positive x direction beyond $x = a$. This phenomenon is called the *tunnel effect* or *leakage through a potential barrier*.

The Schrödinger equation can be written

$$\frac{d^2\psi}{dx^2} = -\frac{8\pi^2 m}{h^2}(E - U)\psi$$

where $U = 0$ except for $0 \leq x \leq a$. We consider the solution in the three regions: A, to the left of the barrier, B within the barrier, and C to the right of the barrier.

In region A, there will be an incident wave and a reflected wave, and from (13.55) we can write

$$\psi_A = A_1 e^{ik_1 x} + B_1 e^{-ik_1 x} \tag{13.68}$$

where

$$k_1^2 = \frac{8\pi^2 mE}{h^2}$$

In region B, where $E < U$, the constant corresponding to k_1 in (13.68) would be an imaginary number, and it is convenient, therefore, to define

$$k_2^2 = \frac{8\pi^2 m(U - E)}{h^2}$$

The solution is then written

$$\psi_B = A_2 e^{k_2 x} + B_2 e^{-k_2 x} \tag{13.69}$$

In region C, there is only an imaginary wave function, with

$$\psi_C = A_3 e^{ik_1 x} \tag{13.70}$$

The problem now is to fit these solutions together smoothly and continuously, since, in accord with one of the basic requirements for a permissible wave function, ψ and also its first derivative must be continuous. Thus,

$$\text{at } x = 0, \qquad \psi_A = \psi_B, \qquad \frac{d\psi_A}{dx} = \frac{d\psi_B}{dx}$$

$$\text{at } x = a, \qquad \psi_B = \psi_C, \qquad \frac{d\psi_B}{dx} = \frac{d\psi_C}{dx} \tag{13.71}$$

There are four conditions, so that all but one of the five arbitrary constants can be determined, and that one is then found from a normalization condition. We use (13.68), (13.69), (13.70), and (13.71) to express four of the constants A_1, B_1, A_2, B_2 in terms of the fifth A_3, with the following results:

$$A_1 = e^{ik_1 a}\left[\cosh k_2 a + \frac{i}{2}\left(\frac{k_2}{k_1} - \frac{k_1}{k_2}\right)\sinh k_2 a\right] \cdot A_3$$

$$B_1 = -\frac{i}{2}e^{ik_1 a}\left(\frac{k_2}{k_1} + \frac{k_1}{k_2}\right)\sinh k_2 a \cdot A_3$$

$$A_2 = \frac{1}{2}e^{ik_1 a}e^{-k_2 a}\left(1 + \frac{ik_1}{k_2}\right) \cdot A_3 \tag{13.72}$$

$$B_2 = \frac{1}{2}e^{ik_1 a}e^{k_2 a}\left(1 - \frac{ik_1}{k_2}\right) \cdot A_3$$

Since the probability of finding a particle in any region is proportional to the square of the amplitude of the wave function, we can determine the probability that the electron passes through the barrier by calculating

$$D = \left| \frac{A_3}{A_1} \right|^2 \tag{13.73}$$

i.e., the probability of penetration is given by the square of the absolute value of the ratio of the amplitudes of the electron waves incident on the barrier and passing through the barrier. From (13.72), (13.73) is found to be

$$D = \frac{1}{1 + \frac{1}{4}\left(\frac{k_2}{k_1} + \frac{k_1}{k_2}\right)^2 \sinh^2 k_2 a} \tag{13.74}$$

We see that there is indeed a finite probability of penetration of the barrier (tunnel effect). Recalling that the hyperbolic sine,

$$\sinh x = \frac{e^x - e^{-x}}{2}$$

we can obtain an approximate form of (13.74) valid when $k_2 a \gg 1$. In this case, $\sinh k_2 a \simeq e^{k_2 a}/2$, and

$$D = \left[\frac{4}{(k_2/k_1) + (k_1/k_2)} \right]^2 e^{-2k_2 a} \tag{13.75}$$

From (13.75) we see that in the limit as $a \rightarrow \infty$ or $(U - E) \rightarrow \infty$, $D \rightarrow 0$, and the barrier is absolutely impermeable, which is the situation for the particle in a box. As an example of (13.75), consider a case in which the barrier width is 1 nm and the top of the barrier is 1 eV above the energy of the incident electron, which is itself 1 eV. Then

$$\frac{k_2}{k_1} = \left(\frac{U - E}{E} \right)^{1/2} = (1)^{1/2} = 1$$

$$D = \left[\frac{4}{(1)^{1/2} + (1)^{1/2}} \right]^2 e^{-2(5.1)}$$

$$D = 1.4 \times 10^{-4}$$

Many important phenomena involve the tunnel effect. An everyday example occurs when an electric circuit is completed by placing two metallic conductors in contact. The current of electrons flows freely across the contact even though the wires are covered by a thin insulating layer of oxide. The electrons readily tunnel through such a barrier—they do not need to surmount the top.

Many electrode processes involve tunneling of electrons through potential barriers at an electrode surface. Thus, the picture given in Fig. 12.13 showing that the electron transfer occurs by way of passage over the top of a potential-energy barrier does not usually represent the real situation. The probability of passage over the top of the barrier of height E_a would be proportional to the Boltzmann factor, $\exp(-E_a/kT)$. The classical probability of overcoming the barrier of 1 eV cited previously would be about 5×10^{-12}. The probability of tunneling through such a barrier as just calculated would be about 1.4×10^{-4}. Thus, the tunnel effect is of primary importance for such electron-transfer reactions. For most ionic transfers tunneling would be negligible, since the mass of a typical ion is 10^5 times greater than that of an electron. The electrodic transfer of protons, however, may involve an appreciable tunnel effect.

There has been great interest also in the contribution of the tunnel effect to thermal reactions involving protons or hydrogen atoms. The correction to the classical theory of reaction kinetics for this specifically quantum mechanical effect is often appreciable. An experimental test of the effect can be made by substitution of a D atom for H. The mass factor $\sqrt{2}$ enters the constants k_2 in the theory and the consequence is an unusually large isotopic effect. In other words, H atoms may often tunnel through barriers that D atoms rarely penetrate.

PROBLEMS

1. Consider an electric circuit consisting of a capacitance C in series with a coil of self-inductance L and zero resistance. If Q is the charge of the capacitor, show that

$$-L\frac{d^2Q}{dt^2} = \frac{Q}{C}$$

   Calculate $Q(t)$ and show that the natural frequency of the circuit is $\omega = 1/\sqrt{LC}$. In the problem of the harmonic oscillator discussed in Section 13.1, what parameters are analogous to the L, Q, and C of the electrical problem?

2. What is the average energy $\bar{\epsilon}$ of a harmonic oscillator of frequency 10^{13} s^{-1} at 0 K, 10 K, 100 K, 1000 K, and 10 000 K? What is $\bar{\epsilon}/kT$ at each temperature? Include the zero-point energy $\frac{1}{2}h\nu_0$.

3. The fundamental vibration frequency of N_2 corresponds to $\sigma = 2360$ cm^{-1}. What fraction of N_2 molecules possess no vibrational energy except their zero-point energy at 500 K?

4. Show that if a progressive wave in one dimension represented by $u_i = a\,e^{i(\omega t - \mathcal{K}x)}$ is superimposed on a wave produced by its total reflection $u_i = a\,e^{i(\omega t + \mathcal{K}x)}$, the resultant vibration at any point is $u = 2a\cos\mathcal{K}x\,e^{i\omega t}$. Sketch the resultant waveform $u(x)$ at several different times $t = 0$, $\tau/4$, $\tau/2$, where τ is the period of the vibration.

5. What is the velocity of an electron in the ground state of the H atom according to the Bohr theory? What would be the de Broglie wavelength of an electron with this velocity?

6. Calculate the kinetic energy of an electron driven from the surface of potassium (work function 2.26 eV) by incident light of wavelength 350 nm. What retarding potential Φ would need to be applied in the apparatus in Fig. 13.6 to prevent the collection of these photoelectrons?

7. A positive and a negative electron can form a short-lived complex called *positronium*. Assume that the Bohr theory of the hydrogen atom could be applied to positronium, and calculate (a) its ionization energy, (b) the energy level of its first excited state, and (c) its radius a_0 in the ground state.

8. Calculate the wavelengths of the H_α line in the Balmer series of ^{1}H, ^{2}H, and ^{3}H.

9. For an electron confined in a one-dimensional box 10 nm long, calculate the number of energy levels that lie between 9 and 10 eV.

10. For a particle moving in the one-dimensional potential well of Fig. 13.14, show that the mean value of x is $\bar{x} = a/2$, and the mean square deviation $\overline{(x - \bar{x})^2} = (a^2/12)[1 - (6/\pi^2 n^2)]$. Show that as n becomes very large, this value agrees with the classical one.

11. A particle of mass m moves in one dimension between $x = a$ and $x = b$, in which region a solution of the Schrödinger equation is $\psi = A/x$, where A is a normalization constant. (a) Calculate A. (b) Show that the average value of x is $\bar{x} = ab/(b-a) \ln (b/a)$.

12. The wave function for a particle moving between 0 and a in a one-dimensional box is $\psi = (2/a)^{1/2} \sin (n\pi x/a)$. Calculate the probability that the particle be found in the middle third of the box (i.e., between $x = a/3$ and $2a/3$) for $n = 1, 2, 3$.

13. What are the wavelengths of (a) a photon, and (b) an electron, each having kinetic energy of 1 eV?

14. The absolute threshold of the dark-adapted human eye for perception of light at 510 nm was measured as 3.5×10^{-17} J, at the surface of the cornea. To how many quanta does this threshold correspond?

15. The diameter of a typical small nucleus is about 10^{-15} m. Suppose that an electron was placed in a one-dimensional infinite potential well 10^{-15} m wide. What would its lowest energy level be? From this result, what would you conclude about the existence of electrons within the nucleus? What about the emission of electrons (β rays) from nuclei?

16. A particle of mass m is inside a spherical container of radius R. Estimate from the uncertainty principle its minimum possible kinetic energy. Apply the result to an electron in a sphere of radius $R = 10^{-10}$ m, $R = 10^{-15}$ m. Compare the latter result with that from Problem 15.

17. Apply the Heisenberg Uncertainty Principle to estimate the kinetic energy of the electron in a hydrogen atom at a distance r from the nucleus. Then calculate the equilibrium distance r_e by minimizing the total energy E, kinetic + potential, with respect to r. Compare E calculated in this way with the experimental value.

18. Repeat the calculation of Problem 17 for the two electrons in a He atom, taking into account the interelectronic repulsion to show that the minimum energy E occurs where $r_1 = r_2 = \frac{1}{7}(h^2/\pi^2 me^2)$ and $E = -\frac{49}{4}(\pi^2 me^4/h^2)$.

19. Suppose that an electron moving in one dimension with a kinetic energy of 9.5 eV meets a barrier of height 10 eV and width 1.0 nm. Estimate the probability that the electron will tunnel through the barrier. (This situation is similar to that of an electron tunneling through a thin oxide layer at an intermetallic contact.)

20. Solve the Schrödinger equation to determine the energy levels and wave functions of a particle in the potential well shown:

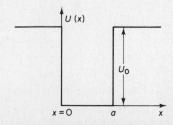

$$x \leq 0, \quad U = U_0; \qquad 0 \leq x \leq a, \quad U = 0; \qquad x \geq a, \quad U = U_0$$

[*Hint:* Solve for ψ in each of the three regions and evaluate the constants by the requirement that ψ and $d\psi/dx$, and hence $(1/\psi)(d\psi/dx)$, must be continuous at the boundaries.]*

21. From Planck's Law, Eq. (13.32), compute the Stefan–Boltzmann equation for the total density of black-body radiation as a function of temperature

$$E(T) = \int_0^\infty E(\nu)\, d\nu = \frac{8\pi^5}{15} \frac{k^4 T^4}{h^3 c^3}$$

22. In the Compton effect, a photon of wavelength λ makes a collision with an electron and is scattered with an altered wavelength λ'. If the electron is initially at rest,

$$\lambda' = \lambda + \frac{2h}{m_e c} \sin^2 \frac{\theta}{2}$$

where m_e is the rest mass of the electron and θ is the angle between the incident and the scattered photon. Derive the equation from the Laws of Conservation of Energy and Momentum. Discuss the relation of the Compton effect to the Heisenberg uncertainty principle.

23. For the de Broglie waves representing particles, the wavelength λ depends upon the frequency ν (i.e., the waves show dispersion). Derive the relation

$$\nu = c\left(\frac{1}{\lambda_c^2} + \frac{1}{\lambda^2}\right)^{1/2}$$

where $\lambda_c = h/m_e c$. You should recall that the mass of the electron depends on its speed v as

$$m = m_e \left(1 - \frac{v^2}{c^2}\right)^{-1/2}$$

24. The group velocity of a wave is defined by $v_g = d\nu/d(1/\lambda)$. Show that the group velocity of a de Broglie particle wave is equal to the ordinary velocity of the particle v.

25. Show that for a particle in a one-dimensional box of length a,

$$\int_0^a \psi_1(x)\psi_2(x)\, dx = 0$$

where ψ_1 and ψ_2 are any two eigenfunctions for two different energy levels. When an integral of the product of two functions vanishes, the functions are said to be *orthogonal*.

*A solution to a similar problem is given by D. ter Haar, *Selected Problems in Quantum Mechanics* (London: Macmillan & Co., Ltd., 1964).

14

Quantum Mechanics and Atomic Structure

Here is this quite beautiful theory, perhaps one of the most perfect, most accurate, and most lovely that man has discovered. We have external proof, but above all internal proof, that it has only a finite range, that it does not describe everything that it pretends to describe. The range is enormous, but internally the theory is telling us, "Do not take me absolutely or seriously. I have some relation to a world that you are not talking about when you are talking about me."

J. Robert Oppenheimer
(1957)*

In the last chapter, we outlined the developments in the history of science that led to the creation of quantum mechanics. The theory was applied to a few systems that illustrated some important consequences of the wave properties of electrons. In this chapter we shall state the minimum number of basic postulates required for the formulation of quantum mechanics. This formulation will then be applied to the few problems that actually yield exact analytic solutions of the Schrödinger equation. Finally, we shall discuss methods that will provide approximate, but often highly accurate solutions for a greater variety of problems of chemical interest.

1. Postulates of Quantum Mechanics

The postulates† stated in this section will need to be extended later by additional postulates that refer to the fundamental property of spin. Everything considered, it is easier to state the postulates for spinless particles, and then later to make the additions required to treat the spin. This procedure does detract from the elegance of the formulation as based on postulates, but it still seems preferable in a first discussion of the subject.

The postulates upon which we can base a logical development of quantum

---

*Physics Today, 10, 12 (1957).

†We may take a scientific *postulate* to be a statement from which it is possible to derive other quantitative statements that are in agreement with the results of physical observations.

mechanics will therefore be stated for a single particle without spin. The generalization to systems of two or more particles will be straightforward. Also, we shall state the postulates for a system in one dimension (one degree of freedom), specified by a coordinate x. Extension to a three-dimensional case is not difficult.

Postulate I:

The *physical state* of a particle at time t is described as fully as possible by a complex wave function $\Psi(x, t)$.

Postulate II:

The wave function $\Psi(x, t)$ and its first and second derivatives $\partial\Psi(x, t)/\partial x$, $\partial^2\Psi(x, t)/\partial x^2$, must be continuous, finite, and single valued for all values of x.

Postulate III:

Any quantity that is *physically observable* can be represented in quantum mechanics by a *Hermitian operator*. A Hermitian operator is a linear operator $\hat{F}$ that satisfies the condition,

$$\int \psi_1^* \hat{F} \psi_2 \, dx = \int \psi_2 (\hat{F}\psi_1)^* \, dx \tag{14.1}$$

for any pair of functions ψ_1, ψ_2 which represent physical states of the particle.

Postulate IV:

The allowable results of an observation of the quantity represented by $\hat{F}$ are any of the eigenvalues f_i of $\hat{F}$, where

$$\hat{F}\psi_i = f_i\psi_i$$

If ψ_i is an eigenfunction of F with eigenvalue f_i, then a measurement of F is certain to yield the value f_i.

Postulate V:

The average or *expectation value* $\langle F \rangle$ of any observable F, which corresponds to an operator $\hat{F}$, is calculated from the formula

$$\bar{F} \equiv \langle F \rangle = \int_{-\infty}^{\infty} \psi^* \hat{F} \psi \, dx \tag{14.2}$$

This formulation assumes that the wave function is *normalized*, i.e.,

$$\int_{-\infty}^{\infty} \psi^* \psi \, dx = 1 \tag{14.3}$$

(where ψ^* is the complex conjugate of ψ, formed by replacing i with $-i$ wherever it occurs in the ψ function).*

*Normalization is necessary to interpret ψ as a *probability amplitude*, as discussed in Section 13.18. The integral of the probability of finding the particle between x and $x + dx$, taken over all space, must be unity, i.e., the particle is *certain* to be somewhere. In scattering problems, however, the wave functions are not square integrable and one deals with the *flux* of probability, which is finite, rather than the probability itself, which diverges.

Postulate VI:

A quantum mechanical operator corresponding to a physical quantity is constructed by writing down the classical expression in terms of the variables x, p_x, t, E, and converting that expression to an operator by means of the following rules:

Classical variable	Quantum mechanical operator	Expression for operator	Operation
x	$\hat{x}$	x	Multiply by x
p_x	$\hat{p}_x$	$\dfrac{\hbar}{i}\dfrac{\partial}{\partial x}$	Take derivative with respect to x and multiply by $\hbar/i$
t	$\hat{t}$	t	Multiply by t
E	$\hat{E}$	$-\dfrac{\hbar}{i}\dfrac{\partial}{\partial t}$	Take derivative with respect to t and multiply by $-\hbar/i$

Postulate VII:

The wave function $\Psi(x, t)$ is a solution of the time-dependent Schrödinger equation,

$$\hat{H}(x, t)\Psi(x, t) = \frac{i\hbar\,\partial\Psi(x,t)}{\partial t} \qquad (14.4)$$

where $\hat{H}$ is the *Hamiltonian operator*.

The Hamiltonian operator is obtained from a classical Hamiltonian expressed in Cartesian coordinates by means of the correspondence rules given in Postulate VI. For the case of a particle subject only to conservative force fields,* the classical Hamiltonian is simply the sum of the kinetic and potential energies,

$$H = \frac{p_x^2}{2m} + U(x, t)$$

Thus, from postulate VI,

$$\hat{H} = -\frac{\hbar^2}{2m}\frac{\partial^2}{\partial x^2} + U(x, t) \qquad (14.5)$$

Substitution of this $\hat{H}$ into (14.4) gives the one-dimensional, time-dependent Schrödinger equation.

2. Discussion of Operators

The postulates were first stated as briefly as possible to allow them to stand out in sharp relief. Some discussion and examples may now be helpful.

---

*For the case of a charged particle is a magnetic field, see J. Griffith, *The Theory of Transition Metal Ions* (London: Cambridge University Press, 1961), p. 432.

The concept of an *operator* is fundamental in quantum mechanics. As the name indicates, an operator is an instruction to carry out a mathematical operation upon a function, which is called the *operand*. For example, in the expression, $(d/dx)f(x)$, the operator is d/dx and the operand is $f(x)$. If $f(x) = x^2$,

$$\frac{d}{dx}f(x) = \frac{d}{dx}x^2 = 2x$$

In the expression $x \cdot f(x)$ we can consider x to be the operator that tells us to multiply $f(x)$ by x. We can write the product of two operators, $\hat{O}_1$ and $\hat{O}_2$ as $\hat{O} = \hat{O}_1\hat{O}_2$. The product operator tells us to perform first the operation O_2 on the operand, and then to perform the operation O_1 on the result. Consider, for instance,

$$\hat{O}_2 = \frac{d}{dx}, \qquad \hat{O}_1 = x, \qquad f(x) = x^2$$

Then,

$$\hat{O}_1\hat{O}_2 f(x) = x\frac{d}{dx}x^2 = 2x^2$$

It is important to note that

$$\hat{O}_2\hat{O}_1 f(x) = \frac{d}{dx}x \cdot x^2 = 3x^2$$

Thus $\hat{O}_2\hat{O}_1 \neq \hat{O}_1\hat{O}_2$. These operators do not *commute*—the order in which they appear makes a difference in the product. Some pairs of operators commute and some do not (noncommuting operators).

An operator $\hat{O}$ is said to be *linear* when for any two functions f and g,

$$\hat{O}(\lambda f + \mu g) = \lambda(\hat{O}f) + \mu(\hat{O}g) \tag{14.6}$$

where λ and μ are arbitrary numbers, either complex or real. For example, d^2/dx^2 is a linear operator, but an operator SQ, which gives the command "take the square of following function," would not be linear.

If we associate physical quantities with linear operators, the expectation value of the quantity, as given by (14.2), obviously must be real, since it is measured with some physical apparatus. If $\langle F \rangle$ is to be real, it must equal its complex conjugate, $\langle F \rangle = \langle F \rangle^*$. The complex conjugate of $\langle F \rangle$, by definition, is obtained by taking the complex conjugate of each part of the integral in (14.2),

$$\langle F \rangle = \langle F \rangle^* = \int_{-\infty}^{\infty} \psi \hat{F}^* \psi^* \, dx \tag{14.7}$$

Hence, from (14.2) and (14.7),

$$\langle F \rangle = \int_{-\infty}^{\infty} \psi^* \hat{F}\psi \, dx = \langle F \rangle^* = \int_{-\infty}^{\infty} \psi(\hat{F}\psi)^* \, dx \tag{14.8}$$

From the definition of a *Hermitian operator* in (14.1), it is evident that the operator $\hat{F}$ in (14.8) is Hermitian. Hence, a sufficient condition that the expectation value be real is that the operator be Hermitian.

If, for a function f and an operator $\hat{O}$, we have

$$\hat{O}f = cf \tag{14.9}$$

where c is a number, then f is called an *eigenfunction* of the operator $\hat{O}$ and c is called an *eigenvalue* of the operator $\hat{O}$. (For Hermitian operators, the eigenvalues c must be real numbers.) The terms *eigenfunction* and *eigenvalue* were introduced in the previous chapter in connection with solutions of differential equations with boundary-value conditions. If $\hat{O}$ is a differential operator, Eq. (14.9) is an expression for the differential equation in operator form, and the problem of finding the eigenfunctions and eigenvalues in (14.9) is mathematically equivalent to the solution of the differential equation and boundary-value problem.

3.   Generalization to Three Dimensions

The postulates of quantum mechanics have been stated for a single particle, having only one degree of freedom (motion in one dimension). In the generalization to three dimensions, $\Psi(x, t) \longrightarrow \Psi(x, y, z, t)$, and the Hamiltonian operator becomes

$$\hat{H} = -\frac{\hbar^2}{2m}\left(\frac{\partial^2}{\partial x^2} + \frac{\partial^2}{\partial y^2} + \frac{\partial^2}{\partial z^2}\right) + U(x, y, z, t)$$

The Laplacian operator

$$\nabla^2 \equiv \frac{\partial^2}{\partial x^2} + \frac{\partial^2}{\partial y^2} + \frac{\partial^2}{\partial z^2}$$

encountered previously in electrostatic theory (p. 448) is usually read as "del squared."

The Schrödinger equation in three dimensions is, therefore,

$$-\frac{\hbar^2}{2m}\nabla^2\Psi + U(x, y, z, t)\Psi = i\hbar\frac{\partial\Psi}{\partial t} \tag{14.10}$$

If the potential U is not a function of time, we can readily separate the variables in this equation, as

$$\Psi(x, y, z, t) = \psi(x, y, z)\,e^{-iEt/\hbar}$$

The time-independent Schrödinger equation then becomes

$$-\frac{\hbar^2}{2m}\nabla^2\psi + U\psi = E\psi \tag{14.11}$$

$$\hat{H}\psi = E\psi \tag{14.12}$$

where the separation constant E can be interpreted as a stationary energy value for the system (as a consequence of Postulate IV). Equation (14.12) has exactly the form shown in (14.9), so that ψ is an eigenfunction, $\hat{H}$ a Hermitian operator, and E an eigenvalue of the system. In the remainder of this chapter, we shall consider solutions of Eq. (14.12) for time-independent eigenfunctions ψ. In Chapter 16, however, we shall return to the time-dependent wave equation (14.10) to solve the problem of the rate of transition between stationary states.

4. Harmonic Oscillator

By suitable transformation of variables, all problems that lead to exact solutions of the Schrödinger equation can be reduced to the same mathematical problem. From the point of view of physical theory, however, it is more enlightening to treat each problem separately. The soluble problems are then: the harmonic oscillator, the rigid rotor, the hydrogen atom (motion of a particle in a coulombic force field), and the hydrogen molecular ion (H_2^+) (motion of particle in combined coulombic fields of two nuclei). There are also a few other special potential functions that allow exact solutions to be obtained.

The one-dimensional harmonic oscillator problem is especially interesting because it is sufficiently difficult to exemplify most points of interest, yet simple enough to allow the mathematical details to be presented in full.

As shown on p. 145, the potential energy

$$U(x) = \tfrac{1}{2}\kappa x^2$$

and from (13.3)

$$\kappa = 4\pi^2 \mu v_0^2 \tag{14.13}$$

where μ is the reduced mass and v_0 the fundamental vibration frequency. The Schrödinger equation (14.11) then becomes

$$\frac{d^2\psi}{dx^2} + \frac{8\pi^2\mu}{h^2}(E - U)\psi = 0 \tag{14.14}$$

For convenience, new parameters are introduced

$$\alpha^4 = \frac{\hbar^2}{\kappa\mu}, \qquad \epsilon = \frac{2\alpha^2\mu E}{\hbar^2} \tag{14.15}$$

Thus, (14.14) becomes

$$\alpha^2 \frac{d^2\psi}{dx^2} + \left(\epsilon - \frac{x^2}{\alpha^2}\right)\psi = 0 \tag{14.16}$$

We next transform the independent variable x into a new variable y by means of

$$x = \alpha y$$

Using the fact that the operator

$$\frac{d^2}{dx^2} = \alpha^{-2}\frac{d^2}{dy^2}$$

we have

$$\frac{d^2\psi}{dy^2} + (\epsilon - y^2)\psi = 0 \tag{14.17}$$

This is an example of a linear differential equation of the second order. A lot of interesting mathematics is concerned with the solution of such equations.*

---

*H. Jeffreys and B. S. Jeffreys, *Methods of Mathematical Physics*, 2nd ed. (London: Cambridge University Press, 1950), Chap. 16.

The theoretical discussion is based on the number and types of *singular points* of the equation. A singular point is any point that is not an *ordinary point*. In Eq. (14.17), for example, an ordinary point $y = y_1$ is any point for which ψ and $d\psi/dy$ can take any pair of values without causing $d^2\psi/dy^2$ to go to infinity. An important property of a linear differential equation is that its singular points are fixed. The theory then shows that at or near any ordinary point, the general solution of the equation can be written as a power series expansion about that point, whose radius of convergence is the distance to the nearest singularity.

In our Eq. (14.17), $y = \infty$ is a singular point because we cannot allow ψ to take any value at $y = \infty$ and still require that $d^2\psi/dy^2$ not go to ∞. In fact, we must require that $\psi = 0$ at $y = \infty$. We therefore choose a function that will make ψ satisfy this condition at $y = \pm\infty$ and multiply this function by a power series that will allow us to solve the equation over the domain $-\infty < y < \infty$.

When y becomes very large, (14.17) reduces to

$$\frac{d^2\psi}{dy^2} - y^2\psi = 0 \tag{14.18}$$

In the limit as $y \to \pm\infty$, (14.18) has the asymptotic solution

$$\psi = e^{\pm y^2/2} \tag{14.19}$$

Since the positive exponential does not behave properly, it is discarded, and we try to find a solution of the original Eq. (14.17) of the form

$$\psi = \mathscr{H}(y)\, e^{-y^2/2} \tag{14.20}$$

When (14.20) is introduced into (14.17), we obtain the differential equation that must be satisfied by $\mathscr{H}(y)$,

$$\frac{d^2\mathscr{H}}{dy^2} - 2y\frac{d\mathscr{H}}{dy} + (\epsilon - 1)\mathscr{H} = 0 \tag{14.21}$$

For Eq. (14.21), $y = 0$ is a regular point so that we can express $\mathscr{H}(y)$ as a power series in y,

$$\mathscr{H}(y) = \sum_v a_v y^v \equiv a_0 + a_1 y + a_2 y^2 + a_3 y^3 + \cdots \tag{14.22}$$

From (14.22),

$$\frac{d\mathscr{H}}{dy} = \sum_v v a_v y^{v-1} \equiv a_1 + 2a_2 y + 3a_3 y^2 + \cdots$$

$$\frac{d^2\mathscr{H}}{dy^2} = \sum_v v(v-1) a_v y^{v-2} \equiv 1 \cdot 2 a_2 + 2 \cdot 3 a_3 y + \cdots$$

Substituting these expressions and (14.22) into (14.21), and arranging in order of ascending powers of y, gives

$$[1 \cdot 2a_2 + (\epsilon - 1)a_0] + [2 \cdot 3a_3 + (\epsilon - 1 - 2)a_1]y$$
$$+ [3 \cdot 4a_4 + (\epsilon - 1 - 2 \cdot 2)a_2]y^2 + \cdots = 0$$

Since y as an independent variable can be assigned any value, for this series to vanish for all values of y, it is necessary that each term vanish:

$$v = 0, \quad 1 \cdot 2a_2 + (\epsilon - 1)a_0 = 0$$
$$v = 1, \quad 2 \cdot 3a_3 + (\epsilon - 1 - 2)a_1 = 0$$
$$v = 2, \quad 3 \cdot 4a_4 + (\epsilon - 1 - 2 \cdot 2)a_2 = 0$$
$$v = 3, \quad 4 \cdot 5a_5 + (\epsilon - 1 - 2 \cdot 3)a_3 = 0, \text{ etc.}$$

We can see that the general rule being followed is that for the vth coefficient (of y^v)

$$(v + 1)(v + 2)a_{v+2} + (\epsilon - 1 - 2v)a_v = 0$$

$$a_{v+2} = -\frac{(\epsilon - 2v - 1)}{(v + 1)(v + 2)}a_v \tag{14.23}$$

The expression (14.23) is an example of a *recursion formula*. If we know a_0 and a_1, (14.23) allows us to calculate all the other coefficients in the power series. The values of a_0 and a_1 are the two arbitrary constants that always occur in the solution of an ordinary differential equation of the second order.

The solution of (14.17) is now at hand, but does it obey the boundary conditions that $\psi \rightarrow 0$ as $y \rightarrow \infty$? We can see that in general it will not do so, because the infinite series in (14.22) would go to ∞ as e^{y^2} and thus would overpower the $e^{-y^2/2}$ factor in (14.20).* To fit the boundary condition, we can terminate the series at some finite number of terms; then the $e^{-y^2/2}$ factor will ensure that $\psi \rightarrow 0$ as $y \rightarrow \infty$. The termination of the series after the vth term can be brought about by selecting the energy parameter ϵ in (14.23) in such a way that the numerator goes to zero for $v = v$ an integer. This condition is therefore $\epsilon - 2v - 1 = 0$, or

$$\epsilon = 2v + 1 \tag{14.24}$$

This condition will terminate either the series with v odd or that with v even but not both. Hence, if v is odd, we set $a_0 = 0$, and if v is even, $a_1 = 0$.

Equation (14.24) is a typical eigenvalue condition. It shows that suitable wave functions ψ cannot be found for any arbitrary values of the energy, but only for certain discrete values given by the condition (14.24). When α is introduced from (14.15), (14.24) becomes

$$E = (v + \tfrac{1}{2})h\nu_0 \tag{14.25}$$

Equation (14.25) is the quantum mechanical expression for the energy levels of a one-dimensional harmonic oscillator. The quantum number v occurs mathematically as a result of the boundary-value condition on the solution of the Schrödinger equation. The energy levels are plotted in Fig. 14.1(a), superimposed on the potential-energy curve given by (4.39).

*Compare the series for $\mathcal{H}(y)$ with that for e^{y^2}:

$$e^{y^2} = 1 + y^2 + \frac{y^4}{2!} + \cdots + \frac{y^v}{(v/2)!} + \frac{y^{v+2}}{[(v/2 + 1)]!} + \cdots$$

The higher terms in this series differ from those for $\mathcal{H}(y)$ in (14.21) simply by a multiplicative constant.

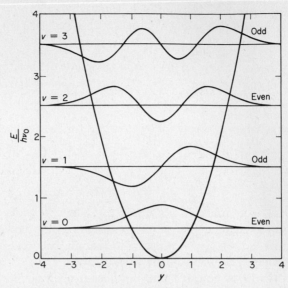

FIG. 14.1(a) The potential function and wave functions for an harmonic oscillator. The scale for the energy levels is given on the right in units of $E/h\nu_0$. The amplitudes of the $\psi(y)$ functions are plotted so that $\psi_v(y)$ is normalized for each energy level. Recall that $y = x/\alpha$ where $\alpha = (h^2/\kappa\mu)^{1/4}$. [H. L. Strauss, *Quantum Mechanics: An Introduction* (Englewood Cliffs, N.J.: Prentice-Hall, Inc., 1968), p. 57.]

5. Harmonic Oscillator Wave Functions

The wave functions ψ_v corresponding to the energy levels (14.25) specified by the quantum number v are the eigenfunctions of the harmonic oscillator problem. Table 14.1 gives the first few polynomials $\mathscr{H}_v(y)$ as obtained from (14.21) and (14.23). The eigenfunctions are then

$$\psi_v = N_v\, e^{-y^2/2}\, \mathscr{H}_v(y) \tag{14.26}$$

where N_v is the appropriate normalization factor obtained from the condition that

$$\int_{-\infty}^{\infty} \psi_v^*(x)\psi_v(x)\, dx = 1, \qquad N_v = \left(\frac{\alpha}{\pi^{1/2}\, 2^v v!}\right)^{1/2}$$

The functions $\psi_v(y)$ are plotted in Fig. 14.1(a).

The polynomials $\mathscr{H}_v(y)$ were known before the advent of quantum mechanics as the *Hermite polynomials*, which occurred in the solution of a differential equation similar to the Hermite equation. The polynomials can be readily obtained by another definition,

$$\mathscr{H}_v(y) = (-1)^v\, e^{y^2}\, \frac{d^v e^{-y^2}}{dy^v} \tag{14.27}$$

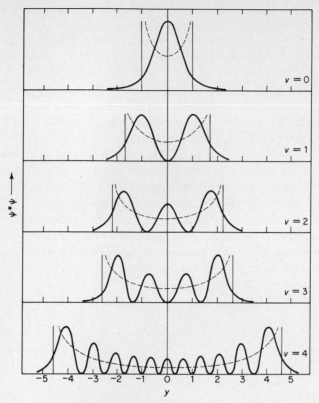

FIG. 14.1(b) Some sketches of the probability density functions for the harmonic oscillator. The dotted curve in each sketch is the probability density function for the classical oscillator with the same physical constants and the same energy. [Chalmers W. Sherwin, *Introduction to Quantum Mechanics* (New York: Holt, Rinehart & Winston, Inc., 1959), p. 53.]

TABLE 14.1
The First Few Hermite Polynomials $\mathscr{H}_v(y)$

$$\mathscr{H}_0 = 1$$
$$\mathscr{H}_1 = 2y$$
$$\mathscr{H}_2 = 4y^2 - 2$$
$$\mathscr{H}_3 = 8y^3 - 12y$$
$$\mathscr{H}_4 = 16y^4 - 48y^2 + 12$$
$$\mathscr{H}_5 = 32y^5 - 160y^3 + 120y$$

We recall from Section 13.18 the interpretation of ψ_v as a probability amplitude, such that $\psi^*\psi \, dx$ gives the probability that the particle be found between x and $x + dx$. In Fig. 14.1(b), we have plotted $\psi_v^*\psi_v$ for several values of v. These curves thus display the probability that the oscillating mass point be found at a distance $x (= \alpha y)$ from its equilibrium position.

In classical mechanics, an oscillator is most likely to be found at the extreme points of its vibration, since here the kinetic energy goes to zero as the motion is reversed. The quantum mechanical result is strikingly different for low values of v. For $v = 0$, for example, the most probable position is at the equilibrium position, $y = 0$. Of course, the zero-point vibration has no classical analog at all, since classically the lowest energy state would entail complete cessation of vibration. We show, however, the classical distribution for an oscillator with the same total energy $E = \frac{1}{2}hv$. The cases $v = 1, 2, 3$, and classical counterparts are shown. Finally, we show the case $v = 4$. For a quantum number this high, the quantum picture is beginning to approach the classical behavior, provided an average is taken over the high frequency oscillations of the wave functions. Such an asymptotic approach of quantum mechanical behavior to classical behavior at large quantum numbers is an example of the *correspondence principle*.

The spacial distributions of ψ are used in calculating the probability of transition from one state to a higher or lower state with consequent absorption or emission of an energy quantum hv. By means of the Franck–Condon principle, discussed in Section 17.17, one can then deduce the results of such spectral transitions.

6. Partition Function and Thermodynamics of Harmonic Oscillator

An important application of the harmonic oscillator model is the statistical mechanical calculation of the partition function f_v shown in Table 5.4. The thermodynamic functions calculated from f_v can be used for the vibrational modes of both molecules and crystals.

From Eqs. (5.28) and (14.24), the partition function for an individual oscillator is

$$z_v = \sum_v \exp \frac{-(v + \frac{1}{2})hv}{kT} = \exp \frac{-hv}{2kT} \sum_v \exp \frac{-vhv}{kT}$$

$$z_v = \frac{\exp(-hv/2kT)}{1 - \exp(-hv/kT)} \tag{14.28}$$

The total vibrational partition function of a molecule or crystal is the product of terms like (14.28), one for each of the normal modes of vibration,

$$z_v = \prod_j z_{v,j} \tag{14.29}$$

For the purposes of tabulation and facility in calculations, the vibrational contributions can be put into more convenient forms. The vibrational energy per mole, from (14.24), (5.37), and (14.28), is

$$U_m = RT^2 \frac{\partial \ln z}{\partial T} = L\frac{hv}{2} + \frac{Lhv \, e^{-hv/kT}}{1 - e^{-hv/kT}}$$

Now $Lh\nu/2$ is the zero-point energy per mole U_{m0}, whence, writing $h\nu/kT = x$, we have

$$\frac{U_m - U_{m0}}{T} = \frac{Rx\, e^{-x}}{1 - e^{-x}} \tag{14.30}$$

The vibrational heat capacity per mole is

$$\left(\frac{\partial U_m}{\partial T}\right)_V = C_{Vm} = \frac{Rx^2}{2(\cosh x - 1)} \tag{14.31}$$

From (5.44), since for the vibrational contribution* $A = G$,

$$\frac{G_m - U_{m0}}{T} = R \ln (1 - e^{-x}) \tag{14.32}$$

Finally, the vibrational contribution to the entropy is

$$S = \frac{U - U_0}{T} - \frac{G - U_0}{T} \tag{14.33}$$

An excellent tabulation of these functions has been given by J. G. Aston.† An abridged set of these values is given in Table 14.2. If the vibration frequency can be

TABLE 14.2
Molar Thermodynamic Functions of a Harmonic Oscillator (Energy Units in Joules)

$x = \dfrac{h\nu}{kT}$	C_V	$\dfrac{(U - U_0)}{T}$	$\dfrac{-(G - U_0)}{T}$	$x = \dfrac{h\nu}{kT}$	C_V	$\dfrac{(U - U_0)}{T}$	$\dfrac{-(G - U_0)}{T}$
0.10	8.305	7.912	19.56	1.70	6.573	3.159	1.677
0.15	8.297	7.707	16.39	1.80	6.393	2.958	1.502
0.20	8.289	7.510	14.20	1.90	6.209	2.778	1.347
0.25	8.272	7.318	12.55	2.00	6.021	2.603	1.209
0.30	8.251	7.130	11.23	2.20	5.640	2.279	0.976
0.35	8.230	6.945	10.14	2.40	5.255	1.991	0.7907
0.40	8.205	6.761	9.230	2.60	4.870	1.734	0.6418
0.45	8.176	6.586	8.439	2.80	4.494	1.507	0.5213
0.50	8.142	6.410	7.753	3.00	4.125	1.307	0.4246
0.60	8.071	6.067	6.615	3.50	3.270	0.9062	0.2552
0.70	7.983	5.740	5.708	4.00	2.528	0.6204	0.1535
0.80	7.883	5.427	4.962	4.50	1.913	0.4204	0.0933
0.90	7.774	5.125	4.339	5.00	1.420	0.2820	0.0556
1.00	7.657	4.841	3.816	5.50	1.036	0.1878	0.0338
1.10	7.523	4.565	3.366	6.00	0.7455	0.1238	0.0209
1.20	7.385	4.301	2.982	6.50	0.5296	0.0815	0.0125
1.30	7.234	4.049	2.645	7.00	0.3723	0.0531	0.0075
1.40	7.079	3.810	2.355	8.00	0.1786	0.0221	0.0025
1.50	6.916	3.582	2.099	9.00	0.0832	0.0092	0.0016
1.60	6.745	3.365	1.875	10.00	0.0376	0.0037	0.0004

*This is evident from (5.45), since z_v is not a function of V, $P = 0$, $G = A + PV = A$.
†In *Treatise on Physical Chemistry*, Vol. 1, 3rd ed., ed. by H. S. Taylor and S. Glasstone, (Princeton, N.J.: D. Van Nostrand Co., Inc., 1942), p. 655.

obtained from spectroscopic observations, such tables can be used to calculate the vibrational contributions to energy, entropy, free energy, and heat capacity. The application of these results to the problem of the heat capacity of solids is described in Section 18.33. For precise work, correction must be made for the anharmonicity of the vibrations.[*]

Let us calculate the vibrational contribution to the entropy of F_2 at 298.2 K. The fundamental vibration frequency is at $\sigma = 892.1$ cm^{-1}. Hence,

$$x = \frac{h\nu}{kT} = \frac{hc\sigma}{kT} = \frac{(6.62 \times 10^{-27})(3.00 \times 10^{10})(892.1)}{(1.38 \times 10^{-16})(298.2)} = 4.305$$

From Table 14.2 and Eq. (14.33),

$$S_m = \frac{U - U_0}{T} - \frac{G - U_0}{T} = 0.5004 + 0.1167 = 0.6171 \text{ J} \cdot \text{K}^{-1} \cdot \text{mol}^{-1}$$

7. Rigid Diatomic Rotor

The problem of the rigid rotor (discussed classically in Section 4.19) leads to the same mathematical formulation as the problem of the angular dependency of the wave function for the electron in the hydrogen atom. We shall treat the case of the rigid rotor here and use the result later in discussing the problem of the hydrogen atom.

As pointed out on p. 144, the motion of two particles of masses m_1 and m_2, which are joined by a rigid connection of length R so that their center of mass always remains at rest, is equivalent to the motion of a single particle of reduced mass μ at a distance R from the origin of coordinates. For such a purely rotational motion there is no potential energy, so that the Schrödinger equation (14.11) becomes, with $U = 0$,

$$-\frac{\hbar^2}{2\mu}\nabla^2\psi = E\psi \tag{14.34}$$

By means of (10.46), we can express (14.34) in polar coordinates and apply the condition that $r = R$, a constant, to obtain

$$\left(\frac{\partial^2}{\partial\theta^2} + \frac{\cos\theta}{\sin\theta}\frac{\partial}{\partial\theta} + \frac{1}{\sin^2\theta}\frac{\partial^2}{\partial\phi^2}\right)\psi(\theta, \phi) + \beta\psi(\theta, \phi) = 0 \tag{14.35}$$

where $\beta = 2\mu R^2 E/\hbar^2$.

Proceeding as in Section 13.20, we separate the variables by a substitution,

$$\psi(\theta, \phi) = \Theta(\theta)\Phi(\phi)$$

which yields two ordinary differential equations,

$$\frac{d^2\Phi}{d\phi^2} + m_l^2\Phi = 0 \tag{14.36}$$

$$\frac{d^2\Theta}{d\theta^2} + \frac{\cos\theta}{\sin\theta}\frac{d\Theta}{d\theta} + \left(\beta - \frac{m_l^2}{\sin^2\theta}\right)\Theta = 0 \tag{14.37}$$

[*]G. N. Lewis and M. Randall, *Thermodynamics*, 2nd ed. (revised by K. S. Pitzer and L. Brewer) (New York: McGraw-Hill Book Company, 1961), p. 430. See Section 17.10.

At this point, m_l is an arbitrary separation constant. As was seen previously for (13.52), the solutions of (14.36) are

$$\Phi(\phi) = \exp(im_l\phi) \tag{14.38}$$

where m_l can take both positive and negative values. We now find that the allowed values of m_l are restricted by the requirements that the wave function and its derivative must be everywhere single valued, finite, and continuous. In the case of the functions (14.38), the requirement that $\Phi(\phi)$ be single valued demands that

$$\Phi(\phi) = \Phi(\phi + 2\pi)$$
$$\exp(im_l\phi) = \exp[im_l(\phi + 2\pi)]$$

so that

$$\exp(im_l 2\pi) = 1$$

The only allowed values of m_l are, therefore,

$$m_l = 0, \pm 1, \pm 2, \pm 3, \text{ etc.} \tag{14.39}$$

The constant m_l has become a *quantum number*.

We next turn our attention to Eq. (14.37) for $\Theta(\theta)$, and introduce the transformation of variables,

$$s = \cos\theta$$
$$g(s) = \Theta(\cos\theta)$$

Since

$$\frac{d\Theta}{d\theta} = -\sin\theta\,\frac{dg}{ds}$$

$$\frac{d^2\Theta}{d\theta^2} = \sin^2\theta\,\frac{d^2g}{ds^2} - \cos\theta\,\frac{dg}{ds}$$

Eq. (14.37) becomes

$$(1 - s^2)\frac{d^2g}{ds^2} - 2s\frac{dg}{ds} + \left(\beta - \frac{m_l^2}{1 - s^2}\right)g = 0 \tag{14.40}$$

We shall not give the details of the solution of this equation, which can be found in any textbook of quantum mechanics.* It is a well-known equation, the solutions of which are the *associated Legendre polynomials*, $P_l^{m_l}(s)$, where the parameter l is related to β by

$$\beta = l(l + 1) \tag{14.41}$$

The solutions to (14.40) in the form of an infinite series usually go to infinity as $s \to 1$ whenever $m_l \neq 0$. As in the case of the Hermite polynomials in the harmonic-oscillator problem, we can prevent the solution going to ∞ only by terminating the series to give a polynomial with a finite number of terms. The condition for such polynomial solutions is that l be zero or a positive integer, such that

*For example, L. Pauling and E. B. Wilson, *Introduction to Quantum Mechanics* (New York: McGraw-Hill Book Company, 1935), p. 118.

$l \geq |m_l|$. The eigenvalue condition on the energy parameter in (14.41) accordingly becomes

$$E_l = l(l+1)\frac{\hbar^2}{2\mu R^2}, \qquad l = 0, 1, 2, 3, \ldots \tag{14.42}$$

The eigenfunctions are

$$\psi_{l,m_l}(\theta, \phi) \equiv Y_{l,m_l}(\theta, \phi) = P_l^{|m_l|}(\cos\theta)\exp(im_l\phi), \tag{14.43}$$

with

$$m_l = -l, -l+1 \cdots 0, 1 \cdots l$$

From (14.43), we note that for each energy level, specified by the value l in (14.42), there will be $2l+1$ different eigenfunctions specified by the allowed values of m_l for the given l. The energy levels of the rigid rotor therefore have a degeneracy of $2l+1$.

In the case of diatomic and linear molecules having only one moment of inertia, $I = \mu R^2$, and Eq. (14.42) is usually written in terms of a rotational quantum number J (instead of l), as

$$E = J(J+1)\frac{\hbar^2}{2I} \tag{14.44}$$

This formula was used in Section 5.15 to calculate the rotational partition function for linear molecules. It will also be applied in Chapter 17 to discuss the energy levels that yield the rotational spectra of molecules.

TABLE 14.3
Surface Spherical Harmonics

			Spherical harmonics*	
l m_l	$P_l^{m_l}(s)$		In polar coordinates	In Cartesian coordinates
0 0	1		$f_{00} = 1$	$s \quad = 1$
1 0	s		$f_{10} = \cos\theta$	$p_z \quad = z/R$
1 1	$(1-s^2)^{1/2}$		$f_{11} = \begin{cases} \sin\theta\sin\phi \\ \sin\theta\cos\phi \end{cases}$	$p_y \quad = y/R$ $p_x \quad = x/R$
2 0	$\frac{1}{2}(3s^2-1)$		$f_{20} = 3\cos^2\theta - 1$	$d_{z^2} \quad = (3z^2 - R^2)/R^2$
2 1	$3s(1-s^2)^{1/2}$		$f_{21} = \begin{cases} \sin\theta\cos\theta\sin\phi \\ \sin\theta\cos\theta\cos\phi \end{cases}$	$d_{yz} \quad = yz/R^2$ $d_{xz} \quad = xz/R^2$
2 2	$3(1-s^2)$		$f_{22} = \begin{cases} \sin^2\theta\sin^2\phi \\ \sin^2\theta\cos 2\phi \end{cases}$	$d_{xy} \quad = xy/R^2$ $d_{x^2-y^2} = (x^2 - y^2)/R^2$
3 0	$\frac{1}{2}(5s^2-3s)$		$f_{30} = 5\cos^3\theta - 3\cos\theta$	$f_{z^3} \quad = (5z^3 - 3R^2z)/R^3$
3 1	$\frac{3}{2}(1-s)^{1/2}(5s^2-1)$	$f_{31} = \begin{cases} \sin\theta(5\cos^2\theta - 1)\sin\phi \\ \sin\theta(5\cos^2\theta - 1)\cos\phi \end{cases}$	$f_{yz^2} \quad = y(5z^2 - R^2)/R^3$ $f_{xz^2} \quad = x(5z^2 - R^2)/R^3$
3 2	$15(1-s^2)s$		$f_{32} = \begin{cases} \sin^2\theta\cos\theta\sin 2\phi \\ \sin^2\theta\cos\theta\cos 2\phi \end{cases}$	$f_{xyz} \quad = xyz/R^3$ $f_{z(x^2-y^2)} = z(x^2 - y^2)/R^3$
3 3	$15(1-s^2)^{3/2}$		$f_{33} = \begin{cases} \sin^3\theta\sin 3\phi \\ \sin^3\theta\cos 3\phi \end{cases}$	$f_{y^3} \quad = y(y^2 - 3x^2)/R^3$ $f_{x^3} \quad = x(x^2 - 3y^2)/R^3$

*The f_{l,m_l} are the real forms of Y_{l,m_l}, and $R = (x^2 + y^2 + z^2)^{1/2}$.

The functions $Y_{l,m_l}(\theta, \phi)$ are called *surface spherical harmonics*. They arise in the solutions of many problems of interest, in both classical physics and quantum mechanics. An example is the problem of *waves on a flooded planet*.* Suppose the earth were a perfect sphere completely covered with water to a uniform depth. The waves on the surface of this idealized ocean could be represented by surface spherical harmonics. The functions are summarized in Table 14.3 in both polar coordinates and Cartesian coordinates.

8. Partition Function and Thermodynamics of Diatomic Rigid Rotor

The discrete energy levels of a linear rigid rotor were given in (14.44). If the moment of inertia I is sufficiently high, these energy levels become so *closely spaced* that the $\Delta\epsilon$ between adjacent levels is much less than kT, even at temperatures of a few kelvin. This condition is, in fact, realized for all diatomic molecules except H_2, HD, and D_2. For F_2, $I = 25.3 \times 10^{-40}$ g·cm^2; for N_2, 13.8×10^{-40}; but for H_2, $I = 0.47 \times 10^{-40}$. These values are calculated from the interatomic distances and the masses of the atoms.

The *multiplicity* of the rotational levels requires some consideration. The number of ways of distributing J quanta of rotational energy between *two* axes of rotation equals $2J + 1$, for in every case except $J = 0$ there are *two* possible alternatives for each added quantum. The statistical weight g of a rotational level J is therefore $2J + 1$.

The rotational partition function now becomes, from (5.52),

$$z_r = \sum (2J + 1) \exp \frac{-J(J + 1)\hbar^2}{2IkT} \tag{14.45}$$

Replacing the summation by an integration, since the levels are closely spaced compared with kT, we obtain

$$z_r = \int_0^\infty (2J + 1) \exp \left[\frac{-J(J + 1)\hbar^2}{2IkT} \right] dJ$$
$$z_r = \frac{2IkT}{\hbar^2} \tag{14.46}$$

In homonuclear diatomic molecules ($^{14}N^{14}N$, $^{35}Cl^{35}Cl$, etc.), odd and even J values are allowed, depending on the symmetry properties of the molecular eigenfunctions. In heteronuclear diatomic molecules ($^{14}N^{15}N$, HCl, NO, etc.), there are no restrictions on the allowed J's. A symmetry number σ is therefore introduced, which is either $\sigma = 1$ (heteronuclear) or $\sigma = 2$ (homonuclear). Thus, (14.46) becomes

$$z_r = \frac{2IkT}{\sigma\hbar^2} \tag{14.47}$$

*Discussed in detail by W. Kauzmann, *Quantum Chemistry* (New York: Academic Press, Inc., 1957), pp. 83–99.

As an example of the application of this equation, consider the calculation of the rotational contribution to molar entropy. From (5.37) and (5.42),

$$S_r = RT \frac{\partial \ln z_r}{\partial T} + k \ln z_r^L = R + R \ln z_r = R + R \ln \frac{2IkT}{\sigma \hbar^2}$$

Note that the rotational energy is simply RT in accordance with the Equipartition Principle.

9. The Hydrogen Atom

If we neglect the translational motion of the atom as a whole and the motion of the atomic nucleus, we can reduce the problem of the hydrogen atom to that of a single electron of mass m in a coulombic field. The motion of the nucleus can be taken into account by using the reduced mass μ of (4.37) instead of m. The problem is similar to that of a particle in a three-dimensional box, except that now there is spherical symmetry. Also, instead of steep walls and zero potential energy within, there is a gradual rise in potential with distance from the nucleus: at $r = \infty$, $U = 0$; at $r = 0$, $U = -\infty$. The potential energy of the electron in the field of the nucleus of charge Ze is given by $U = -Ze^2/r$, which is plotted in Fig. 14.2.

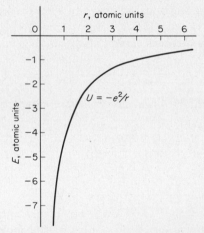

FIG. 14.2 Coulombic potential energy of negative electron in field of positive proton.

The Schrödinger equation therefore becomes

$$\frac{\partial^2 \psi}{\partial x^2} + \frac{\partial^2 \psi}{\partial y^2} + \frac{\partial^2 \psi}{\partial z^2} + \frac{2\mu}{\hbar^2}\left(E + \frac{Ze^2}{r}\right)\psi = 0 \qquad (14.48)$$

For convenience in notation, we can convert to atomic units of distance and energy, in terms of which the equation takes the form

$$\nabla^2 \psi + 2\left(E + \frac{Z}{r}\right)\psi = 0 \qquad (14.49)$$

The spherical symmetry of the potential-energy function suggests that the equation may be solved most readily in spherical polar coordinates, r, θ, and ϕ, which are shown in Fig. 14.3. The coordinate r measures the radial distance from the origin; θ is a colatitude, and ϕ a longitude. Since the electron is moving in three dimensions, three coordinates suffice to describe its position at any time.

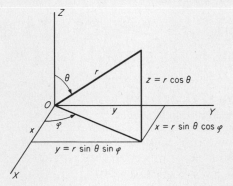

FIG. 14.3 Spherical polar coordinates.

When the Laplacian in (14.49) is transformed into polar coordinates, the equation becomes

$$\frac{1}{r^2}\frac{\partial}{\partial r}\left(r^2\frac{\partial\psi}{\partial r}\right) + \frac{1}{r^2\sin^2\theta}\frac{\partial^2\psi}{\partial\phi^2} + \frac{1}{r^2\sin\theta}\frac{\partial}{\partial\theta}\left(\sin\theta\frac{\partial\psi}{\partial\theta}\right) + 2\left(E + \frac{Z}{r}\right)\psi = 0 \tag{14.50}$$

The variables in (14.50) can be separated because the potential is a function of r alone. Let us substitute

$$\psi(r, \theta, \phi) = R(r)\Theta(\theta)\Phi(\phi) = R(r)Y(\theta, \phi)$$

That is, the wave function is a product of three functions, the first of which depends only on r, the second only on θ, and the third only on ϕ.

We find equations for the separated angular part $Y(\theta, \phi)$ and radial part $R(r)$:

$$\frac{1}{\sin\theta}\frac{\partial}{\partial\theta}\left(\sin\theta\frac{\partial Y}{\partial\theta}\right) + \frac{1}{\sin^2\theta}\frac{\partial^2 Y}{\partial\phi^2} + l(l + 1)Y = 0 \tag{14.51}$$

$$\frac{1}{r^2}\frac{d}{dr}\left(r^2\frac{dR}{dr}\right) + \left(2E + \frac{2Z}{r} - \frac{l(l + 1)}{r^2}\right)R = 0 \tag{14.52}$$

where the separation constant has been written as $l(l + 1)$. The equation for the angular function $Y(\theta, \phi)$ is identical with that found in (14.35) for the rigid-rotor problem. Thus, the angular parts of the wave functions for the hydrogen-atom problem are already at hand in Table 14.3.

The radial equation (14.52) is readily solved by a method similar to that previously used. The resulting solution is based on the *associated Laguerre polynomials*. The function

$$L_r(\rho) = \frac{e^\rho\, d^r(\rho^r e^{-\rho})}{d\rho^r} \tag{14.53}$$

gives the Laguerre polynomial of degree r. If $L_r(\rho)$ is differentiated s times with respect to ρ, one obtains the *associated Laguerre polynomial* of order s and degree $r - s$,

$$L_r^s(\rho) \equiv \frac{d^s L_r(\rho)}{d\rho^s} \tag{14.54}$$

When the $R(r)$ function is normalized, it has the form

$$R_{nl}(r) = \left[\frac{(n - l - 1)!}{2n[(n + l)!]^3}\right]^{1/2} \left(\frac{2Z}{na_0}\right)^{l+(3/2)} r^l \, e^{-Zr/na_0} \, L_{n+l}^{2l+1}\left[\left(\frac{2Z}{na_0}\right)r\right] \tag{14.55}$$

Table 14.4 lists the complete hydrogenlike wave functions for $n = 1$ and $n = 2$.

TABLE 14.4
Normalized Hydrogenlike Wave Functions

	K shell
$n = 1, l = 0, m_l = 0$	$\psi(1s) = \dfrac{1}{\sqrt{\pi}}\left(\dfrac{Z}{a_0}\right)^{3/2} e^{-Zr/a_0}$
	L shell
$n = 2, l = 0, m_l = 0$	$\psi(2s) = \dfrac{1}{4\sqrt{2\pi}}\left(\dfrac{Z}{a_0}\right)^{3/2}\left(2 - \dfrac{Zr}{a_0}\right) e^{-Zr/2a_0}$
$n = 2, l = 1, m_l = 0$	$\psi(2p_z) = \dfrac{1}{4\sqrt{2\pi}}\left(\dfrac{Z}{a_0}\right)^{3/2}\dfrac{Zr}{a_0} e^{-Zr/2a_0} \cos\theta$
$n = 2, l = 1, m_l = \pm 1$[a]	$\psi(2p_x) = \dfrac{1}{4\sqrt{2\pi}}\left(\dfrac{Z}{a_0}\right)^{3/2}\dfrac{Zr}{a_0} e^{-Zr/2a_0} \sin\theta \cos\phi$
	$\psi(2p_y) = \dfrac{1}{4\sqrt{2\pi}}\left(\dfrac{Z}{a_0}\right)^{3/2}\dfrac{Zr}{a_0} e^{-Zr/2a_0} \sin\theta \sin\phi$

[a] The functions here are real linear combinations of the $m_l = +1$ and $m_l = -1$ wave functions (see p. 640).

The energy levels in atomic units are

$$E_n = \frac{-Z^2}{2n^2} \tag{14.56}$$

or, in standard physical units (SI)

$$E_n = -\frac{\mu e^4 Z^2}{8_{\epsilon 0}{}^2 h^2 n^2} \tag{14.57}$$

where μ is the reduced mass of the nucleus and electron. This expression for E_n is identical with that from the old quantum theory of Bohr.

10.   Angular Momentum

The definition in classical mechanics of the angular momentum is recalled in Fig. 14.4. The angular momentum $\mathbf{L}$ of the particle of mass m at the end of vector $\mathbf{r}$

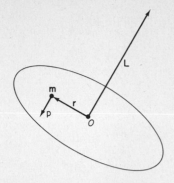

FIG. 14.4 The angular momentum **L** of particle of mass m with respect to point O is defined as

$$\mathbf{L} \equiv \mathbf{r} \times \mathbf{p} \equiv \mathbf{r} \times m\mathbf{v}$$

where **p** is the linear momentum of the particle and **r** is the radius vector from O to the particle. The vector **L** is normal to the plane defined by **r** and **p**.

from the fixed point O is defined as

$$\mathbf{L} \equiv \mathbf{r} \times \mathbf{p} \equiv \mathbf{r} \times m\mathbf{v} \tag{14.58}$$

where **p** is the linear momentum and **v** the velocity of the particle.

In Cartesian coordinates,* (14.58) becomes

$$\begin{aligned} L_x &= yp_z - zp_y \\ L_y &= zp_x - xp_z \\ L_z &= xp_y - yp_x \end{aligned} \tag{14.59}$$

Application of Postulate VI from p. 616 converts the angular momentum in (14.58) to an operator,

$$\hat{\mathbf{L}} = -i\hbar(\mathbf{r} \times \nabla) \tag{14.60}$$

Or, for the components in (14.59),

$$\begin{aligned} \hat{L}_x &= -i\hbar\left(y\frac{\partial}{\partial z} - z\frac{\partial}{\partial y} \right) \\ \hat{L}_y &= -i\hbar\left(z\frac{\partial}{\partial x} - x\frac{\partial}{\partial z} \right) \\ \hat{L}_z &= -i\hbar\left(x\frac{\partial}{\partial y} - y\frac{\partial}{\partial x} \right) \end{aligned} \tag{14.61}$$

Since the eigenfunctions are usually given in spherical polar coordinates, it is convenient to transform the operators in (14.61) into this system:

---

*In terms of unit vectors **i**, **j**, **k** directed along the x, y, z axes, respectively, any vector **A** can be written in terms of its Cartesian components as $\mathbf{A} = A_x\mathbf{i} + A_y\mathbf{j} + A_z\mathbf{k}$. Hence,

$$\mathbf{r} \times \mathbf{p} = (x\mathbf{i} + y\mathbf{i} + z\mathbf{k}) \times (p_x\mathbf{i} + p_y\mathbf{j} + p_z\mathbf{k})$$

Since the vector product obeys the distributive law of multiplication,

$$\mathbf{A} \times (\mathbf{B} + \mathbf{C}) = \mathbf{A} \times \mathbf{B} + \mathbf{A} \times \mathbf{C}$$

the product $\mathbf{r} \times \mathbf{p}$ becomes $\mathbf{r} \times \mathbf{p} = \mathbf{i}(yp_z - zp_y) + \mathbf{j}(zp_x - xp_z) + \mathbf{k}(xp_y - yp_x)$. [See M. L. Boas, *Mathematical Methods in the Physical Sciences* (New York: John Wiley & Sons, Inc., 1966), p. 203.]

$$\hat{L}_x = i\hbar\left(\cot\theta\cos\phi\,\frac{\partial}{\partial\phi} + \sin\phi\,\frac{\partial}{\partial\theta}\right)$$

$$\hat{L}_y = i\hbar\left(\cot\theta\sin\phi\,\frac{\partial}{\partial\phi} - \cos\phi\,\frac{\partial}{\partial\theta}\right) \tag{14.62}$$

$$\hat{L}_z = -i\hbar\left(\frac{\partial}{\partial\phi}\right)$$

The operator

$$\hat{L}^2 = \hat{L}_x^2 + \hat{L}_y^2 + \hat{L}_z^2$$

becomes

$$\hat{L}^2 = -\hbar^2\left[\frac{1}{\sin\theta}\frac{\partial}{\partial\theta}\left(\sin\theta\,\frac{\partial}{\partial\theta}\right) + \frac{1}{\sin^2\theta}\frac{\partial^2}{\partial\phi^2}\right] \tag{14.63}$$

We should note that this expression is identical with the angular part of the Laplacian operator in spherical coordinates, a useful result.

It is now easy to show that the hydrogenlike wave functions are eigenfunctions for the operators $\hat{L}_z$ and $\hat{L}^2$. Therefore, each eigenfunction corresponds to a definite measurable value of the total angular momentum and of the z component of the angular momentum.*

$$\hat{L}^2\psi = l(l+1)\hbar^2\psi \tag{14.64}$$

$$\hat{L}_z\psi = m_l\hbar\psi \tag{14.65}$$

11. Angular Momentum and Magnetic Moment

Solutions of the Schrödinger equation, for an electron in a defined potential, yield probability functions for the velocities of the electron as well as for its position. A moving electron is an electric current and every electric current creates a magnetic field. If an electron located at the end of a vector **r** from the origin, which may be a stationary nucleus, is moving with a velocity **v**, the magnetic induction at the origin is

$$\mathbf{B(r)} = -\frac{\mu_0}{4\pi}\frac{e(\mathbf{r}\times\mathbf{v})}{r^3} = -\frac{\mu_0}{4\pi}\frac{e\mathbf{L}}{mr^3} \tag{14.66}$$

since $m\mathbf{r}\times\mathbf{v} = \mathbf{L}$, the angular momentum. The permeability of vacuum is $\mu_0 = 4\pi\times10^{-7}$ J·s²·C⁻²·m⁻¹. In SI units, where charge is in coulombs and distance in meters, **B** is measured in a unit kg·s⁻²·A⁻¹ called the *tesla* (T), equal to 10^4 gauss (G) in old electromagnetic units.

As viewed from far away, the magnetic induction of a moving electron is equivalent to that of a tiny bar magnet with magnetic moment $\mathbf{p}_m$. The magnitude of the magnetic field of such a magnet would be

$$B(r) = -\frac{\mu_0 p_m}{2\pi r^3} \tag{14.67}$$

---

*We shall leave the calculation of (14.64) and (14.65) from (14.62) and (14.63) as an exercise for the student.

From (14.66) and (14.67), the ratio of the magnetic moment to angular momentum, called the *magnetogyric ratio* γ, is

$$\gamma = \frac{\mathbf{p}_m}{\mathbf{L}} = \frac{e}{2m}$$

The vectors $\mathbf{p}_m$ and $\mathbf{L}$ are parallel, directed along an axis normal to the plane of the current loop (Fig. 14.5).

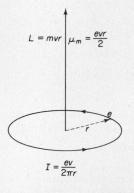

$$L = mvr \quad \mu_m = \frac{evr}{2}$$

$$I = \frac{ev}{2\pi r}$$

FIG. 14.5 Viewed from a distance which is long compared to the radius r of the current loop, the magnetic field due to the current is equivalent to that of a magnetic dipole of moment $\boldsymbol{\mu}$ directed parallel to the angular momentum vector $\mathbf{L}$.

For the case of the electron in the hydrogen atom, the angular momentum can have only quantized values, $\sqrt{l(l+1)}\hbar$, and its component in the direction of a field is restricted to values $m_l\hbar$. The physical nature of the coupling of the orbital angular momentum with the external magnetic field $\mathbf{B}'$ is the magnetic interaction of $\mathbf{p}_m$ with $\mathbf{B}'$, which has a potential energy

$$U = -\mathbf{p}_m \cdot \mathbf{B}' = -p_m B_z' \cos\theta \qquad (14.68)$$

where θ is the angle between the field direction z and the magnetic moment.

The allowed values of the component of magnetic moment in the field direction are, therefore,

$$p_{m,z} = \frac{m_l \hbar e}{2m}$$

There is, therefore, a natural unit of magnetic moment,

$$\mu_B = \frac{e\hbar}{2m} \qquad (14.69)$$

which is called the *Bohr magneton*. In SI units,

$$\mu_B = (9.7232 \pm 0.0006) \times 10^{-24}\ \text{J}\cdot\text{T}^{-1}\ (\text{or m}^2\cdot\text{A})^*$$

12. The Quantum Numbers

The eigenfunctions $\psi(n, l, m_l)$ for a single electron in the field of a nucleus are specified by three quantum numbers, as would be expected for a three-dimensional boundary-value problem in wave mechanics.

$^*\mu_B = 9.2732 \times 10^{-21}\ \text{erg}\cdot\text{gauss}^{-1}.$

The *principal quantum number* n is the successor to the n introduced by Bohr in his theory of the hydrogen atom. The total number of nodes in the wave function is equal to $n - 1$.* These nodes may be either in the radial function $R(r)$, or in the azimuthal function $\Theta(\theta)$.

The quantum number l is called the *azimuthal quantum number* or *angular momentum quantum number*. It is equal to the number of nodes in $\Theta(\theta)$, i.e., to the number of nodal surfaces passing through the origin.† Since the total number of nodes is $n - 1$, the allowed values of l are from 0 to $n - 1$. When $l = 0$, there are no nodes in the function $\Theta(\theta)$ and the wave function is spherically symmetrical about the central nucleus. The electron has an angular momentum $\mathbf{L}$, which is quantized in such a way that its magnitude is

$$|\,\mathbf{L}\,| = \sqrt{l(l + 1)}(\hbar) \tag{14.70}$$

States with $l = 0$ therefore have zero angular momentum.

States with $l = 0, 1, 2, 3$ are denoted as s, p, d, and f states, respectively. In designating a state, the principal quantum number is prefixed to the letter denoting the azimuthal quantum number. For example, $n = 1$, $l = 0$ is a $1s$ state; $n = 2$, $l = 1$ is a $2p$ state; etc.

The quantum number m_l is called the *magnetic quantum number*. The electron in the unperturbed hydrogen atom, in states with $l \neq 0$, has a certain angular momentum given by (14.64). The z component of angular momentum is restricted to $m_l\hbar$. That is, we have a vector of classical length $\sqrt{l(l + 1)}\hbar$ with z component $m_l\hbar$ but with zero average x and y components. This situation allows us to picture a *precession* of the angular momentum vector $\mathbf{L}$ about the z axis, giving zero x and y components on the average.

If the hydrogen atom is placed in a magnetic field directed along the z axis, a definite direction in space is physically established by the field, and the angular momentum vector precesses about this field direction. The solutions of the Schrödinger equation are such that not every orientation between the angular momentum vector and the field direction is allowed. The only allowed directions are those at which the components of the angular momentum along the z axis have certain quantized values given by (14.65) as $L_z = m_l\hbar$.

This behavior is illustrated in Fig. 14.6 for the case in which the azimuthal number $l = 2$. The magnetic number m_l can then have the values $-2, -1, 0, 1, 2$. For any value of l, which specifies the total angular momentum, there are $2l + 1$ values of m_l, which specify the allowed components of the angular momentum in the field direction.

The energy of the precessional motion is quantized. The allowed energy levels are spaced such that $\epsilon = m_l h\nu$, where ν is the frequency of precession of the angular momentum vector in the magnetic field. This is the *Larmor frequency*,

---

*If we consider that there is a nodal surface in the radial function at $r = \infty$, the total number of nodes equals n.

†In case $m_l = l$, the nodal surface becomes a nodal line.

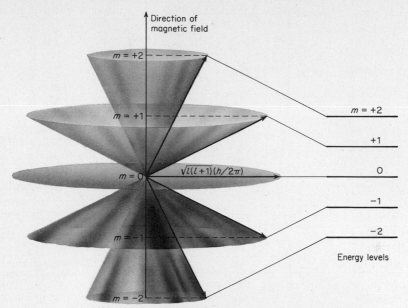

FIG. 14.6 Quantization of the components of angular momentum in a magnetic field **B** for the case $l = 2$.

$$v = \frac{eB}{4\pi m} \qquad (14.71)$$

These energy levels are indicated in Fig. 14.6.

13. The Radial Wave Functions

Figure 14.7(a) is a plot of the radial parts of the wave functions for a few values of n and l. These plots show clearly the nodes in the waves, and the fact that the number of nodes in the radial function equals $n - l - 1$. The amplitude ψ of the electron wave can be positive or negative. The probability of finding the electron in the region between r and $r + dr$ is proportional to $\psi^*\psi = |\psi|^2$, the square of the absolute value of the amplitude. Often we need to know the probability that the electron is at a given distance r from the nucleus, irrespective of direction—in other words, the probability that the electron lies between two spheres, of radii r and $r + dr$. The volume of this spherical shell is $4\pi r^2\, dr$. Hence, the probability of finding the electron somewhere within this shell is $4\pi r^2\psi^*\psi\, dr$. The function $4\pi r^2\psi^*\psi$ is called the *radial distribution function* $D(r)$. It has been plotted in Fig. 14.7(b) for the values of n and l previously used in Fig. 14.7(a).

Consider first the case of $n = 1$, $l = 0$, called a 1s state of the electron in the hydrogen atom. This is the lowest or *ground state* of hydrogen. In the old Bohr theory, the electron in this state revolved in a circular orbit of radius $a_0 =$

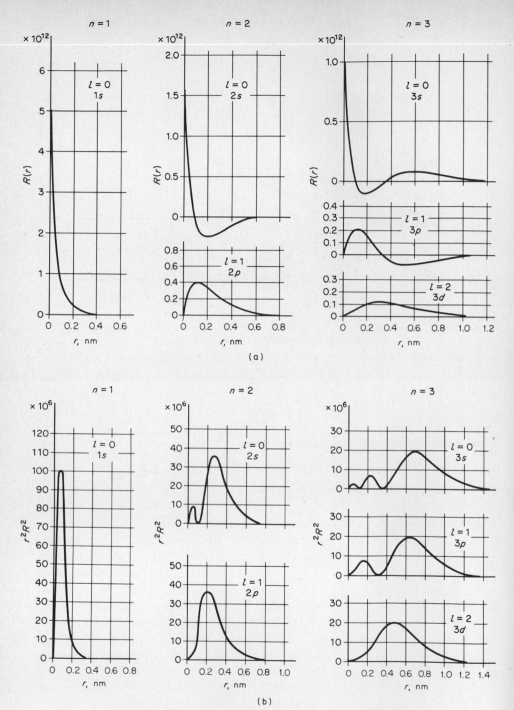

FIG. 14.7 (a) Radial part of wave functions for hydrogen atom; (b) radial distribution functions—giving probability of finding an electron at a given distance r from nucleus after averaging over its angular variables. [After G. Herzberg, *Atomic Spectra* (New York: Dover Publications, Inc., 1944).]

0.0529 nm. The quantum mechanical result indicates that the electron has a certain probability of being anywhere from $r = 10^{-15}$ m, right in the center of the nucleus, to $r = 10^{20}$ m, out in the Milky Way. Nevertheless, the position of *maximum probability* for the electron does correspond to the value $r = a_0$, and the chance that the electron is very far away from the nucleus is comfortingly small. For example, what is the probability that the electron is at a distance of 10 a_0 or 0.529 nm? From Table 14.4, $\psi_{1s} = (\pi a_0^3)^{-1/2} e^{-r/a_0}$, so that the relative probability $(r^2 \psi^2)$ compared to that of finding the electron at a_0 is $10^2 (e^{-10}/e^{-1})^2 = 1.52 \times 10^{-6}$. This is a small but far from negligible number. Another way of expressing the result is to say that one out of about 7×10^5 hydrogen atoms at any instant might be expected to have its electron at a distance 10 a_0 from the nucleus. If an ordinary hydrogen atom were the size of a golf ball, this exceptional hydrogen atom would be the size of a basketball. Of course, it does not stay so big for long, and if we took our snapshot an instant later, we might find the basketball back to its usual size while another golf ball had shrunk to the size of a pea. Thus the hydrogen atom of wave mechanics is not a static rigid structure of fixed size.

In the 2s state, $n = 2$, $l = 0$, we find, in addition to the main peak in the radial distribution function at $r = 5.2\ a_0$, another little peak in the probability at a much smaller value, $r = 0.8\ a_0$. This effect, called *penetration*, gives an important new insight into the behavior of electrons in atomic shells. It shows that in certain states the electron spends a small proportion of its time very close to the nucleus. For a given value of n, penetration is greater the smaller is the value of l, the azimuthal quantum number. Because an electron close to the nucleus experiences a high electrostatic attraction, penetration has a considerable effect in lowering the energy of an electron and thus tends to increase the stability of states in which it occurs.

A wave function for a single electron is called an orbital. The old theory spoke of the *orbits* of electrons revolving like tiny planets about the nucleus as a sun. In the new theory, there are no orbits and all the information about the position of an electron is summarized in its orbital ψ.

14. Angular Dependence of Hydrogen Orbitals

Orbitals with $l = 0$ are called s orbitals and are always spherically symmetric. In this case, ψ is a function of r alone, and does not depend on θ or ϕ. Orbitals with $l = 1$, called p orbitals, have a marked directional character because the ψ function depends on θ and ϕ. The d orbitals, with $l = 2$, are also directional, with rather complicated angular dependences.

In the angular part of the wave functions for the hydrogen atom, the function $\Phi(\phi)$ appears in the complex form

$$\Phi_{m_l}(\phi) = \frac{1}{\sqrt{2\pi}}\, e^{im_l \phi} \tag{14.72}$$

where m_l can take the values $0, \pm 1, \pm 2, \pm 3, \ldots, \pm l$.

For states with $l = 1$ (p orbitals), m_l can be -1, 0, $+1$. For states with $l = 2$ (d orbitals), m_l can be -2, -1, 0, $+1$, $+2$. To display the angular dependence of these orbitals, it is helpful to form new *real* eigenfunctions by linear combinations of the complex ones in (14.72). (The general superposition property of solutions of linear differential equations ensures that such linear combinations of solutions will themselves be solutions.) Table 14.5 shows the linear combinations of the $\Phi(\phi)$ functions, which are the basis for the usual discussion of the angular dependence. The subscripts on the orbitals in complex form denote the values of the quantum number m_l to which they correspond. The subscripts on the real linear-combination orbitals denote the directional properties of the orbitals (or, more precisely, how they transform under rotational operations).

TABLE 14.5
The Functions $\Phi(\phi)$ in the Hydrogenlike Atomic Orbitals

Complex forms	Real linear combinations
p orbitals	
$p_{+1} = \dfrac{1}{\sqrt{2\pi}} e^{i\phi}$	$p_x = \dfrac{1}{\sqrt{2}}(p_1 + p_{-1})$
$p_0 = \dfrac{1}{\sqrt{2\pi}}$	$p_z = p_0$
$p_{-1} = \dfrac{1}{\sqrt{2\pi}} e^{-i\phi}$	$p_y = \dfrac{-i}{\sqrt{2}}(p_1 - p_{-1})$
d orbitals	
$d_{+2} = \dfrac{1}{\sqrt{2\pi}} e^{i2\phi}$	$d_{z^2} = d_0$
$d_{+1} = \dfrac{1}{\sqrt{2\pi}} e^{i\phi}$	$d_{xz} = \dfrac{1}{\sqrt{2}}(d_{+1} + d_{-1})$
$d_0 = \dfrac{1}{\sqrt{2\pi}}$	$d_{yz} = \dfrac{-i}{\sqrt{2}}(d_{+1} - d_{-1})$
$d_{-1} = \dfrac{1}{\sqrt{2\pi}} e^{-i\phi}$	$d_{xy} = \dfrac{-i}{\sqrt{2}}(d_{+2} + d_{-2})$
$d_{-2} = \dfrac{1}{\sqrt{2\pi}} e^{-i2\phi}$	$d_{x^2-y^2} = \dfrac{1}{\sqrt{2}}(d_{+2} - d_{-2})$

Let us consider a couple of examples. The angular part of an orbital denoted $p_0 = p_z$, from Tables 14.3 and 14.5, is

$$\Theta(\theta)\Phi(\phi) = \frac{1}{\sqrt{2\pi}} \cos\theta$$

The transformations of Cartesian to spherical coordinates, shown in Fig. 14.3, are

$$x = r \sin\theta \cos\phi$$
$$y = r \sin\theta \sin\phi$$
$$z = r \cos\theta$$

Hence,

$$p_0 = p_z = \frac{1}{\sqrt{2\pi}} \frac{1}{r} z$$

We see at once why this orbital is denoted as p_z: it has the directional properties of the z coordinate.

If we examine the angular part of p_x, we find

$$p_x = \frac{1}{\sqrt{2}} \sin\theta \left(\frac{1}{\sqrt{2\pi}}\right)(e^{i\phi} + e^{-i\phi})$$

Since

$$\frac{e^{i\phi} + e^{-i\phi}}{2} = \cos\phi$$

$$p_x = \frac{1}{\sqrt{\pi}} \sin\theta \cos\phi = \frac{1}{\sqrt{\pi}} \frac{1}{r} x$$

By similar transformations, we can readily verify the assignments of the subscripts to the other orbitals in Table 14.5.

There are several ways to illustrate the angular dependence of the orbitals. We could plot $\Theta\Phi(\theta, \phi)$ on two separate graphs in polar coordinates. An example of such a plot of $\Theta_{11}(\theta)$ ($l = 1$, $m_l = 1$) is shown in Fig. 14.8, together with $\Theta_{11}^2(\theta)$. The functions plotted, from Table 14.3, are simply $\sin\theta$ and $\sin^2\theta$.

One of the most pictorial ways to display the angular dependence of orbitals is to draw in three-dimensional perspective a representation of the surfaces $\Theta(\theta)$

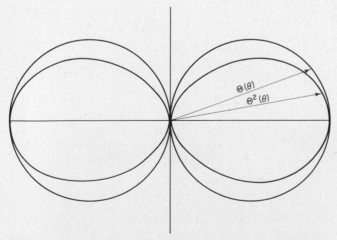

FIG. 14.8 The angular dependence of the wave function $\Theta_{11}(\theta)$ and its square Θ_{11}^2. These are simply graphs of $\sin\theta$ and $\sin^2\theta$ in polar coordinates, the amplitude of the functions at any point being equal to the distance from the origin. These plots display the angular dependence of a p orbital and its square, the latter being proportional to electron density.

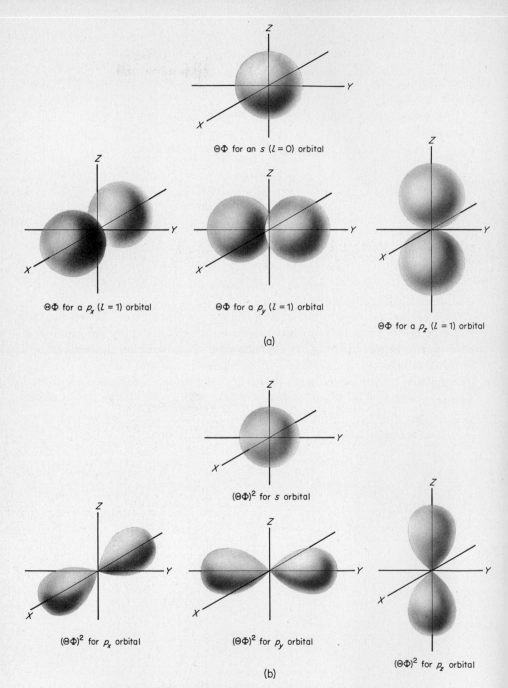

FIG. 14.9 Polar graphs: (a) the absolute values of $\Theta\Phi$, the angular part of the hydrogen-atom wave functions for $l = 0(s)$ and $l = 1(p)$ orbitals; (b) $(\Theta\Phi)^2$ which are proportional to the electron densities.

$\Phi(\phi)$, the spherical surface harmonics. Such surfaces are shown in Fig. 14.9 for the orbitals with $l = 0$ (*s* orbitals) and $l = 1$ (*p* orbitals). We also show $\Theta^2\Phi^2$ since the squares of the angular wave functions determine the electron densities. To obtain the actual amplitude of the wave functions, we must multiply the angular parts $\Theta(\theta)\Phi(\phi)$ shown in Fig. 14.9 by the radial parts $R(r)$ shown in Fig. 14.7. In the same way, we can obtain the electron densities by multiplying $\Theta^2\Phi^2$ in Fig. 14.9 by $R^2(r)$. The angular dependence is the same for all values of r. Thus, the solid shapes in Fig. 14.9 do *not* imply that the orbitals are sharply defined in space. They simply depict the angular dependence of the orbitals for any value of r.

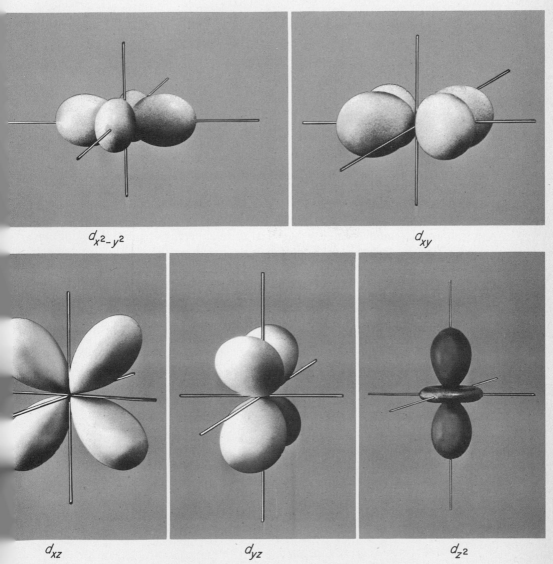

FIG. 14.10 The atomic *d* orbitals (hydrogenlike) [R. G. Pearson, Northwestern University—American Chemical Society].

There are five d orbitals, with $l = 2$ and $m_l = -2, -1, 0, 1, 2$. There are different ways of choosing linearly independent combinations of these orbitals so as to give real functions. The most common solution is that given in Table 14.5. The corresponding surfaces are shown in Fig. 14.10 as three-dimensional models of the angular dependence of $\Theta^2\Phi^2$.

15. The Spinning Electron

In 1921, while doing research in Rutherford's laboratory at Cambridge, Arthur Compton, a young American physicist, had the idea that an electron might possess an intrinsic angular momentum or spin, and thus act as a little magnet. In 1925, Wolfgang Pauli investigated the problem of why lines in the spectra of alkali metals are not single as predicted by the Bohr theory, but actually made up of two closely spaced components. (An example was shown in Fig. 13.7.) He showed that the doublet in the fine structure could be explained if an electron could exist in two distinct states.

G. E. Uhlenbeck and S. Goudsmit of Leiden identified these states as two states of different angular momentum. They suggested, therefore, that the electron possesses an intrinsic angular momentum or *spin*, specified by a quantum number s. By analogy with the relation between the quantum number l and the orbital momentum, the spin angular momentum has a magnitude $S = \sqrt{s(s+1)}\hbar = (\sqrt{3}/2)\hbar$, since s always $= \frac{1}{2}$. Doublets will occur in the alkali-metal spectra if the spin angular momentum can have only two different orientations along any physically established axis, such that the components in the direction of the axis are specified by a quantum number m_s (analogous to m_l), where $m_s = +\frac{1}{2}$ or $-\frac{1}{2}$. These relations are summarized in Fig. 14.11.

The spinning electron acts like a little magnet. The magnetogyric ratio for

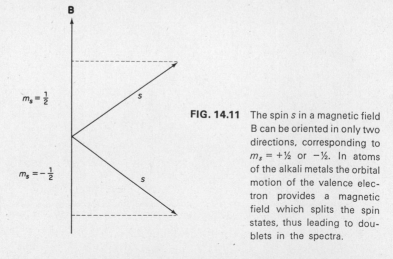

FIG. 14.11 The spin s in a magnetic field B can be oriented in only two directions, corresponding to $m_s = +\frac{1}{2}$ or $-\frac{1}{2}$. In atoms of the alkali metals the orbital motion of the valence electron provides a magnetic field which splits the spin states, thus leading to doublets in the spectra.

electron spin is $\gamma_s = e/m$, just twice* the $\gamma = e/2m$ for the orbital magnetic moment (p. 635). Therefore, the magnetic moment of the electron is

$$\mathbf{P}_m = \gamma_s S = \frac{e}{m} \cdot \frac{\sqrt{3}}{2} \hbar$$

A beautiful experimental demonstration of the fact that the electron acts like a magnet was given by the famous *Stern–Gerlach experiment*. A beam of atoms of an alkali metal, when passed through a strong inhomogeneous magnetic field, is split into two separate beams. Analysis shows that the effect can only be due to the outermost electron in the atom, which may have a spin quantum number $m_s = +\frac{1}{2}$ or $-\frac{1}{2}$. The field splits the atoms into two beams corresponding to the two different spins of their unpaired electrons.

The concept of electron spin first appeared as an extra hypothesis, which needed to be tacked onto the rest of the theory. As a result, four quantum numbers were obtained: n, l, m_l, and m_s. Inasmuch as a quantum number occurs in wave theory for each dimension in which an electron can move, four quantum numbers would correspond to four dimensions. The *fourth dimension* is a phrase made famous by Einstein's relativity theory, in which events are described in a four-dimensional world of three space coordinates and one time coordinate. When Paul Dirac, an English theoretical physicist, worked out a relativistic form of wave mechanics for the electron, he found that the property of spin was a natural consequence of his theory, and no separate spin hypothesis was necessary.[†]

16. Spin Postulates

To include the spin properties of particles in a nonrelativistic quantum mechanics, we must add two new postulates to the seven postulates previously stated for particles without spin.

Postulate VIII:

The operators for spin angular momentum commute and combine in the same way as those for ordinary angular momentum.

Therefore, corresponding with the equations in Section 14.10, we can write equations for spin angular momenta by introducing $\hat{\mathcal{S}}$ wherever we previously had $\hat{L}$.

---

*Relativistic theory confirmed by experiment shows that the factor is $g_e = 2.0023$ instead of exactly 2.

†An article by R. E. Powell, "Relativistic Quantum Chemistry" [*J. Chem. Educ.*, *45*, 558 (1968)] should be consulted for an excellent discussion of the Dirac approach. Powell shows that in the Dirac theory the hydrogenlike orbitals do not have nodes. Thus, if we refer to the quotation at the heading of Chapter 13, we might conclude that *angels* behave like particles *without spin*.

Postulate IX:

For a single electron there are only two possible eigenfunctions of $\hat{\mathscr{S}}^2$ and $\hat{\mathscr{S}}_z$. Their eigenfunctions are called α and β. The eigenvalue equations are

$$\hat{\mathscr{S}}_z\alpha = \tfrac{1}{2}\hbar\alpha, \qquad \hat{\mathscr{S}}_z\beta = -\tfrac{1}{2}\hbar\beta \qquad (14.73)^*$$

and

$$\hat{\mathscr{S}}^2\alpha = \tfrac{1}{2}(\tfrac{1}{2}+1)\hbar^2\alpha, \qquad \hat{\mathscr{S}}^2\beta = \tfrac{1}{2}(\tfrac{1}{2}+1)\hbar^2\beta \qquad (14.74)$$

It is usual to say that an electron represented by spin eigenfunction α has spin $+\tfrac{1}{2}$, and an electron represented by spin eigenfunction β has spin $-\tfrac{1}{2}$. We must understand, however, that the z component of spin angular momentum has a physical significance only if there is some physical way to designate a z axis. Thus, it is customary to depict the vectors for spin angular momentum as in Fig. 14.11, where the z axis is specified by the direction of a magnetic field **B**.

The wave function for an electron can now be represented as the product of an orbital function and a spin function, $\Phi(x, y, z, t)\alpha$ or $\Phi(x, y, z, t)\beta$. Since the spin functions α and β are not functions of the space coordinates x, y, z, the orbital angular momentum operator $\hat{L}$ and the spin angular momentum operator $\hat{\mathscr{S}}$ always commute with each other.

17. The Pauli Exclusion Principle

An exact solution of the Schrödinger wave equation for an atom has been obtained only in the case of the hydrogen atom—a single electron in the field of a positive charge. For some discussions, it is a quite good approximation to assume that electrons in more complex atoms move in the spherically symmetric field of a shielded nucleus. In this *central field approximation*, the allowed stationary states of the electron in the atom can still be classified in terms of the four quantum numbers n, l, m_l, and m_s.

An important rule governs the quantum numbers allowed for electrons in an atom. This is the *Exclusion Principle*, first enunciated by Wolfgang Pauli in a restricted form in 1924 and later given in a more general form. The Pauli Exclusion Principle declares that no two electrons can exist in the same quantum state.[†] In the treatment of atomic structure by the central-field approximation, each electron is assigned to an orbital specified by its four quantum numbers. Thus the Pauli Principle requires that no two electrons in a given atom can have all four quantum numbers, n, l, m_l, and m_s, the same.

Consider in Fig. 14.12 two electrons A and B. Each electron can be denoted by

---

*α and β are not eigenfunctions for $\hat{\mathscr{S}}_x$ and $\hat{\mathscr{S}}_y$, but

$$\hat{\mathscr{S}}_x\alpha = \tfrac{1}{2}\hbar\beta, \qquad \hat{\mathscr{S}}_x\beta = \tfrac{1}{2}\hbar\alpha$$
$$\hat{\mathscr{S}}_y\alpha = \tfrac{1}{2}i\hbar\beta, \qquad \hat{\mathscr{S}}_y\beta = -\tfrac{1}{2}i\hbar\alpha$$

†See the first part of the quotation beginning Chapter 13.

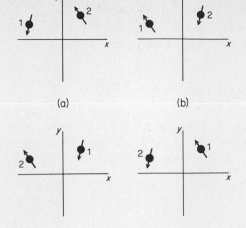

(a)   (b)

(c)   (d)

FIG. 14.12 Exchange of space and spin coordinates for two electrons: (a) original configuration; (b) spin coordinates exchanged; (c) space coordinates exchanged; (d) spin and space coordinates exchanged.

a set of three space coordinates (x, y, z) (of which only x and y are shown here) and a spin coordinate, which can have either of two values, shown by an arrow in the diagram. We can exchange the spacial and/or the spin coordinates of the two electrons, as shown in the diagram. Since the electrons are indistinguishable particles, when we make such an exchange, the wave function ψ of the system must either remain the same ($\psi \rightarrow \psi$) or simply change in sign ($\psi \rightarrow -\psi$). In the first case, we call ψ *symmetric* for the exchange; in the second case, we call ψ *antisymmetric* for the exchange. The total wave function for one electron can be written as a product of a spin part σ (α or β) and a coordinate part ϕ,

$$\psi = \phi(x, y, z)\sigma$$

The general statement of the Pauli Principle, independent of the central field approximation, is as follows: A wave function for a system of electrons must be *antisymmetric* for exchange of the spacial and the spin coordinates of any pair of electrons. (If, therefore, $\phi \rightarrow -\phi$, $\sigma \rightarrow \sigma$; if $\phi \rightarrow \phi$, $\sigma \rightarrow -\sigma$, so that $\psi \rightarrow -\psi$, always.)

The statement of the Pauli Principle in terms of quantum numbers is a special case of the general statement. Consider two electrons 1 and 2, in states specified in the central field model by $n_1, l_1, m_{l_1}, m_{s_1}$, and $n_2, l_2, m_{l_2}, m_{s_2}$. An antisymmetric function would be

$$\psi = \psi_{n_1, l_1, m_{l_1}, m_{s_1}}(1)\, \psi_{n_2, l_2, m_{l_2}, m_{s_2}}(2) - \psi_{n_1, l_1, m_{l_1}, m_{s_1}}(2)\, \psi_{n_2, l_2, m_{l_2}, m_{s_2}}(1)$$

When (1) and (2) are exchanged, $\psi \rightarrow -\psi$, as is necessary. Suppose, however, all four quantum numbers were the same for the two electrons. Then $\psi = 0$, i.e., no such state can exist.

J. C. Slater first pointed out that the antisymmetric wave function for a system of N electrons could most conveniently be written as a determinant:

$$\psi(1, 2, \ldots, N) = \frac{1}{\sqrt{N}} \begin{vmatrix} \psi_a(1) & \psi_a(2) & \cdots & \psi_a(N) \\ \psi_b(1) & \psi_b(2) & \cdots & \psi_b(N) \\ \cdot & & & \\ \cdot & & & \\ \cdot & & & \\ \psi_n(1) & \psi_n(2) & \cdots & \psi_n(N) \end{vmatrix}$$

The subscripts on the ψ's may denote the four quantum numbers specifying an orbital. If we exchange any two columns, we change the sign of the determinant. Hence, the antisymmetry requirement holds. If any set of four quantum numbers becomes the same, for example, $a = b$, then two rows would become the same and the determinant would vanish.

18. Spin–Orbit Interaction

The doublet structure of the atomic spectra of the alkali metals provided the first spectroscopic evidence for electron spin, but the same kind of effect can be seen in the spectrum of atomic hydrogen, provided it is examined with a spectrograph of high resolving power. As an example, consider the H_α line in the Balmer series, the *fine structure* of which is shown in Fig. 14.13. This line arises from transitions

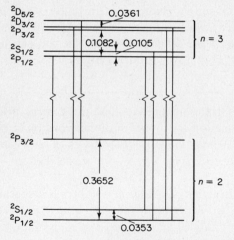

FIG. 14.13 Fine structure of the H_α line in Balmer series of atomic hydrogen. Energy differences are σ in cm^{-1}. The splitting of the states corresponding to different values of j is the result of spin–orbit interaction. The splitting of the $^2S_{1/2}$ and $^2P_{1/2}$ is a result of the Lamb shift, a quantum electrodynamic effect. [See W. E. Lamb and R. C. Retherford, *Phys. Rev.*, *72*, 241 (1947). Also H. G. Kuhn and G. W. Series, *Proc. Roy. Soc. A 202*, 127 (1950).]

between energy levels with $n = 2$ and those with $n = 3$. In the Bohr theory, and in the solutions of the Schrödinger equation for an electron without spin, these energy levels are degenerate, in that several wave functions correspond to each of the levels:

$$n = 2, \qquad l = 0, 1 \qquad 2s, 2p$$
$$n = 3, \qquad l = 0, 1, 2 \qquad 3s, 3p, 3d$$

In the conventional notation, an atomic energy level or term is represented by a capital letter corresponding to the magnitude of $\mathbf{L} = \Sigma\,\mathbf{l}_i$, where $\mathbf{l}_i$ are the l values of the individual electrons. When $L = 0, 1, 2, 3, \ldots$, the term symbol is S, P, D, F, ...

In the case of the H atom, there is only one electron and $L = l$, but the term symbols are used for uniformity. Thus, the terms would be

$$n = 2, \qquad \text{S, P}$$
$$n = 3, \qquad \text{S, P, D}$$

When the electron spin is considered, a new effect occurs which "lifts the degeneracy" of the $n = 2$ and $n = 3$ states, in other words, causes the terms S, P, and D to differ slightly in energy. This effect, called *spin–orbit interaction*, is a magnetic interaction between the magnetic moment due to spin and the magnetic moment due to orbital angular momentum. The interaction is usually written as a contribution to the Hamiltonian of the form

$$H_{s.o.} = \xi(r)\,\mathbf{l}\cdot\mathbf{s} \tag{14.75}$$

where $\mathbf{l}$ designates the vector orbital angular momentum and $\mathbf{s}$ the spin angular momentum.

The spin–orbit interaction arises as follows. The electron with its spin moves with velocity $\mathbf{v}$ in the electrostatic field $\mathbf{E}$ of the shielded nucleus. If the electron were stationary and the nucleus moving around it, it would be obvious that the electron would experience a magnetic field. It makes no difference which charge is considered to be moving, the effective magnetic field is still $\mathbf{B}' = (\mathbf{v}/c \times \mathbf{E})$. The interaction of the electron spin magnetic moment $\boldsymbol{\mu}$ with this field is $\boldsymbol{\mu}\cdot\mathbf{B}'$. For a spherically symmetrical electric potential Φ, $\mathbf{E} = -(\partial\Phi/\partial r)(\mathbf{r}/r^2)$, and thus $\mathbf{B}'$ is proportional to $\mathbf{r} \times m\mathbf{v} = \mathbf{L}$. Then, since $\boldsymbol{\mu}$ is proportional to $\mathbf{s}$, the form of the interaction is $\lambda\,\mathbf{l}\cdot\mathbf{s}$.

The result of the interaction for the electron in a hydrogen atom is that l and s are coupled magnetically to give a new inner quantum number $j = l \pm s$, with $s = \frac{1}{2}$. Consequently, for example, the P terms for $n = 2$ split into two, corresponding to $j = l + s = \frac{3}{2}$ and $j = l - s = \frac{1}{2}$. The j value is written as a subscript to the term symbol. We obtain the following terms:

$$n = 2, \quad \text{S}_{1/2}, \text{P}_{1/2}, \text{P}_{3/2}$$
$$n = 3, \quad \text{S}_{1/2}, \text{P}_{1/2}, \text{P}_{3/2}, \text{D}_{3/2}, \text{D}_{5/2}$$

The fine structure shown in Fig. 14.13 is thus satisfactorily explained.

19. Spectrum of Helium

The atomic spectrum of helium proved to be unexpectedly complicated. After much work, the various lines were sorted out and assigned to transitions between pairs of energy levels designated by their term symbols. The resulting term diagram is shown in Fig. 14.14. The significance of the symbols will be explained as we proceed.

The terms are divided into two distinct sets, and spectral lines occur only by

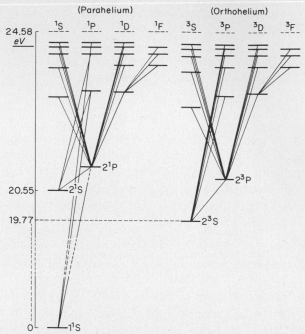

FIG. 14.14 Energy-level diagram of the helium atom with its two separate systems (singlet and triplet term systems). [Wolfgang Finkelnburg, *Structure of Matter* (New York: Academic Press, Inc., 1964), p. 100.]

transitions between terms in the same set. So definite was this separation between two sets of terms, that earlier workers believed they were dealing with two distinct kinds of helium, which they called *parahelium* and *orthohelium*. In modern notation, however, one set is said to consist of *singlets* and the other of *triplets*.

To a first approximation, we can specify a wave function ψ for each electron in the helium atom, i.e., an atomic orbital, by means of the same set of four quantum numbers that were found in the solution of the Schrödinger equation for the hydrogen atom, n, l, m_l, m_s. Of course, we no longer can use the solutions found for the H atom, and no exact solution of this type is available for the He atom itself. Nevertheless, we could imagine one electron in He to be gradually moved either into the nucleus or to infinity, and at no stage in such an imaginary process would we find an abrupt change in the wave function for the remaining electron. For this reason, a one-to-one correspondence can be drawn between the exact hydrogenlike orbitals and some approximate helium orbitals (one-electron wave functions). Thus, one speaks freely of $1s$, $2s$, $2p$, etc. orbitals in helium and more complex atoms, even though the exact form of the wave function may not be known, and, as we shall see later, even the orbital model itself may break down when quantitative calculations are attempted.

The ground state of the He atom has the electron configuration $1s^2$. The two electrons have quantum numbers as follows:

$$n = 1, \quad l = 0, \quad m_l = 0, \quad m_s = \tfrac{1}{2}$$
$$n = 1, \quad l = 0, \quad m_l = 0, \quad m_s = -\tfrac{1}{2}$$

In accord with the Pauli Principle, no two electrons can have the same set of four quantum numbers.

The term symbol for the ground state is ^1S. The general notation for the term symbol is $^{2\mathscr{S}+1}L$. The value of L, which specifies the total orbital angular momentum of all the electrons, is obtained from the vector sum of the l_i, which specify the orbital angular momenta of individual electrons. According as $L = 0, 1, 2, 3$, etc., the state is called S, P, D, F, etc. The left-hand superscript gives the *multiplicity** of the term as $2\mathscr{S} + 1$ where $\mathscr{S}$ is the total spin, specified by the addition of the individual m_s values. In the case of the ground state of helium, $L = 0$ and $\mathscr{S} = 0$, and hence the state is ^1S.

The lowest excited states of He are those in which one electron is in an orbital with principal quantum number $n = 2$. The possible different configurations are $1s^1 2s^1$ and $1s^1 2p^1$. In the case of the H atom, the energy levels calculated from the Schrödinger equation depend only on the value of n, and not on that of l. For atoms with more than one electron, however, the one-electron energy levels will depend strongly on the values of l. The term symbol designates the value of L, so that the excited states will be S states ($L = 0$) and P states ($L = 1$). In the case of the states for He, the S terms always lie below the P terms of the same principal quantum number.

The existence of the singlet and triplet states of helium is clearly the result of the fact that the two electron spins can be either antiparallel ($\mathscr{S} = 0$) or parallel ($\mathscr{S} = 1$). Thus, except for the ground state, in which the Pauli Principle excludes the $\mathscr{S} = 1$ state, each term will be split into a singlet and a triplet. We note from the term diagram that the triplet states (for given values of n and L) always lie lower than the singlets. For example, for $n = 2$, ^3S is 6422 cm^{-1} lower than ^1S.

What is the reason for this large splitting of terms which have the same electron configuration and same L values, but which differ in their total spin? Let us state emphatically that this energy difference is *not* due to any magnetic interaction between the magnetic moments of the spins. Such a magnetic interaction does occur, but it is negligibly small compared to the observed difference in energies of the $1s^1 2s^1$ ^1S and $1s^1 2s^1$ ^3S states. The term splitting is due to a difference in the electrostatic interactions in the system consisting of the $+2$ He nucleus and the two electrons. The electrostatic energies of the two states can be written

$$^1S, \qquad E = F_0 + G_0$$
$$^3S, \qquad E = F_0 - G_0$$

F_0 and G_0 representing integrals obtained in the quantum mechanical calculation of the energy of the system. We call F_0 the *coulombic integral* and G_0 the *exchange integral*. For most states, it would be difficult to calculate these energy integrals, but the spectroscopic data provide precise experimental values.

The exchange energy is a specifically quantum mechanical effect but we can

*It is read as "singlet," "doublet," "triplet," etc.

try to give a qualitative interpretation. In the ³S state, the two electrons have the same spin. Since they are in different orbitals, $1s$ and $2s$, no contradiction of the Pauli Principle is implied. Nevertheless, the two electrons with the same spin must stay away from each other. The orbital notation is simply a shorthand way of describing the positions and velocities of the electrons. If the $1s$ electron and the $2s$ electron with the same spin ever "tried to get into the same region of space," the absolute prohibition of the Pauli Principle would still operate and tend to keep them apart. Since they are kept apart, their repulsive electrostatic energy is small. On the other hand, in the ¹S state, where the two electrons have opposite spins, there is no Pauli Principle prohibition to keep them out of the same region of space. Consequently, the electrostatic repulsion will be markedly higher in the singlet state than in the triplet. Thus we can see that the singlet-triplet splitting is an electrostatic interaction, but fundamentally caused by quantum mechanical rules. It is a nonclassical interaction, and thus difficult to visualize in a physical picture. In summary, the necessary antisymmetry of the wave function *and* the different forms of the singlet and triplet *spin* functions cause the spacial parts of the singlet and triplet functions to differ (resulting in different charge distributions).

We have now explained the general structure of the term diagram of helium by means of a strong electrostatic coulombic interaction which splits terms of different L and a strong electrostatic exchange interaction which further splits terms of the same L but different $\mathscr{S}$.

We have not yet included the spin–orbit interaction, which was found to cause fine structure in the spectrum of the H atom. As might be expected, the spin–orbit interactions also occur in He, and cause a fine structure, which, however, is too small to include in the term diagram of Fig. 14.14.

The total L and total $\mathscr{S}$ for a term can combine to yield a new *inner quantum number J*, and states of different J value will be split by the spin–orbit interaction. In the case of the ¹S states of He, since $L = 0$ and $\mathscr{S} = 0$, J can only have the value 0. For the ³S state with $L = 0$ and $\mathscr{S} = 1$, J must be 1. Similarly, ¹P with $L = 1$ and $\mathscr{S} = 0$ can have only $J = 1$. For ³P, however, J can be 2, 1, or 0. (Figure 14.15 shows the result as a vector addition of L and $\mathscr{S}$.)

The correlation diagram for the first group of excited states in atomic He is summarized in Fig. 14.16. After all the internal interactions have been considered,

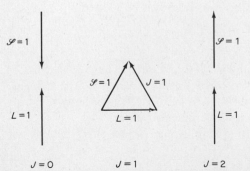

FIG. 14.15   Example of vector addition of L and $\mathscr{S}$ to give resultant J.

the effect of an external magnetic field is shown. When a directional axis is established by the field, the total angular momentum, denoted by the value of J, can assume only those directions relative to the field that have components $M_J\hbar$ in the field direction. This relation of M_J to J is exactly similar to the relation of m_l to l that was shown in Fig. 14.6.

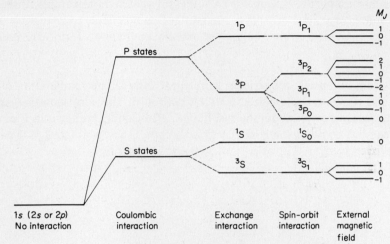

FIG. 14.16   A schematic diagram (not drawn to correct energy scale) to show the way in which an excited configuration of the helium atom with one electron raised from $n = 1$ to $n = 2$ is split into distinct energy states by internal electrostatic and magnetic interactions, and finally by an external magnetic field. When all the degeneracies have been lifted there are 16 different energy levels as required by the eight different orbitals for an excited $n = 2$ electron with spin.

20. Vector Model of the Atom

The physical picture that has been presented for the various interactions between two electrons in the helium atom can be extended to atoms with any number of electrons. A systematic method of considering the interactions is provided by the *vector model* of the atom.

We have seen how an angular momentum vector interacts with an external field and precesses around its direction. In a similar way, the angular momentum vectors of two different electrons within an atom can couple together to form a resultant total angular momentum, and each of the individual vectors then precesses around the resultant vector. We must consider, however, the two different kinds of angular momentum, the intrinsic angular momentum or spin of the electron and the orbital angular momentum due to the motion of the electron about the nucleus.

The way in which these angular momenta interact is summarized in the scheme called *Russell–Saunders coupling.**

(a) The individual spins s_i combine to form a resultant spin

$$\sum_i \vec{s_i} = \vec{\mathscr{S}}$$

The resultant must have integral of half-integral values. For example, three spins of $+\frac{1}{2}$ could give $\mathscr{S} = \frac{3}{2}$ or $\frac{1}{2}$. Two spins of $+\frac{1}{2}$ could couple to give $\mathscr{S} = 1$ or 0.

(b) The individual orbital angular momenta combine to form a resultant L:

$$\sum_i \vec{l_i} = \vec{L}$$

The quantum number L is restricted to integral values. The combination of the l_i can be considered as a quantized vector addition of the corresponding angular momenta, which precess about the resultant L. This model is shown in Fig. 14.17. As an example, consider the configuration $2p^1 3p^1$. The $l_1 = 1$ and $l_2 = 1$ can add vectorially to give $L = 0$, 1, or 2—i.e., S, P and D states.

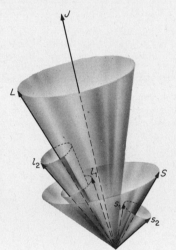

FIG. 14.17 An example of Russell–Saunders coupling. The l_1 and l_2 combine to form a resultant L. The s_1 and s_2 form a resultant $\mathscr{S}$. L and $\mathscr{S}$ precess about their resultant J.

(c) The two resultants L and $\mathscr{S}$ represent the total orbital and the total spin angular momenta of all the electrons in the atom. The spin and orbital angular momentum vectors correspond to magnetic field vectors which exert magnetic forces on each other and may be coupled to form a resultant J called the *inner quantum number*. This gives the total resultant angular momentum of all the electrons in the atom, which is quantized in values $\sqrt{J(J + 1)}\hbar$.

In the absence of an external field, the total angular momentum of the atom must be constant. Hence, L and $\mathscr{S}$ precess around their resultant J, as shown in Fig. 14.17.

*The Russell–Saunders scheme applies to the lighter atoms, but begins to break down in the heavier atoms. In these, the large nuclear charge leads to a strong coupling between the s_i and l_i of each electron due to spin–orbit interaction to give a resultant j_i.

The various energy levels or spectral terms of an atom are denoted by symbols based on this model. The symbol can be expressed as $^{2\mathscr{S}+1}L_J$. Thus, the symbol $^2S_{1/2}$ would denote a state with $2\mathscr{S} + 1 = 2$, or $\mathscr{S} = \frac{1}{2}$, $L = 0$, $J = \frac{1}{2}$. When $L > \mathscr{S}$, the multiplicity gives the number of J values for the given state.

Let us apply the vector model to the lithium atom and hence to the interpretation of its spectrum. The electron configuration of Li in the ground state is $1s^2\,2s^1$. Closed shells (e.g., $1s^2$ in Li) with all states of a given n occupied, have zero angular momentum and spin, and need not be considered in the assignment of term values. The $2s$ electron has $l = 0$ and $s = \frac{1}{2}$. Hence, $L = 0$, $\mathscr{S} = \frac{1}{2}$, $J = \frac{1}{2}$. The ground state in lithium is therefore $1s^2\,2s - {}^2S_{1/2}$. Suppose the optical electron is excited to give the configuration $1s^2\,2p - {}^2P_{3/2,\,1/2}$. These states differ slightly in energy and hence the line corresponding to a transition between the ground state ^2S and the first higher state ^2P is a closely spaced doublet. It was the occurrence of such doublets in the spectra of the alkali metals that first led to the recognition of electron spin. The absorption spectrum of potassium vapor was shown in Fig. 13.7.

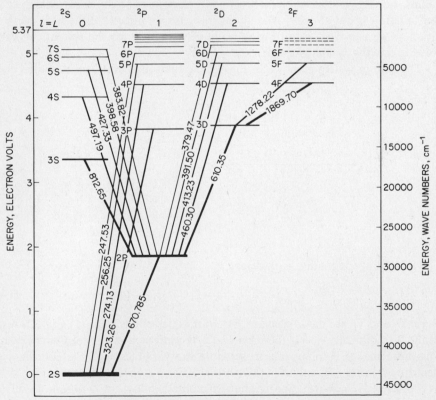

FIG. 14.18   Energy level diagram of the Li atom. The wavelengths (in nanometres) of the spectral lines are written on the connecting lines representing the transitions. Doublet structure is not included. Some unobserved levels are indicated by dashed lines.

The various energy levels of the lithium atom are summarized in Fig. 14.18. Transitions are permitted between these levels only if they obey certain *selection rules*. These are

$$\Delta J = 0, \pm 1$$
$$\Delta L = \pm 1$$

Some of the lower transitions are shown on the diagram, each one corresponding to a line observed in the spectrum of lithium.

21.   Atomic Orbitals and Energies—The Variation Method

Quantum mechanics gives an exact solution for the hydrogen atom. The calculated energy levels and electron distributions are both exact. Any experimental measurement of these quantities can hope only to be nearly as good as the theoretical values. For the next atom, helium with two electrons and a nuclear charge of $+2$, the situation is not so rosy. Already in this case we must face the hard truth that although we can write down the Schrödinger equation for the system, we cannot solve it exactly by analytic methods.

The helium system is shown in Fig. 14.19. In atomic units, the potential energy is

$$U = -\frac{2}{r_1} - \frac{2}{r_2} + \frac{1}{r_{12}} \tag{14.76}$$

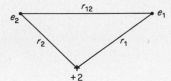

FIG. 14.19   Coordinates for the helium atom.

The Schrödinger equation is therefore

$$\left[\nabla_1^2 + \nabla_2^2 + 2\left(E + \frac{2}{r_1} + \frac{2}{r_2} - \frac{1}{r_{12}} \right) \right] \psi = 0 \tag{14.77}$$

Here, ∇_1^2 and ∇_2^2 are the Laplacians for coordinates of electrons (1) and (2), respectively. The difficulty lies in the term $1/r_{12}$, the interaction between two electrons. This term makes it impossible to separate the variables, i.e., the coordinates of electrons (1) and (2).

Fortunately, powerful approximation methods allow us to obtain useful (and in some cases practically exact) solutions in cases like this. We shall first describe the *Variation Method*. It is important for every student of chemistry to understand this mathematical method, because it lies at the heart of modern theoretical approaches to atomic and molecular structure.

The Schrödinger equation in the operator form of (13.51) is

$$\hat{H}\psi = E\psi$$

We multiply each side of the equation by ψ^*, and integrate over all space, to obtain

$$\int \psi^* \hat{H} \psi \, d\tau = \int \psi^* E \psi \, d\tau$$

Since E is a constant, it can be extracted from the integral as

$$E = \frac{\int \psi^* \hat{H} \psi \, d\tau}{\int \psi^* \psi \, d\tau} \tag{14.78}$$

This expression gives the energy of the system in terms of the correct wave function ψ, which is the solution to the Schrödinger equation. (If you like, you can test this formula on the hydrogen atom by substituting $\psi_{1s} = \pi^{-1/2} e^{-r}$.)

But what use is Eq. (14.78) if we do not know the correct ψ? Suppose we estimate an approximate $\psi^{(1)}$ based on our knowledge of what a reasonable electron distribution might be. Substitution of this trial $\psi^{(1)}$ into (14.78) yields

$$E^{(1)} = \frac{\int \psi^{*(1)} \hat{H} \psi^{(1)} \, d\tau}{\int \psi^{*(1)} \psi^{(1)} \, d\tau}$$

The variation principle states that for any estimated $\psi^{(1)}$, $E^{(1)} \geq E$. The energy calculated from the trial $\psi^{(1)}$ can never be less than the true energy E.[†] The principle applies only for the lowest state of a given symmetry type. The allowable trial $\psi^{(1)}$ is subject to "constraints" due to symmetry and the Pauli Principle.

[†]The variation method is not limited to the Schrödinger equation and was in fact invented by Rayleigh and Ritz for vibration problems. A proof of the variation theorem (as given by H. Shull) goes as follows: Take a complete set of eigenfunction ψ_i of $\hat{H}$, with $\hat{H}\psi_i = E_i \psi_i$. Consider the expectation value for the energy of an arbitrary normalized function Φ in the space spanned by the eigenfunctions of $\hat{H}$. Then one can represent Φ as

$$\Phi = \sum_{i=0}^{\infty} c_i \psi_i \quad \text{with} \quad \int \Phi^* \, \Phi \, d\tau = 1$$

Compute

$$J = \int \Phi^* \hat{H} \Phi \, d\tau$$

$$= \int (\sum c_i \psi_i^*) \hat{H} (\sum c_i \psi_j) \, d\tau$$

$$= \sum c_i^2 E_i \tag{A}$$

because

$$\int \psi_i^* \psi_j \, d\tau = 0 \quad \text{for } i \neq j$$

Now arrange the E_i in a monotonic nondecreasing sequence, $E_0 \leq E_1 \leq E_2 \leq \cdots$ Then we can replace E_i in each term of the sum (A) by E_0 with the assurance that we have never increased the value of the sum, but may have decreased it. Therefore,

$$J = \sum c_i^2 E_i \geq \sum c_i^2 E_0 = E_0 \sum c_i^2$$

From the normalization condition of Φ, $\sum c_i^2 = 1$, so that

$$J \geq E_0$$

Hence the Variation Principle is proved—that the expectation value of E as computed from Φ is an upper bound to the true ground state energy E_0.

The procedure of the Variation Method is now apparent. We must keep trying new estimated ψ's until there is no further change in the calculated energy (or until we are satisfied that the particular kind of trial function we are using has reached the limit of its capabilities). Of course, systematic mathematical methods are available for getting the best trial ψ of any particular form. We might think of the variation method as a systematic effort to discover the electron distribution chosen by Nature for a particular atom or molecule. The ground state distribution is naturally the one with the *lowest possible energy*.

22. The Helium Atom

Armed with the Variation Method, we return to an attack on the helium atom. If we simply neglected the effect of one electron on the motion of the other we could assume that each electron moved in the field of an He^{2+} ion and had a hydrogenlike atomic orbital.* If the proper spin function is included, we would have for the ^1S ground state:

$$^1S: \quad \psi = e^{-Zr_1}\, e^{-Zr_2} \cdot \tfrac{1}{2}[\alpha(1)\beta(2) - \alpha(2)\beta(1)] \tag{14.79}$$

where Z is the nuclear charge ($Z = 2$). The energy calculated from (14.79) and (14.78) is $E^{(1)} = -74.81$ eV, compared to the experimental -78.99 eV.

The discrepancy indicates that the effect of the interaction of the two electrons is important and cannot be neglected. The entire distribution of electron density is altered by the interaction between the two electrons.

We try next the wave function

$$\psi^{(2)} = e^{-Z'r_1}\, e^{-Z'r_2} \tag{14.80}$$

This is like the first trial function except that Z' is a variable parameter, which is adjusted until the minimum energy is found.† This use of an adjustable Z' is common in the variation treatment of atoms and molecules. The effect of changing Z' is to cause the electron distribution to expand (if $Z' < Z$) or to contract (if $Z' > Z$). We therefore call this operation the *adjustment of the scale factor*. The minimum energy occurs when $Z' = Z - (\tfrac{5}{16}) = \tfrac{27}{16}$. The corresponding energy is $E^{(2)} = -77.47$ eV.

We can interpret the wave function of (14.81) with $Z' < 2$ as an indication that each electron partially screens the nucleus from the other one, reducing the effective nuclear charge from $+2$ to $+\tfrac{27}{16}$. One might think that the lower effective charge would cause a higher instead of a lower energy, since the lower is the effective nuclear charge, the less negative is the energy of the electron in the field of the nucleus. The answer is that the lower effective charge must cause a *lower*

---

*The wave function for electron (1) is $1s(1) = e^{-Zr_1}$, that for electron (2) is $1s(2) = e^{-Zr_2}$. The function that expresses the probability for simultaneously finding electron (1) in $1s$ (1) and electron (2) in $1s$ (2) is the product.

†The solution of this problem is worked out in detail in L. Pauling and E. B. Wilson, *Introduction to Quantum Mechanics* (New York: McGraw-Hill Book Company, 1935), pp.184–185 (highly recommended).

kinetic energy for the electron, which more than compensates for the higher potential energy. The lower effective charge leads to an expansion of the "electron cloud" about the nucleus, i.e., the electron has more space in which to move. As pointed out in Section 13.20 for the particle in a box, such a delocalization causes a lowering of the kinetic energy.

It is evident that we have not achieved the best stability for this helium system simply by including the scale factor. As a matter of fact, we have probably allowed the trial helium atom to expand too much. What we should really like to do is allow it to expand somewhat less while including an expression that would explicitly tell one electron to stay away from the other as much as possible. We therefore try a wave function of the form,

$$\psi^{(3)} = (1 + br_{12})\, e^{-Z'r_1}\, e^{-Z'r_2} \tag{14.81}$$

Notice that this function becomes greater as r_{12} increases. Hence, it is weighted in the right direction. The values of the parameters that minimize the energy are $b = 0.364$ and $Z' = 2 - 0.151$. The energy is -78.64 eV, close to the experimental -78.99 eV. Hylleraas in 1930 used a variation function with 14 parameters to obtain a calculated energy in exact agreement with the experimental value.

23. Heavier Atoms—
The Self-Consistent Field

As the atomic number Z, and hence the number of electrons, increases, the application of quantum mechanics to the atom becomes more difficult. Some special cases that permit simplified treatments are inert gas atoms and ions with closed electronic shells, atoms like the alkali metals with one electron outside a closed shell, and those like the halogens with one hole in an otherwise complete shell.

Most theoretical calculations on multielectron atoms are based on a method developed by Douglas Hartree,* a type of variation treatment called the *method of the self-consistent field*. Hartree made the approximation that each electron moves in a spherically symmetrical field, which is the sum of the field due to the nucleus and a spherically symmetrical averaged field due to all the other electrons. This approximation has the great advantage that so long as an electron has a potential energy with spherical symmetry, $U(r_j)$, we can separate the Schrödinger equation for all the N electrons in the atom into N equations, one for each separate electron. Thus, it is possible to calculate one-electron wave functions (orbitals) and to describe them in terms of a set of four quantum numbers, n, l, m_l, m_s. Of course, the orbitals will not have the same form as those of the hydrogen atom, and one should never imagine that the pictures given for hydrogenlike orbitals are valid also for electrons in heavier atoms.

According to Hartree, therefore, the Schrödinger equation for the N-electron atom can be obtained from a Hamiltonian of the form

---

*Douglas Hartree, *The Calculation of Atomic Structures* (New York: John Wiley & Sons, Inc., 1957). Other important contributions to the theory were made by V. Fock and J. C. Slater.

$$H = \sum_{j=1}^{N} H_j = \sum_{j=1}^{N} \left[\frac{p_j^2}{2m} + U_j(r_j) \right] \tag{14.82}$$

The problem thus reduces to the solution of N problems for individual electrons. Since

$$\hat{H}_j \psi(r_j) = E_j \psi(r_j)$$

the N electron wave function is the product of the orbitals,

$$\psi_N(r_1 \cdots r_N) = \psi_1(r_1)\psi_2(r_2) \cdots \psi_N(r_N) \tag{14.83}$$

In this simplified formulation, there is no direct electron–electron interaction. Each electron sees only the average potential due to all the other electrons. The contribution of an electron k to such an average potential can be calculated from its orbital ψ_k. Since $\psi_k^* \psi_k \, d\tau_k$ is the probability of finding electron k in a region of space $d\tau_k = dx_k \, dy_k \, dz_k$, the charge density contributed by electron k to that region $d\tau_k$ is $-e\psi_k^* \psi_k \, d\tau_k$. The electrostatic potential that electron j sees due to the electron k is

$$U_{jk}(r_{jk}) = \int \psi_k^* \psi_k \frac{e^2}{r_{jk}} \, d\tau_k \tag{14.84}$$

The total potential energy of electron j is then

$$U_j(r_j) = -\frac{Ze}{r_j} + \sum_{j \neq k} \int \frac{\psi_k^* \psi_k}{r_{jk}} e^2 \, d\tau_k \tag{14.85}$$

The first term is the interaction with the nucleus, and the summation term is the interaction with all the other electrons.

The Hartree method for carrying out the variation procedure is now as follows:

1. Select any set of N starting ψ_j's for the N electrons in the atom. Call them $\psi_1^{(0)}, \psi_2^{(0)}, \ldots, \psi_N^{(0)}$.
2. Calculate from (14.84) the potential energy of the first electron from the set of $N - 1$ orbitals $\psi_2^{(0)}$ to $\psi_N^{(0)}$.
3. Numerically integrate the Schrödinger equation for the first electron to get a new value of ψ_1, called $\psi_1^{(1)}$.
4. With the new $\psi_1^{(1)}$ and all the old $\psi^{(0)}$'s except $\psi_2^{(0)}$, calculate the potential for the second electron, to get a new $\psi_2^{(1)}$.
5. Repeat this procedure until a whole new set of orbitals $\psi_j^{(1)}$ has been calculated for all the N electrons.
6. Then start the procedure again with step (2) using the $\psi_j^{(1)}$'s to calculate a new $\psi_1^{(2)}$, etc.
7. Continue running through the entire procedure until there is no further change in the potential energy calculated for any of the electrons.

The potential energy of the *self-consistent field* has now been determined and can be used to solve the Schrödinger equation with the Hamiltonian of (14.82). The wave function for the atom is obtained as a product of one-electron wave functions of form (14.83), which we can denote as the Hartree function ψ_H. One can calculate the energy of the atom from

$$E = \frac{\int \psi_H^* \hat{H} \psi_H \, d\tau}{\int \psi_H^* \psi_H \, d\tau}$$

One can also calculate a variety of other atomic properties from ψ_H. All these calculations can actually be made quite rapidly on a modern computer.

In the method as given originally by Hartree, the N-electron wave function was simply a product of one-electron wave functions. In 1930, however, Fock pointed out that most of the effects of electron spin could be taken into account by using, instead of products, Slater determinants, as shown on p. 648. In this way, the electron wave functions would be antisymmetric as required by the Pauli Principle. The self-consistent field method calculated with the antisymmetric determinantal wave functions is called the *Hartree–Fock method*.

The Hartree energies include only the coulombic terms F_{ik}, whereas the Hartree–Fock energies include also the exchange energies G_{ik}. For incomplete electron shells, it is necessary to use more than one determinantal function to obtain functions for which the total orbital angular momentum L and total spin $\mathscr{S}$ are quantized. The m_l and m_s of individual electrons are no longer good quantum numbers, and thus the concept of an individual orbital for each electron cannot be maintained. Consider, for example, the configuration $1s \, 2s \, {}^3S$ of the He atom. The Slater wave function would be the sum of two determinantal functions:

$$\psi = \frac{1}{\sqrt{2}} \left\{ \frac{1}{\sqrt{2}} \begin{vmatrix} 1s\alpha(1) & 2s\beta(1) \\ 1s\alpha(2) & 2s\beta(2) \end{vmatrix} + \frac{1}{\sqrt{2}} \begin{vmatrix} 1s\beta(1) & 2s\alpha(1) \\ 1s\beta(2) & 2s\alpha(2) \end{vmatrix} \right\}$$

The two arrangements, which differ only in that the spins are flipped (i.e., the m_s are assigned differently), must both be included in the wave function with equal weights, since there is no physical reason to choose one rather than the other.

The Hartree–Fock theory has given good agreement with the results of experimental measurements of electron densities in atoms as obtained by X-ray and electron diffraction. Figure 14.20 shows a comparison of the experimental and theoretical values for argon.

The difference between the true energy of an atom or molecule and the Hartree–Fock energy arises from three sources: (a) relativistic terms that are important for inner-shell electrons due to their high velocities, but which have little direct effect on chemical behavior; (b) the *correlation energy* due to interactions between electrons that cause the electrostatic field that they experience to differ from the averaged Hartree–Fock field; (c) magnetic interactions. The correlation energy is, in fact, very important for the chemistry of atoms and molecules, since the energies involved (typically 1 or 2 eV per pair of valence electrons of opposite spin) are exactly in the range that govern chemical reactivities. Also, correlation energies are the main source of intermolecular forces of the London or dispersion type. We have seen in the discussion of the helium atom how correlation energies can be

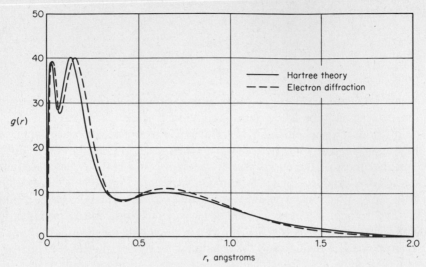

FIG. 14.20 Experimental radial distribution of electrons in argon from electron diffraction compared with the quantum mechanical calculation. [L. S. Bartell and L. O. Brockway, *Phys. Rev.*, *90*, 833 (1953).]

treated by direct introduction of terms including the interelectronic distance r_{ij} into the Hamiltonian.*

24. Atomic Energy Levels—
Periodic Table

The explanation of the structure of the Periodic Table has been one of the greatest achievements in the history of chemistry. We now see clearly that the Periodic Table is the result of two causes. First is the Pauli Exclusion Principle, which states that no two electrons in an atom can occupy the same orbital specified by quantum numbers n, l, m_l, and m_s. Second is the order of the energy levels of the orbitals, which can be predicted quantitatively by the *central field model*. We arrange the different orbitals, each specified by its set of four quantum numbers, in the order of increasing energy and then feed the electrons one by one into the lowest open orbitals, until all the electrons, equal in number to the nuclear charge Z of the atom, are safely accommodated. This process was called by Pauli the *Aufbau Prinzip (Building Principle)*.

Figure 14.21 shows the energy levels of the atomic orbitals calculated as a function of Z by the method of the self-consistent field.† In the limit of low Z,

---

*We shall not discuss correlation energies further at this point, but refer the interested reader to the discussion "Beyond Orbitals, The Correlation Problem," given by R. S. Berry in his excellent resource paper "Atomic Orbitals," *J. Chem. Educ.*, *43*, 283 (1966).

†R. Latter, *Phys. Rev.*, *99*, 510 (1955). More recent work on all the atoms has been reviewed by F. Herman, *Atomic Structure Calculations* (Englewood Cliffs, N.J.: Prentice-Hall, Inc., 1963).

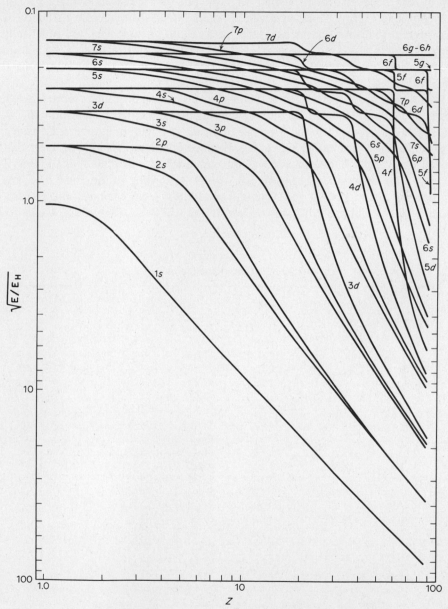

FIG. 14.21 Energy levels of atomic orbitals calculated as functions of nuclear charge (atomic number) [redrawn by M. Kasha from paper of R. Latter]. E is given in units of E_H, the energy of the hydrogen atom in the ground state, 13.6 eV.

all orbitals having the same principal quantum number n fall together at the same energy, because there are too few electrons to cause splitting of the levels. In the limit of high Z, inner orbitals having the same principal quantum number n again fall together in energy. This convergence occurs because the nuclear attraction is now so great that interactions between electrons in the same shell are practically overwhelmed. These are the energy levels observed in X-ray spectra. A hole in the shell with $n = 1$ gives the K series; a hole in the shell with $n = 2$ give the L series; etc.

At intermediate Z, the sequence of energy levels can become much more tangled. This is the region in which interelectronic interactions, such as penetration effects, can lead to departures from the sequence of principal quantum numbers. Consider, for example, the $3d$ orbital. Owing to the shielding of the nucleus by inner electrons, a $3d$ electron experiences an almost constant nuclear attraction until about $Z = 20$ (Ca). Its energy then begins to fall with Z. According to the calculation, the $3d$ crosses the $4s$ at $Z = 28$ (Ni). Actually, we know from chemical and spectroscopic evidence that the crossover occurs at about $Z = 21$ (Sc). Compared with

TABLE 14.6
Periodic System of the Elements

								$(6s, 4f, 5d)$	$(7s, 5f, 6d)$	
								$(6p)$	$(7p)$	
					$(4s, 3d)$		$(5s, 4d)$	55 Cs ——	87 Fr	s
					$(4p)$		$(5p)$	56 Ba ——	88 Ra	s^2
					19 K ——		37 Rb	57 La ——	89 Ac	s^2d
					20 Ca ——		38 Sr	——	——	s^2df^k
					21 Sc ——		39 Y ——	71 Lu	....	$s^2d(f^{14})$
		$(2s)$		$(3s)$	22 Ti ——		40 Zr ——	72 Hf	....	s^2d^2
		$(2p)$		$(3p)$	23 V ——		41 Nb ——	73 Ta	....	s^2d^3
		3 Li ——		11 Na	24 Cr ——		42 Mo ——	74 W	....	s^2d^4
$(1s)$		4 Be ——		12 Mg	25 Mn ——		43 Tc ——	75 Re	....	s^2d^5
1 H		5 B ——		13 Al	26 Fe ——		44 Ru ——	76 Os	....	s^2d^6
2 He		6 C ——		14 Si	27 Co ——		45 Rh ——	77 Ir	....	s^2d^7
		7 N ——		15 P	28 Ni ——		46 Pd ——	78 Pt	....	sd^9
		8 O ——		16 S	29 Cu ——		47 Ag ——	79 Au	....	sd^{10}
		9 F ——		17 Cl	30 Zn ——		48 Cd ——	80 Hg	....	s^2
		10 Ne ——		18 Ar	31 Ga ——		49 In ——	81 Tl	....	p
					32 Ge ——		50 Sn ——	82 Pb	....	p^2
					33 As ——		51 Sb ——	83 Bi	....	p^3
					34 Se ——		52 Te ——	84 Po	....	p^4
					35 Br ——		53 I ——	85 At	....	p^5
					36 Kr ——		54 Xe ——	86 Rn	....	p^6

| 58 | 59 | 60 | 61 | 62 | 63 | 64 | 65 | 66 | 67 | 68 | 69 | 70 | |
| Ce | Pr | Nd | Pm | Sm | Eu | Gd | Tb | Dy | Ho | Er | Tm | Yb | $\left.\right\} 6s^25df^k$ |

| 90 | 91 | 92 | 93 | 94 | 95 | 96 | 97 | 98 | 99 | 100 | 101 | 102 | |
| Th | Pa | U | Np | Pu | Am | Cm | Bk | Cf | Es | Fm | Md | No | $\left.\right\} 7s^26df^k$ |

| $k = 1$ | 2 | 3 | 4 | 5 | 6 | 7 | 8 | 9 | 10 | 11 | 12 | 13 | |

experiment, all the calculated curves in Fig. 14.21 are displaced somewhat toward higher Z.

Table 14.6 shows a concise form of the Periodic Table in which the orbital configurations of the electrons are specified.

25. Perturbation Method

In mathematical physics, we often have a situation in which an exact solution for a differential equation describing a system of interest cannot be obtained, although a solution for a somewhat simplified system may be available. For example, we have the solution for the Schrödinger equation for the hydrogen atom in the absence of any fields, but we might wish to know the effect of an electric or magnetic field on the eigenvalues and eigenfunctions of the field-free system. *Perturbation theory* is a general mathematical method for dealing with such problems. The method will usually give good results with a minimum of computation when the disturbance (perturbation) of the original system is small. We have already used results from perturbation theory in previous discussions without giving proper credit to the method.

In the Schrödinger equation

$$\hat{H}\psi = E\psi \tag{14.86}$$

the Hamiltonian is written as a series expansion,

$$\hat{H} = \hat{H}^0 + \lambda\hat{H}' + \lambda^2\hat{H}'' + \cdots \tag{14.87}$$

where λ is the *perturbation parameter*. As $\lambda \rightarrow 0$, the equation for the unperturbed system is obtained

$$\hat{H}^0\psi^0 = E^0\psi^0$$

for which one assumes that exact solutions are available. The terms, $\lambda\hat{H}' + \lambda^2\hat{H}'' + \cdots$, which should be small compared with $\hat{H}^0$, are called the *perturbations*.

Because the perturbation is small, we can expand both the wave functions and the energies in terms of λ, as

$$\psi_j = \psi_j^0 + \lambda\psi_j^{0'} + \lambda^2\psi_j^{0''} + \cdots$$
$$E_j = E_j^0 + \lambda E_j' + \lambda^2 E_j'' + \cdots \tag{14.88}$$

We now substitute the expressions for $\hat{H}$, ψ, and E into (14.86), and, after collecting coefficients of the same powers of λ, obtain

$$(\hat{H}^0\psi_j^0 - E_j^0\psi_j^0) + (\hat{H}^0\psi_j' + \hat{H}'\psi_j^0 - E_j^0\psi_j' - E_j'\psi_j^0)\lambda +$$
$$(\hat{H}^0\psi_j'' + \hat{H}'\psi_j' + \hat{H}''\psi_j^0 - E_j^0\psi_j'' - E_j'\psi_j' - E_j''\psi_j^0)\lambda^2 + \cdots = 0 \tag{14.89}$$

Since the terms are independent of one another and (14.89) must be valid for an arbitrary choice of λ, each coefficient of λ must vanish separately. Thus, for the first-order perturbation equation, we obtain

$$(\hat{H}^0 - E_j^0)\psi_j' = (E_j' - \hat{H}')\psi_j^0 \tag{14.90}$$

The unknown function ψ'_j can be expanded in terms of the complete set of known functions ψ^0_j, which form the spectrum of $\hat{H}^0$,

$$\psi'_j = \sum_l a_{lj}\psi^0_l \tag{14.91}$$

Substituting this expansion for ψ'_j into (14.90) gives

$$\sum a_{lj}(E^0_l - E^0_j)\psi^0_l = (E'_j - \hat{H}')\psi^0_j \tag{14.92}$$

We multiply (14.90) by ψ^*_j and integrate over all space. Since $\int \psi^{0*}\psi^0_l\, d\tau$ vanishes except for $l = j$, and for this value $E^0_l - E^0_j = 0$, the integral of the left side of (14.92) vanishes, so that

$$0 = \int \psi^{0*}_j(E'_j - \hat{H}')\psi^0_j\, d\tau$$

Hence, since E'_j is a constant, the first-order correction to the energy is

$$\lambda E'_j = \lambda \int \psi^{0*}_j \hat{H}'\psi^0_j\, d\tau \tag{14.93}$$

This result shows that the first-order expression for the perturbation energy is equal to the average of the perturbation function $\lambda H'$ over the unperturbed state of the system. A simplified notation is used for integrals of the type in (14.93):

$$H'_{jk} = \int \psi^{0*}_j \hat{H}'\psi^0\, d\tau \tag{14.94}$$

We shall not give here the mathematical procedures for finding the first-order perturbed wave function $[\psi'_j$ from (14.90)] and the second-order correction to the energy $\lambda^2 H''$. These derivations can be found in any standard textbook on quantum mechanics.*

26. Perturbation of a Degenerate State

An important application of perturbation theory occurs in case two or more unperturbed eigenfunctions have the same unperturbed eigenvalue for the energy. We then say that the state has an n-fold degeneracy, where n is the number of distinct eigenfunctions corresponding to the same energy. The effect of a perturbation of such a system is often to "lift the degeneracy" and to split the energy level into several distinct energy levels. We saw examples of this effect in the discussion of Fig. 14.16.

We shall examine the case in which there are two zero-order eigenfunctions $\psi^{(0)}_1$ and $\psi^{(0)}_2$ belonging to the same energy state $E^{(0)}$. The effect of a first-order perturbation will be investigated. We cannot use the method outlined in Section 14.25 because (14.88) was based on the idea that the perturbed wave function was only slightly different from a *single* unperturbed function $\psi^{(0)}_k$. Now, however, we have two entirely different zero-order wave functions, $\psi^{(0)}_1$ and $\psi^{(0)}_2$. As the

*For example, H. L. Strauss, *Quantum Mechanics, An Introduction* (Englewood Cliffs. N.J.: Prentice-Hall, Inc., 1968), p. 102, or I. N. Levine, *Quantum Chemistry*, Vol. 1 (Boston: Allyn and Bacon, Inc., 1970), p. 210.

perturbation parameter $\lambda \rightarrow 0$, the solution to the perturbed Schrödinger equation must approach a solution to the unperturbed equation, but it will not in general approach either $\psi_1^{(0)}$ or $\psi_2^{(0)}$. The most general solutions to the unperturbed equation will be some linear combinations of $\psi_1^{(0)}$ and $\psi_2^{(0)}$,

$$\chi_1^{(0)} = a_{11}\psi_1^{(0)} + a_{12}\psi_2^{(0)}$$
$$\chi_2^{(0)} = a_{21}\psi_1^{(0)} + a_{22}\psi_2^{(0)} \tag{14.95}$$

Now the perturbed wave functions can be written

$$\psi_1 = \chi_1^{(0)} + \lambda\psi_1' + \lambda^2\psi_1'' + \cdots$$
$$\psi_2 = \chi_2^{(0)} + \lambda\psi_2' + \lambda^2\psi_2'' + \cdots \tag{14.96}$$

We see that as $\lambda \rightarrow 0$, the perturbed wave function reduces to one of the correct zero-order wave functions [yet to be determined precisely by calculation of the coefficients a_{ij} in (14.95)].

We shall not go through the subsequent mathematical development* here, but simply state the final result that the effect of the perturbation is to split the twofold degenerate state into two states.

PROBLEMS

1. Which of the following operators is Hermitian: d/dx, $i\, d/dx$, d^2/dx^2?

2. Show that $[\hat{L}_x\hat{L}_y] \overset{\text{def}}{=} \hat{L}_x\hat{L}_y - \hat{L}_y\hat{L}_x = i\hbar L_z$, where the $\hat{L}_i$ are operators for components of angular momentum. What can you conclude about the possibility of simultaneously measuring two different components of the angular momentum of a particle?

3. Show that $[\hat{L}^2\hat{L}_i] = 0$ (cf. definitions in Problem 2). Is it possible to measure at the same time the length of the angular momentum vector and any particular one of its components?

4. The wave function for the state of lowest energy of a one-dimensional harmonic oscillator is $\psi = Ae^{-Bx^2}$, where A is a normalization constant and $B = (\mu\kappa)^{1/2}/2\hbar$. The potential energy is $U = \frac{1}{2}\kappa x^2$. Derive the total energy E by substituting ψ into the Schrödinger equation.

5. From Table 14.1 and Eq. (14.26), compute the wave functions for a linear oscillator for the values $v = 0, 1, 2, 3, 4$. Plot the functions and mark the domain of y between the points of inflection within which the ψ is oscillatory. Show that the kinetic energy is negative within this region by drawing the potential energy $U(y)$ and indicating the values of the total energy for $v = 0, 1, 2, 3, 4$.

6. If ψ_1 and ψ_2 are wave functions for a degenerate state of energy E, prove that any linear combination $c_1\psi_1 + c_2\psi_2$ is also a wave function.

7. Show that $3\cos^2\theta - 1$ is an eigenfunction of the operator $-\hbar^2(\partial^2/\partial\theta^2 + \cot\theta\,\partial/\partial\theta)$ with the eigenvalue $6\hbar^2$. What physical interpretation can you give to the operator?

---

*It can be found in any textbook on quantum mechanics—e.g., L. Pauling and E. B. Wilson, *Introduction to Quantum Mechanics* (New York: McGraw-Hill Book Company, 1935), pp. 166–179.

8. Prove the following recursion formulas for the Hermite polynomials:

$$\frac{d\mathscr{H}_v}{dy} = 2v\mathscr{H}_{v-1}$$

$$\mathscr{H}_{v+1} - 2y\mathscr{H}_v + 2v\mathscr{H}_{v-1} = 0$$

Show that functions $\mathscr{H}_v(y)$ obeying these formulas must satisfy the differential equation (14.21).

9. The potential energy $U = D(1 - e^{-ax})^2$ was introduced by P. M. Morse as a good approximation to the actual potential energy of diatomic molecules [see Eq. (17.37)]. Plot the function as U/D vs. ax. Solve the Schrödinger equation to determine the energy levels of a particle of mass μ in this potential.*

10. The moment of inertia of H_2 is $I = 0.470 \times 10^{-40}$ g·cm². In pure para-H_2, the spins of the protons are antiparallel and only even values of J are allowed; in pure ortho-H_2, the spins of the protons are parallel and only odd values of J are allowed. From Eq. (14.45), calculate the rotational partition functions of para- and ortho-H_2 as functions of T from 0 to 500 K. Hence, calculate the rotational contributions to the heat capacities C_V and plot vs. T. Comment briefly on results found. How would you compute the C_V for the equilibrium mixture of para- and ortho-H_2? Why can you not use Eq. (14.46) for this problem?

11. (a) Show that the $1s$ and $2s$ hydrogenlike wavefunctions in Table 14.4 are each normalized—i.e., the probability density integrated over all space equals unity. (b) Show that the $1s$ and $2s$ wavefunctions are orthogonal to each other; i.e., $\int \psi_{1s}\psi_{2s}\, d\tau = 0$, where the integral is over all space.

12. The observed ionization energy for the reaction $O^{7+} \longrightarrow O^{8+} + e$ is 867.09 eV. Compare this value with that calculated from quantum theory.

13. Show that the average value of the distance between the nucleus and the electron in a hydrogen atom when l has its maximum value $n - 1$ is

$$\langle r \rangle = n\left(n + \frac{1}{2}\right)a_0$$

where n, l are quantum numbers, a_0 the Bohr radius, and Z the nuclear charge. Calculate $\langle r \rangle$ for the $1s$, $2p$, and $3d$ states of the hydrogen atom.

14. Solve the Schrödinger equation for the motion of a particle of mass m in a spherical volume of radius a in which the potential $V(r) = 0$ for $r < a$ and $V(r) = \infty$ for $r > a$. The eigenfunctions will have the form $\psi(r, \theta, \varphi) = R(r) \exp(\pm i\, m_l\varphi)\, P_l\, m_l(\theta)$. (Try the substitution $R(r) = r^{-1/2}\chi(r)$). Discuss the eigenfunctions, quantum numbers, and energy levels.

15. (a) By means of vector diagrams, find the terms that can arise from one p and one d electron. (b) Show that three equivalent p electrons (as in N) can yield terms ²P, ²D, ⁴S.

16. Show that a system with an odd number of electrons always has even multiplets (doublets, quartets, etc.), and a system with an even number of electrons always has odd multiplets (singlets, triplets, etc.).

---

*The solution to this problem is outlined in *Selected Problems in Quantum Mechanics*, ed. by D. ter Haar (London: Macmillan & Co., Ltd., 1964), p. 108.

17. The term symbol for vanadium in the ground state is ^4F. What is the total spin angular momentum and the total orbital angular momentum?

18. In ions of the first transition series, paramagnetism is due almost entirely to the unpaired spins, being approximately equal to $\mathbf{p}_m = 2\sqrt{\mathscr{S}(\mathscr{S}+1)}$. On this basis, estimate $\mathbf{p}_m$ for V^{3+}, Mn^{2+}, Co^{2+}, and Cu^+ in units of Bohr magnetons.

19. Apply the variational method to the problem of the helium atom using the effective nuclear charge Z' as the variational parameter, and the trial wave function $\psi_I = (Z'^3/\pi a_0^3)e^{-Z'r_1/a_0}e^{-Z'r_2/a_0}$. Show that

$$E = \int \psi_I \hat{H}\psi_I \, dt = \left[-2Z'^2 + \frac{5}{4}Z' + 4Z'(Z'-Z)\right]\frac{2\pi^2 me^4}{h^2}$$

and then minimize E by $\partial E/\partial Z' = 0$.*

20. The lowest energy term of an atom can be found by means of two rules due to Hund. (1) For a given configuration of electrons, the term of lowest energy has the largest value of $\mathscr{S}$, and the largest value of L that is possible for this $\mathscr{S}$. (2) The state of lowest energy corresponds to $J = |L - \mathscr{S}|$ if the unfilled shell is less than half full, and to $J = L + \mathscr{S}$ if it is more than half full. Determine the ground state term for the following atoms: O, Cl, V, Ce.

21. Calculate the magnetic field strength at the center of a hydrogen atom due to the orbital motion of an electron in the $2p$ state. The magnetic field vector will be along the z axis, so that the field strength is found from

$$\langle B_z \rangle = \frac{-e}{\mu c}\int \psi^* \frac{\hat{L}_z}{r^3}\psi \, d\tau$$

and $\hat{L}_z\psi = m_l \hbar\psi$.

*L. Pauling and E. B. Wilson, *Introduction to Quantum Mechanics* (New York: McGraw-Hill Book Company, 1935), p. 184.

15

The Chemical Bond

Compare a concept with a style of painting. Is even our style of painting just arbitrary? Can we choose one at pleasure? (That of the Egyptians, for example). Or is it a mere question of pretty and ugly?

Ludwig Wittgenstein
(1953)*

The electrical discoveries at the beginning of the nineteenth century strongly influenced the concept of the chemical bond. Indeed, Berzelius proposed in 1812 that all chemical combinations were caused by electrostatic attractions. As it turned out 115 years later, this theory happened to be true, although not in the sense supposed by its originator. It did much to postpone the acceptance of diatomic structures for the common gaseous elements, such as H_2, N_2, and O_2. Most organic compounds also fitted poorly into the electrostatic scheme, but until 1828 it was widely believed that these were held together by *vital forces*, arising by virtue of their formation from living things. In that year, Wöhler's synthesis of urea from ammonium cyanate destroyed this distinction between organic and inorganic compounds, and the vital forces slowly retreated to their present precarious refuge in living cells.†

1. Valence Theory

Two classes of compounds came to be distinguished, with an assortment of uncomfortably intermediate specimens. *Polar compounds*, of which NaCl was a prime example, could be described as structures of positive and negative ions held together by coulombic attraction. The nature of the chemical bond in *nonpolar compounds*, such as CH_4, was obscure. Nevertheless, the relations of valence with the periodic table, which were demonstrated by Mendeleyev, emphasized the

Philosophical Investigations (Oxford: Basil Blackwell, 1953), p. 230.
†For an introduction to contemporary vitalism, see E. Wigner, "The Probability of the Existence of a Self-Reproducing Unit," in *Symmetries and Reflections* (Bloomington: Indiana University Press, 1967).

remarkable fact that the valence of an element in a definitely polar compound was usually the same a that in a definitely nonpolar compound, e.g., O in K_2O and $(C_2H_5)_2O$.

In 1904, Abegg pointed out the rule of eight: To many elements in the periodic table he could assign a negative valence and a positive valence the sum of which was eight—e.g., Cl in LiCl and Cl_2O_7, N in NH_3 and N_2O_5. Drude suggested that positive valence was the number of loosely bound electrons an atom could give away, whereas negative valence was the number of electrons an atom could accept.

Once Moseley had clearly established the concept of atomic number (1913), further progress was possible, for then the number of electrons in an atom became known. The special stability of a complete outer octet of electrons was at once noted—for example, Ne, $2 + 8$ electrons; Ar, $2 + 8 + 8$ electrons.

In 1916, W. Kossel made an important contribution to the theory of the electrovalent bond, and in the same year G. N. Lewis proposed a theory for the nonpolar bond. Kossel explained the formation of stable ions by a tendency of atoms to gain or lose electrons until they achieved an inert-gas configuration. Thus, argon has a completed octet of electrons. Potassium has $2 + 8 + 8 + 1$, and it tends to lose the outer electron, becoming the positively charged K^+ ion with the argon configuration. Chlorine has $2 + 8 + 7$ electrons and tends to gain an electron, becoming Cl^- with the argon configuration. If an atom of Cl approaches one of K, the K donates an electron to Cl, and the resulting ions combine as K^+Cl^-, the atoms displaying their valences of one. Lewis proposed that bonds in nonpolar compounds resulted from the sharing of pairs of electrons between atoms in such a way as to form stable octets to the greatest possible extent. Thus, carbon has an atomic number of 6, i.e., 6 electrons, or 4 less than the stable neon configuration. It can share four pairs of electrons with four hydrogen atoms. Each pair of shared electrons constitutes a single *covalent bond*. The Lewis theory explained why the covalence and electrovalence of an atom are usually identical, for an atom usually accepts one electron for each covalent bond that it forms. The number of pairs of shared electrons in a bond is called the *bond order*.

2. The Ionic Bond

The simplest type of molecular structure to understand is formed from two atoms one of which is strongly electropositive (low ionization potential) and the other, strongly electronegative (high electron affinity), for example, sodium and chlorine. In crystalline sodium chloride, one should not speak of an NaCl molecule since the stable arrangement is a three-dimensional crystal structure of Na^+ and Cl^- ions. In the vapor, however, a true NaCl molecule exists, in which bonding is due mainly to electrostatic attraction between Na^+ and Cl^- ions. Spectra of this molecule are observed in the vapor of sodium chloride.

The attractive force between two ions with charges Q_1 and Q_2 can be represented at moderate distances of separation by a coulombic force Q_1Q_2/r^2 or by a potential $U = -Q_1Q_2/r$. If the ions are brought so close together that their electron clouds begin to overlap, a mutual repulsion between the positively charged nuclei becomes evident. Born and Mayer suggested a repulsive potential having the form $U = b\,e^{-r/a}$, where a and b are constants.

The net potential for two ions is therefore

$$U = \frac{-Q_1Q_2}{r} + b\,e^{-r/a} \qquad (15.1)$$

This function is plotted in Fig. 15.1 for NaCl; the minimum in the curve represents the stable internuclear separation for the molecule. Note, however, that at large separations, Na + Cl is a more stable system than $Na^+ + Cl^-$, and thus we find that the molecule NaCl dissociates into atoms.

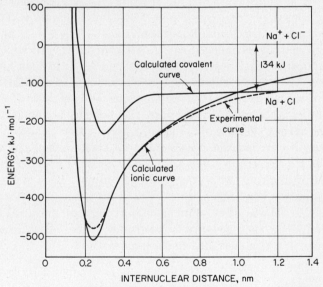

FIG. 15.1 Potential energy of NaCl as a function of internuclear separation. The ionic curve was calculated from Eq. (15.1).

The alkali-halide molecules have been carefully studied since they provide excellent data for detailed examination of theoretical models. Some of their experimental properties are collected in Table 15.1. The chemical bond in these molecules is never purely ionic. In particular the smaller positive ions tend to distort the electronic charge distribution of the larger negative ions, an effect that increases the electron density in the region between the two nuclei. We might say that such bonds acquire a *partial covalent character*.

TABLE 15.1
Experimental Properties of Alkali-Halide Molecules*

Molecule	Equilibrium internuclear distance, r_e nm	Fundamental vibration, σ (cm^{-1})	Dipole moment, μ (D)†	Dissociation energy, D_e (kJ·mol^{-1})
LiF	0.15639	910.34	6.3248	577
LiCl	0.20207	641.	7.1289	469
LiBr	0.21704	563.	7.268	423
LiI	0.23919	498.	6.25	351
NaF	0.19260	536.1	8.1558	477
Na^{35}Cl	0.23609	364.6	9.0020	406
Na^{79}Br	0.25020	298.5	9.1183	360
NaI	0.27114	259.2	9.2357	331
KF	0.21716	426.0	8.5926	490
K^{35}Cl	0.26668	279.8	10.269	423
K^{79}Br	0.28028	219.17	10.628	377
KI	0.30478	186.53	11.05	335
RbF	0.22704	373.3	8.5465	485
Rb^{35}Cl	0.27869	223.3	10.515	414
Rb^{79}Br	0.29447	169.46		377
RbI	0.31768	138.51		318
CsF	0.23455	352.6	7.8839	498
Cs^{35}Cl	0.29064	214.2	10.387	444
Cs^{79}Br	0.30722	149.50		406
CsI	0.33152	119.20	12.1	343

*After M. Karplus and R. N. Porter, *Atoms and Molecules* (New York: W. A. Benjamin, 1970), p. 263.
†See Section 15.15.

3. The Hydrogen Molecule Ion

The classical example of a covalent bond is found in the hydrogen molecule H_2, a system of two protons and two electrons, but a stable molecule exists which is simpler yet, the hydrogen molecule ion, a system of two protons and one electron. We cannot, of course, expect to isolate this charged species and it forms no stable salts $H_2^+ X^-$. In electric discharges through hydrogen gas, however, H_2^+ occurs in high concentration and its spectra and kinetic properties can be studied quite readily. The dissociation energy, $H_2^+ \longrightarrow H^+ + H$ amounts to 2.78 eV and the bond distance H — H is 0.106 nm, almost exactly twice the Bohr radius a_0.

From a theoretical viewpoint, H_2^+ is important because the Schrödinger equation can be separated and solved for this system, and the results of various approximate calculations, such as the variation method, can thus be compared with the exact solution. The coordinates used for the discussion of H_2^+ are shown in Fig. 15.2. Since this is obviously a three-body problem, no general analytic solution is possible.

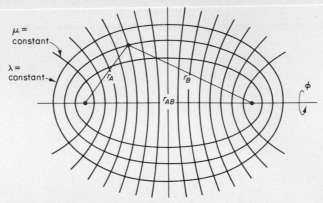

FIG. 15.2   Spheroidal coordinates for the two-center problem of H_2^+.
λ is constant on ellipsoids of revolution, μ on hyperboloids.

A stable molecule consists of electrons in rapid motion about a number of nuclei, while these nuclei themselves execute a more ponderous oscillatory motion about their mean positions.* The electrons flying about the nuclei might be compared to flies buzzing about a family of elephants, although the simile breaks down as soon as one recalls the sizes of nuclei. In any event, when we deal with a system of nuclei and electrons, we can call upon an important basic principle: The motions of nuclei in ordinary molecular vibrations are so slow compared to the motions of electrons that it is possible to calculate the electronic states on the assumption that the nuclei are held in fixed positions. This is the *Born–Oppenheimer approximation*, which is basic to most quantum mechanical calculations of the properties of molecules. As a consequence of the Born–Oppenheimer approximation, it is possible to fix the internuclear distances and thus to calculate the stationary-state wave functions and energy for the electrons in this system of fixed nuclear charges. In the case of the H_2^+ molecule in Fig. 15.2, the distance r_{AB} is held fixed, and only r_A and r_B are varied. Once this problem is solved, a new fixed value of r_{AB} can be taken, and the problem of the electronic motion solved for this new internuclear distance. The internuclear distance r_{AB} is thus a constant parameter in any one calculation.

The energy E of the system for a set of values of r_{AB} can be plotted against r_{AB} to yield what is usually called the *potential-energy curve* of the system. For H_2, the curve is shown in Fig. 15.4 . The force between the nuclei is given by the derivative of the effective potential at any separation, $F = -(\partial E/\partial r_{AB})$. Note that the energy E contains both the kinetic and the potential energies of the electrons plus the potential energy of the nuclei. Yet so far as the nuclear motions are concerned, E can be regarded as the effective potential energy of interaction. The Schrödinger equation for the H_2^+ problem is

$$\nabla^2\psi + \frac{8\pi^2 m}{h^2}\left(E + \frac{e^2}{r_A} + \frac{e^2}{r_B} - \frac{e^2}{r_{AB}}\right)\psi = 0 \tag{15.2}$$

where m is the mass of the electron. This equation can be separated if the indepen-

---

*C. A. Coulson, *Chem. Brit.*, **4**, 113 (1968).

dent variables are transformed to a system of spheroidal coordinates, λ, μ, ϕ, as shown in Fig. 15.2. Thus,

$$\psi(\lambda, \mu, \phi) = L(\lambda)M(\mu)\Phi(\phi) \tag{15.3}$$

By methods similar to those shown for the rigid-rotor and central-field problems, we can obtain solutions to the three ordinary differential equations that result when (15.3) is substituted into (15.2) (of course, after transformation of ∇^2, r_A, and r_B to the new coordinates). We shall not give the rather involved mathematical details of this analysis here, but simply use the results to show in Fig. 15.3 the electron

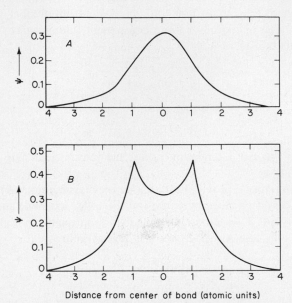

FIG. 15.3 The exact wave function of the ground state of the H_2^+ molecule ion. (A) Values of the wave function along a line normal to the H–H bond and passing through the midpoint of the bond. (B) Values of the wave function along a line passing through the two protons. The protons are located at the two peaks.

density distribution calculated for the H_2^+ molecule.

Figure 15.3 depicts the wave function ψ in two different sections, one along the line passing through the two protons and one along a line perpendicular to that internuclear line at its midpoint. We see that the maxima in ψ lie at the protons, but that there is a considerable amplitude of ψ concentrated in the regions between the two protons. Consequently, since the density of electronic charge is proportional to ψ^2, there is a considerable concentration of negative charge in the region between the two positive protons. The reason for the quite strong bond in H_2^+ is therefore evident. The bond arises from the electrostatic potential energy of interaction between the positive protons and the negative electron. It is a one-electron bond, but the single electron does its job surprisingly well, yielding 268.2 kJ bond energy.

The exact quantum mechanical theory of H_2^+ gives an exact picture of the chemical bond in this molecule. Although other molecules may have more nuclei and more electrons, the fundamental picture will not change—each bond will be due to the electrostatic potential energy of an electron density distribution concentrated near positive nuclei. There are no mysterious forces of any kind involved in holding together the atoms in a molecule.

An important result from the exact theory for H_2^+ comes from the $\Phi(\phi)$ part of the wave function. The wave function must have cylindrical symmetry, so that $\Phi(\phi) = C\,e^{i\lambda\phi}$, where C is a constant and λ is a quantum number that can take values 0, 1, 2, etc. The operator for the z component (along internuclear axis) of the angular momentum is

$$\hat{L}_z = \frac{\hbar}{i}\frac{\partial}{\partial\phi}$$

so that $\Phi(\phi)$ is an eigenfunction for $\hat{L}_z$, with

$$\hat{L}_z\Phi(\phi) = \lambda\hbar\Phi(\phi)$$

The values of the angular momentum about the internuclear axis are therefore quantized in units of $\hbar$.

The quantum number λ provides a basis for the classification of the molecular orbitals of the H_2^+ system, and, by extension, of other diatomic molecules. The notation is similar to that for atomic orbitals based on quantum number l, except that for the molecular case, Greek letters are used as follows:

$$\lambda = 0 \quad 1 \quad 2 \ldots$$
$$\text{orbital:} \quad \sigma \quad \pi \quad \delta \ldots$$

A second important designation of the molecular orbitals in H_2^+ (and, by extension, other homonuclear diatomic molecules) is their symmetry with respect to inversion through the midpoint of the two identical nuclei. As Fig. 15.3 showed, the ground state orbital is symmetrical (German, *gerade*) with respect to such inversion. It is therefore designated as the $1\sigma_g$ orbital. The first excited state is unsymmetrical with respect to this inversion (German, *ungerade*), i.e., it changes sign ($\psi \rightarrow -\psi,\ -\psi \rightarrow \psi$). This orbital is therefore called the $1\sigma_u$ orbital.

4.   Simple Variation Theory of H_2^+

Even though we have the exact solutions for the H_2^+ problem, it is still most instructive to work through in detail a simple variational treatment of this molecule. When we consider more complicated molecules, we shall rely mainly on the variation method, and its application to H_2^+ has the dual advantage of being the most simple example and of allowing a comparison between the approximate and the exact results. The following discussion is based on an analysis given by Linus Pauling.

We take as the variation function a linear combination of the two normalized hydrogen $1s$ atomic orbitals centered on protons a and b, as given on p. 632.

$$\psi = c_1 \psi_{1sa} + c_2 \psi_{1sb} \tag{15.4}$$

From (14.78),

$$E = \frac{\int \psi^* \hat{H} \psi \, d\tau}{\int \psi^* \psi \, d\tau} \tag{15.5}$$

where $\hat{H}$ was given in (15.2).

We introduce the notation

$$H_{aa} = H_{bb} = \int \psi_a^* \hat{H} \psi_a \, d\tau = \int \psi_b^* \hat{H} \psi_b \, d\tau$$

$$H_{ab} = H_{ba} = \int \psi_a^* \hat{H} \psi_b \, d\tau = \int \psi_b^* \hat{H} \psi_a \, d\tau \tag{15.6}$$

$$S = \int \psi_a^* \psi_b \, d\tau$$

Thus, (15.5) becomes

$$E = \frac{c_1^2 H_{aa} + 2c_1 c_2 H_{ab} + c_2^2 H_{bb}}{c_1^2 + 2c_1 c_2 S + c_2^2} \tag{15.7}$$

To minimize the energy with respect to c_1 and c_2, we set derivatives of E with respect to these coefficients equal to zero:

$$\frac{\partial E}{\partial c_1} = 0 = c_1(H_{aa} - E) + c_2(H_{ab} - SE)$$

$$\frac{\partial E}{\partial c_2} = 0 = c_1(H_{ab} - SE) + c_2(H_{bb} - E) \tag{15.8}$$

These are simultaneous linear homogeneous equations. If we tried to solve them in the usual way, by setting up the determinant of the coefficients and dividing it into the same determinant in which a given column was replaced by the constant terms (Cramer's rule), we would get only the trivial solutions, $c_1 = c_2 = 0$. Only in case the determinant of the coefficients itself equals zero can we obtain nontrivial solutions, and then only for certain values of E, which are the *eigenvalues* of this problem. We therefore write the condition for obtaining nontrivial solutions as the vanishing of the determinant of the coefficients of the set of linear homogeneous equations.

$$\begin{vmatrix} H_{aa} - E & H_{ab} - SE \\ H_{ab} - SE & H_{aa} - E \end{vmatrix} = 0 \tag{15.9}$$

In this case, the resulting equation is a quadratic in E. In the general case of N simultaneous equations, it would be an equation of Nth degree in E. An equation of this type is called a *secular equation.** The solutions of (15.9) are

$$E_g = \frac{H_{aa} + H_{ab}}{1 + S}$$

$$E_u = \frac{H_{aa} - H_{ab}}{1 - S}$$

(15.10)

When these eigenvalues are substituted back into (15.8), the equations can be solved for the ratio c_1/c_2, giving (as, indeed, is evident from symmetry)

$$\frac{c_1}{c_2} = \pm 1$$

$$\psi_g = c_1(\psi_{1sa} + \psi_{1sb})$$
$$\psi_u = c_1(\psi_{1sa} - \psi_{1sb})$$

The remaining constant is eliminated by the normalization conditions (which apply, because for each molecular orbital the probability of finding the electron somewhere in space must be unity).

$$\int \psi_g^2 \, d\tau = 1, \qquad \int \psi_u^2 \, d\tau = 1$$

$$c_1^2 \left[\int \psi_{1sa}^2 \, d\tau \pm 2 \int \psi_{1sa}\psi_{1sb} \, d\tau + \int \psi_{1sb}^2 \, d\tau \right] = 1$$

$$c_1^2[1 \pm 2S + 1] = 1$$

$$c_1 = \frac{1}{\sqrt{2 \pm 2S}}$$

The two wave functions are, therefore,

$$\psi_g = \frac{1}{\sqrt{2 + 2S}}(\psi_{1sa} + \psi_{1sb})$$

$$\psi_u = \frac{1}{\sqrt{2 - 2S}}(\psi_{1sa} - \psi_{1sb})$$

(15.11)

It is evident that these are the approximate wave functions corresponding to the $1\sigma_g$ and the $1\sigma_u$ functions obtained from the exact solution.

The integrals H_{aa}, H_{ab}, and S are evaluated as follows. Part of the Hamiltonian in (15.2) is the same as that for the hydrogen atom. Thus,

$$-\frac{h^2}{8\pi^2 m} \nabla^2 \psi_{1sa} - \frac{e^2}{r_a}\psi_{1sa} = E_H \psi_{1sa}$$

where E_H is the energy of the H atom. Then,

$$H_{aa} = \int \psi_{1sa}\left(E_H - \frac{e^2}{r_B} + \frac{e^2}{r_{AB}}\right)\psi_{1sa} \, d\tau = E_H + J + \frac{e^2}{a_0 D}$$

where

$$J = \int \psi_{1sa}\left(-\frac{e^2}{r_B}\right)\psi_{1sa} \, d\tau = \frac{e^2}{a_0}\left[-\frac{1}{D} + e^{-2D}\left(1 + \frac{1}{D}\right)\right]$$

*From Latin *saeculum*, generation, age. The term *secular perturbation* was first introduced into celestial mechanics to describe perturbations causing slow, cumulative effects on orbits.

with

$$D = \frac{r_{AB}}{a_0}$$

Similarly, we find

$$H_{ab} = \int \psi_{1sb}\left(E_H - \frac{e^2}{r_B} + \frac{e^2}{r_{AB}}\right)\psi_{1sa}\, d\tau = SE_H + K + \frac{Se^2}{a_0 D}$$

where

$$S = e^{-D}\left(1 + D + \frac{D^2}{3}\right)$$

$$K = \int \psi_{1sb}\left(-\frac{e^2}{r_B}\right)\psi_{1sa}\, d\tau = -\frac{e^2}{a_0}e^{-D}(1 + D)$$

Thus, we finally obtain for the energies,

$$E_g = E_H + \frac{e^2}{a_0 D} + \frac{J + K}{1 + S}$$

$$E_u = E_H + \frac{e^2}{a_0 D} + \frac{J - K}{1 - S} \tag{15.12}$$

The integral J is called the *coulomb integral* and it gives the coulombic interaction between nucleus a and an electron in the $1s$ orbital of nucleus b (or vice versa). The integral K is called the *resonance* or *exchange integral*, because both wave functions ψ_{1sa} and ψ_{1sb} occur in it. This simple variation treatment gives a dissociation energy of 1.77 eV for H_2^+ as compared to the exact value of 2.78 eV, and an equilibrium distance of 0.132 nm (exact, 0.106 nm).

5. The Covalent Bond in H_2

The most important application of quantum mechanics to chemistry has been its explanation of the nature of the covalent bond. G.N. Lewis, in 1918, declared that this bond consists of a shared pair of electrons. In 1927, W. Heitler and F. London applied quantum mechanics to give the first quantitative theory of the bond.

If two H atoms are brought together, the system consists of two protons and two electrons. If the atoms are far apart, their mutual interaction is effectively nil. In other words, the energy of interaction $U \rightarrow 0$ as the internuclear distance $R \rightarrow \infty$. At the other extreme, if the two atoms are forced closely together, there is a large repulsive force between the positive nuclei, so that as $R \rightarrow 0$, $U \rightarrow \infty$. Experimentally we know that two hydrogen atoms can unite to form a stable hydrogen molecule, whose dissociation energy is 458.1 kJ·mol⁻¹. The equilibrium internuclear separation in the molecule is 0.0740 nm. These facts about the interaction of two H atoms are summarized in the potential-energy curve of Fig. 15.4.

The system of two protons and two electrons is shown in Fig. 15.5, with coordinates appropriately labeled. This system is quite similar to that for the helium atom, shown in Fig. 14.19. The difference is that we now have two nuclei with

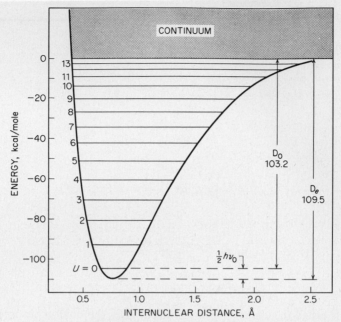

FIG. 15.4 Potential-energy curve for hydrogen molecule. The vibrational energy levels are shown.

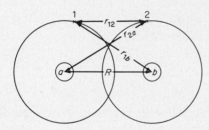

FIG. 15.5 Interaction of two hydrogen atoms.

charges of $+1$ each, instead of one nucleus with charge $+2$. Thus, instead of the expression in (14.76), the potential energy becomes now (in atomic units):

$$U = r_{12}^{-1} - r_{1a}^{-1} - r_{2b}^{-1} - r_{1b}^{-1} - r_{2a}^{-1} + R^{-1}$$

The terms in the potential energy are the following:

$$
\begin{aligned}
U_1 &= r_{12}^{-1} & \text{electron 1 repels electron 2} \\
U_2 &= -r_{1a}^{-1} & \text{electron 1 attracts nucleus } a \\
U_3 &= -r_{2b}^{-1} & \text{electron 2 attracts nucleus } b \\
U_4 &= -r_{1b}^{-1} & \text{electron 1 attracts nucleus } b \\
U_5 &= -r_{2a}^{-1} & \text{electron 2 attracts nucleus } a \\
U_6 &= R^{-1} & \text{nucleus } a \text{ repels nucleus } b
\end{aligned}
$$

(15.13)

In these atomic units, the Schrödinger equation is

$$\{\nabla_1^2 + \nabla_2^2 + (E - U)\}\psi = 0 \tag{15.14}$$

where ∇_1^2 and ∇_2^2 refer to coordinates of electrons 1 and 2, respectively.

The main difficulty with this problem is caused by the term r_{12}^{-1}. If it were not for the presence of this term, the equation would be separable in spheroidal coordinates, as was the H_2^+ problem. We can feel reasonably confident, therefore, that most of what we have learned about the covalent bond in H_2^+ also applies in the case of H_2. In general terms, the bond occurs as a consequence of a concentration of negative charge between the nuclei. The facts that the bond energy is 458.1 kJ in H_2 as against 268.2 kJ in H_2^+, and that the bond distance is quite a bit closer in H_2, 0.0740 nm as compared to 0.106 nm in H_2^+, clearly indicate that two electrons are better than one for holding together two protons.

To compare the theoretical energy with the experimental curve of Fig. 15.4, we must calculate the energy of the system for a number of different values of the internuclear distance R. The repulsion between the nuclei always contributes a term $U_6 = R^{-1}$. To calculate the energy of the electrons, we apply the variation method, as in the cases of the He atom and the H_2^+ molecule.

To begin the calculation, we must make some reasonable choice as to the form of the wave function for each electron in the molecule, i.e., the *molecular orbital*. What shall we take as the first-order approximation for a molecular orbital in the H_2 molecule? If we pull the nuclei far apart, one electron will go with each nucleus, and the system can be represented as the sum of two H atoms. The molecular orbital accordingly would become the sum of two $1s$ atomic orbitals for H atoms, one centered on nucleus a and the other on nucleus b,

$$\psi^{(1)} = 1s_a(1) + 1s_b(1) \tag{15.15}$$

Exactly the same form* was used in the variation treatment of H_2^+. In atomic units, $1s_a(1) = \exp(-r_{1a})$ and $1s_b(1) = \exp(-r_{1b})$; hence,

$$\psi^{(1)} = \exp(-r_{1a}) + \exp(-r_{1b})$$

Equation (15.15) is an example of the formation of a molecular orbital as a linear combination of atomic orbitals. It is called an M.O.–L.C.A.O. approximation.

Our first trial wave function for the H_2 molecule can therefore be written as

$$\psi_{\text{MO}}^{(1)} = [1s_a(1) + 1s_b(1)][1s_a(2) + 1s_b(2)] \tag{15.16}$$

Both electrons have been placed in the same molecular orbital, which was formed as the sum of two $1s$ atomic orbitals. Two electrons can go into such an orbital, in accord with the Pauli Exclusion Principle, provided they have antiparallel spins.

To calculate the energy, we substitute $\psi^{(1)}$ from (15.16) into (15.5). The calculation is essentially similar to that given for H_2^+, but we shall not give the details.† The calculated dissociation energy $D_e(H_2 \rightarrow 2H)$ is 258.6 kJ, and the calculated equilibrium internuclear distance r_e is 0.0850 nm. The experimental values are 458.1 kJ and 0.0747 nm. The quantitative agreement is therefore poor, but the fact

---

*For convenience, we have used the notation $1s_a \equiv \psi_{1sa}$.

†C. A. Coulson, *Proc. Cambridge Phil. Soc.*, **34**, 204 (1938), gives details of a M.O. calculation for H_2.

that the calculation yields a very stable molecule indicates that the model must be fairly reasonable.

The next step, as in the treatment of the He atom, is to introduce the scale factor. The molecular orbital then becomes

$$\psi^{(2)} = \exp(-Zr_a) + \exp(-Zr_b) \tag{15.17}$$

The lowest energy is found when $Z = 1.197$, and gives $D_e = 334.7$ kJ with $r_e = 0.0732$ nm. In this case, unlike that of the He atom, the effective charge is greater than the charge on the individual nucleus. The electrons are thus squeezed into a smaller volume, closer to the nuclei, and their potential energy is lowered. True, the kinetic energy must be raised, but the total energy is still lowered by pulling the electrons closer to the nuclei. The resulting improvement in D_e, however, is rather disappointing. The source of the trouble is obvious. We have not included properly the effect of the interaction between the two electrons. In fact, the M.O. of Eq. (15.17) places both electrons on the same nucleus for a considerable proportion of the time, and thus greatly overestimates the interelectronic repulsion energy.

One way to write a wave function that keeps electrons away from each other is to include what is called a *configuration interaction*. The M.O. of (15.15) is an L.C.A.O. formed entirely of $1s$ orbitals of atomic hydrogen. If terms from $2s$ and other higher states are also included, the electrons can find some additional places to go to get out of each other's way. This treatment at best improves D_e to 386.2 kJ.

So far we have not used any wave functions that include explicitly the interelectronic distance r_{12}. Even the most complicated functions, if they neglect this factor, will not yield a value of D_e greater than 410 kJ, about 10% below the experimental value. Once we include such r_{12} terms, however, the energy again begins to improve. James and Coolidge devised a complicated function of the form

$$\psi = e^{-\delta(\mu_1 + \mu_2)} \sum_{m,n,j,k,p} C_{mnjkp} [\mu_1^m \mu_2^n v_1^j v_2^k + \mu_1^n \mu_2^m \mu_1^k v_2^j] r_{12}^p \tag{15.18}$$

The exponents m, n, j, k, p are integers and $\delta = 0.75$. The C_{mnjkp} are variable parameters. They finally tried an expression with 13 terms, including 5 terms in r_{12}. The calculated $D_e = 455.2$ kJ was within 2.9 kJ of the experimental value. Kolos and Roothaan[*] extended the calculation to 50 terms, obtaining $D_e = 457.8$ kJ and $r_e = 0.0741$ nm. Figure 15.6 shows how the electrons tend to avoid each other in the wave functions used by these workers.

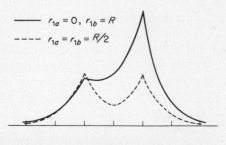

FIG. 15.6 Probability density of the second electron along the axis for two fixed locations of the first electron.

[*]*Rev. Mod. Phys.*, 32, 226 (1960).

A wave function like that of Kolos and Roothaan must represent a spacial distribution of electronic charge practically identical with that which actually occurs in the H_2 molecule. Yet the function is so complicated that one cannot hope to obtain any simple physical interpretation of its terms. *Simple physical pictures of the covalent bond are possible only at the level of simple approximate wave functions.* The cause of the bonding is clear—the electron density builds up around two nuclei and the resultant electrostatic potential energy stabilizes the system—but the bond itself as a simple conceptual model has disappeared into the impenetrable complexity of the wave function.

One of the most graphic ways to present the results of quantum mechanical calculations on simple molecules is to draw the calculated electron densities in the

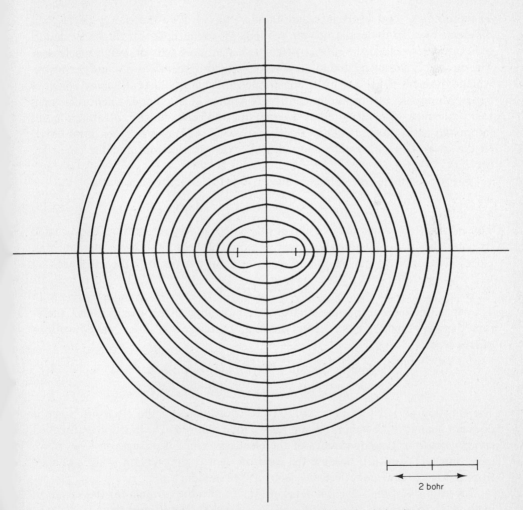

2 bohr

FIG. 15.7 Total electron density in the hydrogen molecule in plane of internuclear axis. Adjacent contour lines differ in electron density by a factor of two.

form of a contour map. A.C. Wahl* has published results of computer-calculated electron-density maps for a number of molecules. The map for H_2 is shown in Fig. 15.7 for the plane passing through the two nuclei. Adjacent contour lines differ in electron density by a factor of two. Thus, they depict a geometric series of electron densities, and in this respect differ from the arithmetic series of contour lines used in geographic maps. The wave functions used to calculate Fig. 15.7 are not quite so good as the Kolos–Roothaan functions, but are adequate to show in detail the electron distribution in a simple diatomic molecule.

6.   The Valence-Bond Method

The procedure used by Heitler and London to calculate the energy of H_2 was different from the molecular-orbital method just outlined. It was the first example of what is now called the *valence-bond method* in molecular quantum mechanics. The method is closely related to the classical structure theory of organic chemistry, which considers molecules to be constructed of *atoms* held together by *chemical bonds*. More precisely, the atoms contribute some of their outer or valence electrons to form bonds with other atoms, so that the molecule consists of atomic cores (atoms that may have provided valence electrons) and bonds between these cores. In the case of H_2, each atom provides one valence electron, and the cores are protons.

The approximate wave function chosen by Heitler and London was

$$\psi_{VB}^{(1)} = 1s_a(1)1s_b(2) + 1s_a(2)1s_b(1) \tag{15.19}$$

The argument by which they arrived at this choice is very interesting. They started by considering two H atoms, each with its own electron. The wave function that places electron 1 on nucleus a and electron 2 on nucleus b is then the product

$$\phi_1 = 1s_a(1)1s_b(2)$$

They then pointed out that the electrons are indistinguishable particles and, therefore, the wave function must express this fact. There are two simple wave functions of this type:

$$\psi_s = 1s_a(1)1s_b(2) + 1s_a(2)1s_b(1) = \phi_1 + \phi_2$$
$$\psi_a = 1s_a(1)1s_b(2) - 1s_a(2)1s_b(1) = \phi_1 - \phi_2 \tag{15.20}$$

The function ψ_s is called *symmetric* in the coordinates of the electrons, since it does not change if the indices (1) and (2) are interchanged. The function ψ_a is called *antisymmetric* in the coordinates of the electrons, since it changes to $-\psi_a$ if we interchange (1) and (2). Because the electron density depends on ψ^2, the change from ψ to $-\psi$ does not affect the electron distribution.

Substitution of the wave functions of (15.20) into the formula for the variation method, (15.5), gives

*A. C. Wahl, "Molecular Orbital Densities: Pictorial Studies," *Science, 151*, 961 (1966).

$$E = \frac{\int (\phi_1 \pm \phi_2)\hat{H}(\phi_1 \pm \phi_2)\, d\tau}{\int (\phi_1 \pm \phi_2)^2\, d\tau} \tag{15.21}$$

7. The Effect of Electron Spins

So far we have not included the spin properties of the electrons, which we must do to obtain a correct wave function. The electron spin quantum number m_s, with allowed values of either $+\frac{1}{2}$ or $-\frac{1}{2}$, determines the magnitude and direction of the spin. We introduce two spin functions α and β corresponding to $m_s = +\frac{1}{2}$ and $m_s = -\frac{1}{2}$. For the two-electron system, there are then four possible complete spin functions:

Spin function	Electron 1	Electron 2
$\alpha(1)\alpha(2)$	$+\frac{1}{2}$	$+\frac{1}{2}$
$\alpha(1)\beta(2)$	$+\frac{1}{2}$	$-\frac{1}{2}$
$\beta(1)\alpha(2)$	$-\frac{1}{2}$	$+\frac{1}{2}$
$\beta(1)\beta(2)$	$-\frac{1}{2}$	$-\frac{1}{2}$

When the spins have the same direction, they are said to be *parallel;* when they have opposite directions, they are *antiparallel.*

The fact that the electrons are indistinguishable forces us again to choose either symmetric or antisymmetric linear combinations for the two-electron system. There are three possible symmetric spin functions:

$$\alpha(1)\alpha(2)$$
$$\beta(1)\beta(2)$$
$$\alpha(1)\beta(2) + \alpha(2)\beta(1)$$

There is one antisymmetric spin function:

$$\alpha(1)\beta(2) - \alpha(2)\beta(1)$$

The possible complete wave functions for the H–H system are obtained by combining these four possible spin functions with the two possible orbital wave functions. This leads to eight functions in all.

At this point in the argument, the Pauli Exclusion Principle enters in an important way. "Every allowable wave function for a system of two or more electrons must be antisymmetric for the simultaneous interchange of the position and spin coordinates of any pair of electrons." As a consequence, the only allowable wave functions are those made up either of symmetric orbitals and antisymmetric spins or of antisymmetric orbitals and symmetric spins. There are four such combinations for the H—H system:

Orbital	Spin	Total spin	Term
$a(1)b(2) + a(2)b(1)$	$\alpha(1)\beta(2) - \alpha(2)\beta(1)$	0 (singlet)	$^1\Sigma$
$a(1)b(2) - a(2)b(1)$	$\begin{cases} \alpha(1)\alpha(2) \\ \beta(1)\beta(2) \\ \alpha(1)\beta(2) + \alpha(2)\beta(1) \end{cases}$	1 (triplet)	$^3\Sigma$

The term symbol Σ expresses the fact that the molecular state has an angular momentum of zero about the internuclear axis, since it is made up of two atomic S terms.* The multiplicity of the term, or number of wave functions corresponding with it, is added as a left-hand superscript. This multiplicity is always $2\mathscr{S} + 1$ where $\mathscr{S}$ is the total spin.

8.   Results of the Heitler–London Method

When we calculate the energy from (15.21) and the valence-bond wave functions of (15.20), we find that the function $\phi_1 + \phi_2$ (i.e., the one for the $^1\Sigma$ state) leads to a deep minimum in the potential-energy curve, but the function $\phi_1 - \phi_2$ (for the $^3\Sigma$ state) leads to a repulsion at all values of R. The results are shown in Fig. 15.8 together with the experimental curve.

In this formulation, the covalent bond is formed between atoms that share a pair of electrons with opposite spins. Only when the spins are antiparallel can the two electron waves be in phase in the region between the nuclei and reinforce each other so as to cause an attractive interaction. If the spins are parallel, there is a

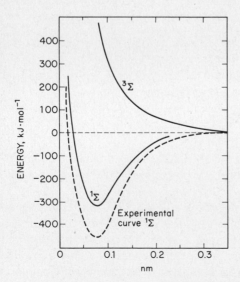

FIG. 15.8 Results of Heitler–London treatment of the H_2 molecule.

*The quantum numbers λ_i of individual electrons add vectorially to yield a resultant Λ, and accordingly as $\Lambda = 0, 1, 2 \ldots$, the state is designated as $\Sigma, \Pi, \Delta, \ldots$.

destructive interference between the electron waves in the region between the two protons. As a result, the electron density between the protons is greatly decreased, and there is a strong net repulsion between the relatively bare positive charges. If two H atoms are brought together, there is only one chance in four that they will attract each other, because the stable state is a singlet and the repulsive state is a triplet.

The calculated binding energy for the simple valence-bond treatment is 303.3 kJ and the calculated r_e is 0.080 nm. This result is actually a little better than that from the simplest molecular-orbital treatment. The reason is that the valence bond wave function does not give so much weight to configurations with both electrons on the same nucleus. Unfortunately, the physical picture of the simple Heitler–London theory is not correct. It would indicate that the binding energy is due mostly to the greater space available for the electrons and hence to a lowering of the *kinetic energy* of the system. In reality, as we saw before, the bond is primarily due to a lowered electrostatic *potential energy*.

The next step is therefore to apply the scale factor to the Heitler–London functions, by using the modified atomic orbitals, $\exp(-Z'r_a)$. With $Z' = 1.166$, the electron distribution is pulled in more closely. The kinetic energy rises, but the potential energy falls more drastically. The calculated bond energy is then 363.2 kJ at $r_e = 0.0743$ nm. As with the molecular-orbital method, really accurate results will require the introduction of terms for the electron correlation. Ultimately, the V.B. and M.O. calculations converge to the same values for the electron density contours, energies, and other calculated properties of H_2.

9. Comparison of M.O. and V.B. Methods

The molecular-orbital and the valence-bond methods are the two basic approaches to the quantum theory of molecules. How do they compare?

The V.B. treatment starts with individual atoms and considers the interaction between them. For two atoms a and b with two electrons (1) and (2), a possible wave function is $\psi_1 = a(1)b(2)$; equally possible is $\psi_2 = a(2)b(1)$, inasmuch as the electrons are indistinguishable. Then the valence-bond wave function is

$$\psi_{VB} = a(1)b(2) + a(2)b(1)$$

The M.O. treatment of the molecule starts with the two nuclei. If $a(1)$ is a wave function for electron (1) on nucleus a, and $b(1)$ is a wave function for electron (1) on nucleus b, the wave function for the single electron moving in the field of the two nuclei can be written as an L. C. A. O., $\psi_1 = c_1a(1) + c_2b(1)$. Similarly, for the second electron, $\psi_2 = c_1a(2) + c_2b(2)$. The combined wave function is the product of these two, or

$$\psi_{MO} = \psi_1\psi_2 = c_1^2a(1)a(2) + c_2^2b(1)b(2) + c_1c_2[a(1)b(2) + a(2)b(1)]$$

Comparing ψ_{VB} with ψ_{MO}, we see that ψ_{MO} gives a large weight to configurations that place both electrons on the same nucleus. In a molecule AB, these are the ionic structures A^+B^- and A^-B^+. The ψ_{VB} neglects these ionic terms. Actually, for most molecules, the simple M.O. considerably overestimates the ionic terms, whereas the simple V.B. considerably underestimates them. The true ionic contribution is usually some compromise between these two extremes, but the mathematical treatment of such a compromise is more difficult. It is necessary to add further terms to the expressions for the wave functions—for instance, ionic terms to the V.B. functions.

10. Chemistry and Mechanics

"The underlying physical laws necessary for the mathematical theory of a large part of physics and the whole of chemistry are thus completely known, and the difficulty is only that the exact application of these laws leads to equations much too complicated to be soluble." This much-quoted statement was written by Dirac in 1929, about three years after the discovery of quantum mechanics, but it remains just as true and exasperating today. Despite the development of high speed computers, which can perform in a few minutes what used to be a year's computation, "it looks as if somewhere around 20 electrons there is an upper limit to the size of a molecule for which accurate calculations are ever likely to become practicable."*

The nature of the problem is illustrated by the methane molecule, CH_4. There are five nuclei and ten electrons. The exact Schrödinger equation for this system would be a partial differential equation in 45 variables. Even with the Born–Oppenheimer approximation, 30 variables remain for the electronic motions. The symmetry of the molecule permits a further simplification of the problem. Nevertheless, in a problem like this, a complete molecular-orbital treatment becomes too complex to handle with present computing facilities. Usually, therefore, the theoretician is content to take the experimental data on the equilibrium positions of the nuclei and to calculate the energy and wave functions for this particular configuration.

We can hope to obtain exact information from quantum mechanical calculations about electron distributions in light atoms and molecules. Such information should give us a much deeper understanding of the nature of the factors responsible for molecular structures. We can then try to translate this quantum mechanical information into concepts such as the nature of chemical bonds, electronegativities, classes of excited states, and so on, which can help us to interpret the chemical behaviors of more complex molecules.

Einstein once wrote the "*Being* is always something which is mentally constructed by us. The nature of such constructs does not lie in their derivation from what is given by the senses. Such a type of derivation is nowhere to be had. The

*C. A. Coulson, "Present State of Molecular Structure Calculations," *Rev. Mod. Phys.*, *32*, 170 (1960). "Ever," however, is a long time, and most other experts would not agree with Coulson. A more encouraging perspective is given by E. Clementi, "Chemistry and Computers," *Intl. J. Quantum Chem.* 1S, 307 (1967).

justification of these constructs, which represent 'reality' for us, lies above all in their quality of making intelligible what is sensually given." The *chemical bond* is a good example of a specifically *chemical construct*, which makes intelligible the results of chemical experimentation and makes possible the design of new chemical experiments. For a chemist engaged in working out a synthesis of chlorophyll, the chemical bond will be more "real" than molecular wave functions, which are comprehensible only as mathematical expressions, such as, for example, the James and Coolidge wave functions for H_2.

Out of the undifferentiated continuum of experience, man crystallizes certain entities. Whether he creates them or discovers them is a paradoxical question. Does nature imitate art or art depict nature? The elementary individual entity is always incompatible in some sense with the continuous field from which it was drawn. Kant expressed this with regard to the atom by saying that the notion of an indivisible atom and the intuition of space are incompatible. Weizsäcker* expressed it by saying that chemistry and mechanics are *complementary*. If we push the mechanics as far as we can, the chemistry disappears. If we push the chemistry to its farthest reaches, the mechanics disappears. In exemplary terms, the James and Coolidge calculation of the H_2 structure is mechanics without chemistry, the synthesis of chlorophyll or the elucidation of the structure of insulin is chemistry without mechanics. To say "chemistry is applied mathematics" is something like saying "poetry is applied music."

11.   Molecular Orbitals for Homonuclear Diatomic Molecules

The V.B. method was based on the chemical concept that, in some sense, atoms exist within molecules, and that the structure of a molecule can be interpreted in terms of its constituent atoms and the *bonds* between them. The M.O. method would like to discard the idea of atoms within molecules, and start with the bare positive nuclei arrayed in definite positions in space. The total number of electrons would be fed one-by-one into this electrostatic field. The M.O. theory is more physical than chemical in its view of a molecular structure, which it sees not as atoms connected by bonds, but as an electronic pudding of varying density, interspersed with some positive nuclear plums.

An orbital is a *one-electron wave function*, i.e., a function of the coordinates of only one electron, for example $\psi(x_1, y_1, z_1)$. If a molecule contains more than one electron, we must recognize that the orbital treatment is only a first approximation to the exact wave function, which for a molecule with N electrons would be a function of the coordinates of all these electrons or $\psi(x_1 y_1 z_1, x_2 y_2 z_2, \ldots, x_N y_N z_N)$. A second approximation to the N-electron wave functions would use two-electron functions called *geminals*,† instead of orbitals. For example, the function of the form $\psi(x_1 y_1 z_1 x_2 y_2 z_2)$ would be a geminal. The concept of the *geminal* emphasizes

*C. F. Weizsäcker, *The World View of Physics* (Chicago: University of Chicago Press, 1952).
†This felicitous name was introduced by H. Shull, *J. Chem. Phys.*, *30*, 1405 (1959).

the significance of pairs of electrons in determining molecular structures. It seems intuitively reasonable to use geminals as a good quantum mechanical approach to the bond structures that have been so useful to the chemist. Further theoretical work will be needed, however, before it can clearly be established whether or not geminals should supplant the more familiar orbitals in calculations of experimental quantities by molecular quantum mechanics.

Just as the electrons in an atom can be assigned to definite atomic orbitals characterized by quantum numbers, n, l, m_l, and occupy the lowest levels consistent with the Pauli Principle, so the electrons in a molecule can be assigned to definite molecular orbitals, and at most two electrons can occupy any particular molecular orbital.

Consider the *molecular orbitals of the hydrogen molecule*. If H_2 is pulled apart, it separates into two hydrogen atoms, H_a and H_b, each with a single $1s$ atomic orbital. If the process is reversed and the hydrogen atoms are squeezed together, these atomic orbitals coalesce into the molecular orbital occupied by the electrons in H_2. A molecular orbital, therefore, can be constructed from a linear combination of atomic orbitals (L. C. A. O.). Thus,

$$\psi = c_1(1s_a) + c_2(1s_b) \tag{15.22}$$

Since the molecules are completely symmetrical, c_1 must equal $\pm c_2$. Thus the two possible molecular orbitals from the $1s$ atomic orbitals are (except for normalization factors),

$$\psi_g = 1s_a + 1s_b$$
$$\psi_u = 1s_a - 1s_b \tag{15.23}$$

These molecular orbitals are shown schematically in Fig. 15.9(a). The $1s$ atomic orbitals are spherically symmetrical (see Section 14.14). If the two orbitals that overlap have the same phase, the resultant is ψ_g, which corresponds to a concentration of electronic charge density between the nuclei. If the two atomic orbitals are opposite in phase, the resultant is ψ_u, which corresponds to a depletion of electronic charge density between the nuclei. Both these molecular orbitals are completely symmetrical about the *internuclear axis;* the angular momentum about the axis is zero, $\lambda = 0$, so that they are σ *orbitals*. The first one is designated a $1s\sigma_g$ orbital. It is called a *bonding orbital*, because the concentration of charge between the nuclei bonds them together. The second one is designated $1s\sigma_u$, and it is called an *antibonding orbital*, corresponding to a net repulsion, because there is no shielding between the positively charged nuclei.

The molecular orbitals we have described are those for H_2, but we can use the same description for other homonuclear diatomic molecules, such as N_2 or Li_2. In a similar way, atomic orbitals for heavier atoms were derived from the exact theory of the hydrogen atom.

The building-up principle for the molecules is now like that for atoms. The nuclei are fixed and the electrons are added one-by-one to the available molecular orbitals of lowest energy. The Pauli Exclusion Principle requires that only two electrons with antiparallel spins can enter any orbital.

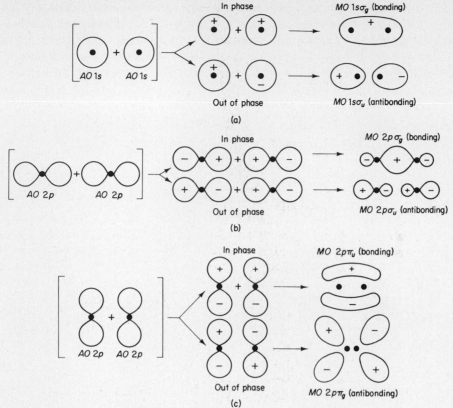

FIG. 15.9 Formation of molecular orbitals by linear combinations of atomic orbitals. Note the + and − signs which indicate the signs (phases) of the wave functions.

In the case of H_2, the two electrons enter the $1s\sigma_g$ orbital. The configuration is $(1s\sigma_g)^2$, and it corresponds to a single electron pair bond between the H atoms.

The next possible molecule would be one with three electrons, He_2^+. This has the configuration $(1s\sigma_g)^2(1s\sigma_u)^1$. There are two bonding electrons and one antibonding electron, so that a net bonding is to be expected. The molecule has, in fact, been observed spectroscopically; it has a dissociation energy of 3.0 eV.

If two helium atoms are brought together, the result is $(1s\sigma_g)^2(1s\sigma_u)^2$. Since there are two bonding and two antibonding electrons, there is no tendency to form a stable He_2 molecule.

The next possible atomic orbitals are the $2s$, and these behave just like the $1s$, providing $2s\sigma_g$ and $2s\sigma_u$ molecular orbitals with accommodations for four more electrons. If we bring together two lithium atoms with three electrons each, the molecule Li_2 is formed. Thus,

$$Li[1s^2 2s^1] + Li[1s^2 2s^1] \longrightarrow Li_2[(1s\sigma_g)^2(1s\sigma_u)^2(2s\sigma_g)^2]$$

Actually, only the outer-shell or valence electrons need be considered, and the molecular orbitals of inner K-shell electrons need not be explicitly designated. The Li$_2$ configuration is therefore written as $[KK(2s\sigma_g)^2]$. The molecule has a dissociation energy of 1.14 eV. The electron density maps for Li$_2$ are drawn in Fig. 15.10.

The hypothetical molecule Be$_2$, with eight electrons, does not occur, since the configuration would be $[KK(2s\sigma_g)^2(2s\sigma_u)^2]$ with no net bonding electrons.

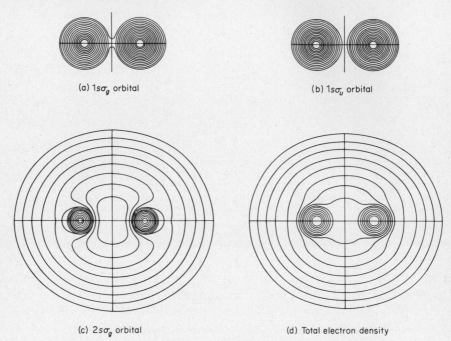

(a) $1s\sigma_g$ orbital   (b) $1s\sigma_u$ orbital

(c) $2s\sigma_g$ orbital   (d) Total electron density

FIG. 15.10   Contour maps of electron density in individual orbitals and of total electron density for Li$_2$ molecule. Adjacent contour lines vary by a factor of two. The outermost contour is an electron density of 6.1×10^{-5} e$^-$/(bohr)^3. The internuclear distance is the experimental value 0.2672 nm. [A. C. Wahl. For discussion, see *Science, 151,* 961 (1966).]

The next atomic orbitals are the $2p$'s shown in Fig. 15.9. There are three of these, p_x, p_y, p_z, mutually perpendicular and with characteristic wasp-waisted figures. The most stable M.O. that can be formed from the atomic p orbitals is one with the maximum overlap along the internuclear axis. This M.O. is shown in Fig. 15.9(b). This bonding orbital and the corresponding antibonding orbital can be written,

$$\psi_g = \psi_a(2p_x) + \psi_b(2p_x), \quad 2p\sigma_g$$
$$\psi_u = \psi_a(2p_x) - \psi_b(2p_x), \quad 2p\sigma_u$$

These orbitals have the same symmetry about the internuclear axis as the σ orbitals formed from atomic s orbitals; thus, they also have zero angular momentum about the axis, so that $\lambda = 0$ and they are σ orbitals.

The molecular orbitals formed from the p_y and p_z atomic orbitals have a distinctly different form, as shown in Fig. 15.9(c). As the nuclei are brought together, the sides of the p_y or p_z orbitals coalesce, and lead to two streamers of charge density, one above and one below the internuclear axis. These orbitals have an angular momentum of one unit, so that $\lambda = 1$ and they are π orbitals.

12. Correlation Diagram

We can obtain a good understanding of the relative energy levels of molecular orbitals by means of the model of the *united atom*. Imagine that we start with two H atoms both in the same quantum state (1s, for example) and squeeze them gradually together until they coalesce to form the united atom of helium. For this imaginary process, we must assume that internuclear repulsion is ignored completely. In this way, we can correlate the initial atomic orbitals in the hydrogen atoms with the final atomic orbitals in the helium atom. The molecular orbitals of the H_2 molecule will lie somewhere on the line joining these two extremes, at the correct internuclear distance for H_2.

Such a *correlation diagram* is shown in Fig. 15.11. We can usually see how the orbitals of the isolated atoms must correlate with those of the united atom by looking at the symmetry properties of the orbitals. For example, suppose that the nuclei A and B in the $1s\sigma_u$ orbital shown in Fig. 15.9 are squeezed together. It is evident that the result would be an orbital with the typical p shape, and the lowest p orbital of the united atom would be the $2p$ of helium. The *noncrossing rule* is also useful in establishing the correlation diagram: as the internuclear distance is varied, two curves for orbitals of the same symmetry cannot cross. Thus, for example, a σ_g can never cross a σ_g, but a σ_g can cross a σ_u. We can tell whether a molecular orbital is bonding or antibonding by examining the correlation diagram to see whether the energy rises or falls as the atoms are brought together to form this orbital. In the diagram of Fig. 15.11, the following are bonding orbitals: $1s\sigma_g$, $2s\sigma_g$, $2p\sigma_g$, $2p\pi_u$, $3s\sigma_g$, $3p\sigma_g$, $3p\pi_u$, $3d\sigma_g$, $3d\pi_u$, $3d\delta_g$.

Extensive calculations of the orbital (one-electron) energies have been made by the self-consistent field method, with the results shown in Table 15.2. Beginning with N_2, where the $3s\sigma_g$ orbital is occupied, its energy falls relative to that of the $2p\pi_u$ orbital, and for O_2 and F_2, the energy of the σ orbital lies below that of the π. This effect is similar to the *penetration* discussed for atomic orbitals of low azimuthal quantum number l (p. 639).

The configurations of other homonuclear diatomic molecules can now be described simply by adding electrons to the available orbitals of lowest energy.

The formation of N_2 proceeds as follows:

$$N[1s^2 2s^2 2p^3] + N[1s^2 2s^2 2p^3] \longrightarrow N_2[KK(2s\sigma_g)^2(2s\sigma_u)^2(2p\pi_u)^4(2p\sigma_g)^2]$$

Since there are six net bonding electrons, it can be said that there is a triple bond between the two N's. One of these bonds is a σ bond; the other two are π bonds at right angles to each other. (It is also possible, however, to form three identical

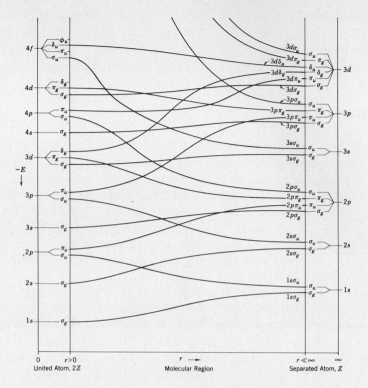

FIG. 15.11 Correlation diagram to show how molecular orbitals are formed from atomic orbitals of separated atoms and correlate with atomic orbitals of the united atom formed as nuclei come together [M. Kasha, *Molecular Electronic Structure*].

equivalent bond orbitals, which can be considered to have a mixture of σ and π characters. This is just another linear combination that some people like better.)

Molecular oxygen is an interesting case:

$$O[1s^2 2s^2 2p^4] + O[1s^2 2s^2 2p^4] \longrightarrow O_2[KK(2s\sigma_g)^2(2s\sigma_u)^2(2p\sigma_g)^2(2p\pi_u)^4(2p\pi_g)^2]$$

There are four net bonding electrons, or a double bond consisting of a σ and a π bond. A single bond is usually a σ bond, but a double bond is not formed from two σ bonds, but from a σ plus a π. In O_2, the $2p\pi_g$ orbitals, which can hold four electrons, are only half filled. Because of electrostatic repulsion between the electrons, the most stable state will be that in which the electrons are assigned as $(2p_y\pi_g)^1(2p_z\pi_g)^1$. The total spin of O_2 is therefore $\mathscr{S} = 1$, and its multiplicity is $2\mathscr{S} + 1 = 3$. The ground state of oxygen is $^3\Sigma$. Because of its unpaired electron spins, O_2 is paramagnetic (Section 17.25).

In the molecular-orbital method, all electrons outside closed shells make a

TABLE 15.2
One-Electron Energies for Homonuclear Diatomic
Molecules, Computed by Molecular Orbital
Self-Consistent Field Method (Energies in
Rydbergs)[a]

Mol.	Ref.	$1\sigma_g$	$1\sigma_u$	$2\sigma_g$	$2\sigma_u$	$1\pi_u$	$3\sigma_u$	$1\pi_g$	$3\sigma_g$
Li_2	1	−4.8806	−4.8802	−0.3604	0.0580	0.1282	0.1834	0.3206	0.7230
	2	−4.8710	−4.8705	−0.3627	0.0551		0.2158		0.7918
Be_2	2	−9.4187	−9.4184	−0.8512	−0.4416		0.1110		1.0882
B_2	3	−15.3530	−15.3514	−1.3552	−0.6990	−0.6824	0.0172		1.2898
C_2	2	−22.6775	−22.6739	−2.0567	−0.9662	−0.8407	−0.0444	0.5295	2.2598
N_2	4	−31.4438	−31.4396	−2.9054	−1.4612	−1.1595	−1.0892	0.5459	2.2054
	2	−31.2911	−31.2885	−2.8421	−1.4274	−1.6908	−1.1110	0.6004	2.2454
O_2	5	−41.1902	−41.1854	−3.0438	−1.9564	−1.0998	−1.1128	−0.7888	1.4784
F_2	2	−52.7191	−52.7187	−3.2517	−2.7225	−1.2159	−1.0922	−0.9489	0.6848

[a]One Rydberg unit = 0.5 hartree (atomic units) = 2.17971×10^{-18}J. Occupied orbitals are to the left of lines; orbitals enclosed by lines are half occupied. Table references follow:
[1]E. Ishiguro, K. Kayama, M. Kotani, and Y. Mizuno, *J. Phys. Soc. Japan, 12*, 1355 (1957).
[2]B. J. Ransil, *Rev. Mod. Phys., 32*, 239, 245 (1960).
[3]A. A. Padgett and V. Griffing, *J. Chem. Phys., 30*, 1286 (1959).
[4]C. W. Seherr, *J. Chem. Phys., 23*, 569 (1955).
[5]M. Kotani, Y. Mizuno, K. Kayama, and E. Ishiguro, *J. Phys. Soc. Japan, 12*, 707 (1957).
[From J. C. Slater, *Quantum Theory of Matter* (New York: McGraw-Hill Book Company, 1968), p. 451.]

contribution to the binding energy of the molecule. The shared electron-pair bond is not particularly emphasized. The way in which the excess of bonding over antibonding electrons determines the tightness of bonding may be seen by reference to the molecules in Table 15.3.

TABLE 15.3
Properties of Homonuclear Diatomic Molecules

Molecule	Bond dissociation energy D_0 kJ·mol⁻¹	Internuclear separation r_e nm	Fundamental vibration frequency ν_0 s⁻¹ × 10⁻¹³	Ground state
B_2	347	0.1589	3.152	$^3\Sigma$
C_2	531	0.1312	4.921	$^1\Sigma$
N_2	711	0.1098	7.074	$^1\Sigma$
O_2	494	0.1207	4.738	$^3\Sigma$
F_2	289	0.1418	2.67	$^1\Sigma$
Na_2	72	0.3078	0.477	$^1\Sigma$
P_2	485	0.1894	2.340	$^1\Sigma$
S_2	347	0.1889	2.176	$^3\Sigma$
Cl_2	239	0.1988	1.694	$^1\Sigma$

13. Heteronuclear Diatomic Molecules

When the two nuclei of a diatomic molecule are different, some of the symmetry of the homonuclear case is lost. There is no longer a center of symmetry between the nuclei so that the $g - u$ designation of the orbitals does not apply. The cylindrical symmetry about the internuclear axis remains, so that λ is still a good quantum number, and the molecular orbitals are still designated as σ, π, δ, etc.

The simplest heteronuclear diatomic molecule is HeH⁺, consisting of two dissimilar nuclei and two electrons. The Schrödinger equation cannot be separated for this system but refined approximate calculations have been made by the variation method. The calculated dissociation energy of the molecule, 182.8 kJ, agrees with the experimental value, as does the bond distance, 0.143 nm. As would be expected, the $Z = 2$ charge of the He nucleus attracts electrons more strongly than does the $Z = 1$ charge of the proton.*

The simplest uncharged heteronuclear diatomic molecule is LiH, which has been the subject of extensive and successful calculations, although there are now four electrons to consider. The molecular-orbital model is still quite adequate, especially in providing a way to designate the higher electronic states of the molecule for spectroscopic purposes. The usual approach is again to write the M.O. as an L.C.A.O. The results are shown schematically in Fig. 15.12, which indicates that the lowest M.O. designated σ^b is formed mainly as a combination of the $1s$

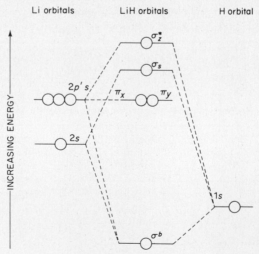

FIG. 15.12 Relative orbital energies in LiH.

*The calculations by B. Anex [*J. Chem. Phys.*, 7, 1651 (1963)] indicate that at the equilibrium internuclear distance of 0.0741 nm, the center of negative charge is at 0.0132 nm and the center of positive charge at 0.0247 nm from the helium nucleus.

orbital of H and the $2s$ orbital of Li with some admixture of the $2p$ orbital of Li that is directed along the bond axis. Thus,

$$\sigma^b = c_1(1s_a) + c_2(2s_b) + c_3(2p_b)$$

The first two electrons in LiH enter the essentially atomic $1s$ orbital of Li, and the next two are placed in the σ^b orbitals to give the configuration $(\sigma^b)^2$. Since the spins are paired and $\Lambda = l_1 + l_2 = 0$, the ground state is $^1\Sigma$.

The calculated electron distribution in LiH is shown in Fig. 15.13. There is

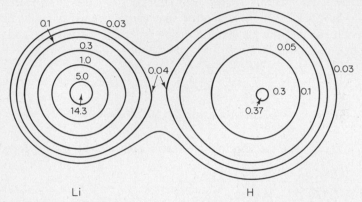

FIG. 15.13 Total electronic density distribution in lithium hydride. [P. Politzer and R. E. Brown, *J. Chem. Phys.*, *45*, 451 (1966).]

a net separation of charge in the ground state, so that the LiH molecule is a dipole. This dipole can be represented as a charge $+\delta$ localized at the Li nucleus and a charge $-\delta$ at the H nucleus. The electronegativity of the H atom is less than that of the Li, so that the atoms do not share the pair of σ_b electrons equally. Such a situation is often described by saying that the bond in LiH has a *partial ionic character*.

14. Electronegativity

Ever since the time of Berzelius, the concept of *electronegativity* has been helpful to chemists. It is not at all easy, however, to provide an operational definition that corresponds with the intuitive day-to-day usage of the concept. For example, Pauling introduced electronegativity as "the power of an atom in a molecule to attract electrons to itself." Although the word *power* is used here in some metaphorical sense, the emphasis on *atoms in molecules* indicates that electronegativity is a *bond property* and certainly not a property of isolated atoms.

Thus when Pauling devised a numerical electronegativity scale, it was based upon *bond energies*. He defined Δ' as the difference in energy between a bond A—B and the geometric mean of the energies of bonds A—A and B—B,

$$\Delta' = D(A—B) - \{D(A—A)\,D(B—B)\}^{1/2} \tag{15.24}$$

The quantity Δ' was observed to increase with the electronegativity difference between A and B. Pauling was able to fit the available data to an empirical expression,

$$\Delta'(A\text{—}B) = 30(X_A - X_B)^2 \tag{15.25}$$

where $X_A - X_B$ is the difference in electronegativity, and $\Delta'(A\text{—}B)$ is expressed in kilocalories. The 30 is merely an arbitrary factor to bring the electronegativity scale to a convenient numerical range. Note that Pauling electronegativities have the dimension of (energy)$^{1/2}$ and units of (kcal/30)$^{1/2}$.

A different and more physical approach to the concept of electronegativity was made by Mulliken.* He suggested that electronegativity could be measured by the arithmetic mean of the first ionization potential of an atom I and its electron affinity A. Thus,

$$M \longrightarrow M^+ + e \qquad I_M$$
$$e + M \longrightarrow M^- \qquad A_M$$
$$\text{Electronegativity} = \frac{I_M + A_M}{2}$$

To take into account the relation of electronegativity to bond properties, the I

TABLE 15.4
Average Electronegativities on the Pauling Scale Determined from Thermochemical Data[a]

I[b]	II	III	II	II	II	II	II	II	II	I	II	III	IV	III	II	I
H 2.20																
Li 0.98	Be 1.57											B 2.04	C 2.55	N 3.04	O 3.44	F 3.98
Na 0.93	Mg 1.31											Al 1.61	Si 1.90	P 2.19	S 2.58	Cl 3.16
K 0.82	Ca 1.00	Sc 1.36	Ti 1.54	V 1.63	Cr 1.66	Mn 1.55	Fe 1.83	Co 1.91	Ni 1.90	Cu 1.65	Zn 1.81	Ga 2.01	Ge 2.01	As 2.18	Se 2.55	Br 2.96
Rb 0.82	Sr 0.95	Y 1.22	Zr 1.33		Mo 2.16			Rh 2.28	Pd 2.20	Ag 1.93	Cd 1.69	In 1.78	Sn 1.96	Sb 2.05	Te	I 2.66
Cs 0.79	Ba 0.89	La 1.10	Hf		W 2.36			Ir 2.20	Pt 2.28	Au 2.54	Hg 2.00	Tl 2.04	Pb 2.33	Bi 2.02		
		Ce 1.12	Pr 1.13	Nd 1.14	Pm	Sm 1.17	Eu	Gd 1.20	Tb	Dy 1.22	Ho 1.23	Er 1.24	Tm 1.25	Yb	Lu 1.27	
					U 1.38	Np 1.36	Pu 1.28									

[a]A. L. Allred, *J. Inorg. Nucl. Chem.*, **17**, 215 (1961).
[b]The oxidation state is specified at the top of each group.

*R. S. Mulliken, *J. Chem. Phys.*, **2**, 782 (1934).

and A used in Mulliken's definition must refer to the *valence states* of the atoms. For example, in the case of a carbon atom, in which promotion of an electron $(2s^2 2p^2 \longrightarrow 2s 2p^3)$ to a higher orbital is necessary before four covalent bonds can be formed, the valence state would be ^5S, whereas the ground state is ^2P. The dimensions of the Mulliken electronegativities are those of (energy), so that they cannot be directly compared with the Pauling values. Nevertheless, the numerical values of Mulliken's absolute electronegativities are roughly proportional to Pauling's numbers, probably by accident.*

A recent summary of electronegativities on the Pauling scale is given in Table 15.4. These values have been found to correlate with a great variety of systematic data on chemical bonds—for example, nuclear quadrupole coupling constants, diamagnetic proton shielding in NMR studies, and frequencies of charge transfer spectra in molecules with metal ligands.

15. Dipole Moments

When a bond is formed between two atoms that differ in electronegativity, there is an accumulation of negative charge on the more electronegative atom, leaving a positive charge on the more electropositive one. The bond then constitutes an electric dipole, which is by definition an equal positive and negative charge, $\pm Q$, separated by a distance r. A dipole, as in Fig. 15.14(a), is characterized by its *dipole moment* $\boldsymbol{\mu}$, a vector having the magnitude Qr and the direction of the line joining the negative to the positive charge. A dipole consisting of charges $\pm e$ (4.80×10^{-10} esu) separated by a distance of 0.1 nm would have a moment of 4.80×10^{-18} esu·cm. The unit 10^{-18} esu·cm is called the *debye*† (D).

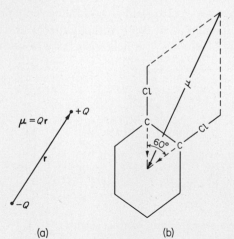

FIG. 15.14   (a) Definition of dipole moment; (b) vector addition of bond dipole moments in orthodichlorobenzene.

(a)   (b)

*R. Ferreira, "Electronegativity and Bonding," *Advan. Chem. Phys.*, *13*, 55 (1967).

†The classic work of Peter Debye, *Polar Molecules* (New York: Dover Publ. Co., 1945), should be read by all serious students.

If a polyatomic molecule contains two or more dipoles in different bonds, the net dipole moment of the molecule is the resultant of the vector addition of the individual bond moments. An example is shown in Fig. 15.14(b).

The experimental study of dipoles in molecules has provided a good deal of information about the nature of heteronuclear bonds. At this point, we shall therefore turn to this experimental approach to the chemical bond and return, later, to a theoretical discussion of more complex molecules.

16.   Polarization of Dielectrics

To discuss the determination of dipole moments, we should review some aspects of the theory of dielectrics. Consider a parallel-plate capacitor with the region between the plates evacuated, and let the charge per unit area on one plate be $+\sigma$ and $-\sigma$ on the other. The electric field within the capacitor is then normal to the plates and has the magnitude* $E_0 = \epsilon_0^{-1}\sigma$. The capacitance is

$$C_0 = \frac{Q}{\Delta\Phi} = \frac{\sigma A}{\epsilon_0^{-1}\sigma d} = \frac{\epsilon_0 A}{d} \tag{15.26}$$

where A is the area of the plates, d the distance, and $\Delta\Phi$ the potential difference between them, and $\epsilon_0 \approx (1/36\pi)\cdot 10^{-9}$ F·m^{-1}.

Now consider the space between the plates to be filled with some substance of negligible electric conductivity. Under the influence of small fields, electrons move quite freely through conductors, whereas in insulators or *dielectrics*, these fields displace the electrons only slightly from their equilibrium positions. An electric field acting on a dielectric thus causes a separation of positive and negative charges. The field is said to *polarize* the dielectric. This *polarization* is shown pictorially in Fig. 15.15(a). The polarization can occur in two ways: by the *induction effect* and by the *orientation effect*. An electric field always induces dipoles in molecules whether or not they contain dipoles to begin with. If the dielectric does contain molecules that are permanent dipoles, the field also tends to aline

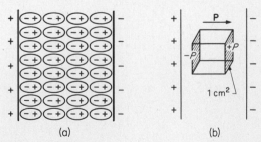

(a)   (b)

FIG. 15.15   (a) Polarization of a dielectric; (b) definition of the polarization vector, **P**.

*See, for example, G. P. Harnwell, *Electricity and Magnetism* (New York: McGraw-Hill Book Company, 1949), p. 26.

these dipoles along its own direction. The random thermal motions of the molecules oppose this orienting action. Our main interest is in the permanent dipoles, but before these can be studied, effects due to the induced dipoles must be clearly distinguished.

When a dielectric is introduced between the plates of a capacitor, the capacitance is increased by a factor ϵ, called the *dielectric constant*. Thus, if C_0 is the capacitance with a vacuum, the capacitance with a dielectric is $C = \epsilon C_0$. Since the charges on the capacitor plates are unchanged, the field normal to the plates is reduced by the factor ϵ, so that $E = E_0/\epsilon$.

The reason the field is reduced is clear from the picture of the polarized dielectric: All the induced dipoles are alined so as to produce an overall dipole moment that cuts down the field strength. Consider in Fig. 15.15(b) a unit cube of dielectric between the capacitor plates, and define a vector quantity **P** called the *dielectric polarization*, which is the dipole moment per unit volume. Then the effect of the polarization is equivalent to that which would be produced by a charge per unit area of $+P$ on one face and $-P$ on the other face of the cube. The field in the dielectric is now determined by the net charge density on the plates, so that

$$\epsilon_0 E = (\sigma - P) \tag{15.27}$$

A new vector has been defined, called the *electric displacement* **D**, which depends only on the charge σ, according to $\mathbf{D} = \sigma/\epsilon_0$. From (15.27),

$$\mathbf{D} = \mathbf{E} + \frac{\mathbf{P}}{\epsilon_0} \quad \text{and} \quad \frac{\mathbf{D}}{\mathbf{E}} = \epsilon \tag{15.28}$$

It follows that

$$\epsilon - 1 = \frac{\mathbf{P}}{\epsilon_0 \mathbf{E}} \tag{15.29}$$

In a vacuum, where $\epsilon = 1$ and $\mathbf{P} = 0$, $\mathbf{D} = \mathbf{E}$.

17. Induced Polarization

The polarization is the sum of two terms, $\mathbf{P} = \mathbf{P}_d + \mathbf{P}_o$. The *induced* or *distortion polarization* is $\mathbf{P}_d$, caused by the separation of positive and negative charges due to the action of the electric field on the dielectric. The *orientation polarization* is $\mathbf{P}_o$, due to the preferential alinements of permanent dipoles in the direction of the electric field.

To compute $\mathbf{P}_d$, we must consider the magnitude of the dipole moment **m** induced in a molecule by the field acting on it. We can assume that this induced moment is proportional to the strength of the local field **F**, and is in the same direction as the field, so that

$$\mathbf{m} = \alpha \mathbf{F} \tag{15.30}$$

The proportionality factor α is called the *polarizability*.* It is the induced moment

---

*For the general case, in which the direction of the induced moment is not the same as that of the field, we must write $\mathbf{m} = \tilde{\alpha}\mathbf{F}$, where $\tilde{\alpha}$ is a *tensor*.

per unit field strength. Note that α/ϵ_0 has the dimensions of volume, since $Qr/(Q/\epsilon_0 r^2) = \epsilon_0 r^3$. The polarizability of a hydrogen atom is $4.5\ a_0^3\epsilon_0$, which is close to that for the volume of a sphere of radius equal to that of the Bohr orbit, $\frac{4}{3}\pi a_0^3\epsilon_0 = 4.19 a_0^3\epsilon_0$. The polarizability of an atom or ion is a good measure of its volume.

In the case of a gas at low pressure, the molecules are so far apart that they do not exert appreciable electrical forces on one another. In this case, the field that polarizes a molecule [$\mathbf{F}$ in (15.30)] is simply the external field $\mathbf{E}$.

$$\mathbf{F} = \mathbf{E} \quad \text{(gas at low pressure)} \tag{15.31}$$

If M is the molar mass, L the Avogadro number, and ρ the density, the number of molecules in unit volume is $L\rho/M$. Hence, the distortion polarization becomes

$$\mathbf{P}_d = \frac{L\rho}{M}\mathbf{m} = \frac{L\rho}{M}\alpha\mathbf{E}$$

and, from (15.29), the dielectric constant of the dilute gas is

$$\epsilon = 1 + \frac{L\rho\alpha}{\epsilon_0 M} \tag{15.32}$$

We can therefore calculate immediately the induced polarization in a gas.

If the dielectric is not a dilute gas, we must consider the influence of the surrounding molecules to estimate the field that acts to polarize a given molecule. This difficult problem has not yet been completely solved, but approximate formulas have been obtained for various special cases. For gases at high pressures, nonpolar liquids, and dilute solutions of polar solutes in nonpolar solvents, the effective $\mathbf{F}$ is usually taken to be*

$$\mathbf{F} = \mathbf{E} + \frac{\mathbf{P}}{3\epsilon_0} \tag{15.33}$$

It follows that $\mathbf{m} = \alpha[\mathbf{E} + (\mathbf{P}/3\epsilon_0)]$, and instead of (15.32), we obtain

$$\frac{\epsilon - 1}{\epsilon + 2}\frac{M}{\rho} = \frac{L\alpha}{3\epsilon_0} = P_M \tag{15.34}$$

The quantity P_M is sometimes called the *molar polarizability*. So far, it includes only the contribution from induced dipoles, and a term due to permanent dipoles must be added to give the total molar polarizability. Equation (15.34) was first derived by O. F. Mossotti in 1850.

18.   Determination of the Dipole Moment

When a molecule is placed in an electric field, there will always be an induced dipole moment. It is evoked almost instantaneously in the direction of the field. It is independent of the temperature, since if the position of the molecule is disturbed by thermal collisions, the dipole is immediately induced again in the field direction. The contribution to the polarization caused by the permanent dipoles, however, is

*A derivation is given by J. C. Slater and N. H. Frank, *Introduction to Theoretical Physics* (New York: McGraw-Hill Book Company, 1933), p. 278; also Y. K. Syrkin and M. E. Dyatkina, *The Structure of Molecules* (New York: Interscience Publishers, 1950), p. 471.

less at higher temperatures, because the random thermal collisions of the molecules oppose the tendency of their dipoles to line up in the electric field.

It is necessary to calculate the average component of a permanent dipole in the field direction as a function of the temperature. Consider a dipole with random orientation. If there is no field, all orientations are equally probable. This fact can be expressed by saying that the number of dipole moments directed within a solid angle $d\omega$ is simply $A\,d\omega$, where A is a constant depending on the number of molecules under observation.

If a dipole moment $\boldsymbol{\mu}$ is oriented at angle θ to a field of strength $\mathbf{F}$ its potential energy* is $U = -\mu F \cos\theta$. According to the Boltzmann equation, the number of molecules oriented within the solid angle $d\omega$ is then

$$A \exp\frac{-U}{kT}\,d\omega = A \exp\frac{\mu F \cos\theta}{kT}\,d\omega$$

The average value of the dipole moment in the direction of the field, by analogy with (4.24), can be written

$$\bar{m} = \frac{\int A \exp(\mu F \cos\theta/kT)\,\mu\cos\theta\,d\omega}{\int A \exp(\mu F \cos\theta/kT)\,d\omega}$$

where the integrals are taken over all possible orientations. To evaluate this expression, let $\mu F/kT = x$, $\cos\theta = y$; then $d\omega = 2\pi \sin\theta\,d\theta = 2\pi\,dy$. Thus,

$$\frac{\bar{m}}{\mu} = \frac{\int_{-1}^{+1} e^{xy}y\,dy}{\int_{-1}^{+1} e^{xy}\,dy}$$

Since

$$\int_{-1}^{+1} e^{xy}\,dy = \frac{e^x - e^{-x}}{x}$$

and

$$\int_{-1}^{+1} e^{xy}y\,dy = \frac{e^x + e^{-x}}{x} + \frac{e^x - e^{-x}}{x^2}$$

then

$$\frac{\bar{m}}{\mu} = \frac{e^x + e^{-x}}{e^x - e^{-x}} - \frac{1}{x} = \coth x - \frac{1}{x} \equiv \mathscr{L}(x)$$

Here, $\mathscr{L}(x)$ is called the *Langevin function*, in honor of the inventor of this treatment.

In most cases, $x = \mu F/kT$ is a small fraction,† so that on expanding $\mathscr{L}(x)$

*G. P. Harnwell, *Electricity and Magnetism* (New York: McGraw-Hill Book Company, 1949), p. 64.

†Values of μ range around 10^{-18} (esu cm). If a capacitor with 1 cm between plates is charged to 3000 V, $\mu F = 10^{-18}[(3 \times 10^3)/(3 \times 10^2)] = 10^{-17}$ erg, compared to $kT = 10^{-14}$ erg at room temperature, so that $\mathscr{L}(x) = 3.33 \times 10^{-4}$ and the average moment in the field direction $\bar{m}$ is only this small fraction of the permanent moment μ.

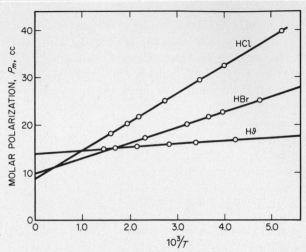

FIG. 15.16 Application of the Debye equation (15.36) to the molar
polarizabilities of the hydrogen halides.

TABLE 15.5
Magnitudes of Dipole Moments

Compound	Moment (D)	Compound	Moment (D)
HCN	2.93	CH_3F	1.81
HCl	1.03	CH_3Cl	1.87
HBr	0.78	CH_3Br	1.80
HI	0.38	CH_3I	1.64
H_2O	1.85	C_2H_5Cl	2.05
H_2S	0.95	$n\text{-}C_3H_7Cl$	2.10
NH_3	1.49	$i\text{-}C_3H_7Cl$	2.15
SO_2	1.61	CHF_3	1.61
CO_2	0.00	CH_2Cl_2	1.58
CO	0.12	$CH{\equiv}CCl$	0.44
NO	0.16	CH_3COCH_3	2.85
KF	8.62	CH_3OH	1.69
KCl	10.48	C_2H_5OH	1.69
KBr	10.41	C_6H_5OH	1.70
LiH	5.883	$C_6H_5NO_2$	4.08
B_2H_6	0.00	CH_3NO_2	3.50
H_2O_2	2.20	$C_6H_5CH_3$	0.37

in a power series, only the first term need be retained, leaving $\mathscr{L}(x) = x/3$, or

$$\bar{m} = \frac{\mu^2}{3kT} F \qquad (15.35)$$

The orientation polarizability due to permanent dipoles is now added to the induced polarizability. Instead of (15.34), the total molar polarizability is therefore

$$\frac{\epsilon - 1}{\epsilon + 2}\frac{M}{\rho} = P_M = \frac{L}{3\epsilon_0}\left(\alpha + \frac{\mu^2}{3kT}\right) \tag{15.36}$$

This equation was first derived by P. Debye.

It is now possible to evaluate both α and μ from the intercept and slope of P_M vs. $1/T$ plots, as shown in Fig. 15.16. The experimental data are values over a range of temperatures of the dielectric constant ϵ. The ϵ can be calculated from the measured capacitance of a parallel-plate capacitor in which the vapor or solution under investigation is the dielectric between the plates. Some values for dipole moments are collected in Table 15.5.*

19. Dipole Moments and Molecular Structure

Dipole moments provide two kinds of information about molecular structure: (1) the extent to which a bond is permanently polarized; (2) an insight into the geometry of the molecule, especially the angles between the bonds. Only a few typical applications will be mentioned.†

Carbon dioxide has no dipole moment in its ground state, despite the difference in electronegativity between carbon and oxygen. We can conclude that the molecule is linear, O—C—O; the two C—O bond moments, which must exist owing to the difference in electronegativity of the atoms, exactly cancel each other by vector addition.

On the other hand, water has a moment of 1.85 D, and must have a triangular structure. It has been estimated that each O—H bond has a moment of 1.60 D, and the bond angle is therefore about 105°, as shown by a vector diagram.

Consider the substituted benzene derivatives:

$\mu = 0$ 1.70   0   0   1.40   1.64

The zero moments of *p*-dichloro- and sym-trichlorobenzene indicate that benzene is planar and that the C—Cl bond moments are directed in the plane of the ring,

*Accurate dipole moments can also be obtained from an analysis of the effect of electric fields on molecular spectra (Stark effect) and from the *electric resonance method* applied to molecular beams.

†Many interesting examples are given by (1) V. I. Minkin and O. A. Osipov, *Dipole Moments in Organic Chemistry* (New York: Plenum Press, 1970) and (2) J. W. Smith, *Electric Dipole Moments* (London: Butterworth, 1955).

thereby adding to zero. The moment of *p*-di-OH benzene, on the other hand, shows that the O—H bonds are not in the plane of the ring, but directed at an angle to it, thus providing a net moment.

The dipole moment of ethyl chloride (2.05 D) is considerably larger than that of chlorobenzene (1.70 D). We might have expected that the electronegative Cl atom would draw electrons away from the π orbitals of benzene and thus place a higher effective negative charge on the Cl by an inductive effect. There must evidently be some more powerful influence that tends to decrease the negative charge on the aromatic Cl. We are thus led to consider resonance structures such as:

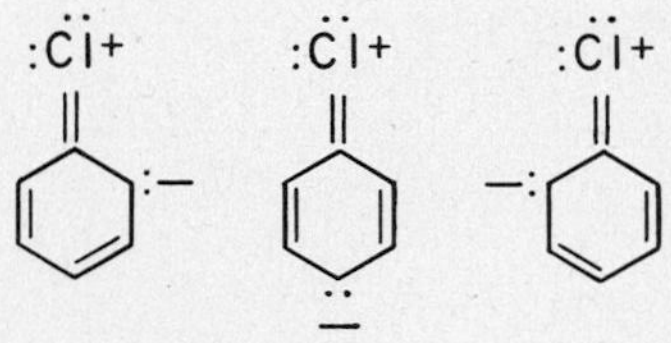

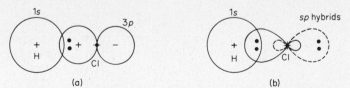

The internuclear distance in HCl is 0.126 nm. If the structure were H^+Cl^-, the dipole moment would be

$$\mu = (1.26)(4.80) = 6.05 \text{ D}$$

The actual moment of 1.03 suggests therefore that the ionic character of the bond is equivalent to a separation of charges of about $e/6$.

How does this separation of charge arise? The most naïve interpretation would be that the different electronegativities of the H and Cl atoms cause the center of negative charge of the *pair of electrons* in the H—Cl bond to be displaced toward the Cl. This model is shown in Fig. 15.17(a). The bond is shown as the pairing of

FIG. 15.17 Two models for the origin of the dipole moment in HCl.

the 1*s* electron of the H with one of the 3*p* electrons of the Cl. In such a model, the remaining nonbonding electrons in Cl are symmetrically disposed about the Cl nucleus and make zero net contributions to the dipole. We know, however, that a greater overlap of the bonding orbitals and hence a stronger bond can be obtained if we first form hybrid *sp* orbitals with the chlorine 3*s* and $3p_x$. One of these hybrid orbitals then overlaps the hydrogen 1*s* orbital to form the bonding orbital that holds the electron pair. Figure 15.17(b) shows this model. Now, however, the pair of electrons in the nonbonding *sp* hybrid orbital are not symmetrically disposed about the Cl nucleus and thus must make a substantial contribution to the dipole moment $H^+ \leftarrow Cl^-$. It is highly probable that such *atomic dipoles* contribute to the total dipole moment in many cases, so that we cannot explain dipole moments entirely in terms of the displacements of bonding electrons.

A measured dipole moment provides an excellent test of the correctness of the electron distribution in a molecule as computed from quantum mechanics. The general equation (14.2) is used to calculate the dipole moment from the best wave functions obtainable. The *dipole moment operator* can be written

$$\mu = \sum_{l=1}^{N} e_l(x_l\mathbf{i} + y_l\mathbf{j} + z_l\mathbf{k})$$

The summation is taken over all the nuclei and electrons in the molecule where e_l is the charge of the lth particle and $\mathbf{i}, \mathbf{j}, \mathbf{k}$ are unit vectors.

20. Polyatomic Molecules

As the numbers of nuclei and electrons in a molecule increase, the computational difficulties in applying quantum mechanics to calculate properties of the molecule soon exceed the range of even the largest and fastest modern computers. It is still possible to think about a molecule in terms of nonlocalized molecular orbitals, by placing the nuclei in fixed positions and adding the electrons to the resultant array of positive charges. It is also possible to write an approximate M.O. in the form of an L.C.A.O., but the number of atomic orbitals used in the *basis set* rapidly increases. The number of integrals that must be evaluated in a M.O.-L.C.A.O. calculation goes up as the fourth power of the number of functions in the basis set. Thus, for a molecule such as ethane, C_2H_6, about 10^6 integrals would need to be evaluated. With more than two nuclei, these integrals become *multicenter integrals* of considerable difficulty even for the best computer programs. For highly symmetrical molecules, such as CH_4 or NH_3, the problem can be simplified by means of powerful theorems from group theory. Nevertheless, severe approximations are still necessary, and as long as they need to be introduced, it is generally most reasonable that they be based on the pre-existing chemical knowledge. In other words, we can think about molecules most effectively in terms of chemical bonds.

The important conceptual advantages of chemical bonds can be maintained by introducing *bond orbitals* or *localized molecular orbitals*. For example, in the water molecule, the atomic arbitals that take part in bond formation are the $1s$ orbitals of the two hydrogens and the $2p_x$ and $2p_y$ of oxygen. The stable structure will be that in which there is maximum overlap of these orbitals. Instead of making the molecular orbital by a linear combination of all four atomic orbitals, we can take them in pairs to form two localized molecular orbitals corresponding to the two O—H bonds.

$$\psi_{\mathrm{I}} = 1s(H_a) + 2p_x(O)$$
$$\psi_{\mathrm{II}} = 1s(H_b) + 2p_y(O)$$

The formation of these bond orbitals is depicted schematically in Fig. 15.18. A pair of electrons of opposite spin is placed in each of these orbitals.

The observed valence angle in H_2O is not exactly $90°$ but actually $105°$. The difference can be ascribed in part* to the polar nature of the bond; the electrons are drawn toward the oxygen, and the residual positive charge on the hydrogens causes their mutual repulsion. In H_2S, the bond is less polar and the angle is $92°$. Note the straightforward fashion in which directed valence has been explained [a posteriori] in terms of the shapes of the atomic orbitals.

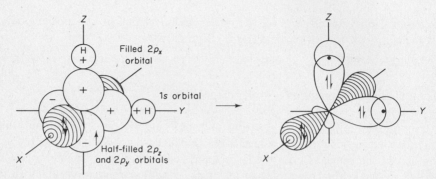

FIG. 15.18   Formation of a molecular orbital for H_2O by overlap of $2p$ orbitals of oxygen and $1s$ orbitals of hydrogens.

A prime example of directed valence is the tetrahedral orientation of the bonds formed by carbon in aliphatic compounds. These bonds are explained by the formation of *hybrid orbitals*. The ground state of the carbon atom is $1s^2 2s^2 2p^2$. There are two uncoupled electrons $2p_x$, $2p_y$, and we might therefore expect carbon to be bivalent. To display the valence of four, the carbon atom must have four electrons with uncoupled spins. The simplest way to obtain this condition is to *promote* one of the $2s$ electrons into the $2p$ state, and to have all the resulting $2p$ electrons with uncoupled spins. Then the outer configuration would be $2s2p^3$, with $2s^1 2p_x^1 2p_y^1$-$2p_z^1$. This excitation requires an energy of about 272 kJ·mol^{-1}, but the binding energy of the four bonds more than compensates for the promotion energy, and carbon is normally quadrivalent rather than bivalent.

If these four $2s2p^3$ orbitals of carbon were coupled with the $1s$ orbitals of hydrogens to yield the methane molecule, we might think at first that three of the bonds should be different from the remaining one. Actually, of course, the symmetry of the molecule is such that all the bonds must be exactly the same. It is possible to form sets of four *hybrid orbitals*, which are linear combinations of the s and p orbitals. One set consists of hybrid orbitals spacially directed from the carbon atom toward the corners of a regular tetrahedron. These are called *tetrahedral orbitals*, t_1, t_2, t_3, t_4. They are shown in Table 15.6. The tetrahedral orbitals are

---

*A more detailed theory shows that the $2s$ electrons of the oxygen also take part in the bonding, forming hybrid orbitals such as those discussed for carbon.

exceptionally stable since they allow the electron pairs to avoid one another to the greatest possible extent. The hybrid t orbitals of the carbon then combine with the $1s$ orbitals of four hydrogen atoms to form a set of four localized molecular orbitals for methane, $\psi_1 = c_1(t_1) + c_2(1s_1)$, etc.

In addition to the tetrahedral hybrids, the orbitals of carbon can be hybridized in other ways. The so-called *trigonal hybrids sp^2* mix $2s$, $2p_x$, and $2p_y$ to form three orbitals at an angle of 120°. These hybrids are also shown in Table 15.6. The fourth A.O., $2p_z$, is perpendicular to the plane of the others. This kind of hybridization is used in ethylene. The double bond consists of an sp^2 hybrid σ bond with a π bond formed by overlapping $2p_z$ orbitals.

If one $2s$ is mixed with a $2p$, we obtain the *digonal hybrids sp*. These are also shown in Table 15.6. This kind of hybridization is used in acetylene.

Table 15.7 summarizes some properties of C—H bonds having different types of hybridization.

TABLE 15.6
Types of Hybridization of s and p Orbitals[a]

Digonal Hybridization

(Two hybrids, D_1 and D_2, pointing along the z axis in opposite directions.)

$D_1 = (1/\sqrt{2})(s + p_z)$
$D_2 = (1/\sqrt{2})(s - p_z)$

Polar representation of digonal hybridization

Trigonal Hybridization

(Three hybrids, T_1, T_2, and T_3 pointing along the xy plane; the first points along the x axis, the other two along the directions forming an angle of 120° with the x axis.)

$T_1 = (1/\sqrt{3})s + (\sqrt{2}/\sqrt{3})p_x$
$T_2 = (1/\sqrt{3})s - (1/\sqrt{6})p_x + (1/\sqrt{2})p_y$
$T_3 = (1/\sqrt{3})s - (1/\sqrt{6})p_x - (1/\sqrt{2})p_y$

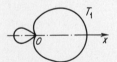

Polar representation of trigonal hybridization

Tetrahedral Hybridization

(Four hybrids, t_1, t_2, t_3, and t_4 pointing toward the vertices of a regular tetrahedron centered at the origin of the coordinates; the first hybrid points along the axis of the triad of x, y, z.)

$t_1 = (1/2)(s + p_x + p_y + p_z)$
$t_2 = (1/2)(s + p_x - p_y - p_z)$
$t_3 = (1/2)(s - p_x + p_y - p_z)$
$t_4 = (1/2)(s - p_x - p_y + p_z)$

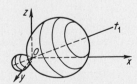

Polar representation of tetrahedral hybridization

[a] R. Daudel, R. Lefebvre, C. Moser, *Quantum Chemistry* (New York: Interscience Publishers, 1959).

TABLE 15.7
Properties of C—H Bonds

Hybridization	Example	Bond length nm	Stretching force constant, N·m⁻¹	Bond energy kJ
sp	acetylene	0.1060	6.937×10^2	506
sp^2	ethylene	0.1069	6.126×10^2	443
sp^3	methane	0.1090	5.387×10^2	431
p	CH radical	0.1120	4.490×10^2	330

The use of hybrid orbitals is by no means restricted to carbon atoms. The hybrid orbital will almost always make it possible to form a stronger covalent bond between two atoms, since the strongly directed character of the orbital will allow the best overlap of the orbital of one atom with the orbital of its partner. Figure 15.19 shows how different hybrid bonds can provide different angles

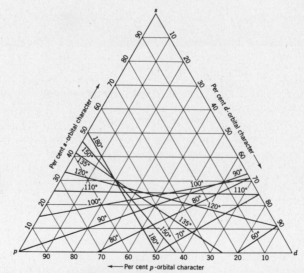

FIG. 15.19 Directional properties of hybrid orbitals from *s*, *p*, and *d* atomic orbitals. [M. Kasha, *Molecular Electronic Structure*, to be published.]

between the bonds. We see, for example, that the 105° bond angle in H_2O could be produced by mixing about 20% *s* character and 80% *p* character. This hybrid would provide stronger bonds for H_2O than the pure *p* bonds shown in Fig. 15.18.

As an example of the use of Fig. 15.19, let us predict the bond angle of the hybrids d^2sp^3, which are important in coordination compounds of the transition metals. For $33\frac{1}{3}\%$ *d*, $16\frac{2}{3}\%$ *s*, and 50% *p*, the diagram shows that the bond angles are 90 and 180°, i.e., the configuation of a regular octahedron. An example is shown in Fig. 15.20.

The "explanation" of directed valence and bond angles in terms of hybrid orbitals has been a quite popular model among chemists since the publication in 1939 of the first edition of *The Nature of the Chemical Bond* by Linus Pauling. Yet

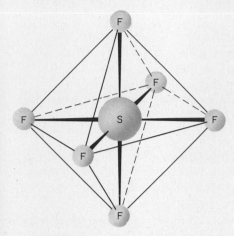

FIG. 15.20 The octahedral molecule SF_6. The sulfur atom provides six hybrid d^2sp^3 orbitals.

one should not overemphasize the quantitative value of the hybridization model. Other general concepts have been proposed that appear to be equally satisfactory in correlating and "explaining" experimental data on directed valence.

In 1957, R. J. Gillespie and R. S. Nyholm* showed that bond angles in inorganic compounds could be explained by considering the most favorable location of all the electron pairs in a valence shell about the central atom, including especially *lone pairs* that were not shared in a bond with another atom. The actual structure would usually be the one in which electrostatic repulsion between the electron pairs was minimized, consistent always with the requirements of the Pauli Principle. Thus two electrons of the same spin have a maximum probability of being at opposite sides of the central nucleus, three electrons at the corners of an equilateral triangle, four electrons at the corners of a tetrahedron, six electrons at the corners of an octahedron.

As an example of these ideas, consider the molecules $SnCl_4$, $SbCl_3$, and $TeCl_2$, shown in Fig. 15.21. These all have an essentially tetrahedral structure, but one of the corners in $SbCl_3$ and two of the corners in $TeCl_2$ are occupied by lone pairs. The bond angle in $SbCl_3$ is decreased to 99.5° from the exact 109.5° of the tetrahedron, in accord with the Gillespie–Nyholm principle that *nonbonding electron pairs repel adjacent electron pairs more strongly than do bonding electron pairs.*

The Gillespie–Nyholm concepts work quite well so long as the central atom is large. When the central atom is small, however, the effects of nonbonded interactions between the atoms surrounding the central atom become important. The

---

Quart. Rev. London, **11**, 339 (1957). See also R. J. Gillespie, *J. Chem. Educ.*, **40**, 295 (1963).

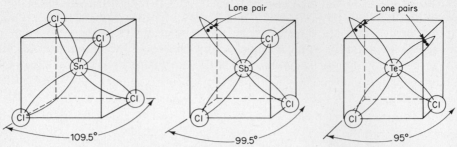

FIG. 15.21 Structures of $SnCl_4$, $SbCl_3$ and $TeCl_2$ illustrate effect of electron repulsion in determination of geometry of molecules in accord with the principles of Gillespie and Nyholm.

structure is then determined to a considerable extent by simple packing requirements, based on repulsions of the van der Waals type between nonbonded atoms. L. S. Bartell has provided much convincing evidence that such effects must be included in any explanation of observed bond angles.*

Eventually, these various qualitative models of molecular geometry may be understood in terms of exact quantum mechanical calculations of molecular structures, but because of the difficulties of the computations, progress toward this goal may be slow. Meanwhile, the simple models will continue to fill their traditional role in the correlation of chemical data and the planning of new experiments.

21. Bond Distances, Bond Angles, Electron Densities

The chemical literature today contains a large amount of information about the structures of polyatomic molecules. The most important source of such data is the spectroscopy of molecules, which will be discussed in Chapter 17. Diffraction patterns, obtained with X rays, electrons, or neutrons, can be analyzed to give detailed structures. Chapter 18 will deal with the theory and method of X-ray diffraction. Because of their good penetrating power, X rays are especially suitable for diffraction by crystals. For diffraction studies with gases, electron beams are generally more useful because the negative electrons are strongly scattered by the electrons and nuclei of molecules. Neutron diffraction is useful for determining the positions of hydrogen atoms, which do not diffract X rays or electrons strongly enough to provide accurate information on positions.

In principle, X-ray and electron scattering can yield direct experimental information about the electron density distribution in a molecule. Figure 14.20 showed the first example of an experimental measurement of electron density in an atom,

J. Chem. Educ., **45**, 754 (1968).

the case of argon. More refined electron diffraction techniques are now being applied to study experimentally the distribution of electron density in molecules and especially in chemical bonds. Relative intensity measurements of scattered electrons are already at the 0.1 % level of accuracy. When the 0.01 % level is reached, we may expect to see good resolution of the electron densities in bonds between the lighter elements. Figure 15.10 showed theoretical electron densities computed by the Hartree–Fock method for a simple molecule. Electron diffraction gives experimental results of comparable accuracy. Such experimental electron densities will soon provide means for stricter tests of the computations. Whereas a dipole moment measurement gives essentially one point in an electron density function, the scattering data give the function along the chemical bond, clearly a much more detailed picture.

22. Electron Diffraction of Gases

The wavelength of 40 kV electrons is 6.0 pm, about one-twentieth the magnitude of interatomic distances in molecules, so that diffraction effects will occur. In Section 13.4, diffraction by a set of slits was discussed in terms of the Huygens construction. In the same way, if a collection of atoms at fixed distances apart (i.e., a molecule) is placed in a beam of radiation, each atom can be regarded as a new source of spherical wavelets. From the interference pattern produced by these wavelets, the spacial arrangement of the scattering centers can be determined. The type of pattern found is shown in Fig. 15.22.

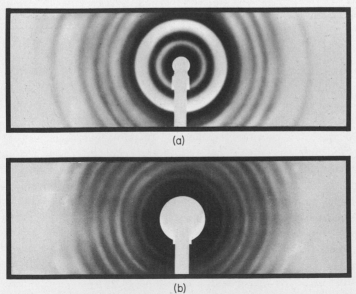

(a)

(b)

FIG. 15.22 Electron diffraction patterns from gases: (a) coronene; (b) phosphorus trichloride [Otto Bastiansen, Norges Tekniske Høgskole, Trondheim].

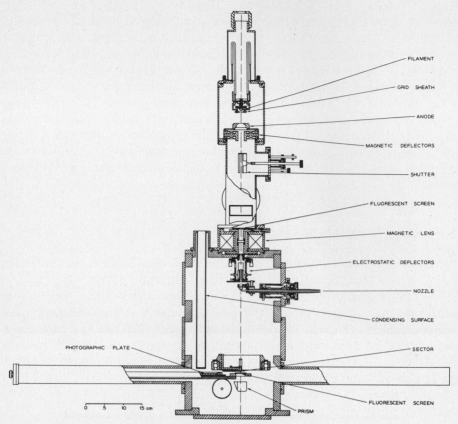

FILAMENT

GRID   SHEATH

ANODE

MAGNETIC   DEFLECTORS

SHUTTER

FLUORESCENT   SCREEN

MAGNETIC   LENS

ELECTROSTATIC   DEFLECTORS

NOZZLE

CONDENSING   SURFACE

PHOTOGRAPHIC   PLATE

SECTOR

FLUORESCENT   SCREEN

PRISM

0   5   10   15 cm

FIG. 15.23 Apparatus for electron diffraction of gases. [L. S. Bartell, University of Michigan.]

An experimental apparatus for electron diffraction of gases is illustrated in Fig. 15.23. The rotating heart-shaped sector is placed in front of the photographic plate to provide an exposure time that increases with the angle of scattering, and thus to compensate for the sharp decrease of scattering intensity with angle.

The electron beam traverses a collection of many gas molecules, oriented at random to its direction. Maxima and minima occur in the diffraction pattern despite the random orientation of the molecules, because the scattering centers occur as groups of atoms with the same definite fixed arrangement within every molecule. Diffraction by gases was treated theoretically (for X rays) by Debye in 1915, but electron diffraction experiments were not carried out until the work of Wierl in 1930.

We can show the essential features of the diffraction theory by considering the simplest case, that of a diatomic molecule. Consider in Fig. 15.24 the molecule AB, with atom A at the origin and B a distance r away. The orientation of the molecule AB is specified by the angles α and ϕ. AP is the projection of AB onto the XY plane. The incident electron beam enters parallel to the Y axis and diffraction

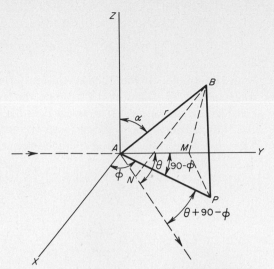

FIG. 15.24 Scattering of electrons by a diatomic molecule.

occurs through an angle θ. The interference between the waves scattered from A and B depends on the difference between the lengths of the paths they traverse. The calculation of the path difference δ requires that we recognize points on the diffracted and undiffracted beams which are in phase with each other. Hence, drop from B a perpendicular BN onto the diffracted direction, and a perpendicular BM onto the undiffracted direction. Now M and N are in phase and the path difference $\delta = AN - AM$.

Since PM is $\perp AY$ and PN is $\perp$ the diffracted beam,

$$\delta = AN - AM = AP \cos(\theta + \phi - 90) - AP \cos(90 - \phi)$$

But $AP = r \sin \alpha$, so that

$$\delta = r \sin \alpha \, [\sin(\theta + \phi) - \sin \phi]$$

$$\delta = 2r \sin \alpha \cos\left(\phi + \frac{\theta}{2}\right) \sin \frac{\theta}{2}$$

To add waves that differ in phase and amplitude, it is convenient to represent them in the complex plane and to add vectorially.* The difference in phase between the two scattered waves is $(2\pi/\lambda)\delta$. We shall assume for simplicity that the atoms A and B are identical. Then the resultant amplitude at P is $A = A_0 + A_0 \, e^{2\pi i \delta/\lambda}$. A_0, called the *atomic form factor for electron scattering*,† depends on the nuclear charge of the atom. The intensity of radiation is proportional to the square of the amplitude or, in this case, to A^*A, the amplitude times its complex conjugate. Thus,

---

*See Courant and Robbins, *What Is Mathematics?* (New York: Oxford University Press, 1941), p. 94.

†Whereas X rays are scattered primarily by the electrons in atoms, fast electrons are scattered primarily by the nuclei.

$$I \sim A^*A = A_0^2(1 + e^{-2\pi i \delta/\lambda})(1 + e^{2\pi i \delta/\lambda})$$
$$= A_0^2(2 + e^{-2\pi i \delta/\lambda} + e^{2\pi i \delta/\lambda})$$
$$= 2A_0^2\left(1 + \cos\frac{2\pi\delta}{\lambda}\right) = 4A_0^2 \cos^2\frac{\pi\delta}{\lambda}$$

To obtain the required formula for the intensity of scattering of a randomly oriented group of molecules, it is necessary to average the expression for the intensity at one particular orientation (α, ϕ) over all possible orientations. The differential element of solid angle is $\sin \alpha \, d\alpha \, d\phi$, and the total solid angle of the sphere around AB is 4π. Hence, the required average intensity becomes

$$I_{av} \sim \frac{4A_0^2}{4\pi} \int_0^{2\pi} \int_0^{\pi} \cos^2\left[2\pi\frac{r}{\lambda} \sin\frac{\theta}{2} \sin\alpha \cos\left(\phi + \frac{\theta}{2}\right)\right] \sin\alpha \, d\alpha \, d\phi$$

On integration,†

$$I_{av} = 2A_0^2\left(1 + \frac{\sin sr}{sr}\right) \tag{15.37}$$

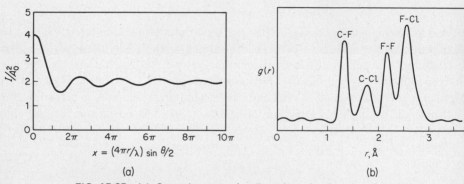

FIG. 15.25 (a) Scattering curve for diatomic molecule—plot of Eq. (15.37); (b) radial distribution function for the compound CF_3Cl.

†Let

$$I_{av} = \frac{A_0^2}{\pi} \int_0^{\pi} \int_0^{2\pi} \cos^2 (A \cos \beta) \, d\beta \sin \alpha \, d\alpha$$

where

$$A = \frac{2\pi r}{\lambda} \sin\frac{\theta}{2} \sin\alpha \quad \text{and} \quad \beta = \phi + \frac{\theta}{2}$$

Since $\cos^2 \beta = (1 + \cos 2\beta)/2$,

$$I_{av} = \frac{A_0^2}{\pi} \int_0^{\pi} \int_0^{2\pi} \left[\frac{d\beta}{2} + \cos (2 A \cos \beta) \, d\beta\right] \sin \alpha \, d\alpha$$

The integral can be evaluated by a series expansion of the cosine

$$\left[\cos x = 1 - \frac{x^2}{2!} + \frac{x^4}{4!} - \cdots\right]$$

followed by a term-by-term integration first over β and then over α.

where

$$s = \frac{4\pi}{\lambda} \sin \frac{\theta}{2}$$

In Fig. 15.25(a), I/A_0^2 is plotted against s, and the maxima and minima in the intensity are clearly evident.

In a more complex molecule with atoms j, k (having scattering factors A_j, A_k) a distance r_{jk} apart, the resultant intensity would be

$$I(\theta) = \sum_j \sum_k A_j A_k \frac{\sin s r_{jk}}{s r_{jk}} \tag{15.38}$$

This is called the *Wierl equation*. The summation must be carried out over all pairs of atoms in the molecule.

23. Interpretation of Electron-Diffraction Pictures

In electron diffraction, scattering by the nuclei provides information about the geometrical structure of molecules. Scattering by the diffusely distributed electrons is less intense, and the observed scattering curves can be corrected for it. The general incoherent background scattering can also be eliminated from the total scattering before the data are used for the determination of structure. One thus uses a molecular scattering $M(s) = (I_t/I_b) - 1$, where I_t is the total intensity and I_b is a suitably chosen background.

This M_s can be related to a radial distribution function $g(r)$, which gives the probability that a nucleus j will be found at a distance r from a nucleus k in the molecule.

$$g(r) = \int_0^{s(\max)} s M(s) \exp(-bs^2) \sin sr \, ds \tag{15.39}$$

Thus $g(r)$ will have a maximum for each value of r corresponding to an internuclear distance in the molecule. The integral is taken from $s = 0$ to the maximum angle of scattering measured. The factor $\exp(-bs^2)$ is a weighting factor introduced to improve the convergence of the integral. The computation of these integrals can now be done quickly on a computer.* An example of a radial distribution function is shown in Fig. 15.25(b) for the molecule CF_3Cl, but peaks are not always so completely resolved. Bond angles in a molecule can also be computed if enough neighboring internuclear distances are known.

Electron diffraction data can also be analyzed by a direct comparison between the experimental $M(s)$ curve and curves computed from the Wierl equation for selected values of molecular parameters (distances and angles).

Some results of electron diffraction studies are collected in Table 15.8. As molecules become more complicated, it becomes increasingly difficult to determine an exact structure, since usually only a dozen or so maxima are visible,

---

*R. A. Bonham and L. S. Bartell, *J. Chem. Phys.*, *31*, 702 (1959).

TABLE 15.8
The Electron Diffraction of Gas Molecules

Diatomic molecules			
Molecule	Bond distance (nm)	Molecule	Bond distance (nm)
NaCl	0.0251	N_2	0.1095
NaBr	0.0264	F_2	0.1435
NaI	0.0290	Cl_2	0.2009
KCl	0.0279	Br_2	0.2289
RbCl	0.0289	I_2	0.2660

Polyatomic Molecules			
Molecule	Configuration	Bond	Bond distance (nm)
$CdCl_2$	Linear	Cd—Cl	0.2235
$HgCl_2$	Linear	Hg—Cl	0.227
BCl_3	Planar	B—Cl	0.173
SiF_4	Tetrahedral	Si—F	0.155
$SiCl_4$	Tetrahedral	Si—Cl	0.201
P_4	Tetrahedral	P—P	0.221
Cl_2O	Bent, $111 \pm 1°$	Cl—O	0.170
SO_2	Bent, 120°	S—O	0.143
CH_2F_2	C_{2v}	C—F	0.1360
CO_2	Linear	C—O	0.1162
C_6H_6	Planar	C—C	0.1393
		C—H	0.108

which obviously will not permit the exact calculation of more than five or six parameters. Each distinct interatomic distance or bond angle constitutes a parameter. It is possible, however, from measurements on simple compounds, to obtain reliable values of bond distances and angles, which may be used to estimate the structures of more complex molecules.

24.   Nonlocalized Molecular Orbitals—Benzene

It is not always possible to assign the electrons in molecules to molecular orbitals localized between two nuclei. Important examples of *delocalization* are found in conjugated and aromatic hydrocarbons.

In the case of benzene, the carbon atomic orbitals are first prepared as trigonal sp^2 hybrids and then brought together with the hydrogens. These localized σ orbitals lie in a plane, as shown in Fig. 15.26(a). The atomic p orbitals extend their lobes above and below the plane [Fig. 15.26(b)], and when they overlap, they form delocalized molecular π orbitals, above and below the plane of the ring. These orbitals hold six mobile delocalized electrons. The shapes of the three lowest π orbitals are shown in Fig. 15.26(c).

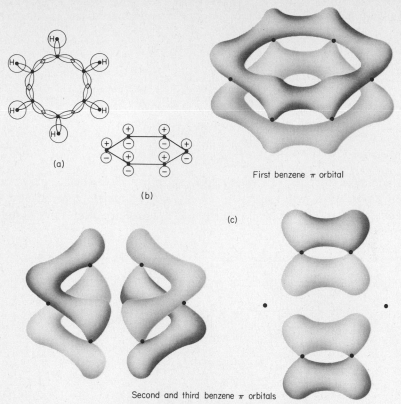

First benzene π orbital

(a)

(b)

(c)

Second and third benzene π orbitals

FIG. 15.26 Molecular orbitals for benzene. (a) Overlap of $sp^2\sigma$
orbitals. (b) The p_z orbitals which overlap to yield π
orbitals. (c) Representation of the three lowest π orbitals.

We can write a π molecular orbital in benzene as a linear combination of six
atomic p orbitals:

$$\psi = c_1\psi_1 + c_2\psi_2 + c_3\psi_3 + c_4\psi_4 + c_5\psi_5 + c_6\psi_6 \qquad (15.40)$$

This ψ is a wave function expressing the property that an electron placed in it can
move around the benzene ring. We calculate the ground state by variation of the
coefficients c_1, c_2, etc., until we find the ψ functions that give the lowest energy.

We substitute the wave function of (15.40) into the basic energy formula of the
variation method, (14.78), to obtain*

$$E = \frac{\int (\sum c_j\psi_j)\hat{H}(\sum c_j\psi_j)\, d\tau}{\int (\sum c_j\psi_j)^2\, d\tau} \qquad (15.41)$$

*To become familiar with the mathematical operations of this theory, you should work
through an example without the summation signs, taking $\psi = c_1\psi + c_2\psi_2$. We use in (15.41)
the real form of wave functions ψ_j.

The summations run from $j = 1$ to $j = 6$. For convenience, we use the notation

$$H_{jk} = \int \psi_j \hat{H} \psi_k \, d\tau$$

$$S_{jk} = \int \psi_j \psi_k \, d\tau \tag{15.42}$$

We can show that $S_{jk} = S_{kj}$ and $H_{jk} = H_{kj}$. Then (15.41) can be written as

$$E = \frac{\sum_j \sum_k c_j c_k H_{jk}}{\sum_j \sum_k c_j c_k S_{jk}} \tag{15.43}$$

To minimize the energy E with respect to the variable coefficients c_i, we set the derivatives of E with respect to the c_i, $\partial E / \partial c_i$, equal to zero, and solve the resultant set of (six) simultaneous equations for the values of c_i. For example, the differentiation of

$$E \sum_j \sum_k c_j c_k S_{jk} = \sum_j \sum_k c_j c_k H_{jk}$$

with respect to c_i, with $\partial E / \partial c_i = 0$, gives

$$E \sum_k c_k S_{ik} + E \sum_j c_j S_{ji} = \sum_k c_k H_{ik} + \sum_j c_j H_{ji}$$

Since $S_{ij} = S_{ji}$ and $H_{ij} = H_{ji}$, this becomes

$$E \sum_j c_j S_{ji} = \sum_j c_j H_{ji}$$

or

$$\sum_j c_j (H_{ji} - E S_{ji}) = 0 \tag{15.44}$$

We have a set of six equations like this as the value of i runs from 1 to 6, corresponding to the process of minimizing the energy with respect to each of the six coefficients c_i. The six simultaneous equations are as follows:

$$c_1(H_{11} - S_{11}E) + c_2(H_{21} - S_{21}E) + \cdots + c_6(H_{61} - S_{61}E) = 0$$
$$c_1(H_{12} - S_{12}E) + c_2(H_{22} - S_{22}E) + \cdots + c_6(H_{62} - S_{62}E) = 0$$
$$\vdots \qquad\qquad \vdots \qquad\qquad \vdots \qquad\qquad \vdots$$
$$c_1(H_{16} - S_{16}E) + c_2(H_{26} - S_{26}E) + \cdots + c_6(H_{66} - S_{66}E) = 0$$

These are *linear homogeneous equations*. As shown on p. 677, the condition for nontrivial solutions is the vanishing of the determinant of the coefficients:

$$
\begin{vmatrix}
H_{11} - S_{11}E & H_{21} - S_{21}E \cdots\cdots\cdots\cdots H_{61} - S_{61}E \\
H_{12} - S_{12}E & H_{22} - S_{22}E \cdots\cdots\cdots\cdots H_{62} - S_{62}E \\
H_{13} - S_{13}E & H_{23} - S_{23}E \cdots\cdots\cdots\cdots H_{63} - S_{63}E \\
H_{14} - S_{14}E & H_{24} - S_{24}E \cdots\cdots\cdots\cdots H_{64} - S_{64}E \\
H_{15} - S_{15}E & H_{25} - S_{25}E \cdots\cdots\cdots\cdots H_{65} - S_{65}E \\
H_{16} - S_{16}E & H_{26} - S_{26}E \cdots\cdots\cdots\cdots H_{66} - S_{66}E
\end{vmatrix} = 0 \tag{15.45}
$$

The secular equation (15.45), when multiplied out, is an equation of the sixth degree in E, and therefore possesses six roots.

Because of the difficulties in evaluating the integrals in the secular equation, an approximate treatment, developed by E. Hückel, has been widely used.* The approximations are the following:

1. $H_{jj} = \alpha$, the *coulombic integral* for all j.
2. $H_{jk} = \beta$, the *resonance integral* for atoms that are bonded together.
3. $H_{jk} = 0$ for atoms that are not bonded together.
4. $S_{jj} = 1$.
5. $S_{jk} = 0$ for $j \neq k$.

With these approximations, the secular equation is greatly simplified, and it becomes

$$
\begin{vmatrix}
\alpha - E & \beta & 0 & 0 & 0 & \beta \\
\beta & \alpha - E & \beta & 0 & 0 & 0 \\
0 & \beta & \alpha - E & \beta & 0 & 0 \\
0 & 0 & \beta & \alpha - E & \beta & 0 \\
0 & 0 & 0 & \beta & \alpha - E & \beta \\
\beta & 0 & 0 & 0 & \beta & \alpha - E
\end{vmatrix} = 0
$$

The roots of this equation of the sixth degree† are

$$E_1 = \alpha + 2\beta, \quad E_{2,3} = \alpha + \beta \text{ (twice)}, \quad E_{4,5} = \alpha - \beta \text{ (twice)}, \quad E_6 = \alpha - 2\beta$$

Since β is negative, they are given in order of ascending energy. When the values of E are put back into the system of linear equations, they can be solved for the coefficients c_j, and thus we can get explicit expressions for the molecular orbitals (wave functions).

It turns out that the lowest M.O. is

$$\psi_A = (6)^{-1/2}(\psi_1 + \psi_2 + \psi_3 + \psi_4 + \psi_5 + \psi_6)$$

This orbital, which can hold two electrons with antiparallel spins, is shown in Fig. 15.25(c). There are two next lowest molecular orbitals, ψ_B and ψ'_B, which together

---

*A. Streitwieser, *Molecular Orbital Theory for Organic Chemists* (New York: John Wiley & Sons, Inc., 1961) gives a complete account of the applications of the Hückel method.

†Divide the determinant by β and let $\epsilon = (\alpha - E)/\beta$, to obtain

$$
\begin{vmatrix}
\epsilon & 1 & 0 & 0 & 0 & 1 \\
1 & \epsilon & 1 & 0 & 0 & 0 \\
0 & 1 & \epsilon & 1 & 0 & 0 \\
0 & 0 & 1 & \epsilon & 1 & 0 \\
0 & 0 & 0 & 1 & \epsilon & 1 \\
1 & 0 & 0 & 0 & 1 & \epsilon
\end{vmatrix} = 0
$$

which, on expansion, yields

$$(\epsilon + 1)^2(\epsilon - 1)^2(\epsilon + 2)(\epsilon - 2) = 0$$

giving the roots shown.

hold four electrons. The six π electrons of benzene occupy these three orbitals of low energy, so that the exceptional stability of the structure is explained by the theory. The total energy would be $E = 2(\alpha + 2\beta) + 4(\alpha + \beta) = 6\alpha + 8\beta$. If the structure of benzene consisted of three localized single bonds and three localized double bonds (i.e., a single Kekule structure), the energy of the six π electrons in the ground state would be $6(\alpha + \beta) = 6\alpha + 6\beta$. Therefore, the resonance energy of benzene is -2β. The experimental value of the resonance energy as estimated from thermochemical data is 150 kJ, so that $\beta = -75$ kJ.

25. Ligand Field Theory*

As the theory of valence was developed during the nineteenth century, chemists recognized many compounds to which the usual rules did not apply. In 1893, Alfred Werner suggested that in so-called *complex compounds*, including salt hydrates, metal ammines, and double salts, an element might display a secondary valence in addition to its normal valence. The secondary valence bonds were also directed in space, so that geometrical and optical isomerisms would be possible. For example, there are no fewer than nine distinct compounds with the empirical formula $Co(NH_3)_3(NO_2)_3$.

In 1916, G. N. Lewis included Werner's complex compounds in his electron theory of valence, with the suggestion that each bond consists of a pair of electrons, *both* of which are donated by a group coordinated to the metal ion. Therefore, this bond was called a *dative bond*, and written with an arrow to indicate the direction of electron transfer. For instance,

$$
\left[
\begin{array}{c}
NH_3 \\
H_3N \searrow \downarrow \nearrow NH_3 \\
Co \\
H_3N \nearrow \uparrow \nwarrow NH_3 \\
NH_3
\end{array}
\right]^{3+} (Cl^-)_3
$$

One way to describe these bonds is to construct hybrid orbitals from the available atomic orbitals and to examine their stereochemistry. The bare cobaltic ion Co^{3+} has an electron configuration:

| | | | | | | | |
| 1s | 2s | 2p | 3s | 3p | 3d | 4s | 4p |

If we take two 3d, one 4s, and three 4p orbitals, we can make six hybrid orbitals d^2sp^3 directed to the corners of a regular octahedron. The valence bond model of the hexamminocobaltic ion would use these six hybrid orbitals to accommodate six pairs of electrons from the six NH_3 molecules. The strong covalent bonds so

*F. A. Cotton, "Ligand Field Theory," *J. Chem. Educ.*, **41**, 466 (1964).

formed would have considerable ionic character, but in this model they would not differ in kind from the d^2sp^3 octahedral hybrids used for the SF_6 molecule shown in Fig. 15.20.

A different way of looking at these structures starts with the idea that the bonding is essentially due to electrostatic interaction between the positive central ion and the dipoles or ions of the coordinated groups. The groups attached to the central ion are called *ligands*. These ligands create an electrostatic field around the central ion, which produces an additional bonding effect called the *ligand field stabilization energy* (L. F. S. E.). The L. F. S. E. arises as a consequence of the action of the field of the ligands on the *d* orbitals of the central ion. In the simple *crystal field* model first given by Bethe,* the ligands were assumed to behave as point negative charges.

We can see the effects of the ligands in a qualitative way from the properties of the *d* orbitals, as shown in Fig. 14.11. In the absence of an external field, the five $3d$ orbitals are all equal in energy (fivefold degeneracy). As soon as these orbitals are placed in an electrostatic ligand field that is not spherically symmetrical, their energies are no longer exactly equal. Those orbitals that point in the direction of the ligands are raised in energy owing to the repulsion between the negative charge density in the *d* orbitals of the central atom and that in the ligands. This effect is shown schematically in Fig. 15.27 for the case of an octahedral field (e.g., six NH_3 at the corners of a regular octahedron). We see that the $d_{x^2-y^2}$ and d_{z^2} orbitals are raised in energy, whereas the d_{xy}, d_{xz}, and d_{yz} are lowered. The splitting of the *d* orbitals in the octahedral field can also be deduced directly from group theory. As shown in Fig. 15.27, however, we can see from the models the physical reason for the observed splitting. The notation for the orbitals in the ligand field is based on their symmetry properties from group theory, as described in the next chapter.

The electron configuration of the complex will depend on the magnitude of the splitting produced by the ligand field. The case of $(FeF_6)^{3-}$ is typical of the effect of a weak field, with a splitting of about 10 000 cm^{-1}, whereas the case of $[Fe(CN)_6]^{3-}$ shows the effect of a strong field, with a splitting about five times as great. In accord with Hund's rule, the *d* electrons tend to occupy orbitals that allow them to keep their spins parallel, because in this way the electrostatic repulsive energy is lowered. On the other hand, if the splitting of the *d* levels by the ligand field is too great, this electrostatic effect will not be large enough to compensate for the energy needed to promote the electrons into the higher orbitals. We see examples of two different situations in Fig. 15.27. The five *d* electrons in $(FeF_6)^{3-}$ have the configuration $t_{2g}^3 e_g^2$ with completely uncoupled spins. In $[Fe(CN)_6]^{3-}$, the configuration is $(t_{2g})^5$, with four spins coupled. Each uncoupled spin acts as a little magnet, so that we can distinguish these configurations experimentally by measuring the magnetic susceptibility of the complexes.

The simple electrostatic model of Bethe is not satisfactory for quantitative calculations, because there is a considerable bonding of a more covalent type between a typical central atom and its ligands.

---

*H. Bethe, *Ann. Physik* [5], *3*, 135 (1929).

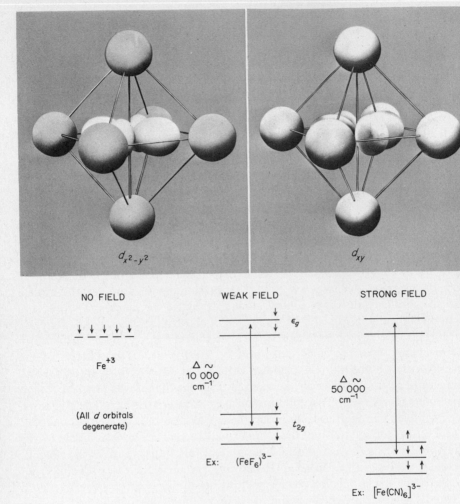

FIG. 15.27 Effect of octahedral ligand field on d orbitals. The electron density in the $d_{x^2-y^2}$ orbital of the central atom interacts more strongly with the ligand field than does the electron density in the d_{xy} orbital [R. G. Pearson].

26. Other Symmetries

The weak field case leads to *high spin complexes*, and the strong field case to *low spin complexes*. The situation of the spins in octahedral complexes of the various transition metals are summarized in Table 15.9.

Figure 15.28 shows how the five d orbitals are split by fields of various symmetries. Of course, the extent of the splitting depends on the intensity of the electrostatic field of the ligands. Sometimes a situation occurs in which one d orbital

TABLE 15.9
d Electrons in Octahedral Complexes

N, Number of Unpaired Spins,
p_m, Predicted Spin-Only Magnetic Moment in Bohr Magnetons

Number of d electrons	Arrangement in weak ligand field		N	p_m	Arrangement in strong ligand field		N	p_m
	t_{2g}	e_g			t_{2g}	e_g		
1	↑	—	1	1.73	↑	—	1	1.73
2	↑ ↑	—	2	2.83	↑ ↑	—	2	2.83
3	↑ ↑ ↑	—	3	3.87	↑ ↑ ↑	—	3	3.87
4	↑ ↑ ↑	↑	4	4.90	↑↓ ↑ ↑	—	2	2.83
5	↑ ↑ ↑	↑ ↑	5	5.92	↑↓ ↑↓ ↑	—	1	1.73
6	↑↓ ↑ ↑	↑ ↑	4	4.90	↑↓ ↑↓ ↑↓	—	0	0
7	↑↓ ↑↓ ↑	↑ ↑	3	3.87	↑↓ ↑↓ ↑↓	↑	1	1.73
8	↑↓ ↑↓ ↑↓	↑ ↑	2	2.83	↑↓ ↑↓ ↑↓	↑ ↑	2	2.83
9	↑↓ ↑↓ ↑↓	↑↓ ↑	1	1.73	↑↓ ↑↓ ↑↓	↑↓ ↑	1	1.73

Examples

	Weak field – high spin	Strong field – low spin
d^4	$CrSO_4$	$K_2Mn(CN)_6$
d^5	$[Fe(H_2O)_6]^{3+}$	$K_3Fe(CN)_6$
d^6	$[Co(H_2O)_6]^{3+}$	$K_4Fe(CN)_6$
d^7	$[Co(H_2O)_6]^{2+}$	Co^{2+}–Phthalocyanine

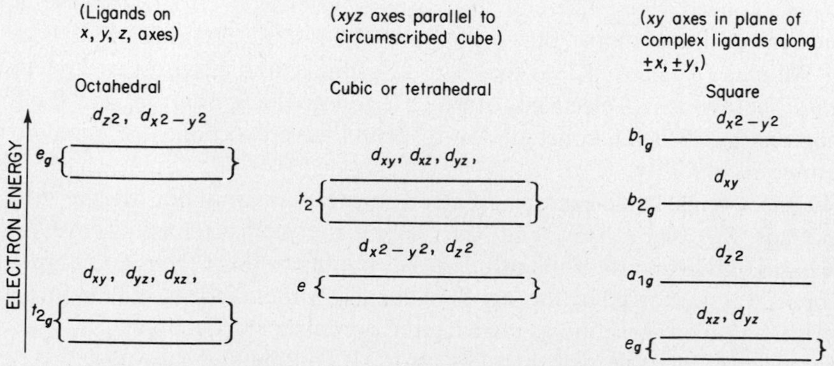

FIG. 15.28 The energy levels of d orbitals in ligand fields with various symmetries. Orbitals within a pair of braces are at same energy.

is filled with a pair of electrons but another d orbital of equal energy is only half filled. This configuration would be degenerate. For example, the electrons in the ion Cu^{2+} might be assigned as

$$\text{either} \quad t_{2g}^6\, d_{z^2}^2\, d_{x^2-y^2}^1$$

$$\text{or} \quad t_{2g}^6\, d_{z^2}^1\, d_{x^2-y^2}^2$$

In such a case, there may be a change in the geometry of the complex, which serves to break the degeneracy, producing a lower and an upper level somewhat different in energy. This change is called the *Jahn–Teller effect*. In the case of Cu^{2+} complexes, we often find that the regular octahedral arrangement does not occur, but instead an irregular octahedral structure appears, in which four ligands lie in a square about the Cu^{2+} and two others are at a greater distance above and below the central ion.

27.   Electron–Excess Compounds

A number of interesting chemical compounds appear to have too many electrons, in the sense that it is difficult to write electron-pair structures for them. A common example is the tri-iodide ion I_3^-. Both I_2 and I^- have achieved completely filled valence orbitals, yet they combine vigorously to form I_3^-. In this case, a satisfactory solution of the problem is provided by simple M.O. considerations. Consider a p_x atomic orbital from each iodine atom, p_x', p_x'', p_x'''. The linear combination of these three atomic orbitals will yield three σ type molecular orbitals, as follows:

$$\sigma_1 = p_x' + (p_x'' + p_x''')$$

$$\sigma_N = p_x'' - p_x'''$$

$$\sigma_2^* = p_x' - (p_x'' + p_x''')$$

These orbitals are shown pictorially in Fig. 15.29. The σ_1 is a bonding orbital extending over all three iodine nuclei, σ_2^* is antibonding, and σ_N is a nonbonding orbital, since it does not involve the central atom and the atomic orbitals are quite distinct from each other. There are $3 \times 7 = 21$ valence electrons from the $5s$ and $5p$ electrons of three I atoms, plus the extra electron from the single negative charge of I_3^-. We place 6 electrons into the three $5s$ orbitals, and 12 into the three p_y and three p_z. There remain 4 electrons, of which 2 go into the bonding σ_1, and 2 into the nonbonding σ_N. We can conclude the I_3^- would have two bonds of $\frac{1}{2}$ order, thus explaining its stability.

The most remarkable example of electron-excess compounds are the inert gas compounds, e.g., XeF_2, XeF_4, and XeF_6. For many years after the discovery of the noble gases, it was an article of faith that they could not form chemical compounds. This belief was reinforced by the school of chemistry that emphasized the formation of an inert-gas configuration as marking the complete satisfaction of all the combining power of an atom. Nevertheless, in 1933, Pauling had suggested that xenon should form a compound XeF_6, and Yost at the California Institute of Technology probably prepared some of the compound but just missed discovering it because

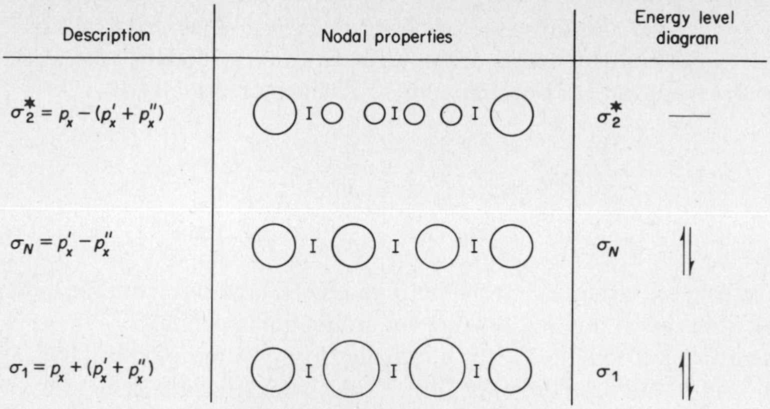

Description	Nodal properties	Energy level diagram
$\sigma_2^* = p_x - (p_x' + p_x'')$	⭕ I 〇 〇 I〇 〇 I ⭕	σ_2^* ——
$\sigma_N = p_x' - p_x''$	⭕ I 〇 I 〇 I ⭕	σ_N ⇅
$\sigma_1 = p_x + (p_x' + p_x'')$	⭕ I 〇 I 〇 I 〇	σ_1 ⇅

FIG. 15.29 Axial molecular orbitals for tri-iodide ion, I_3^-. A pair of electrons goes into the bonding σ_1 orbital and a pair into the nonbonding σ_N. The antibonding σ_2^* is empty. [After G. C. Pimentel and R. D. Spratley, *Chemical Bonding Clarified Through Quantum Mechanics* (San Francisco: Holden-Day, 1969).]

of a minor departure from what later turned out to be a suitable procedure. The dissociation reactions are endothermic for all three of the xenon fluorides, the Xe–F bond energy being about 125 kJ in all cases. The structures are: XeF_2, linear symmetrical; XeF_4, square planar; and XeF_6, probably distorted octahedral. A molecular orbital description of XeF_2 is analogous to that of I_3^-, except that in this case the p orbitals used are not all the same, but two from F and one from Xe.

28. Hydrogen Bonds

One of the firm tenets of classical valence theory was that hydrogen has a valence of one, and can form only one covalent bond by filling its orbital with a pair of electrons of opposite spins, or form one ionic bond as the H^+ or H^- ion. This conclusion, however, is not true, and life itself, as we know it, depends on the ability of the hydrogen atom to act as a link between two other atoms.

If sodium fluoride is dissolved in aqueous hydrofluoric acid, the principal anion formed is not F^- but HF_2^-,

$$F_{(aq)}^- + HF \longrightarrow HF_{2(aq)}^-, \qquad \Delta H = -155 \text{ kJ}$$

The salts $NaHF_2$, KHF_2, etc., are stable crystalline solids. The structure of HF_2^- is linear and symmetric

$$(F - H - F)^-$$

with the F — H distance 0.113 nm compared to 0.092 nm in HF gas. Other hydrogen-bihalide ions are known, but the bonds are much weaker than in the fluoride case.

The molecular orbital explanation of $(HF_2)^-$ is analogous to that for $(I_3)^-$. These are both electron-excess compounds. In the case of HF_2^-, the three σ bonds are formed from the $1s$ H orbital and the $2p$ fluorine orbitals:

$$\sigma_1 = 1s_H + (2p'_{xF} + 2p''_{xF})$$
$$\sigma_N = 2p'_{xF} - 2p''_{xF}$$
$$\sigma_2^* = 1s_H - (2p'_{xF} + 2p''_{xF})$$

The four electrons (one each from the H and the 2F and one from the minus charge) are placed in the σ_1 and σ_N bonding and nonbonding orbitals.

The hydrogen bond in HF_2^- is unusually strong. A more typical hydrogen-bond strength is found in the dimer of formic acid, which has the structure

$$
\begin{array}{ccc}
& O\!-\!H \ldots O & \\
H\!-\!C & & C\!-\!H \\
& O \ldots H\!-\!O &
\end{array}
$$

These hydrogen bonds have energies of 20 kJ each. Similar bonds occur in general between hydrogen and the electronegative elements N, O, F, of small atomic volumes. Besides intermolecular H bonds, like those in formic acid, there are intramolecular H bonds, for example, as found in salicylaldehyde,

$$
\begin{array}{c}
O\!-\!H \\
\vdots \\
O \\
HC
\end{array}
$$

In this case, the existence of the H bond was indicated by the fact that salicylaldehyde failed to display the strong infrared band at about 7000 cm^{-1} characteristic of the first overtone of the O—H stretching frequency at 3500 cm^{-1}.

Hydrogen bonds extended through three dimensions can cause the formation of polymeric structures, as in water and ice (cf. Section 19.2). They can hold molecules together in certain definite arrays or orientations. An important example of this kind of function is found in the structure proposed for the nucleic acids by Watson and Crick,* shown in Fig. 15.30. The two intertwined helical ribbons represent sugar phosphate chains. The bars binding them together represent pairs of purine and pyrimidine bases, one from each chain, held together in a specific orientation by hydrogen bonds.

*J. D. Watson and F. H. C. Crick, *Nature,* 17, 964 (1953).

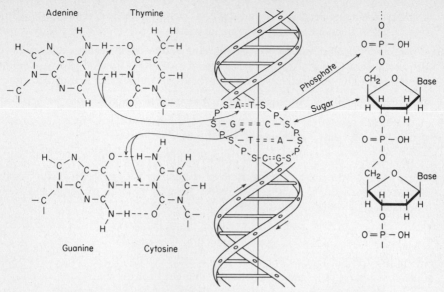

FIG. 15.30 The model of Watson and Crick for the structure of deoxyribonucleic acid (DNA). A double helix of sugar phosphate chains is held together by hydrogen bonded pairs of bases.

PROBLEMS

1. Derive the expressions for the integrals J and K given in Section 15.4.

2. Assume that $S = 0$, and use Eq. (15.12) to plot E_g and E_u against r_{AB}. By calculation or by consultation of reference sources, evaluate S and show what correction the use of the correct $S \neq 0$ makes to your approximate curves for E_g and E_u.

3. Table 2.7 lists single bond energies and Table 15.4 lists electronegativities. Are the values in these tables consistent with the Pauling definition of electronegativity given in Eq. (15.25)?

4. The dissociation energy for the H_2 molecule is $D_0 = 4.46$ eV and its zero-point energy $h\nu_0$ is 0.26 eV. Calculate D_0 and $h\nu_0$ for D_2, T_2, HD, and HT. State any assumptions made.

5. Assign the 13 electrons in BO to appropriate bonding and antibonding orbitals. What is the bond order of the bond in BO?

6. The dipole moment of HBr is 0.780 D, and its dielectric constant is 1.00313 at 273 K and 1 atm. Calculate the polarizability of the HBr molecule.

7. The density of $SiHBr_3$ is 2.690 g·cm^{-3} at 25°C, its refractive index is 1.578, and its dielectric constant is 3.570. Estimate its dipole moment in debyes.

8. The dielectric constant of gaseous SO_2 is 1.00993 at 273 K and 1.00569 at 373 K, at $P = 1$ atm. Calculate the dipole moment of SO_2. Assume ideal gas behavior.

9. Compare the molecules OF, OF$^-$, and OF$^+$, discussing molecular orbitals, bond order, bond lengths, bond energies, and paramagnetism.

10. What kind of hybridization will explain the following molecular geometries: (a) SF_6, octahedral; (b) $Au(CN)_4^-$, square planar; (c) NO_3^-, trigonal planar. Comment upon the question of whether you believe the explanation given is truly convincing.

11. Describe in detail the bonding in CO_2. (Reference to literature may be necessary.)

12. A molecular orbital for the pi bonding in the allyl radical, CH_2CHCH_2, may be constructed as a linear combination of atomic p orbitals as

$$\psi = c_1\psi_1 + c_2\psi_2 + c_3\psi_3$$

where atom 2 is the center carbon atom of the chain. The simple Hückel molecular-orbital energy, with normalized atomic orbitals and considering exchange and overlap integrals only between bonded atoms, is

$$E = \frac{(c_1^2 + c_2^2 + c_3^2)\alpha + 2(c_1c_2 + c_2c_3)\beta}{c_1^2 + c_2^2 + c_3^2 + 2(c_1c_2 + c_2c_3)S}$$

where $\alpha = H_{11} = H_{22} = H_{33}$, $\beta = H_{12} = H_{23}$.
(a) What are H_{ii}, H_{ij}, and S? (Write defining equations.)
(b) For the case $S = 0$, obtain the equations for c_1, c_2, c_3 which minimize the energy.
(c) Set up the determinant for the energy as a function of α and β.
(d) The solutions of the secular equation in (c) are $E_1 = \alpha + \sqrt{2}\beta$, $E_2 = \alpha$, $E_3 = \alpha - \sqrt{2}\beta$. What is the total energy for the case of $C_3H_5^-$, where there are four pi electrons?

13. The orbitals for sp^2 hybridization are

$$\psi_1 = \left(\frac{1}{\sqrt{3}}\right)s + \left(\frac{\sqrt{2}}{\sqrt{3}}\right)p_x$$

$$\psi_2 = \left(\frac{1}{\sqrt{3}}\right)s - \left(\frac{1}{\sqrt{6}}\right)p_x + \left(\frac{1}{\sqrt{2}}\right)p_y$$

$$\psi_3 = \left(\frac{1}{\sqrt{3}}\right)s + c_2p_x + c_3p_y$$

Find the constants c_2 and c_3 so as to obtain three normalized orthogonal orbitals.

14. In an electron-diffraction study of the structure of CS_2, the pictures showed four strong diffraction maxima, designated as 1, 2, 3, 4. Each strong maximum was closely followed by a weak maximum ($1a$, $2a$, $3a$, $4a$) and a deep minimum ($2'$, $3'$, $4'$). By means of the Wierl equation (15.38), show that this electron-diffraction diagram can be explained if there are two different internuclear distances in the molecule, one exactly twice the other. When 40-kV electrons were used, the values of s at which the maxima and minima appeared were

1	1a	2'	2	2a	3'	3	3a	4'	4
4.713	6.312	7.623	8.698	10.63	11.63	12.65	14.58	15.54	16.81

(a) Assuming CS_2 is linear, plot the theoretical intensity I_{av} [Eq. (15.38)] vs. sr, making the approximation that the scattering factor is proportional to the atomic number Z, and locate the values of sr that give maxima and minima.
(b) Calculate the C—S bond distance by averaging the values obtained from each of the observed maxima and minima.

(c) Qualitatively, how would the theoretical intensity curve I_{av} vs. *sr* change on going to CO_2 and to CSe_2?*

15. By means of the concept of hybrid bond orbitals, predict the configuration of the methyl radical CH_3. Summarize recent experimental work on the structure of CH_3.†

16. Cyclo-octatetraene $(CH)_8$ has an eight-membered puckered ring containing four short C—C distances and four longer ones. Describe the hybridization and molecular orbitals used in forming this ring. Contrast the bonding with that in benzene. How would you explain the fact that cyclo-octatetraene is not an aromatic ring compound like benzene?

17. Suppose that the molecules listed in Table 15.1 had purely ionic bonds in their ground states at equilibrium. Calculate the dipole moments that they should have on such a model. Compare the results with the experimental dipole moments. Discuss the differences and their trends for the different alkali-halide molecules.

18. For the molecules listed in Table 15.1, calculate the force constants κ corresponding to the ground-state vibrations. Can you correlate these force constants with the internuclear distances and the ionic characters of the bonds?

*P. Cross and L. O. Brockway, *J. Chem. Phys.*, **3**, 1821 (1935).

†See G. Herzberg, *The Spectra and Structures of Simple Free Radicals* (Ithaca, N.Y.: Cornell University Press, 1971).

16

Symmetry and Group Theory

Tyger, Tyger, burning bright
In the forests of the night,
What Immortal hand & eye
Dare frame thy fearful symmetry?

William Blake
(1793)

The purpose of this short chapter is to introduce some ideas and mathematical techniques that are essential for understanding the structures and properties of molecules and crystals. Symmetry is a concept that dwells with beauty, but in the space available we cannot explore the mathematical formalism deeply enough to reveal much of its complementary beauty. We shall therefore give a schematic account of the subject, based on a few illustrative examples, and hope that the reader will refer to some of the excellent detailed (yet still elementary) discussions that are now available.*

1. Symmetry Operations

Symmetry is described in terms of certain *symmetry operations*, which transform a spacial arrangement into an arrangement that is visually indistinguishable from the original one. For example, consider the equilateral triangle Δ in Fig. 16.1. If we exclude the possibility of removing it from its plane (and thereby turning it over), there are six different operations that produce a coincident triangle from the original. These operations are shown in Fig. 16.1 by means of their effects on an

---

*(1) F. A. Cotton, *Chemical Applications of Group Theory* (New York: Interscience Publishers, 1971).
(2) D. Schonland, *Molecular Symmetry: An Introduction to Group Theory and Its Uses in Chemistry* (Princeton, N.J.: D. Van Nostrand Co., Inc., 1965).
(3) R. McWeeny, *Symmetry, An Introduction to Group Theory and Its Applications* (London: Pergamon Press, 1963).

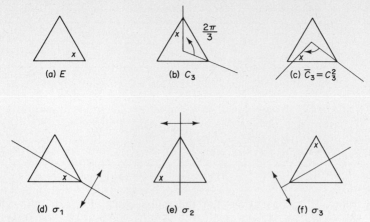

FIG. 16.1 Symmetry operations of the equilateral triangle—point group C_{3v}.

arbitrary general point denoted by X in the triangle. The operations can be listed as follows:

(a) E is the *identity operation*, which leaves each point unchanged.

(b) C_3 is a threefold rotation about an axis passing perpendicularly through the midpoint of Δ. This operation rotates a representative point through an angle of $2\pi/3$ (120°) in a positive (counterclockwise) direction.

(c) $\bar{C}_3$ is a threefold rotation about an axis passing perpendicularly through the midpoint of Δ. This operation rotates a representative point through an angle of $2\pi/3$ (120°) in a negative (clockwise) direction. (This operation could also be written as C_3^2, a positive rotation of $4\pi/3$.)

(d) σ_1 is a reflection in a mirror plane. This operation takes a representative point to an equal distance the other side of the line bisecting the right-hand basal vertex of Δ.

(e) σ_2 is an operation like σ_1, but involves reflection in the line bisecting the upper vertex of Δ.

(f) σ_3 is an operation again like σ_1, but it involves reflection in the line bisecting the left-hand basal vertex of Δ.

We shall encounter later a few additional types of symmetry element, but first we shall examine some properties of the set of elements just listed for the equilateral triangle. The product AB of two symmetry operations is defined to mean that operation B is first performed, and then operation A is performed upon the result. If we examine all the products AB between elements in the set of six symmetry operations for Δ, we find that the product of any two operations always gives the same effect as one of the operations in the original set. This result can be seen in the *multiplication table* for the set of operations, given in Table 16.1.

TABLE 16.1
Multiplication Table of the Symmetry Group C$_{3v}$

			Operation C			
	E	C_3	$\bar{C}_3 = C_3^2$	σ_1	σ_2	σ_3
E	E	C_3	$\bar{C}_3$	σ_1	σ_2	σ_3
C_3	C_3	$\bar{C}_3$	E	σ_3	σ_1	σ_2
$\bar{C}_3 = C_3^2$	$\bar{C}_3$	E	C_3	σ_2	σ_3	σ_1
σ_1	σ_1	σ_2	σ_3	E	C_3	$\bar{C}_3$
σ_2	σ_2	σ_3	σ_1	$\bar{C}_3$	E	C_3
σ_3	σ_3	σ_1	σ_2	C_3	$\bar{C}_3$	E

Operation R labels the rows.

The table follows the convention that the intersection of a row with a column gives the product RC of the row element R and the column element C. As can be seen from the table, RC does not necessarily equal CR, although it may. When it does, we say that R and C *commute*. Thus, $C_3\bar{C}_3 = \bar{C}_3C_3 = E$, whereas $\sigma_1 C_3 = \sigma_2$ and $C_3\sigma_1 = \sigma_3$.

2.   Definition of a Group

A set of elements constitutes a group if its members satisfy the following conditions:

1. The operation of taking the product of two elements having been defined, the product AB of any two elements of the set is itself a member C of the set, $AB = C$.

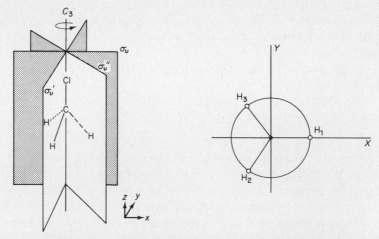

FIG. 16.2   The symmetry elements of CH$_3$Cl, point group **C$_{3v}$**. (The coordinate system is drawn in a way that will be convenient for later treatments.)

2. The set includes an identity element E such that for every member A, $EA = AE = A$.
3. Each element A has an inverse A^{-1}, which is also a member of the set, where $A^{-1}A = AA^{-1} = E$.
4. The associative law of multiplication holds, $A(BC) = (AB)C$.

Examination of the multiplication table of the symmetry operations of Δ shows that these operations form a group. Each symmetry element of the equilateral triangle leaves the midpoint of Δ invariant. A group of symmetry operations for which there is a point that is left invariant under each operation is called a *point group*. The particular point group chosen in our example is denoted as the group $\mathbf{C_{3v}}$, following a notation to be explained later.

Any collection of the elements of a group that by themselves form a group is called a *subgroup* of the original group. From the multiplication table (Table 16.1), we can see that E, C_3, $\bar{C}_3$ form a subgroup of $\mathbf{C_{3v}}$.

The symmetry groups of molecular structures are point groups. Examples of molecules belonging to point group $\mathbf{C_{3v}}$ are ammonia and methyl chloride. The latter is shown in Fig. 16.2, together with the symmetry elements of $\mathbf{C_{3v}}$.

3. Further Symmetry Operations

In addition to the n-fold rotation axes C_n and mirror planes σ, two other types of symmetry operation can occur in molecules.

1. In a rotary reflection or improper rotation, a representative point is rotated through an angle of $2\pi/n$ about some axis and then reflected in a mirror plane σ_h perpendicular to that axis. The order of these operations makes no difference. The operation of rotary reflection can thus be written symbolically as $S_n = \sigma_h C_n$. An example of a molecule with an S_4 axis is methane, as shown in Fig. 16.3.
2. Inversion in a center of symmetry O is denoted by i. The operation takes the representative point P into a point P', equidistant from O, and the direction of OP' is an extension of the line PO. An inversion center is

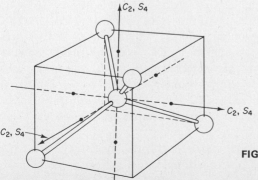

FIG. 16.3 The C_2 axes of methane coincide with S_4 axes.

equivalent to a twofold rotary reflection axis, $i = S_2$. Figure 16.4 shows an example of a molecule with a center of symmetry.

We have now enumerated all the symmetry elements that can occur in molecules. Many different groups of these elements can be formed; these are the molecular point groups. If we consider the equilibrium positions of the nuclei in a molecule, we can assign every molecule to a definite point group.

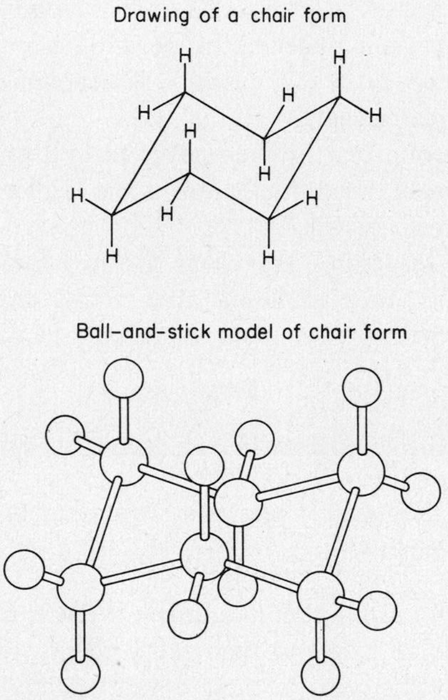

Drawing of a chair form

Ball-and-stick model of chair form

FIG. 16.4 Molecule with a center of symmetry—the chair form of cyclohexane.

4. Molecular Point Groups

We shall now enumerate the various kinds of molecular point groups, using a notation originated by Schoenflies. There is another notation favored by crystallographers but we shall postpone its introduction until Chapter 18, *The Solid State.*

1. The groups with no axis of symmetry are C_s (only a single plane of symmetry), C_i (only a center of symmetry), and C_1 (no nontrivial symmetry elements, the group having only the one element E).
2. Groups having just one axis of n-fold symmetry are denoted by C_n. Examples of molecules belonging to groups C_2 and C_3 are shown in Fig. 16.5(a).
3. Groups with only a single axis of rotary reflection of even order $2n$ are denoted by S_{2n}.
4. Groups with an axis of rotation and a horizontal symmetry plane perpendicular to it are denoted C_{nh}. An example of a molecule belonging to C_{2h} is shown in Fig. 16.5(b).

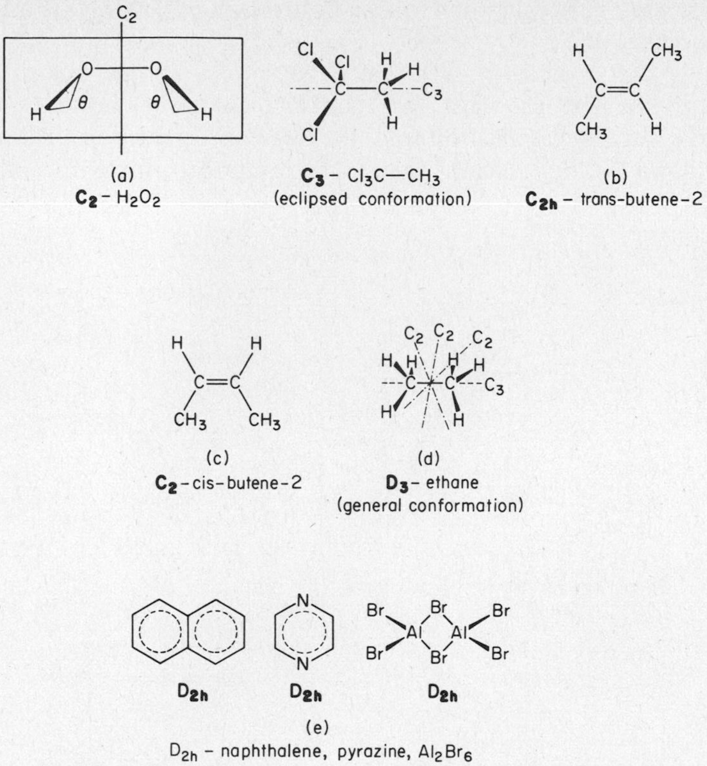

(a)

$C_2 - H_2O_2$

$C_3 - Cl_3C—CH_3$
(eclipsed conformation)

(b)

C_{2h} – trans-butene-2

(c)

C_2–cis-butene-2

(d)

D_3– ethane
(general conformation)

D_{2h} D_{2h} D_{2h}

(e)

D_{2h} – naphthalene, pyrazine, Al_2Br_6

FIG. 16.5 Examples of molecules in various point groups.

5. If a group contains in addition to C_n a vertical symmetry plane containing the axis of rotation, the operation of C_n when $n > 2$ will produce additional equivalent vertical symmetry planes. The resulting group is called $\mathbf{C_{nv}}$; its elements are the rotations of C_n and the reflections σ_v in the vertical planes. When n is even, there are $n/2$ planes, and when n is odd, n planes. An example of a $\mathbf{C_{2v}}$ molecule is *cis*-butene-2 shown in Fig. 16.5(c), and an example of $\mathbf{C_{3v}}$ was previously given in Fig. 16.2.

6. We next consider groups that have more than one symmetry axis. If we add to C_n a twofold axis perpendicular to the principal axis, we obtain the groups $\mathbf{D_n}$, where $n = 2, 3, 4, 5, 6. \ldots$ The operation of C_n on the twofold axis yields n twofold axes perpendicular to C_n. Thus, the group $\mathbf{D_2}$ would have the symmetry operations of rotations of 180° about three mutually perpendicular axes. An example for $\mathbf{D_3}$ would be the molecule ethane in the general conformation in which one CH_3 group was neither exactly eclipsed nor exactly staggered with respect to the other, Fig. 16.5(d).

7. In the groups $\mathbf{D_{nh}}$, the horizontal plane that contains the twofold axis of $\mathbf{D_n}$ is a plane of symmetry. Several examples of $\mathbf{D_{2h}}$ molecules are shown in Fig. 16.5(e).

8. In the groups $\mathbf{D_{nd}}$, n vertical planes of symmetry are added to the elements of $\mathbf{D_n}$. These planes intersect in the principal axis and bisect the angles

between adjacent twofold axes. Two rotary reflection axes of type S_{2n} also occur in $\mathbf{D_n}$.

9. We next consider groups that have more than one axis of symmetry of order $n > 2$. The most important of these groups are those derived from the regular tetrahedron and the regular octahedron. These figures are shown in Fig. 16.6(a). Their pure rotational groups (i.e., the groups of

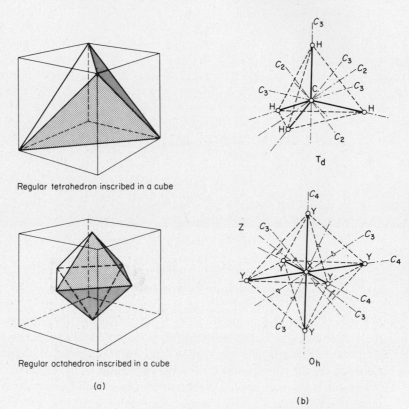

Regular tetrahedron inscribed in a cube

Regular octahedron inscribed in a cube

(a)

(b)

FIG. 16.6   (a) Relations of regular tetrahedron and regular octahedron to the cube. (b) Structures displaying complete tetrahedral ($\mathbf{T_d}$) and complete octahedral ($\mathbf{O_h}$) symmetry.

operations of their rotational symmetry axes) are denoted as $\mathbf{T}$ and $\mathbf{O}$, respectively. The group of the regular octahedron is also the group of the cube, since a regular octahedron can be inscribed in a cube, and clearly possesses all the same symmetry elements. Thus, $\mathbf{O}$ is also the pure rotation group of the cube. A regular tetrahedron can also be inscribed in a cube, but it has a lower symmetry because the corners of the cube are no longer all equivalent. Thus, $\mathbf{T}$ must be a subgroup of $\mathbf{O}$.

In the group $\mathbf{T}$, as can be seen from Fig. 16.6(a), there is a set of four threefold axes, C_3, directed along the body diagonals of the cube, and a set of three twofold axes, C_2, perpendicular to each cube face at its midpoint. In group $\mathbf{O}$, the twofold axes become fourfold axes, C_4, and a set of six additional twofold axes, C_2, appears.

The regular tetrahedron and octahedron also contain mirror planes. Thus, the complete symmetry group of the regular tetrahedron, called $\mathbf{T_d}$, consists of the rotational elements of $\mathbf{T}$ plus six mirror planes σ plus six elements S_4. $\mathbf{T_d}$ is an important point group for molecular structures, examples being CH_4, P_4, CCl_4, and a number of tetrahedral complex ions.

When all the nine mirror planes of the cube are added to $\mathbf{O}$, one obtains the group $\mathbf{O_h}$, important in molecular structure, examples being SF_6 (see Fig. 15.20), $PtCl_6$, and numerous octahedral coordination compounds. The mirror planes lead also to the additional symmetry elements: 6 S_4, 8 S_6, and i.

There are two more regular polyhedra, the regular icosahedron and regular dodecahedron. Both these figures belong to point group $\mathbf{I_h}$. Only a few molecules are known having this symmetry group, notably the icosahedral $B_{12}H_{12}^{2-}$ ion shown in Fig. 16.7.

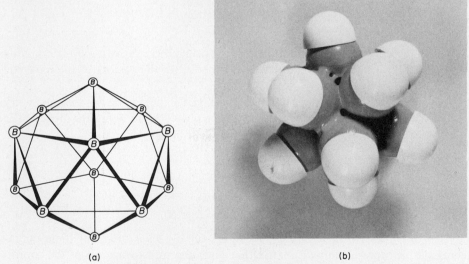

(a)   (b)

FIG. 16.7 (a) Arrangement of the boron atoms in the icosahedral [$\mathbf{I_h}$]$B_{12}H_{12}^{2-}$ ion. (b) Model of the $B_{12}H_{12}^{2-}$ ion.

5. Transformations of Vectors by Symmetry Operations

The physical representation of symmetry elements as rotations, reflections, etc., introduces many of the properties of the molecular point groups, but further development of the theory requires some abstract mathematical formulation of the transformations that are produced by the symmetry operations.

With reference to a set of Cartesian axes X, Y, Z, the coordinates of each atom i in the molecule can be specified by x_i, y_i, z_i. If these coordinates are given for each atom, the position and orientation of the molecule will be exactly specified. The set of three coordinates $x_i y_i z_i$ defines the position of the vector from the origin to the particular atom *i*. This vector can be written as

$$\begin{pmatrix} x_i \\ y_i \\ z_i \end{pmatrix}$$

If the molecule is subjected to a symmetry operation, the coordinates $x_i y_i z_i$ will be transformed to some new values x_i' y_i' z_i'. The transformation of coordinates can always be written as a set of linear equations of the general form:

$$x_i' = a_{11}x_i + a_{12}y_i + a_{13}z_i$$
$$y_i' = a_{21}x_i + a_{22}y_i + a_{23}z_i$$
$$z_i' = a_{31}x_i + a_{32}y_i + a_{33}z_i$$

The usual notation for such a linear transformation is:

$$\begin{pmatrix} x_i' \\ y_i' \\ z_i' \end{pmatrix} = \begin{pmatrix} a_{11} & a_{12} & a_{13} \\ a_{21} & a_{22} & a_{23} \\ a_{31} & a_{32} & a_{33} \end{pmatrix} \begin{pmatrix} x_i \\ y_i \\ z_i \end{pmatrix}$$

The new vector is obtained by multiplying the original vector by the *transformation matrix*. The transformation matrix is the same for all the vectors of the molecule and would apply to a general vector (x, y, z).

As an example, consider the molecule CH_3Cl, belonging to the symmetry group C_{3v}. In Fig. 16.2, it is oriented with the C—Cl bond along the Z axis. The relative coordinates of the three H atoms are as follows:

	x	y	z
H_1	1	0	0
H_2	$-1/2$	$\sqrt{3}/2$	0
H_3	$-1/2$	$-\sqrt{3}/2$	0

The operations of the group C_{3v} are $E, C_3, C_3^2, \sigma_1, \sigma_2, \sigma_3$. Each of these operations can be represented by a matrix that gives the transformation of the coordinates of H_1, H_2, H_3 produced by the symmetry operation.

The matrix that represents the unit operation E will obviously be

$$M(E) = \begin{pmatrix} 1 & 0 & 0 \\ 0 & 1 & 0 \\ 0 & 0 & 1 \end{pmatrix}$$

We can find the matrix $M(C_3)$ that represents C_3 by considering the operation of a rotation about the threefold axis on the coordinates of each of the three H atoms, H_1, H_2 and H_3. The coordinates of H_1 are originally $x = 1, y = 0, z = 0$. From Fig. 16.2, it is evident that the rotation of 120° produced by C_3 shifts the coordinates of H_1 to those of H_2, those of H_2 to those of H_3, and those of H_3 to those of H_1. A little consideration* shows that the matrix representing C_3 is therefore

$$M(C_3) = \begin{pmatrix} -1/2 & -\sqrt{3}/2 & 0 \\ \sqrt{3}/2 & -1/2 & 0 \\ 0 & 0 & 1 \end{pmatrix}$$

The matrix representing $M(\bar{C}_3) = M(C_3^2)$ is now readily derived by matrix multiplication of $M(C_3)$ by itself:

$$M(C_3^2) = \begin{pmatrix} -1/2 & -\sqrt{3}/2 & 0 \\ \sqrt{3}/2 & -1/2 & 0 \\ 0 & 0 & 1 \end{pmatrix}\begin{pmatrix} -1/2 & -\sqrt{3}/2 & 0 \\ \sqrt{3}/2 & -1/2 & 0 \\ 0 & 0 & 1 \end{pmatrix}$$

$$= \begin{pmatrix} -1/2 & \sqrt{3}/2 & 0 \\ -\sqrt{3}/2 & -1/2 & 0 \\ 0 & 0 & 1 \end{pmatrix}$$

The matrix representing σ_1 can be calculated by the same method that yielded

*By considering the vectors to H_1, H_2, and H_3 separately and consecutively, the matrix is readily derived. Thus, the vector to H_1 is initially

$$\begin{pmatrix} 1 \\ 0 \\ 0 \end{pmatrix}$$

The matrix $M(C_3)$ representing the rotation shifts these coordinates to

$$\begin{pmatrix} -1/2 \\ \sqrt{3}/2 \\ 0 \end{pmatrix}$$

so that

$$\begin{pmatrix} -1/2 \\ \sqrt{3}/2 \\ 0 \end{pmatrix} = M(C_3)\begin{pmatrix} 1 \\ 0 \\ 0 \end{pmatrix}$$

Therefore, the matrix $M(C_3)$ must have

$$\begin{pmatrix} -1/2 \\ \sqrt{3}/2 \\ 0 \end{pmatrix}$$

as its first row. Similarly,

$$\begin{pmatrix} -1/2 \\ -\sqrt{3}/2 \\ 0 \end{pmatrix} = M(C_3)\begin{pmatrix} -1/2 \\ \sqrt{3}/2 \\ 0 \end{pmatrix}$$

and

$$\begin{pmatrix} 1 \\ 0 \\ 0 \end{pmatrix} = M(C_3)\begin{pmatrix} -1/2 \\ -\sqrt{3}/2 \\ 0 \end{pmatrix}$$

$M(C_3)$. Then the matrix representations of σ_2 and σ_3 are calculated by matrix multiplication with reference to Table 16.1. We thus obtain the group of matrices in Table 16.2, which provide a *representation* of the symmetry group C_{3v}. The vectors to H_1, H_2, and H_3 are said to provide the *basis* for this representation. It is evident that the representation in Table 16.2 is not the only possible one, and, in fact, any group of matrices that followed the same multiplication table as the symmetry operations of C_{3v} would provide a representation of that group.

TABLE 16.2
Matrix Representation of the Group C_{3v} Derived from
the Basis of Vectors to Hydrogen Atoms in CH_3Cl

$$E = \begin{pmatrix} 1 & 0 & 0 \\ 0 & 1 & 0 \\ 0 & 0 & 1 \end{pmatrix} \quad C_3 = \begin{pmatrix} -1/2 & -\sqrt{3}/2 & 0 \\ \sqrt{3}/2 & -1/2 & 0 \\ 0 & 0 & 1 \end{pmatrix} \quad \bar{C}_3 = \begin{pmatrix} -1/2 & \sqrt{3}/2 & 0 \\ -\sqrt{3}/2 & -1/2 & 0 \\ 0 & 0 & 1 \end{pmatrix}$$

$$\sigma_1 = \begin{pmatrix} 1 & 0 & 0 \\ 0 & -1 & 0 \\ 0 & 0 & 1 \end{pmatrix} \quad \sigma_2 = \begin{pmatrix} -1/2 & -\sqrt{3}/2 & 0 \\ -\sqrt{3}/2 & 1/2 & 0 \\ 0 & 0 & 1 \end{pmatrix} \quad \sigma_3 = \begin{pmatrix} -1/2 & \sqrt{3}/2 & 0 \\ \sqrt{3}/2 & 1/2 & 0 \\ 0 & 0 & 1 \end{pmatrix}$$

6.   Irreducible Representations

For any group it is possible to select many different bases, and from each of these to derive a representation of the group by a group of matrices. However, some of these representations have a specially fundamental importance. We shall now explain which these are and derive some of their properties.

If we consider the particular representation of C_{3v} that was given in Table 16.2, we find that all the matrices are of the simple form:

$$M(R) = \begin{pmatrix} a_{11} & a_{12} & 0 \\ a_{21} & a_{22} & 0 \\ 0 & 0 & 1 \end{pmatrix}$$

This matrix can be written symbolically as

$$M(R) = \left(\begin{array}{c|c} M'(R) & 0 \\ \hline 0 & M''(R) \end{array} \right)$$

In such a case, we say that the matrix $M(R)$ is the *direct sum* of the matrices $M'(R)$ and $M''(R)$. It is easy to see that each set of matrices $M'(R)$ or $M''(R)$ separately furnishes a representation of the group. A representation is often denoted by the symbol Γ, and it is customary to write

$$\Gamma = \Gamma_3 + \Gamma_1$$

where Γ_3 is the representation given by $M'(R)$ and Γ_1 that given by $M''(R)$. We say that the representation of C_{3v} (given in Table 16.2) has been reduced to the sum of a two-dimensional representation Γ_3 and a one-dimensional representation Γ_1.

Examination of the group of matrices that comprise Γ_3 suggests that it cannot be reduced any further, i.e., cannot be set equal to the direct sum of two one-dimensional representations. This is, in fact, the case. Thus, both Γ_3 and Γ_1 are *irreducible representations* (I.R.).

Group theory provides a number of interesting and important theorems regarding irreducible representations and their properties, but we shall not explore these here, fascinating as they will prove to the serious student. As one would expect, all the symmetry groups have been intensively studied and all their various irreducible representations are known, as are various methods for breaking up general representations into direct sums of irreducible representations. For our present purposes, we can use these results as long as we understand their meaning. The mathematical derivations from group theory can be studied later.

It is not even necessary for many applications to molecular structure, quantum mechanics, and spectroscopy to use the I.R.'s directly, since often only their *characters* χ will be necessary. The character of an element in a representation is the trace of the matrix for that element. The trace of a matrix is the sum of its diagonal terms. This definition will be made clear if we consider the two-dimensional I.R., Γ_3, already derived for C_{3v}. This I.R. is shown in Table 16.3 with its characters. For the other irreducible representation of C_{3v}, Γ_1, all the characters are obviously equal to 1.

TABLE 16.3
The Irreducible Representation Γ_3 of
C_{3v} and Its Characters χ

$$E = \begin{pmatrix} 1 & 0 \\ 0 & 1 \end{pmatrix} \quad C_3 = \begin{pmatrix} -1/2 & -\sqrt{3}/2 \\ \sqrt{3}/2 & -1/2 \end{pmatrix} \quad \bar{C}_3 = \begin{pmatrix} -1/2 & \sqrt{3}/2 \\ \sqrt{3}/2 & -1/2 \end{pmatrix}$$

$$\chi(E) = 2 \qquad \chi(C_3) = -1 \qquad \chi(\bar{C}_3) = -1$$

$$\sigma_1 = \begin{pmatrix} 1 & 0 \\ 0 & -1 \end{pmatrix} \quad \sigma_2 = \begin{pmatrix} -1/2 & -\sqrt{3}/2 \\ \sqrt{3}/2 & 1/2 \end{pmatrix} \quad \sigma_3 = \begin{pmatrix} -1/2 & \sqrt{3}/2 \\ \sqrt{3}/2 & 1/2 \end{pmatrix}$$

$$\chi(\sigma_1) = 0 \qquad \chi(\sigma_2) = 0 \qquad \chi(\sigma_3) = 0$$

The question naturally arises as to whether there are any more irreducible representations of C_{3v} beside the two we have found, Γ_3 and Γ_1. There is a general theorem* of group theory which states that a group of g elements will have only a finite number k of different irreducible representations, and that their dimensions n_i satisfy the equation

$$n_1^2 + n_1^2 + \cdots + n_k^2 = g \tag{16.1}$$

Since each n_1 is a positive integer, k cannot be greater than g, and will often be

---

Proof: R. McWeeny, *Symmetry, An Introduction to Group Theory and Its Applications* (London: Pergamon Press, 1963), p. 125.

much less. In the case of C_{3v}, $g = 6$, and we have found two I.R.'s already, one with $n_1 = 1$ and one with $n_3 = 2$. Thus, from (16.1),

$$1^2 + n_2^2 + 4 = 6 \tag{16.2}$$

and there is obviously one I.R. with $n_2 = 1$ yet to be found.

If we had taken a more general basis for the representation of C_{3v}, we would have obtained the other I.R. also. For example, suppose at each H atom we placed a set of three mutually perpendicular vectors, and determined the matrix of the transformation of these three vectors. This procedure would yield 9×9 matrices for the representation, and these matrices on reduction would yield all three of the I.R.'s (some of them more than once).

In the character table for C_{3v}, Table 16.4, we have included the two I.R.'s we found as well as the missing Γ_2.

TABLE 16.4
Character Table for Group C_{3v}

		E	$2C_3$	3σ
Γ_1	A_1	1	1	1
Γ_2	A_2	1	1	-1
Γ_3	E	2	-1	0

We note two features about this character table that will hold true in all cases. The number of I.R.'s is equal to the number of physically distinct kinds of symmetry operation. All operations of the same kind are said to form a *class*, and each element in a given class always has the same character.

The symbols for the I.R.'s that are in most common use were originally given by Mulliken. For C_{3v} they are shown in the table as A_1, A_2, E. They have been generally adopted despite the awkward use of E for two different entities. The symbol A or B denotes a one-dimensional I.R.; E denotes a two-dimensional I.R.; and T denotes a three-dimensional I.R. For a one-dimensional I.R., A is used when the character for rotation about the principal symmetry axis (C_3 in our example) is $+1$, and B is used when this character is -1. The subscripts 1 and 2 designate symmetry or antisymmetry with respect to a C_2 normal to the principal axis, or, if such a C_2 is lacking, with respect to a vertical plane of symmetry. A subscript g or u is used to denote a representation for which the character for an inversion through a center of symmetry is $+1$ (g) or -1 (u), if this operation occurs in the group.

The irreducible representations are of great value in classifying molecular vibrations, orbitals, quantum states, and all the other properties of molecules and crystals that are closely linked with the symmetry of the structures. The character tables are given as appendices to the books listed on p. 732. We shall give one additional example here, the character table for O_h, the symmetry group of the regular octahedron, which is so important in coordination chemistry. The designations of the molecular orbitals used in ligand-field theory, as discussed in Section 15.26, are based on the I.R.'s of the symmetry group of the ligand field (see Table 16.5).

TABLE 16.5
Character Table for Group O_h

O_h	E	$8C_3$	$6C'_2$	$6C_4$	$3C_2(=C_4^2)$	i	$6S_4$	$8S_6$	$3\sigma_h$	σ_d
A_{1g}	1	1	1	1	1	1	1	1	1	1
A_{2g}	1	1	-1	-1	1	1	-1	1	1	-1
E_g	2	-1	0	0	2	2	0	-1	2	0
T_{1g}	3	0	-1	1	-1	3	1	0	-1	-1
T_{2g}	3	0	1	-1	-1	3	-1	0	-1	1
A_{1u}	1	1	1	1	1	-1	-1	-1	-1	-1
A_{2u}	1	1	-1	-1	1	-1	1	-1	-1	1
E_u	2	-1	0	0	2	-2	0	1	-2	0
T_{1u}	3	0	-1	1	-1	-3	-1	0	1	1
T_{2u}	3	0	1	-1	-1	-3	1	0	1	-1

PROBLEMS

1. For each molecule at the left, select the correct point group from the list at the right. Note that some point groups are represented by more than one molecule.

N_2	C_{2v}
CO	C_{4v}
Anthracene ($C_{14}H_{10}$)	$C_{\infty v}$
p-Dichlorobenzene	D_{2h}
SF_5Cl (octahedral)	D_{3h}
Cyclopropane (ignore hydrogens)	D_{3d}
H_2O	$D_{\infty h}$
CO_3^{2-} ion (planar)	
C_2H_6 (trans)	
C_2H_6 (cis)	

2. In the point group C_{2v}, the symmetry operations are E, C_2, and two σ_v perpendicular to each other. Write down the multiplication table for this group in a form similar to that of Table 16.1.

3. Let the C_2 axis of group C_{2v} coincide with the Z axis of a set of cartesian coordinates, and let σ_v be in the xz plane and σ'_v be in the yz plane. Show that the matrices that represent the transformation of a general point x, y, z, by the symmetry operations are

$$E: \begin{bmatrix} 1 & 0 & 0 \\ 0 & 1 & 0 \\ 0 & 0 & 1 \end{bmatrix} \qquad C_2: \begin{bmatrix} -1 & 0 & 0 \\ 0 & -1 & 0 \\ 0 & 0 & 1 \end{bmatrix}$$

$$\sigma_v: \begin{bmatrix} 1 & 0 & 0 \\ 0 & -1 & 0 \\ 0 & 0 & 1 \end{bmatrix} \qquad \sigma_v': \begin{bmatrix} -1 & 0 & 0 \\ 0 & 1 & 0 \\ 0 & 0 & 1 \end{bmatrix}$$

4. If A and X are two elements of a group, then $X^{-1}AX$ will be some element B of the group, and A and B are said to be *conjugate* to each other. A complete set of elements that are conjugate to one another is called a *class* of the group. By means of these definitions, derive the classes of the groups C_{2v} and C_{3v}.

5. The number of irreducible representations of a group is equal to the number of classes in the group. From the result in Problem 4, how many irreducible representations [I.R.] are there for group C_{2v}? How many of these I.R.'s can be obtained from the matrices in Problem 3? How would you obtain the remaining I.R.? Write down the complete character table for C_{2v}, using the correct symbols for the I.R.'s.

6. List the symmetry elements and derive the point group for each of the following molecules: CCl_4, $CHCl_3$, CH_2Cl_2, CH_3Cl, C_2H_5Cl (in various conformations).

7. Prove that a molecule with a center of symmetry cannot have a dipole moment. Are there any other symmetry elements, the existence of which in a molecule would preclude the existence of a dipole movement? Explain.

8. The symmetry operations of point groups are often represented by conventional stereographic projections, for example:

C_{2v} C_{3h}

Draw similar projections for D_{2h} and C_{3v}.

9. Any three p orbitals $xf(r)$, $yf(r)$, $zf(r)$ form an irreducible representation of the group O_h. Write down a matrix representing one each of the various symmetry operations shown in Table 16.4. Show that the p orbitals transform according to T_{1u}.

10. In transition-metal chemistry the behavior of the five d orbitals in an octahedrally symmetric environment is very important. Three of the d orbitals are $d_{xy} = xy\,g(r)$, $d_{yz} = yz\,g(r)$, $d_{xz} = xz\,g(r)$. By considering the effect of various symmetry operations, and using the character table of O_h, show that these transform according to the irreducible representation T_{2g}. The remaining two d orbitals* are $d_{z^2} = (1/2\sqrt{3})(2z^2 - x^2 - y^2)\,g(r)$ and $d_{x^2-y^2} = \frac{1}{2}(x^2 - y^2)\,g(r)$ and form another irreducible representation of O_h. Write down matrices representing the operation i and one C_3 rotation and deduce that these two d orbitals transform according to E_g.

11. A representation Γ of a group G is the direct sum of two other representations Γ_1 and Γ_2. Show that for any element A of G the character χ of A in the representation Γ is the sum of the characters of A in Γ_1 and Γ_2 separately.

*J. S. Griffith, *The Theory of Transition Metal Ions* (London: Cambridge Univ. Press, 1961), p. 226.

17

Spectroscopy and Photochemistry

Do not all fix'd Bodies when heated beyond a certain degree, emit Light and shine, and is not this Emission performed by the vibrating Motions of their parts? . . . As for instance; sea Water in a raging Storm; Quicksilver agitated in vacuo; the Back of a cat, or Neck of a Horse obliquely struck or rubbed in a dark place; Wood, Flesh and Fish while they putrefy; Vapours arising from putrefy'd Waters, usually called Ignes Fatui; stacks of moist Hay or Corn growing hot by fermentation; Glow-worms and the Eyes of some Animals by vital motions; The vulgar Phosphorus agitated by the attrition of any Body, or by the acid Particles of the Air; Ambar and some Diamonds by striking, pressing or rubbing them; Iron hammer'd very nimbly till it become so hot as to kindle Sulphur thrown upon it; the Axle trees of Chariots taking fire by the rapid rotation of the Wheels; and some Liquors mix'd with one another whose Particles come together with an Impetus, as Oil of Vitriol distilled from its weight of Nitre, and then mix'd with twice its weight of Oil of Anniseeds.

Isaac Newton
(1718)*

Spectroscopy measures the interaction of substances with electromagnetic radiation. The frequencies v of absorption or emission of radiation provide data on energy levels through the relation $\Delta E = hv$. By means of theoretical interpretations of the energy levels, usually based on quantum mechanics, detailed information can be derived concerning the structures of the molecules or crystals that provide the spectra.

Spectroscopy has another great interest for the chemist because it is basic to the field of photochemistry, the study of chemical reactions initiated by absorption of electromagnetic radiation. Analysis of the kinetics of photochemical reactions requires a good understanding of the physical processes that can occur when light is absorbed by chemical reactants.

Opticks, or A Treatise of the Reflections, Refractions, Inflections and Colours of Light, 2nd ed. (London: W. & G. Innys, 1718), Query 8.

1. Molecular Spectra

Table 17.1 summarizes the types of molecular spectroscopy of major interest to the chemist. Except for dissociation continua, the spectra of atoms consist of sharp lines, and those of molecules appear to be made up of bands in which a densely packed line structure is sometimes revealed under high resolving power.

TABLE 17.1
Types of Optical Spectra[a]

Type of spectros-copy	Range of energies[b]			Type of molecular energy	Information obtained
	Frequency (Hz)	(cm^{-1})	kJ·mol^{-1}		
Microwave	10^9–10^{11}	0.03–3	4×10^{-4}– 4×10^{-2}	Rotation of heavy molecules	Interatomic distances, dipole moments, nuclear interactions
Far infrared	10^{11}–10^{13}	3–300	4×10^{-2}–4	Rotation of light molecules, vibrations of heavy molecules	Interatomic distances, bond force constants
Infrared	10^{13}–10^{14}	300–3000	4–40	Vibrations of light molecules, vibration–rotation	Interatomic distances, bond force constants, molecular charge distributions
Raman	10^{11}–10^{14}	3–3000	4×10^{-2}– 40	Pure rotation, or vibration–rotation	Interatomic distances, bond force constants, molecular charge distributions
Visible, ultra-violet	10^{14}–10^{16}	3×10^3– 3×10^5	40–4000	Electronic transitions	All above properties plus bond dissociation energies

[a]After J. L. Hollenberg, "Energy States of Molecules," *J. Chem. Educ.*, *47*, 2 (1970).
[b]Compare these values with the average thermal kinetic energy per degree of freedom. $\frac{1}{2}RT \cong 1.3$ kJ·mol^{-1}, or ~ 100 cm^{-1}.

The energy levels responsible for atomic spectra represent different allowed states for the orbital electrons. Likewise, in a molecule, absorption or emission of energy can occur in transitions between different energy levels of the electrons. Such levels would be associated, for example, with the different molecular orbitals discussed in Section 15.11. In addition, however, a molecule can change its energy level in two other possible ways, which do not occur in atoms: by changes in the vibrational energy of the molecule and by changes in the rotational energy of the molecule. These internal energies, like the electronic energy, are quantized, so that the molecule can exist only in certain discrete levels of vibrational and rotational energy.

In the theory of molecular spectra it is customary, as a good first approximation, to consider that the energy of a molecule can be expressed simply as the sum of electronic, vibrational, and rotational contributions,

$$E = E_{\text{elec}} + E_{\text{vib}} + E_{\text{rot}} \qquad (17.1)$$

This separation of the energy into three distinct categories is not strictly correct. The occurrence of electronic energy as a separate term is essentially the Born–Oppenheimer approximation, discussed in Section 15.3. Rotational and vibrational energies cannot be strictly separated from each other because the atoms in a rapidly rotating molecule are pushed apart by centrifugal forces, which thereby affect the character of the vibrations. Nevertheless, the approximation of (17.1) suffices to explain many of the observed characteristics of molecular spectra.

The separations between electronic energy levels are usually much larger than those between vibrational energy levels, which, in turn, are much larger than those between rotational levels. The type of energy level diagram that results is shown in Fig. 17.1, which depicts several of the electronic energy levels of the CO molecule. For the ground electronic state* $X[^1\Sigma^+]$, we also show the vibrational energy

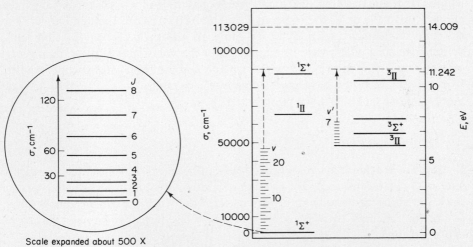

Scale expanded about 500 X

FIG. 17.1 Energy-level diagram of the CO molecule. Several singlet and several triplet electronic states are shown. The vibrational levels associated with lowest singlet and lowest triplet are shown. Actually those levels would continue toward higher energies until the molecule dissociated into a ^3P C atom and a ^3P O atom, at an energy level of 11.242 eV. The energy of ionization, CO $\longrightarrow$ CO$^+$ + e, is 14.009 eV. The left of the diagram, on an expanded scale, shows the rotational levels associated with the lowest ($v = 0$) vibrational level.

*The ground electronic state of a molecule is conventionally denoted by X, and various excited states as A, B, C, etc.

levels, and on an enlarged scale we then show the rotational levels for the lowest ($v = 0$) vibrational level. Associated with each electronic level is a similar set of vibrational levels, each of which is in turn associated with a set of rotational levels. The very small energy differences between successive rotational levels is responsible for the band structure of molecular spectra.

Transitions between different electronic levels give rise to spectra in the visible or ultraviolet region, called *electronic spectra*. Transitions between vibrational levels within the same electronic state are responsible for spectra in the near infrared (< 20 μm), called *vibration–rotation spectra*. Finally, spectra are observed in the far infrared (> 20 μm), arising from transitions between rotational levels belonging to the same vibrational level; these are called *pure rotation spectra*.

In Chapter 9 on chemical kinetics, we considered only thermal reactions. In these the energy for the climb to the activated state comes from the random thermal energies of the reacting molecules. In Chapter 12 on electrodics, we showed how activation energy could be provided by the action of electric fields on ions. A third way to provide the energy of activation is to bring the reactant molecule into collision with quanta of electromagnetic energy (photons). *Photochemistry* is the science of the chemical effects of light, where *light* includes the infrared and ultraviolet, as well as the visible regions of the spectrum, i.e., the range of wavelengths from about 100 to 1000 nm. The energies of quanta in this range vary from about 1 to 10 eV, or 90 to 900 kJ·mol^{-1}. These energies are comparable with the strengths of chemical bonds. Thus, if a molecule absorbs a photon of visible light, definite chemical effects may be expected, yet the collision is still a rather gentle one, and the effects usually follow paths made familiar from spectroscopic studies. In particular, there is almost never enough energy in a single quantum to activate more than one molecule in the primary step.

2. Light Absorption

Light incident upon a macroscopic system can be reflected, transmitted, refracted and scattered, or absorbed. The fraction of incident light absorbed depends on the thickness of the medium that is traversed. The Law of Absorption, originally stated in 1729 in a memoir by P. Bouguer, was later rediscovered by Lambert. It can be expressed as

$$-\frac{dI}{I} = b \, dx \tag{17.2}$$

where I is the intensity of light at a distance x from its entry into the medium, and b is called the [Napierian] *absorption coefficient*. On integration with the boundary condition $I = I_0$ at $x = 0$, we obtain

$$I = I_0 e^{-bx}$$

Thus,

$$\ln \frac{I}{I_0} = \ln \mathscr{T} = -bx \tag{17.3}$$

where $\mathscr{T}$ is the *internal transmittance*.

Workers in this field prefer to use decadic logarithms, so that we also define

$$\log \frac{I}{I_0} = \log \mathscr{T} = -ax \tag{17.4}$$

where a is the [*linear*] *absorption coefficient*. When one speaks simply of the *absorption coefficient*, this a is the quantity that is meant.

In 1852, Beer showed, for many solutions of absorbing compounds in practically transparent solvents, that the coefficient a is proportional to the concentration of solute c. Thus, Beer's Law is

$$\log \frac{I}{I_0} = -\epsilon c x \tag{17.5}$$

where c is the molar concentration and ϵ is called the *molar absorption coefficient*. These absorption laws are the bases for various spectrophotometric methods of analysis. They are obeyed strictly only for monochromatic light.

A device that measures the total amount of incident radiation is called an *actinometer*. This measurement, *actinometry*, is a necessary part of any quantitative study of photochemical reactions. An important type of actinometer is the thermopile, which consists of a number of thermocouples connected in series, with their hot junctions imbedded at a blackened surface that absorbs almost all the incident light and converts it into heat. Calibrated lamps of known energy output are available from the National Bureau of Standards. The emf developed by the thermopile is measured first with the standard lamp, then with the source of radiation of unknown intensity. The reaction vessel is mounted between the thermopile and the light, and the radiation absorbed by the reacting system is measured by the difference between readings with the vessel filled and empty.

3. Quantum Mechanics of Light Absorption

In previous chapters, we discussed applications of quantum mechanics to the structures of atoms and molecules in stationary states, in which the systems do not change with time. If an atom or molecule absorbs or emits quanta of radiation, however, it changes with time from one stationary state to another. Certain general relations can be derived, which will apply to such changes and will govern the intensities of all kinds of spectra.

The method used is called *time-dependent perturbation theory*. We consider the atom or molecule in a stationary state as an unperturbed system and investigate theoretically what happens when it is perturbed by an electromagnetic field that varies with time, for example, the periodic electromagnetic field of a ray of light passing through the system.

In the absence of any perturbation, the atom or molecule is represented by a wave equation

$$(i\hbar)\frac{\partial \Psi}{\partial t} = \hat{H}_0 \Psi \tag{17.6}$$

where $\hat{H}_0$ is the Hamiltonian operator (p. 616) of the unperturbed system. The solution of this equation gives a set of allowed wave functions corresponding to allowed energy levels E_n of the stationary states, with

$$\Psi_n^0(q, t) = \psi_n(q) \exp\left(\frac{iE_n t}{\hbar}\right) \tag{17.7}$$

where $\psi_n(q)$ is a function of the space coordinates, which are all represented formally by q, and is independent of the time.

In the presence of the perturbing field, there will be an additional term $U(q, t)$ in the Hamiltonian, so that the wave equation thus becomes

$$(i\hbar)\frac{\partial \Psi}{\partial t} = (\hat{H}_0 + \hat{U})\Psi \tag{17.8}$$

Let us suppose that initially ($t = 0$) the state of the system is given by $\Psi(q, 0)$. Equation (17.8) then allows us to calculate the wave function at any later time t, $\Psi(q, t)$. We expand $\Psi(q, t)$ as a series in terms of $\Psi_n^0(q, t)$, the time-dependent wave functions of the unperturbed system:

$$\Psi(q, t) = \sum_n a_n(t)\Psi_n^0(q, t) \tag{17.9}$$

To solve for the coefficients a_n, we substitute (17.9) into (17.8). Making use of (17.6), we obtain

$$(i\hbar) \sum \frac{da_n}{dt}\Psi_n^0(q, t) = \hat{U}\Psi(q, t)$$

We multiply each side by $\Psi_m^{0*}(q, t)$, the complex conjugate of Ψ_m^0, and integrate over all q. In the summation on the left side, all terms with $m \neq n$ then vanish as a result of orthogonality of the wave functions, and when $m = n$, the integral equals unity as result of normalization of the wave functions. Thus,

$$i\hbar \frac{da_n}{dt} = \int \Psi_n^{0*}(q, t)\hat{U}\Psi(q, t)\, dq \tag{17.10}$$

Note how this formula can express the transition between one stationary state and another. Suppose that the system is initially in a state $\Psi_1(q, t)$, so that all the coefficients a_n are equal to zero except a_1. As a result of absorption of light of suitable frequency, the system might make a transition to a state $\Psi_2(q, t)$, so that now $a_2 = 1$ and all the other coefficients a_n except a_2 equal zero.

We now assume that the perturbation U is so small that the change in the wave function is small, and, therefore, $\Psi(q, t)$ can be replaced by its unperturbed value, $\Psi_0^0(q, t)$, to yield

$$i\hbar \frac{da_n}{dt} = \int \Psi_n^{0*}(q, t)\hat{U}\Psi_0^0(q, t)\, dq \tag{17.11}$$

Introducing the notation

$$\int \psi_n^*(q)\hat{U}\psi_0(q)\, dq \equiv U_{n0} \tag{17.12}$$

we obtain, by means of (17.7),

$$i\hbar \frac{da_n}{dt} = U_{n0} \exp\left[\frac{i(E_0 - E_n)t}{\hbar}\right] \tag{17.13}$$

This equation is integrated over the time variable, to give

$$a_n(t) = (i\hbar)^{-1} \int_0^t U_{n0} \exp\left[\frac{i(E_0 - E_n)t'}{\hbar}\right] dt' \qquad (17.14)$$

This valuable equation, to a good approximation, shows how the coefficients in (17.9) vary with time, and thus how the system changes with time from one state to another.

We can derive at once an important consequence from (17.14). If the perturbing potential is caused by electromagnetic radiation (e.g., a beam of light of frequency v passing through the sample), the perturbation U will vary with time in accord with an expression such as

$$U(q, t) = F(q)(e^{2\pi i v t} + e^{-2\pi i v t}) \qquad (17.15)$$

The electric and magnetic vectors of the wave of radiation are propagated as traveling waves of frequency v and amplitude proportional to some function $F(q)$. Since the integral in (17.10) is over only the spacial coordinates, the factor U_{n0} must preserve the same time dependence as given for U in (17.15). The integral in (17.14) will therefore contain terms of the form

$$\int e^{2\pi i v t} e^{2\pi i (E_0 - E_n)t/h} dt$$

The two harmonic factors in the integral will always destroy each other by interference, and the integral will go to zero, therefore, unless

$$v = \frac{E_0 - E_n}{h} \qquad (17.16)$$

This is exactly the Bohr condition for a spectral transition, and the present theory therefore shows that in order for a_n to increase, i.e., in order for a transition to occur under the influence of a perturbing potential due to electromagnetic radiation, the basic Bohr condition of (17.16) must be obeyed. Once this condition is met, the intensity of the transition will be governed by the value of the quantity U_{n0}, which is called the *transition moment matrix element*.

Since a_n determines the amplitude of a term n in the wave functions of (17.7), the probability that at time t the system is in the state n is given by

$$P_n = |a_n(t)|^2$$

The transition probability per unit time is

$$\frac{P_n}{t} = \frac{|a_n|^2}{t} = \hbar^{-2} |U_{n0}|^2 \qquad (17.17)$$

4.   The Einstein Coefficients

Consider in Fig. 17.2 two states of a molecule designated by m and n. If the molecule is placed in a beam of electromagnetic radiation, it can undergo transitions between these states under the influence of the perturbing field of the incoming radiation. The number of molecules going from n to m by absorption of a quantum hv_{nm} will be proportional to the number N_n in the state n, and to the density of radiation at

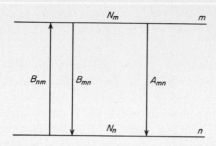

FIG. 17.2   Transitions between two states of a molecule, $E_m - E_n = h\nu_{nm}$.

that frequency $\rho(\nu_{nm})$. Thus,

$$\text{Number of molecules passing from } n \text{ to } m = B_{nm}N_n\rho(\nu_{nm})$$

The B_{nm} is called the *Einstein transition probability for absorption*. The perturbing beam of radiation will cause transitions from m to n such that

$$\begin{matrix}\text{Number of molecules passing from } m \\ \text{to } n \text{ caused by perturbing radiation}\end{matrix} = B_{mn}N_m\rho(\nu_{nm})$$

In addition to the transition from m to n induced by the field of the beam of radiation, there will be a spontaneous emission, with an Einstein transition probability for spontaneous emission given by A_{mn}. We cannot, in fact, calculate this A_{mn} by simple radiation theory, but we can calculate the relation between A_{mn} and B_{mn}, and hence get A_{mn} indirectly. In a steady state, the number of molecules passing from n to m equals the number passing from m to n. Thus,

$$N_m[A_{mn} + \rho(\nu_{mn})B_{mn}] = N_nB_{nm}\rho(\nu_{nm})$$

We know that $B_{mn} = B_{nm}$ from the general property of the matrix element†
$U_{nm} = U_{mn}^*$. Furthermore, from the Boltzmann relation,

$$\frac{N_m}{N_n} = \exp\left(-\frac{h\nu_{nm}}{kT}\right)$$

Planck's radiation law, (13.32), is

$$\rho(\nu) = \frac{8\pi h\nu^3}{c^3}(e^{h\nu/kT} - 1)^{-1}$$

It follows that

$$\frac{A_{mn}}{B_{mn}} = \frac{8\pi h\nu_{nm}^3}{c^3} \tag{17.18}$$

Thus we can calculate A_{mn} from B_{mn}. The Dirac theory of radiation allows us also to calculate A_{mn} directly.

Let us consider a collection of molecules in which some have been raised to

†For any solutions ψ_n, ψ_m of the Schrödinger equation and any Hermitian operator $\hat{M}$,

$$\int \psi_n^*\hat{M}\psi_m \, d\tau = \int \psi_m^*\hat{M}^*\psi_n \, d\tau$$

This *Hermitian property* of the matrix elements is the quantum mechanical foundation of the principle of microscopic reversibility. See Norman Davidson, *Statistical Mechanics* (New York: McGraw-Hill Book Company, 1962), pp. 222–238.

excited states and the radiation or other source of excitation has then been turned off. The Einstein coefficient for spontaneous emission A_{mn} is similar to the rate constant of a first-order reaction,

$$-\frac{dN_m}{dt} = A_{mn}N_m$$

If N_m^0 is the number of molecules initially in the upper state at $t = 0$, at any time t,

$$N_m = N_m^0\, e^{-A_{mn}t}$$

After a time $\tau = 1/A_{mn}$, the number of molecules in the excited state will have declined to $1/e$ of its initial value. For allowed transitions τ is usually of the order of 10^{-8} s. For some excited states, however, τ is much higher ($\sim 10^{-3}$ s). These are called *metastable states*.

Let us now see how the value of the *transition probability*, U_{n0} in (17.12), is determined. Consider the case of a molecule subjected to an electric field E_x directed along the X axis. Each electron in the molecule will experience a force $-E_x e$, which corresponds to a potential energy $E_x e x$ ($F = -\partial U/\partial x$). For all the electrons in the molecule, therefore, this perturbing potential will be

$$U = E_x \sum_j e x_j$$

where the summation is made over all the x coordinates of the electrons, x_j. Note that $-e x_j = \mu_{xj}$, the contribution of the j electron to the x component of the dipole moment of the molecule. The contribution to U_{n0} is

$$\int \psi_n(q)(E_x \sum -e x_j)\psi_0(q)\, dq = \mu_x(n0)E_x \qquad (17.19)$$

The transition probability depends on $\boldsymbol{\mu}_x(n0)$, which is called the *transition moment*. The final result for B_{nm} is derived from (17.17):

$$B_{nm} = \frac{8\pi^3}{3h^2}[\mu_x^2(mn) + \mu_y^2(mn) + \mu_z^2(mn)] \qquad (17.20)$$

where the squares of the three different components of the transition moment are added.

From (17.20), we derive various *selection rules* for spectral transitions. If we are interested in pure rotational spectra, we insert the rotational wave function into (17.20); for vibrational spectra, we use the vibrational wave functions; for electronic spectra, we use wave functions for different electronic states. Even when no transition is allowed by this electric-dipole mechanism, however, other smaller terms may contribute to the perturbing potential and lead to transitions with much lower intensities. Such effects can be quadrupole interactions, or magnetic-dipole interactions with the magnetic field of the light wave.

5. Rotational Levels—Far-Infrared Spectra

The quantum mechanical theory of the rigid linear rotor was treated in Section 14.7. To the quite good approximation that the rotational energy can be treated as independent of vibrations, linear molecules (including all diatomic molecules) will conform to this theory. The moment of inertia of a rigid molecule about an axis is defined as

$$I = \sum m_i r_i^2 \tag{17.21}$$

where m_i is the mass of the ith atom and r_i is its distance from the axis. In a linear molecule, the rotation axis is normal to the internuclear axis of the molecule and passes through the center of mass. Linear molecules belong to point group $\mathbf{D}_{\infty h}$ if they have a plane of symmetry normal to the internuclear axis (e.g., $^{12}C^{16}O_2$) and to point group $\mathbf{C}_{\infty v}$ in the absence of such a plane (e.g., OCS).

As an example of the calculation of a moment of inertia, consider $^{16}O^{12}C^{32}S$ in Fig. 17.3. First consider the center of the S atom to be arbitrarily at the origin

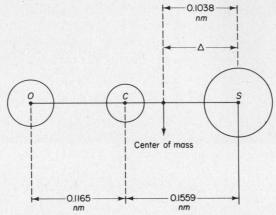

FIG. 17.3 Dimensions of the OCS molecule used in calculating the moment of inertia.

of coordinates $r' = 0$. If the position of the center of mass is at $r' = \Delta$, we can take new coordinates r that have the center of mass as the origin, so that $r' = r - \Delta$. We find Δ from the condition $\sum m_i r_i = 0$, which gives

$$-m_S\Delta + m_O(r_{CS} + r_{CO} - \Delta) + m_C(r_{CS} - \Delta) = 0$$

$$\Delta = \frac{m_O(r_{CS} + r_{CO}) + m_C r_{CS}}{m_S + m_C + m_O}$$

The moment of inertia,

$$I = \sum m_i r_i^2 = m_S\Delta^2 + m_O(r_{CS} + r_{CO} - \Delta)^2 + m_C(r_{CS} - \Delta)^2$$

On substitution of the value for Δ and rearrangement, this gives

$$I = \frac{m_C m_O r_{CO}^2 + m_C m_S r_{CS}^2 + m_O m_S (r_{CO} + r_{CS})^2}{m_C + m_O + m_S} \tag{17.22}$$

Introducing the numerical values shown in Fig. 17.3, we find

$$I = 1.384 \times 10^{-45} \text{ kg·m}^2$$

The energy levels of the rigid rotor are given by (14.44) as

$$E_r = \frac{h^2 J(J+1)}{2I} = BhcJ(J+1) \qquad (17.23)$$

where $B = h/4\pi cI$, the *rotational constant*, is usually given in units of cm^{-1}. The quantum number J can have only integral values. As shown in Section 14.7, the wave functions are

$$\Psi_r^{J,\,k}(\theta, \phi) = P_J^{|k|}(\cos\theta)e^{ik\phi}$$

For each value of J there are $2J + 1$ wave functions characterized by the allowed values of k, which run from $-J$ to $+J$. Each rotational energy level specified by J therefore has a degeneracy of $2J + 1$.

The value of J gives the allowed values of the rotational angular momentum L,

$$L = (h)\sqrt{J(J+1)}$$

This expression is exactly similar to the one for the orbital angular momentum of an electron in the hydrogen atom, as specified by the quantum number l.

To investigate whether a rotating molecule can absorb or emit quanta in passing from one rotational level to another, we use the general theoretical expression for the transition moment, (17.19). Substituting the rotational wave functions, we find for a transition between states J', k', and J'', k'',

$$\mathbf{\mu}(J'k', J''k'') = \int \psi_r^{J'k'}(\mu_x + \mu_y + \mu_z)\psi_r^{J''k''}\, d\tau \qquad (17.24)$$

Let us consider, for example, the z component of the dipole moment (the argument holds for all components):

$$\mu_z = ez = er\cos\theta = \mu^0\cos\theta$$

where μ^0 is the magnitude of the dipole moment. Thus,

$$\mu_z(J'k'J''k'') = \mu^0 \int \psi_r^{J'k'}\cos\theta\,\psi_r^{J''k''}\, d\tau \qquad (17.25)$$

It is clear that unless the permanent dipole moment μ^0 differs from zero, the transition probability will be zero, and no emission or absorption of radiation can occur. We thus have proved that *in order to have a pure rotation spectrum, a rigid rotor must have a permanent dipole moment*. For example, HCl displays an absorption spectrum in the far infrared, but N_2 does not.

If we substitute the rotational wave functions into (17.25) and evaluate the transition probability, we find that μ_z vanishes except for the case in which the quantum number J changes by one unit, $\Delta J = \pm 1$.

An expression for ΔE for the rigid rotor is readily derived from (17.23). For two levels with quantum numbers J and J',

$$\Delta E = h\nu = hcB[J'(J'+1) - J(J+1)] \qquad (17.26)$$

Since

$$\nu = \frac{\Delta E}{h} \quad \text{and} \quad J' - J = 1$$

then
$$v = 2BcJ' \tag{17.27}$$

The spacing between energy levels increases linearly with J, as shown in Fig. 17.4. Absorption spectra arise from transitions from each of these levels to the next higher one. By means of a spectrograph of good resolving power, the absorption band will be seen to consist of a series of lines with an equidistant spacing Δv given by (17.27),
$$\Delta v = v - v' = 2Bc$$

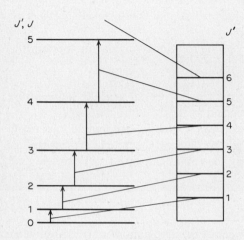

FIG. 17.4 Spacing of rotational energy levels and the lines of a pure rotational spectrum from transitions between levels J and levels $J' = J + 1$.

6. Internuclear Distances from Rotation Spectra

Rotation spectra can provide accurate values of moments of inertia, and hence of internuclear distances and shapes of molecules. Let us consider the example of HCl.

Absorption by HCl has been observed in the far infrared, around $\lambda = 50$ nm or $\sigma = 200$ cm^{-1}. The spacing between successive lines is $\Delta\sigma = 20.1$ to 20.7 cm^{-1}. Analysis shows that the transition from $J = 0$ to $J = 1$ corresponds to a wave number of $\sigma = 1/\lambda = 20.6$ cm^{-1}. The frequency is therefore

$$v = \frac{c}{\lambda} = (3.00 \times 10^{10})(20.6) = 6.20 \times 10^{11} \text{ s}^{-1}$$

The first rotational level, $J = 1$, lies at an energy of
$$hv = (6.20 \times 10^{11})(6.62 \times 10^{-34}) = 4.10 \times 10^{-22} \text{ J}$$
From (17.23) for $(J' = 1) \leftarrow (J = 0)$,

$$\Delta E = \frac{\hbar^2}{I} = 4.10 \times 10^{-22} \text{ J}$$

so that

$$I = 2.72 \times 10^{-47} \text{ kg} \cdot \text{m}^2 = 2.72 \times 10^{-40} \text{ g} \cdot \text{cm}^2$$

Since $I = \mu r_e^2$, where μ is the reduced mass, we can now determine the equilibrium internuclear distance r_e for $H^{35}Cl$:

$$\mu = \frac{(35.0)1}{35.0 + 1} \times \frac{1}{6.02 \times 10^{23}} = 1.61 \times 10^{-24} \text{ g}$$

Hence

$$r_e = \left(\frac{2.72 \times 10^{-47}}{1.61 \times 10^{-27}}\right)^{1/2} = 1.30 \times 10^{-10} \text{ m} = 0.130 \text{ nm}$$

7. Rotational Spectra of Polyatomic Molecules

The rotational energy levels of linear polyatomic molecules are given by (17.23). The moment of inertia, however, may depend on two or more different internuclear distances. Consider, for example, the molecule OCS. The moment of inertia depends on two internuclear distances, O—C and C—S. Clearly, we cannot calculate these two parameters from a single moment of inertia, obtained from the rotational spectrum.

In such cases, the method of isotopic substitution is used. The internuclear distances are determined by the electrostatic configuration of the nuclei and electrons in the structure; to a high degree of approximation, they do not depend on the masses of the nuclei. Thus, isotopic substitution gives new moments of inertia but unchanged internuclear distances. For example, the OCS molecule was investigated in two different isotopic compositions giving the following rotational constants:

$$^{16}O^{12}C^{32}S, \qquad B = 0.202864 \text{ cm}^{-1}$$
$$^{16}O^{12}C^{34}S, \qquad B = 0.197910 \text{ cm}^{-1}$$

The moment of inertia of the linear triatomic molecule was given in (17.22). Substituting the experimental $I = \hbar/4\pi c B$, we obtain two equations with two unknowns r_{CO} and r_{CS}. Although the equations are not linear in the r's, they can be solved by successive approximations or other methods to yield r_{CO} and r_{CS}. The results of a set of such calculations with data from pairs of isotopically substituted OCS molecules are shown in Table 17.2. Internuclear distances obtained in this way are not exactly concordant between pairs, owing to the fact that there is actually a small variation of r_e with the nuclear masses. Can you draw a potential-energy diagram to show how such a variation arises?

For nonlinear molecules, the formulas for the rotational energy levels become more complex, and the spectra, correspondingly, become more difficult to unravel. If a molecule has three moments of inertia, of which two are equal, it is called a *symmetric top*. An example is CH_3Cl, shown in Fig. 17.5. The rotational energy levels of the symmetric top as a rigid rotor are

$$E_r = \frac{\hbar^2}{2I_b}J(J+1) + \frac{\hbar^2}{2}\left(\frac{1}{I_a} - \frac{1}{I_b}\right)K^2 \tag{17.28}$$

Since K specifies the z component of the total angular momentum specified by J,

TABLE 17.2
Internuclear Distances from Pure
Rotation Spectra of OCS by the
Method of Isotopic Substitution

Pairs of molecules						Internuclear distances	
O — C — S vs. O — C — S						C—O nm	C—S nm
16	12	32	16	12	34	0.1165	0.1558
16	12	32	16	13	32	0.1163	0.1559
16	12	34	16	13	34	0.1163	0.1559
16	12	32	18	12	32	0.1155	0.1565

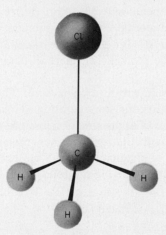

FIG. 17.5 Methyl chloride. A typical symmetric-top molecule.

the allowed values of K are

$$K = J, J - 1, J - 2, \ldots, -J$$

All states with $|K| > 0$ are thus doubly degenerate. When $A > B$, we have a prolate symmetric top (e.g., CH_3Cl); when $A < B$, we have an oblate symmetric top (e.g., BCl_3). As in the case of linear molecules, an infrared spectrum is observed for symmetric tops only when the molecule has a permanent dipole moment. If the principal axis of the top is an axis of symmetry, the dipole moment must, of course, lie along the axis of symmetry. (Why?) In this case, the selection rules are

$$\Delta K = 0, \qquad \Delta J = 0 \quad \text{or} \quad \pm 1$$

Thus, only neighboring levels with the same K can combine with each other. From (17.28), therefore, the lines in the rotation spectrum will be at

$$v = h^{-1}[E_r(J', K') - E_r(J'', K'')] = 2Bc(J'' + 1) = 2BcJ'$$

which is exactly the same formula as that for a linear molecule.

A molecule with three different moments of inertia is an *asymmetric top*. Its rotational energy levels do not follow any simple general formula, but in many cases one of the moments may be much less than the other two, giving a relatively simple spectrum at low resolution.

The model of the molecule as a rigid rotor is an approximation. In a rapidly rotating molecule, there is always a tendency for bonds to stretch by a centrifugal effect. As a result, the moment of inertia increases at higher rotational energies, causing the energy levels to be somewhat closer together than for the corresponding rigid rotor. The equation for the levels corrected for this effect becomes

$$E_r = B_0 hcJ(J + 1) - C_0 hc[J(J + 1)]^2 \qquad (17.29)$$

where C_0 is called the *centrifugal distortion constant*. It is about $10^{-4} B_0$. In the analysis of the symmetric top, we assumed that the molecule was a rigid rotor. If the stretching of bonds due to centrifugal forces is considered, the rotational levels with different K values are split and consequently the individual lines show a fine structure.

8. Microwave Spectroscopy

The moment of inertia of $^{16}O^{12}C^{32}S$ was found on p. 756 to be $I = 138.4 \times 10^{-47}$ kg·m². Its rotational constant is, therefore,

$$B = \frac{\hbar}{4\pi cI} = 0.20286 \text{ cm}^{-1}$$

Spectra at this frequency could be observed with excruciating difficulty in the far infrared, but in practice they are easily measured with microwave spectrometers.

The wavelengths of microwaves are in a range of about 1 to 10 mm. In ordinary absorption spectroscopy, the source of radiation is usually a hot filament or a high pressure gaseous-discharge tube, giving in either case a wide distribution of wavelengths. This radiation is passed through the absorber and the intensity of the transmitted portion is measured at different wavelengths, after separation by means of a grating or prism. In microwave spectroscopy, the source is monochromatic, at a well defined single wavelength, which can, however, be rapidly varied. The frequency of an electronically controlled oscillator can be swept over the frequency range of the wave guide cell. After passage through the cell, which contains the substance under investigation, the microwave beam is picked up by a receiver, and after suitable amplification is fed to a cathode ray oscillograph acting as detector or recorder. The resolving power of this arrangement is 10^5 times that of the best infrared grating spectrometer, so that frequency measurements can be made to seven significant figures. The setup of a typical microwave spectrograph is shown in Fig. 17.6.

One of the most useful extensions of the microwave technique is the inclusion in the cell of a metallic septum by means of which an electric field can be applied to the gas while the spectrum is being scanned. The splitting of quantized energy levels and consequent splitting of spectral lines by an electric field is called the *Stark effect*. The electric field splits the levels specified by J into $J + 1$ levels specified by $M_J = 0, 1, 2, \ldots, J$. The selection rule for M_J depends on the orientation of the Stark electric field to the electric vector of the microwave electric field. If

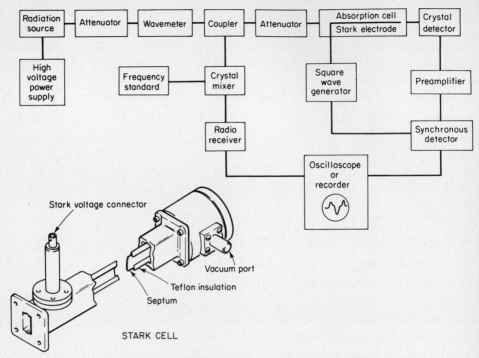

FIG. 17.6 Schematic diagram of a typical microwave spectrometer with Stark modulation. A typical microwave absorption cell would have a cross section of 4 mm × 10 mm and a length of 2 m. The frequency range is from 8 to 40 GHz, and the square wave Stark modulation is 0 to 2000 V at 33.333 kHz.

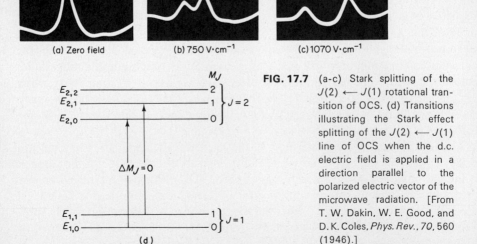

(a) Zero field (b) 750 V·cm⁻¹ (c) 1070 V·cm⁻¹

FIG. 17.7 (a-c) Stark splitting of the $J(2) \leftarrow J(1)$ rotational transition of OCS. (d) Transitions illustrating the Stark effect splitting of the $J(2) \leftarrow J(1)$ line of OCS when the d.c. electric field is applied in a direction parallel to the polarized electric vector of the microwave radiation. [From T. W. Dakin, W. E. Good, and D. K. Coles, *Phys. Rev.*, 70, 560 (1946).]

these electric fields are parallel, $\Delta M_J = 0$; if they are perpendicular, $\Delta M_J = \pm 1$. Figure 17.7 shows the Stark effect in the microwave spectrum of OCS, together with the energy level diagram that explains the observed splitting. The rotational energy level of a molecule in an electric field depends on the dipole moment of the molecule, and the Stark effect in microwave spectra is one of the best methods for measuring dipole moments.

Among the interesting applications of microwave spectroscopy is the study of rotational isomerism. As an example, consider the molecule n-propylchloride, $CH_3—CH_2—CH_2—Cl$. Two *rotational isomers* of this compound are the trans and gauche conformations, shown in projection in Fig. 17.8(a). The microwave spec-

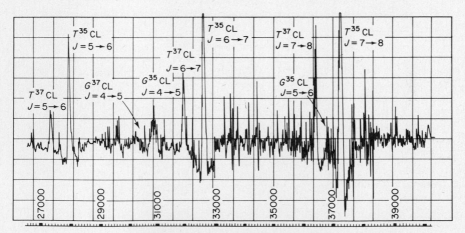

(a) Rotational isomers of the n-propyl halides

(b) Broadband spectrum of n- propyl chloride

FIG. 17.8 Microwave spectrum of n-propyl chloride, showing the trans and gauche forms in various rotational transitions, and the isotope effect due to ^{35}Cl and ^{37}Cl in the molecules. The transition frequencies fit the simple formula $\nu(J) = B^*(J + 1)$ where J is the initial value of the principal rotational quantum number, and B^* is the rotational constant for a "pseudosymmetric top." Thus $B = B + C$, the sum of the two smallest rotational constants for the near prolate rotor, where $B = \hbar^2/2I_b$, $C = \hbar^2/2I_c$. One cannot determine I_a from these spectra. [Hewlett-Packard, Inc., Palo Alto, California.]

trum from17 to 39 GHz is shown in Fig. 17.8(b). Note the sharp rotational lines of the trans isomer and the somewhat broader lines of the gauche isomer. The isotope effect due to different moments of inertia of molecules with ^{35}Cl and ^{37}Cl is also clearly evident.

Pure rotational transitions in heavier molecules are inaccessible to ordinary infrared spectroscopy because, in accord with (17.23), the large moments of inertia would lead to energy-level differences corresponding to excessively long wavelengths. Microwave techniques have made this region readily accessible. With the moments of inertia obtained from microwave spectra, we can calculate internuclear distances to at least ± 0.2 pm. A few examples are shown in Table 17.3.*

TABLE 17.3
Bond Lengths and Angles from Microwave Spectra

Linear Molecules

Molecule	Bond	r_e (nm)	Bond	r_e (nm)
ClCN	C—Cl	0.1629	C—N	0.1163
BrCN	C—Br	0.1790	C—N	0.1159
HCN	C—H	0.1064	C—N	0.1156
OCS	C—O	0.1161	C—S	0.1560
NNO	N—N	0.1126	N—O	0.1191

Symmetric Tops

Molecule	Bond angle	Bond	r_e (nm)	Bond	r_e (nm)
CHCl$_3$	Cl—C—Cl 110°24′	C—H	0.1073	C—Cl	0.1767
CH$_3$F	H—C—H 110° 0°	C—H	0.1109	C—F	0.1385
SiH$_3$Br	H—Si—H 111°20′	Si—H	0.157	Si—Br	0.2209

9. Internal Rotations

For certain polyatomic molecules, the separation of the internal degrees of freedom into vibrations and rotations is not valid. Let us compare, for example, ethylene and ethane, $CH_2{=}CH_2$ and $CH_3{-}CH_3$. The orientation of the two methylene groups in C_2H_4 is fixed by the double bond, so that there is a torsional or twisting vibration about the bond but no complete rotation. In ethane, however, there is an *internal rotation* of the methyl groups about the single bond. "The carbon to carbon single bond in ethane acts like a well-greased axle, about which the two

---

*A comprehensive compilation of microwave spectra, including measured frequencies, assigned molecular species, assigned quantum numbers, and molecular constants determined from these data has been published in 5 volumes as National Bureau of Standards Monograph 70, *Microwave Spectral Tables* (Washington, D.C.: U.S. Govt. Printing Office, 1964–1969).

groups rotate freely."* Thus, one of the vibrational degrees of freedom is lost, becoming an *internal rotation*. Such a rotation would not be difficult to treat if it were completely free and unrestricted, but usually there are potential-energy barriers, which must be overcome before rotation occurs.

Let us consider first the case of free internal rotation, which corresponds to a potential barrier much less than kT. The energy levels are

$$E = \frac{\hbar^2 K^2}{2I_r} \tag{17.30}$$

where K is a quantum number and I_r is the *reduced moment of inertia*. For the case in which the two parts of the molecule constitute coaxial symmetrical tops, such as CH_3—CCl_3,

$$I_r = \frac{I_A I_B}{I_A + I_B}$$

where I_A and I_B are the moments of the tops about the common axis of internal rotation. For other structures, I_r is more complicated.† From (14.46), the partition function for free internal rotation is

$$z_{fr} = \frac{1}{\sigma'} \int_{-\infty}^{+\infty} \exp\left(-\frac{K^2 h^2}{8\pi^2 k T I_r}\right) dK$$

$$z_{fr} = \frac{1}{\sigma'}\left(\frac{8\pi^2 k T I_r}{h^2}\right)^{1/2} \tag{17.31}$$

An example of a free internal rotation was found in cadmium dimethyl,

$$H_3C\text{—}Cd\text{—}CH_3$$

The various contributions to the entropy were computed as follows, at 298 K:

translation and rotation	253.80 J·K⁻¹·mol⁻¹
vibration	36.65 J·K⁻¹·mol⁻¹
free internal rotation	12.26 J·K⁻¹·mol⁻¹
total statistical $S_{298}^{\ominus}$	302.71 J·K⁻¹·mol⁻¹

The Third Law entropy was 302.92; the excellent agreement with the statistical value confirms the hypothesis of free internal rotation in this case.

When the entropy calculated on the assumption of free rotation deviates from the Third Law value, we must consider the possibility of *restricted internal rotation*, with a potential energy barrier $> kT$. Consider, for example, the case of ethane, shown in Fig. 17.9(a) as viewed along the C—C bond axis. The position shown is that of minimum potential energy, $U = 0$; when the CH_3 group is rotated through 60°, we have the H atoms alined and a position of maximum potential energy, $U = U_0$. The variation of U with angle ϕ is shown in Fig. 17.9(b).

*E. B. Wilson, *Science*, **162**, 59 (1968).

†G. N. Lewis and M. Randall, *Thermodynamics*, rev. by K. S. Pitzer and L. Brewer (New York: McGraw-Hill Book Company, 1961), p. 440.

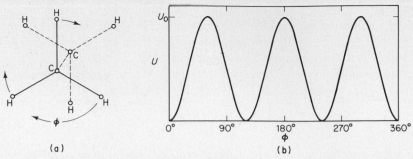

FIG. 17.9   (a) Orientation of CH_3 groups in CH_3–CH_3; (b) potential energy as a function of CH_3 orientation.

The potential-energy curve is represented by

$$U = \tfrac{1}{2}U_0(1 - \cos \sigma'\phi) \tag{17.32}$$

Some barriers to internal rotation are listed in Table 17.4.

TABLE 17.4
Potential Energy Barriers to Internal Rotation About Single Bonds

Molecule	Bond	Barrier (kJ·mol^{-1})
$CH_3)_2O$	C—O	10.8
CH_3—CH_3	C—C	11.5
CH_3—CCl_3	C—C	11.3
$CH_3)_2S$	C—S	8.79
$CH_3)_3SiH$	C—Si	7.66
CH_3SiH_3	C—Si	6.99
$CH_3)_2SiH_2$	C—Si	6.95
CH_3SH	C—S	5.36
CH_3—COF	C—C	4.35
CH_3—OH	C—C	4.48
CH_3—CHO	C—C	4.90
CH_3—CFO	C—C	4.35
CH_3—CH_2·CH_3	C—C	14.9
CH_3—CH=CH_2	C—C	8.28
CH_3—CH_2·Cl	C—C	15.4

There is no simple formula for the partition function for restricted internal rotation. It is necessary to solve the quantum mechanical problem for the energy levels and then to carry out the summation for z. The results have been summarized in tables.*

Microwave spectroscopy has led to major advances in the study of internal rota-

*An excellent set of tables is given by Pitzer and Brewer, *Thermodynamics*, p. 441.

tion.* For internal rotations of methyl groups in simple molecules, some of the pure rotational transitions are split into doublets, usually with spacings of a few megahertz. From this splitting, the barrier height can be determined to an accuracy of better than 5%. Some of these accurate barrier heights are included in Table 17.4. These data present a great challenge, since it should be possible to calculate them as the theoretical techniques of molecular quantum mechanics become sufficiently powerful.†

10. Vibrational Energy Levels and Spectra

Whereas molecular spectra due to transitions between rotational energy levels occur in the far infrared or microwave region, the spectra due to transitions between vibrational energy levels occur in the near infrared. Since each vibrational level is associated with a set of rotational levels, these spectra appear as *bands*, which at high resolution reveal a fine structure of closely packed lines corresponding to the separate rotational levels. Molecular models that include vibrational plus rotational energy may be called *molecular vibrotors*. One can hardly overestimate the importance of these infrared spectra, not only in the development of theories of molecular structure, but also as research tools in organic and biological chemistry.

In Chapter 14, the quantum mechanical problem of the one-dimensional harmonic oscillator was solved in detail. The energy levels were found to be

$$E_v = (v + \tfrac{1}{2})hv_0.$$

Actually, the harmonic oscillator is not a good model for molecular vibrations except at low energy levels, near the bottom of the potential-energy curve. The restoring force in harmonic vibrations is directly proportional to the displacement r. The potential-energy curve is therefore a parabola and dissociation of a harmonic oscillator can never take place, no matter how great the amplitude of vibration. We know very well, however, that the restoring force must actually become weaker as the displacement increases, and for large enough amplitude of vibration the molecule must fly apart. Potential-energy curves for real molecules therefore look like the one in Fig. 15.4, which shows the vibrational energy levels of the H_2 molecule.

Two heats of dissociation may be defined by reference to a curve such as Fig. 15.4. The spectroscopic heat of dissociation D_e is the height from the asymptote to the minimum. The chemical heat of dissociation D_0 is measured from the ground state of the molecule, at $v = 0$, to the onset of dissociation. Therefore,

$$D_e = D_0 + \tfrac{1}{2}hv_0 \tag{17.33}$$

A potential curve like that in Fig. 15.4 corresponds to the model of the anharmonic oscillator. The energy levels corresponding to an anharmonic potential-

---

*E. B. Wilson, *Science*, *162*, 59 (1968).

†A good example of the theoretical approach is the calculation by R. M. Pitzer of the barrier to internal rotation in ethane [*J. Chem. Phys.*, *39*, 1995 (1963)].

energy curve can be expressed as a power series in $(v + \frac{1}{2})$,

$$E_v = h\nu_0[(v + \tfrac{1}{2}) - x_e(v + \tfrac{1}{2})^2 + y_e(v + \tfrac{1}{2})^3 - \cdots]$$

Considering only the first anharmonic term, with *anharmonicity constant* x_e, we have

$$E_v = h\nu_0[(v + \tfrac{1}{2}) - x_e(v + \tfrac{1}{2})^2] \tag{17.34}$$

The energy levels are not evenly spaced, but lie more closely together as the quantum number increases. This fact is illustrated in the levels superimposed on the curve in Fig. 15.4. Since a set of closely packed rotational levels is associated with each of these vibrational levels, it is sometimes possible to determine with great precision the energy level just before the onset of the continuum, and hence to calculate the heat of dissociation of the molecule. The data necessary for such a determination are usually found in long progressions of vibration-rotation bands in electronic absorption or emission spectra (Section 17.19).

11. Vibration-Rotation Spectra of Diatomic Molecules

A diatomic molecule has only one degree of vibrational freedom and thus only one fundamental vibration frequency ν_0. To absorb or emit quanta $h\nu_0$ of vibrational energy, the molecule must possess a permanent dipole moment, since otherwise the transition probability in (17.19) must vanish. Thus, molecules such as NO and HCl display a spectrum in the near infrared due to transitions between vibrational energy levels, but molecules such as H_2 and Cl_2 display no infrared spectra in the gaseous state.

The selection rule governing vibrational transition is $\Delta v = \pm 1$ for the harmonic oscillator, but for a real (anharmonic) oscillator there will be *overtone transitions* with $\Delta v = \pm 2, \pm 3$, etc., although these will have much lower intensities than the *fundamentals* with $\Delta v = \pm 1$.

The vibrational spectrum, however, arises from transitions between definite rotational-energy levels belonging to definite vibrational levels. Therefore, it is a vibration-rotation spectrum. The expression for an energy level in the approximation corresponding to harmonic oscillator and rigid rotor is

$$E_{vr} = (v + \tfrac{1}{2})h\nu_0 + BhcJ(J + 1) \tag{17.35}$$

For a transition between an upper level v', J', and a lower level v'', J'',

$$E_{vr} = (v' - v'')h\nu_0 + B'hcJ'(J' + 1) - B''hcJ''(J'' + 1) \tag{17.36}$$

We must use different rotational constants B' and B'' for the upper and lower states, because the moment of inertia of the molecule is not exactly the same in different vibrational states.

The selection rules for transitions between the vibration-rotation levels of (17.36) are $\Delta v = \pm 1$, $\Delta J = \pm 1$. In the exceptional case that the diatomic molecule has $\Lambda \neq 0$ (i.e., has an angular momentum about the internuclear axis), we can also have $\Delta J = 0$. The best known molecule of this type is NO, which has a $^2\Pi$ ground state.*

A vibration-rotation band may display three branches corresponding to the three cases:

$$J' - J'' = \Delta J = +1, \quad R \text{ branch}$$

$$J' - J'' = \Delta J = -1, \quad P \text{ branch}$$

$$J' - J'' = \Delta J = \quad 0, \quad Q \text{ branch}$$

As an example, the fundamental infrared absorption spectrum of HCl at 2886 cm^{-1} is shown in Fig. 17.10. This band arises from transitions between $v'' = 0$ and $v' = 1$. The appearance of the band at low resolution is shown in (a). The rotational fine structure is not resolved, but we can see the P and R branches. The band is shown in (b) at high resolution. The arrangement of energy levels is shown in (c). Each line corresponds to a distinct value of J' or J''. Note that the lines of greatest intensity do not correspond to $J = 0$ but to J about 4. Why is this so?

For the $v(1 \leftarrow 0)$ band in HCl, we have

$$v = c\sigma = (3 \times 10^{10} \text{ cm} \cdot \text{s}^{-1})(2886 \text{ cm}^{-1}) = 8.65 \times 10^{13} \text{ Hz}$$

as the fundamental vibration frequency. This is about 100 times the rotation frequency found from the far-infrared spectrum.

The force constant of a harmonic oscillator with this frequency, from (14.13), would be

$$\kappa = 4\pi^2 v^2 \mu = 4.81 \times 10^2 \text{ N} \cdot \text{m}^{-1}$$

If the chemical bond is thought of as a spring, the force constant is a measure of its tightness. How would you calculate the classical amplitude of a vibration for the harmonic oscillator model?

Potential-energy curves of the type shown in Fig. 15.4 are so useful in chemical discussions that much effort has been devoted to obtaining convenient mathematical expressions for them. An empirical function that fits fairly well† was suggested by P. M. Morse:

$$U(r - r_e) = D_e[1 - \exp - \beta(r - r_e)]^2 \tag{17.37}$$

Here β is a constant, given in terms of molecular parameters as

$$\beta = \pi v_0 \left(\frac{2\mu}{D_e}\right)^{1/2}$$

where μ is the reduced mass. When the Morse function is used as the potential energy in the Schrödinger equation, the energy levels that are obtained for the oscillator correspond with those in (17.34).

---

*Other examples with $^2\Pi$ ground states are SnF, PO, CH, OH, HCl+; with $^3\Pi$, TiO, C_2, BN.

†D. Steele, E. R. Lippincott, and J. T. Vanderslice, *Rev. Mod. Phys.*, *34*, 239 (1962), compare several potential-energy functions with the experimental data.

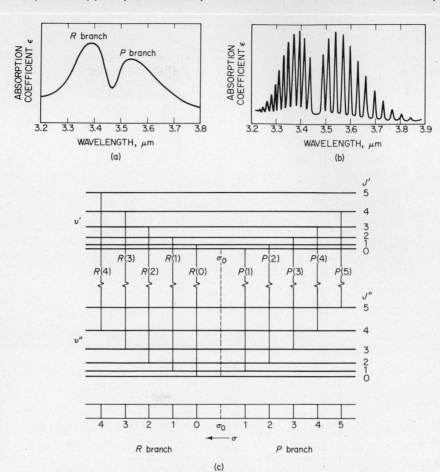

FIG. 17.10 (a) Fundamental vibration band of HCl in near infrared at low resolution. (b) Fundamental vibration band of HCl with resolved rotation spectrum. (c) Schematic energy levels showing structure of P and R branches. Note that the separation between $v'' = 0$ and $v' = 1$ levels is much greater than shown on diagram, a fact indicated by the jagged portions of the transition lines. The *band origin* is designated σ_0.

12. Infrared Spectrum of Carbon Dioxide

A polyatomic molecule need not have a *permanent* dipole moment to have a vibrational spectrum in the infrared, but any vibration that emits or absorbs radiation must cause a *changing dipole moment*. For example, CO_2 is a linear molecule with no permanent dipole. There are $3n - 5 = 4$ normal modes of vibration, as shown in Fig. 17.11. The symmetrical stretching vibration v_1 cannot cause a changing dipole moment, and therefore this vibration is said to be *inactive in the infrared*. The twofold degenerate bending vibration v_2 causes a changing dipole moment, and is

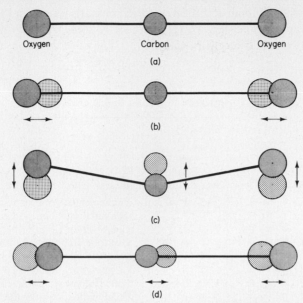

FIG. 17.11   The carbon dioxide molecule (a) is linear and symmetric. It has four degrees of vibrational freedom. In the symmetric stretching mode of vibration (b) the atoms of the molecule vibrate along the internuclear axis in a symmetric manner. In the bending mode (c) the vibration of the atoms is perpendicular to the internuclear axis. This mode is doubly degenerate and there is a similar bending vibration normal to the plane of the paper. In the asymmetric stretching mode (d) the atoms vibrate along the internuclear axis in an asymmetric manner.

The vibrational state of the molecule is described by three quantum numbers, v_1, v_2, and v_3, written as $(v_1 v_2 v_3)$, where v_1 denotes the number of vibrational quanta in the symmetric stretching mode, v_2 the number of vibrational quanta in the bending mode, and v_3 the number of vibrational quanta in the asymmetric stretching mode.

therefore active in the infrared, giving rise to a fundamental absorption band at 667 cm^{-1}. The antisymmetric stretching mode v_3 also causes a changing dipole and is observed in the fundamental absorption band at 2349 cm^{-1}. Notice that the stretch frequency is much higher than the bending one. This is a general result because it is easier to distort a molecule by bending than by stretching, and the force constant is lower for a bending mode.

In addition to the fundamental absorption bands, many combination and overtone bands occur in the infrared spectrum of CO_2, but these have lower intensities than the fundamental. (Their occurrence shows the failure of the harmonic oscillator selection rules.) The analysis of an infrared spectrum consists in sorting out the bands and correctly assigning the fundamental frequencies to their particular normal modes. As we shall see, some vibrations inactive in the infrared region may be active in the Raman spectrum, so that comparisons of data from these two methods facilitate the analysis. The assignment of some of the bands in the CO_2 spectrum is shown in Table 17.5.

TABLE 17.5
The Infrared Vibration Bands of CO_2[a]

Wavelength (μm)	Vibrational quantum numbers ($v_1\ v_2\ v_3$) Initial state	Final state	Assignment of bands
14.986	0 0 0	0 1 0	Fundamental vibration δ
10.4	1 0 0	0 0 1	Combination vibration $v(a) - v(s)$
9.40	0 2 0	0 0 1	Combination vibration $v(a) - 2\delta$
5.174	0 0 0	0 3 0	Overtone vibration 3δ
4.816	0 0 0	1 1 0	Combination vibration $v(s) + \delta$
4.68	0 1 0	2 0 0	Combination vibration $2v(s) - \delta$
4.25	0 0 0	0 0 1	Fundamental vibration $v(a)$
2.27	0 0 0	0 2 1	Combination vibration $2\delta + v(a)$
2.006	0 0 0	1 2 1	Combination vibration $v(s) + 2\delta + v(a)$
1.957	0 0 0	2 0 1	Combination vibration $2v(s) + v(a)$
1.574	0 0 0	2 2 1	Combination vibration $2v(s) + 2\delta + v(a)$
1.433	0 0 0	0 0 3	Overtone vibration $3v(a)$
0.8698	0 0 0	0 0 5	Overtone vibration $5v(a)$

[a]Data from *Landolt-Bornstein Tables*, 6th ed. (Berlin: 1951).

13. Lasers

The word *laser* is an acronym for "Light Amplification by Stimulated Emission of Radiation." The basis of laser action is the stimulated emission measured by the Einstein coefficient B_{mn}, which was discussed in Section 17.3. Figure 17.12

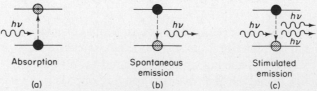

Absorption	Spontaneous emission	Stimulated emission
(a)	(b)	(c)

FIG. 17.12 The stimulated emission of photons $h\nu$ is the basis of laser action. In (a) a molecule in the ground state absorbs a photon and is raised to an excited state. In (b) the molecule is shown to emit a photon spontaneously and to revert to the ground state. In (c) the incoming photon strikes a molecule that is already in an excited state and stimulates the molecule to revert to the ground state as it emits a second photon.

shows in a schematic way how a light wave of characteristic frequency v can enter a medium and produce an electromagnetic perturbation of molecules in an excited state that will cause them to emit radiation at frequency v exactly in phase with the incident radiation. Stimulated emission is the exact reverse of absorption. In a classical picture, when absorption takes place from a light wave passing through a medium, the transmitted light wave has lower amplitude but an unchanged frequency and phase. If stimulated emission occurs, the transmitted wave has a higher amplitude but unchanged frequency and phase.

To obtain an amplification of light through stimulated emission, we must have a *population inversion* among the molecules in the medium, so that there are more molecules in the upper state of the transition than in the lower state. Many types of laser have been studied since the original discovery by C. H. Townes at Columbia University in 1954.* Sometimes the population inversion is achieved through *optical pumping*, sometimes by electrical discharges, and, most interestingly for the chemist, sometimes by a chemical reaction that produces molecules in an excited state.

We shall illustrate the principles of the laser by describing one particular example. Progress has been so rapid that any general survey would not be practical.†
The carbon dioxide laser is based upon the vibrational energy levels of the CO_2 molecule as shown in Fig. 17.13. The three quantum numbers v_1, v_2, v_3 refer to the symmetric stretching mode, the bending mode, and the asymmetric stretching mode, respectively. These normal modes are shown in Fig. 17.11. The rotational fine structure is not included in Fig. 17.13, but each of the vibrational levels is actually associated with a set of closely spaced rotational levels. Figure 17.13 also includes the first vibrational level of N_2, which is at nearly the same energy as the (001) level of CO_2.

The (001) level of CO_2 is ideal as the upper level for a laser operation. It can be excited directly by passing an electric discharge through CO_2, but even more effectively if N_2 is mixed with CO_2. The N_2 is excited by the electric discharge and transfers its energy to the (001) excitation of CO_2. Since N_2 is a homonuclear diatomic molecule (no dipole moment), it cannot undergo vibrational transitions by absorption or emission of radiation. Hence, the N_2 excitation is not lost through radiation but rather transferred in collisions with CO_2 molecules. Because of the close match of energy levels, these are called *resonant collisions*. The first condition for laser operation is therefore achieved when an electric discharge is passed through a mixture of CO_2 and N_2, viz., a population inversion that produces a high concentration of CO_2 molecules in the (001) vibrational state.

Two radiative transitions are possible from (001), as shown in Fig. 17.13, either to (100) with emission of infrared radiation at 10.6 μm, or to (020) with emission at 9.6 μm. The 10.6-μm transitions are stronger than the 9.6-μm by a factor of about 10. The CO_2 molecules in the lower levels become de-excited by

---

*Townes discovered the phenomenon in the microwave region of the spectrum and hence called it *Maser* action. The first laser with visible light was achieved by Maiman in 1960.

†A beautiful introductory book is available: *Lasers and Light*, ed. by A. L. Schawlow (San Francisco: W. H. Freeman Co., 1969).

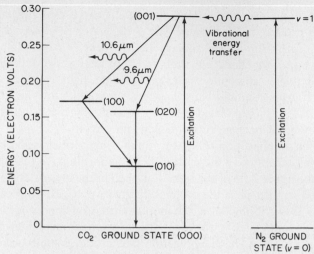

FIG. 17.13 Energy levels concerned in the operation of the CO_2 laser. Addition of nitrogen gas to a carbon dioxide laser results in the selective excitation of the carbon dioxide molecules to the upper laser level. Since nitrogen is a diatomic molecule it has only one degree of vibrational freedom; hence one vibrational quantum number (v) completely describes its vibrational energy levels. Nitrogen molecules can be efficiently excited from the $v = 0$ level to the $v = 1$ level by electron impact in a low pressure nitrogen discharge. Since the energy of excitation of the $N_2(v = 1)$ molecule nearly equals the energy of excitation of the $CO_2(001)$ molecule, an efficient transfer of vibrational energy takes place from the nitrogen to the carbon dioxide in collisions between $N_2(v = 1)$ molecules and $CO_2(000)$ molecules. In such a collision the nitrogen molecule returns from the $v = 1$ level to its ground state by losing one quantum of its vibrational energy, thereby exciting the carbon dioxide molecule from its ground state to the 001 level. The carbon dioxide molecule can then radiatively decay to either the 100 level or the 020 level, in the process emitting infrared light at 10.6 or 9.6 microns, respectively. [From *Lasers and Light*, ed. by A. L. Schawlow (San Francisco: W. H. Freeman & Co., 1969). ©1969 by *Scientific American*, all rights reserved.]

nonresonant collisions, which convert vibrational energy into translational kinetic energy.

The power output of the CO_2 laser can actually be made to occur almost exclusively in a single P branch transition of the 10.6-μm band, usually in the $P(J = 20)$ transition at 10.59 μm. Typical CO_2 lasers are about 2 m long with a continuous power output up to 150 W. Much larger CO_2 lasers have been constructed, with power up to 30 kW. These are awesome devices, in which some properties of the zap guns of "Star Ship Enterprise"* have already been realized. The beam can drill a hole through 1 cm of stainless steel plate in a few seconds. The possibilities for communication and other uses are just beginning to be realized.

14. Normal Modes of Vibration

The model of the harmonic oscillator is not suitable for representing higher vibrational levels and the quantitative details of infrared spectra. Nevertheless, the model is very useful, indeed essential, in the treatment of the mathematical problem of the vibrations of polyatomic molecules. Only in the approximation of linear restoring forces (Hooke's Law $\rightarrow$ harmonic oscillator) does the analytical mechanics of the problem become at all tractable.

Consider a molecule with N nuclei. The location of each nucleus in space can be described by a set of three coordinates x_i', y_i', z_i', so that $3N$ coordinates are required for all the nuclei in the molecule. The positions of the nuclei in the molecule oscillate about certain equilibrium sites. The oscillations of each nucleus can be described in terms of a set of displacements from equilibrium, x_i, y_i, z_i. Thus, the actual Cartesian coordinates of a nucleus with respect to some origin are replaced by the change in these coordinates with respect to some equilibrium value.

In the theory of small vibrations, we assume that the restoring force associated with the displacement of a given coordinate is linearly proportional to each of the other displacements. For example, the restoring force for x_1, the displacement of the x coordinate of nucleus (1) from its equilibrium position, would be

$$F_x^1 = -\kappa_{xx}^{11}x_1 - \kappa_{xy}^{11}y_1 - \kappa_{xz}^{11}z_1 - \kappa_{xx}^{12}x_2 - \kappa_{xy}^{12}y_2 - \cdots - \kappa_{xz}^{1N}z_N \qquad (17.38)$$

The κ's are the force constants. When three (for x, y, z) such extended Hooke's Law equations are written for each nucleus, we get a set of simultaneous dynamical equations for the motions of all the nuclei. In this formulation, each displacement of a nucleus j contributes to the restoring force on every nucleus i (including, of course, itself).

Under certain conditions, the nuclei in the molecule can move in a simple harmonic motion, in which each nucleus undergoes a simple harmonic vibration with the same frequency v and always in phase with every other nucleus. These are the *normal modes of vibration*. For example, the displacement from its equilibrium site of nucleus i would be,

$$x_i = x_{i0} \cos{(2\pi vt + \phi)} \qquad (17.39)$$

where x_{i0} is the amplitude and ϕ the phase. By differentiating twice with respect to time, the acceleration corresponding to (17.39) is $a_i = \ddot{x}_i = -4\pi^2 v^2 x_i$. From $F = ma$,

$$F_i = m_i \ddot{x}_i = -4\pi^2 m_i v^2 x_i$$

*Flagship of the intergalactic fleet in a North American science fiction TV program of the late 1960's.

When such an expression for the force is introduced into the generalized Hooke's Law equations (17.38), one obtains a system of $3N$ homogeneous linear equations:

$$-4\pi^2\nu^2 m_1 x_1 = -\kappa_{xx}^{11}x_1 - \kappa_{xy}^{11}y_1 - \cdots - \kappa_{xz}^{1N}z_N$$

$$-4\pi^2\nu^2 m_1 y_1 = -\kappa_{yx}^{11}x_1 - \kappa_{yy}^{11}y_1 - \cdots - \kappa_{yz}^{1N}z_N$$

$$\vdots \qquad \vdots \qquad \vdots \qquad \vdots \qquad \text{(17.38a)}$$

$$-4\pi^2\nu^2 m_N z_N = -\kappa_{zx}^{N1}x_1 - \kappa_{zy}^{N1}y_1 - \cdots - \kappa_{zz}^{NN}z_N$$

As we saw in a similar situation in Section 14.24 such a system of equations has a nontrivial solution only if the determinant of the coefficients vanishes.

$$\begin{vmatrix} (\kappa_{xx}^{11} - 4\pi^2\nu^2 m_1)\kappa_{xy}^{11} & & \cdots \kappa_{xz}^{1N} \\ \kappa_{yx}^{11} & (\kappa_{yy}^{11} - 4\pi^2\nu^2 m_i) \cdots \kappa_{yz}^{1N} \\ \vdots & \vdots & \vdots \\ \kappa_{zx}^{1N} & \kappa_{zy}^{1N} & \cdots (\kappa_{zz}^{NN} - 4\pi^2\nu^2 m_N) \end{vmatrix} = 0 \qquad \text{(17.40)}$$

This is an equation of degree $3N$ for the square of the frequencies ν^2, a type of secular equation. The solutions for ν^2 would give the values for the frequencies of all the normal modes of vibration. Actually, six of the roots are equal to zero (five for a linear molecule), corresponding to the degrees of freedom of translation and rotation. To solve (17.40) for the numerical values of the frequencies, we would have to know all the force constants, but such information is rarely available. Thus, although (17.40) is an important and interesting equation, in that it displays the mathematical origin of the normal-mode frequencies, it is not of much practical use in calculations.

In fact, the practical problem is usually just the opposite: given the frequencies of a set of normal vibrations from spectroscopic measurements, assign the frequencies to the normal vibrations and derive a set of force constants. Inasmuch as the force constants outnumber the vibration frequencies, this problem evidently cannot be solved on the basis of (17.40) alone. Isotopic substitution may provide additional data, and relations between the force constants can be obtained from models for the interactions between atoms. For moderately complicated molecules ($N > 4$), some drastic reduction in the number of force constants is also necessary. Several models are available. One of the most convenient for the chemist is to divide the displacements of the nuclei into those caused by (1) stretching of chemical bonds, (2) bending of bonds (alteration of bond angles), (3) twisting of bonds (torsional deformation), and (4) a limited number of nonbonded interactions between atoms that come quite close together in the molecule. These nonbonded interactions within molecules are similar to the van der Waals interactions between molecules, which will be discussed in Chapter 19. Let us assume that in one way or another the normal-mode frequencies have been determined. They can then be substituted back into Eqs. (17.38a), and these can be solved for the *ratios of the displacements of the nuclei*. These ratios give the *normal coordinates* for the vibrations, and show how the nuclei move in each of the normal modes of vibration.

15. Symmetry and Normal Vibrations

For practical purposes, it is not usually necessary to work through the calculation of normal coordinates. The form of the normal vibrations can often be deduced from simple symmetry considerations, based on the character table for the point group to which the molecule belongs. If a normal vibration is nondegenerate— i.e., if its particular frequency v belongs to only one normal vibration—the symmetry requirement can be seen quite simply as follows. The total energy of the vibrating molecule cannot change as the result of any symmetry operation performed on it. The energy can be expressed in terms of the normal vibration coordinates q_i as

$$E = \sum \tfrac{1}{2}m_i\ddot{q}_i^2 + \sum \tfrac{1}{2}\kappa_i q_i^2$$

Thus, each q_i must be either symmetric ($q_i \rightarrow q_i$) or antisymmetric ($q_i \rightarrow -q_i$) with respect to each symmetry operation. As an example, Fig. 17.14 shows the effect

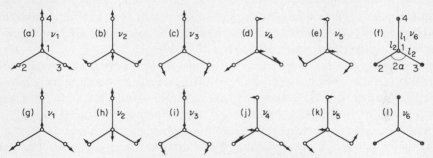

FIG. 17.14 Normal vibrations of an ABC_2 molecule and their behavior for a reflection at the plane of symmetry through AB perpendicular to the plane of the molecule. Motions perpendicular to the plane of the paper are indicated by + or − signs in the circles representing nuclei. [From G. Herzberg, *Molecular Spectra and Molecular Structure* (Princeton, N. J.: D. Van Nostrand Co., Inc., 1945).]

of a symmetry plane on one of the normal vibrations of an ABC_2 molecule, such as H_2CO. If the normal vibration v_i is degenerate, the symmetry problem is somewhat more involved, because now a linear combination of normal modes must be taken, and the effect of the symmetry operation is usually more than just a change in sign.

From the discussion of group representations in the previous chapter, it is evident that each normal mode can serve as the basis for a representation of the symmetry group of the molecule. The representation can be reduced to a set of irreducible representations. Then the normal mode (or, in a degenerate case, a linear combination of normal modes) transforms under the symmetry operations exactly as required by the characters of the irreducible representation (I. R.) to which it belongs.

As an example, consider the molecule H_2CO, whose $(3N - 6)$ normal vibrations are shown in Fig. 17.14. The point group is C_{2v} with the following character table:

C_{2v}	E	C_2	$\sigma_v(xz)$	$\sigma_v(yz)$
A_1	1	1	1	1
A_2	1	1	-1	-1
B_1	1	-1	1	-1
B_2	1	-1	-1	1

All the I.R.'s are one dimensional; therefore, all the normal vibrations must be nondegenerate. The three normal vibrations v_1, v_2, v_3 are totally symmetric and belong to species A_1.* If we take the plane of the molecule to be the xz plane, vibrations v_4 and v_5 belong to species B_1, whereas v_6 belongs to species B_2, being antisymmetric with respect to reflection $\sigma_v'(xz)$ in the plane of the molecule.

Many other examples of the symmetry of normal vibrations can be found in Herzberg's book.† We do not wish to give more details here, but only to establish the following important basic principle: *The character table of the I.R.'s of the point group leads to the classification of the normal vibrations.*

A fundamental vibrational transition occurs when a particular normal mode is excited from its ground state with $v_i = 0$ to its first excited state with $v_i = 1$, while all the other normal modes remain unexcited. In terms of the vibrational wave function, the excitation of a fundamental can be represented as $\psi_{v_i}^1 \leftarrow \psi_{v_i}^0$. In physical terms, the requirement for excitation of a fundamental is that the vibration of the molecule results in an oscillating dipole moment, which can interact with the electric vector of the electromagnetic field. From (17.19), the mathematical requirement is that the *transition moment* be different from zero. In terms of the dipole-moment vector, one of the following integrals must, therefore, be different from zero:

$$\int \psi_v^0 x \psi_v^1 \, d\tau, \qquad \int \psi_v^0 y \psi_v^1 \, d\tau, \qquad \int \psi_v^0 z \psi_v^1 \, d\tau \qquad (17.41)$$

From Section 14.5, we know that the wave functions for the harmonic oscillator are

$$\psi_v = N_v e^{-q^2/2\alpha^2} \mathscr{H}_v \left(\frac{q}{\alpha} \right) \qquad (17.42)$$

where $\mathscr{H}_v$ are the Hermite polynomials and $\alpha^2 = h/2\pi\mu v_0$. For the ground state, $v = 0$, $\mathscr{H}_v = 1$, and hence, $\psi_v^0 = N_0 e^{-q^2/2\alpha^2}$.

In the case of a normal mode of vibration, q represents one of the *normal*

---

*The symmetry of the vibrations is usually designated by lower case symbols, such as a_1, b_1, b_2, etc.

†G. Herzberg, *Infrared and Raman Spectra* (Princeton, N.J.: D. Van Nostrand Co., Inc., 1945).

coordinates. For any normal mode, therefore, any symmetry operation of the molecule can only change q by a factor of ± 1, and must accordingly leave ψ_v^0 unchanged.* We therefore can state the rule: *All the ground state wave functions for the normal vibrations are bases for the totally symmetric representation of the point group of the molecule.*

The first excited wave function, from (17.42), has the form

$$\psi_v^1 = N_v(2x)\, e^{-q^2/2\alpha^2}$$

Since the exponential term is totally symmetric, the wave function for the first excited state itself has the symmetry of q, the normal coordinate; therefore, a wave function for the first excited state ($v = 1$) always has the symmetry of the normal coordinate itself. Hence, the assignment of the normal modes to their group theoretical representation automatically gives the symmetry of the first excited state of each normal mode.

An integral of the product of two functions, $\int f_A f_B\, d\tau$, will vanish unless the integral or some term in it is completely symmetrical. The condition for the nonvanishing of one of the integrals in (17.41) is, therefore, that the product of x, y, or z (as the case may be) with ψ_v^1 must belong to the totally symmetric representation of the group. In other words, ψ_v^1 must belong to the same representation as one of the Cartesian coordinates x, y, or z. The character tables for the various

Character Table of Group C_{2h}

C_{2h}	E	C	i	σ_h	
A_g	1	1	1	1	x^2, y^2, z^2, xy
B_g	1	-1	1	-1	xz, yz
A_u	1	1	-1	-1	z
B_u	1	-1	-1	1	x, y

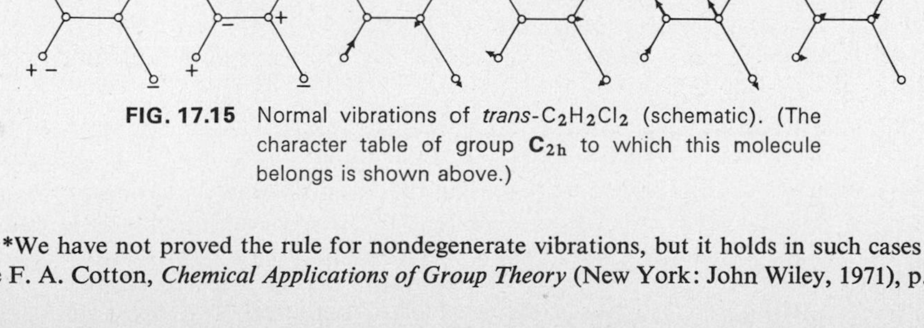

FIG. 17.15   Normal vibrations of *trans*-$C_2H_2Cl_2$ (schematic). (The character table of group C_{2h} to which this molecule belongs is shown above.)

---

*We have not proved the rule for nondegenerate vibrations, but it holds in such cases also. See F. A. Cotton, *Chemical Applications of Group Theory* (New York: John Wiley, 1971), p. 262.

groups usually list the Cartesian coordinates with the representations to which they belong.

Let us take as an example the C_{2v} molecule, CH_2O, shown in Fig. 17.14 with the character table given on p. 778. Which of the normal vibrations will be active? This is a fundamental question for the interpretation of the infrared spectrum. From the character table, we can assign the normal vibrations and the Cartesian coordinates as follows:

$$\begin{array}{ccc} v_1\ A_1 & v_4\ B_1 & x\ B_1 \\ v_2\ A_1 & v_5\ B_1 & y\ B_2 \\ v_3\ A_1 & v_6\ B_2 & z\ A_1 \end{array}$$

In this case, we find a match between each of the normal-mode representations and one of the dipole-moment representations (x, y, z); therefore, *all* the normal modes of formaldehyde are active as fundamentals in the infrared spectrum.

As a second example, let us consider the molecule *trans*-$C_2H_2Cl_2$ shown in Fig. 17.15. The symmetries of the normal modes are obtained by reference to the character table for the group C_{2h}, to which this molecule belongs. The normal modes are thus assigned as follows:

$$\begin{array}{cc} v_1, v_2, v_3, v_4, v_5\ A_g & x\ B_u \\ v_6, v_7 \qquad\qquad A_u & y\ B_u \\ v_8 \qquad\qquad B_g & z\ A_u \\ v_9, v_{10}, v_{11}, v_{12}\ B_u \end{array}$$

FIG. 17.16 The Raman Spectrum of Gaseous Oxygen. (a) Rotation-vibration band excited by Hg line at 435.80 nm. The head of the Q branch occurs at 467.51 nm so that the frequency of the pure vibrational transition $v = 0 \longrightarrow v = 1$ would be $\Delta\sigma_0 = 10^7(\lambda_1^{-1} - \lambda_2^{-1}) = 10^7$ $(435.8^{-1} - 467.51^{-1}) = 1556.25$ cm^{-1}. The iron emission spectrum has been superimposed to serve as a calibration of the wavelength scale. The branches designated as O and S correspond to the Raman selection rules $\Delta J = +2$ and $\Delta J = -2$, and the Q branch, as for infrared spectra, is $\Delta J = 0$. In $^{16}O_2$ only the odd rotational levels are populated. [See Herzberg, *Molecular Spectra and Molecular Structure*, Part I, 2nd ed., pp. 130–135 (New York: D. Van Nostrand & Co., 1950).] (b) Pure rotational spectrum of O_2 excited by the mercury line at 404.7 nm. The Stokes lines are to the long-wavelength side of the exciting lines and the anti-Stokes lines to the shorter-wavelength side. The selection rules are $\Delta J = \pm 2$. The lower J value for the transition is noted for each line. Actually the electronic ground state of O_2 is a triplet ($^3\Sigma_g^-$) but the triplet structure is not resolved in these spectra and thus the quantum number J has been retained. [For a discussion of this point see I. N. Levine, *Quantum Chemistry*, Vol. 2, *Molecular Spectroscopy* (Boston: Allyn and Bacon, 1970), pp. 181–187.] The lines called "ghosts" have nothing to do with the oxygen, but are caused by interference effects due to irregularities in the diffraction grating of the spectrograph. [Alfons Weber, Fordham University].

The comparison shows that the two A_u vibrations (v_6, v_7) and the four B_u vibrations ($v_9, v_{10}, v_{11}, v_{12}$) will be active as fundamentals in the infrared spectrum. This prediction is confirmed by the experimental spectra.

16. Raman Spectra

When a beam of light passes through a medium, a certain amount is absorbed, a certain amount transmitted, and a certain amount scattered. The scattered light can be studied by observations perpendicular to the direction of the incident beam. Most of the light is scattered without change in wavelength (Rayleigh scattering), but there is in addition a small amount of light scattered with an altered wavelength. If the incident light is monochromatic, e.g., an isolated atomic spectral line, the scattered spectrum will exhibit a number of lines displaced from the original wavelength. An example is shown in Fig. 17.16. This effect was first observed by C. V. Raman and K. S. Krishnan in 1928.

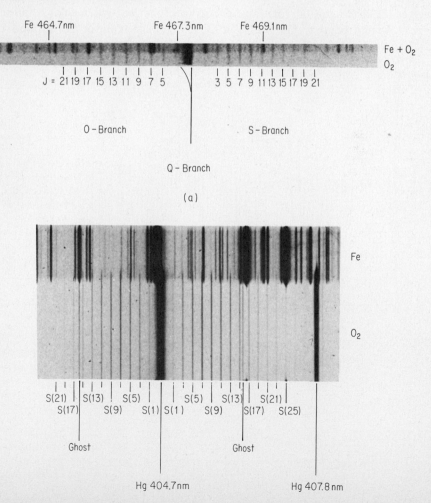

The origin of the Raman effect can be seen as follows. A quantum hv of incident light strikes a molecule. If it is scattered *elastically* its energy is not changed and the scattered light has the same frequency as the incident light. If it is scattered *inelastically*, it can give up energy to the molecule or take up energy from the molecule. This energy exchanged must naturally be in quanta hv', where $hv' = E_1 - E_2$ is the energy difference between two stationary states E_1 and E_2 of the molecule, for example, two definite vibrational energy levels. The frequency of the radiation that has undergone Raman scattering will therefore be

$$v'' = v \pm v'$$

The Raman frequency v' is completely independent of the frequency of the incident light v. We can observe pure rotational and vibration-rotational Raman spectra, which are the counterparts of the absorption spectra in the far and near infrared. The Raman spectra, however, are studied with light sources in the visible or ultraviolet. In many cases, the Raman and infrared spectra of a molecule complement each other, since vibrations and rotations that are not observable in the infrared may be active in the Raman. Thus, the spectrum of oxygen in Fig. 17.16 gives the spacing of the rotational energy levels in the fundamental rotation-vibration Raman band of O_2, although O_2 has no spectrum in the infrared because it has no dipole moment.

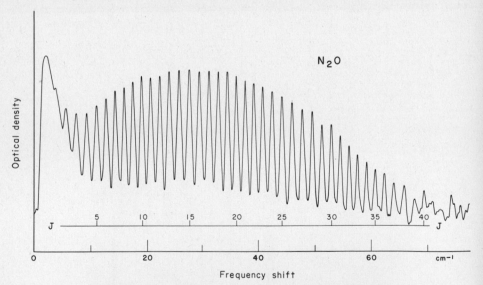

FIG. 17.17 Example of rotational Raman spectrum. A microphotometer curve of a photographic plate of one branch of the spectrum of N_2O. The rotational quantum numbers J are shown [Boris Stoicheff, National Research Council, Ottawa].

One should understand the distinction between *Raman scattering* and *fluorescence*. In both case, a light quantum is produced that has a frequency different from that of the incident quantum. In fluorescence, the system first absorbs the quantum $h\nu$ and then re-emits a quantum $h\nu''$; the incident light must therefore be at an absorption frequency. In Raman scattering, the incident light can be at any frequency.

Raman scattering occurs as a result of the *dipole moment induced in the molecule* by incident light. The induced dipole μ depends on the strength of the electric field F through Eq. (15.30),

$$\mu = \alpha F$$

where α is the polarizability. The transition probability for the Raman effect, from (17.17), therefore becomes

$$|\mu|^{nm} = |F| \int \psi_n^* \alpha \psi_m \, d\tau \tag{17.43}$$

If α is constant, this integral vanishes because of the orthogonality of the wave functions. In order that a vibration or rotation be active in the Raman, therefore, the polarizability must change during the rotation or vibration.

The polarizability changes during rotation of any nonspherical molecule, so that rotational Raman spectra can be obtained from most molecules, giving detailed data on rotational energy levels to supplement those from microwave spectra. Figure 17.17 shows the rotational Raman spectrum of N_2O extending out to values of the rotational quantum number of about $J = 40$.

17. Selection Rules for Raman Spectra

The induced dipole moment μ and the field F that induces it are both vectors, and the relation between them is

$$\mu = \tilde{\alpha} F \tag{17.44}$$

where the quantity $\tilde{\alpha}$ is a *tensor*. When both μ and F are represented in terms of their x, y, and z components, the relation (17.44) is written out as

$$\begin{aligned}
\mu_x &= \alpha_{xx} F_x + \alpha_{xy} F_y + \alpha_{xz} F_z \\
\mu_y &= \alpha_{yx} F_x + \alpha_{yy} F_y + \alpha_{yz} F_z \\
\mu_z &= \alpha_{zx} F_x + \alpha_{zy} F_y + \alpha_{zz} F_z
\end{aligned} \tag{17.45}$$

The nine components of the polarizability tensor are reduced to six by the requirement that $\alpha_{xy} = \alpha_{yx}$, $\alpha_{yz} = \alpha_{zy}$, $\alpha_{xz} = \alpha_{zx}$. The six remaining components, $(\alpha_{xx}, \alpha_{xy}, \alpha_{xz}, \alpha_{yy}, \alpha_{yz}, \alpha_{zz})$ determine the magnitude of the components of the induced dipole moment. It is evident from (17.43) that a given normal vibration can appear in the Raman spectrum only if at least one of the six components of the polarizability tensor differs from zero.

By analogy with the selection rule for infrared spectra, a Raman transition between two vibational levels v' and v'' is allowed only if the product $(\psi_{v'})(\psi_{v''})$ has the same symmetry species as at least one of the components of the polarizability tensor. It is easy to find the symmetry species of the polarizability components for nondegenerate species. Take, for example, the moment in the x direction induced by a field in the y direction,

$$\mu_x = \alpha_{xy} F_y$$

Since μ_x and F_y can either change sign or remain invariant under a symmetry operation, and μ_x behaves like a translation x, and F_y behaves like a translation y, it is evident that α_{xy} behaves like the product xy. This rule also applies to degenerate vibrations. Therefore, the symmetry species of the components α_{xy}, α_{xz}, etc., are obtained immediately as the products of pairs of x, y, etc.

Let us apply this rule to the two examples previously discussed in regard to infrared spectra. For point group C_{2v} (case of formaldehyde), the products are assigned to irreducible representations as follows:

$$xx \quad A_1 \qquad yy \quad A_1 \qquad zz \quad A_1$$
$$xy \quad A_2 \qquad yz \quad B_2$$
$$xz \quad B_1$$

Since at least one of these symmetry species matches each of the normal mode species on p. 780, we conclude that all the normal modes are active in the Raman spectrum in fundamental transitions. For the case of *trans*-$C_2H_4Cl_2$ (C_{2h}), we find the following assignments of the products:

$$xx \quad A_g \qquad yy \quad A_g \qquad zz \quad A_g$$
$$xy \quad A_g \qquad yz \quad B_g$$
$$xz \quad B_g$$

Now only the A_g and B_g vibrational species are active in the Raman, ($v_1, v_2, v_3,$ $v_4,$ and v_6). Note that the species active in the Raman are just those that are inactive in the infrared. It is, in fact, a general rule that for molecules with a center of symmetry (e.g., D_{2h} molecules), vibrations active in the infrared must be inactive in the Raman, and vice versa. The combination of infrared and Raman data in the analysis of the vibrational properties of molecules is therefore a technique of major importance.

18. Molecular Data from Spectroscopy

Table 17.6 is a collection of data derived from spectroscopic observations on a number of molecules.

TABLE 17.6
Spectroscopic Data on the Properties of Molecules[a]

Diatomic molecules				
Molecule	Equilibrium internuclear separation, r_e (nm)	Energy of dissociation, D_0 (eV)	Fundamental vibration σ (cm^{-1})	Moment of inertia (kg·m^2 × 10^{-47})
Cl_2	0.1989	2.481	564.9	114.8
CO	0.11284	9.144	2168	14.48
H_2	0.07414	4.777	4405	0.459
HD	0.07413	4.513	3817	0.611
D_2	0.07417	4.556	3119	0.918
HBr	0.1414	3.60	2650	3.30
HCl^{35}	0.1275	4.431	2989	2.71
I_2	0.2667	1.542	214.4	748
Li_2	0.2672	1.14	351.3	41.6
N_2	0.1095	7.384	2360	13.94
NaCl^{35}	0.251	4.25	380	145.3
NH	0.1038	3.4	3300	1.68
O_2	0.12076	5.082	1580	19.34
OH	0.0971	4.3	3728	1.48

Triatomic molecules								
Molecule A-B-C	Internuclear separation (nm)		Bond angle (deg)	Moments of inertia (kg·m^2 × 10^{-47})			Fundamental vibrations σ (cm^{-1})		
	r_{xy}	r_{yz}		I_A	I_B	I_C	σ_1	σ_2	σ_3
O=C=O	0.1162	0.1162	180		71.67		1320	668	2350
H—O—H	0.096	0.096	105	1.024	1.920	2.947	3652	1595	3756
D—O—D	0.096	0.096	105	1.790	3.812	5.752	2666	1179	2784
H—S—H	0.135	0.135	92	2.667	3.076	5.845	2611	1290	2684
O=S=O	0.140	0.140	120	12.3	73.2	85.5	1151	524	1361
N=N=O	0.115	0.123	180		66.9		1285	589	2224

[a]From G. Herzberg, *Molecular Spectra and Molecular Structure*, Vols. I and II (Princeton, N.J.: D. Van Nostrand Co., Inc., 1950).

19. Electronic Band Spectra

The energy differences ΔE between electronic states in a molecule are in general much larger than those between successive vibrational levels. Thus, the band spectra due to transitions between two different electronic states are observed in the visible or ultraviolet region. The ΔE between molecular electronic levels is usually of the same order of magnitude as that between atomic energy levels, ranging therefore from 1 to 10 eV.

Figure 17.18 represents the ground state of a molecule (curve *A*), and two distinctly different possibilities for an excited state. In one (curve *B*), there is a minimum in the potential-energy curve *B*, so that this represents a stable excited state for the molecule. In curve *C*, there is no minimum, and the state is unstable for all internuclear separations.

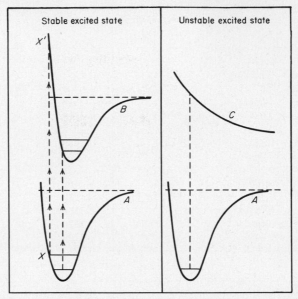

FIG. 17.18 Transitions between electronic levels in molecules. The representation of transitions by vertical lines is based on the Franck-Condon Principle, which indicates that the electronic transition probability is greatest when the change in internuclear distance in the transition is least.

A transition from ground state to unstable state would be followed immediately by dissociation of the molecule. Such a transition gives rise to a continuous absorption band. Transitions between different vibrational levels of two stable electronic states also lead to bands in the spectrum, but in this case the band can be analyzed into closely packed lines corresponding to the different upper and lower vibrational and rotational levels.

A rule known as the *Franck–Condon Principle* helps us to understand electronic transitions. An electron jump takes place much more quickly than the period of vibration of the atomic nuclei ($\sim 10^{-13}$ s), which are heavy and sluggish compared with electrons. Therefore, the positions and velocities of the nuclei are almost unchanged during an electronic transition.* Thus, we can represent a transition by a vertical line drawn between two potential-energy curves in Fig. 17.18. The

*It may be noted that the vertical line for an electronic transition is drawn from a point on the lower curve corresponding to the midpoint in the internuclear vibration. This is done because the maximum in ψ in the ground state lies at the midpoint of the vibration. This is not true in higher vibrational states, for which the maximum probability lies closer to the extremes of the vibration. Classical theory predicts a maximum probability at the extremes of the vibration.

Franck–Condon Principle shows how transitions between stable electronic states may sometimes lead to dissociation. For example, in curve A of Fig. 17.18, the transition XX' goes in the upper state to a vibrational level that lies above the asymptote to the potential-energy curve. Such a transition leads to dissociation of the molecule during the period of one vibration.

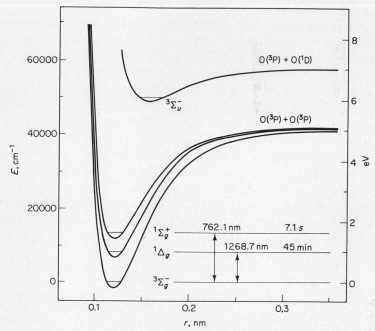

FIG. 17.19 Potential energy curves for the electronic ground state of O_2, two low-lying excited singlet states, and a higher stable excited triplet state. Vibrational dissociation of the three lower states would yield oxygen atoms in their ^3P ground states, but dissociation from the $^3\Sigma_u^-$ triplet would yield one ^3P oxygen atom and one oxygen atom in an excited state ^1D. [After Kasha and Khan.]

A good example of a system for which several important electronically excited states have been studied in detail is the oxygen molecule, as shown in Fig. 17.19. The ground state of O_2 is $^3\Sigma_g^-$, the molecular orbital representation of which was discussed in Section 15.12. There are two low-lying singlet states, $^1\Delta_g$ and $^1\Sigma_g^+$. Dissociation of O_2 from any of these states would yield two oxygen atoms in their ground states, ^3P. The lowest singlet states of molecular oxygen are especially interesting and important because they have such long half-lives (7s and 2700s) for radiative decay, as a consequence of the fact that transition from the singlet back to the triplet with emission of radiation is forbidden by a strong selection rule.

For many years it has been known that a weak red chemiluminescence at about 633 nm accompanies the reaction

$$H_2O_2 + OCl^- \longrightarrow O_2 + Cl^- + H_2O$$

Primarily as a result of work by Kasha and coworkers,* it was recognized that the chemiluminescence involved a bimolecular collision of two molecules of singlet O_2, with the emission of a quantum of red light,

$$(^1\Delta_g)(^1\Delta_g) \longrightarrow (^3\Sigma_g^-)(^3\Sigma_g^-) + h\nu \ (633.4 \text{ nm})$$

Singlet oxygen may play a role in various biological oxidations, in radiation effects on tissues, and in the formation of smog due to photo-oxidations of organic compounds in the atmosphere.

If a molecule dissociates from an excited electronic state, the fragments formed (atoms in the case of a diatomic molecule) are not always in their ground states. To obtain the energy of dissociation into atoms in the ground states, we must therefore subtract any excitation energies of the atoms. For example, in the ultraviolet absorption spectrum of oxygen there is a series of bands corresponding to transitions from the ground state to the $^3\Sigma_u^-$ excited state shown in Fig. 17.19. These bands converge to the onset of a continuum at 175.9 nm, equivalent to 7.05 eV. The dissociation yields a normal atom (^3P state) and an excited atom (^1D state). The atomic spectrum of oxygen shows that this ^1D state lies 1.97 eV above the ground state. Thus, the heat of dissociation of molecular oxygen into two normal atoms $[O_2 \rightarrow 2O \ (^3P)]$ is $7.05 - 1.97 = 5.08$ eV, or 490 kJ·mol^{-1}.

The vibrational and rotational fine structure of electronic bands in the visible and ultraviolet region of the spectrum can provide detailed information about the

FIG. 17.20   An example of an electronic band emission spectrum at high resolution. The spectrum shows the 0,0 band (i.e., the vibrational quantum number $v = 0$ in both lower and upper state) of the $^2\Sigma \longrightarrow {}^2\Sigma$ transition of the diatomic molecule SiN in the violet region of the spectrum. The numbering system refers to the total angular momentum quantum number J. As explained in Section 17.11, when $\Delta J = +1$ we have an R branch, and when $\Delta J = -1$, a P branch. In this spectrum, however, both the states are doublets, i.e., $2\mathscr{S} + 1 = 2$ and the spin $\mathscr{S} = 1/2$, arising from a single unpaired electron outside of closed shells. The nuclear rotational quantum number K combines with the spin quantum number $\mathscr{S}$ to yield $J = K \pm \mathscr{S}$. The case $J = K + 1/2$ gives the P_1 and R_1 branches and the case $J = K - 1/2$ gives the P_2 and R_2 branches. Thus each line is split into a doublet, a result called *spin doubling*. The *band gap* of approximately $4B$ at the origin ν_0 (where B is the rotational constant defined on p. 757) is an obvious feature of the spectrum and confirms the fact that transitions between $K' = 0$ and $K'' = 0$ are forbidden. The wavelengths at the top of the spectrum refer to standard thorium lines and are given in Ångstroms (1 Å = 0.1 nm). [This beautiful spectrum was kindly provided by Thomas Dunn of the University of Michigan.]

*The paper by M. Kasha and A. U. Khan, "The Physics, Chemistry and Biology of Singlet Molecular Oxygen" (Conference N.Y. Acad. Sci., Oct. 1969), provides a fascinating review of this subject.

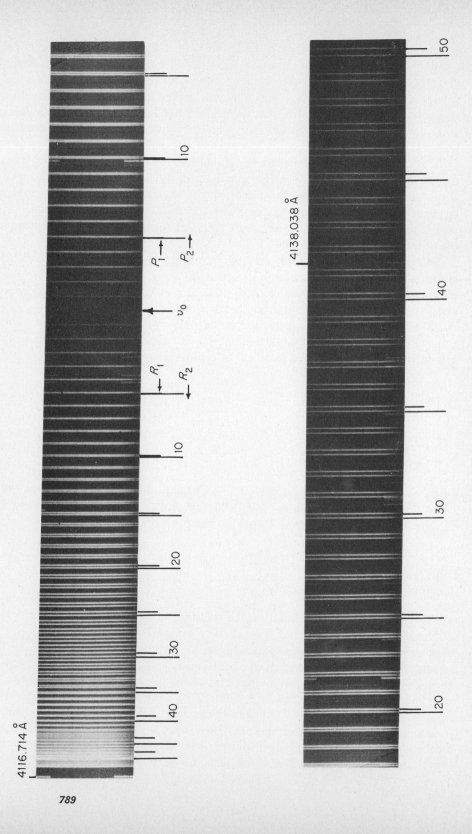

structure of the molecule in its ground state and in electronically excited states. These electronic spectra are more complex and difficult to analyze than infrared and Raman spectra. On the other hand, it is especially important that they be understood by chemists, since they determine the paths followed by photochemical reactions. Figure 17.20 shows one vibrational band in the electronic emission spectrum of silicon nitride (SiN) vapor. The individual rotational lines have been well resolved.

The range of wavelengths from the red end of the visible spectrum at 0.8 μm to the violet at 0.4 μm corresponds to a range of energies from 145 to 290 kJ·mol^{-1}. Somewhat more energy than this is necessary to excite an electron from a typical covalent bond in a small molecule. Most stable compounds of low molecular weight are therefore colorless, although many of them display an absorption in the near ultraviolet.

Molecules containing an unpaired electron (e.g., NO_2, ClO_2, triphenylmethyl) are usually colored. Chromophoric groups, such as $-NO_2$, $-C=O$, or $-N=N$, often cause a compound to be colored, since they contain electrons in π orbitals, and have excited orbitals at relatively low levels.

In an excited electronic state, a molecule may have a quite different shape from that in the ground state. For example, the dimensions of acetylene in the ground state and first excited state are as follows:

	Normal	*Excited*
C—C length	0.1208 nm	0.1385 nm
C—H length	0.1058 nm	0.1080 nm
C—C—H angle	180°	120°

In the ground state, H—C≡C—H is linear; in the excited state, it has a bent trans conformation, with longer bond distances. In this electronic excitation, an electron has been removed from a π orbital and placed in a higher σ orbital.

20. Reaction Paths of Electronically Excited Molecules

The interest of the chemist in an excited molecule often begins exactly where that of the physicist ends. Beyond spectroscopy lies the vast, complex, and vitally important field of photochemistry. Life as we know it depends upon the utilization of solar energy through processes of photosynthesis, but despite long and devoted efforts by many scientists, the mechanisms of photosynthesis are far from being well understood. Visual reception depends upon photochemical reactions of the visual pigments, again a field of intensive research that has so far yielded only a fragmentary understanding of the mechanisms involved. Quite apart from such biochemical applications, photochemistry is at present a subject pursued with enthusiasm by many organic and inorganic chemists, who have discovered that intricate syntheses and rearrangements can be produced by light, which would be difficult or impossible by ordinary thermal reactions. The physical chemist has been particularly interested by the theoretical study of how the energy of excitation of a molecule is redistributed among its internal degrees of freedom, both by intramolecular processes and by collisions or longer range interactions with other molecules. Photochemical kinetics must deal with an array of molecules in special excited states, free radicals and transient intermediates, that can make even the most complex kinetics of thermal reactions seem simple in comparison.

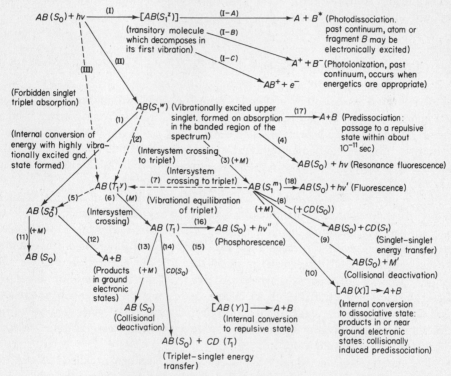

FIG. 17.21 The reaction paths of an electronically excited simple
molecule.

At the risk of dismaying the beginner, but to give some flavor of the complexity of the subject, we show in Fig. 17.21 a summary diagram to illustrate the possible reaction paths of electronically excited simple molecules.* Let us hasten to say that no single individual molecular species would take part in all these reactions; they represent, however, the gamut of possibilities that must be considered by the photochemist.

The molecule AB is assumed to be in a singlet ground state S_0. Absorption of a

*This diagram is taken from the comprehensive monograph *Photochemistry*, by J. G. Calvert and J. N. Pitts (New York: John Wiley & Sons, Inc., 1966), p. 196.

quantum $h\nu$ can yield an excited singlet S_1^z that decomposes in its first vibration along one of a number of possible paths. The absorption spectrum corresponding to $(S_1^z \leftarrow S_0)$ would therefore be diffuse. Alternatively, the absorption $(S_1^w \leftarrow S_0)$ might yield a vibrationally excited but stable singlet, corresponding to a banded region of the spectrum. A forbidden but in some cases still significant singlet to triplet absorption $(T_1^y \leftarrow S_0)$ might yield a molecule in an excited triplet state. Following the initial absorption step, various internal transfers, re-emissions of quanta, and chemical reactions might occur, as detailed in the summary diagram. In the subsequent sections, we shall consider some of these processes in more detail and give some illustrative examples, but first we shall present a brief historical account of basic principles of photochemistry.

21. Some Photochemical Principles

In 1818, Grotthuss and Draper stated a principle which we may call the *Principle of Photochemical Activation: Only the light that is absorbed by a substance is effective in producing a photochemical change.* In earlier days, the distinction between scattering processes and quantum transitions was not understood, so that this activation principle was helpful. It now appears to be an almost self-evident starting point for any discussion of photochemical reactions.

Stark in 1908 and Einstein in 1912 applied the concept of the quantum of energy to photochemical reactions of molecules. They stated a principle which we may call the *Principle of Quantum Activation: In the primary step of a photochemical process, one molecule is activated by one absorbed quantum of radiation.* It is essential to distinguish clearly between the primary step of light absorption and the subsequent processes of chemical reaction. An activated molecule does not necessarily undergo reaction; on the other hand, in some cases one activated molecule through a chain mechanism may cause the reaction of many other molecules. Thus the principle of quantum activation should *never* be interpreted to mean that one molecule *reacts* for each quantum absorbed.*

The energy $E = Lh\nu$, where L is the Avogadro Number, is called *one einstein*. The value of the einstein depends on the wavelength. For orange light with $\lambda = 0.6 \, \mu$m,

$$E = \frac{6.02 \times 10^{23} \times 6.62 \times 10^{-27} \times 3.0 \times 10^{10}}{0.6 \times 10^{-4} \times 10^7 \times 10^3} = 189 \text{ kJ}$$

This energy is enough to break certain rather weak covalent bonds, but for C—C bonds and others with energies higher than 300 kJ, radiation in the ultraviolet region is required.

*The validity of the Stark–Einstein principle depends upon the facts that the lifetimes of excited states are usually short and the intensity of illumination is usually quite low. With the development of lasers and other light sources of high intensity, it has become readily possible to observe primary photochemical processes in which more than one quantum is absorbed by a given molecule.

The *quantum yield* Φ of a photochemical reaction is the number of molecules of reactant consumed or product formed per quantum of light absorbed. When we consider in more detail the mechanisms of photochemical activation, we shall need to define other more specialized *quantum yields*.

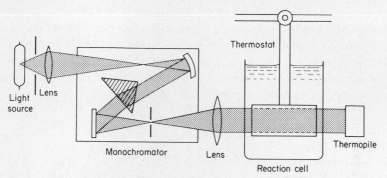

FIG. 17.22   Arrangement for photochemical investigations.

An experimental arrangement for a photochemical study is shown in Fig. 17.22. The light from the source passes through a monochromator, which yields a narrow band of wavelengths in the desired region. The monochromatic light passes through the reaction cell, and the part that is not absorbed strikes an actinometer such as a thermopile.

Consider, for example, an experiment on the photolysis of gaseous HI with light of 253.7 nm wavelength.

$$2HI \longrightarrow H_2 + I_2$$

It was found that absorption of 307 J of energy decomposed 1.30×10^{-3} mol of HI. The energy of the 253.7 nm quantum is $h\nu = (6.62 \times 10^{-34})(3.0 \times 10^{10}/ 2.537 \times 10^{-5}) = 7.84 \times 10^{-19}$ J. The HI has therefore absorbed $307/7.84 \times 10^{-19} = 3.92 \times 10^{20}$ quanta. One *einstein* is defined as $L = 6.02 \times 10^{23}$ quanta, so that $3.92 \times 10^{20}/6.02 \times 10^{23} = 6.22 \times 10^{-4}$ einstein were absorbed. The quantum yield Φ is the number of moles reacted per einstein abosorbed. Hence, $\Phi = 1.30/0.652 = 1.99$ for the photolysis of the HI.

Photosensitization occurs when a molecule that absorbs a quantum then reacts with a molecule of a different species. One of the most reproducible photochemical reactions is the decomposition of oxalic acid photosensitized by uranyl salts. The uranyl ion UO_2^{2+} absorbs radiation from 250 to 450 nm, becoming an excited ion $(UO_2^{2+})^{\star}$, which transfers its energy to a molecule of oxalic acid, causing it to decompose. This reaction has a quantum yield of 0.50.

$$UO_2^{2+} + h\nu \longrightarrow (UO_2^{2+})^{\star}$$

$$(UO_2^{2+})^{\star} + (COOH)_2 \longrightarrow UO_2^{2+} + CO_2 + CO + H_2O$$

The oxalic acid concentration is readily measured by titration with permanganate. A quartz vessel filled with the uranyl-oxalate mixture can be used instead of a thermopile to measure the number of quanta absorbed, from the oxalic acid decomposed and the known quantum yield.

22. Bipartition of Molecular Excitation

In most molecules, the absorption of a quantum leads to a transition from a singlet ground state to a singlet excited state. There will usually be a triplet state somewhat below this excited singlet. In the electronic excitation, one electron from an electron pair bond is excited to a higher state. If the excited electron has a spin antiparallel to that of its mate, the excited state is a singlet, but if the spin of the excited electron is parallel to that of its mate, the state is a triplet.

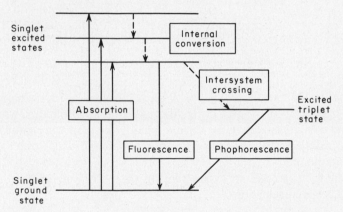

FIG. 17.23 The molecular energy levels concerned in photochemical processes are represented by a generalized *Jablonski diagram.*

A typical situation is shown in Fig. 17.23. We can interpret the initial processes in photochemical activation in terms of this pattern of states. Immediately after the primary quantum jump, a series of extremely rapid events takes place, *before* any photochemical reaction or emission of luminescent radiation can occur. First, there is *internal conversion:* no matter which upper singlet state has been reached in the primary quantum jump, there is often a rapid radiationless transfer of energy to the lowest excited singlet state. Second, there is *intersystem crossing:* there is a competition between the lowest excited singlet and triplet states, and some of the energy is transferred to the triplet by a radiationless process. *Each quantum of radiation absorbed by a normal polyatomic molecule has a probability of bipartition between the lowest excited singlet and triplet states of the molecule.* The basic model for excitation in a polyatomic molecule therefore involves three important states: the ground singlet, the first excited singlet, and the first excited triplet.

As shown in Fig. 17.21, the excited molecule formed in the primary step may re-emit a quantum $h\nu$ of either the same or a different frequency. This emission is either *fluorescence* or *phosphorescence*. According to the nomenclature introduced by G. N. Lewis, a radiative transition between two states of the same multiplicity (usually two singlets) is called *fluorescence*, and a radiative transition between two states of different multiplicity (usually triplet to singlet) is called *phosphorescence*.*

Table 17.7 gives some quantitative data on the spectroscopic properties of aromatic compounds dispersed in an alcohol–ether glass at 77 K. Under these conditions, deactivation by collisions with other molecules is prevented and the phosphorescent spectra are readily observed.

TABLE 17.7
Spectroscopic Properties of Aromatic Compounds
Dispersed in an Alcohol–Ether Glass at 77 K[a]

Compound	Lowest triplet σ (cm^{-1})	Lowest singlet σ (cm^{-1})	Lifetime of phosphorescence τ (s)	Quantum Yield	
				phospho-rescence Φ_p	fluores-cence Φ_f
Benzaldehyde	24950	26750	1.5×10^{-3}	0.49	0.00
Benzophenone	24250	26000	4.7×10^{-3}	0.74	0.00
Acetophenone	25750	27500	2.3×10^{-3}	0.62	0.00
Phenanthrene	21700	28900	3.3	0.14	0.12
Naphthalene	21250	31750	2.3	0.03	0.29
Biphenyl	23000	33500	3.1	0.17	0.21
Decadeuterobiphenyl	23100	33650	11.3	0.34	0.18

[a]See V. L. Ermolaev, *Soviet Phys.*, *80*, 333 (1963), for review and references.

23. Secondary Photochemical Processes—Fluorescence

Fluorescence is an *emission* of electromagnetic radiation. It should not be confused with the *scattering* of radiation, i.e., the *Rayleigh scattering* without change in wavelength, and the *Raman scattering* with change in wavelength. The natural lifetime of an excited state in a molecule undisturbed by collisions is usually about 10^{-8} s, but it can lie anywhere in the range from 10^{-9} to a few seconds. At a pressure of 1 atm, a molecule experiences about 100 collisions in 10^{-8} s. Consequently, excited molecules in most gaseous systems at ordinary pressures usually lose their energies by collision before they have a chance to fluoresce. The fluorescence is said to be *quenched*. In some such systems, fluorescence can be observed if the pressure is sufficiently reduced.

An example is the fluorescence of NO_2 excited by light of wavelength $\lambda = 436$ nm. The absorption spectrum of NO_2 is shown in Fig. 17.24(a). There are many

*The origin of phosphorescence in a metastable state below the lowest excited singlet level was first suggested by Jablonski in 1935.

sharp bands in the visible region of the spectrum, causing the deep brown color of the gas. Starting at about 370 nm in the ultraviolet, however, the bands become more diffuse, and by 330 nm only continuous nonbanded absorption is observed. Below 395 nm, the absorbed quanta would have sufficient energy to dissociate NO_2 by breaking a N—O bond,

$$NO_2 + h\nu \longrightarrow NO\,(^2\Pi) + O\,(^3P)$$

When excited at 436 nm, NO_2 displays a marked fluorescence, which is strongly dependent on pressure. In Fig. 17.24, the relative quantum yield for the fluorescence

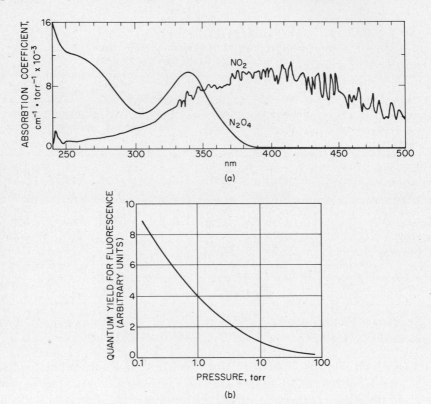

FIG. 17.24 (a) Absorption spectrum of nitrogen dioxide, NO_2, and dinitrogen tetroxide, N_2O_4, at 25°C corrected to pure compound spectra. [From T. C. Hall and F. E. Blacet, *J. Chem. Phys., 20,* 1945 (1952).] (b) Relative quantum yield for fluorescence of NO_2.

is plotted against the pressure, and the increased quenching at higher pressures is evident. Actually, this case is rather unusual since the lifetime of the excited state is 10^{-5} s, instead of about 10^{-8} s.

A kinetic expression for the quenching of fluorescence is obtained by considering the two parallel processes for an excited atom or molecule $M^\star$:

Fluorescence $\qquad M^\star \xrightarrow{k_1} M + h\nu$

Quenching $M^\star + Q \xrightarrow{k_2} M + Q + \text{kinetic energy}$

The total rate of deactivation is

$$\frac{-d[M^\star]}{dt} = k_1[M^\star] + k_2[M^\star][Q]$$

If the intensity of absorbed light is I_0, and the fluorescent intensity is I, the fraction of excited molecules that fluoresce is the *fluorescence yield* y_f,

$$y_f = \frac{I}{I_0} = \frac{k_1[M^\star]}{k_1[M^\star] + k_2[M^\star][Q]} = \frac{1}{1 + (k_2/k_1)[Q]} \qquad (17.46)$$

If k_1 is known from an independent determination of the lifetime τ of the excited state in the absence of quencher Q ($k_1 = \tau^{-1}$), we can evaluate k_2, the specific rate of the quenching process. It is usual to express the results in terms of a *quenching cross section* σ_Q. This is calculated from Eq. (4.55) for the number of binary collisions Z_{12} in a gas. It is the value that must be taken for the cross section $\pi d_{12}^2/4$ in order that the value of k_2 calculated by simple collision theory should exactly equal the experimental value. From (9.54), with $E = 0$ since there is no activation energy for the process of energy transfer, we therefore obtain

$$\sigma_Q = \frac{1}{4L}\left(\frac{\pi\mu}{8kT}\right)^{1/2} k_2$$

An example of the results obtained is shown in Table 17.8, which deals with the quenching of mercury resonance radiation ($^3P_1 \rightarrow {}^1S_0$) by added gases. The great effectiveness of hydrogen and some of the hydrocarbons is due to dissociative reactions, such as

$$\text{Hg}^\star + \text{H}_2 \longrightarrow \text{HgH} + \text{H}, \qquad \text{Hg}^\star + \text{RH} \longrightarrow \text{HgH} + \text{R}$$

TABLE 17.8
Effective Cross Sections for Quenching Mercury Phosphorescence

Gas	$10^{16}\,\sigma_Q - cm^2$	Gas	$10^{16}\,\sigma_Q - cm^2$
O_2	13.9	CO_2	2.48
H_2	6.07	PH_3	26.2
CO	4.07	CH_4	0.06
NH_3	2.94	$n\text{-}C_7H_{16}$	24.0

24. Secondary Photochemical Processes—Chain Reactions

If a molecule is dissociated as a consequence of absorbing a quantum of radiation, extensive secondary reactions may occur, since the fragments are often highly reactive atoms or radicals. Sometimes, also, the products of the primary fission process are in excited states, as so-called *hot atoms* or *hot radicals*.

For example, if a mixture of chlorine and hydrogen is exposed to light in the continuous region of the absorption spectrum of chlorine ($\lambda < 480$ nm), a rapid formation of hydrogen chloride ensues. The quantum yield Φ is 10^4 to 10^6. In 1918, Nernst explained the high value of Φ in terms of a long reaction chain. The first step is dissociation of the chlorine molecule,

$$(1) \qquad Cl_2 + hv \longrightarrow 2Cl, \qquad \Phi_1 I_a$$

which is followed by

$$(2) \qquad Cl + H_2 \longrightarrow HCl + H, \qquad k_2$$
$$(3) \qquad H + Cl_2 \longrightarrow HCl + Cl, \qquad k_3$$
$$(4) \qquad Cl \longrightarrow \tfrac{1}{2}Cl_2 \quad \text{(on wall)}, \qquad k_4$$

If we set up the steady state expression for [Cl] and [H] in the usual way, (Section 9.34), we obtain for the rate of HCl production,

$$\frac{d[HCl]}{dt} = k_2[Cl][H_2] + k_3[H][Cl_2]$$
$$= \frac{2k_2\Phi_1 I_a}{k_4}[H_2]$$

Instead of reaction (4), the chain ending step might be a recombination of chlorine atoms in the gas phase with cooperation of a third body M to carry away excess energy.

$$(5) \qquad Cl + Cl\,(+\,M) \longrightarrow Cl_2\,(+\,M), \qquad k_5$$

In this case, the calculated rate expression would be

$$\frac{d[HCl]}{dt} = k_2[H_2]\left[\frac{\Phi_1 I_a}{k_5[M]}\right]^{1/2}$$

It is likely that reactions (4) and (5) both contribute to the ending of the chain under most experimental conditions, since in experiments with pure H_2 and Cl_2 the rate depends on I_a^n, with n somewhere between $\tfrac{1}{2}$ and 1. The reaction is sensitive to traces of impurities, especially oxygen, which acts as an inhibitor by removing H atoms:

$$H + O_2\,(+\,M) \longrightarrow HO_2\,(+\,M)$$

In contrast with the high quantum yield of the $H_2 + Cl_2$ reaction are the low yields in the photochemical decompositions (*photolyses*) of alkyl iodides. These compounds have a region of continuous absorption in the near ultraviolet, which leads to a break into alkyl radicals and iodine atoms. For example,

$$CH_3I + hv \longrightarrow CH_3 + I$$

The quantum yield of the photolysis is only about 10^{-2}. The reason for the low Φ is that the most likely secondary reaction is a recombination,

$$CH_3 + I \longrightarrow CH_3I$$

Only a few radicals react with another molecule of alkyl iodide,

$$CH_3 + CH_3I \longrightarrow CH_4 + CH_2I$$

25. Flash Photolysis

The technique called *flash photolysis* is especially useful in the study of atoms and radicals that have only a short lifetime before reacting. A powerful flash of light, with energy up to 10^5 J and a duration of about 10^{-4} s, is obtained by discharging a bank of capacitors through an inert gas, such as argon or krypton. The reactants are in a vessel alined parallel with the lamp, and at the instant of the flash, an extensive photolysis occurs. The primary products of the photolysis, usually radicals and atoms, are produced at concentrations much higher than those in an experiment with continuous illumination at low intensities. A good method for following the subsequent reactions of the radicals is to make a continuous record of their absorption spectra.

The apparatus designed by Bair* for high resolution kinetic spectroscopy is shown in schematic form in Fig. 17.25. In the pioneering work of Norrish and

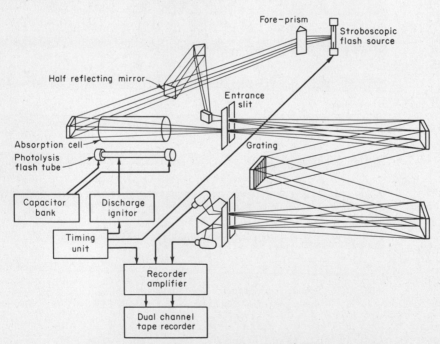

FIG. 17.25 High resolution kinetic spectroscopy by a multiple flash technique [E. J. Bair, Indiana University]. Following the flash photolysis, the concentrations of species in the reaction vessel are followed by dual-beam absorption spectroscopy at high spectral resolution and at an accurate sequence of time intervals as short as 200 μs.

*H. H. Kramer, M. H. Hanes, E. J. Bair, *J. Opt. Soc. Am.*, *51*, 775 (1961).

Porter,* the spectra were recorded photographically at intervals following the flash. The first flash was followed by a second flash—the *specflash*—triggered electronically at precise short intervals after the photoflash. The specflash was arranged to photograph the absorption spectra of the products at successive times measured in the range of microseconds to milliseconds. Each point in a reaction sequence thus required a separate experiment. An example of the data obtained is shown in Fig. 17.26 from the work of G. Porter and F. J. Wright† on the formation

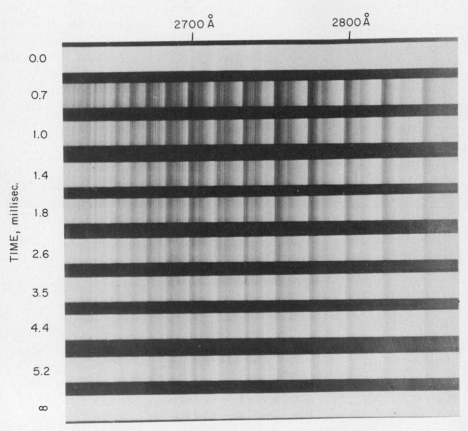

FIG. 17.26 Sequence of spectra of ClO after flash photolysis of a chlorine–oxygen mixture, showing bimolecular decay.

and decay of ClO in mixtures of Cl_2 and O_2. This system is completely reversible; only the chloromonoxy radical ClO is formed in the photoflash, and it reacts completely to $Cl_2 + O_2$ after the flash. The mechanism of formation of ClO is most likely through an intermediate complex ClOO, as

*R. G. W. Norrish and G. Porter, *Nature*, *164*, 685 (1950); G. Porter, *Proc. Roy. Soc. London*, *Ser. A*, *200*, 284 (1950); O. Oldenburg, *J. Chem. Phys.*, *3*, 266 (1935).
  †*Discussions Faraday Soc.*, *14*, 23 (1953).

$$Cl_2 + h\nu \longrightarrow 2Cl$$
$$Cl + O_2 \longrightarrow ClOO$$
$$ClOO + Cl \longrightarrow ClO + ClO$$

The decay of ClO followed ideal bimolecular kinetics,

$$\frac{-d[ClO]}{dt} = k_2[ClO]^2$$

with

$$k_2 = 4.8 \times 10^7 \text{ dm}^3 \cdot \text{mol}^{-1} \cdot \text{s}^{-1} \quad \text{at 298 K}$$

26.   Photolysis in Liquids

When a molecule in a gas absorbs a quantum and splits into two radicals, there is little chance that these particular radicals will recombine with each other. By the time they are separated by one mean free path, the chance that any particular pair will meet again is negligible. In a liquid, on the other hand, radicals formed in a dissociation reaction are surrounded by a cage of closely packed molecules. In many cases, the radicals will recombine within a time not much greater than 10^{-13} s, a typical vibration period. The radicals may diffuse away from each other but even then the chance is high that the same pair soon meet again and recombine. For example, Noyes[*] has estimated that for dissociation of iodine in hexane solution, the probability that the original pair recombines is about 0.5. This kind of recombination of a pair that came originally from the same parent molecule is called *geminate recombination*.

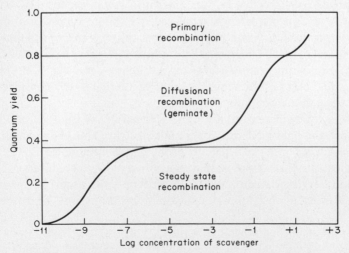

FIG. 17.27 Quantum yield for photolysis in liquid solution in the presence of a scavenger for radicals or atoms.

[*]R. Noyes, *J. Am. Chem. Soc.*, *18*, 999 (1950).

The course of the reaction can be influenced by the addition to the solution of a *scavenger* or radical trap, that is, a substance combining readily with free radicals. Scavengers frequently used have included iodine, vinyl monomers, and diphenyl-picrylhydrazyl (DPPH).

The typical mechanism for a photolysis in a liquid in the presence of a scavenger S can thus be written as follows:

(1) $AB + h\nu \longrightarrow (A + B)$	dissociation
(2) $(A + B) \longrightarrow AB$	geminate recombination
(3) $(A + B) \longrightarrow A + B$	separation by diffusion
(4) $A + S \longrightarrow AS$	scavenging of radicals
$\quad\;\; B + S \longrightarrow BS$	
(5) $A + B \longrightarrow AB$	recombination
(6) $A + X \longrightarrow$ product, $2A \longrightarrow A_2$	reaction of radicals
$\quad\;\; B + X \longrightarrow$ product, $2B \longrightarrow B_2$	

27.   Energy Transfer in Condensed Systems

In the gas phase, the distance between molecules is usually so large that an electronically excited molecule can transfer energy to another molecule only during the close approaches of intermolecular collisions. In a condensed phase, however, energy can be transferred between molecules over considerable distances by various special mechanisms.

The initial discovery of such long-range energy transfer was reported in 1924 in exploratory work by Jean Perrin on the depolarization of fluorescence. When a dye in solution was irradiated with polarized light, the fluorescent light emitted was also polarized provided the dye was present in low concentration, but as the concentration of dye in the solution increased, the extent of polarization of the fluorescent light decreased. Perrin concluded that the molecule of dye that emitted the quantum of depolarized fluorescent light was not the same as the molecule that had absorbed the exciting quantum of polarized light. A transfer of electronic excitation energy had occurred from one molecule to another over a considerable distance, up to 10 nm, in the solution.

In later work, two different solutes were included in the solution, and one was found to absorb the light while the other emitted light. This phenomenon, called *sensitized fluorescence*, provided definite proof of the intermolecular transfer of energy of excitation. A clear-cut example was observed in a solution of 1-chloro-anthracene and perylene, where most of the energy was absorbed by 1-chloro-anthracene, whereas almost all the fluorescent emission was characteristic of perylene.*

The theory of such long-range intermolecular transfers of energy was worked out principally by T. Förster from 1948 onward.† The transfer depends upon the

*E. J. Bowen and R. Livingston, *J. Am. Chem. Soc.*, **76**, 5300 (1954).
†T. Förster, *Radiation Res.*, *Suppl. 2*, 326 (1960).

overlap between the emission band of the donor and the absorption band of the acceptor. The excited donor interacts with the unexcited receptor through a dipole–dipole mechanism similar to that responsible for the London dispersion forces, to be discussed in Section 19.7. The interaction potential thus varies as r^{-6}, where r is the intermolecular distance. This theory indicates that the mean transfer time between pairs is in the range 10^{-11} to 10^{-8} s.

28. Photosynthesis in Plants

It now appears likely that the initial steps in the process of photosynthesis involve transfers of energy of the Perrin–Förster type, similar to those observed in sensitized fluorescence. Green plants, as well as certain protozoa and bacteria, can carry out photochemical reactions having as a net result the conversion of solar radiant energy to chemical free energy stored in the form of carbohydrates and other products. It has been estimated that the annual photosynthetic yield of organic matter over the surface of the earth is at least 10^{13} kg. The photosynthetic reaction is usually considered to consist of two stages, the reduction of water with liberation of oxygen,

$$H_2O \longrightarrow 2H\!-\! +\tfrac{1}{2}O_2$$

and the use of the activated hydrogen to reduce carbon dioxide,

$$2CO_2 + 4H\!-\! \longrightarrow 2(CH_2O)$$

It is not implied that the hydrogen ever exists as free atoms or that the carbohydrate unit (CH_2O) ever exists as free formaldehyde. It is known, however, from studies with $H_2^{18}O$, that all the O_2 liberated does come from the water.

The dominant pigment in green plants and algae is chlorophyll a, the structure of which is shown in Fig. 17.28(a). Chlorophyll b occurs at a much lower concentration. Many other pigments also absorb light in photosynthetic systems, but they appear to function as auxiliary "harvesters of light" and do not have roles in the photosynthetic reactions. Before the primary chemical step occurs in photosynthesis, the excitation energy must be transferred to a particular active site or radiation trap. This harvesting and concentration of the light quanta is shown in pictorial form in Fig. 17.28(b). The exact nature of the catalytic center that traps the quanta is not yet known: it may be chlorophyll a bound in a certain way to protein or lipid, it may be a particular triplet state of the chlorophyll, or it may even be an as yet unidentified derivative of the basic chlorophyll structure.

In any event, owing to the efficiency of the mechanism for harvesting the quanta of excitation energy, the overall efficiency of photosynthesis is astonishingly high. The reaction, $CO_2 + H_2O \rightarrow (CH_2O) + O_2$, is endothermic by about 4.8 eV, whereas a quantum of red light (680 nm) corresponds to an energy of about 1.8 eV. Thus at least 2.6 quanta would be required for each unit of reaction. Actually, the usual overall quantum yield appears to be about $\Phi = 0.125$, i.e., 8 quanta per unit of reaction.

The absorption spectra of chlorophyll a and chlorophyll b in ether solution

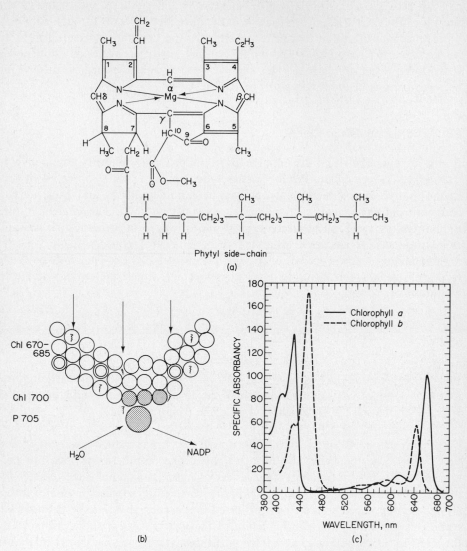

FIG. 17.28 Aspects of function of chlorophyll in photosynthesis. (a) Structure of chlorophyll *a*. In chlorophyll *b*, the CH_3 at position (3) is replaced by CHO. (b) A scheme for funneling of energy absorbed by chlorophyll (Chl 670–685) and other pigments (doubled circles) of plastids into sinks or traps (Chl 700–P705) involved in energy conversion necessary for electron transfer from H_2O to NADP. (c) Absorption spectra of chlorophylls *a* and *b*. [After F. P. Zscheile and C. L. Comar, *Botanical Gazette, 102*, 463 (1941).] Ordinate: specific absorption coefficients $\times 10^{-3}$; abscissa: wavelength (nm). To obtain molar absorption coefficients, multiply by 902.5 for chlorophyll *a*, by 907.5 for chlorophyll *b*.

are drawn in Fig. 17.28(c). The absorption bands centered at 680 nm for chlorophyll *a* and 644 nm for chlorophyll *b* are the ones of importance for photosynthesis. In the plant, all the chlorophylls are dispersed in special organelles called *chloroplasts*, and within the chloroplasts, the pigment molecules are closely associated with membranous structures called *grana*. An electron microscopic picture of a fixed section of chloroplast is shown in Fig. 17.29. It is believed that the regular array of chlorophyll molecules in the grana is conducive to intermolecular energy transfer. Thus the absorption of a quantum *hv* in any molecule is followed immediately by the intermolecular migration of the excitation energy, which shifts from molecule to molecule until it is caught by the active center previously mentioned. It is significant that the quantum yield for fluorescence from chlorophyll in the living cell is very low, but whatever fluorescence does occur is strongly depolarized. These observations lend strong support to the Förster mechanism as the method used to harvest the quanta of radiant energy.

Intensive research on photosynthesis is in progress, and the reader should consult current reviews for information on the details of utilization of the trapped excitation energy in the primary steps of the photosensitized reactions that release O_2 and accept hydrogen into the synthetic cycle.

Elucidation of the mechanism of the first steps in photosynthesis is one of the two greatest challenges to research workers in photochemistry. The other one is to discover how the visual receptor cells in the eye can produce an electric signal of 5 to 10 mV following absorption of a single quantum of light, a performance matched only by the most powerful manmade photomultiplier devices. We know that the primary photochemical step in vision is the absorption of a quantum by a visual pigment *rhodopsin*, which consists of the 11-cis form of vitamin A aldehyde bound to a membrane. The essential reaction is the conversion of the cis compound to the all-trans configuration. We do not know, however, how the aldehyde is bound to the membrane, or how changes in the membrane associated with the cis–trans isomerization lead to the output signal from the receptor cell.

Having briefly mentioned this problem in the hope that some reader may be inspired to solve it, we must now turn our attention to quite different kinds of spectroscopy. If a molecule is placed in a magnetic or electric field, some of the ordinary energy levels may be split, radiative transitions becoming possible between the newly separated levels. The most important spectra of this kind are those measured by the techniques of electron paramagnetic resonance (EPR) and nuclear magnetic resonance (NMR).

29. Magnetic Properties of Molecules

The theory of the magnetic properties of molecules resembles in many ways that for the electric polarization discussed in Section 15.16. Thus, a molecule can have a permanent magnetic moment and also a moment induced by a magnetic field. Magnetic phenomena are discussed in terms of two field vectors—**H** the magnetic field strength, and **B** the magnetic induction or magnetic flux density. When a mag-

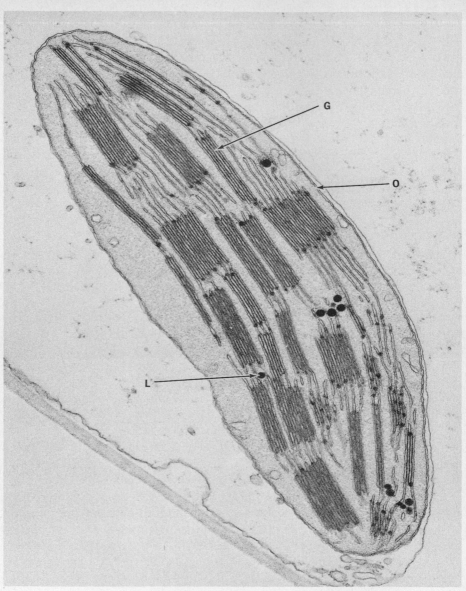

FIG. 17.29(a) A lettuce chloroplast visualized by osmium tetroxide staining and thin sectioning. The letter *G* marks a stack of grana membranes seen in cross section. These membranes contain chlorophyll, other pigments, and various proteins which function as electron carriers. The letter *O* marks the outer chloroplast membrane which functions in part to keep the enzymes needed for carbon dioxide fixation contained within the chloroplast. Nonfunctional lipid droplets are denoted by the letter *L*. The magnification is 33 150X. The electron micrograph was supplied by R. A. Dilley, C. J. Arntzen, and R. Fellows.

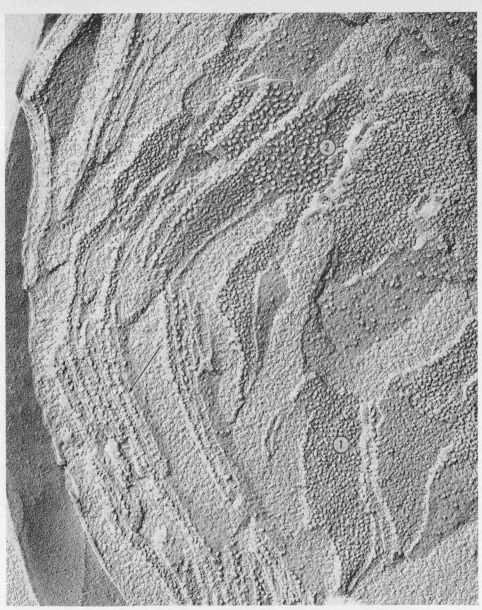

FIG. 17.29(b) A spinach chloroplast visualized by the freeze-etching technique. In this preparation, the carbon–platinum replica of the frozen-fractured chloroplast shows surface or tangential views of the interior of grana membranes (1 and 2). A barely visible cross section of several grana membranes is shown (arrow) and can be compared to those seen in Fig. 17.29(a). The particles, 1 and 2, found on the interior of grana membranes are probably part of the photochemical and/or enzymatic apparatus of the membrane. The magnification is 69 550X. [R. A. Dilley].

netic field acts upon a material system, the observable effects depend upon the magnetic flux density **B**. It is therefore incorrect to use **H** in such a context.* In the SI system, the unit of **B** is $kg \cdot s^{-2} \cdot A^{-1}$, called the *tesla* (T). It is equivalent to 10^4 gauss in electromagnetic units.

The magnetic analog of the electrical polarization **P** (Section 15.16) is the magnetization **M**, the magnetic moment per unit volume, and

$$\mathbf{M} = \frac{\mathbf{B}}{\mu_0} - \mathbf{H} \tag{17.48}$$

where μ_0 is the *permeability of vacuum*.

The permeability μ is the magnetic counterpart of the electrical *permittivity* ϵ, and

$$\mathbf{B} = \mu \mathbf{H}$$

Hence, (17.48) becomes

$$\mathbf{M} = \mathbf{H}\left[\frac{\mu}{\mu_0} - 1\right] \tag{17.49}$$

and the *magnetic susceptibility*,

$$\chi = \frac{\mathbf{M}}{\mathbf{H}} = \frac{\mu}{\mu_0} - 1 \tag{17.50}$$

The *molar susceptibility* would be

$$\chi_m = \frac{M}{\rho} \chi$$

where M is the molar mass and ρ the density.

The magnetic analog of (15.36) is

$$\chi_M = L\left(\alpha_m + \frac{p_m^2}{3kT}\right) \tag{17.51}$$

where α_m is the induced magnetic moment and p_m is the permanent magnetic dipole moment. As in the electrical case, the two contributions can be separated by measurements of the temperature dependence of χ_M. Pierre Curie showed in 1895 that for gases, solutions, and some crystals,

$$\chi_M = D + \frac{C}{T} \tag{17.52}$$

where D and C are constants. The theoretical analysis by Langevin in 1905 led to (17.51) and hence to the value $C = Lp_m^2/3k$ for the *Curie constant*.

An important difference from the electrical case now appears in that χ_M can be either negative or positive. If χ_M is negative, the medium is called *diamagnetic*. If χ_M is positive, it is called *paramagnetic*. Examples of these two types of magnetic behavior are shown in Fig. 17.30, with a representation of the paths of magnetic lines of force passing through the two kinds of substances. The magnetic field in

*A clear analysis of the vectors describing an electromagnetic field is given by A. Sommerfeld, *Lectures on Theoretical Physics*, Vol. 3, *Electrodynamics* (New York: Academic Press, 1953).

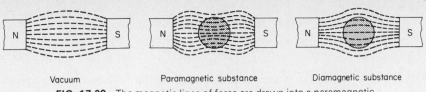

Vacuum   Paramagnetic substance   Diamagnetic substance

FIG. 17.30 The magnetic lines of force are drawn into a paramagnetic substance, so that the field in it is greater than in free space. The lines of force are pushed out of a diamagnetic substance, so that the field in it is less than in free space.

a diamagnetic medium is less than in a vacuum; in a paramagnetic medium, it is greater than in a vacuum.

An experimental measurement of susceptibility can be made with the *magnetic balance*. The specimen is suspended from a balance beam so that it is half inside and half outside the region between the poles of a strong magnet (about 0.5 T). When the magnet is turned on, a paramagnetic sample is drawn into the field region, whereas a diamagnetic sample is pushed out. The force necessary to restore the original balance point is

$$mg = \frac{(\chi_1 - \chi_2)}{2} AB^2 \tag{17.53}$$

where A is the cross-sectional area of the specimen, χ_1 its susceptibility, χ_2 the susceptibility of the surrounding atmosphere, and B the magnitude of the magnetic induction.

Diamagnetism is the counterpart of the electrical distortion polarization. The effect is exhibited by all substances and is independent of temperature. A simple interpretation is obtained if one imagines electrons moving about nuclei to be like currents in a wire. If a magnetic field is applied, the velocity of the electrons is changed, inducing a magnetic field, which in accord with Lenz's Law is opposed in direction to the applied field. The diamagnetic susceptibility is therefore always negative. In the case of an electric current in a wire, if the applied field is kept constant, the induced field quickly dies out owing to the resistance. Inside an atom, however, there is no resistance to the electronic current, so that the opposing field persists as long as the external magnetic field is maintained.

30. Paramagnetism

When paramagnetism occurs, the paramagnetic susceptibility χ is usually 10^3 to 10^4 times the diamagnetic susceptibility, so that the small diamagnetic effect is overwhelmed. Paramagnetism is related to the orbital angular momenta and spins of the electrons in a substance.

An electron revolving in an orbit is like an electric current in a loop of wire. The magnitude of the magnetic moment of a current loop is defined as $p_m = AI$ where A is the area of the loop and I is the current. If the current is due to an electron of charge $-e$ and mass m_e moving with velocity v, $I = -ev/2\pi r$, so that $p_m = -(ev/2\pi r)\pi r^2 = -evr/2$, where r is the radius of the loop. The angular momentum of the electron of mass m_e is $L = m_e vr$. Hence

$$\frac{\mathbf{P_m}}{\mathbf{L}} = \gamma = \frac{-e}{2m_e} \tag{17.54}$$

The ratio of magnetic moment to angular momentum is the magnetogyric *ratio* γ; since $\mathbf{p_m}$ and $\mathbf{L}$ are antiparallel vectors, directed along an axis normal to the plane of the current loop, we can take the ratio of their magnitudes as in (17.54).

The angular momentum can have only quantized values $\sqrt{l(l+1)}(\hbar)$. If the atom is in an external field, the frequency of precession of the angular momentum vector is also quantized, so that the component of $\mathbf{L}$ in the field direction is restricted to values $m_l \hbar$. The allowed values of the component in the field direction of the orbital magnetic moment of the electron are

$$p_{m,z} = -m_l \left(\frac{e\hbar}{2m_e} \right) = -m_l \mu_B \tag{17.55}$$

We see, therefore, that there is a natural unit of magnetic moment, $\mu_B = e\hbar/2m_e$. It is called the *Bohr magneton*. In SI units, its value is 9.2732×10^{-24} J·T^{-1} (equivalent to 9.2732×10^{-21} erg·G^{-1}).

As we have seen, the electron also has an intrinsic angular momentum or spin, which gives rise to a corresponding magnetic moment. A spinning electron is a little magnet. For the electron spin, however, $\gamma = -e/m_e$, twice the value for orbital motion. The spin angular momentum is quantized in units of $\sqrt{s(s+1)}(\hbar)$, where $s = \frac{1}{2}$. In an external field, the components of the spin angular momenturm are also quantized, and the allowed values in the direction of the field are $\pm\frac{1}{2}\hbar$. The corresponding magnetic moment in the field direction of one unpaired electron spin is $(e/m_e)(\frac{1}{2}\hbar) = e\hbar/2m_e$, one Bohr magneton. The quantization of the spin components is shown in Fig. 14.11.

In the case of the magnetic moment of molecules, only the contributions due to spin are important. Within a molecule there are strong internal electric fields, directed along the chemical bonds. In a diatomic molecule, for example, such a field is directed along the internuclear axis. This internal field holds the orbital angular momenta of the electrons in fixed orientations. They cannot line up with an external magnetic field, and thus the contribution they would otherwise make to susceptibility is ineffective. It is said to be *quenched*. Only the effect due to electron spin remains, and this is not influenced by the internal field. Thus a measurement of the permanent magnetic moment of a molecule can tell us how many unpaired spins are in its structure. The magnetic moment for n unpaired spins is $\sqrt{n(n+2)}$ Bohr magnetons.

31. Nuclear Properties and Molecular Structure

An experimental investigation of the structure of a molecule seeks first of all to learn how the nuclei of the atoms in the molecule are arranged in space. This information specifies the bond distances and bond angles in the structure. We want

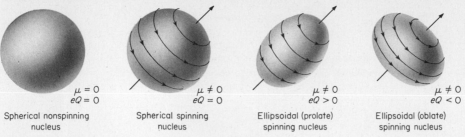

$$\begin{array}{cccc} \mu = 0 & \mu \neq 0 & \mu \neq 0 & \mu \neq 0 \\ eQ = 0 & eQ = 0 & eQ > 0 & eQ < 0 \end{array}$$

| Spherical nonspinning nucleus | Spherical spinning nucleus | Ellipsoidal (prolate) spinning nucleus | Ellipsoidal (oblate) spinning nucleus |

FIG. 17.31   Varieties of nuclei classified according to their magnetic dipole moments p_m and quadrupole moments eQ.

to know also how the electrons are distributed among the nuclei. The distribution of electronic charges tells us the nature of the bonding and should ultimately explain the chemical reactivity of the molecule. Measurements of dipole moments, magnetic susceptibility, optical spectra, and X-ray and electron diffraction yield information about electronic structure. These methods are all based on the interaction of the molecule with an external probe consisting of some kind of electromagnetic field. In most cases, however, these probes are not delicate enough to reveal the finer details of the electron distribution.

Since 1945, important breakthroughs have been achieved in the study of the electronic structure of molecules. The idea is *to use the nuclei themselves as probes* to reveal the electron distribution that surrounds them. Nuclei are not unfeeling point charges. They have definite properties of their own which make them sensitive to the electromagnetic environment in which they may be placed.

The significant properties of the nuclei are their *magnetic moments* and their *electric quadrupole moments*. A nucleus may possess an intrinsic *nuclear spin* and thus act as a little magnet with magnetic moment $\mathbf{p}_m$. Although a nucleus cannot have an intrinsic dipole moment, it may have a quadrupole moment eQ. If a nucleus has a quadrupole moment, the distribution of charge in the nucleus must depart from perfect sphericity. We can represent such nuclei as ellipsoids.

We thus classify the nuclei as shown in Fig. 17.31. Nuclei with quadrupole moments and/or magnetic moments different from zero can act as delicate probes that report on the electromagnetic field in their surroundings. The properties of a number of important nuclei are summarized in Table 17.9.

32.   Nuclear Paramagnetism

All nuclei with odd mass numbers have spins, designated by a quantum number *I*, the value of which is an odd multiple of $\frac{1}{2}$. Nuclei with even mass numbers have spins that are even multiples of $\frac{1}{2}$. Since the nuclei have positive charges, a nucleus with spin must also have a magnetic moment, which is always parallel to the spin angular momentum vector and has the magnitude

TABLE 17.9

Properties of Representative Nuclei[a]

Nuclide	Natural abundance (%)	Spin I ($h/2\pi$)	Magnetic moment p_m (nuclear magnetons)	Quadrupole moment ($e \times 10^{-24}$ cm²)	NMR frequency (MHz at field of 1 T)
^{1}H	99.9844	$\frac{1}{2}$	2.79270		42.577
^{2}H	0.0156	1	0.85738	2.77×10^{-3}	6.536
^{10}B	18.83	3	1.8006	1.11×10^{-2}	4.575
^{11}B	81.17	$\frac{3}{2}$	2.6880	3.55×10^{-2}	13.660
^{13}C	1.108	$\frac{1}{2}$	0.70216		10.705
^{14}N	99.635	1	0.40357	$2 \quad \times 10^{-2}$	3.076
^{15}N	0.365	$\frac{1}{2}$	-0.28304		4.315
^{17}O	0.07	$\frac{5}{2}$	-1.8930	-4.0×10^{-3}	5.772
^{19}F	100	$\frac{1}{2}$	2.6273		40.055
^{31}P	100	$\frac{1}{2}$	1.1305		17.235
^{33}S	0.74	$\frac{3}{2}$	0.64272	-6.4×10^{-2}	3.266
^{39}K	93.08	$\frac{3}{2}$	0.39094		1.987

[a]A complete table is given by J. A. Pople, W. G. Schneider and H. J. Bernstein, *High Resolution Nuclear Magnetic Resonance* (New York: McGraw-Hill Book Company, 1959), p. 480.

$$p_N = \gamma_N I h \tag{17.56}$$

where γ is the *nuclear magnetogyric ratio*.

It was first predicted that the magnetic moment of the proton would be one *nuclear magneton* μ_N, since it has spin $I = \frac{1}{2}$. If m_p is the mass of a proton,

$$\mu_N = \frac{eh}{2m_p} = 5.0493 \times 10^{-27} \text{ J} \cdot \text{T}^{-1}$$

Actually, however, the proton was found to have a magnetic moment of 2.7927 μ_N. Since m_p is almost 2000 times the electronic mass m_e, the nuclear magneton is about 2000 times less than the electronic or Bohr magneton, $\mu_B = eh/2m_e$. The magnitude of the magnetic moment of a nucleus is often written as

$$p_N = g_N \mu_N I \tag{17.57}$$

where g_N is called the *nuclear g factor*.

Nuclear magnetism was first revealed in the *hyperfine structure* of spectral lines. As an example, consider the hydrogen atom, a proton with one orbital electron. The nucleus can have a spin $M_I = \pm\frac{1}{2}$ and the electron can have a spin $m_s = \pm\frac{1}{2}$. The nuclear and electron spins can be either parallel or antiparallel to each other, and these two different alinements differ slightly in energy, the parallel state being higher. Thus, the ground state of the hydrogen atom is a closely spaced doublet, and this splitting can be observed in the atomic spectrum of hydrogen, if a spectrograph of high resolving power is employed. The spacing between the two levels, $\Delta E = h\nu$, corresponds to a frequency ν of 1402 MHz.

After the prediction of the astrophysicist van der Hulst, an intense emission of radiation at this frequency was observed from clouds of interstellar dust. The study of this phenomenon as well as other sources of 1420 MHz radiation in the universe form an important part of radioastronomy.

33.   Nuclear Magnetic Resonance

If a nucleus with spin I is placed in a magnetic field, the magnetic moment vector must precess about the field direction, and its component in the direction of the field is restricted to values

$$p_N = M_I g_N \mu_N \tag{17.58}$$

where $M_I = I, I - 1, I - 2, \ldots, -I$. Figure 17.32 shows an example for a

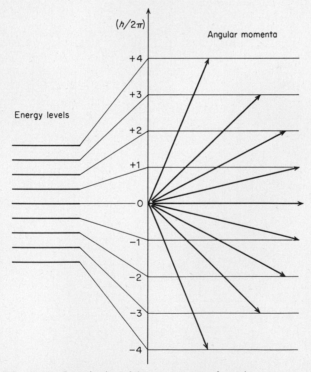

FIG. 17.32   Quantization of the components of angular momentum of nuclear spin for the case I = 4.

nucleus with $I = 4$ (e.g., ^{40}K). In the magnetic field, the states with different values of M_I have slightly different energies. The potential energy of the nucleus when its magnetic moment has a component p_N in the direction of a field $\mathbf{B}_0$ is

$$U_m = -p_N B_0 = -M_I g_N \mu_N B_0 \tag{17.59}$$

The energy spacing between two adjacent levels ($\Delta M_I = \pm 1$) is

$$\Delta E = g_N \mu_N B_0 \tag{17.60}$$

An an example, consider the case of the nucleus ^{19}F for which $g_N = 5.256$. In a field of 1 T,

$$\Delta E = (5.246)(5.0493 \times 10^{-27}) = 2.653 \times 10^{-26} \text{ J}$$

The frequency of radiation for this ΔE would in the shortwave region at

$$\nu = \frac{\Delta E}{h} = 19.85 \times 10^6 \text{ Hz} = 19.85 \text{ MHz}$$

In 1946, E. M. Purcell and F. Bloch independently developed the method of *nuclear magnetic resonance* to measure these transition frequencies. The principle of this method is shown in Fig. 17.33. The field B_0 of the magnet is variable from

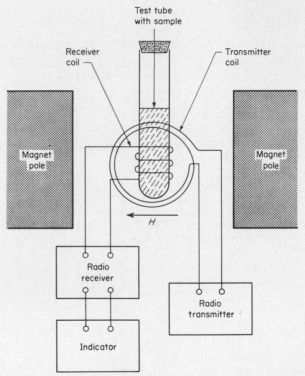

FIG. 17.33 Simplified apparatus for the basic nuclear-magnetic-resonance experiment [Varian Associates].

0 to 1.0 T.* This field produces an equidistant splitting of the nuclear energy levels that correspond to different values of M_I. The low power radiofrequency transmitter operates at, for example, 60 MHz. It causes a small oscillating electromagnetic field to be applied to the sample. This field induces transitions between the energy levels, by a resonance effect, when the frequency of the oscillating field equals the transition frequency ν. When such transitions occur in the sample, the resultant

*One often refers colloquially to the absolute magnitude B_0 of a field $\mathbf{B}_0$ as "the field B_0."

oscillation in the field induces a voltage oscillation in the receiver coil, which can be amplified and detected. In the instrument shown, the magnetic field of the large magnet is fixed and the radiofrequency of the oscillator is fixed; the resonance condition is achieved by superimposing a small variable *sweep field* on the field of the large magnet.

Figure 17.34 shows an oscillographic trace of the voltage variations over a small range of magnetic sweep field, $\pm 200 \ \mu$T from the fixed field of 0.7 T, with ethanol (C_2H_5OH) as the sample. Note that each different kind of proton in the molecule CH_3CH_2OH appears at a distinct value of B_0. The reason for this splitting is that the different protons in the molecule have slightly different magnetic environments, and hence slightly different resonant frequencies. The areas under the peaks are in the ratio 3:2:1, corresponding to the relative numbers of protons in the different environments. Each peak also has a fine structure.

When the frequency of the oscillatory field of the transmitting coil in Fig. 17.33 is the same as the natural precession frequency of the nuclear magnetic moment in the strong external field, energy is absorbed from the field. In other words, quanta of microwave radiation are absorbed as the nuclear magnetic quantum number M_I increases by one unit. To have continuous absorption of energy from the oscillating field, however, there must be some effective mechanism by which the nuclear magnet can lose this energy and return from the excited state to the ground state so as to be able to make another quantum jump with absorption of energy. The resonance effect measures the net absorption of energy (i.e., the difference between the energy absorbed in going from the lower to the upper state and the energy emitted in going from the upper to the lower state). Since at equilibrium there are more systems in the lower state (in accord with the Boltzmann factor $e^{-\Delta E/kT}$), there is a net absorption of energy.

The system can return to the lower state not only by emission of radiation, but also by various radiationless mechanisms, which are called *relaxation processes*. If it were not for these relaxation processes, nuclear resonance would be impossible in practice, since there would be no way to restore the thermal equilibrium which keeps the lower states more populated than the upper ones.

There are two kinds of relaxation to be distinguished. The relaxation that establishes the equilibrium value of the nuclear magnetization along the direction of the external field is called *longitudinal relaxation*. It follows a kinetic law that is first order, in that the rate of relaxation depends on the first power of the excess (over the equilibrium value) of the number of nuclei in the upper state. From (9.5),

$$N - N_e = (N - N_e)_0 \ e^{-k_1 t} = (N - N_e)_0 \ e^{-t/\tau_1}$$

where the reciprocal of the rate constant k_1 is called τ_1, the *longitudinal relaxation time*. This process is also called *spin–lattice relaxation*. It is due to the interaction of the nuclear spins with various fluctuating local fields in the matter surrounding the oriented nuclei. Addition of any paramagnetic substance to the solution can greatly shorten τ_1. Figure 17.35 shows the effect of added paramagnetic ions on the τ_1 of protons in water.

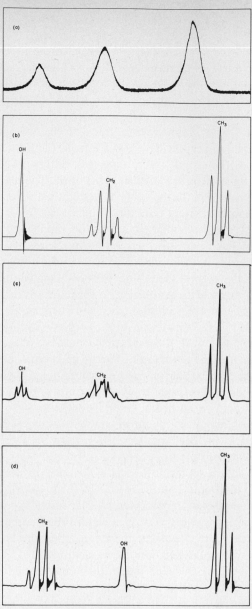

FIG. 17.34 Aspects of·the NMR spectrum of ethanol. (a) Spectrum at low resolution at 40 MHz. Peaks occur for the three nonequivalent protons in $CH_3 \cdot CH_2 \cdot OH$ and the peak areas are in the ratio $3 : 2 : 1$. (b) Spectrum of ethanol in presence of a trace of acid. The spin–spin splitting due to interaction between CH_2 and CH_3 protons is observed but the OH proton exchanges so rapidly between molecules that no splitting of its resonance peak is observable. (c) Spectrum at high resolution of carefully purified dry ethanol. The OH proton peak is now split into a triplet by interaction with the two adjacent CH_2 protons, and the quartet of resonances for the CH_2 protons has become a rather poorly resolved octet. (d) Spectrum of 6% C_2H_5OH in $CDCl_3$. The solvent is not an acceptor of H-bonds and hence the H-bonds that occur in pure ethanol are mostly destroyed. Consequently the chemical shift of the OH proton is greatly decreased. The H-bond acts to shield the proton from the external field. The OH peak now actually occurs downfield from the CH_2 peaks.

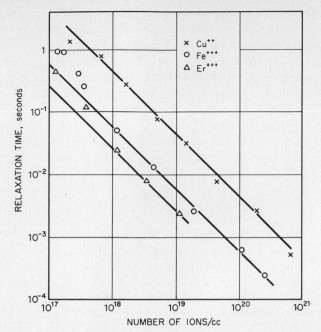

FIG. 17.35 Relaxation time τ_1 for protons in water that contains paramagnetic ions, measured at 29 MHz.

The other kind of relaxation process is called *transverse relaxation* (time τ_2). If the nuclei precessing about a field direction are in phase with one another, there will be a net component of magnetic moment in a plane XY normal to the axis Z of the magnetic field. Any disturbing field that tends to destroy this phase coherence will cause relaxation of the XY component of magnetic moment. One such process is *spin–spin relaxation*, in which a nucleus in a higher spin state transfers energy to a neighboring nucleus by exchanging spins with it.

Through the Heisenberg uncertainty principle of (13.45), the width ΔE of an emission line is related to the mean lifetime Δt of the excited state:

$$\Delta E \cdot \Delta t = \frac{\hbar}{2}$$

$$\Delta v = \frac{\Delta E}{h} = \frac{1}{4\pi \, \Delta t}$$

34. Chemical Shifts and Spin–Spin Splitting

The environment surrounding the nucleus has a small but definitely measurable effect on the field sensed by the nucleus. On this fact depends the usefulness of NMR in studying the structure of molecules and the nature of chemical bonds. The electrons surrounding a nucleus are acted upon by the external field to produce an induced diamagnetism which partially shields the nucleus. This shielding effect amounts to only about 10 parts per million of the external field, but the precision of NMR measurements is so great that the effect is readily measured to within 1%. The result is called the *chemical shift*. We saw in Fig. 17.34 an example in which the frequency of proton resonance in CH_3CH_2OH was somewhat different for each of the three different kinds of H atom in the structure. Because the chemical shift is due to the diamagnetism induced by the external field, its absolute magnitude depends on the strength of the field.

The chemical shift is expressed relative to that of some standard substance. In the case of proton resonance, the standard is usually water. We define the chemical shift parameter δ as

$$\delta = \frac{B_0(\text{sample}) - B_0(\text{reference})}{B_0(\text{reference})} \times 10^6 \qquad (17.61)$$

Some examples of δ for different kinds of H atoms in organic molecules are given in Table 17.10. The chemical shift parameter itself can be useful in determining

TABLE 17.10

Chemical Shifts of Proton Resonances in Parts Per Million Relative to Water as Zero[a]

Group	δ	Group	δ
$-SO_3H$	-6.7 ± 0.3	H_2O	(0.00)
$-CO_2H$	-6.4 ± 0.8	$-OCH_3$	$+1.6 \pm 0.3$
RCHO	-4.7 ± 0.3	$\equiv C-H$	$+2.4 \pm 0.4$
$RCONH_2$	-2.9	RSH	$+3.3$
ArOH	-2.3 ± 0.3	$-CH_2-$	$+3.5 \pm 0.5$
ArH	-1.9 ± 1.0	RNH_2	$+3.6 \pm 0.7$
$=CH_2$	-0.6 ± 0.7	$\equiv C-CH_3$	$+4.1 \pm 0.6$
ROH	-0.1 ± 0.7	R_2NH	$+4.4 \pm 0.1$

[a]After J. D. Roberts, *Nuclear Magnetic Resonance* (New York: McGraw-Hill Book Company, 1959). Data from H. S. Gutowsky, L. H. Meyer, and A. Saika, *J. Am. Chem. Soc.*, 75, 4567 (1953).

an unknown molecular structure. The proton NMR spectrum of a molecule provides us a sort of catalog of the different locations of hydrogen atoms in the structure.

When the NMR spectrum shown in Fig. 17.34(a) was studied at higher resolution, the peak for CH_2 was split into four lines and that for CH_3 into three lines. The spectrum obtained at high resolution was shown in Fig. 17.34(b). The effect which splits the CH_2 and CH_3 peaks into multiplets is not a chemical shift. This conclusion is proved by the fact that the observed splitting does not depend on the strength of the applied field. The effect is caused by the interaction of the nuclear spins (magnetic moments) of one set of equivalent protons with those of another set. It is therefore called *spin–spin splitting*.

Consider first the two protons A and B of the CH_2 group. How can the spins of protons A and B be arranged in relation to the spin of a given proton in the CH_3 group? There are four possible ways:

1. spin parallel to both A and B;
2. spin antiparallel to both A and B;
3. spin parallel to A, antiparallel to B;
4. spin antiparallel to A, parallel to B.

The last two arrangements clearly have the same energy. The result of the spin-spin interaction is that each proton in the CH_3 group can feel three slightly different magnetic fields, depending on its relation to the spins of the CH_2 protons. The resonance signal thus splits into three components, as shown in Fig. 17.36. In the same way, a proton in the CH_2 group feels slightly different fields depending on how its spin is related to those of the CH_3 group. There are four energetically different arrangements possible, as shown in Fig. 17.36.

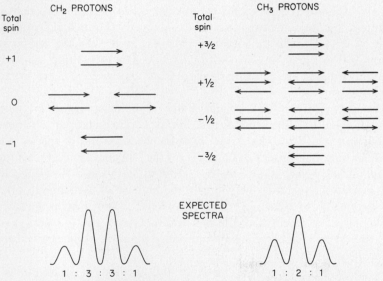

FIG. 17.36 Spin–spin splittings of nmr spectrum of $CH_3 \cdot CH_2OH$. Each proton in CH_2 "senses" four different spin arrangements of CH_3 and each proton of CH_3 "senses" three different spin arrangements of CH_2.

What will be the effects of the interaction of the CH_3 and CH_2 protons with the proton in the OH group? We should expect each of the lines shown in Fig. 17.34(b) to be split into a doublet, and the CH_2 should be split more than the CH_3 since the former is closer to the OH group. These predictions are confirmed by the experimental spectrum in Fig. 17.34(c). We also see the spectrum of the OH proton, which is split into a (121) triplet by interaction with the CH_2. If a trace of acid is added to the alcohol, a dramatic change occurs in the spectrum [Fig. 17.34(b)]. The OH triplet becomes a singlet and the splitting of CH_2 and CH_3 by the OH disappears. The H^+ acid acts as a catalyst for the rapid exchange of protons between the OH groups of different molecules. Thus, the lifetime of an OH in any given conformation becomes too short to permit the spin–spin interaction with other nuclei in the molecule.

35.   Chemical Exchange in NMR

One of the most interesting applications of NMR spectroscopy is in the measurement of the rates of certain fast reactions. This method can be applied only when the mean lifetime τ_R of the reactant is comparable in magnitude to one of the relaxation times τ_1 or τ_2 of the nucleus being monitored by NMR. The rates of a number of reactions, however, may fall into such a range, including conformation changes, internal rotations, and proton transfers.

As an example, consider the spectra of 4,4-dimethylaminopyrimidine shown in Fig. 17.37 at a series of different temperatures. Unless there is rapid rotation about the N—C bond to the ring, the two methyl groups are nonequivalent. At low temperatures, therefore, where rotation is inhibited, two distinct peaks occur, separated by 17.8 Hz, the difference in chemical shifts. As the temperature is raised from 228 to 240 K, the peaks broaden and then merge into a broad single peak. As the temperature is further raised from 240 to 265 K, the broad peak sharpens progressively.

The broadening of the peaks, which is due to the reaction that exchanges the nonequivalent sites, i.e., the internal rotation, is a consequence of the Heisenberg uncertainty principle. For a reaction $A \rightleftharpoons B$, the mean lifetime of species A is $\tau_R = k_1^{-1}$, where k_1 is the first-order rate constant for the exchange. Thus, the width of the NMR peak at half-height is $\Delta v = k_1/\pi$. Before using this relation to calculate k_1, we must subtract the natural line width in the absence of exchange to obtain the broadening due only to exchange. In the other limit, at higher exchange rates when the individual peaks have fused, the line width is $\Delta v = \frac{1}{4}(v_A - v_B)(\pi k_1)^{-1}$ where v_A and v_B are the individual frequencies of peaks A and B, as determined from the measurements at low temperatures.

Many applications have been made of this method. It is beginning to be used to study such problems as the interaction of substrate molecules with active centers of enzymes. To be really useful in this area, however, the sensitivity of NMR spectroscopy will need to be increased by about a factor of 10, so that biochemical compounds can be studied at the low concentrations of physiological interest.

36.   Electron Paramagnetic Resonance

The energy of a magnetic moment $\mathbf{p}_m$ in a magnetic field $\mathbf{B}_0$ is

$$E = -\mathbf{p}_m \cdot \mathbf{B}_0 \tag{17.62}$$

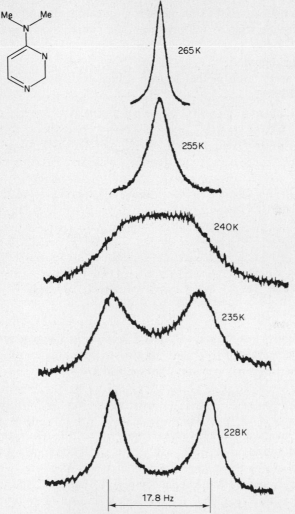

FIG. 17.37 Spectra of the methyl groups of 4,4-dimethylaminopyrimi-
dine at 100 MHz and various temperatures. This series
shows a sequence of spectra essentially of the form of
Fig. 17.34(a), as there is no appreciable coupling between
the methyl protons and any other nuclei. (We are grateful
to A. R. Katritzky and G. J. T. Tiddy for this spectrum.)
[From Ruth M. Lynden-Bell and Robin K. Harris, *Nuclear
Magnetic Resonance Spectroscopy* (London: Nelson,
1970).]

If the field is directed along the z axis of a system of Cartesian coordinates (the orientation always taken by convention), Eq. (17.62) becomes

$$E = -g\mu_B M_z B_0 \tag{17.63}$$

where M_z is the component of the spin angular momentum along the z axis. For the case of a single free electron, as we have seen, the simple theory gave $g = 2$, but actually $g = 2.00229$. For an unpaired electron in a molecule or crystal, g will differ somewhat from this figure.

The way the energy levels of the electron spin states are split by the magnetic field is shown in Fig. 17.38.

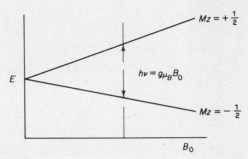

FIG. 17.38   Energies of electron spin states in a magnetic field as function of the flux density B_0.

If the system under investigation is placed in a cell in a magnetic field, we can determine $h\nu$ by techniques of resonance spectroscopy similar in principle to those used in NMR. The frequency per unit applied field is

$$\frac{\nu}{B_0} = \frac{g\mu_B}{h} = \frac{(2.00229)(9.273 \times 10^{-24})}{6.62 \times 10^{-34}} = 2.8 \times 10^{10} \text{ Hz} \cdot \text{T}^{-1}$$

Readily available magnets for EPR work range up to 5 T. Thus, the absorption frequencies would range up to 150×10^9 Hz ($= 150$ GHz). We therefore note that EPR spectra are in the microwave region of the spectrum, whereas NMR spectra are in the shortwave radiofrequency region. For EPR spectroscopy, it is customary to use a fixed frequency for the source and to vary the magnetic field until a resonant condition is met. The radiation sources are usually Klystron tubes, invented in 1939 by Russell, Varian, and Hansen, and since intensively developed for use in radar installations. The tubes have a limited range of frequency tuning and usually operate in the X band (8 to 12 GHz) or K band (27 to 35 GHz). Commercial EPR apparatus has been built around these microwave sources.

To yield an EPR spectrum, a substance must have one or more unpaired spins. The two principal applications of EPR have been in the study of free radicals in solution and paramagnetic ions in the crystalline state. Photochemical intermediates in the triplet state are often followed by EPR spectra. The method of *spin labeling*, whereby an organic molecule with an unpaired electron is attached to a membrane structure, may provide unusual data about the properties of membranes from analysis of the fine structure and line widths of the EPR spectrum of the molecule used as a probe.

PROBLEMS

1. *Scrapie* is a neurological disease of sheep that appears to be caused by an infectious agent smaller than a typical virus. By irradiation with nearly monochromatic ultraviolet light, the action spectrum for inactivation of the scrapie agent was measured.* The peak in this action spectrum does not coincide with that for inactivation of typical RNA or DNA viruses. Discuss the interpretation of this action spectrum and its significance for our understanding of the nature of the scrapie agent.

2. Sketch the absorption spectrum of a molecule such as CH_3Cl on the following scale. Above the scale, label the type of process that causes the absorption of radiation: (a) vibrational; (b) electronic; (c) rotational transitions; (d) ionization and molecular decomposition; (e) predissociation. Indicate below the scale the positions of the various spectral regions: (i) radiofrequency; (ii) visible; (iii) microwave; (iv) ultraviolet; (v) infrared.

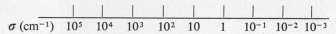

$$\sigma \ (cm^{-1}) \quad 10^5 \quad 10^4 \quad 10^3 \quad 10^2 \quad 10 \quad 1 \quad 10^{-1} \quad 10^{-2} \quad 10^{-3}$$

3. If the length of a molecule such as H—Cl were multiplied by a factor of 10 without changing its mass, how would its infrared and microwave spectrum be affected?

4. The frequency of the O—H stretching in CH_3OH is at $\sigma = 3300 \ cm^{-1}$. Predict σ for the O—D stretch in CH_3OD.

5. The bond length in BO is 0.1205 nm. Calculate the rotational contribution to the entropy of BO(g) at 298 K.

6. Calculate the wavelengths of the first eight lines in the pure rotational spectrum of ClF. The bond length is 0.1630 nm and the centrifugal distortion constant $C_0 = 10^{-4}B_0$. Draw the calculated spectrum.

7. Calculate the ratio of the number N_J of molecules in a sample of HI at (a) 300 K and (b) 1000 K having rotational quantum numbers $J = 5$ and $J = 0$ ($I = 4.31 \times 10^{-47} \ kg \cdot m^2$). (c) Calculate the values of J at the maximum N_J at these two temperatures.

8. The rotational spectrum of HF has lines 41.9 cm^{-1} apart. Calculate the moment of inertia and the bond length of this molecule.

9. When acetylene is irradiated with the 435.83 nm mercury line, a Raman line attributed to the symmetrical stretching vibrations is observed at 476.85 nm. Calculate the fundamental frequency for this vibration.

*R. Latarjet, *et al.*, *Nature*, *227*, 1341 (1970).

10. The following lines were observed in the vibration-rotational bands of the infrared spectrum of HI, in decreasing order of σ in cm^{-1}: 2288.5, 2277.5, 2266.1, 2254.3, 2242.2, 2216.7, 2203.6, 2190.0, 2176.0, 2161.7, 2147.5, 2132.7, 2117.7. Calculate (a) the rotational constants in the lowest and first excited vibrational states; (b) the moments of inertia in the two states; (c) the internuclear distances in the two states; (d) the origin of the fundamental band; and (e) the force constant of the H—I bond. (f) Plot the observed spectrum and the energy levels from which it arises.

11. A ^{39}K nucleus has a spin $I = \frac{3}{2}$ and a nuclear g factor of 0.2606. (a) Draw a diagram to show all possible orientations of the magnetic moment of a ^{39}K nucleus in a magnetic field. (b) Calculate the transition frequency from one of these orientations to an adjacent one in a field of 0.100 T.

12. Sketch the proton NMR spectrum of $CHCl_2$—CH_2Cl: (a) at low resolution, giving relative intensities of peaks; (b) at resolution adequate to show the spin–spin splitting, indicating relative intensities of peaks.

13. Sketch the proton NMR spectrum at high resolution from CH_3CHBr_2, indicating the origin of the peaks.

14. An NMR spectrometer operating at a frequency of 60 MHz gives proton spectra at a field of 1.4092 T. At what field would the ^{11}B spectrum be observed at 60 MHz? The nuclear spin of ^{11}B is $\frac{3}{2}$ and $g_B = 1.7920$.

15. The ion Ti^{3+} has one unpaired electron. Calculate the contribution of the spin of this electron (a) to the magnetic moment of the ion and (b) to the molar magnetic susceptibility of Ti^{3+} at 298 K.

16. A photographic film is exposed for 10^{-3} s to a 100-W incandescent light at a distance of 10 m. If 5% of the power is emitted as visible light to which film is sensitive (say, at 600 nm), estimate the number of silver atoms that will be produced in an AgBr grain 10 μm in diameter. Mention any assumptions used and their reasonableness.

17. The quantum yield for inactivation of the $T1$ bacteriophage at 260 nm is $\Phi = 3 \times 10^{-4}$. If the molar mass of this species is $M = 10^7$g, estimate the target cross section of the phage at which absorption of the quantum has a lethal effect.

18. The dissociation energy D_0 of H_2 is 4.46 eV and its zero-point energy $\frac{1}{2}h\nu_0$ is 0.26 eV. Estimate D_0 and $\frac{1}{2}h\nu_0$ for D_2 and T_2.

19. By means of the concept of hybrid bond orbitals, predict the structure of CH_3. Discuss the probable microwave and infrared spectra of this molecule.

20. In the microwave spectrum of $^{12}C^{16}O$, the $J = 0 \longrightarrow 1$ rotational transition has been measured at 115 271.204 MHz. Calculate the value of the moment of inertia and the internuclear separation r_e in CO. Discuss the factors limiting the accuracy of r_e as compared to that of the spectroscopic measurement.

21. Qualitatively, how will the molar absorption coefficient ϵ compare in infrared and visible regions? Discuss the physical factors that determine the differences.

22. Pictured are some normal modes of vibration of several molecules. Indicate which are active as infrared fundamentals.

	Molecule	*Mode*	

(a)   [F, H / C=C / H, F structure]   C=C stretch   "rocking"

(b)   [F, F / C=C / H, H structure]   C=C stretch   out-of-plane twist

(c) CH_3CF_3 torsion of C—C bond   C—C stretch

23. The following are two equivalent resonance structures of the singly ionized anion of a common type of organic dye molecule:

$$R \diagdown \overset{+}{N}{=}CH{-}CH{=}CH{-}CH{=}CH{-}N \diagup R$$
$$R \diagup \qquad\qquad\qquad\qquad\qquad \diagdown R$$

$$R \diagdown N{-}CH{=}CH{-}CH{=}CH{-}CH{=}\overset{+}{N} \diagup R$$
$$R \diagup \qquad\qquad\qquad\qquad\qquad \diagdown R$$

(R may be various organic groups: CH_3, for example). Note that the two structures differ only in the positions of the double bonds, much like the resonance forms of benzene. When two such structures exist, involving a series of conjugated double bonds, it is a good approximation to assume that the two π electrons associated with each double bond are free to move the length of the molecule in a field of constant potential energy but are prevented from leaving the molecule by an infinite potential energy barrier. This is similar to the quantum mechanical problem of a particle in a box where the potential energy is zero inside the box and rises to infinity at the walls. The length of the box here is a, the distance between the two nitrogen atoms. (a) Derive the formula $E_n = n^2h^2/8ma^2$ for the energy levels of an electron in such a box of length a (one dimensional), starting from either the Schrödinger equation or de Broglie's relation. (b) From the result of part (a), draw an energy level diagram showing the five lowest energy states with their respective quantum numbers n, and sketch the associated wave functions. (c) The six π electrons occupy the lowest energy states, subject to the Pauli Exclusion Principle, which requires that no two electrons have the same values of both n and the spin quantum number, s ($s = +\frac{1}{2}$ or $-\frac{1}{2}$). Indicate on your energy level diagram where the electrons would be. (d) Light absorption raises an electron from the highest filled energy level to the lowest unfilled level. Compute the energy difference between these levels and the wavelength in nanometres of the absorbed light. For this you need to know a, which is 0.84 nm. (e) How much will the wavelength increase for a dye containing one additional vinylene group, —CH=CH—, for which a is increased by about 0.14 nm?

24. The fluorine NMR spectrum of F_2BrC—CCl_2Br is shown at 40 MHz as a function of temperature. Explain qualitatively what is happening and correlate the lines observed at 193 K with the three rotational isomers of the molecule. Calculate an upper limit to the average rate of rotation about the C—C bond at 193 and 213 K.

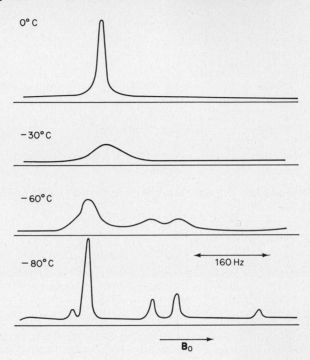

25. The following microwave absorption lines were observed for NaCl vapor at 800°C:

Frequency (MHz)	Rough relative intensity
26051.1	1.0
25847.6	0.6
25666.5	0.3_5
25493.9	0.3_0
25473.9	0.2_3
25307.5	0.1_8
25120.3	0.1_0

The probable error in these frequencies is ± 0.75 MHz (about ten times worse than usual in microwave spectroscopy, owing to difficulties associated with the high temperature). (a) Assign rotational (J) and vibrational (v) quantum numbers to these transitions. (b) Determine from these lines the values of B_v and r_v for each vibrational state, and derive B_e and r_e, for both isotopic molecules. Handy conversion factor: $B(\text{MHz}) = (h/8\pi^2 I) = (5.05531 \times 10^3)/I(\text{amu} \cdot \text{nm}^2)$. Atomic masses are as follows:

Na 22.997139 g·mol⁻¹, 100% abundance
³⁵Cl 34.97993 g·mol⁻¹, 75.4% abundance
³⁷Cl 36.97754 g·mol⁻¹, 24.6% abundance

26. The moment of inertia of a molecule of cyclo-octasulfur S_8, is 2314×10^{-40} g·cm² relative to the principal symmetry axis of the crown-configuration molecule. Suppose these molecules are rotating freely about this axis in a certain crystalline form of sulfur. Calculate the contribution of this free rotation to the entropy at 25°C. The molecule reaches equivalent orientations four times in one rotation.

27. The ground state of NO is a doublet ($g = 2$). Another doublet lies at 1490 J·mol⁻¹ above the ground state. (a) Calculate the electronic partition function of NO at 25°C. (b) Calculate the ratio of the number of molecules in the excited state to the number of molecules in the ground state.

28. The infrared spectrum of N_2O shows fundamental bands for three normal modes of vibration. Assuming the structure is linear, how does the spectrum distinguish between N—N—O and N—O—N? Sketch the normal modes.

29. Sketch the expected ³¹P NMR spectrum for HPF_2 for the following types of spin–spin coupling: (a) $J(P - F) > J(P - H)$; (b) $J(P - H) > J(P - F)$, where J is the spin–spin coupling between the designated nuclei.

30. Draw reasonable potential-energy curves for ground and excited states that on the basis of the Franck–Condon principle will provide a reasonable explanation of the following observations. (a) A molecule has a $v' = 0$ to $v'' = 0$ transition as the most intense band in its emission spectrum. (b) A molecule has a broad structureless absorption band but its emission spectrum is quite sharp and has fine structure.

31. It is found that 1.6 torr of N_2 reduces the fluorescent intensity of sodium resonance radiation at 200°C to half its value in the absence of added gas. If the natural lifetime of an excited Na atom is 10^{-7} s, calculate the quenching cross section of the N_2 molecule.

32. The kinetics of recombination of bromine atoms was measured following a flash photolysis of gaseous Br_2.* For Br atoms in argon, the rate followed the equation $d[Br_2]/dt = k_3[Br]^2[Ar]$. The values of k_3 were: 300 K, 3.7; 333 K, 2.7; 348 K, 2.3. All are in units of $10^9 \times$ dm⁶·mol⁻²·s⁻¹. Estimate the activation energy for the reaction and discuss possible interpretations of the value obtained.

*R. L. Strong, J. C. W. Chien, P. E. Graf, and J. E. Willard, *J. Chem. Phys.*, **26**, 1287 (1957).

18

The Solid State

Textbooks & Heaven only are Ideal
Solidity is an imperfect state.
Within the cracked and dislocated Real
Nonstoichiometric crystals dominate.
Stray Atoms sully and precipitate;
Strange holes, excitons, wander loose; because
of Dangling Bonds, a chemical Substrate
Corrodes and catalyzes—surface Flaws
Help Epitaxial Growth to fix adsorptive claws.

John Updike
(1968)*

It has been said that the beauty of crystals lies in the planeness of their faces. Crystallography began when relations between the plane faces were subjected to scientific measurement. In 1669, Niels Stensen, Professor of Anatomy at Copenhagen and Vicar Apostolic of the North, compared the interfacial angles in a collection of quartz crystals. An interfacial angle is defined as the angle between lines drawn normal to a pair of faces. He found that the corresponding angles in different crystals were always the same. After the invention of the contact goniometer in 1780, this conclusion was checked and extended to other substances, and the constancy of interfacial angles has been called the *first law of crystallography*.†

1. The Growth and Form of Crystals

A crystal grows from solution or melt by the deposition onto its faces of molecules or ions from the liquid. If molecules deposit preferentially on a certain face, that face will not extend rapidly in area, compared with faces at angles to it on which deposition is less frequent. The faces with the largest area are therefore those on which added molecules deposit most slowly.

---

*"The Dance of the Solids," from *Midpoint and Other Poems* (New York: Alfred A. Knopf, 1968), p. 20.
†An interesting history of crystallography is *Origins of the Science of Crystals* by J. G. Burke (Berkeley: Univ. of California Press, 1966).

Sometimes an altered rate of deposition can completely change the form, or *habit*, of a crystal. Sodium chloride grows from water solution as cubes, but from 15% aqueous urea as octahedra. It is believed that urea is preferentially adsorbed on the octahedral faces, preventing deposition of sodium and chloride ions, and therefore causing these faces to develop rapidly in area.

Some of the most ornate crystals are formed by *dendritic growth*. The snow crystals in Fig. 18.1 all display the symmetry of ice, but with many variations on the hexagonal theme. In such dendritic growth, a definite crystallographic axis

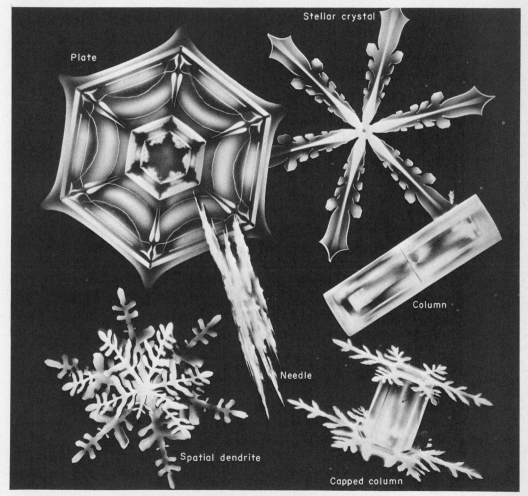

FIG. 18.1   Six typical habits of snow crystals. The different forms are characteristic of growth at different altitudes, the major determinant factor being the temperature. This drawing was made by H. Wimmer for an article "Wintry Art in Snow" by John A. Day, *Natural History, 71,* 24 (1962) [American Museum of Natural History, New York].

(a)

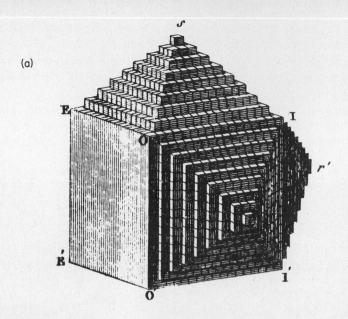

(b)

FIG. 18.2 (a) Model of a crystal structure proposed by René Haüy in *Traité élémentaire de Physique,* Vol. 1 [Paris: De L'Imprimerie de Delance et Leseur, 1803]; (b) a crystal of the rhombic type of tobacco necrosis virus in which the molecular order is unusually good. Magnification 42 000X [Ralph W. G. Wyckoff and L. W. Labaw, National Institutes of Health, Bethesda, Md.].

coincides with the axis of each growing branch; the tips are positions from which heat is most rapidly removed from the crystal.*

Already in 1665, Robert Hooke† had speculated on the reason for the regular forms of crystals, and decided that they were the consequence of a regular packing of small spherical particles.

> ... So I think, had I time and opportunity, I could make probable that all these regular Figures that are so conspicuously *various* and *curious*, and do so adorn and beautify such multitudes of bodies ... arise only from three ot four several positions or postures of *Globular* particles.... And this I have *ad oculum* demonstrated with a company of bullets, and some few other very simple bodies, so that there was not any regular Figure, which I have hitherto met withal ... that I could not with the composition of bullets or globules, and one or two other bodies, imitate, and even almost by shaking them together.

In 1784, René Just Haüy, Professor of Humanities at the University of Paris, proposed that the regular external form of crystals was the result of a regular internal arrangement of little cubes or polyhedra, which he called the *molécules intégrantes* of the substance. A model of a crystal structure drawn by Haüy is shown in Fig. 18.2(a). Figure 18.2(b) is an electron micrograph of a crystal of tobacco necrosis virus.‡ Thus, after 174 years, the Haüy model was confirmed by direct observation.

2. Crystal Planes and Directions

The faces of crystals, and also planes within crystals, can be characterized by means of a set of three noncoplanar axes. Consider in Fig. 18.3(a) three axes having lengths a, b, and c, which are cut by the plane ABC, making intercepts OA, OB and OC. If a, b, c are chosen as unit lengths, the lengths of the intercepts may be expressed as OA/a, OB/b, OC/c. The reciprocals of these lengths will then be a/OA, b/OB, c/OC. Now it has been established that it is always possible to find a set of axes on which the reciprocal intercepts of crystal faces are small whole numbers. Thus, if h, k, l are small integers,

$$\frac{a}{OA} = h, \qquad \frac{b}{OB} = k, \qquad \frac{c}{OC} = l$$

This is equivalent to the *Law of Rational Intercepts*, first enunciated by Haüy. The use of the reciprocal intercepts (hkl) as indices to define the crystal faces was proposed by W. H. Miller in 1839. If a face is parallel to an axis, the intercept is at ∞, and the Miller index becomes $1/\infty$ or 0. The notation is also applicable to planes drawn within the crystal. As an illustration of the Miller indices, some of the planes in a cubic crystal are shown in Fig. 18.3(b).

*Bruce Chalmers, in *Growth and Perfection of Crystals* (New York: John Wiley & Sons, Inc., 1958), pp. 291–302.

†*Micrographia, or Some physiological Descriptions of Minute Bodies made by Magnifying glasses with observations and Inquiries thereupon* (London: Jo. Martyn and Ja. Allestry, 1665).

‡*J. Ultrastruct. Res.*, 2, 8 (1958).

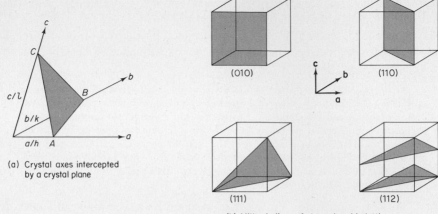

(a) Crystal axes intercepted by a crystal plane

(b) Miller indices of planes in cubic lattice

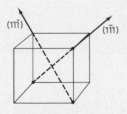

(c) Example of notation for directions in a crystal structure

FIG. 18.3   Planes in crystals are designated by the reciprocals of their intercepts with properly chosen crystal axes. Directions in crystals are designated by the components of the direction vector along the crystal axes. (a) Crystal axes intercepted by a crystal plane. (b) Miller indices of planes in cubic lattice. (c) Example of notation for directions in a crystal structure.

   In the Miller index for a crystal face, only the ratio $h: k: l$ is significant. Thus (420) would be the same face as (210). For planes within crystals, multiplication of the Miller index by an integer would change the interplanar spacing. Thus the *planes* 420 would include all the planes 210, and in addition a set of planes midway between them. According to current crystallographic notation, (hkl) refers to a crystal face and hkl (without parentheses) to a set of planes. Curly brackets are used to designate *all the equivalent faces* of a crystal, i.e, a *form* of a crystal. For example, we would say that cubic sodium chloride has the {100} form.
   The direction of a line in a crystal is denoted with square brackets by [uvw]. We place the origin of coordinates at one point on the line and then [uvw] is the direction from the origin to a point on the line ($u\mathbf{a} + v\mathbf{b} + w\mathbf{c}$) where $\mathbf{a}$, $\mathbf{b}$, and $\mathbf{c}$

are the crystallographic axes. The directions $[\bar{1}11]$ and $[1\bar{1}1]$ are shown in a cubic crystal in Fig. 18.3(c).

3. Crystal Systems

According to the set of axes used to represent their faces, crystals may be divided into seven systems, summarized in Table 18.1. They range from the completely general set of three unequal axes (a, b, c) at three unequal angles (α, β, γ) of the triclinic system, to the highly symmetrical set of three equal axes at right angles of the cubic system.

TABLE 18.1
The Seven Crystal Systems

System	Axes	Angles	Example
Cubic	$a = b = c$	$\alpha = \beta = \gamma = 90°$	Rock salt
Tetragonal	$a = b, c$	$\alpha = \beta = \gamma = 90°$	White tin
Orthorhombic	$a; b; c$	$\alpha = \beta = \gamma = 90°$	Rhombic sulfur
Monoclinic	$a; b; c$	$\alpha = \gamma = 90°; \beta$	Monoclinic sulfur
Rhombohedral	$a = b = c$	$\alpha = \beta = \gamma$	Calcite
Hexagonal	$a = b; c$	$\alpha = \beta = 90°; \gamma = 120°$	Graphite
Triclinic	$a; b; c$	$\alpha; \beta; \gamma$	Potassium dichromate

4. Lattices and Crystal Structures

Instead of considering, as Haüy did, a crystal that is made of elementary material units, it is helpful to introduce a geometrical idealization, consisting only of a regular array of points in space, which is called a *lattice*. An example in two dimensions is shown in Fig. 18.4.

The lattice points can be connected by a regular network of lines in various ways. Thus the lattice is broken up into a number of *unit cells*. Some examples are shown in the figure. Each cell requires two vectors, **a** and **b**, for its description.

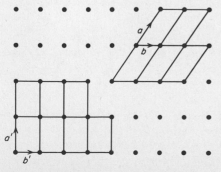

FIG. 18.4 Two-dimensional lattice with unit cells.

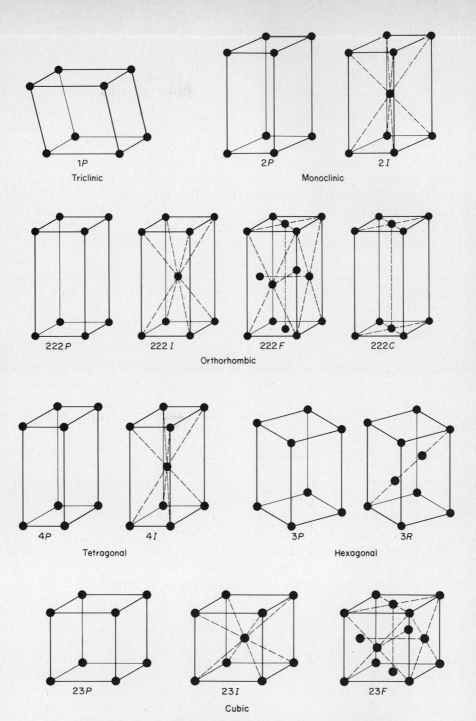

1P
Triclinic

2P   2I
Monoclinic

222P   222I   222F   222C
Orthorhombic

4P   4I   3P   3R
Tetragonal   Hexagonal

23P   23I   23F
Cubic

FIG. 18.5 The fourteen Bravais lattices. The symbols that designate a lattice include numbers denoting symmetry axes and letters P (primitive), I (body centered), F (face centered), C (end centered), R (rhombohedral).

A three-dimensional *space lattice* can be similarly divided into unit cells described by three vectors.

If each point in a space lattice is replaced by an identical atom or group of atoms there is obtained a *crystal structure*. The lattice is an array of points; in the crystal structure each point is replaced by a material unit. The positions of atoms in unit cells are denoted by coordinates which are fractions of the unit cell dimensions. For example, if a unit cell has sides a, b, c, an atom at $(\frac{1}{2}, \frac{1}{4}, \frac{1}{2})$ would be at $(a/2, b/4, c/2)$ relative to the origin $(0, 0, 0)$ at a corner of the cell.

In 1848, A. Bravais showed that all possible space lattices could be assigned to one of only 14 types.* The 14 Bravais lattices are shown in Fig. 18.5. They give the allowed different translational relations between points in an infinitely extended regular three-dimensional array. The choice of the 14 lattices is somewhat arbitrary, since in certain cases alternative descriptions are possible.

5. Symmetry Properties

When one studies an actual crystal, some of the faces may be so poorly developed that it is difficult or impossible to see its full symmetry. It becomes necessary to consider an ideal crystal in which all the faces of the same kind are developed to the same extent. The symmetry of a crystal is evident not only in the development of its faces, however, but also in all its physical properties, e.g., electrical and thermal conductivity, piezoelectric effect, and refractive index.

Closer consideration of the crystal systems or the Bravais lattices reveals a curious and important fact. The systems allow different symmetry elements in crystals, but the only kinds of axis that occur are C_2, C_3, C_4, or C_6. We never see a crystal with a C_5, C_7, or a C_8, or indeed any other axial symmetry except the four already mentioned. As we saw in Chapter 16, fivefold, sevenfold, and other axes can occur perfectly well in isolated molecules. Ferrocene, for example, is a well known molecule with a C_5 symmetry axis. Why do C_5 axes occur in molecules but not in crystals? The reason for this restriction is that it is impossible to fill all of space with figures of fivefold symmetry. We can see this result even in two dimensions. It is possible to tile a floor with parallelograms $[C_2]$, equilateral triangles $[C_3]$, squares $[C_4]$, or regular hexagons $[C_6]$. It is impossible to tile a floor with regular pentagons, heptagons, etc., without leaving gaps in the tiling. The fact that actual crystals never display any axial symmetry elements other than C_2, C_3, C_4, C_6, therefore leads to the following conclusion: Crystals must be constructed of regular subunits that fill all of space in a definite geometric array. The regular crystal forms observed in nature are outward manifestations of inner regularities in structure. The significance of the 14 Bravais lattices is now more evident—they are the 14 possible ways of arranging points in a regular array so as to fill all space.

---

*A lattice that contains body-, face-, or end-centered points can always be reduced to one that does not (*primitive lattice*). Thus the face-centered cubic can be reduced to a primitive rhombohedral. The centered lattices are chosen when possible because of their higher symmetry.

The symmetry elements in crystals are therefore restricted to the following: m, C_2, C_3, C_4, C_6, i, E. The crystallographic point groups are restricted to groups that can be formed from these symmetry elements. There are exactly 32 crystallographic point groups, which specify the 32 *crystal classes*. They are summarized in Table 18.2.

TABLE 18.2

Crystal Classes and Systems

System	Crystal classes[†]						
Triclinic	$C_1 = 1$	$C_s = \bar{1}$					
Monoclinic	$C_2 = 2$	$C_s = m = \bar{2}$	$C_{2h} = 2/m$				
Orthorhombic				$C_{2v} = 2mm$	$D_2 = 222$	$D_{2h} = mmm$	
Rhombohedral	$C_3 = 3$	$C_{3i} = \bar{3}$		$C_{3v} = 3mm$	$D_3 = 32$	$C_{3d} = \bar{3}m$	
Tetragonal	$C_4 = 4$	$S_4 = \bar{4}$	$C_{4h} = 4/m$	$D_{2d} = \bar{4}2m$	$C_{4v} = 4mm$	$D_4 = 42$	$D_{4h} = 4/mmm$
Hexagonal	$C_6 = 6$	$C_{3h} = \bar{6}$	$C_{6h} = 6/m$	$D_{3h} = \bar{6}m$	$C_{6v} = 6mm$	$D_6 = 62$	$D_{6h} = 6/mmm$
Cubic	$T = 23$		$T_h = m3$	$T_d = \bar{4}3m$		$O = 43$	$O_h = m3m$

[†]Crystal classes are given in Schoenflies and International (Hermann-Mauguin) notations. The latter notation gives the set of symmetry elements which determine the point group. The symbol for a crystal class is derived by writing down the symmetry elements associated with three important directions in the crystal.

Monoclinic: the axis normal to the other two.

Orthorhombic: the three axes a, b, c.

Tetragonal: the c axis; the a axis; an axis $\perp c$ and 45° to a.

Hexagonal: the c axis, the a axis $\perp c$, and an axis $\perp c$ and 30° to a.

Rhombohedral: the body diagonal and an axis normal to it. (Or principal and secondary axes of hexagonal system.)

Cubic: the directions [001], [111], and [110].

When a mirror plane is normal to a rotation axis it is written as the denominator of a fraction. For example: $4/mmm$ is the holohedral class of the tetragonal system having a 4 along the c axis with an m normal to it, and m's normal to each the other two standard directions.

Although the Schoenflies symbols introduced in Chapter 16 are still used, crystallographers prefer new international symbols introduced by Hermann and Mauguin,* and these symbols are included in Table 18.2.

All crystals necessarily fall into one of the *seven systems*, but there are several *classes* in each system. Only one of these, called the *holohedral class*, possesses the complete symmetry of the system. For example, consider two crystals belonging to the cubic system, rock salt (NaCl) and iron pyrites (FeS_2). Crystalline rock salt, Fig. 18.6, possesses the full symmetry of the cube: three fourfold axes, four threefold axes, six twofold axes, three mirror planes perpendicular to the fourfold

*W. F. de Jong, *General Crystallography* (San Francisco: W. H. Freeman, Inc., 1959), pp. 8–21. The International symbols for the symmetry elements of crystals are as follows:

Identity element (no symmetry)	1	Fourfold rotation axis (rotor)	4
Mirror plane of symmetry	m	Fourfold rotatory inverter	$\bar{4}$
Twofold rotation axis (rotor)	2	Sixfold rotation axis (rotor)	6
Threefold rotation axis (rotor)	3	Sixfold rotatory inverter	$\bar{6}$
Threefold rotatory inverter	$\bar{3}$	Center of symmetry (inverter)	$\bar{1}$

axes, six mirror planes perpendicular to the twofold axes, and a center of inversion. The crystal belongs to point group **O$_h$**. The cubic crystals of pyrites might at first

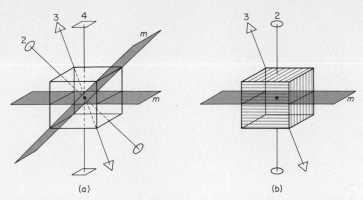

FIG. 18.6 Symmetry elements of two cubic crystals: (a) rock salt **O$_h$**; (b) pyrites, **T$_h$**.

seem to possess all these symmetry elements too. Closer examination reveals, however, that the pyrites crystals have characteristic striations on their faces, as shown in the picture, so that all the faces are not equivalent. These crystals therefore do not possess the six twofold axes with the six planes normal to them, the fourfold axes have been reduced to twofold axes, and the 3 have become $\bar{3}$. The point group is **T$_h$**.

In other cases, such departures from full symmetry are only revealed, as far as external appearance goes, by the orientation of etch figures formed on the surfaces. Sometimes the phenomenon of pyroelectricity provides a useful symmetry test. When a crystal with no center of symmetry is heated, a difference in potential develops between its faces. This can be observed by the resultant electrostatic attraction between individual crystals.

All these differences in symmetry are caused by the fact that the full symmetry of the point lattice has been modified in the crystal structure, as a result of replacing the geometrical points by groups of atoms. Since these groups need not have so high a symmetry as the original lattice, classes of lower than holohedral symmetry can arise within each system.

6. Space Groups

The crystal classes are the various groups of symmetry operations of finite figures, e.g., actual crystals. They are made up of operations by symmetry elements that leave at least one point in the crystal invariant. This is why they are called *point groups*.

In a crystal structure, considered as an infinitely extended pattern in space, new types of symmetry operation are admissible, which leave no point invariant.

These are called *space operations*. The new symmetry operations involve *translations* in addition to rotations and reflections. Clearly only an infinitely extended pattern can have a space operation (translation) as a symmetry operation. The new symmetry elements that result from combination of translation with the symmetry operations of the point groups are *glide planes* and *screw axes*. Examples of their operations are shown in Fig. 18.7. The operation produced by a glide plane is a

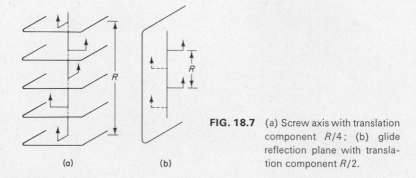

(a)   (b)

FIG. 18.7 (a) Screw axis with translation component $R/4$; (b) glide reflection plane with translation component $R/2$.

reflection across the plane followed by a specified translation parallel to the plane. In different kinds of glide planes, the translation can be half an axial length, half a face diagonal, or one-fourth a face diagonal. A screw axis combines rotation about an axis with a translation in the direction of the axis for a distance equal to some fraction of the lattice spacing in that direction.

The possible groups of symmetry operations of infinite figures are called *space groups*. They may be considered to arise from combinations of the 14 Bravais lattices with the 32 point groups.* A space group may be visualized as a sort of crystallographic kaleidoscope. If one structural unit is introduced into the unit cell, the operations of the space group immediately generate the entire crystal structure, just as the mirrors of a kaleidoscope produce a symmetrical pattern from a few bits of colored paper. The space group expresses the totality of the symmetry properties of a crystal structure, and mere external form or bulk properties do not suffice for its determination. One must make a determination of the inner structure of the crystal, and this is made possible by the methods of X-ray diffraction.

7. X-Ray Crystallography

A group of physicists at the University of Munich in 1912 was interested in both crystallography and the behavior of X rays. P. P. Ewald and A. Sommerfeld were studying the passage of light waves through crystals. At a colloquium discussing

---

*A good example of the construction of space groups is given by Lawrence Bragg, *The Crystalline State* (London: G. Bell & Sons, Ltd., 1933), p. 82. The space-group notation is described in *International Tables for the Determination of Crystal Structure*, Vol. 1. There are exactly 230 possible crystallographic space groups.

some of this work, Max von Laue pointed out that if the wavelength of the radiation became as small as the distance between atoms in the crystals, a diffraction pattern should result. There was some evidence that X rays might have a wavelength in this range, and W. Friedrich agreed to make an experimental test. An X-ray beam passed through a crystal of copper sulfate gave a definite diffraction pattern. Figure 18.8 shows a modern example of an X-ray diffraction pattern

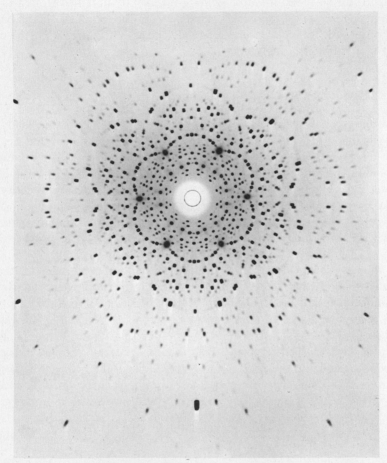

FIG. 18.8   X-ray diffraction by the Laue method of a crystal of beryl.
[Eastman Kodak Research Laboratories.]

obtained by the Laue method. The wave properties of X rays were thus definitely established and the new science of X-ray crystallography began.

The condition for diffraction maxima from a one-dimensional array of scattering centers was given on p. 579. For a three-dimensional array,

$$a(\cos \alpha - \cos \alpha_0) = h\lambda$$
$$b(\cos \beta - \cos \beta_0) = k\lambda \qquad (18.1)$$
$$c(\cos \gamma - \cos \gamma_0) = l\lambda$$

In these equations, α_0, β_0, γ_0 are the incident angles of the X-ray beam with the rows of scattering centers parallel to each of the three axes of the crystal, and α, β, γ are the corresponding angles of diffraction.

If monochromatic X rays are used, there would be only a slight chance that the orientation of the crystal would be such as to yield diffraction maxima. The Laue method, however, uses a continuous spectrum of X radiation with a wide range of wavelengths, the so-called *white radiation*, conveniently obtained from a tungsten target at high voltages. In this case, at least some of the radiation should be at the proper wavelength to experience interference effects, no matter what the crystal-to-beam orientation.

8. The Bragg Analysis of X-ray Diffraction

When the news of the Munich work reached England, it was immediately taken up by William Bragg and his son Lawrence, who had been working on a corpuscular theory of X rays. Lawrence Bragg, using photographs of the Laue type, analyzed the structures of NaCl, KCl, and ZnS (1912, 1913). In the meantime (1913), the elder Bragg devised a spectrometer that measured the intensity of an X-ray beam by the amount of ionization it produced, and he found that the characteristic X-ray line spectrum could be isolated and used for crystallographic work. Thus, the Bragg method uses a monochromatic (single wavelength) beam of X rays.

The Braggs developed a treatment of X-ray scattering by a crystal that was much easier to apply than the Laue theory, although the two are essentially equivalent. It was shown that the scattering of X rays could be represented as a "reflection" by successive planes of atoms in the crystal. Consider, in Fig. 18.9, a set of parallel planes in the crystal structure and a beam of X rays incident at an angle θ. Some of the rays will be "reflected" from the upper layer of atoms, the angle of reflection being equal to the angle of incidence. Some of the rays will be absorbed, and some will be "reflected" from the second layer, and so on with successive layers. All the waves "reflected" by a single crystal plane will be in phase. Only

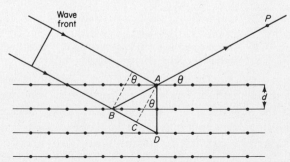

FIG. 18.9   Derivation of the diffraction condition for X rays based on the equivalence of X-ray scattering from a set of planes to a "reflection" from successive planes.

under certain strict conditions will the waves "reflected" by different underlying planes be in phase with one another. The condition is that the difference in length of path between the waves scattered from successive planes must be an integral number of wavelengths, $n\lambda$. If we consider the "reflected" waves at the point P, this path difference for the first two planes is $\delta = \overline{AB} - \overline{BC}$. Since, $\overline{AB} = \overline{BD}$ and $\delta = \overline{CD}$ and $\overline{CD} = \overline{AD} \sin \theta$, therefore, $\delta = 2d \sin \theta$. The condition for reinforcement or Bragg "reflection" is thus

$$n\lambda = 2d \sin \theta \tag{18.2}$$

According to this viewpoint, there are different *orders* of "reflection" specified by the values $n = 1, 2, 3, \ldots$. The second-order diffraction maxima from 100 planes may then be regarded as a "reflection" due to a set of planes 200 with half the spacing of the 100 planes.

The Bragg equation indicates that for any given wavelength of X rays there is a lower limit to the spacings that can give observable diffraction spectra. Since the maximum value of $\sin \theta$ is 1, this limit is given by

$$d_{\min} = \frac{n\lambda}{2 \sin \theta_{\max}} = \frac{\lambda}{2}$$

9. Proof of Bragg Reflection

The Bragg equation is so fundamental that we cannot leave its proof resting on the unsupported assertion that X-ray scattering from a regular structure behaves like reflection from a set of parallel planes. We shall therefore derive the Laue conditions in vector form and show how they lead to the Bragg formula.

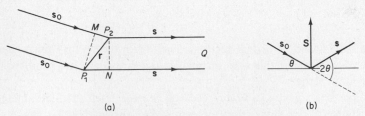

(a)   (b)

FIG. 18.10 Constructions for derivation of Bragg relation. (a) To show that the difference in path of X rays scattered from P_1 and P_2 is $\mathbf{r} \cdot (\mathbf{s} - \mathbf{s_0})$. (b) To show that $\mathbf{s}$ is related to $\mathbf{s_0}$ by a reflection in a plane that bisects the angle between the two vectors.

Consider first in Fig. 18.10(a) the X-ray scattering from any two lattice points P_1, P_2. (Imagine, if you like, that on each lattice point there is an electron which scatters the X rays.) Let $\mathbf{s_0}$ be a unit vector normal to the incident plane wave and $\mathbf{s}$ be a unit vector normal to the scattered plane wave. Suppose the scattered rays are detected at a point Q, which is at a distance from P_1 and P_2 very large compared to r, the distance $P_1 P_2$. Then $P_1 Q$ and $P_2 Q$ can be taken as parallel. The path difference is

$$\delta = P_1 N - P_2 M$$

$$\delta = \mathbf{r} \cdot \mathbf{s} - \mathbf{r} \cdot \mathbf{s}_0 = \mathbf{r} \cdot (\mathbf{s} - \mathbf{s}_0) = \mathbf{r} \cdot \mathbf{S}$$

The vector $\mathbf{r}$ between two lattice points can always be expressed as $(m\mathbf{a} + n\mathbf{b} + p\mathbf{c})$ where m, n, p are integers and $\mathbf{a, b, c}$ are the crystal axes.

Figure 18.10(b) represents the relation between $\mathbf{s}_0$, $\mathbf{s}$ and their difference $\mathbf{S} = \mathbf{s} - \mathbf{s}_0$. We note that $\mathbf{S}$ is in the direction of the normal to the plane that would result if $\mathbf{s}_0$ were reflected into $\mathbf{s}$ through an angle 2θ. Thus, the relation between the incident and scattered wavefronts can be represented by the geometrical equivalent of a reflection from a plane bisecting the two vectors.

In order that the waves scattered by P_1 and P_2 be in phase, the path difference δ must equal an integral number N of wavelengths λ.

$$\delta = \mathbf{r} \cdot \mathbf{S} = (m\mathbf{a} + n\mathbf{b} + p\mathbf{c}) \cdot \mathbf{S} = N\lambda$$

Since this expression must be true when m, n, or p change by integral values, each of the products must also equal an integral number of wavelengths. Thus,

$$\mathbf{a} \cdot \mathbf{S} = h\lambda$$
$$\mathbf{b} \cdot \mathbf{S} = k\lambda \qquad (18.3)$$
$$\mathbf{c} \cdot \mathbf{S} = l\lambda$$

These are the Laue equations in vector form.

Bragg identified the integers h, k, l in the Laue equations with the Miller indices of a lattice plane. We know $\mathbf{S}$ is the direction of the bisector of normal and incident beams. From (18.3),

$$\left(\frac{\mathbf{a}}{h} - \frac{\mathbf{b}}{k} \right) \cdot \mathbf{S} = 0$$

$$\left(\frac{\mathbf{a}}{h} - \frac{\mathbf{c}}{l} \right) \cdot \mathbf{S} = 0$$

When the scalar product of two vectors equals zero, the vectors are perpendicular. Therefore, $\mathbf{S}$ is perpendicular to the plane hkl, since it is perpendicular to two vectors in that plane.

Recalling that $\mathbf{s}$ and $\mathbf{s}_0$ are *unit* vectors, we see from (18.3) that the perpendicular distance of the plane from the origin, i.e., the interplanar spacing, is

$$d = \frac{(\mathbf{a}/h) \cdot \mathbf{S}}{|\mathbf{S}|} = \frac{\lambda}{|\mathbf{S}|} = \frac{\lambda}{2 \sin \theta}$$

which completes the proof of the Bragg relation.

10. Fourier Transforms and Reciprocal Lattices

Any periodic function $f(x)$ of period L can be represented by a Fourier series,

$$f(x) = \sum_n a_n \, e^{2\pi i n x / L} \qquad (18.4)$$

The coefficients a_n are given by

$$a_n = \frac{1}{L} \int_{-L/2}^{+L/2} f(x)e^{-2\pi inx/L}\, dx$$

As $L \to \infty$, the sum in (18.4) can be replaced by an integral to give

$$f(x) = \frac{1}{2\pi} \int_{-\infty}^{+\infty} g(\xi)e^{ix\xi}\, d\xi$$

where

$$g(\xi) = \int_{-\infty}^{+\infty} f(x)e^{-ix\xi}\, dx$$

The functions $f(x)$ and $g(\xi)$ are a pair of Fourier transforms. We say, for example, that $g(\xi)$ is the Fourier transform of $f(x)$ and vice versa.

We can readily extend the definitions to three dimensions to give

$$f(x, y, z) = \frac{1}{2\pi} \iiint_{-\infty}^{+\infty} g(\xi, \eta, \zeta)e^{i(x\xi+y\eta+z\zeta)}\, d\xi\, d\eta\, d\zeta \tag{18.5}$$

and

$$g(\xi, \eta, \zeta) = \iiint_{-\infty}^{+\infty} f(x, y, z)e^{-i(x\xi+y\eta+z\zeta)}\, dx\, dy\, dz \tag{18.6}$$

We can now think of x, y, z and ξ, η, ζ as being the coordinates of two three-dimensional spaces. The space of x, y, z is defined by three standard vectors, **a, b, c**, such that any vector **r** drawn from the origin of coordinates is equal to

$$\mathbf{r} = x\mathbf{a} + y\mathbf{b} + z\mathbf{c} \tag{18.7}$$

Similarly, the space of ξ, η, ζ can be defined by three standard vectors **A, B, C**, with any vector **R** drawn from its origin of coordinates equal to

$$\mathbf{R} = \xi\mathbf{A} + \eta\mathbf{B} + \zeta\mathbf{C} \tag{18.7a}$$

A volume element in xyz space would be

$$dv = \mathbf{a} \cdot (\mathbf{b} \times \mathbf{c})\, dx\, dy\, dz$$

and in $\xi\eta\zeta$ space,

$$dV = \mathbf{A} \cdot (\mathbf{B} \times \mathbf{C})\, d\xi\, d\eta\, d\zeta$$

Suppose we wish to represent $f(x, y, z)$ in (18.5) in vector form as $f(\mathbf{r})$. Such a representation will require the following relation between the standard vectors of xyz space and those of $\xi\eta\zeta$ space:

$$\mathbf{a} \cdot \mathbf{A} = \mathbf{b} \cdot \mathbf{B} = \mathbf{c} \cdot \mathbf{C} = 1$$

$$\mathbf{a} \cdot \mathbf{B} = \mathbf{a} \cdot \mathbf{C} = \mathbf{b} \cdot \mathbf{A} = \mathbf{b} \cdot \mathbf{C} = \mathbf{c} \cdot \mathbf{A} = \mathbf{c} \cdot \mathbf{B} = 0$$

If the vectors **a, b, c** have the dimension of length, the vectors **A, B, C** have the dimension of reciprocal length. Thus, we call the space of xyz *physical space* and the space of $\xi\eta\zeta$ *reciprocal space*. Functions in reciprocal space are Fourier transforms of functions in physical space, and vice versa.

Corresponding to each of the 14 Bravais lattices in physical space, there will be a reciprocal lattice. The points **p** of a space lattice are given by

$$\mathbf{p} = m\mathbf{a} + n\mathbf{b} + p\mathbf{c}$$

where m, n, p are integers. [Note that this is simply a special case of (18.7), applied to points on a lattice.] The points **P** of the reciprocal lattice are given by

$$\mathbf{P} = h\mathbf{A} + k\mathbf{B} + l\mathbf{C} \tag{18.8}$$

The scalar product **P** • **p** is

$$\mathbf{P} \cdot \mathbf{p} = (h\mathbf{A} + k\mathbf{B} + l\mathbf{C}) \cdot (m\mathbf{a} + n\mathbf{b} + p\mathbf{c}) = (hm + kn + lp) = N$$

where N is an integer. Therefore,

$$\exp(2\pi i \mathbf{P} \cdot \mathbf{p}) = 1 \tag{18.9}$$

This result shows that reciprocal space might well be called *Fourier space*. The significance of Fourier space (reciprocal space) is that whereas a crystal structure exists in real space, an X-ray *diffraction diagram* exists in Fourier space (reciprocal space). Each point on an X-ray diffraction diagram corresponds to a set of integers *hkl* that specify a vector in Fourier space. Thus, each point in the reciprocal lattice corresponds to a set of planes *hkl* in the real lattice. If we could examine a crystal structure with a sufficiently powerful microscope, we would see the real structure. What we obtain from the pattern of diffracted X rays is a Fourier transform of the real structure.

11.   Structures of Sodium and Potassium Chlorides

Among the first crystals to be studied by the Bragg method were sodium and potassium chlorides. A single crystal was mounted on the spectrometer, as shown in Fig. 18.11(a), so that the X-ray beam was incident on one of the important crystal faces, (100), (110), or (111). The apparatus was arranged so that the scattered beam entered the ionization chamber, which was filled with methyl bromide. The intensity of the beam was measured by the charge collected by an electrometer.

The experimental data are shown plotted in Fig. 18.12 as intensity of scattered beam vs. twice the angle of incidence of beam to crystal. As the crystal is rotated, successive intensity maxima flash out as the angles conforming to the Bragg condition (18.2) are reached. In these first experiments, monochromatic X radiation was obtained from an X-ray tube having a palladium target.

At the beginning of these experiments, Bragg knew neither the wavelength of the X rays nor the structures of the crystals. He did know, from external form, that both NaCl and KCl could be based on one of the cubic lattices—simple, body centered, or face centered. By comparing the spacings calculated from the X-ray data with those expected for these lattices, he could assign the correct lattice to each crystal.

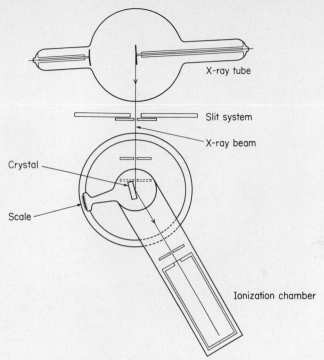

FIG. 18.11(a)   Bragg X-ray spectrometer.

The distance between planes hkl in a cubic lattice is*

$$d_{hkl} = \frac{a_0}{(h^2 + k^2 + l^2)^{1/2}}$$ (18.10)

When this is combined with the Bragg equation, we obtain

$$\sin^2 \theta = \frac{\lambda^2}{4a_0^2} (h^2 + k^2 + l^2)$$

Thus, each value of $\sin \theta$ observed for a diffraction maximum can be *indexed* by assigning to it the value of hkl for a set of planes meeting the Bragg condition. For a simple cubic lattice, values of $h^2 + k^2 + l^2$ correspond to sets of planes hkl as follows:

hkl	100	110	111	200	210	211	220	221	300	etc.
$h^2 + k^2 + l^2$	1	2	3	4	5	6	8	9	9	etc.

*The equation of the plane hkl is $hx + ky + lz = a_0$. The distance from any point (x_1, y_1, z_1) to the plane is

$$d = \frac{hx_1 + ky_1 + lz_1 - a_0}{\sqrt{h^2 + k^2 + l^2}}$$

Hence, when the point is at the origin, we find $d = a_0/\sqrt{h^2 + k^2 + l^2}$.

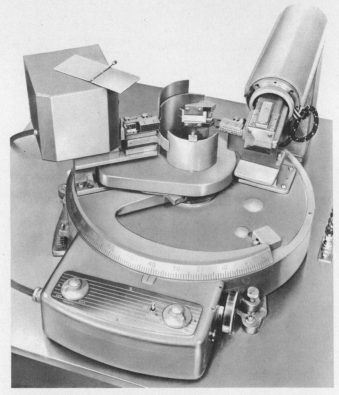

FIG. 18.11(b)   A modern X-ray spectrometer. The head of the X-ray
tube is at right. The diffracted beam is measured by
the counter tube at left. [General Electric Company.]

If the observed X-ray pattern from a simple cubic crystal is plotted as intensity
vs. $\sin^2 \theta$, we should obtain a series of six equidistant maxima. Then the seventh
would be missing, since there is no set of integers hkl such that $h^2 + k^2 + l^2 = 7$.
There would then follow seven more equidistant maxima with the fifteenth missing, seven more with the twenty-third missing, four more with the twenty-eighth
missing, etc.

In Fig. 18.13(a), we see the 100, 110, and 111 planes for a simple cubic lattice.
A structure could be based on this lattice by placing an atom at each lattice point.
If an X-ray beam entered such a structure at the Bragg angle, $\theta = \sin^{-1}(\lambda/2a)$,
the rays scattered from one 100 plane would be exactly in phase with the rays
from successive 100 planes. The strong scattered beam would be called the "first-
order reflection from the 100 planes." A similar result is obtained for the 110
and 111 planes. With a structure based on a simple cubic lattice, we should obtain
a diffraction maximum from each set of planes hkl, since for any given hkl all the
atoms are included in the planes.

Figure 18.13(b) shows a structure based on a body-centered cubic lattice.

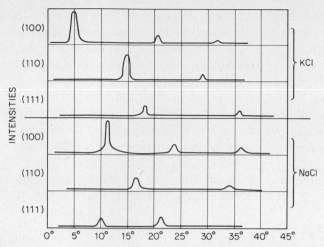

FIG. 18.12 Bragg spectrometer data, I vs. 2θ, taken from cubic crystals of NaCl and KCl with Pd–K_α X rays.

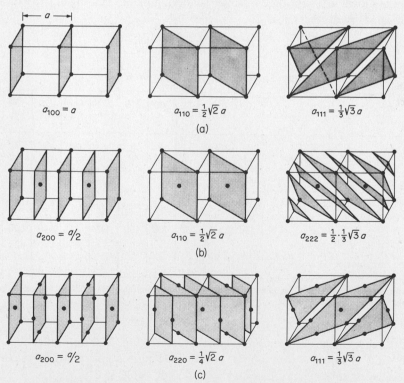

$a_{100} = a$ $\qquad$ $a_{110} = \frac{1}{2}\sqrt{2}\,a$ $\qquad$ $a_{111} = \frac{1}{3}\sqrt{3}\,a$

(a)

$a_{200} = a/2$ $\qquad$ $a_{110} = \frac{1}{2}\sqrt{2}\,a$ $\qquad$ $a_{222} = \frac{1}{2}\cdot\frac{1}{3}\sqrt{3}\,a$

(b)

$a_{200} = a/2$ $\qquad$ $a_{220} = \frac{1}{4}\sqrt{2}\,a$ $\qquad$ $a_{111} = \frac{1}{3}\sqrt{3}\,a$

(c)

FIG. 18.13 Spacings in cubic lattices: (a) simple cubic; (b) body-centered cubic; (c) face-centered cubic.

The 110 planes, as in the simple cubic case, pass through all the lattice points, and thus a strong first-order 110 reflection occurs. In the case of the 100 planes, however, we find a different situation. Exactly midway between any two 100 planes, there lies another layer of atoms. When X rays scattered from the 100 planes are in phase and reinforce one another, the rays scattered by the interleaved atomic planes will be retarded by half a wavelength, and hence will be exactly out of phase with the others. The observed intensity will therefore be the difference between the scattering from the two sets of planes. If the atoms are all the same and thus have identical scattering powers, the resultant intensity will be reduced to zero by the destructive interference between the scattering from the two interleaved sets of planes.

The second-order diffraction from the 100 planes, occurring at the Bragg angle with $n = 2$ in (18.2), can equally well be expressed as the scattering from a set of planes, called the 200 planes, with just half the spacing of the 100 planes. In the body-centered cubic structure, all the atoms lie in these 200 planes, so that all scattering is in phase, and a strong scattered beam is obtained. The same situation holds for the 111 planes: the first-order 111 will be extinguished, but the second-order 111, i.e., the 222 planes, will give strong scattering. If we examine successive planes *hkl* in this way, we find for the body-centered cubic structure the results shown in Table 18.3, in which scattering planes missing due to extinction are indicated by dotted lines.

TABLE 18.3
Calculated and Observed Diffraction Maxima[a]

hkl	100	110	111	200	210	211	—	220	300 221	310
$h^2 + k^2 + l^2$	1	2	3	4	5	6	(7)	8	9	10
simple cubic	|	|	|	|	|	|	—	|	|	|
body-centered cubic	:	|	:	|	:	|	—	|	:	|
face-centered cubic	:	:	|	|	:	:	—	|	:	:
Sodium chloride	:	:	|	|	:	:	—	|	:	:
	200	220	222	400	420	422	—	440	600 422	620
Potassium chloride	|	|	|	|	|	|	—	|	|	|

[a]A solid vertical line indicates that Bragg "reflection" is observed from the indicated plane; a dotted vertical line indicates that Bragg "reflection" does not occur.

In the case of the face-centered cubic structure, Fig. 18.13(c), reflections from the 100 and 110 planes are missing, and the 111 planes give a strong maximum. The results for subsequent planes are included in Table 18.3.

In the first investigation of NaCl and KCl, the X-ray wavelength was not known, so that the spacings corresponding to the diffraction maxima could not be calculated. The values of sin θ, however, could be used directly. Table 18.3 gives a comparison of the observed maxima with those calculated for the three different cubic lattices. Bragg found the curious result that apparently NaCl was face centered whereas KCl was simple cubic.

The reason why the KCl structure appeared to behave toward X rays like a simple cubic array is that the scattering powers of K$^+$ and Cl$^-$ ions are almost indistinguishable, since they both have an argon configuration with 18 electrons. Since Na$^+$ and Cl$^-$ ions differ in scattering power, however, the face-centered structure of NaCl was revealed. The "reflections" observed from the 111 face of NaCl included a weak peak at about 10°, in addition to the stronger peak at about 20°, corresponding to that observed with KCl.

These results are all explained by the structure shown in Fig. 18.14, which

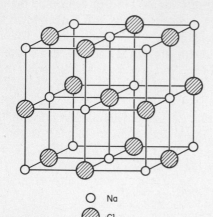

O Na

Cl

FIG. 18.14   Rock salt structure. Both NaCl and KCl have this structure.

consists of a face-centered cubic array of Na$^+$ ions and an interpenetrating face-centered cubic array of Cl$^-$ ions. Each Na$^+$ ion is surrounded by six equidistant Cl$^-$ ions, and each Cl$^-$ ion by six equidistant Na$^+$ ions. The 100 and 110 planes contain equal numbers of both kinds of ions, but the 111 planes consist of either all Na$^+$ or all Cl$^-$ ions. Now if X rays are scattered from the 111 planes in NaCl, whenever scattered rays from successive Na$^+$ planes are exactly in phase, the rays scattered from the interleaved Cl$^-$ planes are retarded by half a wavelength and are therefore exactly out of phase. The first-order 111 maximum is therefore weak in NaCl, because it represents the difference between these two scatterings. In the case of KCl, where the scattering powers are almost the same, the first-order maxima are almost extinguished by interference. Thus, both NaCl and KCl have the same structure, in complete agreement with the experimental X-ray evidence.

Once the NaCl structure was established, it was possible to calculate the wavelength of the X rays used. From the density of crystalline NaCl, $\rho = 2.163$ g·cm^{-3}, the molar volume is $M/\rho = 58.45/2.163 = 27.02$ cm^3·mol^{-1}. Then the volume occupied by each NaCl unit is $27.02/(6.02 \times 10^{23}) = 44.88 \times 10^{-24}$ cm^3. In the unit cell of NaCl, there are eight Na$^+$ ions at the corners of the cube, each shared between eight cubes, and six Na^+ ions at the face centers, each shared between two cubes. Thus, per unit cell, there are $\frac{8}{8} + \frac{6}{2} = 4\ Na^+$ ions. There is an equal number of Cl^- ions, and therefore four NaCl units per unit cell ($Z = 4$). The volume of the unit cell is, therefore,

$$4 \times 44.88 \times 10^{-24} = 179.52 \times 10^{-24}\ cm^3$$

The interplanar spacing for the 200 planes is

$$d_{200} = \tfrac{1}{2}a = \tfrac{1}{2}(179.52 \times 10^{-24})^{1/3} = 2.82 \times 10^{-8}\ cm = 0.282\ nm$$

When this value and the observed diffraction angle were put into the Bragg equation, the wavelength of the $Pd-K_{\alpha_1}$ X ray was calculated as

$$\lambda = 2(0.282) \sin 5°58' = 0.0586\ nm$$

Once measured in this way the wavelength can be used to determine the interplanar spacings in other crystal structures. Conversely, crystals with known spacings can be used to measure the wavelengths of other X-ray lines. The most generally useful target material is copper, with $\lambda = 0.1537\ nm$ (K_{α_1}), a convenient length relative to interatomic distances. When short spacings are of interest, molybdenum (0.0708 nm) is useful, and chromium (0.2285 nm) is often employed for study of longer spacings.

Many investigations of crystal structure have used photographic methods to record the diffraction patterns, but the Bragg spectrometer method is generally applicable. Improved spectrometers are widely used, incorporating a scintillation-crystal detector in place of the electrometer and ionization chamber.

12. The Powder Method

The simplest technique for obtaining X-ray diffraction data is the powder method, first used by P. Debye and P. Scherrer. Instead of a single crystal with a definite orientation to the X-ray beam, a mass of finely divided crystals with random orientations is taken. The experimental arrangement is illustrated in Fig. 18.15(a). The powder is contained in a thin-walled glass capillary, or deposited on a fiber. Polycrystalline metals are studied in the form of fine wires. The sample is rotated in the beam to average as well as possible over the orientations of the crystallites.

Out of many random orientations of the little crystals, there will be some at the Bragg angle for X-ray "reflection" from each set of planes. The direction of the reflected beam is limited only by the requirement that the angle of reflection equal the angle of incidence. Thus, if the incident angle is θ, the reflected beam makes an angle 2θ with the direction of the incident beam, as shown in Fig. 18.15 (b). This angle 2θ may itself be oriented in various directions around the central beam direction, corresponding to the various orientations of the individual crystallites. For each set of planes, therefore, the reflected beams outline a cone of scattered radiation. This cone, intersecting a cylindrical film surrounding the specimen, gives rise to the observed lines. On a flat plate film, the observed pattern

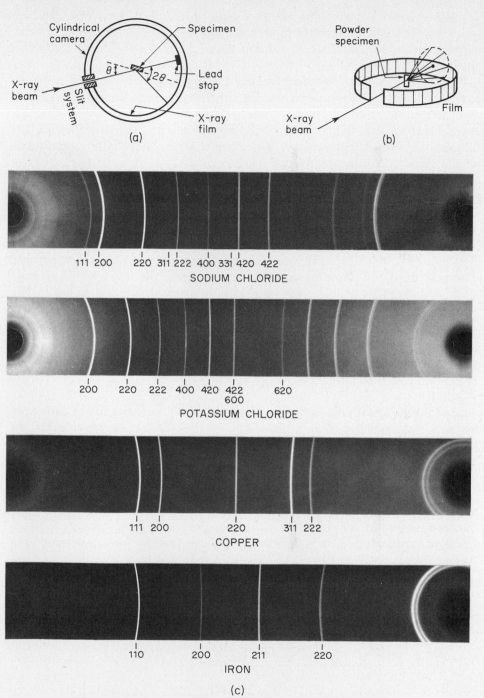

FIG. 18.15 X-ray diffraction by the powder method. [A Lessor, I.B.M. Laboratories.]

consists of a series of concentric circles. Figure 18.15(c) shows the Debye–Scherrer (powder) diagrams obtained from several examples of important types of cubic crystal structures. The observed patterns may be compared with the theoretical predictions in Table 18.3.

After we obtain a powder diagram, our next step is to index the lines, assigning each to the responsible set of planes. The distance x of each line from the central spot is measured carefully, usually by halving the distance between the two reflections on either side of the center. If the film radius is r, the circumference $2\pi r$ corresponds to a scattering angle of 360°. Then, $x/2\pi r = 2\theta/360°$. Thus we calculate θ and, from (18.2), the interplanar spacing.

To index the reflections, we must know the crystal system to which the specimen belongs. This can sometimes be determined by microscopic examination. Powder diagrams of monoclinic, orthorhombic, and triclinic crystals may be almost impossible to index. For the other systems, straightforward methods are available. Once we have found the dimensions of the unit cell by calculation from a few large spacings (100, 110, 111, etc.), we can calculate all the interplanar spacings and compare them with those observed, thus completing the indexing. Then more precise unit-cell dimensions can be calculated from high-index spacings. The general formulas giving the interplanar spacings are derived from analytic geometry.*

13. Rotating Crystal Methods

The rotating single crystal method, with photographic recording of the diffraction pattern, was developed by E. Schiebold about 1919. It has been, in one form or another, the most widely used technique for precise structure investigations.

The crystal, which is preferably small and well formed, perhaps a needle 1 mm long and 0.5 mm wide, is mounted with a definite axis perpendicular to the beam, which bathes it in X radiation. The film may be held in a cylindrical camera. As the crystal is rotated slowly during the course of the exposure, successive planes pass through the orientation corresponding to the Bragg condition. Sometimes only part of the data is recorded on a single film, by oscillating the crystal through some angle smaller than 360°. In the Weissenberg and other moving-camera methods, the film is moved back and forth with a period synchronized

---

*For example,

$$\text{Tetragonal} \qquad d_{hkl}^{-2} = \frac{h^2 + k^2}{a^2} + \frac{l^2}{c^2}$$

$$\text{Orthorhombic} \qquad d_{hkl}^{-2} = \left(\frac{h}{a}\right)^2 + \left(\frac{k}{b}\right)^2 + \left(\frac{l}{c}\right)^2$$

$$\text{Hexagonal} \qquad d_{hkl}^{-2} = \frac{4}{3}\frac{h^2 + hk + k^2}{a^2} + \left(\frac{l}{c}\right)^2$$

For the remaining formulas, see the *International Tables for Determination of Crystal Structures* (1952).

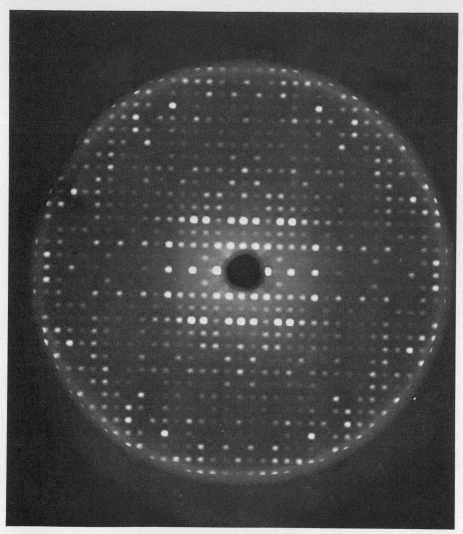

FIG. 18.16 An X-ray picture taken with a precession camera of a monoclinic crystal of lysozyme iodide (courtesy of L. K. Steinrauf, Indiana University). From the positions and intensities of the diffraction maxima in such pictures, the detailed three-dimensional crystal structure, and hence molecular structure, of lysozyme was determined in 1965 by D. C. Philips and his coworkers at the Royal Institution in London. Lysozyme was the second protein and the first enzyme to have its structure determined by X-ray crystallography. The molecule contains 1950 atoms. An account of this work has been given in a beautifully illustrated article by Philips [*Scientific American*, Nov., 1966].

with the rotation of the crystal. Thus, the position of a spot on the film immediately indicates the orientation of the crystal at which the spot was formed.

We cannot give here a detailed interpretation of these several varieties of rotation pictures.* An example is shown in Fig. 18.16, taken from a crystal of an enzyme, lysozyme. The spots are indexed and their intensities are measured. The data thus obtained are the raw materials for the determination of the crystal structure.

14. Crystal-Structure Determinations

The reconstruction of a crystal structure from the intensities of the various X-ray diffraction maxima is analogous to the formation of an image by a microscope. According to Abbe's theory of the microscope, the objective gathers various orders of light rays diffracted by the specimen and resynthesizes them into an image. This synthesis is possible because two conditions are fulfilled in the optical case: the phase relationships between the various orders of diffracted light waves are preserved at all times, and optical glass is available to focus and form an image with radiation having the wavelengths of visible light. Electron beams can be focused with electrostatic and magnetic lenses, but we have no such lenses for forming X-ray images. Also, the way in which the diffraction data are necessarily obtained (one-by-one) means that all the phase relationships are lost.† The essential problem in determining a crystal structure is to regain this lost information in some way or other, and to resynthesize the structure from the amplitudes and phases of the diffracted waves.

We shall return to this problem, but first let us see how the intensities of the various spots in an X-ray picture are governed by the crystal structure.‡ The Bragg relation fixes the angle of scattering in terms of the interplanar spacings, which are determined by the arrangement of points in the crystal lattice. In an actual structure, each lattice point is replaced by a group of atoms. The arrangement and composition of this group are the primary factors that control the intensity of the scattered X rays, once the Bragg condition has been satisfied.

*See Bragg, *op. cit.*, p. 30. Also Bunn, *op. cit.*, p. 137.

†A *hologram* is an optical construct that might be said to be something between a crystal structure and a diffraction pattern. To make a hologram, a three-dimensional structure is illuminated with a strong source of *coherent light* (from a laser). A diffraction pattern, the hologram, is produced as the result of interference between light scattered from the object and the coherent incident light. The pattern can be recorded on a photographic plate from which a picture print is made. If the positive hologram is illuminated with the original coherent background light, the transmitted light can be brought, with a suitable lens, to focus into an image of the original object. In the case of the X-ray diffraction diagram, no coherent background radiation is available.

It has been suggested that the mental image which is formed in the neural network of the brain is a sort of *hologram*. Thus, the process of recalling an event or object from memory would be analogous to the reconstruction of an object from a hologram. If this model is true, our memories would be in a Fourier space similar to that of the X-ray diffraction diagrams of crystal structures.

‡This treatment follows that given by M. J. Buerger in *X-Ray Crystal Structure Analysis*, Ch. 10 (New York: John Wiley & Sons, Inc., 1960).

As an example, consider in Fig. 18.17(a), a structure formed by replacing each point in a body-centered orthorhombic lattice by two atoms (e.g., a diatomic molecule). If a set of planes is drawn through the black atoms, another parallel but slightly displaced set can be drawn through the white atoms. When the Bragg condition is met, as in Fig. 18.7(b), the reflections from all the black atoms are in

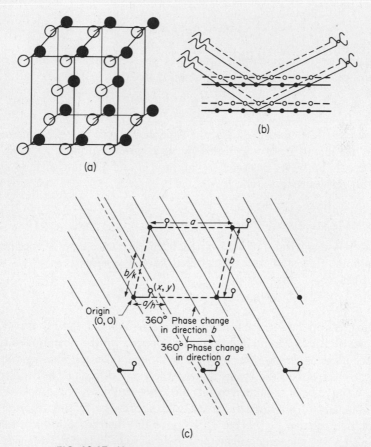

FIG. 18.17 X-ray scattering from a typical structure.

phase, and the reflections from all the white atoms are in phase. The radiation scattered from the blacks is slightly out of phase with that from the whites, so that the resultant amplitude, and therefore intensity, is diminished by interference.

The problem is to obtain a general expression for the phase difference. An enlarged view of a section through the structure is shown in Fig. 18.17(c), with the black atoms at the corners of a unit cell with sides a and b, and the whites at displaced positions. The coordinates of a black atom may be taken as $(0, 0)$ and those of a white as (x, y). A set of planes hk is shown, for which the Bragg condition is fulfilled; these are actually the 32 planes in the figure. Now the spacings

a/h along a and b/k along b correspond to positions from which scattering differs in phase by exactly $360°$ or 2π radians, i.e., scattering from these positions is exactly in phase. The difference in phase between these planes and those going through the white atoms is proportional to the displacement of the white atoms. The phase difference ϕ_x for displacement x in the a direction is $x/(a/h) = \phi_x/2\pi$, or $\phi_x = 2\pi h(x/a)$. The total phase difference for displacement in both a and b directions becomes

$$\phi_x + \phi_y = 2\pi\left(h\frac{x}{a} + k\frac{y}{b}\right)$$

By extension to three dimensions, the total phase change that an atom at (xyz) in the unit cell contributes to the plane (hkl) is

$$\phi = 2\pi\left(\frac{hx}{a} + \frac{ky}{b} + \frac{lz}{c}\right) \tag{18.11}$$

We may recall Section 15.22 that the superposition of waves of different amplitude and phase can be accomplished by vectorial addition. If f_1 and f_2 are the amplitudes of the waves scattered by atoms (1) and (2), and ϕ_1 and ϕ_2 are the phases, the resultant amplitude is $F = f_1 e^{i\phi_1} + f_2 e^{i\phi_2}$. For all the atoms in a unit cell,

$$F = \sum_j f_j e^{i\phi_j} \tag{18.12}$$

When the phase ϕ_j from (18.11) is introduced, an expression is obtained for the resultant amplitude of the waves scattered from the hkl planes by all the atoms in a unit cell:

$$F(hkl) = \sum_j f_j e^{2\pi i(hx_j/a + ky_j/b + lz_j/c)} \tag{18.13}$$

The expression $F(hkl)$ is called the *structure factor* of the crystal. Its value is determined by the exponential terms, which depend on the positions of the atoms, and by the *atomic scattering factors* f_j, which depend on the number and distribution of the electrons in the atoms and on the scattering angle θ. Structure-factor expressions have been tabulated for all the space groups.*

The intensity of scattered radiation is proportional to the absolute value of the amplitude squared, $|F(hkl)|^2$. Thus, once the Bragg condition is satisfied for a set of planes hkl, the structure factor allows us to calculate the intensity of X-ray scattering from hkl. The relation between intensity and structure factor includes a number of physical terms for which explicit formulas are available.†

As an example of the use of the structure factor, let us calculate $F(hkl)$ for the 100 planes in a face-centered cubic structure, e.g., that of gold. In this structure, there are four atoms in the unit cell ($Z = 4$), which may be assigned coordinates $(x/a, y/b, z/c)$ as follows: (000), $(\frac{1}{2}\frac{1}{2}0)$, $(\frac{1}{2}0\frac{1}{2})$ and $(0\frac{1}{2}\frac{1}{2})$. Therefore, from (18.13),

---

*International Tables for the Determination of Crystal Structures (1952). It is usually possible to narrow the choice of space groups to two or three by means of a study of missing reflections (hkl) and comparison with the tables.

†M. J. Buerger, *X-Ray Crystal Structure Analysis* (New York: John Wiley and Sons, 1960). Chaps. 7 and 8.

$$F(100) = f_{Au}(e^{2\pi i \cdot 0} + e^{2\pi i \cdot 1/2} + e^{2\pi i \cdot 1/2} + e^{2\pi i \cdot 0})$$
$$= f_{Au}(2 + 2e^{\pi i}) = 0$$

since

$$e^{\pi i} = \cos \pi + i \sin \pi = -1$$

Thus the structure factor vanishes and the intensity of scattering from the 100 planes is zero. This is almost a trivial case, since inspection of the face-centered cubic structure reveals that there is an equivalent set of planes interleaved midway between the 100 planes, so that the resultant amplitude of the scattered X rays must be reduced to zero by interference. In more complicated instances, however, it is essential to use the structure factor to obtain a quantitative estimation of the scattering intensity expected from any set of planes *hkl* in any postulated crystal structure.

15. Fourier Synthesis of a Crystal Structure

The electrons in crystals are responsible for the scattering of X rays. Thus, it is rather artificial to represent the crystal structure as an array of atoms located at points (x, y, z). A continuous distribution of electron density $\rho(x, y, z)$ would be a more realistic model. The expression for the structure factor in (18.11), given as a sum over discrete atoms, then becomes an integral over the continuous distribution of electronic scattering matter.*

$$F(hkl) = \int_0^a \int_0^b \int_0^c \rho(x, y, z)\, e^{2\pi i(hx/a + ky/b + lz/c)}\, dx\, dy\, dz \qquad (18.14)$$

Inasmuch as the electron density $\rho(x, y, z)$ is a function with the periodicity of the lattice, it can be written as a Fourier series in three dimensions:

$$\rho(x, y, z) = \sum\sum\sum A(pqr)\, e^{+2\pi i(px/a + qy/b + rz/c)} \qquad (18.15)$$

To evaluate the Fourier coefficient $A(pqr)$, we substitute (18.15) into (18.14), obtaining

$$F(hkl) = \int_0^a \int_0^b \int_0^c \sum\sum\sum A(pqr) \exp\left[2\pi i\left(\frac{hx}{a} + \frac{ky}{b} + \frac{lz}{c}\right) \right]$$
$$\times \exp\left[2\pi i\left(\frac{px}{a} + \frac{qy}{b} + \frac{rz}{c}\right) \right] dx\, dy\, dz \qquad (18.16)$$

The integrals of the exponential functions over a complete period always vanish, so that the only term that remains in (18.16) is that for which $p = -h$, $q = -k$, and $r = -l$, which leads to

$$F(khl) = \int_0^a \int_0^b \int_0^c A(\bar{h}\bar{k}\bar{l})\, dx\, dy\, dz = VA(\bar{h}\bar{k}\bar{l})$$

where V is the volume of the unit cell. When this value of the Fourier coefficient is put into (18.15), we obtain

*Equation (18.14) holds for unit cells with perpendicular axes. A slight modification would be necessary for others, involving a change in the variables x, y, z to a new set of axes.

$$\rho(x, y, z) = \frac{1}{V} \sum\sum\sum F(hkl) \exp\left[-2\pi i\left(\frac{hx}{a} + \frac{ky}{b} + \frac{lz}{c}\right)\right] \quad (18.17)$$

The summation is carried out over all values of h, k, l, so that there is one term for each set of planes hkl, and hence for each spot on the X-ray diffraction diagram.

This remarkable equation summarizes the entire problem of a structure determination, since the crystal structure is equivalent to the function $\rho(xyz)$ in (18.17). Positions of individual atoms are peaks in the electron density function ρ, with heights proportional to the atomic numbers (number of electrons). If we knew the $F(hkl)$'s we could immediately plot the crystal structure. All we know, however, are the intensities of the diffraction maxima, and these are proportional to $|F(hkl)|^2$. As stated earlier, we know the amplitudes but we have lost the phases in taking the X-ray pattern.

In one method of solution, we assume a trial structure and calculate the intensities. If the assumed structure is even approximately correct, the most intense observed reflections should have large calculated intensities. We then compute the Fourier series by taking the *observed F's* for these reflections with the *calculated phases*. If we are on the right track, the graph of the Fourier summation

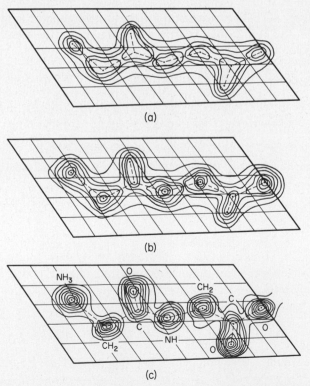

(a)

(b)

(c)

FIG. 18.18 Fourier map of electron density in glycylglycine projected on base of unit cell: (a) 40 terms; (b) 100 terms; (c) 160 terms.

should give new positions of the atoms, from which new F's can be calculated, which allow more of the phases to be correctly determined. As more terms are included in the Fourier synthesis, the resolution of the structure improves, just as the resolution of a microscope improves with objectives that catch more orders of diffracted light. Figure 18.18 shows three Fourier syntheses for the structure of glycylglycine. The first, (a), included 40 terms in the summation and gave a resolution of about 0.25 nm; the third, (c), with 160 terms, gave a resolution of about 0.11 nm.

Sometimes the phases can be readily determined if one atom that is much heavier (i.e., has many more electrons) than any of the other atoms occupies a

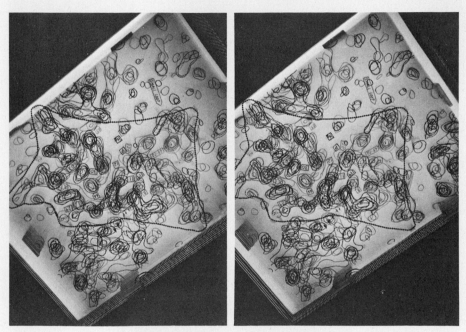

FIG. 18.19 Stereoscopic view of map of electron density in three dimensions of the structure of ribonuclease-S, obtained by Fourier synthesis of X-ray diffraction data. This protein has a molecular weight of 12 000 and consists of 124 amino-acid residues. The amplitudes and phases of 6000 X-ray diffraction maxima were established. Heavy atom derivatives were prepared with the uranyl cation, the tetracyanoplatinate (II) anion and dichloroethylenediamine platinum II. A resolution of about 0.20 nm has been achieved. The stereo effect can usually be obtained by dividing the two pictures with a card and allowing each eye to focus on one of the stereo views. [H. W. Wyckoff, D. Tsernoglu, A. W. Hanson, J. R. Knox, B. Lee, and Frederic M. Richards, Department of Molecular Biophysics and Biochemistry, Yale University, *J. Biol. Chem., 245,* 305 (1970).]

known position in the structure.* The X-ray diffraction diagram now becomes almost a hologram, for the scattering from the heavy atom with its known phase plays the same role in reconstructing the structure as does the coherent background radiation in the hologram. Of course, we still do not have any X-ray lenses for projecting the image of the structure directly, but we can use other ways of making an optical synthesis.

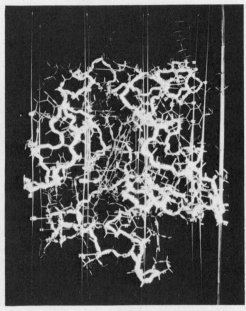

(a)   (b)

FIG. 18.20   (a) A space-filling model at 0.40 nm resolution of the structure of ferricytochrome from horse heart. The crevice that holds the heme molecule (shown in black) is a vertical cleft about 2.1 nm long. The heme is parallel to the cleft and perpendicular to the surface of the apoprotein molecule. The porphyrin rings of the heme extend into the interior of the apoprotein which serves as a hydrophobic pocket for the rings. The heme is covalently linked to two sulfur atoms of the apoprotein by thioether links from the vinyl sidechains of heme to cysteinyl residues in the apoprotein. (b) The structure is shown here at 0.28 nm resolution as a model in which bonds are represented by thin sticks and the main polypeptide chain is outlined with white yarn. The model of the heme is shown in exact relation to the crevice in the apoprotein. [R. E. Dickerson, M. L. Kopka, J. Weinzierl, J. Varnum, D. Eisenberg, and E. Margoliash, *J. Biol. Chem. 242*, 3014 (1967).]

*The coordinates of the heavy atom can usually be determined from a calculation of the Patterson function or vector map of the structure, a Fourier series like that in (18.17) except that $F^2(hkl)$ replaces $F(hkl)$.

The first direct application of the heavy-atom method was to the structure of phthalocyanine by Robertson and Woodward (1940). The molecule has a hole at its center and various metal atoms introduced at that site can form coordinate nitrogen to metal bonds. An important application of the heavy-atom method was the first determination of the structure of penicillin by Dorothy Hodgkin and Barbara Low. X-ray diffraction pictures obtained from Na, K, and Rb salts led to calculation of the phases.

The number of terms that must be included in the Fourier summation to achieve a given resolution increases rapidly with the number of atomic positions that are to be determined in the unit cell. For a protein having a molecular weight of 2×10^4, about 10^4 Fourier terms would be required to achieve a resolution of 0.20 nm. Figure 18.19 is a representation of the three-dimensional Fourier map of the electron density in the enzyme ribonuclease-S.*

Various ways have been employed to depict the three-dimensional structures of proteins by solid models. At a low resolution (~ 0.5 nm), sections of balsa wood can be cut to represent the limits of contours of electron density corresponding to van der Waals radii of the atoms. Figure 18.20(a) is a space-filling model of the structure of ferricytochrome-c at a resolution of 0.40 nm. As resolution improves, the individual atoms in the polypeptide chain can be discerned and high resolution skeletal models can be constructed with a wire framework. Figure 18.20(b) is such a model of the ferricytochrome-c structure at a resolution of 0.28 nm, in which the path of the main polypeptide chain has been delineated with white yarn.

16. Neutron Diffraction

Not only X-ray and electron beams, but also beams of heavier particles may exhibit diffraction patterns when scattered from the regular array of atoms in a crystal. Neutron beams are sometimes especially useful. The wavelength is related to the mass and velocity by the de Broglie equation, $\lambda = h/mv$. Thus a neutron with a speed of 3.9×10^5 cm·s^{-1} (kinetic energy 0.08 eV) would have a wavelength of 0.1nm. The diffraction of X rays is caused by the orbital electrons of the atoms in the material through which they pass; the atomic nuclei contribute practically nothing to the scattering. The diffraction of neutrons, on the other hand, is primarily caused by two other effects: (a) *nuclear scattering* due to interaction of the neutrons with the atomic nuclei, (b) *magnetic scattering* due to interaction of the magnetic moments of the neutrons with permanent magnetic moments of atoms or ions.

In the absence of an external magnetic field, the magnetic moments of atoms in a paramagnetic crystal are arranged at random, so that the magnetic scattering of neutrons by such a crystal is also random. It contributes only a diffuse background to the sharp maxima occurring when the Bragg condition is satisfied for the nuclear scattering. In *ferromagnetic materials*, however, the magnetic moments

*F. M. Richards, *J. Biol. Chem. 242*, 3485 (1967).

are regularly alined, so that the resultant spins of adjacent atoms are parallel, even in the absence of an external field. In *antiferromagnetic materials*, the magnetic moments are also regularly alined, but in such a way that adjacent spins along certain directions are opposed. The neutron diffraction patterns distinguish experimentally between these different magnetic structures and indicate the direction of alinement of spins within the crystal.

For example, manganous oxide, MnO, has the rock salt structure (Fig. 18.14), and is antiferromagnetic. The detailed magnetic structure as revealed by neutron diffraction is shown in Fig. 18.21. The manganous ion Mn^{2+} has the electronic structure $3s^2 3p^6 3d^5$. The five $3d$ electrons are all unpaired, and the resultant magnetic moment is $2\sqrt{\frac{5}{2}(\frac{5}{2} + 1)} = 5.91$ Bohr magnetons. If we consider Mn^{2+} ions in successive 111 planes in the crystal, the resultant spins are oriented so that they are directed either all positively or all negatively along the [111] direction.

Neutron diffraction can also be used to locate hydrogen atoms in crystal

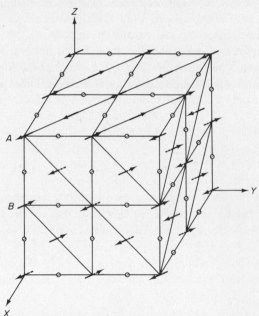

FIG. 18.21   The magnetic structures of the transition-element oxides, such as MnO, indicating the doubled unit-cell when the magnetic moment directions are taken into account. The arrows drawn at the positions of the metal ions indicate the directions of their magnetic moments: the small circles show the intervening oxygen atoms. In each case there are ferromagnetic sheets of atoms parallel to the (111) plane. For MnO, NiO the moments lie in this plane, as shown in the figure, but for FeO they point at right angles to the plane. [G. E. Bacon, *Applications of Neutron Diffraction in Chemistry* (London: Pergamon Press, 1963), p. 110.]

structures. It is usually difficult to locate these by means of X-ray or electron diffraction, because the small scattering power of the hydrogen is overshadowed by that of heavier atoms. The hydrogen nucleus, however, scatters neutrons strongly. Thus it has been possible to work out the structures of such compounds as UH_3 and KHF_2 by neutron diffraction analysis.*

17. Closest Packing of Spheres

Quite a while before the first X-ray structure analyses, some shrewd theories about the arrangement of atoms and molecules in crystals were developed from purely geometrical considerations. For example, from 1883 to 1897, W. Barlow discussed a number of structures based on the packing of spheres.

There are two simple ways in which spheres of the same size can be packed together so as to leave a minimum of unoccupied volume, in each case 25.9% void. These are the hexagonal-closest-packed (hcp) and the cubic-closest-packed (ccp) arrangements depicted in Fig. 18.22. In these closest-packed structures, each sphere is in contact with 12 others, a hexagon of 6 in the same plane, and two triangles of 3, one above and one below. In the hcp structure, the spheres in the upper triangle lie directly over those in the lower triangle; thus the layers repeat in the order *AB, AB, AB, AB*. In the ccp structure, the orientation of the upper triangle is rotated 60° from that of the lower. Thus the repeat order is *ABC, ABC, ABC*.

The ccp structure is based on a face-centered cubic (fcc) unit cell. The closest-packed layers form the 111 planes of this structure, and they are thus stacked normal to the [111] directions. The hcp structure is based on a hexagonal unit cell. The closest-packed layers form the 001 planes, which are stacked normal to the sixfold *c* axis.

The ccp structure is found in the solid state of the inert gases, in crystalline methane, etc.—symmetrical atoms or molecules held together by van der Waals forces. The high temperature forms of solid H_2, N_2, and O_2 occur in hcp structures.

TABLE 18.4
Structures of the Metals

Cubic closest packed (fcc)—(ccp)		Hexagonal closest packed (hcp)		Body-centered cubic (bcc)	
Ag	γFe	αBe	Os	Ba	Mo
Al	Ni	γCa	αRu	αCr	Na
Au	Pb	Cd	βSc	Cs	Ta
αCa	Pt	αCe	αTi	αFe	βTi
βCo	Sr	αCo	αTl	γFe	V
Cu	Th	βCr	Zn	K	βW
		Mg	αZr	Li	βZr

*S. W. Peterson and H. A. Levy, *J. Chem. Phys.*, **20**, 704 (1952).

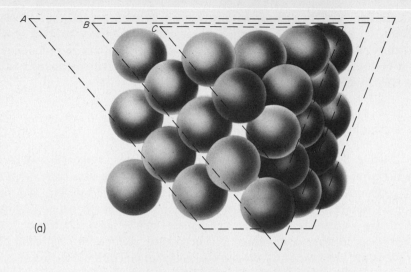

(a)

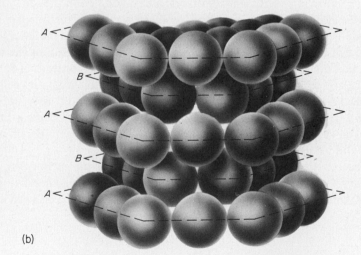

(b)

FIG. 18.22 Structures that arise from the closest packing of uniform spheres. (a) Cubic-closest packing. The closest packed layers are the 111 planes in the face-centered cubic structure. (b) Hexagonal-closest packing. The closest packed layers are the 001 planes [L. V. Azaroff, *Introduction to Solids* (New York: McGraw-Hill Book Company, 1960].

Most typical metals crystallize in the ccp, the hcp, or a body-centered-cubic (bcc) structure. Some examples are collected in Table 18.4. Other metal structures include the following:* the diamond-type cubic of grey tin and germanium; the

*For descriptions, see R. W. G. Wyckoff, *Crystal Structures* (New York: Interscience Publishers, 1963).

face-centered tetragonal, a distorted fcc, of γ manganese and indium; the rhombohedral layered structures of bismuth, arsenic, and antimony; the body-centered
tetragonal of white tin; the simple cubic of polonium. It will be noted that many
of the metals are polymorphic (allotropic), with two or more structures depending
on conditions of temperature and pressure.

The nature of the binding in metal crystals will be discussed later. For the
present, we may think of them as a network of positive metal ions packed primarily according to geometrical requirements, and permeated by mobile electrons,
the so-called *electron gas*.

18. Binding in Crystals

Two different theoretical approaches to the nature of the chemical bond in molecules were described in Chapter 15. In the valence-bond method, the point of
departure is the individual atom. In the molecular-orbital method, the electrons
are not assigned possessively to the individual atoms.

For studying the nature of binding in crystals, two methods closely related to
the two basic models for molecules are available. In one case, the crystal structure
is pictured as an array of regularly spaced atoms, each possessing electrons used
to form bonds with neighboring atoms. These bonds may be ionic, covalent, or
intermediate in type. Extending throughout three dimensions, they hold the
crystal together. The alternative approach is to consider the nuclei at fixed positions in space and gradually to pour the electron cement into the periodic array of
nuclear bricks. Both these methods yield useful and distinctive results, displaying
complementary aspects of the nature of the crystalline state. The first method,
growing out of the valence-bond theory, is the *bond model* of the solid state. The
second method, an extension of the method of molecular orbitals, we shall call,
for reasons to appear later, the *band model* of the solid state. The bond model is
sometimes called the *tight binding approximation*, and the band model, the *collective
electron approximation*.

19. The Bond Model of Solids

If we consider that a solid is held together by chemical bonds, we shall naturally
proceed to classify the bond types. Our classification may not be clear-cut in all
instances, but it is nevertheless useful as a conceptual framework for distinguishing
various types of solid.

(1) *Van der Waals Bonds*. These bonds result from forces between inert atoms
or essentially saturated molecules. The forces are the same as those responsible
for the *a* term in the van der Waals equation. Crystals held together in this way
are sometimes called *molecular crystals*. Examples are nitrogen, carbon tetrachloride, benzene. The molecules tend to pack together as closely as their sizes

and shapes allow. The binding between the molecules in van der Waals structures represents a combination of factors, such as dipole–dipole and dipole–polarization interactions, and the quantum mechanical *dispersion forces*, first elucidated by F. London, which are often the principal component. The theory of these forces will be discussed in the next chapter.

(2) *Ionic Bonds.* This type of bond is familiar from the case of the NaCl molecule in the vapor state (Section 15.2). In the NaCl crystal, the coulombic interaction between oppositely charged ions leads to a regular three-dimensional structure; each positively charged Na^+ ion is surrounded by six negatively charged Cl^- ions, and each Cl^- is surrounded by six Na^+. There are no distinct NaCl molecules. The ionic bond is spherically symmetrical and undirected; an ion will be surrounded by as many oppositely charged ions as can be accommodated spacially, provided that the requirement of overall electrical neutrality is satisfied.

(3) *Covalent Bonds.* These bonds are the result of the sharing of electrons by atoms. When extended through three dimensions, they may lead to a variety

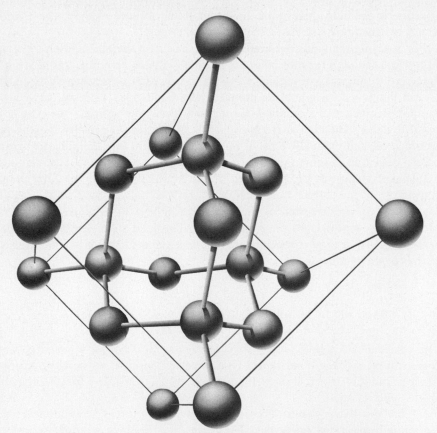

FIG. 18.23 Cubic unit cell of the diamond structure (drawn by F. M. Thayer). [G. H. Wannier, *Solid State Theory,* Cambridge University Press, 1959.] If the four atoms tetrahedrally surrounding a central atom are of a different species, the structure would be that of zinc blende (ZnS).

of crystal structures, depending on the number of electrons available for bond formation.

A good example is the diamond structure in Fig. 18.23. This structure can be based on two interpenetrating face-centered cubic lattices. Each atom on one lattice is surrounded tetrahedrally by four equidistant atoms on the other lattice. This arrangement constitutes a three-dimenstional polymer of carbon atoms joined together by tetrahedrally oriented bonds. Thus the configuration of the carbon bonds in diamond is similar to that in aliphatic compounds, such as ethane. Germanium, silicon, and grey tin also crystallize in the diamond structure.

A similar structure is assumed by compounds such as ZnS (in the zinc-blende structure), AgI, AlP, and SiC. In all these structures, each atom is surrounded by four unlike atoms oriented at the corners of a regular tetrahedron. In every case, the binding is primarily covalent. The structure can occur whenever the number of outer-shell electrons is four times the number of atoms; it is not necessary that each atom provide the same number of valence electrons.

In graphite, the more stable allotrope of carbon, the bonds resemble those in aromatic organic compounds. The structure is shown in Fig. 18.24. Strong bonds operate within each layer of carbon atoms, whereas weaker binding joins the layers—hence, the slippery and flaky nature of graphite. The C—C bond distance within the layers of graphite is 0.134 nm, identical with that in anthracene. As in aromatic hydrocarbons (p. 719), we can distinguish two types of electrons within the graphite structure. The σ electrons are paired to form localized pair (sp^2) bonds, and the π electrons are free to move throughout the planes of the C_6 rings.

Atoms with a valence of only 2 cannot form isotropic three-dimensional structures. Consequently we find interesting structures such as that of selenium (Fig. 18.25) and tellurium, which consist of endless chains of atoms extending through the crystal, the individual chains being held together by much weaker forces. Another way of solving the problem is illustrated by the structure of rhom-

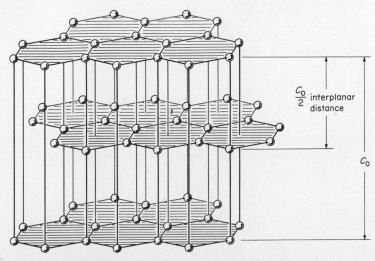

$\dfrac{c_0}{2}$ interplanar distance

c_0

FIG. 18.24 The hexagonal structure of graphite.

bic sulfur, Fig. 18.26. Here there are well defined, puckered, eight-membered rings of sulfur atoms. The bivalence of sulfur is maintained and the S_8 molecules are held together by van der Waals attractions. Elements like arsenic and antimony, which in their compounds display a covalence of 3, tend to crystallize in structures containing layers or sheets of atoms.

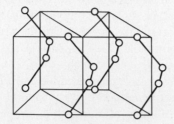

FIG. 18.25 Structure of selenium.

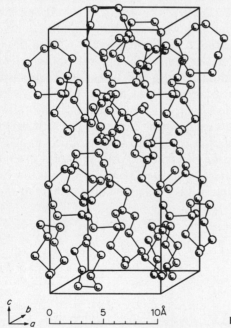

FIG. 18.26 Structure of rhombic sulfur.

(4) *Bonds of Intermediate Type.** In solids, as in individual molecules, such bonds can be considered to result from resonance between covalent and ionic contributions. Alternatively, we may consider the polarization of one ion by an oppositely charged ion. An ion is polarized when its electron distribution is distorted by the presence of an oppositely charged ion. The larger an ion, the more readily is it polarized, and the smaller an ion, the more intense is its electric

*The type of crystal structure can often be predicted from the *ionicity* of the bonds: J. C. Phillips, "Ionicity of Chemical Bonds in Crystals," *Rev. Mod. Phys. 42*, 317 (1970).

field and the greater its polarizing power. Usually, therefore, the larger anions tend to be strongly polarized by the smaller cations. Even apart from the effect of size, cations are less polarizable than anions because their net positive charge tends to hold their electrons in place. The structure of an ion is also important: rare-gas cations, such as K^+, have less polarizing power than transition cations, such as Ag^+, because their positive nuclei are more effectively shielded.

The effect of polarization appears in the structures of the silver halides. AgF, AgCl, and AgBr have the rock salt structure, but as the anion becomes larger, it

FIG. 18.27 The arrangement of molecules in the ice crystal. The orientation of the water molecules as represented in the drawing is arbitrary; there is one proton along each oxygen–oxygen axis, closer to one or the other of the two oxygen atoms. [Linus Pauling, *The Nature of the Chemical Bond* (Ithaca: Cornell University Press, 1960).]

becomes more strongly polarized by the small Ag$^+$ ion. Finally, in AgI the binding has little ionic character and the crystal has the zinc-blende structure. It has been confirmed spectroscopically that crystalline silver iodide is composed of atoms and not ions.

(5) *Hydrogen Bonds*. The hydrogen bond, discussed in Section 15.28, is important in many crystal structures, e.g., inorganic and organic acids, salt hydrates, ice. The structure of ordinary ice is shown in Fig. 18.27. The coordination is similar to that in wurtzite, the hexagonal form of zinc sulfide. Each oxygen is surrounded tetrahedrally by four nearest neighbors at a distance of 0.276 nm. The hydrogen bonds hold the oxygens together, leading to a very open structure. By way of contrast, hydrogen sulfide has a ccp structure, each molecule having 12 nearest neighbors.

(6) *Metallic Bonds*. The metallic bond is closely related to the ordinary covalent electron-pair bond. Each atom in a metal forms covalent bonds by sharing electrons with its nearest neighbors, but the number of orbitals available for bond formation exceeds the number of electron pairs available to fill them. As a result, covalent bonds resonate among the available interatomic positions. In the case of a crystal, this resonance extends throughout the entire structure, thereby producing greatly enhanced stability. The empty orbitals permit a ready flow of electrons under the influence of an applied electric field, leading to metallic conductivity.

20. Electron-Gas Theory of Metals

The collective-electron or band model for solids had its origin in the theory of metals. The cohesive energy of metals, and their high electric and thermal conductivities, remained a mystery until after the discovery of the electron in 1895. Drude in 1905 suggested that a piece of metal is like a box, or three-dimensional potential well, containing a gas of freely mobile electrons. If an external electric field is applied, the negative electrons flow up the gradient of electric potential, i.e., an electric current occurs. Conductivity σ is the ratio of current per unit area $\mathbf{j}$ to field $\mathbf{E}$, $\sigma = \mathbf{j}/\mathbf{E}$. As was seen in the discussion of ionic conductivity, $\sigma = C|Q|u$, where C is the concentration of charge carriers, $|Q|$ the absolute value of their charge, and u their mobility. The conductivity of metals could be explained on Drude's model if it was assumed that all the valence electrons were included in C, and, furthermore, that the mobility u was so high that the electrons could move freely over hundreds of atomic distances without being appreciably deflected by collisions with nuclei or other electrons. In other words, the term *electron gas* was no mere figure of speech; electrons in metals really appeared to have kinetic properties similar to those of gas molecules.

A serious objection to the Drude theory then became manifest. If electrons really behaved like gas molecules, they should take up kinetic energy when the metal is heated. In accord with the principle of equipartition of energy, Section 4.18, the translational energy of a mole of electrons should be $\frac{3}{2}LkT$, where L is the Avogadro Number and k the Boltzmann constant, giving an electronic contribution to the heat capacity of $C_V = (\partial U/\partial T)_V = \frac{3}{2}Lk$ per mole. Experimentally, however, there was no electronic heat capacity of any such magnitude. In fact, the heat capacities of metals at ordinary temperatures were close to the value given by the ancient rule of Dulong and Petit, $C_V = 3Lk$ per mole, which was accounted for completely by the $3L$ degrees of vibrational freedom per atom. At very low temperatures, a small electronic heat capacity could be detected, and in a few metals (e.g., nickel), there appeared to be a detectable increase in the electronic contribution at very high temperatures. But, practically speaking, the electronic heat capacity predicted by the Drude theory was nowhere to be found.

This paradoxical failure of the electron-gas theory was really a failure of statistical mechanics to give the correct C_V. The resolution of the problem, first given by Sommerfeld in 1928, required a deep analysis and ultimate revision of the Boltzmann law of statistical mechanics.

21. Quantum Statistics

We must go back to the point in Chapter 13 where we obtained the solution for the motion of a particle in a three-dimensional box. At that time, we were considering particles without spin and the Pauli Exclusion Principle had not yet been described. Consequently, if we were to distribute N electrons in the energy levels of Fig. 13.15, while neglecting any electrostatic interaction between the electrons, they would on the average occupy the levels in accord with a Boltzmann distribution law,

$$N_i = g_i N_0 \exp\left(-\frac{\epsilon_i}{kT}\right)$$

This expression would give the number of electrons N_i in a level at energy ϵ_i and degeneracy g_i. In particular, at 0 K, $N_i = 0$ for all $i > 0$, and all the electrons would occupy the lowest energy level, $\epsilon_0 = 0$. Now, electrons cannot really behave in such a way, because they are elementary particles of spin $s = \frac{1}{2}$, and the Pauli Principle will allow only $2g_i$ electrons to be placed in any energy level ϵ_i. Thus, even at the absolute zero, there must be a wide spread of occupied energy levels. All the lowest states are filled with pairs of electrons until some maximum energy level E_F is reached. If we draw the distribution function, the probability $p(E)$ of filling a level as a function of energy E of the level, we find the result shown in the dashed curve in Fig. 18.28: $p(E) = 1$ until we reach E_F, after which $p(E) = 0$.

This $p(E)$ is an example of a *Fermi–Dirac distribution function*. It is the function to be expected when elementary particles are distributed in translational

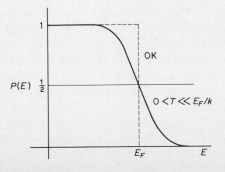

FIG. 18.28 The Fermi-Dirac distribution function for energies of electrons in a metal.

energy levels with the requirement that they obey the Pauli Exclusion Principle. At any temperature above 0 K, some electrons will move into higher energy levels, and at temperatures still small compared with E_F/k, the distribution function will have the appearance of the solid curve in Fig. 18.28.

The mathematical expression for the distribution function is*

$$p(E) = \frac{1}{e^{(E-E_F)/kT} + 1} \tag{18.18}$$

Note that when $p(E) = \frac{1}{2}$, $E = E_F$, where E_F is called the *Fermi energy*. In metals, the Fermi energy acts as an effective cut-off level for allowed energies of electrons.

As long as $E_F \gg kT$, the $p(E)$ function has the general form shown in Fig. 18.28. When $E_F \leq kT$, however, the distribution becomes similar to the classical Boltzmann one. For example, at 1000 K, $kT = 0.086$ eV, and for sodium, $E_F = 3.12$ eV, a typical value for a metal. Thus the electron gas in metals will follow the quantum mechanical (Fermi–Dirac) distribution law. It is now evident why the electron gas does not contribute appreciably to the heat capacity of a metal. The only way an electron could gain energy when the metal is heated would be to move into a higher energy state, but, for the typical electron deep in the Fermi sea, there are no empty levels available, since almost all the higher levels nearby would be already filled with electrons. Only a relatively few electrons at the top of the distribution can find empty levels into which to move. (These electrons occupy what is called the *Maxwellian tail* of the *Fermi–Dirac distribution*.) As a result, at ordinary temperatures, the electronic heat capacity is almost negligible.

22. Cohesive Energy of Metals

The cohesion of metals can be understood in qualitative terms as the consequence an electrostatic attraction between the positive cores of the metal atoms and the negative fluid of mobile electrons. A quantitative treatment of the problem involves solution of the Schrödinger equation for the energy of many electrons in a periodic electric field specified by the crystal structure of the metal.

Consider in Fig. 18.29 a simplified model of a one-dimensional structure. For concreteness, the nuclei are taken to be those of sodium, with a charge of $+11$. The position of each nucleus represents a deep potential energy well for the electrons. If these wells were far apart, the electrons would all fall into fixed positions on the sodium nuclei, giving rise to $1s^2 2s^2 2p^6 3s^1$ configurations, typical of isolated sodium atoms. This is the situation shown in Fig. 18.29(a). In the metal, however, the potential wells are not far apart and are not infinitely deep, and the actual situation is more like that shown in Fig. 18.29(b). The electrons can tunnel through the barriers, an electron on one nucleus slipping through to occupy a position on a neighboring nucleus. We are thus no longer concerned with energy levels of individual sodium atoms but with levels of the crystal as a whole. The

---

*The Fermi–Dirac distribution law can be derived immediately from the *grand partition function*. [L. D. Landau and E. M. Lifshitz, *Statistical Physics* (London: Pergamon Press, 1958), p. 152.] The derivation shows that $E_F = \mu$, the chemical potential of the electron.

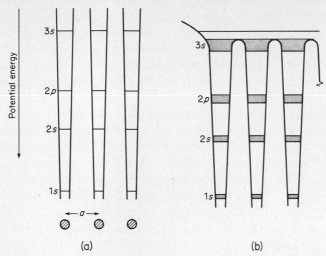

FIG. 18.29   Energy levels in sodium: (a) isolated atoms; (b) section of crystal. The sharp levels in the atom have become bands in the crystal.

Pauli Principle tells us that no more than two electrons can occupy exactly the same energy level. Therefore, once the possibility of electrons moving through the structure is admitted, we can no longer consider the energy levels to be sharply defined. The sharp $1s$ energy level in an isolated sodium atom is broadened in crystalline sodium into a band of closely packed energy levels. A similar situation arises for the other energy levels, each becoming a band of levels as shown in Fig. 18.29(b).

Each atomic orbital contributes one level to a band. In the lower bands ($1s$, $2s$, $2p$), there are just enough levels to accommodate the number of available electrons, so that the bands are completely filled. If an external electric field is applied, the electrons in the filled bands cannot move under its influence, for to be accelerated by the field they would have to move into somewhat higher energy levels. This is impossible for electrons in the interior of a filled band, because all the levels above them are already occupied and the Pauli Principle forbids their accepting additional tenants. Nor can the electrons at the very top of a filled band acquire extra energy, since there are no higher levels for them to move into. Occasionally, it is true, an electron may receive a jolt of energy and be knocked completely out of its band into a higher unoccupied band.

So much for the electrons in the lower bands. The situation is different in the uppermost band, the $3s$, which is only half-filled. An electron in the interior of the $3s$ band still cannot be accelerated because the levels directly above are already filled. Electrons near the top of the filled zone, however, can readily move up into unfilled levels within the band. The topmost band has actually broadened sufficiently to overlap the peaks of the potential-energy barriers, so that electrons in the top upper levels can move quite freely throughout the crystal structure. According to this idealized model, in which the nuclei are always arranged at the

points of a perfectly periodic lattice, no resistance would oppose the flow of an
electric current. The actual resistance is caused by departures from perfect periodicity. One important loss of periodicity results from the thermal vibrations of the
atomic nuclei. These vibrations destroy the perfect resonance between the electronic
energy levels and thus cause a resistance to the flow of electrons. As would be
expected, the resistance from this mechanism increases with temperature. Increased
resistance also results if an alloying constituent is added to a pure metal, and the
regular periodicity of the structure is diminished by the foreign atoms.

At this point, the reader may well be thinking that this is a pretty picture for a
univalent metal like sodium, but what of magnesium with its two $3s$ electrons
and therefore completely filled $3s$ band? Why is it not an insulator instead of a
metal? More detailed calculations show that in such cases, the $3p$ band is low
enough to overlap the top of the $3s$ band, providing a large number of available
empty levels. Actually, this happens for the alkali metals also. The way in which
the $3s$ and $3p$ bands in sodium broaden as the atoms are brought together is shown
in Fig. 18.30. The interatomic distance in sodium at 1 atm pressure and 298 K is

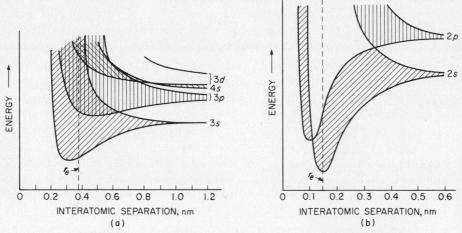

FIG. 18.30 Approximate quantum mechanical calculation of the
formation of energy bands as atoms are brought together
into a crystal: (a) sodium, (b) diamond.

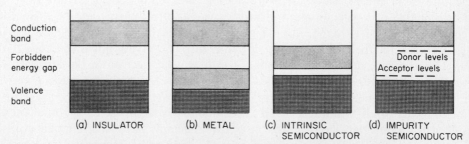

FIG. 18.31 Schematic band models of solids classified according to
electronic properties.

$r_e = 0.38$ nm. At this distance, there is no longer any gap between the $3s$ and $3p$ bands. In the case of diamond, on the other hand, there is a large energy gap between the filled *valence band* and the empty *conduction band* at its $r_e = 0.15$ nm.

Thus, conductors are characterized either by partial filling of bands or by overlapping of the topmost bands. Insulators have completely filled lower bands with a wide energy gap between the topmost filled band and the lowest empty band. These models are represented in Fig. 18.31.

23. Wave Functions for Electrons in Solids

The existence in crystals of bands of allowed electronic energy levels separated by energy gaps in which electronic states are forbidden has been derived (in a qualitative way) from the somewhat similar discrete energy states and energy gaps for electrons in isolated atoms. This approach is called the *tight-binding approximation* when it is worked out in a quantitative way. It is also possible to start with the free electron gas and to derive the existence of energy bands and gaps from the effects of the periodic potential of the atomic cores. From a basic theorem of differential equations,* we can derive the following important result: If $\psi_0(x)$ is a solution of the one-dimensional Schrödinger equation for a free electron, a solution for motion of the electron in a potential $U(x)$ that is periodic with a period a, i.e.,

$$U(x) = U(x - a),$$

can always be obtained in the form

$$\psi(x) = \psi_0(x)\, u(x - a) \tag{18.19}$$

where $u(x - a)$ itself has the same period a as the potential U. In solid-state theory, this result is called *Bloch's theorem* and functions of the form (18.19) are *Bloch functions*. They form the basis of many quantum mechanical calculations of properties of crystals.

A solution for the free electron as a running wave was given in Eq. (13.55) as $\psi_0 = e^{ikx}$, where $k = 2\pi\sigma = 2\pi/\lambda = 2\pi p/h$, where λ is the de Broglie wavelength and p is the momentum of the electron. Then the Bloch function becomes

$$\psi(x) = e^{ikx}\, u(x - a)$$

Since u is periodic with the period a of the potential, it can be expanded in a Fourier series,

$$u(x - a) = \sum A_n\, e^{-2\pi i n x/a} \tag{18.20}$$

If the perturbing potential $U(x - a)$ is small, we can neglect the terms beyond A_0 except when $k = \pi n/a$, so that the Bloch function becomes

$$\psi = A_0\, e^{ikx} + A_n\, e^{ik_n x}$$

where $k_n = k - 2n\pi/a$.

---

*Floquet's theorem. See E. T. Whittaker and G. N. Watson, *Modern Analysis* (London: Cambridge University Press, 1952), p. 412.

Let us plot the energy E of the electron vs. k. For the free electron,

$$E = \frac{p^2}{2m} = \frac{h^2 k^2}{8\pi^2 m} \quad \text{since} \quad p = \frac{hk}{2\pi} \qquad (18.21)$$

This is the equation of a parabola. The effect of the periodic potential is to introduce energy gaps at $k = \pm\pi/a$, $\pm 2\pi/a \pm n\pi/a$. The result is shown in Fig. 18.32. For the lower values of k, the curve of E vs. k coincides with the

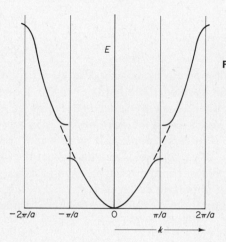

FIG. 18.32　Energy vs. wave number for motion of an electron in the one-dimensional periodic potential of Eq. (18.20). The range of allowed k values from $-\pi/a$ to $+\pi/a$ is the first *Brillouin Zone* for this system. The unbroken parabola represented by following the dashed curve would give the energy of the free electron in accord with Eq. (18.21).

parabola for free electrons. As k approaches $\pm\pi/a$, however, the slope of E vs. k decreases, and at $k = \pm\pi/a$, there is a discontinuity in E over a range of values that are *forbidden energy states* for the electron in the periodic structure. Thus, a band of allowed energy levels is followed by a gap, then another band of allowed energies occurs, followed by another gap, and so on.

The condition for a discontinuity in energy, $k = \pm n\pi/a$, is simply the Bragg condition, $n\lambda = 2d \sin\theta$, since $k = 2\pi/\lambda$, and $a = d \sin\theta$, with $\theta = 90°$ for the one-dimensional case. The discontinuities occur at wavelengths for which electrons incident on the line of atoms would meet the Bragg condition for scattering. Thus, the electrons at these wavelengths cannot pass through the structure but will be "reflected" (in the Bragg sense) back out.

The regions of allowed energy are called *Brillouin zones*. The zones exist in k space, which from (18.9) and (18.21) is clearly reciprocal space or Fourier space. Thus, the Brillouin zones in the one-dimensional case are segments of a one-dimensional Fourier space for the electrons. The first zone extends from $k = -\pi/a$ to $k = +\pi/a$. The second zone is the k space from $-2\pi/a$ to $-\pi/a$ plus that from $\pi/a + 2\pi/a$.

The extension of these concepts to three dimensions is straightforward. The Brillouin zones can be drawn as volumes in k space, which are bounded by planes specified by the condition

$$k_x n_1 + k_y n_2 + k_z n_3 = \frac{\pi}{a}(n_1^2 + n_2^2 + n_3^2)$$

These are just the values of $k(k_x, k_y, k_z)$ for which Bragg reflection would occur.*

24. Semiconductors

Solids are classified on the basis of their electrical conductivity into three types:

1. Conductors or metals, which offer a low resistance to the flow of electrons when a potential difference is applied. The resistivities of metals lie in the range 10^{-6} to 10^{-8} $\Omega \cdot m$ at room temperature, and increase with the temperature.
2. Insulators, which have high resistivities, from 10^8 to 10^{20} $\Omega \cdot m$ at room temperature.
3. Semiconductors, the resistivities of which are intermediate between those of typical metals and typical insulators, and decrease with rising temperature, usually as $e^{\epsilon/kT}$.

The electrical properties of crystals depend on the magnitude of the *band gap* between the filled valence band and the higher empty conduction band. One way to determine the gap is to measure the wavelength at which optical absorption begins in the crystal, the so-called *absorption edge*. The energy at the onset of absorption should correspond to the energy necessary to transfer an electron from the top of the filled valence band to the bottom of the conduction band. Values of the band gap ϵ for crystals with the silicon structure are as follows: C (diamond), 5.2 eV; Si, 1.09 eV; Ge, 0.60 eV; Sn (grey), 0.08 eV.

The ratio of the number of electrons thermally excited to the conduction band to the number in the valence band is given by a Boltzmann factor $e^{-\epsilon/2kT}$. For diamond, ϵ is so high that electrons rarely reach the conduction band by thermal excitation, so that the crystal is a good insulator. In the cases of Si and Ge, however, there will be an appreciable number of *conduction electrons* produced by thermal excitation from the valence band. These crystals are typical *intrinsic semiconductors*.

When an electron in an intrinsic semiconductor jumps into the conduction band, it leaves behind a *hole* in the valence band. Electrons in a completely filled valence band could make no contribution to the conductivity. As soon as holes appear, however, the remaining electrons find some empty states available and thus can contribute to the conductivity. A hole in a band of negative electrons is effectively a point of positive charge.† The jump of an electron into a hole is

---

*For further discussion and examples, see W. J. Moore, *Seven Solid States* (New York: W. A. Benjamin, Inc., 1967).

†A. H. Wilson, *Semiconductors and Metals* (London: Cambridge University Press, 1939), pp. 8–10, gives a simple proof.

equivalent to the jump of a positive charge into the position vacated by the electron. We can therefore treat the motion of electrons in an almost filled band as if the holes were positive charges moving in an almost empty hole band.

Electrons and holes, besides having opposite charges, may have different effective mobilities, u_e and u_h. If C_e and C_h are the concentrations of electrons and holes, the electrical conductivity becomes

$$\kappa = e\,(C_e\,u_e + C_h\,u_h) \tag{18.22}$$

where e is the absolute value of the electronic charge. In pure silicon, we should expect that $C_e = C_h$. In other substances, however, C_e may differ appreciably from C_h. In such cases, we refer to the *majority carriers* and the *minority carriers* of electric current in the semiconductors.

25.   Doping of Semiconductors

We can draw an analogy between an intrinsic semiconductor like silicon and a weakly ionizing solvent like water.

$$H_2O \rightleftharpoons H^+ + OH^-, \qquad K_w = [H^+][OH^-]$$

$$Si \rightleftharpoons h^+ + e^-, \qquad K_i = [h^+][e^-]$$

In pure intrinsic silicon,

$$[h^+] = [e^-] = K_i^{1/2} = A(T)\,e^{-\epsilon/2kT}$$

The solution of a weak base in water is then analogous to the solution in silicon or germanium of atoms (As, Sb, P, etc.) that have more valence electrons than the four of silicon itself.

$$NH_3 + H_2O \longrightarrow NH_4^+ + OH^-$$

$$As + Ge \longrightarrow As^+ + Ge^-$$

The solution of a weak acid in water is analogous to the solution in silicon or germanium of atoms (B, In, etc.) that have fewer valence electrons than silicon.

$$CO_2 + H_2O \longrightarrow HCO_3^- + H^+$$

$$In + Si \longrightarrow In^- + Si^+$$

Figure 18.33 depicts the doping of silicon; the corresponding energy-band diagram was shown in Fig. 18.31.

When [as in Fig. 18.33(a)] a P atom is substituted for a Si atom, four of the valence electrons of P can enter the valence band, but the fifth electron must enter some level of higher energy. In fact, this energy state is only 0.012 eV below the conduction band. Therefore, the bound electrons in the *impurity levels* can readily be thermally excited into the conduction band. The doped semiconductor will have a greatly enhanced conductivity compared with pure intrinsic silicon. A

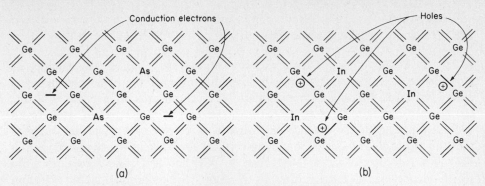

FIG. 18.33   (a) Donors produced by Group V impurities in germanium; (b) acceptors produced by Group III impurities.

semiconductor such as Si doped with P is called *n-type*, because the majority
current carriers are negatively charged (electrons). A dopant such as P or As, which
can give electrons to the conduction band, is called a *donor*, and the extra energy
levels just below the conduction band are *donor levels*.

If the doping atom, boron for example, has fewer valence electrons than silicon,
the schematic structure would look like Fig. 18.33(b). There would be a hole or
missing electron in the tetrahedral bonds about the B atom. Consequently, a new
level occurs within the band gap. In the case of B in Si, these impurity levels are
only 0.01 eV above the top of the valence band. Electrons can readily jump from
the top of the valence band to fill such *acceptor levels*. The positive holes left in
the valence band strongly enhance the electrical conductivity and silicon doped
with boron becomes a *p-type* semiconductor.

Ionization energies of several solutes in Si and Ge are summarized in Table
18.5.

TABLE 18.5
Ionization Energies of Solutes in Silicon and Germanium (in eV)

	In silicon	In germanium
Donors		
Li^+	0.0093	0.033
P^+	0.012	0.045
As^+	0.0127	0.049
Sb^+	0.0096	0.039
Acceptors		
B^-	0.0104	0.045
As^-	0.0102	0.057
In^-	0.0112	0.16
Ca^-	0.04	0.49

26.   Nonstoichiometric Compounds

A departure of a compound from chemical stoichiometry can greatly enhance its conductivity. Zinc oxide, ZnO, usually contains more Zn than O, whereas nickel oxide, NiO, usually contains more O than Ni. Such nonstoichiometry may be due to the incorporation of extra atoms into the crystal at interstitial sites, or to vacancies caused by the absence of atoms from normal sites. Evidence suggests that the excess zinc in ZnO is largely interstitial Zn, whereas the excess oxygen in NiO is largely due to vacant Ni^{2+} sites. In NiO and ZnO, the departures from stoichiometry, even at temperatures up to 1000°C, are quite small, about 0.1 atom percent excess of Zn or O. In other metallic oxides and chalcogenides, however, the departures from stoichiometric composition may be much wider.

We may note that the excess Zn in ZnO acts as a typical donor and leads to *n*-type semiconductivity. In NiO, the Ni^{2+} vacancies, each associated with two Ni^{3+} to preserve electrical neutrality, provide typical acceptor levels leading to *p*-type semiconductivity. (We distinguish in nomenclature between a *vacancy*, the absence of an atom from a normal structural site, and a *hole*, the absence of an electron from a normal band or bond orbital.)

27.   Point Defects

In 1896, the English metallurgist Roberts-Austen showed that gold diffused faster in lead at 300°C than sodium chloride diffused in water at 15°C. This is one example of the surprising ease with which atoms can sometimes move about in the solid state. Rate processes, such as diffusion, sintering, and tarnishing reactions, provide further evidence. It was difficult to believe that atoms could easily move in solids by changing places with one another: the activation energy for such a process would be much too high.

More reasonable mechanisms were suggested by I. Frenkel, in 1926, and by W. Schottky, in 1930. They proposed the first exact models for what are now called *point defects* in crystals. The Frenkel defect is shown in Fig. 18.34(a) as

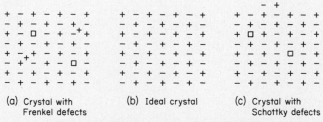

(a) Crystal with Frenkel defects   (b) Ideal crystal   (c) Crystal with Schottky defects

FIG. 18.34   Intrinsic point defects in crystals.

originally proposed for silver chloride. It consists of an Ag^+ ion that has left its lattice site and moved to an interstitial position. The Frenkel defect consists of the vacancy plus the interstitial. The Schottky defect is shown in Fig. 18.34(c), as it occurs in sodium chloride. It consists of a pair of vacancies of opposite signs. Frenkel and Schottky defects are called *intrinsic defects*. They do not alter the exact stoichiometry of the crystal, but they provide a facile mechanism by which atoms can move within the crystal. It is much easier for an atom to move from an occupied site into a vacancy than for two atoms in occupied sites to change places directly.

We can calculate the concentration of the various point defects from simple statistical considerations. It costs an energy U to make a defect, but entropy S is gained owing to the disorder associated with the entropy of mixing of the defects with the normal lattice sites. If n imperfections are distributed among the total of N crystal sites, the entropy of mixing is

$$S = k \ln W = k \ln \frac{N!}{(N - n)!\, n!}$$

If ϵ is the increase in energy per defect, and if we neglect contributions to the energy caused by changes in vibration frequencies in the neighborhood of the defect, we can write for the change in the Helmholtz free energy,

$$\Delta A = \Delta U - T \Delta S$$

$$\Delta A = n\epsilon - kT \ln \frac{N!}{(N - n)!\, n!}$$

At equilibrium,

$$\left(\frac{\partial \Delta A}{\partial n}\right)_T = 0$$

Applying the Stirling formula ($\ln X! = X \ln X - X$), we find

$$\ln \frac{n}{N - n} = -\frac{\epsilon}{kT}$$

and thus if $n \ll N$,

$$n = Ne^{-\epsilon/kT} \tag{18.23}$$

As an example, if ϵ is about 1 eV and T is 1000 K,

$$\frac{n}{N} \approx 10^{-5}$$

For a pair of vacancies, the expression for the number of ways of forming the defect is squared and we get finally, for Schottky defects,

$$n = Ne^{-\epsilon/2kT} \tag{18.24}$$

For Frenkel defects,

$$n = (NN')^{1/2}\, e^{-\epsilon/2kT} \tag{18.25}$$

where N' is the number of interstitial sites available.

28.   Linear Defects: Dislocations

The stress–strain curve for a typical solid is traced in Fig. 18.35. The linear region of the curve, where strain is proportional to stress, represents an *elastic deformation*; if the stress is released in such a region, the solid springs back to its original length. It is true that a low stress maintained for a long time may produce some permanent deformation of a solid, a phenomenon called *creep*. But the rapid

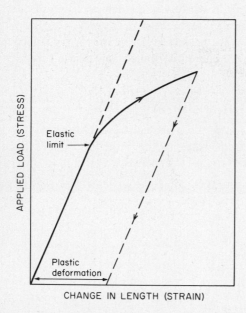

FIG. 18.35 Stress vs. strain behavior of a typical metal rod.

CHANGE IN LENGTH (STRAIN)

irreversible deformation of a solid, which we call *plastic deformation*, begins only when a certain critical stress is exceeded. The stamping of steel sheets into car panels is an example of plastic deformation.

The problem for metals under deformation is not why they are so strong, but why they are so weak. The calculated elastic limit of a perfect crystal would be 10^2 to 10^4 times that actually observed. There must be some imperfections or defects in the actual crystals, which cause them to deform plastically under quite small loads.

A solution to this problem was worked out quite independently in 1934 by Taylor, Orowan, and Polanyi.* Crystals contain defects called *dislocations*. Mott[†] has compared these defects to rucks in carpets. We all have experience of pulling carpets over floors and know that there are two ways of doing it. One can take hold

*G. I. Taylor, *Proc. Roy. Soc. London, Ser. A, 145*, 362 (1934); E. Orowan, *Z. Physik, 86*, 604, 634 (1934); M. Polanyi, *Z. Physik, 89*, 660 (1934). (The idea of the defects, now called *dislocations*, was discussed earlier, especially in the work of Prandtl, Masing, and Dehlinger.)

[†]N. F. Mott, *Atomic Structure and the Strength of Metals* (London: Pergamon Press, 1956).

of one end and tug, or one can make a ruck in one end of the carpet and gently edge it to the other end. For a big heavy carpet, the second way is the way to do it that involves least effort. The dislocation most like a ruck is the *edge dislocation* shown in Fig. 18.36, which represents a model of crystal structure viewed along a dislocation line. The dislocation line is perpendicular to the plane represented by the plane of the paper. The presence of the dislocation allows the crystal to deform readily under the influence of a shear stress. The atoms are displaced in the *slip plane* that includes the dislocation line. Thus the dislocation can move across the crystal from one side to the other, the result being a displacement of the top half

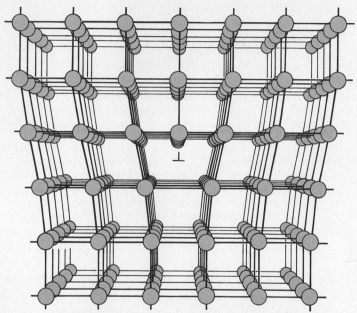

FIG. 18.36 Model to show the structure of an edge dislocation. The dislocation line is normal to the plane of the paper at the point marked ⊥. The deformation is the result of inserting an extra half-plane of atoms in the upper half of the crystal. Interatomic distances in the upper half-crystal are compressed and those in the lower half-crystal are extended. [N. B. Hannay, *Solid State Chemistry* (Englewood Cliffs, N.J.: Prentice-Hall, Inc., 1967).]

of the crystal relative to the bottom half. Figure 18.37 shows a photograph of a dislocation in a macroscopic structure of soap bubbles in two dimensions.

Figure 18.38 is a striking electron micrograph of directly observed dislocations in thin metal films.

The other basic kind of dislocation is the *screw dislocation*. You can visualize this defect by cutting a rubber stopper parallel to its axis, and then pushing on one end so as to create jog at the other end. If you suppose that initially the stopper

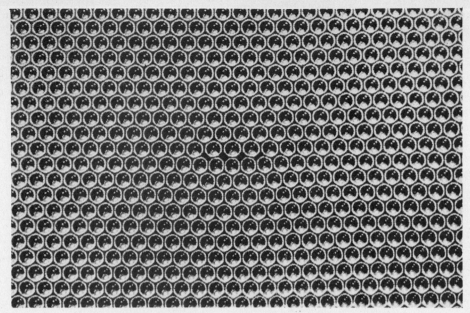

FIG. 18.37 W. L. Bragg and J. F. Nye [*Proc. Roy. Soc.* (London) *A190,* 474 (1947)] showed how many of the properties of crystals could be visualized by means of a two-dimensional raft of bubbles on the surface of an aqueous solution of detergent. This picture (by W. M. Lomer) shows a dislocation in a bubble-raft. The dislocation is most easily seen by turning the page 30° and sighting at a low angle.

contained atoms at regular lattice points, the results of the deformation would be to convert the parallel planes of atoms normal to the axis into a kind of spiral ramp. Such a displacement of the atoms constitutes a screw dislocation; the dislocation line is along the axis of the stopper. A model of the emergence of a screw dislocation at the surface of a crystal is shown in Fig. 18.39.

29. Effects Due to Dislocations

Even a single crystal of metal can be bent into shapes with curved boundaries. The dislocations in the crystal structure make such curvature possible. A mason solves the same problem when he builds an arch out of rectangular bricks. The solution is to match $N + 1$ bricks in one layer against N in the preceding layer.

Cohesive forces between atoms in a crystal offer little resistance to the gliding motion of dislocations, and only a crystal without dislocations would approach its maximum theoretical strength. Sometimes thin metallic whiskers can be grown virtually free of dislocations. Pure iron whiskers have been obtained with tensile strengths up to 1.4×10^8 N·cm^{-2}, compared to a maximum of 3×10^7 for the

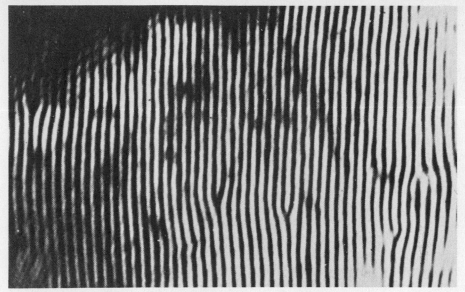

FIG. 18.38 Electronmicrograph of dislocations shown by a Moiré
pattern of overlapping (111) Au and Pd films. The Moiré
spacing is about 3.5 nm and corresponds to an effective
magnification of about 400 000. [D. W. Pashley, J. W.
Menter, and G. A. Bassett.]

strongest steel wire. The (impossible?) dream of metallurgists is to produce structural metals so free of dislocations that they would maintain their theoretical
strengths in practical applications.

The deformation of a metal does not consist simply in the sliding of dislocations
already present—the deformation process, which occurs by means of dislocations,

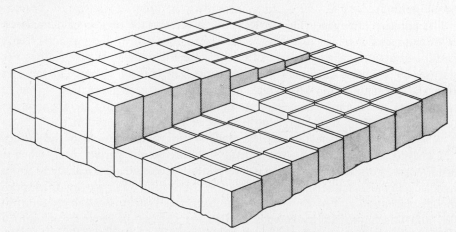

FIG. 18.39 An emergent screw dislocation at the surface of a crystal.
The atoms (or molecules) are represented by small cubes.

itself produces more dislocations. There are many different mechanisms for multiplication of dislocations, most of which are based on the existence within the crystal of an obstacle to the glide of a dislocation.

The simplest dislocation source of this kind was described by Frank and Read in 1950. Dislocations do not usually lie entirely in one slip plane. At each end of a segment of dislocation within a given slip plane there will be a *jog* where it jumps to a neighboring slip plane. A short dislocation segment between jogs is said to be *pinned*, because if a stress is applied, the pinned dislocation cannot glide in the slip plane. The pinned dislocation is the *Frank–Read source*. The only possible motion for the pinned segment is to swell out into an arc. As stress is continued, this arc expands more and more until finally it forms a complete *dislocation loop*. Then a new dislocation arc can begin to grow from the source. Under continuous moderate stress, such a source can emit dislocations like an antenna broadcasting radio waves (figuratively speaking).

We can now begin to see the answer to one of the most basic questions in the science of materials: How can the addition of a small percentage of an alloying element (carbon in iron, copper in aluminum, as examples) so greatly increase the mechanical strength of a metal? Since the metal deforms through movements of dislocations, anything that hinders such movements will increase the resistance of the metal to deformation. A foreign atom introduced into the structure of a metal will tend to reside at a position of minimum free energy. Since positions adjacent to dislocations are sites of lower free energy, foreign atoms will tend to segregate at dislocations. In this way, they tend to stabilize dislocations by locking them into place. If a dislocation were to slide away and leave the foreign atom behind, extra work (free energy) would be required to establish the foreign atom in a normal structural site. Thus, alloying elements can greatly increase the strength of metals by rendering their dislocations much less mobile.

Dislocations also provide preferred sites within crystals for occurrence of chemical reactions and physical changes (such as phase transformation, precipitation or etching).

The point of emergence of a dislocation at the surface of a crystal is also a site of enhanced chemical reactivity, a fact often revealed by the pattern of etch pits formed at the surface. We can often measure the number of dislocations per unit area by counting such etch pits. Densities range from 10^5 m^{-2} in the best silver and germanium crystals (small crystals can be grown practically free of dislocations) to 10^{16} m^{-2} in a severely deformed metal. Figure 18.40 shows an example of a study of dislocations by selective etching and decoration.

When a crystal is etched, the easiest place to remove an atom from the surface is at a dislocation, since here the crystal structure is not perfect. But, when a crystal grows from the vapor or melt, the easiest place to deposit each new atom or molecule is also at the site of a dislocation. In Section 11.5, we saw that spontaneous nucleation of a new phase in a homogeneous system is extremely improbable, and this improbability extends also to the deposition of an atom onto a perfect crystal face. In brief, such a process is improbable because there is a loss in entropy in the condensation process, but there is no compensating decrease in

FIG. 18.40 Screw dislocations and other defects are revealed in cleaved surfaces of graphite by etching and decoration. The crystal is cleaved to a thickness of about 30 nm, etched in CO_2 at 1150 °C for about 40 min. The CO_2 reacts faster at linear than at highly curved surface steps and hence "unravels the dislocation" by selective etching. Short etches in ozone and in a mixture of O_2 and Cl_2 expand surface vacancies. Finally, a minute amount of gold is evaporated onto the surface, and the gold atoms collect preferentially at the surface defects. The slice of crystal is examined by direct transmission electron microscopy. The overall magnification in the picture shown is about 100 000 X. The concentrations of screw dislocations in synthetic pyrolytic graphite crystals ranged from 10^6 to 5×10^8 cm^{-2}, but they were extremely rare in natural crystals. [G.R. Henning, Argonne National Laboratory.]

energy for the first few atoms in a new layer, since they have no neighbors to welcome them. If there is an emergent screw dislocation at the surface, however, a new atom can readily deposit at the edge of this developing screw. Thus the crystal grows most readily in the form of a developing helix. This mechanism was suggested in 1949 by F. C Frank.* Curiously enough, pictures of helical growth of crystals had been published many years previously, but until the dislocation theory was worked out, the significance of the observations was not realized. Figure 18.41 shows such a spiral growth from a pair of dislocations on a crystal of silicon.

30. Ionic Crystals†

The binding in most inorganic crystals is predominantly ionic in character. Therefore, since coulombic forces are undirected, the sizes of the ions play a most important role in determining the final structure. Several attempts have been made to calculate a consistent set of ionic radii, from which the internuclear distances in ionic crystals could be estimated. The first table, giving by V. M. Goldschmidt in 1926, was modified by Pauling. These radii are listed in Table 18.6.

TABLE 18.6

Ionic Crystal Radii[a], nm

Li^+	0.060	Na^+	0.095	K^+	0.133	Rb^+	0.148	Cs^+	0.169
Be^{2+}	0.031	Mg^{2+}	0.065	Ca^{2+}	0.099	Sr^{2+}	0.113	Ba^{2+}	0.135
B^{3+}	0.020	Al^{3+}	0.050	Sc^{3+}	0.081	Y^{3+}	0.093	La^{3+}	0.115
C^{4+}	0.015	Si^{4+}	0.041	Ti^{4+}	0.068	Zr^{4+}	0.080	Ce^{4+}	0.101
O^{2-}	0.140	S^{2-}	0.184	Cr^{6+}	0.052	Mo^{6+}	0.062		
F^-	0.136	Cl^-	0.181	Cu^+	0.096	Ag^+	0.126	Au^+	0.137
				Zn^{2+}	0.074	Cd^{2+}	0.097	Hg^{2+}	0.110
				Se^{2-}	0.198	Te^{2-}	0.221	Tl^{3+}	0.095
				Br^-	0.195	I^-	0.216		

[a] From L. Pauling, *The Nature of the Chemical Bond*, 3rd ed. (Ithaca: Cornell University Press, 1960), p. 514.

First, let us consider ionic crystals having the general formula *CA*. They may be classified according to the *coordination number* of the ions—i.e., the number of ions of opposite charge surrounding a given ion. The CsCl structure, shown in Fig. 18.42(a), has eightfold coordination. The NaCl structure [Fig. 18.42(b)] has sixfold coordination. Although zinc blende [Fig. 18.42(c)] is itself covalent, there are a few ionic crystals, e.g., BeO, with this structure, which has fourfold coordination. The coordination number of a structure is determined primarily by the number of

*F. C. Frank, *Discussions Faraday Soc.*, 5, 48 (1949).

†Students who have not yet surveyed the variety of inorganic crystal structures should peruse the standard work of A. F. Wells, *Structural Inorganic Chemistry*, 3rd ed. (New York: Oxford University Press, 1961).

FIG. 18.41   A double spiral on the surface of a crystal of silicon carbide centered on the site of the emergence of a pair of screw dislocations (300 X). W. F. Knippenberg, *Philips Research Reports, 18,* 161 (1963). A further discussion is given by A. Rabenou, "Chemical Problems in Semiconductor Research," *Endeavour, 26,* 158 (1966).

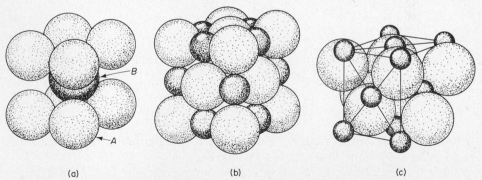

(a)   (b)   (c)

FIG. 18.42   Three important crystal structures of Type *CA.* [After R. W. G. Wyckoff, *Crystal Structures* (New York: John Wiley & Sons, Inc., 1963).] (a) A perspective drawing of the ions associated with a unit cube of the cesiumchloride structure. This structure is based on a simple-cubic lattice in which each lattice point is associated with a CsCl unit, for example Cl^- at (0. 0. 0) and Cs^+ at $(\frac{1}{2}, \frac{1}{2}, \frac{1}{2})$. (b) A perspective drawing of the unit cube of the sodium-chloride structure. The larger spheres are Cl^- ions and the smaller spheres Na^+ ions. This picture may be compared with Fig. 18.14. (c) A perspective drawing showing how Be^{2+} (smaller spheres) and O^{2-} (larger spheres) would pack together in their crystal structure with fourfold coordination. This is the same structure as that of zinc blende (ZnS), in which, however, the bonding is primarily covalent and the atoms would therefore be more nearly the same size. This structure is similar to that of diamond shown in Fig. 18.23.

the larger ions, usually the anions, that can be packed around the smaller ion, usually the cation. It should therefore depend upon the *radius ratio*, of cation to anion, r_C/r_A. A critical radius ratio is obtained when the anions packed around a cation are in contact with both the cation and with one another.

The structures of CA_2 ionic crystals are governed by the same coordination principles. Three common structures are shown in Fig. 18.43. In fluorite, each Ca^{2+} is surrounded by eight F^- ions at the corners of a cube, and each F^- is surrounded by four Ca^{2+} at the corners of a tetrahedron. This is an example of 8:4 coordination. The structure of cassiterite illustrates a 6:3 coordination. The cadmium-iodide structure illustrates the result of a departure from typically ionic binding. The iodide ion is easily polarized, and one can distinguish definite CdI_2 groups forming a layerlike arrangement.

31.   Cohesive Energy of Ionic Crystals

Equation (15.1) was an expression for the energy of interaction of a pair of ions, given as the sum of an electrostatic attraction and a short-range repulsion. The energy of a crystal composed of a regular array of positive and negative ions can be calculated by summing the interaction energy over all the pairs of ions in the array. To simplify the discussion, we shall consider only crystals made up of two kinds of ions, will charge numbers z_i and z_j. The energy of formation of the crystal from gaseous ions at infinite separation is called the crystal energy, ΔU_c. It can be calculated by summing (15.1) over all the pairs of ions in the crystal,

$$\Delta U_c = \sum_{\substack{\text{all pairs} \\ \text{of ions}}} \frac{Q_i Q_j}{4\pi\epsilon_0 r_{ij}} + b \sum_{\substack{\text{all pairs} \\ \text{of ions}}} \exp(-cr_{ij}) \qquad (18.26)$$

In a double sum over i and j, each pair would be counted twice, once as (i, j) and once as (j, i). Thus, (18.26) becomes

$$\Delta U_c = \frac{1}{2} \sum_{i,j}' \frac{Q_i Q_j}{4\pi\epsilon_0 r_{ij}} + \frac{1}{2} b \sum_{i,j}' \exp(-cr_{ij}) \qquad (18.27)$$

where $\sum'$ means that the cases $i = j$ are excluded from the double sum. The symbol r_{ij} denotes the distance from the ion i to the ion j.

The first summation can be visualized by sitting upon any ion i in the structure and adding up the coulombic energies of its interaction with all the other ions. One must take into account the attractions between this ion and the oppositely charged ions coordinated around it, the repulsions between this ion and ions of like sign at somewhat larger separations, then attractions with unlike ions once removed, etc. Therefore, for each ion, the electrostatic interaction will be a sum of terms, alternately plus and minus, and diminishing in magnitude in accord with the r_{ij}^{-1} law. For any given structure, this summation amounts to relating all the different interionic distances to the distance r_0, the shortest distance between oppositely charged ions (strictly speaking, measured at 0 K).

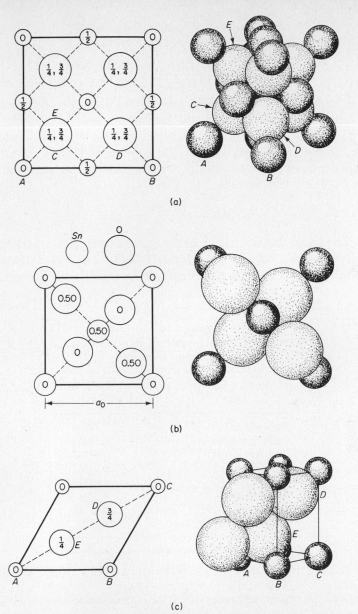

(a)

(b)

(c)

FIG. 18.43 Three important types of CA_2 crystal structures. [After R. W. G. Wyckoff, *Crystal Structures* 2nd. ed. Vol. 1 (New York: John Wiley & Sons, Inc., 1963).]   (a) The structure of fluorite, CaF_2, shown in perspective drawing and projected on the basal 100 plane. The smaller dark spheres are the Ca^{2+} ions.   (b) The structure of cassiterite, SnO_2, also that of rutile, TiO_2, and other oxides of quadrivalent metals. The small darker spheres are the metal ions. The projection is on the 100 plane of the tetragonal unit cell.   (c) The structure of cadmium iodide, CdI_2. The larger spheres represent iodide ions, but because of the partly covalent nature of the bonding the relative sizes of the Cd and the I are really not so different as they are in this purely ionic model. The projection is onto the base of the hexagonal unit cell.

891

Sums of this kind were first calculated by E. Madelung in 1918. When the summation is made, the first term in (18.27) can be written as

$$U_M = \frac{1}{2}e^2 \sum_{i,j}{}' \frac{z_i z_j}{4\pi\epsilon_0 r_{ij}} = -\frac{L\mathcal{M}' z_i z_j e^2}{r_0} \tag{18.28}$$

where U_M is called the *Madelung energy* and $\mathcal{M}'$ is a dimensionless quantity called the *Madelung constant*. The electrostatic potential energy of a single univalent ion located in a particular crystal structure would be $-\mathcal{M}'e^2/r$ and the corresponding energy for one mole of the ionic crystal is given by (18.28).

Madelung constants in the literature are almost always tabulated in terms of a somewhat different equation for U_M,

$$U_M = -\frac{L\mathcal{M} z^2 e^2}{r_0} \tag{18.29}$$

where z is the highest common factor of the charges of the two ions. For example $z = 1$ for NaCl, CaF_2, Al_2O_3; $z = 2$ for ZnS, TiO_2; $z = 3$ for AlN, etc. Table 18.7 gives values of the Madelung constants $\mathcal{M}$ for a number of important crystal structures.*

TABLE 18.7
Madelung Constants Based on r_0, Nearest Cation–Anion Distance[a]

Crystal structure type	Madelung constant $\mathcal{M}$[a]
Rock salt (AB)	1.74756
Cesium chloride (AB)	1.76267
Zinc blende (AB)	1.63805
Wurtzite (AB)	1.64132
Fluorite (AB_2)	5.03878
Rutile (AB_2)	4.7701
Cuprite (A_2B)	4.44248
Corundum (A_2B_3)	25.0312

[a] Note that $\mathcal{M}$ is a dimensionless number. This $\mathcal{M}$ is defined for use in Eq. (18-29).

In terms of (18.29), the crystal energy of (18.27) becomes

$$\Delta U_c = -L\mathcal{M}\frac{e^2}{r} + A\exp\left(-\frac{r}{\rho}\right) \tag{18.30}$$

where A and ρ are constants related to those of the law of repulsion. We can eliminate one of these constants by means of an equilibrium condition and calculate the other from the compressibility of the crystal. This calculation is effective because compressibility is mainly determined by the repulsive forces between the ions.

*A good description of methods for taking these lattice sums is given by C. Kittel, *Introduction to Solid State Physics* (New York: John Wiley & Sons, Inc., 1966), p. 91.

The crystal energy must be a minimum with respect to variation of r, so that $(\partial \Delta U_c / \partial r)_T = 0$. Hence, from (18.30),

$$\left(\frac{\partial \Delta U_c}{\partial r}\right)_T = \frac{L \mathscr{M} e^2}{r^2} - \frac{A \exp(-r/\rho)}{\rho} = 0$$

$$A = \frac{\rho L \mathscr{M} e^2}{r^2} \exp\left(\frac{r}{\rho}\right)$$

$$\Delta U_c = -L \mathscr{M} \frac{e^2}{r}\left(1 - \frac{\rho}{r}\right) \tag{18.31}$$

The compressibility β of the crystal is related to ΔU_c and the volume per mole V_m by

$$\frac{1}{\beta} = V_m \left(\frac{\partial^2 \Delta U_c}{\partial V^2}\right)_T$$

The volume $V_m = L \varphi r^3$, where φ is a factor that can be calculated from the crystal structure. Thus the constant ρ in (18.31) can be calculated from measurements of the compressibility.

32. The Born–Haber Cycle

The crystal energies computed from (18.31) can be compared with experimental values obtained from thermochemical data by means of the *Born–Haber cycle*. We shall consider NaCl as a typical example.

$$
\begin{array}{ccc}
\text{Na}^+(g) + \text{Cl}^-(g) & \xrightarrow{\;-\Delta U_c\;} & \text{NaCl}(c) \\
{\scriptstyle -e}\uparrow I \quad {\scriptstyle +e}\uparrow -A & & \Big\downarrow -\Delta H_f \\
\text{Na}(g) + \text{Cl}(g) & \xleftarrow{\;\Delta H_s + \frac{1}{2}D_0\;} & \text{Na}(c) + \frac{1}{2}\text{Cl}_2(g)
\end{array}
$$

The energetic quantities in this cycle are defined as follows, all per mole:

$\Delta U_c = $ the crystal energy
$\Delta H_f = $ standard enthalpy of formation of NaCl(c)
$\Delta H_s = $ enthalpy of sublimation of Na(c)
$D_0 = $ energy of dissociation of $\text{Cl}_2(g)$ into atoms
$I = $ ionization potential of Na
$A = $ electron affinity of Cl

For the cycle, from the First Law of Thermodynamics,

$$\oint dU = 0 = -\Delta U_c - \Delta H_f + \Delta H_s + \tfrac{1}{2}D_0 + I - A$$

or

$$\Delta U_c = -\Delta H_f + \Delta H_s + \tfrac{1}{2}D_0 + I - A \tag{18.32}$$

All the quantities on the right side of this equation can be evaluated, at least for alkali-halide crystals. The values obtained for crystal energies from (18.32) can therefore be compared with those calculated from (18.31). The ionization potentials I are obtained from atomic spectra, and the dissociation energies D_0 can be accurately determined from molecular spectra. In recent years, spectroscopic methods of measurement have yielded accurate values for electron affinities.*

A summary of data for various crystals is given in Table 18.8. When a calculated crystal energy deviates widely from the value obtained by means of the Born–Haber cycle, large nonionic contributions to the cohesive energy of the crystal are indicated.

TABLE 18.8
Data to Illustrate the Born–Haber Cycle[a]

Crystal	$-\Delta H_f$	I	ΔH_s	D_0	A	ΔU_c [Eq. (18.32)]	ΔU_c [Eq. (18.31)]
NaCl	414	490	109	226	347	779	795
NaBr	377	490	109	192	318	754	757
NaI	322	490	109	142	297	695	715
KCl	435	414	88	226	347	674	724
KBr	406	414	88	192	318	686	695
KI	356	414	88	142	297	632	665
RbCl	439	397	84	226	347	686	695
RbBr	414	397	84	192	318	673	673
RbI	364	397	84	142	297	619	644

[a]The various energy terms in Eq. (18.32) are given in kilojoules per mole, and the experimental crystal energy ΔU_c is computed and compared with the theoretical value from Eq. (18.31).

Although the theory of the cohesive energy of ionic crystals is satisfactory, it is not quite good enough to predict correctly the particular crystal structure that a substance will have in its most stable state. The net energy differences between possible alternative structures are usually small, of the order of 40 kJ or less, and theoretical calculations based upon rigid-sphere ions are not sufficiently precise to denote with certainty the most stable structure. For example, the Madelung energy of NaCl in the CsCl-type structure would be only 10 kJ higher than that in the actual structure, supposing that the same interionic distance was maintained. Thus there is a need for more refined calculations of crystal energies, including terms due to mutual polarization of ions, dipole–quadrupole and quadrupole-quadrupole dispersion terms, and a better treatment of repulsive forces.

33. Statistical Thermodynamics of Crystals—Einstein Model

From an accurate partition function for a crystal, we could calculate all its thermodynamic properties by making use of the general formulas of Chapter 5.

For 1 mol (L atoms) of a monatomic crystal there are $3L$ degrees of freedom. For practical purposes, we consider $3L$ vibrational degrees of freedom, since

*See, e.g., R. S. Berry and C. W. Reimann, *J. Chem. Phys.*, *38*, 1540 (1963).

$3L - 6$ is so close to $3L$. The precise determination of $3L$ normal modes of vibration for such a system would be an impossible task, and it is fortunate that some quite simple approximations give good answers.

First of all, let us suppose that the $3L$ vibrations arise from independent oscillators, and that these are harmonic oscillators, which is a good enough approximation at low temperatures, when the amplitudes are small. The model proposed by Einstein in 1906 assigned the same frequency v to all the oscillators.

The crystalline partition function according to the Einstein model, from (14.28), is

$$Z = e^{-3Lhv/2kT}(1 - e^{-hv/kT})^{-3L} \tag{18.33}$$

It follows immediately that

$$U - U_0 = 3Lhv(e^{hv/kT} - 1)^{-1} \tag{18.34}$$

$$S = 3Lk\left[\frac{hv/kT}{e^{hv/kT} - 1} - \ln(1 - e^{-hv/kT})\right] \tag{18.35}$$

$$G - U_0 = 3LkT \ln(1 - e^{-hv/kT}) \tag{18.36}$$

$$C_V = 3Lk\left(\frac{hv}{2kT} \operatorname{csch} \frac{hv}{2kT}\right)^2 \tag{18.37}$$

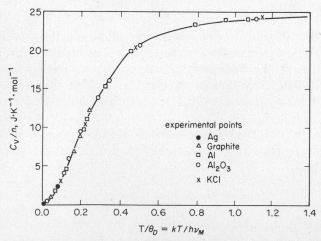

FIG. 18.44 The heat capacity of solids as a function of temperature. The molar heat capacity is divided by n, the number of atoms per mole, and the temperature is divided by the Debye temperature for the solid. All the heat capacity data then fall close to a common curve, illustrating the validity of the Debye analysis of the heat-capacity problem. The solid curve is that calculated from the Debye theory. The Einstein model is not so good as the Debye model, especially at lower temperatures.

The predicted temperature variation of C_V is particularly interesting. We recall that Dulong and Petit, in 1819, noted that the molar heat capacities of the solid elements, especially the metals, were usually around $3R \simeq 25 \text{ J} \cdot \text{K}^{-1}$. Later measurements showed that this figure was merely a limiting value for high temperatures, approached by different elements at different temperatures. If we expand the expression in (18.37) and simplify somewhat,* we obtain

$$C_V = \frac{3R}{1 + \frac{1}{12}(h\nu/kT)^2 + \frac{1}{360}(h\nu/kT)^4 + \cdots}$$

When T is large, this expression reduces to $C_V = 3R$. Over the range of lower T, the curve in Fig. 18.44 is nearly followed, the heat capacity being a universal function of (ν/T). The frequency ν can be determined from one experimental point at low temperatures, and then the entire heat capacity curve can be drawn for the substance. The agreement with the experimental data is quite good. It is clear that the higher the fundamental vibration frequency ν, the larger is the quantum of vibrational energy, and the higher the temperature at which C_V attains the classical value of $3R$. For example, the frequency for diamond is $2.78 \times 10^{13} \text{ s}^{-1}$, but for lead it is only $0.19 \times 10^{13} \text{ s}^{-1}$, so that C_V for diamond is only about 5.4 $\text{J} \cdot \text{K}^{-1} \cdot \text{mol}^{-1}$ at room temperature, but C_V for lead is 25.0. The elements that follow Dulong and Petit's rule at room temperatures are those with relatively low vibration frequencies.

34.   The Debye Model

Instead of a single fundamental frequency, it would be a better model to take a spectrum of vibration frequencies for the crystal. The statistical problem then becomes somewhat more complicated. One possibility is to assume that the frequencies are distributed according to the same law as that given in Section 13.5 for the distribution of frequencies in black-body radiation. This problem was solved by P. Debye.

Instead of using (18.34), we must obtain the energy by averaging over all the possible vibration frequencies ν_i of the solid, from 0 to ν_M, the maximum frequency. This gives

$$U - U_0 = \sum_{i=0}^{M} \frac{h\nu_i}{e^{h\nu_i/kT} - 1} \tag{18.38}$$

Since the frequencies form a virtual continuum, the summation can be replaced by an integration if we use the distribution function for the frequencies found in (13.28) (multiplied by 3 since we now have one longitudinal and two transverse mechanical vibrations, instead of the two transverse oscillations of the radiation case). Thus,

$$dN = f(\nu)\, d\nu = 12\pi \frac{V}{c^3} \nu^2\, d\nu \tag{18.39}$$

*Recalling that $\text{csch } x = 2/(e^x - e^{-x})$, and $e^x = 1 + x + (x^2/2!) + (x^3/3!) + \cdots$.

where c is now the velocity of elastic waves in the crystal. Then Eq. (18.38) becomes

$$U - U_0 = \int_0^{v_M} \frac{hv}{e^{hv/kT} - 1} f(v)\, dv \qquad (18.40)$$

Before substituting (18.39) into (18.40) we eliminate c by means of (13.28), since when $N = 3L$, $v = v_M$, for each direction of vibration, so that

$$3L = \frac{4\pi}{c^3} V v_M^3, \qquad c^3 = \frac{4\pi}{3L} V v_M^3, \qquad dv = \frac{9L}{v_M^3} v^2\, dv \qquad (18.41)$$

Then (18.40) becomes

$$U - U_0 = \frac{9Lh}{v_M^3} \int_0^{v_M} \frac{v^3\, dv}{e^{hv/kT} - 1} \qquad (18.42)$$

By differentiation with respect to T,

$$C_V = \frac{9Lh^2}{kT^2 v_M^3} \int_0^{v_M} \frac{v^4 e^{hv/kT}\, dv}{(e^{hv/kT} - 1)^2} \qquad (18.43)$$

Let us set $x = hv/kT$, whereupon (18.43) becomes

$$C_V = 9Lk\left(\frac{kT}{hv_M}\right)^3 \int_0^{hv_M/kT} \frac{e^x x^4\, dx}{(e^x - 1)^2} \qquad (18.44)$$

The Debye theory predicts that the heat capacity of a solid as a function of temperature should depend only on the characteristic frequency v_M. If the heat capacities of different solids are plotted against kT/hv_M, they should fall on a single curve. Such a plot is shown in Fig. 18.44, and the confirmation of the theory appears to be good. Debye has defined a *characteristic temperature*, $\Theta_D = hv_M/k$, and some of these characteristic temperatures are listed in Table 18.9 for various solids.

TABLE 18.9
Debye Characteristic Temperatures K

Substance	Θ_D	Substance	Θ_D	Substance	Θ_D
Na	159	Be	1000	Al	398
K	100	Mg	290	Ti	350
Cu	315	Ca	230	Pb	88
Ag	215	Zn	235	Pt	225
Au	180	Hg	96	Fe	420
KCl	227	AgCl	183	CaF_2	474
NaCl	281	AgBr	144	FeS_2	645

The application of (18.44) to the limiting cases of high and very low temperatures is of considerable interest. When the temperature becomes large, $e^{hv/kT}$ becomes small, and the equation may readily be shown to reduce to simply $C_V = 3R$, the Dulong and Petit expression. When the temperature becomes low, the integral may be expanded in a power series to shown that

$$C_V = aT^3$$

where a is a constant. This T^3 law holds below about 30 K and is of great use in extrapolating heat capacity data to absolute zero in connection with studies based on the Third Law of Thermodynamics (Section 3.23).

A direct experimental determination of the distribution of vibration frequencies in a solid is now possible from measurements of the scattering of slow neutrons by a single crystal. An example of such a determination for vanadium by Zemlyanov and coworkers is shown in Fig. 18.45. It would appear from such results that the

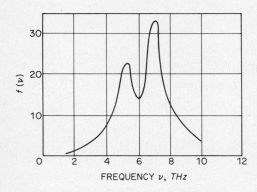

FIG. 18.45 The measurement of $f(v)$ for vanadium by means of incoherent neutron scattering. The frequency is in teraherz (Hz × 10^{12}) and $f(v)$ in arbitrary units. [After P. A. Egelstaff, *Thermal Neutron Scattering* (New York: Academic Press, 1965).]

distribution function used in the Debye theory of heat capacities is only a rough approximation of reality. The whole subject of the vibration spectra of solids, their interpretation in terms of interatomic forces, and their consequences in various thermodynamic properties, is under rather intensive development at the present time.

PROBLEMS

1. Show by means of drawings the directions [111], [100], [110], and [122] in a cubic unit cell.

2. The possible tetragonal Bravais lattices are primitive P and body-centered I. Show clearly that end-centered C is not a distinct type.

3. What is the plane of closest packing in the (a) fcc and (b) bcc structures? What is the relation between the length of cube a_0 and the closest internuclear distance in these structures?

4. Show that a fcc lattice can also be represented by a rhombohedral lattice. Calculate the rhombohedral angle α. If a general position in fcc lattice is (uvw), what would be its coordinates referred to the rhombohedral lattice?

5. Calcium fluoride is face-centered cubic with four CaF_2 groups per unit cell. The (111) reflection with X rays of wavelength 0.1542 nm has $\theta = 14.18°$, $\sin \theta = 0.245$,

$\cos \theta = 0.970$. (a) Calculate the length of the unit cell edge. (b) Calculate the density of the crystal.

6. Tetraborane is monoclinic with $a = 0.868$, $b = 1.014$, $c = 0.578$ nm, $\beta = 105.9°$, $\sin \beta = 0.962$, $\cos \beta = -0.274$. Draw a diagram showing the a and c axes and the orientation of the 101 planes. Calculate the interplanar spacing for 101 based on your diagram.

7. The crystal structure of $(CH_3)_2SO_2$ $(M = 94.13)$ is orthorhombic with $a = 0.736$, $b = 0.804$, $c = 0.734$ nm. There are four molecules in the unit cell. Calculate the density of the crystal.

8. For each of the following Schoenflies symbols (a) for point groups, associate the correct one of the following Hermann–Mauguin symbols (b),
   (a) C_{3v}, D_{2d}, T, O_h, D_4, D_{4d}, S_4
   (b) 23, $\bar{4}$, $m3m$, $\bar{8}2m$, 422, $\bar{4}2m$, $3m$.

9. Magnesium oxide $(M = 40.30)$ is cubic and has a density of 3.620 g·cm^{-3}. An X-ray diffraction diagram of MgO powder has lines at values of $\sin \theta = 0.399$, 0.461, 0.652, 0.764, 0.798, and 0.922. Index the pattern and determine the type of cubic structure. Calculate the wavelength of X rays used. Assume that the number of MgO units per unit cell is the smallest consistent with the structure type.

10. Tridymite, the high temperature form of SiO_2, is hexagonal with $a = 0.503$ nm and $c = 0.822$ nm. The space group is $P6/mmc$, and there are 4 SiO_2 units per unit cell. The Si atoms occupy positions with C_{3v} symmetry:

$$\tfrac{1}{3}, \tfrac{2}{3}, z; \qquad \tfrac{2}{3}, \tfrac{1}{3}, \bar{z}; \qquad \tfrac{2}{3}, \tfrac{1}{3}, \tfrac{1}{2} + z; \qquad \tfrac{1}{3}, \tfrac{2}{3}, \tfrac{1}{2} - z$$

with $z = 0.44$.
The oxygen atoms occupy positions with $D_{3h} - \bar{6}m2$ symmetry:

$$\tfrac{1}{3}, \tfrac{2}{3}, \tfrac{1}{4}; \qquad \tfrac{2}{3}, \tfrac{1}{3}, \tfrac{3}{4}$$

The other six oxygens are in positions of $C_{2h} - 2/m$ symmetry:

$$\tfrac{1}{2}, 0, 0; \qquad 0, \tfrac{1}{2}, 0; \qquad \tfrac{1}{2}, \tfrac{1}{2}, 0; \qquad \tfrac{1}{2}, 0, \tfrac{1}{2}; \qquad 0, \tfrac{1}{2}, \tfrac{1}{2}; \qquad \tfrac{1}{2}, \tfrac{1}{2}, \tfrac{1}{2}$$

(a) Calculate the density of tridymite.
(b) Draw the projection of this structure on the ab plane, showing all of the atoms in at least one unit cell.
(c) Calculate the distance from a silicon atom to each of the two types of oxygen bonded to it.
(d) What are the dimensions of the reciprocal lattice of this structure?

11. The zinc blende structure of ZnS is cubic with Zn at $0, 0, 0$; $\tfrac{1}{2}, \tfrac{1}{2}, 0$; $\tfrac{1}{2}, 0, \tfrac{1}{2}$; $\tfrac{1}{2}, \tfrac{1}{2}, 0$; and S at $\tfrac{1}{4}, \tfrac{1}{4}, \tfrac{1}{4}$; $\tfrac{1}{4}, \tfrac{3}{4}, \tfrac{3}{4}$; $\tfrac{3}{4}, \tfrac{1}{4}, \tfrac{3}{4}$; $\tfrac{3}{4}, \tfrac{3}{4}, \tfrac{1}{4}$. Calculate the structure factor of the lowest angle "reflection" from a ZnS crystal. The atomic scattering factors are $f(Zn) = 24.7$ and $f(S) = 12.3$. The lowest angle reflection is observed at $\theta = 14.3°$ when the wavelength of the X rays is 0.1542 nm. Calculate the dimension a_0 of the unit cell.

12. In the CaF_2 structure, Ca^{2+} ions are at $0, 0, 0$,; $\tfrac{1}{2}, \tfrac{1}{2}, 0$; $\tfrac{1}{2}, 0, \tfrac{1}{2}$; and $0, \tfrac{1}{2}, \tfrac{1}{2}$; and F^- ions are $\tfrac{1}{4}, \tfrac{1}{4}, \tfrac{1}{4}$; $\tfrac{3}{4}, \tfrac{3}{4}, \tfrac{1}{4}$; $\tfrac{3}{4}, \tfrac{1}{4}, \tfrac{3}{4}$; $\tfrac{1}{4}, \tfrac{3}{4}, \tfrac{3}{4}$; $\tfrac{3}{4}, \tfrac{3}{4}, \tfrac{3}{4}$; $\tfrac{1}{4}, \tfrac{1}{4}, \tfrac{3}{4}$; $\tfrac{1}{4}, \tfrac{3}{4}, \tfrac{1}{4}$; and $\tfrac{3}{4}, \tfrac{1}{4}, \tfrac{1}{4}$. Write the structure factor for 311 in terms of the atomic scattering factors. Reduce the formula to simplest terms. That is, in $F(311) = Cf(Ca^{2+}) + Df(F^-)$, evaluate the numbers C and D (real or complex).

13. The diamond structure is face-centered cubic with atoms at $0, 0, 0$; $\frac{1}{2}, \frac{1}{2}, 0$; $\frac{1}{2}, 0, \frac{1}{2}$; $0, \frac{1}{2}, \frac{1}{2}$; $\frac{1}{4}, \frac{1}{4}, \frac{1}{4}$; $\frac{3}{4}, \frac{3}{4}, \frac{1}{4}$; $\frac{3}{4}, \frac{1}{4}, \frac{3}{4}$; $\frac{1}{4}, \frac{3}{4}, \frac{3}{4}$. The unit cell edge length is 0.3570 nm. Calculate the carbon–carbon bond length in diamond.

14. Show that the void volume for structures of closest packed spheres is 25.9% in both ccp and hcp. For spheres of radius 0.100 nm, calculate a_0 for ccp, and a_0, c_0 for hcp.

15. Show graphically that fivefold axes of symmetry are inconsistent with translational symmetry and hence cannot be elements of a space group.

16. A face-centered cubic structure gave the following powder pattern with X rays of wavelength 0.1542 nm:

h k l	$\sin^2 \theta$
1 1 1	0.0526
2 0 0	0.0701
2 2 0	0.1402
3 1 1	0.1928
2 2 2	0.2104

(a) Derive a general formula for $\sin^2 \theta$ of a cubic substance in terms of wavelength, the unit cell dimension, and the Miller indices.
(b) Calculate the interplanar spacing for each of the preceding lines.
(c) From each of these lines, determine a value for the length of the unit cell edge.

17. Suppose that a crystal of NaCl has a concentration of 10^{-3} atom fraction of (a) Schottky and (b) Frenkel defects. If the $a_0 = 0.5629$ nm at 25°C, as in a perfect crystal, calculate the densities of the crystal for (a) and (b).

18. What is the approximate concentration of vacancies in a crystal of silver at its melting point 1234 K if the energy of formation of a vacancy equals one-half the latent heat of sublimation ΔH_s? You can make a reasoned estimate for ΔH_s.

19. Assume that there is one conduction electron per atom in a gold crystal, and that the energy levels available to the electrons are those of a cubic box given in Eq. (13.67). Show that the number of electrons per unit volume in an energy range from E to $E + dE$ is given by the density-of-states function,

$$N(E)\, dE = \frac{\pi}{2} \left(\frac{8m}{h^2} \right)^{3/2} E^{1/2}\, dE$$

Hence, calculate the maximum energy level occupied by the electrons in the gold crystal at 0 K.

20. Calculate the nearest Ni—Ni distances in NiO and Ni at 25°C, given that NiO crystallizes with the NaCl structure (an approximation) with $a_0 = 0.4180$ nm and Ni in the fcc structure with $a_0 = 0.3517$ nm. Comment on the result found.

21. A Debye–Scherrer powder pattern of a cubic crystal with X rays of $\lambda = 0.1539$ nm displayed lines at the following scattering angles:

No. of line	1	2	3	4	5	6	7	8	9
θ (deg)	13.70	15.89	22.75	26.91	28.25	33.15	37.00	37.60	41.95
Intensity	w	vs	s	vw	m	w	w	m	m

where w is *weak*, s is *strong*, m is *medium*, and v is *very*. (a) Index these lines.

(b) Calculate a_0 for the crystal. (c) Identify the crystal. (d) Explain the relative intensities of lines 4 and 5 by means of the structure factor.

22. Cesium chloride has a unit cell length of $a_0 = 0.411$ nm and a crystal energy of 636.8 J·mol^{-1}. Calculate the crystal energy of T1CN which has the CsCl structure with $a_0 = 0.382$ nm. Assume the same value for ρ, the repulsive parameter in Eq. (18.31).

23. The Debye characteristic temperature of copper is $\Theta_D = 315$ K. Calculate the entropy of copper at 25°C, assuming that $\alpha = 4.95 \times 10^{-5}$ K^{-1} and $\beta = 7.50 \times 10^{-7}$ atm^{-1}, both independent of T.

24. Calculate the proton affinity of NH_3 from the following data, i.e., ΔU for $NH_3 + H^+$ $\longrightarrow NH_4^+$. NH_4F crystallizes in the ZnO-type structure, whose Madelung constant is 1.6381 based on r_0 the nearest cation–anion distance, 0.263 nm. The electron affinity of F is 350 kJ·mol^{-1}. The ionization potential of H is 1305 kJ·mol^{-1}. Heats of formation are given in Tables 2.2 and 2.6. The heat of reaction, $\frac{1}{2}N_2 + 2H_2 + \frac{1}{2}F_2 \longrightarrow NH_4F(c)$, is 469 kJ.

19

Intermolecular Forces and the Liquid State

My body floated weightlessly through space, the water took possession of my skin, the clear outlines of marine creatures seemed almost provocative, and economy of movement acquired moral significance. Gravity—I saw it in a flash—was the original sin, committed by the first living beings who left the sea. Redemption would come only when we returned to the ocean as already the sea mammals have done.

J. Y. Cousteau
(1963)*

Gases, at least in the ideal approximation approached at high temperatures and low densities, are characterized by complete randomness on the molecular scale. The ideal crystal, on the other hand, is Nature's most orderly arrangement. Because the extremes of perfect chaos and perfect order are both relatively simple to treat mathematically, the theories of gases and of crystals have advanced at respectable rates. Liquids, however, representing a peculiar compromise between order and disorder, have so far defied a comprehensive theoretical treatment.

In an ideal gas, the molecules move independently of one another and interactions between them are neglected. The energy of the perfect gas is simply the sum of the energies of the individual molecules, their internal energies plus their translational kinetic energies; there is no intermolecular potential energy. It is therefore possible to write down a partition function such as that in Eq. (5.44) from which all the equilibrium properties of the gas are readily derived. In a crystalline solid, translational kinetic energy is usually negligible. The molecules, atoms, or ions vibrate about equilibrium positions to which they are held by strong intermolecular, interatomic, or interionic forces. In this case too, an adequate partition function, such as that in (18.33), can be obtained. In a liquid, on the other hand, the situation is much harder to define. The cohesive forces are sufficiently strong to lead to a condensed state, but not strong enough to prevent a considerable translational energy of the individual molecules. The thermal motions introduce a

*From an introduction to *Sensitive Chaos*, by T. Schwenk (London: Rudolf Steiner Press, 1965).

disorder into the liquid without completely destroying the regularity of its structure. It is therefore difficult to evaluate the partition function for a liquid.

It is sometimes convenient to classify liquids, like crystals, from a rather chemical standpoint, according to the kinds of cohesive forces that hold them together. Thus there are the ionic liquids such as molten salts, metallic liquids consisting of ions and mobile electrons, liquids such as water held together mainly by hydrogen bonds, and finally molecular liquids in which cohesion is due to the van der Waals forces between essentially saturated molecules. Many liquids fall into this last group, and even when other forces are present, the van der Waals contribution may be large. The nature of these forces will be considered later in this chapter.

1. Disorder in the Liquid State

The crystal is energetically a more favorable structure than the liquid to which it melts. It is necessary to add energy, the latent heat of fusion, to effect the melting. The equilibrium situation, however, is determined by the difference in free enthalpy, $\Delta G = \Delta H - T\,\Delta S$. With increasing temperature, the greater randomness of the liquid, and hence its greater entropy, finally make the $T\,\Delta S$ term large enough to overcome the ΔH term, so that the crystal melts when the following condition is reached:

$$T(S_{\text{liq}} - S_{\text{cry}}) = H_{\text{liq}} - H_{\text{cry}}$$

The sharpness of the melting point is noteworthy: there is no continuous gradation of properties between liquid and crystal. The sharp transition is due to the rigorous geometrical requirements that must be fulfilled by a crystal structure. It is not possible to introduce small regions of disorder into a crystal without seriously disturbing the structure over a long range. Two-dimensional models of gaseous, liquid, and crystalline states are illustrated in Fig. 19.1. The picture of the liquid was constructed by J. D. Bernal by introducing around atom A only five other atoms, instead of its normal close-packed coordination of six. Every effort was then made to draw the rest of the circles in the most ordered arrangement possible. The one point of abnormal coordination among some hundred atoms sufficed to produce a long-range disorder similar to that in the liquid state. When the thermal motions destroy the regular structure in one region of a crystal, the irregularity rapidly spreads throughout the entire specimen; disorder in a

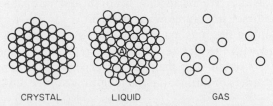

CRYSTAL   LIQUID   GAS

FIG. 19.1 Two-dimensional models of states of matter.

crystal is contagious. We do not mean to imply that crystals are ideally perfect, and admit of no disorder at all. The amount of disorder allowed is usually limited, and when the limit is exceeded, melting occurs.

2.   X-Ray Diffraction of Liquid Structures

The study of the X-ray diffraction of liquids followed the development of the method of Debye and Scherrer for crystalline powders. As the particle size of the powder decreases, the width of the lines in the X-ray pattern increases. With particles around 10 nm in diameter, the lines have become diffuse halos, and with still further decrease in particle size, the diffraction maxima become blurred out altogether.

If a liquid were completely amorphous, i.e., without any regularity of structure, it should also give a continuous scattering of X rays without maxima or minima. This was actually found not to be the case. A typical pattern, that obtained from liquid mercury, is shown in Fig. 19.2, as a microphotometer tracing of the

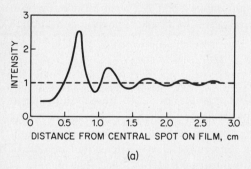
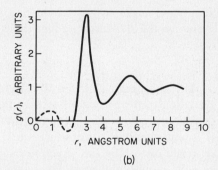

(a)   (b)

FIG. 19.2 X-ray diffraction of liquid mercury: (a) photometric tracing
of diffraction diagram; (b) radial distribution function for
liquid mercury.

photograph. One or two, or sometimes more, intensity maxima appear, having positions that often correspond closely to some of the larger interplanar spacings in the crystal structures. It is interesting that a crystal like bismuth, which has a peculiar and rather loose solid structure, is transformed on melting into a more closely packed structure. We recall that bismuth and water are among the few substances that contract in volume when melted.

The fact that only a few maxima are observed in the diffraction patterns from liquids is in accord with the picture of short-range order and increasing disorder at longer range. To obtain the maxima corresponding to smaller interplanar spacings or higher orders of diffraction, the long-range order of the crystal must be present.

The arrangement of atoms in such a monatomic liquid is described by introducing the *radial distribution function* $g(r)$. Taking the center of one atom as origin, this $g(r)$ gives the probability of finding the center of another atom at the end of a vector of length r drawn from the origin. The chance of finding another atom between a distance r and $r + dr$, irrespective of angular orientation, is therefore $4\pi r^2 g(r)\, dr$ (cf. p. 637). It is now possible to obtain for the intensity of scattered X radiation an expression similar to that in (15.38), except that instead of a summation over individual scattering centers, there is an integration over a continuous distribution of scattering matter, specified by $g(r)$. Thus,

$$I(\theta) \propto \int_0^\infty 4\pi r^2 g(r) \frac{\sin sr}{sr}\, dr \tag{19.1}$$

As before,

$$s = \frac{4\pi}{\lambda} \sin \frac{\theta}{2} \tag{19.2}$$

By an application of Fourier's integral theorem, this integral can be inverted,* yielding

$$4\pi r^2 g(r) \propto \frac{2}{\lambda} \int_0^\infty I(\theta) \frac{\sin sr}{sr}\, d\theta \tag{19.3}$$

With this relationship, we can calculate a radial distribution curve, such as that plotted in Fig. 19.2(b), from an experimental scattering curve, such as that in Fig. 19.2(a).

The regular coordination in the close-packed liquid-mercury structure is clearly evident, but the fact that maxima in the curve are rapidly damped out at larger interatomic distances indicates that the departure from the ordered arrangement becomes greater as one travels outward from any centrally chosen atom. In general, liquid metals have approximately close-packed structures quite similar to those of the solids, with the interatomic spacings expanded by about 5%. The number of nearest neighbors in a close-packed structure is 12. In liquid sodium, each atom is found to have on the average 10 nearest neighbors.

One of the most interesting liquid structures is that of water. The first extensive X-ray diffraction study of liquid water was made by Morgan and Warren in 1938. In more recent measurements, Narten, Danford, and Levy have considerably improved the accuracy of the method and calculated the radial distribution functions over a range of temperatures as shown in Fig. 19.3.† At values of $r < 0.25$ nm, $g(r)$ vanishes, indicating that 0.25 nm is an effective molecular diameter within which no other molecule can approach a given central molecule. At 4°C, $g(r)$ is nearly unity for $r > 0.8$ nm, so that beyond this distance the density of neighboring molecules equals the bulk density. Thus, the local order imposed by a central

---

*See H. Bateman, *Partial Differential Equations of Mathematical Physics* (New York: Dover Publications, 1944), p. 207, for mathematical discussion.

†We follow the discussion of these data given by D. Eisenberg and W. Kauzmann, *The Structure and Properties of Water* (New York: Oxford University Press, 1969).

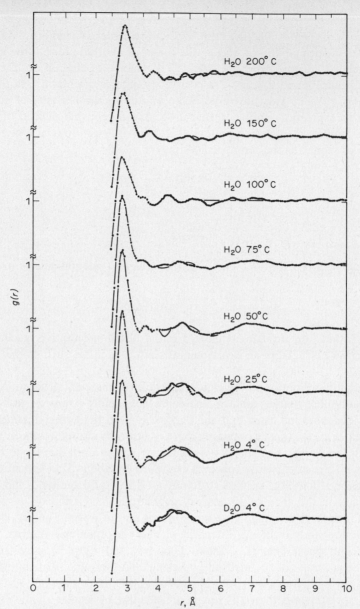

FIG. 19.3 Radial distribution functions, $g(r)$, for liquid H_2O at various temperatures and for liquid D_2O at 4°C. Note that the base line of each curve is one unit above that for the curve below. The points were determined from experiments. The experiments at and below 100°C were carried out at atmospheric pressure; those above 100°C were at the vapor pressure of the sample. [From A. H. Narten, M. D. Danford, and H. A. Levy, *Discussions Faraday Soc., 43,* 97 (1967).]

molecule upon its neighbors does not extend beyond 0.8 nm. At higher temperatures, the short-range order becomes even less evident, and at 200°C it disappears beyond 0.6 nm. There is a sharp peak in $g(r)$ at 0.29 nm due primarily to nearest neighbors of the central molecule, and when $g(r)$ is integrated over the volume element $4\pi r^2\, dr$ in this shell, the number of nearest neighbors is calculated to be 4.4 at all temperatures from 4 to 200°C. These results indicate that the coordination in liquid water is approximately tetrahedral, as we know to be the case in Ice-I (Fig. 18.27). The peaks in the $g(r)$ function for water at 0.45 to 0.53 nm and at 0.64 to 0.78 nm are also in good accord with the tetrahedral arrangement.

The small but distinct peak at 0.35 nm cannot be explained by the tetrahedral structure. The structure of Ice-I has six interstitial sites ("cavity centers") at a distance of 0.348 nm from the central molecule. It was therefore suggested that when ice melts, some of the water molecules move from their tetrahedral sites into these interstitial sites, thus accounting for the peak in $g(r)$ at 0.35 nm. There is a contraction in volume of about 9% when ice melts. The X-ray data indicate that occupancy of the interstitial sites increases from 45% at 4°C to 57% at 200°C.

3. Liquid Crystals

In some substances, the tendency toward an ordered arrangement is so great that the crystalline form does not melt directly to a liquid phase at all, but first passes through an intermediate stage (the *mesomorphic* or *paracrystalline* state), which at a higher temperature undergoes a transition to the liquid state. These intermediate states have been called *liquid crystals*, since they display some of the properties of each of the adjacent states. Thus, some paracrystalline substances flow in a gliding stepwise fashion and form *graded droplets*, having terracelike surfaces; other varieties flow quite freely but are not isotropic, exhibiting interference figures when examined with polarized light. An example is shown in Fig. 19.4.

Liquid crystals tend to occur in compounds the molecules of which are markedly unsymmetrical in shape. For example, in the crystalline state, long-chain molecules may be lined up as shown in Fig. 19.5(a). On raising the temperature, the kinetic energy may become sufficient to disrupt the binding between the ends of the molecules but insufficient to overcome the strong lateral attractions between the long chains. Two types of anisotropic melt might then be obtained, shown in Fig. 19.5(b) and (c). In the smectic (Greek σμηγμα, *soap*) state, the molecules are oriented in well-defined planes. When a stress is applied, one plane glides over another. In the nematic (Greek νημα, *thread*) state, the planar structure is lost, but the orientation is preserved. With some substances, notably the soaps, several different phases, differentiated by optical and flow properties, can be distinguished between typical crystal and typical liquid.

FIG. 19.4 A liquid crystalline solution of a polypeptide in dichloro-acetic acid solution viewed by natural light with a low powered microscope. [C. Robinson, J. C. Ward, and R. B. Beevers, Courtaulds Research Laboratory, Maidenhead, *Discussions Faraday Soc., 25,* 29 (1958)].

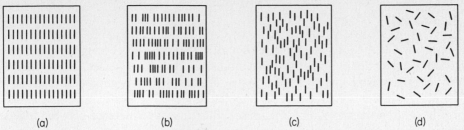

(a)   (b)   (c)   (d)

FIG. 19.5 Possible degrees of order in condensed states of long-chain molecules: (a) crystalline—orientation and periodicity; (b) smectic—orientation and arrangement in equispaced planes, but no periodicity within planes; (c) nematic—orientation without periodicity; (d) isotropic fluid—neither orientation nor periodicity.

A compound frequently studied in its paracrystalline state is p-azoxyanisole,

$$CH_3O\!\!-\!\!\bigcirc\!\!-\!\!\overset{\overset{\displaystyle O}{\uparrow}}{N}\!\!=\!\!N\!\!-\!\!\bigcirc\!\!-\!\!OCH_3$$

The solid form melts at 357 K to a liquid crystal that is stable to 423 K, at which point it undergoes a transition to an isotropic liquid. The compound ethyl p-anisalaminocinnamate,

$$CH_3O\!\!-\!\!\bigcirc\!\!-\!\!CH\!\!=\!\!N\!\!-\!\!\bigcirc\!\!-\!\!CH\!\!=\!\!CH\!\!-\!\!COOC_2H_5$$

passes through three distinct paracrystalline phases between 356 and 412 K. Cholesteryl bromide behaves rather differently.* The solid melts at 367 K to an isotropic liquid, but this liquid can be supercooled to 340 K, where it passes into a metastable liquid-crystalline form.

> Liquid crystals, it is to be noted, are not important for biology and embryology because they manifest certain properties which can be regarded as analogous to those which living systems manifest (models), but because living systems actually are liquid crystals, or, it would be more correct to say, the paracrystalline state undoubtedly exists in living cells. The doubly refracting portions of the striated muscle fibre are, of course, the classical instance of this arrangement, but there are many other equally striking instances, such as cephalopod spermatozoa, or the axons of nerve cells, or cilia, or birefringent phases in molluscan eggs, or in nucleus and cytoplasm of echinoderm eggs.
>
> The paracrystalline state seems the most suited to biological functions, as it combines the fluidity and diffusibility of liquids while preserving the possibilities of internal structure characteristic of crystalline solids.†

*J. Fischer, *Z. Physik Chem.*, *160 A*, 110 (1932).

†Joseph Needham, *Biochemistry and Morphogenesis* (London: Cambridge University Press, 1942), p. 661. The predictions of Needham, made some 30 years ago, about the importance of paracrystalline states in biological systems are borne out by current work on phase changes in cell membranes [*Liquid Crystals and Ordered Fluids*, eds. J. F. Johnson and R. S. Porter (New York: Plenum Press, 1970)]. The behavior of liquid crystals in electric fields may suggest models for electrically excitable membranes. [G. H. Brown, J. W. Doane, and V. D. Neff, *Critical Rev. Solid State Sci. 1*, 303 (1970).]

4. Glasses

The glassy or vitreous state of matter is another example of a compromise between crystalline and liquid properties. The structure of a glass is essentially similar to that of an associated liquid such as water, so that there is a good deal of truth in the old description of glasses as supercooled liquids. The two-dimensional models in Fig. 19.6, given by W. H. Zachariasen, illustrate the difference between a glass and a crystal.

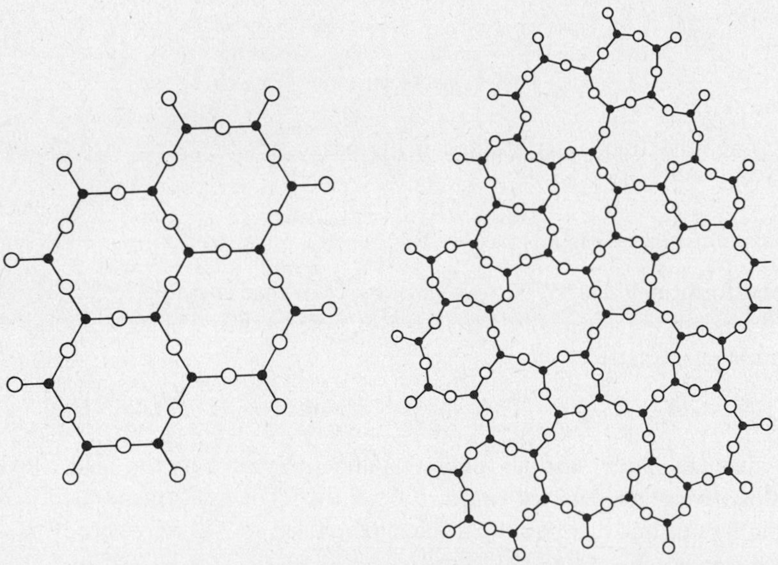

FIG. 19.6   Schematic representation in two dimensions of the difference between the structure of a crystal (left) and a glass (right). Solid circles represent silicon atoms; open circles are oxygen atoms. [After W. H. Zachariasen, *J. Am. Chem. Soc., 54,* 3841 (1932).]

The bonds are the same in both cases, e.g., in silica the strong electrostatic Si—O bonds. Thus, both quartz crystals and vitreous silica are hard and mechanically strong. The bonds in the glass differ considerably in length and therefore in strength. Thus, a glass on heating softens gradually rather than melts sharply, since there is no one temperature at which all the bonds become loosened simultaneously.

The extremely low coefficient of thermal expansion of some glasses, notably vitreous silica, is explicable in terms of a structure such as that in Fig. 19.6. The structure is very open, and as in the case of liquid water, increasing the temperature may allow a closer coordination. To a certain extent, therefore, the structure may "expand into itself." This effect counteracts the normal expansion in interatomic distance with temperature.

5. Melting

Table 19.1 contains data on the melting points, latent heats of melting, latent heats of vaporization, and entropies of melting and vaporization of a number of substances. The heats of melting are much lower than the heats of vaporization. It requires much less energy to convert a crystal to liquid than to vaporize a liquid. The entropies of melting are also considerably lower than the entropies of vapori-

TABLE 19.1
Data on Melting and Vaporization

Substance	Enthalpy of fusion (kJ·mol⁻¹)	Enthalpy of vaporization (kJ·mol⁻¹)	Melting point (K)	Entropy of fusion (J·K⁻¹·mol⁻¹)	Entropy of vaporization (J·K⁻¹·mol⁻¹)
		Metals			
Na	2.64	103	371	7.11	88.3
Al	10.7	283	932	11.4	121
K	2.43	91.6	336	7.20	87.9
Fe	14.9	404	1802	8.24	123
Ag	11.3	290	1234	9.16	116
Pt	22.3	523	2028	11.0	112
Hg	2.43	64.9	234	10.4	103
		Ionic crystals			
NaCl	30.2	766	1073	28.1	456
KCl	26.8	690	1043	25.7	389
AgCl	13.2		728	18.1	
KNO₃	10.8		581	18.5	
BaCl₂	24.1		1232	19.5	
K₂Cr₂O₇	36.7		671	54.7	
		Molecular crystals			
H₂	0.12	0.92	14	8.4	66.1
H₂O	5.98	47.3	273	22.0	126
Ar	1.17	7.87	83	14.1	90.4
NH₃	7.70	29.9	198	38.9	124
C₂H₅OH	4.60	43.5	156	29.7	124
C₆H₆	9.83	34.7	278	35.4	98.3

zation. The latter are quite constant, from 90 to 100 kJ·mol⁻¹ (Trouton's rule). The constancy of the ΔS(fusion) is not so marked. For some classes of substances, however, notably the close-packed metals, the entropies of fusion are remarkably constant.

6. Cohesion of Liquids—Internal Pressure

Whatever the model chosen for the liquid state, the cohesive forces are of primary importance. Ignoring, for the time being, the origin of these forces, we can obtain an estimate of their magnitude from thermodynamic considerations. This estimate is provided by the *internal pressure*.

We recall from Eq. (3.45) that

$$\left(\frac{\partial U}{\partial V}\right)_T = T\left(\frac{\partial P}{\partial T}\right)_V - P$$

In the case of an ideal gas, the internal-pressure term $P_i = (\partial U/\partial V)_T$ is zero because intermolecular forces are absent. In the case of an imperfect gas, $(\partial U/\partial V)_T$ becomes appreciable, and in the case of a liquid it may become much greater than the external pressure.

The internal pressure is the resultant of the forces of attraction and the forces of repulsion between the molecules in a liquid. It therefore depends markedly on the volume V, and thus on the external pressure P. This effect is shown in the following data for diethyl ether at 298 K:

P (atm):	200	800	2000	5300	7260	9200	11 100
P_i (atm):	2790	2840	2530	2020	40	−1590	−4380

For moderate increases in P, P_i decreases only slightly, but as P exceeds 5000 atm, P_i begins to decrease rapidly and goes to large negative values as the liquid is further compressed. This behavior reflects on a larger scale the law of force between individual molecules; at high compressions, the repulsive forces become predominant.

Internal pressures at 1 atm and 298 K are summarized in Table 19.2, taken from a compilation by J. H. Hildebrand. With normal aliphatic hydrocarbons, there is a gradual increase in P_i with the length of the chain. Dipolar liquids tend to have somewhat larger values than nonpolar liquids. The effect of dipole interaction is nevertheless not predominant. As might be expected, water with its strong hydrogen bonds has an exceptionally high internal pressure.

Hildebrand was the first to point out the significance of the internal pressures of liquids in determining solubility relationships. If two liquids have about the same P_i, their solution has little tendency toward positive deviations from Raoult's Law. The solution of two liquids differing considerably in P_i will usually exhibit

TABLE 19.2
Internal Pressures of Liquids (298 K and 1 atm)

Compound	P_i (atm)
Diethyl ether	2370
n-Heptane	2510
n-Octane	2970
Tin tetrachloride	3240
Carbon tetrachloride	3310
Benzene	3640
Chloroform	3660
Carbon bisulfide	3670
Mercury	13 200
Water	20 000

considerable positive deviation from ideality, i.e., a tendency toward lowered mutual solubility.

7. Intermolecular Forces

It should be clearly understood from earlier discussions that all the forces between atoms and molecules are electrostatic in origin. They are ultimately based on Coulomb's Law of the attraction between unlike, and repulsion between like charges. One often speaks of long-range forces and short-range forces. Thus, a force that depends on r^{-2} will be effective over a longer range than one dependent on r^{-7}. All these forces may be represented as gradients of potential energies, $\mathbf{F} = -\text{grad}\ U$. For forces with radial symmetry, i.e., no preferred direction in space, $F = -\partial U(r)/\partial r$. It is often convenient to describe the potential energies rather than the forces themselves. We can then distinguish the following varieties of intermolecular and interionic potential energies:

1. The coulombic energy of interaction between ions with net charges, leading to a long-range attraction, with $U \propto r^{-1}$.
2. The energy of interaction between permanent dipoles, with $U \propto r^{-6}$.
3. The energy of interaction between an ion and a dipole induced by it in another molecule, with $U \propto r^{-4}$.
4. The energy of interaction between a permanent dipole and a dipole induced by it in another molecule with $U \propto r^{-6}$.
5. The forces between neutral atoms or molecules, such as the inert gases, with $U \propto r^{-6}$.
6. The overlap energy arising from the interaction of the positive nuclei and electron cloud of one molecule with those of another. The overlap leads to repulsion at very close intermolecular separations, with $U \propto r^{-9}$ to r^{-12}.

The van der Waals attractions between molecules must arise from interactions belonging to classes 2, 4, and 5.

The first attempt to explain them theoretically was made by W. H. Keesom (1912), based on the interaction between permanent dipoles. Two dipoles in rapid thermal motion may sometimes be oriented so as to attract each other, sometimes so as to repel each other. On the average, they are somewhat closer together in attractive configurations, and consequently there is a net attractive energy. This energy was calculated* to be

$$U_d = -\frac{2\mu_1^2\,\mu_2^2}{3kTr^6} \tag{19.4}$$

where μ is the dipole moment. The observed r^{-6} dependence of the interaction energy is in agreement with deductions from experiment. This theory is, of course, not an adequate general explanation of van der Waals forces, since considerable attractive forces exist between molecules, such as the inert gases, with no vestige of a permanent dipole moment.

Debye, in 1920, extended the dipole theory to take into account the *induction effect*. A permanent dipole induces a dipole in another molecule and a mutual attraction results. This interaction depends on the polarizability α of the molecules, and leads to a formula,

$$U_i = -\frac{\alpha_2\,\mu_1^2 + \alpha_1\,\mu_2^2}{r^6} \tag{19.5}$$

This effect is quite small and does not help us to explain the case of the inert gases.

The primary cause of the attractive forces between neutral molecules is the *dispersion interaction*. An interpretation can be given as follows. In a neutral molecule, such as argon, the positive nucleus is surrounded by a "cloud" of negative charge. Although the time average of this charge distribution is spherically symmetrical, at any instant it will be somewhat distorted. We see this clearly in the case of the neutral hydrogen atom, in which the electron is sometimes on one side of the proton, sometimes on the other. Thus, a snapshot taken of an argon atom would reveal a little dipole with a certain orientation. An instant later, the orientation would be different, and so on, so that over any macroscopic period of time the instantaneous dipole moments would average to zero. We should not think that these snapshot dipoles interact with those of other molecules to produce an attractive potential. This cannot happen, because there will be repulsion just as often as attraction; there is not enough time for the instantaneous dipoles to line up with one another. There is, however, an interaction between snapshot dipoles and the polarization they produce. Each instantaneous argon dipole induces an appropriately oriented dipole moment in neighboring atoms, and these moments interact with the original to produce an instantaneous attraction.

The quantum mechanical treatment of this dispersion interaction was given by F. London in 1930, and the resultant forces are now often called *London forces*, whereas the term *van der Waals forces* is kept for the sum of all the interactions between uncharged molecules. If we suppose that a pair of equivalent oscillators

*J. E. Lennard–Jones, *Proc. Phys. Soc. London*, **43**, 461 (1931).

have a characteristic frequency v_0, the potential of the London interaction is calculated* to be

$$U_L = -\frac{3hv_0\alpha^2}{4r^6} \qquad (19.6)$$

A more useful general formula for the interaction of two different molecules A and B is

$$U_L = -\frac{3I_A I_B}{2(I_A + I_B)}\frac{\alpha_A\alpha_B}{r^6} \qquad (19.7)$$

Here I_A and I_B are the first ionization potentials of the molecules, and α_A and α_B their polarizabilities. Equation (19.7) is related to (19.6) by the substitution of I/h for the characteristic frequency v_0.

The important contributions to the potential energy of intermolecular attraction that we have listed all display an r^{-6} dependence. The different contributions to the intermolecular attractions are shown in Table 19.3 for a number of gases.

The complete expression for the intermolecular energy must include also a repulsive term, the overlap energy, which becomes appreciable at very close distances. Thus, the total interaction can be written

$$U = -Ar^{-6} + Br^{-n} \qquad (19.8)$$

The value of the exponent n is from 9 to 12. Figure 4.4 showed the curves for a number of gases, with $n = 12$. The constants A and B are calculated from the equations of state.

TABLE 19.3
Calculated Contributions to the Intermolecular Potential[a]

Molecule	Dipole moment μ (D)	Polarizability $\alpha \times 10^{30}$ (m³)	Energy hv_0(eV)	Coefficients of r^{-6}		
				Orientation[b] $\frac{2}{3}\mu^4/kT$	Induction[b] $2\mu^2\alpha$	Dispersion[b] $\frac{3}{4}\alpha^2 hv_0$
CO	0.12	1.99	14.3	0.0034	0.057	67.5
HI	0.38	5.4	12	0.35	1.68	382
HBr	0.78	3.58	13.3	6.2	4.05	176
HCl	1.03	2.63	13.7	18.6	5.4	105
NH₃	1.5	2.21	16	84	10	93
H₂O	1.84	1.48	18	190	10	47
He	0	0.20	24.5	0	0	1.2
Ar	0	1.63	15.4	0	0	52
Xe	0	4.00	11.5	0	0	217

[a]J. A. V. Butler, *Ann. Rep. Chem. Soc. London*, 34, 75 (1937).
[b]Units of J·m⁶ × 10⁻⁷⁹.

*For a simple quantum mechanical derivation of Eq. (19.6), see W. Kauzmann, *Quantum Chemistry* (New York: Academic Press, Inc., 1957), p. 507.

8. Equation of State and Intermolecular Forces

The calculation of the equation of state of a substance from a knowledge of the intermolecular forces is a problem of great complexity. The method of attack may be outlined in principle, but the mathematical difficulties have proved so formidable that a solution has been obtained only for a few simplified cases. The calculation of the equation of state reduces to calculating the partition function Z for the system. From Z, the Helmholtz free energy A is immediately derivable, and hence the pressure, $P = -(\partial A/\partial V)_T$.

To determine the partition function, $Z = \sum e^{-E_i/kT}$, the energy levels of the system must be known. In the cases of ideal gases and crystals, we can use energy levels for individual consituents of the system, such as molecules or oscillators, ignoring interactions between them. In the case of liquids, such a simplification is not possible because it is precisely the interaction between different molecules that is responsible for the characteristic properties of a liquid.

Let us consider for simplicity a gas of interacting mass points. Any monatomic gas would be a good example. We shall follow a delightful derivation given by Landau and Lifshitz.* The energy is written as the sum of kinetic and potential energies:

$$E(p, q) = \sum_{j=1}^{N} \left(\frac{p_j^2}{2m} + U(q_j) \right)$$

Here p, q stand for generalized coordinates; for N atoms there would be $3N$ each of p's and q's. The kinetic energy is obtained simply by summing over the independent kinetic energies of all the atoms, but the potential energy is, in general, a function of all the coordinates.

The partition function is taken in the integral form of Eq. (5.54),

$$Z = \int e^{-E(p,q)/kT} \, d\tau \tag{19.9}$$

where $d\tau$ includes all the differentials of momenta and coordinates. The integral in (19.9) is the product of a kinetic energy term and a potential energy term, and the latter can be written as

$$Q = \int \cdots \int e^{-U/kT} \, dV_1 \, dV_2 \cdots dV_N \tag{19.10}$$

where the integral is now taken over the volume elements of each atom.

For a perfect gas, there is no intermolecular potential energy, and $U = 0$, so that Q would become simply V^N. The kinetic energy term is the same for perfect and imperfect gases. Thus, we can write

$$A = -kT \ln Z = A^p - kT \ln \frac{1}{V^N} \int \cdots \int e^{-U/kT} \, dV_1 \cdots dV_N$$

*L. S. Landau and E. M. Lifshitz, *Statistical Physics* (London: Pergamon Press, 1959), p. 219.

where A^p is the Helmholtz free energy for a perfect gas. If we add and subtract 1 from the integrand,

$$A = A^p - kT \ln\left\{\frac{1}{V^N} \int \cdots \int (e^{-U/kT} - 1)\, dV_1 \cdots dV_N + 1\right\} \quad (19.11)$$

We now assume that the gas is of such low density that only encounters between *pairs* of molecules need be considered. We also assume that the sample of gas is so small that only one such collision between pairs is occurring at any instant. (There is no loss in generality in such an assumption because A is an extensive state quantity and doubling the amount of gas simply doubles A). The interacting pair can be chosen from the N atoms in $\frac{1}{2}N(N-1)$ different ways. If the energy of interaction of the pair is U_{12}, the integral in (19.11) becomes

$$\frac{N(N-1)}{2} \int \cdots \int (e^{-U_{12}/kT} - 1)\, dV_1 \cdots dV_N$$

Since U_{12} depends only on the coordinates of two atoms, we can integrate over all the others, obtaining V^{N-2}. Since N is very large, we can replace $N(N-1)$ by N^2, to obtain

$$\frac{N^2 V^{N-2}}{2} \iint (e^{-U_{12}/kT} - 1)\, dV_1\, dV_2$$

We introduce this expression for the integral into (19.11), to find

$$A = A^p - kT \ln\left\{\frac{N^2}{2V^2} \iint (e^{-U_{12}/kT} - 1)\, dV_1\, dV_2 + 1\right\} \quad (19.12)$$

We now use the fact that for $x \ll 1$,

$$\ln(1 + x) \approx x$$

so that (19.12) becomes

$$A = A^p - \frac{kTN^2}{2V^2} \iint (e^{-U_{12}/kT} - 1)\, dV_1\, dV_2 \quad (19.13)$$

Instead of the coordinates of the two atoms, we can introduce the coordinates of their center of mass and their relative coordinates (as on p. 151). Integration over the relative coordinates gives V, so that (19.13) simplifies to

$$A = A^p - \frac{kTN^2}{2V} \int (e^{-U_{12}/kT} - 1)\, dV \quad (19.14)$$

This expression is usually written as

$$A = A^p + \frac{N^2 kT}{V} B(T) \quad (19.15)$$

where

$$B(T) = \tfrac{1}{2} \int (1 - e^{-U_{12}/kT})\, dV \quad (19.16)$$

The pressure is

$$P = -\left(\frac{\partial A}{\partial V}\right)_T \quad (19.17)$$

so that

$$P = \frac{NkT}{V}\left[1 + \frac{NB(T)}{V}\right] \qquad (19.18)$$

Since the perfect gas pressure is $P^p = NkT/V$, this derivation has thus given us an expression for the second virial coefficient $B(T)$ in terms of the potential energy of interaction U_{12} between a pair of molecules. This is a lovely result, but it applies only to gases with small deviations from ideality. For dense gases, and especially for liquids, further development of the theory depends on evaluation of the configuration integral Q for more general interactions. For dense gases, a series development of Q is possible, paralleling the empirical expressions for the third, fourth, and higher virial coefficients. For liquids, however, the series does not converge, and this avenue of theoretical progress is blocked.

9. Theory of Liquids

Faced with the mathematical cul-de-sac of the general configuration integral, theoreticians have been left with three approaches to the theory of the liquid state.

1. The construction of simplified models for which the configuration integral might be evaluated, and comparison of the results from the models with experiment.
2. Attempts to calculate the radial distribution function, from which, for certain simple liquids (liquid argon, for example) the thermodynamic properties could be evaluated.
3. Numerical calculations with large digital computers, using a Monte Carlo method.

In a two-dimensional application of the Monte Carlo method, one starts with a random configuration of molecules, specified by their coordinates, $x_1 y_1$, $x_2 y_2, \ldots, x_j y_j$. The interaction between any pair of molecules is specified by a suitable choice of the intermolecular potential-energy function, $U(r_{ij})$. The computer selects at random one molecule—say, at $x_j y_j$—and moves it to a new position $x'_j = x_j + \alpha\delta x$, $y'_j = y_j + \beta\delta y$, where α and β are randomly selected fractions and δx and δy are the maximum displacements of the coordinates in one move. The computer can now calculate the total intermolecular potential energy by summing $U(r_{ij})$ all the pairs of molecules,

$$U = \sum_{\text{pairs}} U(r_{ij})$$

The average value of the potential energy of the system is then calculated from the general formula [compare Eq. (5.25)],

$$\langle U \rangle = \frac{\sum U \exp(-U/kT)}{\sum \exp(-U/kT)}$$

The calculation is continued until enough configurations are taken to provide a reasonable statistical average. In practice, the computation is modified so as to exclude moves that would cause one molecule to approach so close to another as to lead to a large repulsion.

Long (and expensive) computer times are required by this method. For example, for a system of 108 molecules, 65 000 configurations per hour can be calculated, and a least 50 000 configurations are needed for a satisfactory average.

Computer calculation can also be applied to nonequilibrium configurations of molecules by the method of *molecular dynamics*. In this method, the actual equation of motion is solved numerically for each molecule, by calculating the force **F** on each molecule due to all the other molecules, and integrating the equation $\mathbf{F} = m\mathbf{a}$. Then the trajectories of all the molecules are plotted to give a dynamic picture of how the collection of molecules changes with time.

Figure 19.7 shows the results of calculations by Wainwright and Alder.[*] The bright lines are the computer output of the motions of a pair of molecules displayed on the face of a cathode-ray tube. The upper figure represents a solid, in which the molecules are constrained to vibrate about equilibrium positions. As the temperature is raised, there is a sudden transition to a liquid state, for which the trajectories are as shown in the lower picture. The computations were based on a rigid sphere model of interaction between molecules, and it is interesting to see that even with such a simple model, the molecular dynamics indicate the existence of a sharp phase transition, solid $\rightarrow$ liquid

Since 1933, Eyring and his students have sought a simple model for the structure of liquids which would avoid the intricacies of the general statistical theory.[†] They began by considering the *free volume* of the liquid, the void space that is not occupied by rigid molecular volumes. In a liquid at ordinary temperature and pressure, about 3% of the volume is free volume. We can deduce this figure, for example, from the studies of Bridgman on the compressibility β. As long as compression consists essentially in squeezing the free volume out of the liquid, β remains relatively high. When the free volume is exhausted, β drops precipitously.

The Eyring model for a liquid is shown in Fig. 19.8. The vapor is mostly void space with a few molecules moving at random. The liquid is mostly filled space with a few vacancies moving at random. As the temperature is raised, the concentration of molecules in the vapor increases and the concentration of vacancies in the liquid also increases. Thus, as the vapor density increases, the liquid density decreases until they become equal at the critical point.

The average density ρ_{av} of liquid and vapor in equilibrium should remain approximately constant. Actually, there is a slight linear decrease with temperature, so that

$$\rho_{av} = \rho_0 - aT \tag{19.19}$$

where ρ_0 and a are constants characteristic of each substance. This relation was

---

[*]B. J. Alder and T. E. Wainwright, *J. Chem. Phys.*, *31*, 459 (1959); *Sci. Amer.*, Oct. 1959, 113.

[†]H. Eyring and M. S. Jhon, *Significant Liquid Structures* (New York: John Wiley & Sons, Inc., 1969).

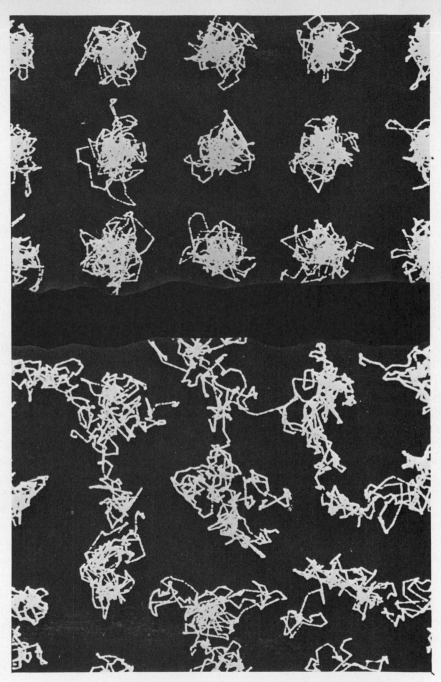

FIG. 19.7 Computer simulation of the transition from solid to liquid based on calculation of trajectories of molecules by method of molecular dynamics. Upper half shows solid behavior; lower half, liquid behavior.

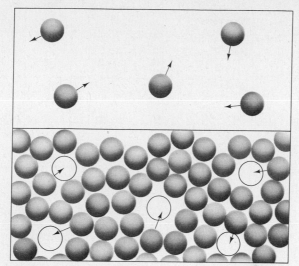

FIG. 19.8 Vacancies in a liquid behave like molecules in a gas. In a liquid, vacancies move among molecules. In a gas, molecules move among vacancies.

discovered by L. Cailletet and E. Mathias in 1886 and is called the *Law of Rectilinear Diameters*. Some examples are shown in Fig. 19.9, where the data for helium, argon, and ether are plotted in terms of reduced variables to bring them onto the same scale.

How is the free volume distributed throughout the liquid? Eyring suggests that the most favorable distribution is in the form of vacancies of approximately molecular size. A molecule adjacent to such a vacancy would have gaslike properties, and hence the introduction of such vacancies would give the highest increase in entropy for a given expense of energy in making free volume. A molecule not next to a vacancy would have solidlike properties.

If we neglect any increase in volume due to vacancies of other than molecular size and assume that vacancies are randomly distributed in the liquid, the fraction of molecules next to a vacancy is $(V_l - V_s)/V_l$, and this is the fraction of gaslike molecules, V_l and V_s being the molar volumes in liquid and solid states. The remaining volume fraction V_s/V_l is solidlike. Eyring and Ree thus suggest that the heat capacity of a monatomic liquid like argon should be represented by the equation

$$C_{Vm} = \left(\frac{V_s}{V_l}\right)3R + \left(\frac{V_l - V_s}{V_l}\right)\frac{3}{2}R \qquad (19.20)$$

The agreement with experiment is encouraging. Other properties of the liquid can be calculated by taking its partition function Z as the product of solidlike and gaslike terms.

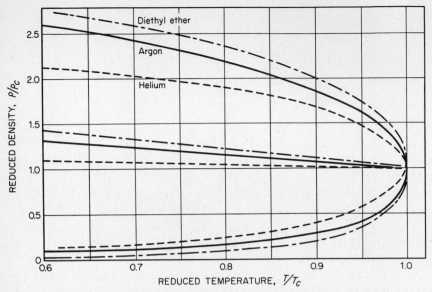

FIG. 19.9 Law of Rectilinear Diameters. The straight lines indicate the average reduced density of the liquid and vapor phases in equilibrium.

10. Flow Properties of Liquids

When we consider the mechanical properties of a substance, be it fluid or solid, it is convenient to divide the applied stresses into two classes, *shear stresses* and *compressive stresses*. These are shown in Fig. 19.10. A stress is a force applied to unit area of the body. A change in the dimensions of the body produced by the action of a stress is called a *strain*. The ratio of stress to strain is an *elastic modulus*.

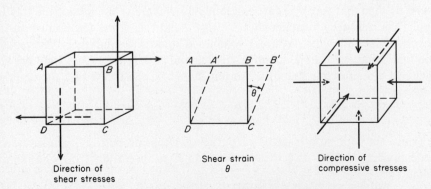

FIG. 19.10 Shear stresses and compressive stresses.

Let us restrict our attention to isotropic bodies. A compressive stress applied uniformly over the surface of such a body (i.e., a pressure) produces a uniform compression. The bulk modulus K is defined as the ratio

$$K = \frac{\text{compressive force per unit area}}{\text{change in volume per unit volume}}$$

$$K = V\left(\frac{\partial P}{\partial V}\right)_T$$

The bulk modulus $K = \beta^{-1}$, where β is the compressibility.

A shear stress produces an angular deformation θ, as shown in Fig. 19.11. The result of the shear is to displace a plane in the body parallel to itself relative to other parallel planes in the body. The shear stress is the force F' per unit area of the *plane displaced*. The shear strain is the ratio of the displacement between any two planes to the distance between the planes. The shear strain is measured by $\tan \theta = x'$.

Perhaps the most typical property of fluids is the fact that they begin to flow appreciably as soon as a shear stress is applied. A solid, on the other hand, apparently supports a considerable shear stress, opposing to it an elastic restoring force proportional to the strain, and given by Hooke's Law, $F' = -\kappa'x'$. In fact, however, even a solid flows somewhat, but usually the stress must be maintained for a long time before the flow is noticeable. This slow flow of solids is called *creep*. Although creep is usually small, its occurrence suggests that the flow properties of liquids and solids differ in degree and not in kind. Under high stresses, creep passes over into the *plastic deformation* of solids, which can be observed in such operations as the rolling, drawing, or forging of metals. These operations proceed by a mechanism based upon the gliding of slip planes.

The fact that liquids flow immediately under a very small shear stress does not necessarily mean that there are no elastic restoring forces within the liquid structure. These forces may exist without having a chance to be effective, owing to the rapidity of the flow process. The skipping of a thin stone on the surface of a pond demonstrates the elasticity of a liquid. An interesting silicone polymer has been exhibited under the name of *bouncing putty*. This curious material is a hybrid of solid and liquid in regard to its flow properties. Rolled into a sphere and thrown at a wall, it bounces back as well as any rubber ball. Set the ball on a table and it gradually collapses into a puddle of viscous putty. Thus, under long-continued stress it flows slowly like a liquid, but under a sudden sharp blow it reacts like a rubber.

It is difficult to measure the reversible or elastic properties of liquids under shear stresses, since they are usually overwhelmed by the irreversible *shear viscosity*. If η is the shear viscosity and κ' is the shear elasticity, Maxwell showed that an instantaneous shear distortion should have a *relaxation time* $\tau = \eta/\kappa'$. For ordinary light liquids, with $\kappa' \approx 10^9$ N$\cdot$m^{-2} (similar to its value in a loosely bound crystal) and $\eta = 10^{-3}$ kg$\cdot$m$^{-1}\cdot$s^{-1}, $\tau \approx 10^{-12}$ s. Thus, the relaxation frequency (10^{12} s^{-1}) would be outside the range of measurement. With polymer solutions, however, η can be much higher, and κ' can be lower, so that the relaxation frequency may fall within the range of ultrasonic frequencies.*

Another interesting problem† arises when we ask whether a liquid subjected to compressive stress displays any irreversible effects due to a *volume viscosity*. In the case of compressive stresses, the elastic response or change in volume is easy to detect, but the viscous effect is transient and hard to detect. The experimental approach is to study the absorption of ultrasonic waves in liquids. The deformation caused by a sound wave is partly a change in volume and partly a shear, but it is possible to separate the contributions to the absorption coefficient due to the different mechanisms. For glycerol, a highly associated liquid, the absorption of ultrasonic waves is about twice the value calculated from shear viscosity alone, so that a large contribution is due to compressive viscosity. In water also, the absorption of ultrasonic waves is due mainly to the high volume viscosity.

11. Viscosity

Some of the hydrodynamic theory of fluid flow was discussed in Section 4.24. We described how the coefficient of viscosity η could be measured from the rate of flow through cylindrical tubes. This is one of the most convenient methods for liquids as well as gases. The equation for an incompressible fluid is suitable for liquids, whereas that for a compressible fluid is used for gases. Thus, from (4.60),

$$\eta = \frac{\pi R^4 \, \Delta P}{8l(dV/dt)}$$

In the *Ostwald viscometer*, one measures the time required for a bulb of liquid to discharge through a capillary under the force of its own weight. It is usual to make relative rather than absolute measurements with these instruments, so that the dimensions of the capillary tube and volume of the bulb need not be known. The time t_0 required for a liquid of known viscosity η_0, usually water, to flow out of the bulb is noted. The time t for the liquid of unknown viscosity is similarly measured. The unknown viscosity is then

$$\eta = \frac{\rho}{\rho_0} \frac{t}{t_0} \eta_0$$

where ρ_0 and ρ are the densities of water and of the other liquid.

Another useful viscometer is the Höppler type, based on the Stokes formula for the frictional resistance of a spherical body falling through a fluid, $F = 6\pi\eta rv$.

$$\eta = \frac{F}{6\pi rv} = \frac{(m - m_0)g}{6\pi rv} \tag{19.21}$$

*W. P. Mason, W. O. Baker, M. T. McSkimin, and J. H. Heiss, *Phys. Rev.*, **75**, 936 (1949).

†Paraphrased from the remarks of H. S. Green, "The Structure of Liquids," *Encyclopedia of Physics* (Berlin: Springer, 1960), p. 92. See also the comprehensive discussions in K. H. Herzfeld and T. Litovitz, *Absorption and Dispersion of Ultrasonic Waves* (New York: Academic Press, 1959).

By measuring the rate of fall in the liquid (terminal velocity v) of metal spheres of known radius r and mass m, the viscosity can be calculated, since the force F is equal to $(m - m_0)g$, where m_0 is the mass of liquid displaced by the ball.

The hydrodynamic theories for the flow of liquids and gases are very similar. The kinetic-molecular mechanisms differ widely, as we might immediately suspect from the difference in the dependence of gas and liquid viscosities on temperature and pressure. In a gas, the viscosity increases with temperature and is practically independent of pressure. In a liquid, the viscosity increases with pressure and decreases with increasing temperature.

The typical dependence of liquid viscosity on temperature was first pointed out by J. deGuzmann Carrancio in 1913. The viscosity coefficient may be written

$$\eta = A \exp\left(\frac{\Delta E_{vis}}{RT}\right)$$

The quantity ΔE_{vis} is the energy barrier that must be overcome before the ele-

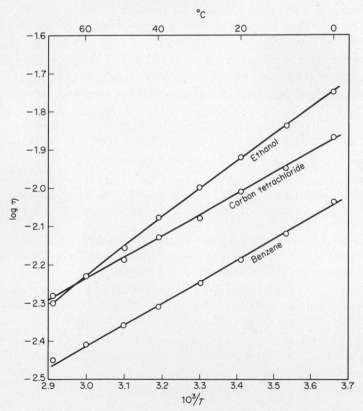

FIG. 19.11 Temperature dependence of viscosity plotted as log η vs. T^{-1}. The data for carbon tetrachloride and benzene fall on straight lines; but the data for ethanol at higher temperatures deviate from a straight line in the direction of lower viscosities.

mentary flow process can occur. It is expressed per mole of liquid. The term $\exp(-\Delta E_{vis}/RT)$ can then be explained as a Boltzmann factor giving the fraction of the molecules having the requisite energy to surmount the barrier. Thus, ΔE_{vis} is an activation energy for viscous flow.

The viscosities of several liquids as functions of temperature are plotted in Fig. 19.11 as $\log \eta$ vs. T^{-1}. In most cases, a good linear relationship is found. The activation energies ΔE_{vis} are usually about $\frac{1}{3}$ to $\frac{1}{4}$ the latent heats of vaporization. Liquids with hydrogen-bonded structures often exhibit a deviation from the linear $\log \eta$ or T^{-1} plot at higher T, in the direction of lowered η. How would you explain this effect?

PROBLEMS

1. The normal vapor pressure P_0 of acetic acid is 20.6 torr at 303 K. Two drops of CH_3COOH (l), each of radius 5.00 nm, are in equilibrium with CH_3COOH (g) in an otherwise empty container. Calculate the pressure of the CH_3COOH vapor. If the two drops coalesce to one drop, what will be the new vapor pressure? What assumptions did you make to obtain a solution?

2. The thermal pressure coefficient $(\partial P/\partial T)_V$ for CCl_4 is 117.5 N·m⁻²·K⁻¹ at 315 K and 10 N·m⁻² pressure. Calculate the internal pressure.

3. The liquid and vapor densities of ethanol in equilibrium at various temperatures are

θ, °C	100	150	200	220	240
ρ(liq), g·cm⁻³	0.7157	0.6489	0.5568	0.4958	0.3825
ρ(vap), g·cm⁻³	0.00351	0.0193	0.0508	0.0854	0.1716

The critical temperature is 243°C. What is the critical volume?

4. The crystal structure of solid krypton is cubic close packed with a distance between nearest neighbors of 0.405 nm at 58 K. The polarizability of a Kr atom is 2.46×10^{-24} cm³, and its first ionization potential, which may be taken as $h\nu_0$ in the dispersion interaction, is 13.93 eV.
   (a) Calculate the dispersion interaction for a pair of Kr atoms 0.405 nm apart.
   (b) Show how the total energy for a mole of solid krypton may be calculated, assuming that only nearest neighbor interactions are important.

5. The flow time for H_2O in a viscometer at 20°C is 120 s. The density of water is 0.997 g·cm⁻³, and its viscosity is 1.002 g·m⁻¹·s⁻¹. The flow time of benzyl alcohol (density 1.042 g·cm⁻³) is 663 s. Calculate the viscosity of benzyl alcohol.

6. At 1050 K, a certain molten glass has a viscosity of 1.46×10^5 kg·m⁻¹·s⁻¹ and a density of 3.54×10^3 kg·m⁻³. How long would a platinum sphere 1.00 cm in diameter take to fall through 1.00 cm of this molten glass?

7. Two forms of the virial equation of state are used:
   (a) $\dfrac{PV_m}{RT} = 1 + \dfrac{B(T)}{V_m} + \dfrac{C(T)}{V_m^2} + \cdots$
   (b) $\dfrac{PV_m}{RT} = 1 + B'(T)P + C'(T)P^2 + \cdots$
   Show that $B' = B/RT$, $C' = (C - B^2)/(RT)^2$.

8. For a gas of rigid spheres of diameter d with an attractive energy $U(r) = -Ar^{-6}$, show that second virial coefficient is*

$$B = \frac{2}{3}\pi L d^3 \left\{ 1 - \sum_{n=1}^{\infty} \frac{3}{n!\,(6n-3)} \left(\frac{A}{d^6 kT} \right)^n \right\}$$

9. Use the expression in Problem 8 to calculate B for argon given that $d = 0.293$ nm and $A = 3 \times 10^{-77}$ J·m^6. Hence, calculate PV_m/RT for argon at 100 atm and 300 K and compare it with the experimental value.†

10. The London attractive energy between two like atoms a distance r apart is

$$E_L = \frac{-\frac{3}{2}\mu^2 \alpha}{(4\pi\epsilon_0)^2 r^6}$$

where α is the polarizability and μ is the value of the instantaneous dipole moment. It can be shown that $\mu^2 = \alpha\,\Delta E/2$, where ΔE is the energy for the transition that contributes the most to the dispersion forces. Calculate E_L for two argon atoms at several values of r, assuming $\alpha = 1.63 \times 10^{-30}$ m^3 and $\Delta E =$ the ionization potential I, 15.4 eV.

11. The following equation for the viscosity of a liquid was derived by Eyring from the activated-complex theory of rate processes:

$$\eta = \frac{L}{V_m}(2\pi m k T)^{1/2} V_f^{1/3} \exp\left(\frac{\Delta H_{\text{vis}}^{\ddagger}}{RT} \right)$$

where V_m is the molar volume of the liquid and V_f is its free volume per molecule. From the data shown in Fig. 19.11, calculate $\Delta H_{\text{vis}}^{\ddagger}$ for CCl_4. By means of reasonable assumptions about the dimensions of the molecules in the liquid state, estimate V_f and hence calculate the pre-exponential factor in the Eyring equation and compare it with the experimental value.

*See J. O. Hirschfelder, C. F. Curtiss, and R. B. Bird, *Molecular Theory of Gases and Liquids* (New York: John Wiley & Sons, Inc., 1954), pp. 131, 158, 166.

†F. Din, ed., *The Thermodynamic Functions of Gases*, Vol. 2 (London: Butterworth & Co., Ltd., 1958).

20

Macromolecules

Just as man, under the regard of the paleontologist, merges anatomically into the mass of mammals who preceded him, so the cell, in the same descending series, both qualitatively and quantitatively merges into the world of chemical structure and visibly converges towards the molecule.

Pierre Teilhard de Chardin
(1955)*

Much of our present understanding of very large molecules is based on pioneering work by Hermann Staudinger and his many coworkers, who published over 700 papers in this field from 1920 to 1950. Staudinger introduced the term *macromolecular* to characterize substances with molecular weight $M_r > 10\,000$. A system is called macromolecular if it shows the qualities (viscosity, light scattering, ultrasedimentation, colligative properties, etc.) characteristic of this special state, as will be outlined in this chapter. Staudinger defined *molecular colloids* (as distinct from *association colloids*) as substances that behaved as colloids in solution owing to the large sizes of their molecules.

Originally, a *polymer* denoted a substance composed of molecules with a formula X_n, made by repetition of n identical structural units X. Combination of two structural units led to *copolymers* $X_n Y_m$ (in contrast to *homopolymers* X_n). Now we often speak also of substances such as proteins and nucleic acids as *polymers* because they contain a large number of similar structural units joined together by the same kind of linkage.

1. Types of Polyreactions

High polymers can be synthesized from small molecules by different types of *polyreactions*.

Le Phénomène Humain (Paris: Edition du Seuil, 1955).

928

Addition polymerization consists in the addition of one molecule to another via a utilization of unsaturated valences. For example, polyethylene is formed by successive addition of $CH_2{=}CH_2$ units to a growing polymer chain. The growth of the chain can be initiated by introduction of a free radical R, which adds to the unsaturated molecule to give a larger radical,

$$R{-} + CH_2{=}CH_2 \longrightarrow R{-}CH_2{-}CH_2{-}$$

which, in turn, adds to another molecule of C_2H_4,

$$R{-}CH_2{-}CH_2{-} + CH_2{=}CH_2 \longrightarrow R{-}CH_2{-}CH_2{-}CH_2{-}CH_2{-}$$

Such addition steps can continue at a very rapid rate during the growth phase of polymerization. Finally, however, the growing radical $R(CH_2)_n{-}CH_2{-}$ may encounter a second radical and the polymerization chain is teminated by recombination or disproportionation.

$$R(CH_2)_n{-}CH_2{-} + R(CH_2)_m{-}CH_2 \longrightarrow R(CH_2)_{m+n+2}R$$
$$\longrightarrow R(CH_2)_nCH_3 + R(CH_2)_{m-1}CH{=}CH_2$$

If two different kinds of monomer, A and B, are used as starting materials, copolymerization can occur, with the possibility of forming a wide variety of copolymers depending on the proportions of A and B in the product. An important industrial example is the copolymer of styrene and butadiene in a ratio of about $1:3$ in the synthetic rubber *SBR*.

Polycondensation is a type of reaction that proceeds through the elimination of a smaller molecule and formation of a bond between two monomers, each of which contains two functional groups so that the reaction can occur repeatedly to yield a macromolecule. An example is the synthesis of the polyamide Nylon 66 by Carothers in 1934:

$$
\begin{array}{lll}
\text{COOH} & \text{NH}_2 & \text{COOH} \\
| & | & | \\
(\text{CH}_2)_4 & (\text{CH}_2)_6 & (\text{CH}_2)_4 \\
| & | & | \\
\text{COOH} \quad + & \text{NH}_2 \quad\longrightarrow & \text{CO} \quad + \quad \text{H}_2\text{O} \\
& & | \\
& & \text{NH} \\
\text{adipic} & \text{hexamethylene} & | \\
\text{acid} & \text{diamine} & (\text{CH}_2)_6 \\
& & | \\
& & \text{NH}_2
\end{array}
$$

The product has functional end groups and condensation can continue and will yield products with $M_r \sim 15\,000$.

The active species in a polymerization reaction is not necessarily a radical, but can be an ion or an active complex formed with a suitable catalyst. In 1963, Karl Ziegler and Giulio Natta shared a Nobel Prize for their work in developing techniques for synthesis of polymers with specified stereoisomerism. The key to these reactions was the discovery of heterogenous catalysts based on mixtures of $Al(C_2H_5)_3$ and $TiCl_4$.

2.   Distribution of Molar Masses*

It is evident that macromolecules formed by any of the polyreactions outlined will not all have the same final molecular mass m. (For example, see Fig. 20.1.) There will be a distribution of the masses of the macromolecules with a probability $W(m)$ that the mass lies between m and $m + dm$. This nonuniformity of the molecular masses causes a number of problems in the interpretation of some of the properties of macromolecules in solution.

In a solution of styrene in benzene, all the molecules of styrene are identical and all methods of determining the molar mass M should yield the same value, within experimental uncertainties. In a solution of polystyrene in benzene, however, the masses of the individual polymer molecules are distributed over a range of values. Consequently different methods of determining the molar mass from properties of the solution can yield different values.

Colligative properties, such as osmotic pressure, depend on the number of molecules in a solution. Thus, the mass per mole calculated from a colligative property is the *number average*, defined by

$$\bar{M}_N = \frac{L \sum N_i m_i}{\sum N_i} \tag{20.1}$$

To obtain $\bar{M}_N$, we add the products of each molecular mass m_i and the number of molecules N_i having that mass and divide by the total number of molecules, finally multiplying by the Avogadro Number L.

In case there is a distribution of molecular masses, the number average tends to emphasize the lower masses. Suppose there were only two kinds of molecule, one with $Lm_i = 100 \text{ g·mol}^{-1}$ and one with $Lm_2 = 10\,000 \text{ g·mol}^{-1}$. The number average would be $\bar{M}_N = 5050 \text{ g·mol}^{-1}$, despite the fact that over 99% of the mass of the substance was in the heavier molecules.

Some experimental determinations of molar mass, for example, those based on the scattering of light, depend on the masses of material in the different fractions. These methods yield a *mass average* molar mass defined by

$$\bar{M}_m = \frac{L \sum N_i m_i m_i}{\sum N_i m_i} = \frac{L \sum N_i m_i^2}{\sum N_i m_i} \tag{20.2}$$

Consider a sample containing 10% by mass of polymer with $Lm_1 = 10\,000$ g·mol^{-1} and 90% by mass of one with $Lm_2 = 100\,000$ g·mol^{-1}. Then,

$$\bar{M}_m = \frac{0.1(10\,000) + 0.9(100\,000)}{1} = 91\,000 \text{ g·mol}^{-1}$$

whereas

$$\bar{M}_N = \frac{0.1(10\,000) + 0.09(100\,000)}{0.19} = 52\,500 \text{ g·mol}^{-1}$$

---

*We distinguish between the molar mass M, the mass divided by the amount of substance, and the relative molecular mass M_r a dimensionless quantity, also called the *molecular weight*. M_r is the *ratio* of the average mass per molecule of the natural isotopic composition of the substance to 1/12 the mass of an atom of the nuclide ^{12}C.

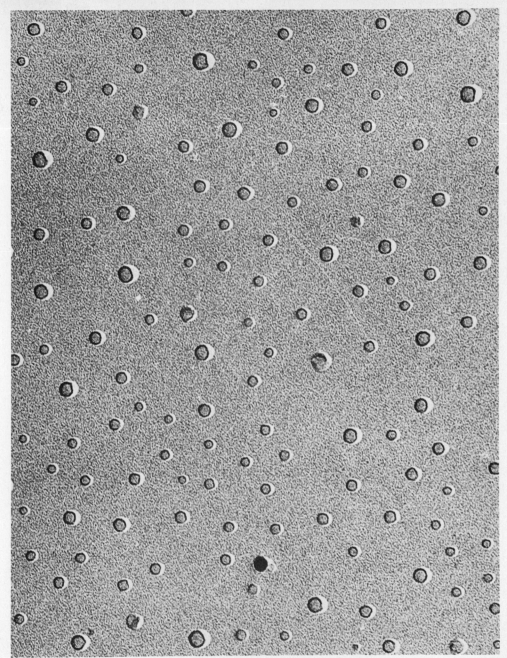

FIG. 20.1 An electron micrograph of molecules of brominated natural rubber, magnification 100 000X. This picture illustrates the distribution of molecular masses that usually occurs in solutions of high polymers, and the consequent necessity of taking a properly weighted average of properties that depend upon molecular mass. [Photograph from Dr. S. Nair, The Rubber Research Institute of Malaya.]

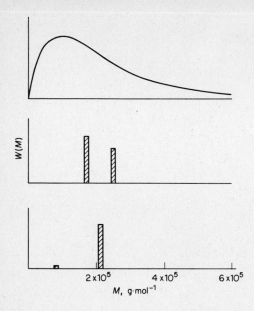

FIG. 20.2 Three different distributions of molar masses, each yielding $\bar{M}_N = 10^5$ and $\bar{M}_m = 2 \times 10^5$ g·mol⁻¹.

Figure 20.2 shows three different distributions of molar masses, each yielding $\bar{M}_N = 10^5$ g·mol⁻¹ and $\bar{M}_m = 2 \times 10^5$ g·mol⁻¹.

We shall now consider several methods for measuring the molecular masses, sizes, and shapes of macromolecules in solution.

3.   Osmotic Pressure

The theory of osmotic pressure, discussed in Section 7.15, applies also to solutions of polymers. Ordinary solutions follow the van't Hoff equation, $\Pi = cRT$, up to a concentration of about 1%, but serious deviations occur in polymer solutions at much lower concentrations. In 1914, Caspari reported that the osmotic pressure of a 1% solution of rubber in benzene was about twice the value calculated from the van't Hoff equation. Similar large deviations have been frequently noted.

The evaluation of data on the osmotic pressure of high polymer solutions presents three principal problems: (1) How can we determine the molar mass when the van't Hoff law fails? (2) Can the data on osmotic pressure be explained in terms of the sizes and shapes of the molecules of polymer? (3) Can data on osmotic pressure provide information on the interactions of polymer molecules with each other and with solvent molecules?

In 1945, McMillan and Mayer* showed that the osmotic pressure of nonelectrolytes could be represented by a power series in the concentrations, exactly like the virial equation for a nonideal gas,

$$\Pi = RT\left(\frac{C}{\bar{M}_N} + BC^2 + DC^3 + \cdots\right) \tag{20.3}$$

*W. G. McMillan and J. E. Mayer, *J. Chem. Phys.*, **13**, 276 (1945).

Here C is the concentration in mass per unit volume (e.g., $g \cdot cm^{-3}$). In fairly dilute solutions, the third term becomes negligible, and

$$\frac{\Pi}{C} = \frac{RT}{\bar{M}_N} + RTBC \tag{20.4}$$

Thus, a plot of Π/C vs. C should give a straight line in the region of low concentration. The intercept of the line extrapolated to $C = 0$ yields $RT/\bar{M}_N$, and hence the number average molar mass $\bar{M}_N$. Figure 20.3 shows osmotic pressures

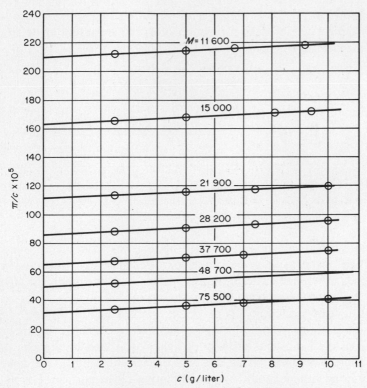

FIG. 20.3 Osmotic pressures of solutions of polypyrrolidones in water at 298 K plotted in accord with Eq. (20.4). The number average molar masses $\bar{M}_N$ are noted on the curves.

of aqueous solutions of polypyrrolidones* plotted according to (20.4). The values of $\bar{M}_N$ range from 11 600 to 75 500 for different fractions.

Various designs of osmometers have been used for these studies. The semipermeable membrane is usually a swollen cellulose film, which is mounted between rigid supports in such a way as to present a large area open to the solutions. In measurements of the osmotic pressure of high polymers, we must try to minimize the Donnan effects discussed in Section 12.18 by establishing conditions under which the

*J. Hengstenberg, *Makromol. Chem.*, **7**, 572 (1951).

charge of the polymer is minimized, or by working with a considerable excess of neutral salt.

4. Light Scattering—
The Rayleigh Law

The scattering of light by a colloidal solution was described by Tyndall in 1871. He passed a beam of white light through a gold sol and observed bluish light scattered normal to the incident beam. A typical arrangement for the study of light scattering is shown in Fig. 20.4. The detector of scattered light is mounted so that the intensity can be measured as a function of θ, the scattering angle from the direction of the incident beam.

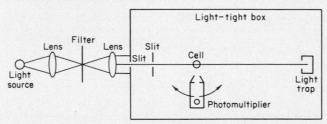

FIG. 20.4 Diagram of the basic construction of a light-scattering apparatus.

If a solution or sol is sufficiently dilute, the total scattered light is simply the sum of the light scattered from individual particles. The light scattered from a particle depends on the area intercepting the beam, and thus upon the square of the effective radius of the particle, a^2.

A theory of light scattering was originally worked out by Rayleigh in 1871 for the case of isotropic particles whose dimensions are small compared to the wavelength λ of the light. The intensity of light I_θ scattered by a single particle through an angle θ depends on the intensity of incident light I_0, the distance r_s from the scattering volume, and the polarizability α of the particle. For unpolarized light, the relation is*

$$R_\theta = \frac{I_\theta \, r_s^2}{I_0} = \frac{8\pi^4 \, \alpha^2}{\lambda^4}(1 + \cos^2 \theta) \tag{20.5}$$

The quantity R_θ is called the *Rayleigh ratio.*

We may note especially the strong inverse dependence of scattered intensity on wavelength. As a consequence, blue light is scattered much more strongly than red light. If we look at the sky by daylight, we see the sunlight scattered by gas molecules and dust particles through various angles θ, and the sky appears blue. At sunset, however, we look directly at the sun, and it appears red because the

*A derivation is given by P. J. Flory, *Principles of Polymer Chemistry* (Ithaca: Cornell University Press, 1953), p. 287. The discussion in P. 934-6 uses electrostatic units. To conform to SI system, α should be replaced by $\alpha/4\pi\epsilon_0$ wherever it occurs.

blue light has been selectively scattered away from the direct beams. Animal colors are often due in part to scattered light,* sometimes modified by interference phenomena.

5. Light Scattering by Macromolecules

The total light scattered in all directions from the incident beam as it traverses a suspension is measured by the *turbidity* τ. If the intensity I_0 of an incident beam is reduced to I on passage through a distance x,

$$\frac{I}{I_0} = \exp(-\tau x)$$

This expression is formally similar to the Bouguer–Lambert Law for the absorption of light.

If N particles of uniform size are randomly distributed at a concentration N/V per unit volume, the Rayleigh ratio for the intensity of light scattered per unit volume of suspension, from (20.5), would be

$$R'_\theta = \frac{I_\theta r_s^2}{I_0} = \frac{8\pi^4 \alpha^2 N}{\lambda^4 V}(1 + \cos^2 \theta) \tag{20.6}$$

The turbidity is related to this Rayleigh ratio at 90° by

$$\tau = \left(\frac{16\pi}{3}\right) R'_{90} \tag{20.7}$$

Light scattering is not measured as a turbidity, but by an actual determination of the intensity of the light at 90° and other angles relative to the incident beam. Results, however, are often reported in terms of the τ calculated from (20.7).

In 1947, Debye extended the Rayleigh theory to solutions of macromolecules. The polarizability α is not a convenient quantity to measure, but by means of the Clausius-Mossotti theory of Section 15.17 it can be related to data on the refractive index **n** of the solution. The excess polarizability α of the solution over that of the solvent is related to the dielectric constants ϵ of the solution and $\epsilon°$ of the solvent by

$$\epsilon - \epsilon° = 4\pi \frac{N}{V}\alpha$$

For the relatively high frequencies of visible light, $\epsilon \approx \mathbf{n}^2$, so that

$$\mathbf{n}^2 - \mathbf{n}_0^2 = 4\pi \frac{N}{V}\alpha$$

$$\alpha = \frac{(\mathbf{n} + \mathbf{n}_0)(\mathbf{n} - \mathbf{n}_0)}{4\pi} \frac{V}{N}$$

The quantity $(\mathbf{n} - \mathbf{n}_0)/C$, where C is the concentration in mass per unit volume,

*C. H. Greenewalt, *Hummingbirds* (New York: Doubleday & Company, Inc., 1960), p. 169.

is called the *refractive index increment* of the solute, and for a linear variation it can be written as (dn/dC). For a dilute solution $(\mathbf{n} + \mathbf{n}_0) \approx 2\mathbf{n}_0$ so that

$$\alpha = \frac{\mathbf{n}_0}{2\pi}\left(\frac{dn}{dC}\right)\frac{M}{L}$$

when M is the molar mass. By substitution into (20.6),

$$R_\theta = \frac{2\pi^2 \mathbf{n}_0^2 (dn/dC)^2}{L\lambda^4} CM(1 + \cos^2 \theta) = KCM(1 + \cos^2 \theta) \qquad (20.8)$$

where K is the defined combination of parameters $2\pi^2 \mathbf{n}_0^2 (dn/dC)^2/L\lambda^4$. Thus a measurement of R_θ can lead to a determination of molar mass M (or the molecular weight by a suitable choice of units).

In more concentrated solutions, (20.8) can be corrected to yield

$$K\frac{C}{R_\theta}(1 + \cos^2 \theta) = \frac{1}{M} + 2BC \qquad (20.9)$$

This equation has the same form as that for the osmotic pressure. If a solution of polymers of differing molecular masses is studied, light scattering yields the mass average molar mass, $\bar{M}_m$,[*] whereas osmotic-pressure data yield the number average $\bar{M}_N$. We can therefore combine data from the two methods to obtain information about the distribution of molecular masses in the polymer solution.

When the dimensions of the scattering particle are no longer small compared to the wavelength λ, the theory of scattering becomes more complicated. We must then consider interference between light waves scattered from different parts of the same particle. The theory for such scattering was worked out by Gustav Mie in 1908 for the case of a spherical particle.[†]

The next extension of the theory is to take into account the sizes *and* shapes of the molecules. These factors are included in the *particle scattering factor* $P(\theta)$. One obtains the equation,[‡]

$$K\frac{C}{R_\theta}(1 + \cos^2 \theta) = \frac{1}{MP(\theta)} + 2BC \qquad (20.10)$$

It is possible to obtain an accurate value of M from data on light scattering, by examining the solutions over a range of concentrations C and angles θ. The data are then extrapolated to zero concentration and zero angle. This graphical extrapolation is called a *Zimm plot* after its inventor. Consider the data obtained by Doty[§] and his coworkers for a fraction of cellulose nitrate dissolved in acetone.

---

[*]From Eq. (20.9) applied to a distribution of polymers with concentrations C_i and molecular weights M_i, the limiting value of $R_\theta/(1 + \cos^2 \theta)$ as $C \longrightarrow 0$ is $K \sum C_i M_i$. Therefore, the limiting value of KC/R_θ is

$$\frac{\sum C_i}{\sum C_i M_i} = \frac{1}{\bar{M}_m}$$

[†]*Ann. Physik* (4), *25*, 377 (1908). See H. C. Van de Hulst, *Light Scattering by Small Particles* (New York: John Wiley & Sons, Inc., 1957).

[‡]B. H. Zimm, *J. Chem. Phys.*, *16*, 1093 (1948).

[§]A. M. Holtzer, H. Benoit, and P. Doty, *J. Phys. Chem.*, *58*, 624 (1954).

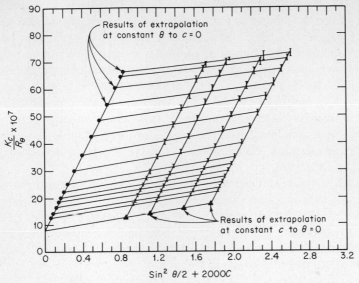

FIG. 20.5(a)   Light scattering of cellulose nitrate in acetone at 25°
presented on a Zimm plot.

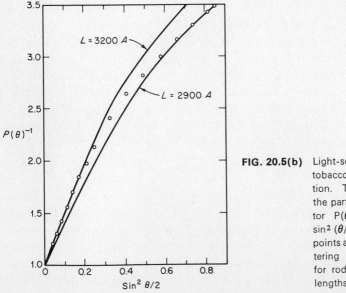

FIG. 20.5(b)   Light-scattering data from
tobacco mosaic virus solu-
tion. The reciprocal of
the particle scattering factor $P(\theta)$ is plotted vs.
$\sin^2 (\theta/2)$.   Experimental
points and theoretical scattering curves are shown
for rods of two different
lengths.

In Fig. 20.5(a), the values of KC/R_θ are plotted against $\sin^2\theta + kC$, where k is an
arbitrary constant (2000 in this case) used to spread out the data. Each set of points
at constant concentration is extrapolated to zero angle and each set of points at
constant angle is extrapolated to zero concentration. The two limiting sets of
points are extrapolated to zero ordinate, where they should intersect at a value

of $1/\bar{M}_m$ free of effects due to interaction between particles (C) or interference between different parts of the same particle (θ). In this case $\bar{M}_m = 400\ 000\ \text{g}\cdot\text{mol}^{-1}$. The value of $\bar{M}_N$ from osmotic pressure was $234\ 000\ \text{g}\cdot\text{mol}^{-1}$. The difference indicates a rather broad distribution of molecular masses in this sample.

In this polymer, each monomer unit has $M_r = 294$ and a length of 0.515 nm. The number of units in a chain of $M_r = 400\ 000$ would be 1350. If completely extended, this chain would be 695 nm long. The value of the rms end-to-end distance, $(\bar{r^2})^{1/2}$, from light scattering was 150 nm. This value is rather large for a molecule of such a short completely extended length, and suggests that this polymer is unusually stiff and extended. Most randomly coiled polymers tend to be much more tightly coiled and compact.

Figure 20.5(b) shows some determinations of $P(\theta)$ from light scattering by tobacco mosaic virus, together with the theoretical curves for rods 290 and 320 nm long. The length of the virus appears to fit a value of 300 ± 5 nm. The $\bar{M}_m$ is calculated to be $39.5 \times 10^3\ \text{kg}\cdot\text{mol}^{-1}$. From the length, mass, and density we can estimate an effective diameter of 15 nm. This work is a good example of the use of light scattering to characterize a macromolecule in solution.

6. Sedimentation Methods:
The Ultracentrifuge

The motion of a particle in a fluid medium under the influence of a gravitational field is determined by the balance between the force of gravitation and the frictional resistance of the medium. If a particle of mass m falls in a medium of density ρ, the gravitational force is $(1 - v_1\rho)mg$, where v_1 is the partial specific volume of a particle (i.e., V_1/M_1, where V_1 is the partial molar volume and M_1 is the molar mass), and hence $v_1\rho mg$ is the weight of fluid displaced by the particle. If dx/dt is the velocity of sedimentation, $f(dx/dt)$ is the frictional force, where f is the frictional coefficient. A terminal steady velocity is reached, at which

$$f\left(\frac{dx}{dt}\right) = (1 - v_1\rho)mg \tag{20.11}$$

We may note the f^{-1} is the velocity under a unit force.

Jean Perrin in 1908 studied with a microscope the sedimentation of gamboge particles in an aqueous suspension under the influence of the earth's gravitational field. The weakness of this field restricted his method to quite massive particles. The development of centrifugal techniques, capable of producing gravitational accelerations up to $3 \times 10^5 g$, is due in large measure to The Svedberg of Uppsala. Beginning about 1923, he devoted great skill to the problem of characterizing macromolecules, especially proteins, by means of centrifugal sedimentation. In a centrifugal field of force, we replace g in (20.11) by $\omega^2 x$, where ω is the angular velocity and x is the distance from the center of rotation. Thus,

$$f\left(\frac{dx}{dt}\right) = (1 - v_1\rho)m\omega^2 x \tag{20.12}$$

The quantity

$$s = \frac{dx/dt}{\omega^2 x} \tag{20.13}$$

is called the *sedimentation constant*. It is the sedimentation rate for a unit centrifugal acceleration. For a given molecular species in a given solvent at a given temperature, s is a characteristic constant. It has often been expressed in Svedberg units, equal to 10^{-13} seconds.

For the case of a spherical particle of radius a, Stokes calculated*

$$f = 6\pi\eta a$$

where η is the coefficient of viscosity of the medium. This is a special case, however, and usually we have no simple formula for f, even if we know the shape of the particle, which usually we do not.

Fortunately, we can eliminate f if we can measure some other property that depends on it. For free diffusion in a dilute solution, the diffusion coefficient is

$$D = \frac{RT}{Lf} \tag{20.14}$$

This relation was derived by Einstein in 1905.† It follows directly from Eq. (10.19) if we replace u_i/Q by f^{-1}.

Assuming that the diffusional f is the same as the centrifugal f, we can eliminate it between (20.12) and (20.14) to obtain

TABLE 20.1
Characteristic Constants of Protein Molecules at 20°C

Proteins	v_1 $(cm^3 \cdot g^{-1})$	s $(10^{-13}\ s)$	D $(cm^2 \cdot s^{-1}) \times 10^7$	M $kg \cdot mol^{-1}$
Myoglobin (beef heart)	0.741	2.04	11.3	16.9
Hemoglobin (horse)	0.749	4.41	6.3	68
Hemoglobin (man)	0.749	4.48	6.9	63
Hemocyanin (octopus)	0.740	49.3	1.65	2800
Serumalbumin (horse)	0.748	4.46	6.1	70
Serumalbumin (man)	0.736	4.67	5.9	72
Serumglobulin (man)	0.718	7.12	4.0	153
Lysozyme (egg yellow)	(0.75)	1.9	11.2	16.4
Edestin	0.744	12.8	3.18	381
Urease (jack bean)	0.73	18.6	3.46	480
Pepsin (pig)	(0.750)	3.3	9.0	35.5
Insulin (beef)	(0.749)	3.58	7.53	46
Botulinus toxin A	0.755	17.3	2.10	810
Tobacco mosaic virus	0.73	185	0.53	31400

*G. G. Stokes, *Trans. Cambridge Phil.*, Ser. *9*, No. 8 (1851). See also R. S. Bradley, *The Phenomena of Fluid Motions* (Reading, Mass.: Addison-Wesley Publ. Co., 1967), Ch. 8.

†A. Einstein, *Ann. Physik*, *17*, 549 (1905); *19*, 371 (1906).

$$M = \frac{RTs}{D(1 - v_1\rho)} \tag{20.15}$$

This equation, derived by Svedberg in 1929, has been the basis of many measurements of molecular weights from sedimentation velocities. For precise determinations, as usual, we should extrapolate s, D, and v_1 to infinite dilution. Table 20.1 summarizes the data on some protein molecules that were studied in this way.

Figures 20.6 and 20.7 illustrate the optical techniques used to follow the sedimentation of macromolecules in an analytical ultracentrifuge.

A second method of using the centrifuge is based on the *sedimentation equilibrium*. When equilibrium is reached, the rate at which a solute is driven outward

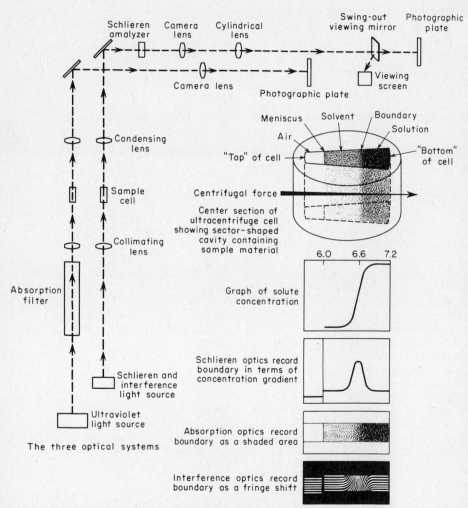

FIG. 20.6 Summary of methods used for study of high polymer molecules in the analytical ultracentrifuge. [Beckman Instruments, Inc.]

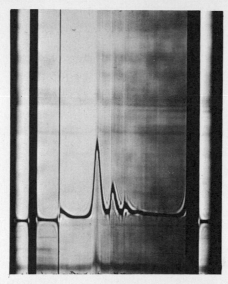

Schlieren pattern (multi-components). Preparation of ribosomes from Escherichia coli. 29,500 rpm

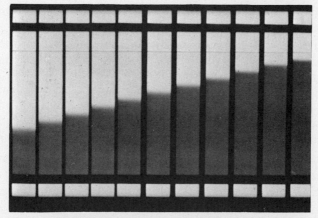

Ultraviolet absorption. Purified calf thymus DNA, 0.004 % in 0.2 M NaCl, pH 7.0. Speed 59,780 rpm. Photos at two minute intervols after reaching speed.

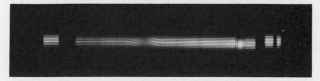

Rayleigh interference pattern. Bushy stunt virus 0.5 gm/ml in buffer at pH 4.1. Speed 14,290 rpm. Time 105 minutes.

FIG. 20.7 Examples of centrifuge patterns using the three different optical systems shown in Fig. 20.6. [Spinco Division, Beckman Instruments, Inc.]

by the centrifugal force just equals the rate at which it diffuses inward under the influence of the concentration gradient. The sedimentation rate is

$$\frac{dn}{dt} = \frac{c\,dx}{dt} = c\omega^2 xm(1 - v_1\rho)\left(\frac{1}{f}\right)$$

The diffusion rate is

$$\frac{dn}{dt} = -\left(\frac{kT}{f}\right)\frac{dc}{dx}$$

At equilibrium, the sum of these rates is zero, and we obtain

$$\frac{dc}{c} = \frac{M(1 - v_1\rho)\omega^2 x\,dx}{RT}$$

Integration between x_1 and x_2 leads to

$$M = \frac{2RT\ln(c_2/c_1)}{(1 - v_1\rho)\omega^2(x_2^2 - x_1^2)} \tag{20.16}$$

The molar mass so obtained is $\bar{M}_m$, the mass average.

The *sedimentation equilibrium* method does not require an independent measurement of D in order to fix the molecular weight, but the time required for equilibrium is so long that the method was not much used with substances having a molar mass greater than 5000 g·mol^{-1}.

The equilibrium method is based on a condition of zero net flux across any plane in the solution normal to the radius of the cell. At the top meniscus and at the bottom of the cell, there can never be any net flux, so that the equilibrium condition must hold at these sections at all times.* Shortly after the centrifuge is brought to speed, therefore, a determination of the concentrations at these special planes can be used to give the equilibrium values. This modification of the equilibrium method increases its useful scope.

If a solution of a substance of low molecular weight is centrifuged, there will be at equilibrium a gradient in density across the cell. If we add to the cell a substance of high molecular weight, it should float in this solution of graded density at the particular position at which its buoyant density equals the density of the solution. If the macromolecular substance is made up of fractions of different molecular weights, each fraction should form a band at a particular plane in the cell. This technique is called *density-gradient ultracentrifugation*.†

This method has been useful in studies of the replication of nucleic acids both *in vivo* and *in vitro*. One of the major problems in biochemistry is the mechanism of genetic control of heredity. The genetic material is DNA, a high polymer formed of units consisting of sugar phosphates coupled with nitrogen bases. A model for the replication of the DNA molecule is shown in Fig. 20.8, as proposed by Watson and Crick.‡ The density-gradient method was applied to this problem as follows.

*W. J. Archibald, *J. Phys. Chem.*, *51*, 1204 (1947).

†M. S. Meselson, F. W. Stahl, and J. Vinograd, *Proc. Natl. Acad. Sci.*, *44*, 671 (1958); *43*, 581 (1957).

‡F. H. C. Crick and J. D. Watson, *Proc. Roy. Soc. London, Ser. A*, *223*, 80 (1954).

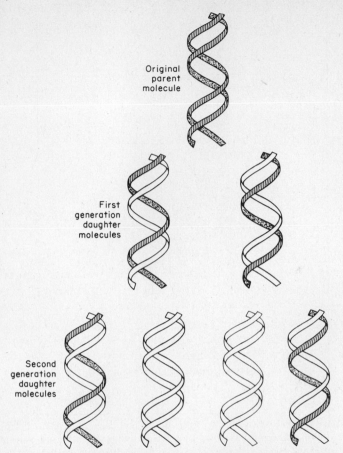

Original
parent
molecule

First
generation
daughter
molecules

Second
generation
daughter
molecules

FIG. 20.8   Illustration of the mechanism of DNA duplication proposed
by Watson and Crick. Each daughter molecule contains
one of the parental chains (black) paired with one new
chain (white). Upon continued duplication, the two original parent chains remain intact, so that there will always
be found two molecules each with one parental chain.
[M. S. Meselson and F. W. Stahl, California Institute of
Technology.]

Bacteria were grown in a medium containing heavy nitrogen ^{15}N. They were then
abruptly shifted to a medium containing only ^{14}N. If the DNA molecules replicate
according to the Watson–Crick model, at the end of the first generation, the ^{15}N
labeled material will be depleted, and mixed ^{14}N and ^{15}N molecules will be formed.
At the second generation, ^{14}N^{15}N and ^{14}N^{14}N molecules should occur. Figure
20.9 shows the results of one of the experiments. The bands observed in the density
gradient in the centrifuge correspond with the predictions from the Watson–
Crick model for replication.

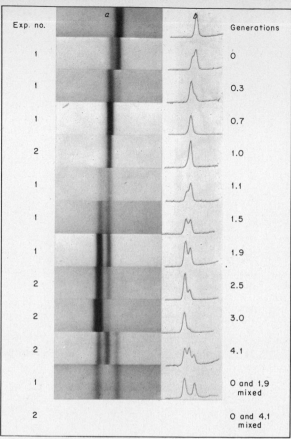

FIG. 20.9 An application of density-gradient centrifugation. Ultraviolet absorption photographs showing DNA bands resulting from density-gradient centrifugation of lysates of bacteria sampled at various times after the addition of an excess of ^{14}N substrates to a growing ^{15}N-labeled culture. Each photograph was taken after 20 hours of centrifugation at 44 770 rpm under the conditions described in the text. The density of the CsCl solution increases to the right. Regions of equal density occupy the same horizontal position on each photograph. The time of sampling is measured from the time of the addition of ^{14}N in units of the generation time. The generation times were estimated from measurements of bacterial growth. b: Microdensitometer tracings of the DNA bands shown in the adjacent photographs. The microdensitometer pen displacement above the base line is directly proportional to the concentration of DNA. The degree of labeling of a species of DNA corresponds to the relative position of its band between the bands of fully labeled and unlabeled DNA shown in the lowermost frame, which serves as a density reference. A test of the conclusion that the DNA in the band of intermediate density is just half-labeled is provided by the frame showing the mixture of generations 0 and 1.9. When allowance is made for the relative amounts of DNA in the three peaks, the peak of intermediate density is found to be centered at $50 \pm 2\%$ of the distance between the ^{14}N and ^{15}N peaks. [M. Meselson and F. W. Stahl.]

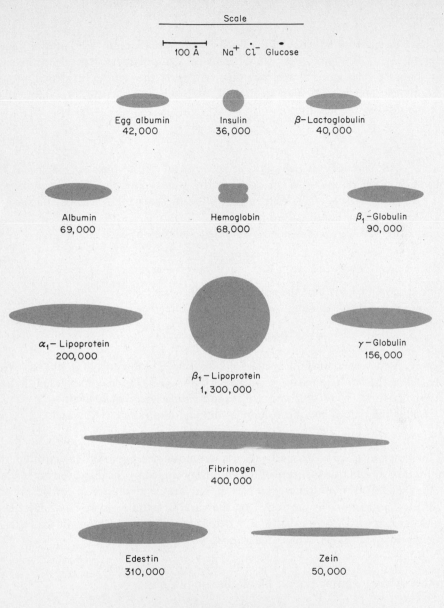

RELATIVE DIMENSIONS OF VARIOUS PROTEINS

FIG. 20.10 Estimated dimensions of various protein molecules as seen in projection. Most of the proteins are represented as ellipsoids of revolution. β-lipoprotein is a sphere [J. L. Oncley, Harvard University]. The molecular weight M_n is given under each name.

7. Viscosity

We have discussed the theory of the viscosity of gases (Section 4.25) and of liquids (Section 19.11). The viscosities of solutions of high polymers depend on the sizes and shapes of the molecules in solution. An early theoretical investigation was made by Einstein in 1906. For the case of rigid spherical particles in a dilute solution, he derived the equation,

$$\lim_{\phi \to 0} \left[\frac{(\eta/\eta_0) - 1}{\phi} \right] = 2.5 \tag{20.17}$$

Here η is the viscosity of the solution and η_0 that of the pure solvent; ϕ is the volume fraction of the solution occupied by solute particles. The fraction η/η_0 is called the *viscosity ratio* and $(\eta/\eta_0) - 1$ represents the fractional increase in viscosity caused by the dissolved particles.

Equation (20.17) was extended by Simha to ellipsoidal particles, to give

$$\lim_{\phi \to 0} \left[\frac{(\eta/\eta_0) - 1}{\phi} \right] = v \tag{20.18}$$

where v is related to the axial ratio of the ellipsoid.

Since it is difficult to measure ϕ directly, we often use the expression

$$[\eta] = \lim_{C \to 0} \left[\frac{(\eta/\eta_0) - 1}{C} \right] \tag{20.19}$$

where C is the concentration in mass per volume. The quantity $[\eta]/100$ has often been called the *intrinsic viscosity*, but the I.U.P.A.C. (1952) recommended that we use $[\eta]$ as defined by (20.19) and call it the *limiting viscosity number*.

If v_1^∞ is the partial specific volume at infinite dilution, $\phi = C v_1^\infty$ and

$$[\eta] = v v_1^\infty$$

In the cases of proteins and other macromolecules hydrated in solution, we cannot take v_1^∞ as a true measure of the volume occupied by the dissolved molecule.*

Figure 20.10 shows some examples of the sizes and shapes of native protein molecules in solution, as derived from the various techniques we have described.

Molar masses M have usually been determined from viscosities on the basis of semiempirical formulas. Thus, Staudinger in 1932 proposed

$$[\eta] = KM$$

A more general relation due to Mark and Houwink is

$$[\eta] = KM^\alpha \tag{20.20}$$

This relation predicts that the logarithm of $[\eta]$ should be a linear function of the logarithm of molar mass M. Some data plotted in this way are shown in Fig. 20.11(a) for polyisobutene in two different solvents. The relation seems to hold well for a given polymer and solvent when M exceeds 30 kg·mol^{-1}. For flexible coils,

---

*J. L. Oncley, *Ann. N. Y. Acad. Sci., 41*, 121 (1941), has provided methods for treating this problem.

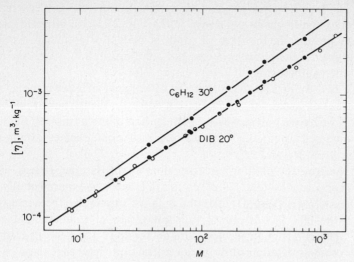

FIG. 20.11(a) Intrinsic viscosity vs. molar mass relationship for poly-isobutene in diisobutene (DIB) at 20° and in cyclo-hexane at 30°C.

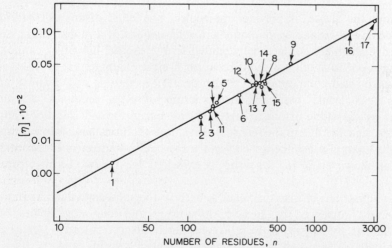

FIG. 20.11(b) Intrinsic viscosities of seventeen proteins in the form of random coils: (a) insulin, (2) ribonuclease, (3) hemoglobin, (4) myoglobin, (5) β-lactoglobin, (6) chymotrypsinogen, (7) pepsinogen, (8) aldolase, (9) bovine serum albumin, (10) glyceraldehyde-3-phos-phate dehydrogenase, (11) methemoglobin, (12) lactate dehydrogenase, (13) enolase, (14) alcohol dehydrogenase, (15) ovalbumin, (16) heavy subunit of myosin, (17) "51A" protein. [A. H. Reisner and J. Rowe, *Nature, 222,* 558 (1969).]

α is close to 2. Once the curve has been determined, therefore, a measurement of η can be used to determine an unknown M. In Fig. 20.11(b), intrinsic viscosities of 17 different proteins in the form of random coils are plotted in accord with (20.20). These are not the native proteins, but solutions in 6 M guanidine chloride with 0.5 M β-mercaptoethanol.

If we have two well fractionated samples of polymer of known molar masses, we can measure their viscosities and evaluate K and α for the particular polymer and solvent used. If we have a polymer with a distribution of molar masses, the viscosity average $\bar{M}_v$, obtained from (20.20), would be a rather complicated average falling somewhere between $\bar{M}_N$ and $\bar{M}_m$.*

For solid spheres, the limiting viscosity number $[\eta]$ is independent of molar mass, i.e., the coefficient α in Eq. (20.20) equals 0. It is found experimentally that all globular proteins, regardless of molar mass, have essentially the same intrinsic viscosity ($\alpha = 0$). This fact is one of the best pieces of evidence that globular proteins are very compact molecules, similar to solid spheres. The structures of crystalline proteins determined by X-ray diffraction show that these particular protein molecules are also compact and globular in the solid state.

8. Stereochemistry of Polymers

In discussing the shape of polymer molecules, one must distinguish carefully between *configurations* and *conformations*. *Configurations* of a molecule are steric arrangements which are symmetrically related to one another but which cannot be changed from one to another without breaking bonds. Examples are cis–trans isomers of unsaturated compounds or cyclopropane derivatives, and D and L configurations in molecules with asymmetric atoms or other centers of chirality. *Conformations* are steric arrangements that can occur through rotations about single bonds, and are due to the fact that certain orientations have lower energies than others. Examples are the gauche and trans conformations in ω-substituted paraffins (as shown on p. 763 for *n*-propyl chloride) and the two forms of cyclohexane.

The conformations of macromolecules, both in the solid state and in solution, can be greatly influenced by specific short-range and long-range interactions between different regions within the molecules. Such effects can lead to symmetries in an otherwise random coil, such as helices, staggered forms, and the like. Of course, the possibilities for such intramolecular interactions will depend on the basic configuration of the molecule concerned. An example in which the conformation is determined by hydrogen bonding is the α-helical structure of polypeptides, discovered by Pauling and Corey in 1951, which is shown in Fig. 20.12. Some polypeptides of individual α-amino acids will form α-helices, but some will not, owing to steric repulsions, electrostatic interactions, or other reasons. In a

---

*C. Tanford, *Physical Chemistry of Macromolecules* (New York: John Wiley & Sons. 1961), p. 111.

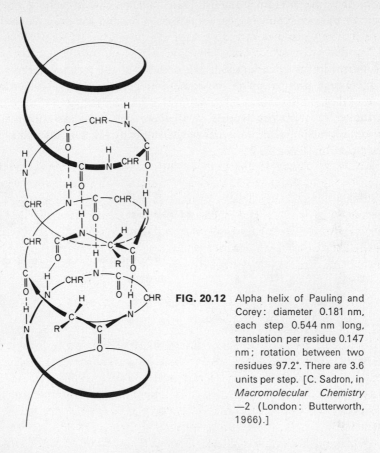

FIG. 20.12 Alpha helix of Pauling and Corey: diameter 0.181 nm, each step 0.544 nm long, translation per residue 0.147 nm; rotation between two residues 97.2°. There are 3.6 units per step. [C. Sadron, in *Macromolecular Chemistry* —2 (London: Butterworth, 1966).]

protein, there is a definite primary structure in the sequence composed of twenty different amino acids, so that depending upon a subtle interplay of various competing factors, some parts of the protein polypeptide chain may be helical whereas other parts form a random coil. As the temperature is raised, the transition, helix → random coil, often occurs over a narrow range of temperatures, so that it appears to be a cooperative phenomenon similar in some respects to the melting of a solid.

The great advance in macromolecular synthesis made by Ziegler and Natta was to devise methods for making polymers of predetermined stereochemical configurations. A classic example was the synthesis of natural rubber, a product of the tree *Hevea brasiliensis*. Rubber is a poly-cis-isoprene. A different type of tree, e.g., *Palaquium oblongifolia*, produces guttapercha, a plastic substance but not an elastomer, which has the structure of poly-trans-isoprene. Until 1955, however, all the methods used to synthesize polyisoprene gave a mixture of cis and trans-configurations. Polymeric mechanisms involving free-radical chain reactions usually have no stereospecificity, and only with the application of the Ziegler catalyst was a route opened toward stereospecific macromolecular synthesis.

Another type of stereoisomerism in polymers is caused by the presence of asymmetric or pseudoasymmetric carbon atoms. Consider for example, polybutadiene:

$$(-CH_2-CH=CH-CH_2-)_n$$

Two mirror-image configurations are possible at the tertiary carbon atom. If all the methylene groups are in the same configuration, the polymer is called *isotactic;* if they alternate in configuration, it is *syndiotactic;* and if they are randomly distributed, it is *atactic.* Models of all these structures are shown in Fig. 20.13 and some of the physical properties of the sterically regular butadiene polymers are summarized in Table 20.2.

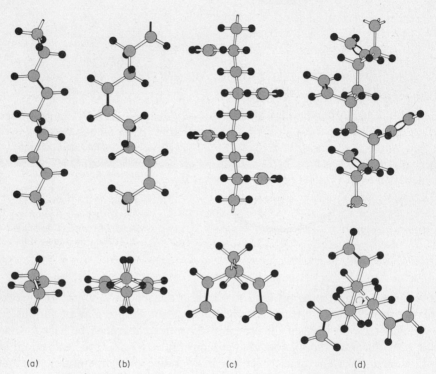

(a)   (b)   (c)   (d)

FIG. 20.13 Lateral and vertical views of the chain conformations of four stereoisomeric polybutadienes: (a) trans-1, 4; (b) cis-1, 4; (c) syndiotactic-1, 2; (d) isotactic-1, 2. [G. Natta, Nobel Lecture, *Angewandte Chemie, 76,* 553 (1964).]

Stereoregularity (e.g., isotaxis and syndiotaxis) usually forces the polymer chain into a certain conformational order (e.g., helix or planar zig-zag), at least in the solid state, in some cases also in the dissolved or molten state. Looking at a single polymer chain, we can consider conformational regularity as one-dimensional crystallinity. Highly interesting is the possibility for one-dimensional polymorphism which consists of defined different chain conformations, or different conformations

TABLE 20.2
Physical Properties of Four Sterically Regular Butadiene Polymers

Polymers (Infrared analyses)	M. Pt. [°C]	Identity period nm	Density [g·cm⁻³]
trans-1,4 (99—100%)	146[a]	$\begin{cases} 0.485 \text{ (Form I)} \\ 0.465 \text{ (Form II)} \end{cases}$	0.97 / 0.93
cis-1,4 (98—99%)	2	0.86	1.01
isotactic-1,2 (99% 1,2-units)	126	0.65	0.96
syndiotactic-1,2 (99% 1,2-units)	156	0.514	0.96

[a] Trans-1.4-polybutadiene exists in two crystalline forms. Form I is stable below 75°C, and Form II is stable between 75°C and the melting point.

of side groups. Examples are polybutene-1, which at normal temperature assumes a 3_1-helix but at higher temperatures a 4_1-helix, and syndiotactic polypropene which can assume a planar zig-zag and the conformation of a double-helix.

Sometimes, the preferred chain conformation is brought about by polar groups within the polymer chain. Good examples are aliphatic polyethers.* In this case, the members with short methylene sequences have helical chains whereas the ones with longer methylene sequences assume a planar zig-zag conformation.

A long-chain polymer molecule can often assume many different conformations in solution. Any theory of the properties of polymer solutions must begin, therefore, with a study of the statistical problem posed by the different orientations of the monomer segments of the polymer chosen. The most simple model of the polymer chain replaces the actual chemical bonds with links permitting perfectly free rotation; it also neglects the fact that no two monomer units can occupy the same location in space.

Consider in Fig. 20.14 a freely jointed polymer chain with one end fixed at the origin. What is the probability W that the other end falls in a volume element $dx\,dy\,dz$ at a distance r from the origin? The problem is like that of the free diffusion of a single molecule with a fixed jump distance a, i.e., a random walk in three dimensions. The solution is the Gaussian distribution,

$$W(x, y, z)\, dx\, dy\, dz = \left(\frac{\beta}{\pi^{1/2}}\right)^3 e^{-\beta^2 r^2}\, dx\, dy\, dz \tag{20.21}$$

The probability that one end of the chain lies at a distance r from the origin, irrespective of direction (cf. the distribution of molecular velocities in Section 4.13), is

$$W(r)\, dr = \left(\frac{\beta}{\pi^{1/2}}\right)^3 e^{-\beta^2 r^2} \cdot 4\pi r^2\, dr \tag{20.22}$$

*H. Tadokoro, Y. Takahashi, Y. Chotani, and H. Kakida, *Makromol. Chem.*, *109*, 96 (1967).

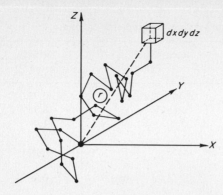

FIG. 20.14 Conformation of a freely rotating polymer chain of 23 links with one end at origin. Analog of a random walk in three dimensions.

Differentiation of (20.22) shows that the maximum in $W(r)$ occurs at a value

$$r = \beta^{-1}$$

Thus β^{-1} is the most probable value of r.

If a is the length of a unit link of the chain, and N is the number of links,

$$\beta^{-1} = a\left(\frac{2N}{3}\right)^{1/2} \tag{20.23}$$

For a zig-zag chain of carbon atoms, with bond angle α and length of the C—C bond l,

$$a = l\left(\frac{1 + \cos \alpha}{1 - \cos \alpha}\right)^{1/2}$$

The chain with freely rotating links is quite an unrealistic model for an actual polymer. It is necessary in further developments to consider the effect of restrictions on the bond angles, and the "free volume effect" that results from the fact that no two segments can fall at the same location. In this way, a reasonable theory for these flexible coil polymers can be developed. Such models would not apply, however, to proteins. As shown in Fig. 20.10, these behave more like rigid spheroids, or sometimes, as in the case of collagen,* rigid rods.

9.   Elasticity of Rubber

The most remarkable property of a rubber or *elastomer* is its ability to sustain a large deformation without breaking and to snap back to its original conformation when the deforming stress is removed. An elastomer is like a liquid in its deformability and like a solid in its elasticity. The properties of elastomers are fundamentally related to their structures of long chainlike polymer molecules. Any long chain polymer can exhibit such properties under the right physical conditions.

The essential change that occurs in stretching an elastomer is shown schematically in Fig. 20.15. In the unstretched rubber, the polymer chains are coiled

*H. Boedtker and P. Doty, *J. Am. Chem. Soc.*, 78, 4267 (1956).

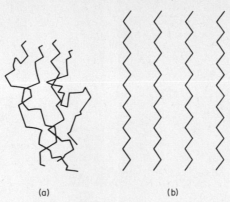

(a)   (b)

FIG. 20.15 Idealized models of chains in rubber: (a) contracted;
(b) stretched.

and twisted more or less at random. In the stretched condition, the chains uncoil
to a considerable extent and tend to become oriented along the direction of
elongation. Thus, stretching causes the randomly oriented chains to assume a
more ordered alinement. The unstretched, disordered configuration is a state
of greater entropy. If the tension is released, the stretched rubber spontaneously
reverts to the unstretched condition.

Electron microscopy of thin films of rubber reveals directly such changes in
the orientations of crystallites, which were originally detected by X-ray diffraction.
Figure 20.16 shows crystalline and amorphous regions in an unstrained film of
natural rubber, and highly oriented crystalline structure in a film that was strained
to 200% before crystallization.

Robert Boyle and his contemporaries used to talk about the "elasticity of a
gas." Although we seldom hear this term today, we should note that the elasticity
of a gas and the elasticity of a rubber band have the same thermodynamic interpretation. Consider a volume of gas compressed by a piston in a cylinder. If the
pressure is released, the piston springs back as the gas expands. The expanded
gas is in a state of higher entropy than the compressed gas. It is in a more disordered
state because each molecule has a larger volume in which to move. A stretched
rubber band is like a compressed gas. If we release the tension on the rubber,
it spontaneously contracts. The contracted state of the rubber band, like the expanded state of the gas, is the state of greater disorder and hence of higher entropy.

From Eq. (3.43) the pressure is

$$P = -\left(\frac{\partial A}{\partial V}\right)_T = T\left(\frac{\partial S}{\partial V}\right)_T - \left(\frac{\partial U}{\partial V}\right)_T \tag{20.24}$$

For a gas, the $(\partial U/\partial V)_T$ term is small and effectively $P = T(\partial S/\partial V)_T$. The pressure
is proportional to T and is determined by the change in entropy with volume. The
analog of (20.24) for a strip of elastomer of length l subject to a tension K is

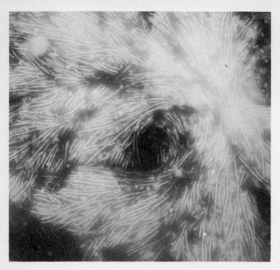 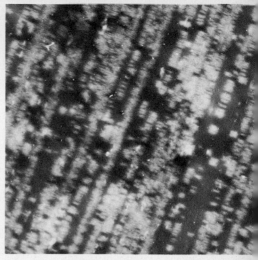

FIG. 20.16 (Left) Crystalline spherulite in unstrained thin film of cross-linked natural rubber. Electron micrograph at 20 000X after staining with osmium tetroxide. Dark regions are amorphous, light regions crystalline. (Right) Crystallization in a thin film of natural rubber strained at 200% before crystallization. The long rows are in the direction of strain and the short filamentous crystals are perpendicular to that direction. Electron micrograph at 10 000X after staining with osmium tetroxide. [E. H. Andrews, Department of Materials, Queen Mary College, London.]

$$-K = T\left(\frac{\partial S}{\partial l}\right)_T - \left(\frac{\partial U}{\partial l}\right)_T$$

It is found experimentally that K is proportional to T, so that, just as in the case of the gas, the term involving the energy is relatively unimportant.

We can now use the statistical theory of polymers to calculate the probabilities W of stretched and unstretched conformations, and then obtain the change in entropy from the Boltzmann relation $S = k \ln W$.*

10. Crystallinity of Polymers

Although they cannot always be obtained as perfect three-dimensional crystals, high polymers often have a high degree of order in the solid state, which can be investigated by X-ray diffraction. For example, much important work has been done on X-ray diffraction by *fibers*, either natural (hair, silk, porcupine quills) or artificial (nylon, polypeptides).

*W. Kuhn and F. Grun, *J. Polymer Sci., 1*, 183 (1946).

In some cases, these fibers have a well defined periodicity parallel to the fiber axis, but a more random organization in other directions. The type of pattern that results can be understood from the case of diffraction by a linear array (p. 579). Consider a set of scattering centers with a repeat distance c along the c axis. If the Bragg condition is fulfilled at angle θ,

$$c \sin \theta = l\lambda$$

where l is an integer. The pattern on the film will consist of a set of layer lines with a spacing a, such that

$$\tan \theta = \frac{a}{R}$$

where R is the distance between specimen and film.

0°C   25°C

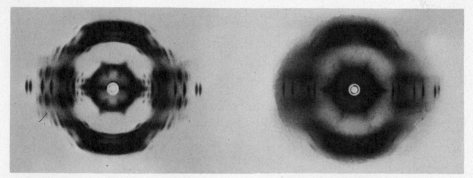

Above 19°C

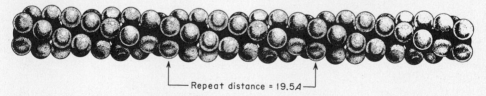

└─── Repeat distance = 19.5Å ──┘

Below 19°C

└─Repeat distance = 16.9Å─┘

FIG. 20.17 X-ray diffraction patterns of oriented fibers of polytetrafluoroethylene above and below the transition at 292 K and models of sections of the polymer to show the transition that occurs. [C. A. Sperati and H. W. Starkweather, Du Pont Experimental Station, Wilmington, Delaware.]

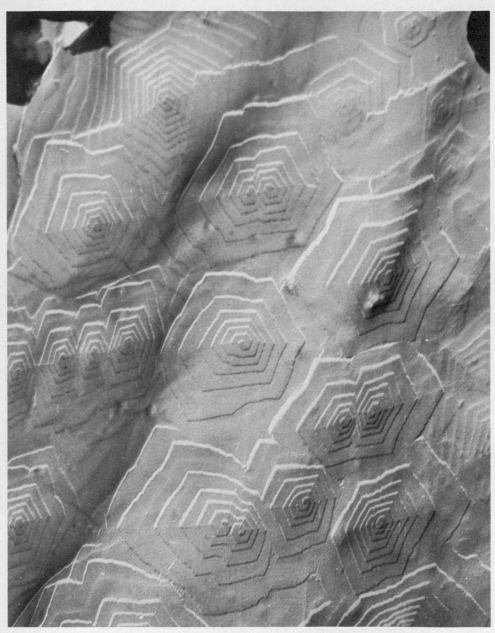

FIG. 20.18 Portion of a single crystal of polyoxymethylene precipitated from solution in cyclohexanol. The crystal surface was shadowed by evaporation of chromium at a glancing angle and examined by electron microscopy at 20 000X. The pattern of spirals suggest that the crystal grows from spiral dislocations.

Figure 20.17 shows an example of such a diagram for fibers of polytetrafluoro-ethylene. This polymer undergoes a transition in the crystalline state at 292 K, which is associated with an expansion of the repeat distance along the polymer chain. Such X-ray fiber diffraction diagrams have been important in elucidat-ing the structures of proteins and nucleic acids. As shown in Chapter 18, however, complete elucidations of protein structures can be achieved by X-ray diffraction analysis of single crystals of complexes of proteins with atoms of heavy metals.

In the case of high polymers, the terms "crystalline" and "amorphous" must usually be regarded as defining extreme conditions that may seldom be met in practical examples. Even a randomly ordered polymer system has a considerably higher order than a liquid, and even single crystals of polyethylene depositing from xylene solution display bent regions due to chain folding that have a considerably lower regularity than the inner regions of a crystal of a typical nonpolymeric compound. Real macromolecular systems often can exist in quite a number of different phases having degrees of order intermediate between those of ideal liquids and ideal crystals.

Neverthelesss, single crystals of high polymers can be grown from solution. The electron micrograph in Fig. 20.18 shows a portion of the surface of a single crystal of polyoxymethylene grown from solution in cyclohexanol.*

PROBLEMS

1. A suspension contains equal numbers of particles with molecular weights 10 000 and 20 000. Calculate $\bar{M}_N$ and $\bar{M}_m$. A suspension contains equal masses of particles with molecular weights 10 000 and 20 000. Calculate $\bar{M}_N$ and $\bar{M}_m$.

2. The following data were obtained for the osmotic pressures of nitrocellulose in acetone at 20°C:

$C(g \cdot dm^{-3})$	1.16	3.66	8.38	19.0
$\Pi(cm\ H_2O)$	0.62	2.56	8.00	25.4

Calculate the limiting value of Π/C and hence $\bar{M}_N$.

3. M. L. Huggins† derived the following equation for the osmotic pressure Π of a poly-mer solution as a function of the concentration of solute, C_2:

$$\frac{\Pi}{C_2} = \frac{RT}{M_2} + \frac{RT\rho_1}{M_1\rho_2^2}\left(\frac{1}{2} - \xi\right)C_2 + \frac{RT\rho_1}{3M_1\rho_2^3}C_2^2 + \cdots$$

where ρ_1 and ρ_2 are the densities of solvent and solute, M_1 is the molar mass of solvent, and M_2 is the number average molar mass of the polymer. The ξ is an interaction constant characteristic of the solution. Data on polystyrene in toluene obtained by Bawn‡ et al. show that when values of the constants are introduced at 25°C the Huggins equation becomes

*P. H. Geil, *Polymer Single Crystals* (New York: John Wiley & Sons, 1963).
†M. L. Huggins, *J. Phys. Chem.*, **46**, 151 (1942).
‡C. Bawn, R. Freeman, and A. Kamaliddin, *Trans. Faraday Soc.*, **46**, 862 (1950).

$$\frac{\Pi}{C_2} = \frac{RT}{M_2} + 2.03 \times 10^5 \left(\frac{1}{2} - \xi\right) C_2 + 6.27 \times 10^4 C_2^2$$

where Π is in $g \cdot cm^{-2}$, C_2 in $g \cdot cm^{-3}$, and $\rho_2 = 1.080\ g \cdot cm^{-3}$. The data for one sample of polystyrene were

$10^3 C_2$	1.55	2.56	2.93	3.80	5.38	7.80	8.68
Π	0.16	0.28	0.32	0.47	0.77	1.36	1.60

Plot these data as a straight line and determine M_2 and ξ.

4. Prove that the mean square end-to-end length of a linear polymer chain with free rotation about the bonds of the chain is $R^2 = Na^2$ where N is the number of bonds of length a. Hence calculate the root mean square end-to-end length of a linear polystyrene molecule having $M = 10^5$.

5. Consider a segment of length a in a freely orienting polymer chain. Under the influence of a force F the segment acquires an orientation in the x direction. By analogy with the theory given in Section 15.18 for the average value of a dipole moment in the field direction, show that the average value of the x component of a is

$$\langle a_x \rangle = a \left[\coth \frac{aF}{kT} - \frac{kT}{aF} \right]$$
$$= a \mathscr{L}\left(\frac{aF}{kT}\right)$$

where $\mathscr{L}$ is the Langevin function. How large a force would be required to elongate the linear polystyrene chain described in Problem 4 by 20%?

6. The relative viscosity of a polymer solution containing 1.00 g of polymer in 100 cm³ is 2.800. A solution half as concentrated has a relative viscosity of 1.800.
   (a) Calculate the intrinsic viscosity. (Assume that a graphical treatment would give a straight line, and calculate the intercept analytically).
   (b) If the appropriate values of K and a in the Mark-Houwink equation are 5.00 $\times$ 10^{-4} and 0.600, calculate the molecular weight of the polymer.

7. For horse hemoglobin in water solution at 20°C, $D = 6.3 \times 10^{-7}\ cm^2 \cdot s^{-1}$, $s = 4.41 \times 10^{-13}$ s, $v = 0.749\ cm^3 \cdot g^{-1}$, $\rho = 0.9982\ g \cdot cm^{-3}$. Calculate the molecular weight.

8. The viscosities relative to pure solvent of a fraction of polystyrene of $\bar{M}_N = 280\ 000$ dissolved in tetralin at 20°C were

conc. (%)	0.01	0.025	0.05	0.10	0.25
η_r	1.05	1.12	1.25	1.59	2.70

Calculate the value of the exponent α in the Staudinger equation. Hence, estimate the relative viscosity of a 0.10 solution of polystyrene with $\bar{M}_N = 500\ 000$ in the same solvent.

9. The diffusion coefficient of the insulin molecule in water at 20°C is $8.2 \times 10^{-7}\ cm^2 \cdot s^{-1}$. Estimate the mean time required for an insulin molecule to diffuse through a distance equal to the diameter of a typical living cell, about 10 μm.

10. Calculate the most probable value of the chain length of normal $C_{20}H_{42}$ given the C—C bond length of 0.15 nm and the bond angle of 109°28'.

11. For a solution of cellulose trinitrate ($M_r = 140\ 000$) in acetone, $dn/dc = 0.105$ $\text{cm}^3 \cdot \text{g}^{-1}$ and $n_0 = 1.3589$. Calculate the ratio of intensities of transmitted to incident light at wavelengths of 400 and 700 nm through 1.00 cm thickness of a solution of the polymer containing 2.00 g per 100 cm^3.

12. Suppose you have a weight on a rubber band so as to keep it under constant tension. If you then heat the rubber band, will the weight rise or fall? Give a thermodynamic answer, then try the experiment.

13. An electrophoretic study of an aqueous solution of protein reveals two protein species, with $M_r = 60\ 000$ and $120\ 000$. In a solution containing 1.76% protein by weight at $25°C$ the amount (n) of the larger protein is 1.56 times that of the smaller one.
    (a) Calculate $\bar{M}_N$ and $\bar{M}_m$ for the proteins in solution.
    (b) Estimate the viscosity of the solution assuming the protein molecules behave as rigid spheres of density 1.290 $\text{g} \cdot \text{cm}^{-3}$.
    (c) Estimate the ratio of sedimentation coefficients s for the two proteins.

Appendix A

A Consistent Set of Physical Constants*

Quantity	Symbol	Value	SI Units
Speed of light	c	2.997925	10^8 m·s^{-1}
Electron charge	$-e$	1.602192	10^{-19} C
Planck constant	h	6.62620	10^{-34} J·s
Avogadro number	L	6.02217	10^{23} mol^{-1}
Electron rest mass	m_e	9.10956	10^{-31} kg
Proton rest mass	m_p	1.67261	10^{-27} kg
Atomic mass unit	amu	1.66053	10^{-27} kg
Faraday constant	F	9.64867	10^4 C·mol^{-1}
Rydberg constant	R_∞	1.0973731	10^7 m^{-1}
Bohr radius	a_0	5.291772	10^{-11} m
Bohr magneton	μ_B	9.27410	10^{-24} J·T^{-1}
Nuclear magneton	μ_N	5.05095	10^{-27} J·T^{-1}
Gas constant	R	8.3143	J·K^{-1}·mol^{-1}
Boltzmann constant	k	1.38062	10^{-23} J·K^{-1}

*These values were obtained from the comprehensive review of fundamental constants by B. N. Taylor, W. H. Parker, and D. N. Langenberg [*Rev. Mod. Phys.*, *41*, 375 (1969)]. They differ somewhat from the set of constants adopted in 1963 by the National Bureau of Standards (NBS Technical News Bulletin, Feb. 1963) and the International Committee on Weights and Measures.

Appendix B

Recommended Unit Prefixes

Multiples and Submultiples	Prefix	Symbol
10^{12}	tera	T
10^9	giga	G
10^6	mega	M
10^3	kilo	k
10^{-1}	deci	d
10^{-2}	centi	c
10^{-3}	milli	m
10^{-6}	micro	μ
10^{-9}	nano	n
10^{-12}	pico	p
10^{-15}	femto	f
10^{-18}	atto	a

Name Index

Subject Index